Basic to Advanced Computer Aided Design Using

NX 11

Modeling, Drafting, Assemblies, and Sheet Metal

A Project Oriented Learning Manual

By:

Stephen M. Samuel, PE

Adam Ericksen, Ph.D.

DESIGN VISIONARIES

Superior Vision Yields Optimal Products

Basic to Advanced Computer Aided Design Using NX 11

ISBN: 978-1-935951-10-0

Published by:

Design Visionaries

7034 Calcaterra Drive

San Jose, CA 95120

info@designviz.com

www.designviz.com

www.nxtutorials.com

Phone: (408) 997-6323

DESIGN VISIONARIES

Superior Vision Yields Optimal Products

Proudly Printed in the United States of America

Published February 2017

Dedication

We dedicate this book to the inventors of the world. Your persistence, your creativity, and your vision are a beacon of hope for a bright future ahead.

About the Authors

Stephen M. Samuel PE, Founder and President of Design Visionaries, has over 25 years of experience developing and using high-end CAD tools and mentoring its users. During a ten-year career at Pratt & Whitney Aircraft, he was responsible for implementing advanced CAD/CAM technology in a design/manufacturing environment. He has trained thousands of engineers in Unigraphics, written self-paced courses in UG Advanced Modeling and Best Practices, and performed design work for numerous Fortune 500 companies. Stephen is the author of distinctive publications on Nastran, UG/NX CAD, Solid Edge, SolidWorks, and Teamcenter Engineering PLM. Stephen holds several US patents and enjoys a life of creativity and intellectual challenge in the city of San Jose, CA. He happily shares his life with three amazing children, his wife and his 84-year-old powerhouse of a mother that lives in a home right next door.

Adam Ericksen, Ph.D., is a Design Engineer and NX Trainer at Design Visionaries, with over ten years of teaching experience. He holds a Ph.D. in Mathematics from the University of Southern California, where he taught mathematics to engineers and other students of science. Adam has contributed to product designs spanning several industries since joining Design Visionaries, and served as an expert NX trainer for a variety of satisfied customers throughout the US.

Acknowledgements

We would like to thank the following people for their tireless efforts. Without the contributions from each of you, this book would be a mere shadow of what it has become.

Jennine Scott, Landon Ritchie, Linda Howlett

Special thanks to Landon Ritchie for the cover design, and John Faria for proofreading.

What readers have to say about our previous books

"Practical Unigraphics NX Modeling for Engineers was extremely effective and much better than [other] textbooks. I would highly recommend this book to other professors and students alike."

Fred Dyen, Director of St. Louis University's Aviation Maintenance Institute (AMI)

"The UG NX textbook is well organized. Its tutorial style of learning is easy for students to utilize. The practice exercises are essential. From my experience in teaching students to use UG software, I have found that this is the best textbook currently on the market for teaching UG NX."

Dr. Pat Spicer, Professor at Western Illinois University

Preface

Dear reader,

Thank you for purchasing *Basic to Advanced Computer Aided Design Using NX 11*, the latest offering in our series of CAD training text books. Design Visionaries is an engineering consulting firm that performs many design projects great and small, including industrial design, product design and engineering analysis. Our customers entrust us with a wide range of designs, including medical devices, aerospace components, heavy machinery, and consumer products.

The methods outlined in this book go beyond an academic use of the software. They are tricks of the trade that come from thousands of hours of actual use of the software to design some of the most difficult products in the world. In addition, Design Visionaries offers world class on-site training which enables us to develop and evolve our training material to provide maximum benefit. Please enjoy this text, and we invite you to log on to our websites – www.designviz.com and nxtutorials.com, where you can download the part files pack that accompanies this book. There are also additional free materials, other advanced materials, products, and goodies.

Thank you,

Stephen Samuel

Adam Ericksen

February 2017

Contents

1 Introduction

This book has been written with an underlying philosophy that comes from years of engineering design experience. Engineers are pretty bright in general, so we've written this book to take advantage of that fact. Our book begins with the basics and examples explained to every last detail. As the book progresses, more and more is left to the reader. We believe this enables faster learning as you won't have to sift through superfluous instructions. We hope you enjoy this material that we've truly poured our hearts into.

Using NX is like playing a piano. In the same way that chords are as important as individual notes, NX commands are far more powerful when used in concert with others. Our book makes an effort to show not only the details of the most important commands, but the powerful combinations that we have used to bring about excellent designs.

For even more tutorials, visit nxtutorials.com. There are some great tutorials on some very useful but esoteric NX functions. There are even some awesome projects with plans already available if you need a quick break from NX and want to get your hands busy building!

1.1 Fundamental Concepts of Solid Modeling

Siemens PLM Software NX 11™ (commonly referred to as *Unigraphics*, *UG*, or *NX*) is a 3D Computer-Aided Design (CAD) software that is used at companies all over the world to design state-of-the-art products. This manual teaches you the basics of modeling, assemblies, drafting, and sheet metal in NX. Each of these functions is organized into a separate *application*. This majority of this book concentrates on what is typically the most complicated task in NX: three-dimensional geometric modeling in the Modeling application.

1. Parametric Models

A *parametric model* has variables (*parameters*) that control its geometry. Parametric models are "smart" models that remember how they were built and "rethink" themselves as their parameters or features change. Most models constructed in NX are parametric.

2. Non-Parametric Models

A *non-parametric model* has none of the parameters or feature-driven geometry; it is sometimes referred to as a "dumb" model. Geometry from non-native files (e.g., STEP) imports into NX as dumb solids. To modify a non-parametric model, it is often necessary to rebuild some of the geometry, or to make use of the *Synchronous Modeling* tools.

3. Hybrid Models

A *hybrid model* is neither completely parametric nor non-parametric; rather it is comprised of both types of geometry. Portions of the geometry may be so complicated that it is more efficient to remove parameters, while some of the geometry is still feature-driven. This is a very flexible and powerful modeling technique.

Building a completely parametric model takes more planning and foresight; however, the extra effort can pay off if you are building multiple versions of a model, or if you can anticipate potential design modifications. The experienced NX user will determine which modeling approach is appropriate for each design project.

4. Associativity

Associativity is an important option that arises on many tools in NX. It allows for parent-child relationships to be established between entities in your model so that if the parent changes, so does the child. Associative links between components in assemblies give rise to fully parametric, smart assemblies. Knowing how and when to employ associativity is key to building a successful design.

5. Curves

Curves are inherently one-dimensional objects, and can be created either in a plane, or in three-dimensional space. Each curve has a set of *control points*. For example, lines have control points at both ends and at the midpoint, while circles have a basic control point at their center. These control points are useful for a variety of tasks, such as constraining new geometry, and making measurements of existing geometry.

6. Sketches

Sketches are by definition planar, parametric features. Sketches are always defined upon a planar entity, with a reference that orients the sketch horizontally or vertically. As with curves, sketch entities also have control points.

The general approach in creating a sketch is to define its orientation, quickly create the curves that define it (do not worry about dimensions when creating geometry), and constrain it by placing geometric constraints and dimensions. At any point after the sketch is constrained, the dimensions that define the sketch can be changed and the model will then update.

Once a sketch is completed, it may be used to create other geometry. A typical use case involves sketching a profile and invoking the **Extrude** command for a specified distance to create a solid body. Sketches may also be used to define complex geometry, such as free-form surfaces.

7. Reference Features

Reference features include **Datum Plane**, **Datum Axis** and **Datum Coordinate System** features, which are used as construction entities and may be thought of as parametric reference geometry. A common use for reference features is to position sketches in order to create meaningful geometric relationships.

8. Design Features

Design features are used to create solids. They are often based on either curves or sketches (e.g., the **Extrude** command). Although design features are parametric in nature, they are only fully parametric if they are created upon parametric geometry, (i.e., solids created from sketches).

9. Surfaces

Like design features, surfaces are often based upon curves or sketches. They are the most powerful and mathematically complex of the geometric tools in NX and are often used when the desired geometry is too complicated to create using design features alone.

10. Feature Operations

Feature operations are used to add detail to the existing geometry. These are great tools and save a lot of time while creating complex geometry. Commonly used feature operations include **Edge Blend**, **Draft**, **Chamfer**, **Pattern Feature**, **Shell**, **Trim**, and the usual *Boolean operations* (**Unite**, **Subtract**, and **Intersect**).

1.2 The NX *".prt"* File

NX uses only one filetype for its core applications – the *".prt"* file. An NX part file is capable of acting as a model, a drawing, an assembly, or storing manufacturing toolpaths – in fact, **one part file can do all four of these things at the same time**! For this reason, there are no files with extensions other than *".prt"* for drawings, assemblies, etc. It is possible to produce geometry in the top-level assembly file – which is for most purposes a bad practice! This can be confusing to new users, but leads to great flexibility in how you choose to organize your CAD data.

One available approach is to organize all design and manufacturing into a single *".prt"* file – this one file could be accessed by the designer, drafter, and NC programmer as necessary. Another approach is to place each of the drawing and manufacturing data into their own *".prt"* file which acts as an assembly with one component – namely, the model. In this scheme, the designer, drafter, and NC programmer each have their own file to own, and can work simultaneously.

1.3 How to Use This Book

Learning CAD is an active process – you won't learn to use NX by just reading this book. Each chapter is written in a tutorial style, building upon the last, so the best way to read this book will be with both NX and the book open, doing the exercises as directed in the text.

- Bullet points indicate instructions – when you see a bulleted item, follow along on the computer.

The first section, *The User Interface (UI)*, consists of Chapters 2 through 4, and while you may be tempted to jump ahead to Chapter 5, we strongly encourage you to work through this foundational material, at least quickly. Those chapters contain a lot of information that you will find necessary to follow instructions later in the book, so we also recommend reviewing those chapters again once you have created some models. Test out the UI again later with more complex geometry than the very-simple examples below.

The next section, *Basic Modeling, Assemblies, and Drafting*, is comprised of Chapters 5 through 12. In this section, we will familiarize you with most of the basic tools needed for creating parts and assembling them. Mastery of the techniques in these chapters will prepare you to make a variety of machined and plastic parts.

The final section of the book, *Intermediate and Advanced Topics*, consists of Chapters 13 through 19. The line between what is "intermediate" and what is "advanced" is blurry, and you will find some topics easier than others throughout these chapters. To fully master some techniques and tools, you will benefit from knowledge of the manufacturing processes needed to produce the parts you design, which is not covered in this book.

As you work through the exercises in the book, you will want to save them, as some are re-used in later exercises. At this time, you will create a directory in which to save these files.

- In Windows, create a folder called *"My NX 11 Files"* in an easy-to-find location – we strongly recommend *"C:\My NX 11 Files"*, as we will refer to this folder throughout this book.
- Visit www.designviz.com/goodies and download the NX 11 part files that accompany this book. Save a copy of this folder in *"C:\My NX 11 Files"* as *"Downloaded Files"*. If for some reason the part files are unavailable for download, please email info@designviz.com and we will get them to you as soon as possible.
- Within *"C:\My NX 11 Files\Downloaded Files"* there is a folder called *"Completed"*. This folder contains the completed part files that you are asked to create in the exercises throughout the book – you can use these files to check your work!

2 The Ribbon and Other Toolbars

Our philosophy in writing this book is that you will only learn the material by *doing*. In this chapter, you will make several customizations to the UI to improve your learning experience, and practice basic file management. Let's get started right away with a tour of the user interface! Begin by launching NX 11.

NX 11.0

When NX is completely loaded, your screen should appear as below. The look and feel of NX 11 is similar to that of many Windows™ based products. Commands are arranged at the top of the screen on a ribbon, and selections are made primarily with the left mouse button. NX is comprised of dozens of distinct applications, each of which offers the user different tools for specific design or manufacturing tasks, many of which are industry-specific. In this book, we will focus on four applications – *Modeling*, *Drafting*, *Sheet Metal*, and *Assemblies*. When you first launch NX, the *Gateway* application appears, as shown below.

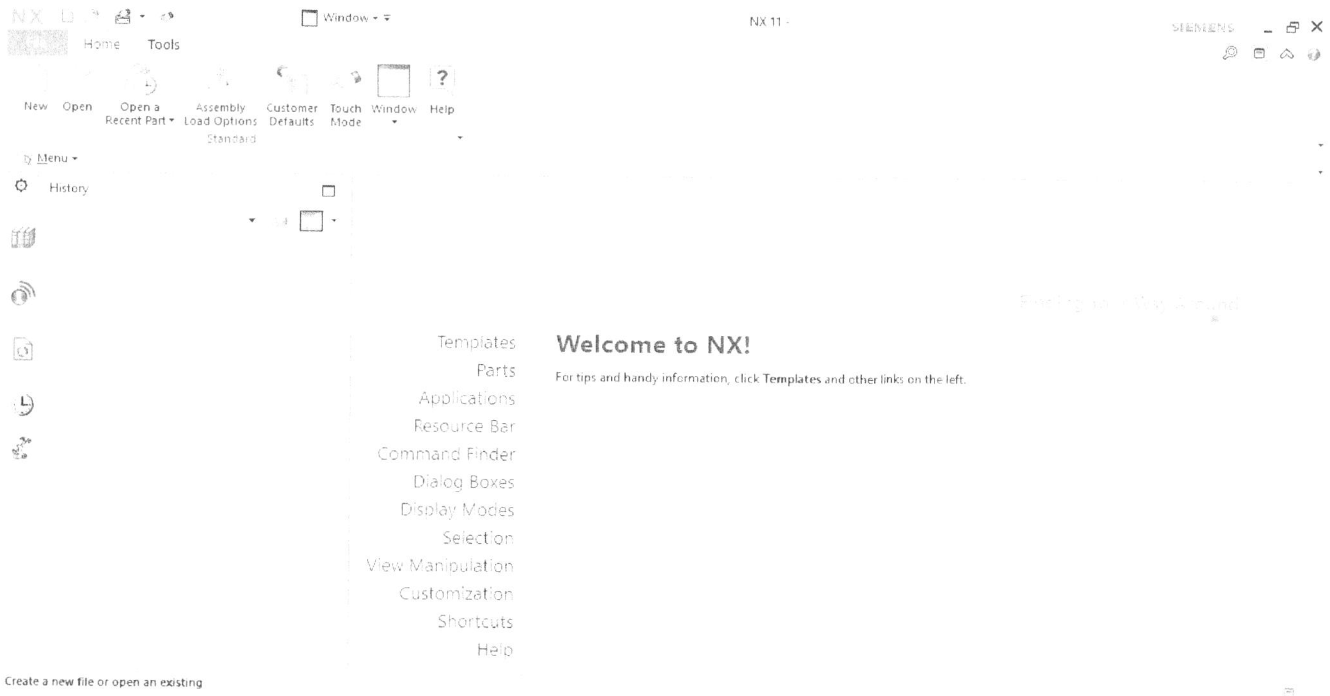

2.1 The Resource Bar

Our tour begins in a roundabout way, on the *Resource Bar*. The Resource Bar is a vertical collection of tabs with an attached palette on the far-left side of the NX window, adjacent to the graphics window.

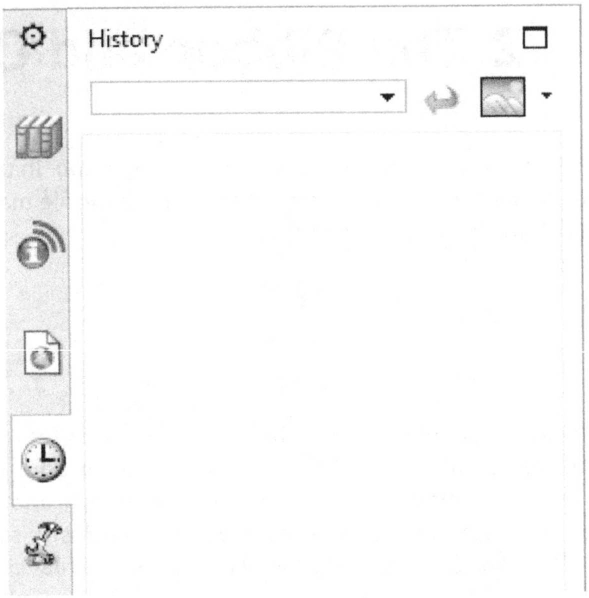

Each of the tabs in the resource bar serves a different important purpose, but the most important within the scope of this book are: the *Assembly Navigator*, the *Constraint Navigator*, the *Part Navigator*, and the *Roles* tab. The *Roles* tab is important when preparing to use NX because the tools available on the various menus in NX vary depending on your choice of role. Find the icon shown below, and click on it.

We will be using the *Advanced* role in what follows. The *Advanced* role is found in the **Content** folder – click on it to enable it at this time.

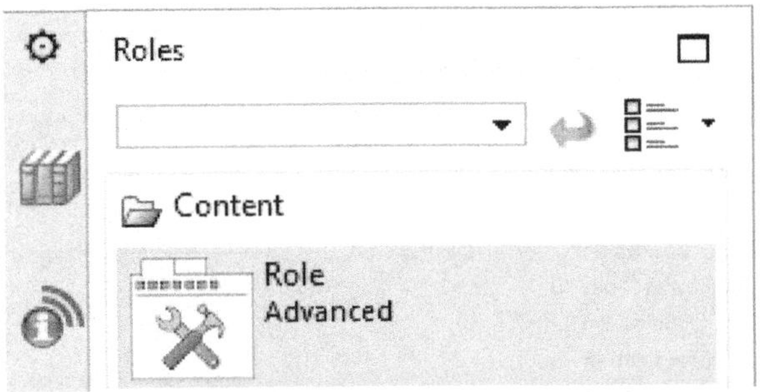

The Part Navigator (currently hidden) will be of particular importance as we begin solid modeling because it contains the *Model History* – a sequential summary of all the parametric operations performed to produce a model. We will study the Part Navigator in depth in Chapter 4, and the Assembly Navigator in Chapter 11.

2.2 The Quick Access Toolbar

The *Quick Access Toolbar* shows icons for many commonly-used file management commands. By default, it includes icons (listed in order as shown below) for **Save**, **Undo/Redo**, **Cut/Copy/Paste**, the **Repeat Command** drop-down menu, **Touch Mode**, the **Switch Window** command, and the **Window** drop-down menu. The **Toolbar Options** drop-down on the far right includes options for additional file management operations, such as **New** and **Open**.

2.3 Managing Part Files

NX handles standard file management tasks (creation of new files, opening of existing files, etc.) in much the same way as most *Windows*-based applications.

2.3.1 Creating New Files

Creating new files in NX is just like creating files in virtually any other Windows-based software!

- To begin select the **New File** command from the far-left side of the ribbon.

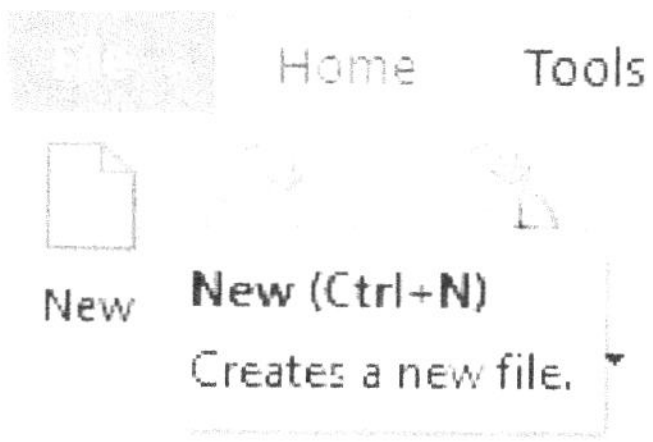

- The **New** dialog box shown below will then appear. Set the **Units** to *Millimeters*, select the **Model** template, and click **OK**. Note the variety of tabs on the **New** dialog box – *Model, Drawing, Simulation*, etc. Each of these will show templates for a specific application in NX and will launch that application when you create a new file from that template.

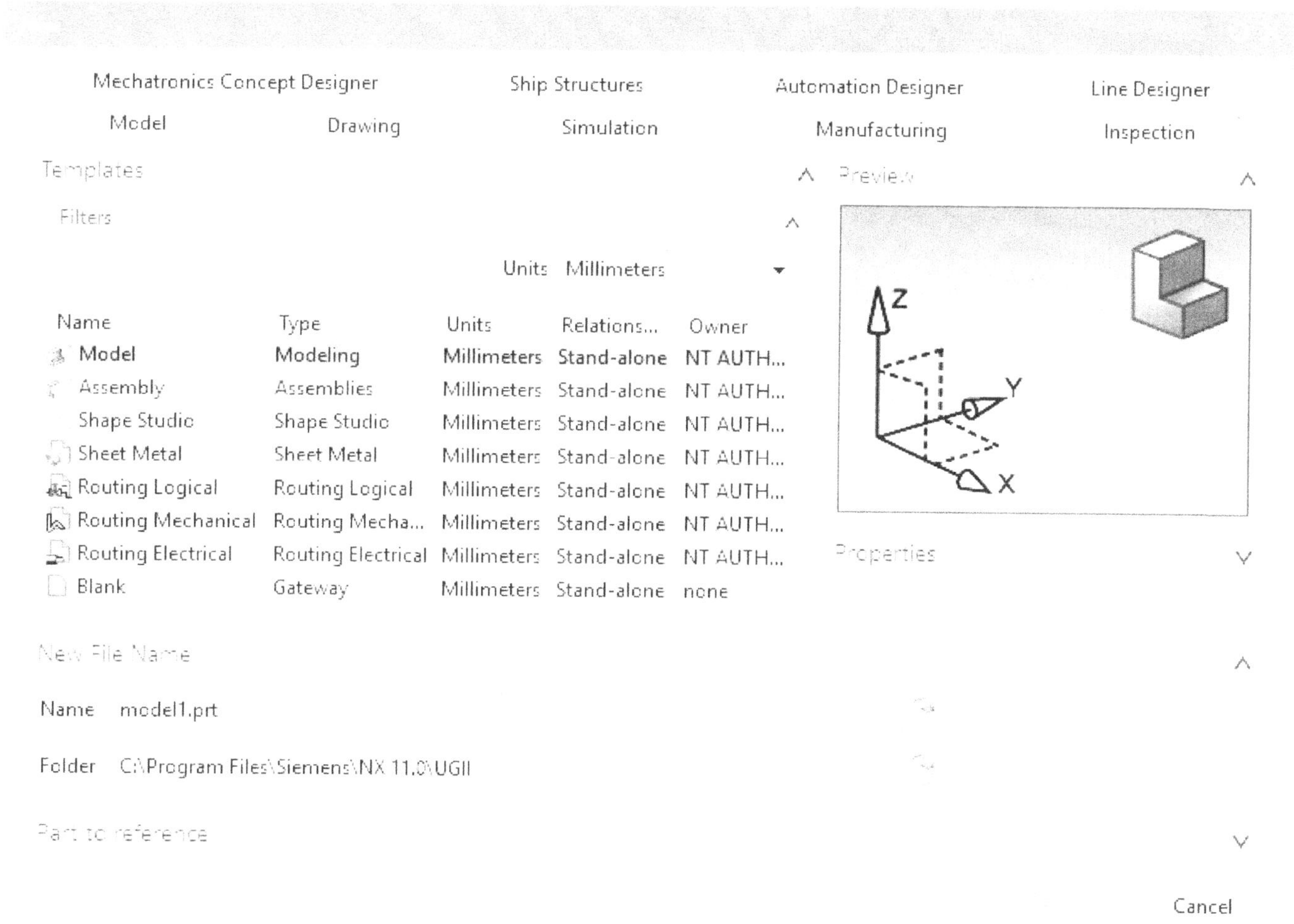

Notes:

1. *The default directory,* C:\Program Files\Siemens\NX 11.0\UGII, *is by default a read-only directory. If you create files to be stored in this directory, upon saving, NX will later prompt you to relocate them.*

2. ***Units** cannot be changed from within NX, so you really should try to get the desired **Units** right at the time of file creation! There is a command-lin utility called* "UG_CONVERT_PART" *in the folder* "C:\Program Files\Siemens\NX 11.0\NXBIN" *that you can run to convert a part from* Millimeters *to* Inches *(or vice versa).*

- When your new file opens, you will find yourself in the **Modeling** application. Note that the Resource Bar on the left has changed, and that in the present setup, the Part Navigator is open.

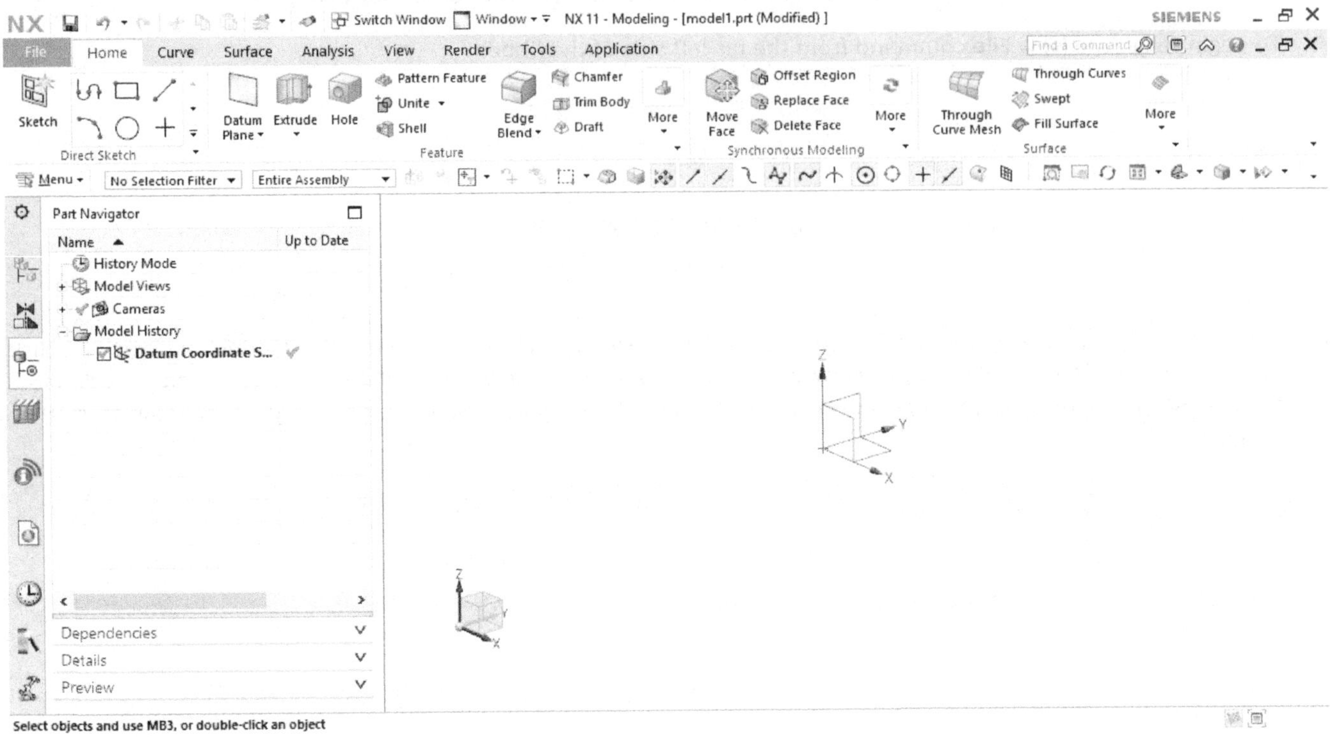

2.3.2 Opening Files

Opening files is straightforward. NX will not allow you to have multiple files of the same name open, even if they are wildly different parts, so be aware that when opening an assembly, any parts open in your session must not have the same name as a component in that assembly, or else the component will fail to load!

- Select **File / Open** to open an existing part file.
- Navigate to *"C:\My NX 11 Files/Downloaded Files"* and select *"Opening Files.prt"*. Click **OK**.
- This file is opened in its own window within NX. You can push the **Restore** button next to the **Close** button along the top of the ribbon to see the window within NX.

- The window can be resized as you please. There isn't much advantage to working with this tiny window, since you can't have multiple windows visible at once!

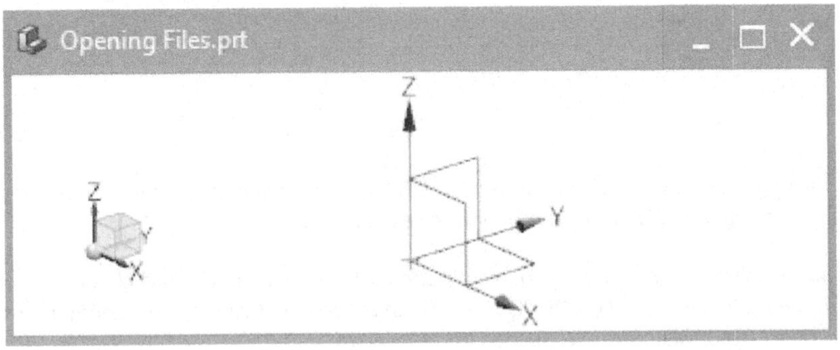

2.3.3 Switching Windows

Once you have multiple parts open, you might like to switch between them!

- The **Window** drop-down along the Quick Access Toolbar can be useful – although, it only displays the 10 most recently accessed files.

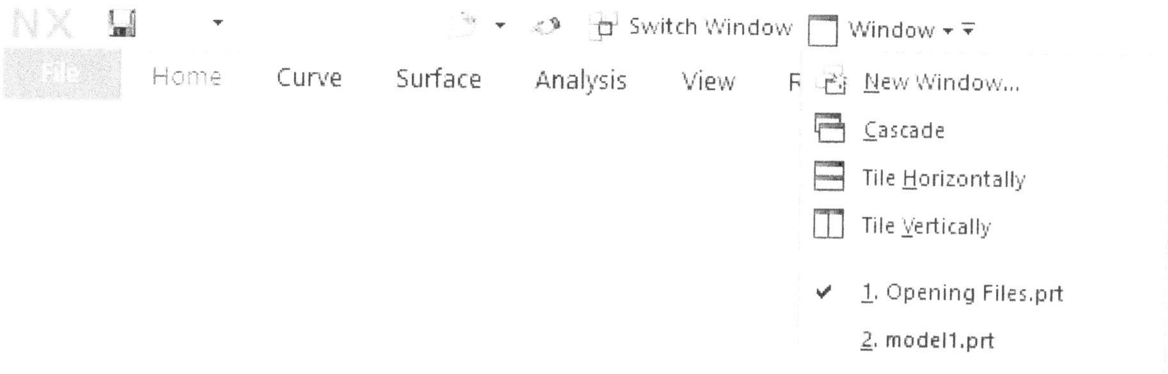

- The **Switch Window** tool, new in NX 11, is a better way to cycle through the windows you have open. Holding **[Ctrl]** and pressing **[Tab]** cycles through the choices.

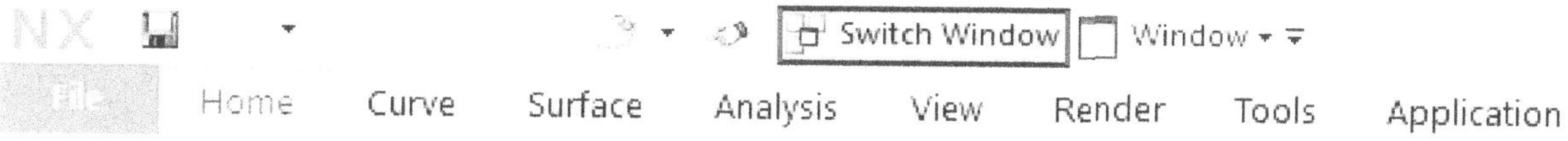

- Use **Switch Window** to cycle between the two parts you have open, ultimately landing on *model1.prt*!

2.3.4 The Work Part vs. The Displayed Part

When you work with a *".prt"* file that is acting as an assembly, there is a distinction between the part that you have open in the present window (the *Displayed Part*) and the part that you are working on and actively editing (the *Work Part*). In short, the Work Part is the *".prt"* file in your assembly in which your operations have their effect. The Part Navigator shows the Model History for the Work Part. The Assembly Navigator shows the structure of the Displayed Part, as an assembly. Since there is no separate file extension for assemblies, it is not always clear whether a part you open will be a part or an assembly. In the exercise below, you will learn how to distinguish parts from assemblies, and to set the Work Part within the Displayed Part.

- Open the file *"The Displayed Part.prt"* from *"C:\My NX 11 Files\Downloaded Files"*.
- In the Graphics Window, you will see two bodies, but the Part Navigator shows only *Datum Coordinate System (0)* in the Model History.

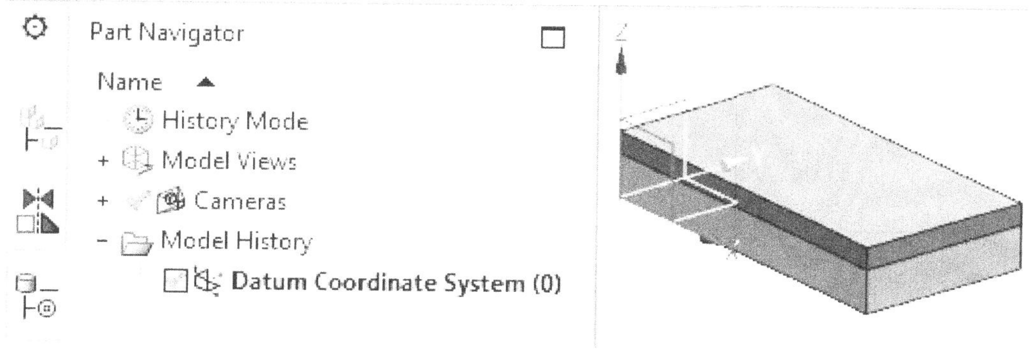

- The two "bodies" that you see are actually components, and the *".prt"* file that you opened is an assembly! Switching to the Assembly Navigator should clear this up.

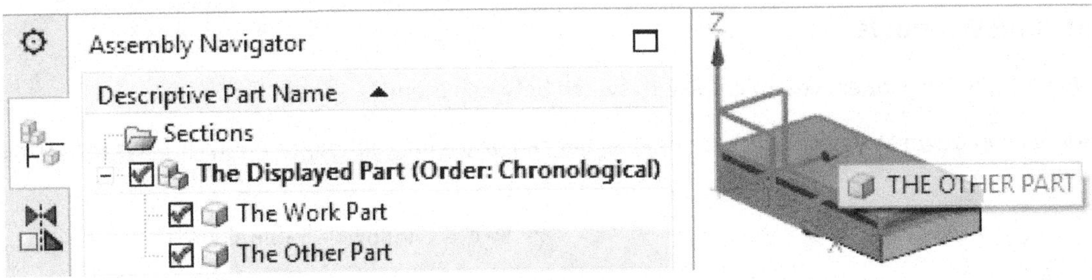

- The datum coordinate system you saw in the Model History is the only feature within *"The Displayed Part.prt"* – regarded as a part. The Part Navigator gives information about the Work Part, which is currently *"The Displayed Part.prt"*.
- Set *"The Work Part.prt"* as the Work Part – to do so, simply double-click on *The Work Part* in the Assembly Navigator. Note that upon making *"The Work Part"* the Work Part, *"The Other Part"* becomes de-emphasized in the Graphics Window.

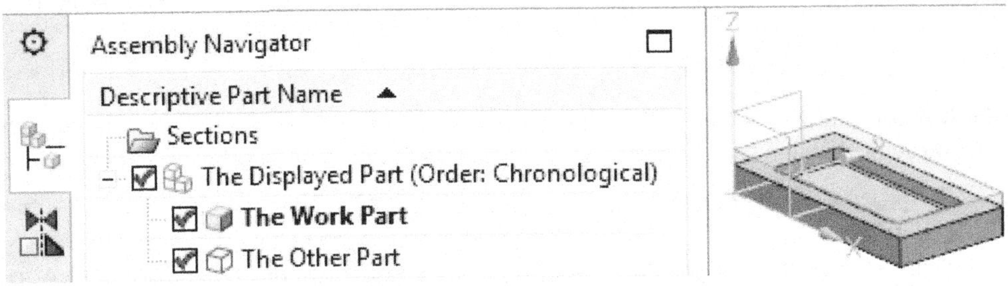

- Also, note that the application window now shows the name of both the Work Part and the Displayed Part.

NX 11 - Modeling - [The Work Part.prt in Assembly The Displayed Part.prt]

- Switch back to the Part Navigator. Now, the Model History for *"The Work Part.prt"* is displayed. You can now select features from *"The Work Part.prt"* and edit them. This is a typical example of a parametric model – every feature is the result of an operation in the Modeling application. You will learn to edit features in the Model History of a parametric model in Chapter 4.

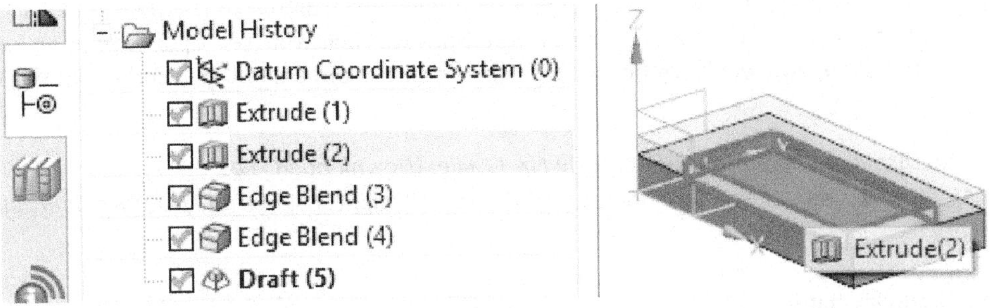

- Switch back to the Assembly Navigator and double-click on *The Other Part* to make *"The Other Part.prt"* the Work Part.

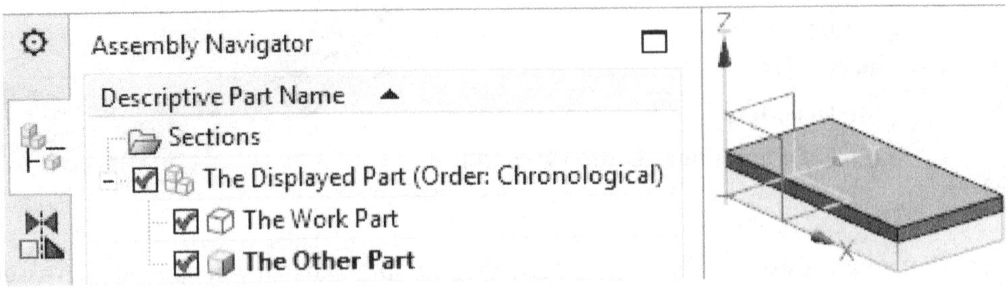

- Switch to the Part Navigator. Observe that the Model History for *The Other Part* is rather uninformative – there is a datum coordinate system, and then a feature called *"Body (1)"*. This model has no memory of how it was made, and no parameters driving its geometry. This is a typical example of a non-parametric model, sometimes called a "dumb" model. Since the body has no features recorded in the Model History, it cannot be modified by the same methods as a parametric model. In Chapter 16, you will learn about the **Synchronous Modeling** commands – an amazing set of tools for making direct edits to models, regardless of whether they are parametric or non-parametric.

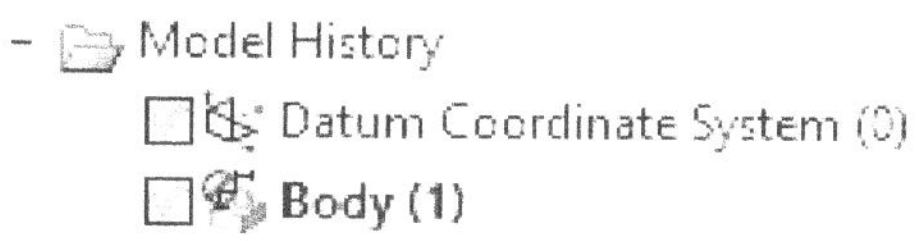

- Switch to the Assembly Navigator. Right-click on *The Work Part* and select **Make Displayed Part**.

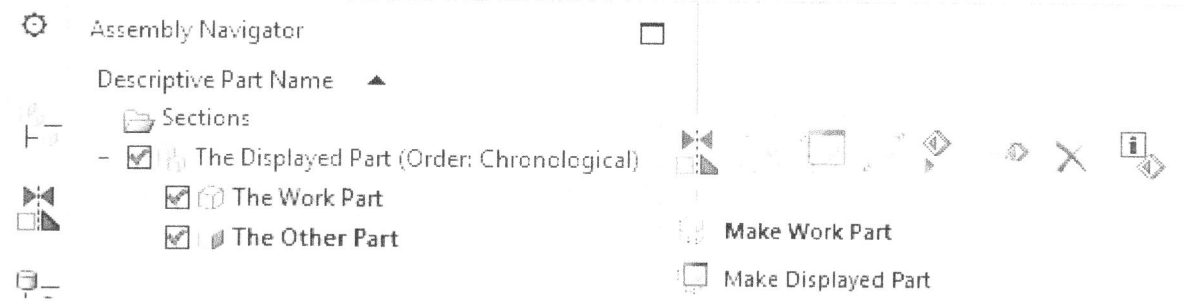

- This opens *"The Work Part.prt"* in its own window; in this window, *"The Work Part.prt"* is the Displayed Part! You can return to the window displaying *"The Displayed Part.prt"* by way of the **Switch Window** command, but in this special scenario, there is a more efficient way to get there. Since *"The Displayed Part.prt"* is a parent assembly for *"The Work Part.prt"*, you can navigate back to *"The Displayed Part.prt"* by way of the Assembly Navigator! Right-click on *The Work Part* in the Assembly Navigator, and choose **Display Parent** / *The Displayed Part.*

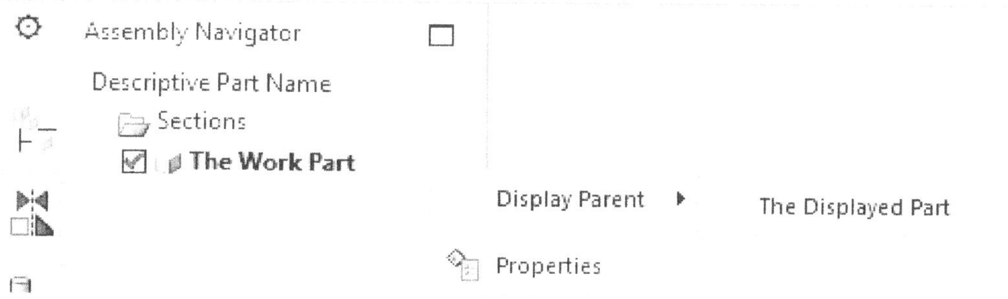

- The technique you just practiced is only possible when the parent assembly is actually open. Other assemblies that might have *"The Work Part.prt"* as a component will not appear in the list here.

2.3.5 Saving Files

All your hard work won't be worth anything if you can't save your files!

- Change the Displayed Part to *"model1.prt"*.
- Select **File / Save / Save**, as shown below. Note that the **Save** command will save the current Work Part and any modified components within.

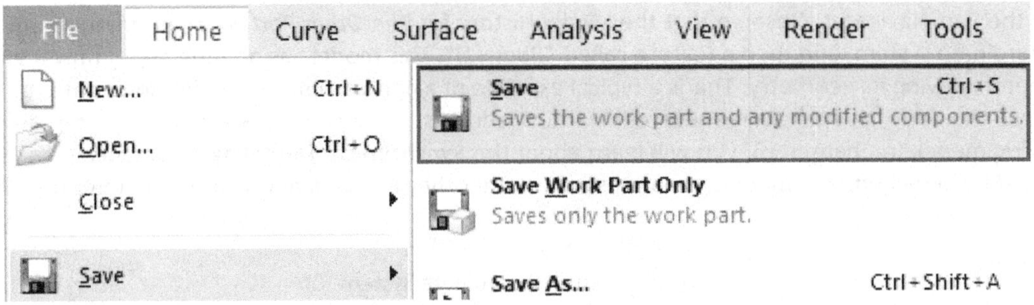

- NX will prompt you to give your file a name – although it already has the name *"model1.prt"*, nothing has been recorded yet. The reason the **Name Parts** dialog appears is that you left the default name of *"model1.prt"*. The directory *"C:\Program Files\Siemens\NX 11.0\UGII"* is read-only by default, so you won't be able to save there. Find the folder *"C:\My NX 11 Files"* that you created in Chapter 1.3 and save your model there!

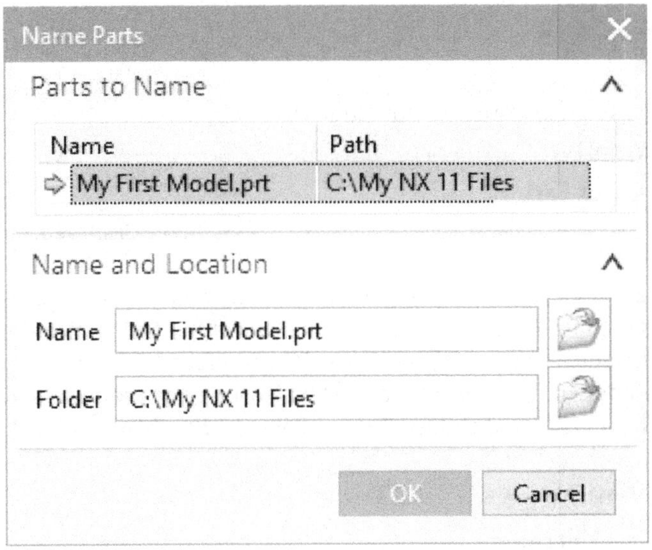

2.3.6 Closing Files

The commands found under **File/ Close** can be used to close individual files or all the files that are currently loaded in a session. Most are self-explanatory. You can also close files individually with the **Close** button on the far-right side of the ribbon.

- At this point, your **Window** drop-down should display four parts.

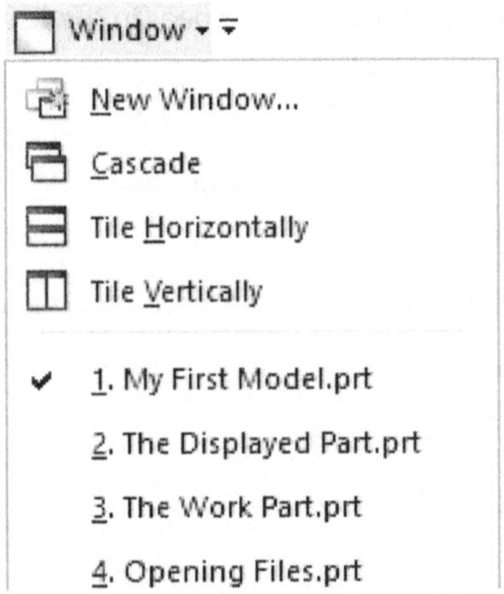

- Select **File / Close / Selected Parts**.

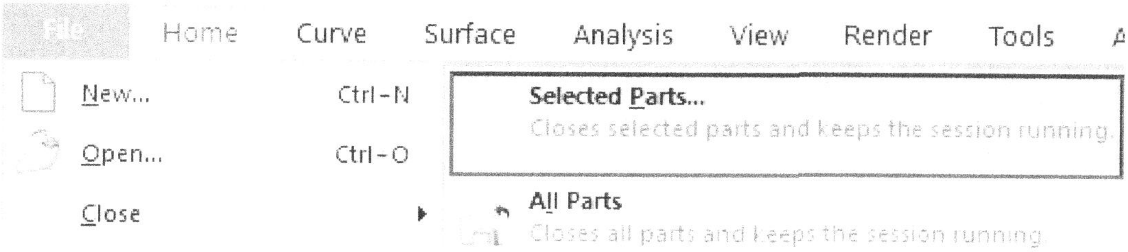

- The **Close Part** dialog that appears will let you close *any* parts open in the session, even if they are not open in their own windows. For instance, *"The Other Part.prt"* was opened when it became the Work Part, but it is not displayed in its own window.
- Choose *"My First Model.prt"* and hold **[Ctrl]** and select *"The Other Part.prt"*. Click **OK**.

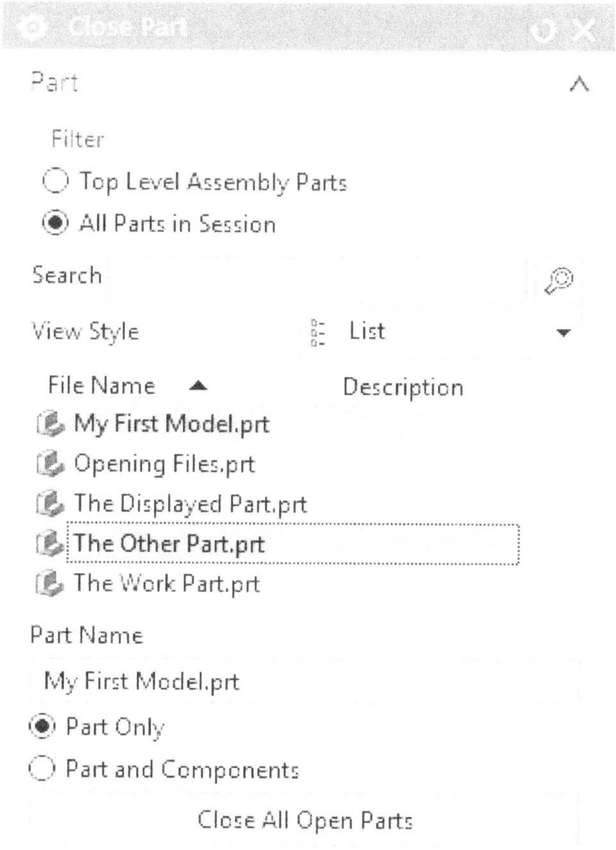

- Use the keyboard shortcut **[Ctrl]+[Tab]** to switch the window to *"The Displayed Part.prt"*. Open the Assembly Navigator. The empty checkbox next to *The Other Part* indicates that it is closed. You can also close components from the Assembly Navigator.

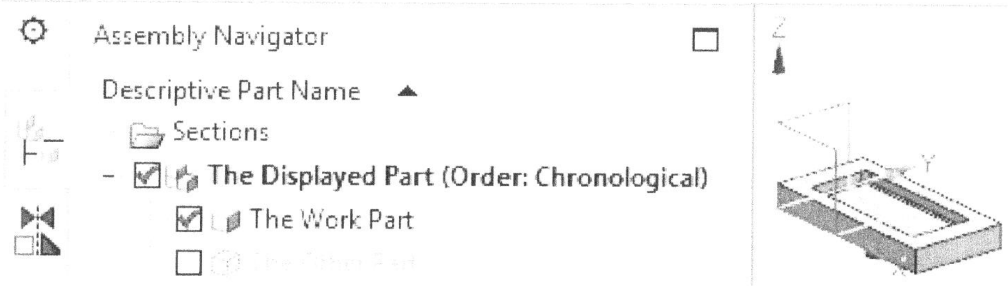

- When you close an assembly, NX will ask if you want to close the components as well. This is relevant when you have certain components open in their own windows. In our example, *"The Work Part.prt"* is open in its own window. **Close**

"The Displayed Part.prt" by clicking on the **Close** button in the top right corner of the window (not the "X" in the very top right, that closes the whole application!)

- When the **Close Part** dialog appears, click **Part Only**.

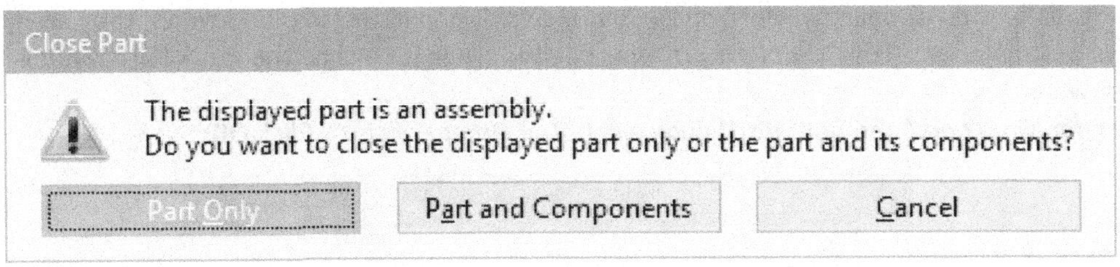

- *"The Work Part.prt"* is still accessible through the **Window** drop-down.

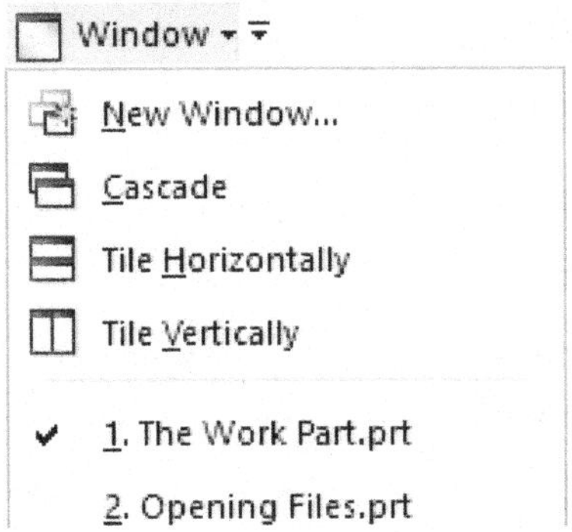

- **Close** *"The Work Part.prt"* by the method of your choosing.

2.3.7 Assembly Load Options

Since assembly components are stored in separate files, file management for assemblies becomes a little more complex. You must consider how each component is saved, as well the assembly itself. If you save the top-level assembly, NX will save all of its modified components – which may lead to versioning problems.

As you begin working with larger assemblies, you will find that fully opening all components and leaving all geometry within those components visible will impact the performance of your computer, sometimes causing NX to run very slowly. There are a variety of tools available in NX to manage component visibility, of which **Assembly Load Options** are the most basic.

When you first open an assembly, the default setting (in an out-of-the-box NX installation) is for the components to be *partially loaded*. This improves system performance when loading large assemblies (in comparison with, say, fully loading all components). If you are in the **Gateway** application, with no part files open, **Assembly Load Options** appears in the **Standard** group on the **Home** tab.

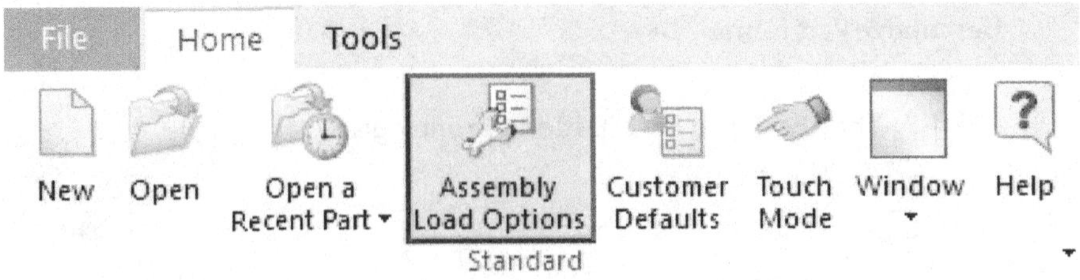

If you have any files open, you can always access **Assembly Load Options** at **File / Assembly Load Options**.

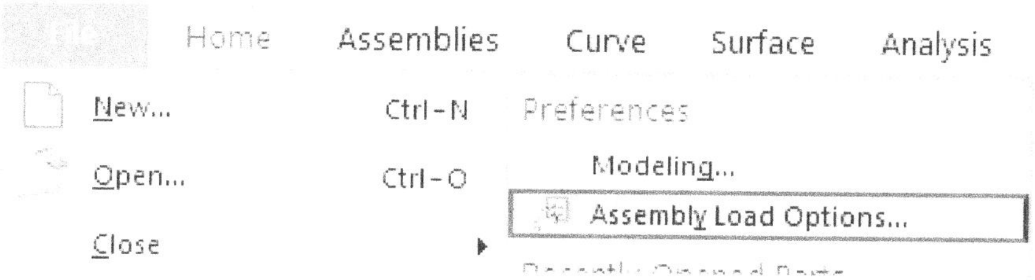

- Open your **Assembly Load Options**. Note that under **Scope**, the default setting has **Use Partial Loading** checked.

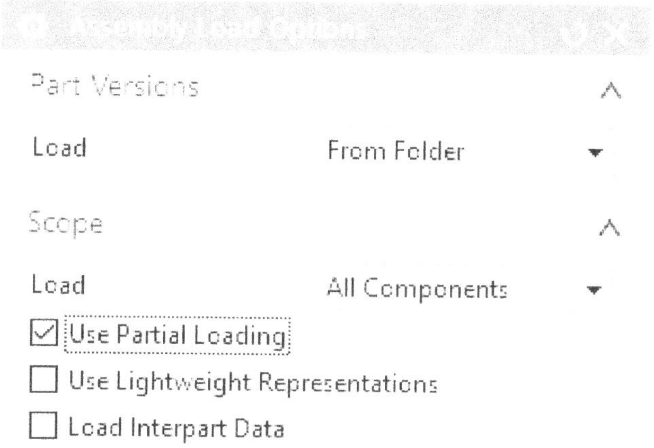

***Note:** The default **Load** option is* "From Folder". *NX will not search in embedded folders when this option is chosen; therefore, the directory with the assembly file must also contain all of the components.*

- Change the **Load** drop-down to *Structure Only*.

- Click **OK** to save your **Assembly Load Options**.
- **Open** *"The Displayed Part.prt"*.
- You will notice that in the Assembly Navigator, the checkbox next to each component is unmarked. The **Load** option *Structure Only* means that only the top-level file of the assembly is loaded and all of the components remain closed when it loads.

- The checkmarks indicate whether a component has been at least partially loaded or not. Check the box next to *The Work Part*. This partially loads *"The Work Part.prt"*.
- Right-click on *The Work Part* in the Assembly Navigator. Note that you can still open *"The Work Part.prt"* fully.

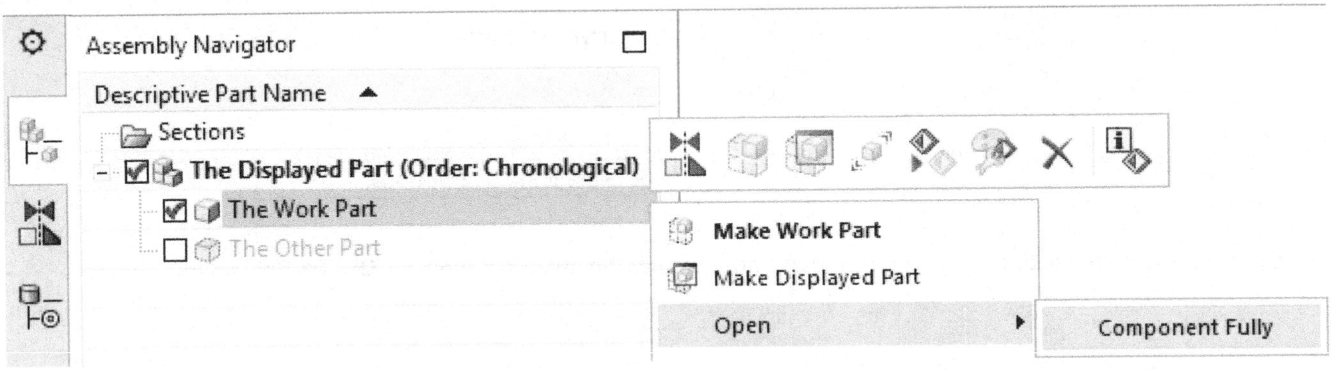

- The distinction between partially and fully loading a component is subtle. For now, it suffices to say that fully loading all components will impact system performance when opening a large assembly.
- Select **Assembly Load Options** again. Reset the **Assembly Load Options** tool to restore the default settings, and click **OK** to save.

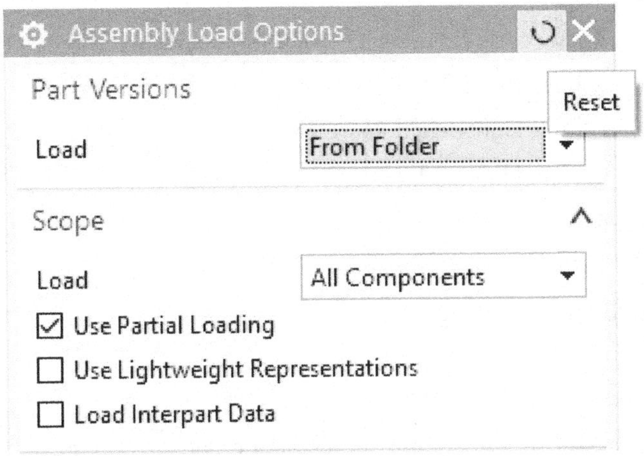

If component part files are moved or stored in different directories, the **From Search Folders** option provides the most flexibility. The option **Add Folder to Search** bring up an expanded menu that allows you to create one or more search directories. Any added search directories are only effective in the current session of NX.

Special Tip: *To search all subdirectories of a specified directory, use an ellipsis (three periods) "..." at the end of the chosen directory pathname. Otherwise, NX only searches in the specified directory.*

2.3.8 Save As

The behavior of the **Save As** command merits a special discussion for components in an assembly. When you use **Save As** to make a copy of a part, each assembly in your NX session that contains that part as a component will replace the old copy with the new one. This is often not the desired outcome, so you must take care to close any assemblies that contain your part before using **Save As** if you don't want all instances of that part to be replaced. For the special scenario in which you wish to replace certain instances of a part (but not all), there is a special **Save As** command called **Make Unique**, which we will study in Chapter 18.3.

- Open *"The Displayed Part.prt"* from *"C:\My NX 11 Files\Downloaded Files"*.
- Make *"The Work Part.prt"* the Displayed Part.
- Select **File / Save / Save As** and name a copy of your part *"The Other Work Part.prt"*. A second **Save As** dialog will appear – close this one without entering a name. The second window was asking you to name a copy of *"The Displayed Part"*. In general, when you perform a **Save As** on a bottom-level component, the **Save As** will work its way up the assembly tree for any assembly that has it open, asking you to rename each parent assembly. You should now see an information window and a dialog called **Save As** asking if you want the **Save As** to continue. Click **Yes**.

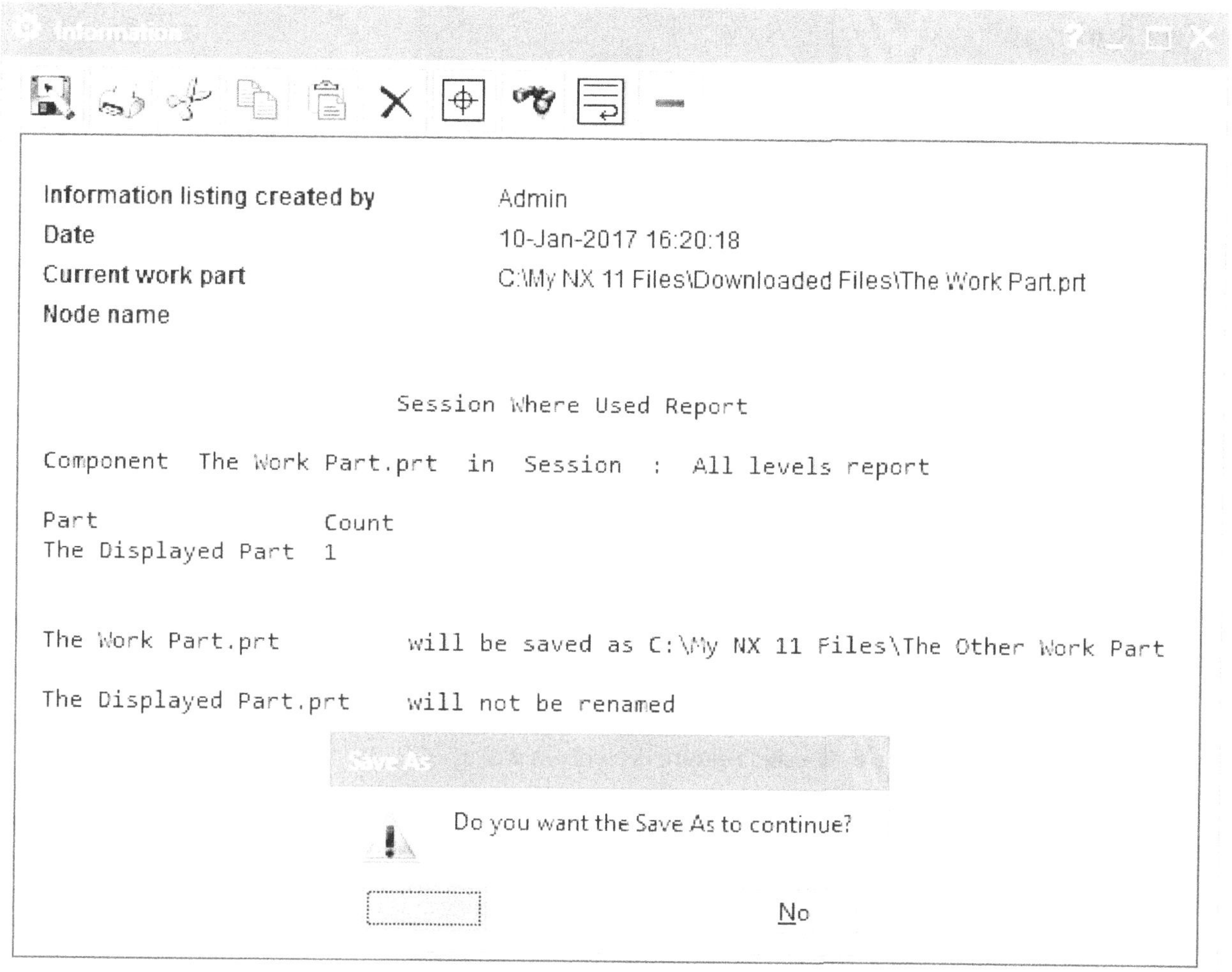

- Close the information window. A new dialog will appear letting you know that *"The Displayed Part.prt"* now references *"The Other Work Part.prt"*, not *"The Work Part.prt"*.

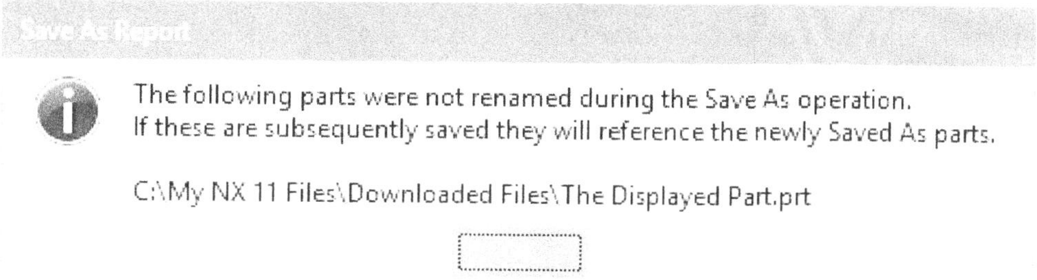

- Make *"The Displayed Part.prt"* the Displayed Part, and you will see that now, *"The Other Work Part"* appears in place of *"The Work Part"* in the assembly tree.

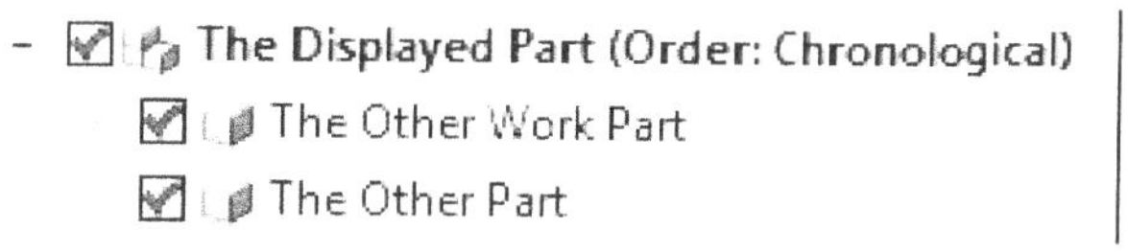

- Close *"The Displayed Part.prt"* and *"The Other Work Part.prt"* so that only *"Opening Files.prt"* remains open. Don't save the changes made to "The Displayed Part.prt".
- Change the Displayed Part to *"Opening Files.prt"*.

2.4 The Ribbon

The user interface for NX is icon-based, with icons organized on a ribbon in the same style as other modern CAD software and *Windows* applications. The ribbon is very customizable so you can easily display the commands that you use the most. All of the functionality is organized into *tabs* on the *ribbon*. Within each tab, there are *groups*, within each group there are *commands*, and within certain groups there are *galleries* and *drop-down menus* with further groups and commands. Each group and tab also has a **Ribbon Options** drop-down menu indicated by a down arrow ▾. Simply click on icons to open the corresponding tools, galleries, or drop-downs.

- This is the **Home** tab.

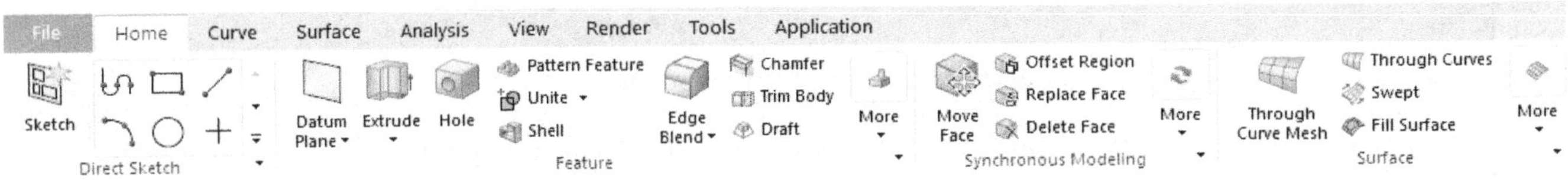

- On the **Home** tab, this is the **Feature** group.

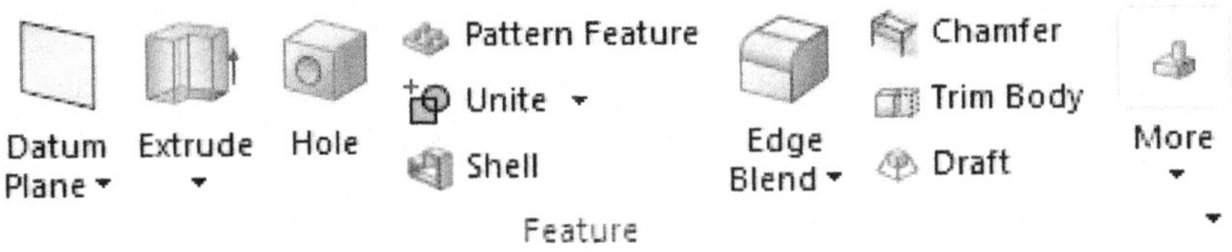

- Within the **Feature** group, this is the **Design Feature** drop-down menu.

- Within the **Design Feature** drop-down menu, this is the **Revolve** command.

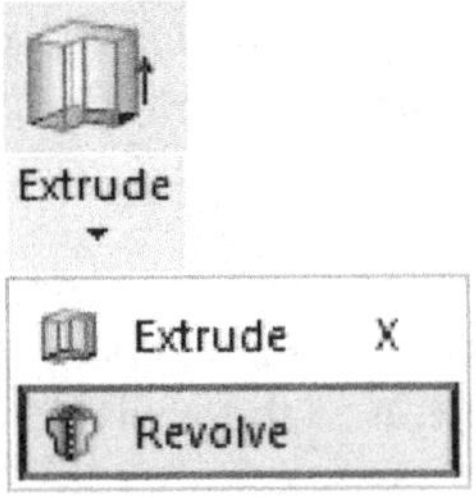

- Galleries come in two forms. For example, this is the **More** gallery on the **Feature** group. Galleries like this are in collapsed form, and behave like drop-down menus.

- This is the **Sketch Curve** gallery on the **Direct Sketch** group. You can expand the **Sketch Curve** gallery by clicking the bottom right button. Galleries in this form are uncollapsed, and allow you to see some of the tools inside.

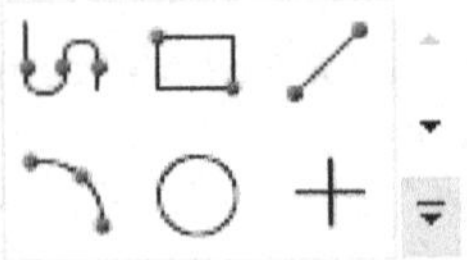

- This is what the **Sketch Curve** gallery looks like when fully expanded.

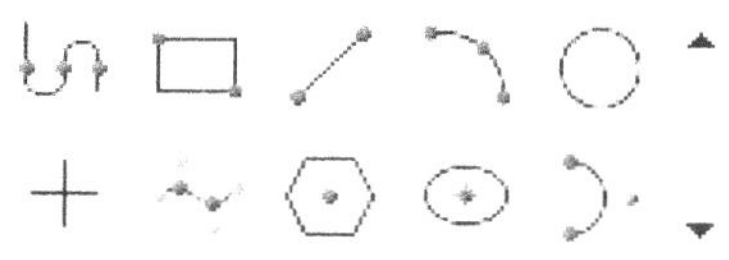

- The **Ribbon Options** drop-down menu (▾) for the **Feature** group determines which galleries, drop-down menus, and commands appear in that group.

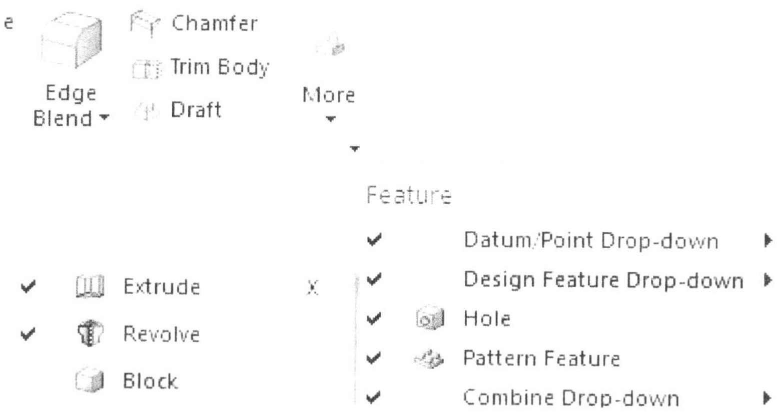

- Tabs can be shown or hidden by right-clicking at the top of the ribbon, and checking or unchecking the corresponding tab name.

2.5 The Top Border Bar

The *Top Border Bar* is located just beneath the ribbon, and by default it comes with three groups: the **Menu**, **Selection Group**, and the **View Group**.

Nearly any command that can be accessed on the ribbon can also be accessed via **Menu**.

For example, the **Extrude** command is found via the menu at **Menu / Insert / Design Feature / Extrude**.

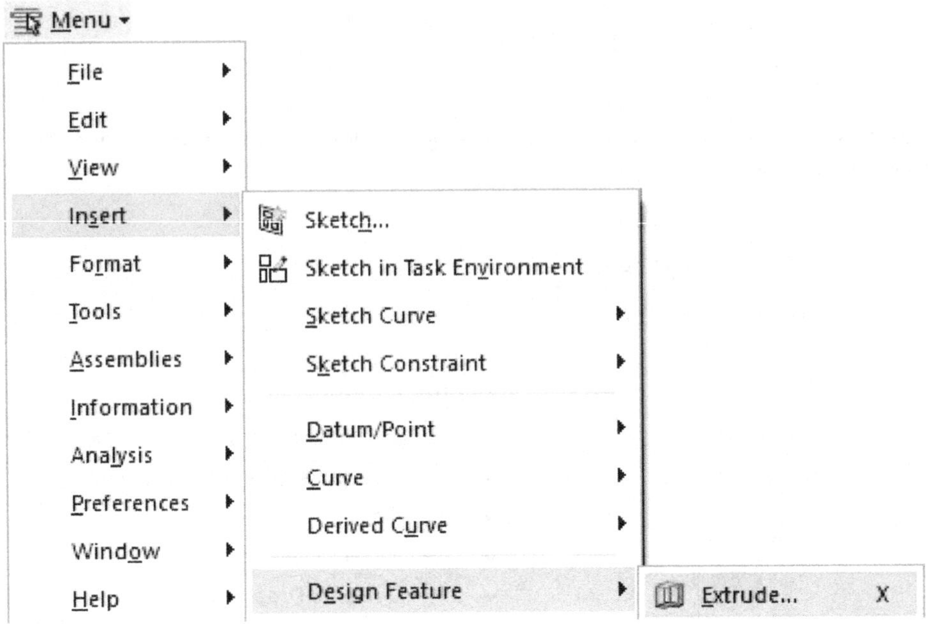

The **Selection Group** is extremely important, as it sets the rules for how your cursor interacts with entities in the graphics window. You will learn about the **Selection Group** in detail in Chapter 3.3.

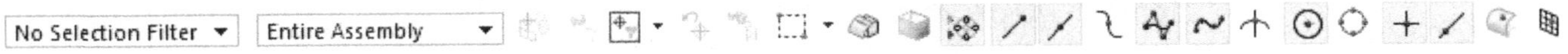

The **View Group** contains commands that control the presentation of geometry within the graphics window – you will learn about these tools throughout Chapter 3.

2.6 Command Finder

The **Command Finder** is a godsend: use it to find commands that you cannot locate in the menus or toolbars. Unfortunately, there is no finder for the **Command Finder** so you will have to find it for it yourself!

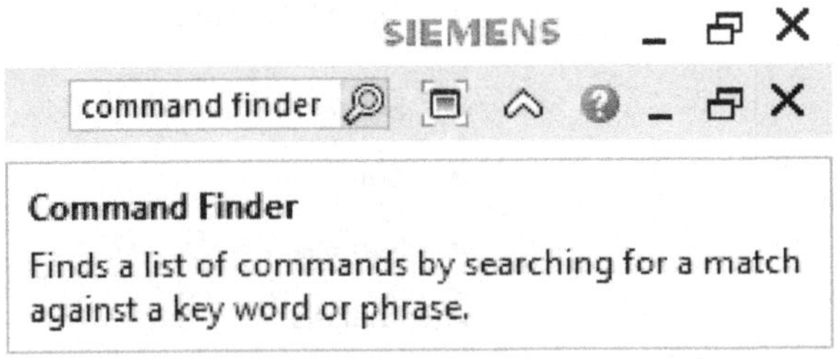

*Hint: **Command Finder** is in the top right corner of the program!*

You can even type in commands that exist in other CAD programs and NX will find them for you. For example, type "loft" into the **Command Finder** (a command from some other program), and **Command Finder** will locate the NX equivalent, **Through Curves**.

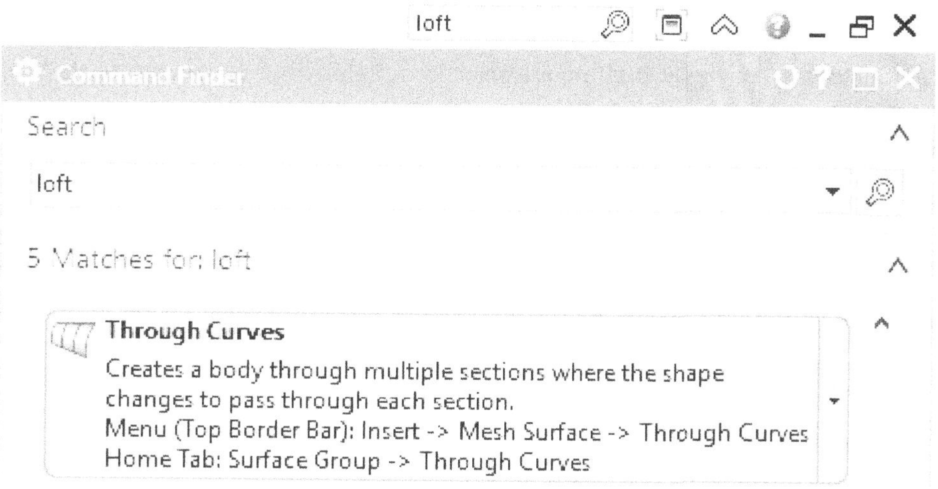

Again - this is a fantastic tool for quickly finding features that are hidden in NX's vast library of capabilities! So no more excuses!

After the search results come back, hover the cursor over the correct result and it will show you exactly where it is located. To use the tool instantly, you can just click on the search result as it appears in the **Command Finder**.

In the early chapters of this book, we will tell you explicitly where commands are on the ribbon, but as the material progresses and becomes more advanced, we will assume more of you, the reader. If we ask you to use a command that we haven't mentioned yet – use the **Command Finder**!

2.7 Dialog Boxes

There are commonalities between most dialog boxes in NX, so in order to prepare you for the menus you will encounter later, in this section we review the anatomy of the **Extrude** dialog box (recall that the **Extrude** command is located on the **Home** tab).

2.7.1 Red Asterisks

Red asterisks indicate inputs on the menu that require selections. Before you can complete the operation, you must provide valid inputs for each red asterisk. Click on the asterisk itself, or the words adjacent (*"Select Curve (0)"* in this case) to add to each input in a dialog box. Usually it is <u>not</u> a good idea to click on any icons to the right of this text, as they often perform special actions.

Once you have made a valid selection the red asterisks will become green check marks. For example, when you make a valid choice of **Section** for the **Extrude** command, you will see something like the image below.

2.7.2 Numerical & Dynamic Inputs

Often there are text entry boxes for required numerical inputs for operations. For instance, on the **Extrude** dialog box, there are two by default – the **Start** and **End Limits**.

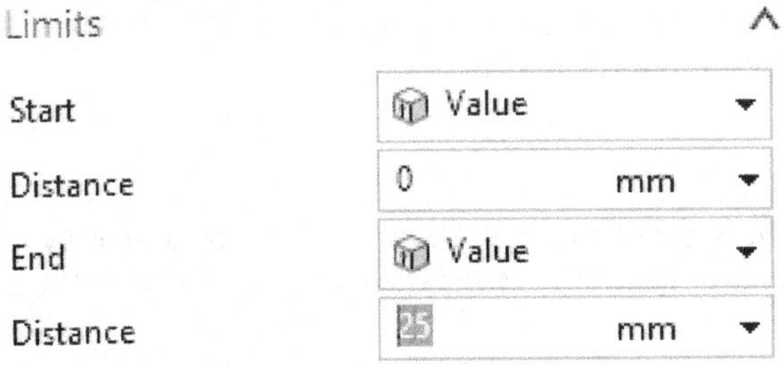

You can modify these by writing over the text in the box, and when you do, very often you will also encounter an arrow and a floating dynamic input box as shown below.

You can either drag the end of the arrow or change the number in the floating text input box. Any time you change a number by typing over it, if you want the dynamic preview of your model to update, you must press **[Enter]** or **[Tab]**.

2.7.3 Cue Display

The *Cue Display* contains important information about what NX is expecting from you. If you are performing an operation and are uncertain what to do next, *read the cue*! The Cue Display is located in the bottom left corner of the graphics window. Its content depends on what part of the menu you are actively working on.

Select planar face to sketch or select section geometry

Select objects to infer vector

2.7.4 Status Display

The *Status Display* often gives a summary of the last operation performed. This display is sometimes a good supplement to the cue, and it can be particularly helpful when picking geometry in the graphics window. If you are ever uncertain about what geometry you just picked, look at the status. The Status Display is only visible when a selection is made or an operation is underway or completed, and it can be found in the bottom center of the graphics window. For instance, when you perform an **Extrude**, the Status Display tells you what kind of body was created, as shown below.

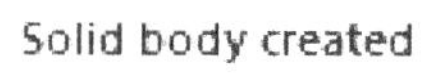

2.7.5 OK, Apply, Cancel, Close

These should be familiar from other Windows-based software. **OK** completes the operation and closes the dialog box. **Apply** completes the operation but leaves open the dialog box – this is helpful when you want to use the command again right away. **Cancel** closes the dialog box without completing the operation, as does the **Close** button in the top right corner.

Note that **OK** and **Apply** are grayed out if there are any red asterisks remaining.

2.7.6 Dialog Options

The **Dialog Options** button is a gear-shaped button found in the top left corner of any dialog box, and allows you to customize the appearance of the dialog box. An important tool within **Dialog Options** is **Help**, which brings up the online documentation for the dialog box!

Another very important option in the **Dialog Options** drop-down is the choice between seeing *Less* and *More*. For example, *Extrude (Less)* shows only **Section**, **Direction**, **Limits**, and **Boolean**.

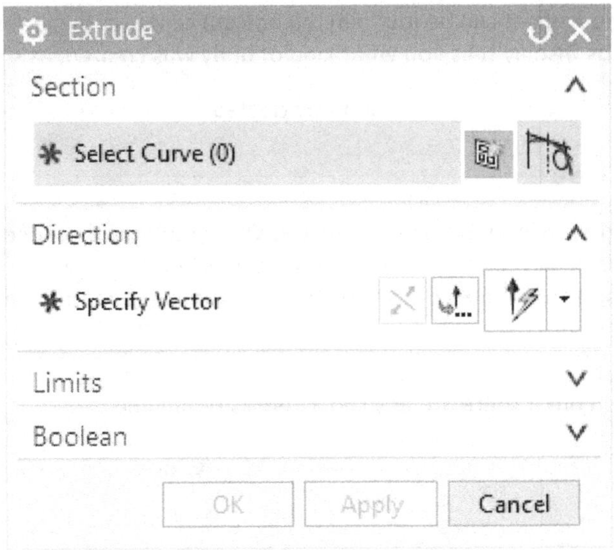

We recommend that you always make sure that your dialogs are showing *More*.

2.7.7 Expanding & Minimizing Sections

Sections of each dialog box can be opened and closed simply by clicking on the header.

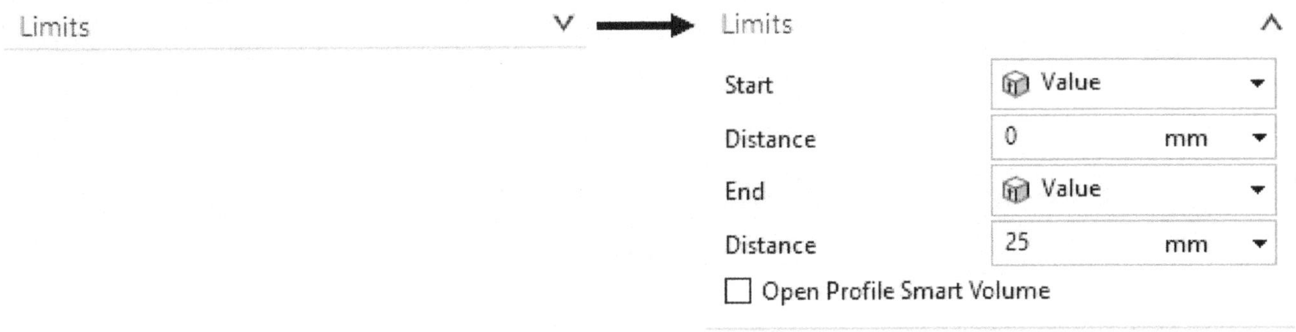

2.7.8 The Reset Button

Dialog boxes in NX have memory, in the sense that after using a tool, the next time you open that tool, the same values that you last entered will be present in each of the inputs. The **Reset** button restores each tool to its default values. It can be especially handy for more complex tools whose dialog boxes have many required inputs and parameters.

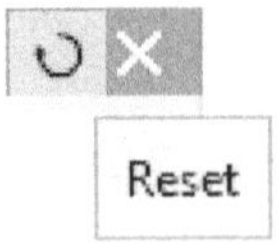

2.7.9 Sketch Section

The **Sketch Section** icon appears on many dialog boxes where section curves are required as input.

The **Sketch Section** icon allows you to build a sketch on-the-fly from within the dialog box of the tool you are using. You can click on the button directly to bring up the **Create Sketch** dialog. You can also make implicit use of the button by selecting a plane, face, or (in certain scenarios) edge, at which point you will then enter the *Sketch Task Environment* on that entity. It is important to note that any sketches created by way of the **Sketch Section** icon will be internal to the feature in question. See Chapter 6.10 for further details.

2.7.10 Reverse Direction

Tools that require a direction vector as input always have a **Reverse Direction** icon. It, well, reverses the direction of the vector once it is created.

After your vector is selected, you will often see a golden arrow in the graphics window.

If you put your cursor over this arrow, it will let you know that you can double-click on it to reverse the direction.

In order to see this vector, the *Specify Vector* input on the dialog must be highlighted in yellow, as shown below.

2.7.11 Construction Geometry on-the-fly

Many dialog boxes have additional icons that allow you to build construction geometry on-the-fly.

The **Extrude** tool has, for example, a **Vector** constructor icon that brings up the **Vector** dialog box, with various options for creating a direction vector for the extrusion.

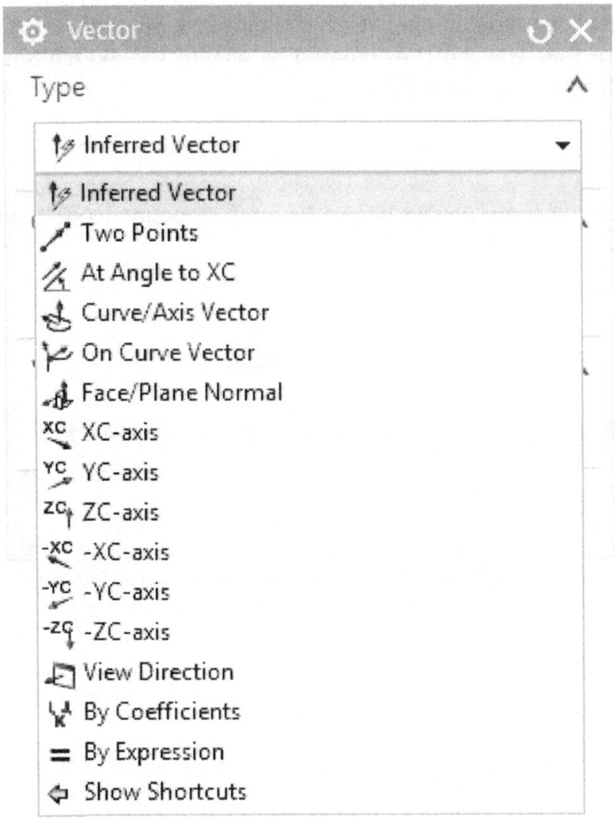

Oftentimes, these constructor icons also have an adjacent drop-down menu that indicates many of the same choices (but usually not all of the choices) from the corresponding dialog box. After gaining familiarity with the icons, you can use this drop-down menu as a shortcut.

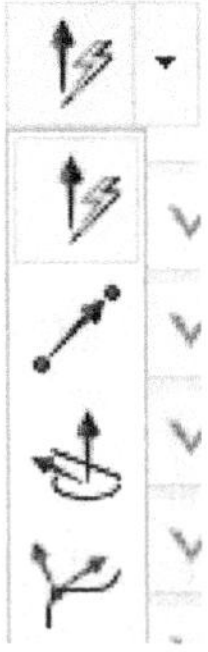

2.7.12 The Middle Mouse Button

The middle button on a three-button mouse has a special role in NX. When a dialog box is open, clicking the middle mouse button will accept the current selection and move on to the next required input. If all required inputs (marked by red asterisks) have been satisfied, the middle mouse button will click either **OK** or **Apply** from the bottom of the dialog, depending on which is highlighted.

- As an example, select the **Sweep Along Guide** tool. Use the **Command Finder** to locate it.
- Note that there are two required inputs for **Sweep Along Guide** – a **Section**, and a **Guide**.

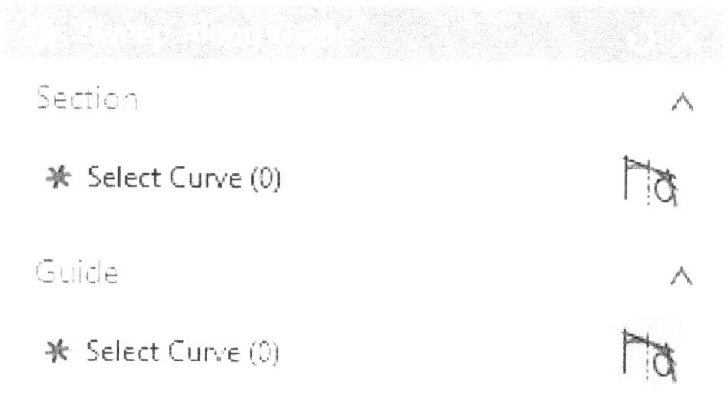

- Without making a selection, click the middle mouse button. This advances the "current selection" to the next required input – the **Guide**.

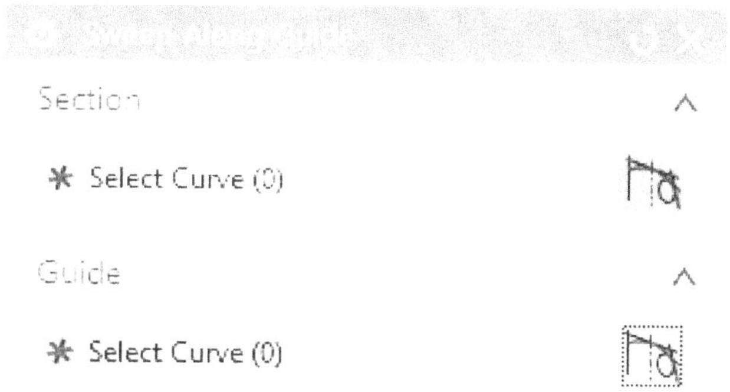

- Clicking the middle mouse button again cycles back to the **Section** – since no selection was made, it is still required!

When a button is highlighted on a dialog, the middle mouse button has a tendency to push that button.

- Select the **Extrude** tool. Click on *"Select Vector"*.

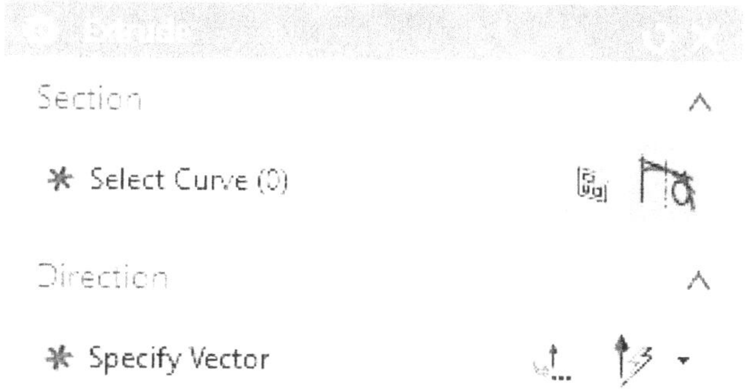

- When you click the middle mouse button, the menu cycles back to the **Section**. Notice that the **Sketch Section** icon is highlighted.

- Now when you click the middle mouse button, it has the effect of pushing the **Sketch Section** icon. This will bring up the **Create Sketch** dialog!

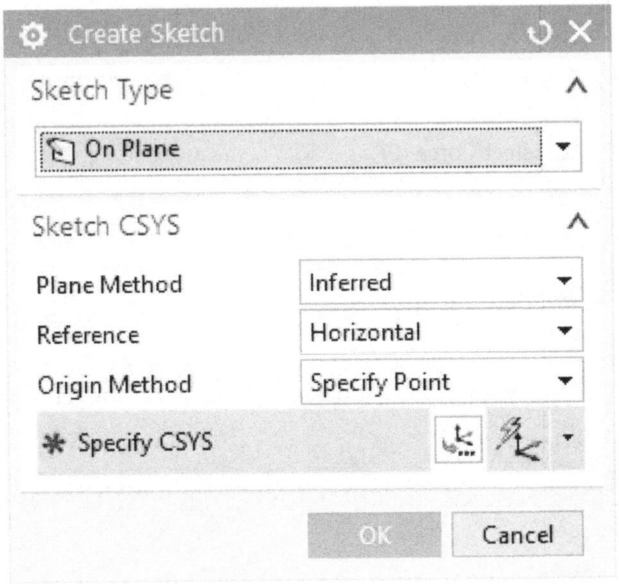

- **Close** the **Create Sketch** dialog, and then the **Extrude** dialog.

2.7.13 Lists

Some tools allow you to select many distinct sets of entities of the same type. These different sets are recorded as items on a list. Adding selections to list items requires its own special attention, as it is very easy to accidentally add multiple sets of entities to a single list item!

Consider the **Edge Blend** tool. The **Edge Blend** tool adds a fillet along an edge where two faces on a body meet. In fact, the tool will allow you to apply multiple fillets to different sets of edges at the same time, each with its own radius. In the **Edge Blend** dialog, these sets of edges are recorded as **List** items.

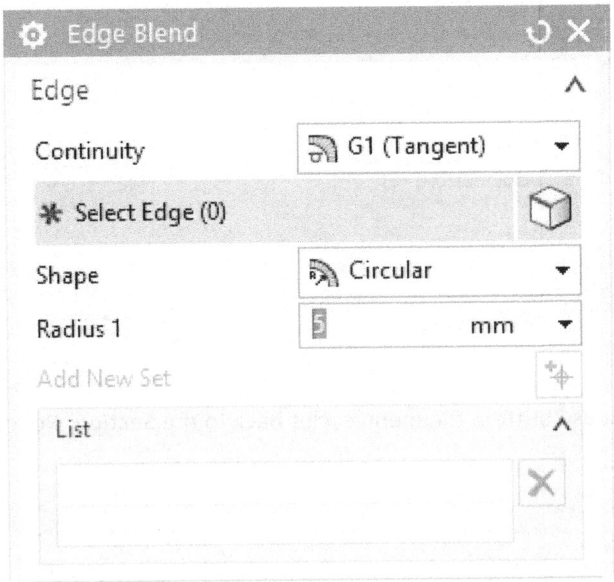

The key step that is required to create a new list item is to push the **Add New Set** button. Upon making a selection for one list item, the **Add New Set** icon will become highlighted, and you can actually just click the middle mouse button to accept and create a new list item!

2.8 Pop-Up Toolbars

There are a number of useful toolbars that appear when you left-click or right-click, either in empty space in the graphics window, or on existing entities in the Graphics Window, Part Navigator, and Assembly Navigator.

The **View Pop-Up Toolbar** appears when you left-click somewhere in the Graphics Window. It shows a variety of viewing, and editing tools, specific to the type of object that you left-clicked on.

The shortcut toolbar looks like this when you left-click on a feature...

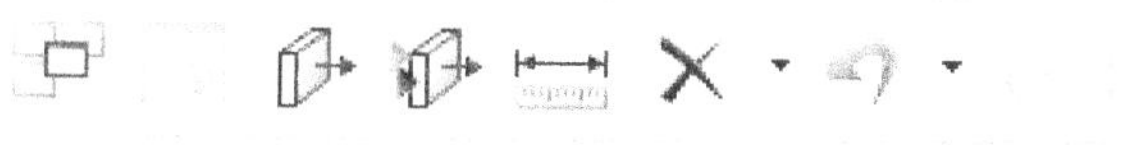

And like this when you left-click on empty space.

The **View Shortcut Menu** appears upon right-clicking empty space in the graphics window, and includes the **Selection Group** for quick access.

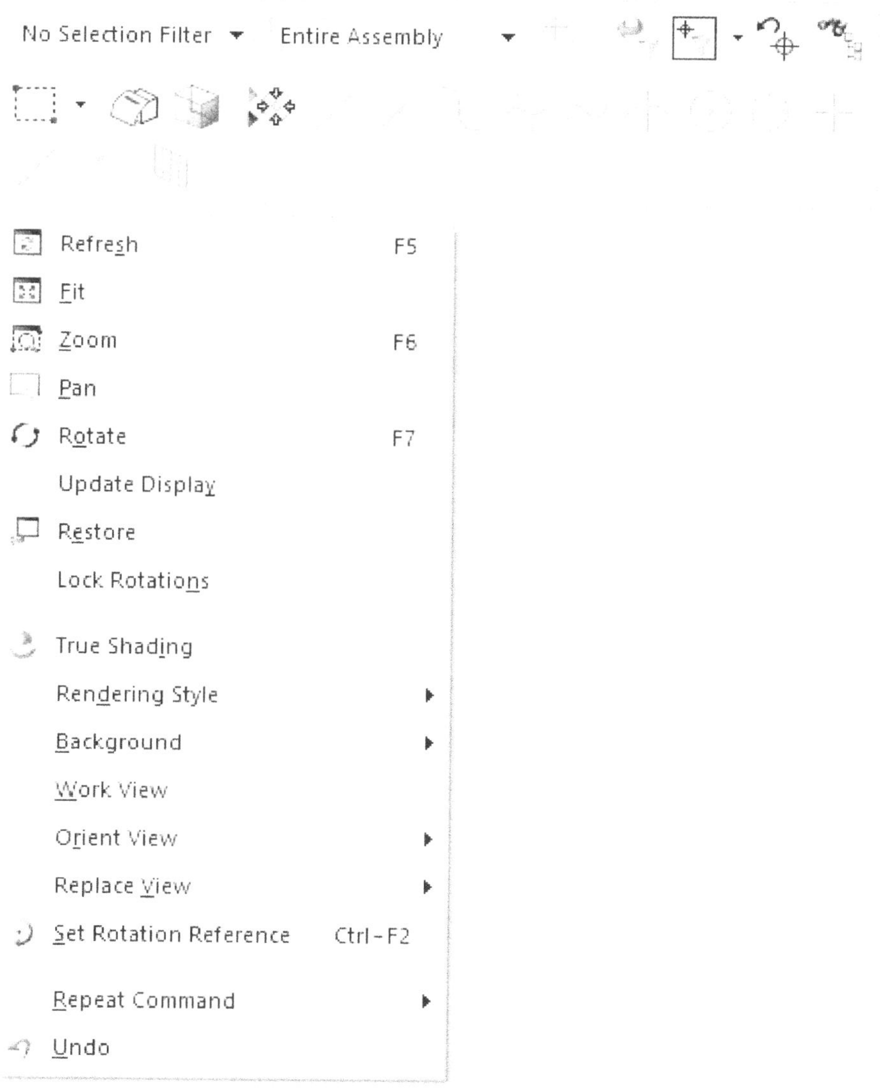

Right-clicking on a piece of geometry combines the shortcut toolbar with a popup menu with the usual editing operations. The **View** subgroup at the bottom of the list expands the menu to show the same commands on the menu above.

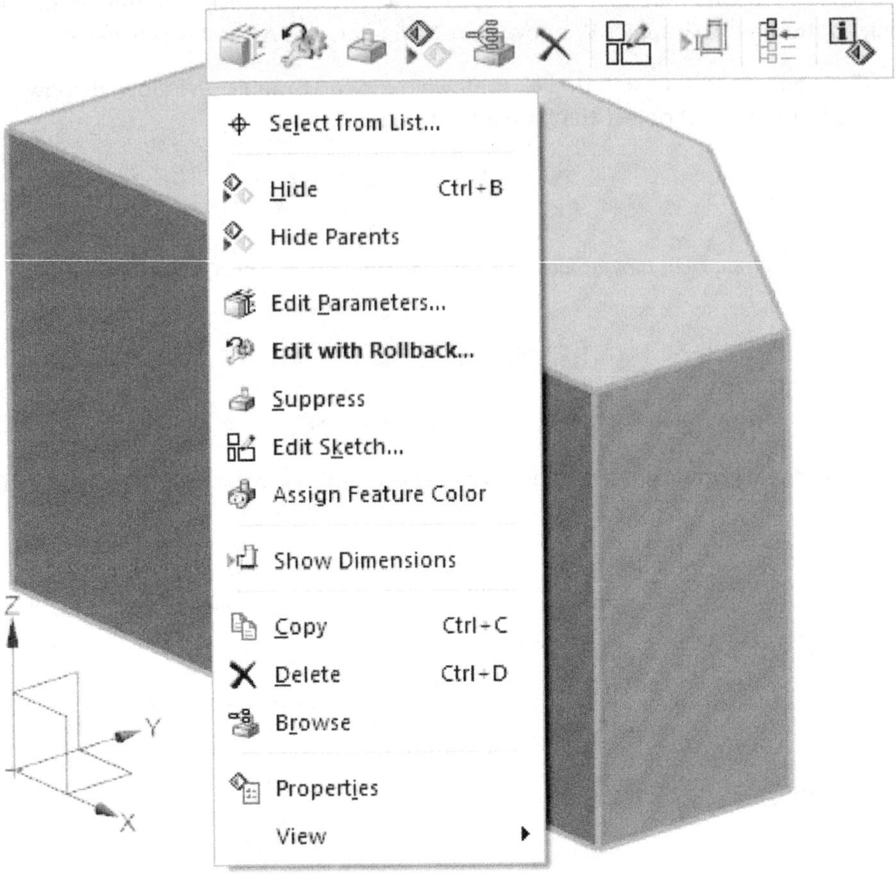

There is also a **View Radial Shortcut Menu** that appears when you click-and-hold the right mouse button in empty space in the graphics window. You will learn about the commands on this menu throughout Chapter 3.

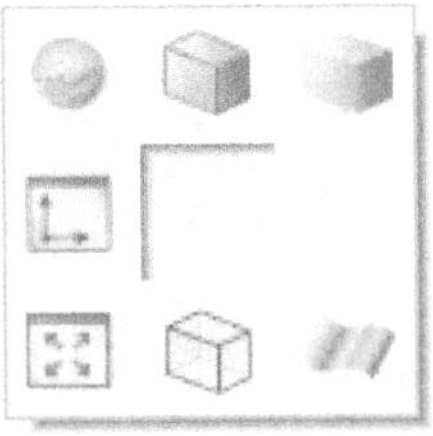

There are three **Application Radial** menus available to you with shortcuts for your favorite commands. These are accessed by holding **[Ctrl] + [Shift]** and clicking a mouse button in empty space in the Graphics Window. These menus can be customized by right-clicking on the ribbon, and selecting the **Shortcuts** tab within the **Customization** dialog box.

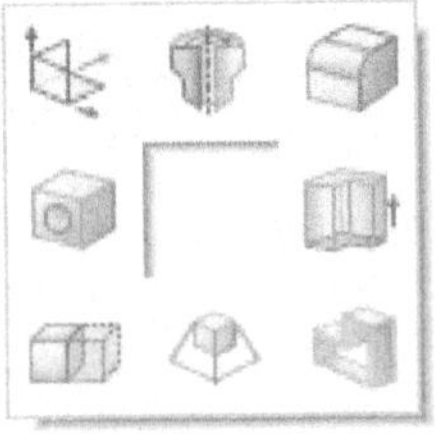

[Ctrl]+[Shift]+MB1

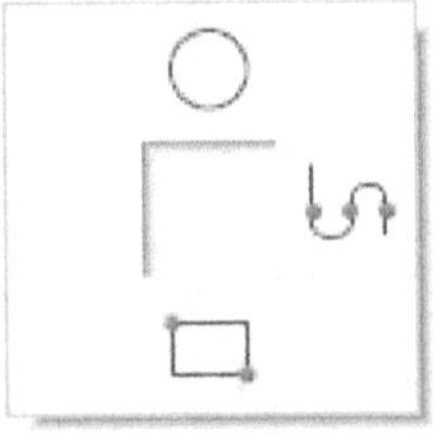

[Ctrl]+[Shift]+MB2

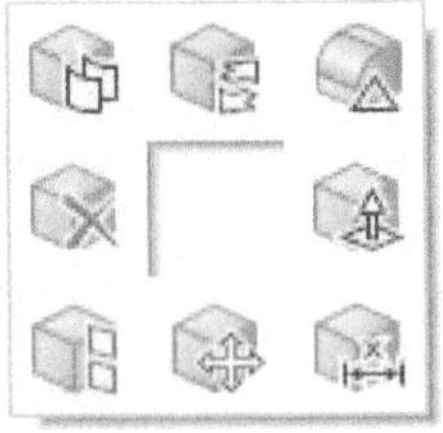

[Ctrl]+[Shift]+MB3

Mastery of the application radial menus can be a real time-saver when modeling. We strongly recommend that you practice using them once you have learned what commands you need most!

2.9 Customization

In this exercise, you will customize the ribbon and save the layout in a new role. There are two sketch tools in NX 11, and the purpose of this exercise is to remove one of them – **Direct Sketch** – from the ribbon, to make way for the other – **Sketch in Task Environment**. In Chapter 6, you will learn to sketch using **Sketch in Task Environment**, so this customization will ensure that you have easy access to the correct sketch tool.

- Begin on the **Home** tab. Click on the **Ribbon Options** down-arrow (▾) on the far-right side of the ribbon, and uncheck **Direct Sketch Group**.

- This removes the **Direct Sketch** group from the far-left side of the ribbon!

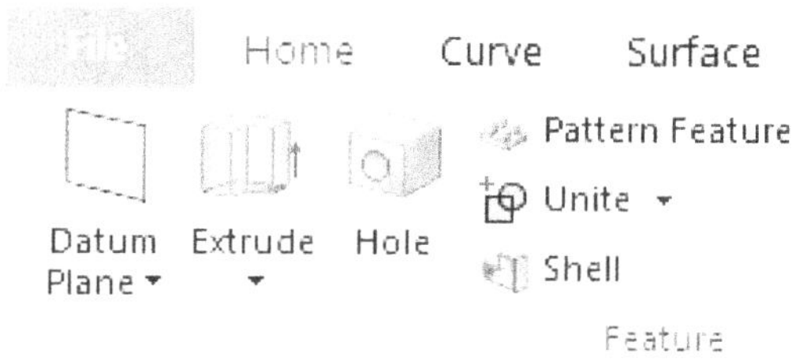

- Right-click on any empty space on the ribbon, and select **Customize** (or, equivalently, use the keyboard shortcut **[Ctrl]+[1]**). Using the **Search** bar in the **Customize** menu, search for *Sketch in Task Environment*.

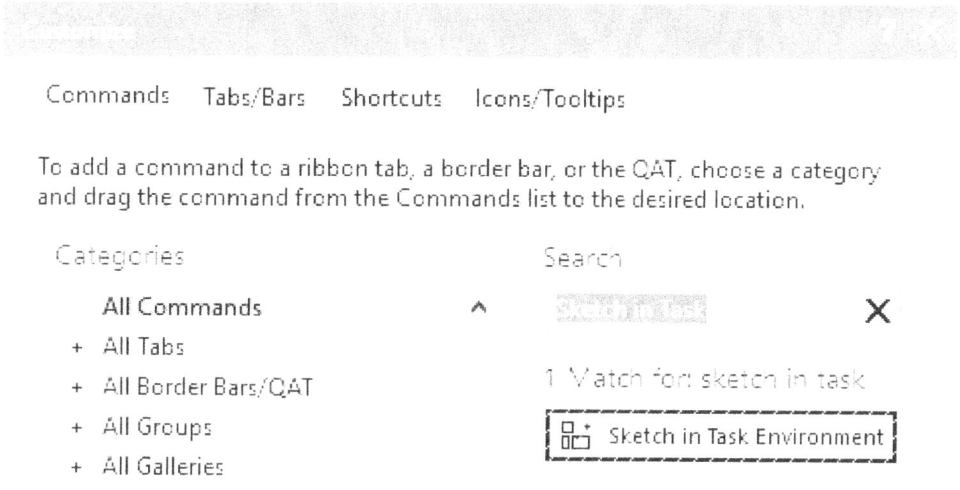

- Click and drag the **Sketch in Task Environment** icon all the way to the far-left side of the **Home** tab.

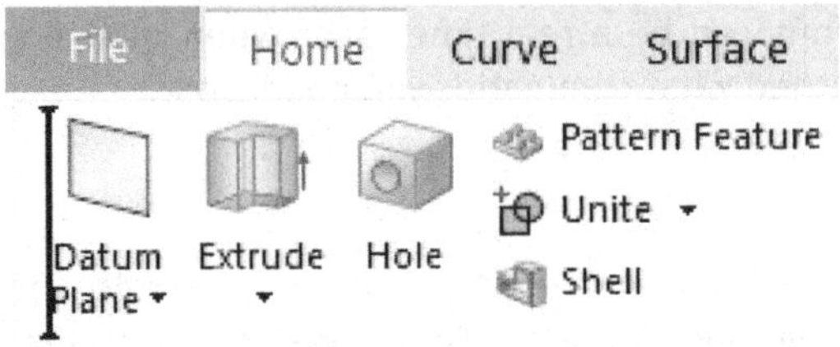

- Release once it is placed as shown below.

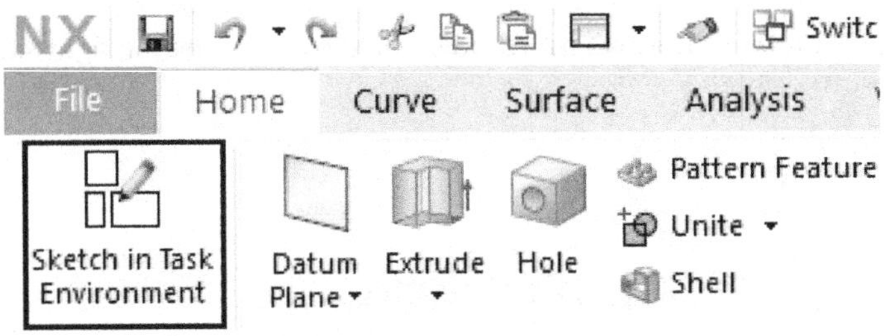

- Next, switch to the **Curve** tab, and click on the far-right **Ribbon Options drop-down arrow** (▾). Uncheck the **Direct Sketch group**, and check **Sketch in Task Environment**.

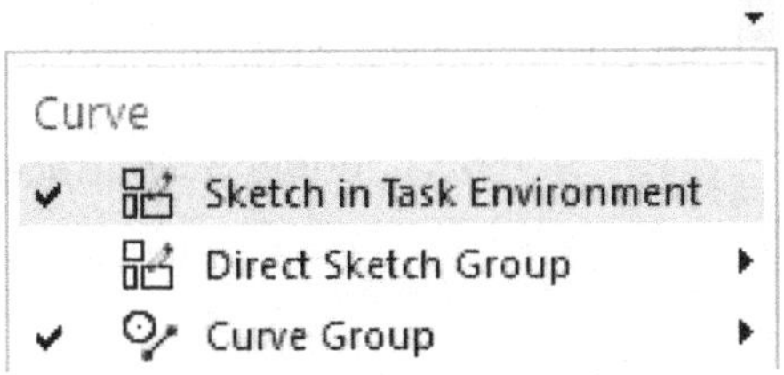

- Select the **Roles** tab from the Resource Bar. Right-click on empty space in the **Roles** navigation panel and select **New User Role**.

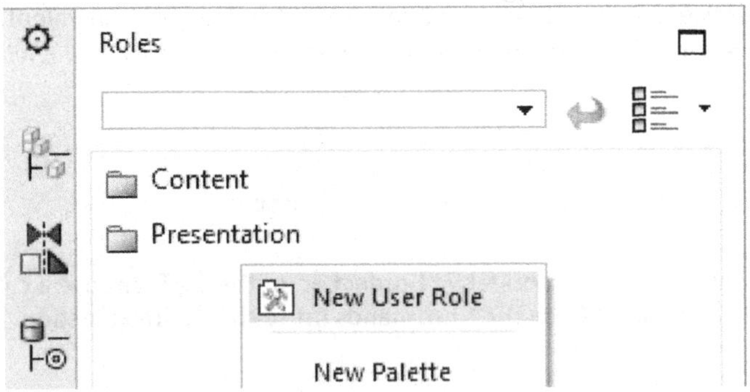

- Name your role "*Learning NX 11*" and give it a brief description as shown below. Set the **Role Type** to *Content and Presentation*, and check the *Include All* checkbox in **Applications**.

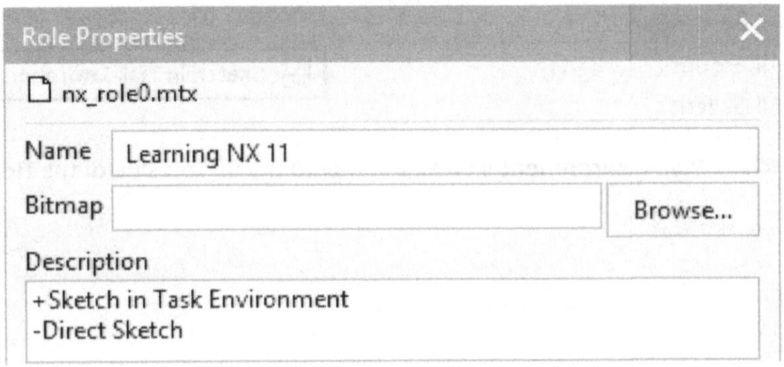

- Having created your own role, you will now see a folder called **User** in the **Roles** navigator, which shows your custom role.

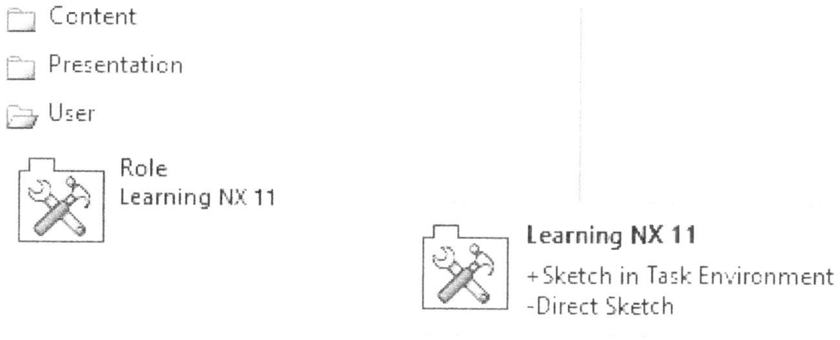

- To activate a User-Defined role, right-click on it and select **Apply**.

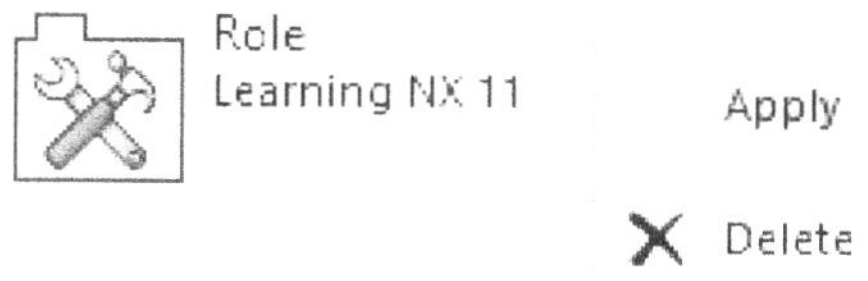

- To update a User-Defined role after making changes to the layout, right-click on it and select **Save Role**.

2.9.1 Keyboard Shortcuts

Another useful feature of the **Customize** dialog is its ability to customize keyboard shortcuts in NX.

- Open the **Customize** menu again with the keyboard shortcut **[Ctrl]+[1]**.
- On the **Customize** dialog, push the **Keyboard** button.

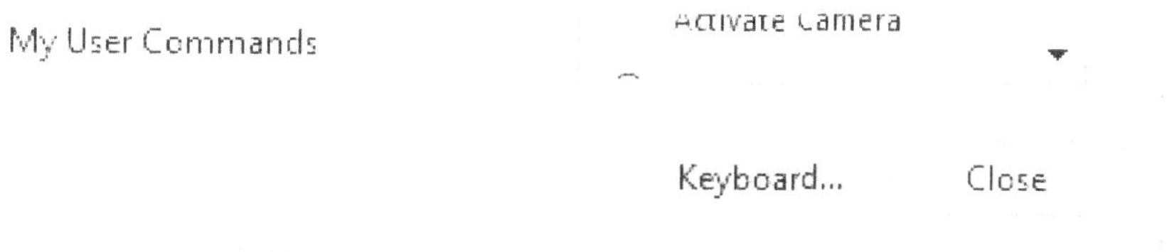

- In the **Customize Keyboard** dialog that appears, select *Selection* from the **Categories** on the left. In the **Commands** list, find and select *Deselect All*. In the **"Press new shortcut key"** text entry field, press **[Shift]+[D]**. Push the **Assign** button to save this change.

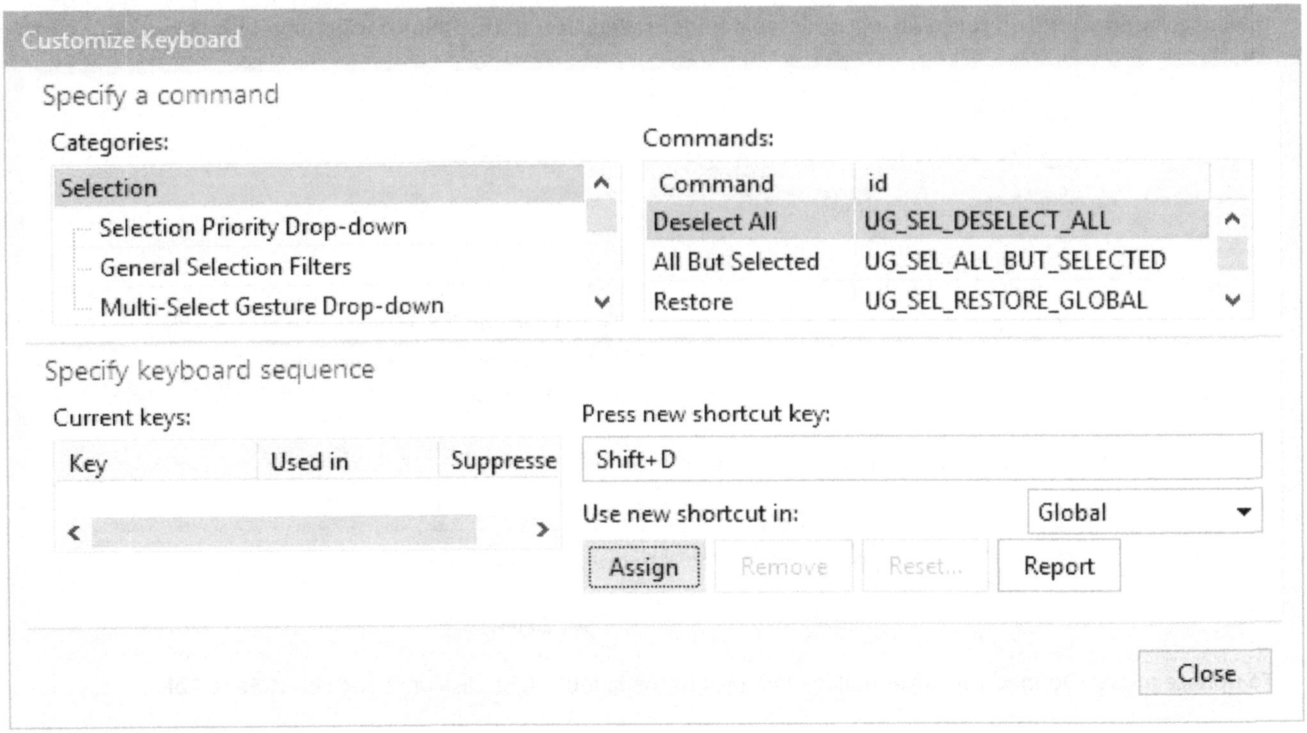

- **Save** the changes to your *"Learning NX 11"* role.

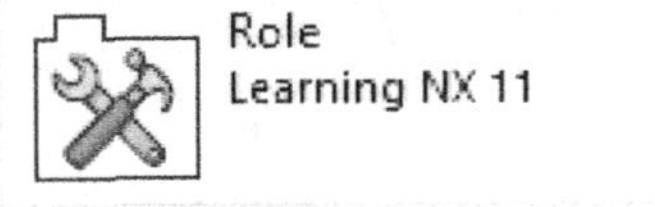

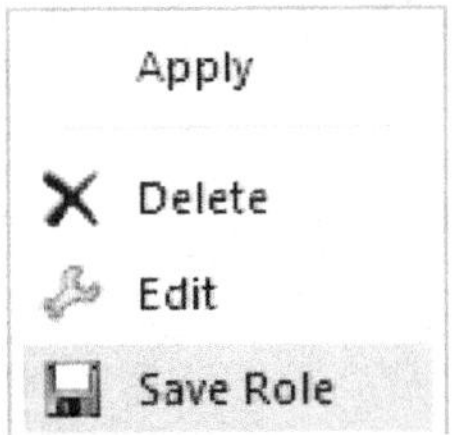

- Now, wherever the **Deselect All** command appears, the keyboard shortcut will be included in the text that accompanies the description. Try searching for **Deselect All** with **Command Finder**!

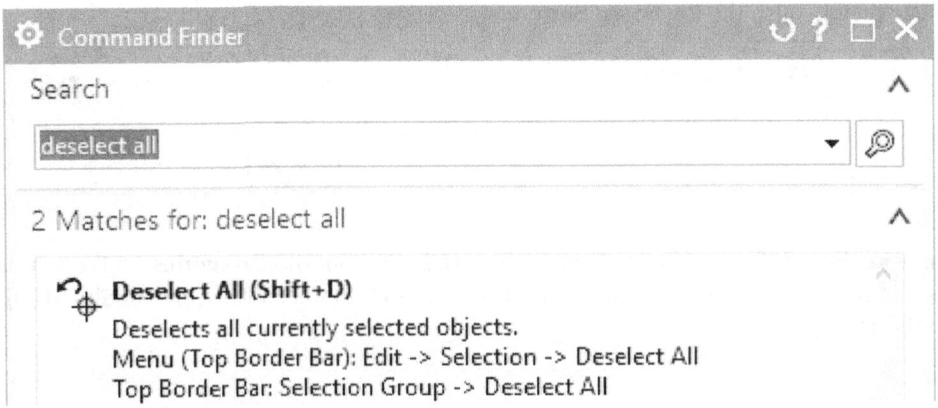

- To see a list of all currently-assigned keyboard shortcuts, select **Menu / Information / Custom Menu Bar / Shortcut Keys**, or push the **Report** button on the **Customize Keyboard** dialog.

2.9.2 Configuring Navigator Columns

At times in this book you will want to see the columns in your Assembly Navigator ordered in a specific way. In the exercise below, you will learn to reorder columns in the Assembly Navigator. The same technique applies to the Part Navigator, and other navigators that you might encounter in your use of NX.

- Switch to the Assembly Navigator, and right-click anywhere in the empty space contained therein. All the way at the bottom of the menu, select **Properties**.

- In the **Assembly Navigator Properties** dialog, select the **Columns** tab. Find *Position* in the menu (you will probably have to scroll down) and check the checkbox to the left. Use the **Move Up** arrow to send it all the way to the top of the list, just behind *Name*.

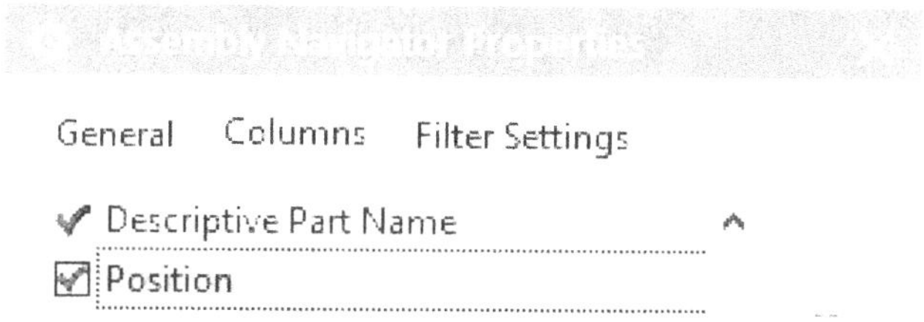

- Click **OK** to save your changes. This reorders the columns in the Assembly Navigator so that, following the *Name* column, you can now see the *Position* column. The *Position* column is a rough indicator of degrees of freedom remaining within a component's immediate parent assembly. You will make use of it in Chapter 11. Since there are no components in *"Opening Files.prt"*, this doesn't show very much!

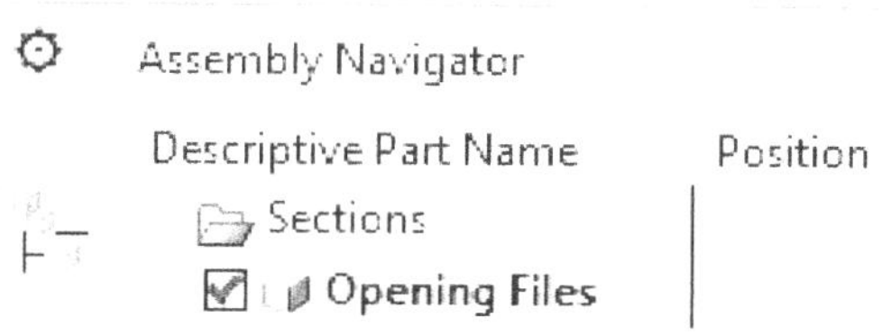

2.10 Preferences and Customer Defaults

You will need one more customization prior to doing the exercises in Chapter 6. There is a setting which chooses whether to use **Direct Sketch** or **Sketch in Task Environment** as the default tool for *editing* sketches. You can manage this setting in the **Modeling Preferences** menu, found at **File / Preferences / Modeling**

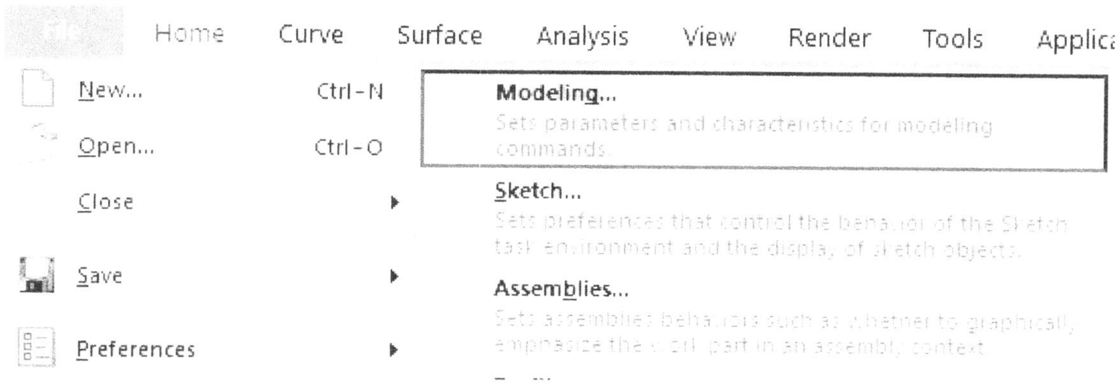

It is critical to understand that all changes made to **Modeling Preferences** are session settings – meaning that when you close NX, they are forgotten. The next time you open NX, the settings in the **Modeling Preferences** menu are loaded from the **Customer Defaults**. This means that to make a change that will persist from session to session, you must change the setting from the **Customer Defaults** menu.

- Select **File / Utilities / Customer Defaults**. Under **Modeling / General**, switch to the **Edit** tab, change your **Edit Sketch Action** to **Task Environment**, as shown below.

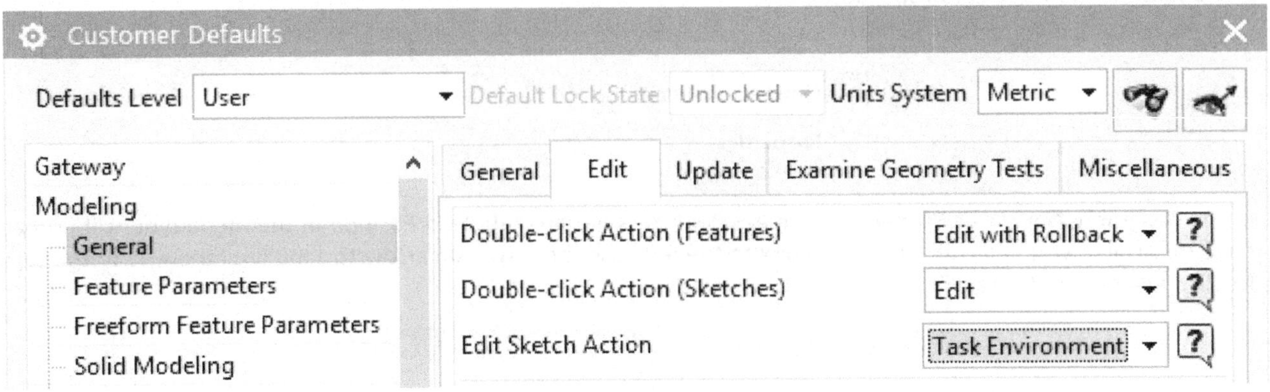

A nice feature of both the **Customer Defaults** menu, as well as every **Preferences** menu, is the built-in **Command Finder**.

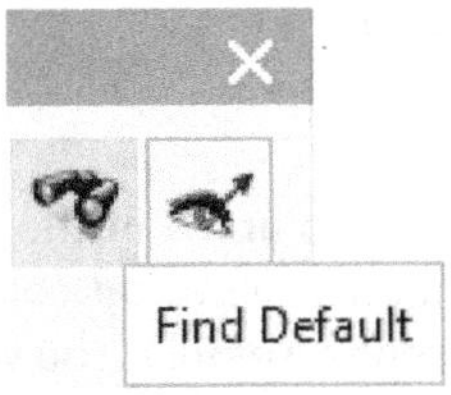

The binoculars icon, wherever it occurs, indicates search functionality. Recognize it and take advantage of it to work your way through menus faster!

Changes made to the **Customer Defaults** menu do not take effect until the next session of NX begins. At this point you should exit NX and reopen it.

3 The Graphics Window

Our introduction to the user interface would be woefully incomplete without an in-depth study of the most important part – the **Graphics Window**!

3.1 Coordinate Systems

When you create a new file using the Modeling template, there are three coordinate systems already in place – the Absolute Coordinate System, the Work Coordinate System, and a Datum Coordinate System. It is easy to confuse these in the Graphics Window, so we begin by describing the various coordinate systems you will encounter as you begin using NX.

3.1.1 The Absolute Coordinate System

The *Absolute Coordinate System* gives a part file its units and scale. The absolute coordinate system is positioned at the point with (x, y, z) coordinates (0, 0, 0). The Absolute Coordinate System cannot be selected or made visible, but you can see its orientation from the triad in the bottom left corner of the graphics window. This triad does not indicate the position of the Absolute Coordinate System, only its orientation.

When working with assemblies, it is often convenient to create all components such that they share a common absolute coordinate system. Each part file has its own Absolute Coordinate System, and so in an assembly, it is possible to make reference to either the Absolute Coordinate System of the Work Part, or the Displayed Part.

3.1.2 The Work Coordinate System

The Work Coordinate System (hereafter referred to as the WCS) is a non-parametric, non-associative coordinate system from which many tools infer an orientation for the operation in question. The WCS is not a selectable entity, though it appears in the graphics window as shown below. You can toggle the display of the WCS on or off with the [W] key.

In the modeling template that comes with a standard installation of NX, the WCS is positioned at the Absolute Coordinate system initially. You can reposition the WCS with the commands found in the **WCS** group in the **More** gallery in the **Utilities** group on the **Tools** tab, as shown below.

A key point about the WCS is that references made to the WCS at the time of a feature's creation are not fully associative, in the sense that, if the WCS changes its position or orientation, features that reference the WCS will be unaffected. If this doesn't totally make sense now, don't worry – in Chapter 5.1, you'll do an exercise that illustrates this behavior fully.

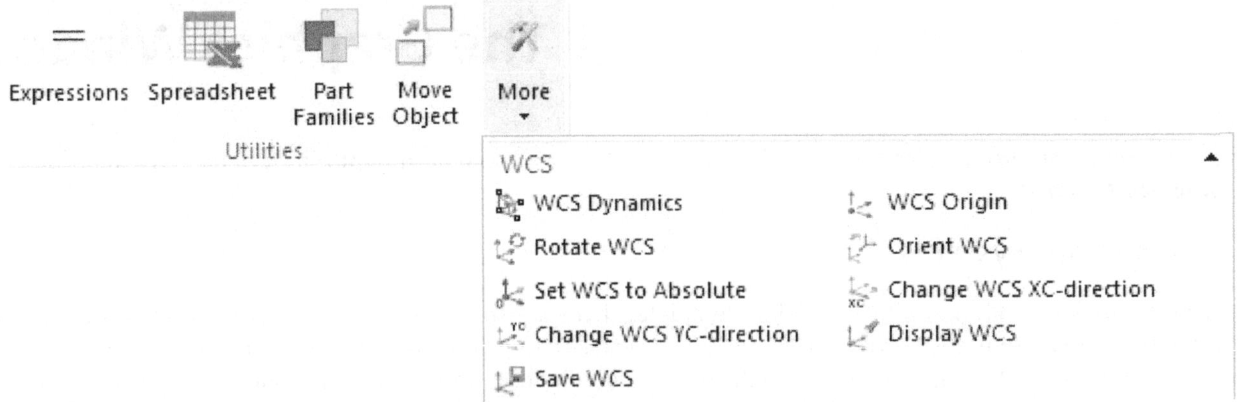

While tools that infer – or use – an orientation _once_ make reference to the WCS, the Absolute Coordinate System gives the part file its overall orientation and sets distances appropriately in the units of the part file. We will see shortly that coordinates for points can be specified in relation to either the WCS or the Absolute Coordinate System.

The Work Coordinate System is a feature of the Displayed Part – there is no "Work Part WCS" for each component in an assembly.

3.1.3 Dynamic Coordinate Systems

Dynamic coordinate systems appear in many places throughout NX, whenever you are positioning an entity with a *Dynamic* method. For example, the **WCS Dynamics** tool allows you to move the WCS interactively by dragging it to new locations in the graphics window. You can activate it from the **Tools** tab as shown above, or by double-clicking on the WCS in the graphics window. The dynamic coordinate system that appears has seven handles, as shown below. These are common to all dynamic coordinate systems.

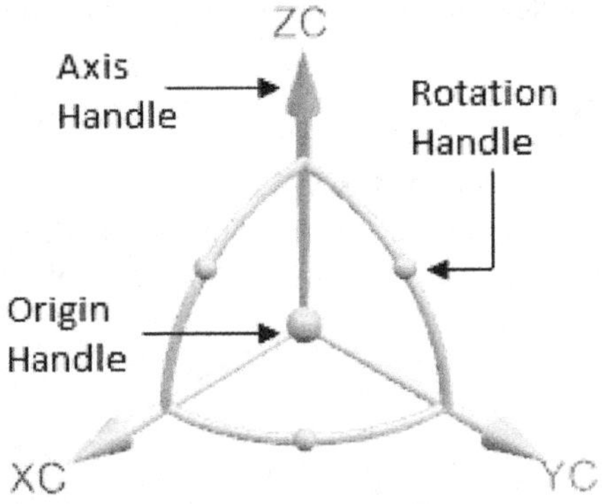

When you click on either an axis handle or a rotation handle, a dynamic dialog will appear – this dialog box allows you to enter specific offset values. To exit **WCS Dynamics**, you can hit the **[Esc]** key or click the **middle mouse button**.

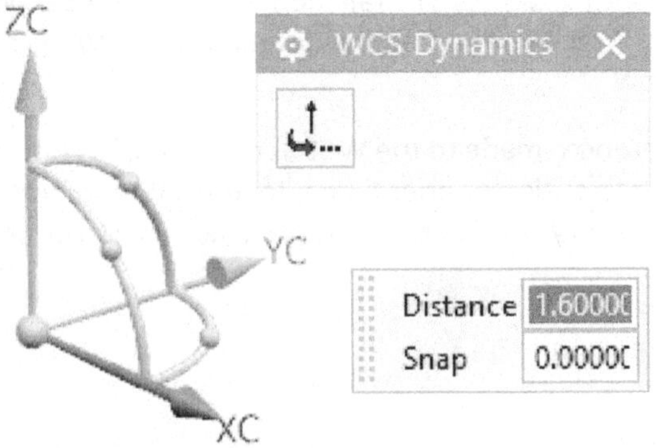

3.1.4 OrientXpress

The *OrientXpress* is a coordinate system which inherits its orientation from the WCS. It appears whenever you are required to input a direction vector and the vector type is set to **Inferred**.

On certain dialogs where a direction vector and an angle are required, the OrientXpress may take on a form similar to what you see below.

3.1.5 Datum Coordinate Systems

Finally, we arrive at the parametric, selectable, associative kind of coordinate system – the **Datum Coordinate System**! When you create a file using the *Modeling* template, it comes with a **Datum Coordinate System** (Datum CSYS) in it, which happens to be aligned with both the WCS and the Absolute Coordinate System.

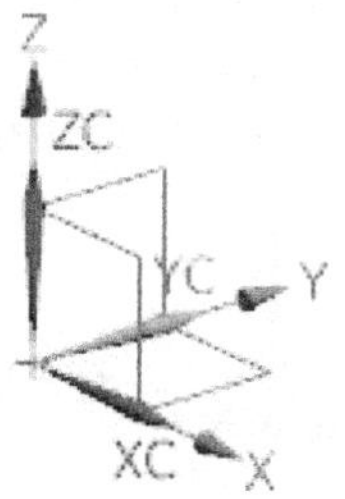

Datum coordinate systems are created using the **Datum CSYS** tool, found in the **Datums** drop-down in the **Feature** group on the **Home** tab.

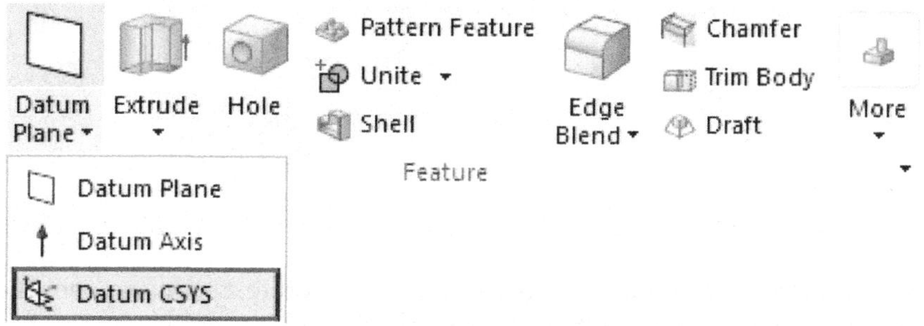

In Chapter 5.1, you will practice making some datum coordinate systems. We would love to have you create some datum coordinate systems right away, but for now it's more important that you get a good grasp on the rest of the UI.

3.1.6 The View Coordinate System

There is one more coordinate system that is useful to keep in mind as you use NX, although it never actually appears in the Graphics Window. You should imagine that the *View Coordinate System* is sitting on your screen at the center of the Graphics Window. It is never explicitly named, but a number of navigation operations and dynamic positioning motions are best understood in relation to this imaginary coordinate system. The XY plane of this coordinate system is on your screen, with X pointing horizontally and Y pointing vertically, and the Z axis points straight out of the screen at the viewer.

3.2 Orienting the View

Most commands in NX are made by using the mouse to navigate between menus and to pick geometry. NX requires the use of a three-button mouse to maximize productivity, especially for navigation. The primary functions of the mouse buttons are illustrated below.

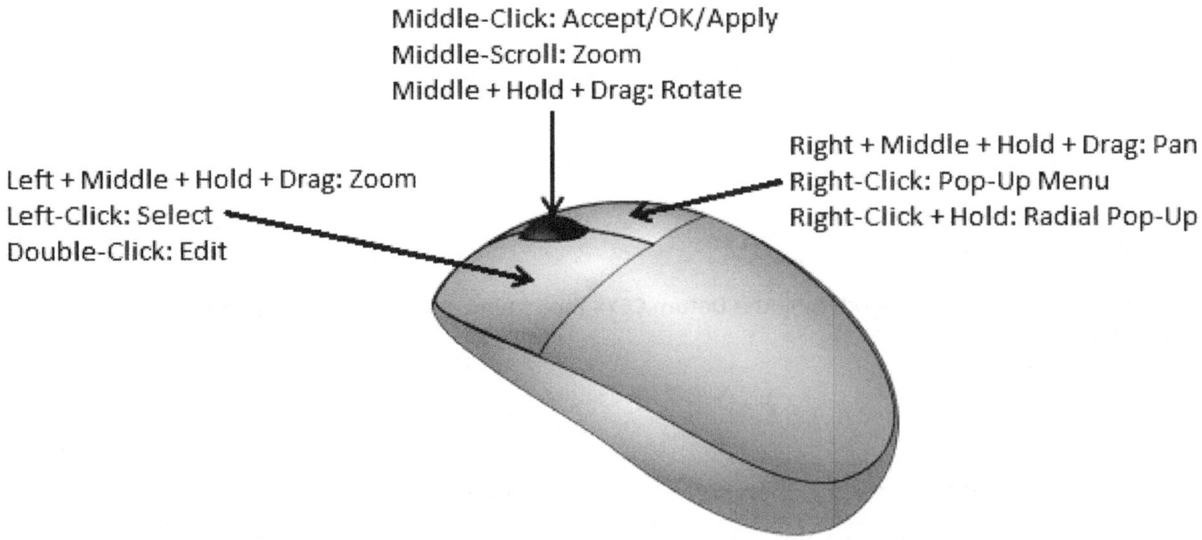

*Note: NX refers to left-click, middle-click, and right-click as **MB1, MB2, MB3**, respectively.*

There are many ways to perform the three basic navigation operations – rotate, zoom, and pan. These tools are found on the **View** group on the Top Border Bar, but we don't recommend the use of **View** group buttons, because it is much easier to use your mouse!

- From the downloaded part files folder, open the file *"The Graphics Window.prt"*.
- This part file contains several bodies, curves, and datums, with Model History as shown below.

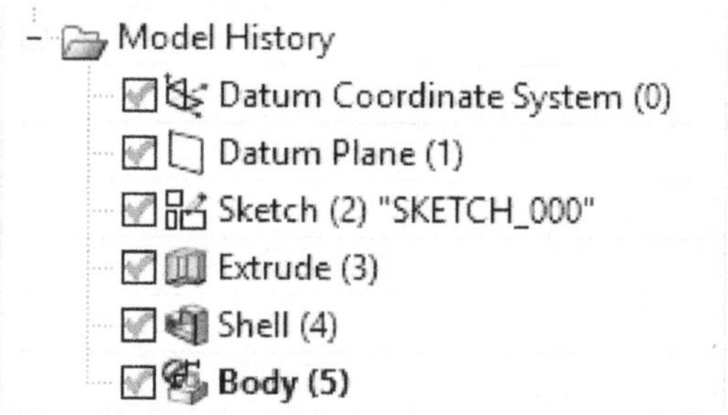

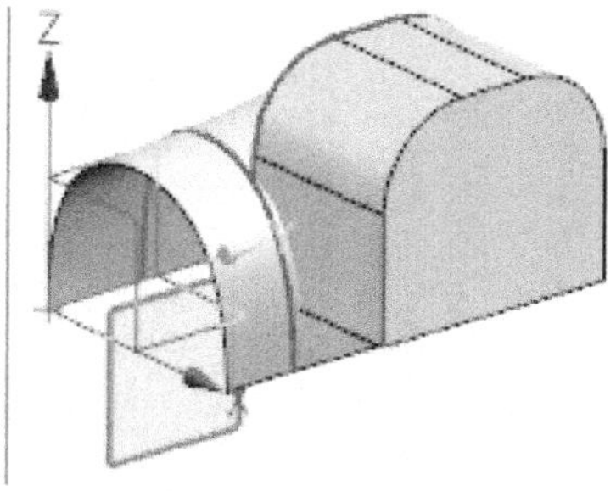

- Follow the instructions below with this part!

3.2.1 Zoom

- Roll the scroll wheel – each click of the scroll wheel zooms in or out by **25%**.
- Drag while pressing and holding the left and middle mouse buttons simultaneously.
- Drag while pressing and holding **[Ctrl]** and the middle mouse button.

3.2.2 Pan

- Drag while pressing and holding the middle and right mouse buttons simultaneously.
- Drag while pressing and holding **[Shift]** and the middle mouse button.

3.2.3 Rotate

- Press and hold the middle mouse button and move the cursor. The rotate icon will appear. Drag in the direction of the rotation you want.
- The following tricks allow for pure rotation about the axes of the *View Coordinate System* described in Chapter 3.1.6:
  - Middle-click-and-hold-and-drag near the right/left edge of the Graphics Window for X-axis rotation.
  - Middle-click-and-hold-and-drag near the bottom edge of the Graphics Window for Y-axis rotation.
  - Middle-click-and-hold-and-drag near the top edge of the Graphics Window for Z-axis rotation.
- The red-blue-green triad in the bottom left corner of the graphics window indicates the orientation of the Absolute Coordinate System, and it can be used for rotation as well. Click on one of the axes to enable pure rotation about that axis. You can enter a precise angle by which to rotate as well. To restore totally free rotation, click on the white sphere at the origin of this coordinate system.

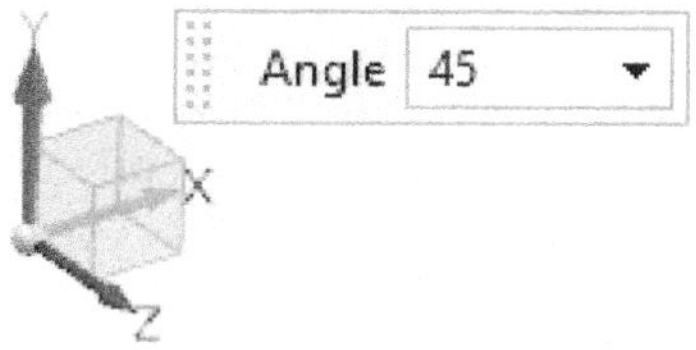

- An important aspect of rotation is the rotation point – by default, NX will set the rotation point at the center of a large box containing everything in your model. If you wish to set a specific rotation point, <u>click and hold</u> the middle mouse button and wait for the concentric circles to appear, as shown below. Once the circles appear, continue holding down the middle mouse button and move your mouse and you will rotate about that point. This technique works best when your cursor is over an existing point, or a point on an edge/face, etc., rather than a free-floating point in space.

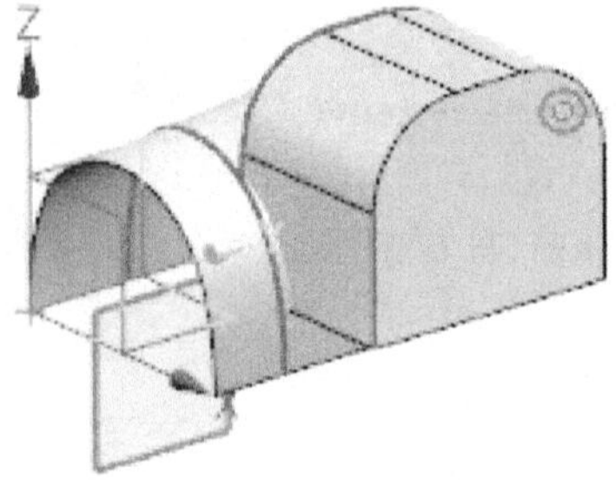

- You can also set a specific rotation point by right-clicking in empty space in the Graphics Window and choosing **"Set Rotation Reference"** from the pop-up view shortcut menu. Once a rotation reference is set, you can clear it by selecting **"Clear Rotation Reference"** from the same menu.

3.2.4 Standard Views

You can orient the view to a standard view of your model – *Top, Front, Left, Right, Back, Bottom, Trimetric,* or *Isometric.* The icons used to obtain each of these views are found on the **View** tab in the **Orientation** group. These views are all defined in relation to the Absolute Coordinate System of your model.

If it's too much hassle to make your way to the **View** tab, you might find that the **Orient View** drop-down menu found in the **View** group on the Top Border Bar is more convenient!

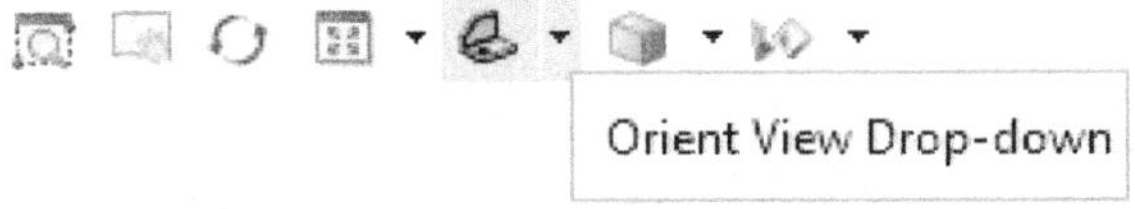

3.2.5 Fit

Fit resizes the model to fill the entire Graphics Window, maintaining the current camera angle. Very useful if you lose your model or if the clipping planes begin to cut through your model.

3.2.6 Snap View

The **Snap View** command adjusts the camera angle so that the View Coordinate System of Chapter 3.1.6 aligns with the nearest coordinate plane of the Absolute Coordinate System. The orientation of the Absolute Coordinate System is indicated by the triad in the bottom left corner of the graphics window.

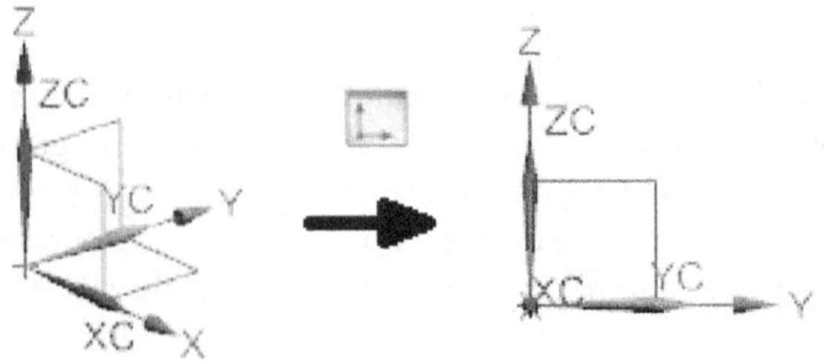

Here, ***Snap View*** *changes the camera angle so that the view is normal to the Y-Z plane.*

The **Snap View** command is not found on the ribbon. You can find it on the **View Radial Pop-Up Menu**, which appears when you click and hold the right mouse button on empty space in the graphics window. It also has a keyboard shortcut of **[F8]**.

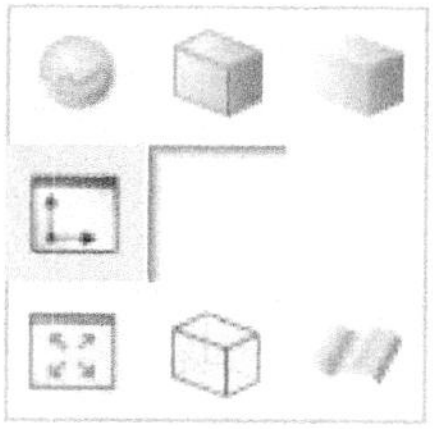

3.3 Selection

In NX, all tools and entities in the model are selected with the left mouse button. The following concerns the selection of entities within the Graphics Window.

3.3.1 Action-Object Selection Order

A point of confusion for new NX users who are familiar with other CAD software is the Action-Object ordering for selecting tools and then objects from the model. In other CAD systems, the user must often select geometry before choosing the desired tool (Object-Action Selection Order), but in NX the order is usually reversed. First you select the tool, and then tell the software what the relevant entities are in the part file.

3.3.2 The Selection Toolbar

The **Selection Group** is on the Top Border Bar (see Chapter 2.5), and the menus and buttons in this group control how you are able to select entities from the Graphics Window.

In this group, there are a number of useful tools for allowing and disallowing certain objects to be selected. We will refer to and explain several of these throughout the chapter.

3.3.3 The Selection Ball and Crosshairs

When selecting geometry, the cursor display will change into a "selection ball" as it is moved into the graphics window, as shown below. Any geometry that falls within the selection ball is a candidate for selection, assuming that it is a valid entity for the active dialog box.

You can adjust the size and behavior of the selection ball using the options found in **Selection Preferences** menu, found at **Menu / Preferences / Selection**.

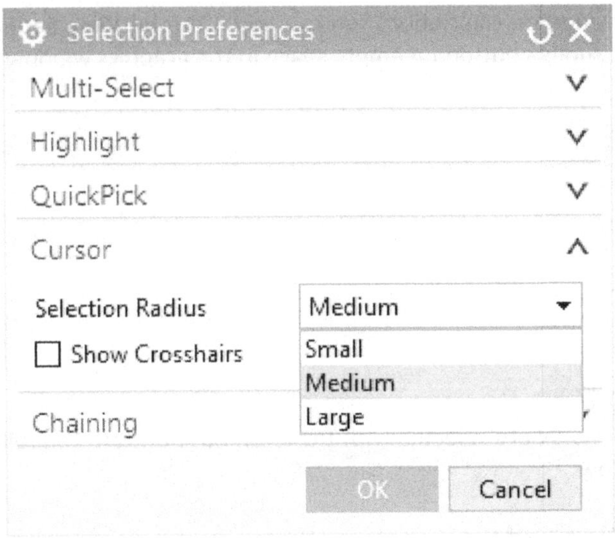

If you wish to enable crosshairs, you can do so by checking the box shown above. Crosshairs are useful if you require horizontal or vertical alignment between certain entities.

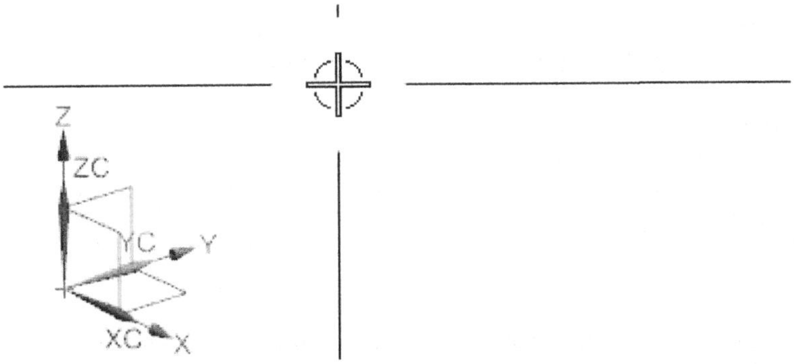

3.3.4 QuickPick

Sometimes it is not possible to position the selection ball such that the geometric entity you are trying to select is unambiguous. In these cases, the **QuickPick** menu helps you to disambiguate and come up with the right selection. You can invoke the **QuickPick** menu in several ways.

- Put your cursor near both of the curves in the model. Wait three seconds until the three white boxes appear next to the selection ball, as shown below.

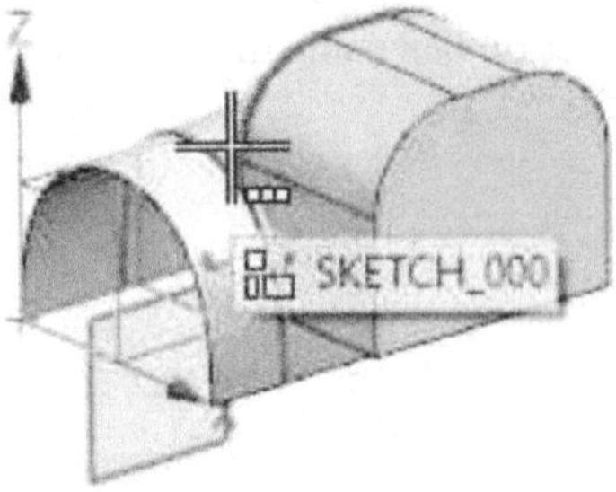

- Without moving your cursor, left-click, and the **QuickPick** dialog box will appear. This allows you to choose among the multiple entities that touched the selection ball at the time of your click. Note that NX will allow you to select faces and edges of bodies, and distinguishes between the feature *"SKETCH_000"* and the curves contained in that sketch.

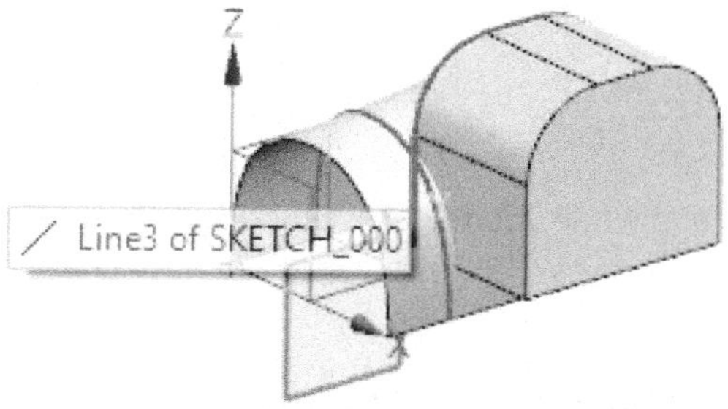

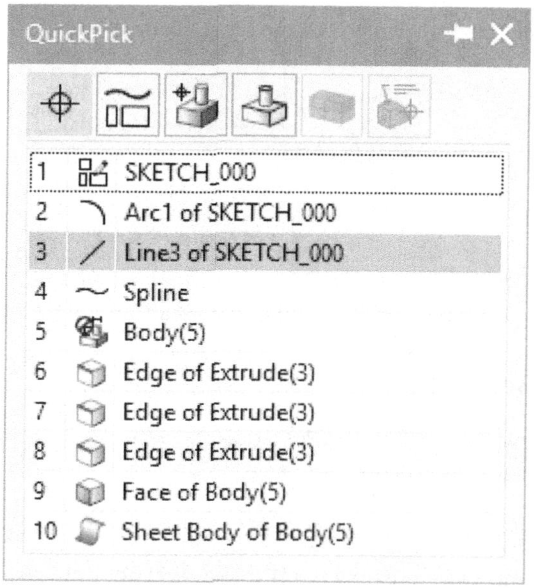

- The icons along the top of the **QuickPick** menu allow you to filter for specific types of entities. This can be very helpful when the list in the **QuickPick** menu is extensive. From left to right, the filters are: construction entities, features, body parts, components, annotation. Cycle through the available buttons to see what kinds of entities are included in each group.

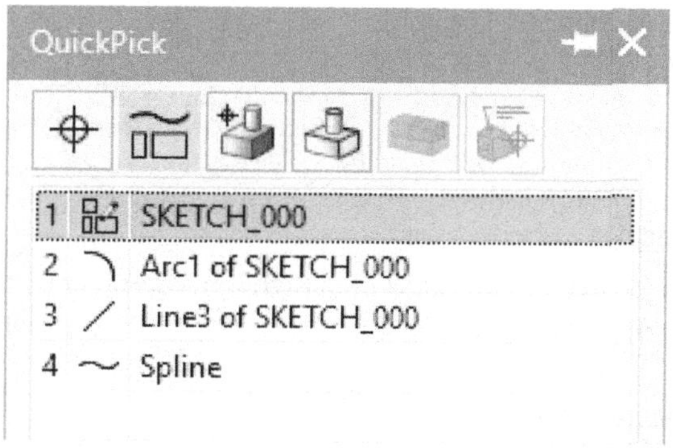

- Close the **QuickPick** menu without making a selection.
- You can control the delay time (waiting period until the three white boxes appear) for **QuickPick** from the **Selection Preferences** menu (**Menu** / **Preferences** / **Selection**).

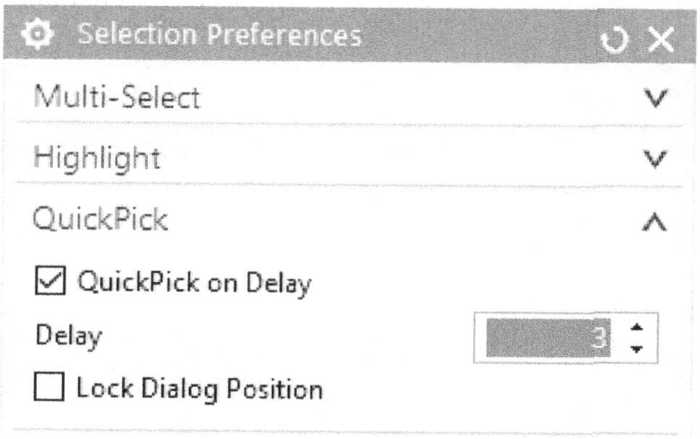

- If you prefer not to wait for the delay, or if you disable it entirely, you can also access the **QuickPick** menu by right-clicking on an object, and choosing **"Select from List"**.

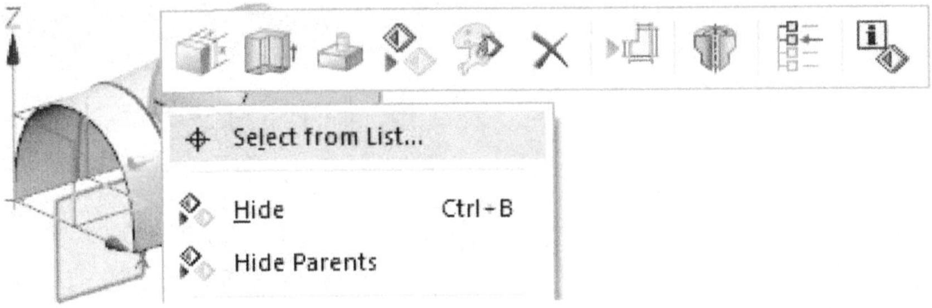

- You can also left-click-and-hold and wait about half a second, and when you release, QuickPick will appear.

3.3.5 Deselection

There are several ways to deselect objects. Make a selection in the Graphics Window, and then practice each of the following deselection techniques.

3.3.5.1 Deselecting All Objects

- The Escape Key – pushing the **[Esc]** key will not only deselect all objects, but close any dialog box you may have open.
- Clicking on empty space in the Part Navigator.
- The **Deselect All** button, found on the Selection Toolbar. Also, recall that in Chapter 2.9.1, you customized the UI so that **Deselect All** has a keyboard shortcut of **[Shift]+[D]**. This keyboard shortcut is enabled as long as you are using the *"Learning NX 11"* role created in that chapter. Try out your keyboard shortcut!

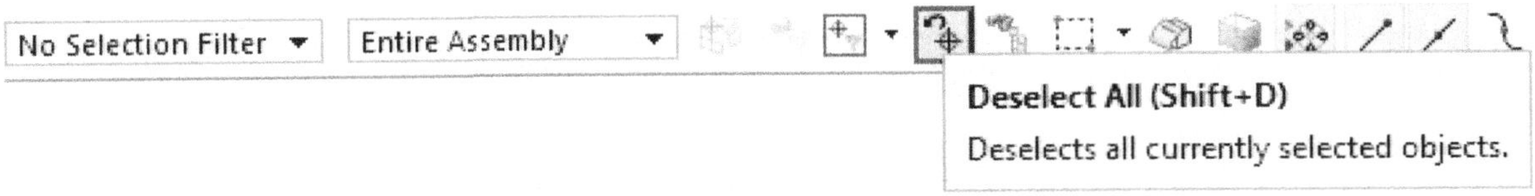

3.3.5.2 Deselecting Individual Objects

- From the Graphics Window, hold **[Shift]** and left-click on a highlighted object to deselect it.
- From the Part Navigator, hold **[Ctrl]** and left-click on a highlighted object to deselect it.

3.3.6 Multiple Selection & Deselection

There are a variety of ways in which you can select or deselect multiple objects in NX.

3.3.6.1 From the Graphics Window

Once you have selected something, to add to your selection, simply click on another entity.

- Begin by clicking the datum plane in the model. To select a datum plane from the Graphics Window, you must put your cursor over the edge of that datum.

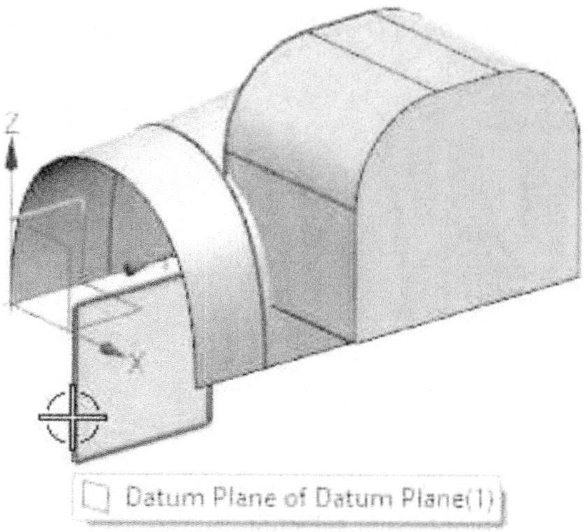

- Note that if your cursor is in the middle of the datum plane, the software will not select that datum plane!

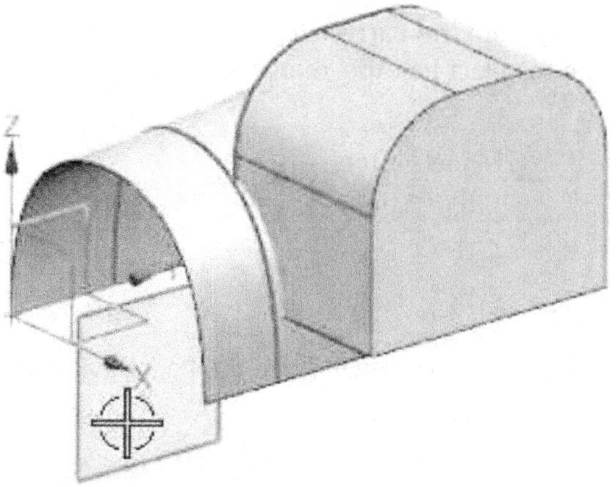

- Use the **QuickPick** menu to select the YZ plane of the datum coordinate system.

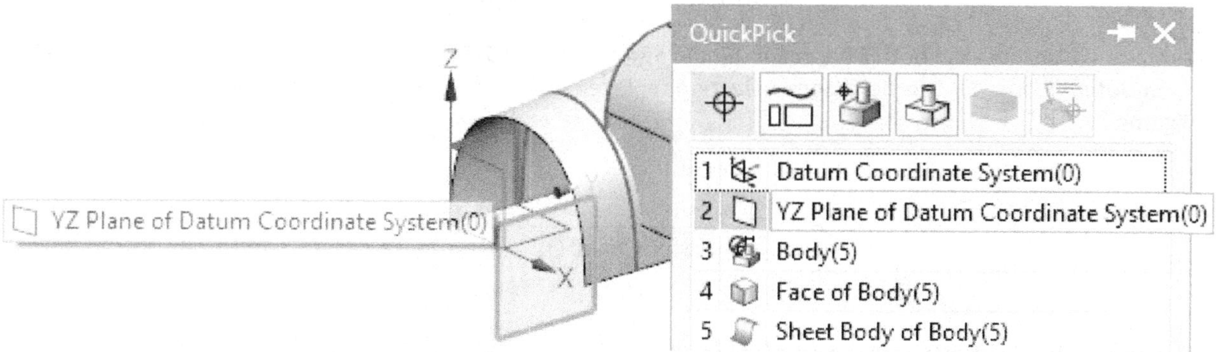

- Note your Status Display.

YZ Plane selected - total 2

- Note that your Part Navigator shows only *Datum Plane (1)* selected.

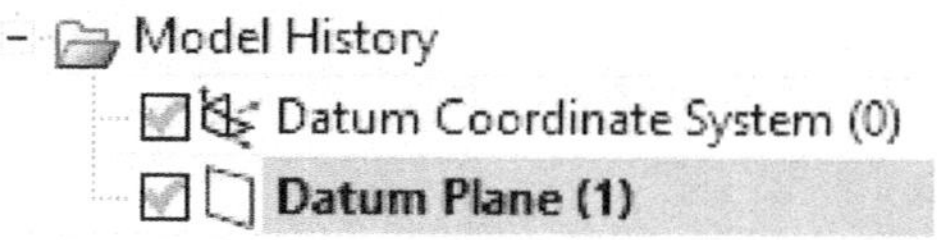

- Hold the **[Shift]** key and click on *Datum Plane (1)* in the graphics window, so that only the YZ plane of the datum coordinate system remains in your selection.

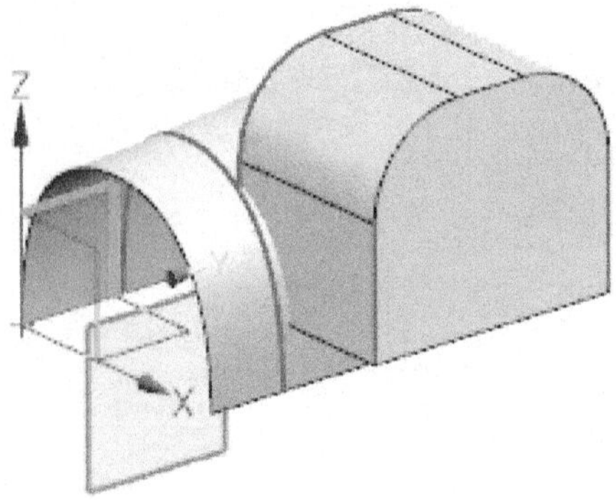

- Hold **[Shift]** and click on the YZ plane to empty your selection.

3.3.6.2 From the Part Navigator

- Click on *Datum Coordinate System (0)* from the Part Navigator. Note that this selects the entire feature, whereas in the Graphics Window we were able to select just the YZ plane. From the Part Navigator, you are only able to select entire features.

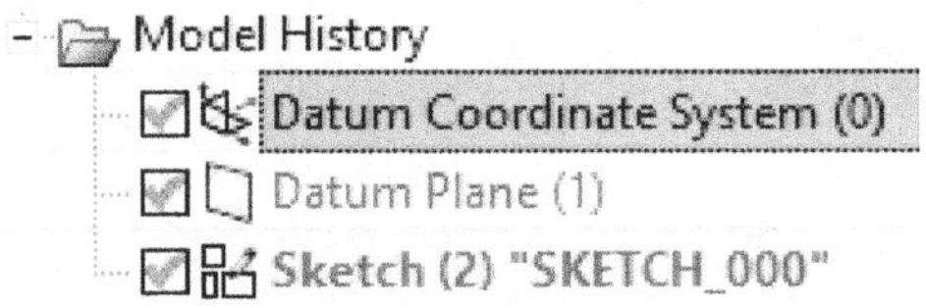

- To add to your selection, hold **[Ctrl]** and click on *Sketch (2)* in the Part Navigator.

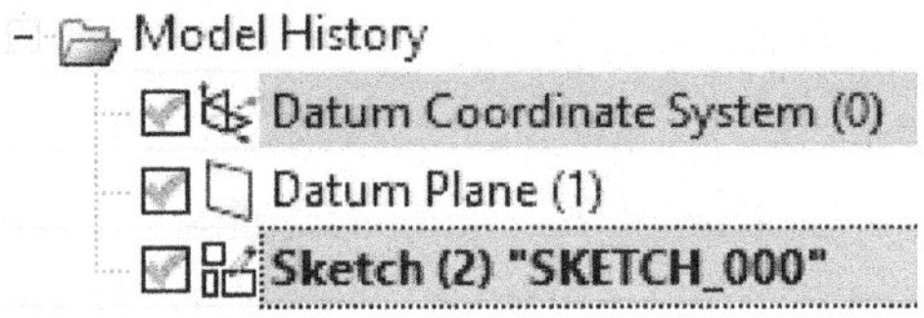

- To simultaneously deselect both *Datum Coordinate System (0)* and *Sketch (2)*, and also select *Datum Plane (1)*, simply click on *Datum Plane (1)*. This behavior – selection replacement – only occurs when making selections from the Part Navigator.

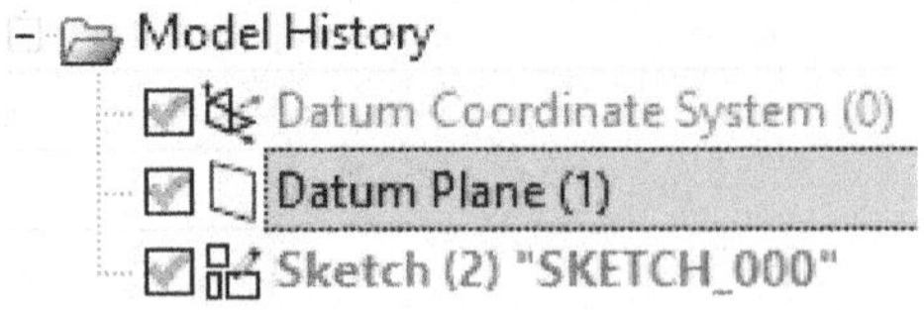

- To deselect all items from within the Part Navigator, just click on any empty space in the Part Navigator.

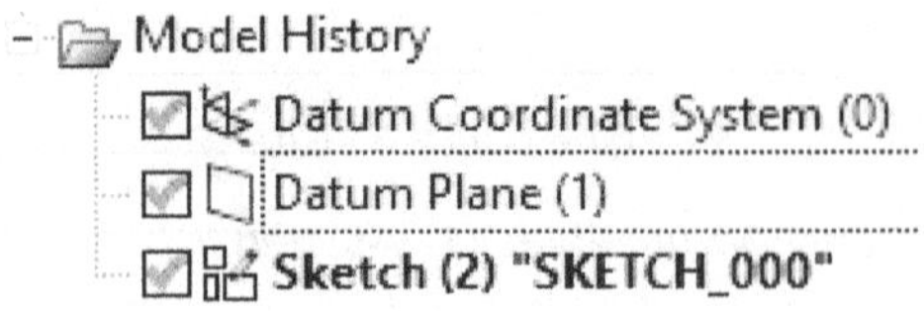

- Click on *Datum Coordinate System (0)* in the Part Navigator.

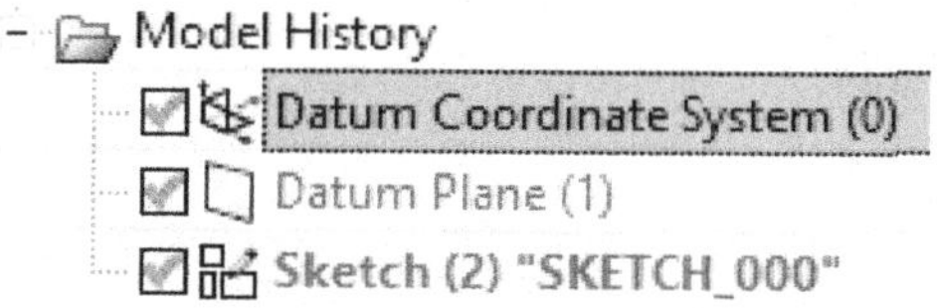

- You can also hold **[Shift]** down and click on another line in the Model History to make a multiple selection. The difference is that it will select all lines between the first and last. Hold **[Shift]** and click on *Sketch (2)* in the Part Navigator.

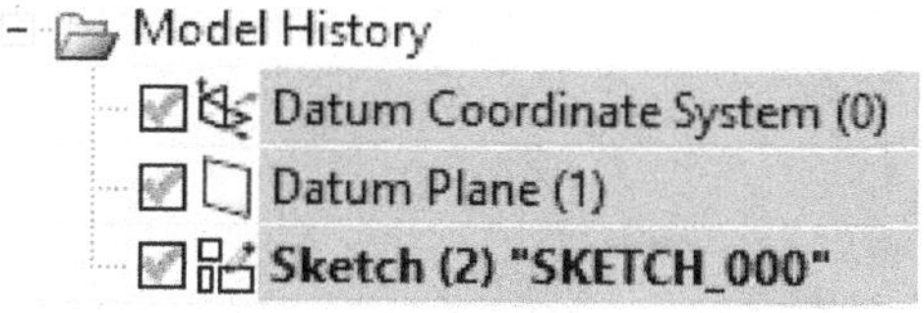

3.3.6.3 Multi-Select Gesture

In the Graphics Window, there is another way to make multiple selections and deselections – upon clicking and holding and dragging, a rectangle will appear. Anything inside this rectangle will be added to your selection, or – if you are holding down **[Shift]** – removed from your selection. The shape of your multi-select gesture tool is determined by the **Multi-Select Gesture** drop-down menu found in the **Selection Group**. You can choose a rectangle, lasso, or circle.

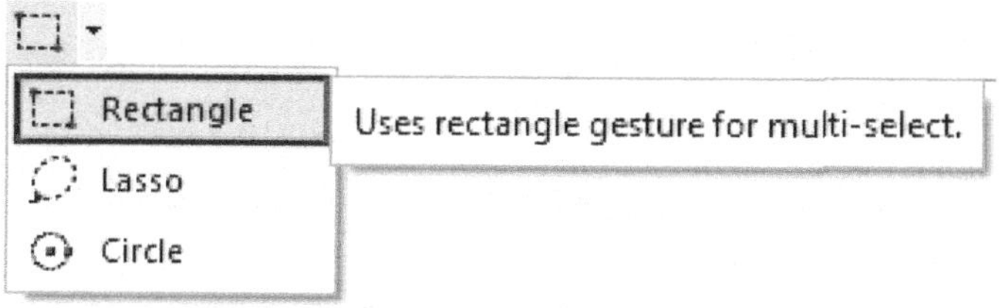

- Hold **[Shift]** and click-and-hold-and-drag a rectangle as shown below.

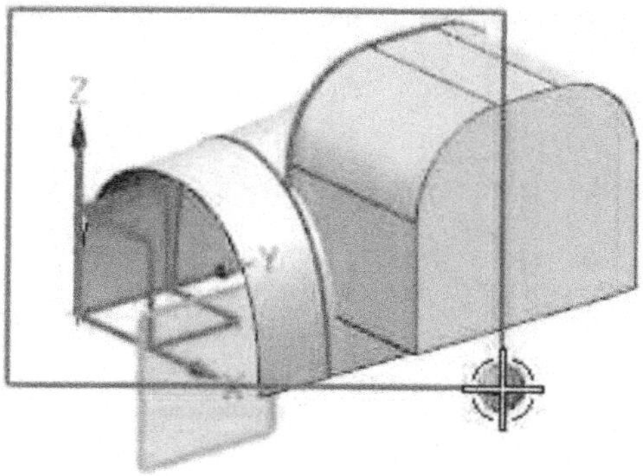

- When you release the left mouse button, you will find that both *Datum Coordinate System (0)* and *Sketch (2)* are no longer selected.

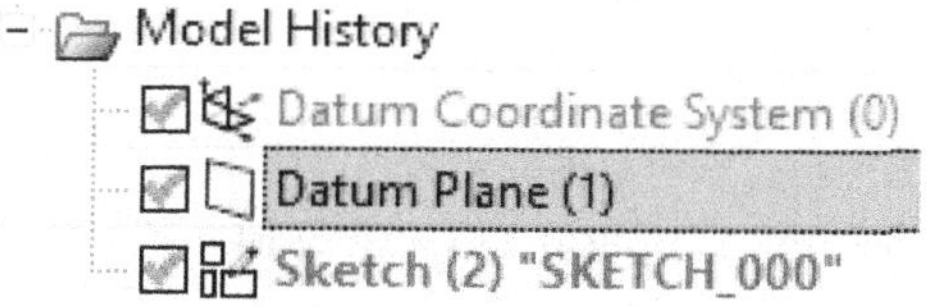

- Change your multi-select gesture to *Lasso*.

- (No longer holding down **[Shift]**) click-and-hold-and-drag the lasso as shown below.

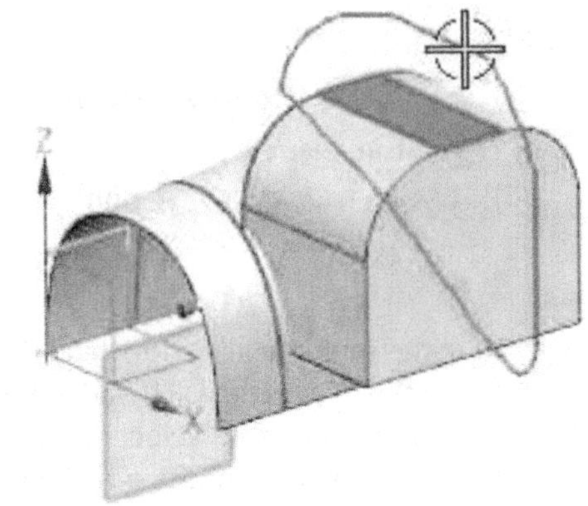

- Change your multi-select gesture to *Circle*.

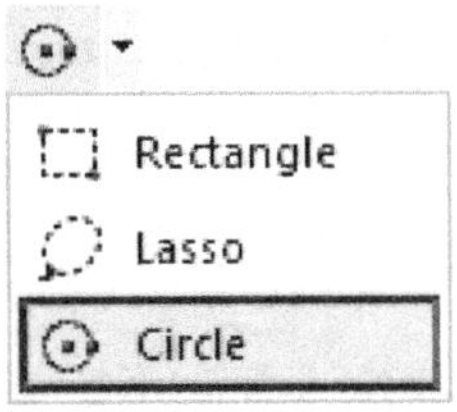

- Click-and-hold-and-drag to select the datum coordinate system and the edge of the sheet body, as shown below.

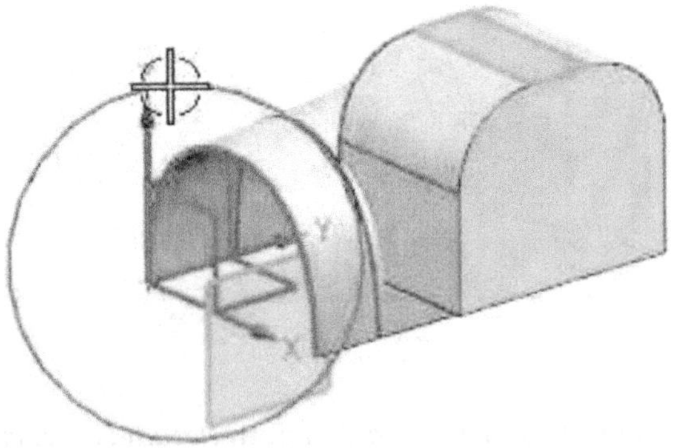

- Now that you are familiar with all three multi-select gestures, you can decide which is best for any given selection or deselection!

3.3.6.4 <u>Keyboard Shortcuts</u>

Familiar keyboard shortcuts from other *Windows* applications also work!

- **[Ctrl]+[A]** (**Select All**) will select all remaining unselected geometry. Note that **Select All** does not select features, but instead faces, bodies, and edges of those features. It does select construction features (datums, sketches).

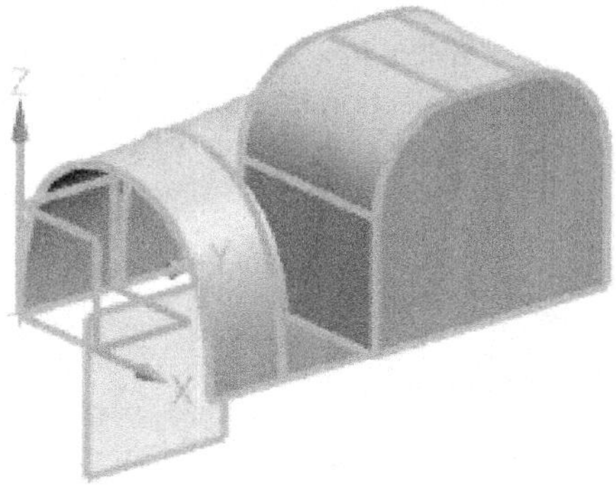

- **[Esc]** will deselect all selected items. Note that if you have a dialog box open, it will also close that!

3.3.7 Selection Drop-Down Menus

In the **Selection Group**, there are three critically important selection drop-down menus – the **Selection Type Filter**, the **Selection Scope**, and the **Selection Intent**. In this section, we study these tools.

3.3.7.1 <u>Selection Type Filter</u>

The **Selection Type Filter** drop-down menu determines *<u>what</u>* you are able to select with your cursor. It is imperative that you check your **Selection Type Filter** before making a selection – often times when you are unable to make a selection, it is due to the **Selection Type Filter** setting!

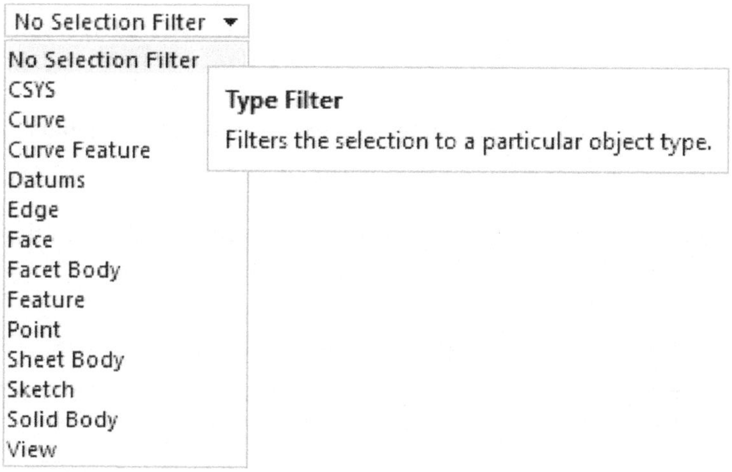

Note that when you have tool open (such as **Extrude**) and the active item on the menu requires a selection (such as **Section**, in the case of **Extrude**), you will only see types in the **Selection Type Filter** that are relevant for that type of selection. The types shown below are the only valid inputs for the **Section** of an **Extrude**.

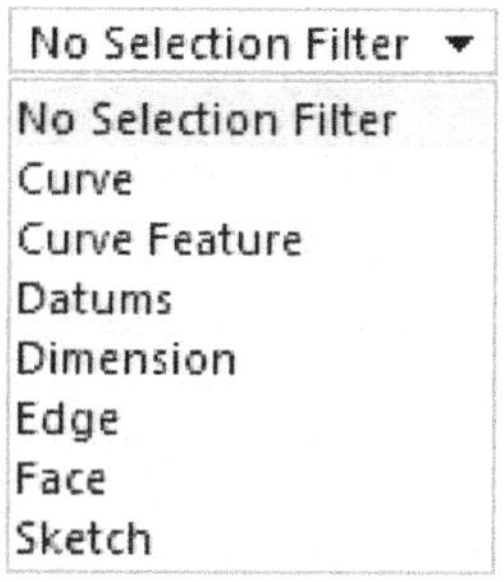

- Set your **Selection Type Filter** to *Datums*.

- Make sure that all geometry in your part file is visible (use **Fit** or switch to the **Trimetric** view if necessary. Use **[Ctrl]+[A]** to select all geometry matching your selection filters.

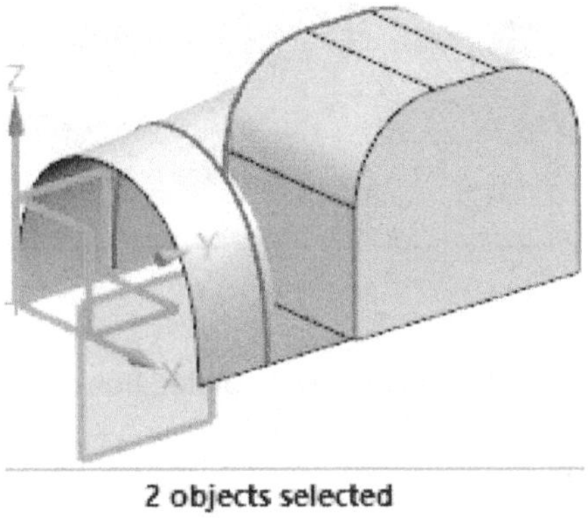

- Your **Select All** command found the two datum features – the datum coordinate system and the datum plane. **Deselect All** via whatever method you want.
- You can manually change the **Selection Type** Filter back to *No Selection Filter*, but there is a shortcut – the **Reset Filters** button. Find it and click on it.

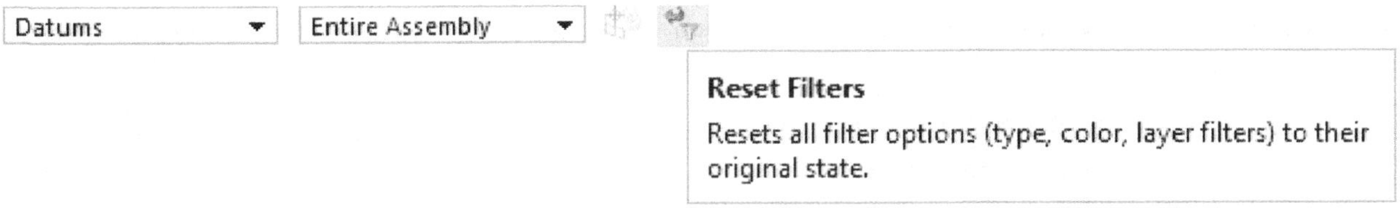

- After using the **Reset Filters** button, your **Selection Type Filter** is restored to the default value – *No Selection Filter*.

3.3.7.2 Selection Scope

The **Selection Scope** drop-down menu determines *where* you are able to select from. In the **Modeling** application, this drop-down menu won't matter much until you start building assemblies.

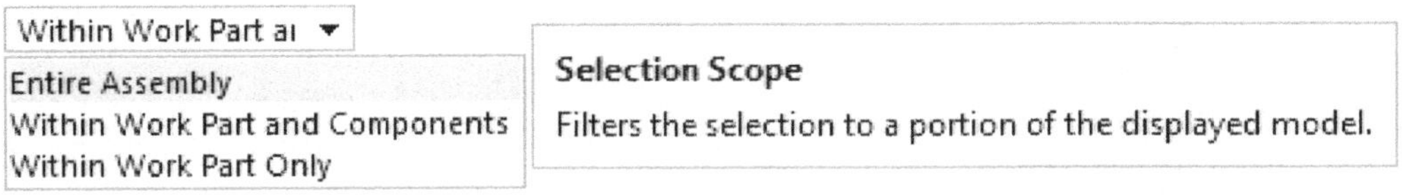

In the early chapters of this book, you *will* want to pay attention to your **Selection Scope** when creating a sketch or using tools in the sketcher.

- Double-click on *Sketch (2)* in the Part Navigator. This opens the sketch for editing. Your Model History will now appear as shown below.

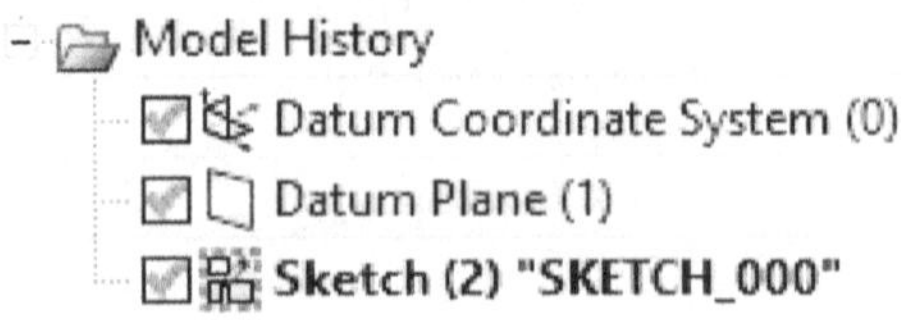

- Note that your **Selection Scope** now includes an option for *Within Active Sketch Only*.

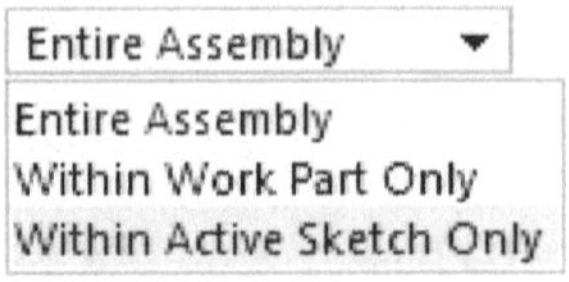

- Use the **Finish Sketch** icon on the far-left side of the ribbon to exit the Sketch Task Environment.

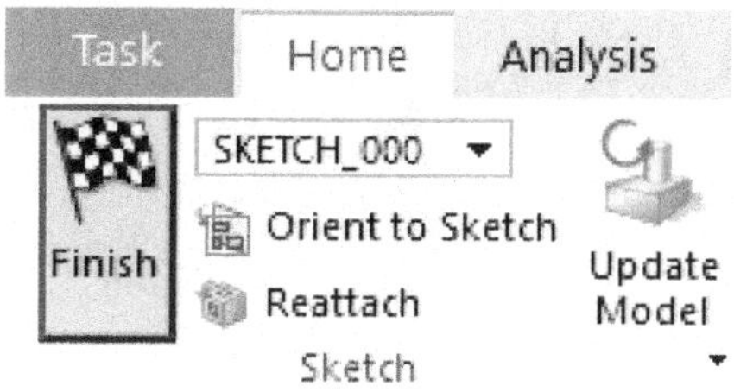

3.3.7.3 <u>Selection Intent</u>

Often, when a tool requires or wants a selection of you, several new icons and/or drop-down menus will appear on the Selection Toolbar to help you further refine your selection intent. The role of the **Selection Intent** drop-down is to determine *<u>how</u>* your selection should continue.

- Double-click on the body of *Extrude (3)*. The **Extrude** dialog will appear, and in the middle of the Selection Toolbar, you should see a new drop-down menu appear. This drop-down is your **(Selection Intent) Curve Rule**.

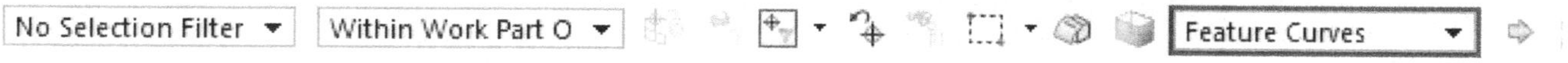

- Use **Deselect All** to clear the selection.
- Change the **Selection Intent** drop-down to *Single Curve*. Note that some of the icons to the right of the drop-down have changed – certain choices here come with their own special options that take the form of buttons on the Selection Toolbar. When you go to pick a curve from the sketch, only that curve is selected.

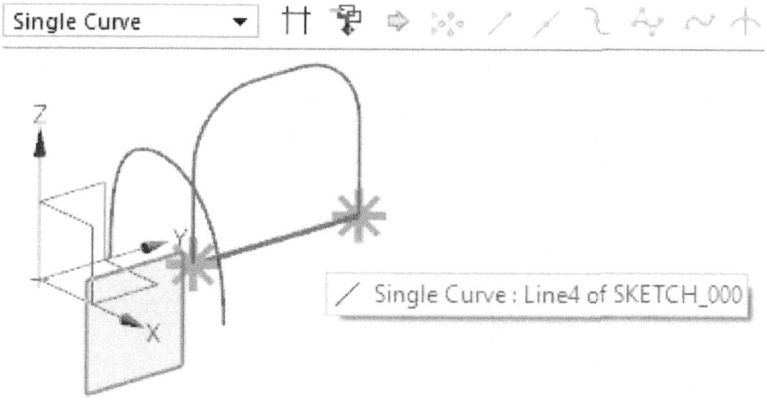

- Change the **Selection Intent** drop-down to *Tangent Curves*. Now when you make a selection, it will continue along tangent curves connected at their ends.

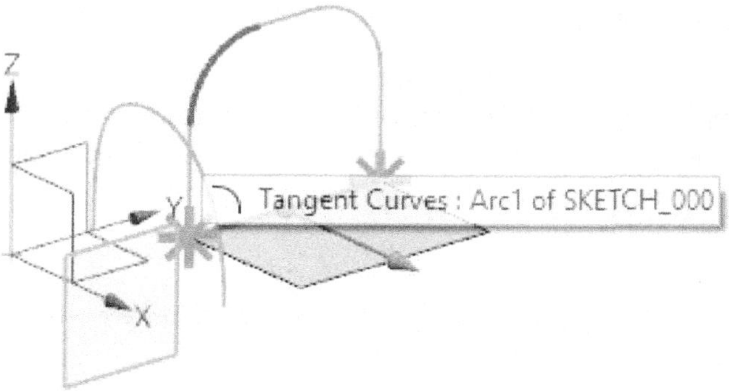

- Use **Deselect All** again to clear your selection.

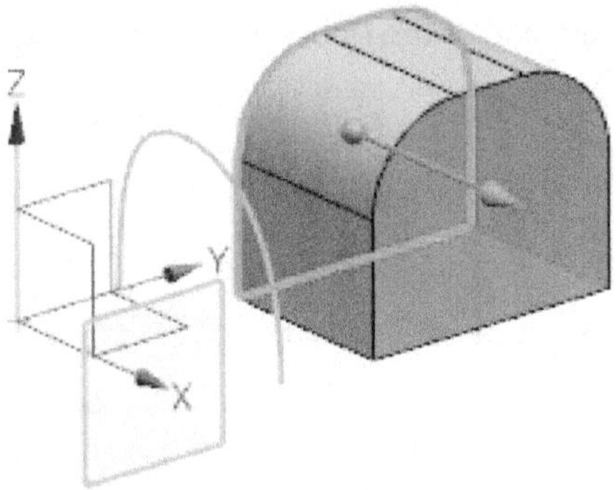

- Set your **Selection Intent** drop-down to *Connected Curves*. Now when you make a selection, it will continue along curves that are connected at their ends, but not necessarily tangent.

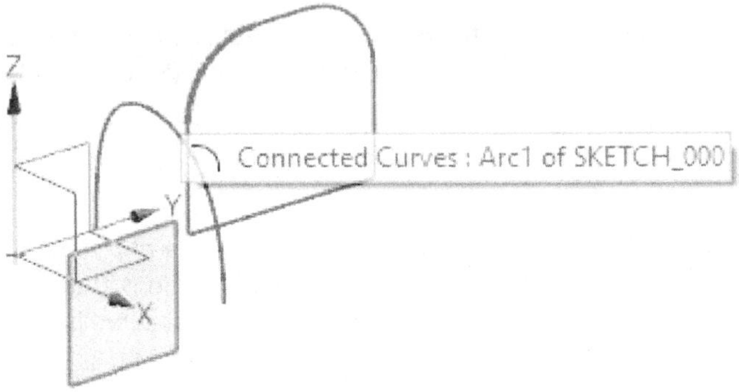

- There is a nice shortcut for changing the **Selection Intent Curve Rule** without using the drop-down: simply right-click on your selection, and on the drop-down that appears, you can select a new appropriate **Selection Intent Curve Rule**.

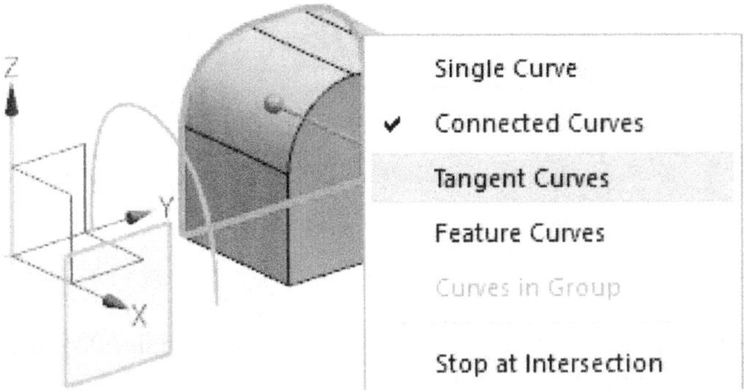

- Close the **Extrude** tool without saving your changes.

As you learn about different modeling tools throughout this book, you will have many opportunities throughout the text to explore different options on these drop-down menus as they appear. For other commands that you will encounter later (e.g., when faces are the required input), it may be called the **(Selection Intent) Face Rule**.

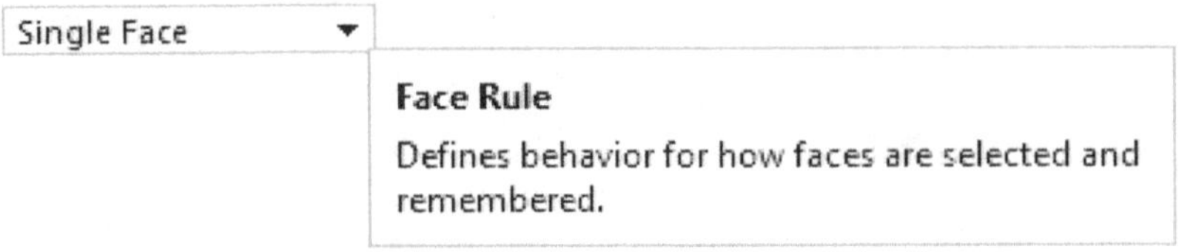

For simplicity, we will refer to any such drop-down menus as the **Selection Intent** drop-down.

3.3.8 Snap Points

Curves in NX have control points that can be selected for various purposes – defining datums, constraints, dimensions, etc. Whether or not these control points are selectable depends on whether the corresponding **Snap Point** icon is enabled on the Selection Toolbar. The **Snap Point** icons are shown below.

The leftmost icon in this group determines whether the entire set of **Snap Point** icons is enabled or disabled.

With **Snap Points** enabled, when you put your cursor on a curve on a control point you should see the corresponding snap point icon appear near your cursor. Let's practice selecting snap points.

- From the Part Navigator, right-click on *"SKETCH_000"* and select **Edit with Rollback**. This opens the sketch for editing as though it were the latest feature in the Model History.

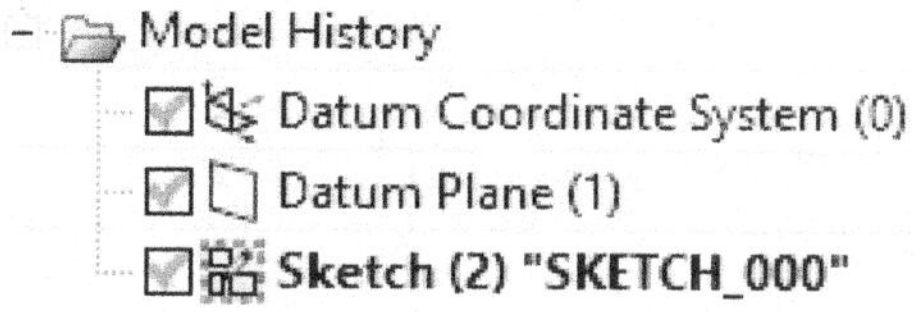

- Put your cursor over the rightmost endpoint of the bottom horizontal line as shown below, and note that the *End Point* snap point icon appears. If it does not, you should make sure it is enabled on the Selection Toolbar.

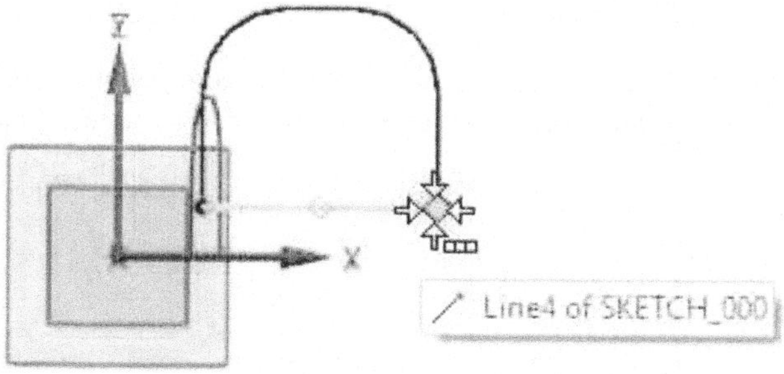

- Disable the *Point on Curve* snap point icon.

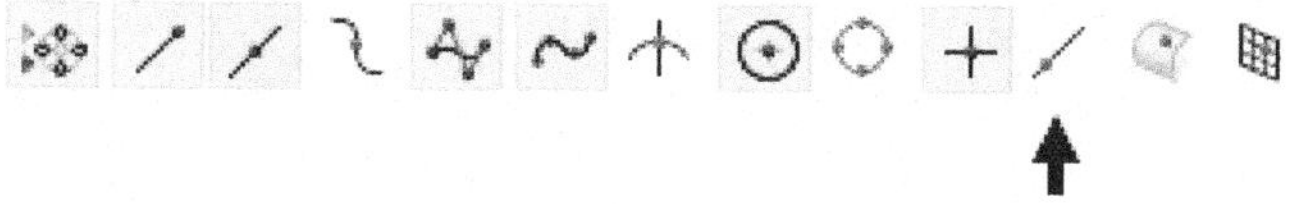

- When you move your cursor away from the end point to another part of the curve, you should no longer see a snap point icon.

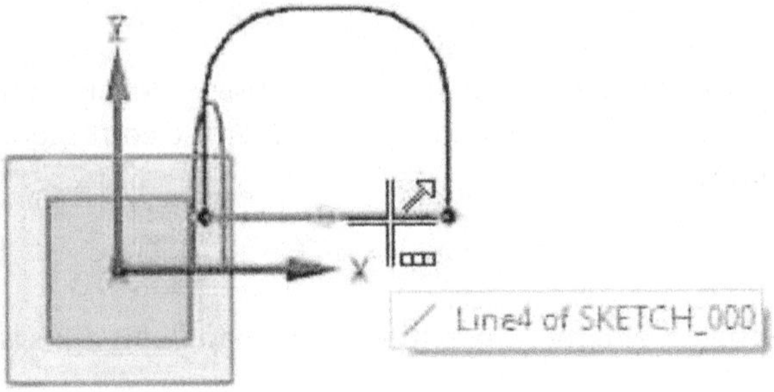

- One snap point icon is particularly important – the **Arc Center** snap point icon.

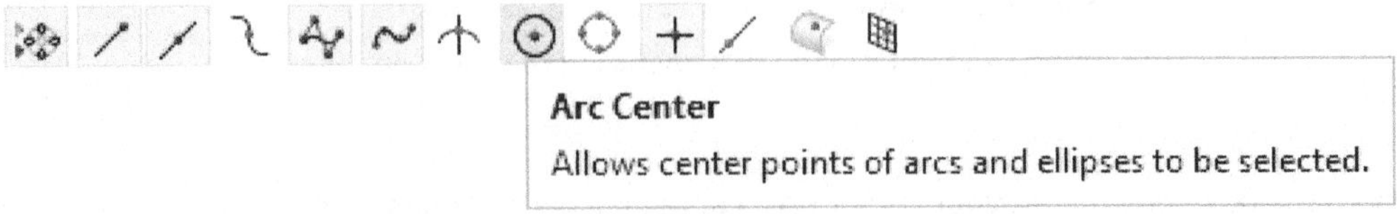

- There is a subtle difference in cursor placement between selecting an arc and its center in NX. If your cursor is on the outside of an arc or circle, but with the selection ball touching the arc, the software infers that you would like to select the arc or circle itself.

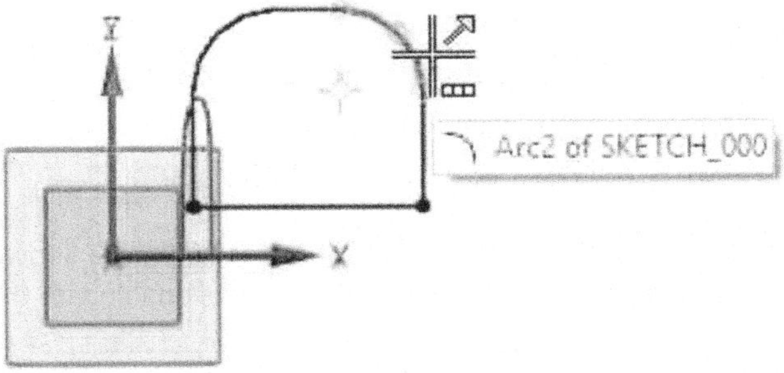

- If your **Arc Center** snap point icon is enabled, and you move your cursor ever-so-slightly to the interior of the arc, again with the Selection Ball touching the curve, the software infers that you would like to select the arc center.

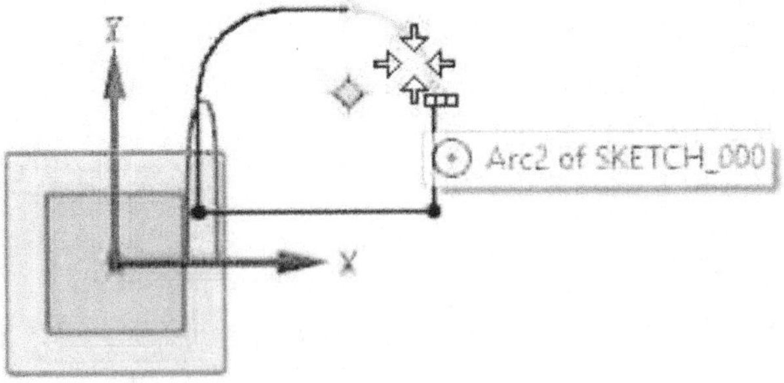

- **Finish** the sketch.

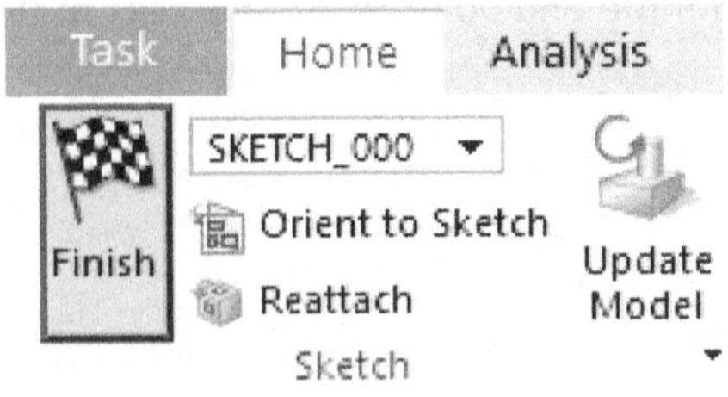

The priority given to arc centers when the *Arc Center* snap point icon is enabled can make it quite difficult to select arcs for other purposes (e.g., applying tangency constraints), so if you need to select several arcs, you might find it useful to disable the *Arc Center* snap point icon!

3.3.9 The Class Selection Menu

The **Class Selection** menu often appears when using certain tools in the Action-Object ordering that may or may not have their own dialog boxes (e.g., **Show**, **Hide**). It allows for greater flexibility than any of the individual selection drop-downs discussed so far, in that you can create compound filters. After making a selection with the **Class Selection** menu, you must click **OK**, at which point NX will either perform the action requested, or take you to another dialog box.

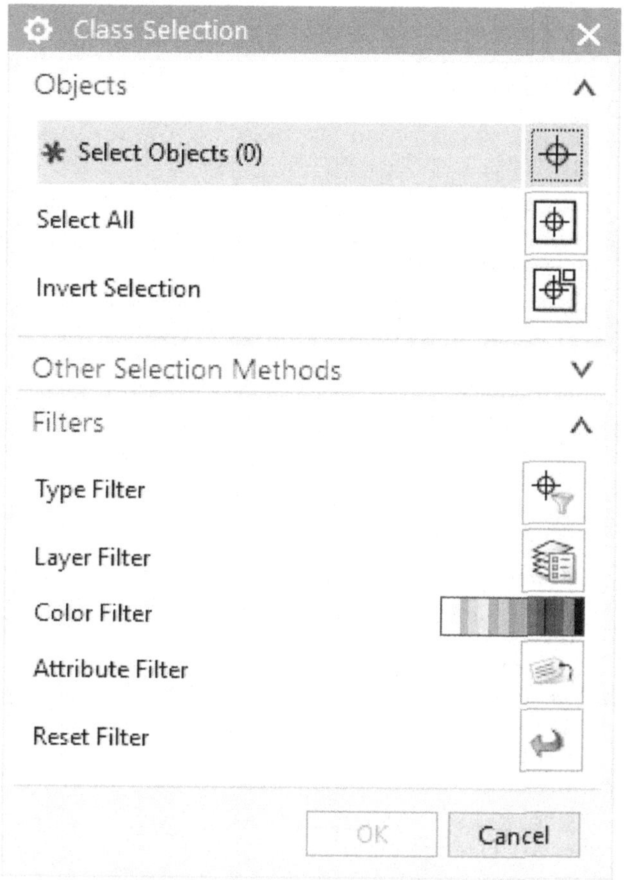

3.3.10 Detailed Filtering

NX offers a refinement of the **Selection Type Filter** that will allow you to filter for objects with specific subtypes. This is called **Detailed Filtering**, and it is available both from the Selection Toolbar, and within the **Class Selection** menu via the **Type Filter** button.

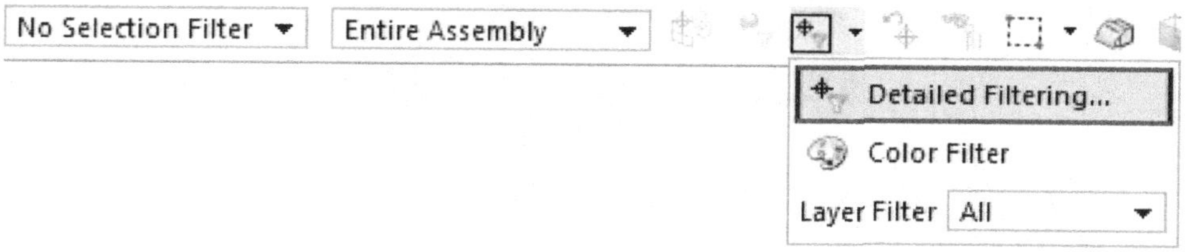

- Open the **Detailed Filtering** menu, and select *Face* from the list of **Types**. Note that you can select multiple **Types** from this list by holding **[Ctrl]** and clicking on them.

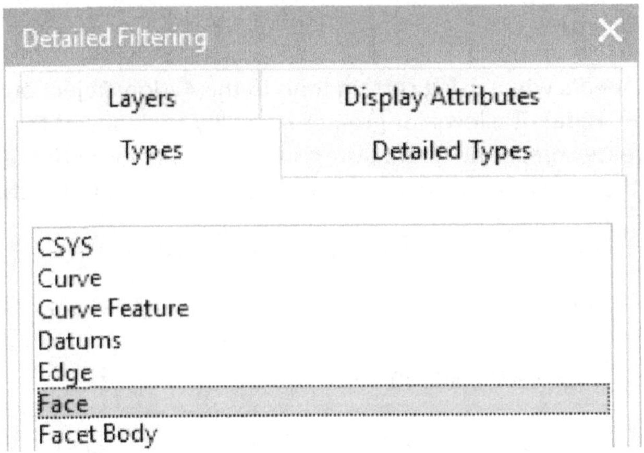

- Select the **Detailed Types** tab, and choose *Plane* from the list.

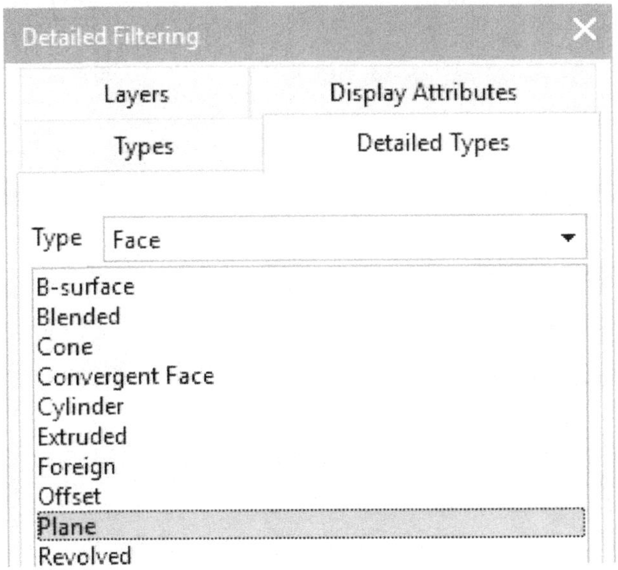

- Close the **Detail Filtering** dialog, and use the keyboard shortcut **[Ctrl]+[A]** to select all planar faces in your model. Note that it finds twelve because the solid body of *Extrude (3)* is hollow!

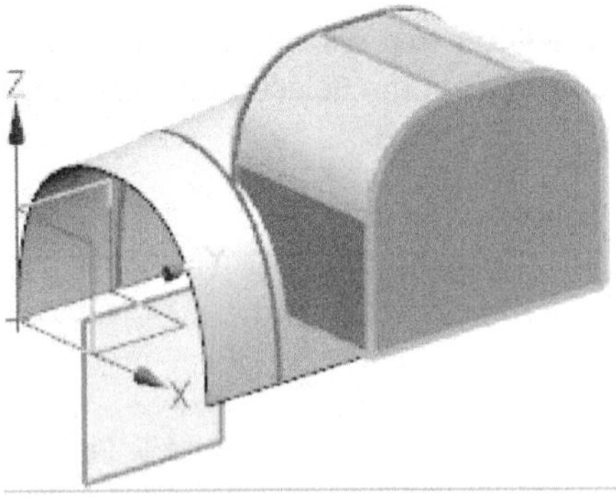

12 objects selected

- **Deselect All**, and use **Reset Filters** to restore your **Selection Type Filter** to its defaults.

3.3.11 Selection Priority

A somewhat subtler aspect of selection is the priority given to different types of entities when you click on them in the Graphics Window. By default, when you left-click in the Graphics Window, the *Top Selection Priority* is given to features – entities that arise from parametric, associative operations, and which populate the Model History.

There is a tool that allows you to designate another type of entity as the top selection priority, although it is not enabled on the Selection Toolbar by default. When it is enabled, the **Selection Priority** drop-down appears on the Selection Toolbar to the left of the **Detailed Filtering** tool.

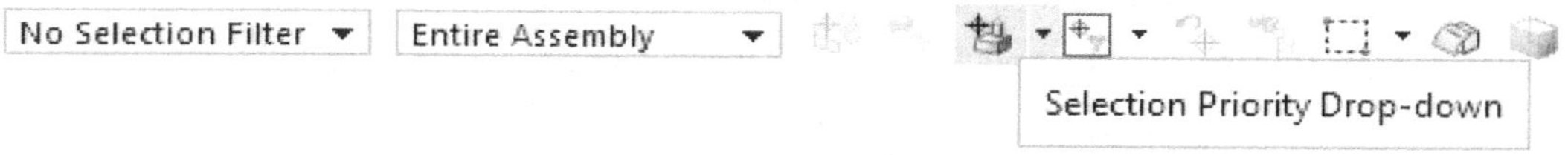

One reason that changing the top selection priority is useful is because there are keyboard shortcuts to reassign the top selection priority. There are no keyboard shortcuts for types in the **Selection Type Filter**, and the **Customize** tool does not allow you to create any such shortcuts either. In this sense, the keyboard shortcuts that reassign the top selection priority are the closest thing in NX to keyboard shortcuts for options in the **Selection Type Filter**.

- To enable the **Selection Priority** drop-down, you must use the **Ribbon Options** drop-down on the far-right side of the Top Border Bar.

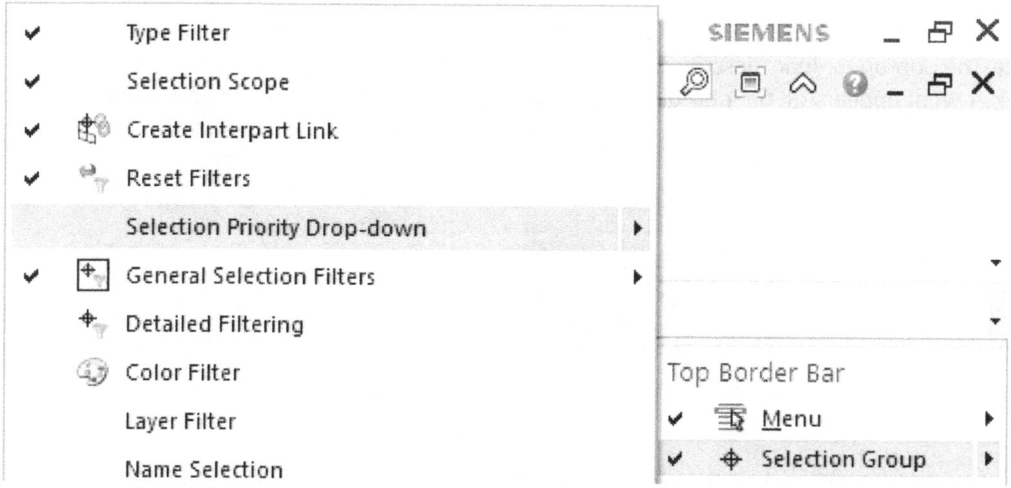

- Bring your cursor over the body of *Extrude (3)* and notice that NX wants to select the feature, *Extrude (3)*.

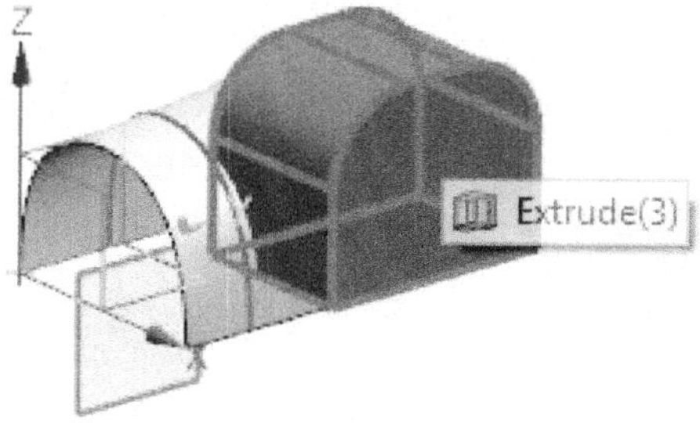

- Change the top selection priority to *Body*. The keyboard shortcuts are a very effective way to switch the top selection priority, so if you memorize them, it really isn't necessary to have this drop-down available on the Selection Toolbar.

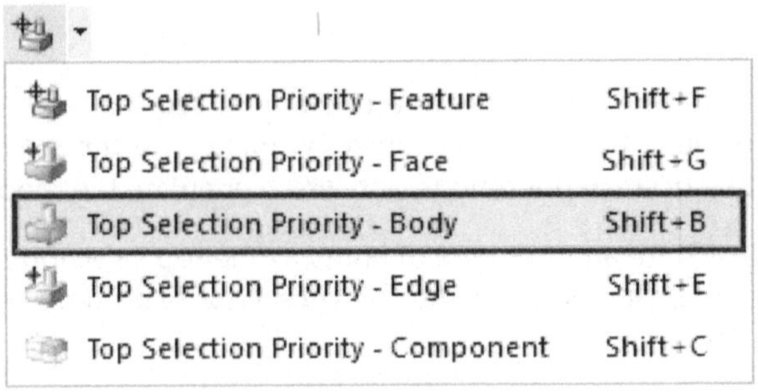

- Now when you put your cursor over the body of *Extrude (3)*, NX wants to select the body!

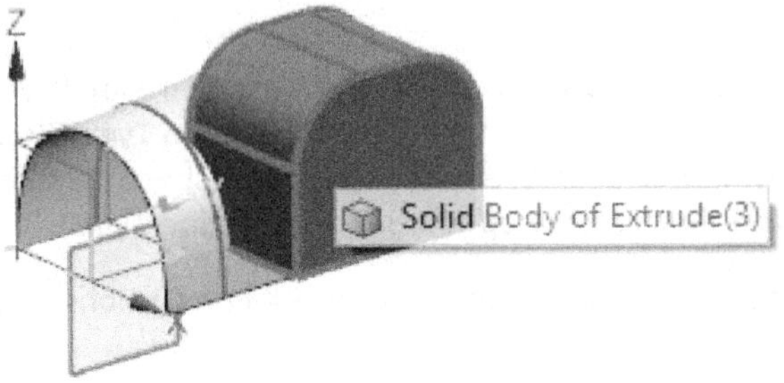

- The distinction might seem subtle or unimportant, but there are advantages to changing the top selection priority. For instance, the pop-up toolbars described in Chapter 2.7 all display type-specific tools. Click on *Solid Body of Extrude (3)* and look at what appears on the pop-up toolbar – these tools only appear when your selection is of type *Body*.

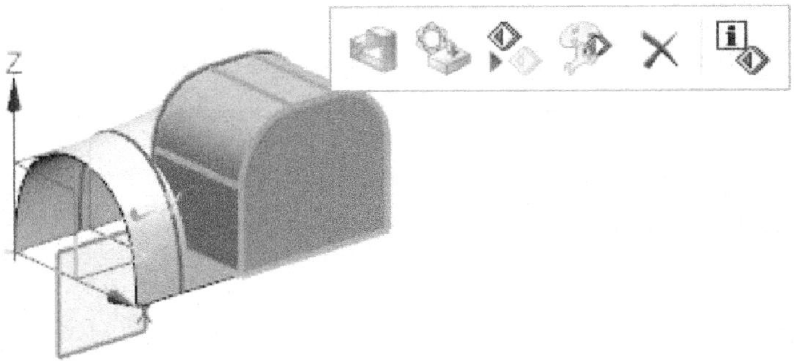

- Set the top selection priority back to *Feature* (**[Shift]+[F]**). Now when you click on *Extrude (3)*, the tools on the pop-up toolbar are all specific to features.

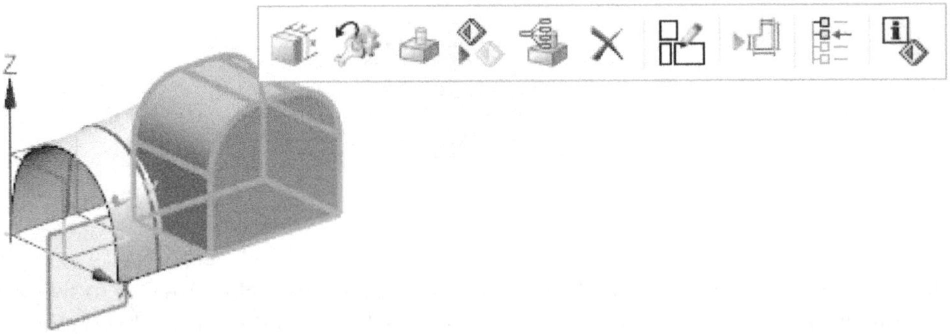

- Set the top selection priority to *Face* (**[Shift]+[G]**).
- Right-click menus are also type-sensitive. Right-click on the front face of the body of *Extrude (3)*. Notice that in the right-click menu, there are a number of tools that let you create objects based on the face – for instance, **Sketch in Task Environment** would position a sketch on the selected face, and **Datum Plane** would position a plane parallel to the selected face.

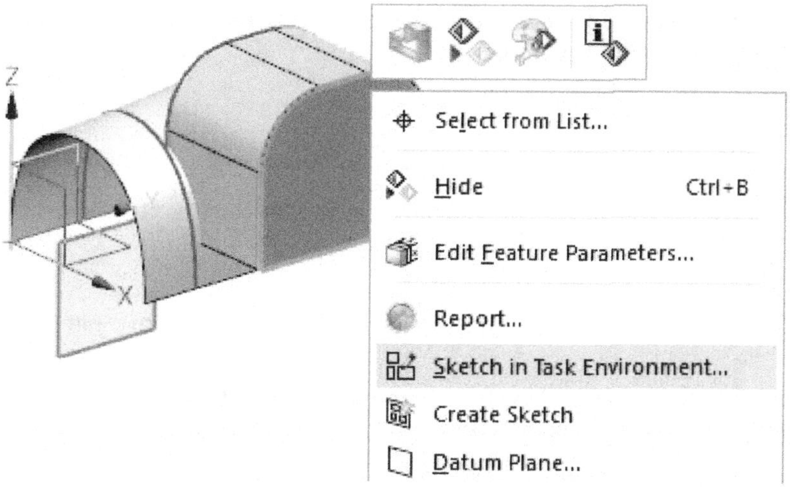

- Make no selection from the right-click menu – **[Esc]** will close it. Set the top selection priority back to *Feature*. Now, when you right-click on *Extrude (3)*, the entries in the right-click menu are all feature-specific.

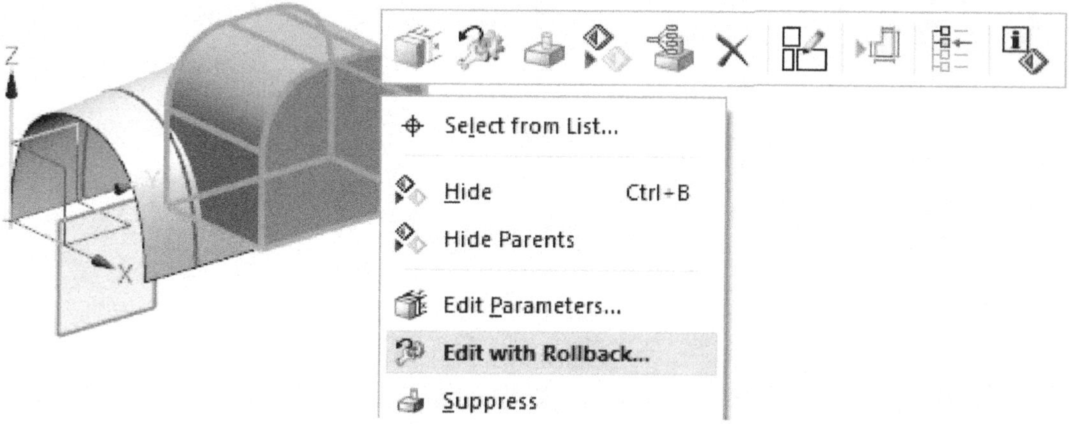

The main advantage that the Selection Priority keyboard shortcuts offer advanced NX users is the ability to perform many actions in the "Object – Action" selection ordering. The "Action – Object" selection ordering described in Chapter 3.3.1 is useful when you are learning NX, and indeed is required for many types of actions. But as you learn the software better you will want to take as many shortcuts as possible to avoid excess clicking. Accessing tools via shortcut menus in the "Object – Action" selection order with the help of the Selection Priority keyboard shortcuts is a trick that every expert user of NX should know.

3.4 Visibility and Display

As your models become more complex, managing visibility and display settings will be extremely helpful.

3.4.1 Show, Hide, and Show and Hide

On the ribbon, you can find the **Show** and **Hide** commands in the **Show Hide Gallery**, in the **Visibility** group, on the **View** tab.

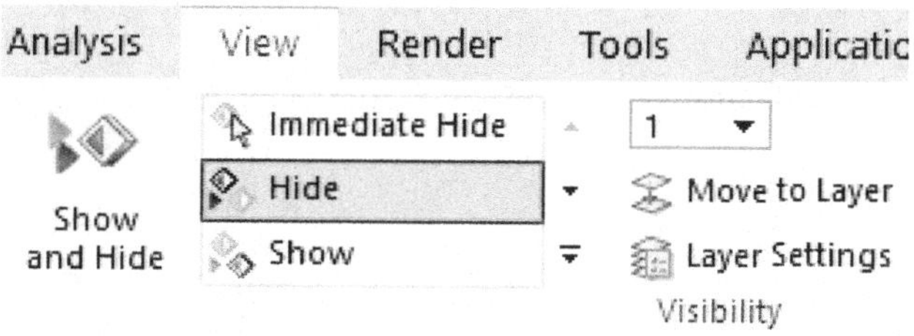

It is usually not convenient to switch to the **View** tab to access these tools. Most can also be found in the **Show/Hide drop-down menu** within the **View** group on the **Top Border Bar**.

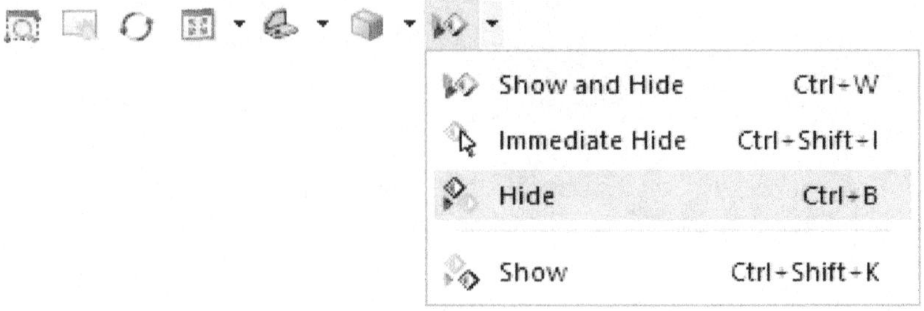

- Select the **Hide** command from either the **View** tab or from the **Top Border Bar** (your choice). Note that when you use the **Hide** command in the "Action – Object" selection ordering, the **Class Selection** menu will appear.

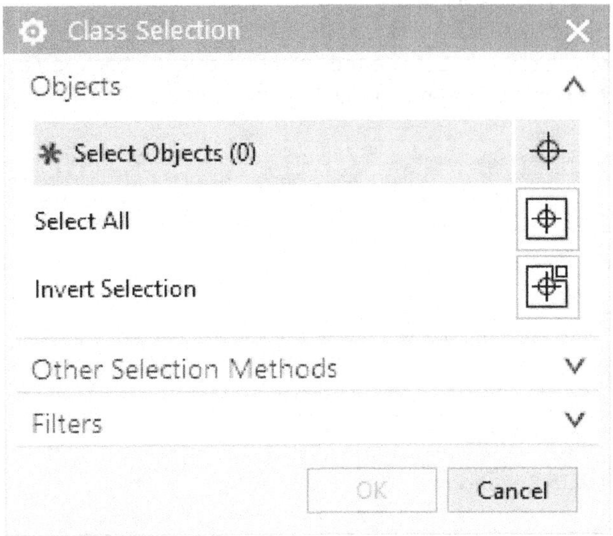

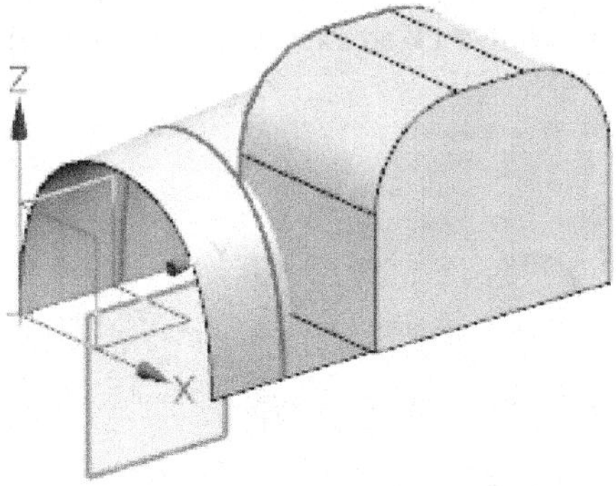

- Your Cue Display (remember, it's in the bottom left corner of the NX application) is your only indication that the **Class Selection** menu has anything to do with the **Hide** command you just asked for!

Select objects to hide

- Set your **Selection Type Filter** to *Datums*, and use the **Select All** button on the **Class Selection** menu.

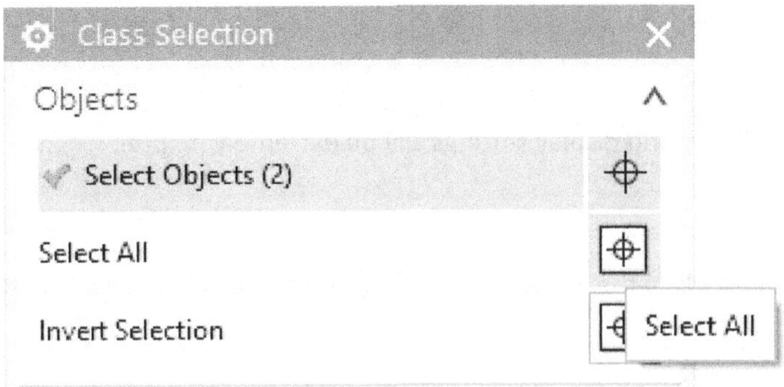

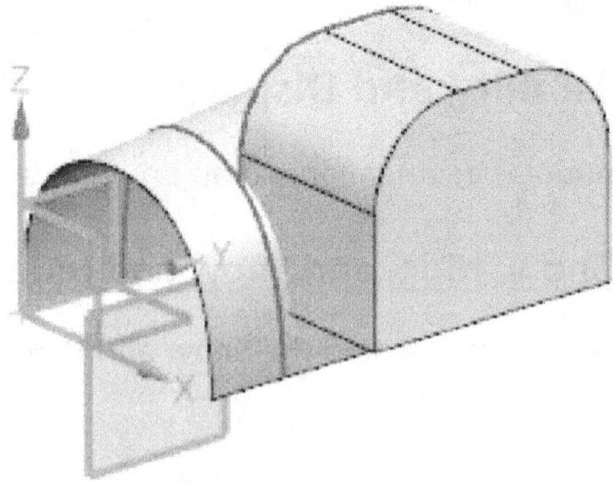

- When you click **OK**, the datum coordinate system and the datum plane will disappear!

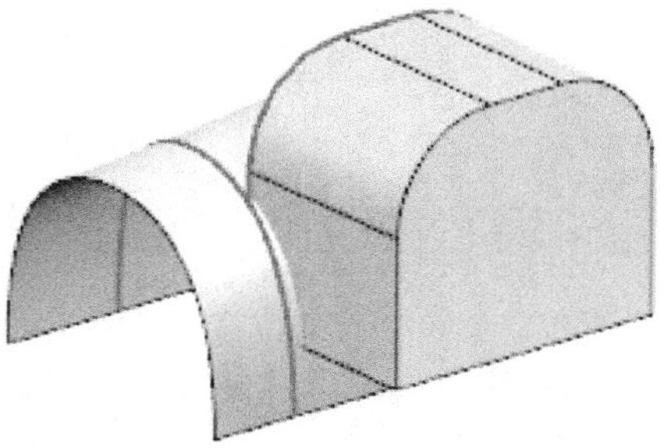

- Next, select the **Show** command from either the **View** tab or the Top Border Bar, your choice. Again, the **Class Selection** menu appears. Note that your visibility settings have inverted – everything that was shown is now hidden, and everything that was hidden is now shown. This lets you make your selection for **Hide** from only those objects that are hidden!

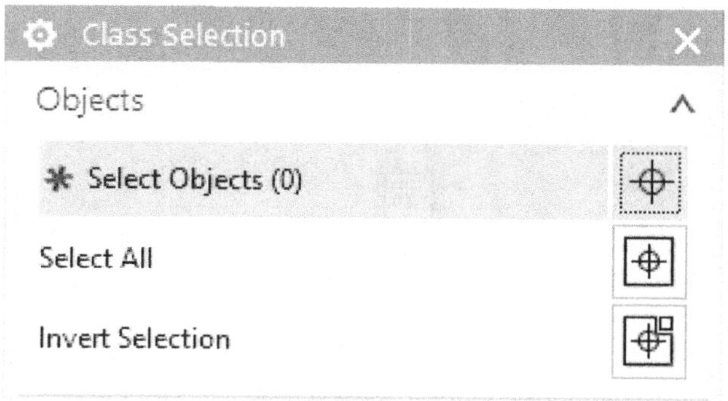

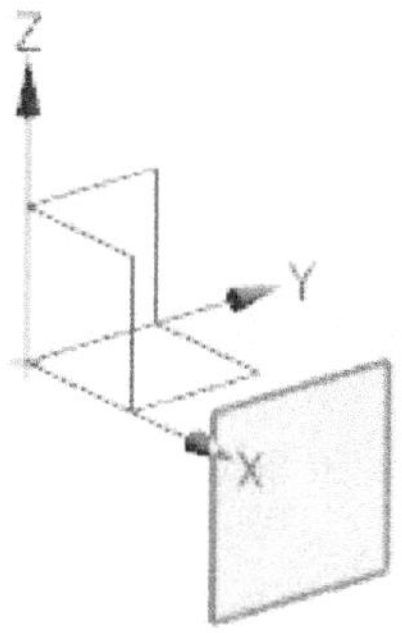

- An advantage that the **Class Selection** menu offers is the ability to set precise filters that go away as soon as the operation is completed. Within the **Filters** section of the **Class Selection** dialog, click on the **Type Filter** icon.

- In the **Select by Type** dialog, choose **Datums**, and click on the **Detail Filtering** button.

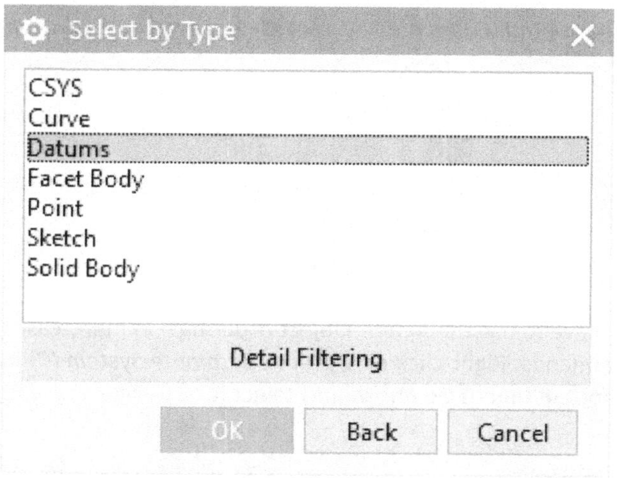

- Within the **Datums** menu that appears, you can choose the subtype **Datum Plane** to refine your **Selection Type Filter** further! Click **OK** when you have made this selection, and click **OK** again when you return to the **Select by Type** menu.

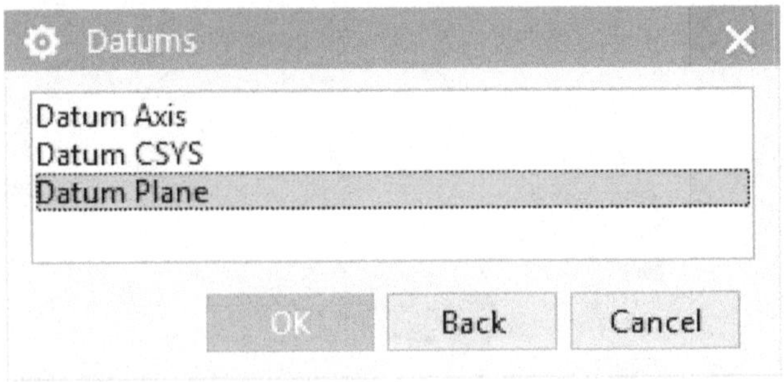

- When you return to the **Class Selection** menu, click on the **Select All** button again. This time, it only grabs your *Datum Plane*!

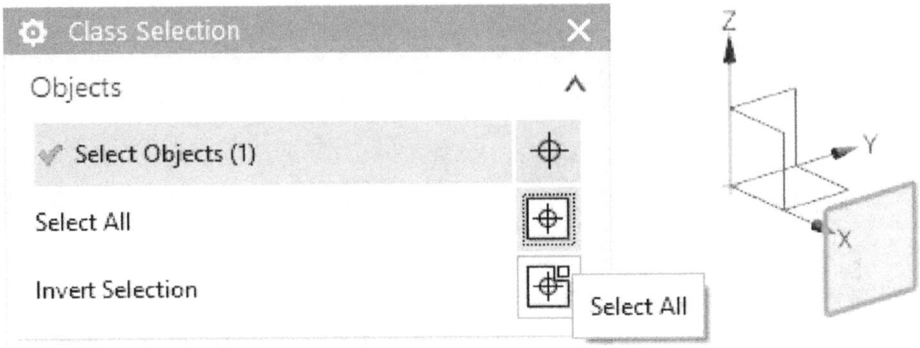

- Pretty nifty, right!? Click **OK** so that the datum plane becomes visible.

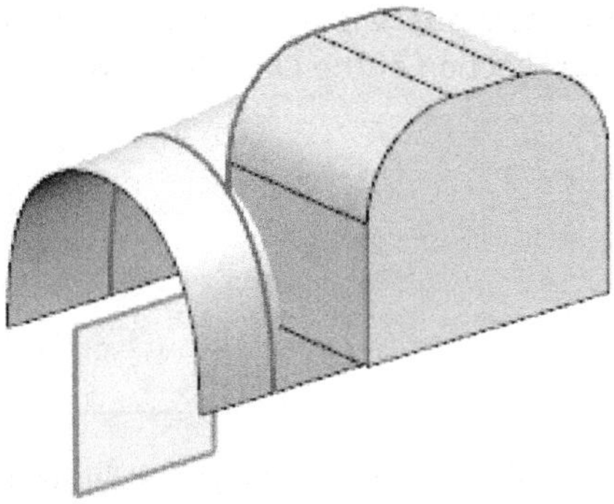

- In your Model History, note that *Datum Coordinate System (0)* is grayed out. This is what happens to features that are hidden from view.

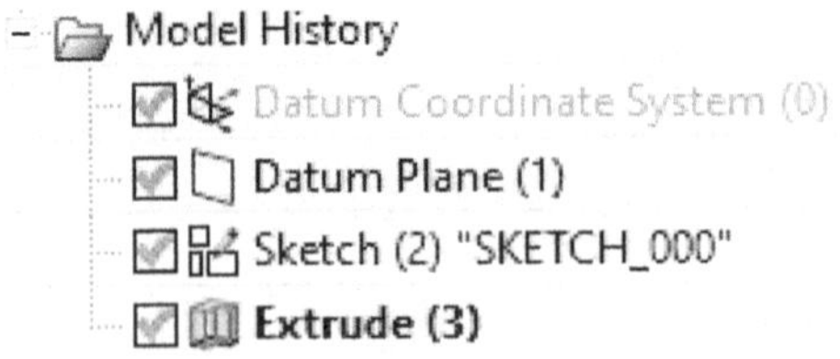

- By the way, you don't always have to use the Action-Object ordering – in many cases, you can right-click on an object and find a list of relevant commands. Right-click on *Datum Coordinate System (0)* in the Part Navigator and you will see **Show** at the top of the shortcut menu! Go ahead and select it.

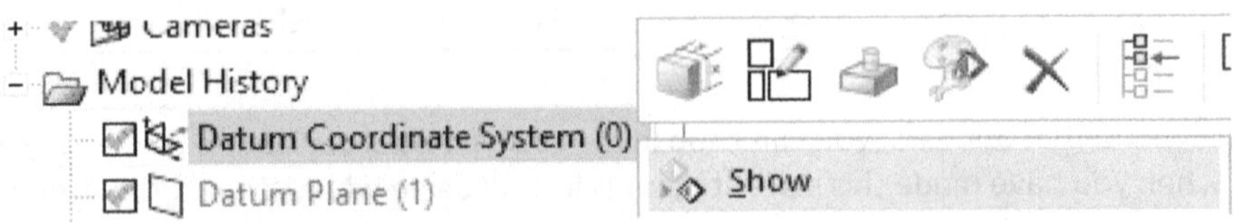

- Now all your features are visible again. Let's use the **Hide** command one more time to get an even better appreciation for what the **Class Selection** menu can do. This time, use the keyboard shortcut **[Ctrl]+[B]** to select the Hide command.
- Click on the **Type Filter** button again.

- In the **Select by Type** menu, click on *Datums*, and then press and hold the **[Ctrl]** key and click on *Sketch*.

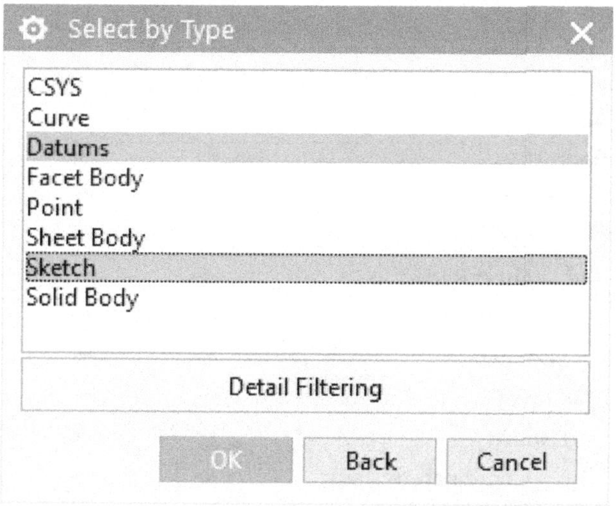

- Click **OK** to return to the **Class Selection** menu. Note that your **Selection Type Filter** drop-down menu is blank – the compound **Type Filter** that you set is not in that list, so it doesn't display in the drop-down!

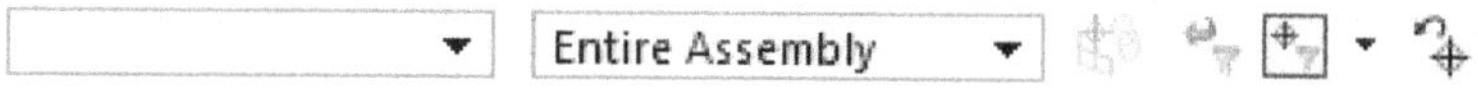

- Use the **Select All** button to select all features satisfying the **Type Filter** criteria.

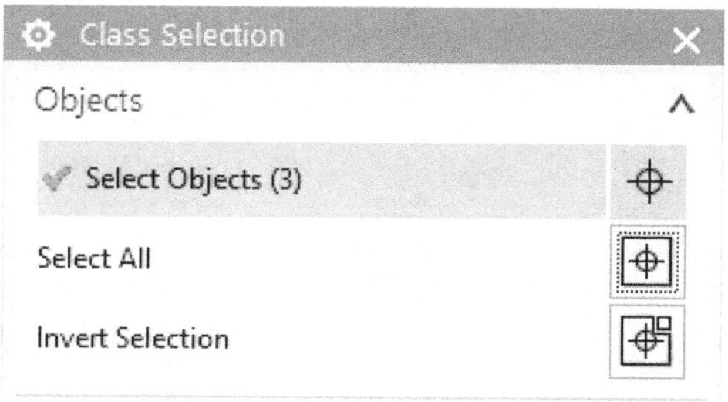

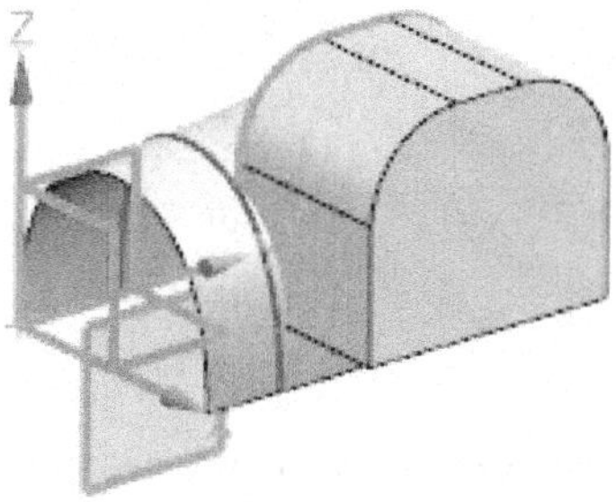

- This time when you click **OK** the sheet body, solid body, and curve remain visible.

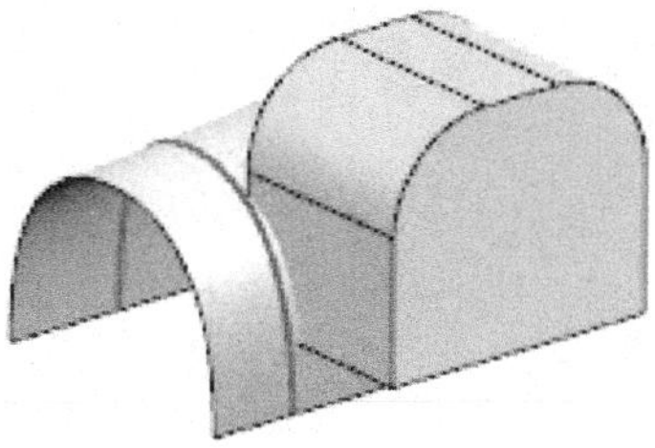

- The **Show** and **Hide** commands are great for selecting individual objects or sets of objects, but the **Show and Hide** command is a much more efficient way to show or hide whole classes of objects. The **Show and Hide** command is found both in the **View** tab, as well as on the **Top Border Bar**.

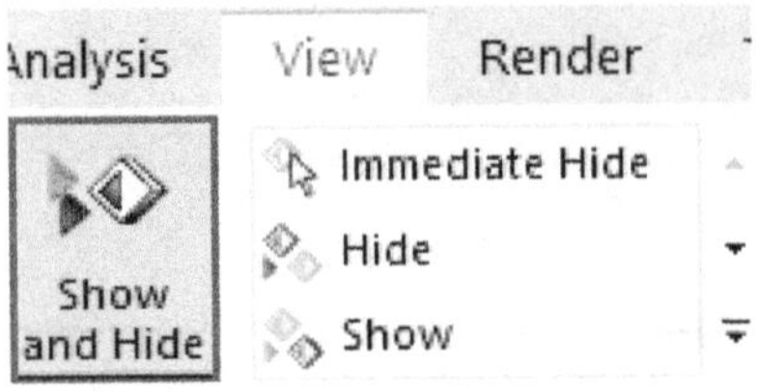

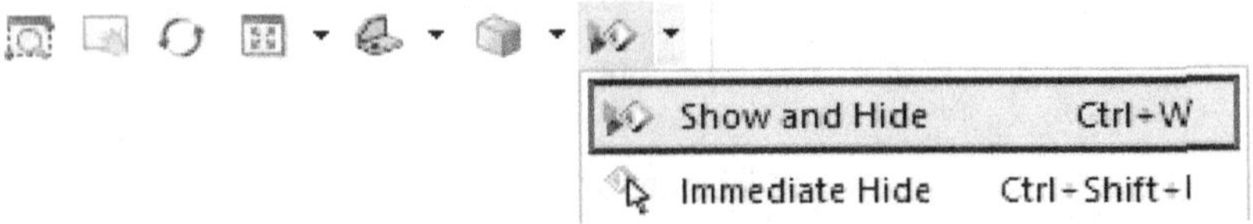

- Select the **Show and Hide** command from either location. The menu will appear as shown below.

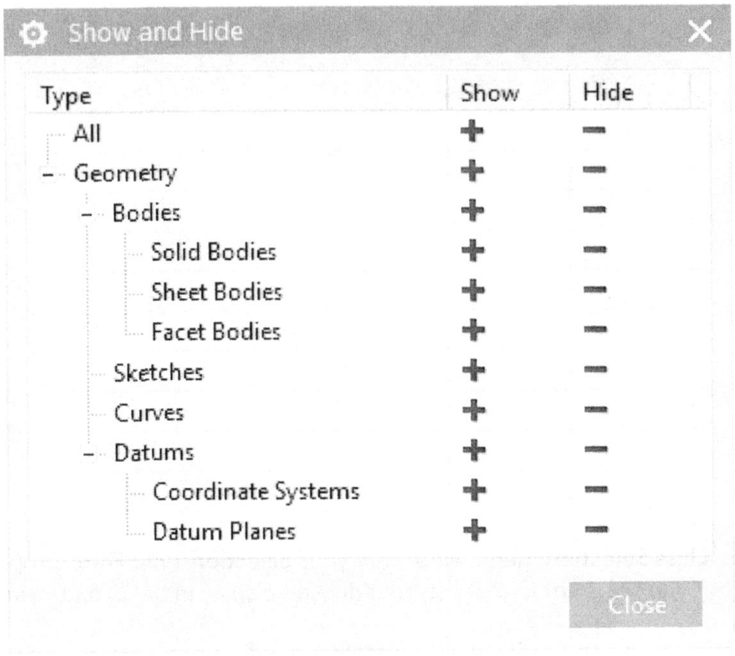

- By using the + and – icons in the menu, you are able to show or hide entire classes of objects. Give it a try! Use **Show All** and then whatever combination of + and – icons you want!

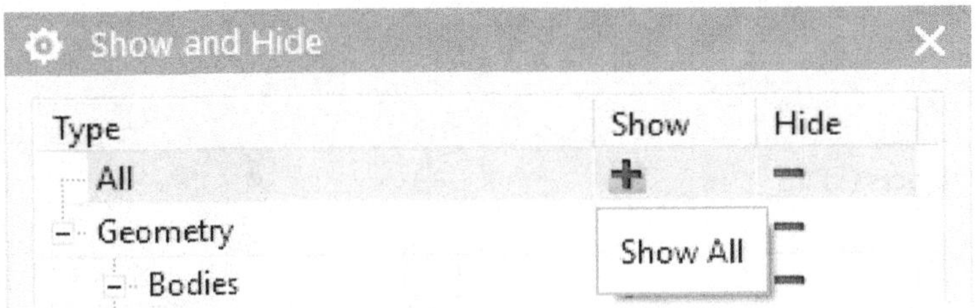

- The **Show and Hide** menu gives you a great way to hide all your construction geometry and only show the solid model that you are building. Use **Hide All**.

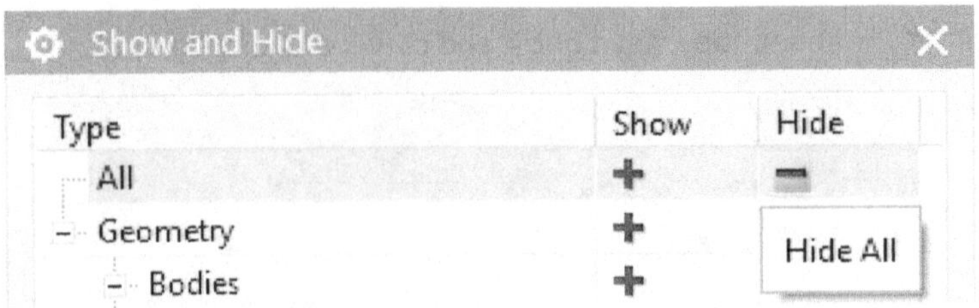

- Then, use **Show Solid Bodies**.

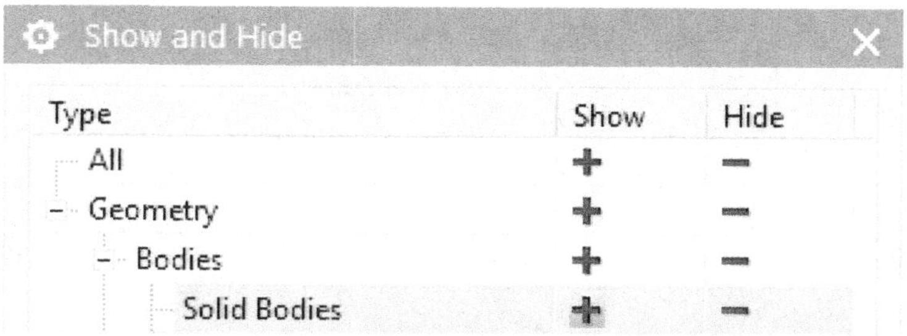

- After this, again, only the solid body of *Extrude (3)* will be visible!

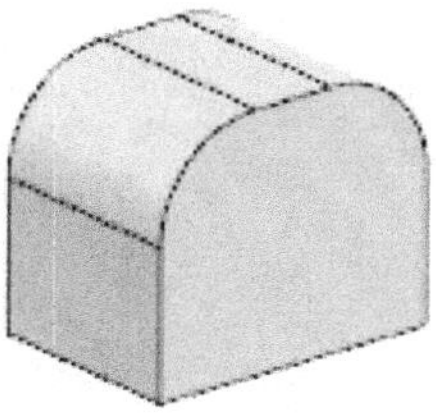

3.4.2 Rendering Styles

At times, you will wish to change the rendering style in the graphics window. Up until now we have been using the *Shaded with Edges* rendering style. Buttons that allow you to switch rendering styles can be found in two locations. First, you can find them in the **Style** group on the **View** tab.

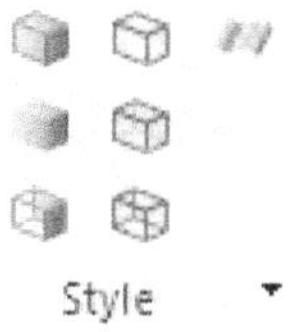

As with other commands from the **View** tab, these can also be found in the **Rendering Style** drop-down menu in the **View** group on the Top Border Bar.

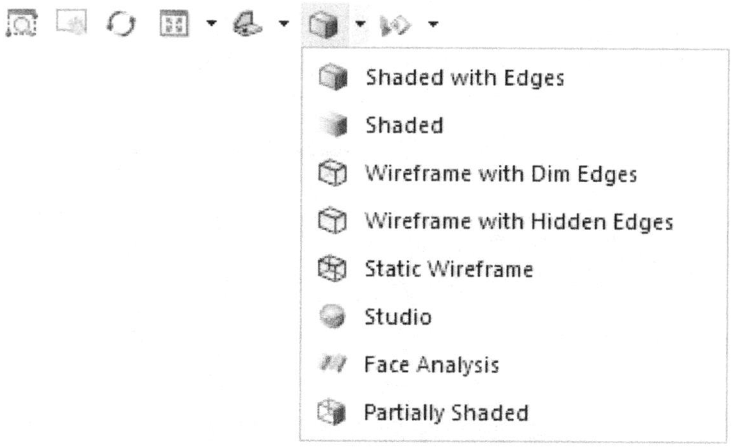

The names are, for the most part, self-explanatory. A few notable exceptions are *Studio*, *Face Analysis*, and *Partially Shaded* – these have special roles in rendering, displaying face analysis objects, and in object display, respectively.

3.4.3 Object Display

The display properties of objects in your part file can be manipulated with the **Edit Object Display** command, found in the **Visualization** group on the **View** tab. It also has a keyboard shortcut of **[Ctrl]+[J]**.

This tool allows you to modify the color, transparency, line widths, and various other aspects of an objects visualization in the graphics window.

- Select the **Edit Object Display** command. Note that this is another example of a tool that brings up the **Class Selection** menu. Select the solid body.

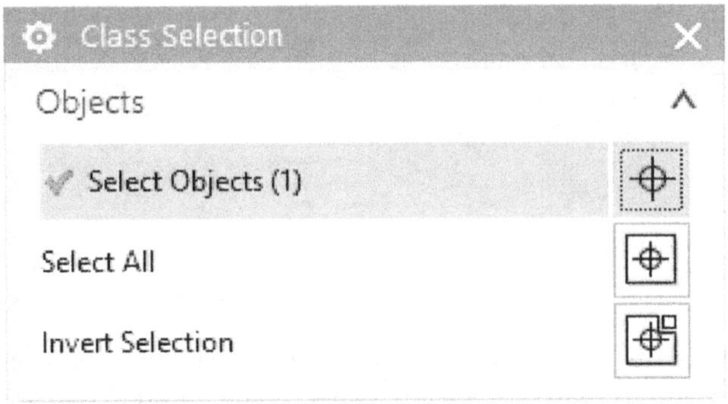

- This time when you click **OK**, NX will bring up a new dialog box – **Edit Object Display**. We will modify the color of the cylinder, by clicking on the **Color** swatch.

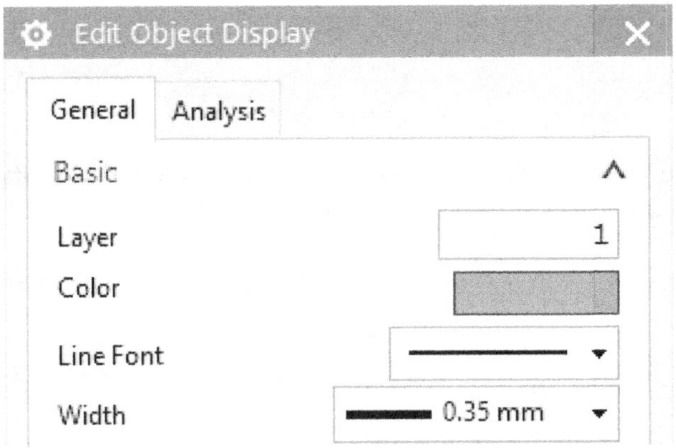

- On the **Color** menu, there is an extensive **Palette**. Choose your favorite color and click **OK**!

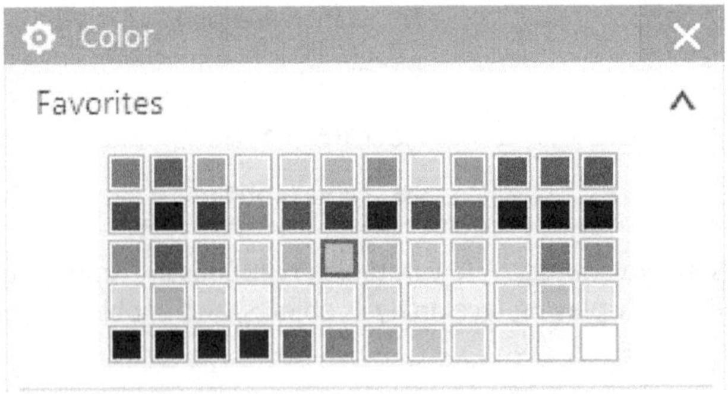

- When you return to the **Edit Object Display** menu, note that there is a translucency slider. Slide the bar to *50* and watch your model become transparent!

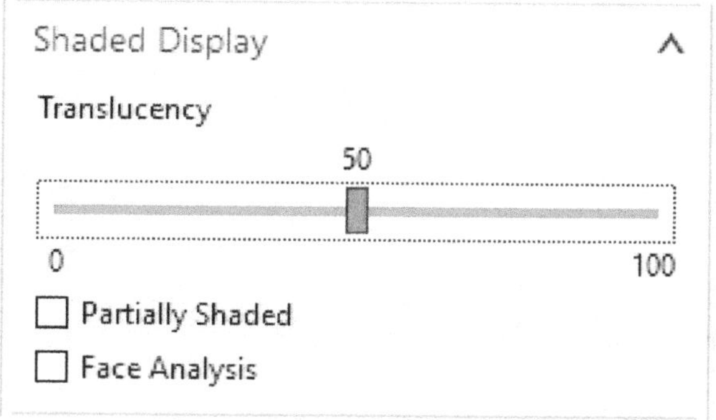

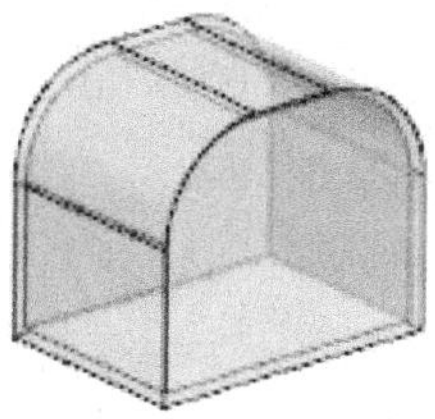

- Click **OK** again to complete the operation! Your cylinder will then take on whatever color you selected.
- Use **[Ctrl]+[Z]** to undo the changes you made to the color and translucency so your model looks the same as ours in the following exercises!

Another noteworthy setting in the **Edit Object Display** dialog is the *"Apply Changes to Owning Part"* checkbox. This can be enabled when you are changing the color of a component in an assembly – check this box when you want the color change to apply to the actual part, and not the instance within the assembly.

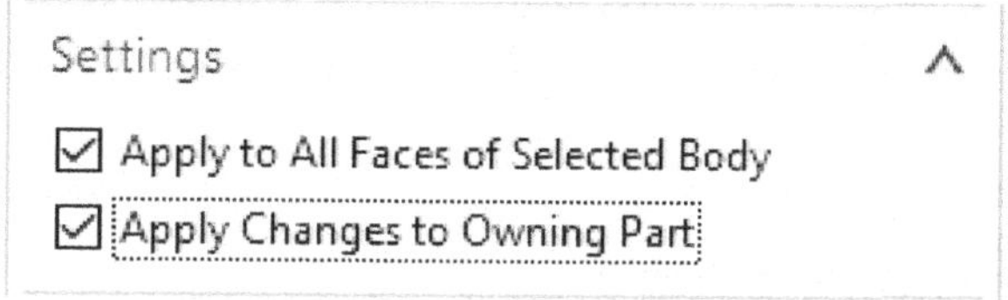

3.4.4 Reference Sets

In an NX assembly, the visible geometry from each component in that assembly is determined by the active reference set for that component. A reference set is simply a collection of entities within a part file. When that part file is brought into an assembly as a component, some reference set in that part file is activated – meaning that only the geometry within that reference set is visible. NX part files have three default reference sets – *Entire Part*, *Empty*, and *Model*. All other reference sets are user-defined.

Reference Set	Contains
Entire Part	All geometry in the part file
Empty	No geometry from the part file
Model	Only bodies from the part file Best Practice: only solid bodies

In the following project, you will activate the different reference sets in the part file *"The Graphics Window.prt"* by viewing it as a component in an assembly, make modifications to the existing reference sets, and create a user-defined reference set. The **Reference Sets** tool is found on the ribbon all the way at the bottom of the **More** gallery on the far-right side of the **Assemblies** tab[1]. The **Assemblies** tab is only visible when the **Assemblies** application is enabled, so for now you will prefer to access the Reference Sets tool via **Menu / Format / Reference Sets**, which is always accessible, regardless of whether the **Assemblies** application is enabled. The **Reference Sets** tool controls the reference sets for the Work Part.

[1] See Chapter 11.

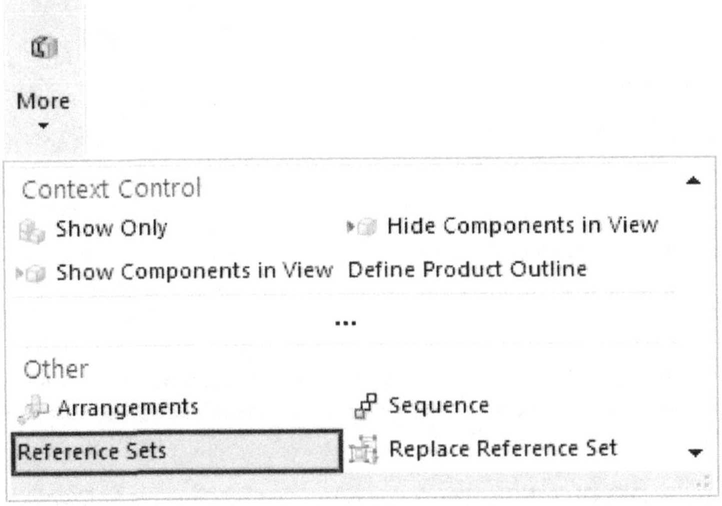

- Open the file *"The Graphics Window – Assembly.prt"* from *"C:\My NX 11 Files\Downloaded Files"*.
- This file contains *"The Graphics Window"* as a component, as shown in the assembly tree structure in the Assembly Navigator. Use the **Assembly Navigator Properties** dialog described in Chapter 2.9.2 to display the **Reference Set** column. As you can see, the active reference set for *"The Graphics Window"* is *Entire Part*.

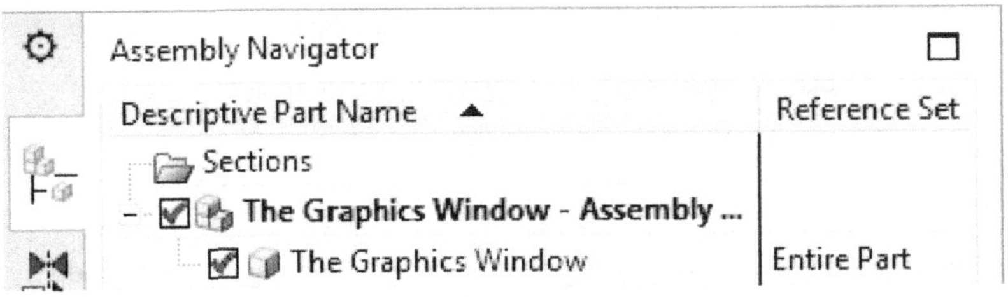

- Right-click on *"The Graphics Window"* in the Assembly Navigator, and select **Replace Reference Set** / *MODEL*.

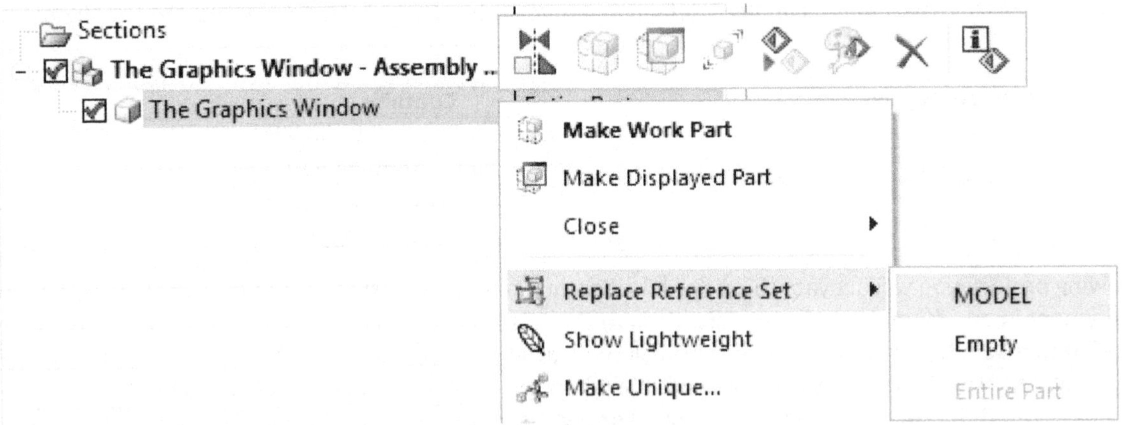

- Now, only the objects in the *Model* reference set in *"The Graphics Window"* are visible in The Graphics Window. These are the solid body of *Extrude (3)* and the sheet body of *Body (5)*.

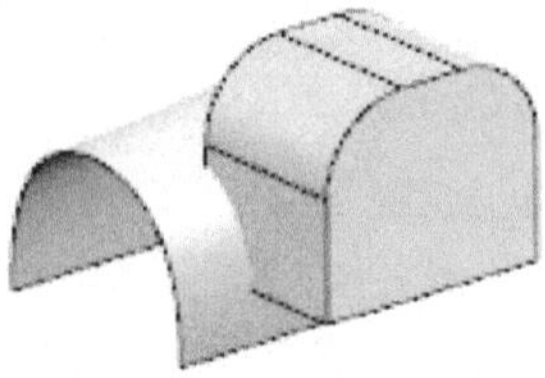

- The best practice is to use the *Model* reference set to show only the solid body of the part in question. Let's learn how to make edits to reference sets. Double-click on *"The Graphics Window"* in the Assembly Navigator to make it the

Work Part. Note that setting *"The Graphics Window"* as the Work Part sets its reference set to *Entire Part* for the time being.

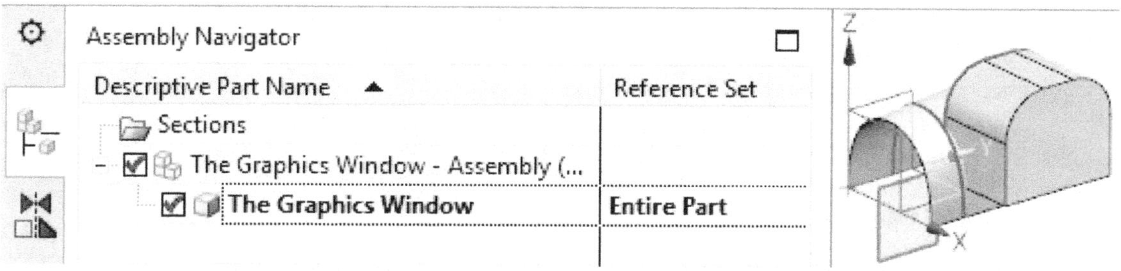

- Select the **Reference Sets** tool and click on *Model* in the list. Currently, *Model* has both a solid body and a sheet body within it.

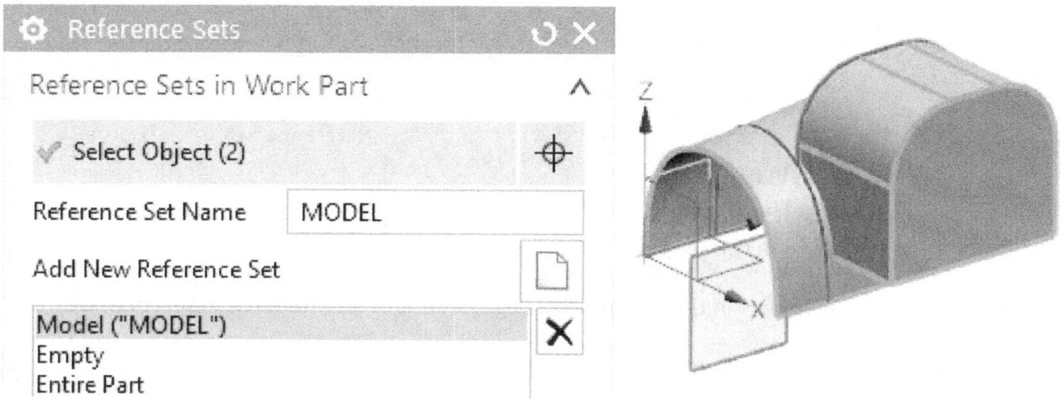

- Deselect the sheet body of *Body (5)* so that only the solid body of *Extrude (3)* remains in the *Model* reference set.

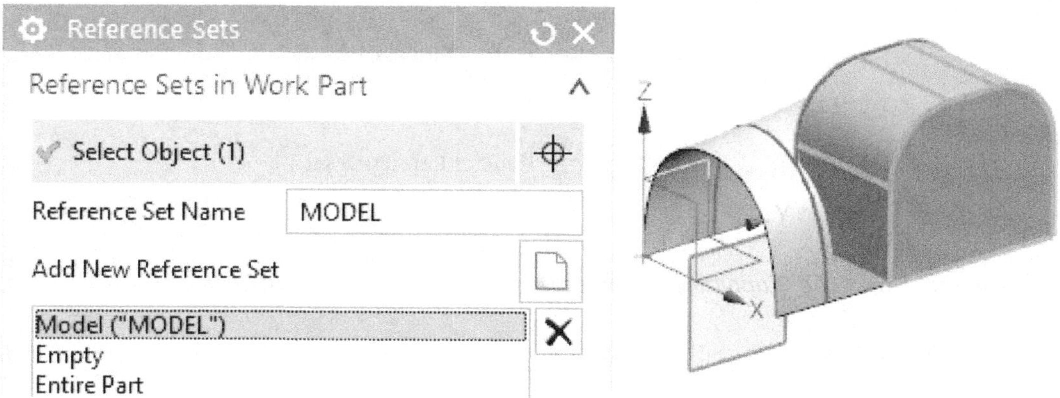

- To set *Model* as the active reference set, you can push the **Set as Current** button below on the dialog.

- Now only the solid body of *Extrude (3)* is visible.

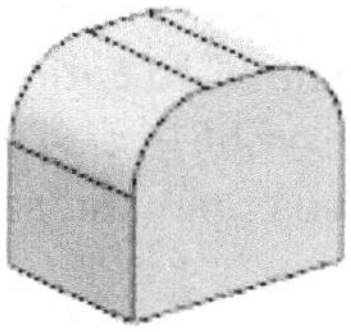

- Select *Entire Part* from the list and push the **Set as Current** button. Now use the **Add New Reference Set** button to create a user-defined reference set, and name it *"DATUMS"*.

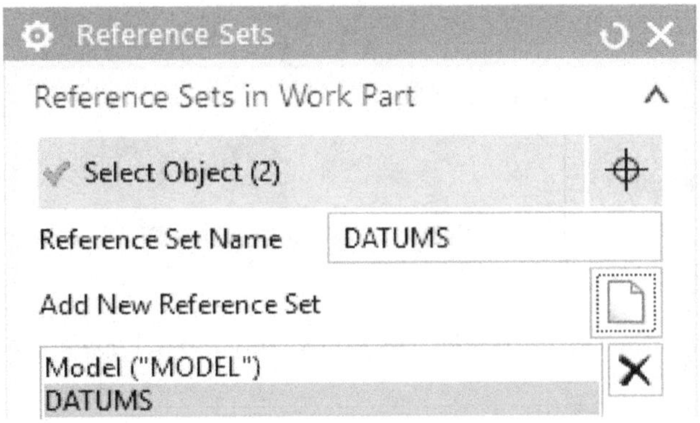

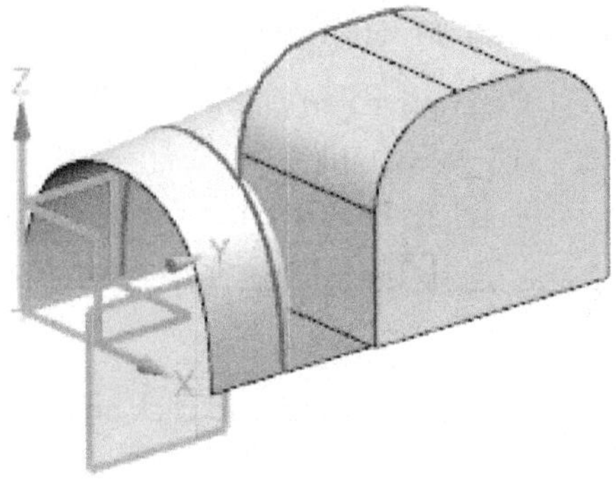

- Set the Displayed Part as the Work Part. Note that the active reference set for *"The Graphics Window"* switches back to *Model*.

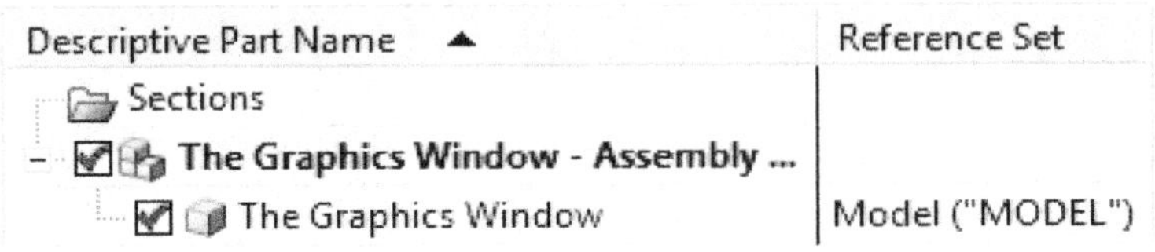

- Right-click on *"The Graphics Window"* in the Assembly Navigator, and select **Replace Reference Set** / *DATUMS*.

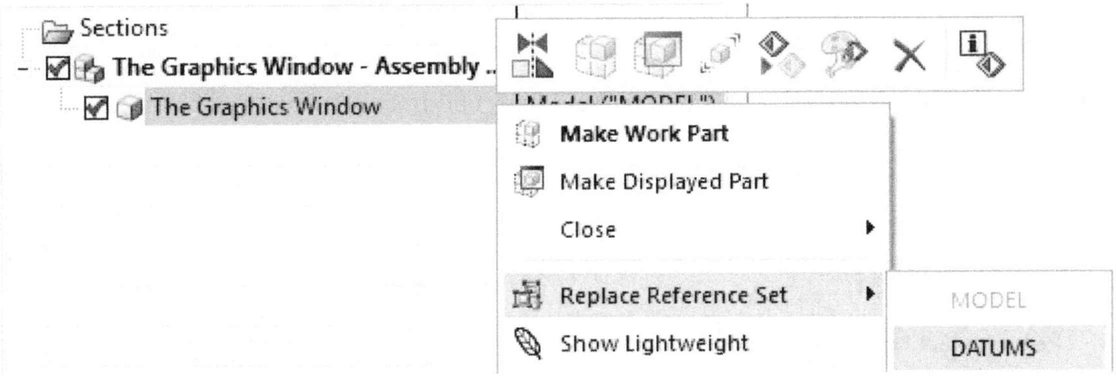

- Now only the datums from *"The Graphics Window"* are visible in the Graphics Window.

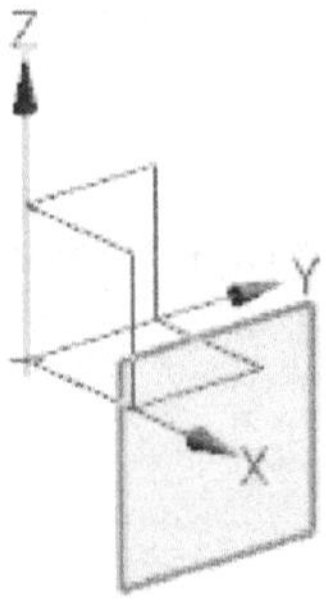

- Set the active reference set for *"The Graphics Window"* to *Entire Part* again.
- **Save** the changes to *"The Graphics Window – Assembly.prt"*. This will also save the changes made to the reference sets in *"The Graphics Window.prt"*, since the **Save** command saves all changes to modified components within the Work Part.

3.4.5 Section Views

Section views are very useful, especially when working with assemblies, for seeing inside of enclosures, or seeing how parts fit to each other. There are two basic tools on the ribbon that manage section views – **Edit Section**, and **Clip Section**. Both are found in the **Visibility** group on the **View** tab.

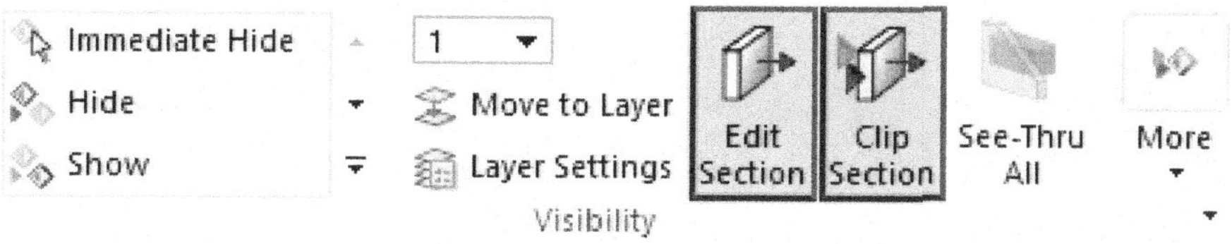

Edit Section is used to create a section view, or modify the *Work Section View*, if one has already been created. **Clip Section** toggles the *Work Section View* on or off.

- Make "The Graphics Window.prt" the Displayed Part.
- Use the **Show and Hide** tool to **Show** *Sheet Bodies*.
- Select the **Edit Section** tool. A dynamic coordinate system will appear that controls the position of the section plane.

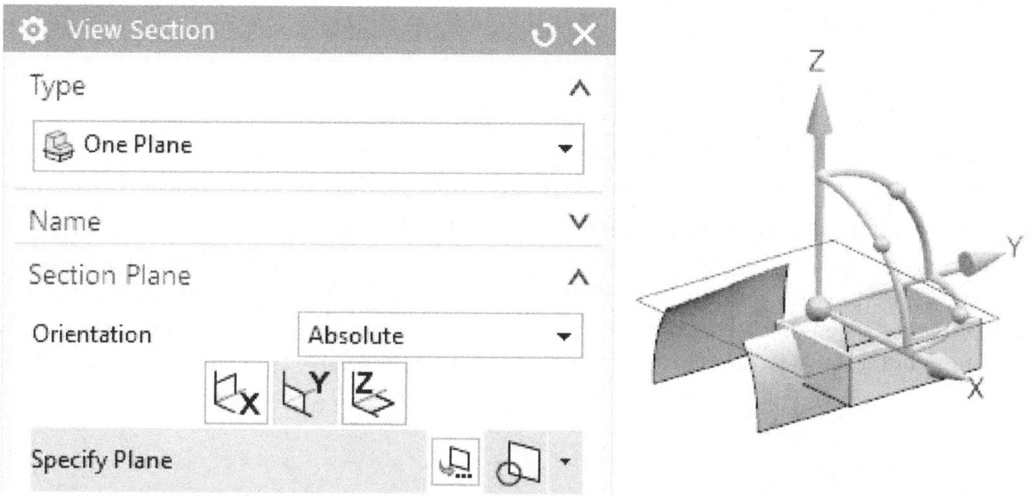

- Push the **Set Plane to Y** icon to quickly reorient the section plane so that it is normal to the Y axis of the absolute coordinate system.
- **Cap Settings** are a very important part of the **View Section** dialog – when the **"Show Cap"** checkbox is marked, solid bodies appear solid. The color of the body along its intersection with the section plane is determined by the **Cap Color**.

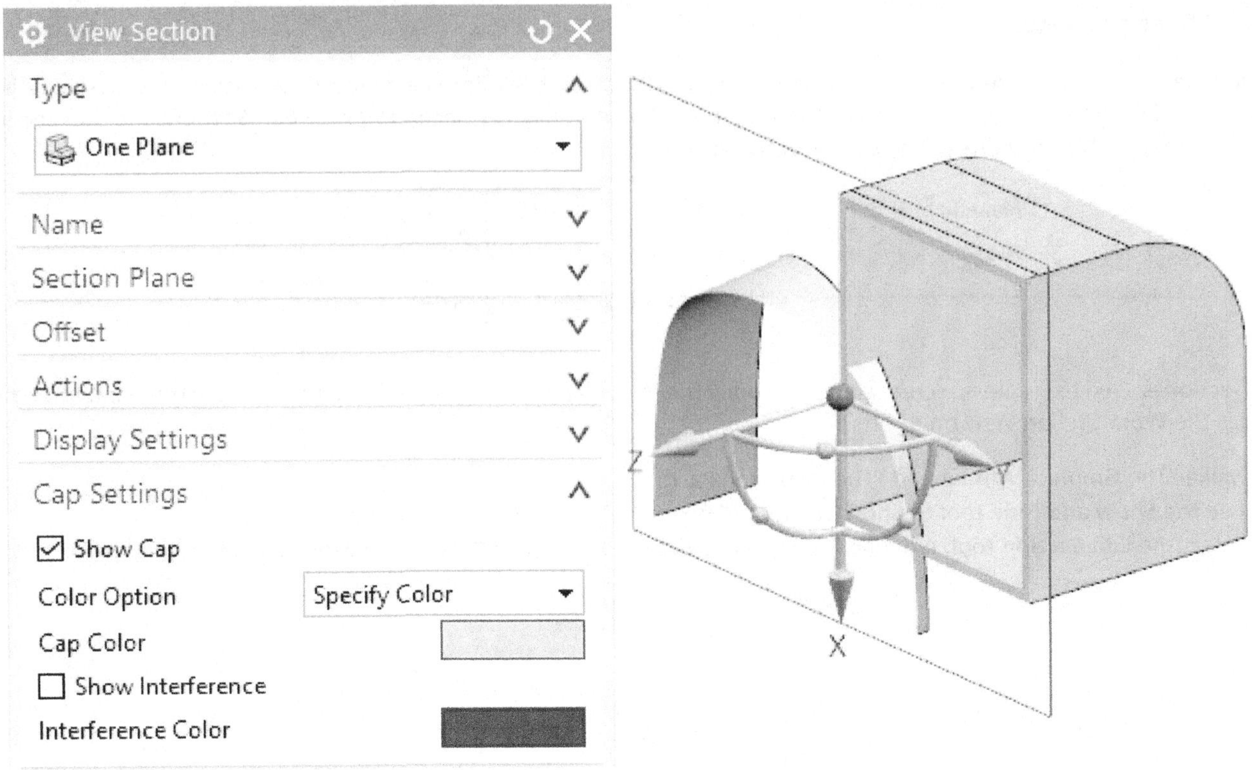

- Uncheck the *"Show Cap"* checkbox. Now your solid bodies appear completely hollow, and you see the inside surfaces. It can be hard to discern where faces meet the section plane when the **Cap** is not displayed. To help this, you can check the *"Show Section Curves Preview"* checkbox in the **Section Curve Settings** part of the dialog. The *"Display Curves while Dragging"* keeps the section curves visible while you manipulate the section plane with the dynamic coordinate system, but it tends to dramatically increase rendering time.

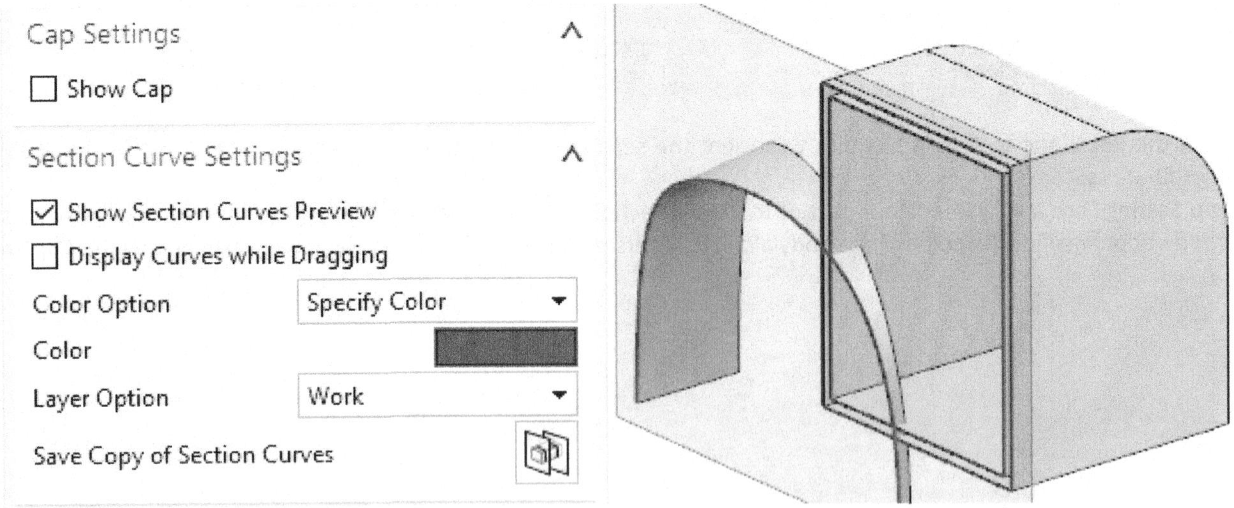

- Turn off section curves and enable the cap again, and rotate the dynamic coordinate system about its y axis by using the spherical rotation handle to set the section plane as shown below.

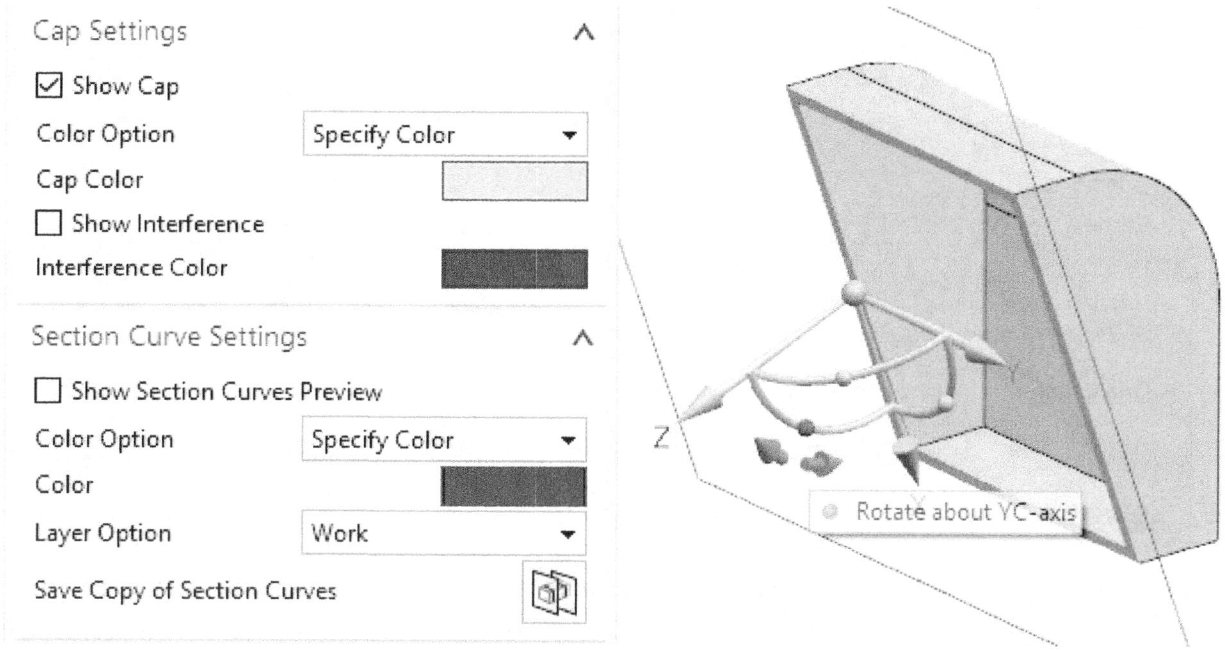

- Click **OK** to save your section view. Now you can **Clip** and **Unclip** the view with the **Clip Section** button.

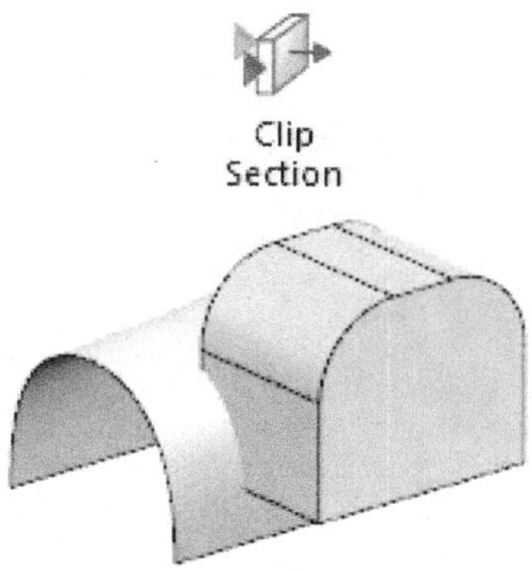

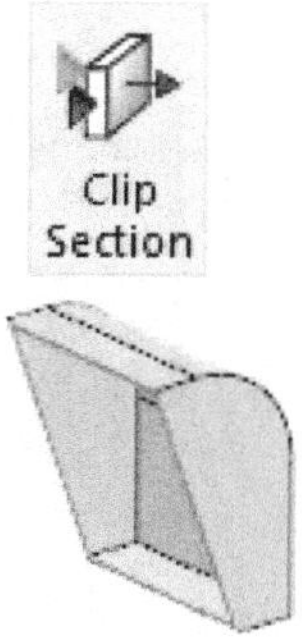

- Sections are managed at the Displayed Part level, and so they are accessible in the Assembly Navigator. Double-clicking on the Work Section here has the same effect as the **Clip Section** button. To create another section, right-click on the *Sections* folder and select **New Section**.

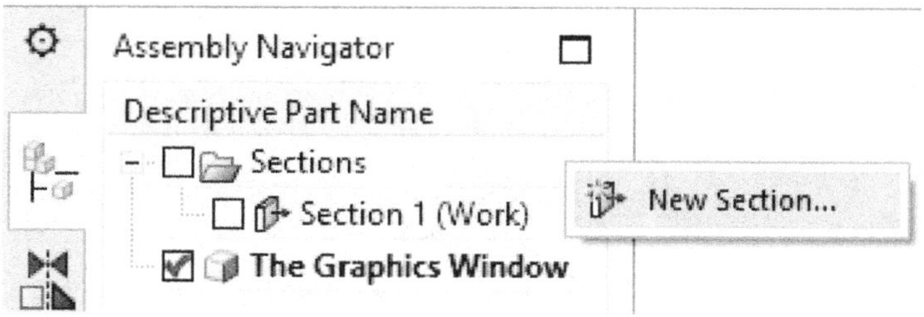

- In addition to the simple planar sections that are usually of interest, you can also produce a "Box" section view which clips your model with six planes. Change the **Type** to *Box* and push the **Set Plane to X** icon to set up these six planes as shown below.

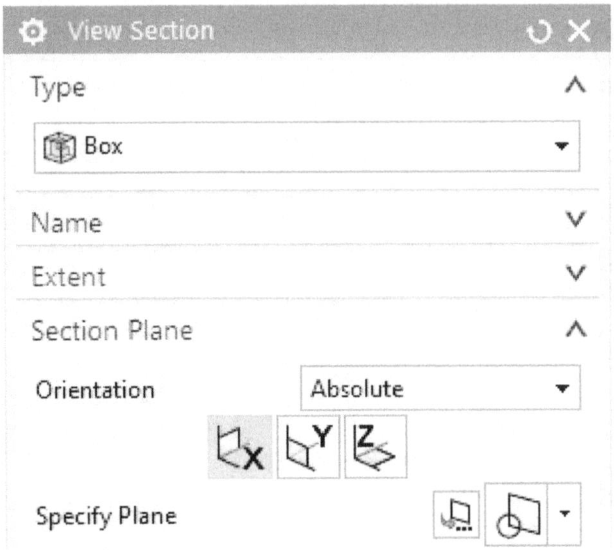

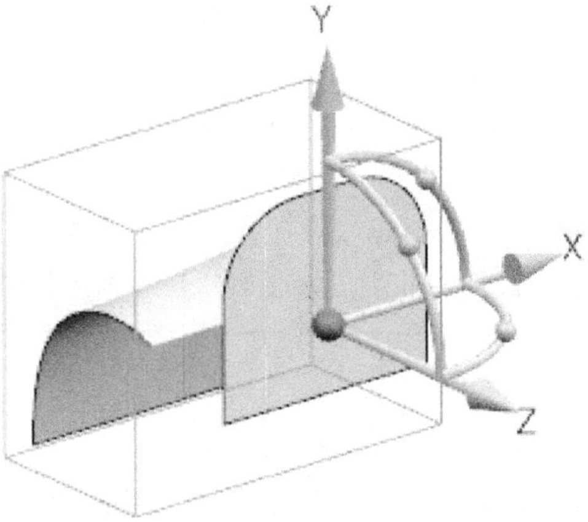

- Drag the Z handle until it appears as shown below. Next, select the left face of the box by positioning your cursor as shown below and clicking.

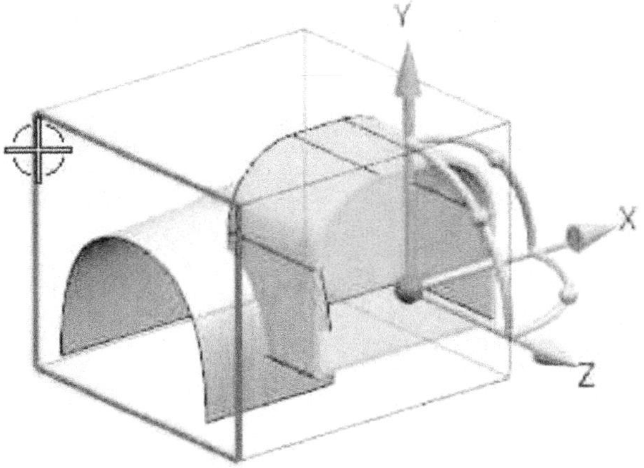

- Now the dynamic coordinate system controls the selected face of the box. Drag the Z handle of the dynamic coordinate system inward until the section appears as shown below.

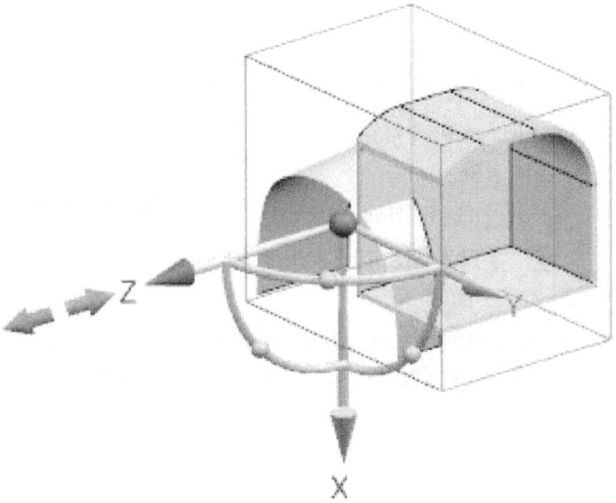

- Click **OK** to save your view. Now you have two sections stored in the *Sections* folder in the Assembly Navigator, and *Section 2* is the Work Section. To make *Section 1* the Work Section, just double-click on it. The checkboxes to the left indicate visibility of the section curves – you can check or uncheck them to make section curves visible or invisible, even for sections other than the Work Section!

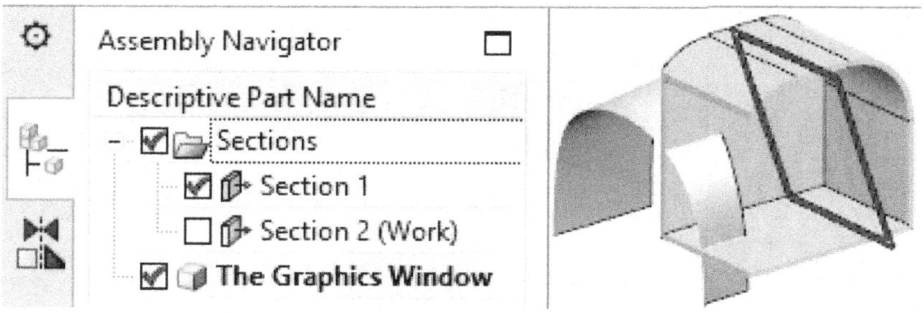

- **Unclip** the Work Section and hide all section curves.

3.4.6 Cleaning Up the Display

There are a number of tools that are useful for cleaning up undesired entities from the Graphics Window. Most are accessible through the View Shortcut Menu (see Chapter 2.8) that appears when you right-click in the empty space in the Graphics Window, but there are many more found in the depths of the View tab. Our aim here is not to cover the various rendering options and tools extensively, but instead to share with you some of the most useful tools for quickly fixing the display in the Graphics Window.

3.4.6.1 Resetting Clipping Planes

The clipping planes are a feature of your camera that determine which portion of the 3D environment should be rendered. Sometimes, in the course of zooming in and out a lot, your clipping planes will cut through your model and produce what appears to be a section view, although you don't have much control over it. The **Fit** command is useful because it resets the clipping planes so that your model doesn't appear cut. Any of the standard view tools described in Chapter 3.2.4 will also reset the clipping planes.

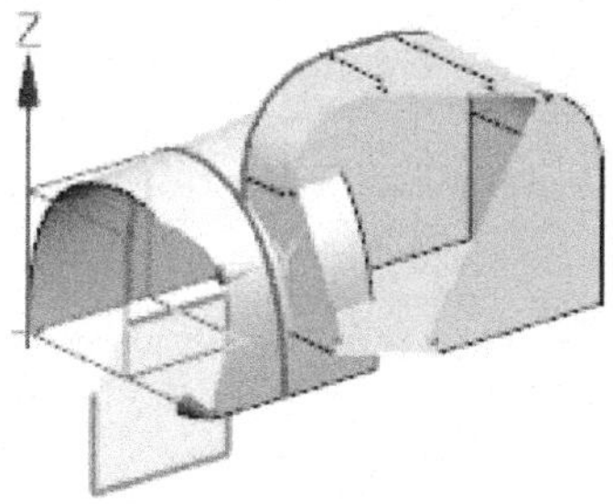

3.4.6.2 Refresh

Certain actions will leave artifacts on the screen – typically requests to display some kind of information in the Graphics Window. For instance, the **Show Dimensions** command displays the parameters of a feature directly in the Graphics Window. These dimensions persist until the Graphics Window is cleaned using **Refresh**.

- Right-click on *Extrude (3)* and select **Show Dimensions**.

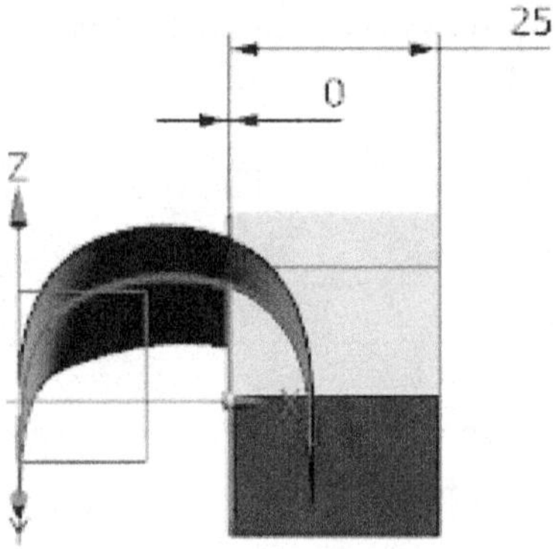

- Right-click in empty space in the Graphics Window and select **Refresh**, or use the keyboard shortcut **[F5]**, and the displayed dimensions vanish!

3.4.6.3 Update Display

Curves are not rendered perfectly in any CAD software – if you zoom in you will see the "faceting" that appears smooth from a distance. When your part or assembly is complicated, you might even see many curves faceted from a distance. To regenerate the rendering, you can use the **Update Display** command, which will recalculate the appropriate size facets for the current perspective.

- Zoom in on the curved edge of the body of *Extrude (3)* and you will see that the rendering quality diminishes – you can see that the arc is approximated by small line segments. Right-click in empty space in the Graphics Window and select **Update Display**.

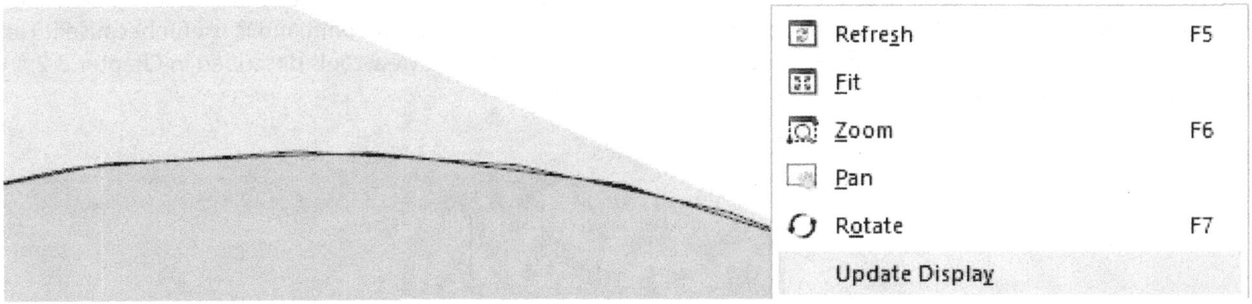

- This improves the rendering for your current camera.

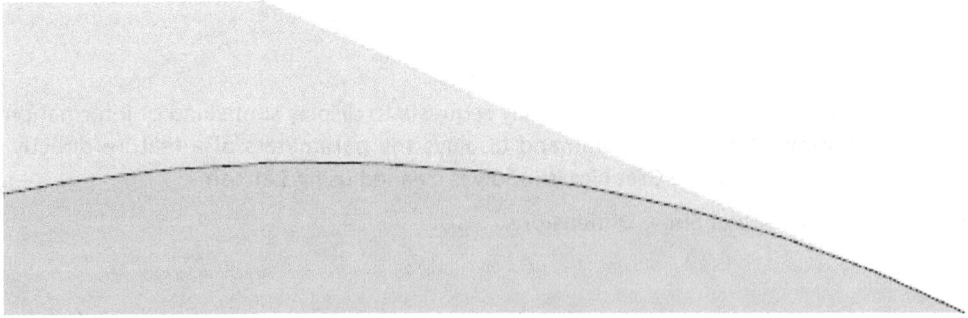

3.5 Layers

Layers in NX are used to organize geometry within your part file or organize components in an assembly. Using layers to segregate geometry is an extremely powerful technique and can be understood fully when one envisions the physical analog of layers: traditional drafting sheets. Imagine you have a number of sheets of see-thru paper on a drafting board. You have 256 pieces of paper—each numbered. Each layer has a tick mark so you can line the sheets up to one another. When you are creating a model that has many different sketches, datum planes and other intermediate geometry, or an assembly with many different components you may place them on different layers. If you then imagine what you might do with the various layers of see-thru paper it gives you great insight as to what you can do with layers and geometry in NX.

3.5.1 Layer Settings

The **Layer Settings** tool allows you to determine whether each of the 256 layers in your part file is visible or selectable. The **Layer Settings** command is found in the **Visibility** group on the **View** tab.

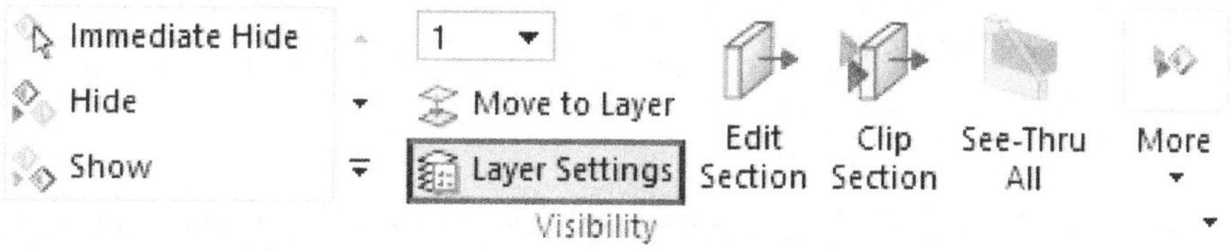

The **Layer Settings** menu appears as shown below for *"The Graphics Window.prt"*.

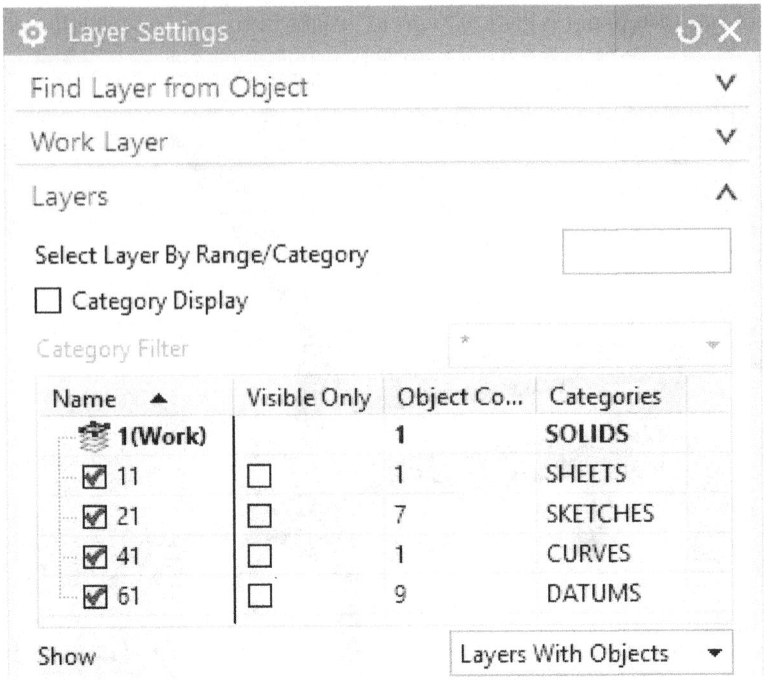

There are several important concepts to be aware of when working with **Layer Settings**.

3.5.1.1 Work Layer

The layer that you place on the top of the stack is the one that you work on. This is called the *Work Layer*. When you author geometry using modeling operations (e.g., **Extrude**), that geometry is placed on the Work Layer.

3.5.1.2 Selectable

Imagine that you have a number of layers on the drafting board underneath the work layer. You may want to temporarily lift the Work Layer and make some edits to the geometry that is on layers below. These are called *Selectable Layers*. Selectable layers are indicated by a red checkmark in the Name column, and they are necessarily visible. The Work Layer is necessarily selectable.

3.5.1.3 Invisible

You may want to hide a layer that may or may not have geometry on it. Imagine rolling it up and putting it in a bin. This is called an *Invisible Layer*.

- Uncheck the box in the **Name** column for Layer 61. This makes the objects on Layer 61 invisible.

3.5.1.4 Visible Only

Imagine a drafting sheet covered in plastic wrap. You will still be able to see what's on it but you will not be able to accidentally change it. The analog in NX is called a Visible Only Layer.

- Check the box in the **Visible Only** column for Layer 11. This makes the objects on Layer 11 visible-only.

3.5.2 Layer Categories

The **Category** column displayed in the **Layer Settings** menu indicates which categories a given layer might belong to. A part file created with the *Modeling* template from an out-of-the-box NX installation will have certain Layer Categories preconfigured, but they are fully customizable using the **Layer Category** tool, found in the **Visibility** group on the **View** tab.

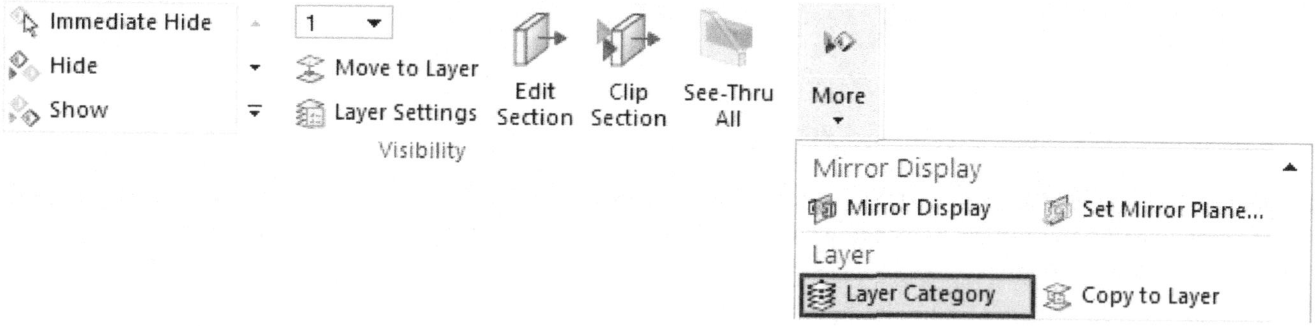

- Select the **Layer Category** tool. In the text entry field, name a new category "My New Layer Category". Push the **Create/Edit** button.

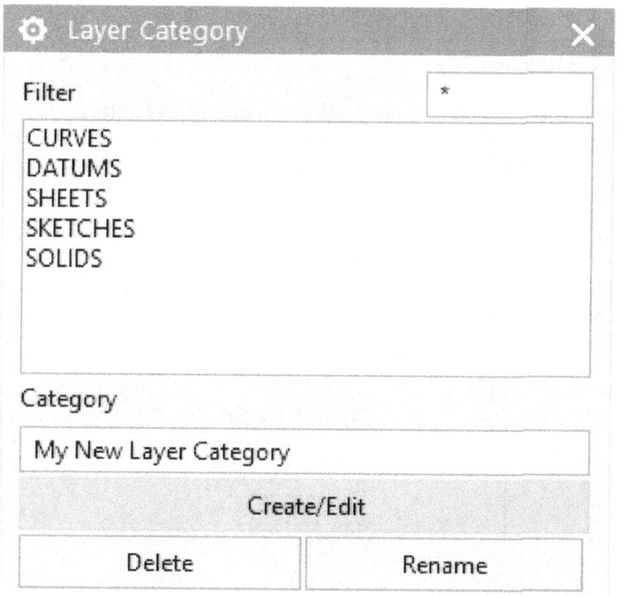

- This brings up a new dialog that allows you to select layers and add or remove them from the category. Select layers *101-110* and push the **Add** button. Click **OK** to save your changes. When you return to the **Layer Category** dialog, click **OK** to save your changes.

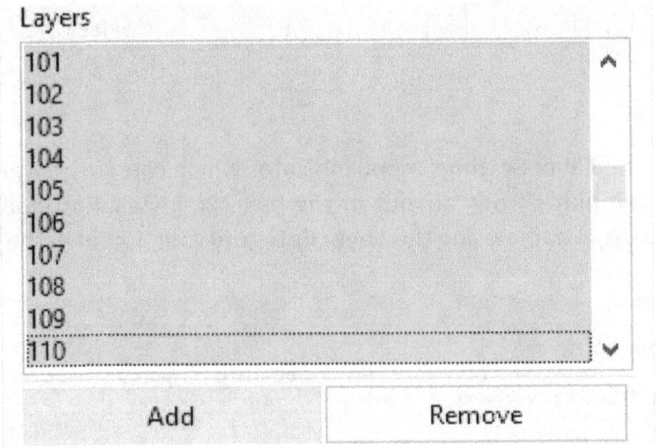

- You can also modify layer categories from within the **Layer Settings** dialog. Check the *Category Display* checkbox to see more clearly which layers belong to which categories. By right-clicking on a layer, you can add it to, or remove it from, a category.

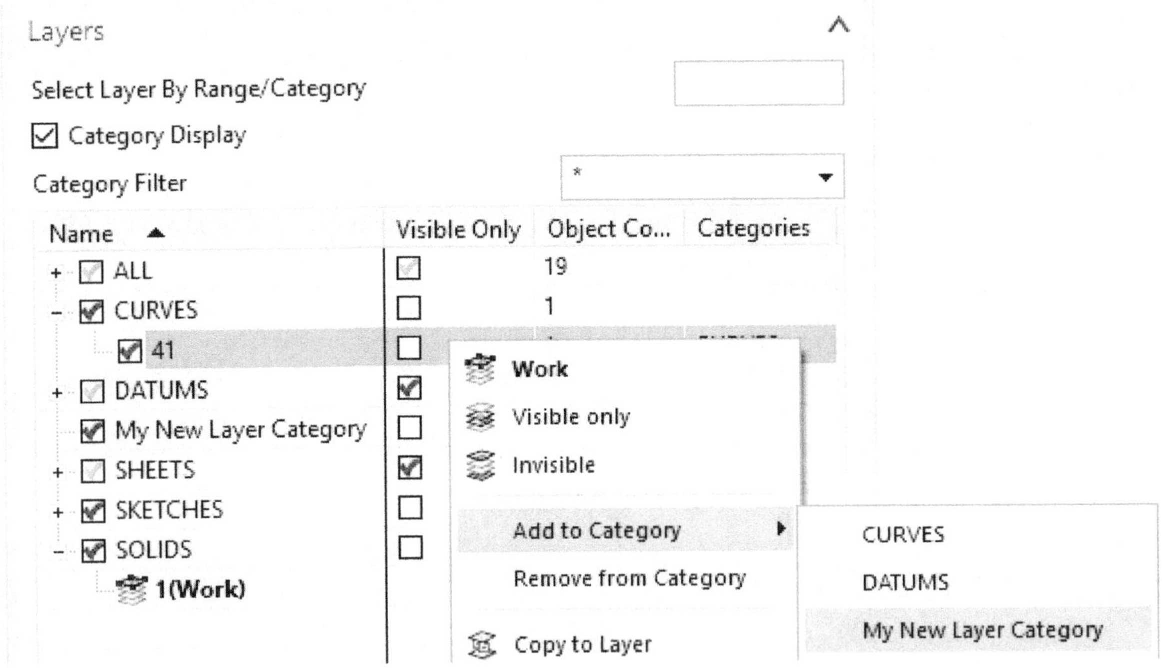

3.5.3 Move to Layer

The **Move to Layer** command moves an entity in the Graphics Window from one layer to another. **Move to Layer** is found in the **Visibility** group on the **View** tab.

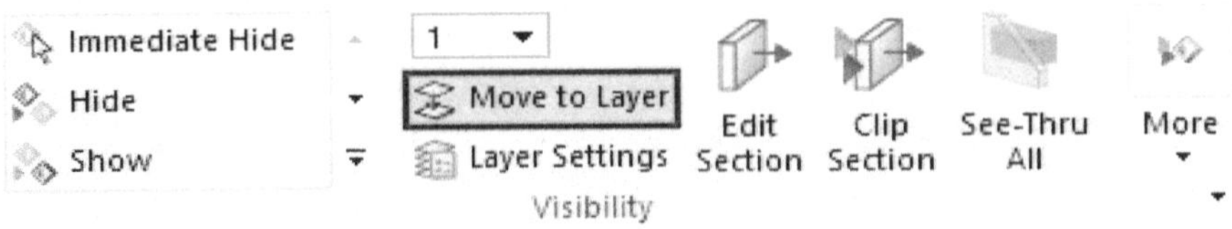

- Select the **Move to Layer** command. The **Class Selection** dialog will appear.
- Set the **Selection Type Filter** to *Sketch*, and use the **Select All** button on the **Class Selection** dialog.

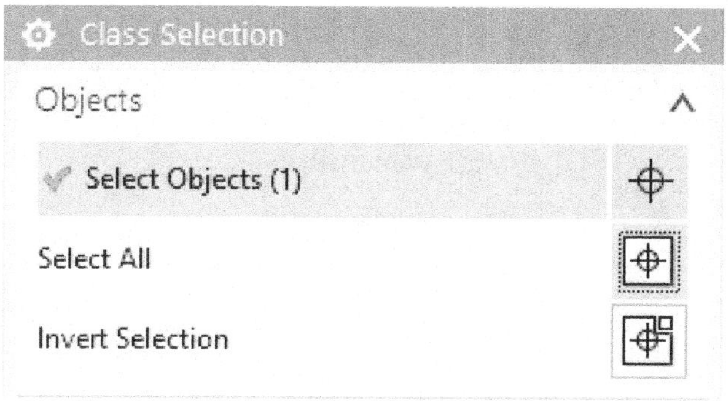

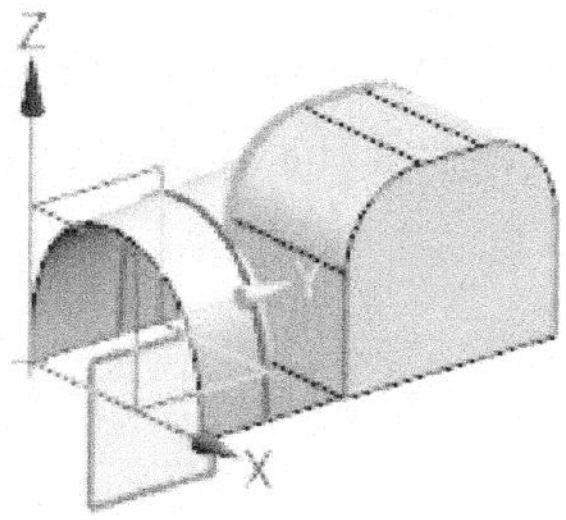

- Click **OK**. The **Layer Move** dialog will appear. Specify a **Destination Layer** of *22*. Click **OK**, and the sketch will be moved.

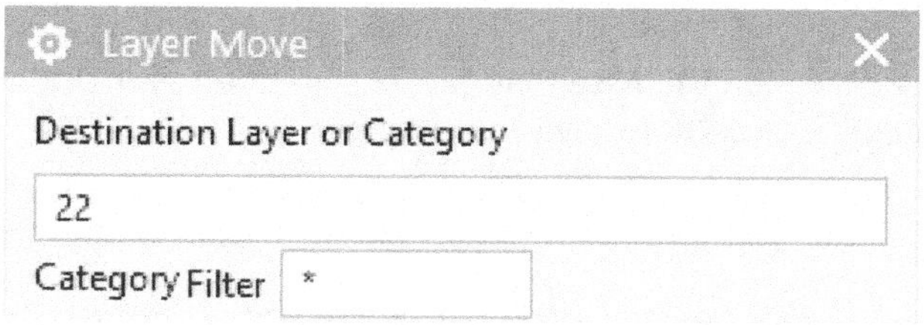

- Now in your **Layer Settings** menu, layer *22* will appear instead of layer *21*, because of the setting *Layers With Objects* in the **Show** drop-down.

Name ▲	Visible Only	Object Count	Categories
1(Work)		**2**	**SOLIDS**
☐ 11	☑	1	SHEETS
☑ 22	☐	7	SKETCHES
☑ 41	☐	1	CURVES
☐ 61	☐	9	DATUMS

- **Save** the changes to the file *"The Graphics Window.prt"*.

3.5.4 Assemblies and Layers

When you bring a *".prt"* file into an assembly as a component, you can choose whether the geometry in that file should segregate onto the same layers in the assembly, or move to different layers. It is important to distinguish between the part file itself, and instances of that part in assemblies – they can be given different layer settings.

- Make *"The Graphics Window – Assembly.prt"* the Displayed Part.
- Select the **Layer Settings** tool, and observe that the layers with objects are exactly the same as those in *"The Graphics Window.prt"*. Also, note that the layer selectability and visibility settings in the assembly are independent of those in the component.

Name ▲	Visible Only	Object Count	Categories
1(Work)		**2**	**SOLIDS**
☑ 11	☐	1	SHEETS
☑ 22	☐	7	SKETCHES
☑ 41	☐	1	CURVES
☑ 61	☐	9	DATUMS

- From the Assembly Navigator, right-click on *"The Graphics Window.prt"* and select **Properties**.

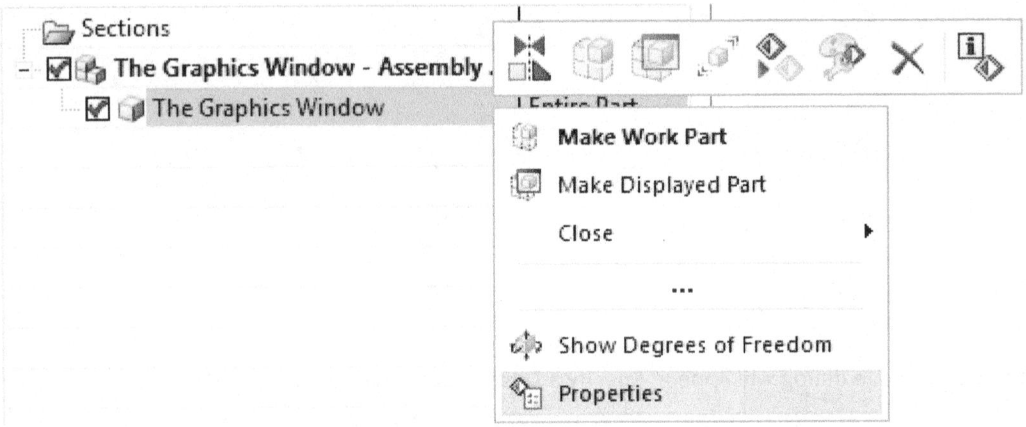

- In the **Component Properties** dialog, on the **Assembly** tab, change the **Layer Option** drop-down to *Specified Layer*. Let the **Layer** default to the value *1*. Click **OK** to save your changes.

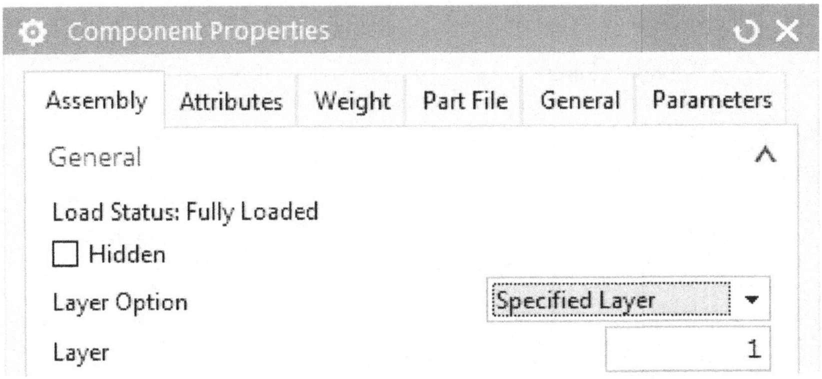

- Select the **Layer Settings** tool again. Now all the geometry in the present instance of *"The Graphics Window.prt"* is on layer *1*.

Name	Visible Only	Object Count	Categories
1(Work)		**20**	**SOLIDS**

- Make *"The Graphics Window.prt"* the Displayed Part.
- Select the **Layer Settings** tool. Note that the **Layer Settings** for *"The Graphics Window.prt"* are unaffected!

Name	Visible Only	Object Count	Categories
1(Work)		**1**	**SOLIDS**
☑ 11	☑	1	SHEETS
☑ 22	☐	7	SKETCHES
☑ 41	☐	1	CURVES
☐ 61	☐	9	DATUMS

- Make *"The Graphics Window – Assembly.prt"* the Displayed Part.
- **Save** your changes to *"The Graphics Window – Assembly.prt"*.

4 Editing Parametric Models

The **Part Navigator** is a powerful, easy-access organization tool for parametric models. It is located in the **Resource Bar** on the left side of the NX interface. We have seen how to do some simple things to entities in the **Model History** within the **Part Navigator** at this point, and in this chapter, we will explore more fully the different ways that you can make edits to your parametric model within the Part Navigator.

- Open *"Editing Parametric Models.prt"* from *"C:\My NX 11 Files\Downloaded Files"*.
- Use the techniques of Chapter 2.9.2 to ensure that the **Layer** column is visible in the Part Navigator, following the **Name** column.

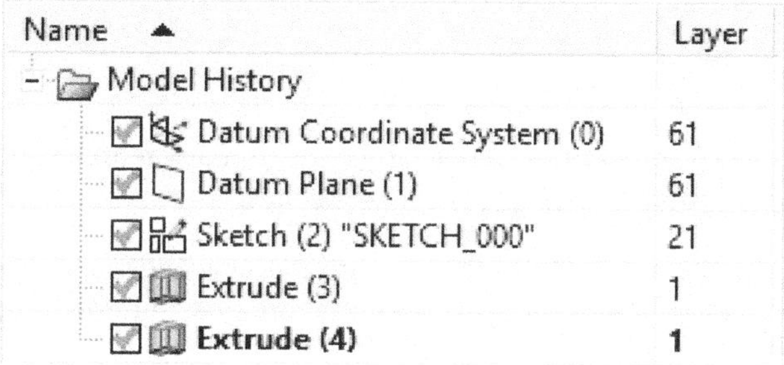

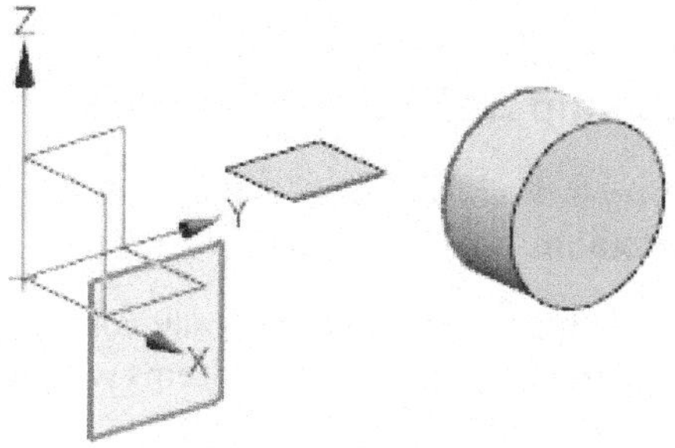

4.1 Dependencies

By default, your **Model History** displays features chronologically, in the order that they were created. This ordering is helpful for seeing how NX builds the model from features, but doesn't offer any insight into the relationships between those features. We begin by describing some tools that allow you to understand the dependencies within a model.

Dependencies are the parent-child relationships created when building up a CAD model. The skill of creating (and controlling) well-behaved parent/child relationships is what separates the expert-level user from the novice. When you make edits to a feature in the Model History, any dependent features can be impacted by your edit, so before making edits, it is worthwhile to explore your tools for understanding dependencies.

In a geometrically complex model, it is often helpful to learn more about the parent/child relationship structure. There are several ways to study this structure, but the most basic way is to make use of color-coding in the Part Navigator. NX color-codes parent and child relationships in the Model History, as shown below. Once a design feature is highlighted (selected) its parents (referenced features) are color-coded pink, and children (dependent features) are color-coded blue.

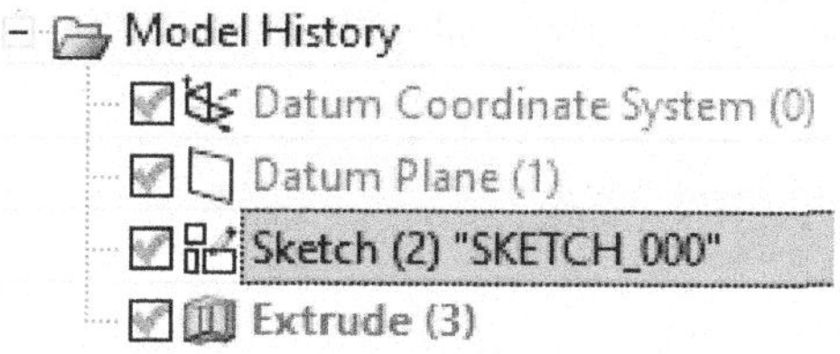

4.1.1 The Dependencies Panel

The parent/child relationships are shown explicitly for any highlighted item in the Model History in the **Dependencies** panel, located at the bottom of the **Part Navigator**.

- Expand the **Dependencies** panel and highlight *Sketch (2)* – the parent/child relationships involving *Sketch (2)* are made completely explicit in the panel, as shown below.

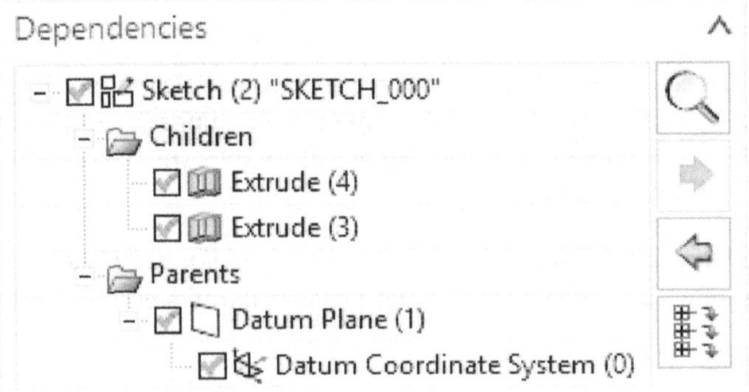

4.1.2 Timestamp Order vs. Reference Sets

The Model History is displayed by default in *Timestamp Order*. The number next to the name of each feature in the Model History is a timestamp, and Timestamp Order shows timestamps in ascending order by default.

There is an alternative to the chronologically-correct Model History in the Part Navigator. You can view a model in terms of its reference sets, where the top-level items within each reference set are the bodies it contains, and the dependencies in that body are exposed.

- To turn off Timestamp Order, right-click in any empty space within the **Part Navigator** (for example, just to the left of *Datum Coordinate System (0)*). On the menu that appears, you will find **Timestamp Order** checked.

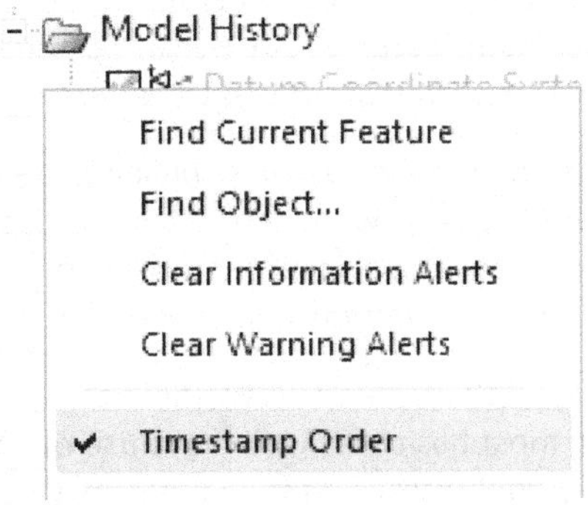

- Uncheck **Timestamp Order** and the Model History will disappear in favor of a "Reference Sets" section, and an "Unused Items" folder. Note that there is an invisible spline on layer *41* that now appears, which was absent from the Model History.

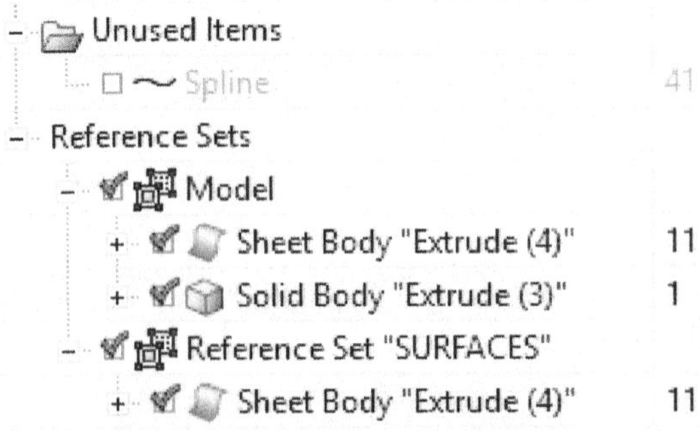

- Within each body, you can use the + buttons to the left of each item you can expand a tree that illustrates the parent features and their dependencies, similar to what you see in the **Dependencies** panel.

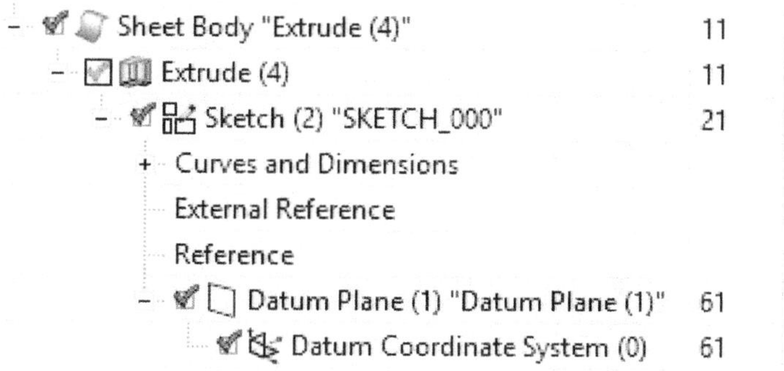

- You can also modify the reference sets within the part from the Part Navigator. Unfortunately, you cannot add new reference sets. Right-click on *Model* and select **Edit**.

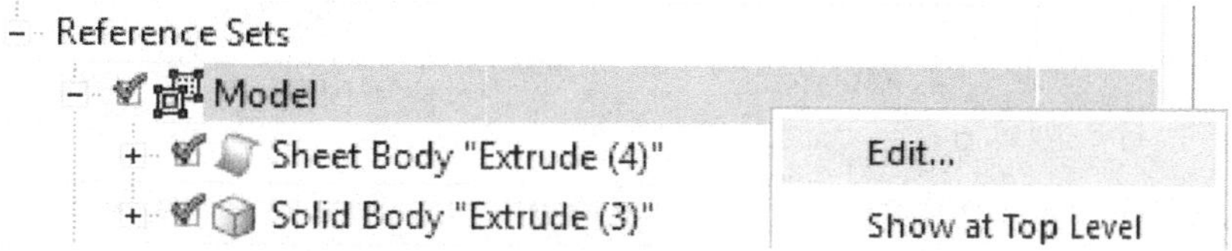

- Deselect the surface from the *Model* reference set so that only the solid body remains.

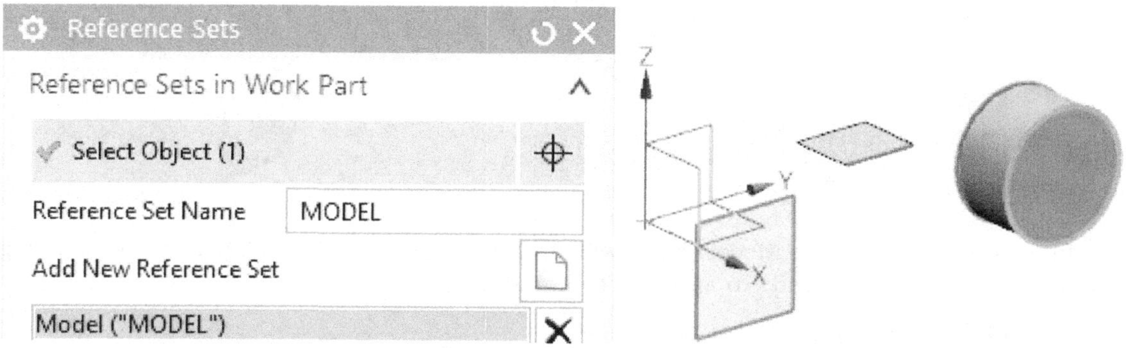

- For the rest of this chapter, we will work in Timestamp Order, so please re-enable it at this time!

4.1.3 Reordering Features

When you *reorder* an operation in NX 11, it is moved to a different place in the model's creation history and is assigned a new timestamp. Reordering features requires attention to the dependencies in the model, as NX will not allow you to reorder a feature prior to one of its parents.

- Select *Extrude (3)*. Note that *Extrude (4)* is not color-coded blue. It is independent of *Extrude (3)*.

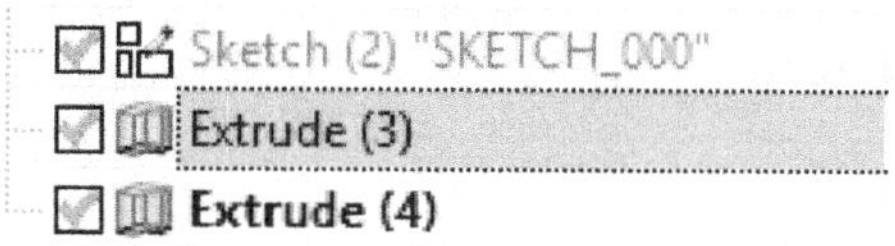

- Since *Extrude (4)* is independent of *Extrude (3)*, it can be reordered before *Extrude (3)* in the Model History. Right-click on *Extrude (4)* and select **Reorder Before** / *Extrude (3)*. Note that *Sketch (2)* is absent from the list because it is a parent.

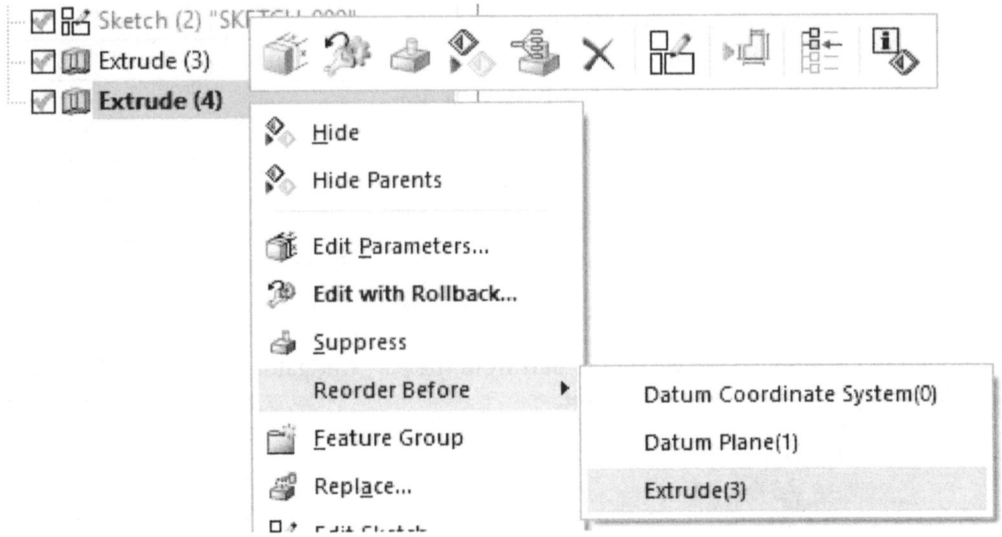

- Now click on *Extrude (3)* in the Part Navigator. Since the timestamps were switched, now *Extrude (3)* corresponds to the newly created sheet body!

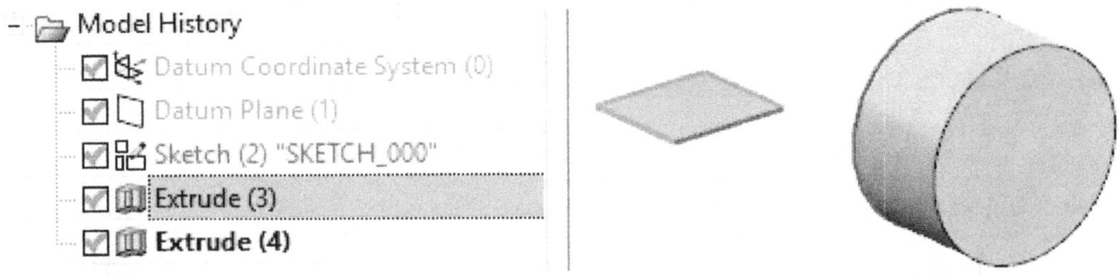

- You can also reorder features in the Model History with drag-and-drop techniques. Click-and-hold-and-drag *Extrude (4)* to the line between *Sketch (2)* and *Extrude (3)*, as shown below.

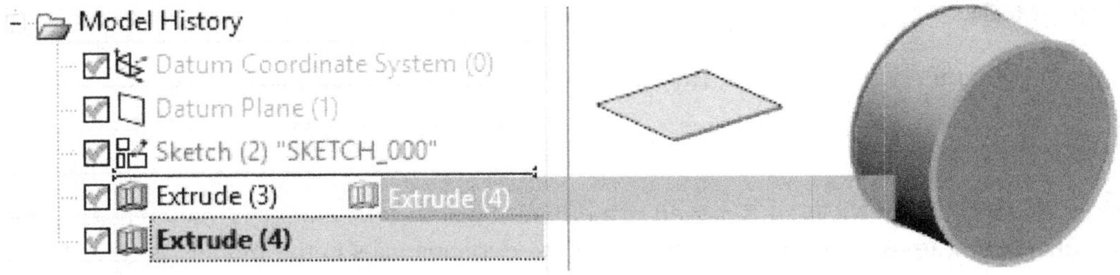

4.2 Deleting, Copying, and Suppressing Features

In these exercises, you will learn how to remove features from the model history permanently (by deleting them), temporarily (by suppressing them), and how to reuse features (by copying and pasting them).

4.2.1 Suppression

The green check marks to the left of each feature in the Model History can be a source of confusion for new NX users – those checkboxes indicate whether a feature is active or suppressed. Suppression has a few things in common with both hiding features and deleting them, but it is the same as neither.

When you **Suppress** an individual feature in NX, it is removed from the model's creation history, but remains present with its timestamp in the history. The model will revert to the state it would have been in if the feature never existed (as with **Delete**), but you can easily unsuppress it.

- Click on the green checkmark to uncheck the box next to *Sketch (2)*. Note that *Extrude (3)* and *Extrude (4)* both become suppressed as well.

- Click on the empty checkbox next to *Extrude (4)*. Since *Sketch (2)* is a parent feature for *Extrude (4)*, it also becomes unsuppressed.

4.2.2 Deleting Features

When you **Delete** an individual feature in NX it is permanently removed from the model's creation history. Consequently, the model will revert to a state it would have been in if the feature had never existed. A deleted feature can never be retrieved unless an **Undo** ([Ctrl]+[Z]) operation is possible.

- Use the keyboard shortcut **[Ctrl]+[D]** to access the **Delete** command. Surprise! The Class Selection menu appears. Select *Datum Plane (1)* from the Part Navigator.

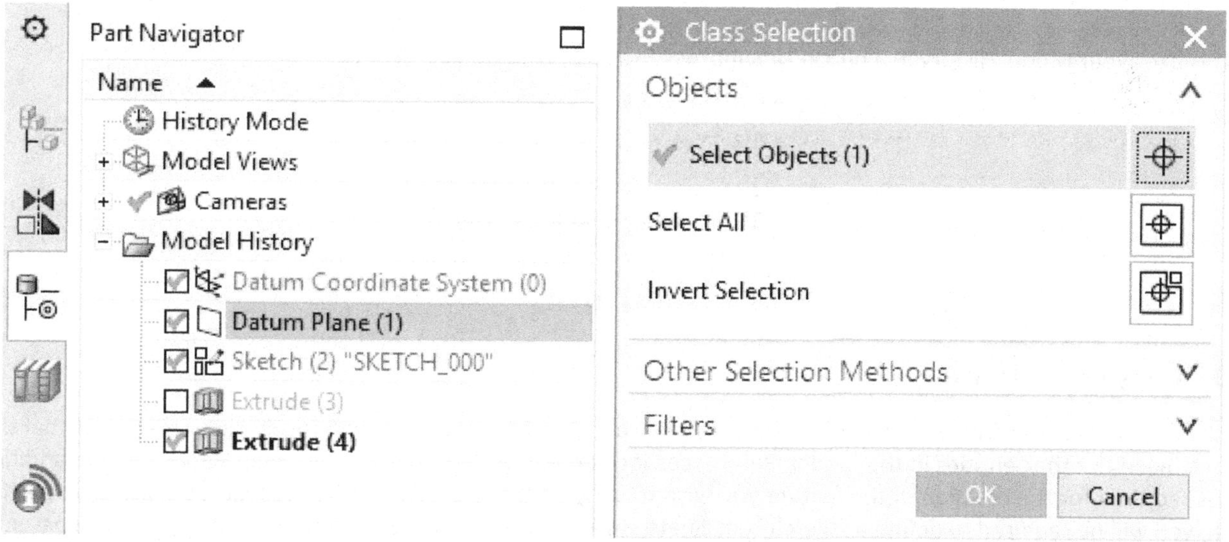

- Upon clicking **OK**, you will find that the **Delete** operation had disastrous consequences for the model – *Sketch (2)* depended on *Datum Plane (1)*, and so when *Datum Plane (1)* was deleted, *Sketch (2)* lost its references! The Model History will show an error.

- You will also encounter an Information Window with an error message in it.

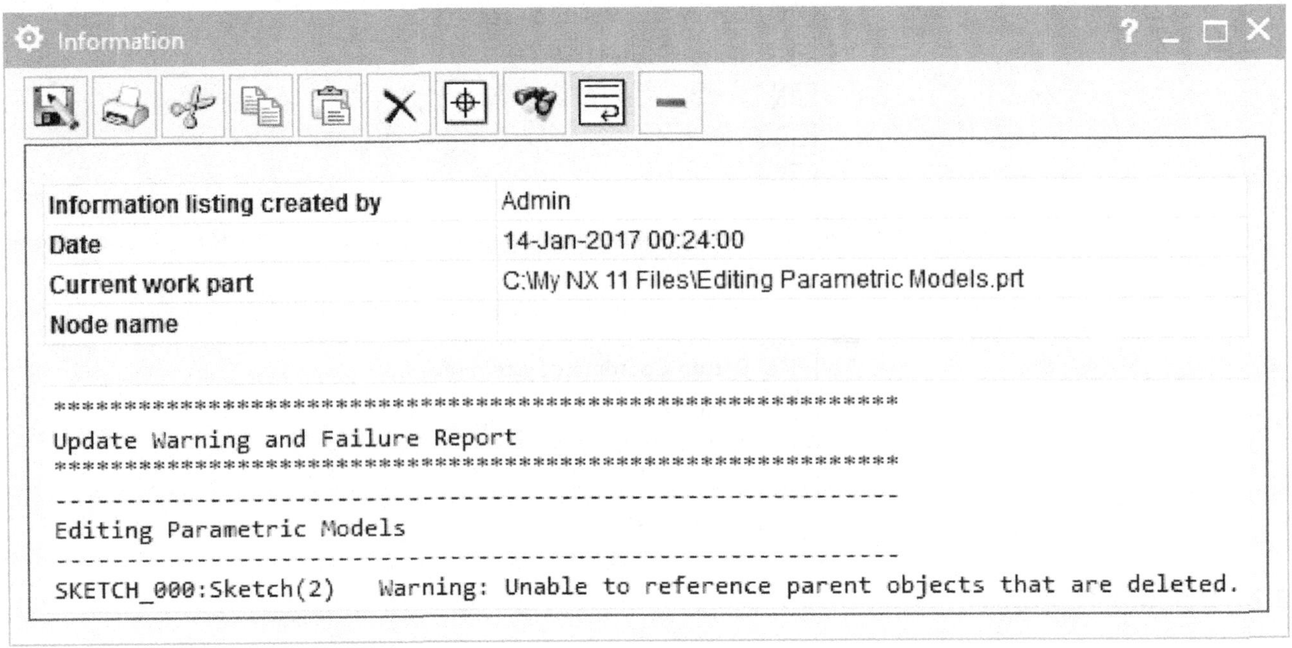

- Oops! The best way to bring back a deleted feature that you didn't intend to delete is with Undo (**[Ctrl]+[Z]**). Close the Information Window and use Undo, either with the keyboard shortcut **[Ctrl]+[Z]**, or from the Quick Access Toolbar.

- This fixes your Model History in a hurry! **Unsuppress** *Extrude (3)* while you're at it.

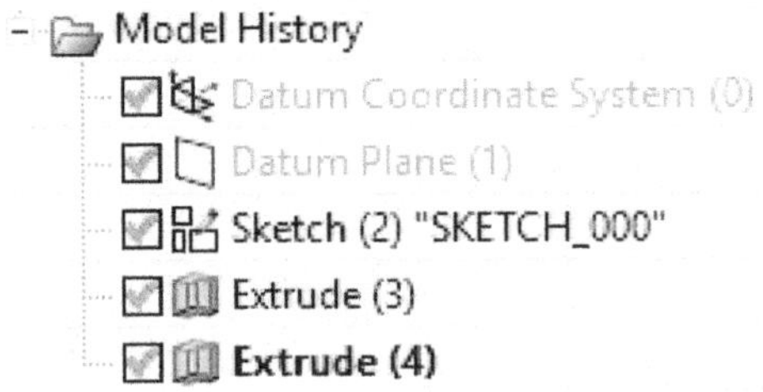

4.2.3 Copying Features

When you reuse features from the model history by copying and pasting them, you must resolve any references that those features make to other entities in the model. References include parent features, but also sometimes include additional entities required for the creation of the feature you wish to copy. In the example below, you will see that, upon pasting a sketch, you will be required to define a "sketch coordinate system" as a reference – which ends up not being a parent!

- Right-click on *Datum Plane (1)* and select **Copy**. You can also select it and use the keyboard shortcut **[Ctrl]+[C]**.

- There is no option to **Paste** when you right-click in the Part Navigator. Use **[Ctrl]+[V]** to **Paste**. You will be prompted to select a face, in relation to which the new plane will be offset, since the old plane was offset from its parent. Choose the XZ plane of the datum coordinate system. Use the **Reverse Direction** button to point in the -Y direction.

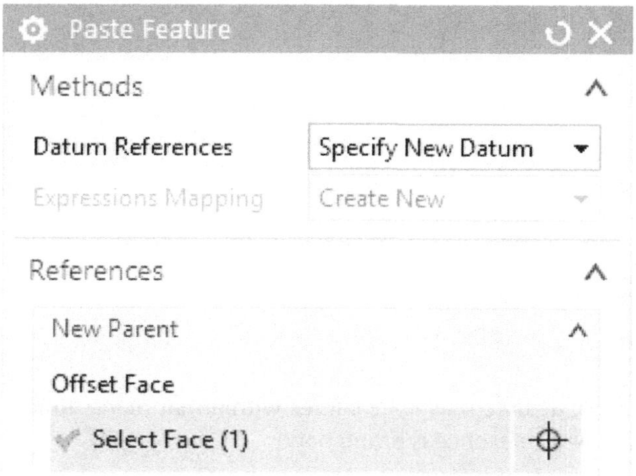

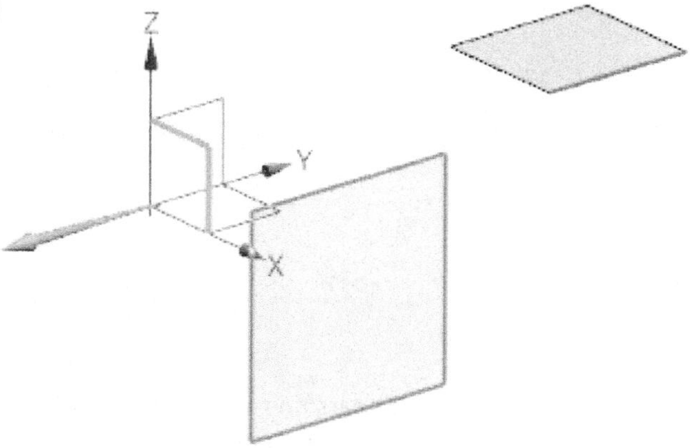

- Next, select *Sketch (2)* and use **[Ctrl]+[C]** to copy it. Use **[Ctrl]+[V]** to paste it. Now, despite the fact that the only parent of *Sketch (2)* is *Datum Plane (1)*, NX is prompting you to specify a coordinate system. Push the CSYS dialog button as shown below.

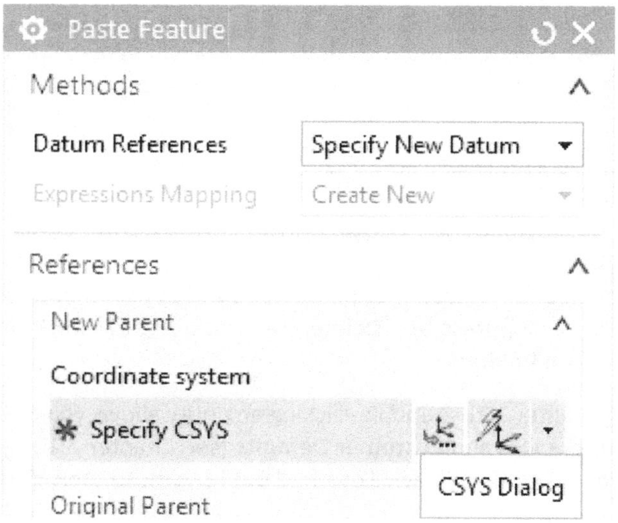

- Use the rotation handle between XC and YC to rotate the dynamic CSYS so that XC is pointing in the -Y direction. Then drag the XC handle to a distance of *50 mm* in the Y direction.

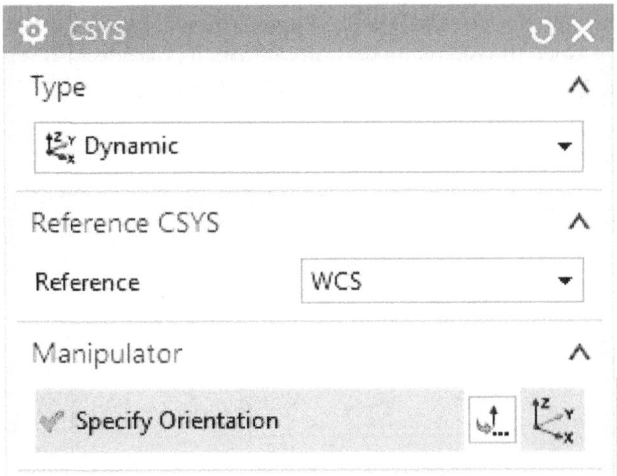

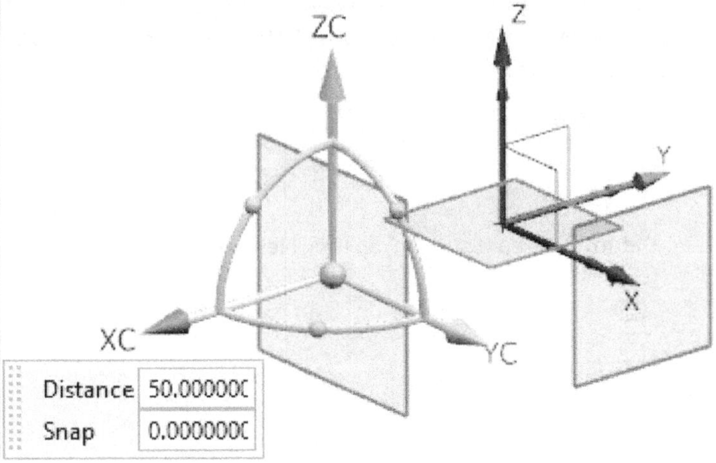

- The newly created sketch bears no relation to *Datum Plane (5)*, despite having its curves within that plane. In fact, it has no parent features – the coordinate system you specified as a reference is brand new!

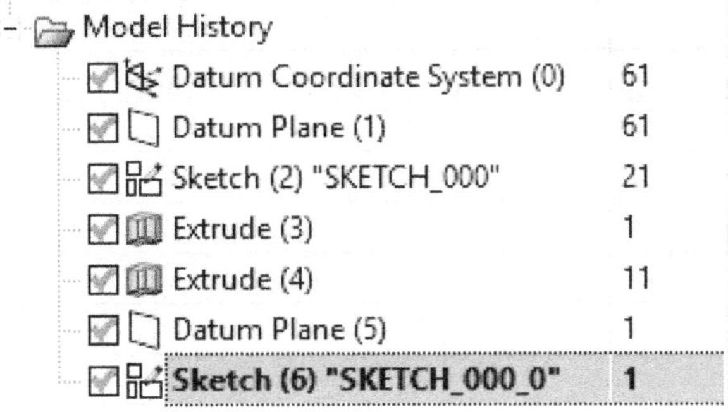

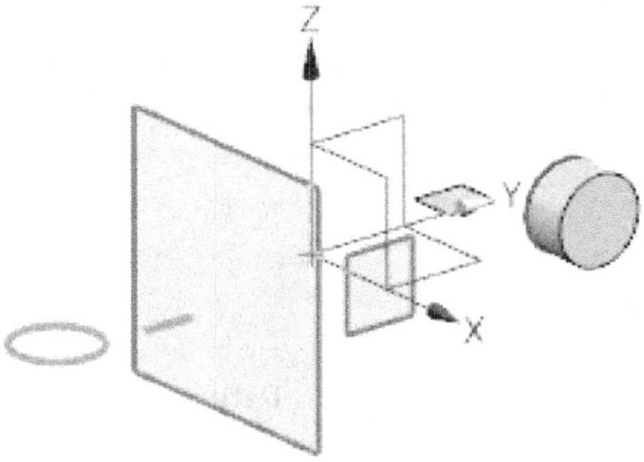

4.3 Edit Tools

Now that you appreciate the importance of respecting dependencies within your model, and have seen the rather extreme tools for editing your Model History – **Suppress** and **Delete** – at work, we will work through the use of **Make Current Feature**, **Edit Parameters** and **Edit with Rollback**.

You have seen at several points in Chapter 3, that double-clicking an entity allows you to edit it, but now we can clarify the exact **Edit Action**. In **Modeling Preferences** and **Customer Defaults** (see Chapter 2.10), you can choose whether double-clicking invokes Edit Parameters, or Edit with Rollback. Let's find out how those tools differ!

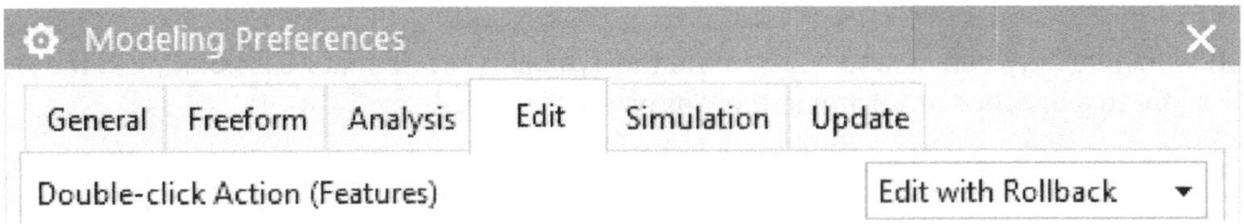

4.3.1 Make Current Feature

Make Current Feature is used to roll back the clock on your model to a specified state. It can be used as a diagnostic tool, to pinpoint the operation that caused a model to go haywire, or to insert features into the Model History with an earlier timestamp.

- Right-click on *Datum Plane (1)* in the Model History, and select **Make Current Feature**.

- This rolls back the clock so that *Datum Plane (1)* is the latest feature. Any new operations will be added to the Model History between *Datum Coordinate System (1)* and *Extrude (2)*.

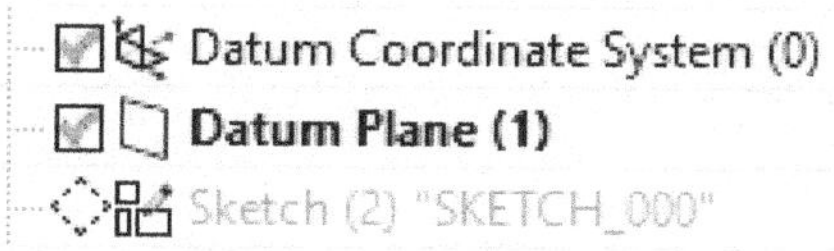

- **Copy** and **Paste** *Datum Plane (1)* using the techniques you learned in Chapter 4.2.3, with the YZ plane as a reference, offset in the -X direction. You will need to use the **Reverse Direction** button again. Note that later timestamps increment up as new features are created.

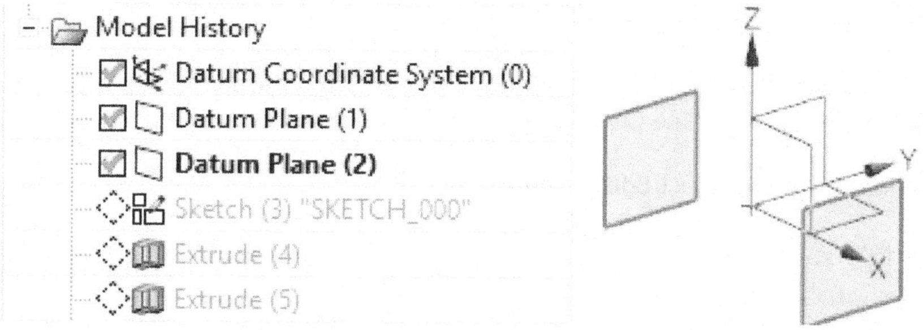

- To roll the clock back to the present version of the model, you can right-click on *Sketch (7)* and select **Make Current Feature**, but this is a good opportunity to practice some keyboard shortcuts. Click on empty space in the Graphics Window to make sure absolutely nothing is selected anywhere, and then press **[Ctrl]+[Shift]+[→]**.

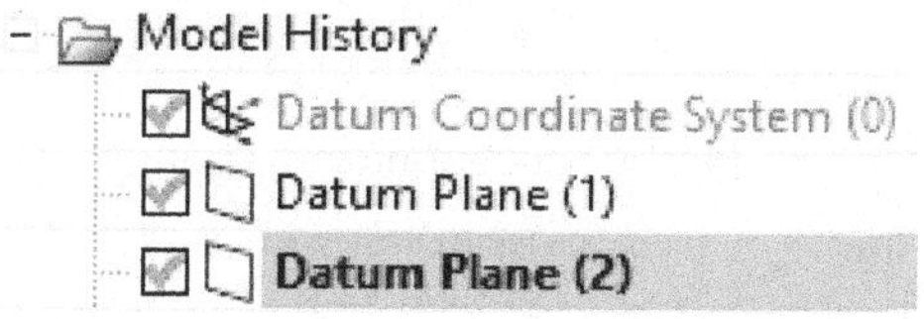

- This makes the next feature in the Model History the current feature. Cool, right? Now use **[Ctrl]+[Shift]+[Home]**.

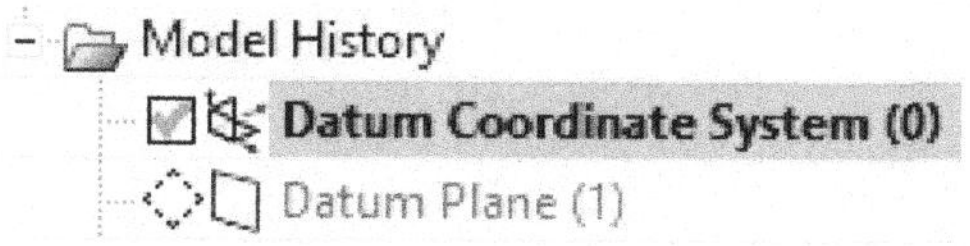

- This rolls the Model History all the way back to the first feature. Now use **[Ctrl]+[Shift]+[End]**.

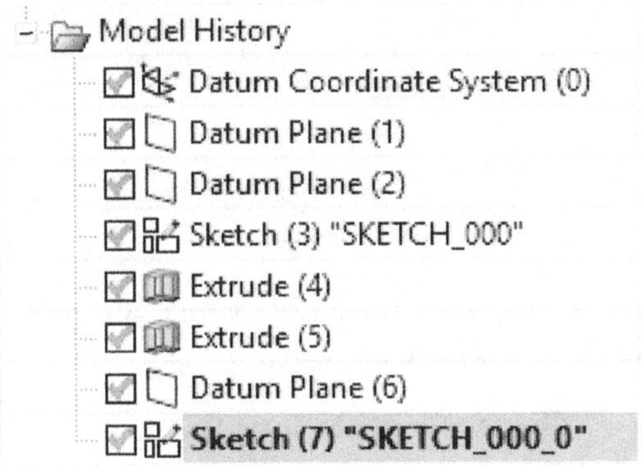

- Can you guess what **[Ctrl]+[Shift]+[←]** does?

4.3.2 Edit Parameters

The **Edit Parameters** tool re-opens the dialog box corresponding to the feature selected, and allows you to adjust the parameters and other inputs, rather than deleting the feature and reusing the tool. This can save you the work of rebuilding all the downstream dependent features if you set up your model well.

Note: for sketches, **Edit Parameters** is a different command, and the **Edit** command for sketches behaves basically as **Edit Parameters** does for non-sketch features. See Chapter 6.2.3 for details on how **Edit Parameters** works for sketches.

- Right-click on *Datum Plane (1)* and select **Edit Parameters**.

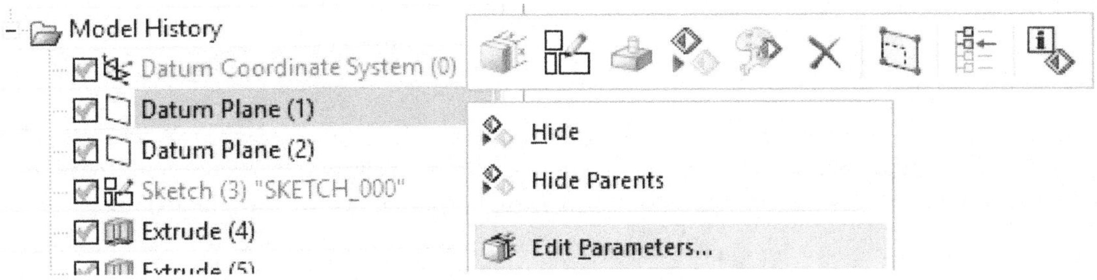

- Change the **Offset Distance** to *25 mm* and click **OK** to commit the change.

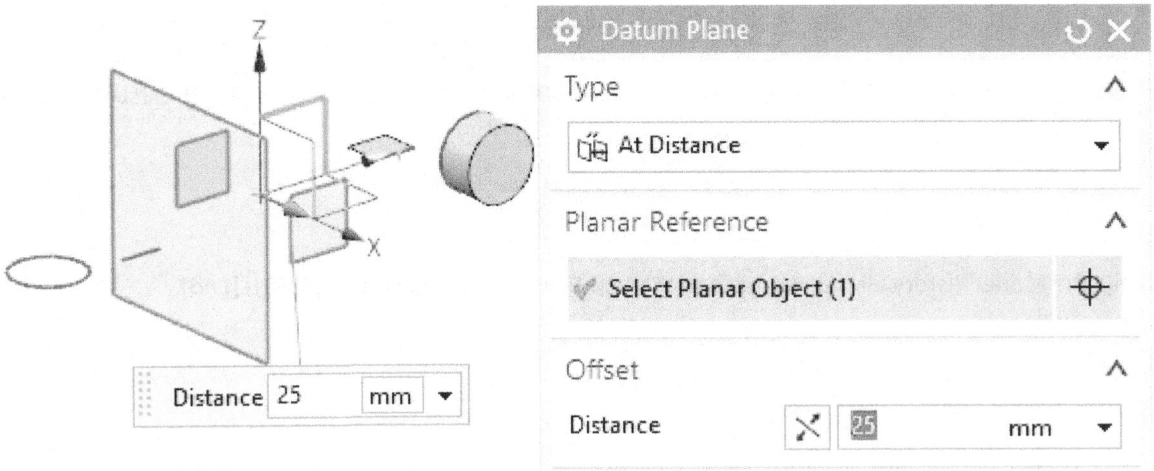

- Click **OK** to commit the change.

4.3.3 Edit with Rollback

Succinctly, **Edit with Rollback = Make Current Feature + Edit Parameters**. At times your dependent features will be covering the feature you wish to edit, and in these scenarios, **Edit with Rollback** will temporarily suppress all features with a later timestamp so that your workspace is free of those later features.

- Right-click on *Datum Plane (2)* and select **Edit with Rollback**.

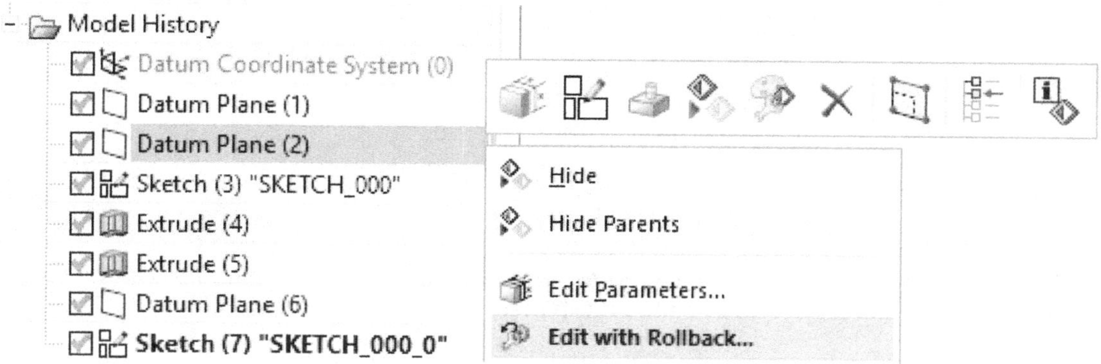

- Note your Model History when the **Datum Plane** dialog box appears. The features following *Datum Plane (2)* in the Model History appear as though **Make Current Feature** rolled the Model History back to *Datum Plane (2)*.

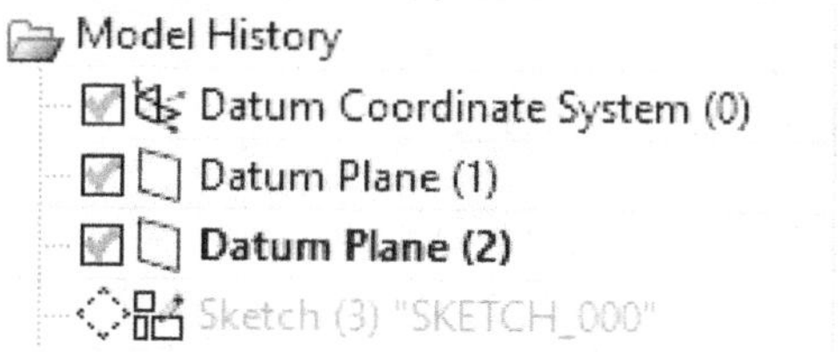

- Change the **Offset Distance** to *10 mm*.

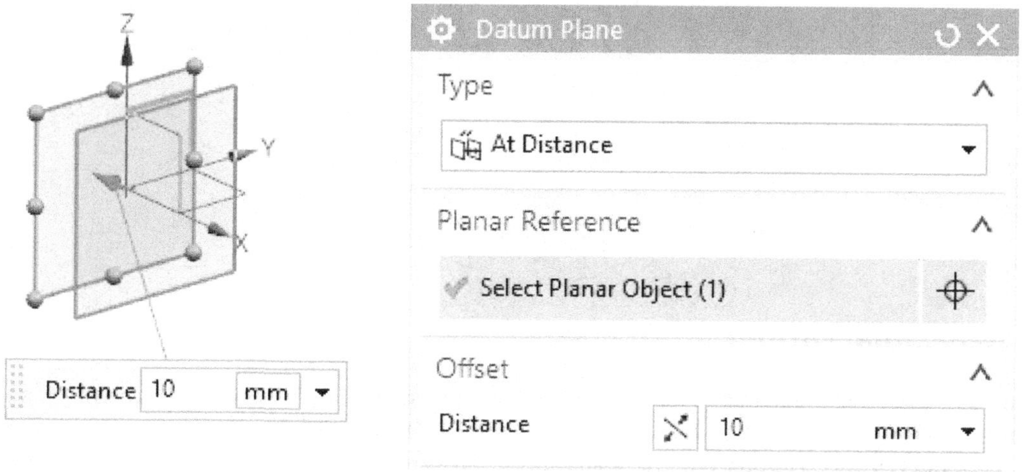

- When you click **OK** to commit the change, your model will be made current again.
- **Save** your file!

4.3.4 The Details Panel

The Part Navigator holds yet another way to make edits to the parameters governing features in your model history – the **Details** panel.

- Expand the **Details** panel and select *Extrude (5)* from the Model History. The **Details** panel will then show all the parameters controlling *Extrude (5)*.

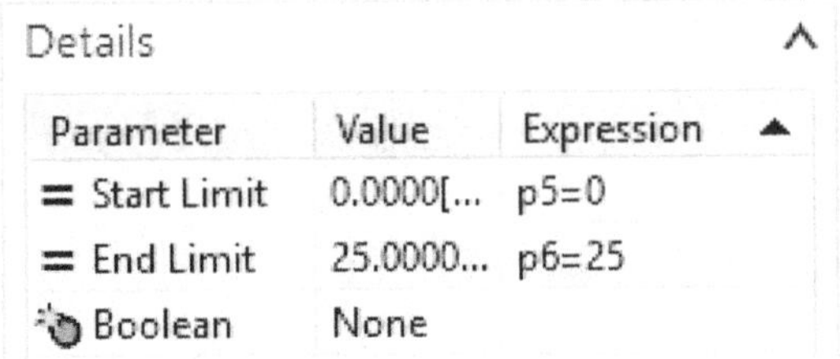

- Double-click on the value for **Start Limit** in the **Expression** column (*p5* in our example). This allows you to change the value at will – go ahead and set it to *10 mm* and press **[Enter]**.

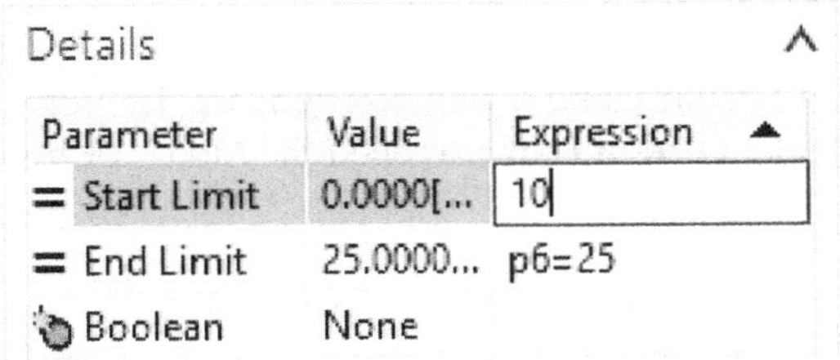

- As you can see, the **Details** panel gives you a fast way to make edits to specific parameters for features without re-opening their dialog boxes!

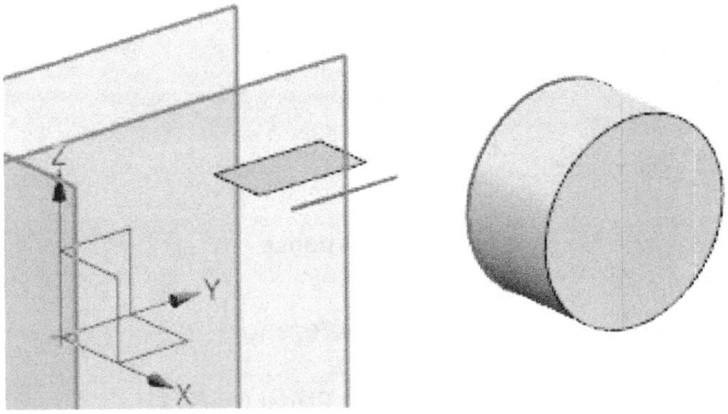

4.4 Non-Associative Entities

The kind of edits that you have made in this chapter are specific to associative features. Associativity is a key concept in NX, and at its heart, it concerns whether an entity has, or remembers its parents. For curves, associativity also entails that they appear in the Model History. Non-parametric solids and surfaces in your model will appear in your Model History, but they can't be converted to parametric entities in the same way that curves can.

- Make *Layer 41* visible. You might recall from Chapter 4.1.2 that when viewing this part in terms of its reference sets, rather than its model history, an invisible spline appeared in the Unused Items folder in the Part Navigator.
- This spline is an example of a non-associative curve. It is not recorded in the Model History, and if other entities were used in its creation, it doesn't remember that.

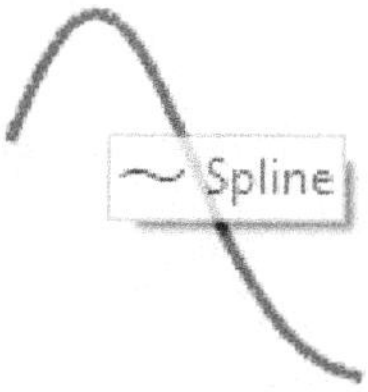

- Double-clicking on this curve edits it by way of the **Studio Spline** tool that you will learn about in detail in Chapter 13. In the **Settings** section of the dialog, there is a checkbox named *Make Associative*. Check it and click **OK** to finish editing. Now the curve appears in the Model History as *Spline (8)*.

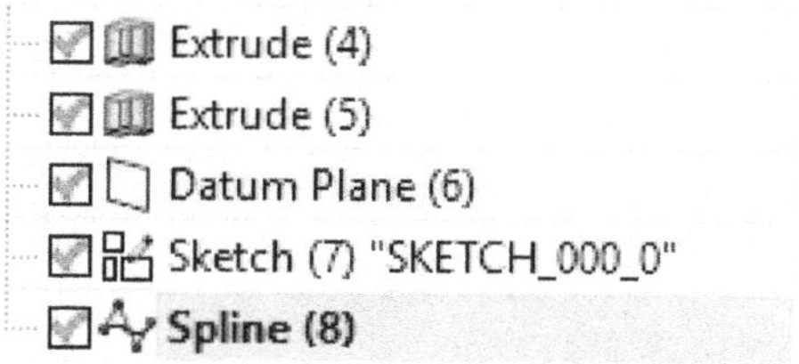

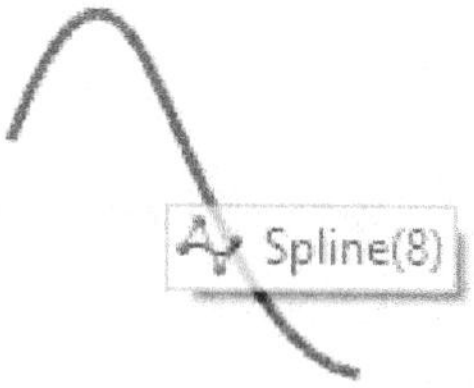

- Most non-associative curves can be made associative in this way – upon editing, they are converted to splines, lines, or arcs, as appropriate.
- **Save** your file!

5 Datums

Datums are the selectable, parametric, associative reference entities that are often required for positioning features on a model. There are four basic types of datums in NX – *datum coordinate systems, datum planes, datum axes, and datum points*. These can be found in the **Datum/Point** drop-down menu in the **Feature** group on the **Home** tab.

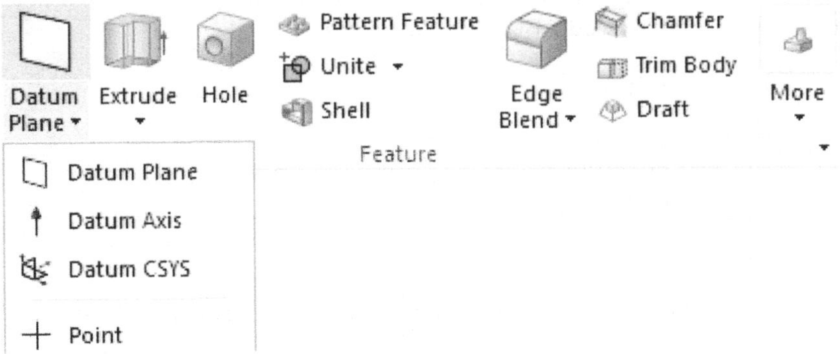

Datums are useful for many other modeling operations such as mirroring, patterning, and positioning holes. Defining a datum often requires a combination of model geometry and other datums. For example, in the figure below, the plane is placed at an angle to the top face. Since the model in the figure below has no straight edges, a datum axis was created to define an axis of rotation.

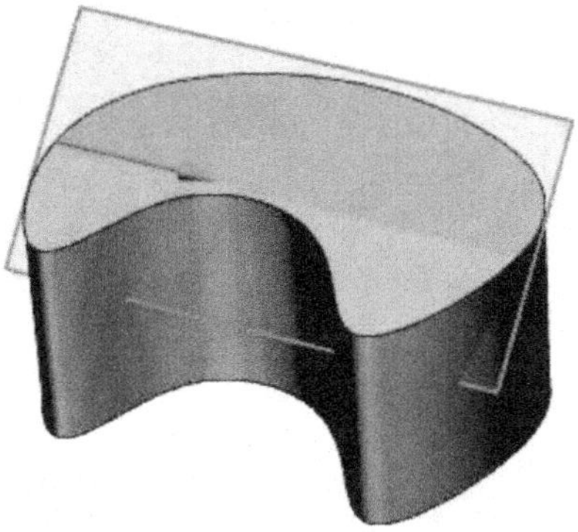

5.1 Datum Coordinate System

Datum Coordinate Systems were first mentioned in Chapter 3.1.5. Now that you have a good understanding of how dependencies work in the Model History in NX, we can make good on the promise made in Chapter 3.1.2 to fully explain the sense in which references to the WCS are not fully associative. You will also get some practice with the **Datum CSYS** tool!

- Create a new file (**Units**: *Millimeters*) named *"Coordinate Systems.prt"* and place it in *"C:\My NX 11 Files"*.
- Select the **Datum CSYS** tool, and change the **Reference** to *Selected CSYS*. Select the existing CSYS, *Datum Coordinate System (0),* as shown below.

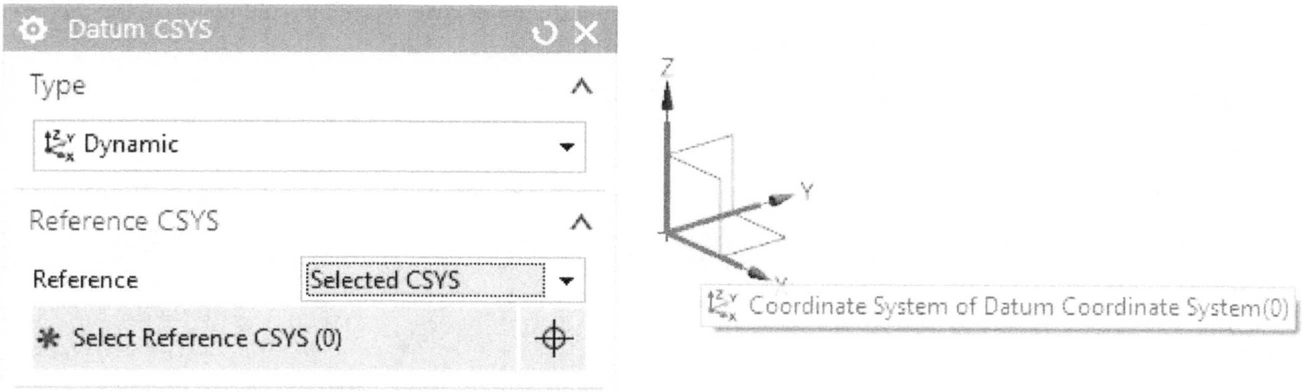

- When the Dynamic CSYS appears, click on its origin, and change the Y and Z Offset values to *30 mm*. Click **OK** to place the new CSYS.

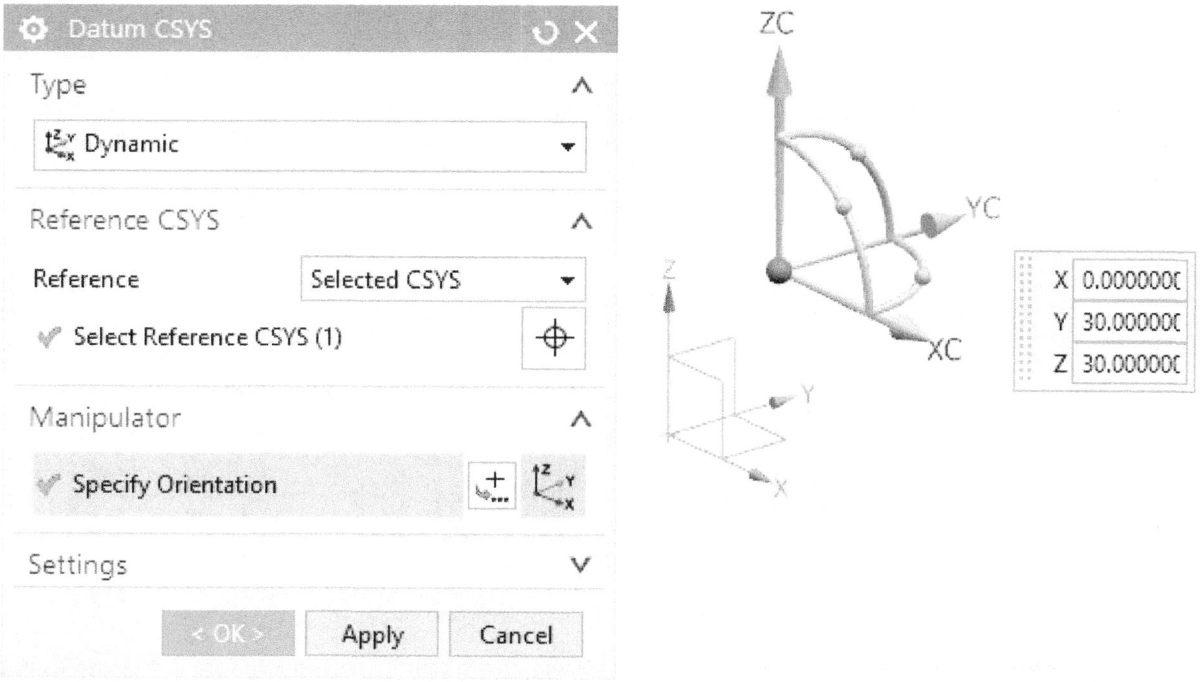

- If the WCS is not visible, first use **[W]** to toggle its display on. Next, activate the **WCS Dynamics** tool. Move the WCS *30 mm* in the XC direction.

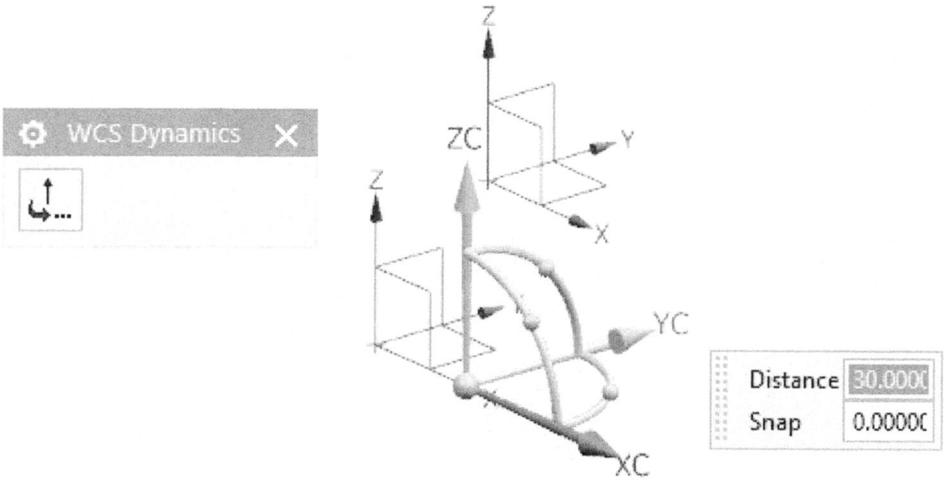

- Use the **Datum CSYS** tool again, this time with *WCS* chosen from the **Reference** drop-down menu. Move the Dynamic CSYS *30 mm* in the YC direction. Click **OK** to create the CSYS.

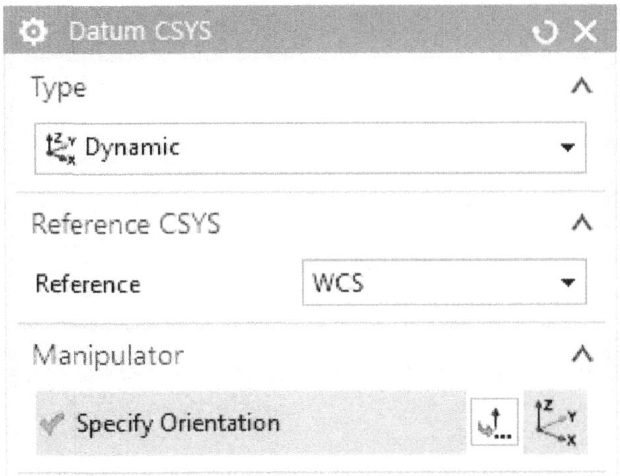

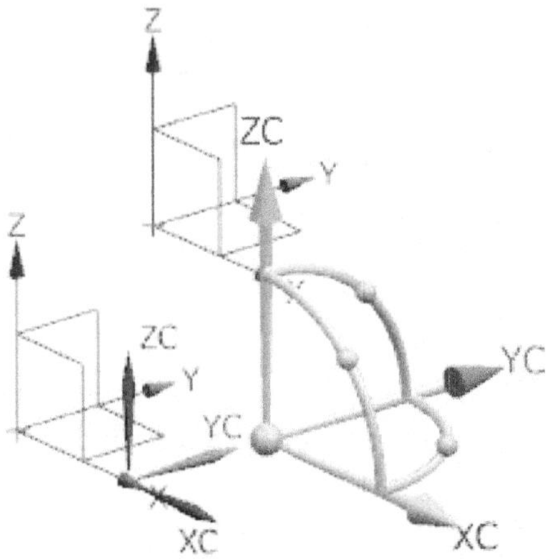

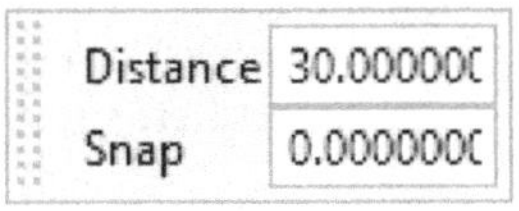

- Next, use the **Set WCS to Absolute** command. Note that *Datum Coordinate System (2)* does <u>not</u> move, even though it was positioned in relation to the WCS. The positioning of *Datum Coordinate System (2)* was determined at the time of creation, and there is no associative relationship between *Datum Coordinate System (2)* and the WCS after its initial placement.

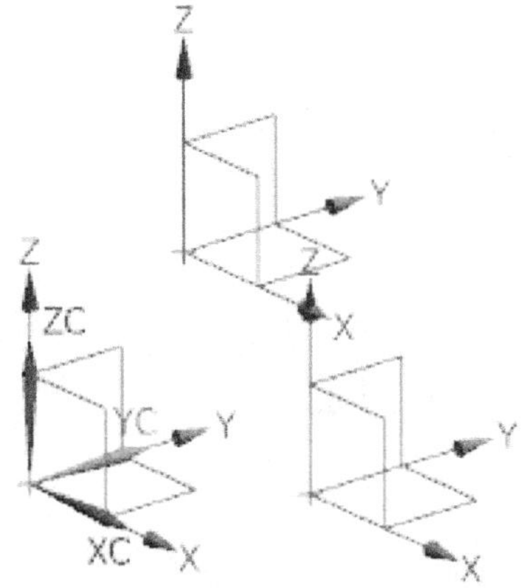

- Double-click *Datum Coordinate System (0)* from the Part Navigator to edit its position. Rotate it *45 degrees* in the YZ plane. When you click **OK** to finish editing the position, note that *Datum Coordinate System (1)* maintains its relative position to *Datum Coordinate System (0)*. There is an associative relationship between the two, that remains even after the initial placement of *Datum Coordinate System (1)*.

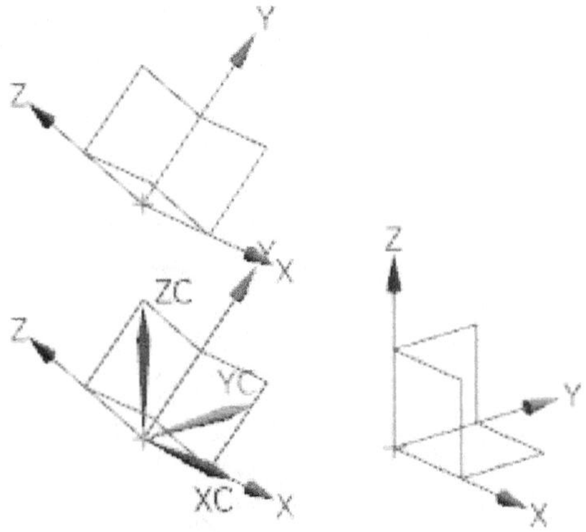

- **Save** your file!

5.2 Datum Plane

Datum planes are very useful for creating sketches, constraining and positioning features, and defining cross-sections for solid models. You may use solid geometric entities (such as faces, vertices, and control points) or other reference entities when defining datum planes.

- Open the part file *"Datums.prt"* from *"C:\My NX 11 Files\Downloaded Files"*.

5.2.1 At Distance

Datum planes can be offset from planar faces, or from other datum planes.

- Open the **Datum Plane** tool and set the **Type** to *At Distance*. Select the top face of the solid. The **Datum Plane** menu will have an **Offset Distance** field in which to enter a **Distance** value and the number of planes to create.

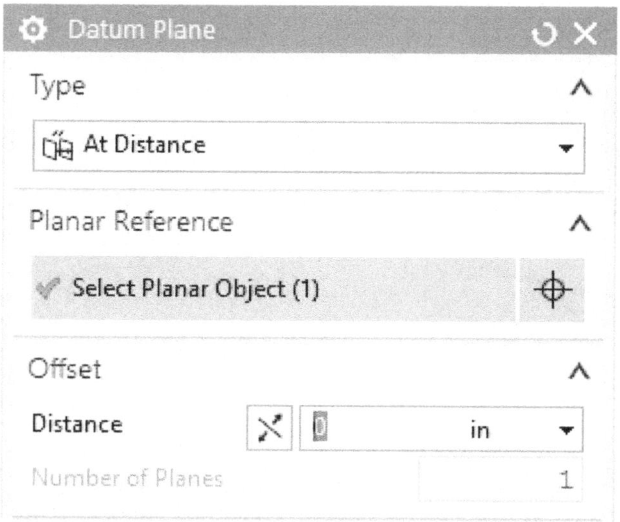

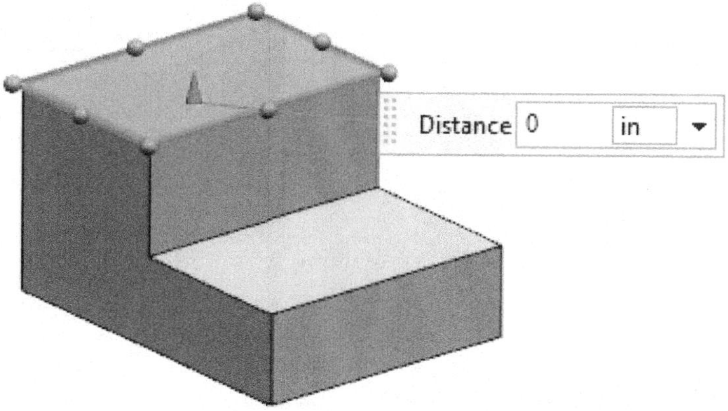

- A preview of the datum plane, along with an arrow appears in the Graphics Window, as shown in the following figure. The arrow indicates the positive offset direction.

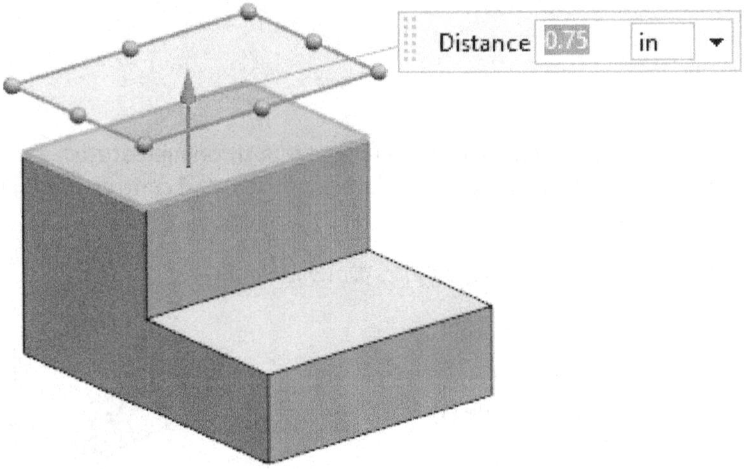

- Enter an **Offset Distance** of *0.75 in* and set the **Number of Planes** field to *1*. Click **Apply** to create your datum plane.
- Create a second datum plane with **Type** set to *At Distance*, but this time use the datum plane that you just created as the **Planar Reference**. Type in an **Offset Distance** value of *.38 in*.

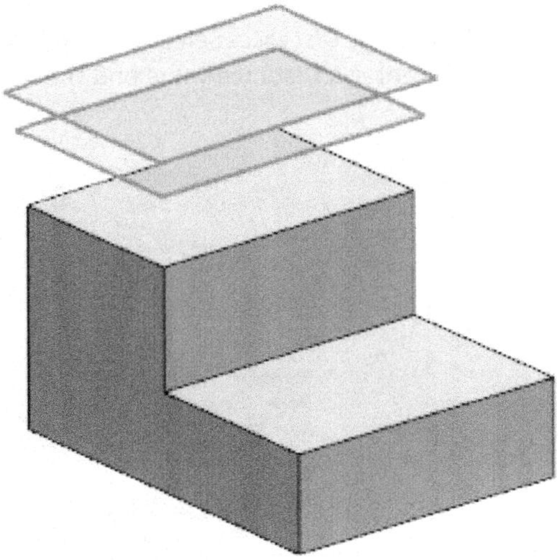

5.2.2 Bisector

A datum plane of type *Bisector* is centered between two faces or other datum planes. If those faces meet along an edge, or those planes meet along an axis, the plane will share that edge or axis, and bisect the angle between those faces or planes.

- Hide the *At Distance* datum planes created previously
- Select the **Datum Plane** tool and set the **Type** to *Bisector*.
- Designate the left face (shown below) as the **First Plane**.

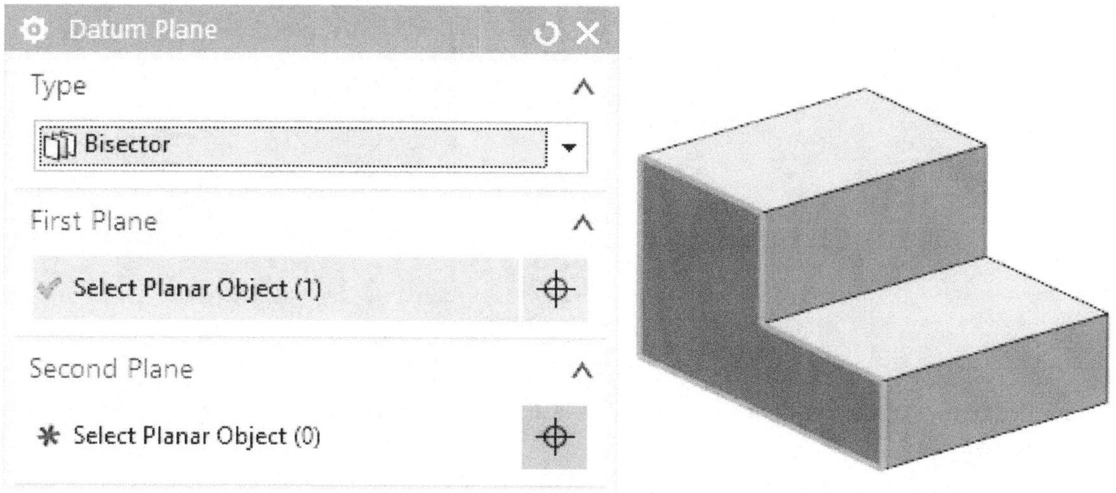

- Rotate the model and select the opposite face. The preview will appear as shown below. Click **OK** to create this new datum plane!

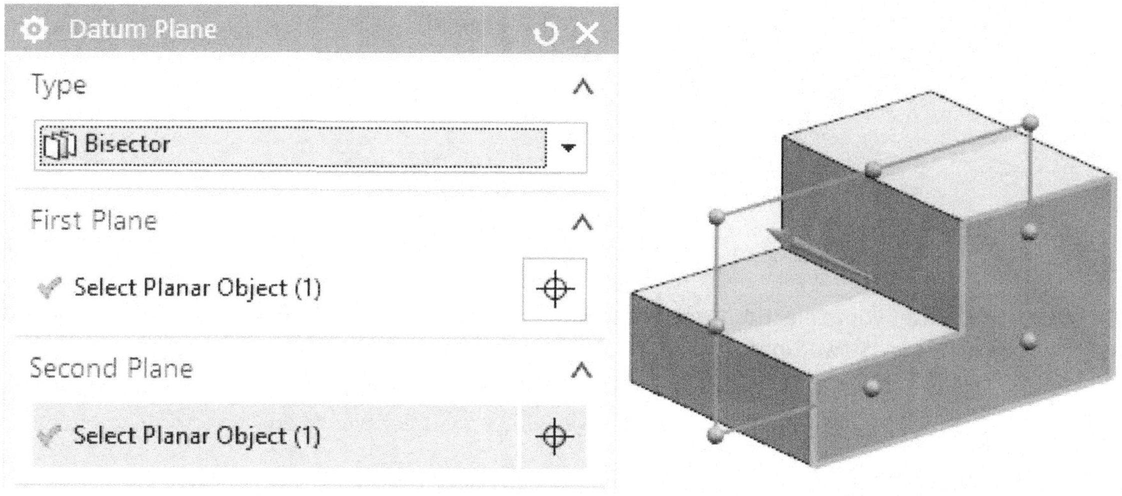

5.2.3 Two Lines

In the **Datum Plane** dialog, the *Inferred* setting for **Type** is often pretty smart – the vast majority of your selections can be made with the **Type** set to *Inferred.* In this exercise, you will learn that you can even preselect geometry before opening the **Datum Plane** tool, in this case to build a *Two Lines* datum plane.

- Hide the *Bisector* datum plane you just created.
- Open the **Datum Plane** tool and **Reset** it so that the **Type** becomes *Inferred*, and then close it.
- Set your **Selection Type Filter** to *Edge* and choose the two edges shown below.

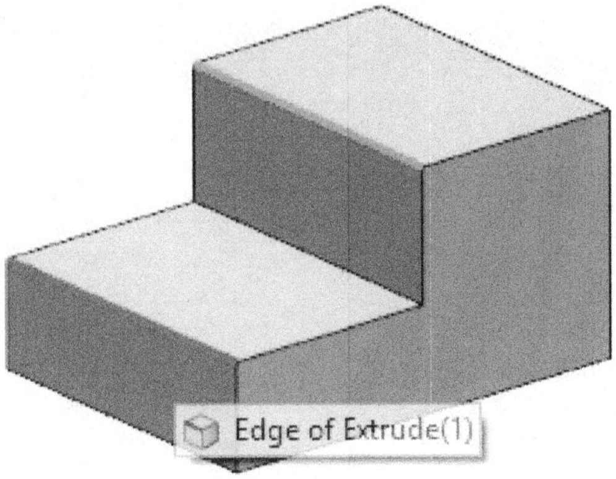

- Now choose the **Datum Plane** tool. Since the **Type** was last set to *Inferred*, it accepts your preselection and produces a plane of type *Two Lines*.

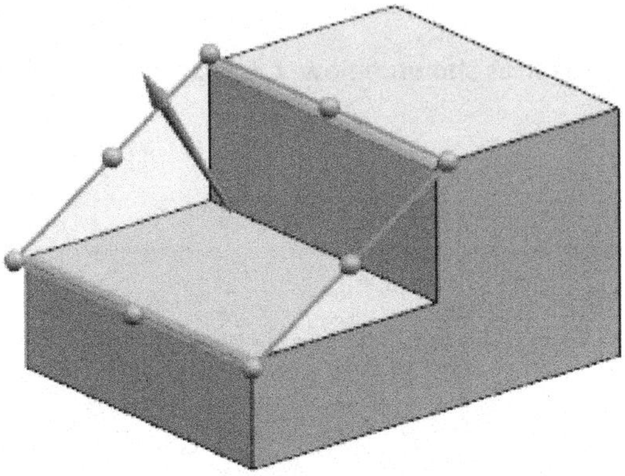

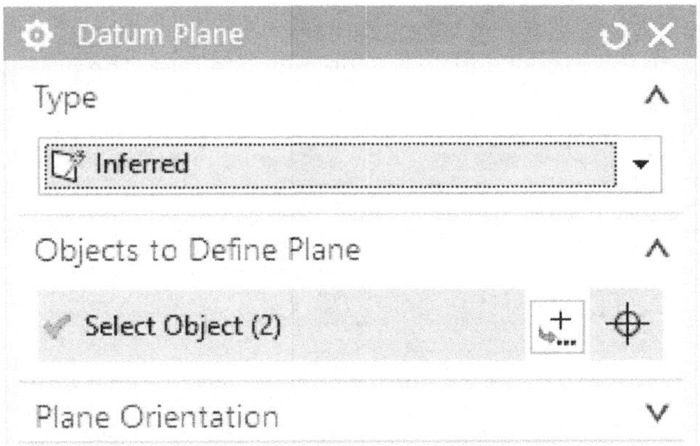

- Click **OK** to create it. Reset your **Selection Type Filter** and then double-click on the datum plane just created. You will see that the **Type** is stored as *Two Lines*.

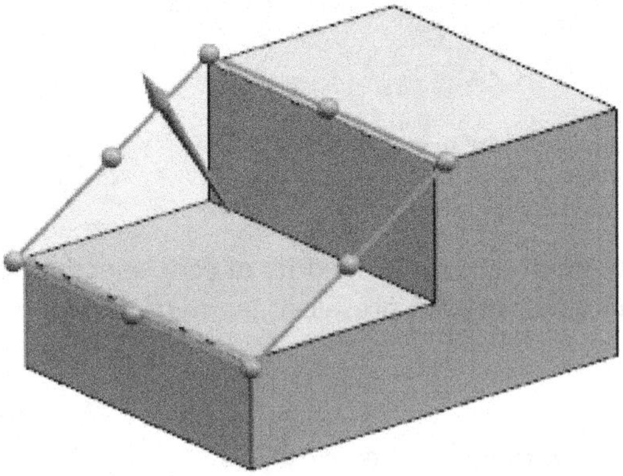

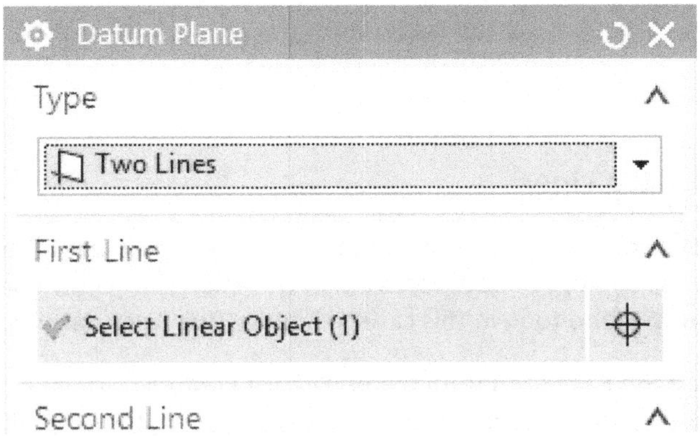

5.2.4 At Angle

Datum planes may be constrained at an angle to a planar face or a datum plane. In order to accomplish this, it must pass through either an edge or a datum axis. The exercise below walks you through the creation of a typical angled datum plane.

- **Hide** the *Two Lines* datum plane you just created.
- Select the **Datum Plane** tool and set the **Type** to *At Angle*.
- Select the top face of your model for the **Planar Reference**, and the edge shown below for the **Through Axis**.

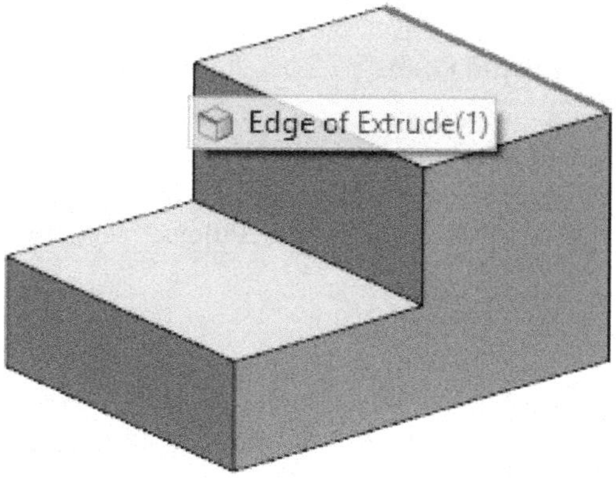

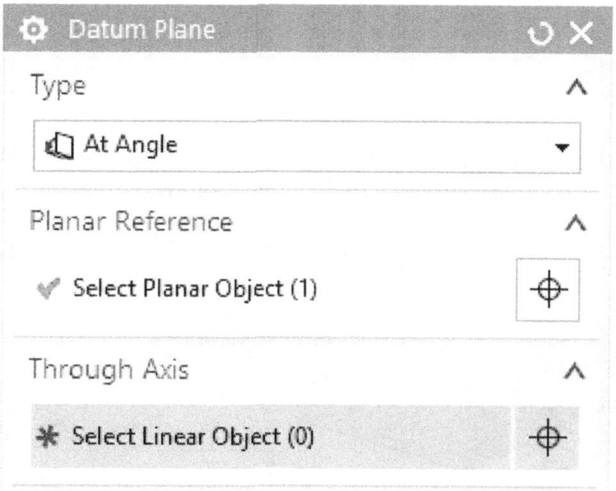

- In the **Angle** value field input *±45 degrees* as necessary to create the datum plane as shown below. Click **OK**.

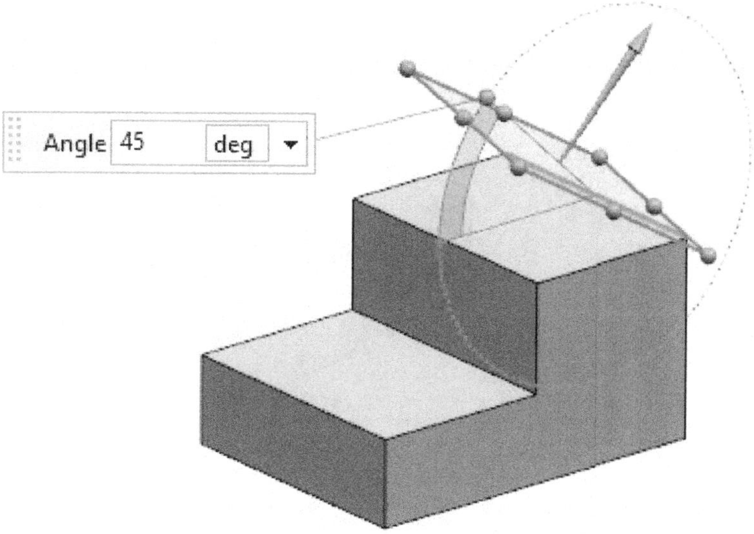

5.2.5 Three Points

Geometrically, it takes three non-collinear points in 3D space to uniquely determine a plane. This is an excellent way to position a datum plane.

- **Hide** the *At Angle* datum plane you just created.
- Select the **Datum Plane** tool and set the **Type** to *Curves and Points*. Set the **Subtype** to *Three Points*. Choose the three points shown below. You may have to ensure that your *End Point* snap point icon is enabled.

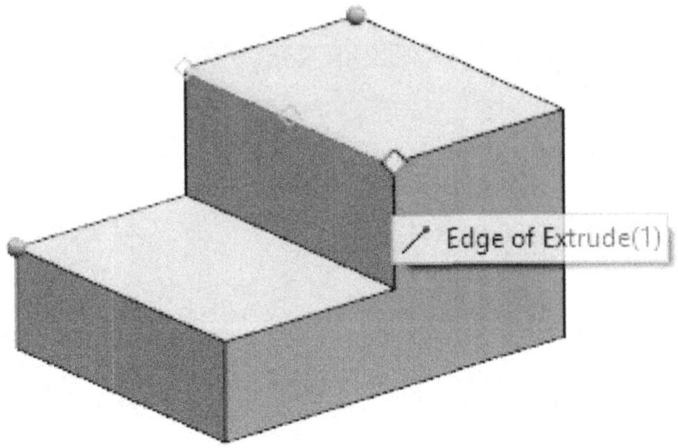

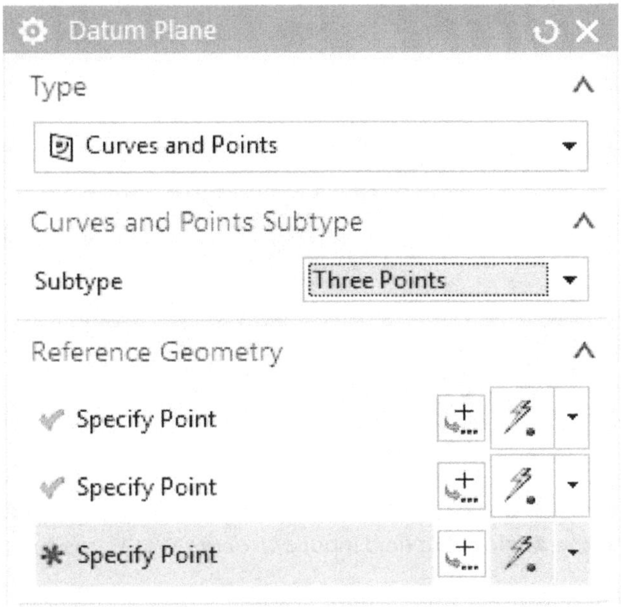

- In fact, the *Inferred* **Type** is smart enough to produce this plane when you click those three points. Click **OK** to create the plane.

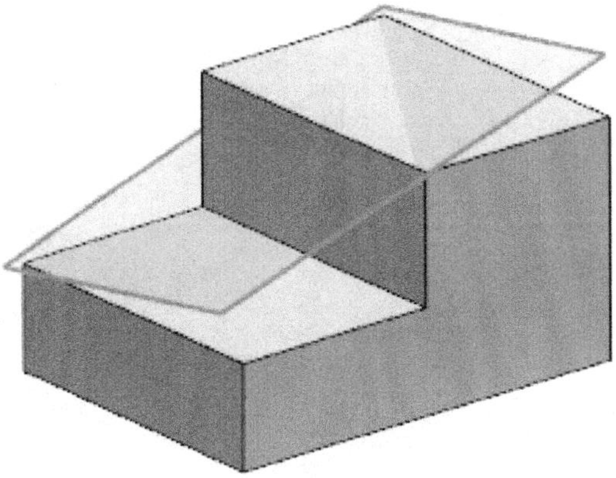

5.2.6 Tangent

Datum planes can be constrained to be either tangent to cylindrical faces or through the center axis. To explore these options, we will make use of a cylindrical boss on top of the model.

- Hide the *Three Points* datum plane that you just created.
- **Unsuppress** *Extrude (2)* from the Model History. You will see a cylindrical boss appear atop the model.

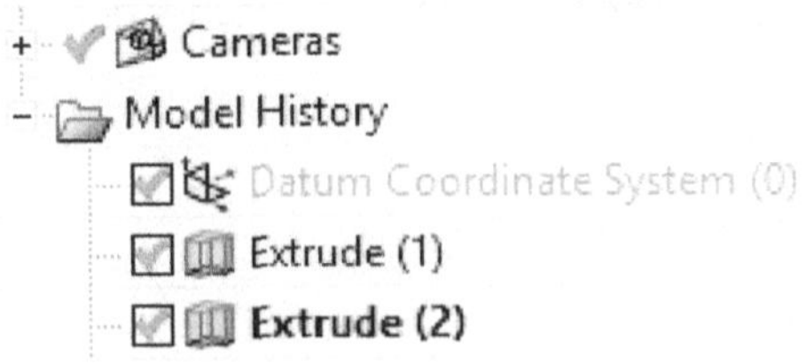

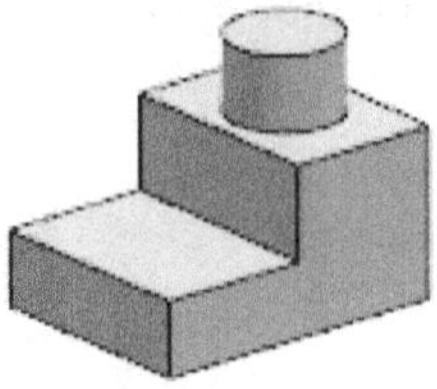

- Select the **Datum Plane** tool and set the **Type** to *Tangent*, and the **Subtype** to *Angle to Plane*. For **Reference Geometry**, choose the cylindrical face, and for the **Planar Object**, choose the planar face shown below. Set the **Angle** to *45 degrees*. Click **OK** to create your datum plane.

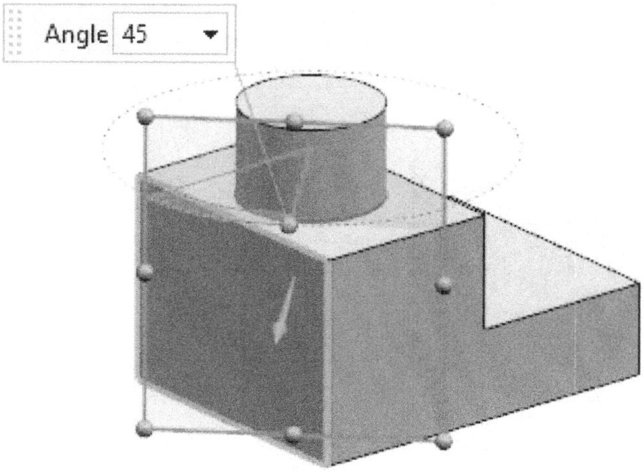

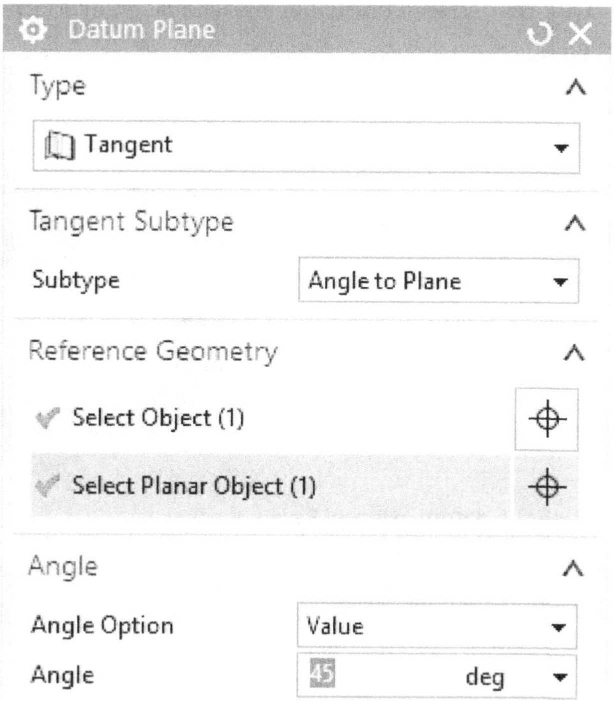

- **Hide** the newly-created datum plane and **Save** your file!

5.3 Datum Axis

Datum Axes are useful for creating angled datum planes and circular patterns, as well as for bodies of revolution. They are defined by specifying constraints and solid geometric entities, just as for *Datum Plane* features.

Typical objects used to define datum axes include edges, two points on a solid body, the axis of a cylinder, or the intersection of two planes.

5.3.1 Intersection

In this exercise, you will create a datum axis where a plane and face intersect.

- Continue working with *"Datums.prt"*. **Show** the bisector datum plane constructed in Chapter 5.2.2.
- Select the **Datum Axis** command. Choose the datum plane and the face as shown below.

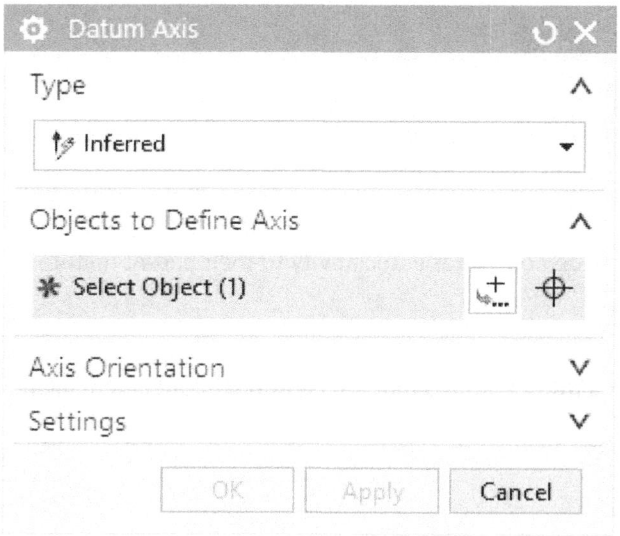

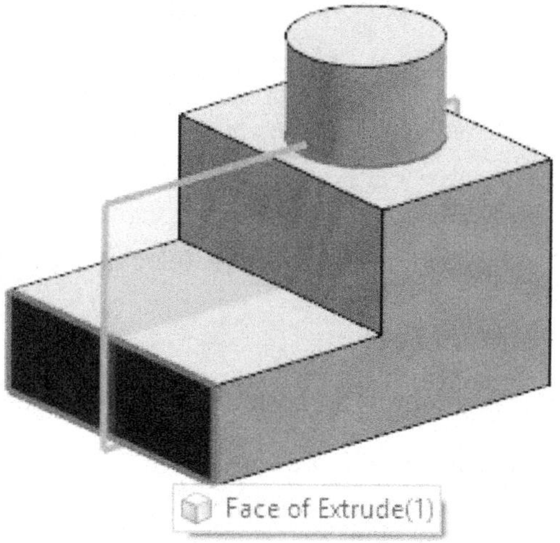

- Click **OK** to create the datum axis, and hide the bisector datum plane. Your model should appear as below.

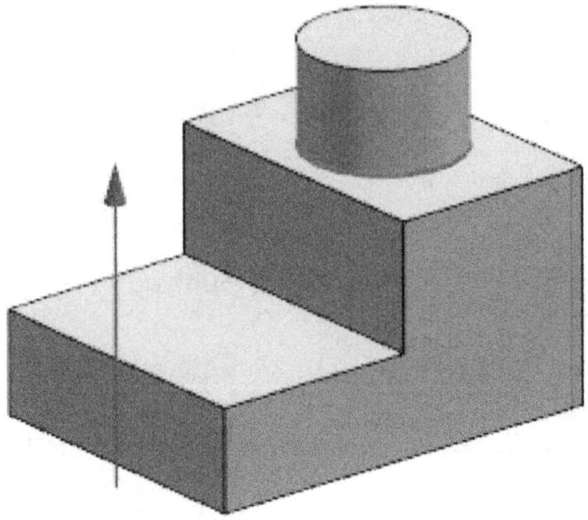

- **Hide** the datum axis and **Save** your file!

The other methods for building datum axes are straightforward – we leave it to you to practice making them and determine which is best for your designs.

5.4 Point

The **Point** tool creates a point entity that is fully associative and parametric in relation to parent features. When the **Type** is set to *Inferred*, notice that the **Reference** only allows you to select the Absolute Coordinate System or the WCS – and not any existing Datum CSYS. As illustrated in Chapter 5.1, references to the Absolute Coordinate System and WCS are not fully associative. The **Point** tool will therefore only produce an associative relationship to a parent feature when the point is built using one of the other methods in the **Type** drop-down.

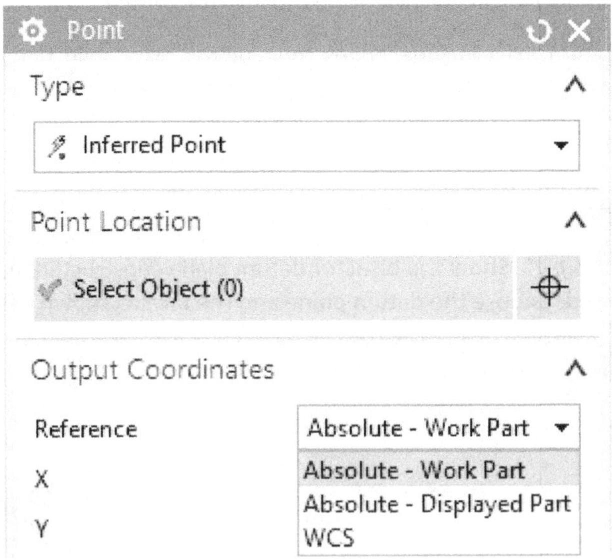

In the following exercises, you will create point entities and demonstrates associativity to their parent features.

5.4.1 Point on Face

Points on faces are useful for positioning holes, bosses, and other features that are centered about a defining point.

- Continue working with *"Datums.prt"*.

- Select the **Datum Point** command, and set the **Type** to *Point on Face*. Click on the front face shown below, and set the **U Parameter** and **V Parameter** to *0.5* and *0.75* respectively, as shown below[2]. Click **OK** to create the point.

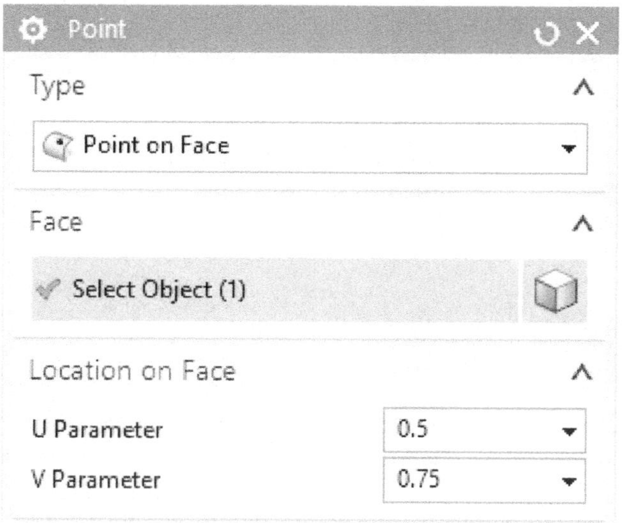

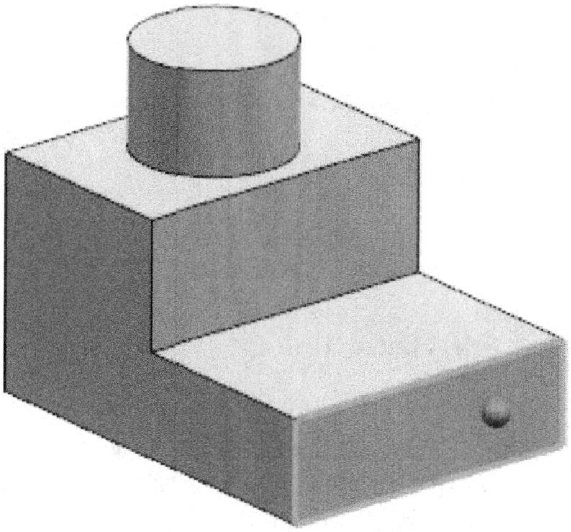

- Edit *Extrude (1)* and change the **End Limit** from *3 in* to *6 in*.
- After editing, you will find that your datum point entity has maintained its correct relative position on the face.

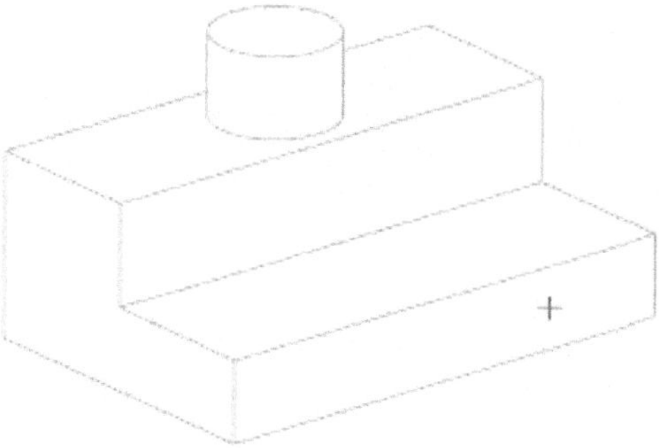

5.4.2 Intersection Point

Points at the intersection of an existing curve or edge, with another curve, edge, face, or datum, are often useful for building curves that intersect properly.

- **Hide** the *Point on Face* point, and **Show** the *Three Points* datum plane constructed in Chapter 5.2.5 (ours is *Datum Plane (8)*).
- Select the **Point** tool and set the **Type** to *Intersection Point*. Select the *Three Points* datum plane as the **Curve, Surface, or Plane**. Choose the edge shown below as the **Curve to Intersect With**. Click **OK** to create the point.

[2] Every face in NX has parameters U and V that vary between 0 and 1. You will learn more about these important parameters in Chapter 19.

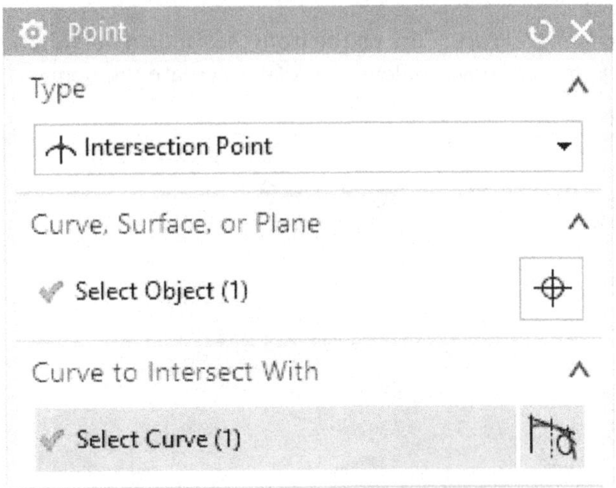

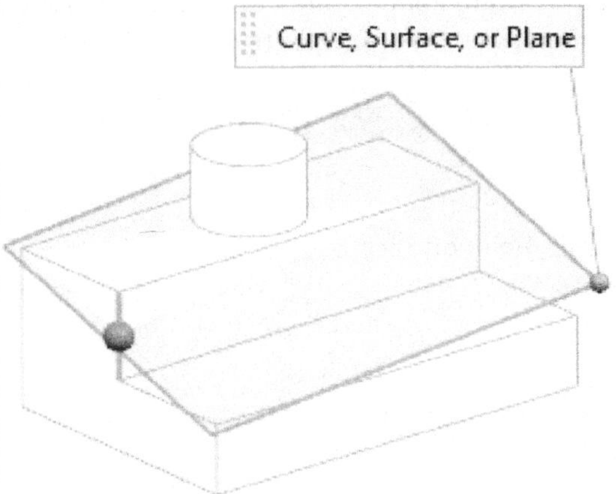

- Select *Extrude (1)* from the Model History and open the **Details** panel within the Part Navigator. Modify the *Vertical Dimension between Line1 and Line1* to have a value of *4.5 in*. Your point remains at the intersection of the edge with the *Three Points* datum plane.

Details

Parameter	Value	Expression
Start Limit	0.0000[in]	p4=0
End Limit	6.0000[in]	p5=6
Vertical Dimension between Line1 and Line1	4.5000[in]	p0=4.5
Horizontal Dimension between Line4 and Line4	2.0000[in]	p1=2
Horizontal Dimension between Line7 and Line7	4.0000[in]	p2=4
Vertical Dimension between Line5 and Line5	1.0000[in]	p3=1

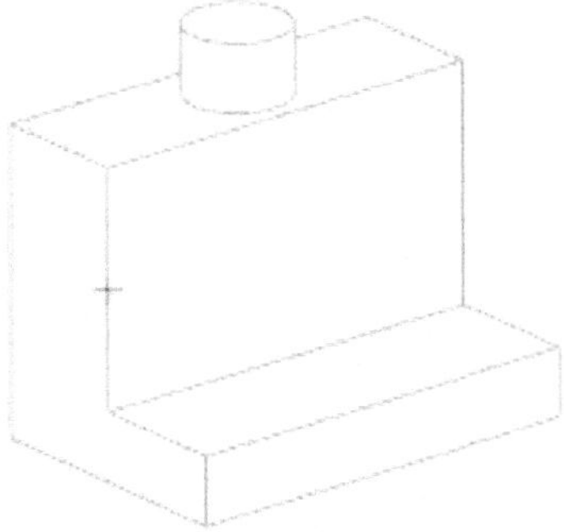

- **Save** your file!

6 Sketches

Most solid models begin with a sketch – a planar set of curves constructed in a special 2D environment. Sketching is arguably one of the more difficult techniques to completely master in NX, but it is well-worth the effort. A single sketch can capture a tremendous amount of design intent and inform the sequencing of subsequent solid modeling tools so that the model is robust and resilient.

6.1 Sketch in Task Environment

There are two sketch tools in NX – **Direct Sketch**, and **Sketch in Task Environment**. It is the opinion of the authors that, between these two, **Sketch in Task Environment** is the superior tool for beginners and experts alike. If you skipped Chapters 2.9 and 2.9.1, you will need to go back to those sections and do the exercises – in those chapters, we customized the ribbon and user preferences of NX so that **Direct Sketch** is disabled entirely.

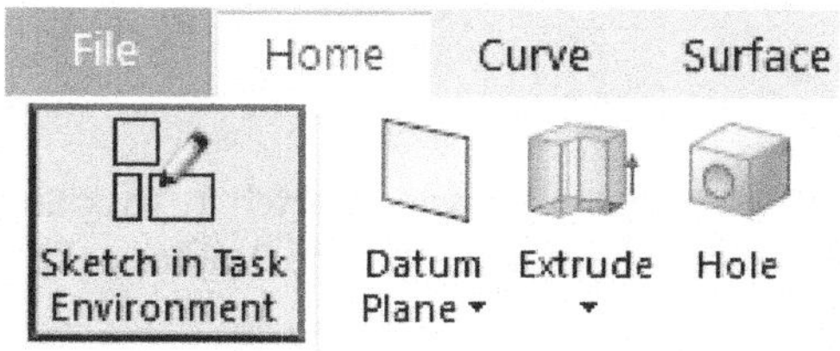

6.1.1 The Sketch Task Environment

The *Sketch Task Environment* is in some sense a sub-application that you can enter in a variety of ways, the most basic of which is to use the **Sketch in Task Environment** tool. You also enter the Sketch Task Environment when creating or modifying a sketch internal to a feature (see Chapter 6.10) or when using **Edit with Rollback** to modify a sketch. In Chapter 2.9.1 you modified the **Edit** action for sketches to also enter the Sketch Task Environment.

- Create a new part file (**Units**: *Inches*) called *"Sketch in Task Environment.prt"*.
- Select the **Sketch in Task Environment** tool.
- The **Create Sketch** dialog will appear. When the **Plane Method** is set to *Inferred*, the tool will by default assume that you want to sketch on the XY plane of the Work Coordinate System (see Chapter 3.1.2). The "required" input *"Specify CSYS"* is actually optional when the **Plane Method** is set to *Inferred*, so at this point you can just click **OK** to continue.

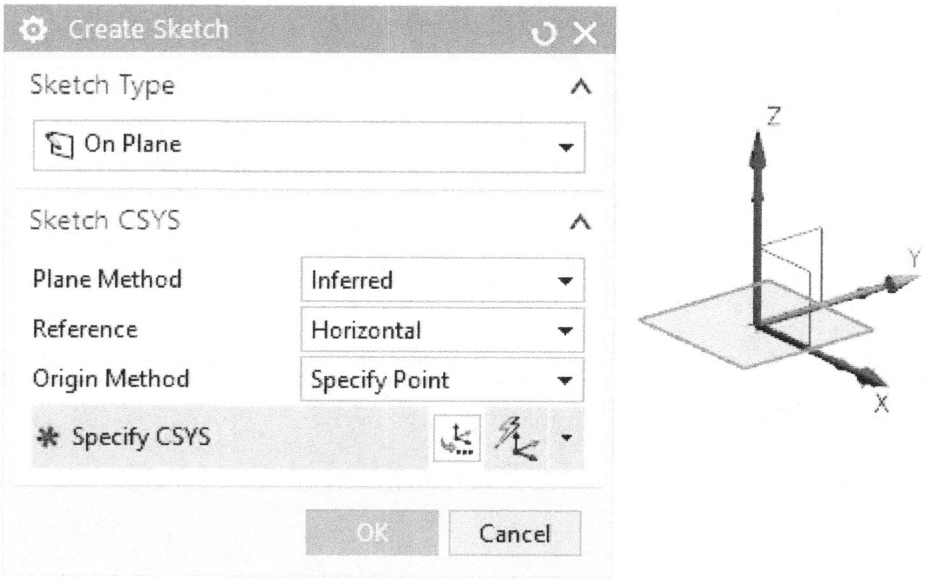

- This brings you into the Sketch Task Environment. Note that the tools on the **Home** tab have changed, and that the ribbon has lost several tabs that were present in the **Modeling** application (e.g., the **Surface** tab). From within the Sketch Task Environment, you cannot access most tools from the Modeling application. Also, note that the usual **File** drop-down at the top has been replaced by the **Task** drop-down, which provides tools for creating and managing sketches within the Sketch Task Environment.

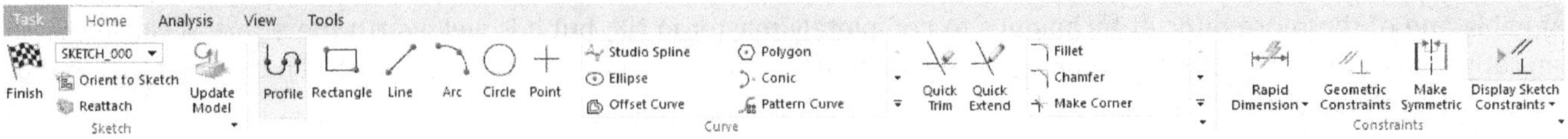

- The Model History shows a new sketch named *"SKETCH_000"*. The blue shading on the icon to the left indicates that *SKETCH_000* is actively being edited in the Sketch Task Environment.

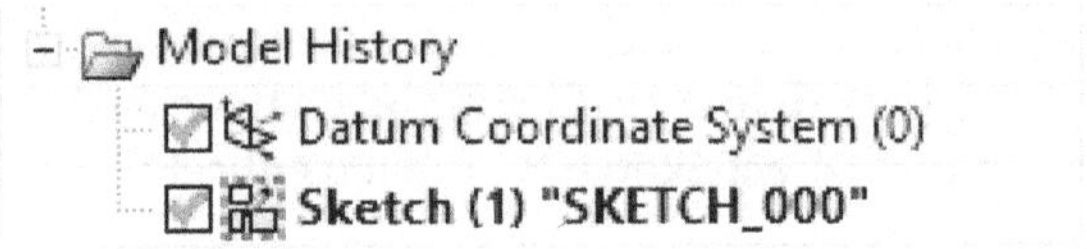

6.1.2 The Profile Tool

By default, when you enter the Sketch Task Environment using **Sketch in Task Environment**, the **Profile** tool is enabled. The **Profile** tool creates chain-connected series of lines and arcs. The **Profile** tool is found in the **Curve** group on the **Home** tab in the sketch task environment.

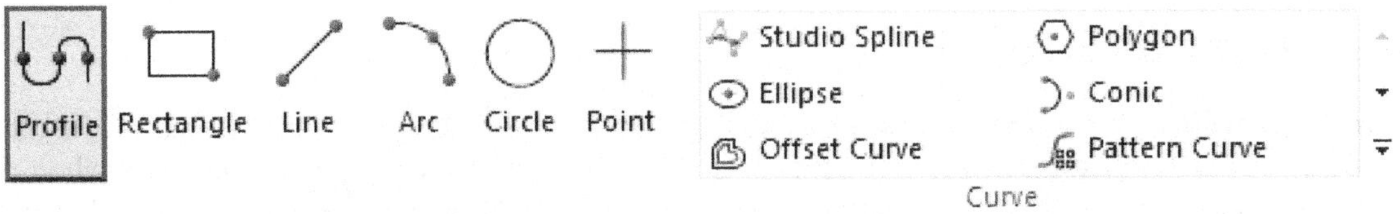

The profile tool has two modes – line and arc. As you can see in the **Object Type** section below, the shaded *line* icon indicates that the **Profile** tool is currently in *line* mode.

- To create a line segment using the profile tool, <u>two clicks</u> are required – one for the start point and one for the end point. Click <u>once</u> at the *Sketch Origin*. Note that the *Sketch Origin* is a different entity than *Point of Datum Coordinate System (0)*.

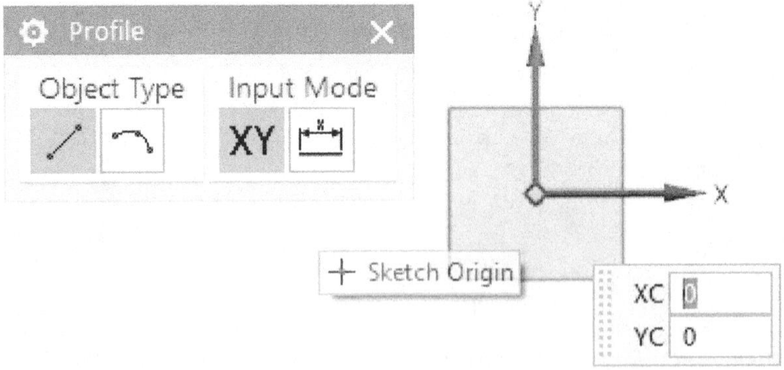

- As you move your cursor vertically, NX previews your line segment and shows the length in a dialog box. Drag to a **Length** of *25* and **Angle** of *90*, as shown below.

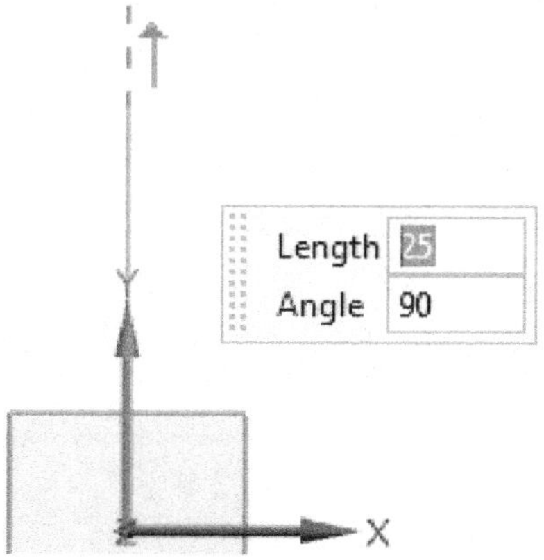

- Click again to complete the line segment. Note that the appearance has changed, and your line has a name.

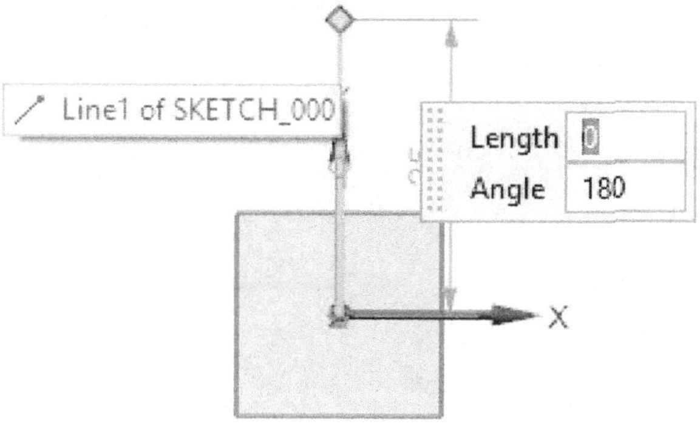

- Move your cursor to the location shown below, and click again. This time it only took one click to produce a line segment! The **Profile** tool uses each endpoint as the start point for the next line or arc, depending on mode. This allows you to sketch complicated contours quickly!

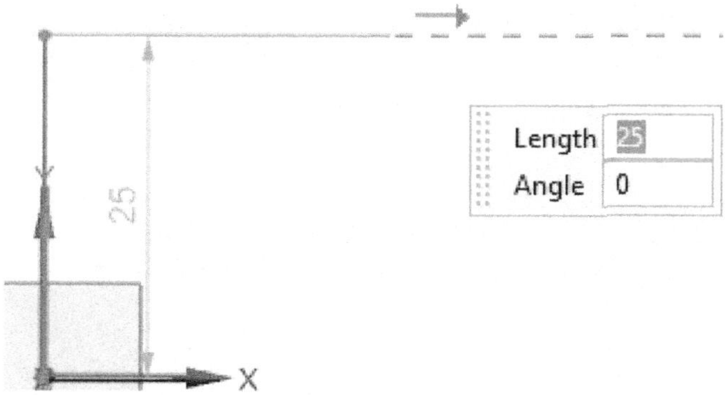

- Type *"25"* into the **Length** field, push **[Enter]**, and move your cursor down and to the right until a dotted horizontal line appears, joining the cursor location to the end of your first (vertical) line segment. Click when your cursor is locked to the dotted line. *Guide lines* like this appear as you sketch to help you align your clicks with existing features of the sketches. It is worthwhile to learn to use these efficiently, as they can dramatically improve the speed with which you produce accurate sketches on the fly.

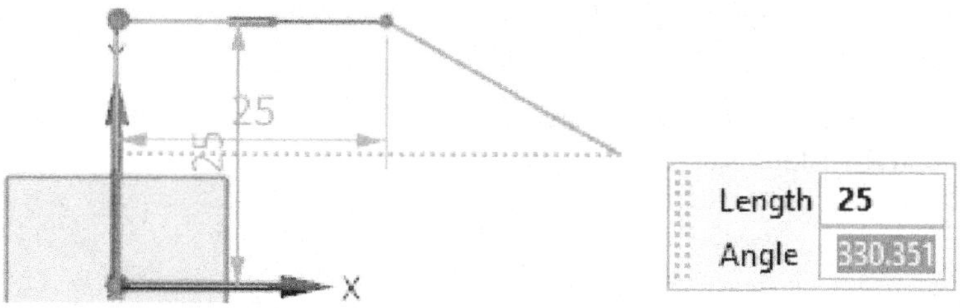

- Notice that typing in a value created a driving dimension for the last line segment! Move your cursor straight down until another horizontal guide line appears. Click once your cursor has locked to the guide line.

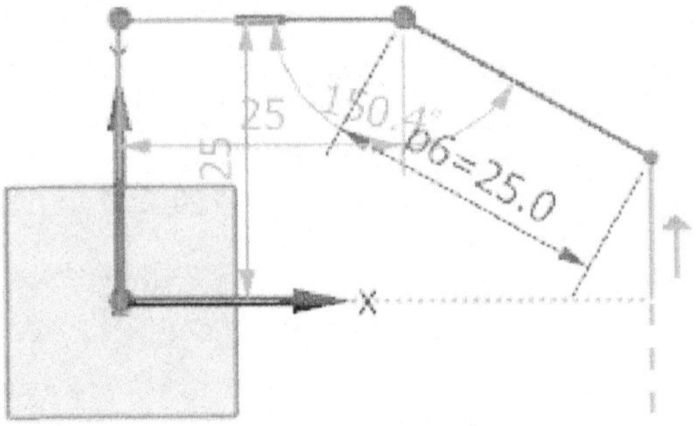

- Finally, move your cursor back to the sketch origin to complete a closed contour. Make sure you select the start point of the first line.

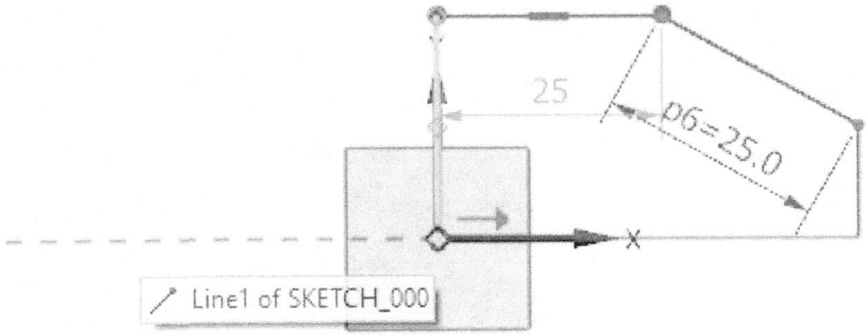

- Congratulations on completing your first sketch! Close the **Profile** tool when you are done.

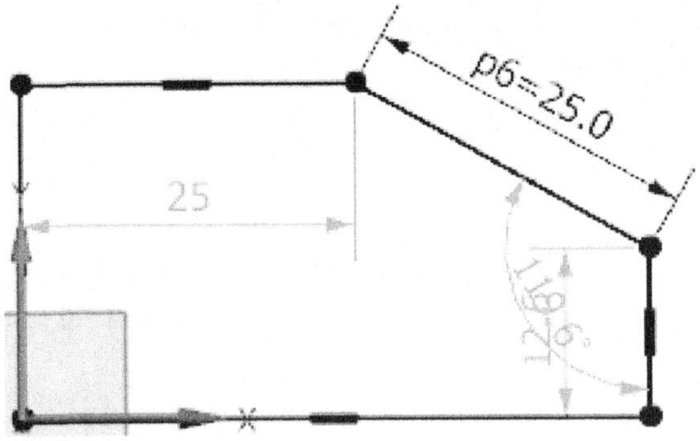

6.1.3 Finish Sketch

The **Finish Sketch** command is used to save the changes made in the Sketch Task Environment to the **Sketch** feature in the Model History. The **Finish Sketch** command is found in the **Sketch** group on the far-left side of the **Home** tab.

- Go ahead and push the **Finish Sketch** button to save your first sketch! You can also use the keyboard shortcut **[Q]**. Your sketch is now saved in the Model History.

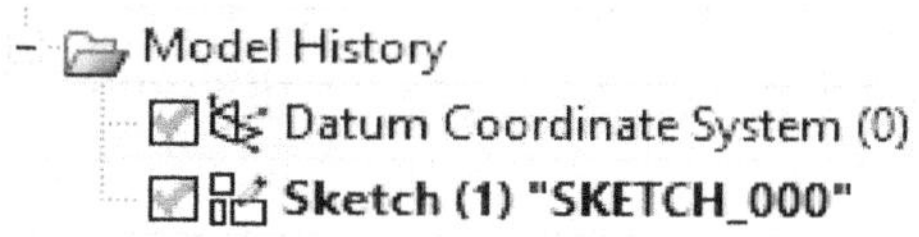

6.2 Editing and Managing Sketches

Before we delve deeper into the Sketch Task Environment, it is worthwhile to learn how to edit and manage your existing sketches.

6.2.1 Dynamically Modifying Sketches

When you have no tools open in the Sketch Task Environment, you can dynamically modify your sketch entities with a click-and-hold-and-drag motion. Be careful to avoid selecting sketch constraints if they are visible in the graphics window – especially the *Horizontal* and *Vertical* constraints attached to lines on your sketch.

- Double-click on *"SKETCH_000"* from the Model History to edit it. With default NX settings, this would bring you into the Direct Sketch environment, but because of the changes made to the **Customer Defaults** in Chapter 2.9.1, this brings you back into the Sketch Task Environment, with no tool selected.
- Drag the top horizontal line in your sketch upward and to the right, and you will see the sketch dynamically update as shown below.

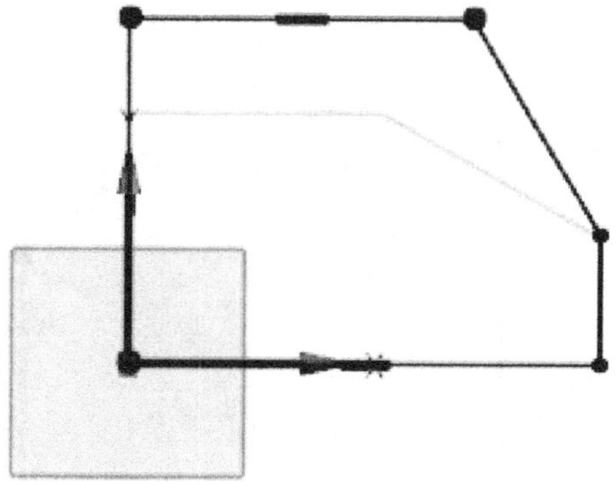

- When you release the left mouse button, the auto dimensions all update. Note that the auto dimensions do not constrain your sketch, but that the one driving dimension keeps the length of the diagonal line fixed at *25 in*. Auto dimensions are driven.

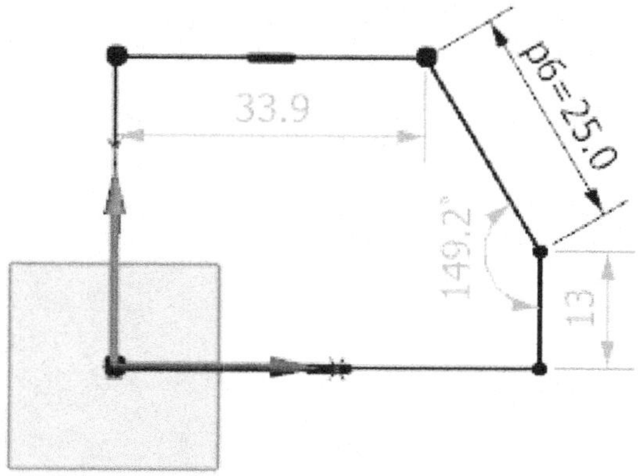

6.2.2 Deleting Sketch Entities

Deleting sketch entities is just like deleting features from a model.

- Sketch constraints are selectable entities and when you are dynamically modifying sketches you should avoid them. To delete a sketch constraint, just select it and delete it! Select the **Horizontal** constraint on the top horizontal line in your sketch.

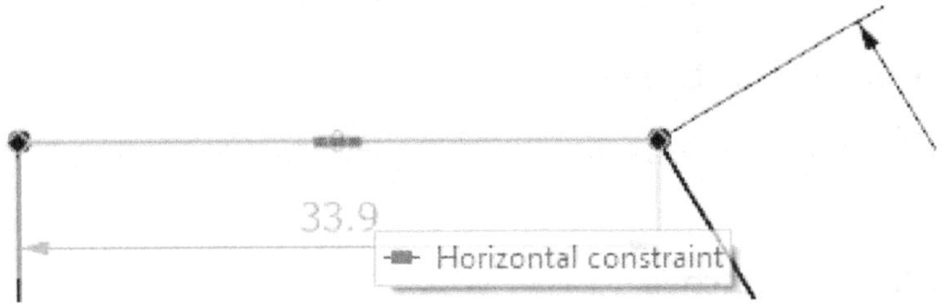

- After selecting a sketch constraint, the **Delete** command will appear on the pop-up toolbar.

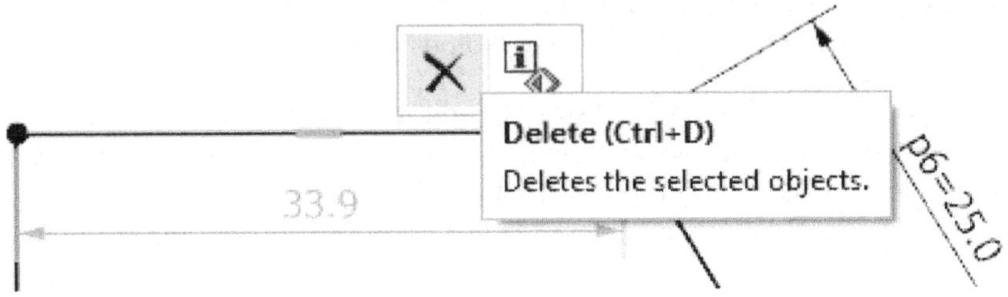

- Now you can dynamically drag the rightmost endpoint of the line upward and it will become angled!

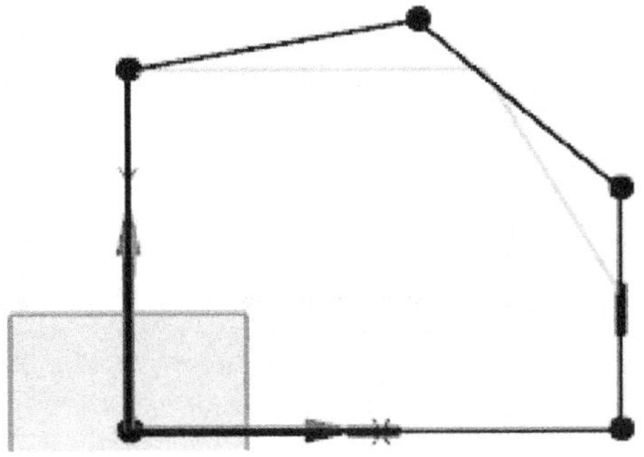

- Sketch curves are deleted in the same way. Click on the previously-horizontal line, and select **Delete** from the pop-up toolbar.

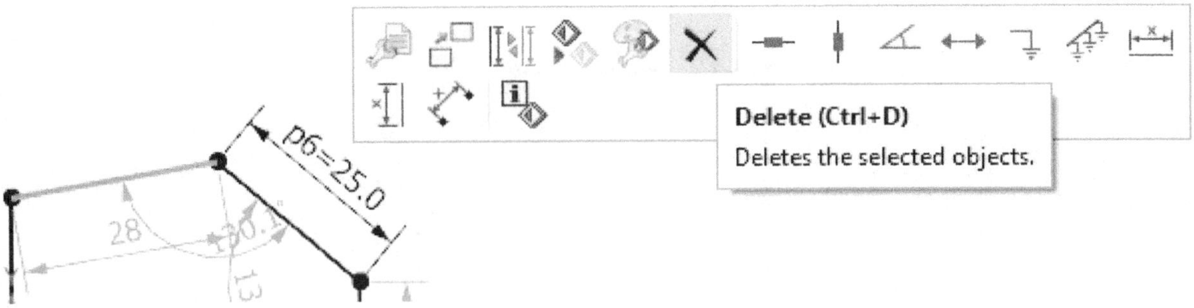

- Your sketch will now appear as shown below.

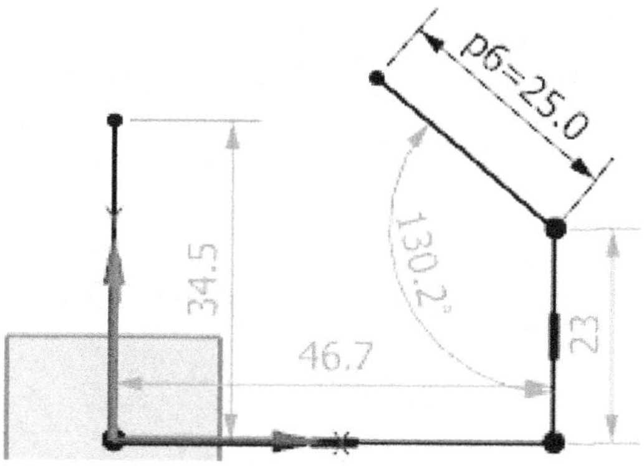

6.2.3 Editing Sketch Curves, Constraints, and Dimensions

Sketch entities are parametric but not fully associative, in the sense that they are smart, and capable of establishing dependencies, but they lack timestamps. Editing is possible, but often will not bring up the original dialog for a sketch object.

- Double-clicking on the leftmost vertical line allows you to edit it, although it brings up an unfamiliar dialog named **Line (Non-Associative)**. Click the **Close** button to exit the **Line** dialog without making any changes. For lines and arcs, double-clicking will bring up the dialog for the corresponding 3D curve tool (as described in Chapter 13).

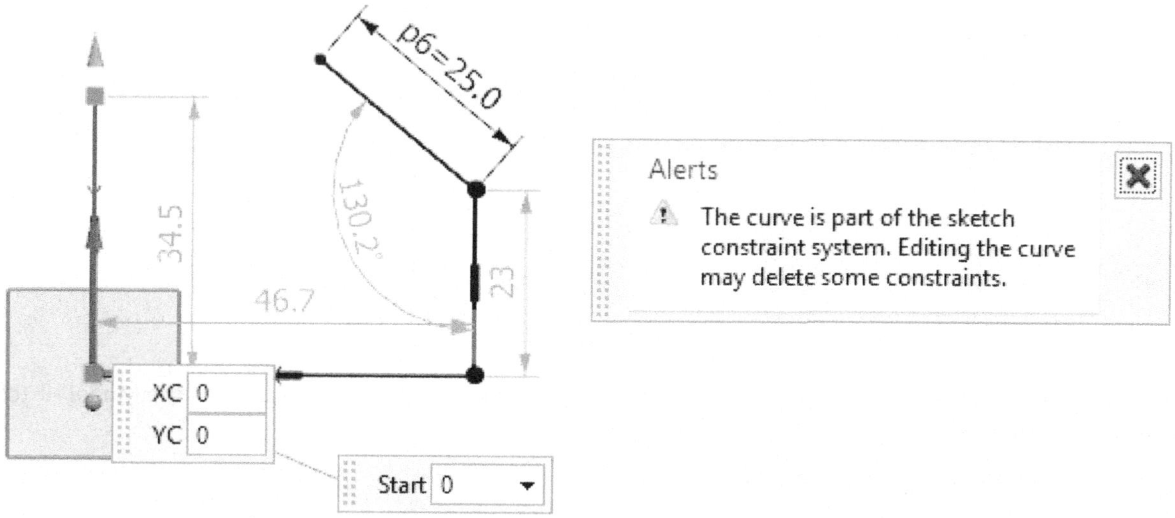

- Double-clicking on a dimension will bring up the dialog for the appropriate type of sketch dimension, and allow you to modify the value. You can also convert driving sketch dimensions to reference by way of the **Reference** checkbox.

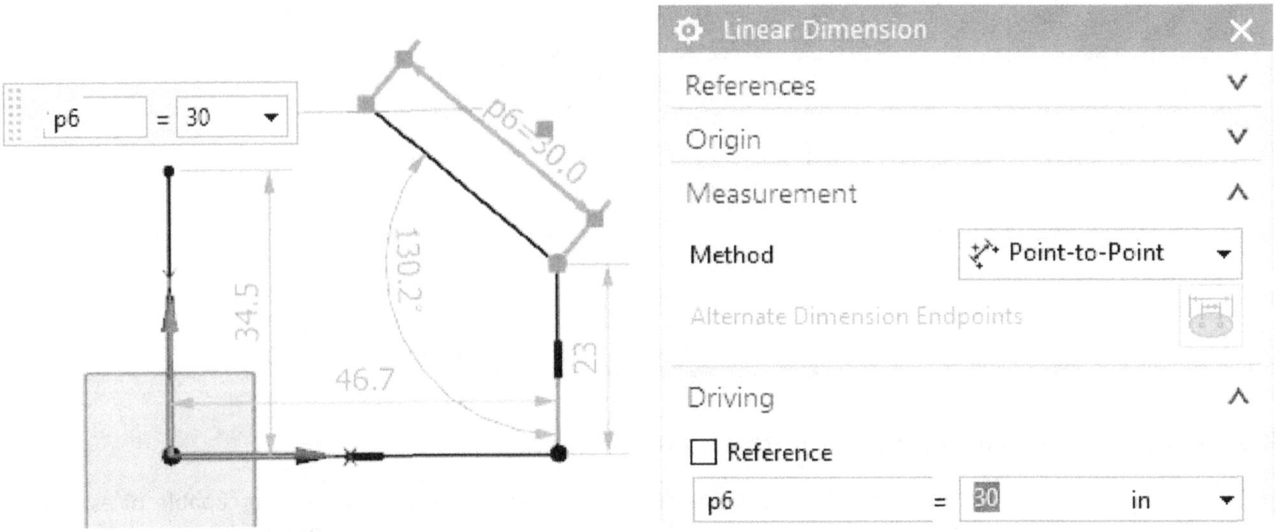

- Double-clicking a sketch auto dimension and accepting the changes in the dialog will convert it into a driving dimension. Double-click on the angular dimension and change the value to *135 degrees* and click **OK** to save the change, and it will become a driving dimension.

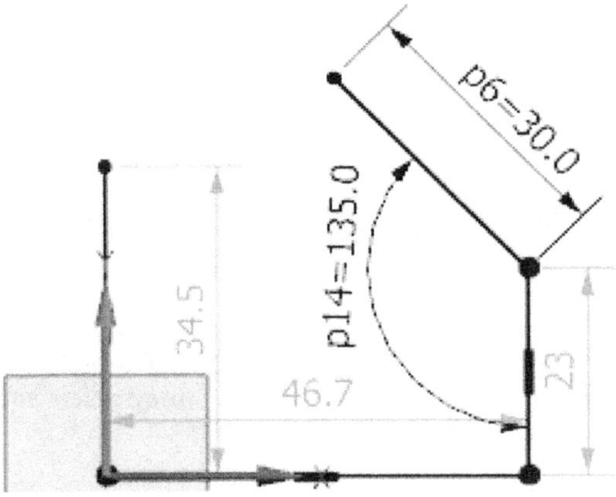

- There is another important way to modify sketch dimensions. Right-click on *"SKETCH_000"* in the Model History, and select **Edit Parameters**.

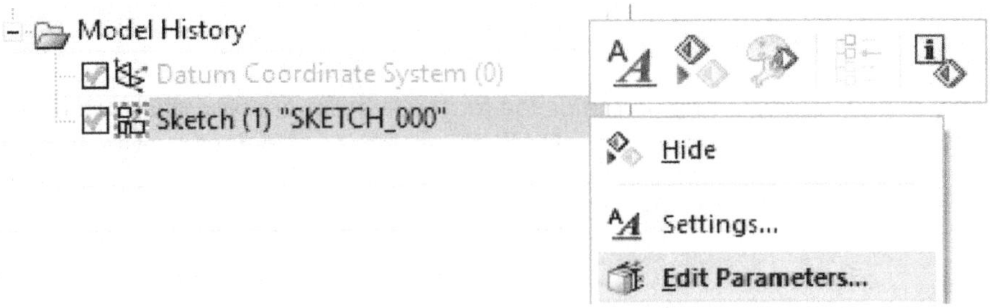

- In the **Sketch Parameters** dialog that appears, you get a list of all driving dimensions in your sketch, and you can individually modify or delete them.

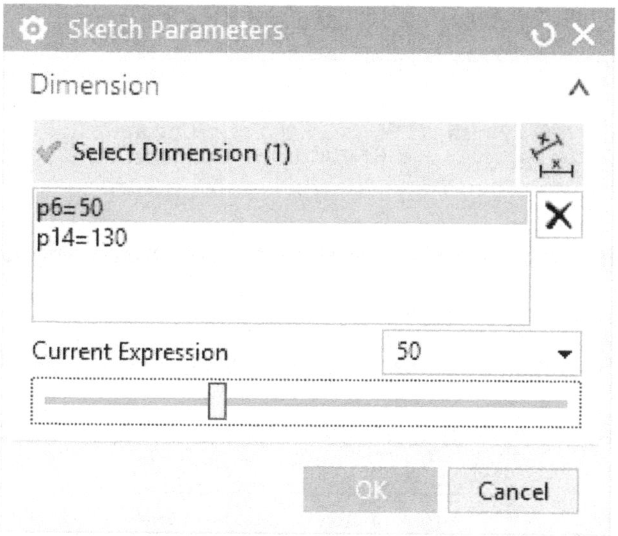

6.2.4 Exit Sketch

The **Exit Sketch** command will discard all changes made while in the Sketch Task Environment. If you are creating a sketch from scratch, this means that the sketch will not be created; if you are editing an existing sketch, only the changes are discarded.

Exit Sketch is found near the bottom of the **Task** drop-down at the top of the ribbon (where the **File** drop-down usually appears). Note that there are some other sketch management tools in the **Task** drop-down, such as **New Sketch**, which will allow you to create a new sketch, potentially with a different sketch coordinate system, without leaving the Sketch Task Environment.

- Well, we really messed up that first sketch. Use the **Exit Sketch** command to discard the changes you made in the last few exercises. This will restore *"SKETCH_000"* to its previous state.

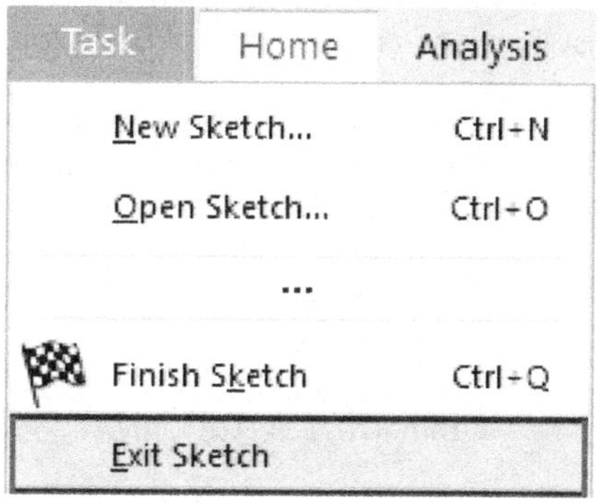

6.2.5 Naming Sketches

From within the Sketch Task Environment, you can rename your sketch.

- Double-click on *"SKETCH_000"* in the Part Navigator, or right-click on it and select **Edit**.
- In the **Sketch** group on the **Home** tab, the drop-down menu that shows *"SKETCH_000"* can be used to rename sketches, and to change the active sketch within the Sketch Task Environment.

- Double-click the name to rename your sketch *"MY FIRST SKETCH"*.

6.2.6 Creating New Sketches from Within the Sketch Task Environment

You can also create a new sketch without leaving the Sketch Task Environment.

- Next, let's create a new sketch right here from within the Sketch Task Environment. From the **Task** drop-down, select **New Sketch**.

- Click **OK** when the **Create Sketch** dialog appears, to create the new sketch also in the XY plane. A new **Sketch** feature named *"SKETCH_000"* will appear in the Model History, and the **Profile** tool will be selected.

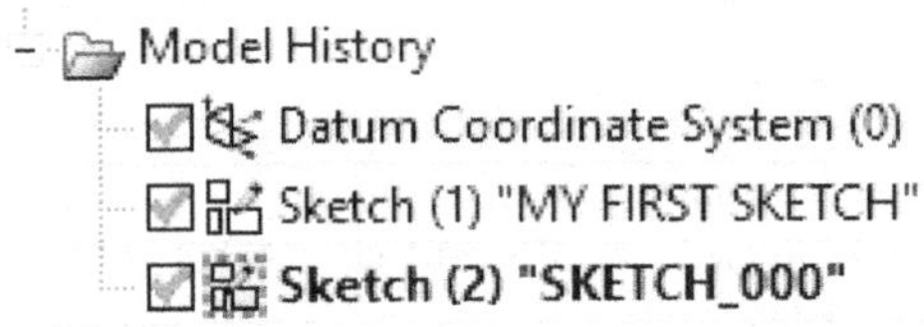

6.2.7 The Profile Tool – Making Arcs

In Chapter 6.1.2, you learned how to use the **Profile** tool to make chain connected series of lines, but it is capable of much more – you can sketch a shape involving just lines and arcs using the **Profile** tool, and quickly too!

- Let's get some practice using the **Profile** tool to create both lines and arcs in this second sketch! First you will want to set your **Selection Scope** drop-down to *Within Active Sketch Only*, so that you don't pick up inferred constraints to objects outside of the active sketch as you create the profile.

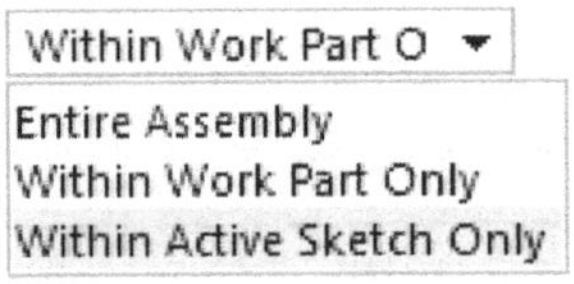

- Click at the point in the sketch plane with coordinates *(10, 15)* to initiate a line.

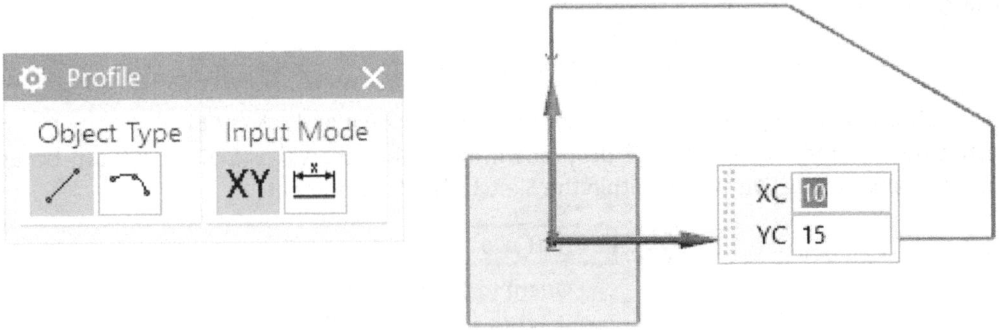

- Move *15 in* to the right and click to create the first line segment.

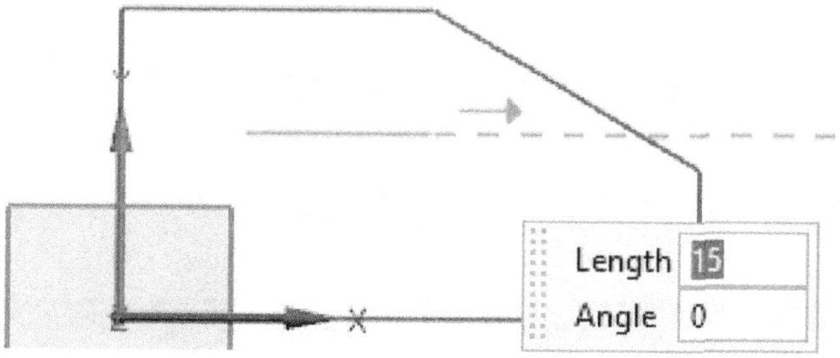

- To switch the **Profile** tool into arc mode, you can click on the arc icon in the **Object Type** section of the **Profile** dialog, but we want to share a shortcut with you that will significantly speed up your use of the tool.

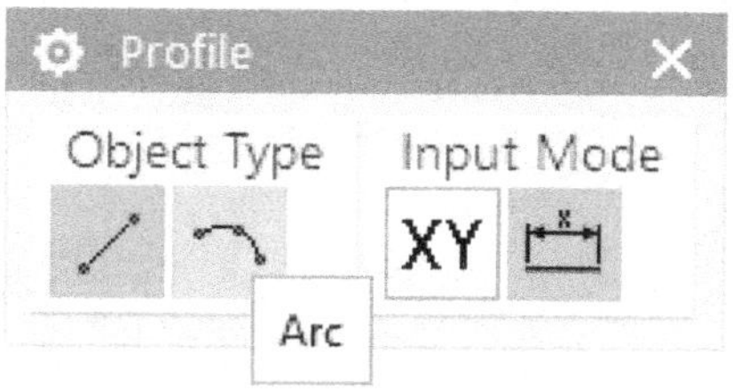

- The shortcut works like this: when the **Profile** tool is open, clicking and holding the left-mouse button anywhere in your sketch and then dragging, even a short distance, will toggle the **Object Type** from *Line* to *Arc*. Do a quick click-and-hold-and-drag-and-release motion to toggle the **Object Type** to *Arc*. In order for the preview to display the arc tangent to the line (as shown below), you must move your cursor out of the line to the right as you perform this dragging motion. If your arc appears perpendicular to the line, move your cursor (without clicking-and-dragging again) back to the endpoint of the line, and move it straight to the right until the preview appears tangent.

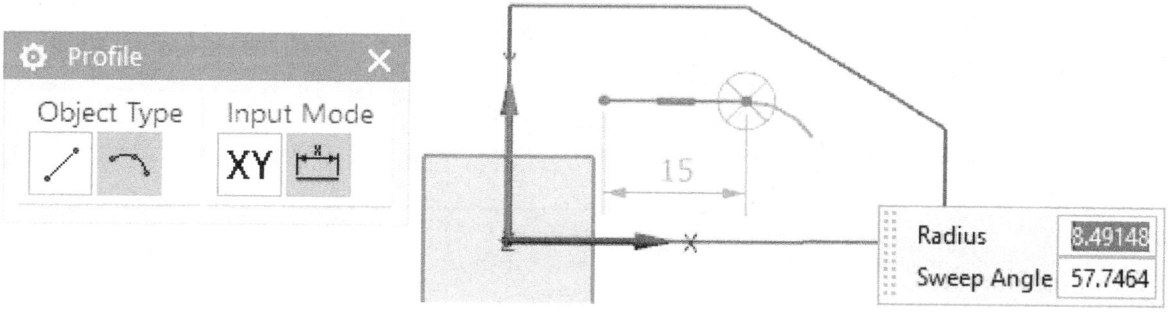

- For the **Radius**, enter a value of *4 in*, and move your cursor until the arc snaps to a **Sweep Angle** of *180 degrees*. Click to finish the arc.

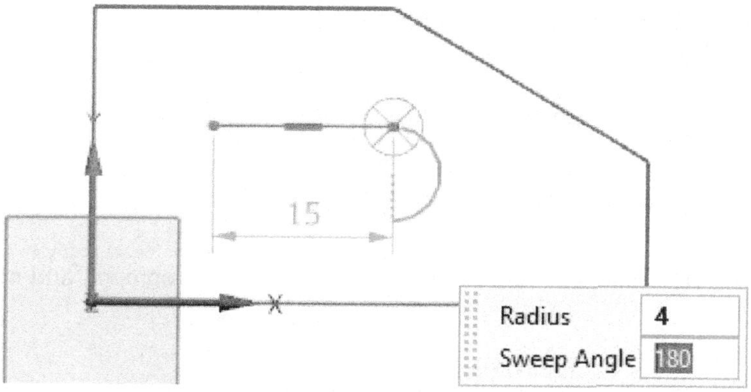

- When you complete the arc, the **Profile** tool will automatically switch back into **Line** mode. Move your cursor to the left by *15 in* and allow the tool to infer a tangency constraint with the arc. Click to create the line.

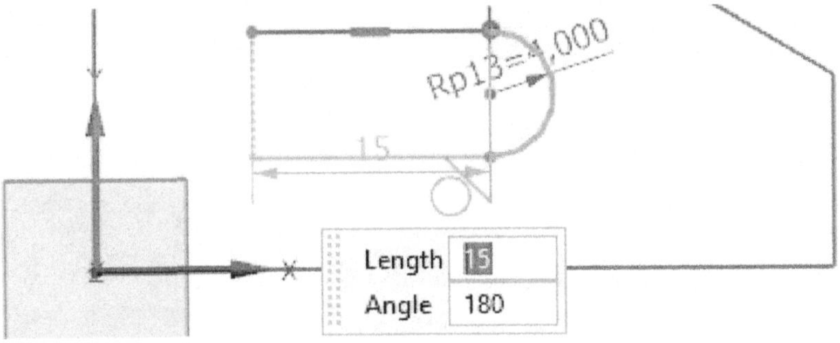

- Click-and-hold-and-drag to create another arc on the left side, and select the start point of the first line as the endpoint for the arc. Again, allow it to infer a tangency constraint.

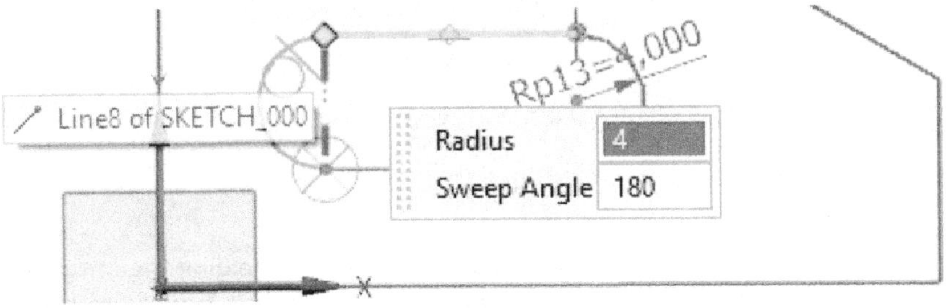

- Your second sketch is now finished!

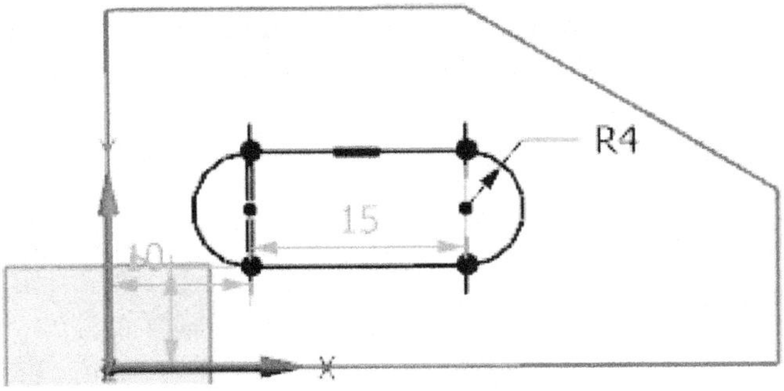

6.2.8 Opening Sketches

When there are multiple sketches in your part file, the naming drop-down can also be used to open a different sketch without leaving the Sketch Task Environment!

- Use the naming drop-down to rename this new sketch *"MY SECOND SKETCH"*. Note that this drop-down also shows *"MY FIRST SKETCH"*. Go ahead and click on it!

- This makes *"MY FIRST SKETCH"* the active sketch in the Sketch Task Environment, and now you can edit its sketch entities directly.

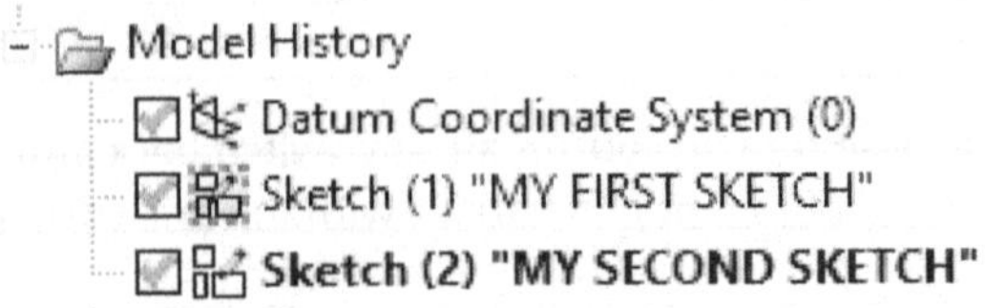

- You can also open another sketch in the Model History by double-clicking on it in the Part Navigator, or right-clicking on it and selecting **Open**. By the way, all of these are the same as selecting **Open Sketch** from the **Task** drop-down.
- Push the **Finish Sketch** button to save all your changes!
- **Save** your file!

6.3 The Sketch Coordinate System

When you create a sketch, that sketch comes with its own internal coordinate system, with axes labeled X, Y, and Z. The horizontal and vertical axes in the sketch are indicated by X and Y, and the Z axis of the sketch coordinate system points at you when you are looking at the sketch plane from a normal view. You can control the X, Y, and Z axes of the sketch coordinate system explicitly at the time of sketch creation – a technique we will practice several times throughout this chapter.

- Continue working with *"Sketch in Task Environment.prt"*.
- **Hide** *"MY FIRST SKETCH"* and *"MY SECOND SKETCH"*.
- Select the **Sketch in Task Environment** tool. When the **Create Sketch** dialog appears, choose the YZ plane for the *"Specify CSYS"* input. To select the datum plane, you will want to put your cursor along its edge or corner – when your cursor is in the middle of the datum plane, your selection ball is unable to find the plane. You will see a red-green-blue triad labeled X, Y, and Z appear on top of the datum coordinate system. This is a preview of the sketch coordinate system.

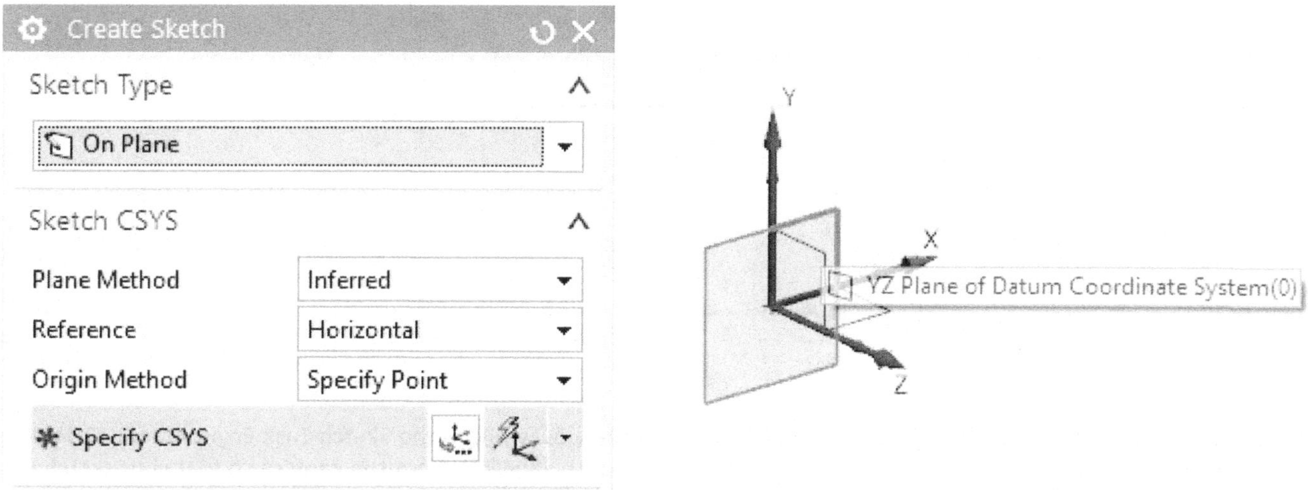

- Once you click on the YZ plane, two gold direction arrows will appear. These indicate the *Plane Normal* and the *Horizontal* direction. You can double-click on them to reverse them. Don't do that for now, just click **OK** to enter the Sketch Task Environment.

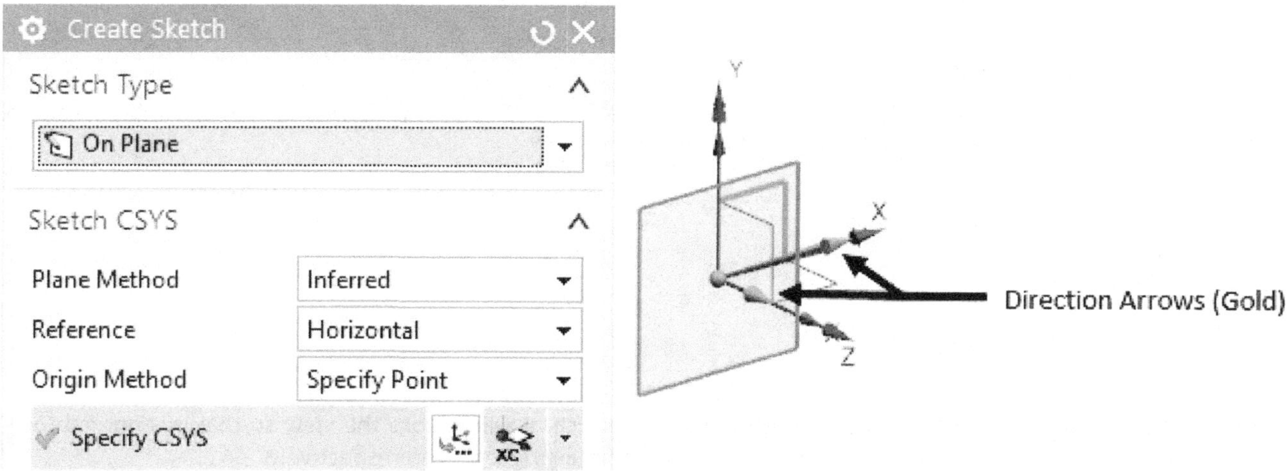

- Once you enter the Sketch Task Environment, you will now see the sketch coordinate system and its labels sitting directly on top of the datum coordinate system, but they are not the same.

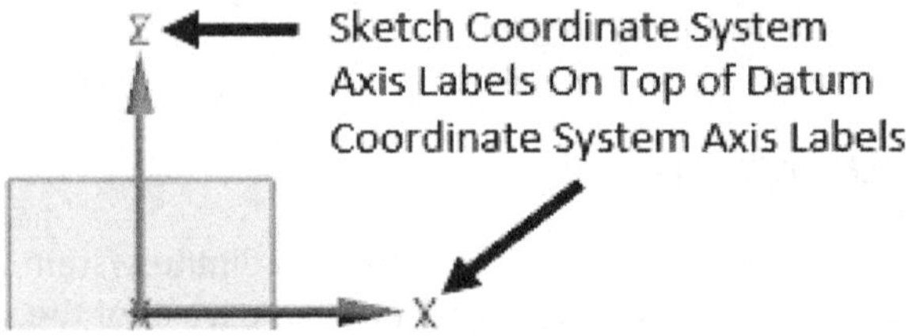

- To avoid confusion, you can simply hide *Datum Coordinate System (0)* from the Model History. Then only the sketch coordinate system is visible.

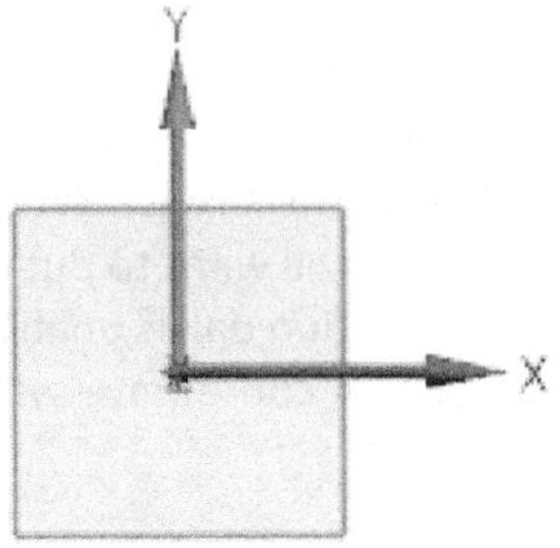

- Use the **Profile** tool to sketch a rectangle, with the bottom left corner attached to the *Sketch Origin*.

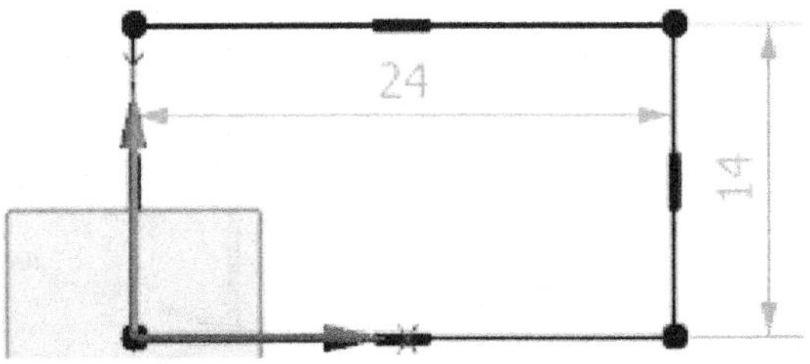

6.3.1 Orient to Sketch

Sometimes you will either intentionally or inadvertently rotate the view while in the Sketch Task Environment. **Orient to Sketch** provides a fast way to restore your view to the sketch plane – it will reorient the camera so that your sketch fills the graphics window, and the sketch plane normal (Z) is pointing at you.

Orient to Sketch is conveniently located on the right side of the **Radial View Pop-up Toolbar** that appears when you click and hold the right mouse button in empty space in the graphics window.

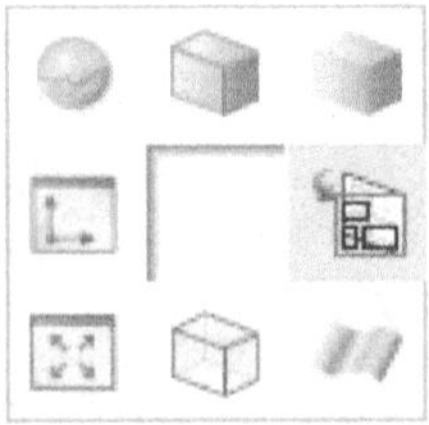

- Rotate the view in your sketch, and then push **Orient to Sketch**. It will restore the view so that it is normal to the sketch plane, with the sketch horizontal and vertical axes pointing right and up respectively.

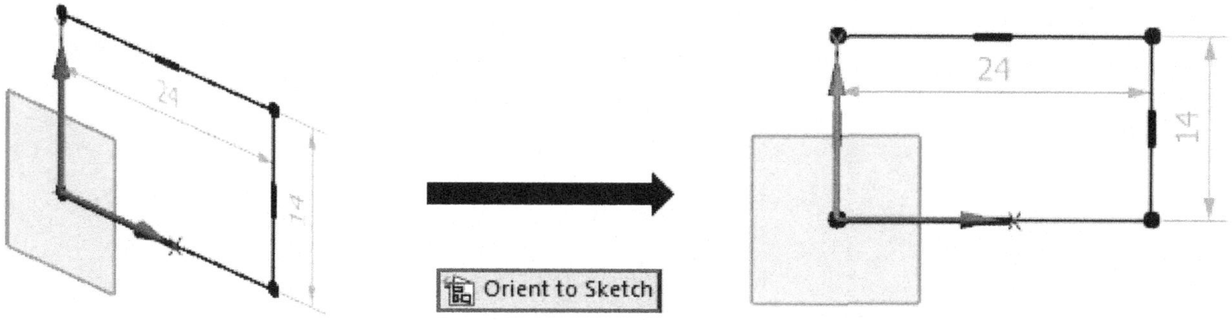

6.3.2 Reattach Sketch

Sometimes as a design evolves, it is necessary to rethink the location of a sketch coordinate system, especially when sketch features are defined purely in relation to the sketch coordinate system, and do not reference geometry outside of the sketch. The **Reattach Sketch** tool allows you to modify the sketch coordinate system freely – you can place a sketch on an entirely different plane, adjust the normal and horizontal directions, and move the origin around!

- **Hide** *"SKETCH_000"* and **Show** both *"MY FIRST SKETCH"* and *"MY SECOND SKETCH"*. Open *"MY SECOND SKETCH"*.
- Select the **Reattach Sketch** tool. In the **Sketch Orientation** section of the dialog, set the **Vector Type** to *Inferred*.

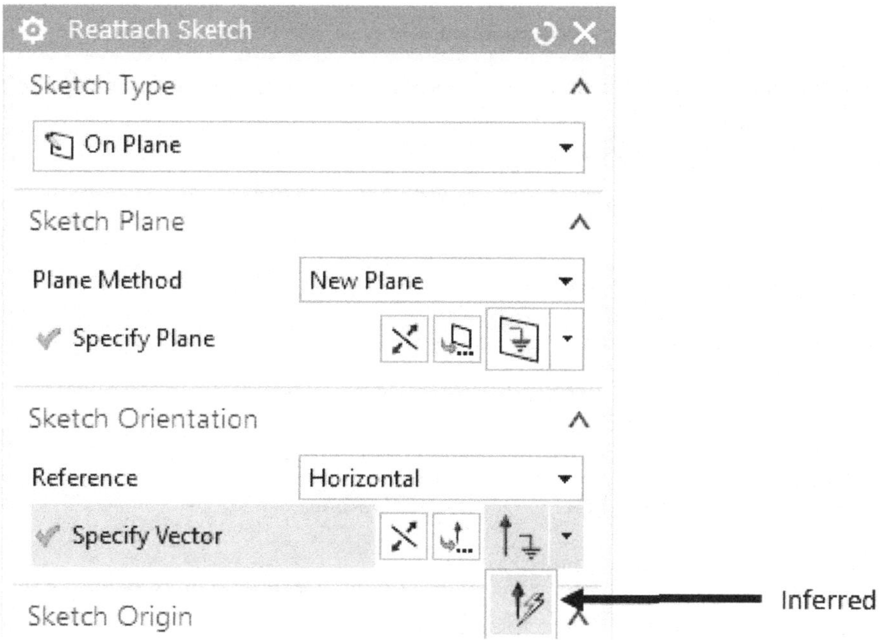

- For the **Reference Vector**, choose the diagonal line from *"MY FIRST SKETCH"*.

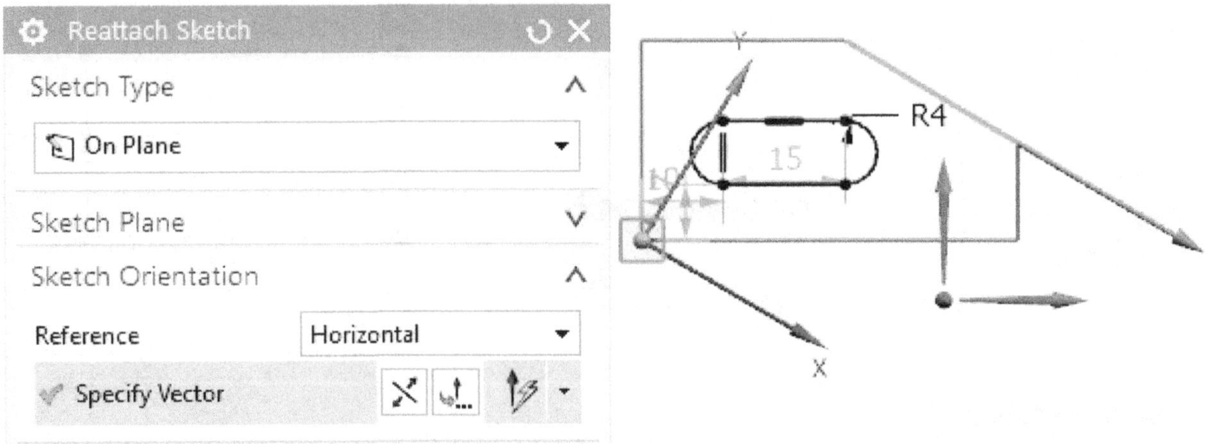

- Click **OK** to commit your changes. The active sketch will rotate so that the slot shape becomes parallel to the diagonal line from *"MY FIRST SKETCH"*. Dynamically drag the slot shape so that it is contained within the contour of *"MY FIRST SKETCH"*, as shown below.

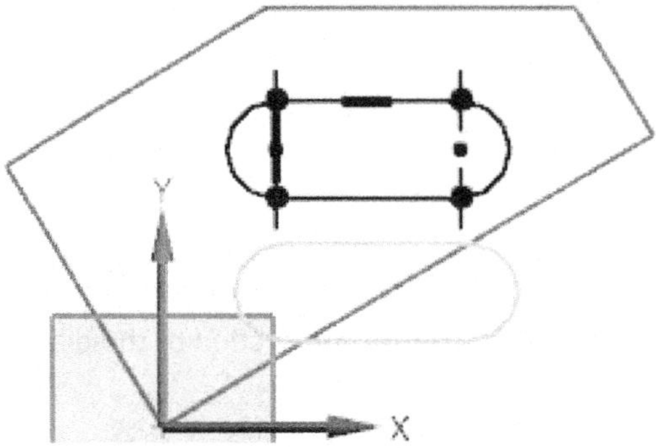

- Select the **Reattach Sketch** tool again. This time, within **Sketch Origin**, change the **Point Type** to *Inferred*.

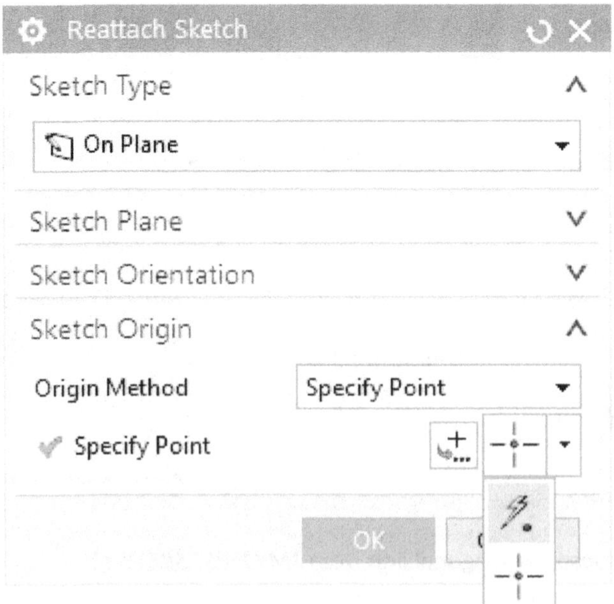

- For the **Origin Point**, choose the bottom right-most corner from *"MY FIRST SKETCH"*.

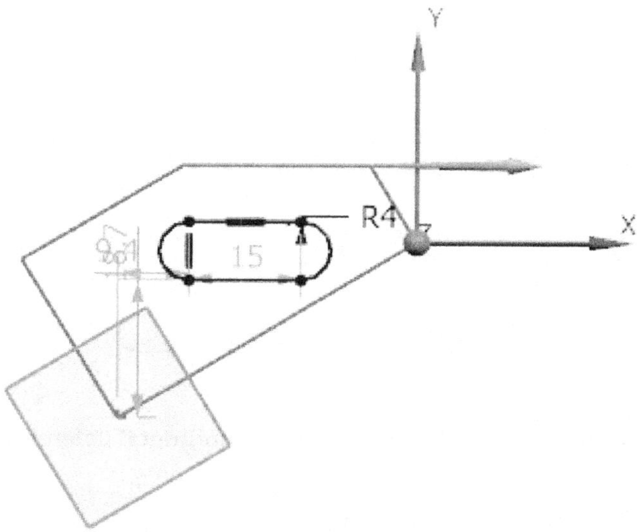

- Click **OK** and the slot shape will translate along the length of the bottom line from *"MY FIRST SKETCH"*.

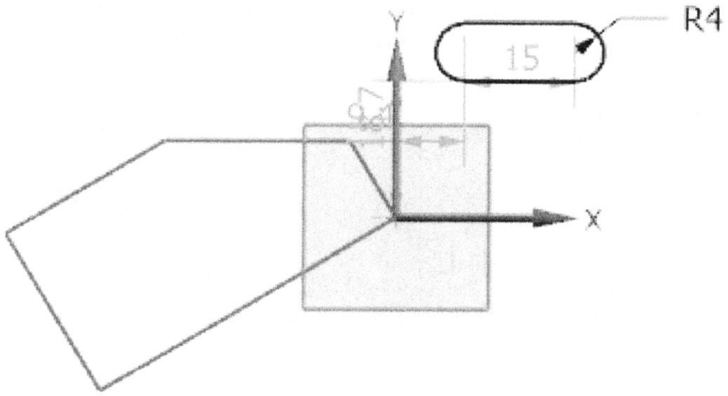

- Select **Reattach Sketch** again. Within the **Sketch Plane** section of the dialog, change the **Plane Type** to *Inferred*.

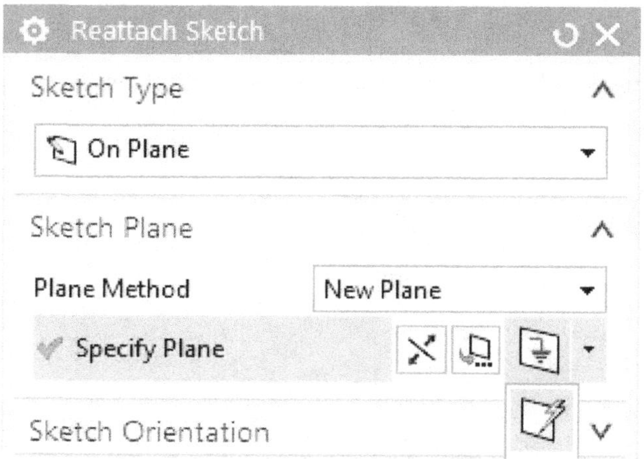

- Select the YZ plane of the datum coordinate system in your model. The selected **Horizontal Reference** will be projected to the new sketch plane – note that it points the opposite direction of the sketch plane's natural horizontal (the Y axis).

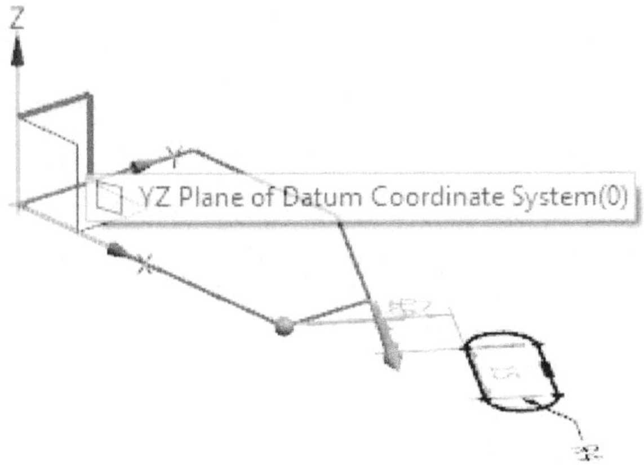

- You will need to use the **Reverse Direction** button to align the **Horizontal Reference** with the positive Y axis of the datum coordinate system.

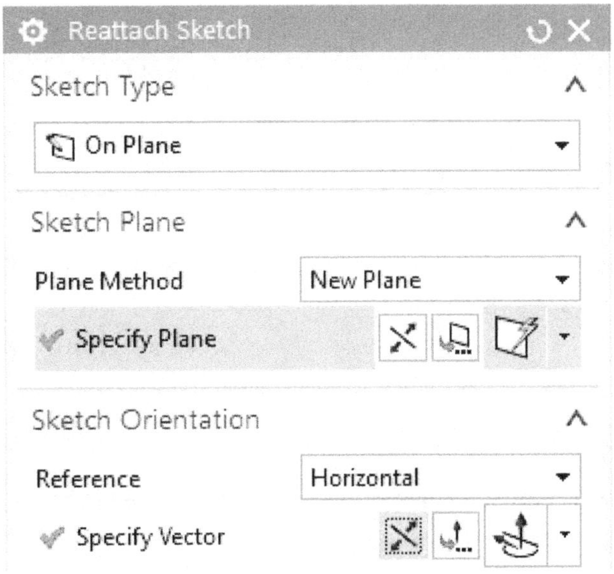

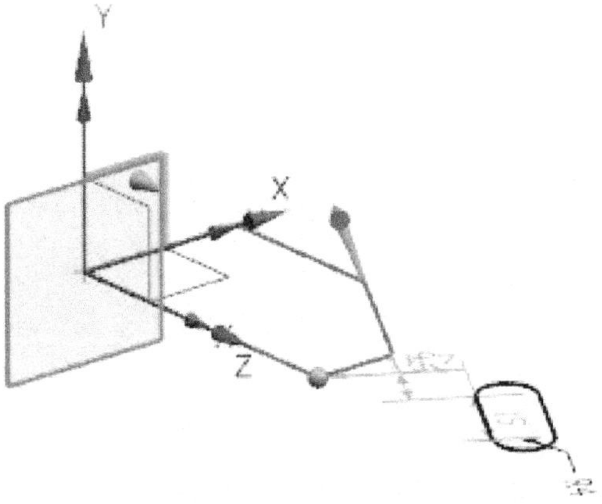

- Click **OK** to commit your changes and the sketch will jump onto the YZ plane!

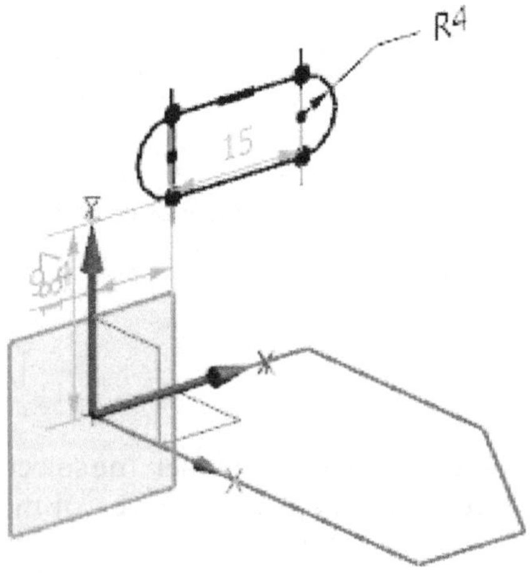

- **Finish** your sketch to return to the **Modeling** environment.
- **Save** your file!

6.3.3 Sketching on a Face

Often you will want to sketch on faces of existing solid and sheet bodies. In doing so, it is important to understand how the sketch coordinate system is attached to that face in order to build a robust parametric model.

- Open the file *"Sketches on Faces and Paths.prt"* from *"C:\My NX 11 Files\Downloaded Files"*.
- In this part file, there is a block with a chamfered edge, which has no parametric history – the entity is *Body (0)* in the Model History.

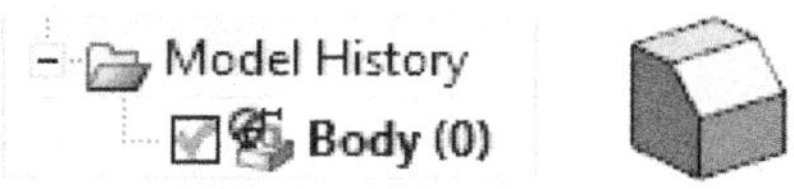

- Select **Sketch in Task Environment** and put your cursor on the chamfered face near the top leftmost corner. You will see a preview of the sketch coordinate system attached to that top leftmost corner.

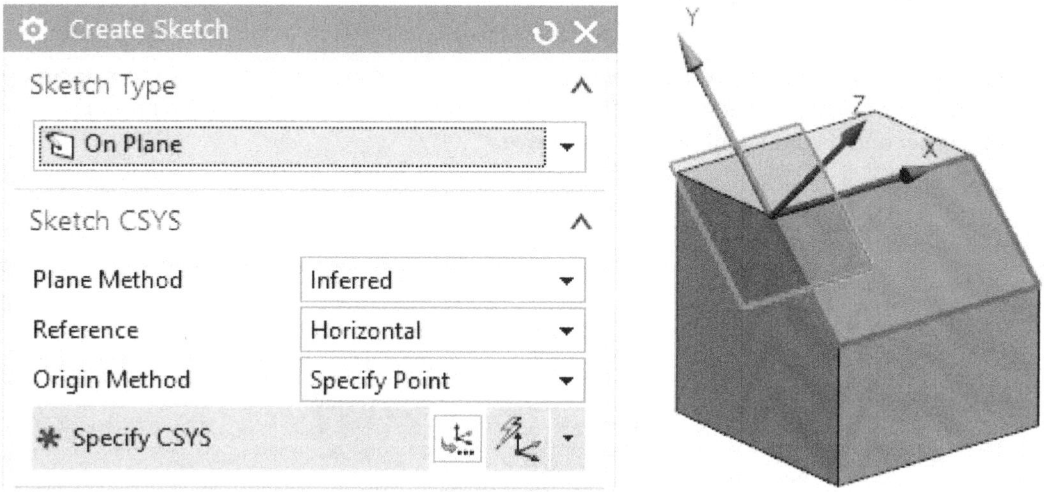

- Without making a selection, move your cursor so that it is still over the chamfered face, but nearest the bottom rightmost corner. The sketch coordinate system will move to that corner.

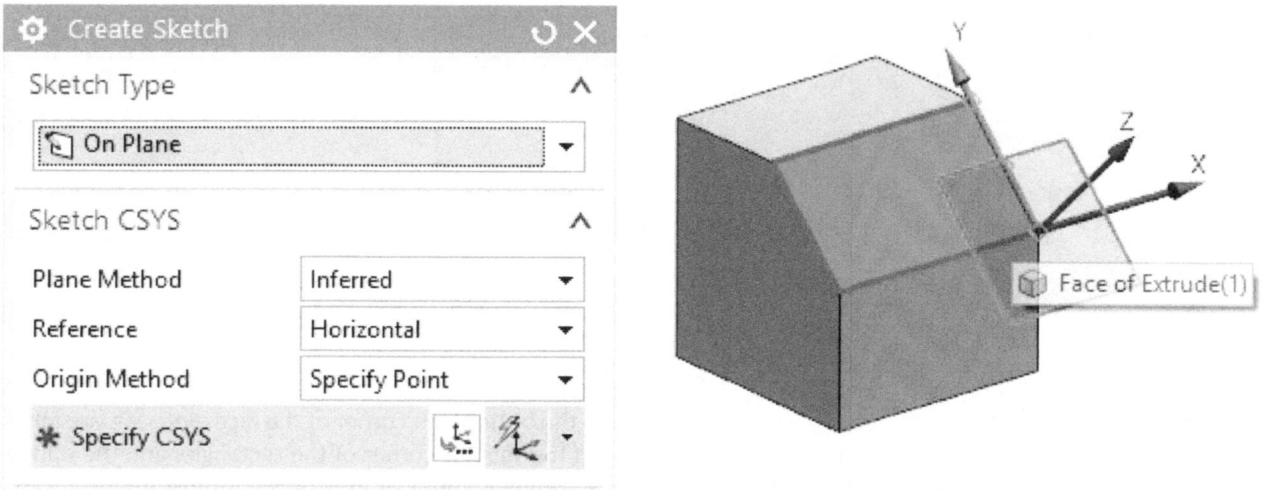

- While the **Plane Method** is set to *Inferred*, there are a number of factors influencing the sketch coordinate system inference. These include snap points, and the orientation of the *View Coordinate System* as described in Chapter 3.1.6. Move your cursor near the midpoint of the edge, and the preview of the sketch coordinate system will move to that midpoint. (This will only happen if your *Midpoint* Snap Point icon is enabled!) Click to position your sketch coordinate system here.

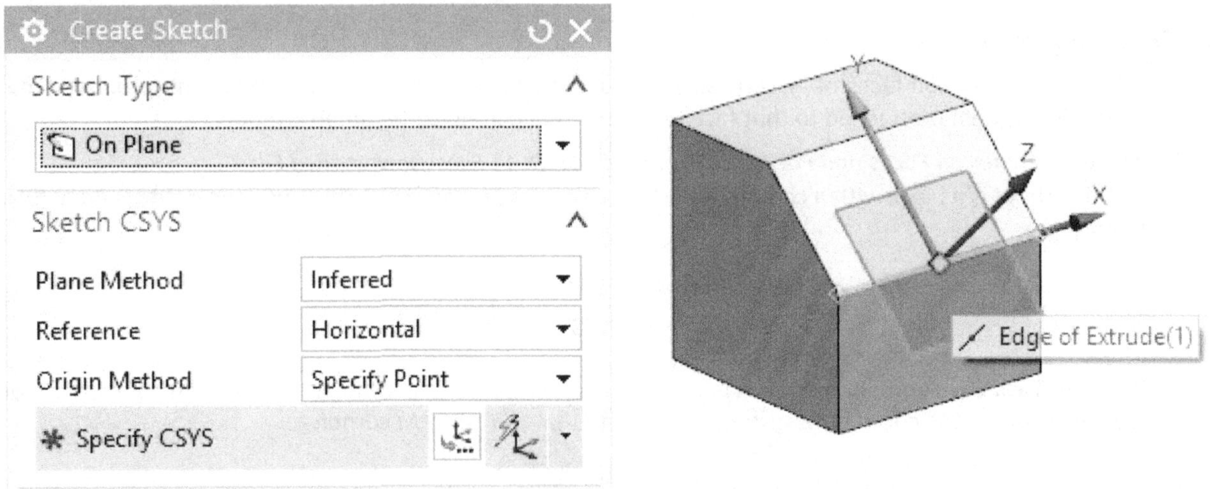

- Once in the Sketch Task Environment, sketch a rectangle using the **Profile** tool starting from the approximate location shown below. Because of where you first click, your rectangle should pick up a **Vertical Alignment** constraint between its top leftmost corner and the midpoint of the edge above.

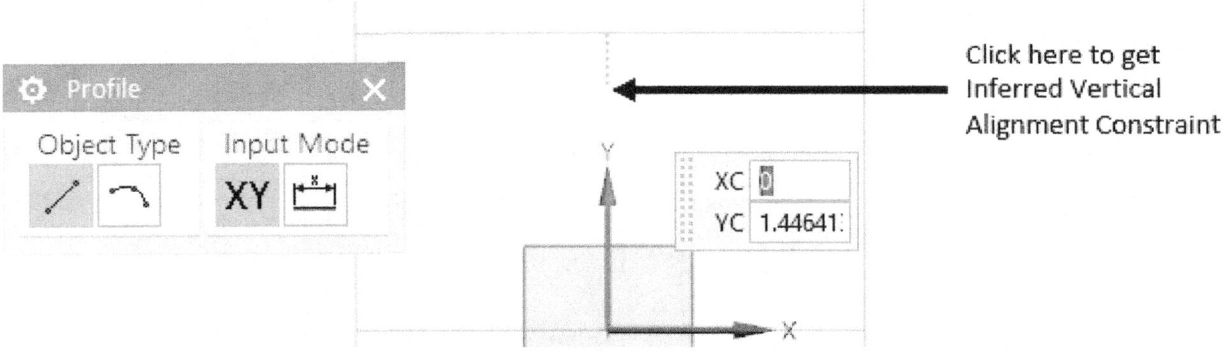

- The exact dimensions don't matter, but try to get them close to those shown below.

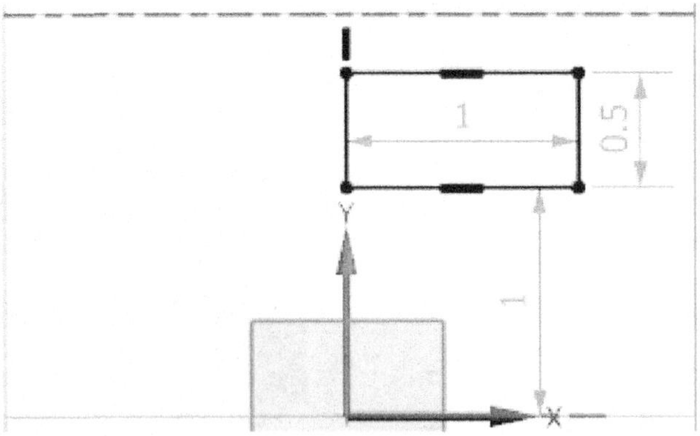

- Use **Reattach Sketch** to move the sketch coordinate system to the bottom left corner of the face. Because you allowed your sketch to create a vertical alignment constraint between the top left corner of the rectangle, and the midpoint of the edge of the block, the rectangle doesn't move. When sketching on faces, it is of the utmost importance that your constraints and dimensions refer to external objects in the way that best captures your design intent.

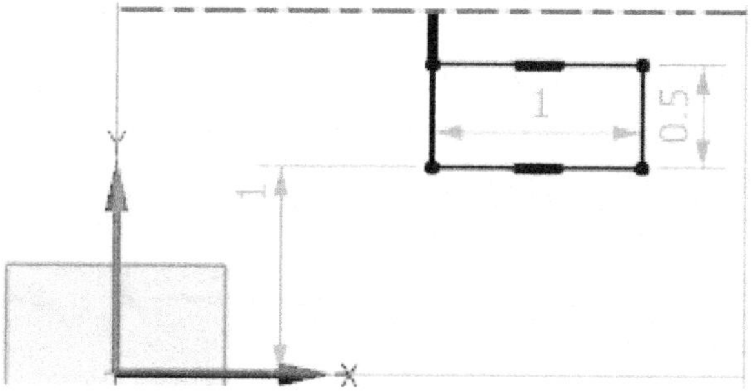

- **Finish** your sketch!

6.3.4 Sketch on Path

For many types of geometry, the best construction method is to sweep a section curve along a drive curve. Good examples are moldings, railings, and bezels. In cases like these it is often a good choice to create your section curves by way of a sketch of **Type** *"On Path"*.

- Create a new sketch, and set the **Sketch Type** to *On Path*. For the **Path**, choose the bottom edge of the chamfered face, as shown below.

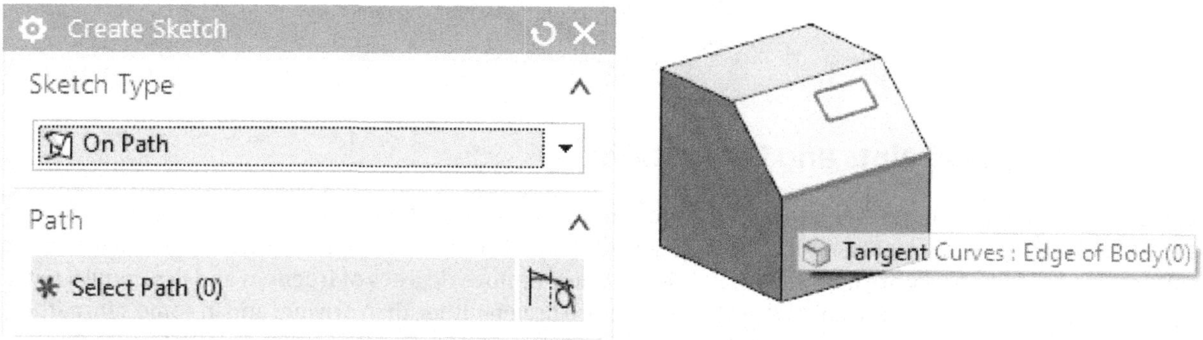

- Setting your sketch coordinate system properly is virtually always required when creating sketches on paths – the orientation of the inferred coordinate system is usually awful. Modify the sketch coordinate system as shown below.

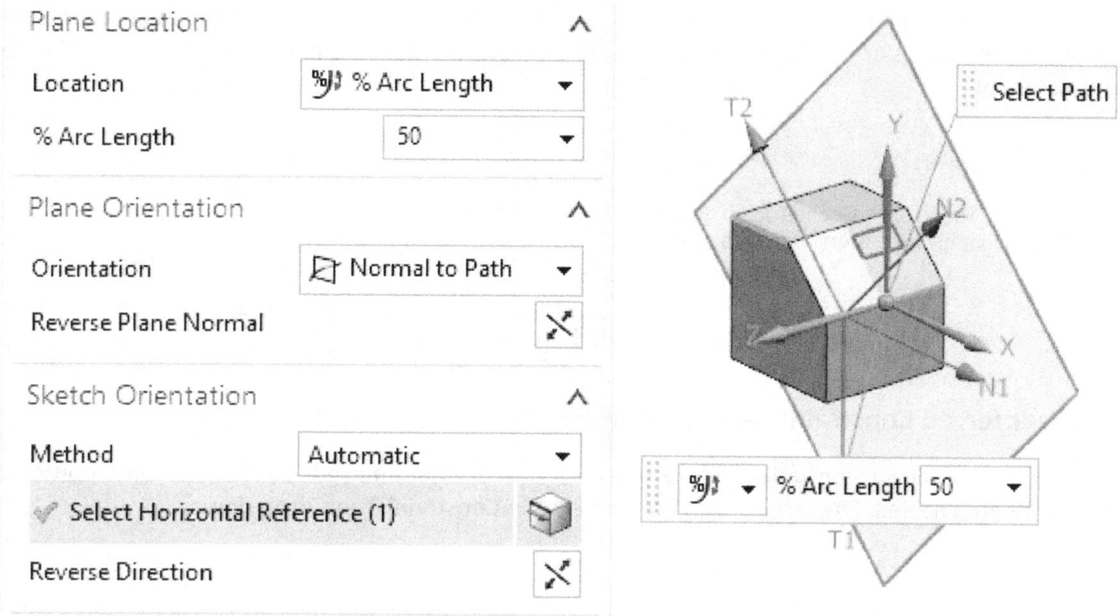

- Once in the Sketch Task Environment, create the rectangle shown below using the **Profile** tool.

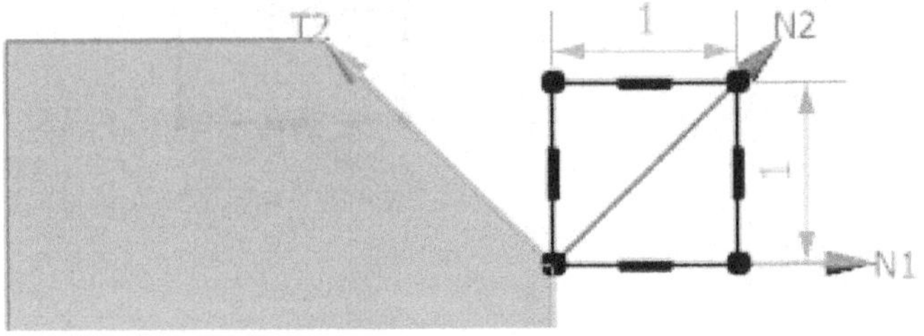

- Finish your sketch, and your model will appear as shown below.

- **Save** your file!

6.4 Automatic Constraints and Dimensions

In the course of creating your first few sketches, NX automatically created certain dimensions and constraints. There are a number of tools available for managing the creation and display of these automatic constraints and dimensions. The constraints created automatically are hard constraints, in that they remove degrees of freedom and can conflict with other constraint and driving dimensions. The automatic dimensions are driven rather than driving, and in some sense, they don't really remove degrees of freedom since you can dynamically modify your sketch even with them in place.

6.4.1 Display Constraints and Auto Dimensions

Sketch constraints and auto dimensions are visible in the Graphics Window if the **Display Sketch Constraints** and **Display Sketch Auto Dimensions** tools are enabled. These tools are found in the **Constraint Tools** drop-down in the **Constraints** group on the **Home** tab.

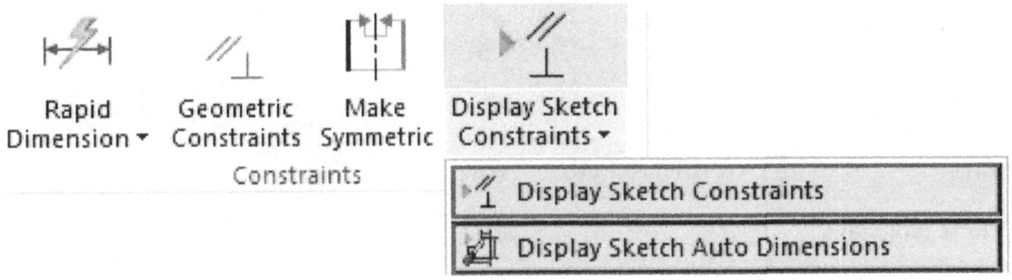

6.4.2 Create Inferred Constraints and Continuous Auto Dimensioning

Auto dimensions and inferred constraints are created if the **Create Inferred Constraints** and **Continuous Auto Dimensioning** tools are enabled. These tools are also found in the **Constraint Tools** drop-down.

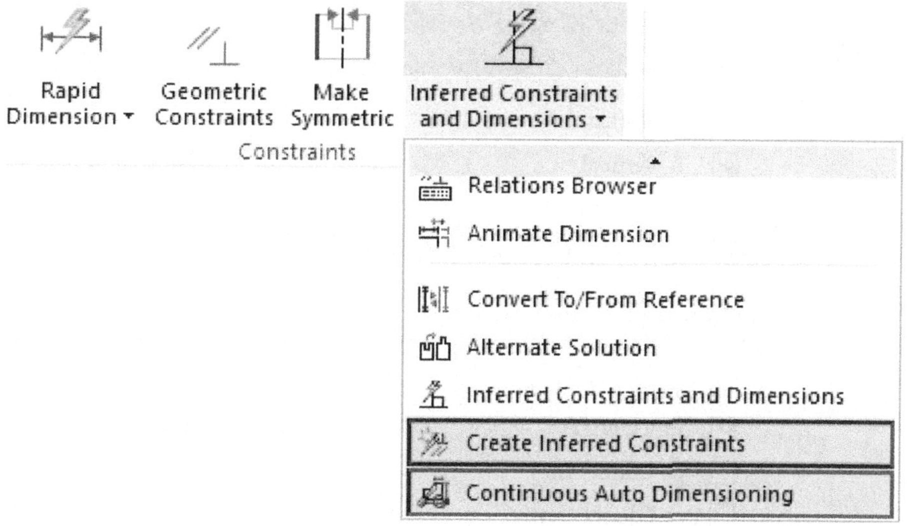

The **Create Inferred Constraints** tool only determines whether future sketch entities will automatically have constraints and auto dimensions attached to them – it does not affect previously placed sketch entities. Disabling **Continuous Auto Dimensioning** does not delete previously-created auto dimensions, but ensures that future sketch entities will not be subjected to auto dimensions. Enabling **Continuous Auto Dimensioning** will create auto dimensions for all unconstrained sketch geometry. Note that your *Status Display* (at the bottom of the Graphics Window) reports the number of auto dimensions.

Sketch is fully constrained with 3 auto dimensions

6.4.3 Inferred Constraints and Dimensions

The **Inferred Constraints and Dimensions** tool allows you to set the logic that dictates which inferred constraints and auto dimensions get created.

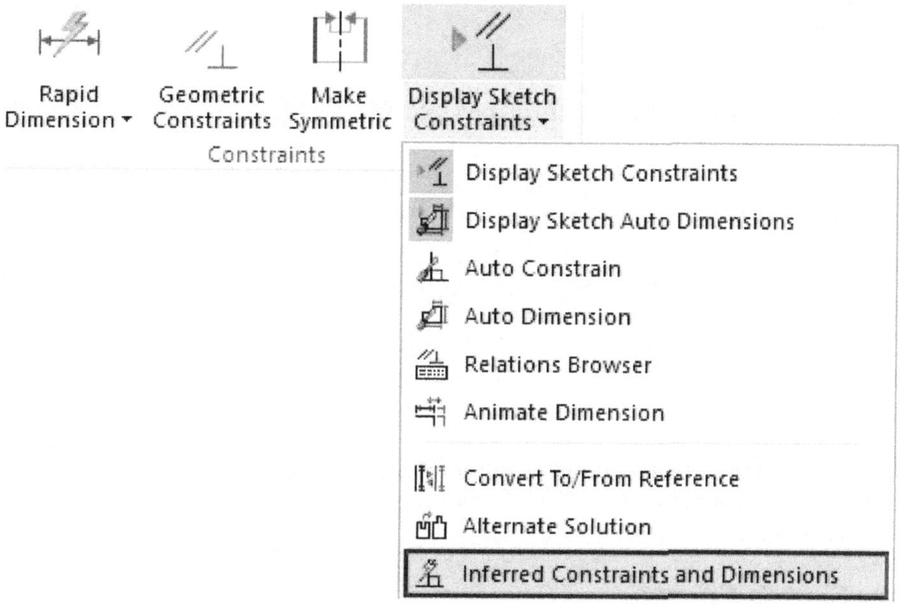

The default settings for inferred constraints are shown below.

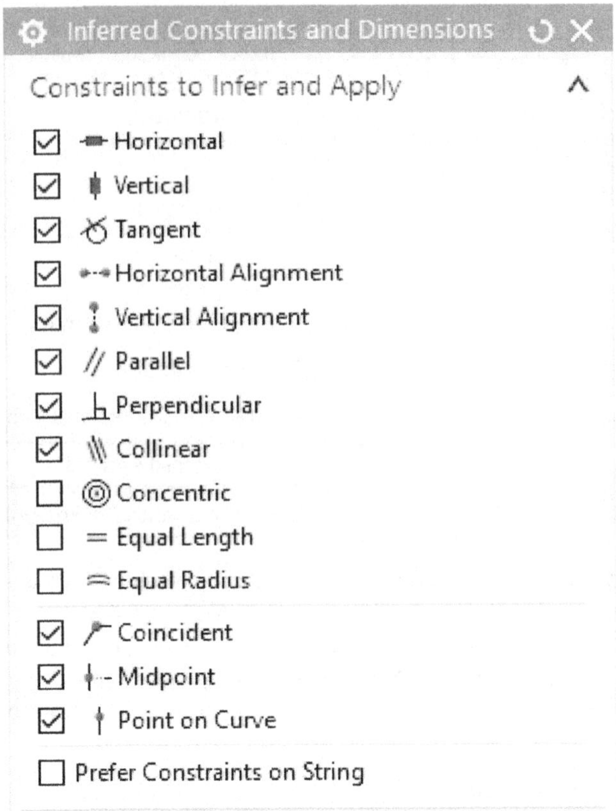

The dimensions placed when **Continuous Auto Dimensioning** is enabled are "weak" or *driven* dimensions, rather than *driving* dimensions – they can be overwritten by other constraints and dimensions and will not conflict with those other constraints. Remember when you typed in the dimension *"25 in"* in creating your first profile? That dimension became a driving dimension thanks to the *Create Dimension for Typed Values* checkbox shown below. The **Auto Dimensioning Rules** shown below can be reordered using the arrows next to the list.

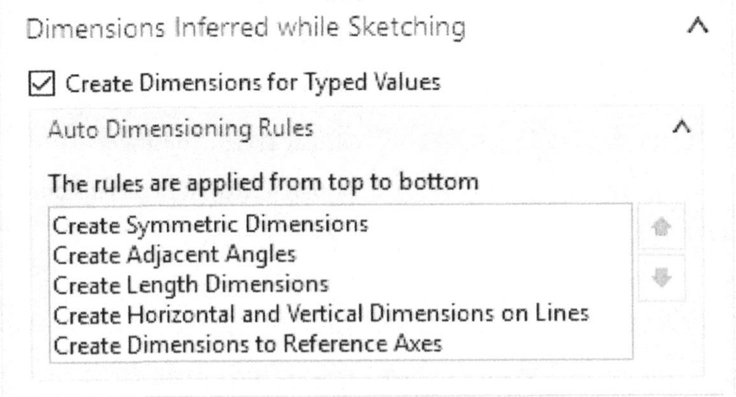

In this book, we will not modify the default settings in the **Inferred Constraints and Dimensions** tool, but we encourage you to try different options and see what you like. Throughout the subsequent exercises, we will hide auto dimensions so as not to clutter the screenshots in this book.

6.5 Elementary Sketch Curves

With some of the basics of the Sketch Task Environment covered, it's time to learn to make some different sketch curves! We'll start with those fundamental tools in the **Curve** group on the **Home** tab that create curves from scratch.

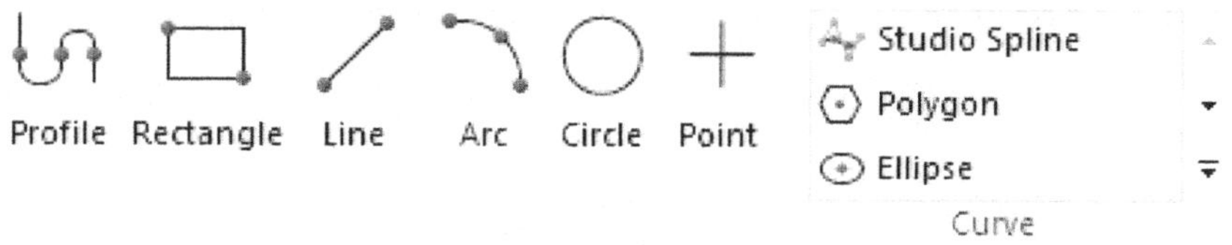

6.5.1 Line

The **Line** tool produces a line segment with just two clicks. Unlike the **Profile** tool, the **Line** tool will not automatically start another line.

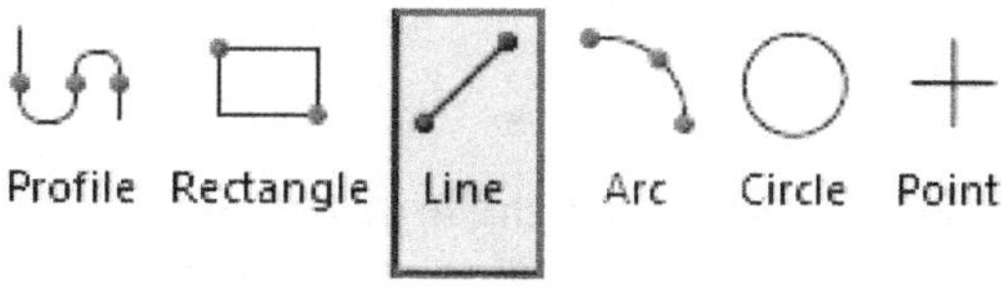

- Create a new part file (**Units**: *Inches*) called *"Elementary Sketch Curves.prt"* and create a sketch on the XY plane.
- Name your sketch *"LINE"*.
- Select the **Line** tool. Click once to determine the start point of the line.

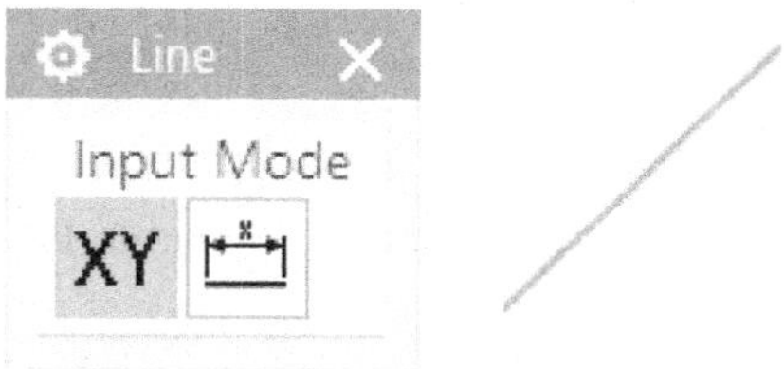

- Click again to finish the line. It's as easy as that!

- **Finish** your sketch.

6.5.2 Arc

The **Arc** tool has two methods for creating arcs – **Arc by 3 Points**, and **Arc by Center and Endpoints**.

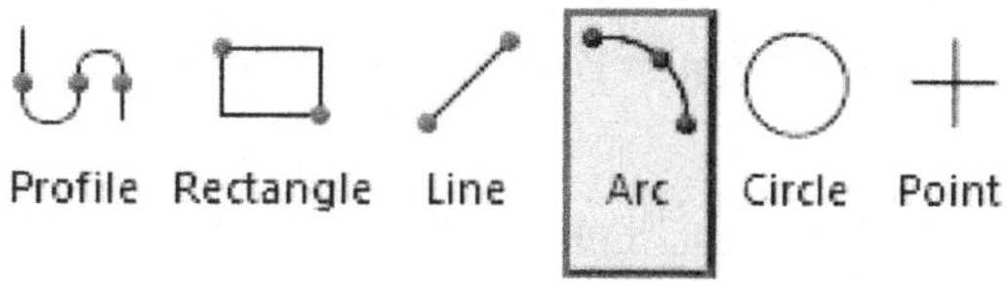

- **Pan** to a different part of the sketch plane, away from the line you created in the last exercise.
- Select the **Arc** tool and set the **Arc Method** to **Arc by 3 Points**. Click once to specify the start point of the arc and move the cursor to the right.

- Click again to specify the end point of the arc, and note that as you move the cursor, the **Radius** varies in the dynamic input box.

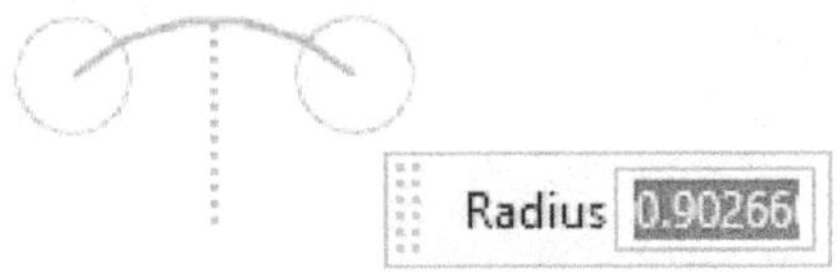

- Click a third time once you are pleased with the radius and the arc will be created.

- **Pan** to a different part of the sketch plane, and change **the Arc Method** to **Arc by Center and Endpoints**. Click once to specify the arc center, and then again to specify the start point of the arc.

- As you move the cursor, the **Radius** and **Sweep Angle** will vary dynamically, and your third click will determine the arc end point.

- **Finish** your sketch.

6.5.3 Circle

The **Circle** sketch tool has two methods for creating circles – *Circle by Center and Diameter*, and *Circle by 3 Points*.

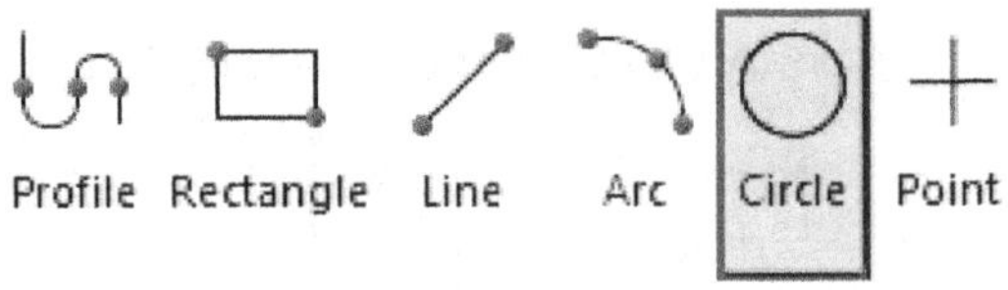

- Create a new sketch on the XY plane and name it *"CIRCLE"*.
- **Circle by Center and Diameter** requires two clicks – the first click indicates the center, and the second click specifies a point on the circle.
- **Circle by 3 Points** requires three clicks – the behavior is similar to the **Arc by 3 Points** method in the **Arc** tool.
- Practice both modes. Get comfortable using them with clicks alone, as well as with the help of the dynamic input boxes!

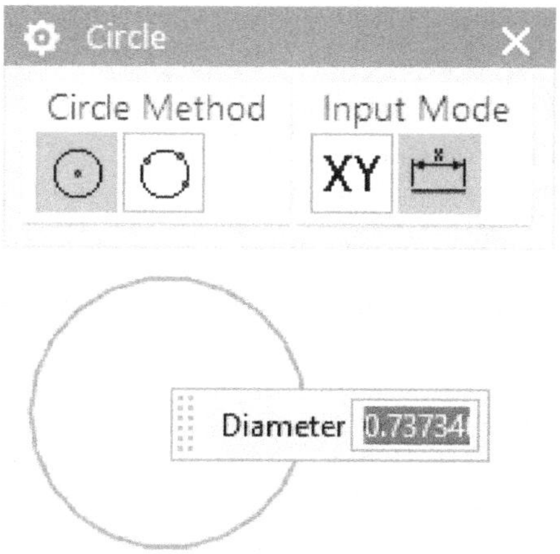

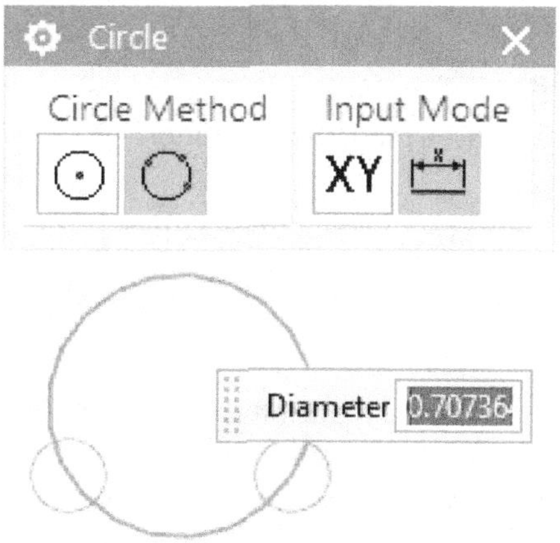

- **Finish** your sketch.

6.5.4 Point

The **Point** tool is completely self-explanatory. Select the tool and click anywhere in the graphics window to place a point on the sketch plane! A useful feature of the sketch **Point** tool is the constructor button that allows you to build a point using the 3D **Point** tool that you learned about in Chapter 5.1.

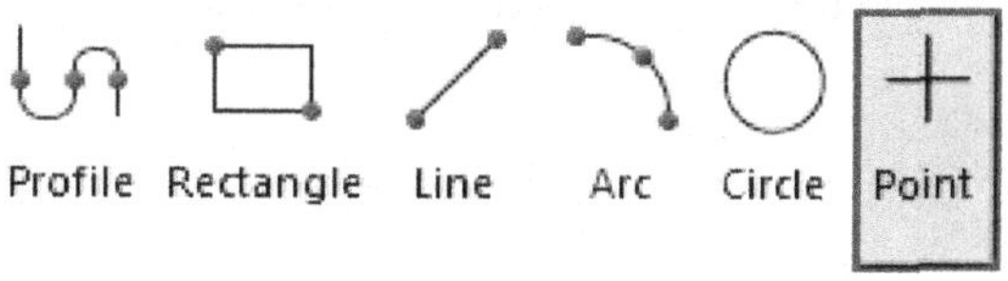

- Create a new sketch on the XY plane and name it *"POINT"*.
- Select the **Point** tool and push the point constructor button.

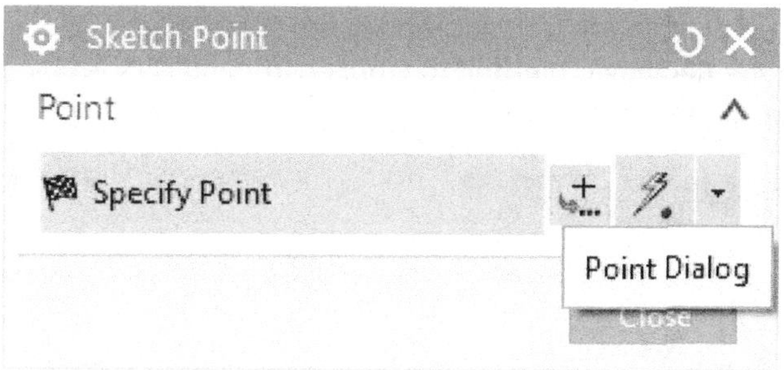

- In the **Point** dialog that appears, set the **Type** to *Between Two Points*. Choose the arc centers of two of the circles you made in the *"CIRCLE"* sketch.

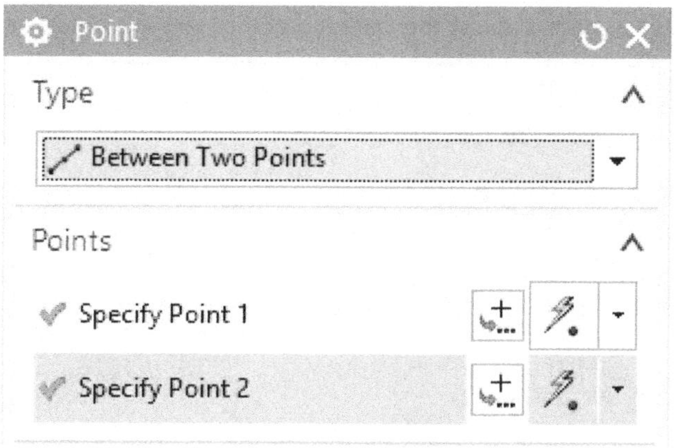

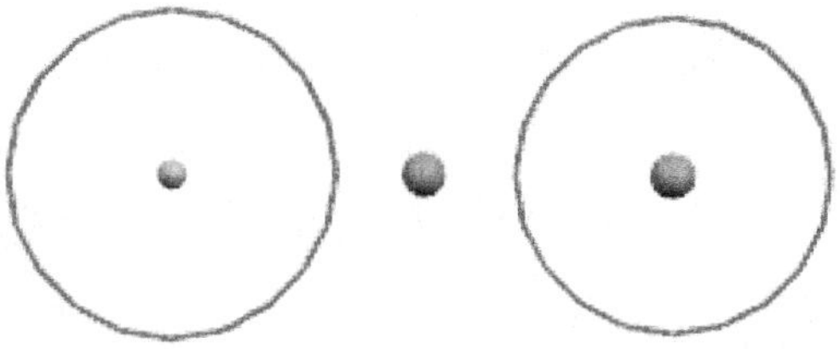

- **Finish** your sketch!

6.5.5 Rectangle

The **Rectangle** tool has three modes of operation - *By 2 Points*, *By 3 Points*, and *From Center*.

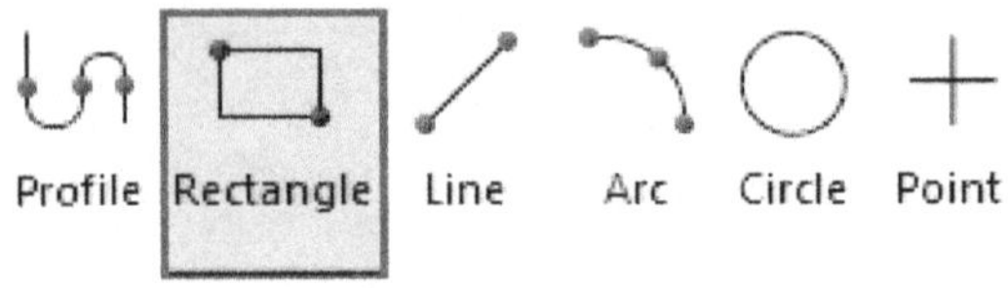

- Create a new sketch on the XY plane and name it *"RECTANGLE"*.
- Select the rectangle tool. The default **Rectangle Method** is *By 2 Points*. This will produce a rectangle aligned with the sketch coordinate system, where your clicks indicate opposite corners.

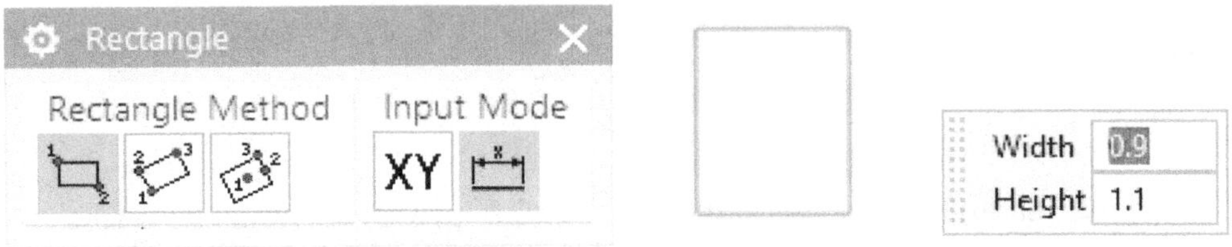

- Set the **Rectangle Method** to *By 3 Points*. With this method, your three clicks specify three of the four corners of the resulting rectangle. A rectangle constructed in this way need not be oriented to the sketch coordinate system.

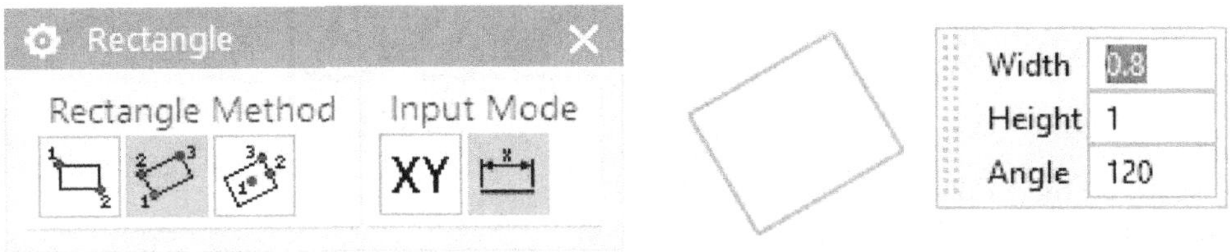

- Use the **Point** tool to place a point at an arbitrary location in the sketch plane.
- Select **Rectangle** tool and set the **Rectangle Method** to *From Center* and click on the point you created. Your next two clicks will orient the rectangle and set its boundaries.

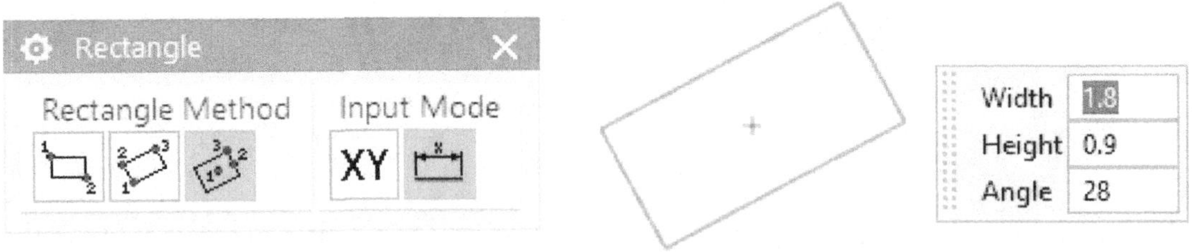

- Notice how that in this case, the rectangle comes with two free **Midpoint** constraints! If you fully constrain the point, these midpoints ensure that the rectangle it will remain centered about the selected point even if it is resized!

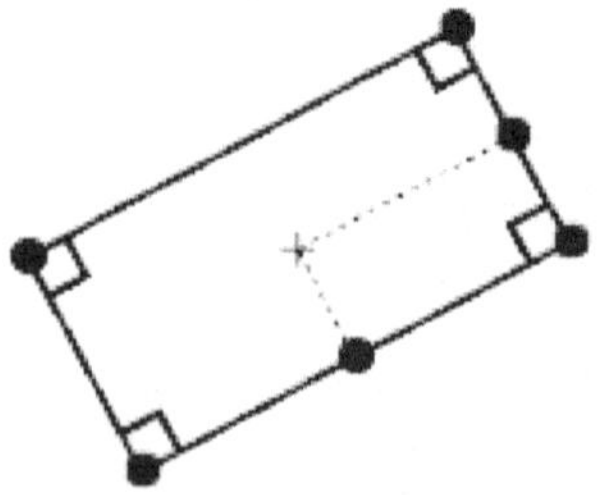

- **Finish** your sketch!

6.5.6 Polygon

The **Polygon** tool is used to construct *regular* polygons – that is, polygons with all side lengths equal, and all angles between those sides equal.

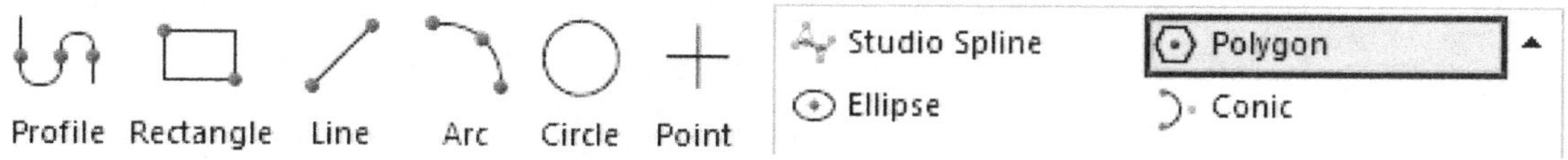

- Create a new sketch on the XY plane and name it *"POLYGON"*.
- The **Polygon** tool works like the **Circle** tool with **Circle Method** set to *Circle By Center and Diameter* – your first click indicates the **Center Point**, and your second click controls the **Size** of your polygon.
- There are three ways to size regular polygons.
  - *Side Length*: the side length
  - *Inscribed Radius*: the radius of a circle in which the polygon is inscribed
  - *Circumscribed Radius*: the radius of a circle circumscribed within the polygon

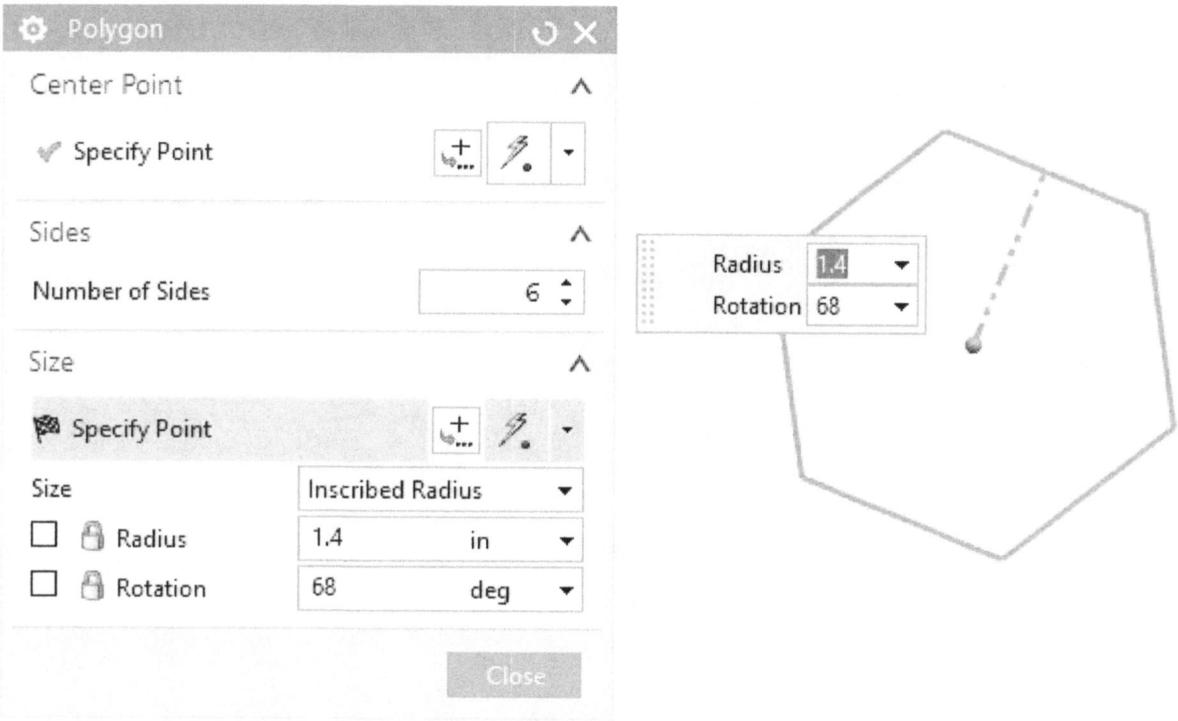

- Change the **Number of Sides** to **6** and choose a sizing method of your choice. Enter values for the radius and rotation, or click to place the polygon without dimensional constraints. To edit the parameters of your polygon, you can double-click on the *Polygon* constraint that appears in the Graphics Window at the center of the polygon, and the **Polygon** dialog box will reappear.

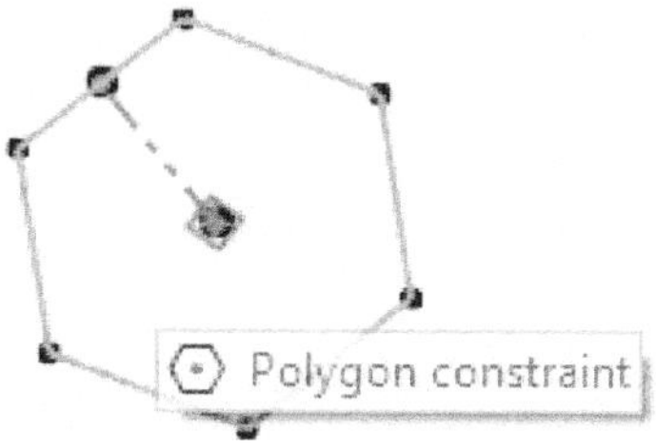

- **Finish** your sketch!

6.5.7 Ellipse

The sketch curves that we have created up to this point have all consisted of lines and arcs. You can do quite a lot with just "circles and squares" but some designs require more sophisticated curves. The **Ellipse** tool produces either a closed ellipse or an open elliptical curve.

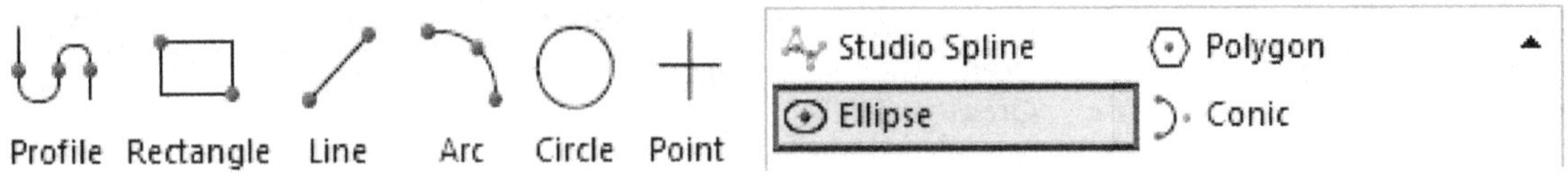

- Create a new sketch on the XY plane and name it *"ELLIPSE"*.
- Select the **Ellipse** tool. It has straightforward controls for the **Major Radius**, **Minor Radius**, and **Rotation Angle**.

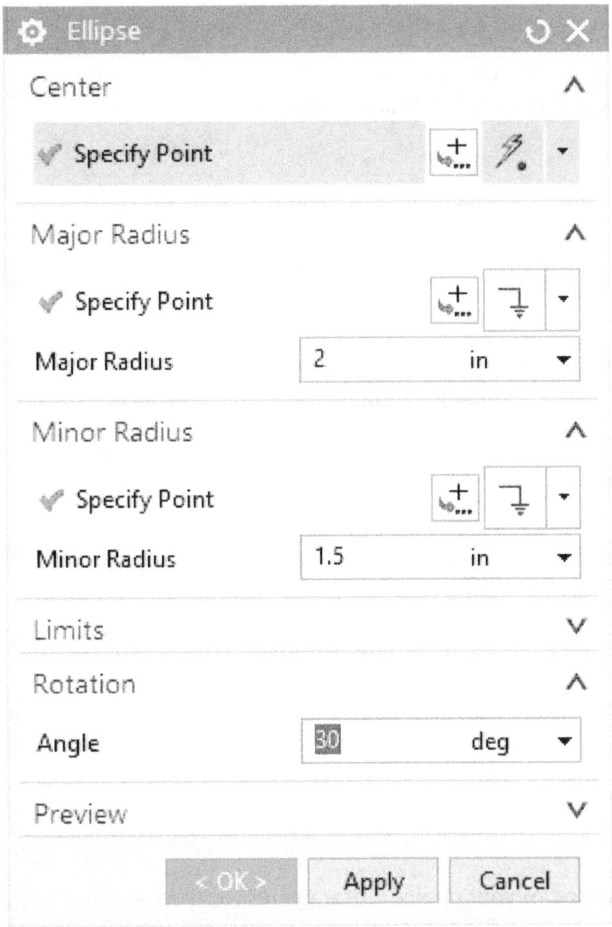

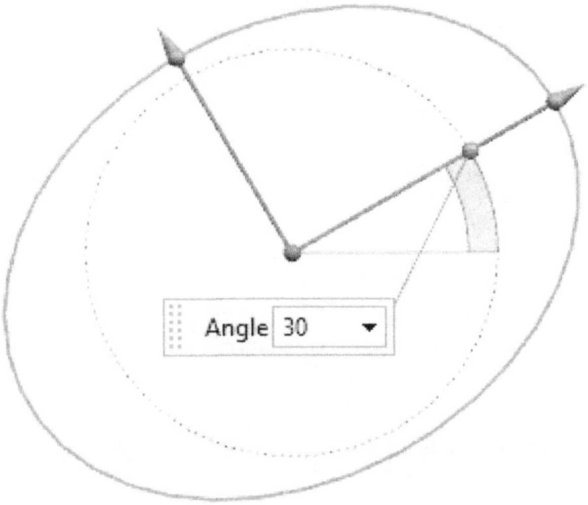

- A slightly annoying aspect of the **Ellipse** tool is that the dimensions that you enter in the dialog are never saved – neither as dimensions, nor parameters. You can double-click on the ellipse after it is created to modify those values, but you cannot drive them associatively.

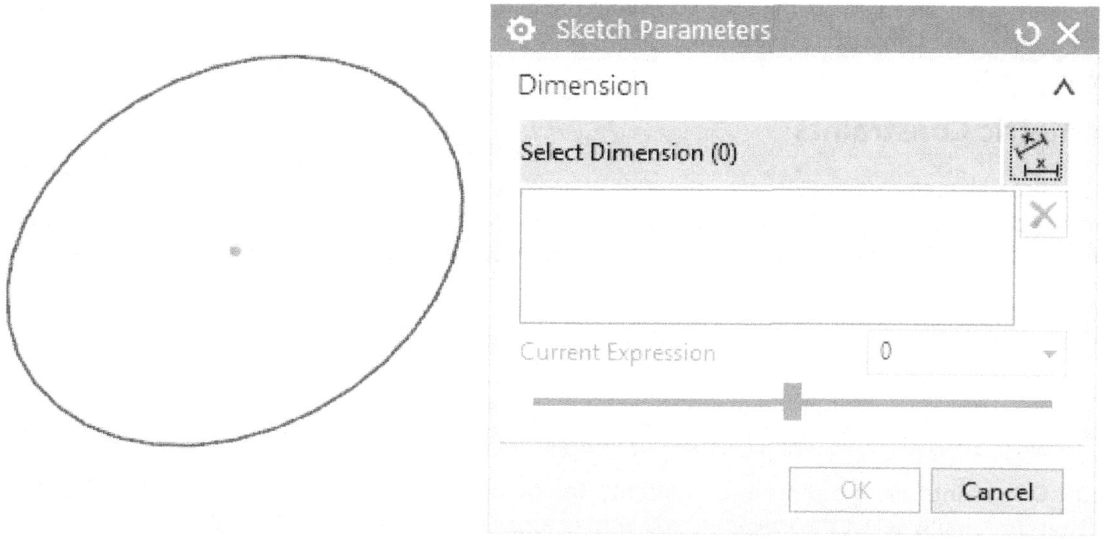

- If you need to control the major or minor radii or diameters, you should create explicit dimensions from end to end. In Chapter 6.7.6, you will practice parameterizing an ellipse.
- **Finish** your sketch!

6.5.8 Conic

Another relatively simple curve that allows you to create many more shapes than just lines and arcs is the conic tool – this allows you to create a curve that represents the intersection of a cone with the sketch plane.

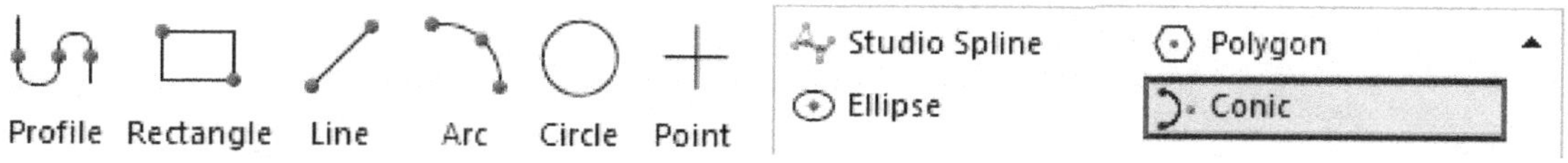

- Create a new sketch on the XY plane and name it *"CONIC"*.
- Select the **Conic** tool. To create your conic, three clicks are required – it behaves similarly to the arc tool when the **Arc Method** is set to *Arc by 3 Points* – your first two clicks indicate the endpoints of the conic, and your third click indicates the control point. The control point does not actually lie on the conic. The parameter **Rho** determines how sharp the minimum radius of the conic is – in some sense, how strongly the control point pulls the curve toward it.

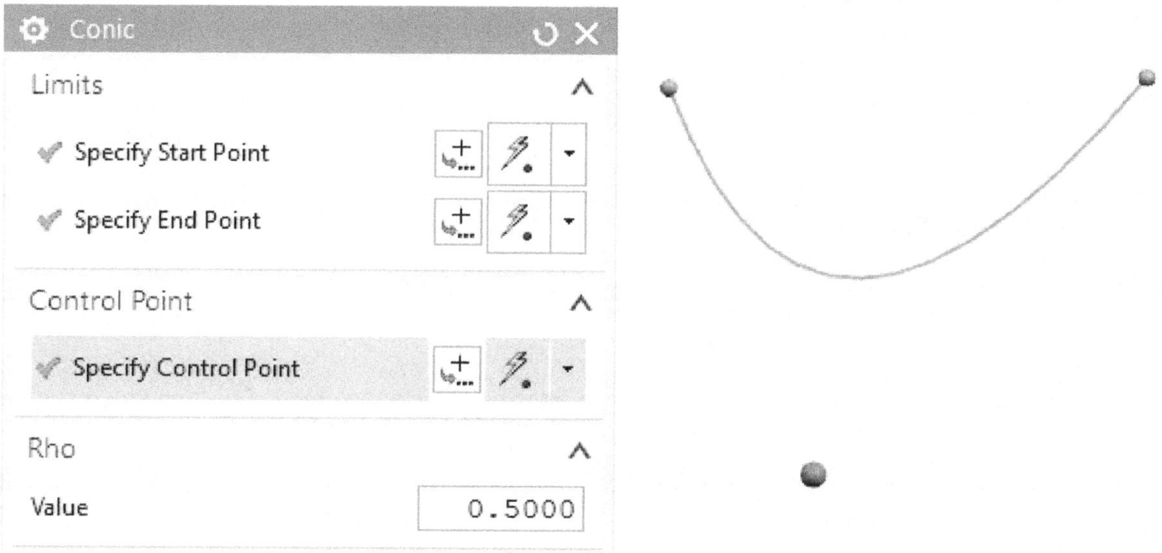

- The **Rho Value** is not saved as a parameter, so as with the **Ellipse** tool, if you need to modify it, you must double-click the curve to make the **Conic** dialog reappear.
- **Finish** your sketch!

6.6 Geometric Constraints

Until now, the constraints that you have encountered in your sketches have been automatically created as a result of the **Create Inferred Constraints** option. To create other constraints that NX is unable to infer, you must use the **Geometric Constraints** tool, found in the **Constraints** group on the right side of the **Home** tab in the Sketch Task Environment.

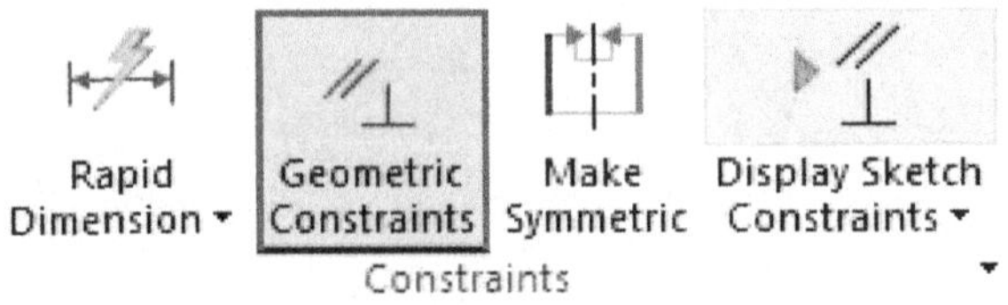

The **Geometric Constraints** dialog box behaves differently from most other dialog boxes in NX – there is no **OK** button and no **Apply** button. You simply select the constraint you wish to apply, then click on the object(s) you wish to constrain. Most constraints require at least one *Object to Constrain*, and exactly one *Object to Constrain to*. Once these are selected, the constraint is generated without the need for you to confirm or save it.

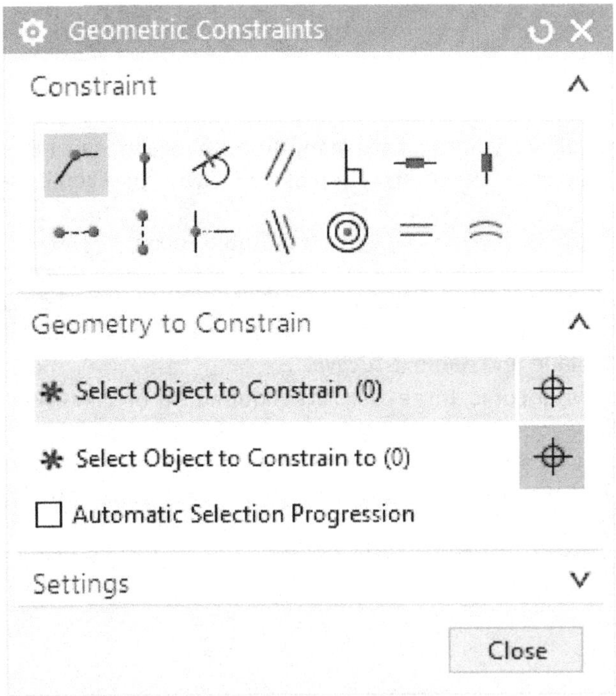

6.6.1 Automatic Selection Progression

Most constraints require both an *Object to Constrain*, and an *Object to Constrain to*, and will in fact allow many objects for the first input. *Automatic Selection Progression* advances the menu automatically after you specify the *Object to Constrain*, so that your next click specifies the *Object to Constrain to*.

When you only need to specify constraints pairwise between selected objects, *Automatic Selection Progression* will save you clicks, but when you want to constrain many objects to one, you will want to disable it.

In the next few exercises, you will be applying constraints pairwise, so we strongly recommend that you enable *Automatic Selection Progression*. If you don't, you will need to manually click on *"Select Object to Constrain to (0)"* after selecting the *"Object to Constrain"*.

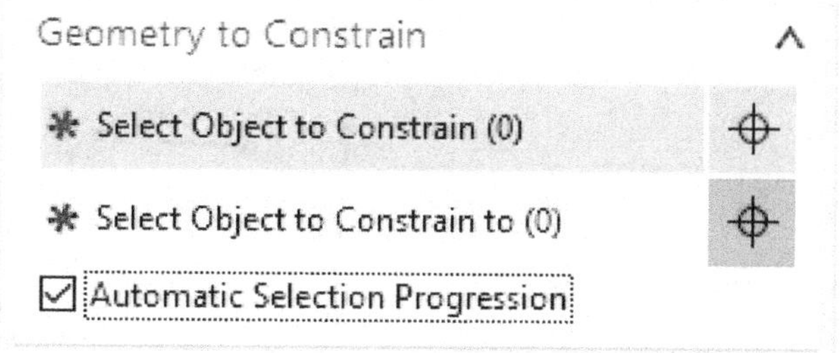

Note that by default, *Automatic Selection Progression* is disabled, and that if you **Reset** the **Geometric Constraints** tool, the checkbox will become unmarked.

For the next few exercises, you will need a file from the downloaded parts folder.

- Open *"Geometric Constraints.prt"* from *"C:\My NX 11 Files\Downloaded Files"*.
- In the Model History, you will find three sketches.

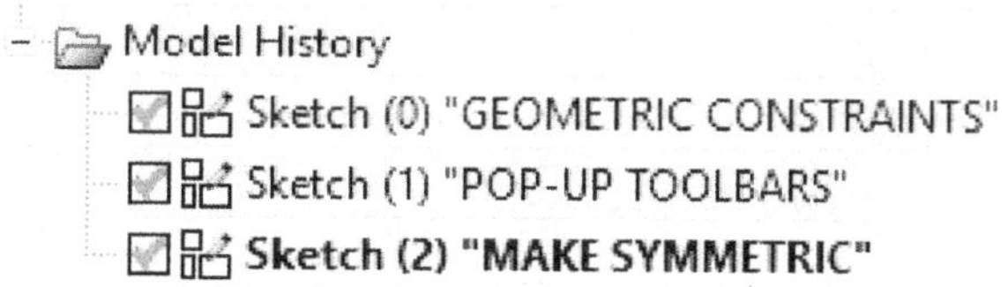

6.6.2 Coincident

The *Coincident* constraint forces two existing points in a sketch to coincide. A typical use case for this constraint is to make the endpoint of one curve coincide with the start point of another. The *Coincident* constraint requires two (or more) points as input.

You will use the *Coincident* constraint to close the gaps found in the sketch in *"Geometric Constraints.prt"*.

- Open the sketch *"GEOMETRIC CONSTRAINTS"* by double-clicking on it.

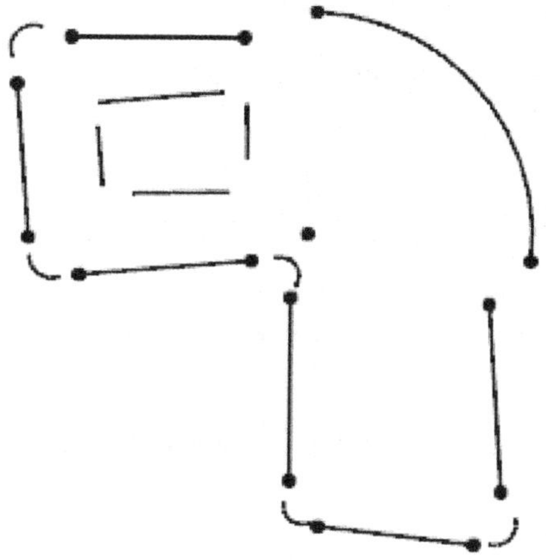

- Select the **Geometric Constraints** tool and choose the *Coincident* constraint. Make sure that **Automatic Selection Progression** is checked on. Select the right endpoint of the horizontal line segment as the *Object to Constrain*, and the left endpoint of the large arc as the *Object to Constrain to*, as shown below.

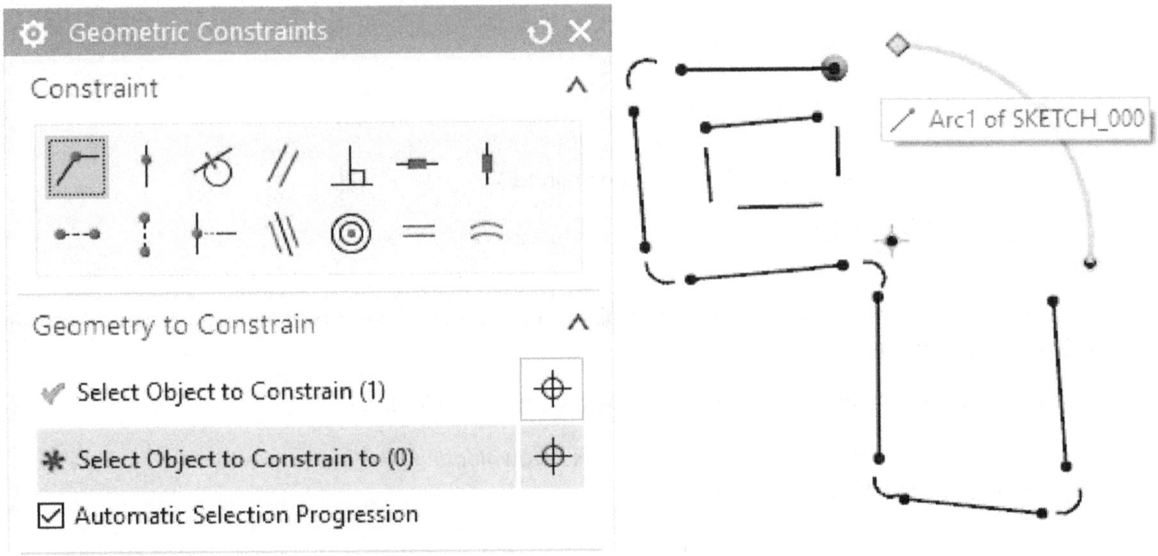

- Do this for each pair of adjacent endpoints until your sketch appears as shown below.

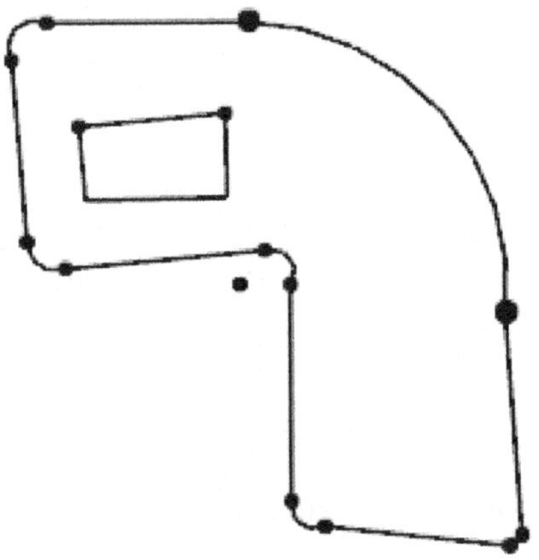

6.6.3 Horizontal & Vertical

The *Horizontal* and *Vertical* constraints are used to force line segments to be parallel to the X and Y axes of the sketch coordinate system, respectively.

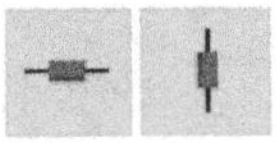

When you sketch a line that is nearly horizontal or vertical, NX will automatically apply a *Horizontal* or *Vertical* geometric constraint to them when the preview is within a default value of *3 degrees* of horizontal or vertical. You can modify the default value in **Sketch Preferences**.

- Select the **Horizontal** constraint, and choose the line atop your sketch that is nearly horizontal.

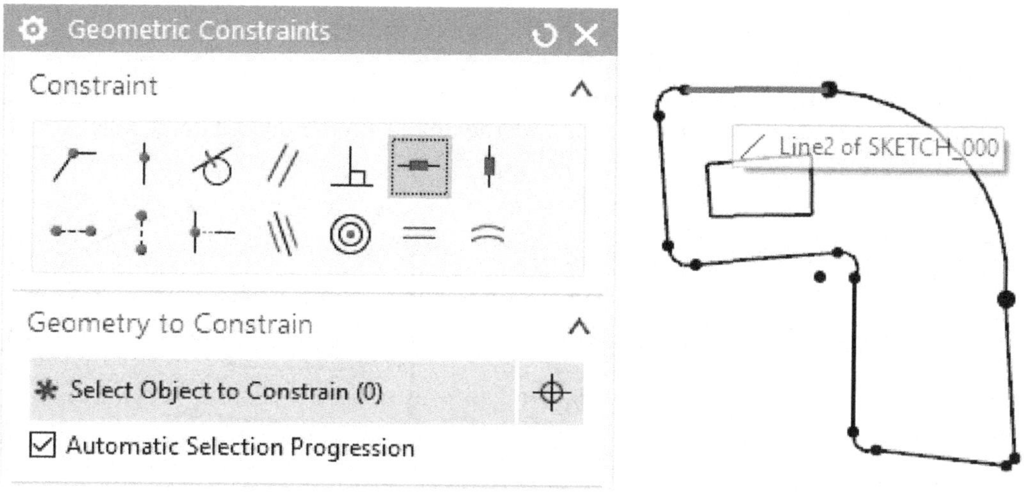

- Select the *Vertical* constraint, and choose the line to the far left of the sketch to make it vertical.

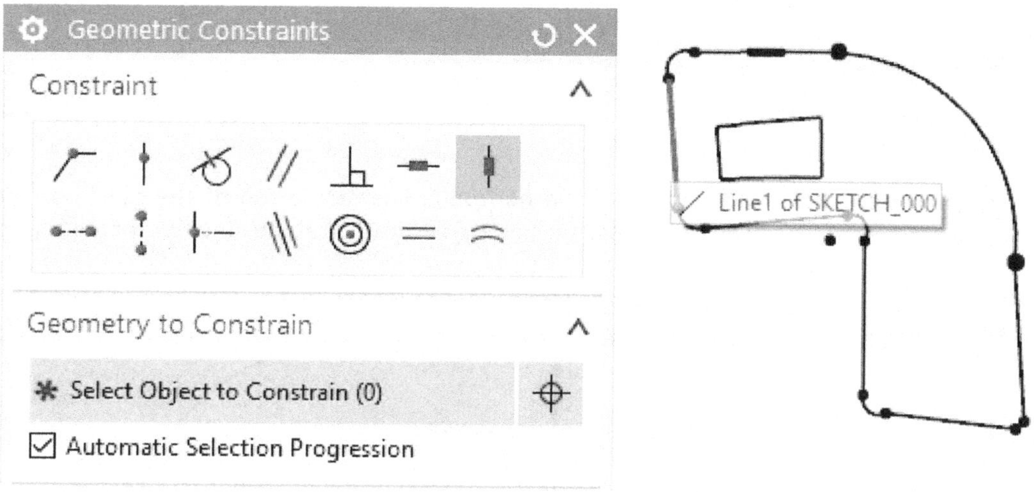

- Apply one more *Horizontal* constraint to the line atop the quadrilateral inside.

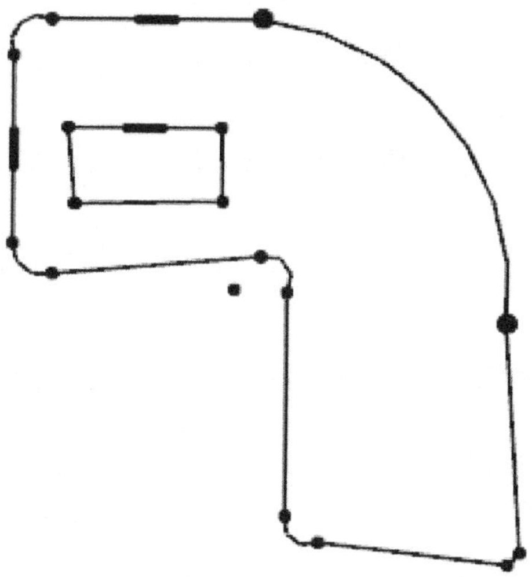

6.6.4 Parallel & Perpendicular

The **Parallel** constraint forces two (or more) lines to be parallel. The **Perpendicular** constraint forces two (or more) line segments to be perpendicular to each other. These tools are pretty straightforward.

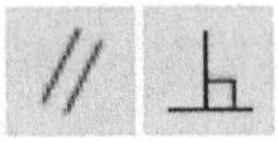

- Choose the *Parallel* constraint, and select the horizontal line at the top of the sketch as the *"Object to Constrain"*. For the *"Object to Constrain to"*, choose the line directly underneath the interior quadrilateral, as shown below.

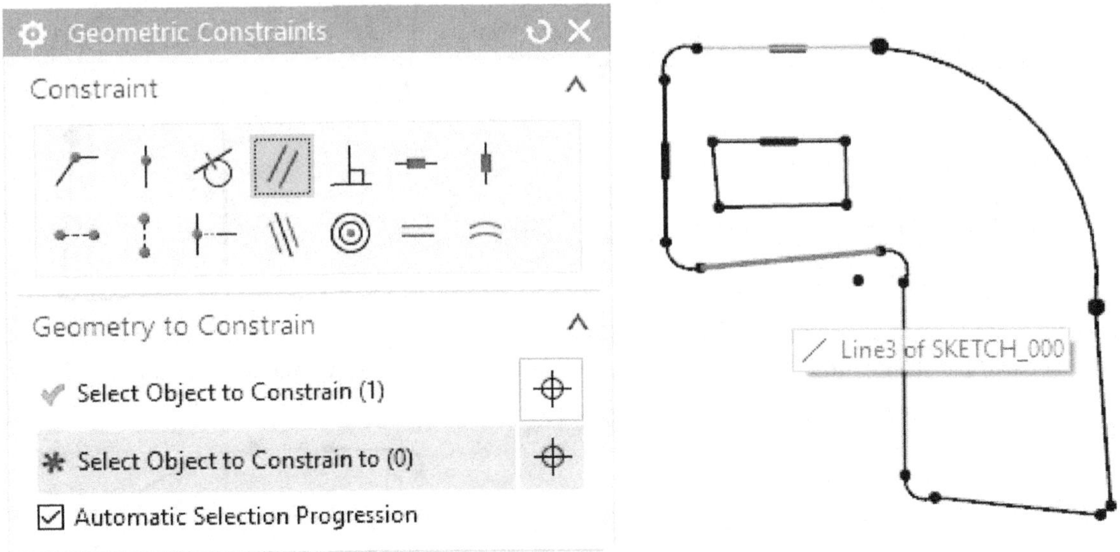

- Choose the *Perpendicular* constraint, and select the horizontal line at the top of the sketch as the *"Object to Constrain"*. For the *"Object to Constrain to"*, choose the (nearly) vertical line at the other end of the large arc, as shown below.

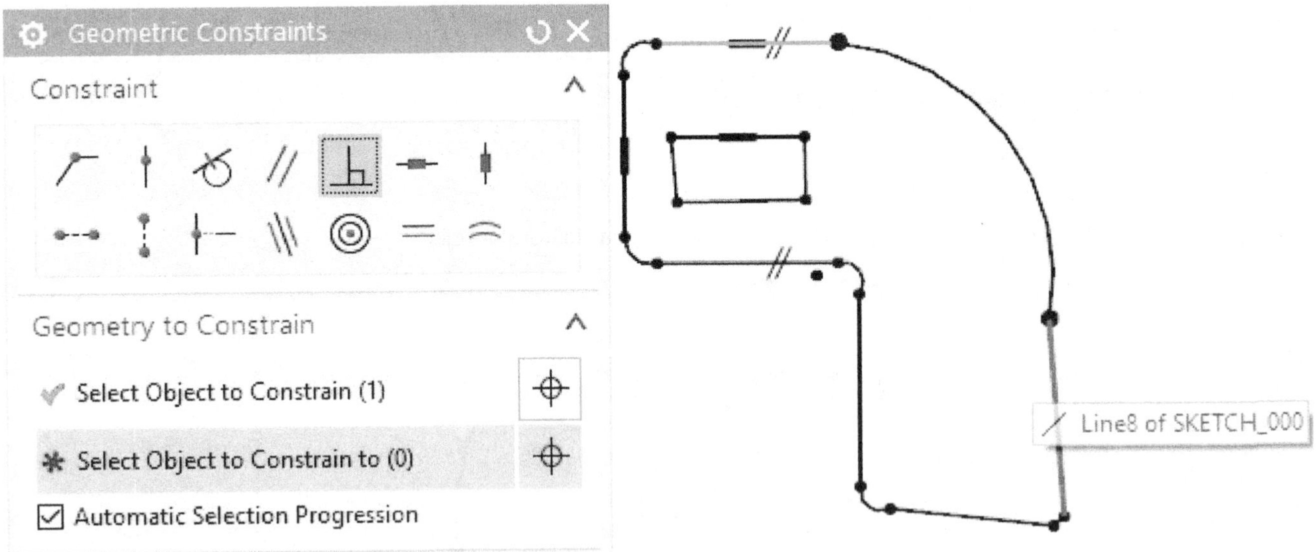

- Add three more perpendicular constraints to the corners of the inside quadrilateral to turn it into a rectangle. Your sketch should now look something like this!

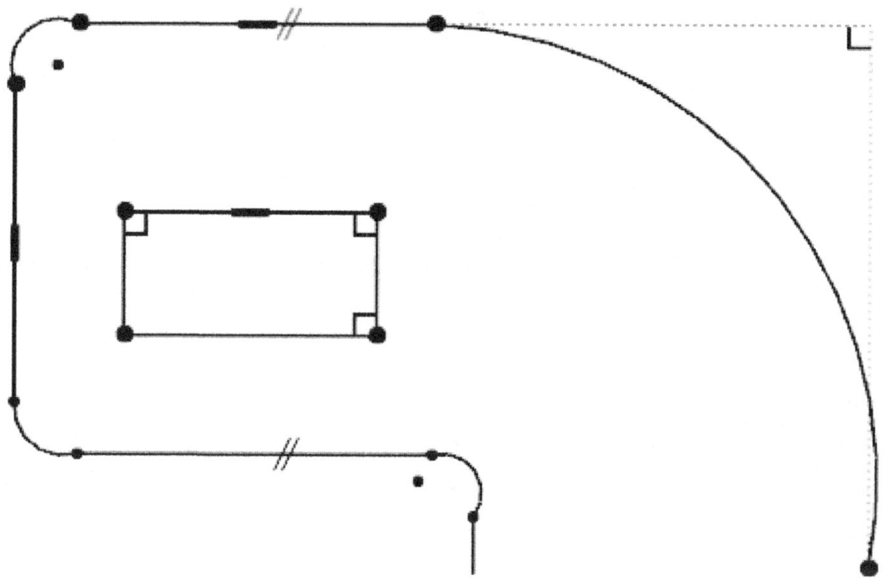

6.6.5 Tangent

The *Tangent* constraint forces tangency between two curves at a point. The appearance of the constraint in the graphics window is slightly unusual – it is a line passing through the coincident point at a 90° angle to the tangent line!

- Select the *Tangent* constraint, and choose the horizontal line atop the sketch as the *"Object to Constrain"*. Select the large arc to the right as the *"Object to Constrain to"*.

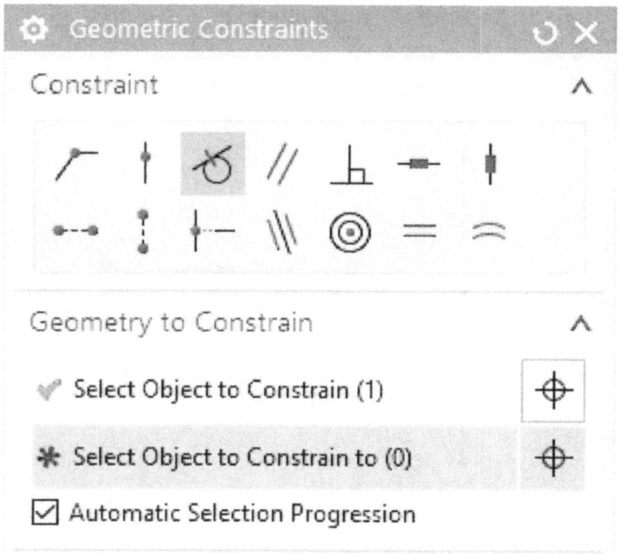

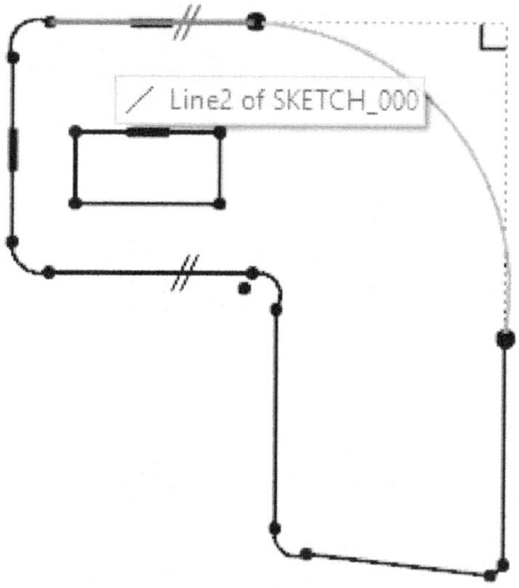

- Apply six more *Tangent* constraints between the three small arcs on the upper left side of the sketch and their neighboring lines.

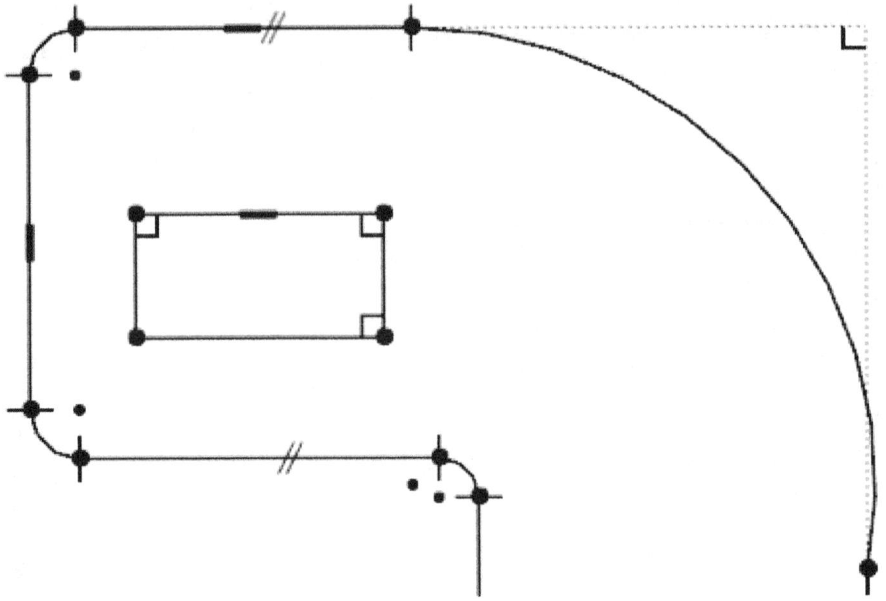

6.6.6 Point on Curve

Point on Curve is possibly a misleading name for a very useful constraint – it should really be thought of as a point constrained to be on the (possibly) infinite extension of a curve. It may be used to align points, endpoints of curves, or arc center points to lines or other curves. The *Point on Curve* constraint requires a single curve and at least one point as input.

- Select the *Point on Curve* constraint, and then choose the left vertical line of the interior rectangle as the *"Object to Constrain"*. For the *"Object to Constraint to"*, choose the arc center of the arc beneath it.

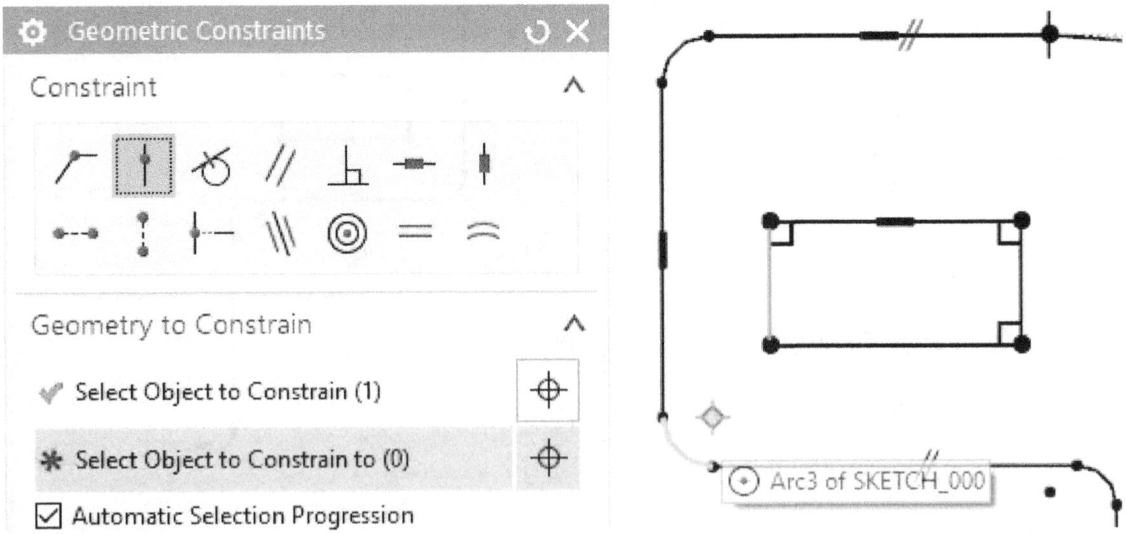

- The *Point on Curve* constraint appears in the graphics window as a dotted line connecting the arc center to the vertical line segment.

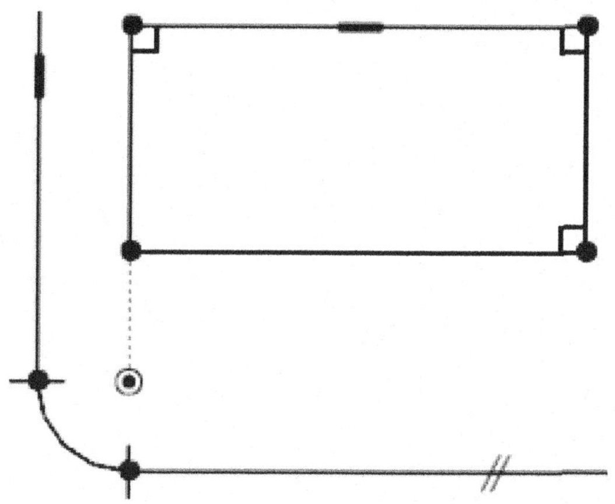

6.6.7 Midpoint

If you are accustomed to other CAD software, you may be surprised by the functionality of the *Midpoint* constraint. The *Midpoint* constraint is extremely useful for centering sketch objects relative to others. The *Midpoint* constraint requires at least one line and at least one point as input.

The graphic below illustrates the effect of the *Midpoint* constraint. Prior to the application of the constraint (right), the point (B) bears no obvious relation to the line (A). After application of the constraint (left), the point moves along a line parallel to (A) until the nearest point to B on A is the midpoint.

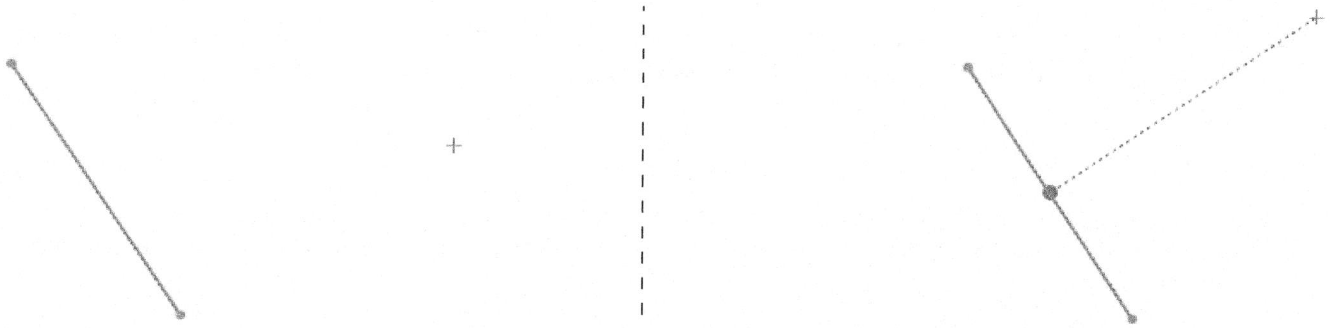

- Select the *Midpoint* constraint, and choose the leftmost vertical line in the sketch as the *"Object to Constrain"*. For the *"Object to Constrain to"*, select the midpoint of the leftmost vertical line of the inside rectangle.

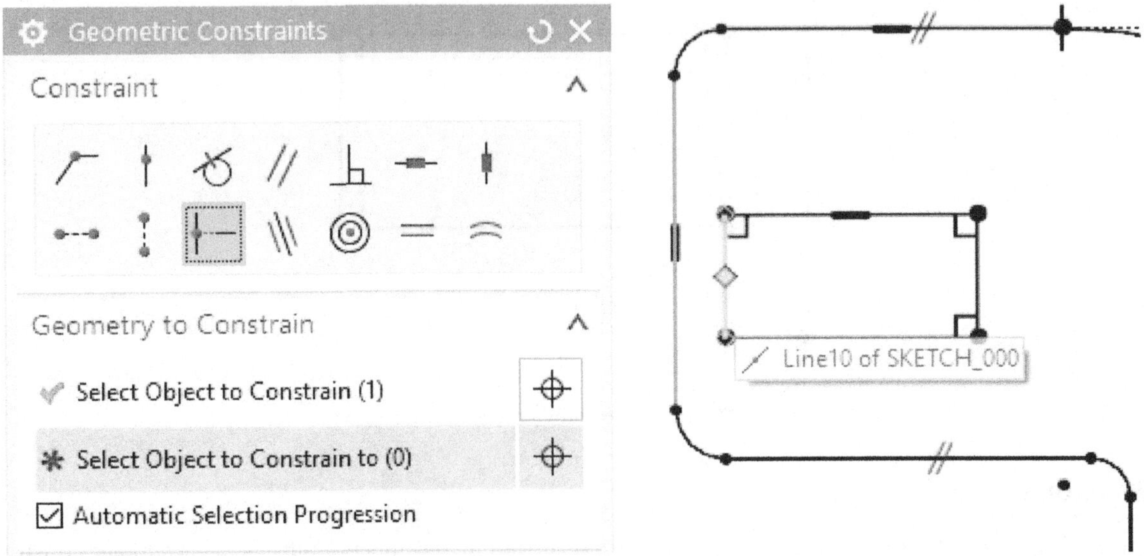

- The *Midpoint* constraint appears in the Graphics Window also as a dotted line.

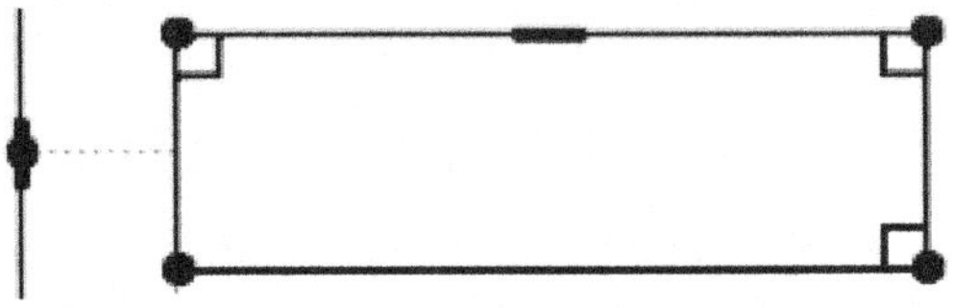

6.6.8 Collinear

The *Collinear* geometric constraint is similar to the *Point on Curve* constraint. The *Collinear* constraint aligns two or more lines with each other, as suggested by the icon. The *Collinear* constraint requires two (or more) lines as input.

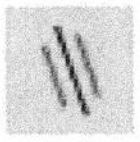

- Select the *Collinear* constraint, and choose the rightmost vertical line of the inside rectangle as the *"Object to Constrain"*. For the *"Object to Constrain to"*, choose the (nearly) vertical line attached to the arc below and to the right of the rectangle.

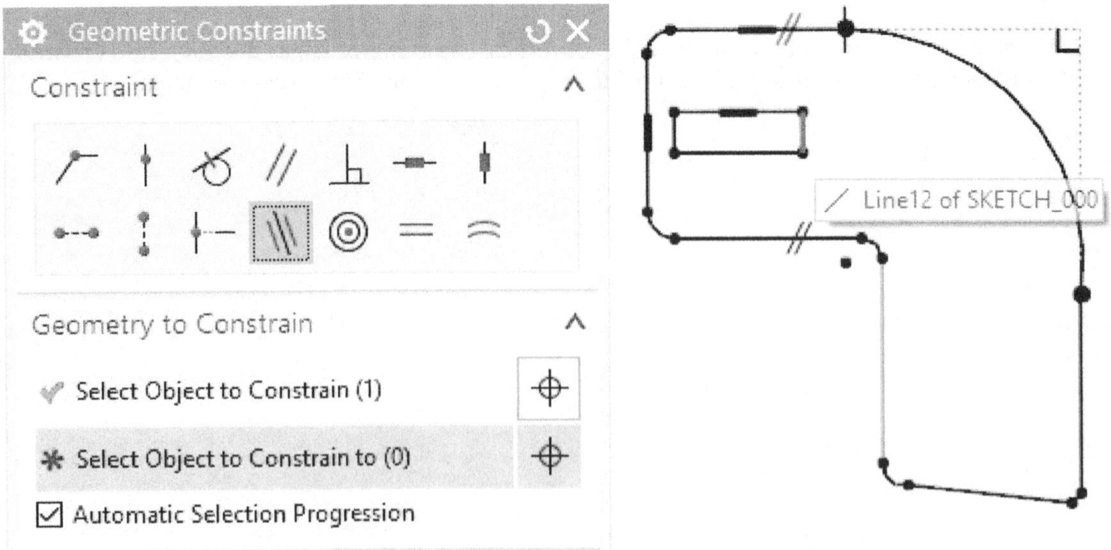

- Like both *Point on Curve* and *Midpoint*, the *Collinear* constraint appears in the Graphics Window as a dotted line.

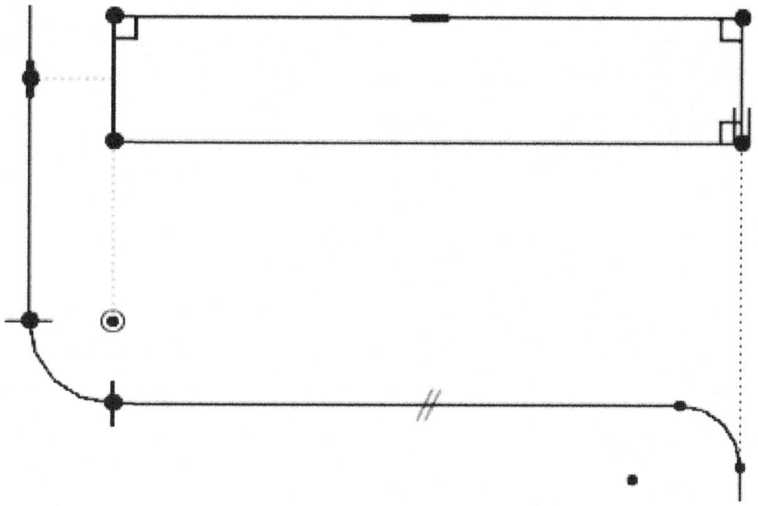

6.6.9 Concentric

The *Concentric* constraint is used to force the arc centers of circles or arcs to coincide – it offers no additional functionality over the *Coincident* constraint. The *Concentric* constraint requires two (or more) arcs or circles as input.

- Select the *Concentric* constraint. Choose the large arc in the top right corner of the shape as the *Object to Constrain*, and the small arc in the interior of the shape as the *Object to Constrain to*.

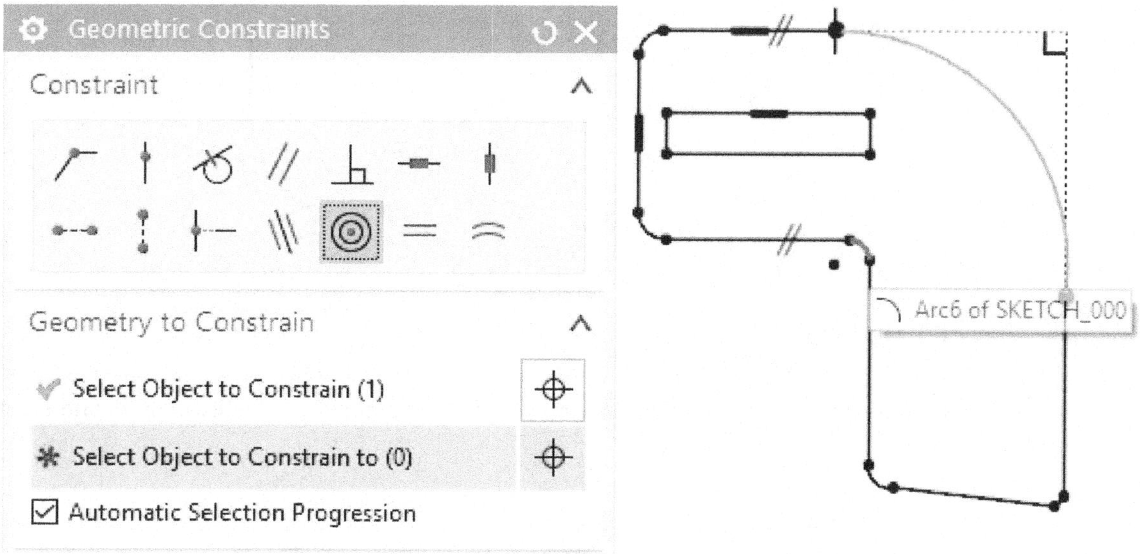

6.6.10 Horizontal Alignment and Vertical Alignment

The *Horizontal Alignment* and *Vertical Alignment* constraints are new in NX 11. They simply align points horizontally or vertically in the sketch.

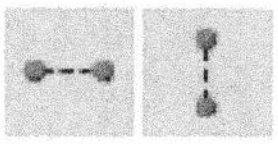

Horizontal Alignment and *Vertical Alignment* constraints can be applied to the control points of arcs to make them tangent to horizontal or vertical lines – you will practice this in the exercise below.

- Select the *Horizontal Alignment* constraint, and choose the arc center of the arc in the bottom right corner of the sketch as the *"Object to Constrain"*. For the *"Object to Constrain to"*, choose the rightmost endpoint of the arc (or the line endpoint with which it is coincident) as shown below.

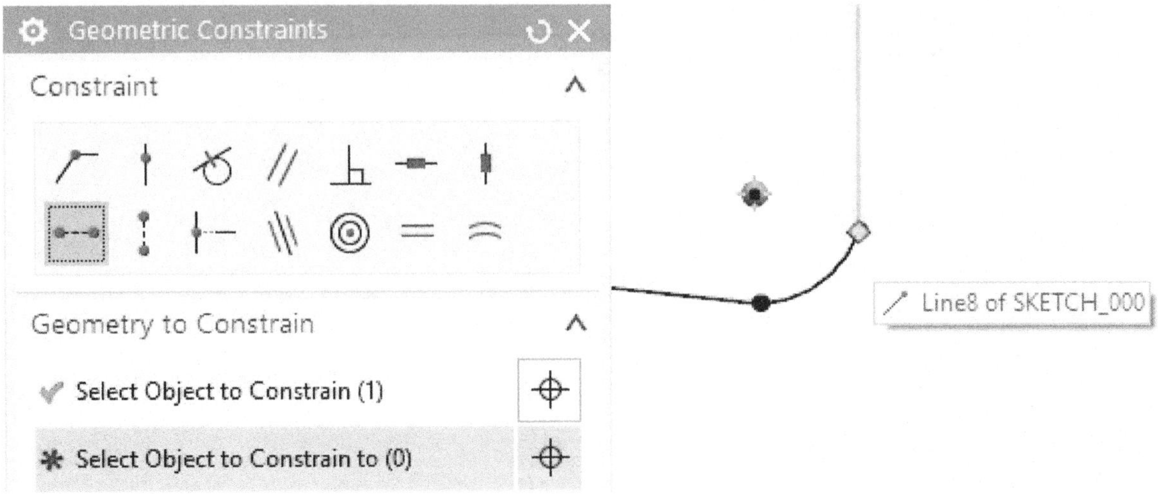

- Select the *Vertical Alignment* constraint, and choose the arc center of the arc in the bottom right corner of the sketch as the *"Object to Constrain"*. For the *"Object to Constrain to"*, choose the leftmost endpoint of the arc (or the line endpoint with which it is coincident) as shown below.

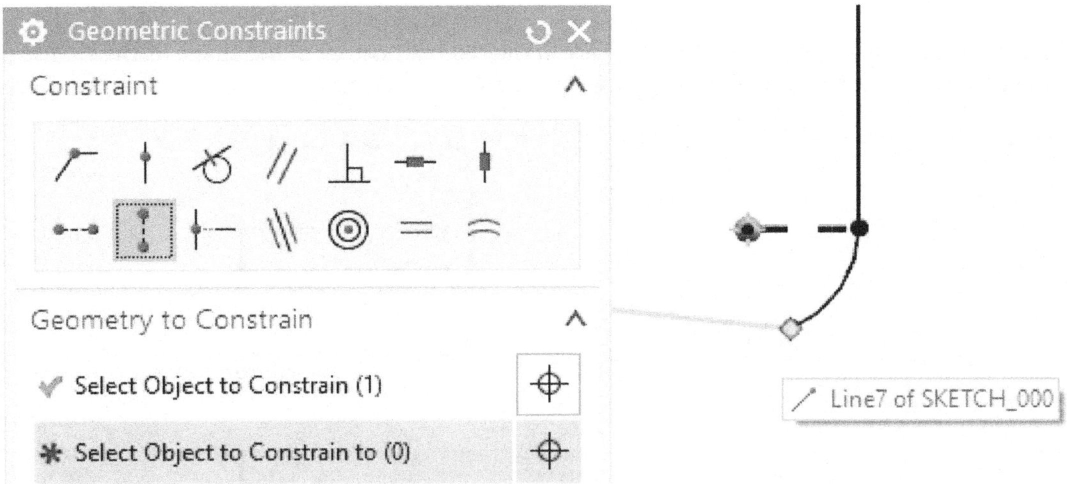

- Add three more *Horizontal Alignment* and *Vertical Alignment* constraints between the arc centers and endpoints of the arcs shown below.

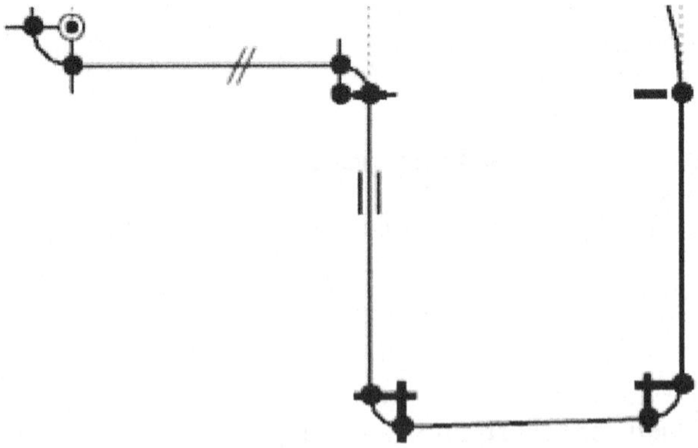

6.6.11 Equal Length and Equal Radius

The *Equal Length* constraint is straightforward – it simply forces two lines to have the same length. Note that it only works on lines, and not arcs or any other kind of sketch curves. The *Equal Length* constraint requires (at least) two line segments as input.

The *Equal Radius* constraint is similar to the *Equal Length* constraint, but it can only be applied to arcs and circles. The *Equal Radius* constraint requires two (or more) arcs or circles as input.

Be sure not to confuse the icon when you use the tool – the *Equal Length* constraint will not allow you to select arcs! The appearance of the constraint in the graphics window is an equals sign, just like that of *Equal Length*.

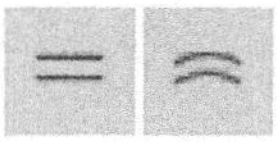

- For the present *Equal Length* constraints, it will be helpful to disable *Automatic Selection Progression*. Choose the *Equal Length* constraint and choose the five line segments shown below as the *"Object to Constrain"*.

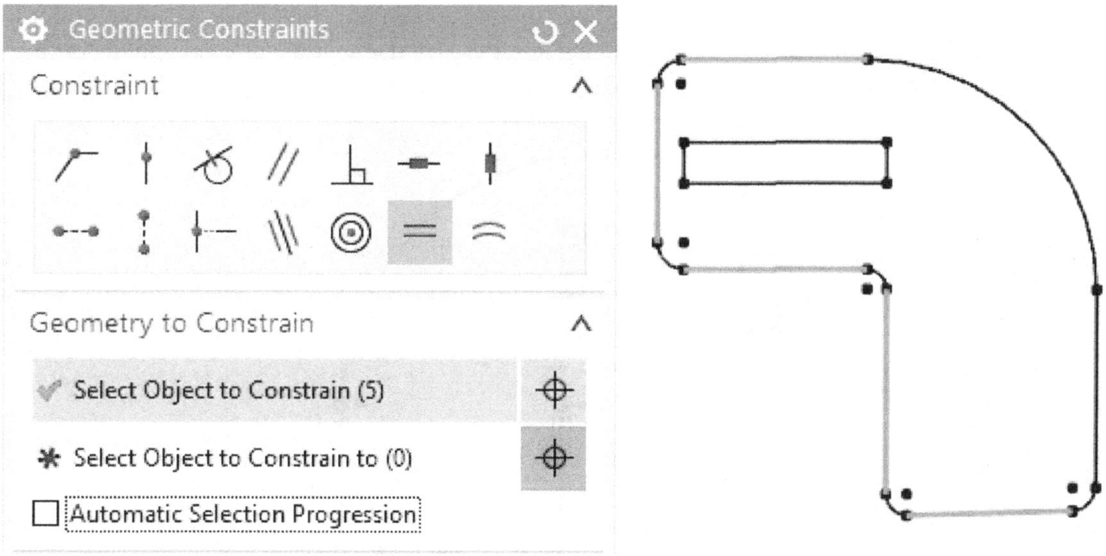

- Since *Automatic Selection Progression* is disabled, you have to manually click *"Select Object to Constrain to"* (or click the middle-mouse button to accept and move on), and then choose the rightmost vertical line segment as the *"Object to Constrain to"*.
- Use the same technique for the *Equal Radius* constraint – choose all but one of the small arcs in the sketch as the *"Object to Constrain"*, then choose the last as the *"Object to Constrain to"*.

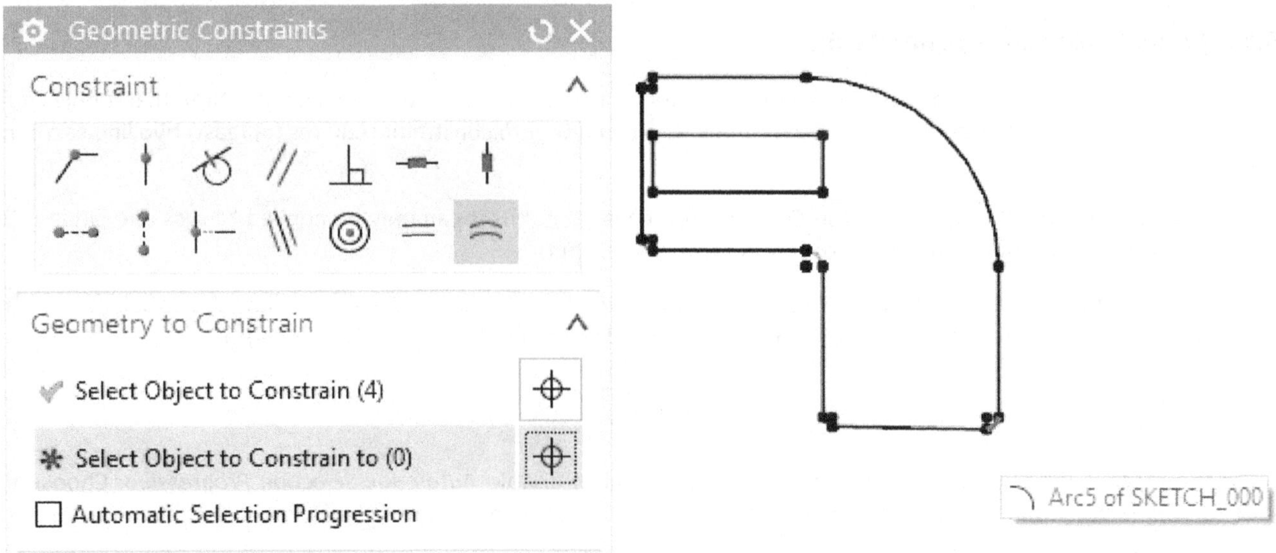

- When you are finished, your sketch will look something like this. Try dynamically dragging points on the sketch to resize the arcs and lines!

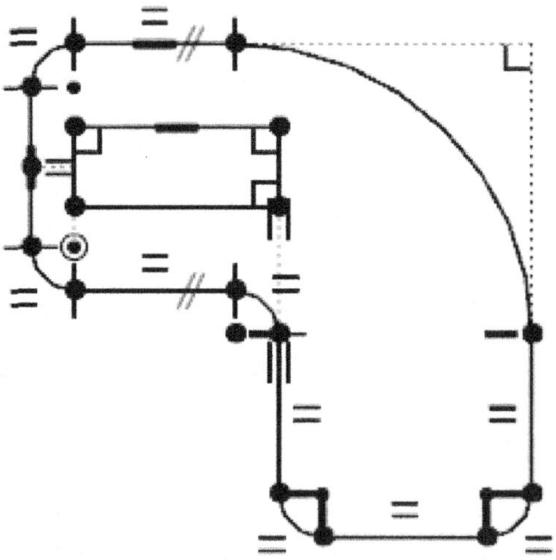

- **Finish** the sketch!

6.6.12 Other Geometric Constraints

There are a number of other, lesser-used constraints that you can enable by clicking their checkboxes within the **Settings** section of the **Geometric Constraints** dialog. We leave it to you to explore these and use them as you find necessary.

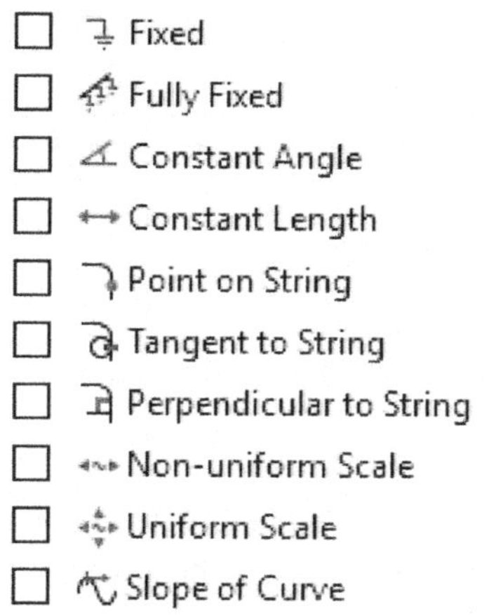

6.6.13 Pop-Up Toolbars

It is not actually necessary to use the **Geometric Constraints** dialog in order to apply constraints to your sketch. Upon selection of appropriate sketch entities, constraints applicable to those sketch entities will appear on the pop-up toolbars, as described in Chapter 2.8. When you are first learning NX, it will be better to use the **Geometric Constraints** tool to get a handle on what constraints are available, and what they accept as inputs, but once you are comfortable, the pop-up toolbars are an excellent shortcut that can speed up your sketch workflow.

- Continue working with *"Geometric Constraints.prt"* and open the sketch called *"POP-UP TOOLBARS"*.

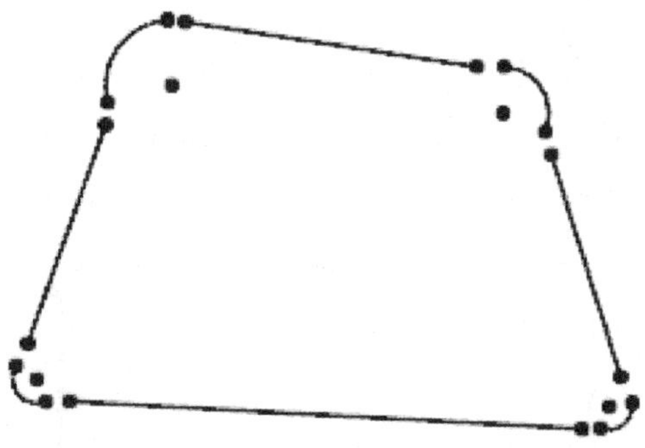

- Select the top line in the sketch, and from the pop-up toolbar that appears, find and push the *Horizontal* constraint button.

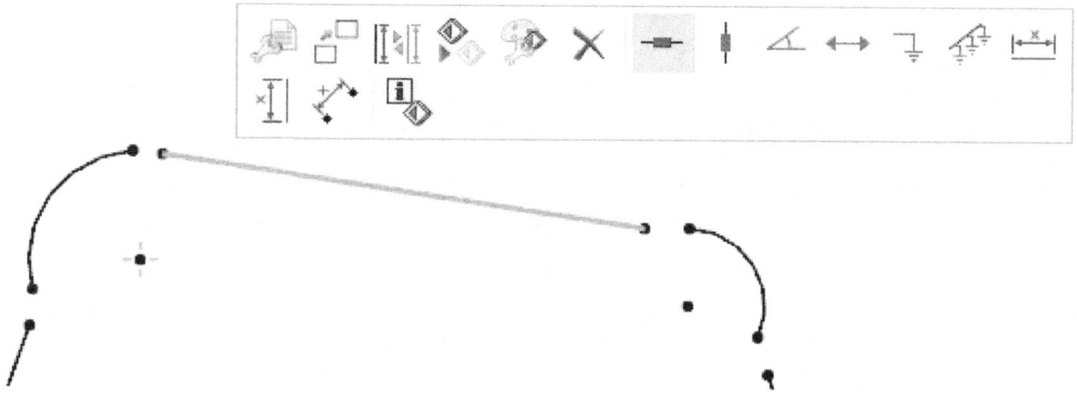

- Select the leftmost endpoint of the (now) horizontal line, and then click on the adjacent endpoint of the arc in the top left corner of the sketch. From the pop-up toolbar, find and select the *Coincident* constraint button.

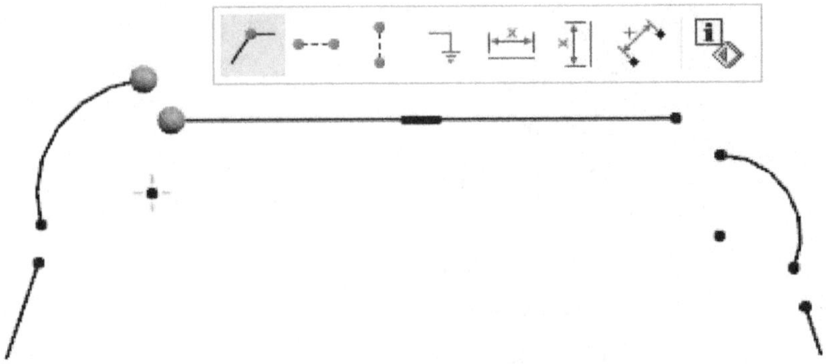

- Click on the horizontal line and then the arc to its left. Remember that (as you learned in Chapter 3.3.8) in order to select the arc and not its arc center, your selection ball must be touching the arc from the outside. From the pop-up toolbar that appears, select the *Tangent* constraint icon.

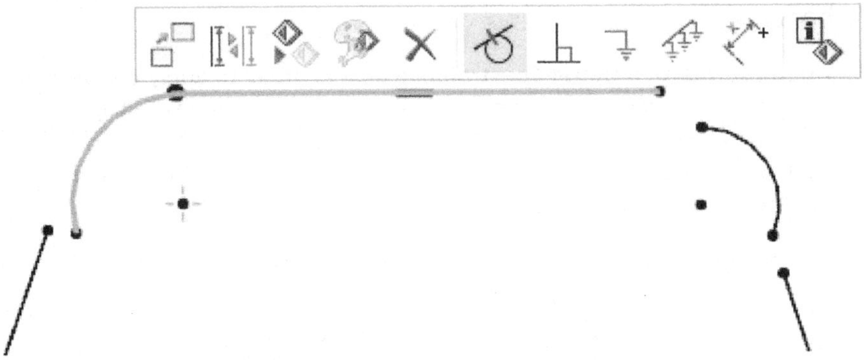

- Continue applying constraints from the pop-up toolbars in this fashion so that the lines are horizontal and vertical, and all arcs are tangent to the lines and have equal radius.

6.6.14 Make Symmetric

The **Make Symmetric** command applies a symmetry sketch constraint to curves in your sketch. The required inputs for **Make Symmetric** are a bit different than those of the other **Geometric Constraints,** so it has its own dialog. The **Make Symmetric** tool is found next to **Geometric Constraints** in the **Constraints** group on the **Home** tab.

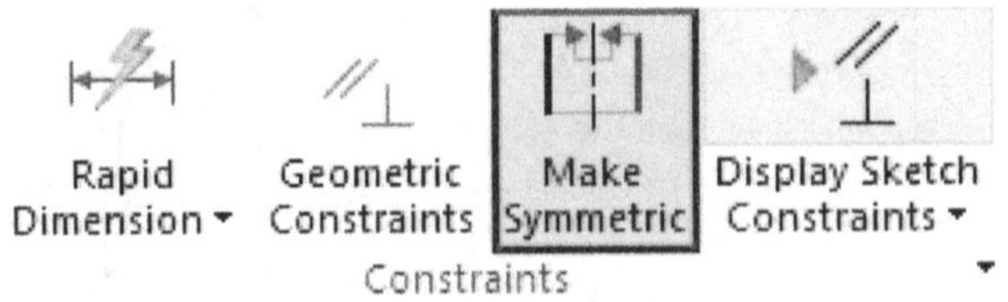

- Continue working with *"Geometric Constraints.prt"* and open the sketch called *"MAKE SYMMETRIC"*.

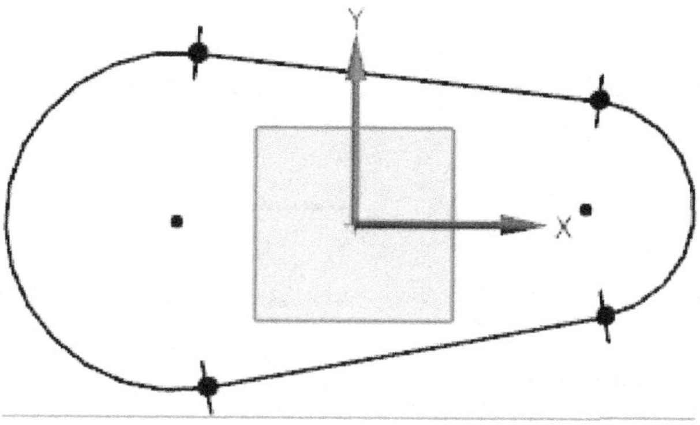

Sketch is fully constrained with 6 auto dimensions

- Select the **Make Symmetric** command, and choose the two arcs in the sketch for the *Primary Object* and *Secondary Object*. Note that your selection is *oriented* – make sure that the golden orientation arrows are pointing toward each other after you make your selection. For the **Symmetry Centerline**, select the Sketch Vertical Axis.

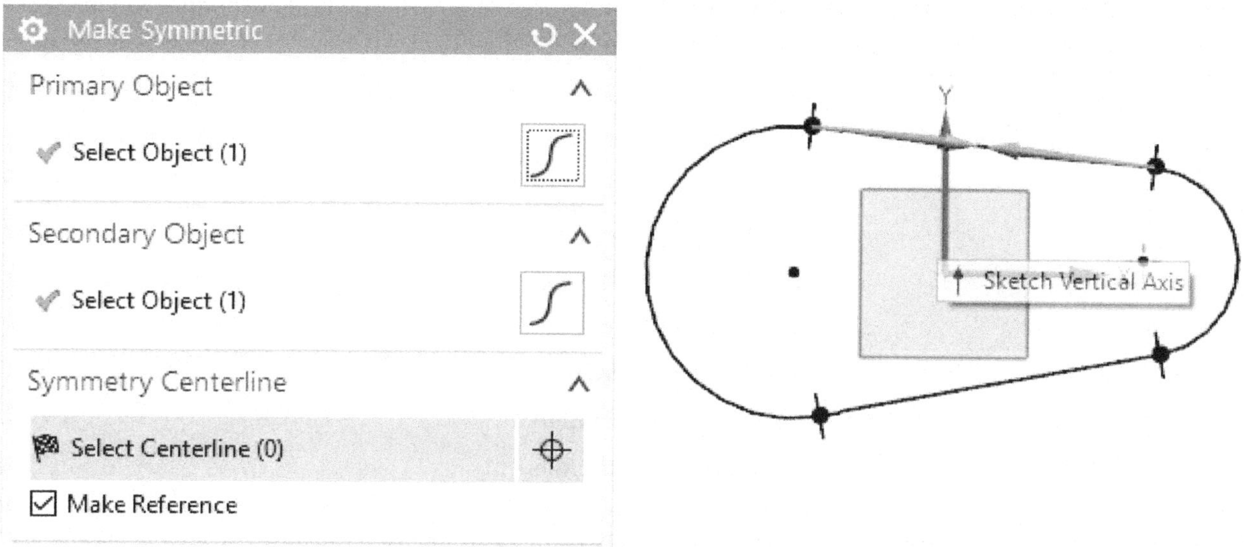

- Note that the one symmetry constraint removes three degrees of freedom! Now, the arc centers and endpoints are horizontally aligned.

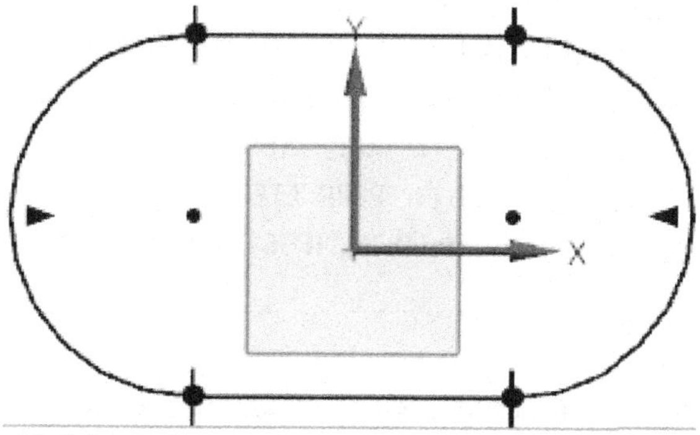

Sketch is fully constrained with 3 auto dimensions

- Apply a *Point on Curve* constraint between the arc center of one arc, and the sketch horizontal axis. The remaining two degrees of freedom can be removed by adding dimensions to the sketch – the topic of the next chapter!

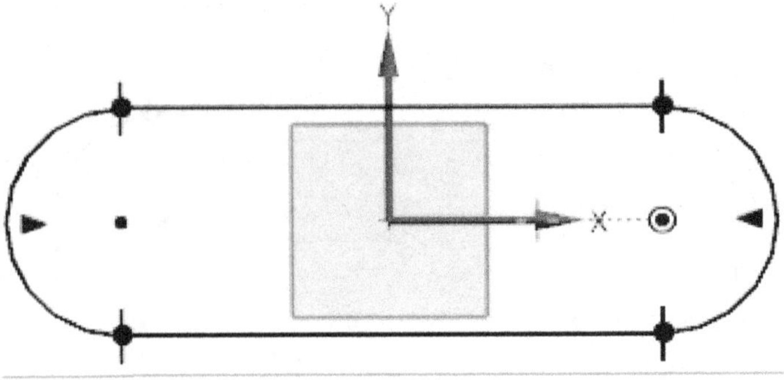

Sketch is fully constrained with 2 auto dimensions

- **Finish** the sketch, and **Save** your file!

6.7 Dimensions

The dimension tools in the **Sketch Task Environment** can be found on the **Dimension** drop-down menu, found in the **Constraints** group on the **Home** tab. Sketch dimensions are parameters that drive the behavior of entities within the sketch and which can be used in formulas outside the sketch. Sketch auto dimensions, in contrast, are driven, and do not give rise to parameters that can be used outside of the sketch.

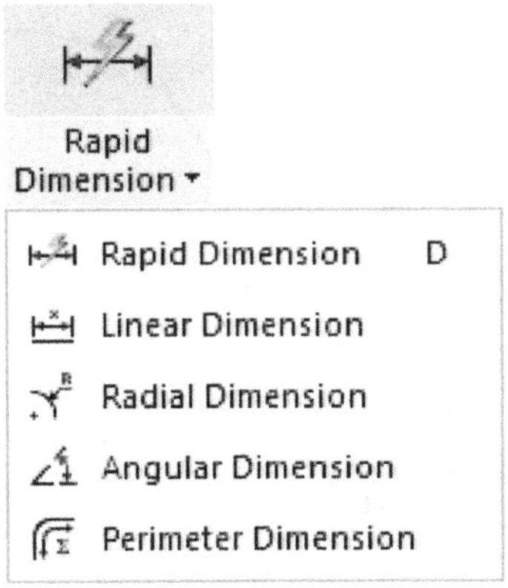

In the exercises below, you will need to make use of some sketches in a part file from the downloaded part files folder.

- Open *"Sketch Dimensions.prt"* from the downloaded part files folder. In the Model History, you will find five sketches.

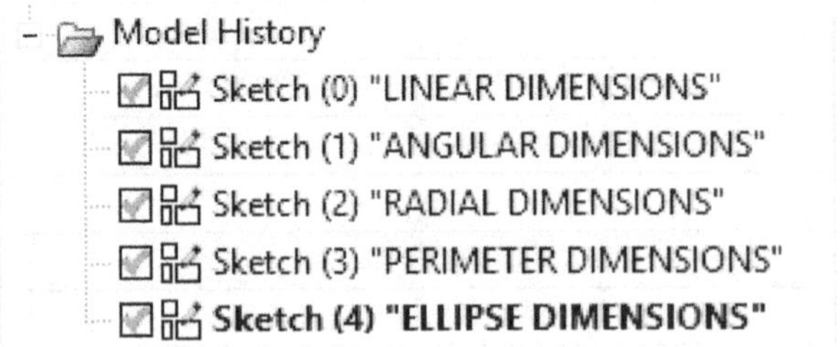

6.7.1 Linear Dimension

The **Linear Dimension** tool has a variety of **Measurement Methods** that produce a linear measurement between two objects.

- Open the sketch *"LINEAR DIMENSIONS"*.

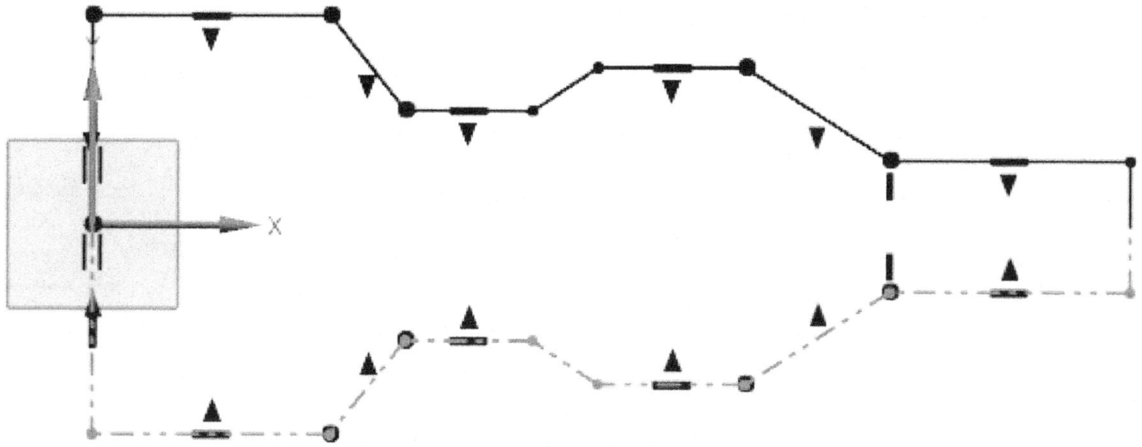

6.7.1.1 Horizontal

The *Horizontal* **Measurement Method** creates a linear dimension between two objects in the sketch horizontal direction. If you click on an existing horizontal line, the tool will infer that you want a dimension between its endpoints.

- Select the **Linear Dimension** tool and set the **Measurement Method** to *Horizontal*. Choose the leftmost and rightmost upper endpoints of the sketch as the *First Object* and *Second Object*. Before you can modify the value, you must click to place the dimension – once it is on the sketch, you modify its value. Enter a value of *35 in*.

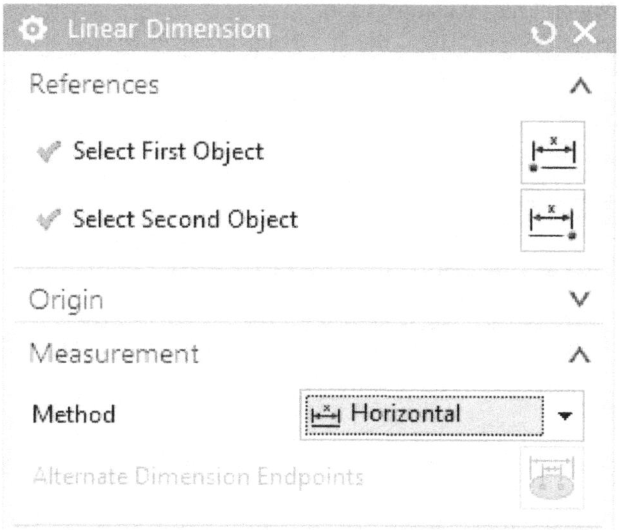

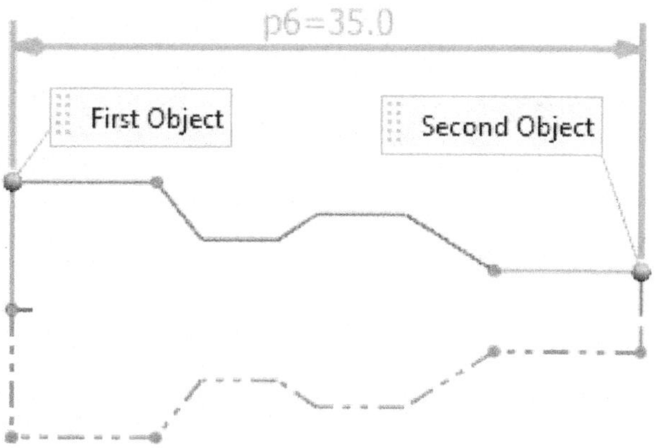

6.7.1.2 Vertical

The *Vertical* **Measurement Method** creates a linear dimension between two objects in the sketch vertical direction. If you click on an existing vertical line, the tool will infer that you want a dimension between its endpoints.

- Select the **Linear Dimension** tool and set the **Measurement Method** to *Vertical*. Choose the endpoints of the diagonal line shown below as the *First Object* and *Second Object*. Enter a value of *4 in*.

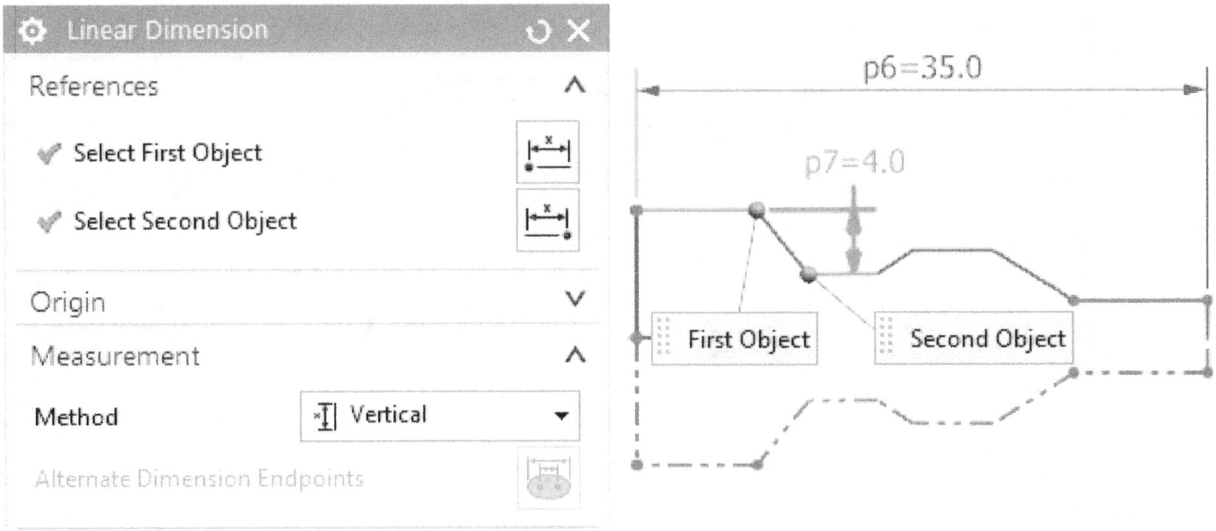

6.7.1.3 Point-to-Point

The *Point-to-Point* **Measurement Method** creates a dimension between two points in a sketch, along the direction of the line joining the two points.

- Select the **Linear Dimension** tool and set the **Measurement Method** to *Point-to-Point.* Choose the endpoints of the diagonal line shown below as the *First Object* and *Second Object*. Enter a value of *5 in*.

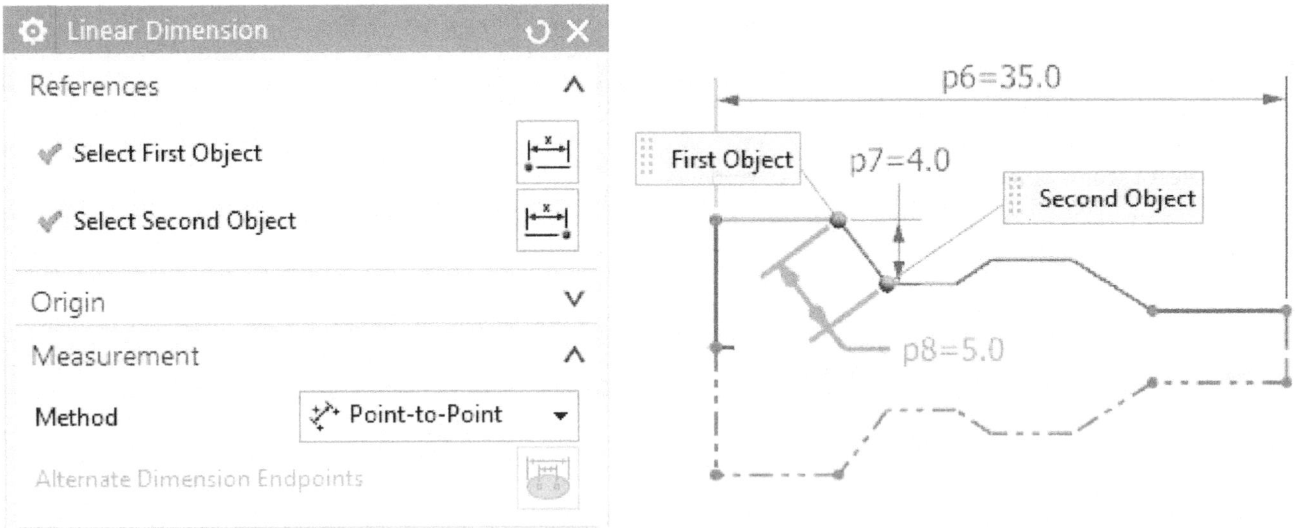

6.7.1.4 Perpendicular

The *Perpendicular* **Measurement Method** creates a dimension between a line and a point in a sketch, along the direction perpendicular to the line.

- Select the **Linear Dimension** tool and set the **Measurement Method** to *Perpendicular.* Choose the diagonal line shown below as the *First Object*, and the start point of the next diagonal line as the *Second Object*, as shown below. Enter a value of *3 in.*

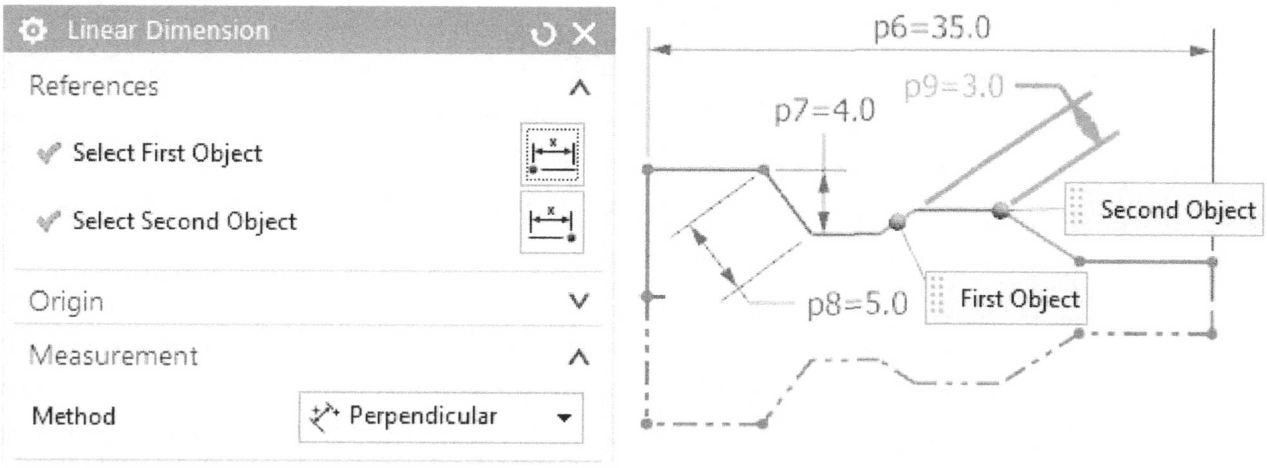

6.7.1.5 Cylindrical

The *Cylindrical* **Measurement Method** creates a dimension between two sketch points or a line and a point, which indicates the diameter of a cylindrical face obtained by revolving the sketch. If a point and a line are selected, the dimension is oriented using the same logic as the *Perpendicular* **Measurement Method**, and if two points are selected, the logic is the same as the *Point-to-Point* **Measurement Method**.

- Select the **Linear Dimension** tool and set the **Measurement Method** to *Cylindrical*. Choose the top left and bottom left endpoints of the sketch as the *First Object* and *Second Object*, as shown below. Enter a value of *16 in.*

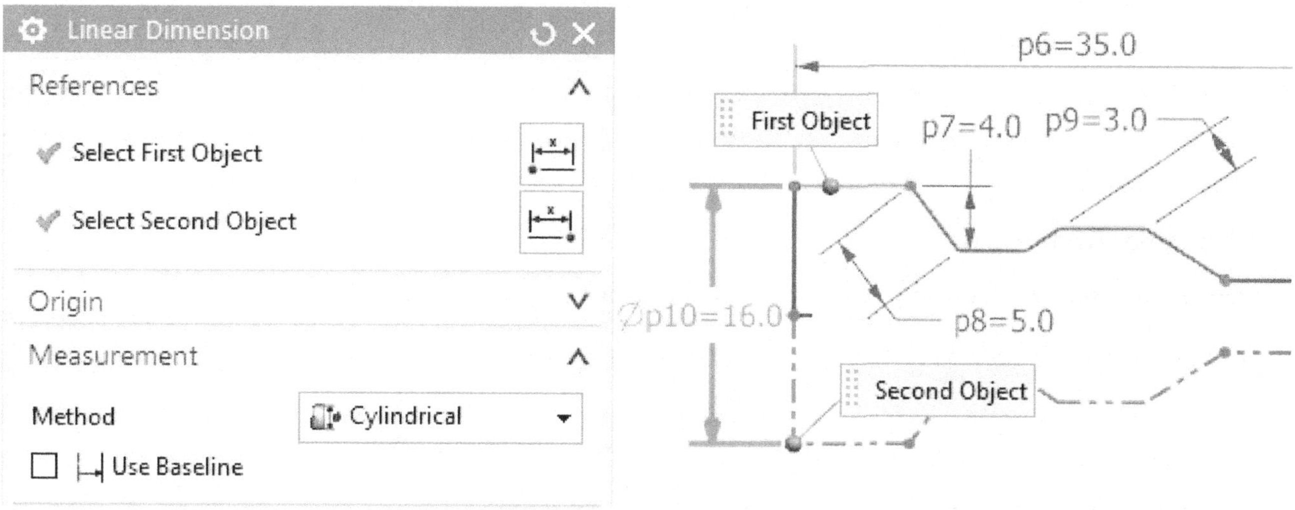

6.7.2 Rapid Dimension

The **Rapid Dimension** tool has all the same **Measurement Methods** as the **Linear Dimension**, **Angular Dimension**, and **Radial Dimension** tools. By default, the **Measurement Method** is set to **Inferred**, which is good enough for most practical purposes. The tool is smart, and depending on how you click it will figure out whatever you want, whether it be a linear dimension, an angular dimension, a perpendicular dimension, a radius or diameter, or anything in between! The only **Measurement Method** not covered by *Inferred* is *Cylindrical* – to place a *Cylindrical* dimension, you must either set the **Measurement Method** beforehand, or edit the dimension and change the type after placing it.

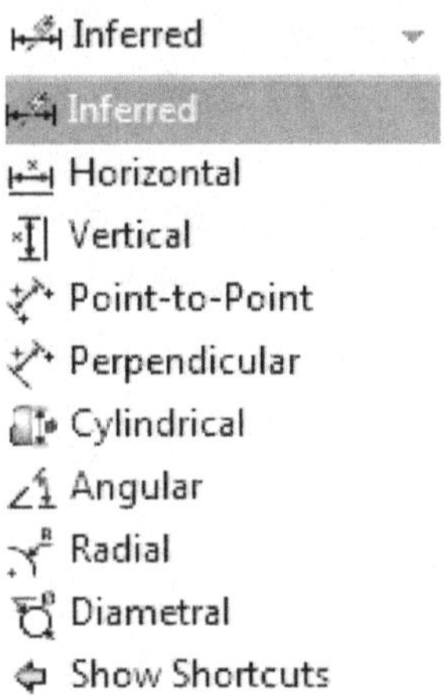

- Use the **Rapid Dimension** tool to add the linear dimensions shown below to your sketch. Can you create them all with the *Inferred* **Measurement Method**?

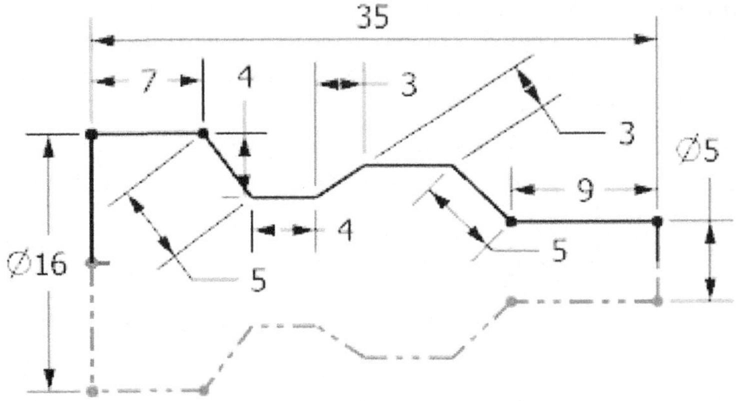

6.7.3 Angular Dimension

The **Angular Dimension** tool is used to constrain the angle between a sketch line and some other linear object.

- Open the sketch *"ANGULAR DIMENSIONS"*. Inside you will find a quadrilateral as shown below.

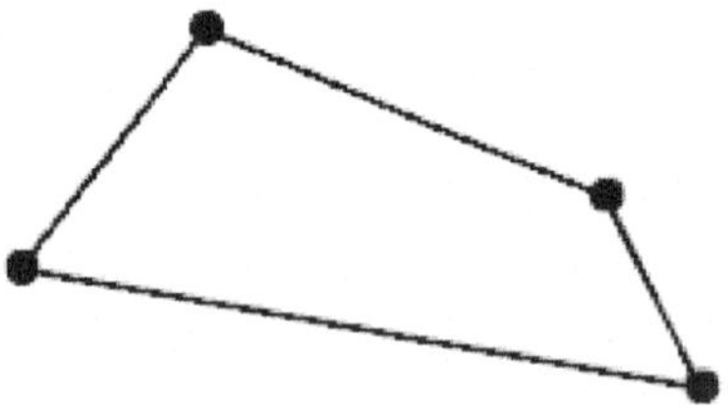

- Select the **Angular Dimension** tool, and choose the two lines shown below as the *First Object* and *Second Object*. Bring your cursor to the interior of the quadrilateral and click to place the dimension. Enter a value of *110 degrees*.

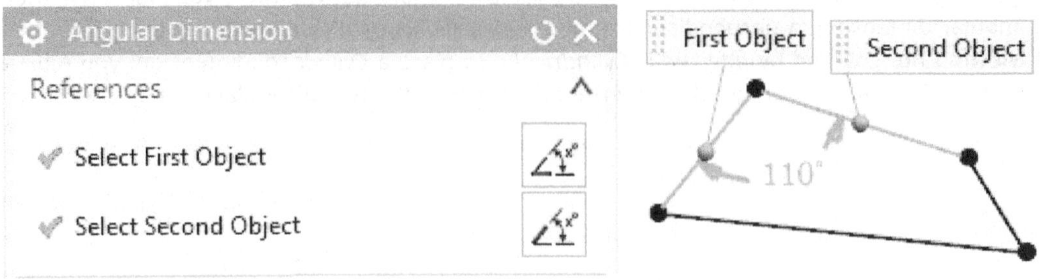

- The exact measurement you control with an **Angular Dimension** depends on where you bring your cursor when you place the dimension on the sketch.

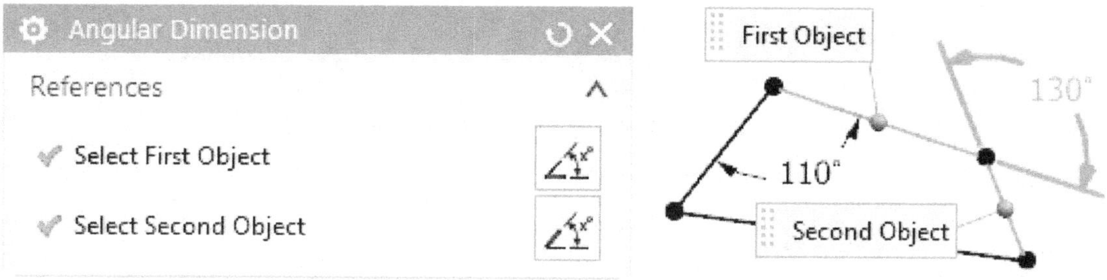

6.7.4 Radial Dimension

The **Radial Dimension** tool is used to create a dimension that controls the radius or diameter of an arc.

- Open the *"RADIAL DIMENSIONS"* sketch. Inside you will find two circles.

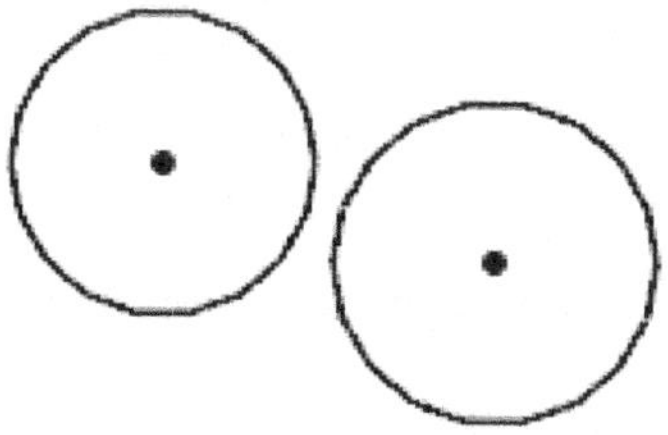

6.7.4.1 Radial

The *Radial* **Measurement Method** produces a dimension controlling the radius of an arc or circle.

- Select the **Radial Dimension** tool and set the **Measurement Method** to *Radial*, and choose the left circle. Enter a radius of *12 in*.

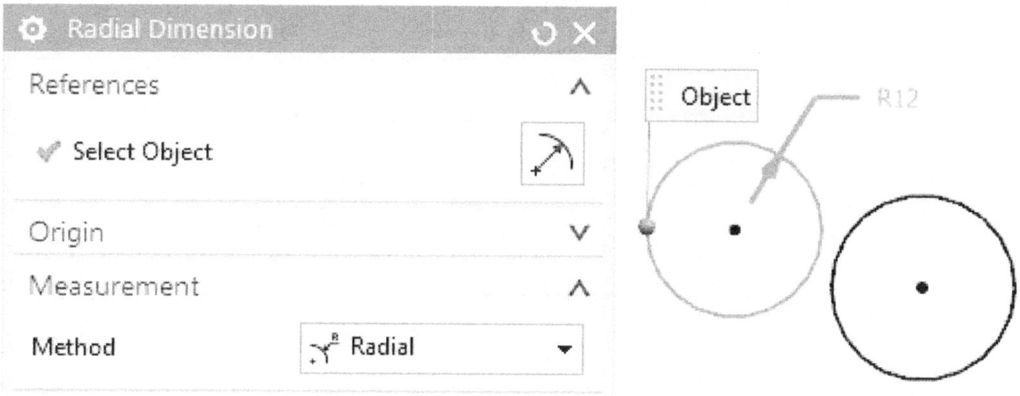

6.7.4.2 Diametral

The *Diametral* **Measurement Method** produces a dimension controlling the radius of an arc or circle.

- Select the **Radial Dimension** tool and set the **Measurement Method** to *Diametral*, and choose the left circle. Enter a diameter of *24 in*.

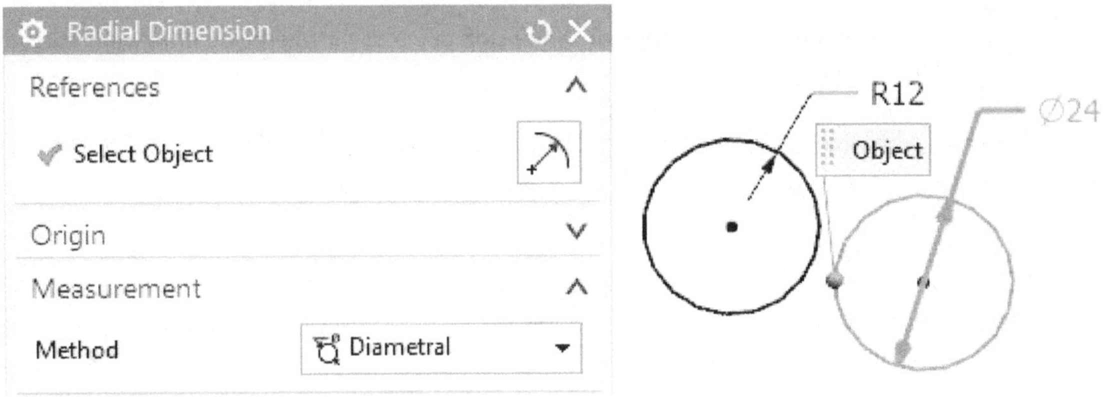

6.7.5 Perimeter Dimension

The **Perimeter Dimension** tool allows you to constrain the total length of a series of lines and arcs. It will not work on ellipses, conics, or splines. The **Perimeter Dimension** tool is found on the **Dimensions** drop-down in the **Constraints** group. Perimeter dimensions are the only kind of sketch dimensions that the **Rapid Dimension** tool cannot produce.

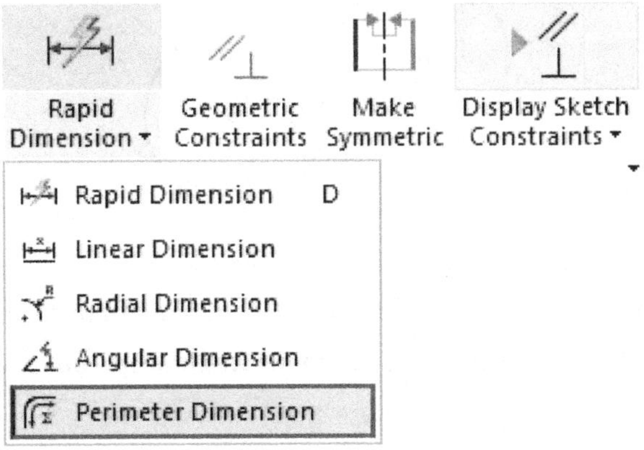

- Open the sketch called *"PERIMETER DIMENSIONS"*.

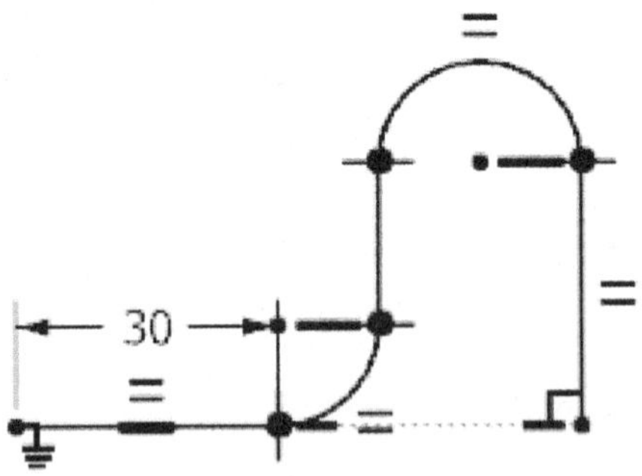

- Select the **Perimeter Dimension** tool and select all the curves in the sketch. Enter a value of *150 in* for the **Distance**.

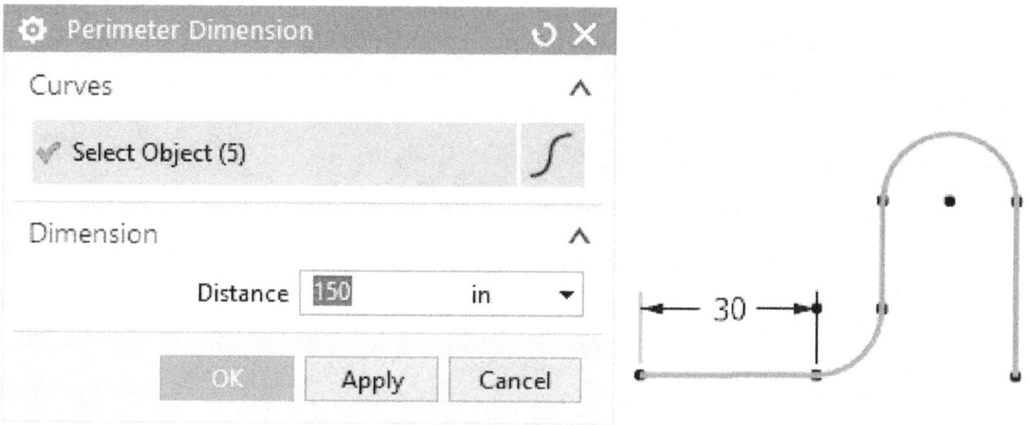

- The **Perimeter Dimension** created is unlike other sketch dimensions, as it does not appear in the Graphics Window. If you want to edit the value of the **Perimeter Dimension** or delete it altogether, the best way to do so is from the **Sketch Parameters** dialog that appears when you right-click on the sketch in the Model History and select **Edit Parameters** (first introduced in Chapter 6.2.3).

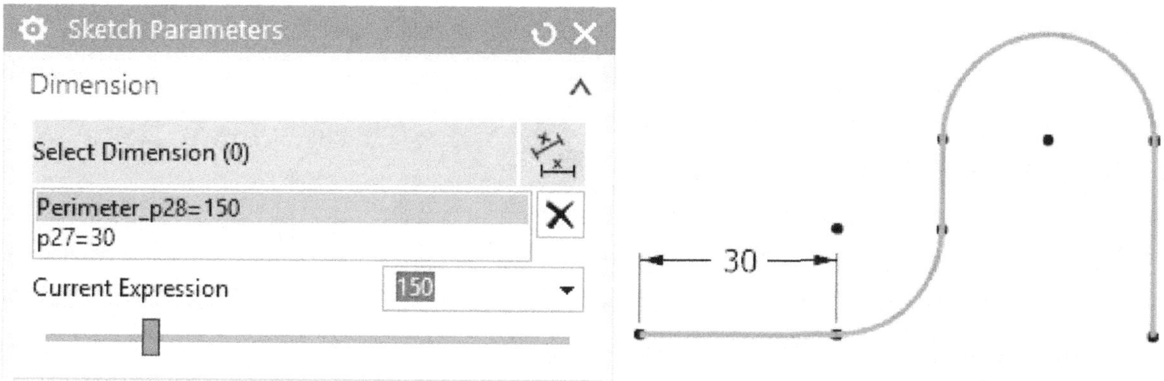

6.7.6 Parameterizing an Ellipse

You might recall from Chapter 6.5.7 that the values you enter into the **Ellipse** tool do not become driving parameters for the ellipse created. If ever need to create a sketch ellipse which is driven parametrically, the technique in the exercise below will prove useful.

- Continue working with *"Sketch Dimensions.prt"* and open the sketch called *"ELLIPSE DIMENSIONS"*.
- It is easy to dimension from the ellipse center to an extreme point along the ellipse semi-major or semi-minor axis. Make sure that you select a point on the ellipse as the *Second Object*.

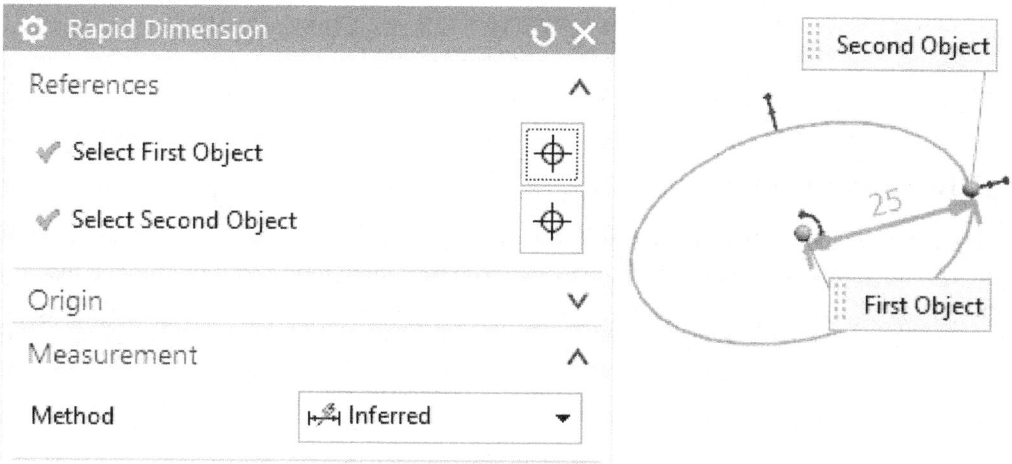

- Once there is a single dimension from the ellipse center to an inferred point on the ellipse, NX will not accept another such dimension. Instead, you will need to dimension to a point positioned along the ellipse's other axis. To create such a point, use the **Sketch Point** tool with the **Point Type** set to *Quadrant Point* and click near the extreme point along the other elliptical axis.

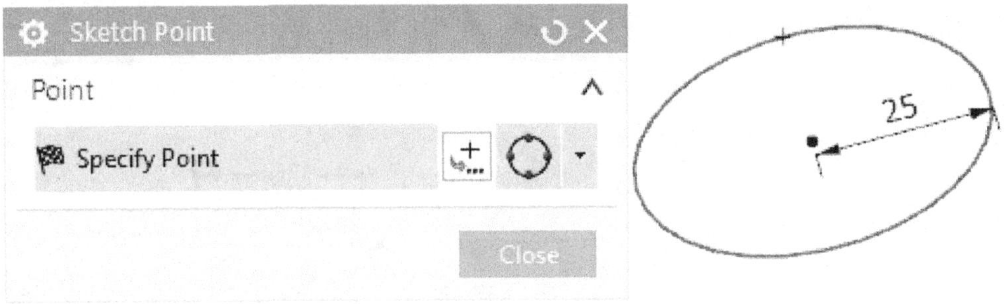

- Now you can create a dimension from the ellipse center to the sketch point entity with the help of the **Measurement Method** *Point-to-Point*.

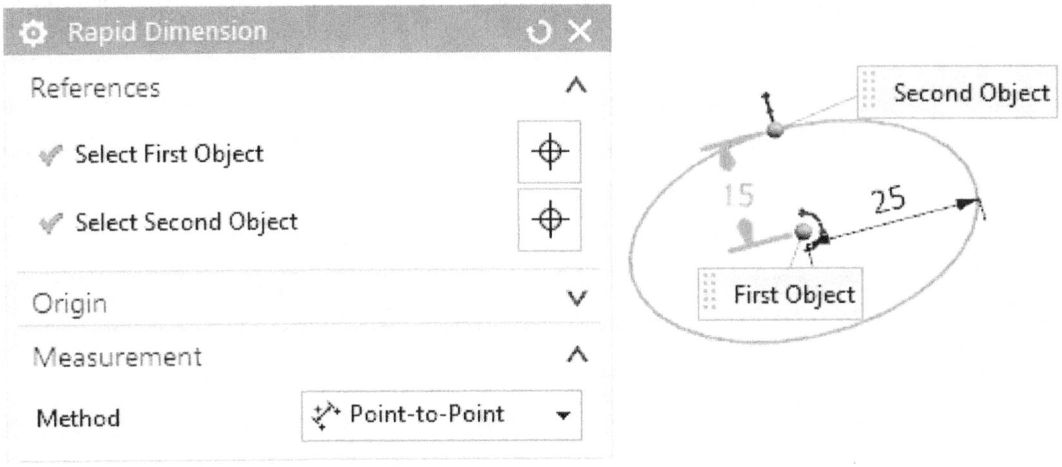

- If you need a parametrically-driven sketch ellipse in a model, you can now write formulas to drive these dimensions!

6.8 Managing Constraints and Dimensions

NX has several tools that help you to manage and edit the effects of existing sketch constraints and dimensions.

- Open the file *"Managing Constraints & Dimensions.prt"* from the downloaded part files folder. In the Model History, you will find three sketches, two of which are suppressed.

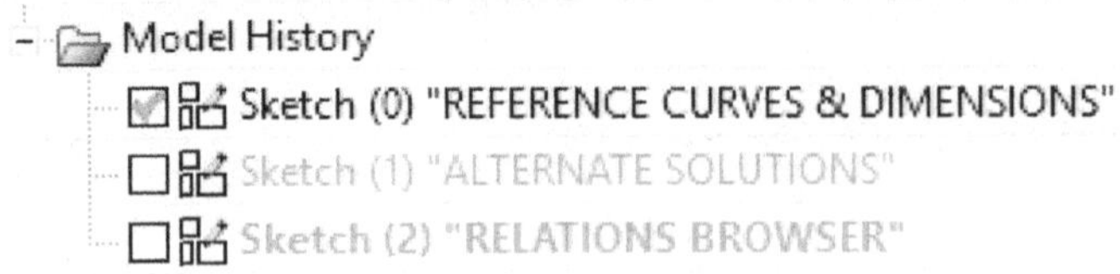

6.8.1 Convert To/From Reference

The **Convert To/From Reference** sketch tool allows the creation of reference curves and dimensions within a sketch. Reference curves show up as dotted lines in your sketch. They can be constrained and dimensioned just as regular sketch curves, but cannot be used by many operations, especially those outside of the sketch. Converting a dimension to reference results in that dimension becoming *driven* by the other constraints and dimensions governing the sketch.

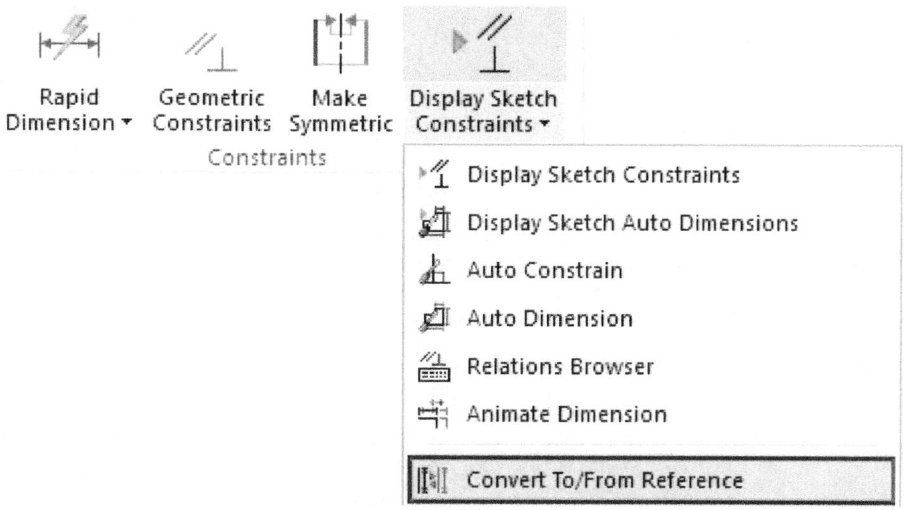

- Open the sketch *"REFERENCE CURVES & DIMENSIONS"*. Inside, you will find a slice of a circle. The intention here is to drive the size of the arc with a perimeter dimension rather than the radial dimension currently given, and for the two lines connecting the endpoints of the arc to its center to become construction lines.

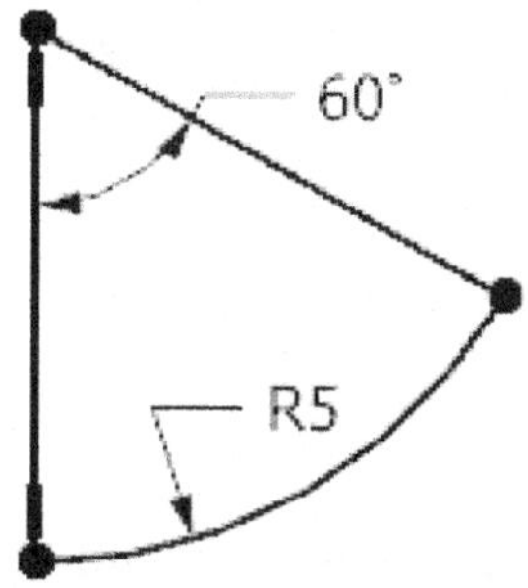

- Open the **Convert To/From Reference** tool, and select the two lines and the radial dimension. Click **OK** to perform the conversion.

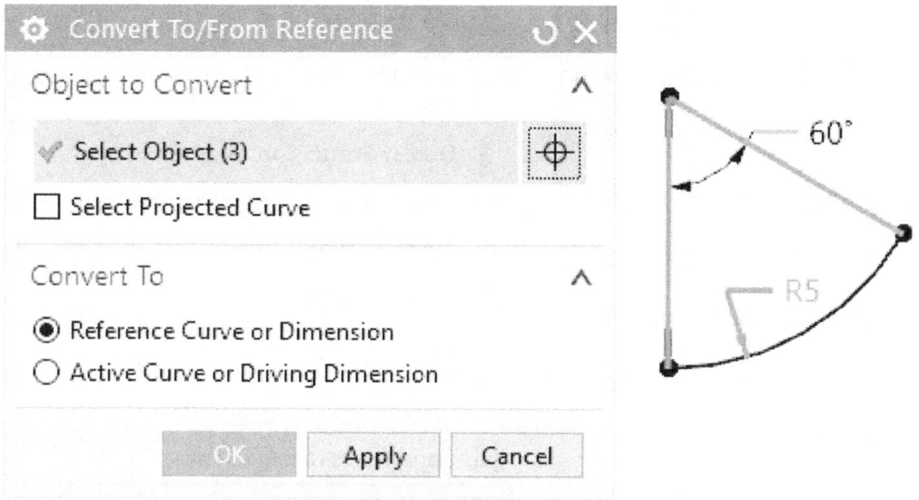

- Apply a **Perimeter Dimension** of *6 in* to the arc. Now the radial dimension simply reports the radius, subject to the other sketch constraints.

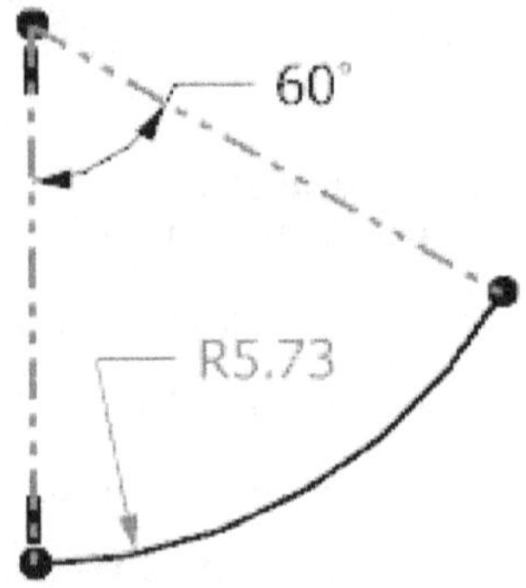

- Oftentimes, it is more convenient to convert sketch entities to reference/active by using the icon on the pop-up toolbar. Try it out!

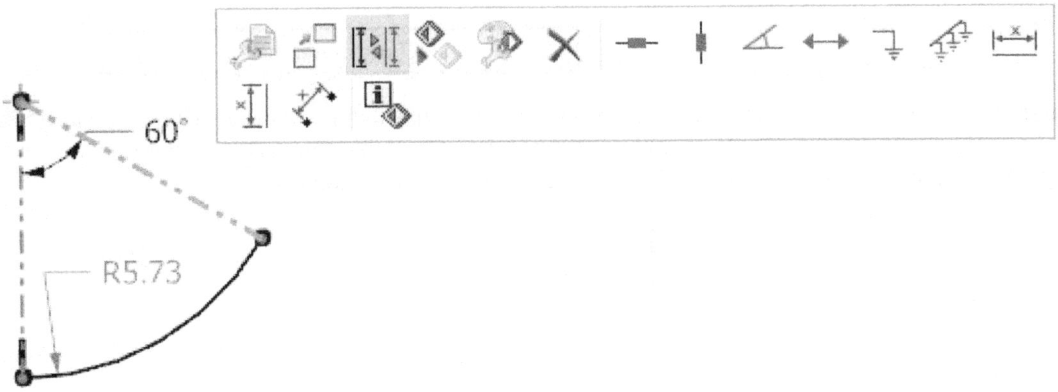

6.8.2 Alternate Solution

Since sketch constraints and dimensions lack timestamps, they are solved simultaneously starting from the current positions of all sketch entities. Sometimes NX will misinterpret your design intent and some dimensions or constraints will need to be reversed, or cycled.

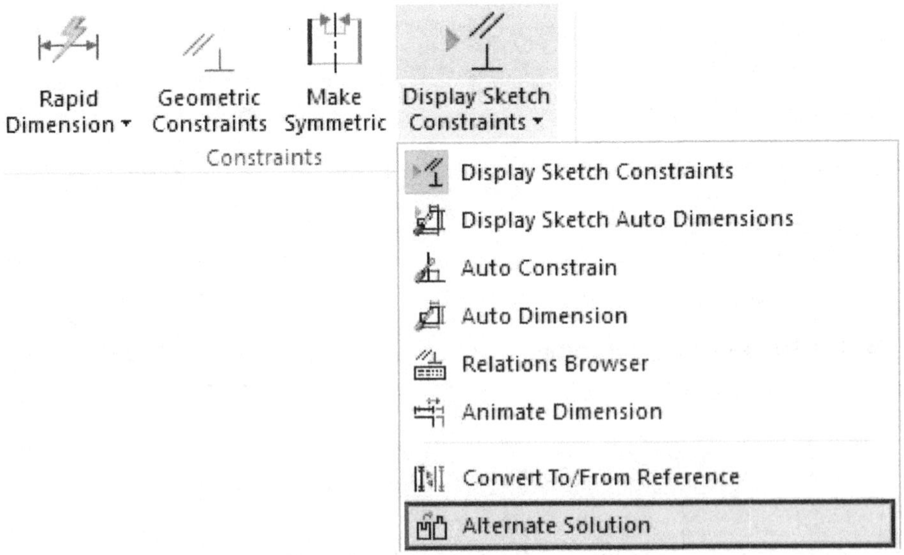

- **Unsuppress** and then open the sketch *"ALTERNATE SOLUTIONS"*.

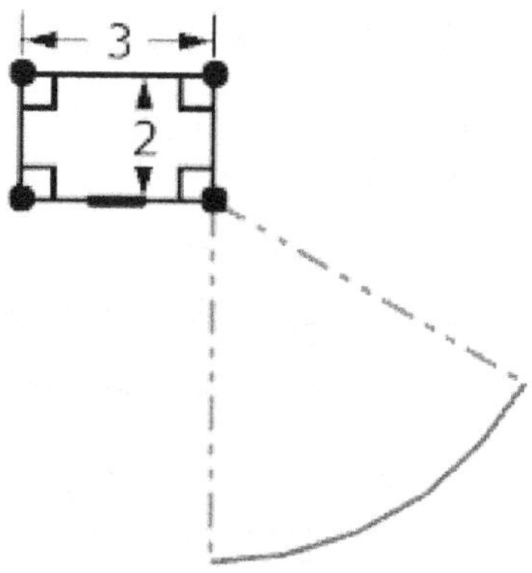

- Select the **Alternate Solution** tool and choose the horizontal dimension of *3 in.*

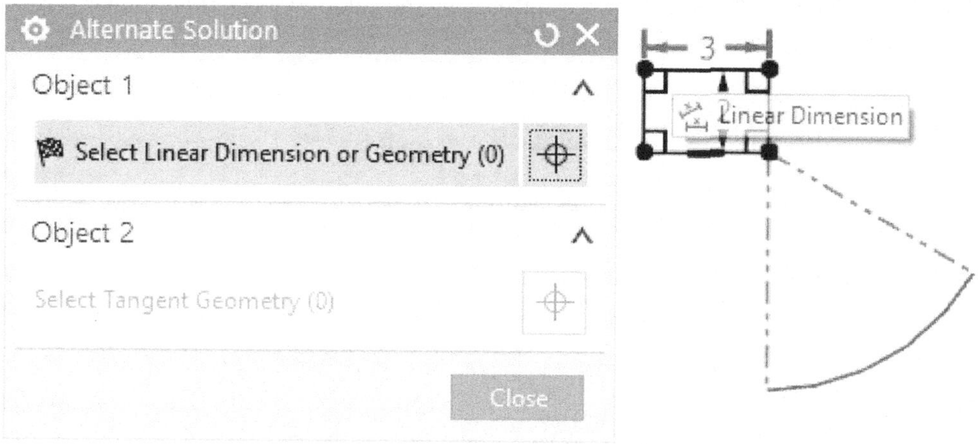

- In light of the existing *Coincident* constraint on the bottom left corner of the rectangle, the only other solution is for the left side of the rectangle to flip to become the right side!

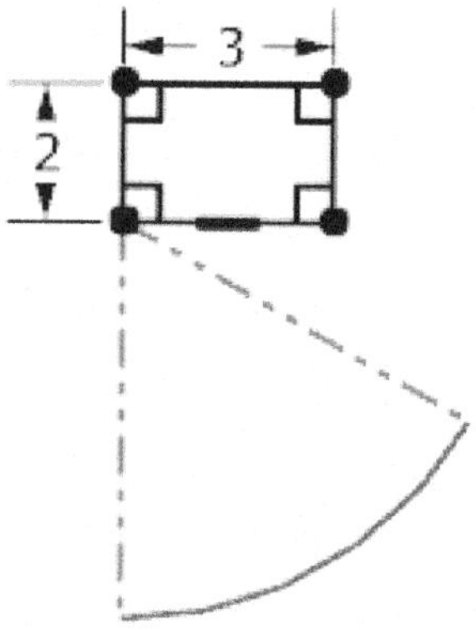

- In practice, you may find more complicated results that need to be alternated, but this simple example illustrates the basic effect of the tool!

6.8.3 Sketch Relations Browser

It's pretty easy to create conflicting sketch constraints and dimensions, especially because of the many inferred constraints NX will create in the sketch environment. When this happens, the fact that sketch entities lack timestamps can make it difficult to figure out which sketch constraints or dimensions are problematic and should be deleted. In older versions of NX, there was a rather clumsy tool called **Show/Remove Constraints** that allowed you to see which constraints conflicted with each other, but it was limited to only geometric constraints.

NX 11 has a new tool called the **Sketch Relations Browser** which enhances the capabilities of the old **Show/Remove Constraints** tool and allows you to deal with problematic sketch dimensions as well.

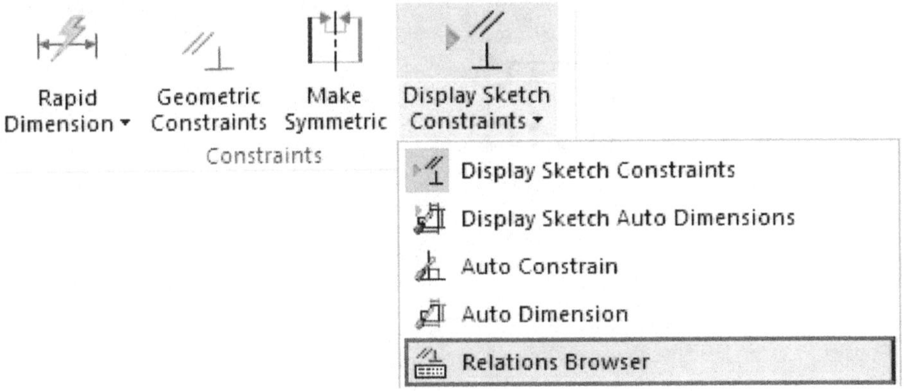

- **Unsuppress** and open the sketch *"RELATIONS BROWSER"*. In this sketch, there are a number of sketch constraints and dimensions that are in conflict with each other, as indicated by the color-coding – they appear in red.

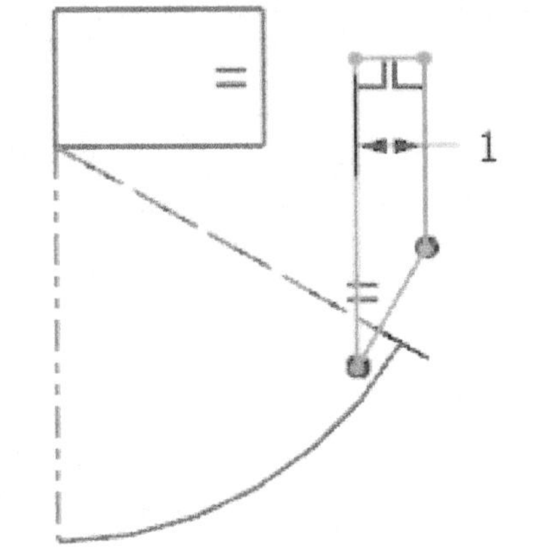

Sketch contains conflicting constraints

- Open the **Sketch Relations Browser**. By default, the **Top-level Node Objects** displayed are *Curves*, and the **Status** column will tell you which ones are subject to problematic constraints.

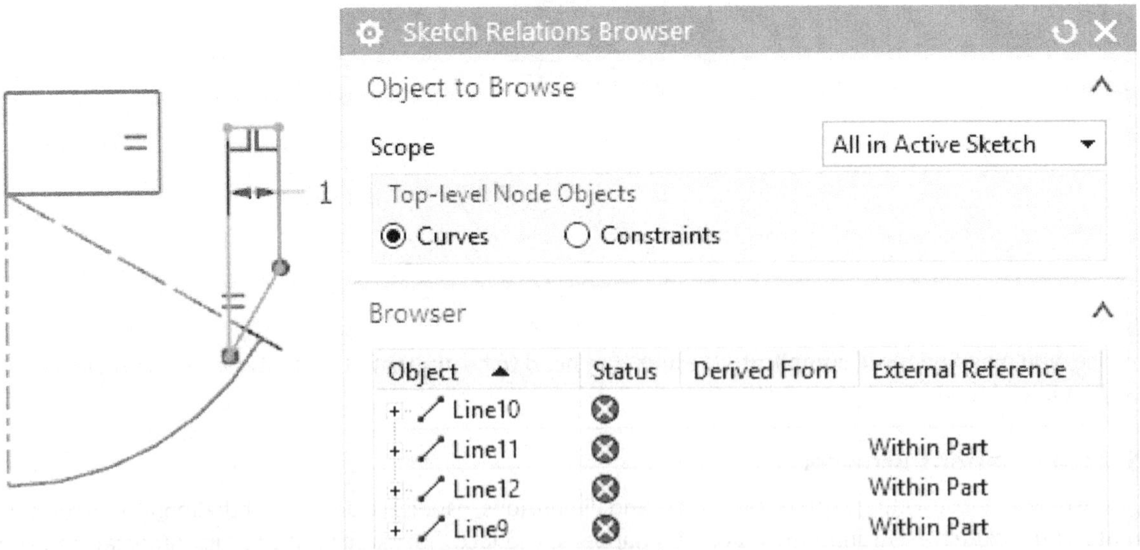

- You can toggle the **Top-level Node Objects** to *Constraints* and the list will reorganize so that constraints appear at the top, and now the **Status** column tells you which conflict with others. Select and right-click on the two *Perpendicular* constraints, and select **Delete**.

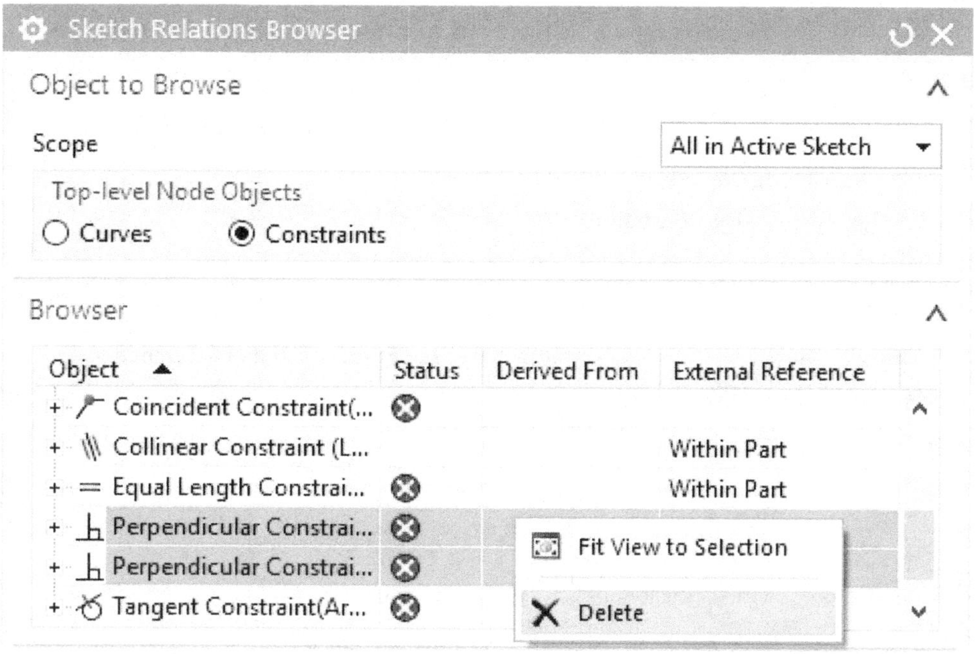

- These constraints conflicted with a coincident constraint and the horizontal dimension. Now that they are gone, your sketch is no longer overconstrained!

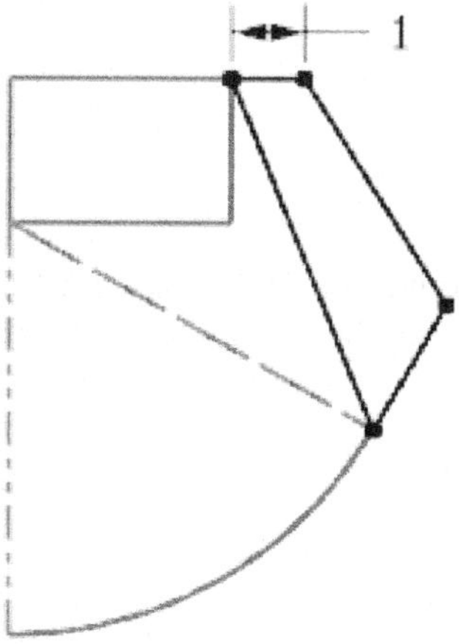

- Another handy feature of the **Sketch Relations Browser** is the ability to convert sketch curves and dimensions to and from reference from within the browser!

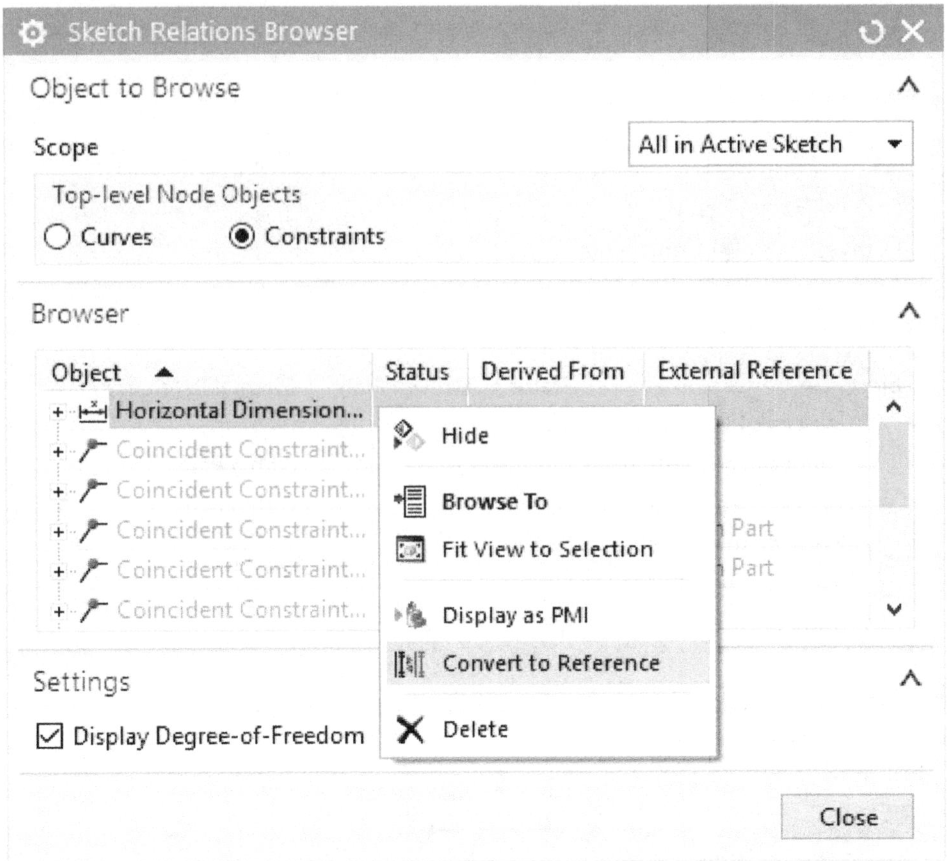

6.9 Derived Sketch Curves

Oftentimes the most efficient way to create a complex sketch is to create curves from curves. The **Curve** group contains a number of tools that create curves derived from other curves, and the majority of these tools create constraints that ensure that if the parent curves change, the derived curves will maintain the correct relationship to the parent curves.

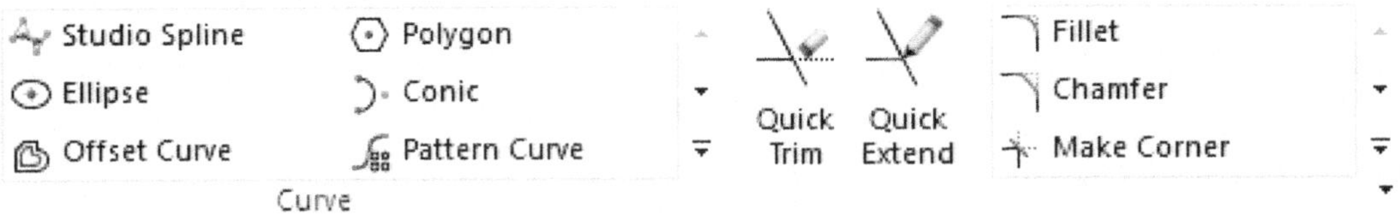

6.9.1 Offset Curve

The **Offset Curve** tool creates a set of curves which are uniformly offset from a set of input curves. The offset curves are created with an "Offset Constraint" which ensures that the offset curves maintain their relationship to the input curves if the input curves change.

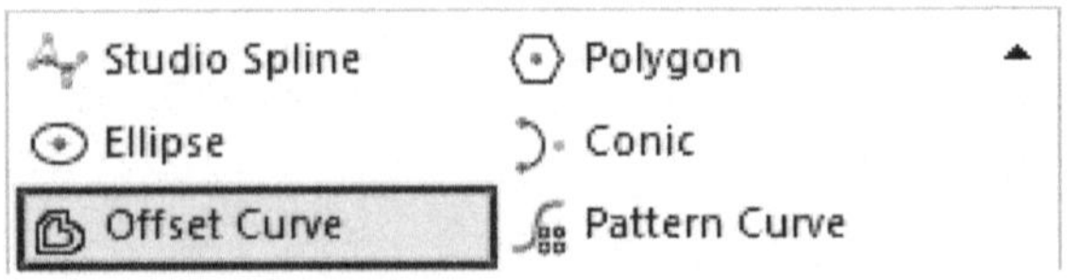

- Create a new part file (**Units**: *Inches*) called *"Derived Sketch Curves.prt"*, and place it in *"C:\My NX 11 Files"*.
- Create a sketch on the XY plane of the datum coordinate system, and position a circle of diameter *6 in* at the sketch origin. The sketch coordinate system is hidden from the screenshots below.
- Select the **Offset Curve** tool, and use it to offset the circle with the parameters shown below.

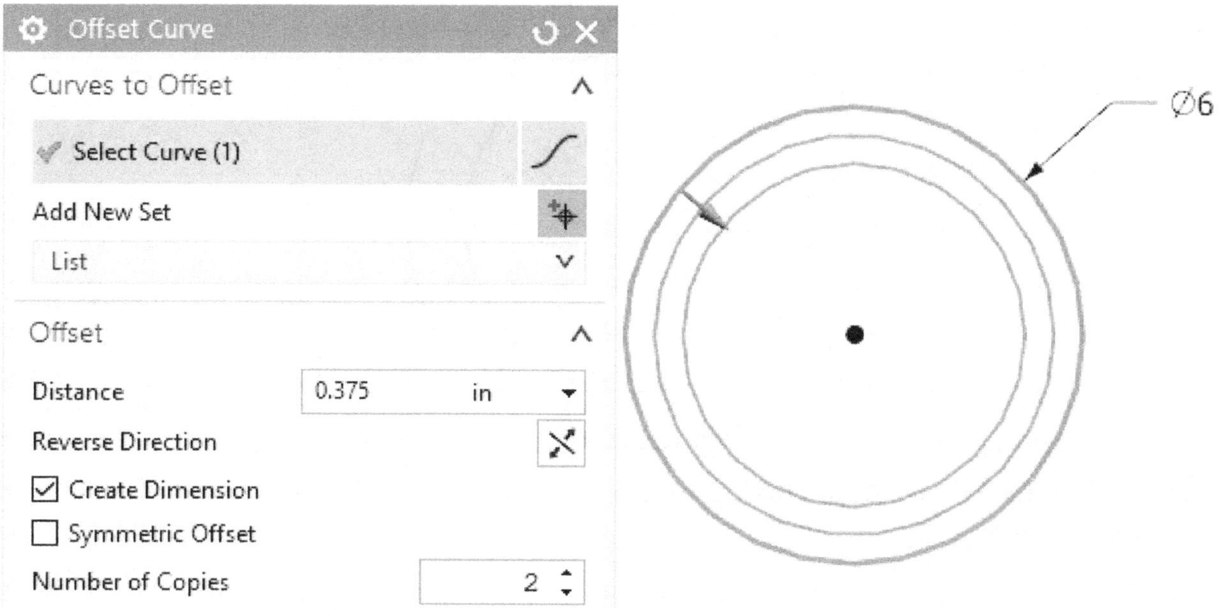

- Because the *"Create Dimension"* checkbox was marked, the two interior offset circles come with offset dimensions. Use **Convert To/From Reference** to convert the middle circle to reference.

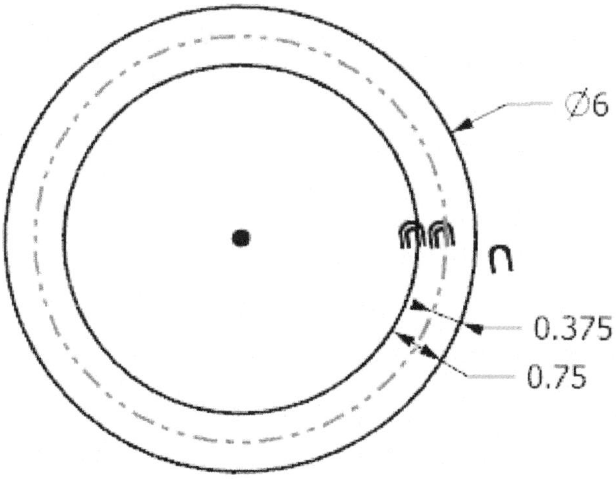

- The "nested U's" constraint symbol is the *Offset Constraint* – you can double-click on this to edit the parameters for the offset curves! If you delete the constraint, the offset curves become unconstrained, but are not deleted.

6.9.2 Pattern Curve

Patterning sketch curves is another great way to create smart sketches. Like **Offset Curve**, **Pattern Curve** will produce copies of an input curve subject to a "Pattern Constraint" which you can double-click to edit the pattern.

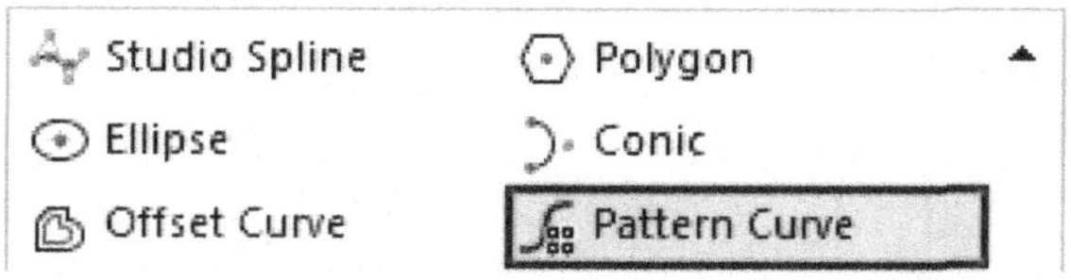

- Position a circle of diameter *0.25 in* with its center constrained to the reference circle and the sketch vertical axis.

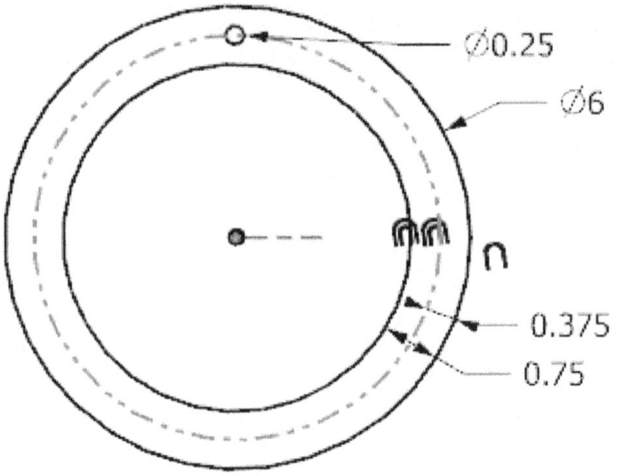

- Select the **Pattern Curve** tool, and use it to pattern the small circle around the center of the larger circles with the parameters shown below.

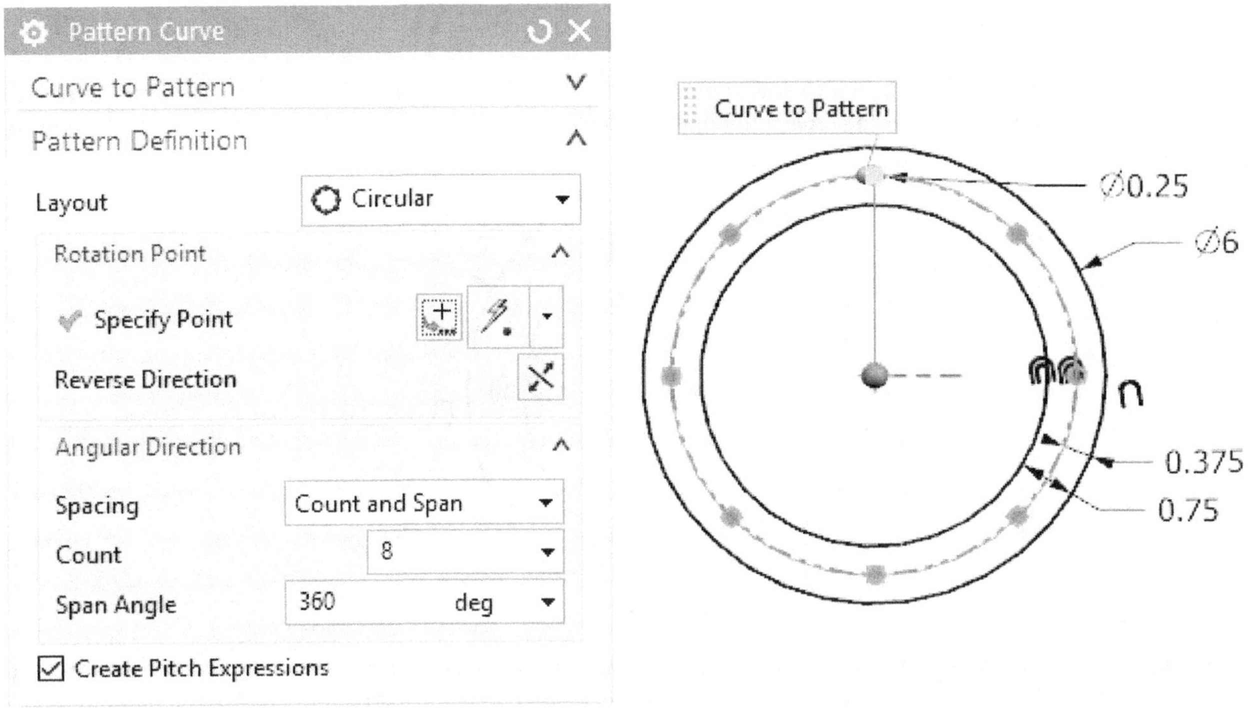

- The *Pattern Constraint* appears as a circular pattern of dots, and you can double-click on it to edit the pattern. Note: if you delete the constraint, the pattern members become unconstrained, but are not deleted.

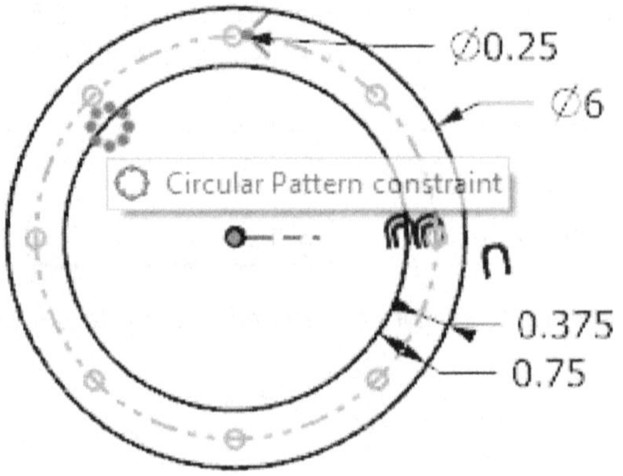

6.9.3 Intersection Point

It often happens that your sketch curves will need to be constrained to curves that pass through the sketch plane transversely. In order to constrain your sketch curves to these other curves, you usually need to create intersection points in the sketch plane, where the transverse curves meet the plane. The **Intersection Point** tool creates these "smart" points, which will follow the parent curves if edits are made.

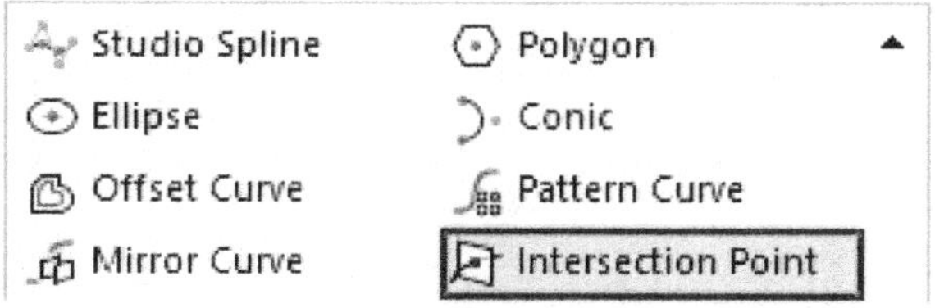

- Create a new sketch in the YZ plane of the datum coordinate system.
- Use the **Intersection Point** tool to create a point where the smallest circle in the first sketch intersects the YZ plane, in the positive Y direction. You might have to use the **Cycle Solutions** button to alternate between the two intersection points.

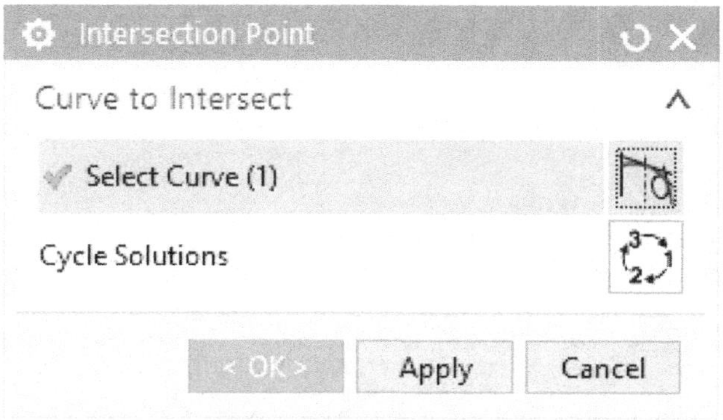

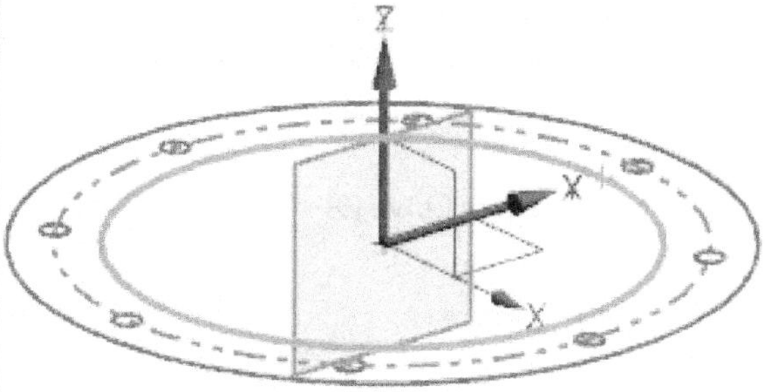

6.9.4 Mirror Curve

Mirroring curves is yet another procedure for reducing the total amount of individual constraints needed in a sketch. Like **Offset Curve** and **Pattern Curve**, the **Mirror Curve** command produces smart copies that are associatively controlled by a "Mirror Constraint".

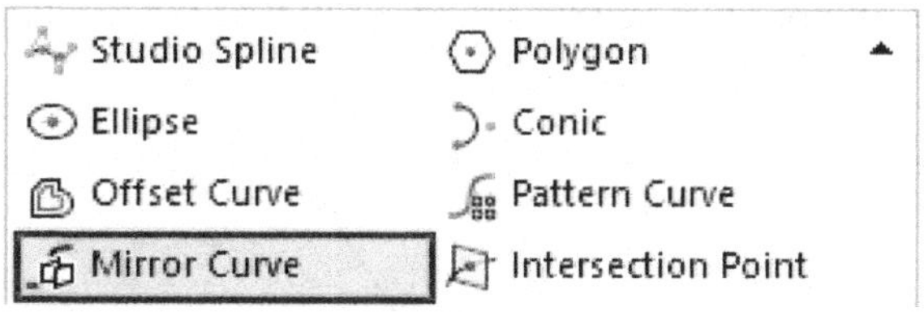

- Continue working on the sketch in the YZ plane, and create the contour shown below using the **Profile** tool. Make sure that the bottom right point is coincident with the *Intersection Point*.

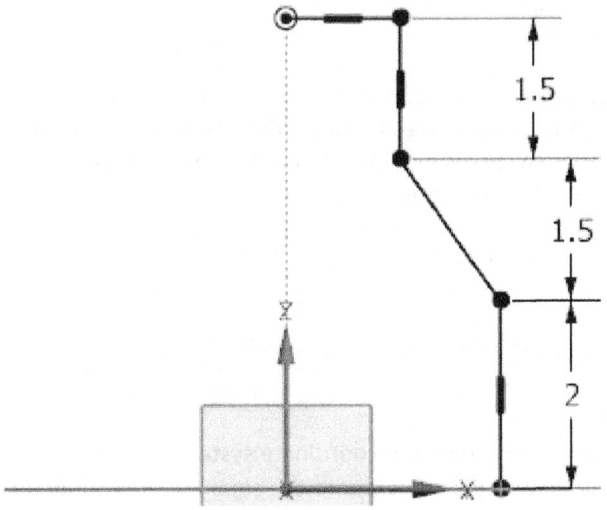

- Select the **Mirror Curve** tool. Mirror the lower three curves about the horizontal line.

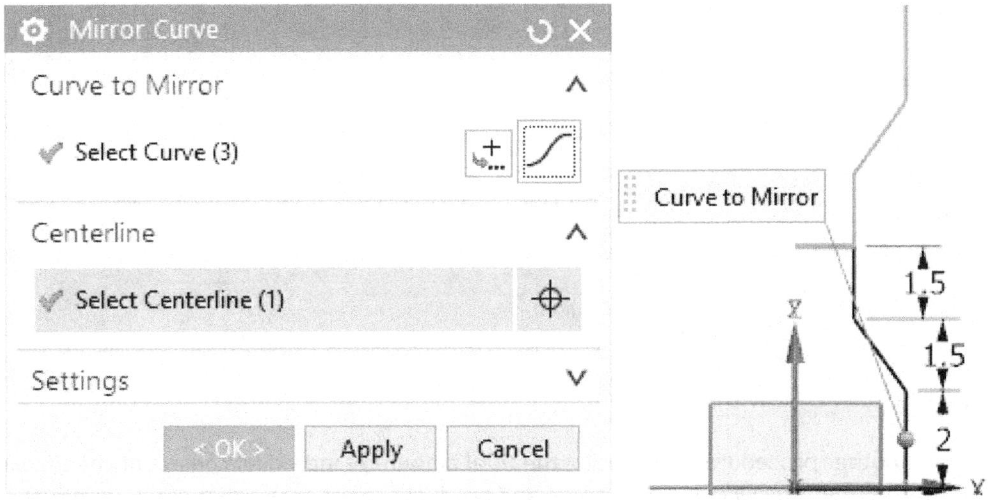

- There is a checkbox in the **Mirror Curve** dialog that ensures that the centerline is converted to reference. As with **Offset Curve** and **Pattern Curve**, this produces a *Mirror Constraint* that can be edited or deleted.

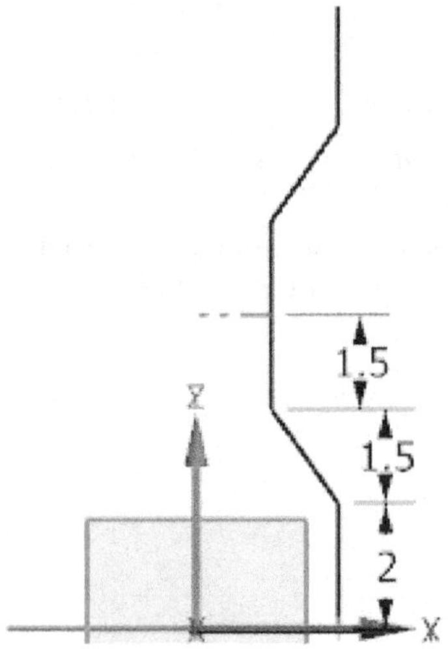

6.9.5 Project Curve

Projecting curves from outside of the sketch plane into the sketch plane so that they become modifiable sketch entities is another common workflow. The **Project Curve** tool can be used to project curves from other sketches, 3D curves, or edges of bodies into the sketch plane, where they become *"recipe curve"*[3] to which other curves can be constrained.

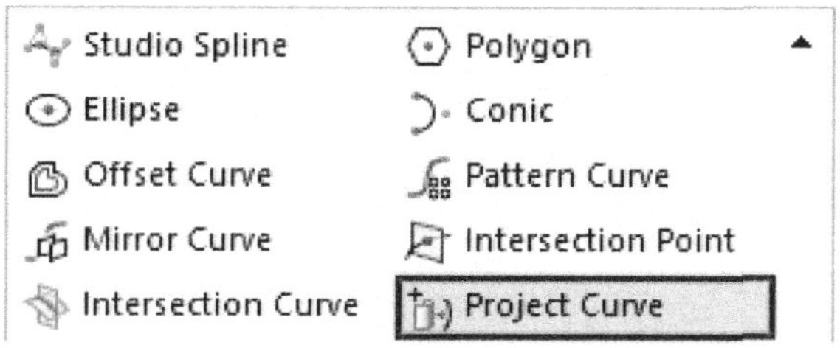

- Create a new sketch at the end of the path constructed in the second sketch.

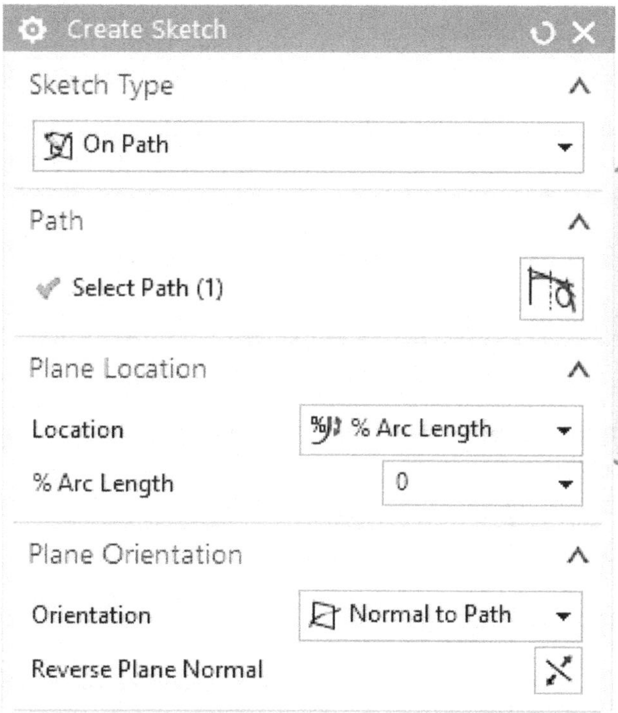

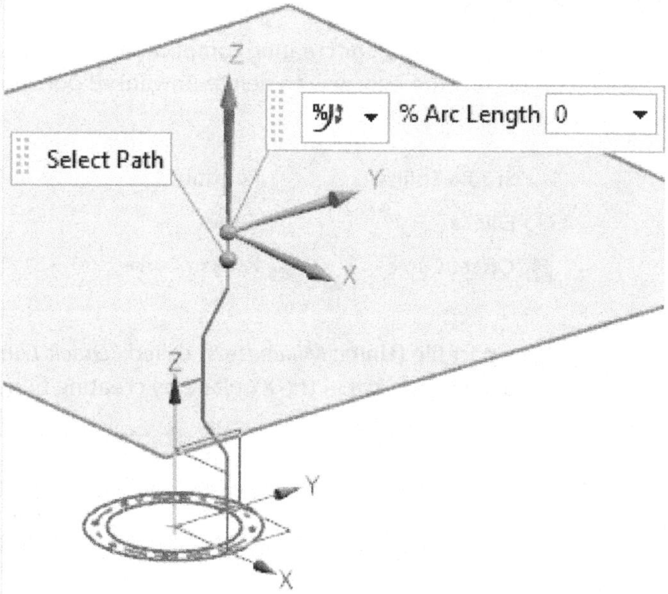

- Select the **Project Curve** tool and select all possible curves from the first sketch. Note that the reference circle cannot be projected.

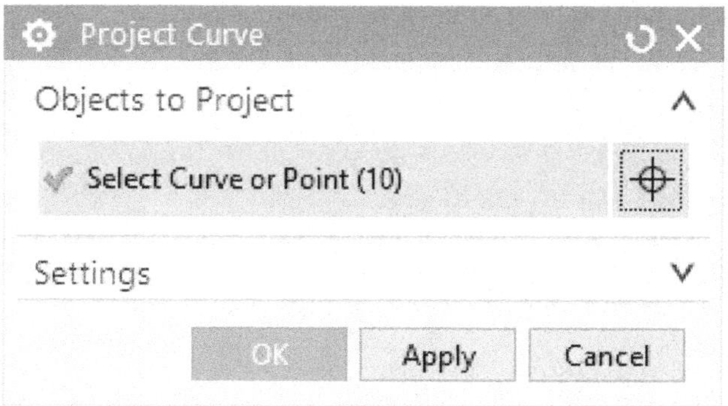

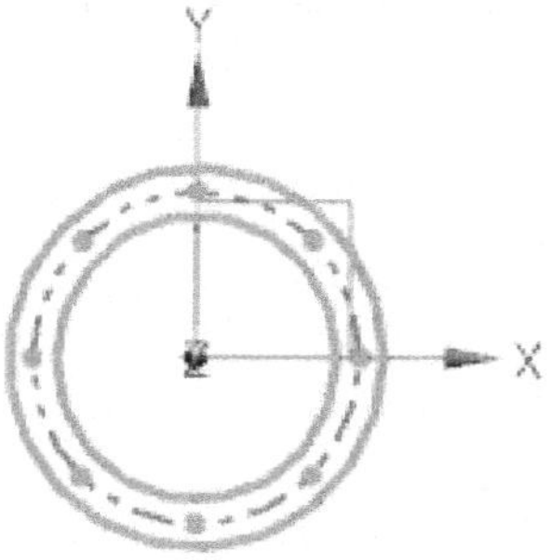

---

[3] *"Recipe curves"* are special sketch curves that result from **Project Curve** and **Intersection Curve** (not used in this text).

- Recipe curves are color-coded so that you can tell them apart from ordinary sketch curves. To edit the projected curves, you can double-click on any one of them and the **Project Curve** dialog will reappear for editing.

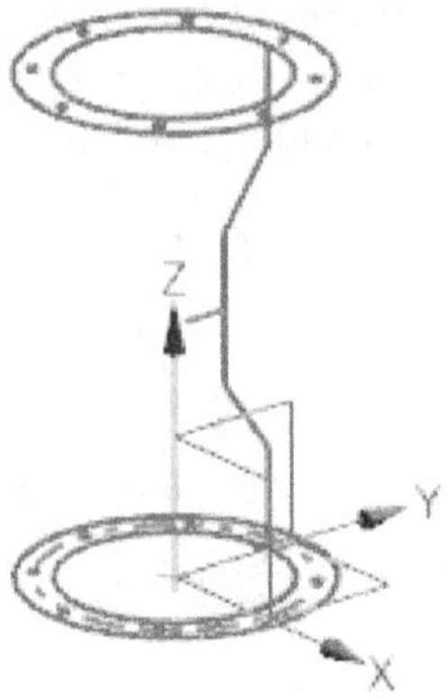

- **Finish** your sketch and **Save** your part file!

6.9.6 Quick Trim

The **Quick Trim** tool is helpful for creating complicated sketches by combining rudimentary sketch operations and deleting the excess curves. It is the only way to delete unwanted portions of sketch curves between their intersections with other sketch curves.

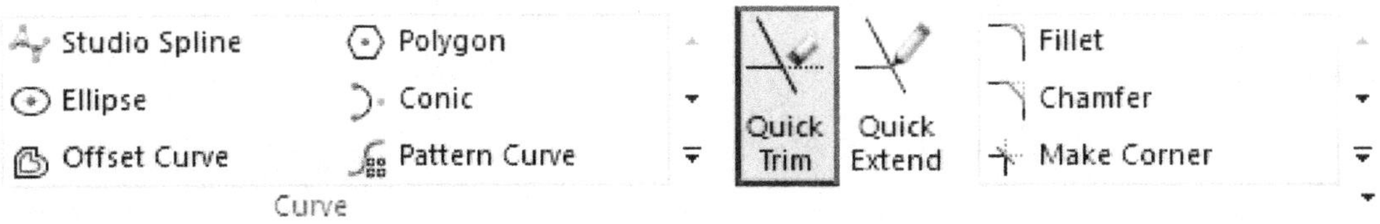

- Create a new part file (**Units**: *Millimeters*) called *"Quick Trim & Fillet.prt"* and place it *"C:\My NX 11 Files"*.
- Create the following sketch in the XY plane by creating two rectangles.

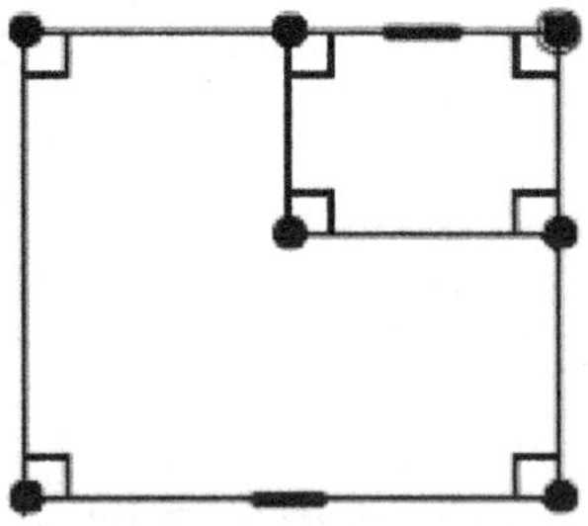

- Select the **Quick Trim** tool. You can click-and-hold-and-drag your cursor through sketch curves to delete portions between intersecting sketch curves. The cursor will become a pencil, as shown below.

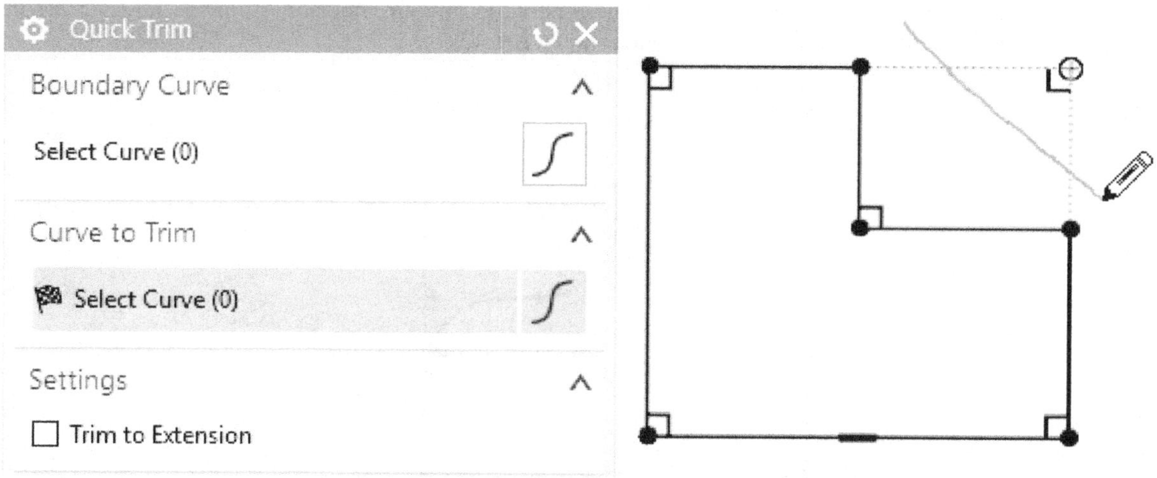

- You can also just click on the portions of curves that you want to delete, but we thought you'd like this neat shortcut!

6.9.7 Fillet

The **Fillet** tool is used to round sharp corners on sketches (much like an actual fillet rounds sharp edges on an object). Where two lines intersect at their endpoints, an arc is placed, tangent to both lines, and the excess portions of the lines are trimmed.

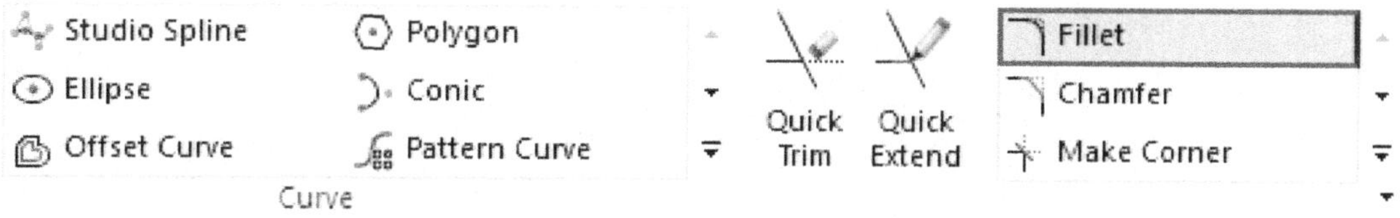

- Select the **Fillet** tool. Like the **Quick Trim** tool, when you click-and-hold-and-drag, the cursor will become a pencil which you can use to draw through the lines you would like to join with an arc. Click and hold and drag as shown below.

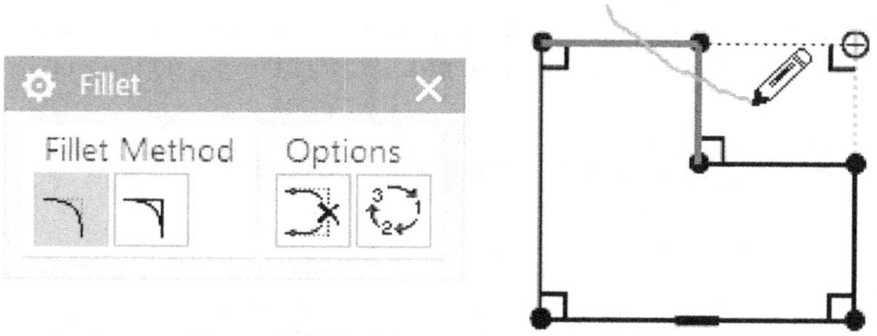

- The other way to use the **Fillet** tool is to sequentially click the lines you wish to blend. Click the vertical line segment and then the horizontal one, and then click again to finalize the arc!

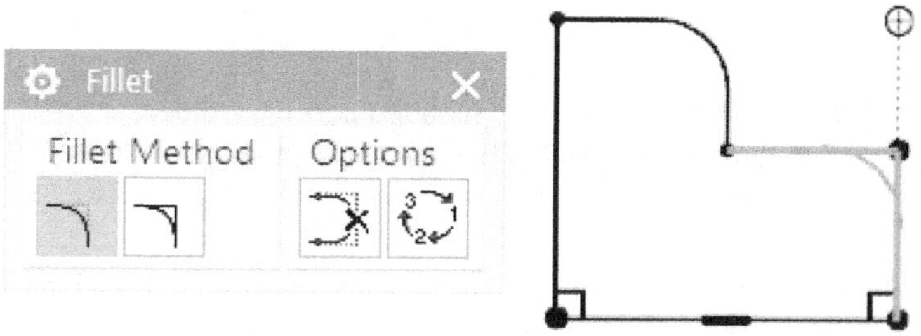

- Round the interior corner using the **Fillet** tool with the technique of your choosing. Apply *Equal Length* constraints and *Equal Radius* constraints, and add the dimensions shown below.

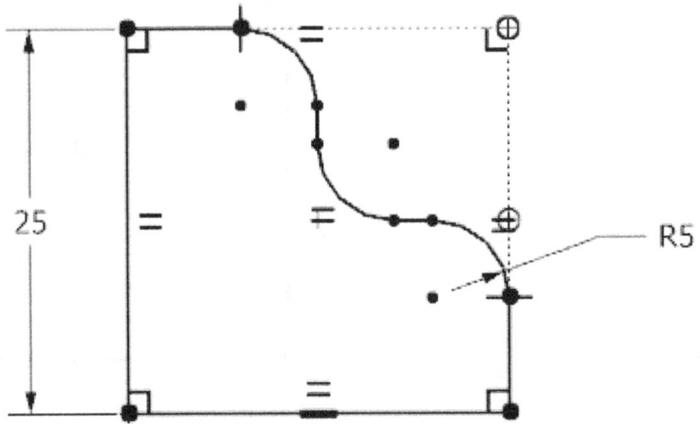

- Use **Mirror Curve** to mirror the outer tangent curves to produce the shape shown below.

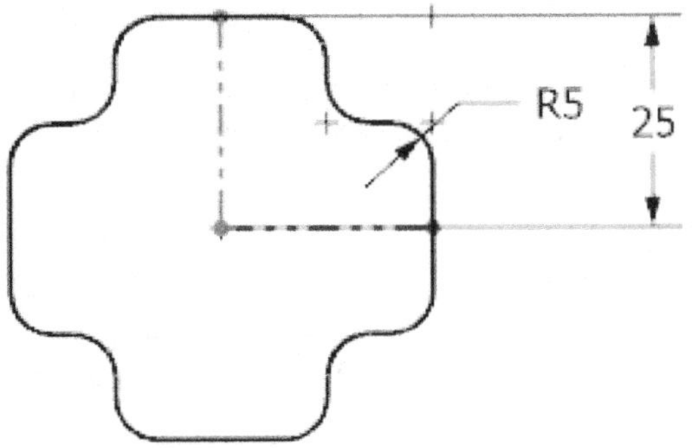

- **Finish** your sketch, and **Save** your file!

6.10 Internal Sketches

Certain features that allow curves from a sketch as their input section curves are capable of storing those sketches internally. There are two basic requirements that a feature must satisfy before a sketch that it is made from can be made internal to that feature:

1. The **Sketch Section** icon must be present on the dialog box for the feature.

2. The sketch must have no other dependent features.

In the exercise below, you will learn how to make sketches internal and external, and how to edit internal sketches.

- **Open** the part file *"Internal Sketches.prt"* from the downloaded part files folder.
- Inside you will find two sketches, and three extruded bodies. The first sketch – *"CIRCLE"* – has *Extrude (1)* as its only dependent feature.

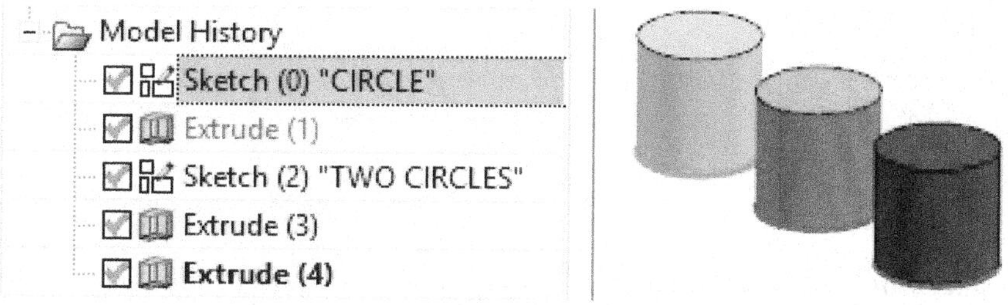

- Right-click on *Extrude (1)* and select **Make Sketch Internal**.

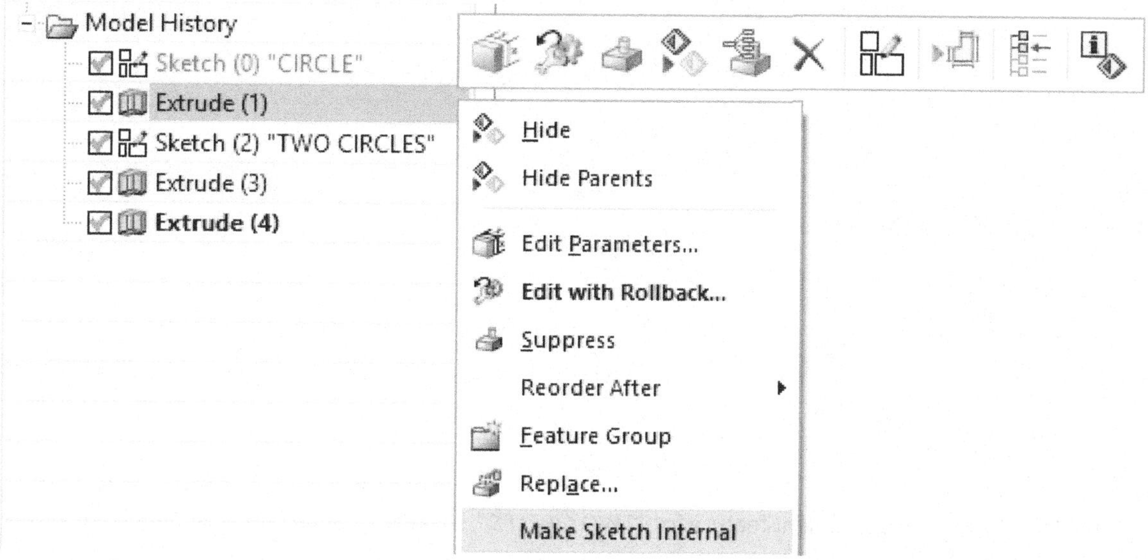

- This simplifies the Model History and automatically hides the sketch *"CIRCLE"* so that you don't have to manage its visibility independently of *Extrude (1)*. To edit the internal sketch, right-click on *Extrude (1)* and select *Edit Sketch*.

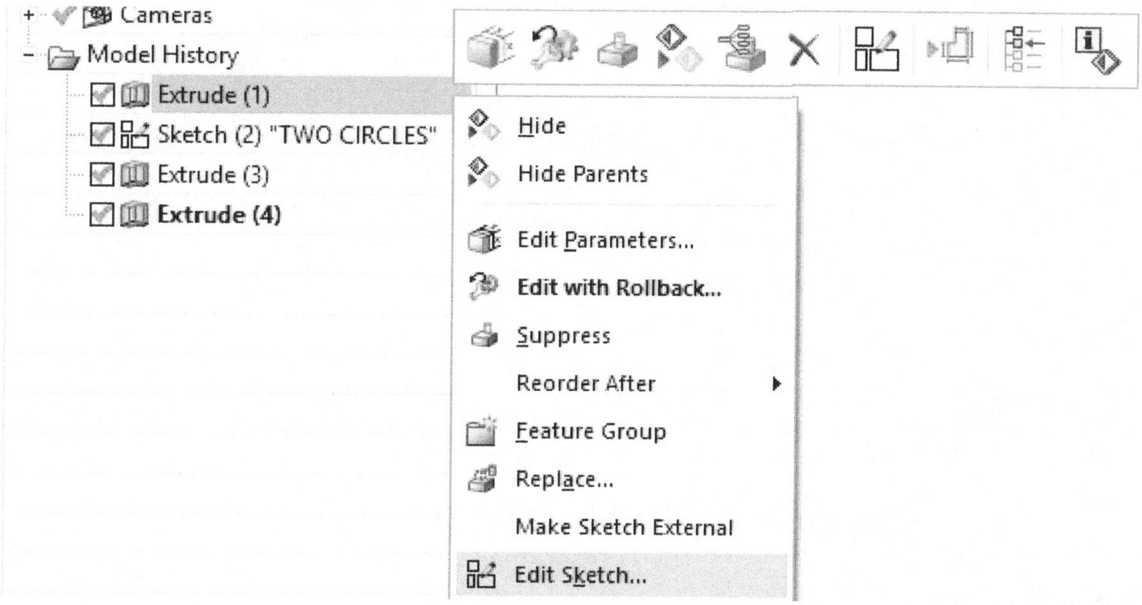

- This brings you into the Sketch Task Environment, with the edit action **Edit with Rollback**. Use **Finish Sketch** or **Exit Sketch** to return to the **Modeling** environment.

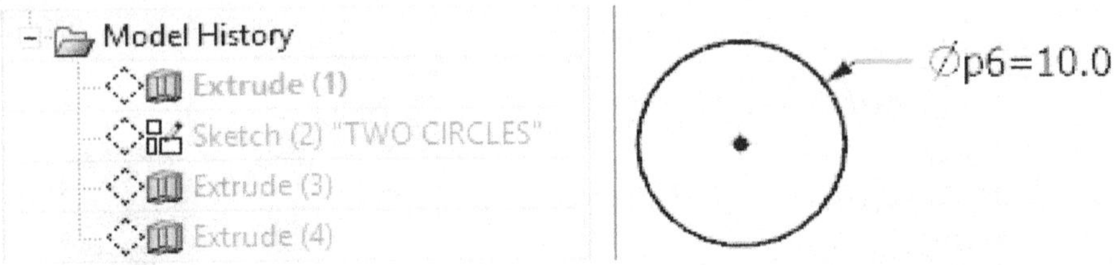

- Note that when you select the sketch *"TWO CIRCLES"*, both *Extrude (3)* and *Extrude (4)* highlight in blue in the Model History – both are dependent features, so the sketch *"TWO CIRCLES"* cannot be made internal to either.

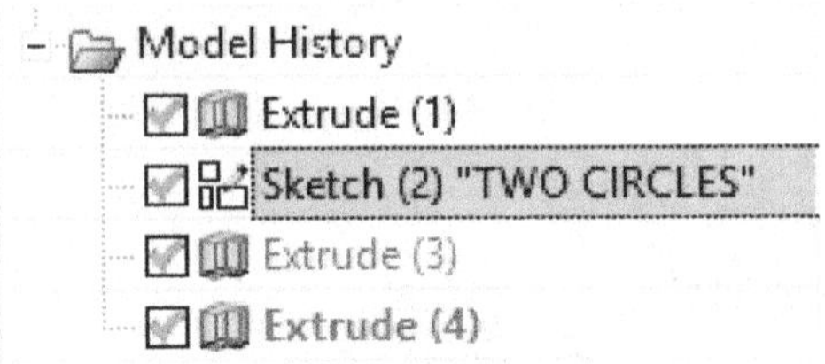

- Right-click on *Extrude (4)* and delete it.

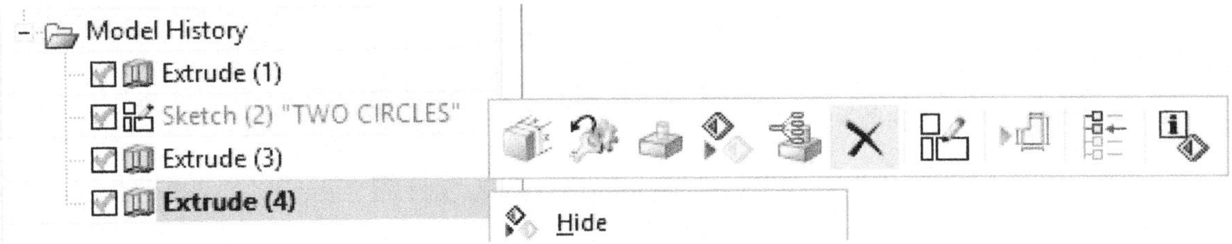

- Now, *Extrude (3)* is the only dependent feature for *"TWO CIRCLES"*, so you can right-click on it and make *"TWO CIRCLES"* internal to *Extrude (3)*.

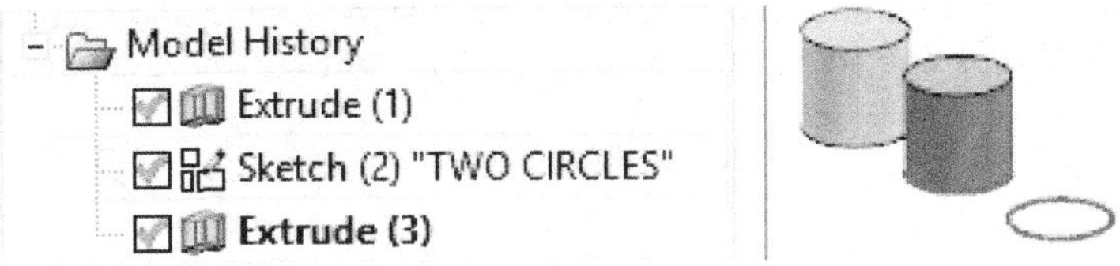

- Note that the entire sketch need not be used by the dependent feature in order for it to be made internal.

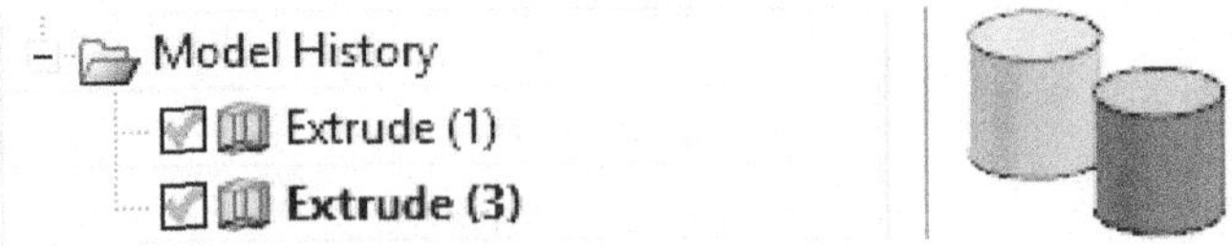

- **Save** your work!

There is a setting in **Modeling Preferences** (on the **General** tab of the **Modeling Preferences** dialog) that will automatically make sketches internal to their child features as possible. If you prefer to work with internal sketches, you might find this setting useful!

6.11 Direct Sketch

We conclude this chapter with a brief lesson on the **Direct Sketch** environment. So far, we have been using the **Sketch in Task Environment** command rather than **Direct Sketch**, for the following reasons.

Direct Sketch tends to confuse new users because:

- The appearance of the ribbon doesn't change – you can be *in* a sketch and not realize it if you navigate away from the **Home** tab.
- The **Direct Sketch** group contains tools that look almost identical to those in the **Curve** group on the **Curve** tab – it is easy to mistakenly grab the 3D **Line** tool instead of the sketch **Line** tool.
- Certain basic tools, such as the **Geometric Constraints** tool, are buried in the **More** gallery in **Direct Sketch** mode – thus requiring more clicks to make a precise sketch.

All that said, **Direct Sketch** is fantastic if you know all the shortcuts and don't want to wait for the system to enter the **Task Environment**. It allows advanced users to work much more efficiently to build a model from scratch, placing sketches and using them to **Extrude**, without even clicking the **Finish** icon!

- Create a new part file (**Units**: *Inches*) called *"Direct Sketch.prt"* and place it in *"C:\My NX 11 Files"*.
- Select the *Advanced* role from the **Roles** tab.
- Click on the **Profile** tool in the **Direct Sketch** group.

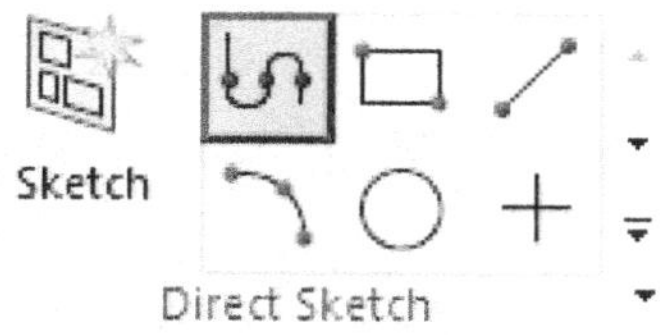

- Note that when the **Profile** tool dialog box appears, a preview of a sketch plane is generated – the sketch plane is determined by the same logic as the *Inferred* **Plane Method** from the **Create Sketch** dialog. If you wish to explicitly define the sketch plane, you can click the **Sketch** icon on the **Profile** dialog.

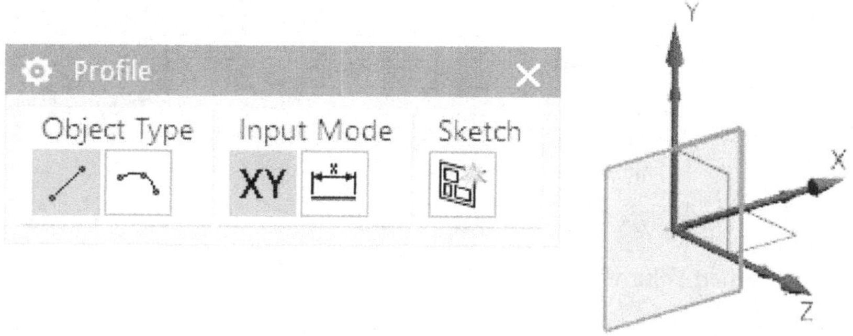

- Click on the YZ plane and click at the origin to begin using the **Profile** tool. Sketch the slot shape shown below.

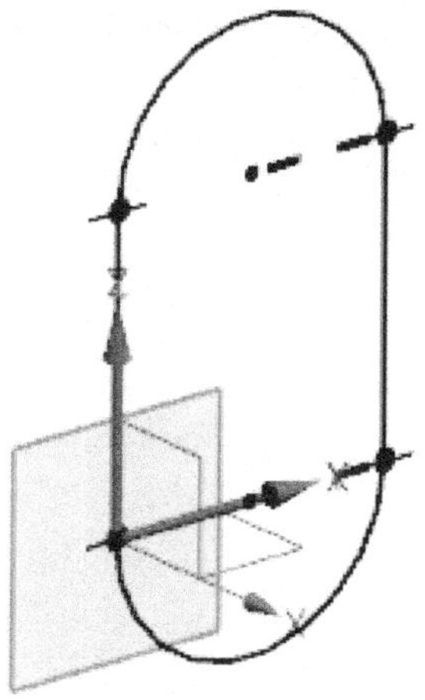

- Notice that, when you click the plane to sketch on, the view is NOT oriented to view the sketch plane that you selected. Orienting the screen to the sketch plane tends to give you a better perspective of the profile they are sketching and an easier ability to sketch horizontal, vertical and perpendicular lines – but for the advanced user, is not strictly necessary.

- Note that the left side of the **Home** tab has expanded into **Direct Sketch** mode displaying more sketch tools. Most tools are found in the **More** drop-down. One advantage of using **Direct Sketch** is that it is not necessary to use **Finish Sketch** before moving on and selecting other Modeling tools.

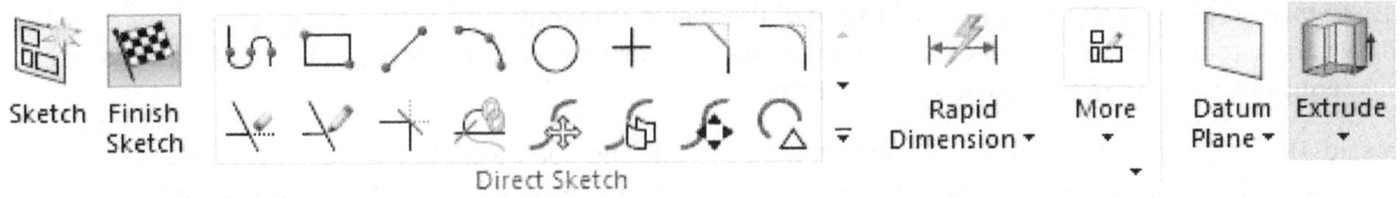

- Select the **Extrude** command. There is no need to select your sketch (this is not typical) – **Extrude** automatically grabs the whole sketch as the **Section**. Click **OK** to create a solid body from your sketch.

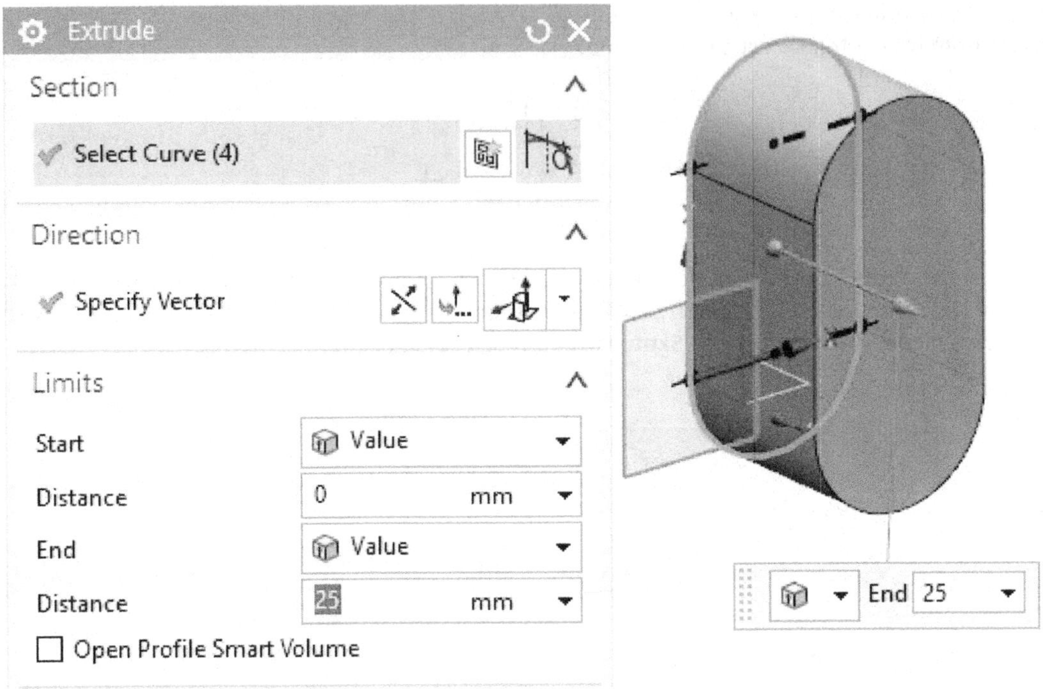

- Since *Extrude (2)* is the only dependent feature of *Sketch (1)*, you can make the sketch internal – go ahead and practice.
- **Save** your file!
- For the rest of the book, we leave it up to you to decide which sketch tool you'd like to use. If you prefer **Sketch in Task Environment**, now is a good time to switch back to your *Learning NX 11* role, but if you like **Direct Sketch**, you can continue with the **Advanced** role. You can also customize your Learning NX 11 role to include both and use that for the duration of the text.

7 Simple Solids and Surfaces

Once you have a sketch, or several sketches, there are four basic types of bodies that you can create from those sketches using solid modeling operations.

- *Extruded* – an extruded body is obtained by sweeping an input section string along a vector, with specified start and end distance limits. The **Extrude** command creates this kind of body in NX.

- *Revolved* – a body of revolution is obtained by sweeping an input section string around an axis, with specified start and end angle limits. The **Revolve** command creates bodies of revolution in NX.

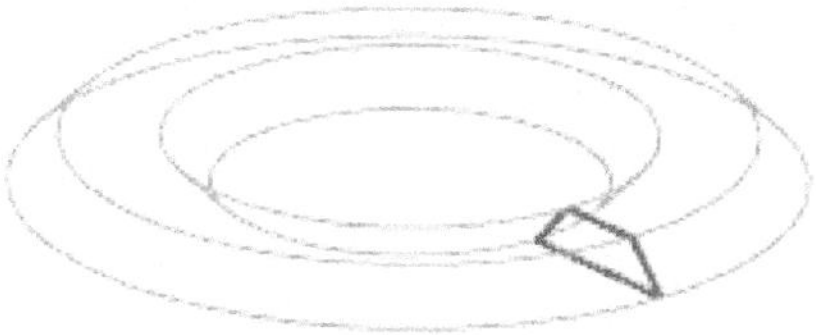

- *Swept* – a swept body is obtained by sweeping the input section string along a guide curve. The **Sweep Along Guide** command produces general swept bodies in NX.

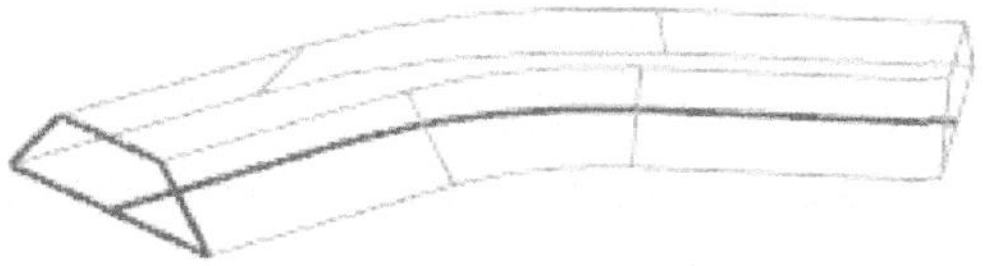

- *Lofted* – lofted bodies are obtained by specifying non-intersecting section strings (not necessarily the same shape!), which the lofted body must pass through. The **Through Curves** command produces general lofted bodies in NX.

More sophisticated shapes, especially surfaces, are built using tools combine multiple sections with multiple guides, and are closely related to sweeps and lofts, but are more complex than either – these are called mesh surfaces in NX.

In addition to these four basic types of bodies shown above, you can also build simple primitive shapes from scratch. The primitive design features in NX include **Block**, **Cylinder**, **Cone**, and **Sphere**. These commands are found in the **More** gallery in the **Feature** group on the **Home** tab. In the exercises below, we illustrate these fundamental tools, as well as a few others for creating simple solids and surfaces.

7.1 Extrude

The **Extrude** command is fundamental to solid modeling. The **Extrude** function takes a set of curves (called the *section string*) and sweeps them along a vector to create a solid or sheet body. If your section string is closed, the result will be a solid body, and if the section string is open, the result will be a sheet body. The **Extrude** command is found in the **Feature** group on the **Home** tab.

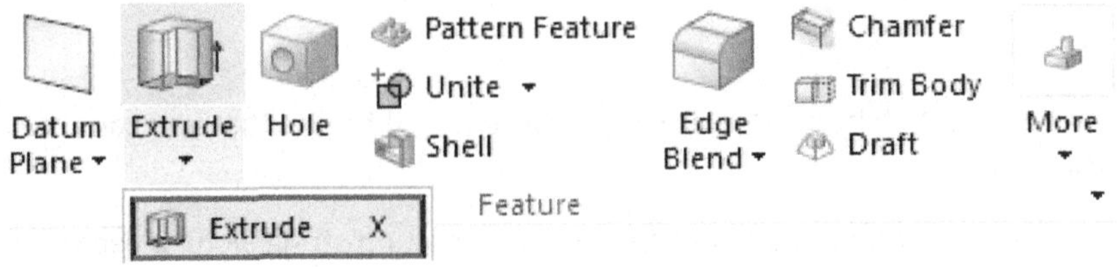

While most applications of the **Extrude** tool involve a planar section string (or at least curves from parallel planes), it is possible for the curves in the string to be twisted in three-dimensional space, although you must take care to ensure that the resulting body will not have self-intersections.

- Create a new file called *"Extrude.prt"* and place it in *"C:\My NX 11 Files"*.
- Select the **Extrude** tool. The **Extrude** dialog box will then appear. The red asterisk in the **Section** field indicates that NX wants you to specify a curve. Since the **Sketch Section** icon is present, if you click on the XY plane of the datum coordinate system, you will enter the Sketch Task Environment on that plane. The **Direction** will automatically become a normal vector to the sketch plane, as well. Go ahead and select the XY plane.

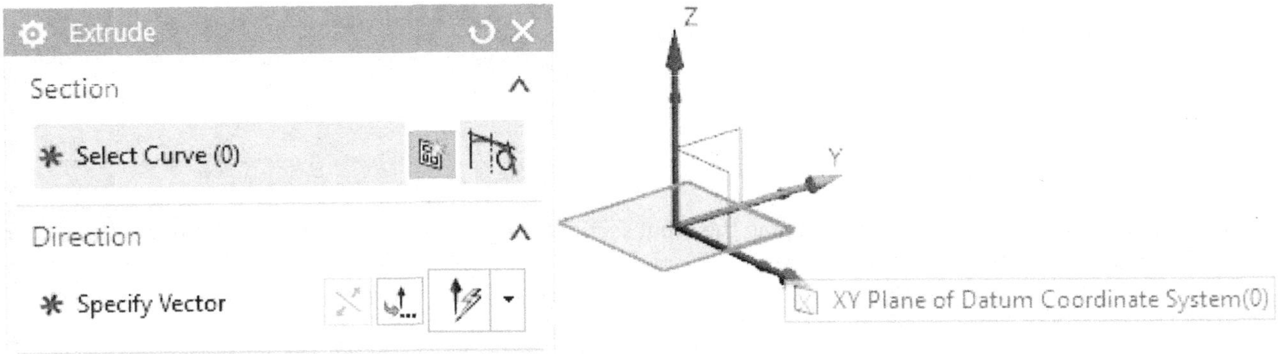

- In the Sketch Task Environment, create the sketch shown below. **Finish** the sketch to return to the **Extrude** dialog.

- When the selection is a closed string, the preview shows a solid body defined by **Start** and **End Limit** values. You can modify these values in the **Extrude** dialog and the preview will update when you push **[Tab]** *or* **[Enter]**.

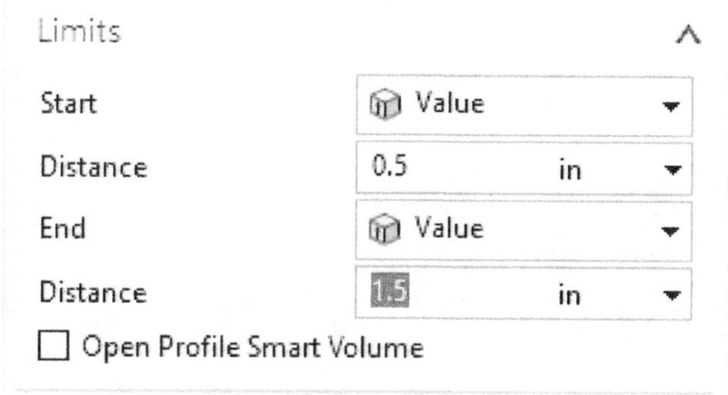

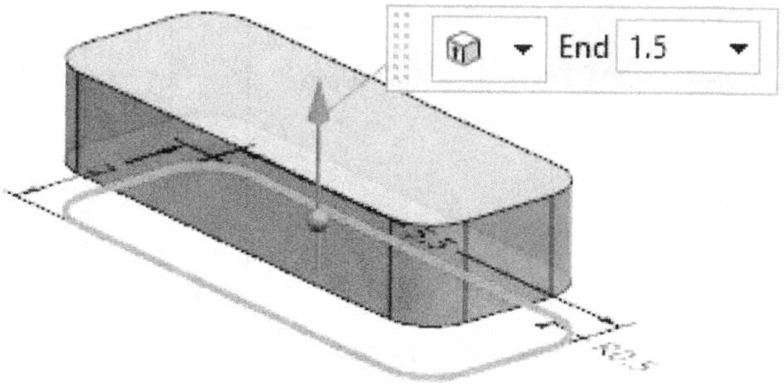

- You can also use the handles to dynamically modify your extruded body – the arrow controls the **End**, and the spherical handle controls the **<u>Start</u>**. Simply click-and-hold-and-drag either handle to change the value dynamically.

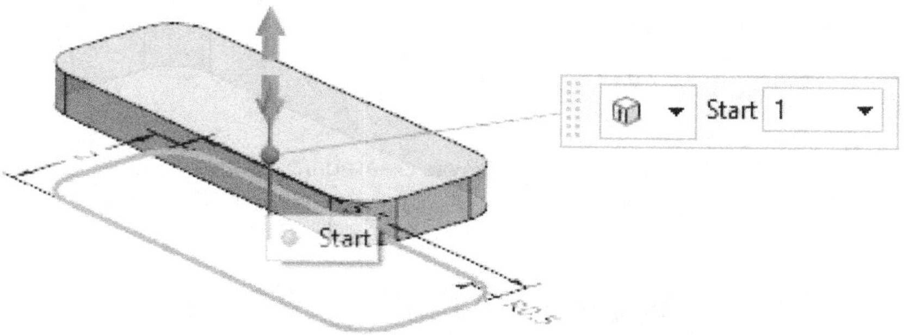

- There are no restrictions on the relation between the **Start** and **End** values – NX doesn't care if you use negative values, or whether the **End** value is greater than the **Start** value, or if they are the same – go wild!

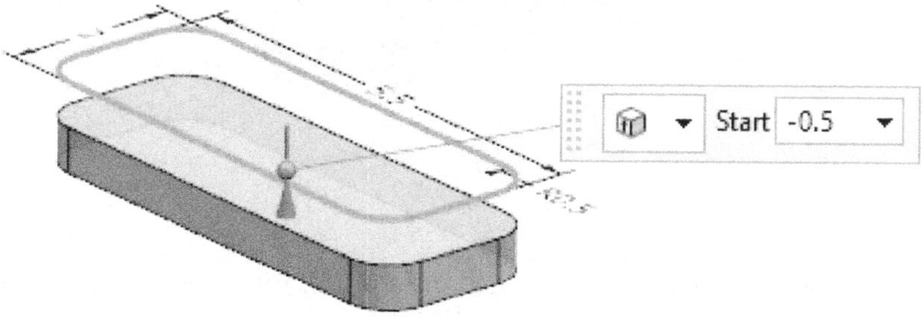

- There are some additional options within the **Extrude** dialog that are very useful. For instance, you can apply a **Draft** angle to the walls of the extruded body. When you specify a **Draft**, the draw direction is necessarily the same as the extrusion direction, and so whether the **Start Limit Value** is greater than the **End Limit Value** or not will impact how the **Angle** parameter is interpreted. Use the **Reverse Direction** button to **Extrude** in the -ZC direction, then set the **Start** to *0 in* and the **End** to *1.5 in* and specify a **Draft Angle** of *5 degrees*.

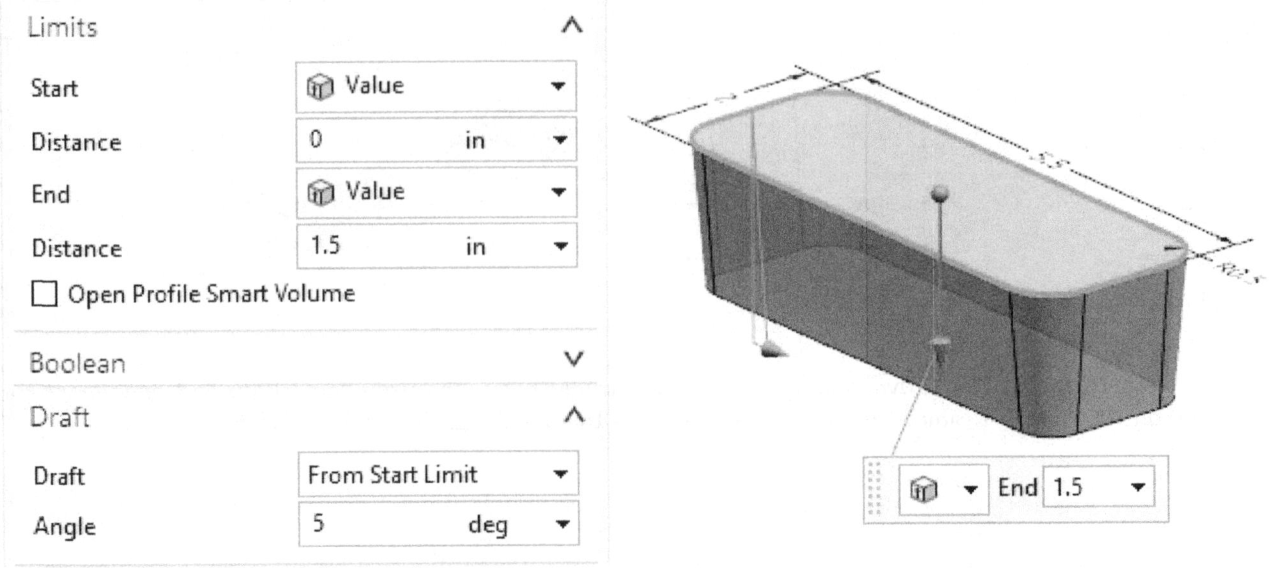

- Click **OK** to create the *Extrude* feature. Since the sketch was created internally to the feature, it appears in neither the Model History, nor the Graphics Window.

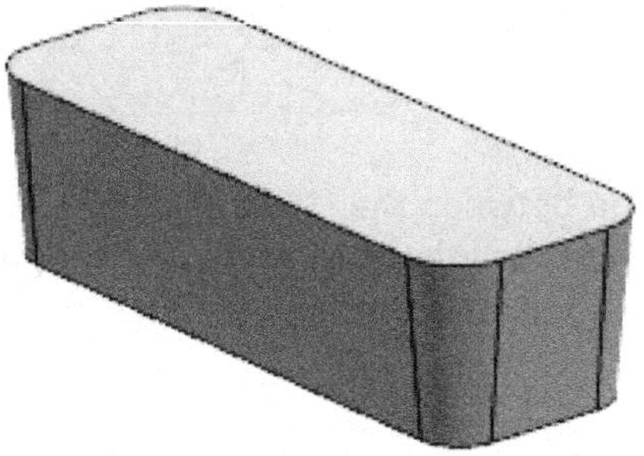

- **Save** your part file!

7.2 Revolve

The **Revolve** tool creates a body of revolution from a section string, which is not required to be closed. The axis of revolution can be a curve, edge, existing axis, or you can specify it by giving a direction vector and a point entity.

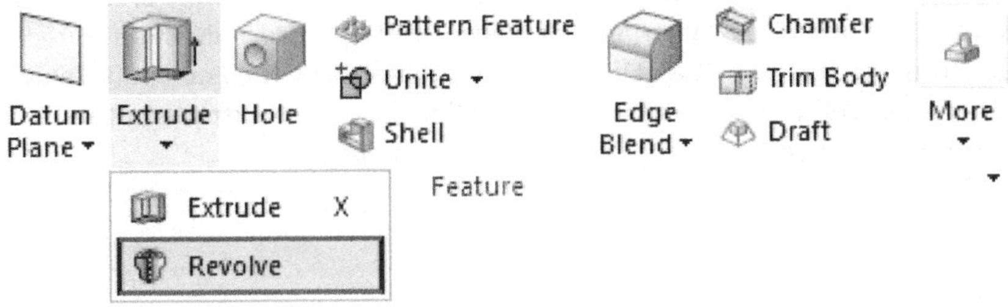

- Create a new file called *"Revolve.prt"* and place it in *"C:\My NX 11 Files"*.
- Create the sketch shown below on the YZ plane of the datum coordinate system.

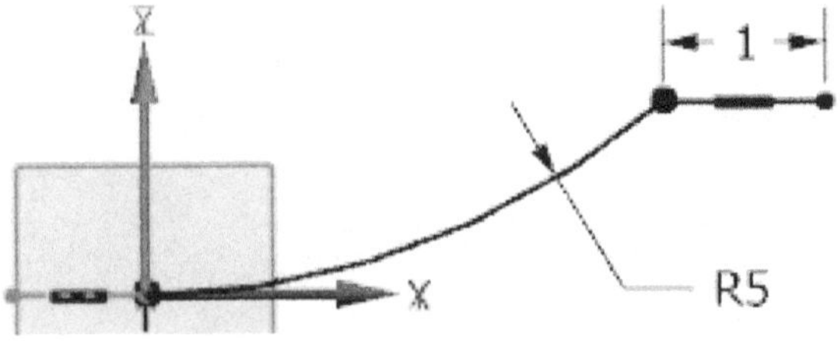

- Select the **Revolve** tool and choose your sketch as the **Section**. For the **Axis**, use the Z axis of the *OrientXpress*, as shown below.

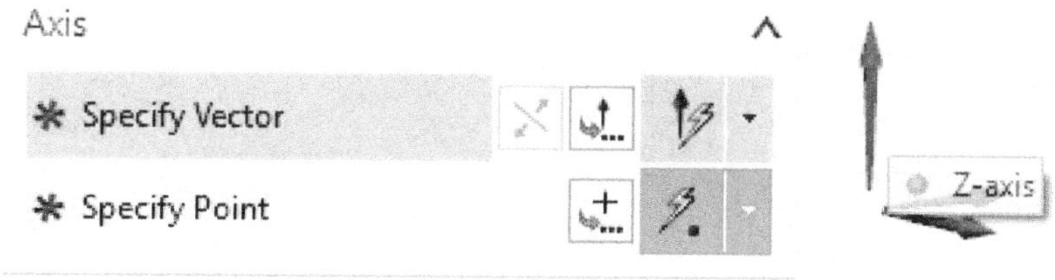

- A common point of confusion for new NX users is the role that the **Point** plays in determining the **Axis**. Notice that after selecting the **Vector**, you still need to specify a **Point**.

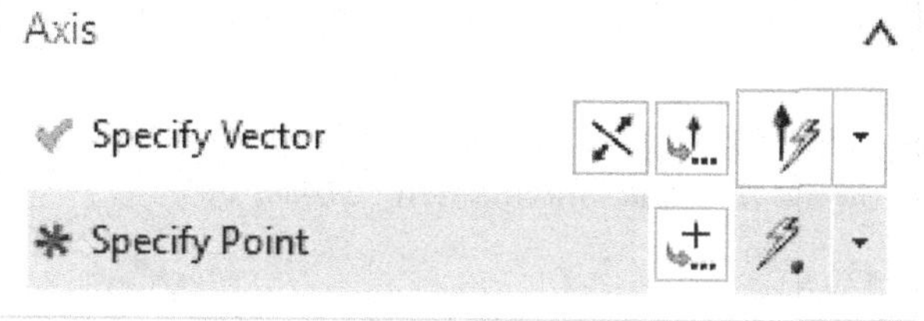

- The vector gives the axis its direction, but it is still not anchored in three-dimensional space – this is the role of the **Point**. Click on the point of the datum coordinate system.

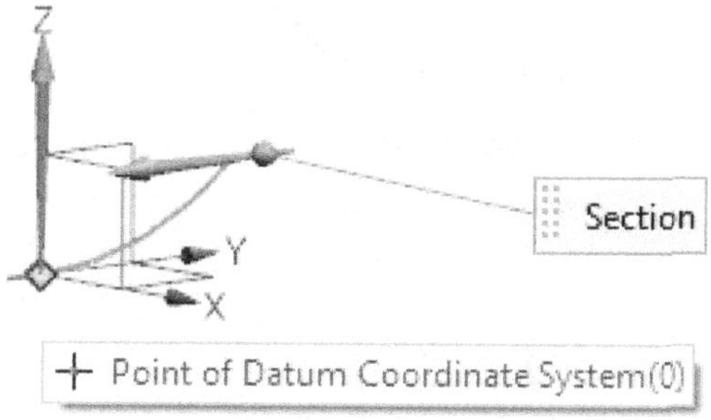

- Now that the vector is required to pass through a specific point, there is enough data to determine an axis! When prompted for the **Vector**, if you select an existing line, edge, or axis, NX will infer a point based on the input.
- Specify a two-sided offset from *0 in* to *0.125 in* so that the body of revolution is a solid of uniform thickness.

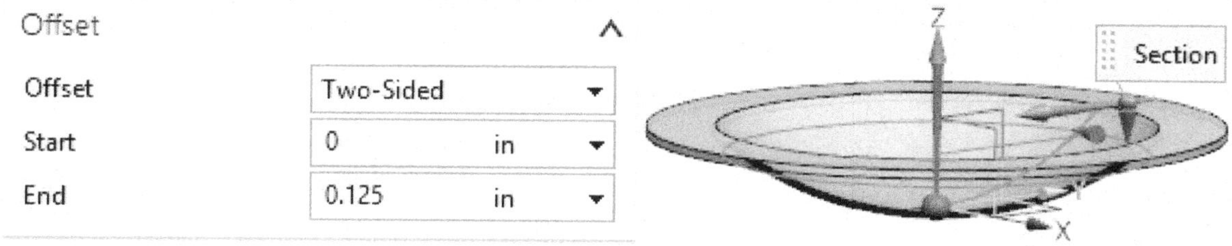

- **Save** your part file!

7.3 Sweep Along Guide

Both **Extrude** and **Revolve** are special cases of sweep operations. There are a number of sweep tools in NX, the most basic and general of which is the **Sweep Along Guide** tool. The **Sweep Along Guide** tool requires that you specify a section string and a guide string, and the resulting body has cross section identical to the section at each plane normal to the guide.

Sweep Along Guide is found in several places on the ribbon, the most convenient of which is in the **More** gallery from the **Surface** group on the **Home** tab.

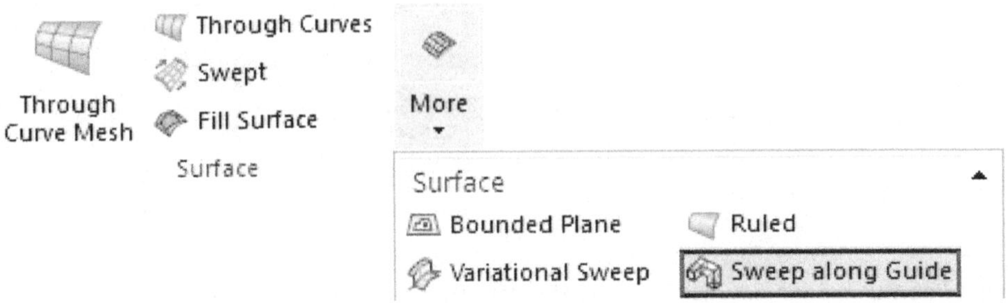

The **Sweep Along Guide** has several noteworthy features not present in other sweep tools: it has a **Boolean** option (see Chapter 8.3), and both the section and guide strings can have sharp corners (most sweeps require tangent-continuous guide strings).

- Create a new file (**Units**: *Inches*) called *"Sweep Along Guide.prt"* and place it in *"C:\My NX 11 Files"*.
- Create the sketch shown below on the XY plane, and name it *"GUIDE"*.

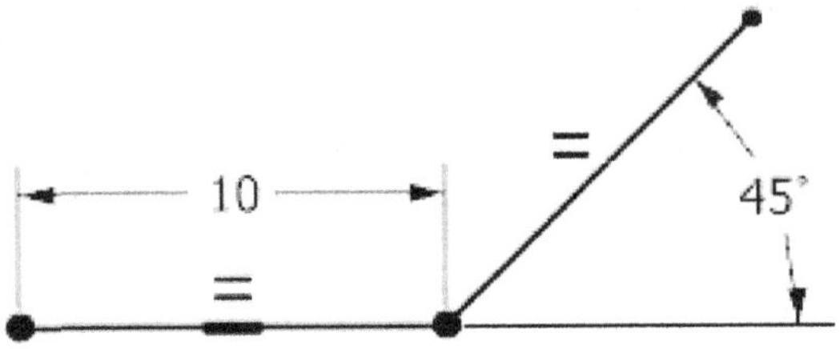

- Create the sketch shown below as a sketch of **Type** *On Path* attached to *"GUIDE"* at *0%* arc length. Name it *"SECTION"*.

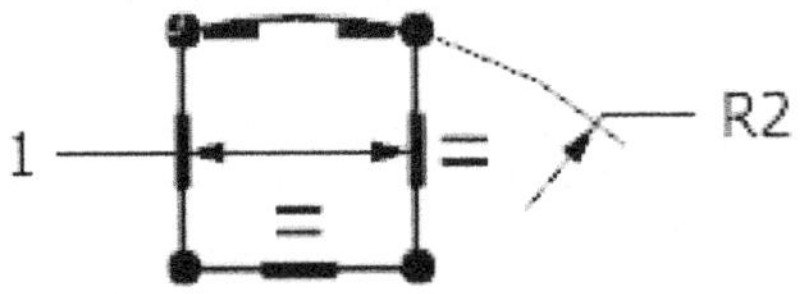

- Select the **Sweep Along Guide** tool, and choose the sketch *"SECTION"* as the **Section**, and the sketch *"GUIDE"* as the **Guide**. Click **OK** to create the solid body shown below.

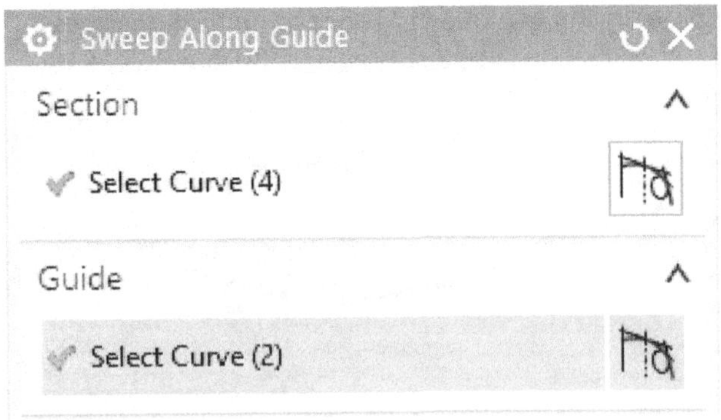

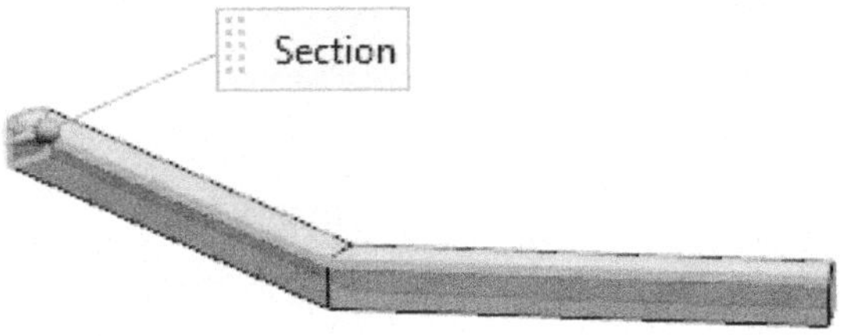

- **Save** your file!

7.4 Tube

The **Tube** command is the only sweep command that does not require a section as input – the section is always a pair of circles, specified by an **Outer Diameter** and **Inner Diameter**. Instead, the **Tube** command asks only for the **Guide** curve. The **Inner Diameter** is allowed to be zero if you wish to model an entirely solid tube. The **Tube** command is found in several places on the ribbon, the most convenient of which is in the **More** gallery from the **Surface** group on the **Home** tab.

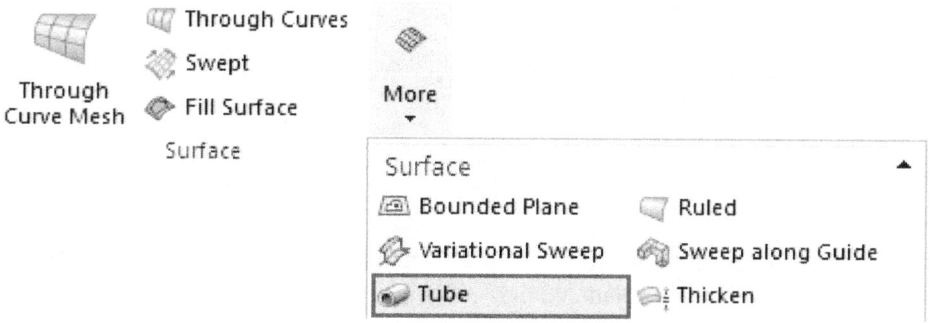

- Create a new file (**Units**: *Inches*) called *"Tube.prt"* and place it in *"C:\My NX 11 Files"*.
- Create the sketch shown below in the XY plane.

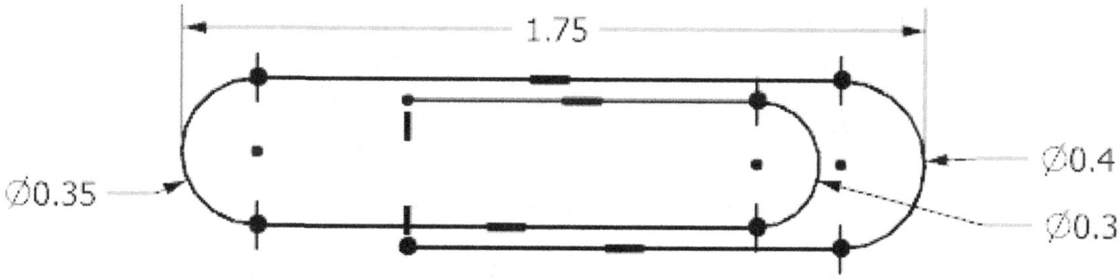

- Select the **Tube** command and choose the sketch as the **Path**. Enter the parameters shown below and click **OK**.

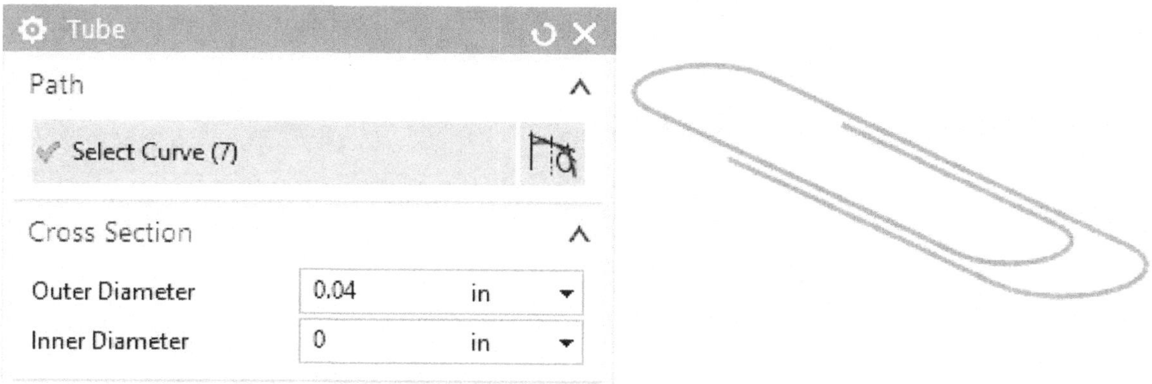

- Your model should look like a paper clip. **Save** your file!

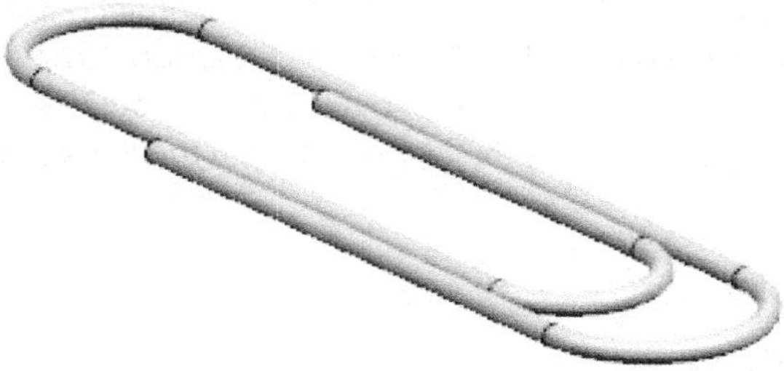

7.5 Bounded Plane

In general, when you have a closed, chain-connected set of curves in three-dimensional space, there is no unique or obvious way to build a surface that has them as the boundary – typically more data is required to define a surface. However, when the curves are coplanar, the planar region enclosed by the curves is an especially simple and useful surface that NX can generate from only the boundary curves. The **Bounded Plane** tool does exactly that, and it is found in the **More** gallery in the **Surface** group on the **Home** tab.

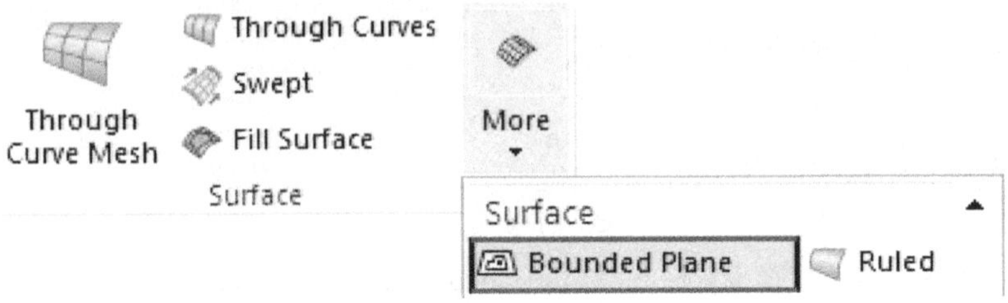

- Create a new file (**Units**: *Inches*) called *"Bounded Plane.prt"* and place it in *"C:\My NX 11 Files"*.
- Create the sketch shown below in the XY plane.

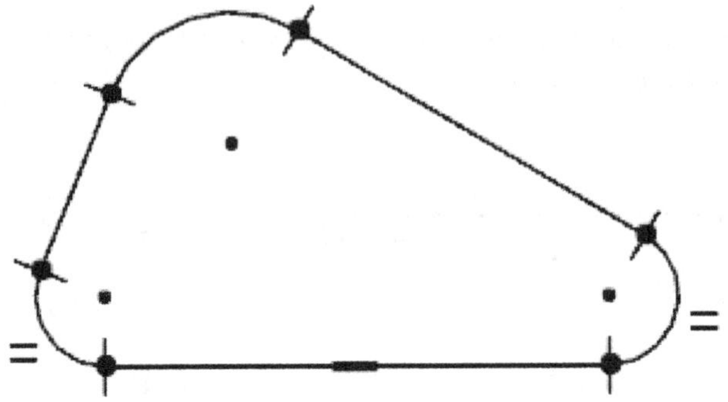

- Select the **Bounded Plane** tool and choose the sketch as the **Planar Section**. Click **OK** to generate the surface.

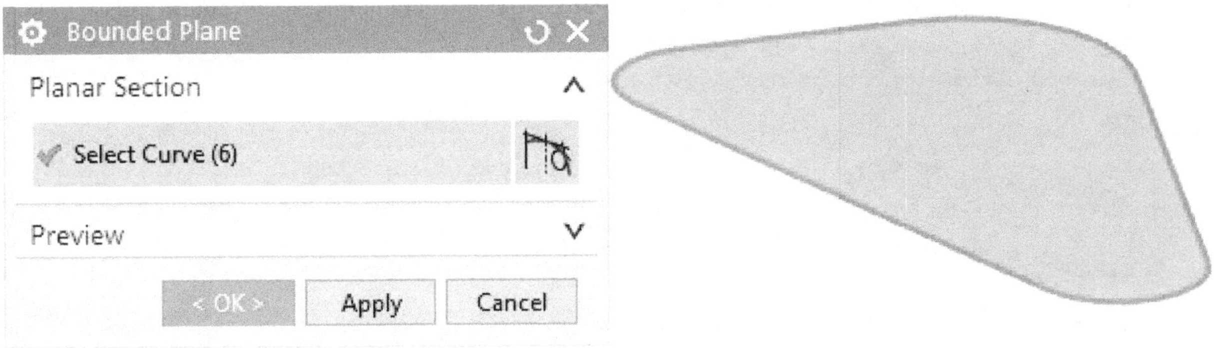

- **Save** your file!

7.6 Ruled Surface

The **Ruled Surface** tool is the simplest example of a loft. It requires two disconnected section strings as input, and it creates a surface or solid body from them by connecting each point on the first section string to each point of the second section string by a straight line. It offers some modest control over how the lines should be drawn (alignment by parameter, arc length, etc.). The **Ruled Surface** command is found in the **More** gallery in the **Surface** group on the **Home** tab.

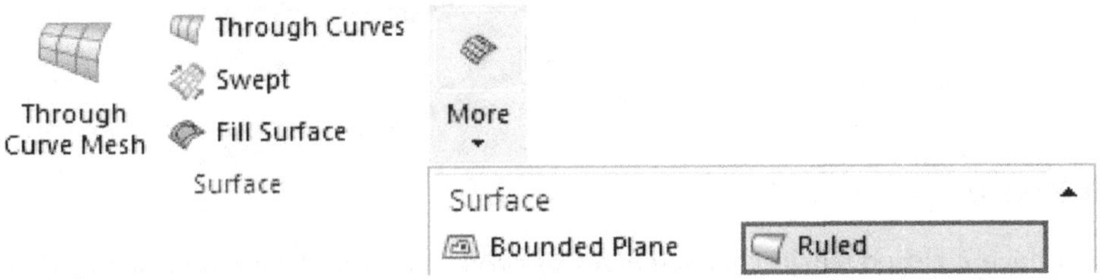

- Create a new file (**Units**: *Inches*) called *"Ruled Surface.prt"* and place it in *"C:\My NX 11 Files"*.
- Create the sketch shown below in the XZ plane. Make sure that the dimensions and constraints in the sketch only make reference to the sketch coordinate system and not to the datum coordinate system.

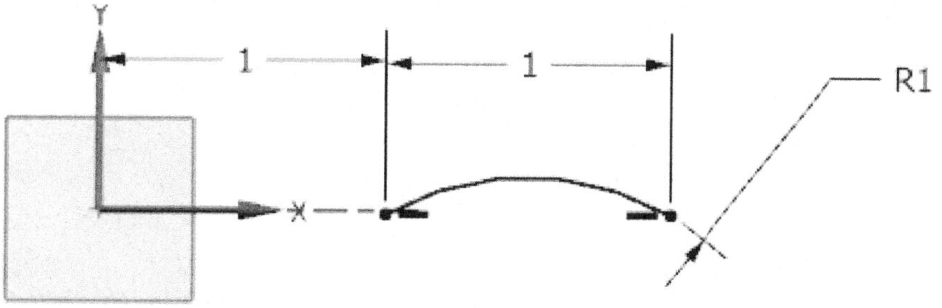

- **Copy** & **Paste** ([Ctrl]+[C], [Ctrl]+[V]) the first sketch onto the YZ plane. If you created your sketch constraints and dimensions only in relation to the sketch coordinate system, this will be easy.

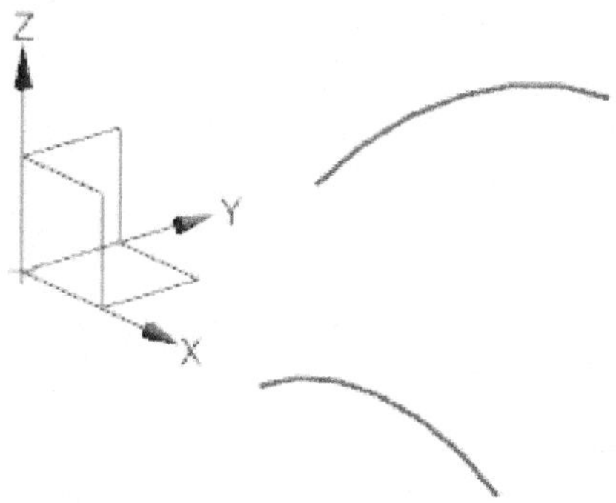

- Select the **Ruled Surface** command and choose the two sketches as **Section String 1** and **Section String 2**. In **Alignment**, make sure the **Preserve Shape** checkbox is marked – this ensures that the cross sections inherit their shape directly from the section strings. The section orientation is sensitive to *where* you click on the defining curves – try to click on the same side of each arc. If your orientations become mismatched, use the **Reverse Direction** button to repair your surface and eliminate self-intersections.

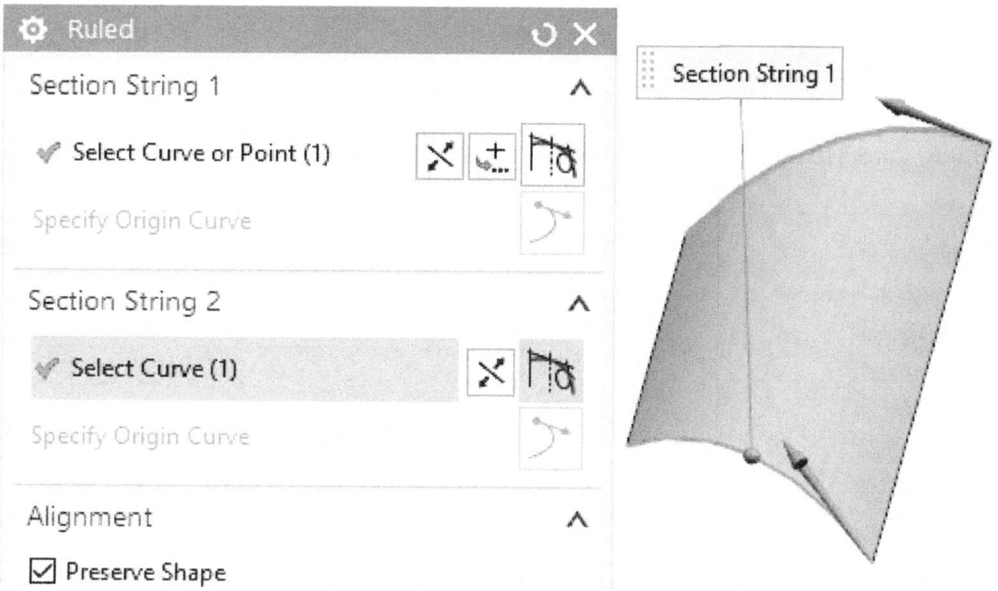

- **Save** your file!

7.7 Through Curves

The **Through Curves** command is the most general tool that enables you to build a body that is *lofted* between two or more disconnected *section strings*. Each section must be a chain-connected series of curves – no section string can have multiple loops. The **Through Curves** command is found in the **Surface** group on the **Surface** tab, as well as in the **Surface** group on the **Home** tab.

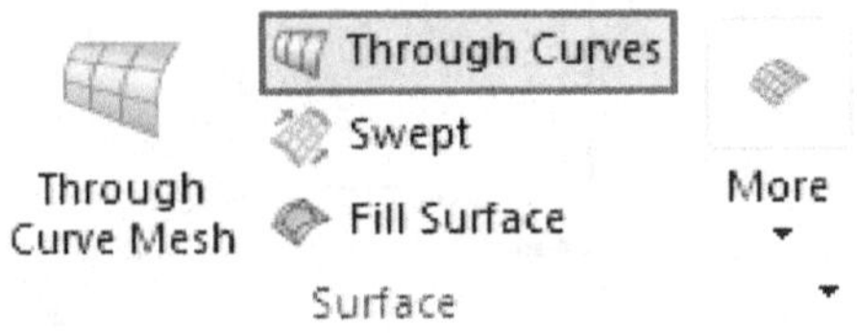

To avoid self-intersection, the orientation of each **Section String** must be consistent. In this exercise, you will define the necessary construction geometry and create a surface using the **Through Curves** tool.

- Create a new file (**Units**: *Inches*) called *"Through Curves – Sheet.prt"* and place it in *"C:\My NX 11 Files"*.
- Create four parallel datum planes, each spaced two inches apart.

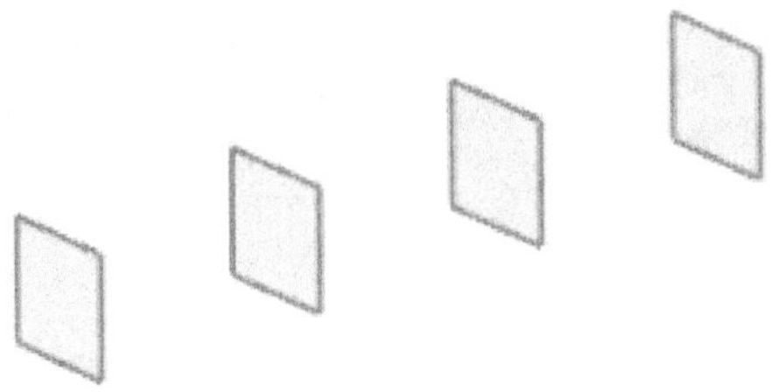

- Create a sketch of an arc on each datum plane, with endpoints horizontally aligned along the sketch horizontal axis, and with arc center along the sketch vertical axis. Each arc should have a linear dimension between its endpoints, equal to its radius. The first sketch is shown below.

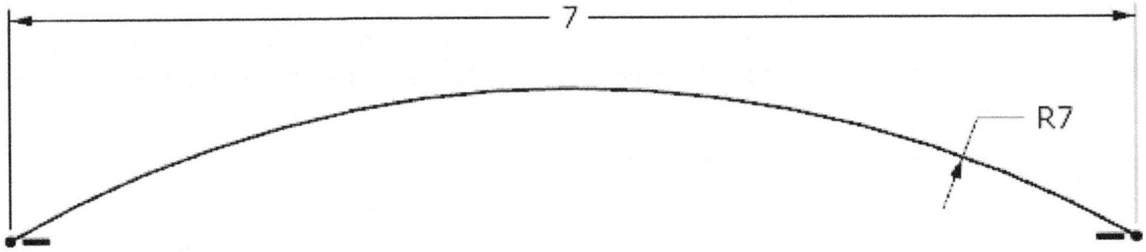

- The radii of the arcs, from left to right, are *7 in, 4 in, 2 in, 1 in.*

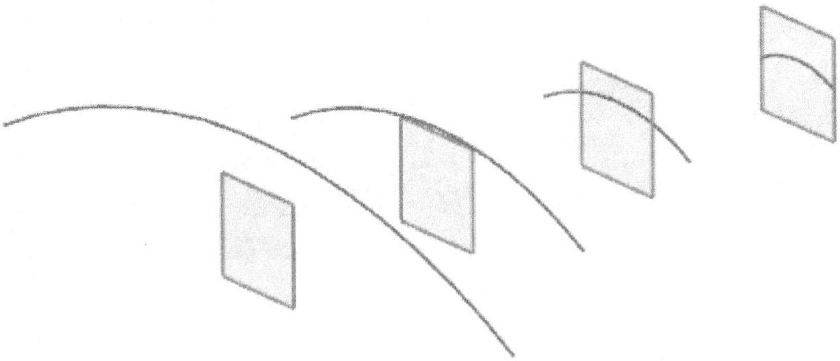

- Open the **Through Curves** command. Select the first arc as **Section 1**. Note the **List** subsection of the dialog; in order to create and specify additional sections, you must push the **Add New Set** button when you are done with each selection.

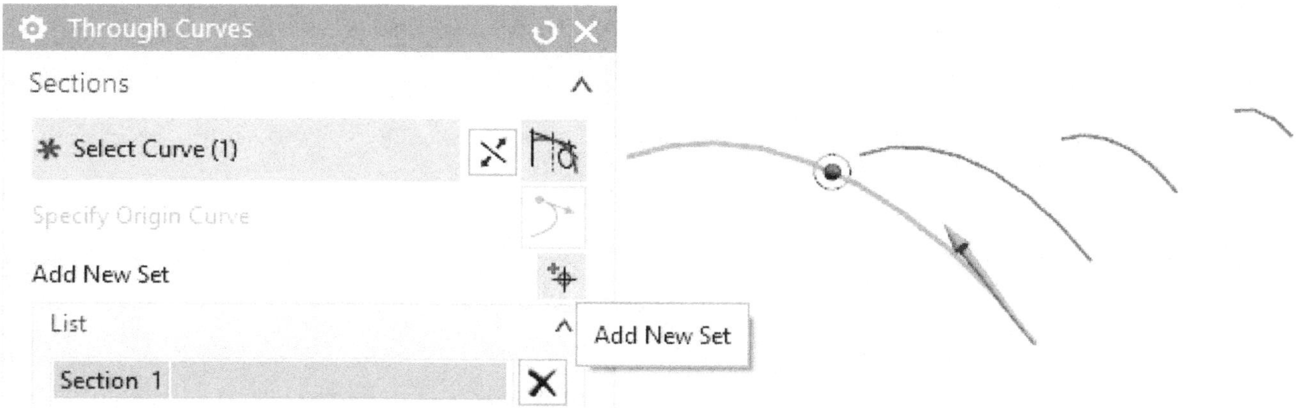

- Push the **Add New Set** button and select the second arc as **Section 2**.

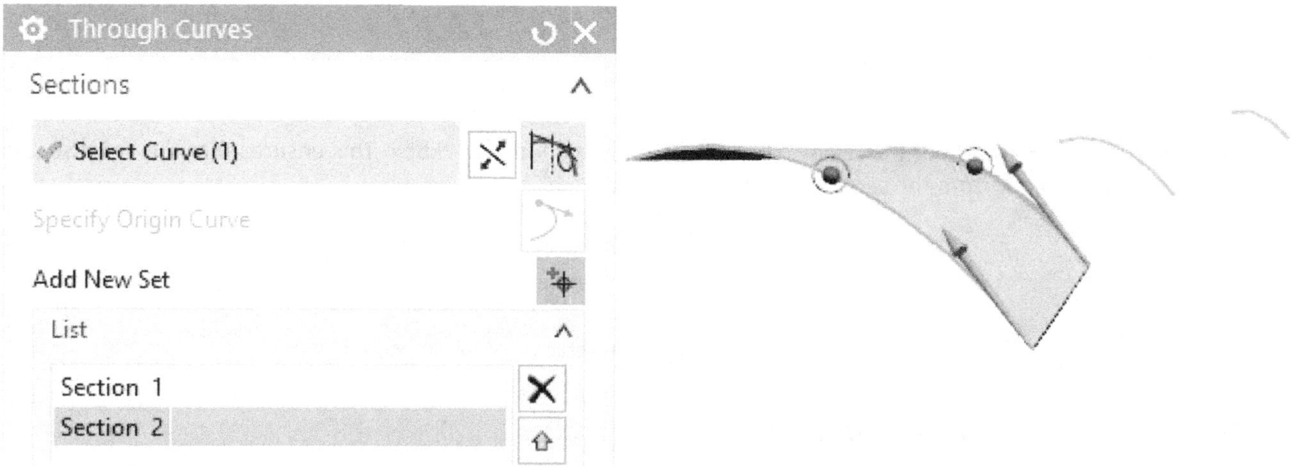

- The section orientation is sensitive to *where* you click on the defining curves – try to click on the same side of each arc. If your orientations become mismatched, use the **Reverse Direction** button to repair your surface and eliminate self-intersections.

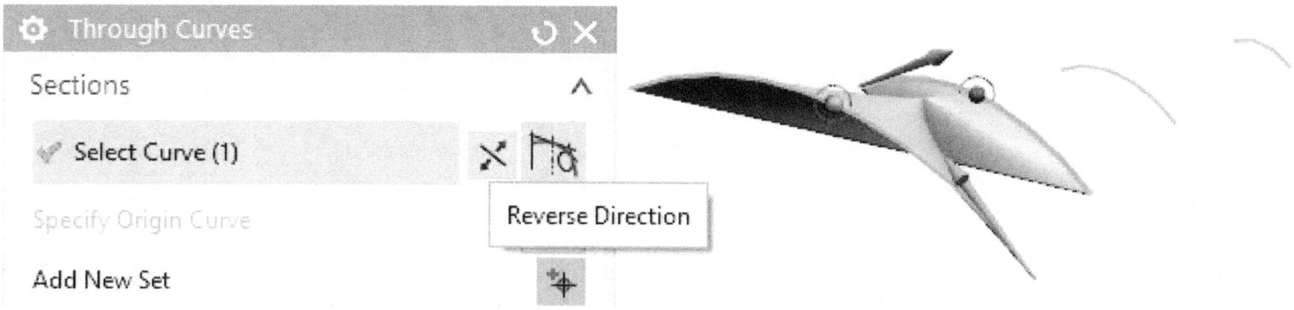

- Continuing from left to right, add two more sections. The order of the sections is extremely important.

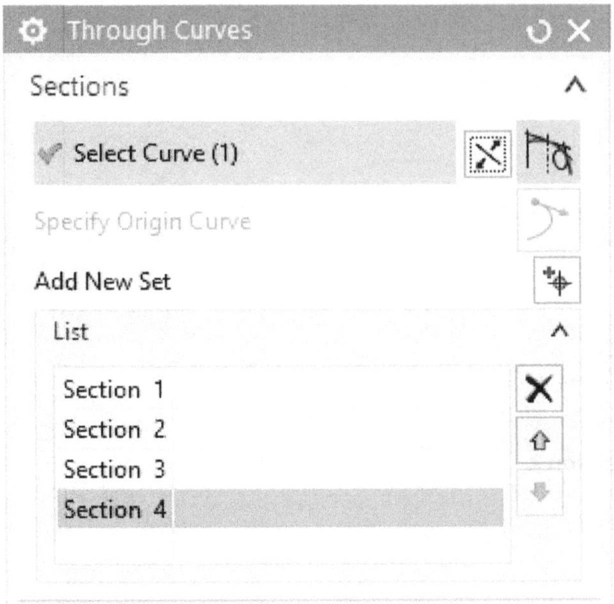

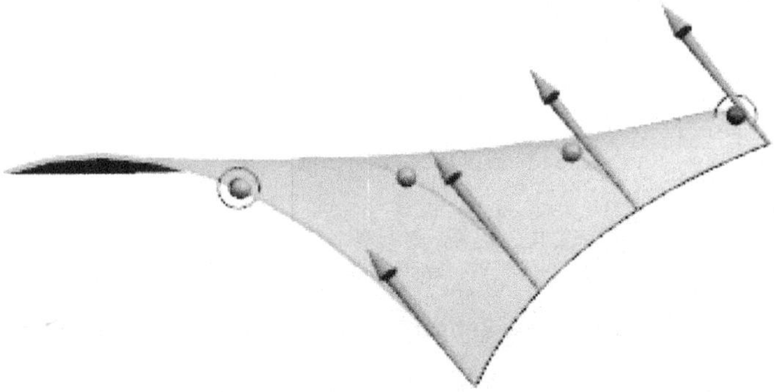

- Within the **Alignment** section of the dialog, check the *Preserve Shape* checkbox. This ensures that the cross section inherits its shape directly from the section curves throughout.

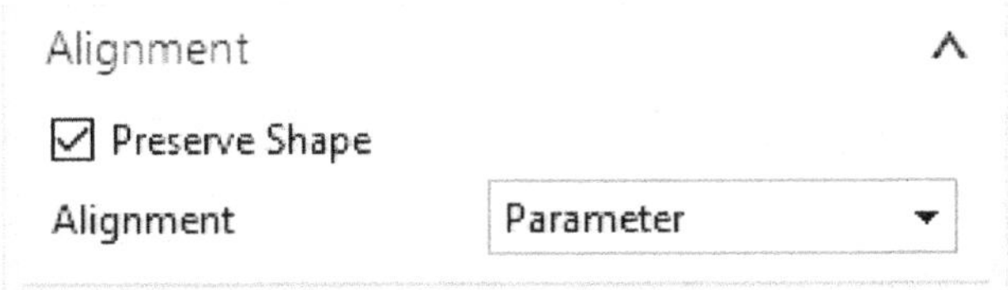

- Click **OK** to create a beautiful freeform surface!

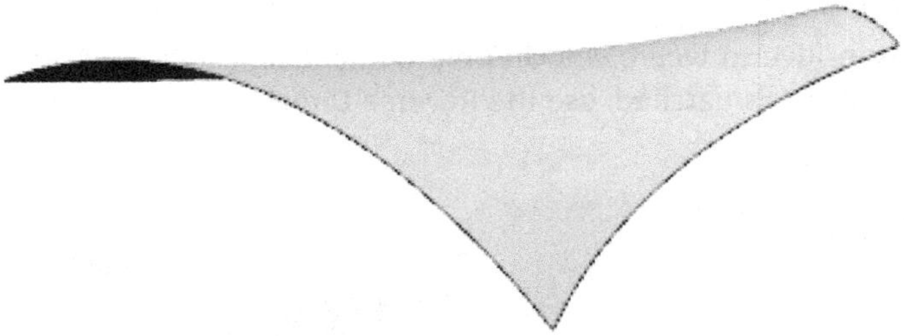

- **Save** your file!

7.8 Thicken

Thicken is a useful tool when you need to create a body of uniform thickness from a surface that you already have. **Thicken** is found in many places on the ribbon, the most convenient of which is the **More** gallery of the **Surface** group on the **Home** tab.

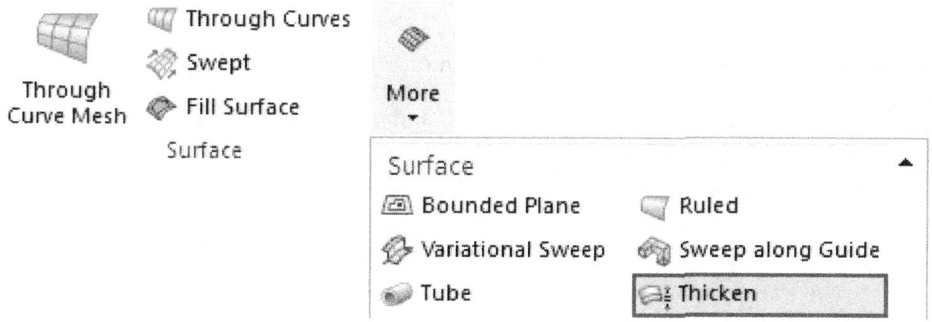

- Continue working with *"Through Curves – Sheet.prt"*.
- Select the **Thicken** tool. There are two **Offset** parameters, and their difference is the overall thickness of the resultant body. Select the sheet body of **Through Curves** as the **Face**, and specify **Offset** values as shown below. This will result in a body of uniform thickness which has the **Through Curves** surface as a mid-surface!

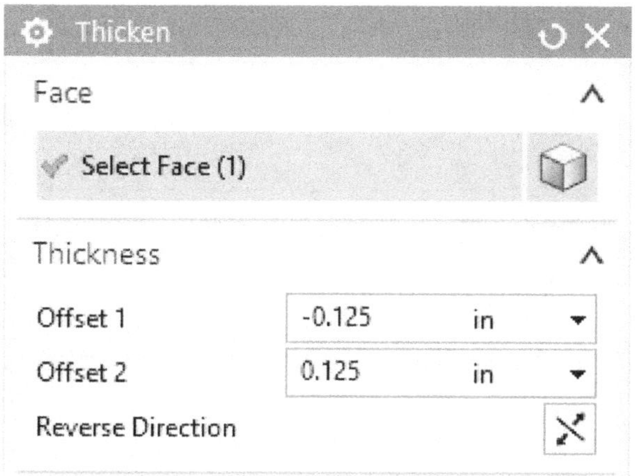

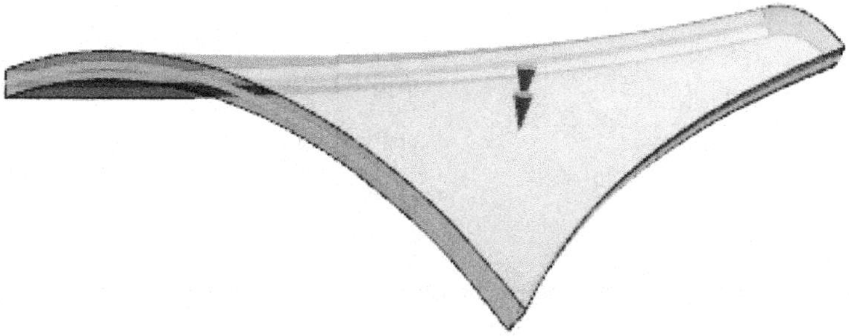

- The sheet body of **Through Curves** is not connected to the solid body created by **Thicken** – if you set your **Selection Type Filter** to *Sheet Body*, you will see clearly that you can select the sheet body separately.

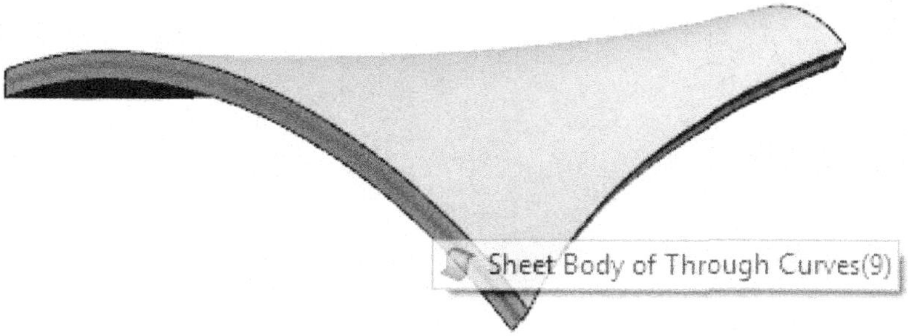

- **Hide** the sheet body from **Through Curves**.
- **Save** your file!

7.9 Primitives

To create elementary solid shapes, you may define **Primitives**. The four primitive commands are found in the **More** gallery in the **Feature** group on the **Home** tab. The tools that produce primitives are **Block**, **Cylinder**, **Cone**, and **Sphere**.

In this chapter, you will create geometry similar to that found in the *"Datums.prt"* part file, with the help of the **Block** tool and the **Cylinder** tool. You will then add a sphere to the part!

7.9.1 Block

In this project, you will create two blocks, one on top of the other.

- Create a new part file (**Units**: *Inches*) called *"Primitives.prt"*, and place it in *"C:\My NX 11 Files"*.
- Select the **Block** tool, set the **Type** to **Origin and Edge Lengths**, and enter the dimensions as shown below. When you are done, click on the **Point Dialog** button to define the origin for the block, as shown below.

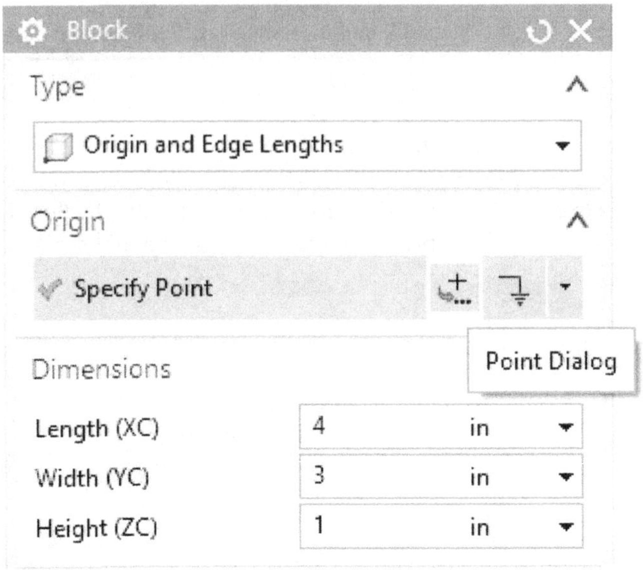

- This brings up the **Point** dialog box. Change the YC coordinate to *-3 in*, as shown below. out the **Output Coordinates** as shown below, and click **OK**.

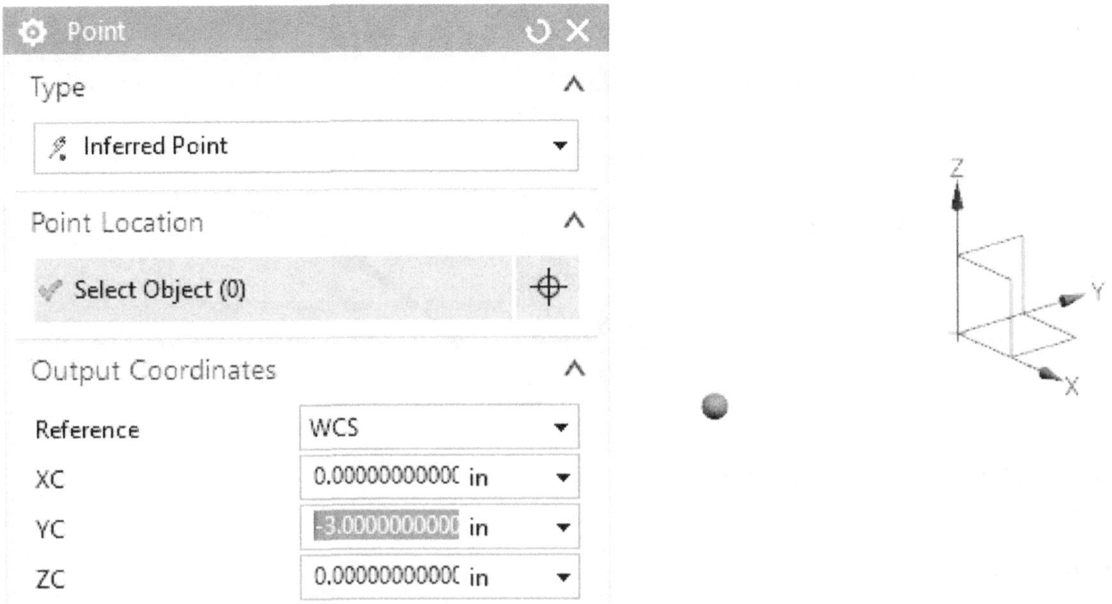

- When you get back to the **Block** dialog box, click **Apply** to produce the first block.
- For the second block, change the **Height (ZC)** value to *1.25 in*, as shown below. For the **Origin**, choose the endpoint shown below.

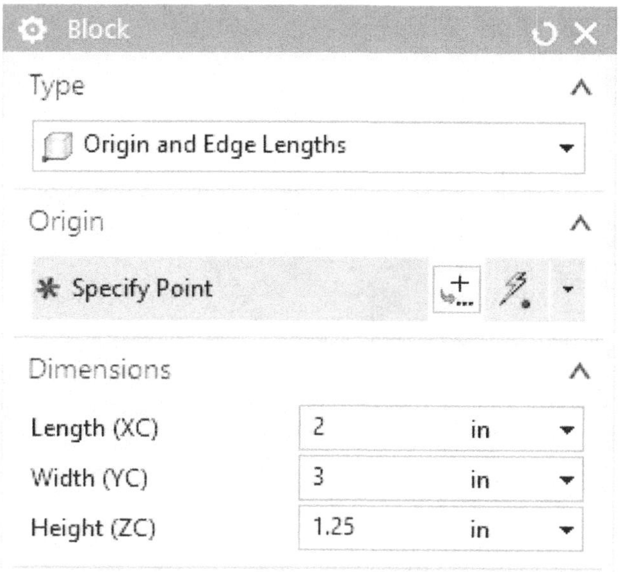

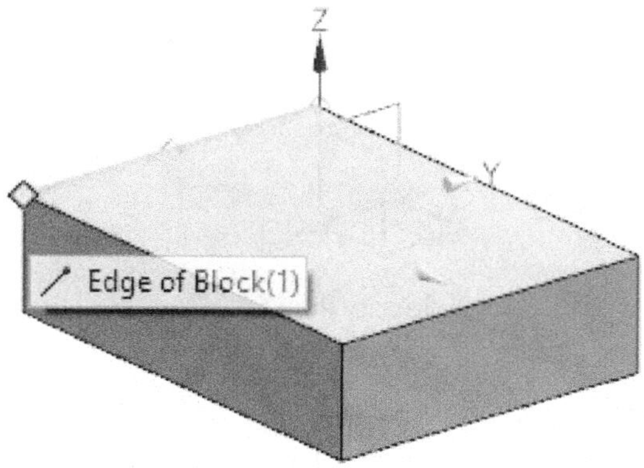

- Hide the *Datum CSYS* feature and **Save** your file.

Question: How did the software choose the XC, YC, and ZC directions for the blocks created so far?

Answer: It infers them from the ***WCS****!*

The position and dimensions of blocks created using the **Block** tool are parametric and associative, but the orientation is not, since it depends on the WCS!

7.9.2 Cylinder

In this project, you will create a **Cylinder** of **Diameter** *0.88 in* and **Height** *2.5 in*.

- Continue working with *"Primitives.prt"*, and select the **Cylinder** tool.
- Set the **Diameter** to *1.5 in*, and the **Height** to *1 in*. After entering these values, click on the **Point Dialog** icon, as shown below.

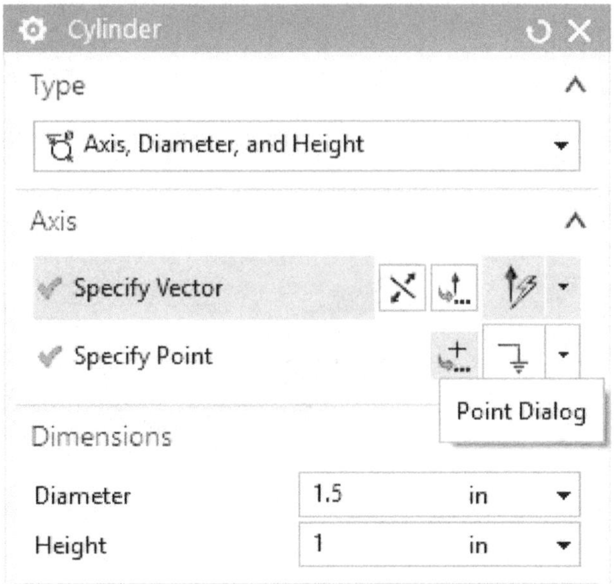

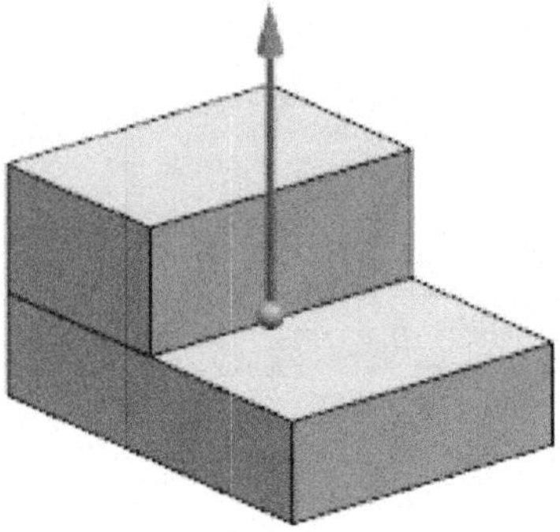

- In the **Point** dialog, set the **Type** to *Point on Face*, and place a point on the top face of the top block with U and V parameters both set to *0.5*. This will center your cylinder on top of the top "step."

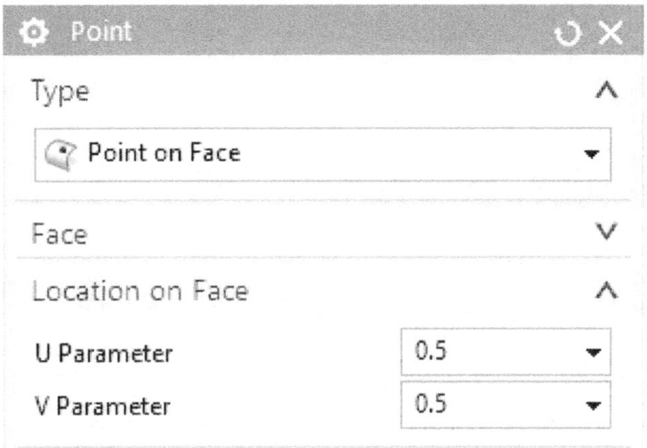

- Click **OK** to position the point, and then **OK** to create the cylinder.

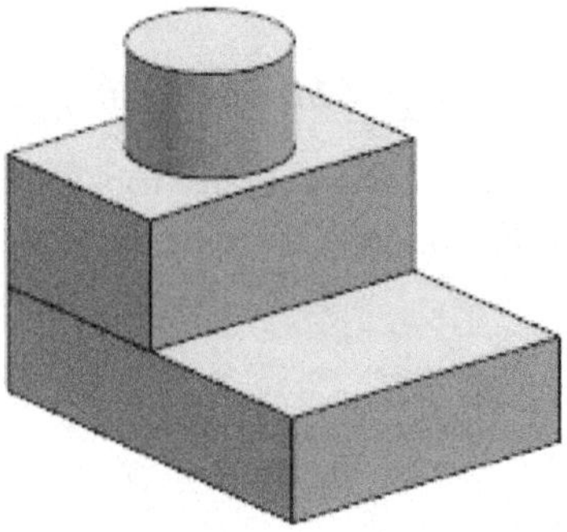

- **Save** your file!

While the cylinder inherited its orientation from the vector in the **Axis** section of the menu, both blocks inherited their orientations from the WCS. As such, their orientation is not fully associative, and for this reason, some users object to the use of primitives. You will have to decide for yourself whether you care – there are plenty of design scenarios in which the orientation simply doesn't matter!

7.9.3 Sphere

In the exercise below, you will add a sphere to the model by way of the **Sphere** tool!

- Continue working with *"Primitives.prt"*, and select the **Sphere** tool.
- Enter a **Diameter** of *1 in* and push the **Point Dialog** button.

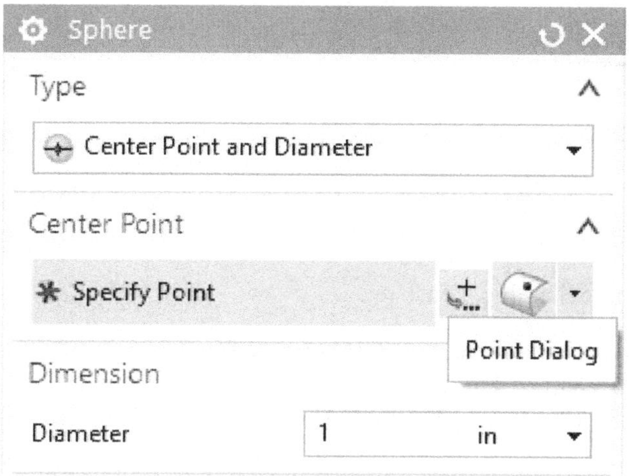

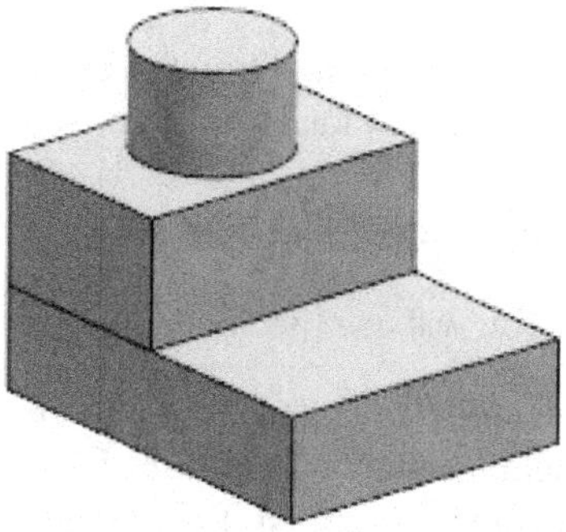

- In the **Point** dialog, set the **Type** to *Point on Face*, and select the top face of the lower block. Enter the parameters shown below. Click **OK** to return to the **Sphere** dialog.

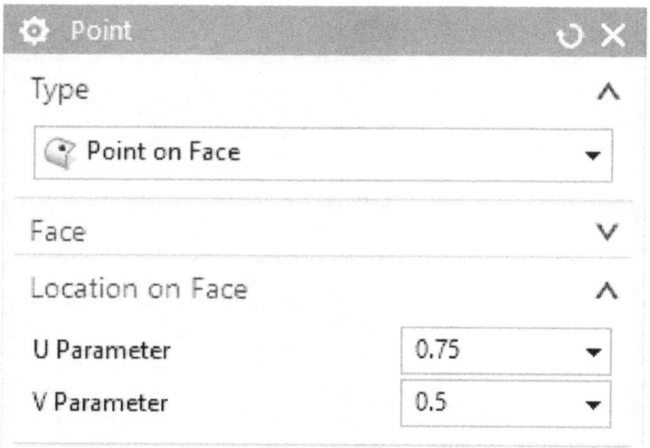

- Click **OK** to create a sphere as shown below.

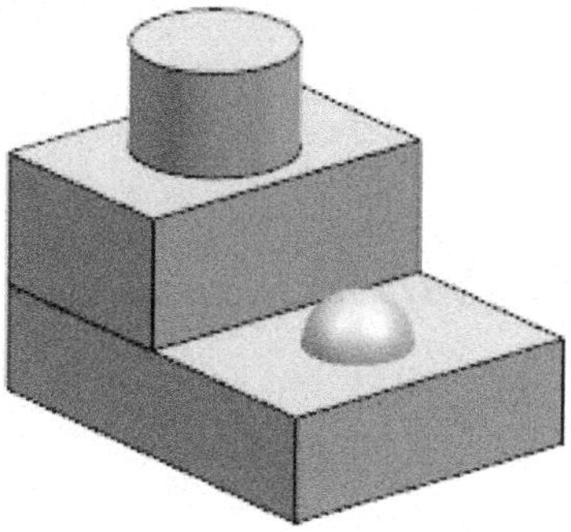

- **Save** your file!

7.10 Shell

Shell, like **Thicken**, is a command from the **Offset/Scale** group. We would be remiss not to introduce this tool early, as it facilitates the creation of cavities and plastic parts which require a uniform thickness throughout.

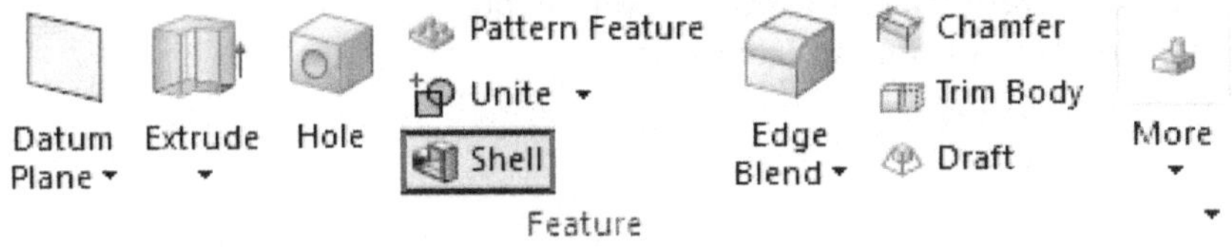

The **Shell** tool removes material from an existing solid body, resulting in a solid body of constant wall thickness. In defining a **Shell** feature, you specify faces that to be removed by the tool – the remaining walls are given the specified wall thickness.

7.10.1 Remove Faces, then Shell

In this exercise, you will use the **Shell** command to model a simple bushing.

- Create a new file (**Units**: *Inches*) called *"Bushing.prt"* and place it in *"C:\My NX 11 Files"*.
- Create a **Cylinder** with a **Diameter** of *3 in* and a **Height** of *4 in*.
- **Extrude** the bottom edge of the cylinder up in the +Z direction with the parameters shown below.

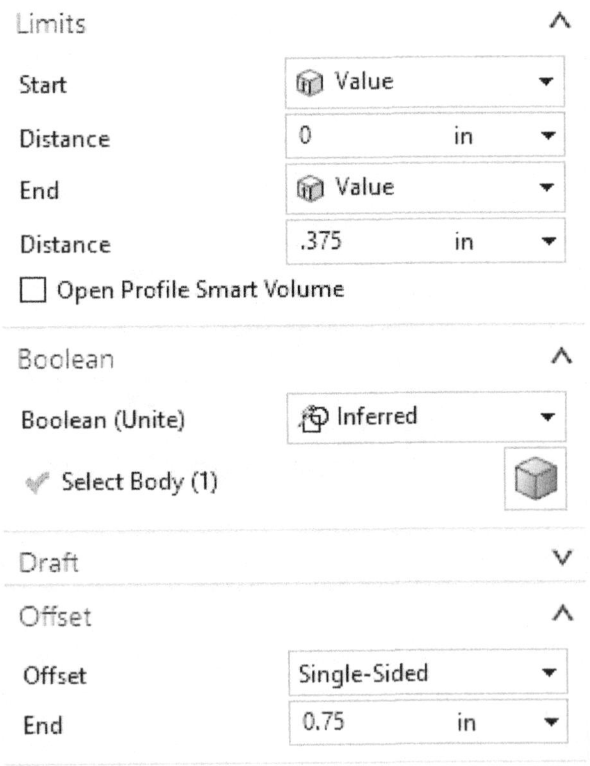

- Select the **Shell** tool, and input a **Thickness** of *0.375 in*. Choose the top and bottom faces of your model.

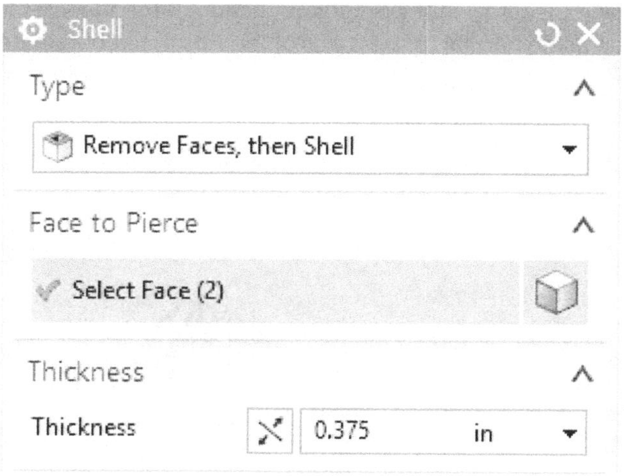

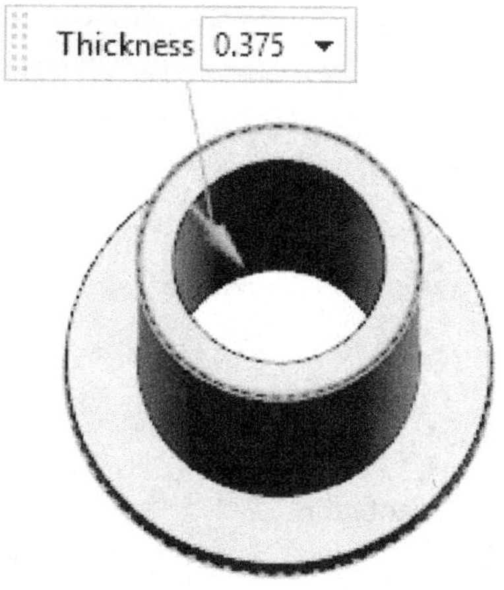

- **Save** your file!

7.10.2 Shell All Faces

In the **Shell** dialog, the **Type** *Shell All Faces* results in only the interior of your solid model being hollowed out – no faces are deleted, but an amount of material equal to the **Thickness** is left on the inside of all body faces.

- Create a new file (**Units**: *Inches*) called *"Shell All Faces.prt"* and place it in *"C:\My NX 11 Files"*.
- Use the **Sphere** command to create a sphere with a **Diameter** of *4 in*.

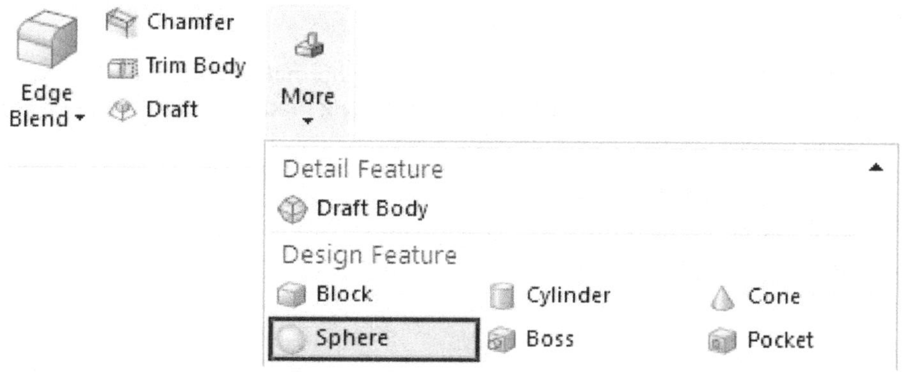

- Select the **Shell** tool and set the **Type** to **Shell All Faces**. Choose the solid body of the sphere as the **Body to Shell**. Set the **Thickness** to *0.25 in*.

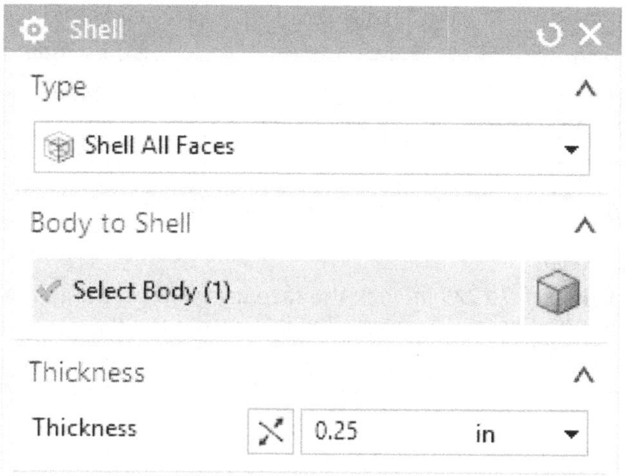

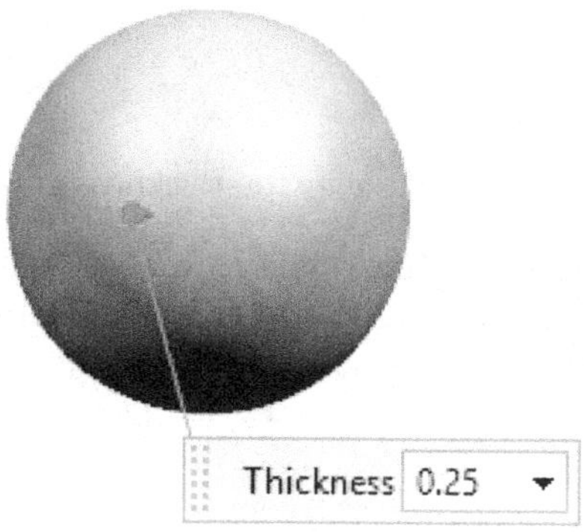

- Use the **Edit Section** command to get a look inside the body. In the **Set Plane** section of the **View Section** dialog, you can use the **Set Plane to X** icon to set the section plane normal to the X axis of the Absolute Coordinate System.

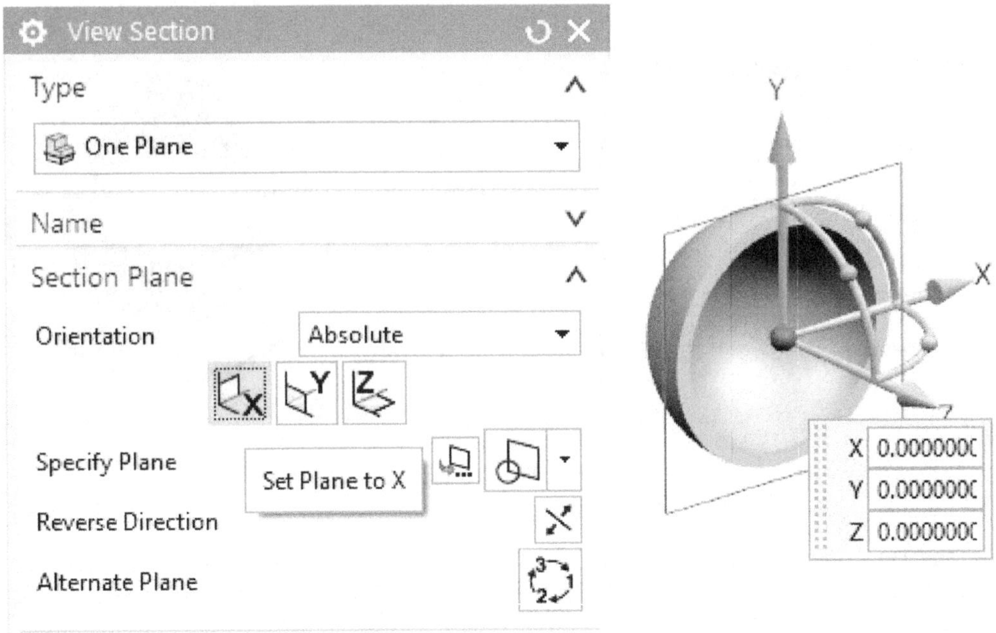

- **Save** your file!

7.11 Offset Face

The **Offset Face** command offsets one or more faces of a body along normal vectors at each point on those faces. The **Offset Face** feature is very useful when making small changes to existing models and can be applied to both parametric and non-parametric models.

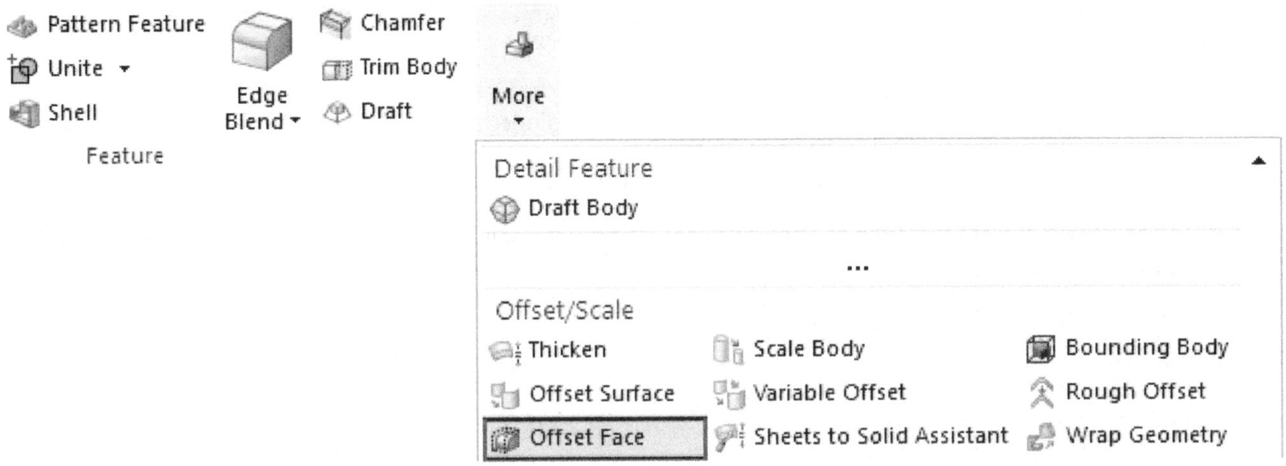

In this exercise, you will modify the thickness of the flange in the model from Chapter 7.10.1.

- Open the part file *"Bushing.prt"* from *"C:\My NX 11 Files"*.
- Select the **Offset Face** tool, and input an **Offset Value** of *0.1875 in*. Pick the face as shown in the following figure. If necessary, use the **Reverse Direction** button so that the offset is removing material from the flange.

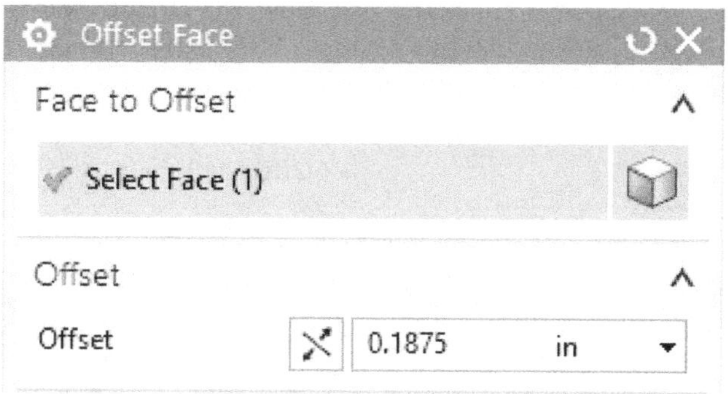

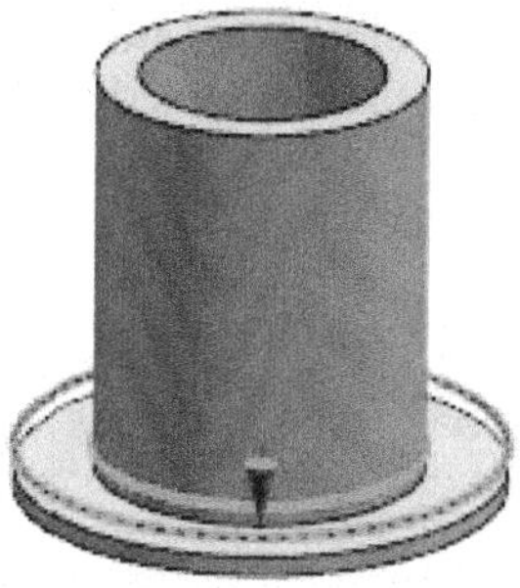

- **Save** your file!

7.12 Bounding Body

The **Bounding Body** command is new in NX 11. It provides a simple means of finding a block or cylinder which contains the entities selected with minimal excess. It is very useful if you need to design a mold, or to model the stock from which a part is to be machined. **Bounding Body** is found in the **Offset/Scale** group in the **More** gallery of the **Feature** group on the **Home** tab.

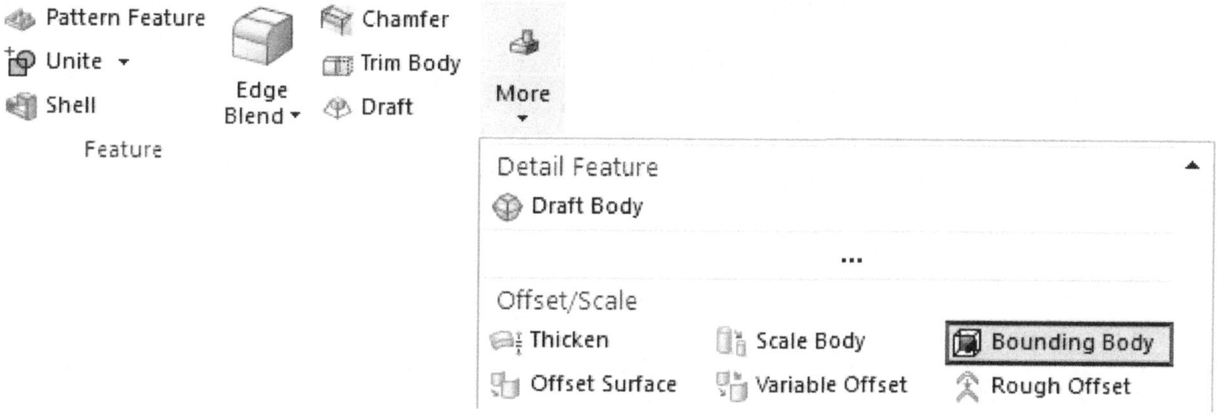

- Continue working with *"Through Curves – Sheet.prt"*.
- Select the **Bounding Body** tool and choose the solid body of *Thicken* as the **Object**.

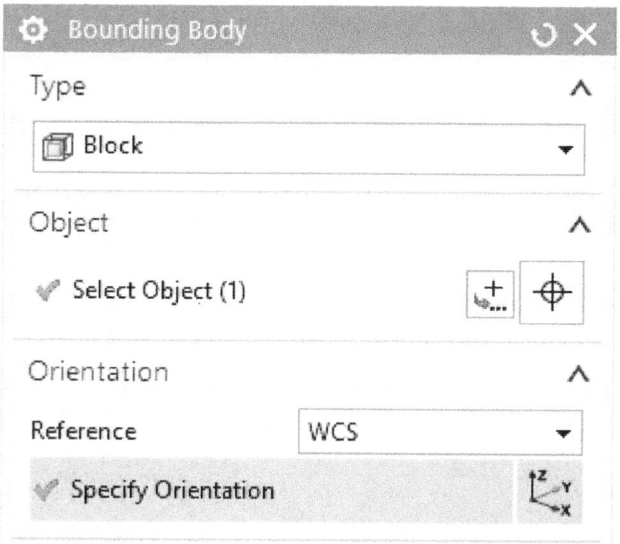

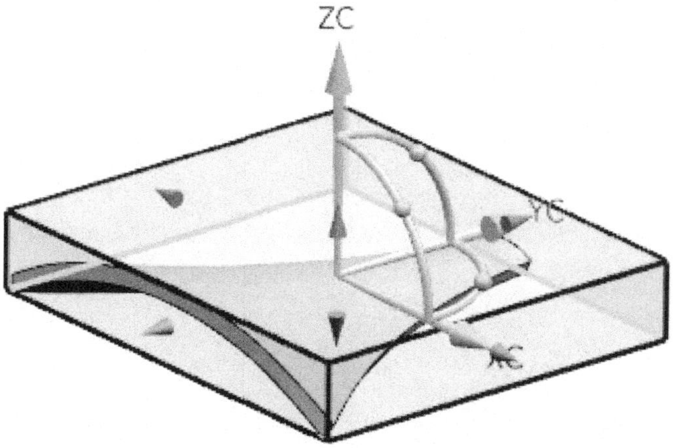

- In **Parameters**, you have the option to set a *Single Offset* – check this box. Now, the **Offset** value determines how far all faces of the bounding body should be offset – you can use this to add a little extra material. By disabling *Single Offset*, you get the ability to specify individual offsets on each of the six faces.

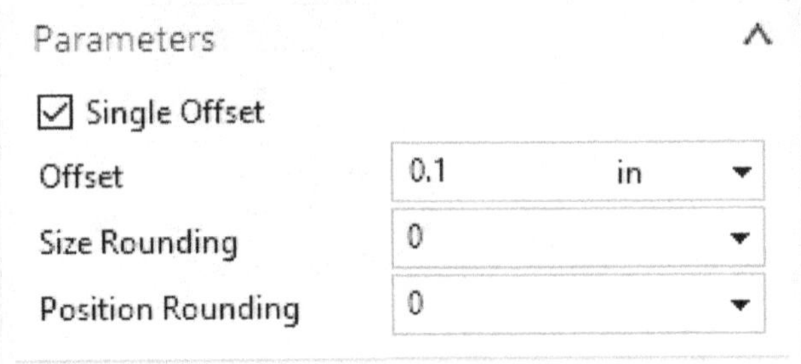

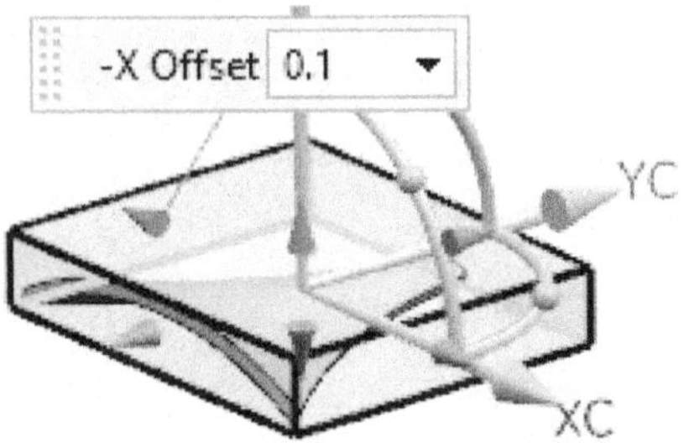

- Click **OK** to create the *Bounding Body* feature. It remains translucent. (You may need to enable translucency on the **View** tab).

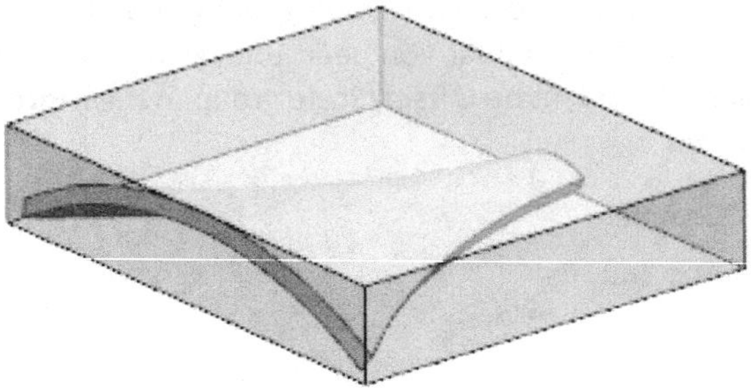

- **Save** your file!

8 Boolean, Combine, and Trim Operations

Building complex solid geometry often requires combining multiple entities. In NX, a **Boolean** operation is a feature operation that takes a **Target** body as input, and combines it with one or more **Tool** bodies. There are three Boolean operations – **Unite**, **Subtract**, and **Intersect**.

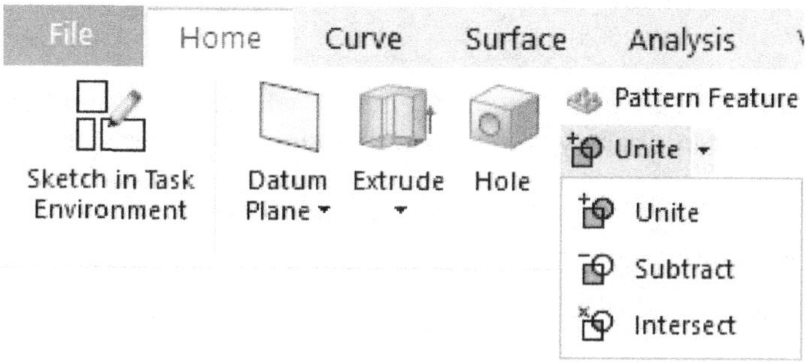

The next figure illustrates the effects of each **Boolean** operations on solids.

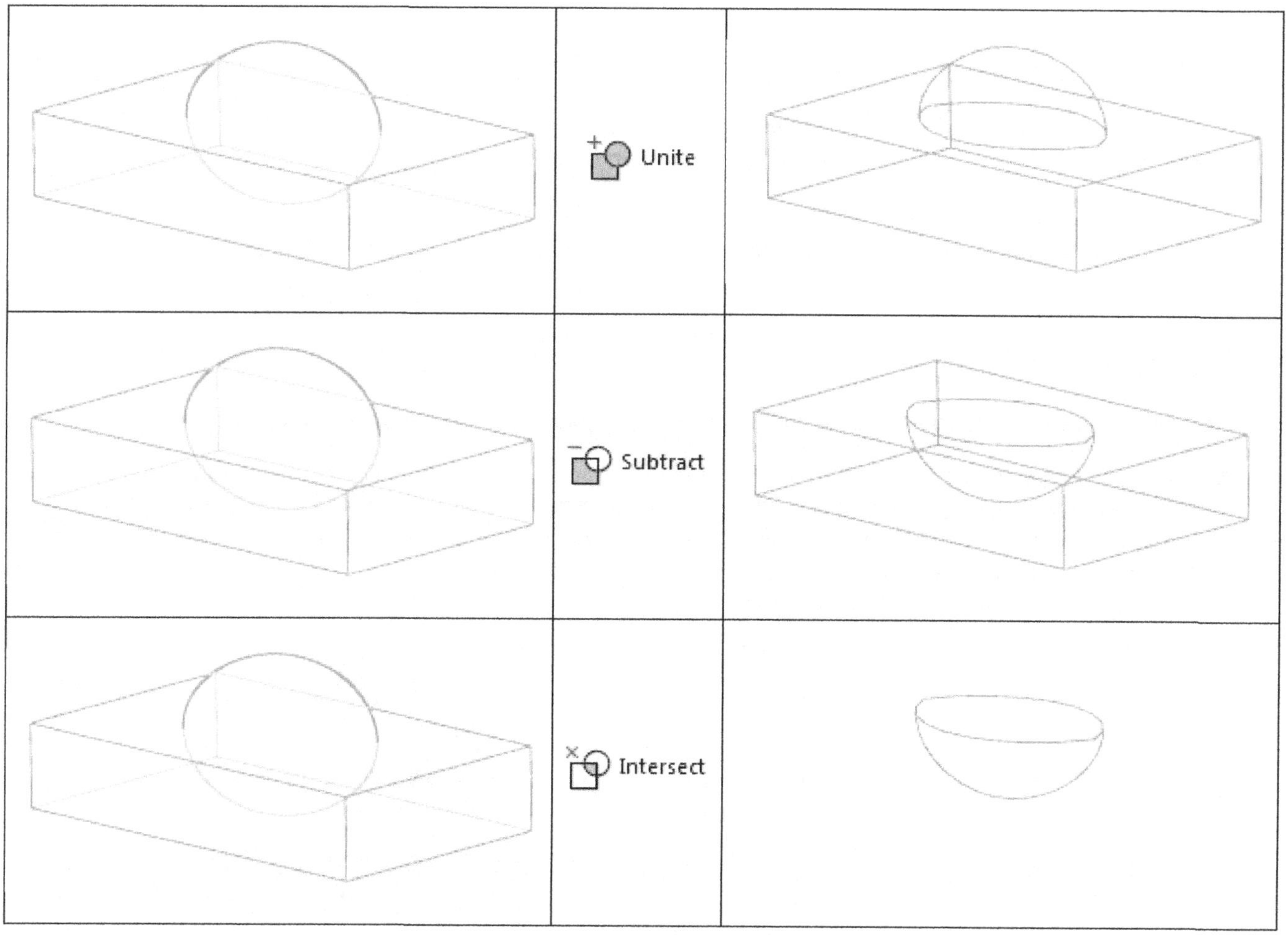

*Note: For **Unite**, **Subtract** and **Intersect**, NX 11 allows you to retain the **Target** and **Tool** body designations. Even after the **Boolean** operation has been completed, it is possible to differentiate between the original two parts.*

Boolean operations are of such fundamental importance that they are included as an option in many solid modeling tools, such as **Extrude**, **Revolve**, **Sweep Along Guide**, **Block**, **Cylinder**, and more. When an internal **Boolean** option is given, it will appear on the dialog box, often with a default setting of *Inferred*.

In addition to **Boolean** operations, there are a number of other useful combination operations in NX found in the **Combine** and **Trim** groups, located within the **More** gallery in the **Feature** group on the **Home** tab.

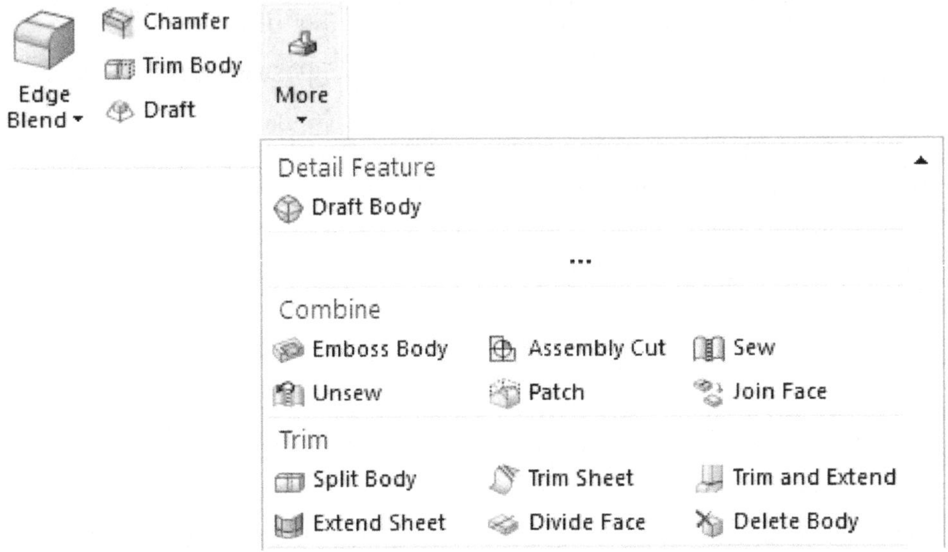

8.1 Unite and Subtract

In this example, you will create combine four bodies into a single solid body using **Unite** and **Subtract** operations.

- Open the file *"Primitives.prt"* that you created in Chapter 7.9 from *"C:\My NX 11 Files"*. If you skipped that exercise, you can find the same geometry in the part file *"Unite and Subtract.prt"* in *"C:\My NX 11 Files\Downloaded Files"*.
- Select the **Unite** command from the ribbon. For the **Target** select the cylinder, and for the **Tool** select the two blocks, as shown below. Click **OK** to combine them into a single solid body.

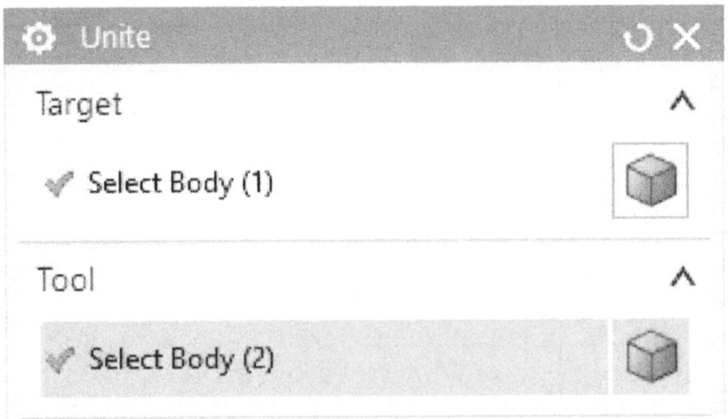

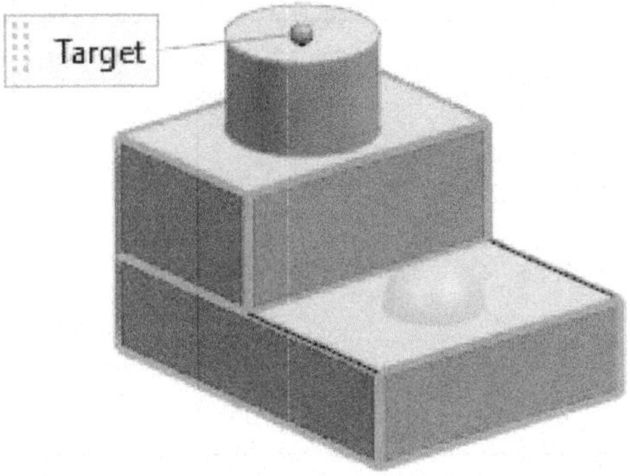

- Next, select the **Subtract** command. Choose the solid body that resulted from the last operation as the **Target**, and the sphere as the **Tool**. There is a checkbox for *Keep Tool* within **Settings** that will allow you to retain the solid body of the **Tool**, but in this case, leave it unchecked so that the **Tool** actually becomes deleted from your model.

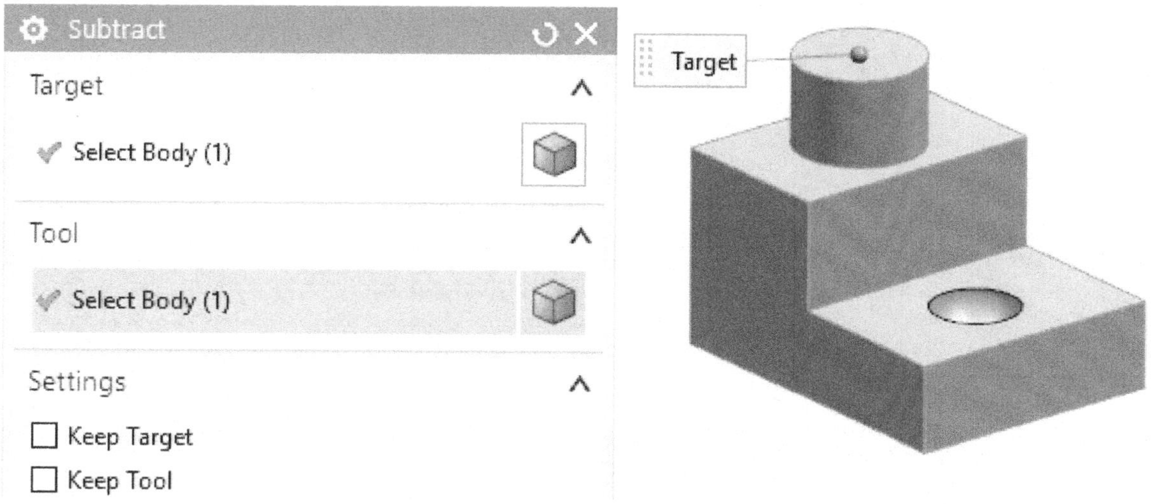

- You now have a single solid body in your model. You may find it confusing that when your **Selection Type Filter** is set to *No Selection Filter*, the *Block* and *Cylinder* appear to be separate! This is because the setting *No Selection Filter* prioritizes features because of the default setting *Top Selection Priority – Feature* described in Chapter 3.3.11. If you set the **Selection Type Filter** to *Solid Body*, the entire body will highlight as one object.

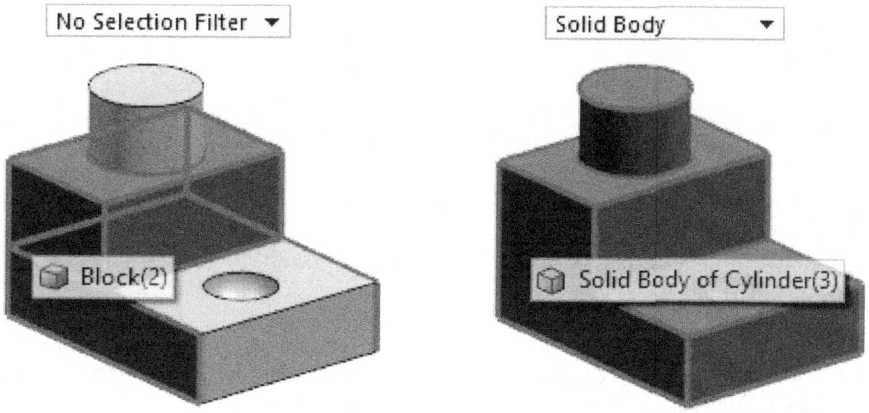

- **Save** your part file!

8.2 Intersect

Intersecting two solids is a great way to make complicated shapes. In this exercise, you will make a tab that must be curved with a radius in one direction. Since this geometry is curved in two planes, it is a good candidate for construction using the **Intersect** operation.

- Create a new file (**Units**: *Inches*) called *"Intersect.prt"* and put it in your folder *"C:\My NX 11 Files"*.
- Create the sketch shown below on the XY plane of *Datum Coordinate System (0)*.

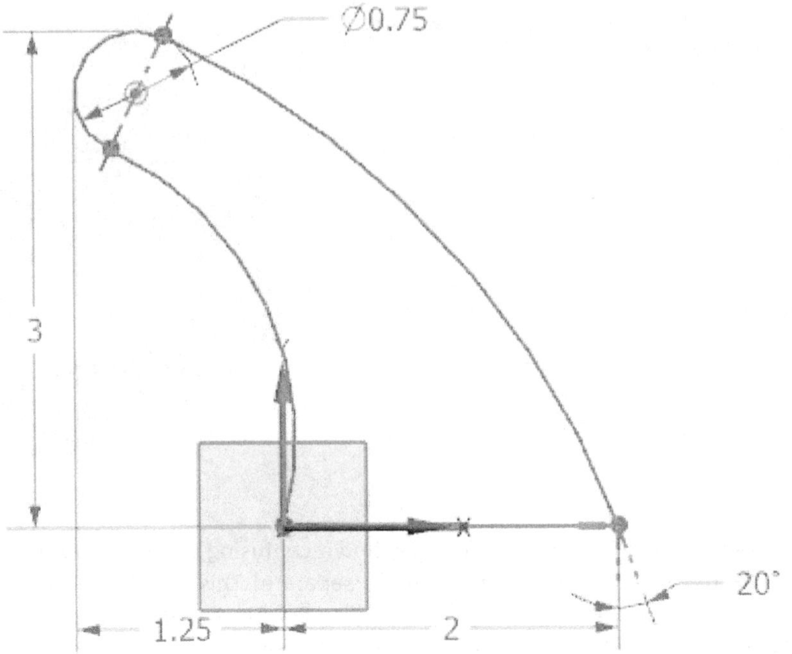

- **Extrude** your sketch symmetrically by *0.5 in*, and then make the sketch internal to the *Extrude* feature. **Hide** your datum coordinate system.

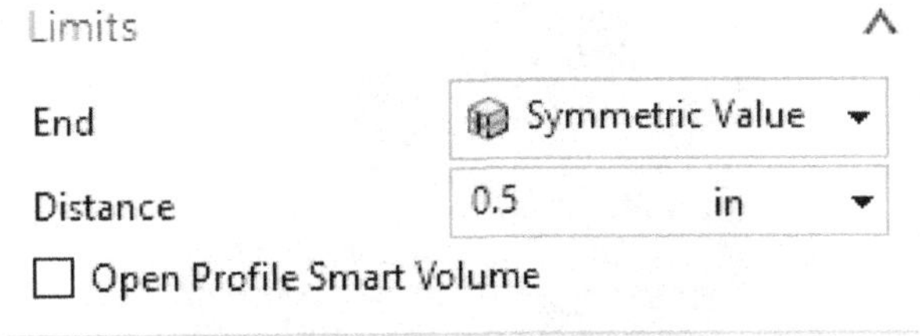

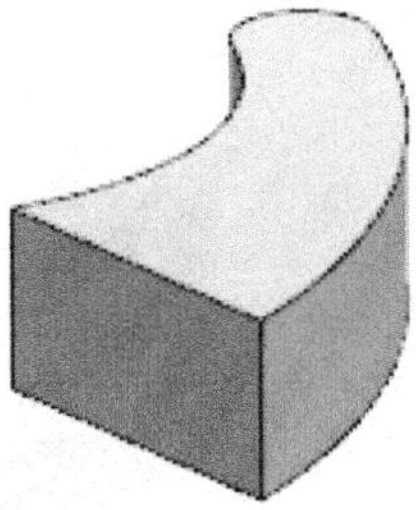

- Create another sketch, this time on the YZ plane of the datum coordinate system. Make sure your sketch coordinate system is oriented as shown below.

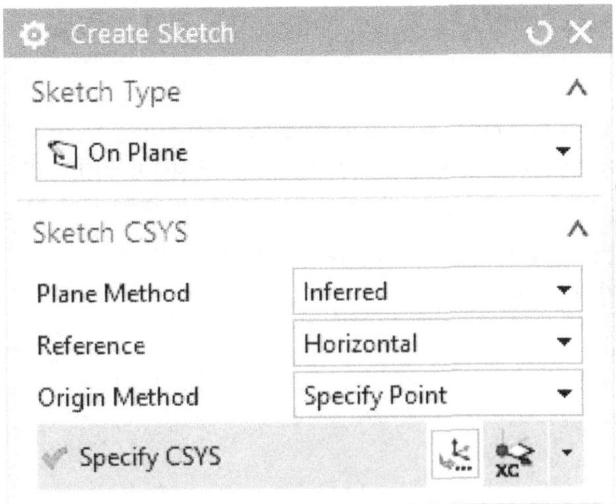

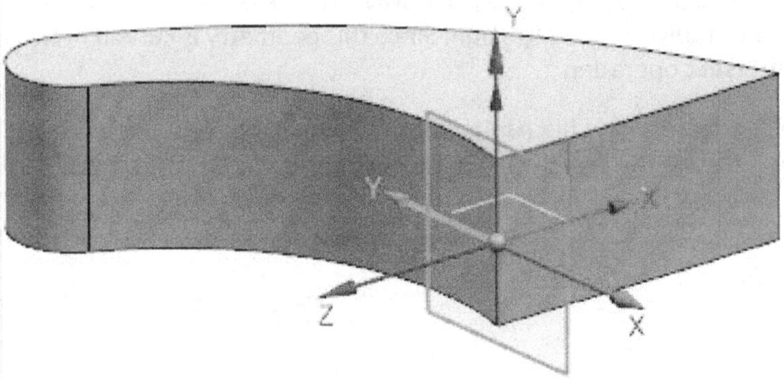

- Sketch the arc shown below. It must be tangent to the horizontal reference line on the left, and the leftmost endpoint should be constrained horizontally to the midpoint of the rightmost vertical edge.

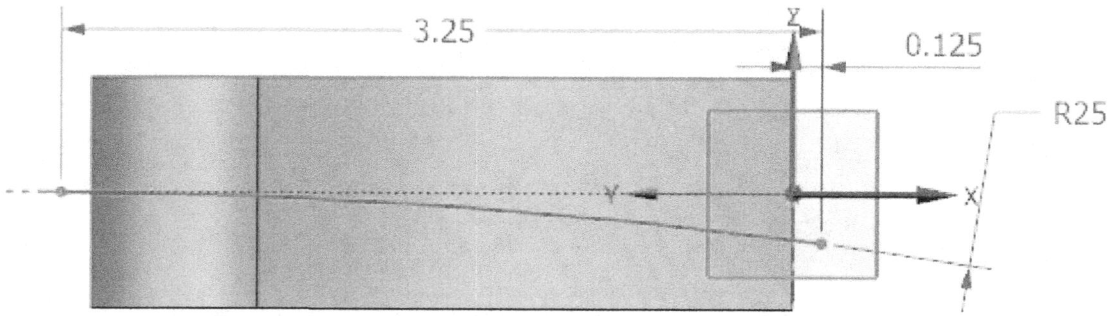

- **Extrude** the arc from *-1.25 in* to *2 in* in the YC direction. In the **Offset** section of the **Extrude** dialog, specify a **Symmetric Offset** of *0.2 in*. Make sure that the **Boolean** is set to *None*!

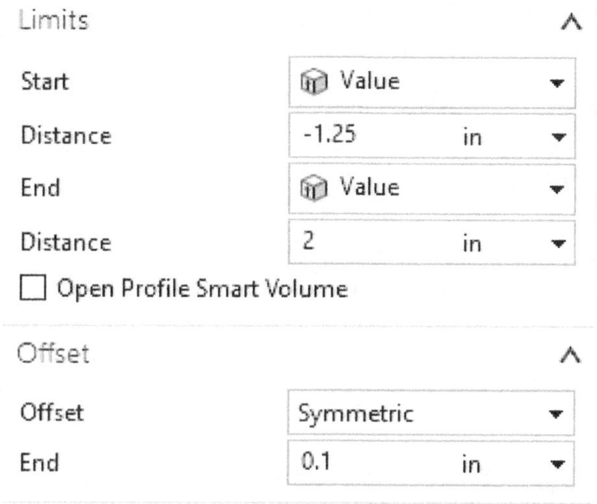

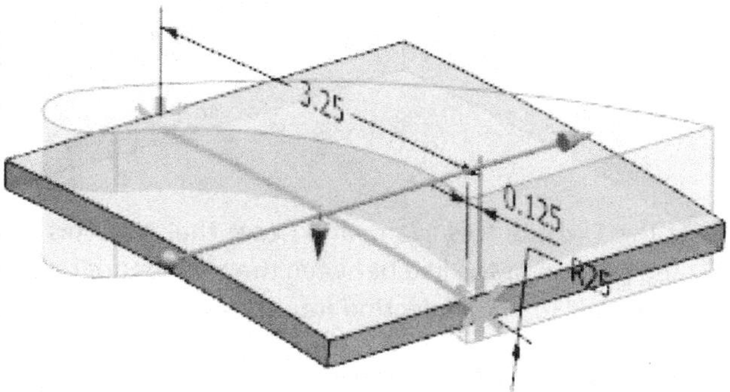

- Make the second sketch internal to the second **Extrude** and then select the **Intersect** tool. Choose one of the two bodies above as the **Target**, and the other as the **Tool**. Your model will appear as below.

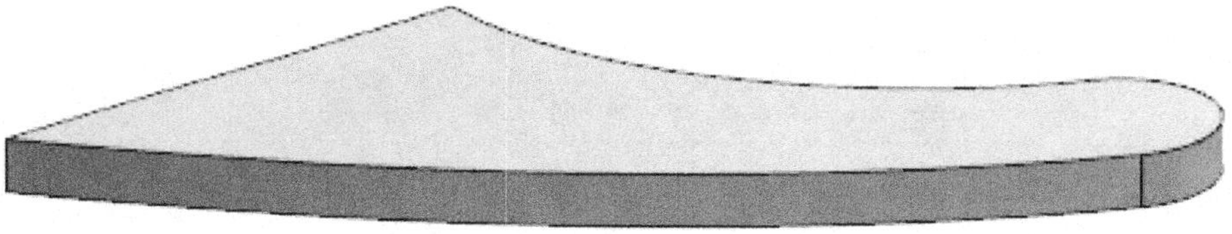

- **Save** your file!

8.3 Internal Booleans

In the previous exercise, you explicitly set the **Boolean** option to *None* when creating the second extruded body. In this exercise, you will learn what to expect from the *Inferred* setting. Usually, the body you are creating passes through another, *Inferred* with result in a *Subtract*, whereas if it protrudes out of another body, *Inferred* will result in *Unite*.

- Create a new file called *"Internal Booleans.prt"*. Create the sketch shown below on the XY plane of the datum coordinate system. Note that there is only a single horizontal line along the bottom of the sketch.

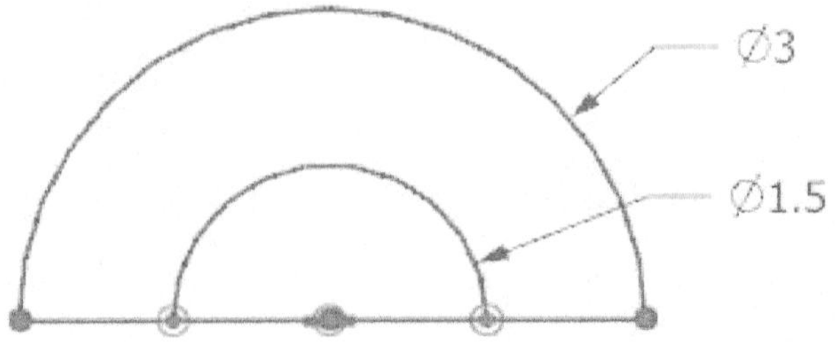

- **Extrude** the outer profile from *0 in* to *0.5 in* in the –Z direction, to produce the solid shown in the following figure. You will probably want to set your **Selection Intent** to *Connected Curves.*

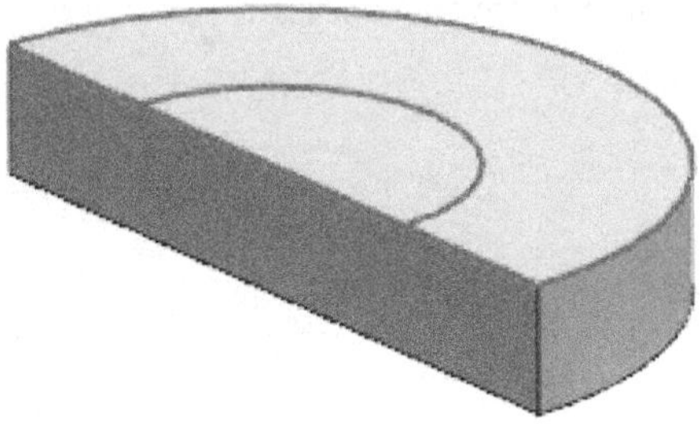

- Select the **Extrude** tool again, and **Reset** the tool. This time you will extrude just the inner profile. In order to select only the portion of the line between the smaller arc's two endpoints, you will need to enable the **Stop at Intersection** tool, adjacent to your **Selection Intent** drop-down.

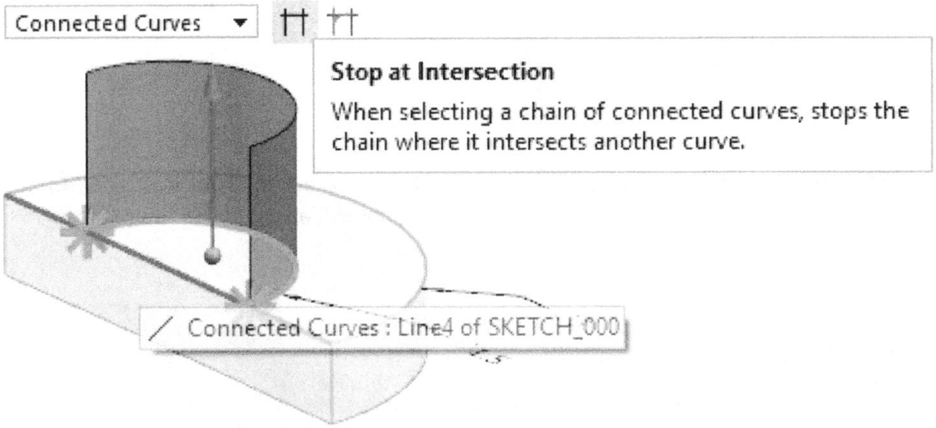

- Even though the whole line highlights when you go to make your selection, thanks to **Stop at Intersection**, you will only select the portion of the line on the arc interior. **Extrude** the inner profile from *0 in* to *1 in*. Note that when the **Boolean** is set to *Inferred* here, the inference is that a *Unite* is desired.

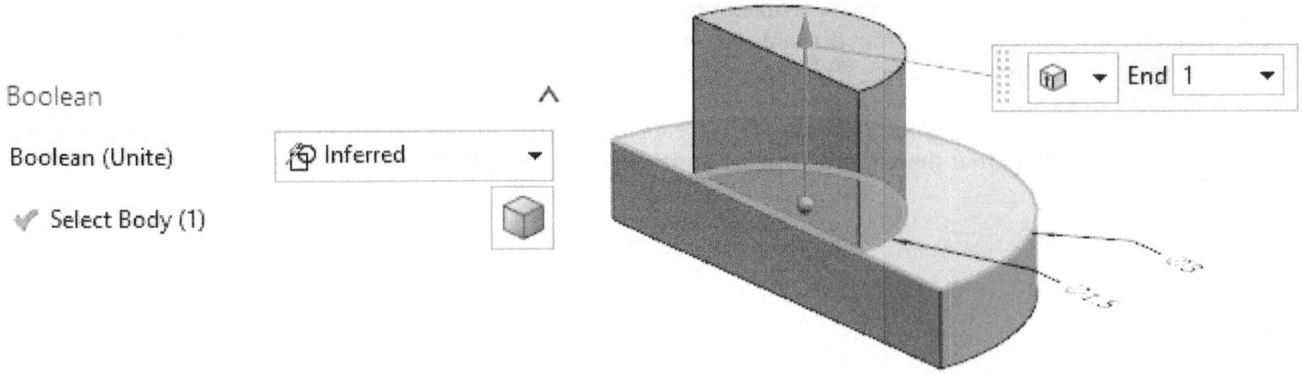

- Create the following sketch on the top face of the solid.

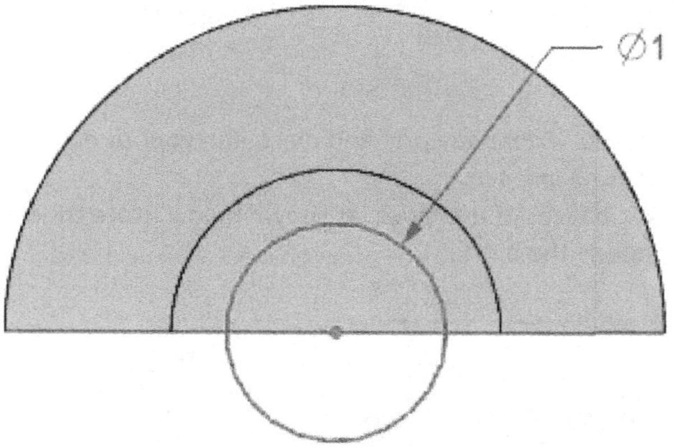

- **Extrude** the sketch you just created downward into the body, and set the **End** to *Through All*, rather than *Value*. Note that this time, the *Inferred* setting gives rise to a **Boolean** of *Subtract*.

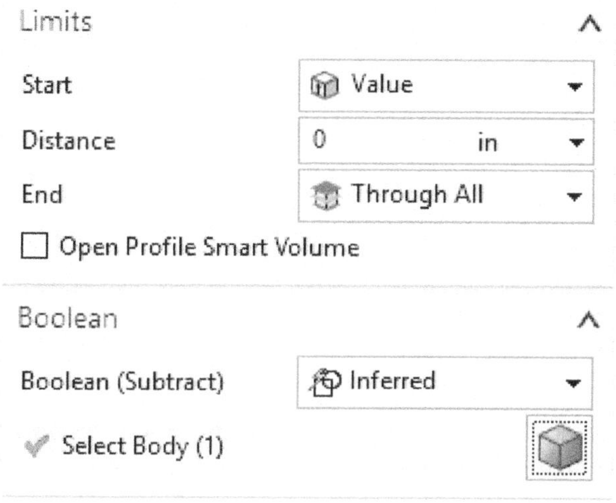

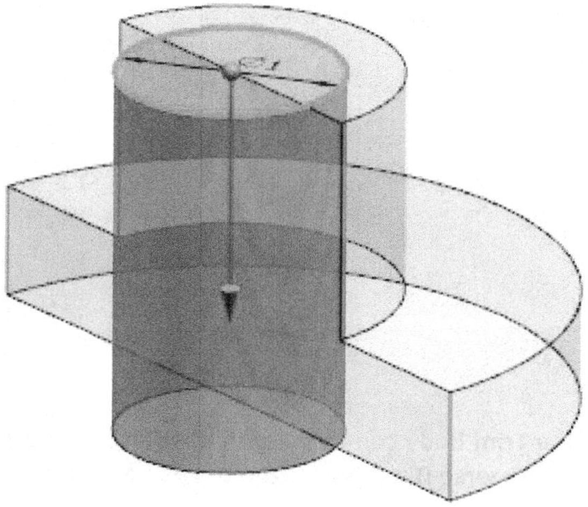

- Click **OK** to finish your solid, and **Save** your file!

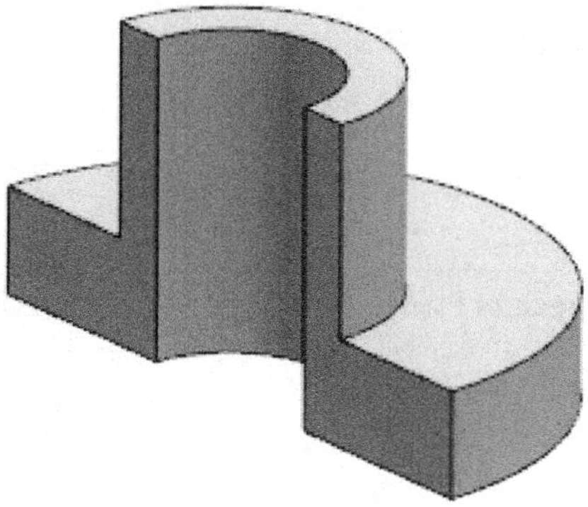

8.4 Trim Body

The **Trim Body** tool is used to remove one half of a body, where the body is divided into two regions using datum plane or a sheet body as a trimming tool. The **Trim Body** command is found in the **Feature** group on the **Home** tab.

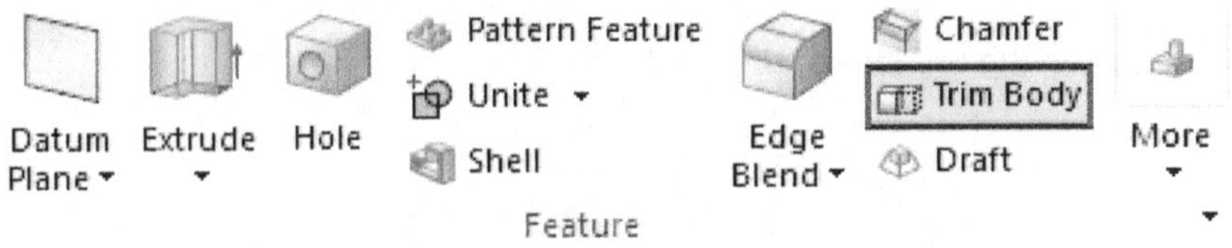

- Create a new file (**Units**: *Inches*) called *"Trim Body.prt"* and place it in your directory *"C:\My NX 11 Files"*.
- Create a **Block** with dimensions *2 in* x *2 in* x *4 in*.
- Sketch an arc on one of the *2 in* x *4 in* faces of the block, as shown below. Note the horizontal alignment of endpoints, and the tangency with the *2 in* edge of the block.

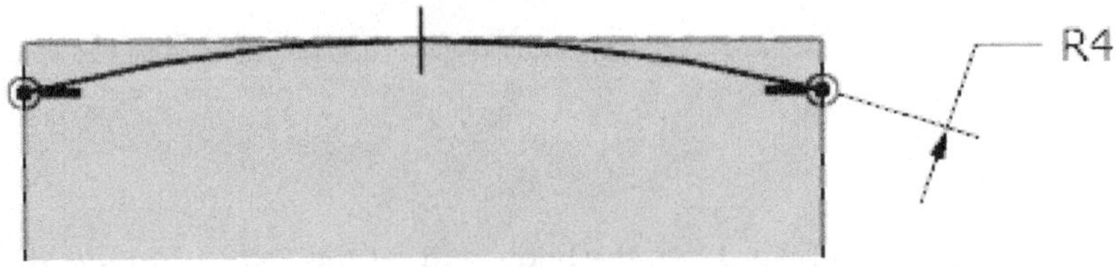

- **Extrude** your sketch from *-0.5 in* to *2.5 in* toward the block, to produce a sheet body as shown below. Make the sketch internal to the sheet.

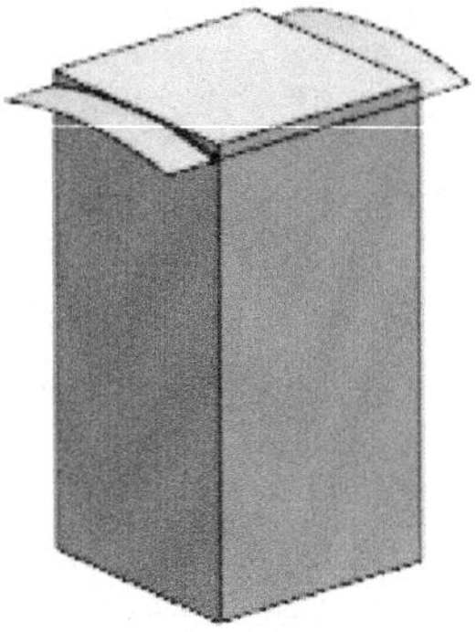

- Use the **Trim Body** command with the block as the **Target** and the sheet just created as the **Tool**. You may need to use the **Reverse Direction** button to ensure that the discarded region is the small one, as shown below.

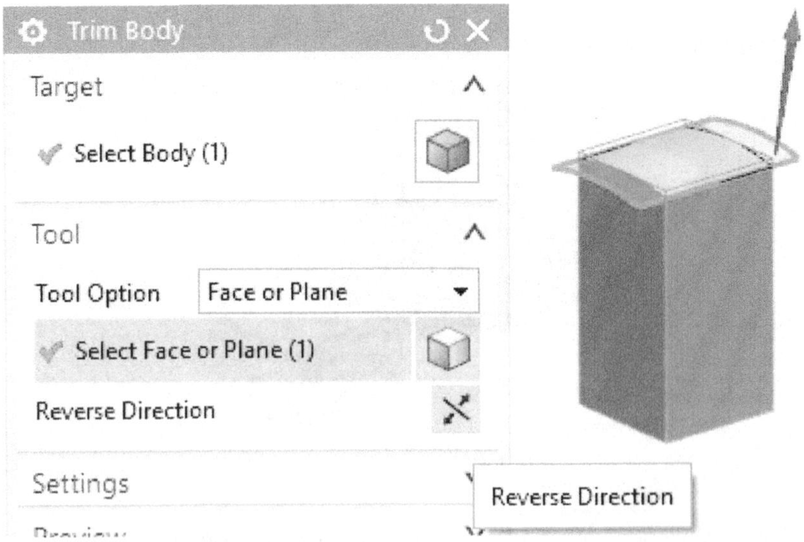

- **Hide** the sheet body. Create an identical sketch on the adjacent *2 in* x *4 in* face. Note that you can constrain the endpoints of the arc to the endpoints of the edges this time!

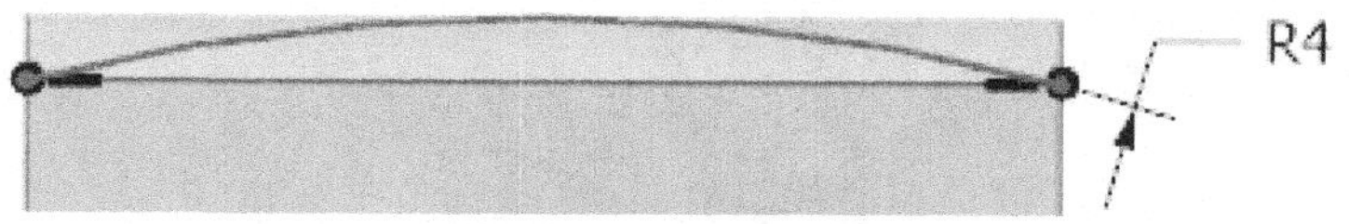

- **Extrude** this sketch to create another sheet body, and use **Trim Body** to trim the solid with the sheet again! Make the sketch internal to the sheet.

- **Save** your file!

8.5 Delete Body

Delete Body is an extremely useful trim operation that we will use in the next few exercises (and throughout the remainder of the book). Quite simply, it deletes a sheet or solid body from your model. You might ask – why use **Delete Body** over **Delete**? The difference is that **Delete Body** is a parametric and fully associative operation that respects dependencies.

This tool is fantastic for cleaning your model of construction geometry that you otherwise would need to hide in order to clearly see your model, but still allows for modifications to that construction geometry, should the need for changes arise.

The **Delete Body** command is found in the **Trim** group – way down at the bottom of the **More** gallery in the **Feature** group on the **Home** tab.

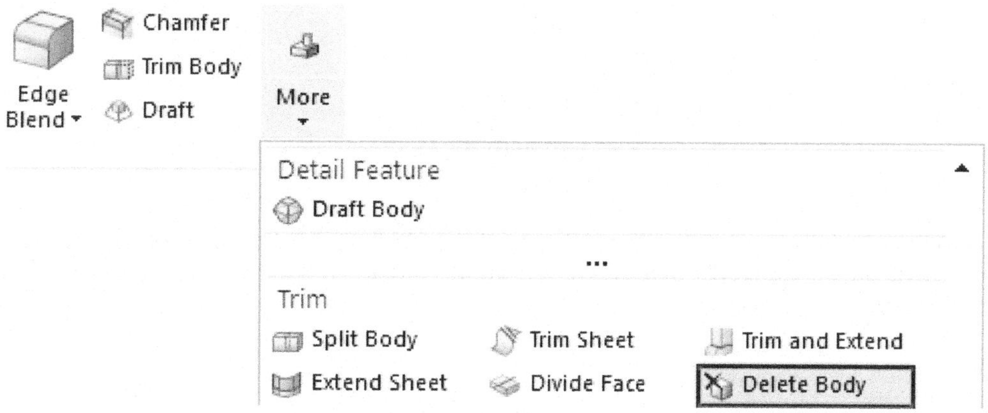

- Continue working with *"Trim Body.prt"*, and **Show** the two sheet bodies in the part file.

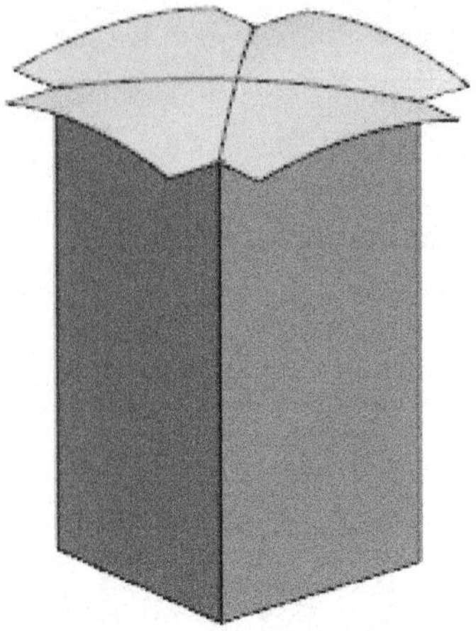

- Use the **Delete Body** command to delete the two sheet bodies.

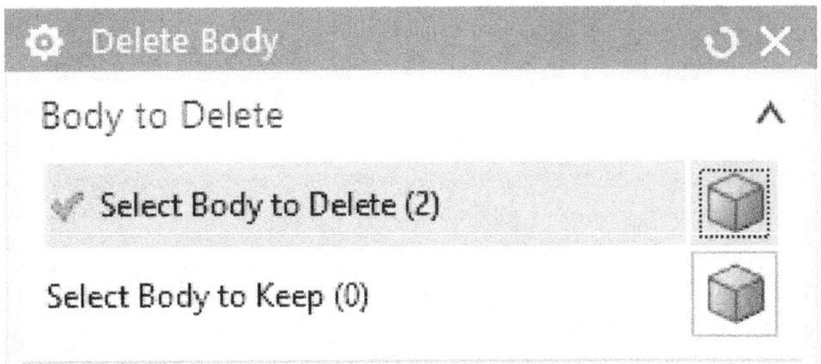

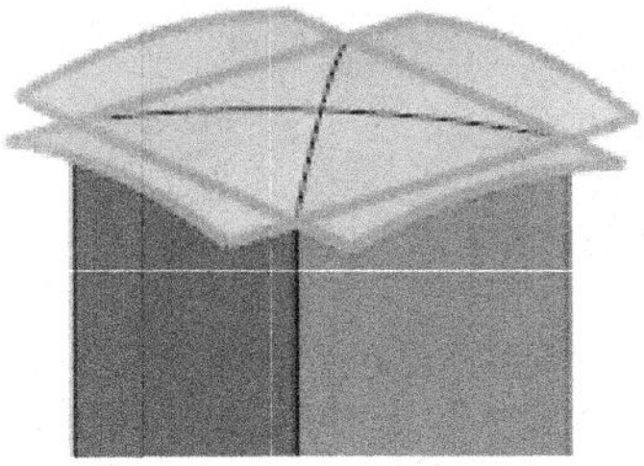

- Upon completing the operation, the sheet bodies will be deleted and in the Model History you will see a new feature called *Delete Body (7)*.

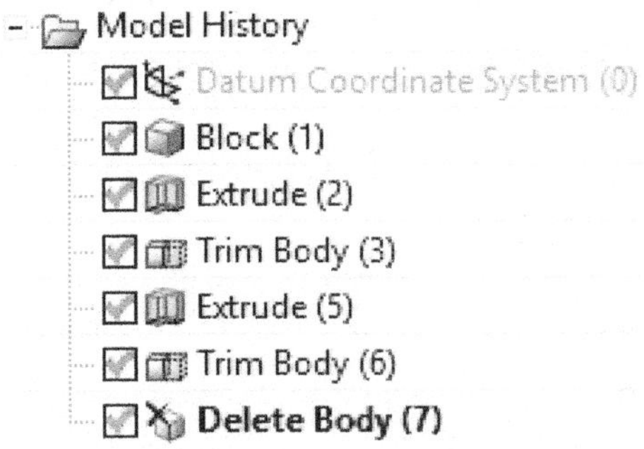

- The beauty of **Delete Body** is that it respects the dependencies between *Extrude (2)* and *Trim Body (3)*, and *Extrude (5)* and *Trim Body (6)*. *Delete Body (7)* can be suppressed or even deleted if you need to recover the sheet bodies. It's a deletable delete! What's not to love?
- **Save** your file!

8.6 Split Body

Split Body is basically the same as **Trim Body**, but it retains the **Tool** instead of discarding it. It is also found in the **Trim** group in the **More** gallery of the **Feature** group on the **Home** tab.

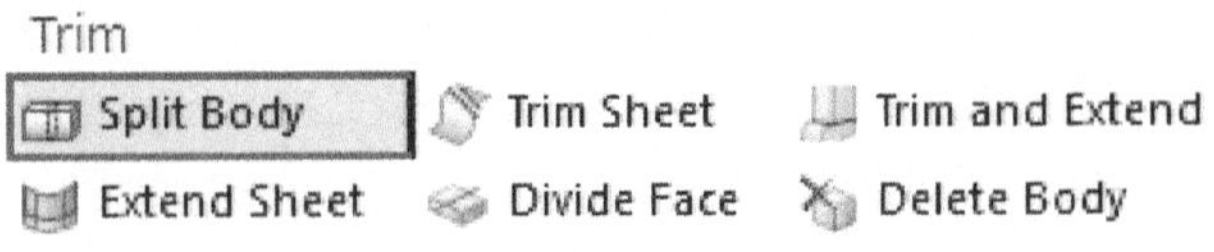

- Create a new part file (**Units**: *Inches*) called *"Split Body.prt"* and place it in your folder *"C:\My NX 11 Files"*.
- Create a **Block** with dimensions *2 in* x *2 in* x *2 in*.

- Use the **Datum Plane** tool with **Type** set to *Inferred*, and select two opposite corners of the block, and the midpoint from the edge between them to construct the datum plane.

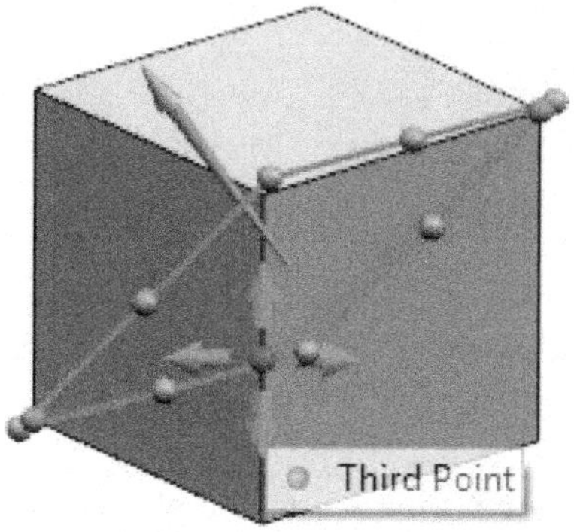

- Select the **Split Body** tool, and choose the block as the **Target** and the datum plane as the **Tool**. The output of your operation will be two separate solid bodies. Give them different colors!

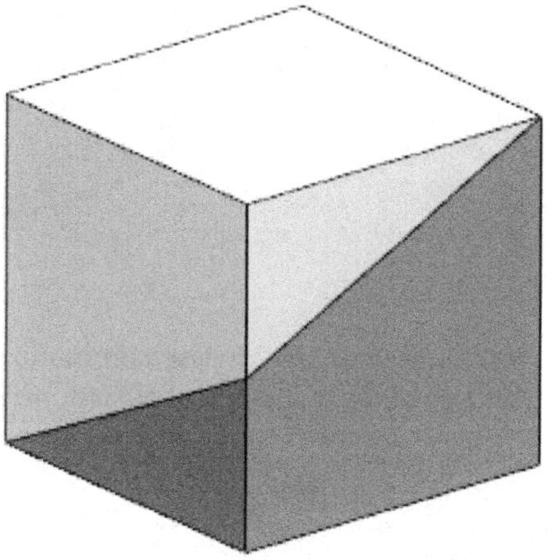

- **Save** your file!

8.7 Sew and Unsew

The **Boolean** and **Trim** operations discussed so far are great for combining solids bodies with each other or with sheet bodies, but to combine sheet bodies with one another, we need a different operation.

The **Sew** command is used to take distinct sheet bodies with shared edges, and merge them into a single sheet body. It doesn't remove edges, but it does result in a larger sheet body that has multiple faces. Conversely, the **Unsew** command is used to remove a face or set of faces from a sheet body. The selected face becomes a separate sheet body from the original.

The **Sew** and **Unsew** commands are found in the **Combine** group, in the **More** gallery of the **Feature** group on the Home tab.

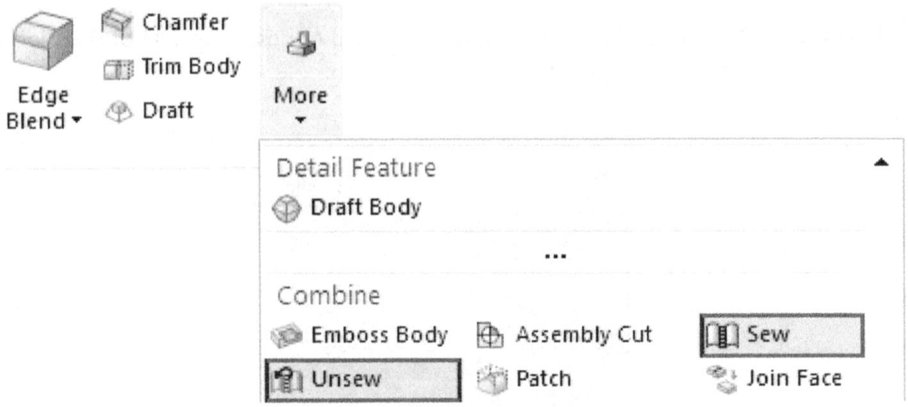

- Create a new file (**Units**: *Inches*) called *"Sew & Unsew.prt"* and place it in your folder *"C:\My NX 11 files"*.
- Create a **Block** with dimensions *2 in* x *2 in* x *0.25 in*. Select the **Unsew** command and use it to turn the face shown below into a separate sheet body.

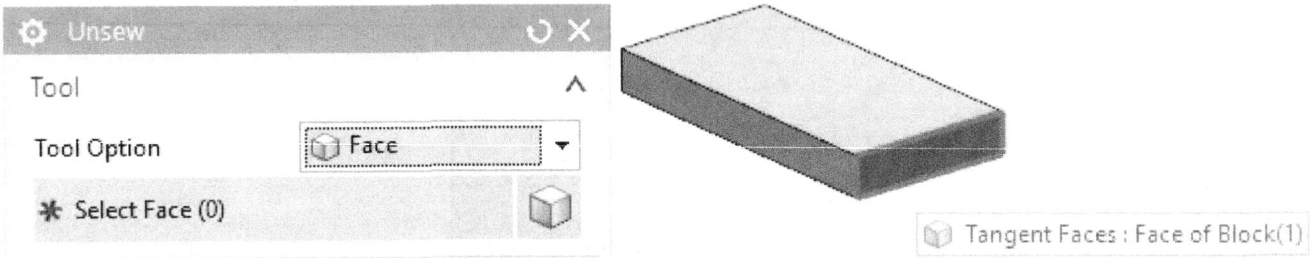

- Use **Delete Body** to then delete that face. Note that the **Unsew** command actually hollowed out the interior of the block – it went from being a solid body to two distinct sheet bodies with no material inside.

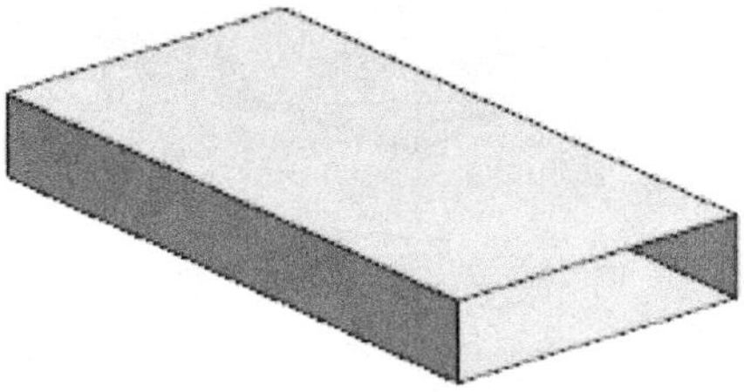

- Sketch an arc of radius *1 in* on the top face of the block as shown below. Its endpoints should be touching the face along the edge of the missing face.

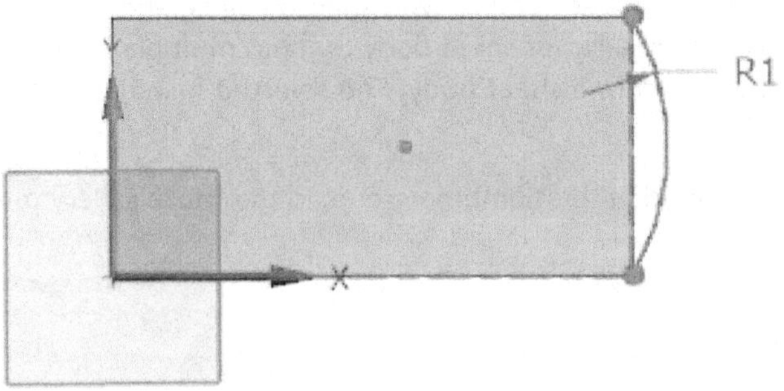

- **Extrude** the sketch in the -ZC direction from *0 in* to *0.25 in* to produce a sheet body as shown below. Make the sketch internal to the sheet body.

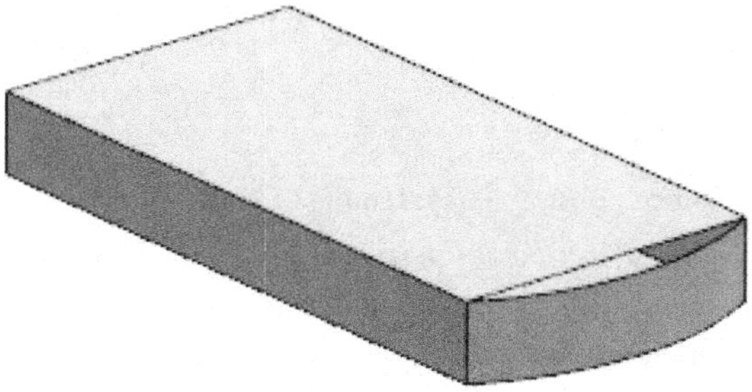

- Select the **Bounded Plane** tool. Set your **Selection Intent** to *Single Curve*, and choose the top edge of the block and the top edge of the curved sheet body as the **Planar Section**, as shown below.

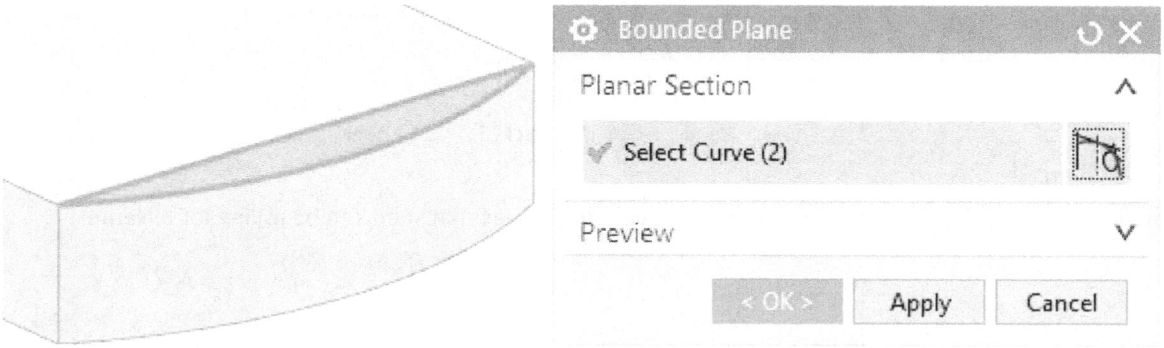

- Flip your model and do the same on the other side! Finally, we are ready to **Sew**!
- With the **Sew** tool, select the sheet body leftover from the block as the **Target**, and the three other sheet bodies (*Extrude (4)*, *Bounded Plane (5)*, and *Bounded Plane (6)* as the **Tool**.

- The result of your **Sew** operation is a solid body. To confirm this, set your **Selection Filter** to *Solid Body* and put your cursor on the body.

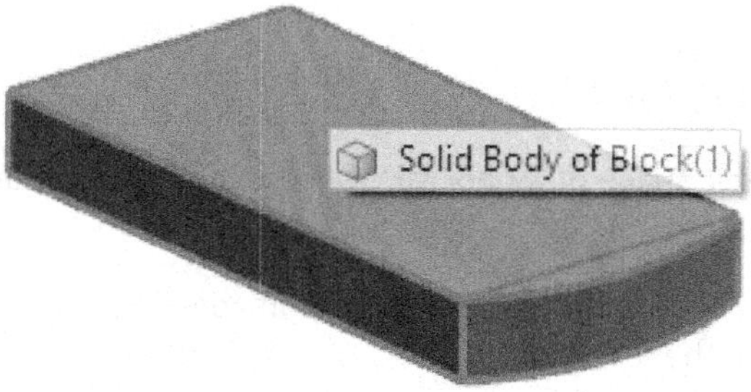

- To remove the smooth edges between the top face of the block and the bounded planes, you can use the **Join Face** command, also found in the **Combine** group.

- From the **Join Faces** dialog, select **On Same Surface**.

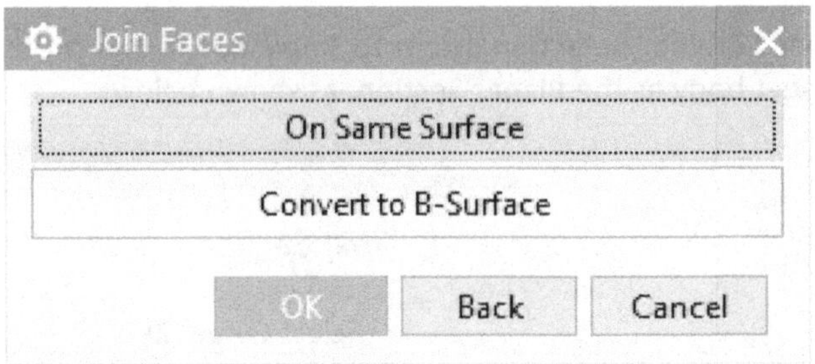

- At this point you will meet a rather cryptic dialog called **Join Faces** that seems to be asking for a **Name**.

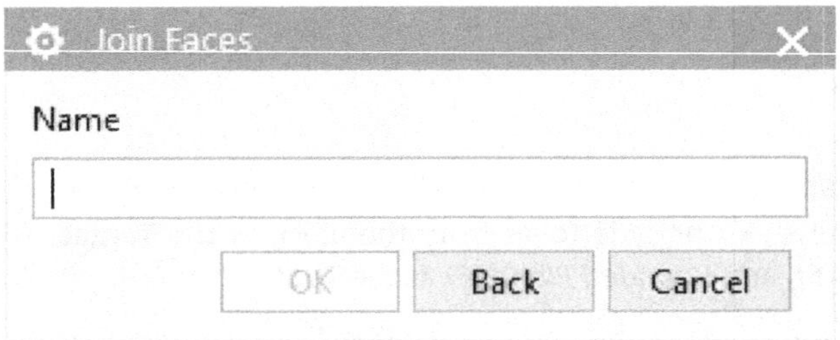

- Ignore the **Name** input, and look at your Cue Display (all the way at the bottom left of your NX window). Always look at the Cue Display when you aren't sure what the software wants from you!

Select body

- Select the solid body. Before you even have a chance to click **OK**, the smooth edges will have vanished between the faces of the block and the former bounded planes. **Join Face** is a rather odd tool, the dialog interface leaves plenty of room for improvement!

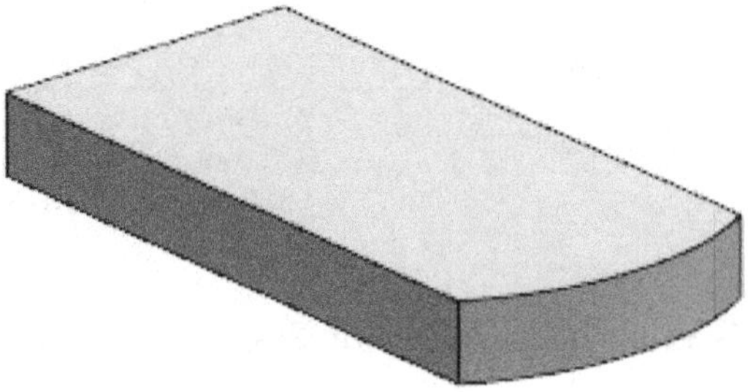

- Click **OK** to save your part file!

8.8 Project – The Pool Rack, Part I

In this exercise, you will create the basic shape of a pool rack with several applications of the **Extrude** tool, each with an appropriate **Boolean** setting.

- Begin by creating a new file called *"Pool Rack.prt"* and place it in your folder *C:\My NX 11 Files*.
- Create the sketch shown below on the X-Y plane.

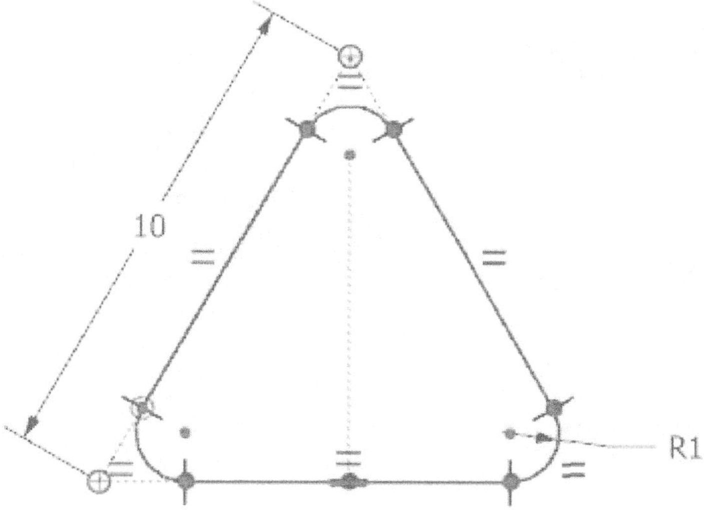

- **Extrude** your sketch in the +ZC direction from **0 in** to **1.5 in**. Specify a **Draft** *From Start Limit* of **2 degrees**, and a *Two-Sided* **Offset** from **0 in** to **0.25 in**.

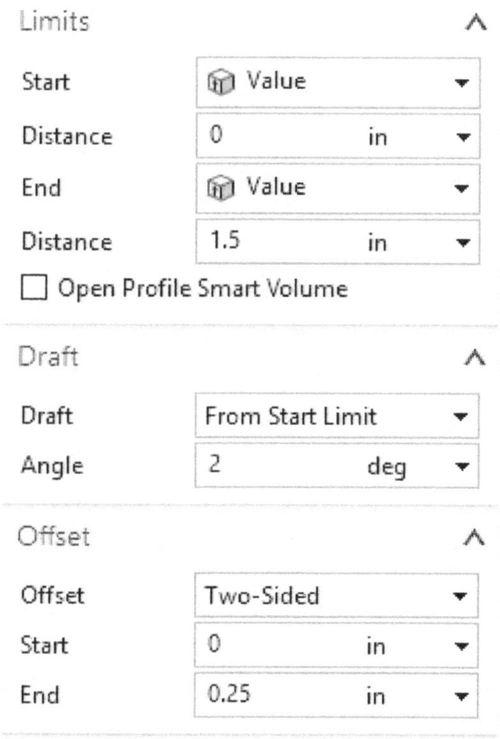

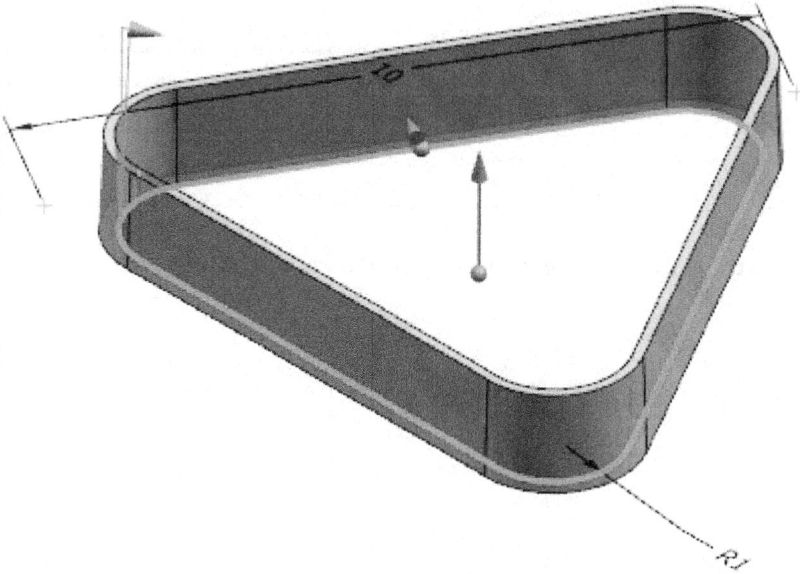

- Click **OK** to make your solid, then make the sketch internal to the extrusion.
- Next, **Extrude** the bottom interior edge of the solid from *0 in* to *0.125 in* in the +ZC direction. Make sure your **Selection Intent** is set to *Tangent Curves* to facilitate the selection. Set the parameters in your **Extrude** dialog box as shown below.

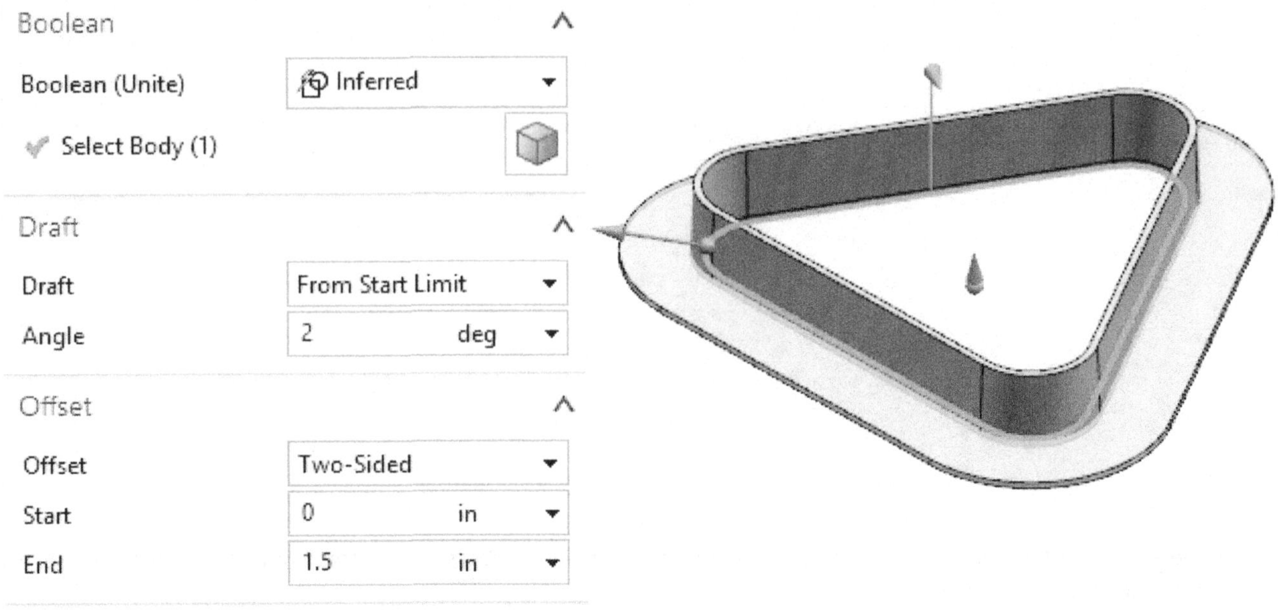

- Create one more sketch, this time on the top face of the last extrusion. Sketch a rectangle centered as shown below.

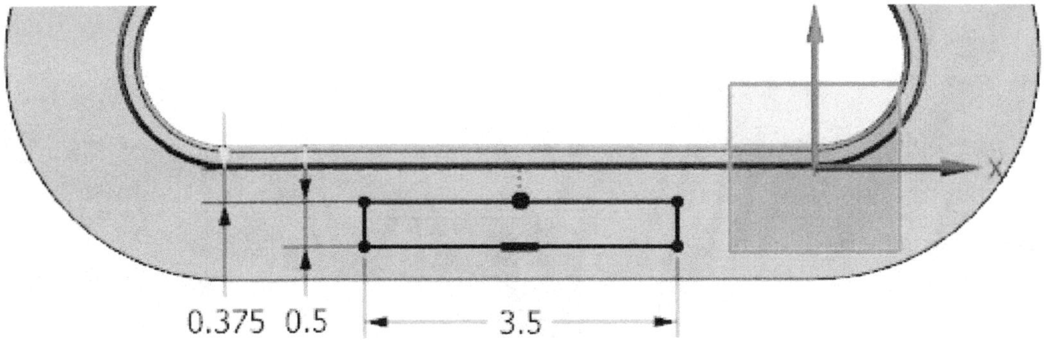

- **Extrude** this sketch back into the solid body with the settings as shown below.

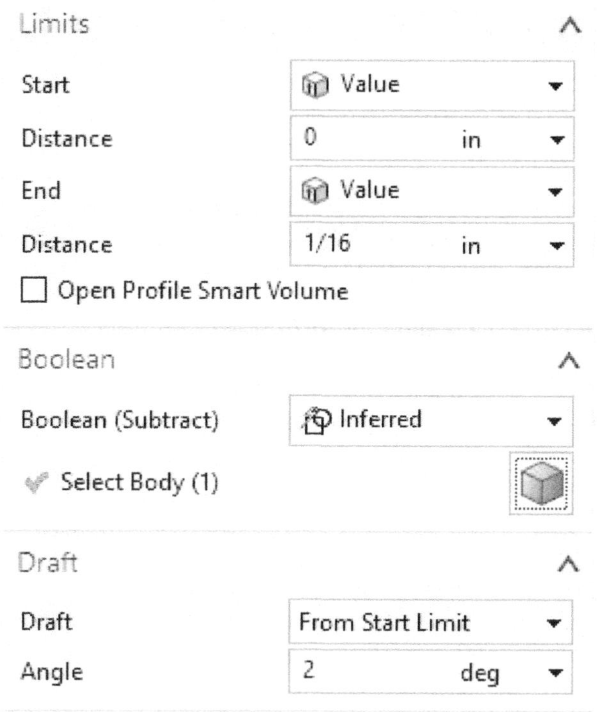

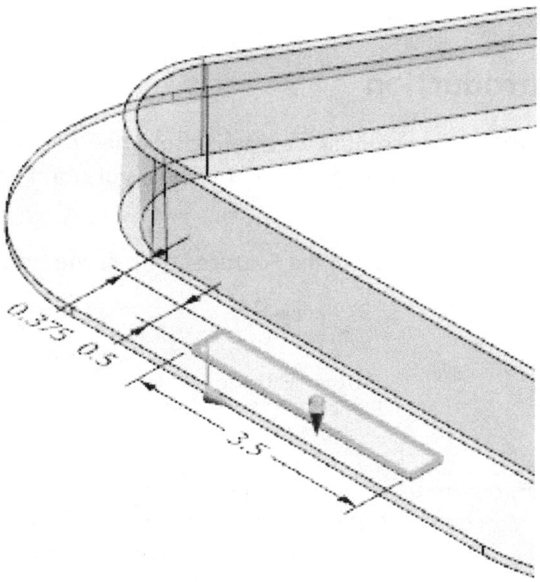

- **Save** your part file! You'll use it again it later.

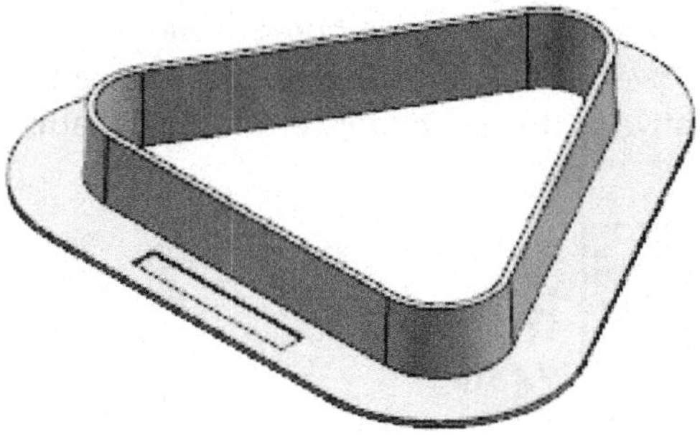

9 Design Features, Detail Features, and More

9.1 Introduction

Design Features are essential for creating solid geometry. Unlike sketches and curves, design features allow you to create solid geometry directly; you can create holes, slots, and other geometries that result from manufacturing and machining operations.

Design features are found in the **Feature** group of the **Home** tab or via **Menu / Insert / Design Feature**.

Detail Features are used to treat edges and corners – for instance, fillets and blends. Detail features are also found in the **Feature** group on the **Home** tab. In this chapter, you will learn about Design Features and Detail Features of fundamental importance in solid modeling, as well as some other tools that don't fit into those groups neatly, but are otherwise also essential for creating the vast majority of simple parts.

9.2 Hole

Although you can create a hole simply by subtracting a cylinder from a solid, use of the **Hole** tool allows you to create fully parametric *Hole* features in a variety of shapes – counterbored, countersunk, threaded, and more. One important advantage that using the **Hole** tool has over subtracting a cylinder of the right size is that the hole callout on the part drawing can easily be made fully associative to the parameters of the **Hole** feature created. The **Hole** tool is found in the **Feature** group on the **Home** tab.

In the exercises that follow, we will practice positioning holes of various types.

9.2.1 General

The *General* hole type is appropriate for holes sized to your exact measurements. Simply specify the actual diameter.

- Create a new part file (**Units**: *Millimeters*) called *"Holes.prt"* and place it in *"C:\My NX 11 Files"*.
- Begin with a **Block** that is *100 mm* x *80 mm* x *10 mm*.
- Select the **Hole** tool, and set the **Type** to *General Hole*. Note the presence of the **Sketch Section** icon – this means that when you click on the top face of the sketcher as shown below, you will enter the Sketch Task Environment and create an internal sketch.

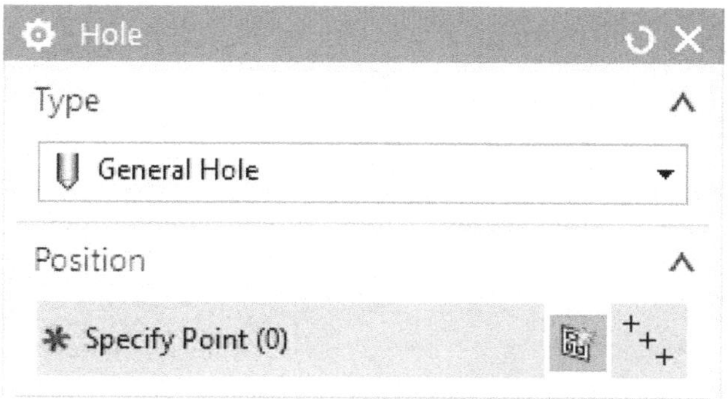

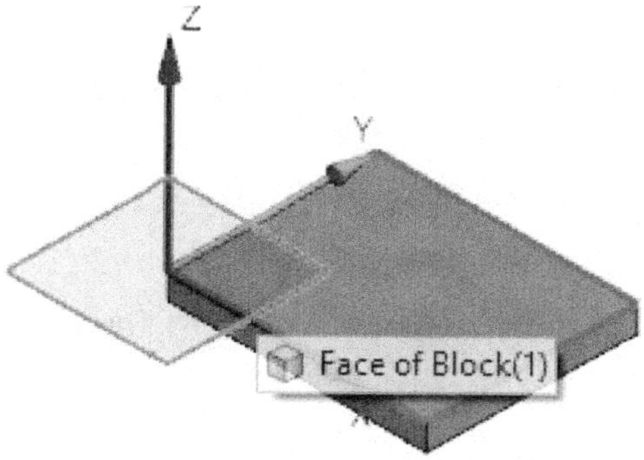

- When you arrive in the Sketch Task Environment, the **Sketch Point** tool is loaded, and a point has been placed at the exact position you clicked on the top face of the block.

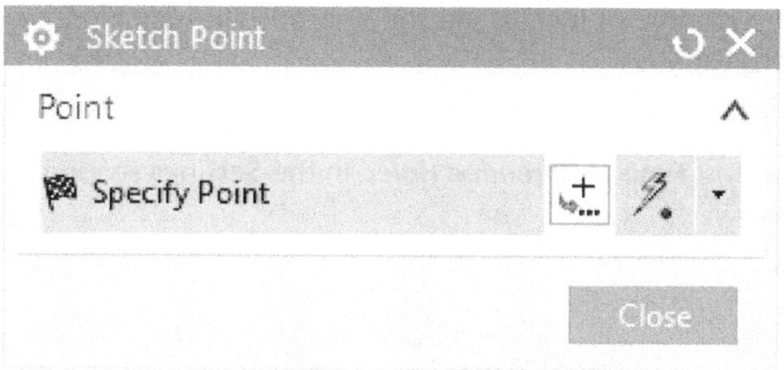

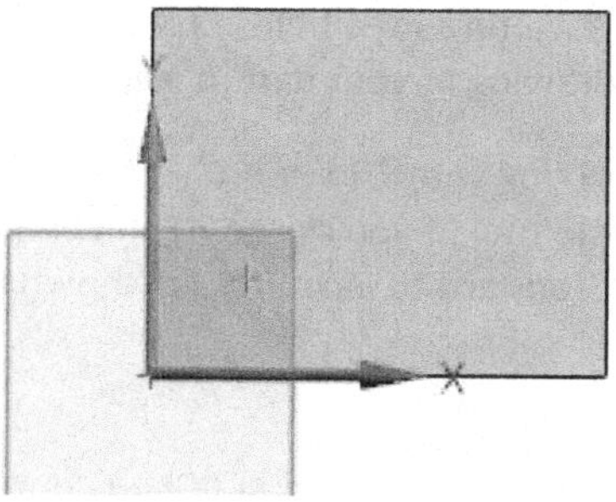

- Position your hole using the **Rapid Dimension** tool in the sketcher, as shown below.

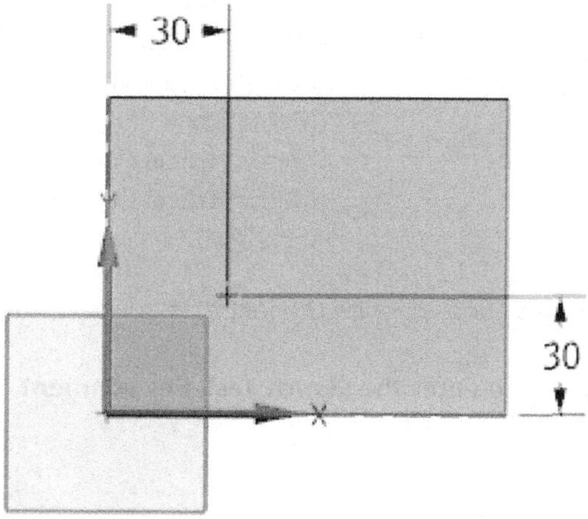

*Note: You can create more than one point at a time, making it possible to create multiple holes with a single application of the **Hole** tool.*

- Once you have positioned your point, click **Finish**. This will create the internal sketch and return you to the **Hole** dialog box.
- Make sure the **Form** is *Simple* and set the **Depth Limit** to *Through Body*. Click **Apply** when you are done.

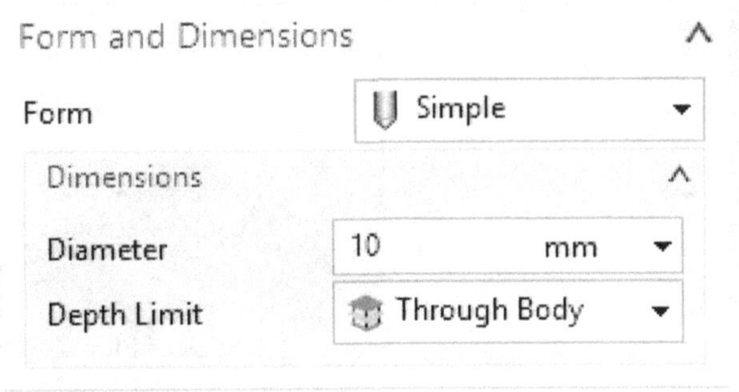

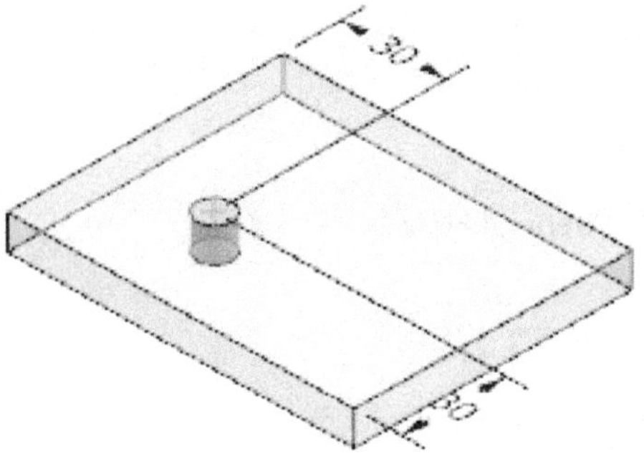

- **Save** your file!

9.2.2 Threaded

The ability to create threaded holes spares you the trouble of looking up the appropriate major/minor diameters, etc. when making a drawing of your part. It's an excellent tool!

- Continue working with *"Holes.prt"*.
- Open the **Hole** tool (if you closed it) and change the **Type** to *Threaded Hole*. In the **Settings** section of the dialog box, change the **Standard** to *Inch UNC*, as shown below.

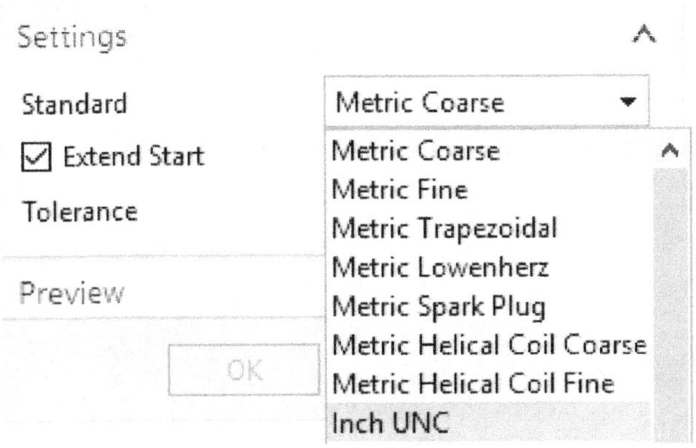

- This is a fantastic option within the **Hole** tool – despite the fact that our parts units are metric, we are not constrained to only metric thread sizes.
- Click on the top face of the block again to enter the **Sketch Task Environment**. Position the resulting point as shown below.

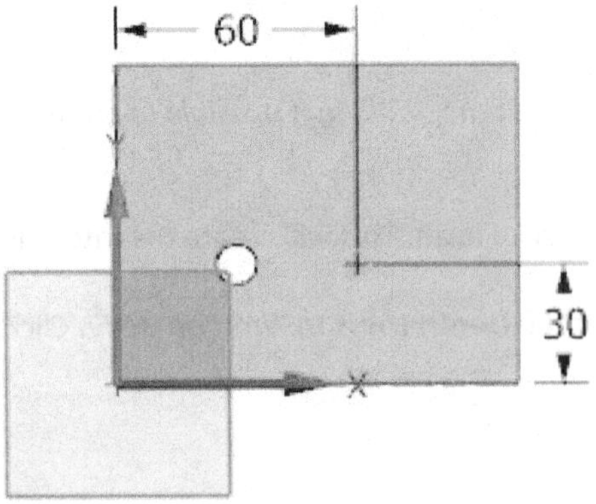

- Set the parameters for your threaded hole as shown in the figure below.

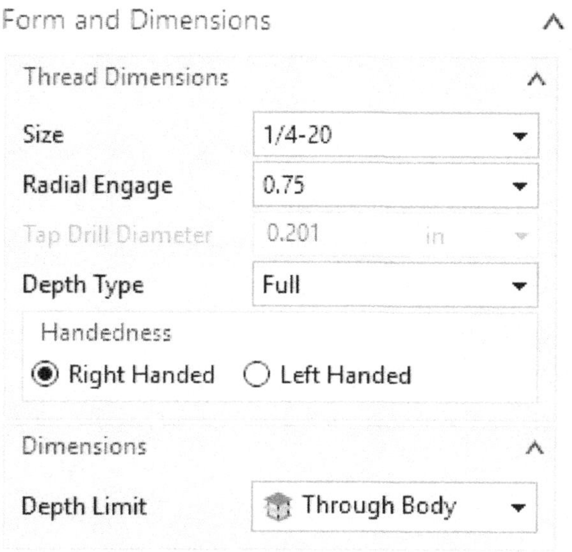

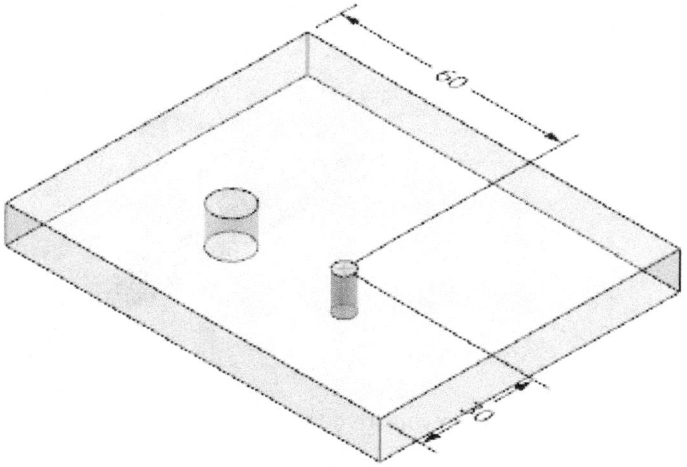

- Click **OK** to create the threaded hole. You will notice that the threads are only symbolic. Not showing the threads saves memory and loading time. It is possible to add detailed threads using the **Thread** tool. We will not use the **Thread** tool here, but encourage you to try it if you are curious!

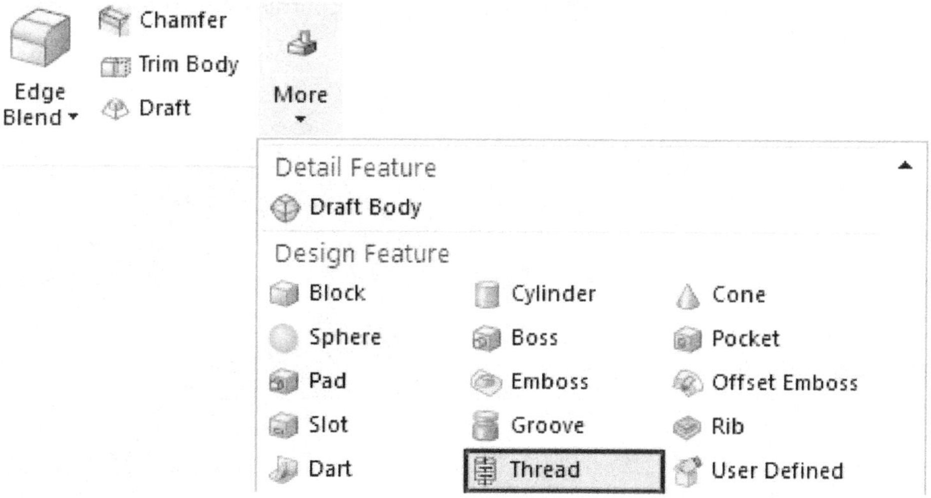

- **Save** your file!

9.2.3 Screw Clearance

The *Screw Clearance Hole* type is useful for producing holes that have standard amounts of clearance to fit threaded holes of standard sizes.

- Continue working with *"Holes.prt"*.
- Select the **Hole** tool, change the **Type** to *Screw Clearance Hole*, set the **Form** to *Countersunk*, and the **Screw Size** to *M10*. Position the hole anywhere you'd like on the top face of the block. Your model should appear as below!

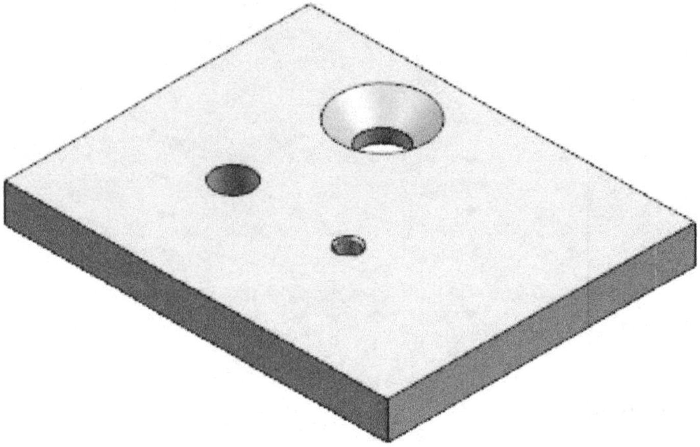

- **Save** your file when you are done.

9.2.4 Depth Limits

Depth Limits are options for controlling the depth of your hole either with a parameter value, or geometrically rather than numerically. We will walk through the different **Depth Limit** options in this exercise.

- Create a new file called *"Hole Depth Limits.prt"* and place it in *"C:\My NX 11 Files"*.
- Sketch a *200 mm* x *100 mm* rectangle in the XY plane of the datum coordinate system, and **Extrude** it *30 mm* in the +ZC, with a *Two-Sided* **Offset** from *0 mm* to *3 mm*.

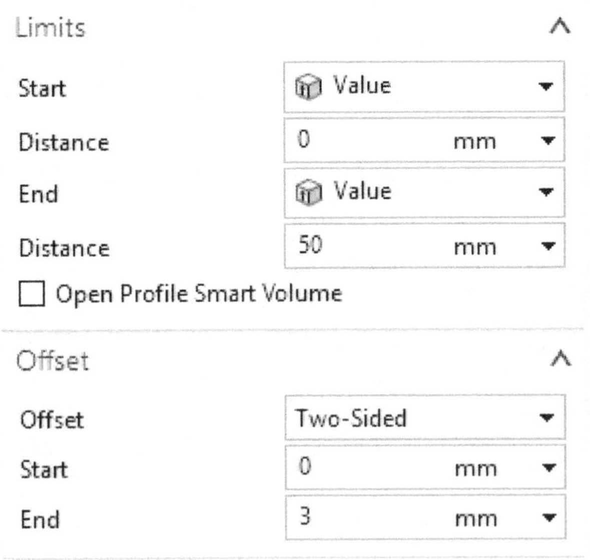

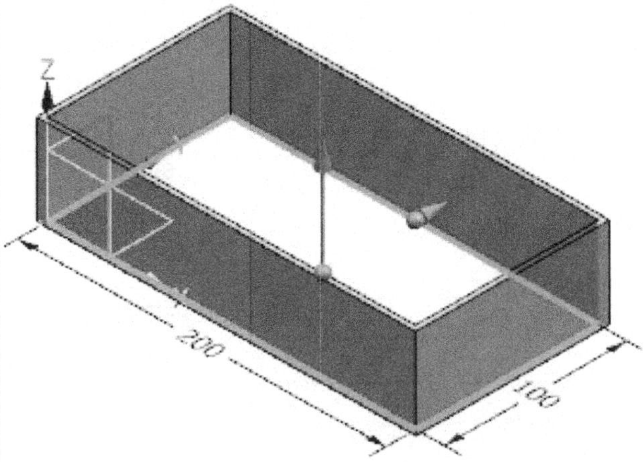

- Select the **Hole** tool and **Reset** it. This should the **Type** to *General Hole*. Select the face shown below.

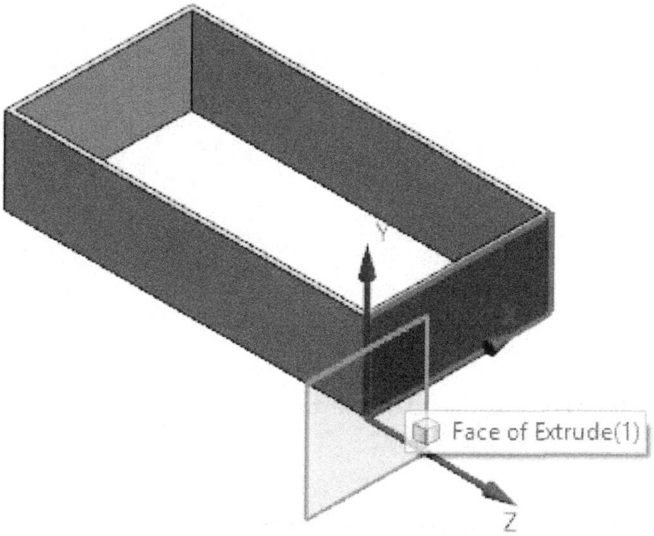

- Use two *Midpoint* constraints to center your point on the face, and then **Finish** the sketch.

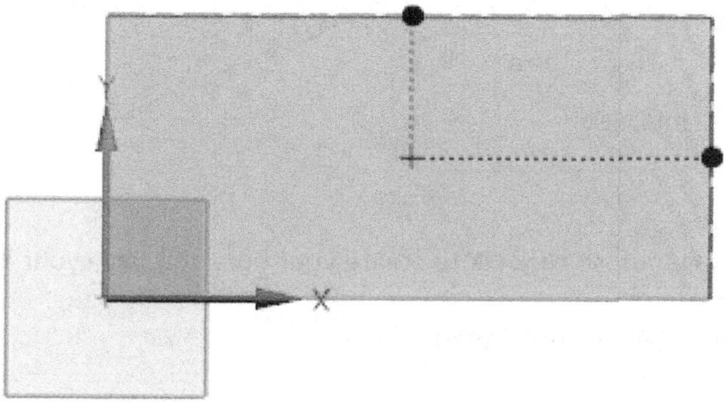

- When you return to the **Hole** tool, make sure the **Depth Limit** is set to *Value* and enter a **Depth** of *50 mm*. Your preview should appear as below.

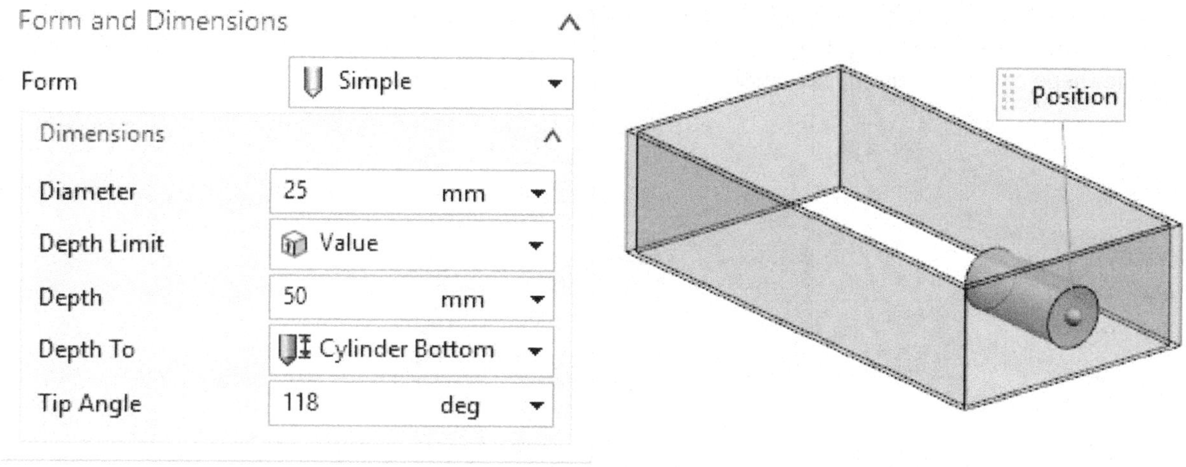

- Now change the **Depth Limit** to *Until Selected*. Select opposite inner face of the body and then you will see that the hole goes to that face as in the following figure.

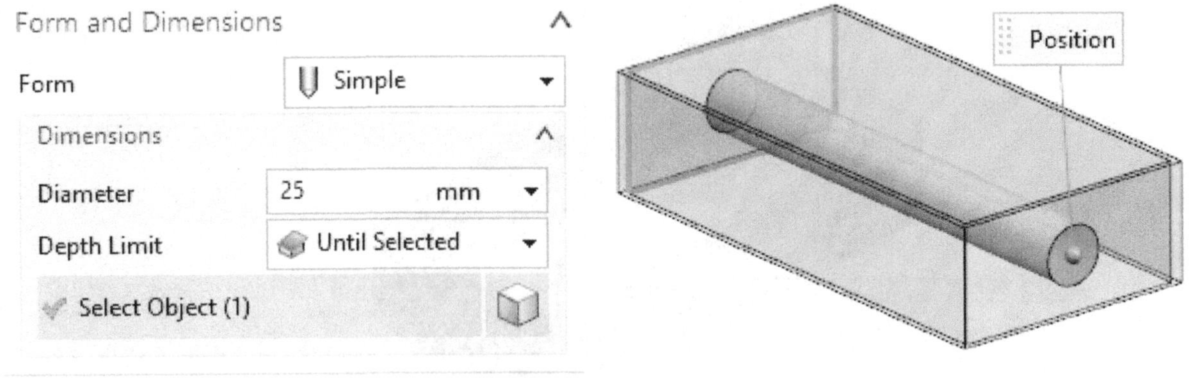

- Finally, try **Until Next**. This will extend the hole to the next face it reaches, as shown in the following figure.

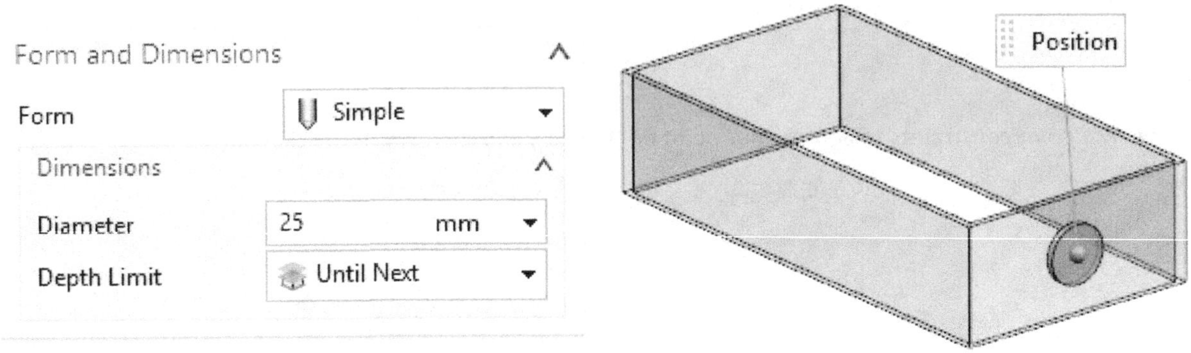

- The other options are rather obvious, so click **OK** to create your hole and **Save** your file!

9.2.5 Two Holes Centered at the Same Point

Occasionally you will wish to create two holes centered at the same point, similar to a counterbored hole, but where the bottom face of the larger hole is in the shape of the drill tip, and not flat. This comes up in manifold design, where to reduce backflow through a very long narrow hole, you might want to lead a larger hole into the smaller hole.

- Open the file called *"Two Holes Centered at the Same Point.prt"* from *"C:\My NX 11 Files\Downloaded Files"*.
- In this part, you will find a body with a simple hole in it. There is also a section view already created – **Clip** the section to see the shape of the hole already created.

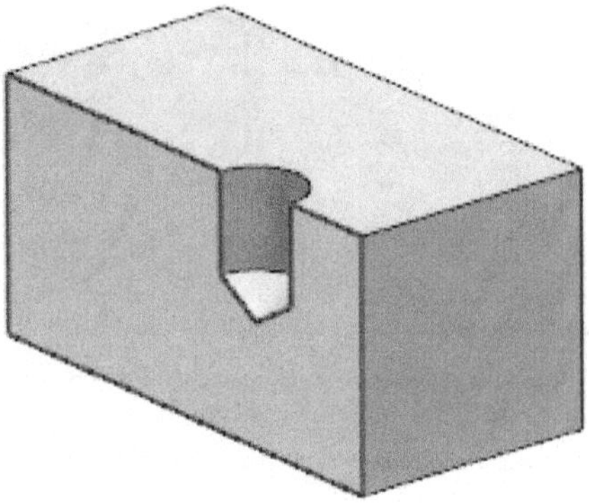

- Open the **Hole** tool, set the **Type** to *General*. For the **Position**, choose the arc center of the top edge of the existing *Simple Hole* feature.

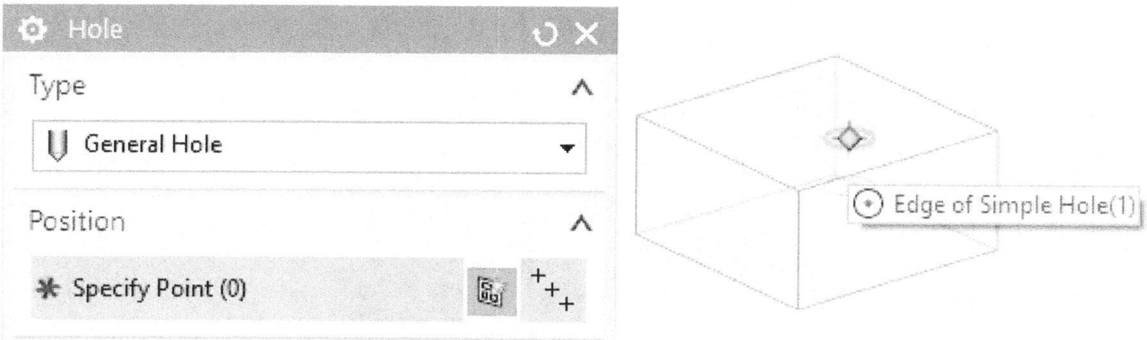

- This results in an error because the **Direction** is set to *Along Face Normal* by default. The point isn't actually attached to a face, so NX cannot infer a direction vector.

- Change the **Hole Direction** to *Along Vector* and NX will be able to generate a preview. Choose the cylindrical face of the existing *Simple Hole* feature as the direction vector, and enter the parameters shown below. Click **OK** to create the **Hole** feature.

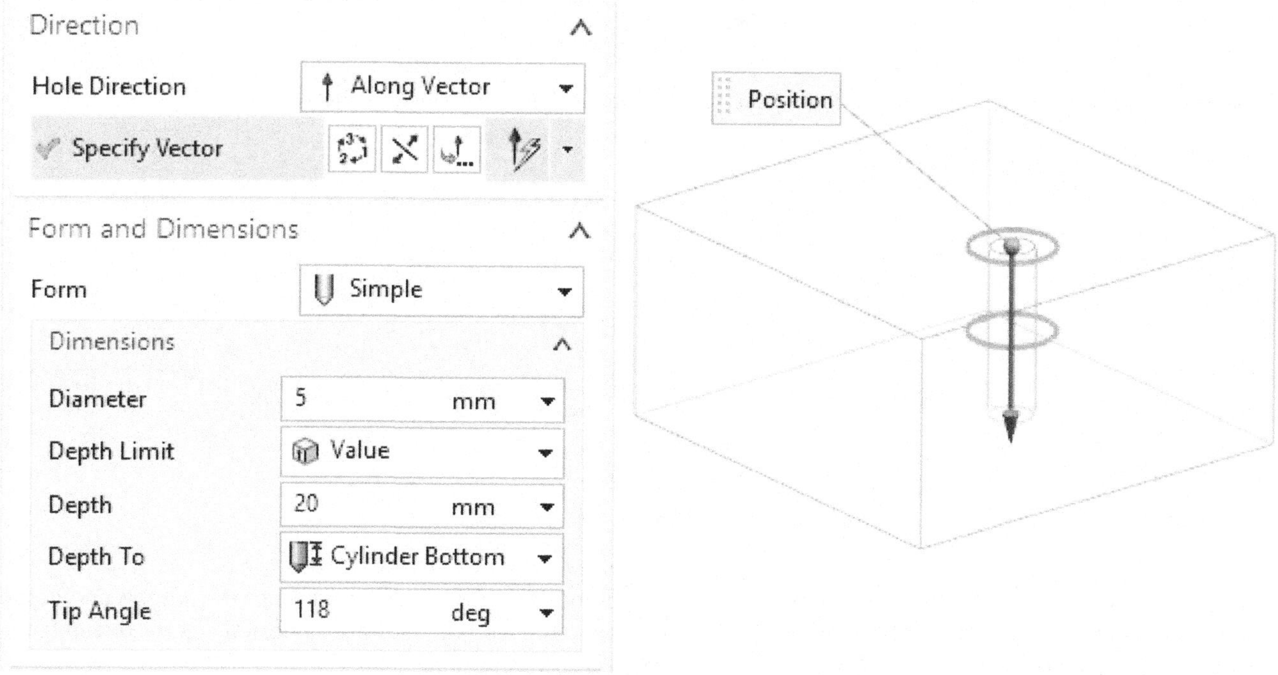

- **Clip** the Work Section to see the shape of your new hole.

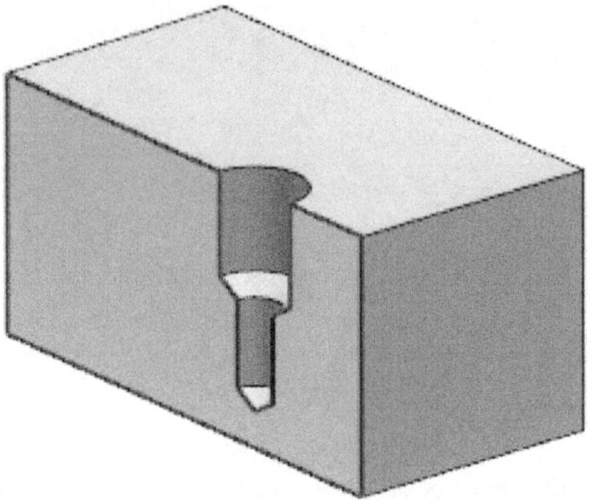

- **Save** your file!

9.2.6 Hole on a Curved Surface

If you wish to construct a hole on a curved surface, it is usually easiest to do so with the help of a *Point* feature of type *Point on Face* – you can't sketch on a curved face.

- Open the part file *"Hole on a Curved Face.prt"* from *"C:\My NX 11 Files\Downloaded Files"*.
- In this part file, you will find a cylindrical body.

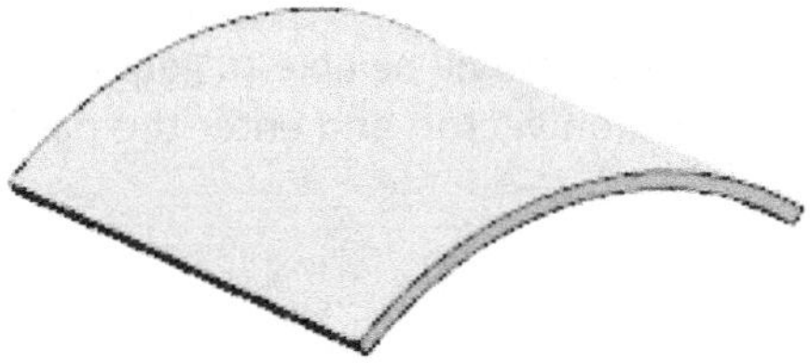

- Use the **Point** tool with **Type** set to *Point on Face* to position a point on the face as shown below.

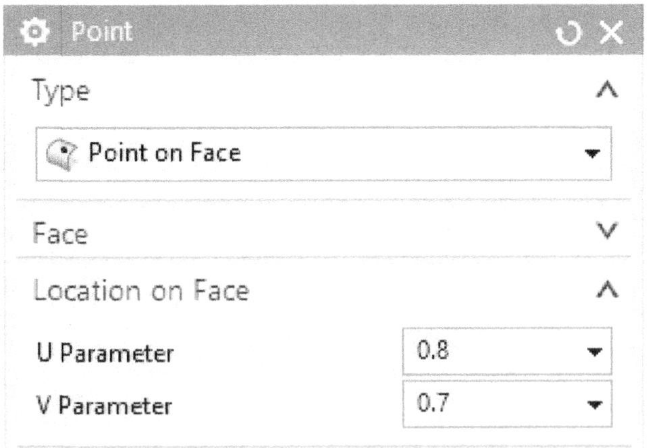

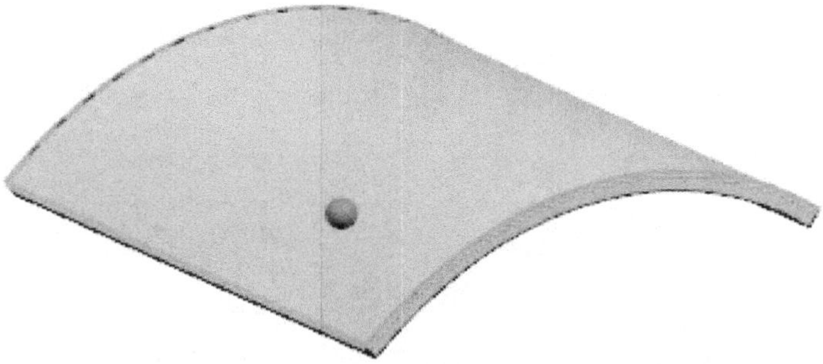

- Open the **Hole** tool and with **Type** set to *General*, select the point entity just constructed. Since the point is understood to be on the curved face, the **Hole Direction** *Normal to Face* automatically calculates a normal vector to the face at that point. Set the parameters shown below and click **OK** to create your hole.

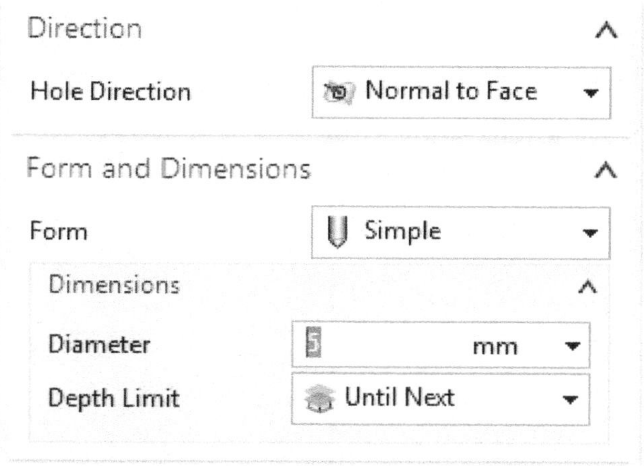

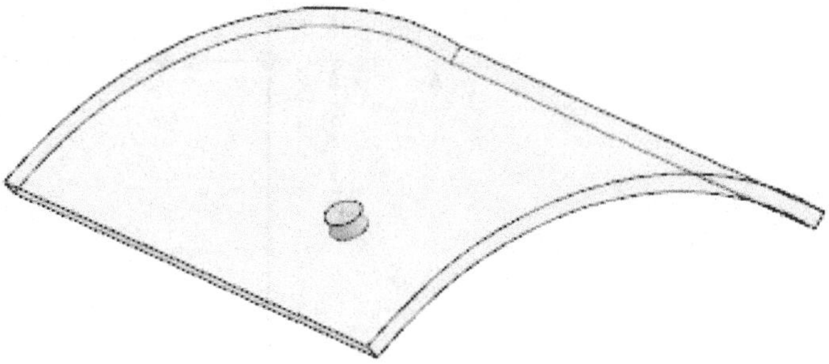

- **Save** your file!

9.3 Blends

Edge Blend and **Face Blend** are used to replace sharp edges between faces with rounded faces. These rounded faces are usually known as fillets. There are a variety of other more sophisticated blend tools that we will address in Chapter 15.

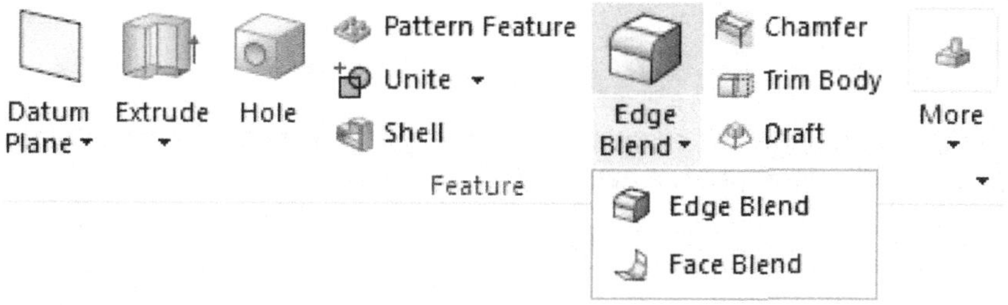

The order in which you apply blends can make a big difference on complicated models that often have intersecting blends. In general, you should apply the largest radius and most critical blends first and move on to the smaller radius and less critical blends last. However, experience teaches that some experimentation is often required to complete the blending of a solid model. Often times it pays off to do the blending after the part design is completed.

9.3.1 Edge Blend

The simplest way to use the **Edge Blend** tool is to specify the radius of the cross-section of the blend face. A good way to predict the result is to think of a ball with the specified radius being rolled across the surfaces, blending the faces as in the figure below. Wherever the ball touches the corner is where the blend surface will be created.

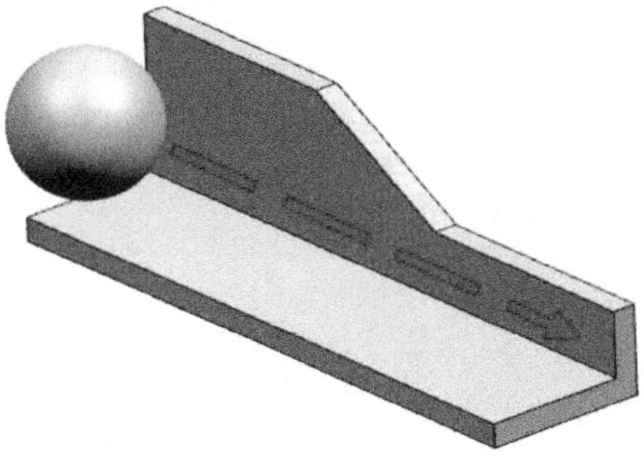

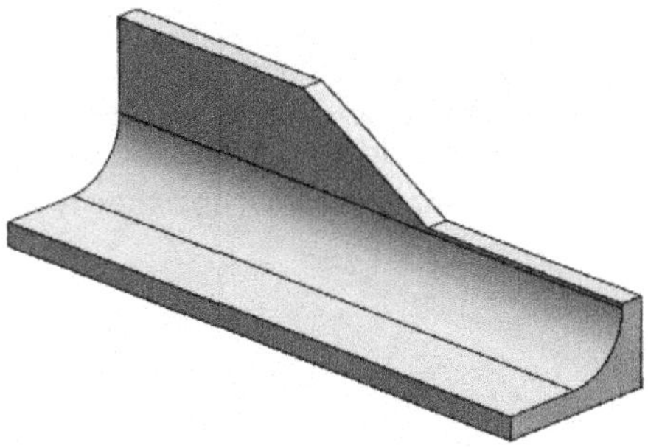

- Create a new part file (**Units**: *Inches*) called *"Blends.prt"* and place it in *"C:\My NX 11 Files"*.
- Create the sketch shown below in the XY plane and **Extrude** it *2 in* in the +Z direction.

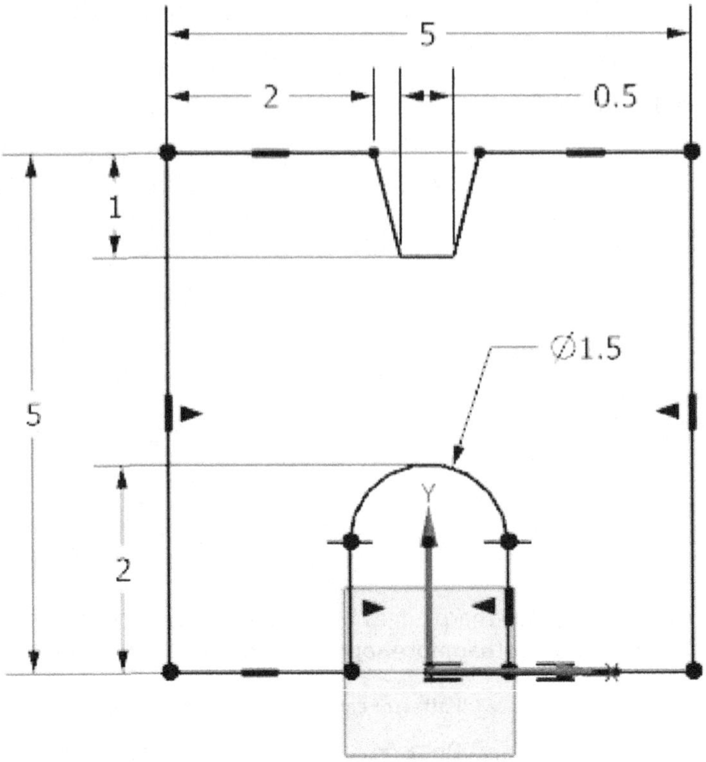

- Select the **Edge Blend** tool, and select the edges shown in the following figure. Set the **Blend Face Continuity** to *G1 (Tangent)*, the **Shape** to *Circular*, and specify a **Radius** of *0.5 in*.

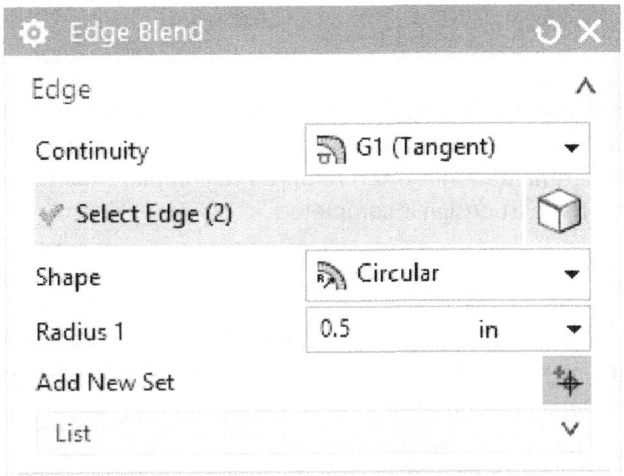

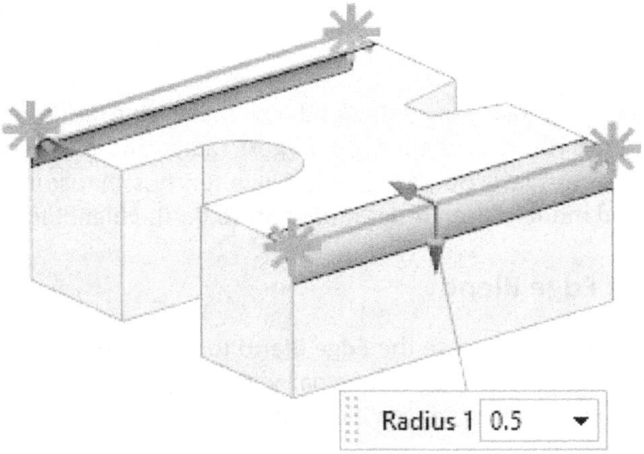

- Expand the **List** subsection of the menu, and click the **Add New Set** button. This allows you to apply different blend radii to different sets of edges and have the software infer how they should meet.
- Select the four edges shown below, and specify a **Radius** of *0.25 in* for this set.

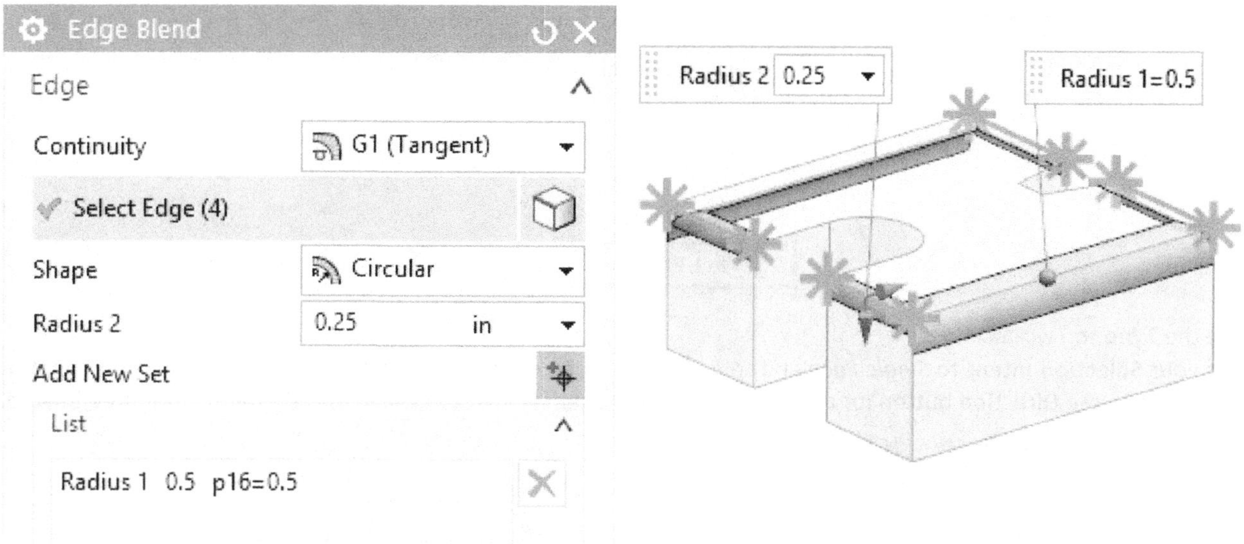

- Click the **Add New Set** button again, and select the edges shown in the following figure. Specify a **Radius** of **0.125 in**.

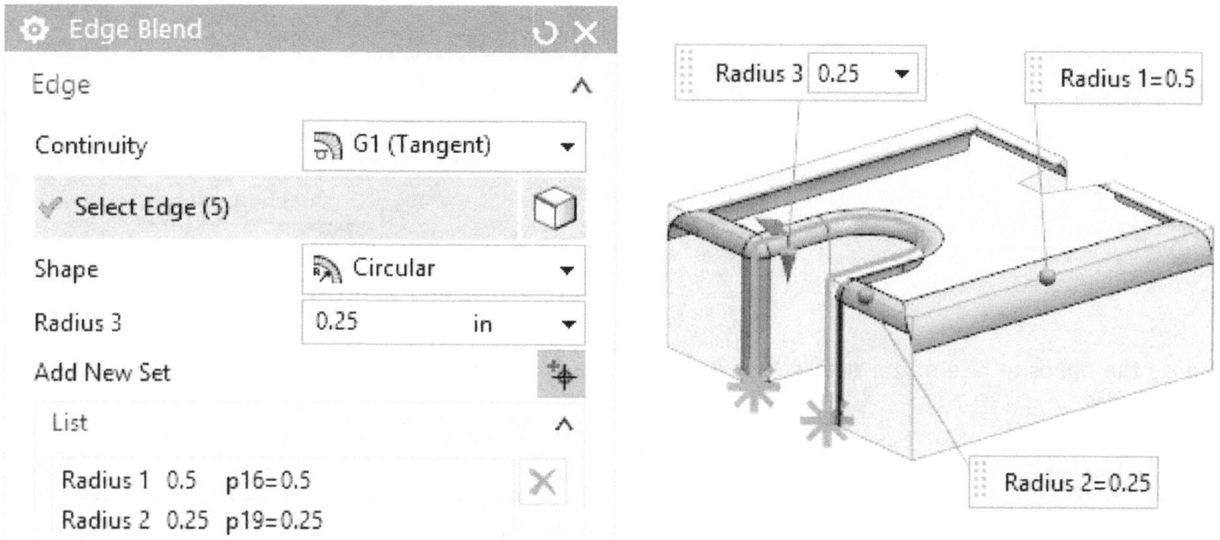

- Finish the blend by clicking **OK**. **Save** your file when you are done!

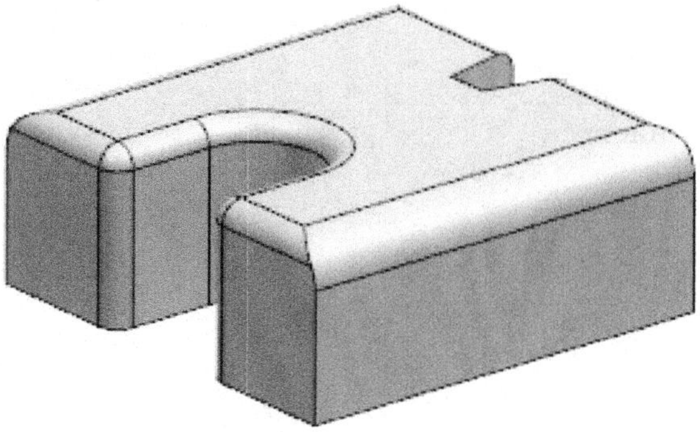

9.3.2 Face Blend

A **Face Blend** can be created between two sets of faces of solid or sheet bodies that are not necessarily adjacent. Continue working with *"Blends.prt"*.

- Select the **Face Blend** tool.

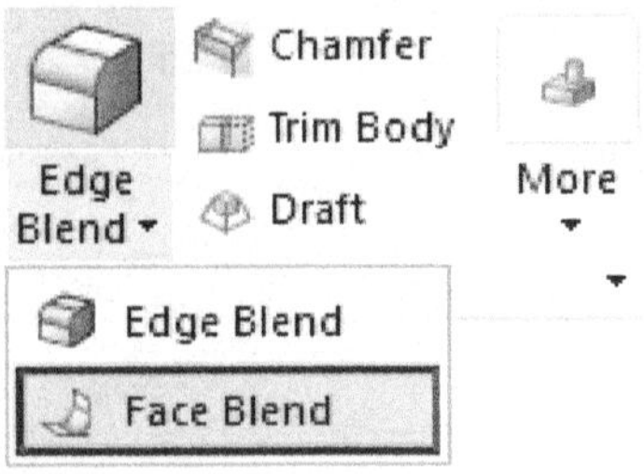

- Set the **Type** to *Two-face*.
- Set your **Selection Intent** to *Single Face* and select the face shown in the following figure as *Face 1*. You might need use the **Reverse Direction** button (or double-click on the golden arrow) to ensure that the face is oriented inward.

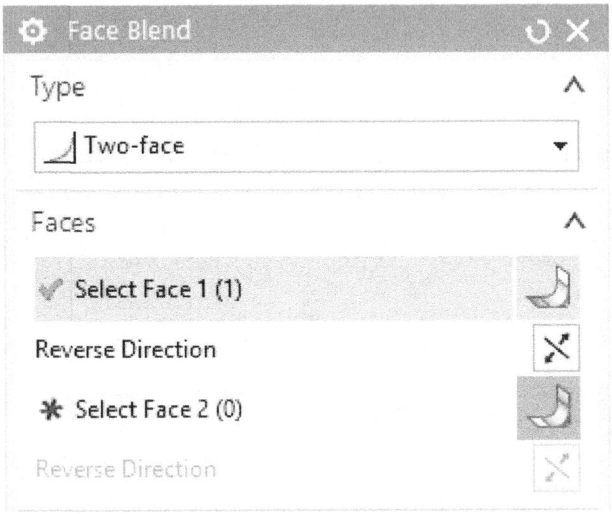

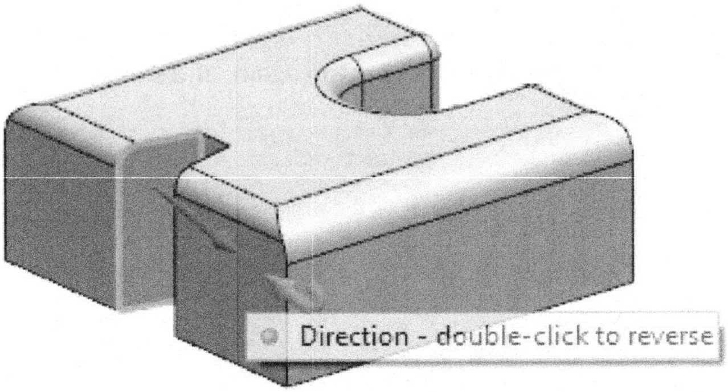

- Choose the opposite face of the slot as *Face 2*. Within the **Cross Section** part of the menu, set the parameters as shown below.

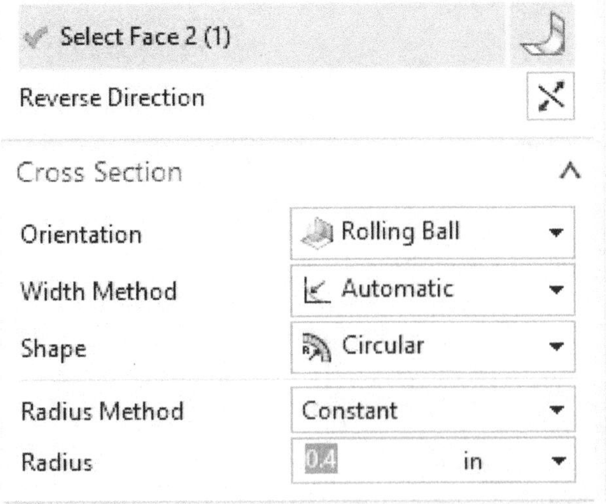

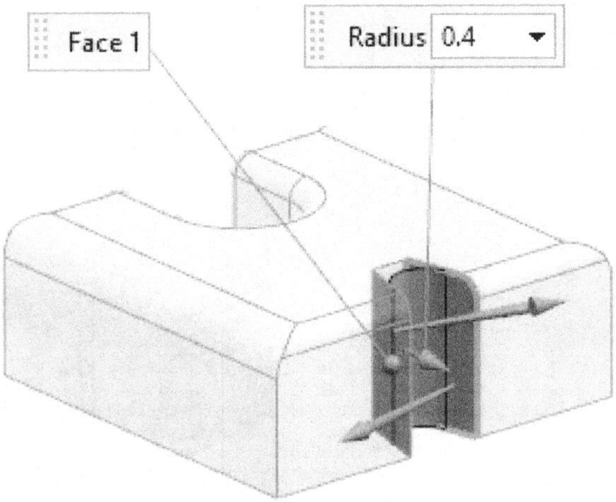

- The **Face Blend** tool rounds the selected faces of the slot despite the fact that they are not touching, and replaces the planar face between them with a single smooth transition.

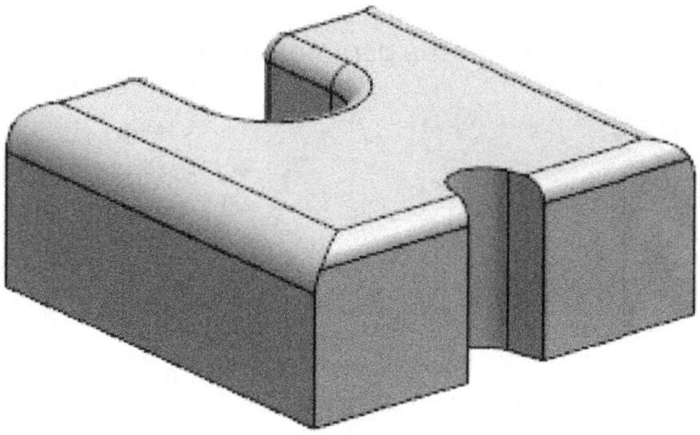

9.4 Chamfer

The **Chamfer** tool produces a flat surface made by cutting off an edge or corner or by adding a flat surface to a corner at an angle – like an **Edge Blend**, it is used to improve sharp corners either for manufacturability, handling, or strength.

Consider the solid body shown below. You could easily make this body by extruding a sketch of the entire top view outline including the radius and chamfer details. But by using the **Blend** and **Chamfer** features, you improve the readability of the Model History, and leave open the option of removing the chamfers or blends, if required later.

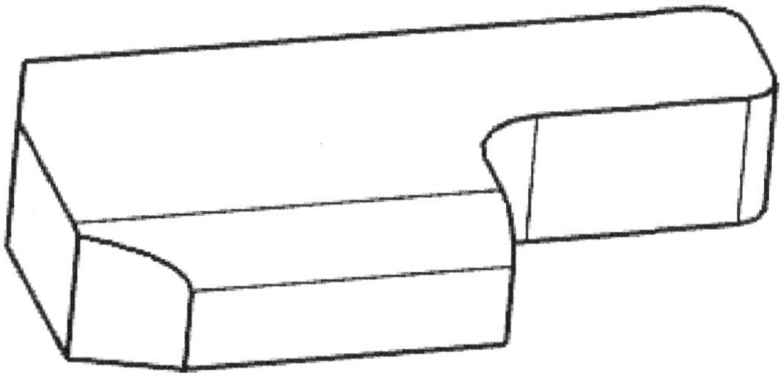

- Create a new file (**Units**: *Inches*) called *"Chamfer.prt"* and place it in *"C:\My NX 11 Files"*.
- Create the following sketch on the XY plane.

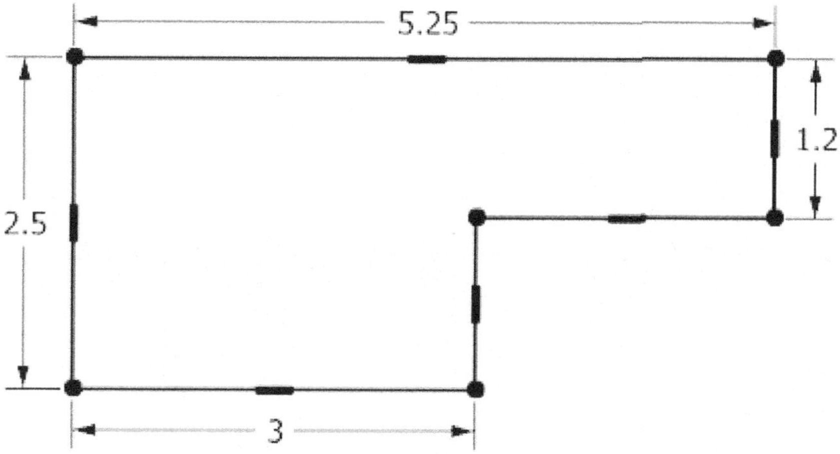

- **Extrude** by *1 in* in the +Z direction
- Select the **Chamfer** tool. Select the edges and set the parameters as shown below

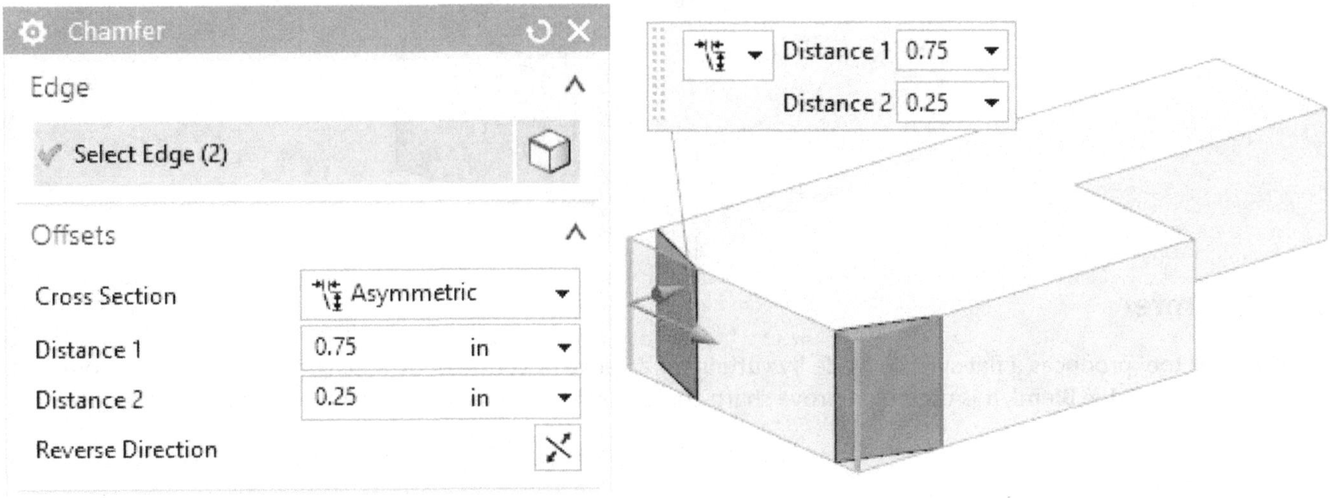

- Select the **Edge Blend** tool, and choose the edges and radii as shown below. Note that Radius 2 applies to both of the rightmost vertical edges.

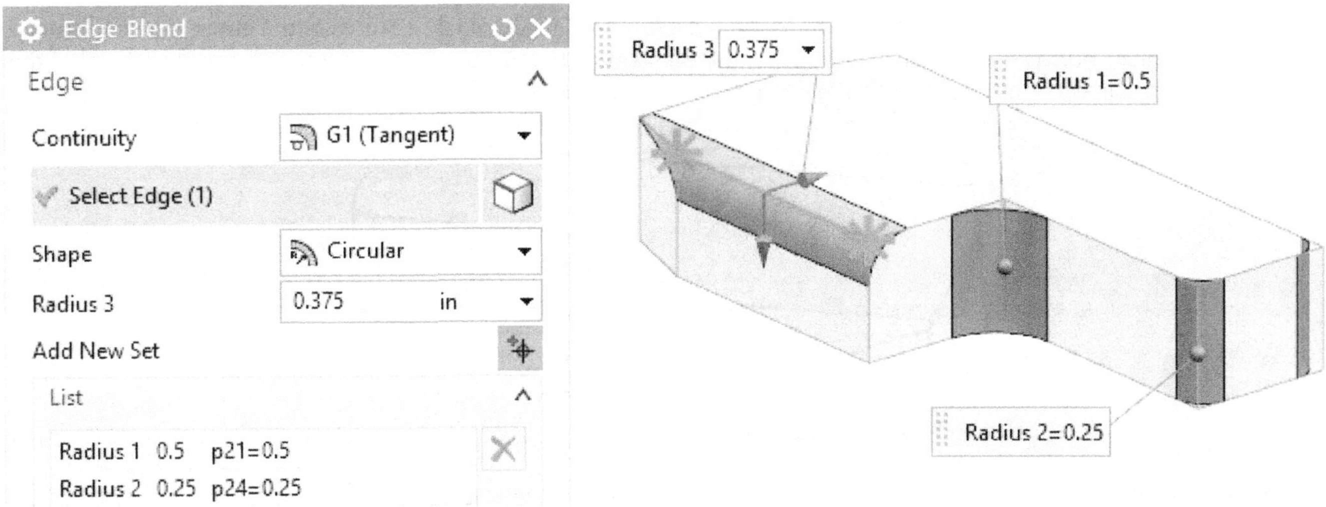

- Your model will appear as below. **Save** your file!

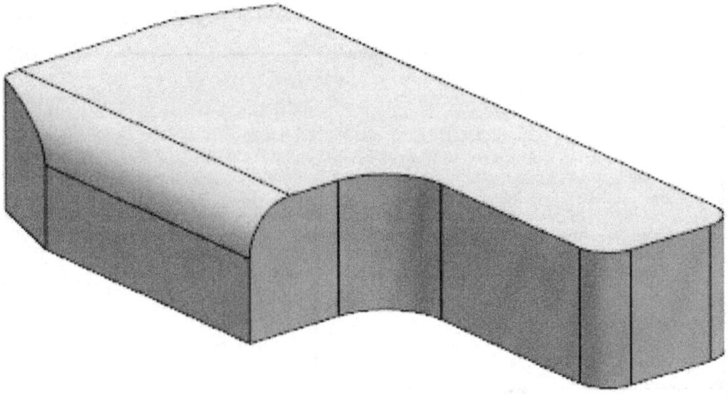

9.5 Draft

Draft is used to apply an angle to the faces of pre-existing solid geometry. You have already learned that a draft angle can be applied to the faces of an extruded body at the same time that it is created. The standalone **Draft** command however, can be used to apply a draft angle to some or just a few of the faces of a solid, relative to some direction vector. In the exercises below, you will explore some of these options.

Consider the solid body shown below. This geometry is typical of a cast part with a non-planar parting line, which in this case is the bottom surface. You will need to apply drafts to the left, right, front, and rear surfaces.

This type of draft can be created using **Draft** tool with **Type** set to *Edges*. It is similar to the *From Plane*, but instead of specifying a plane you select edges that remain fixed when the draft is applied.

- Create a new part file (**Units:** *Inches*) called *"Draft.prt"*, and place it in *"C:\My NX 11 Files"*.
- Create the sketch shown in the following figure on the XZ plane, and **Extrude** it from *0 in* to *4 in* in the +YC direction.

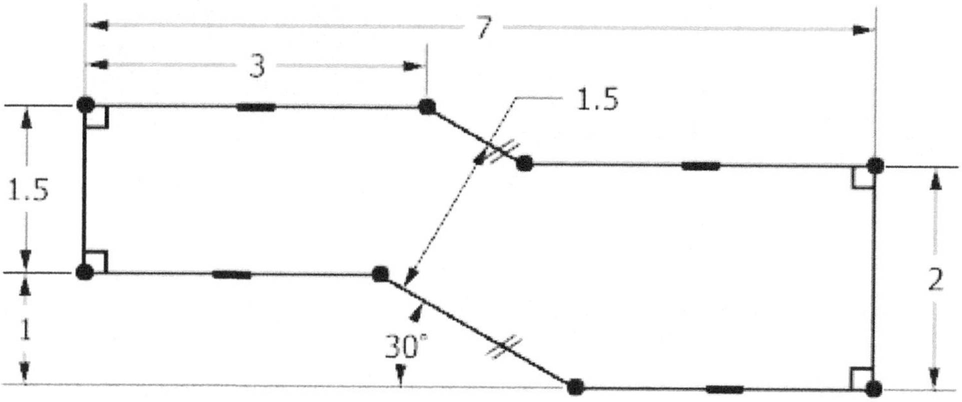

- Select the **Draft** command and set the **Type** to *Edges*. Let +ZC be the **Draw Direction**, and select the edges shown below as the **Stationary Edges**.

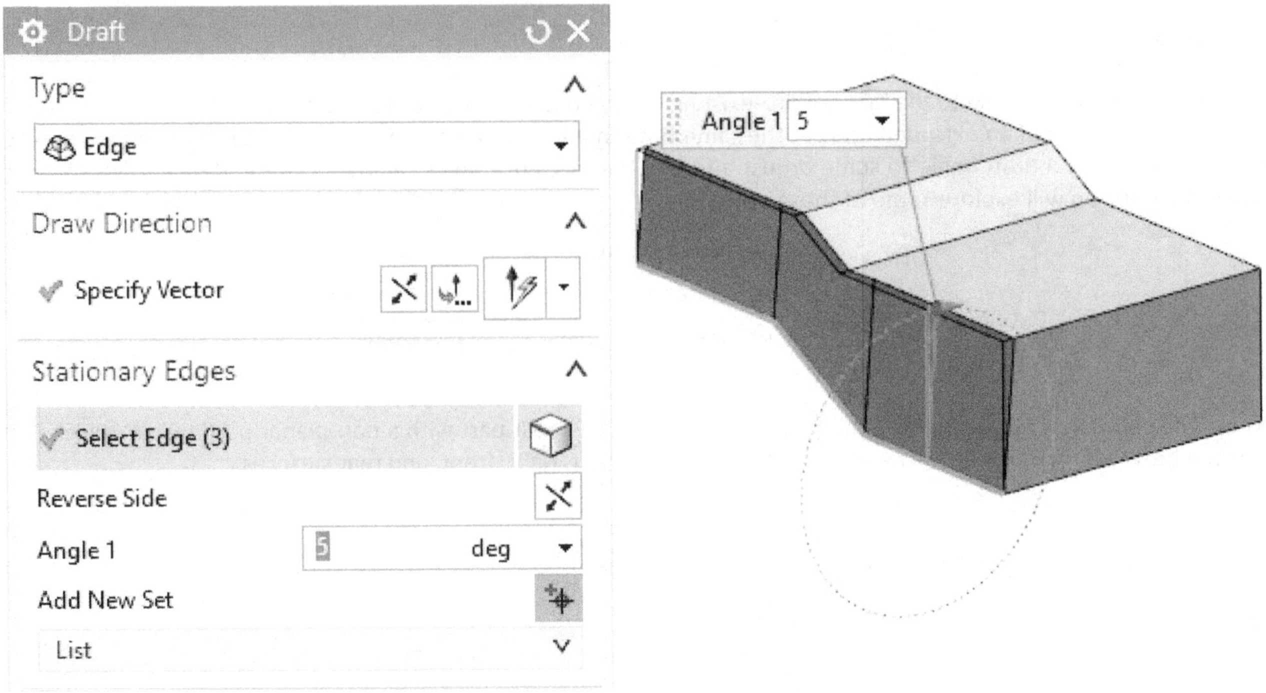

- Rotate your model 180 degrees and select the corresponding edges from the opposite face.

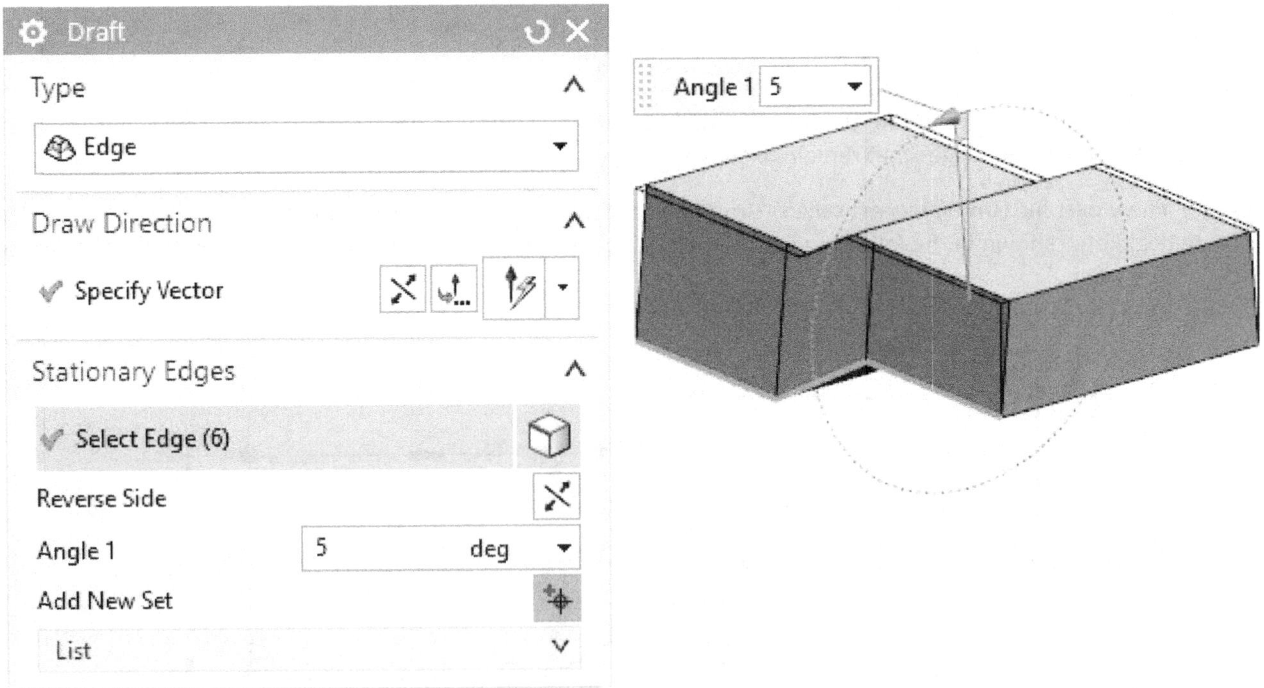

- Click **OK** to apply the **Draft** command.
- **Save** your file!

9.6 Pocket

The **Pocket** tool allows you to easily create rectangular, circular, or custom-shaped cuts in a solid body (shown below). Like holes, they require a placement plane. The **Slot** and **Pad** tools are very similar to the **Pocket** tool – if you understand how to use one, you should be able to employ them all.

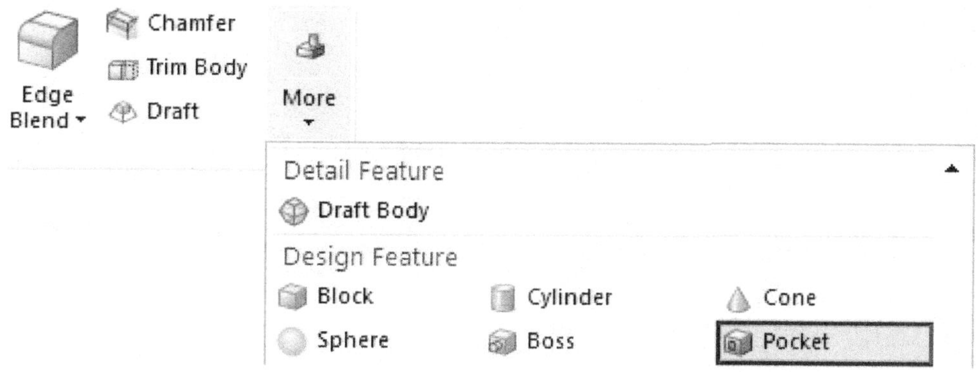

- Create a new file (**Units**: *Inches*) called *"Pocket.prt"* and place it in *"C:\My NX 11 Files"*.
- Create a **Block** with dimensions *4 in* x *7 in* x *2 in*.
- Select the **Pocket** tool, and choose **Rectangular**.
- When the **Rectangular Pocket** menu appears, click the top face of the block as shown below.

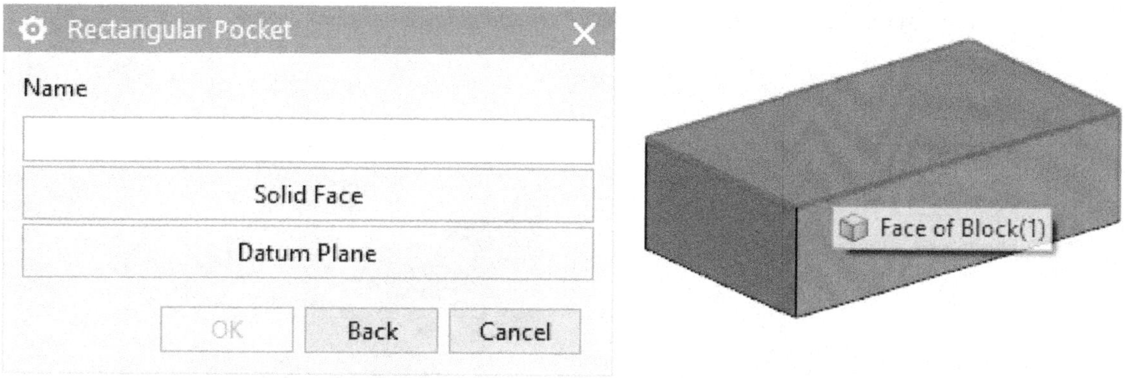

- When the **Horizontal Reference** menu appears, select one of the *7 inch* edges of the selected face.

- On the **Rectangular Pocket** dialog, enter the parameters shown below.

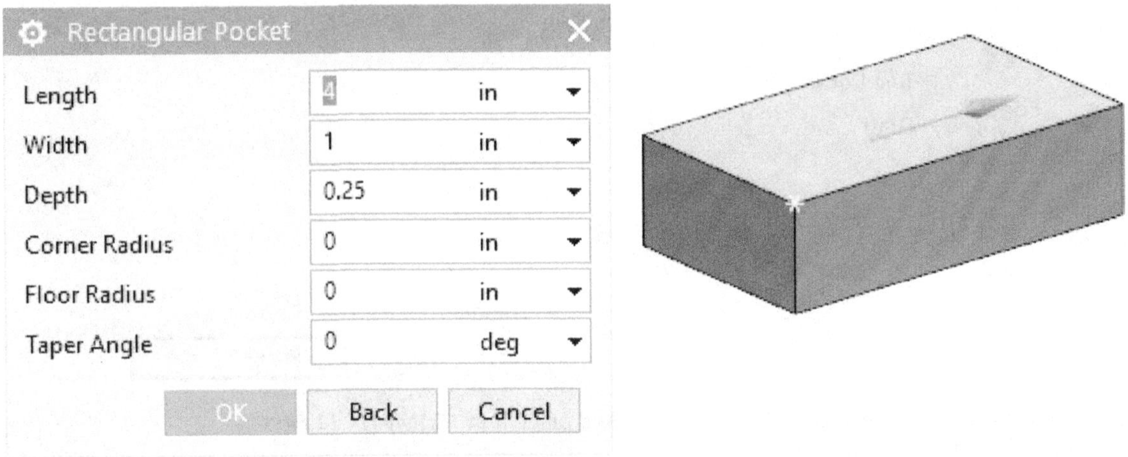

- Choose **OK**. A preview will be generated, centered where you clicked on the face. The next step is to position the pocket on the face. We will do so by first specifying perpendicular dimensions from the edges of the top face to the edges of the pocket. From the **Positioning** menu, select *Perpendicular*, as shown below.

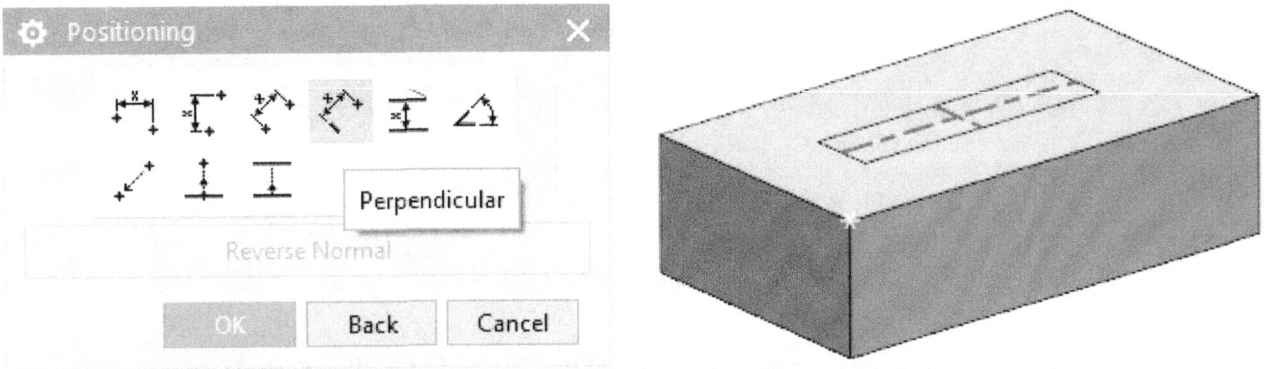

- This brings up a rather cryptic dialog box named **Perpendicular**. Choose the edge shown below.

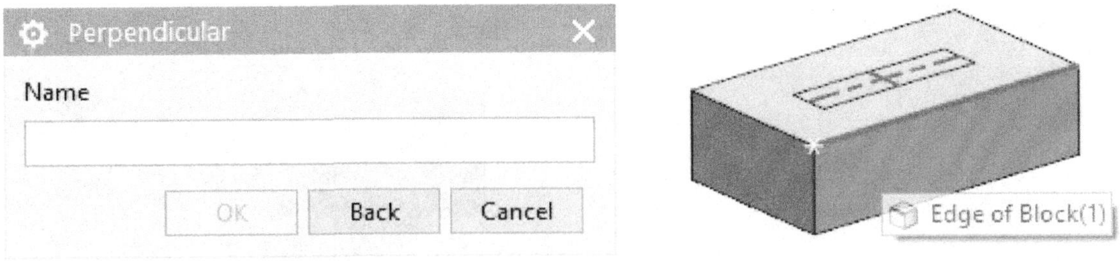

- The **Perpendicular** dialog box now appears different, and the Cue Display reads *"Select tool edge"*. Now select the nearest edge of the pocket to the edge you selected in the last step.

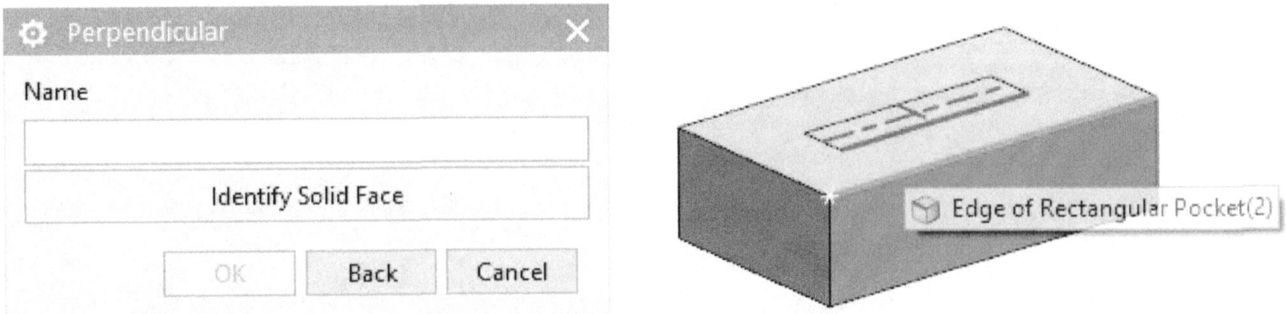

- The **Create Expression** dialog box will appear in the graphics window, as shown below. Enter a value of *1 in*.

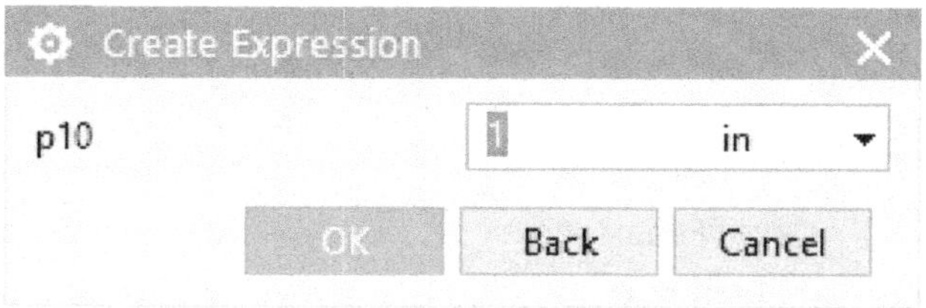

- Repeat the last few steps to create a perpendicular dimension of *1.5 inches* between the *4-inch* edge of the top face and the *1 inch* edge of the pocket nearest to it. Note: if your pocket was initially positioned as in the images above, it will be necessary to enter a value of *-1.5 in*. You will now see two dimensions in the graphics window.
- From the **Positioning** dialog box, select the **Angular** method. This will be used to rotate the pocket relative to the *7-inch* edge selected for the first perpendicular dimension.

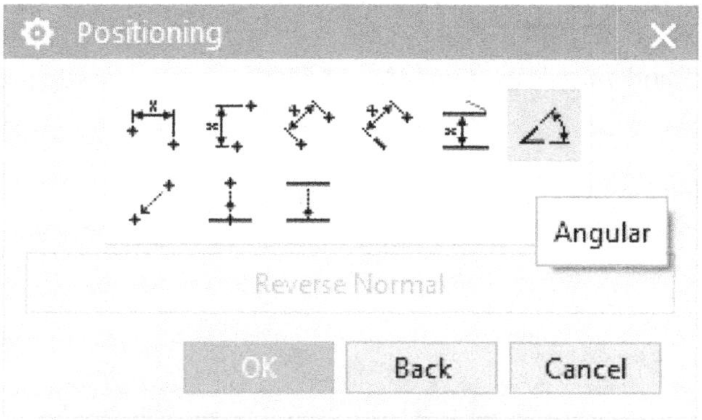

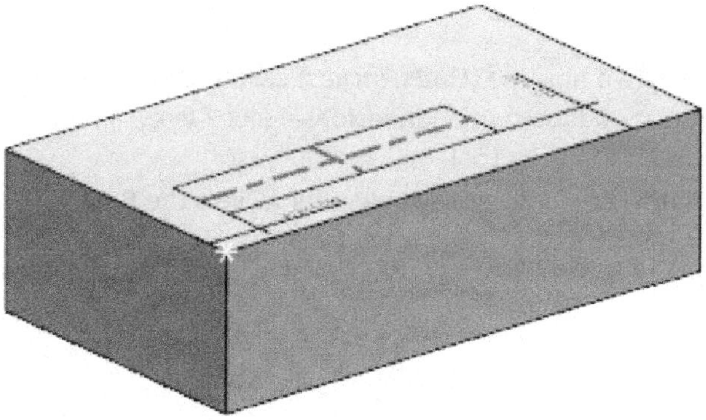

- As with the perpendicular dimensions just created, you must specify an edge of the target body and then an edge of the tool body. What is not obvious from the menu is that *where* you click on the edges determines the orientation of the angle. Click on the right side (+X direction) of the *7-inch* edge of the target above, and then click on the right side (+X direction) of the nearest *4-inch* edge of the tool. Give the resulting angular dimension a value of *20 degrees*.

While creating angular dimensions, it is important to pay attention to which end of the target and tool edges are selected – it is easy to get the opposite angle. Another common mistake is canceling from the **Positioning** menu. Selecting **Cancel** in this menu at any time completely exits the **Pocket** tool.

*Note: **Pad** command is very similar to the **Pocket** feature, but rather than removing material, it adds it. To create a **Pad**, follow the instructions provided for the **Slot**, including selection of a **Horizontal Reference**. A **Corner Radius** and **Taper Angle** may also be specified for a **Rectangular Pad**.*

9.7 Boss and Groove

The **Boss** tool produces a cylindrical extrusion attached to an existing solid body. The **Groove** tool subtracts an annulus of material from a cylindrical face on a body. In this project, we will apply the **Groove** tool to the cylindrical face of a boss.

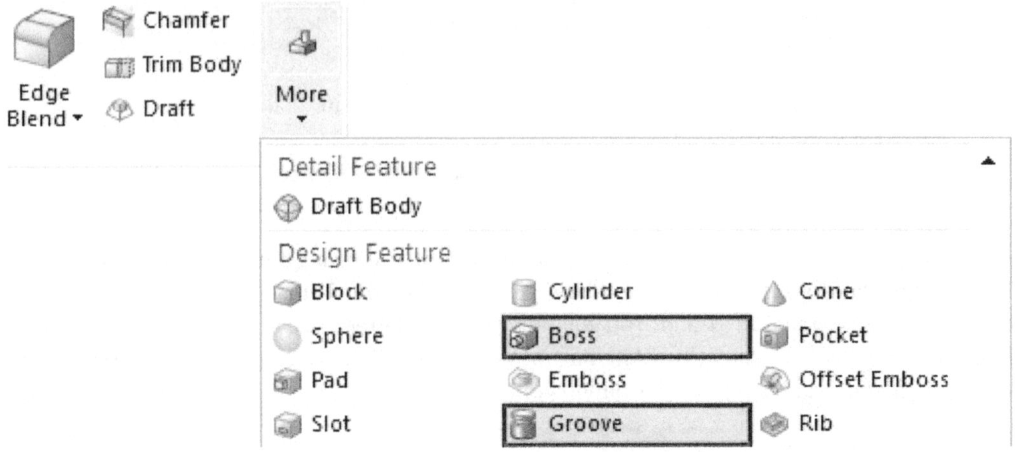

- Create a new file (**Units**: *Inches*) called *"Boss & Groove.prt"* and place it in *"C:\My NX 11 Files"*.
- Create a **Block** with dimensions *4 in* x *7 in* x *2 in*.
- Select the **Boss** tool.
- Enter the parameters shown below into the **Boss** dialog, and choose the top face of the block.

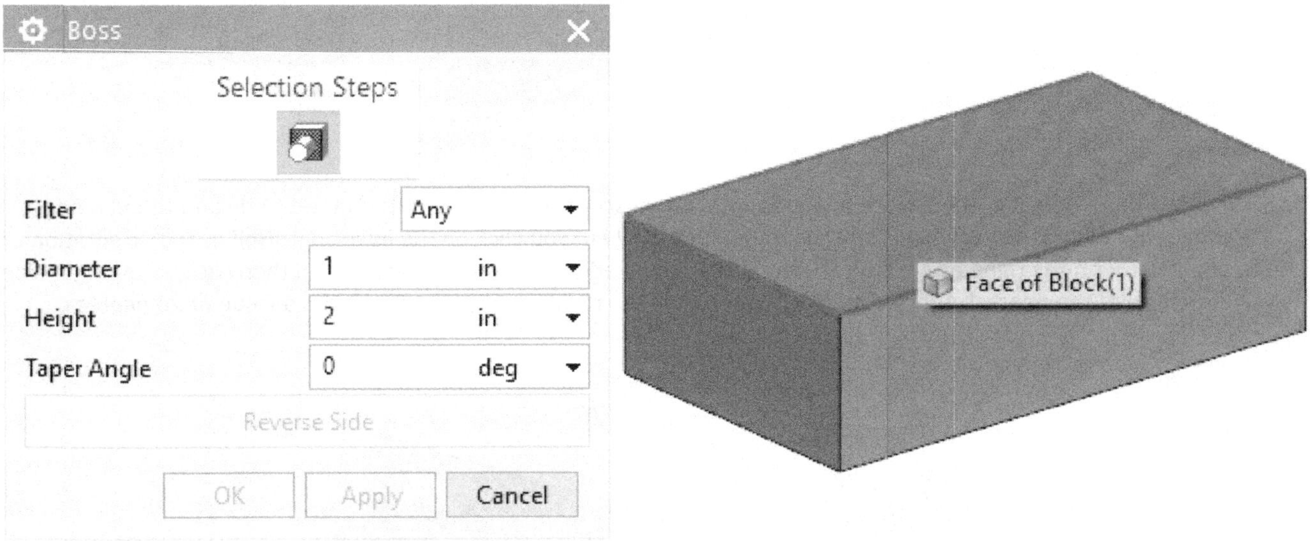

- Use two perpendicular dimensions to position the boss in the center of the top face of the block – *2 inches* and *3.5 inches* will suffice. Click **Apply** on the **Positioning** dialog box rather than **OK** so that you can apply multiple dimensions.

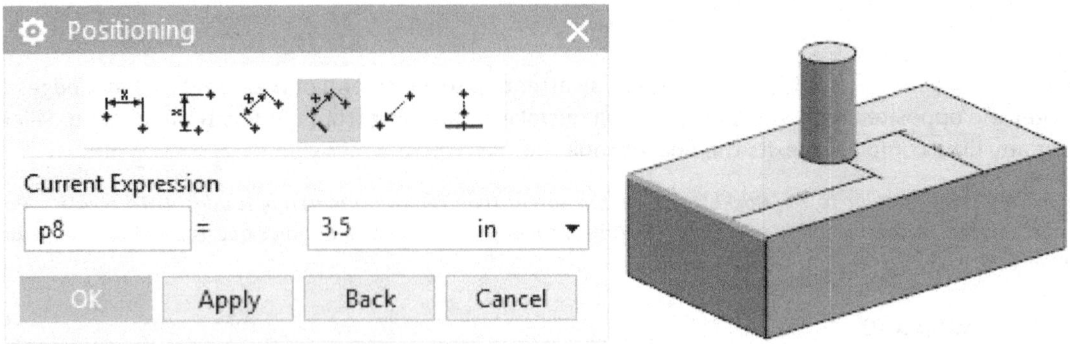

- Select the **Groove** tool, and choose **Rectangular**.

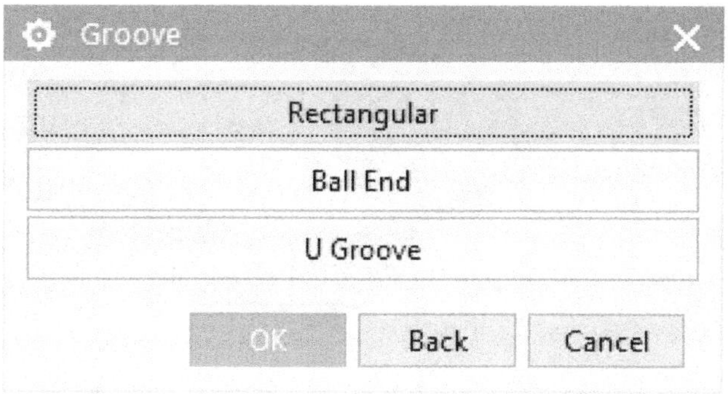

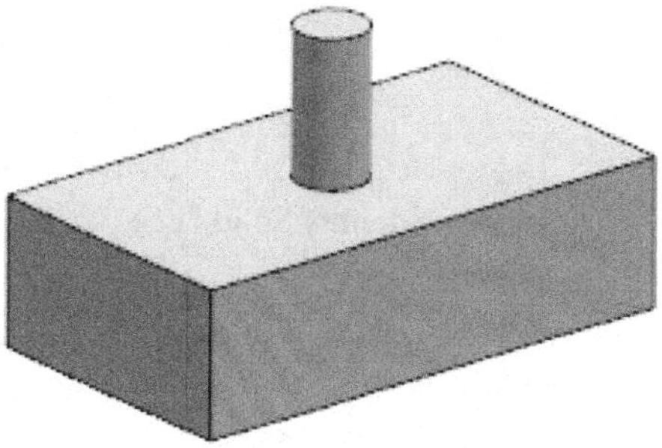

- Pick the cylindrical face of the boss. Enter the values shown below into the **Rectangular Groove** dialog box. The **Groove Diameter** is the inner diameter of the subtraction annulus. The width is the thickness of this annulus.

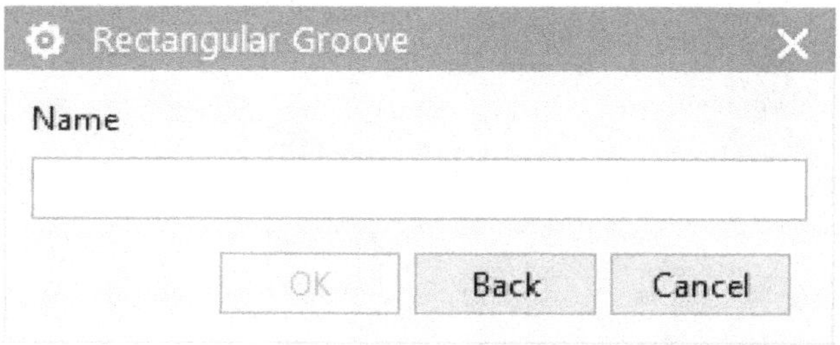

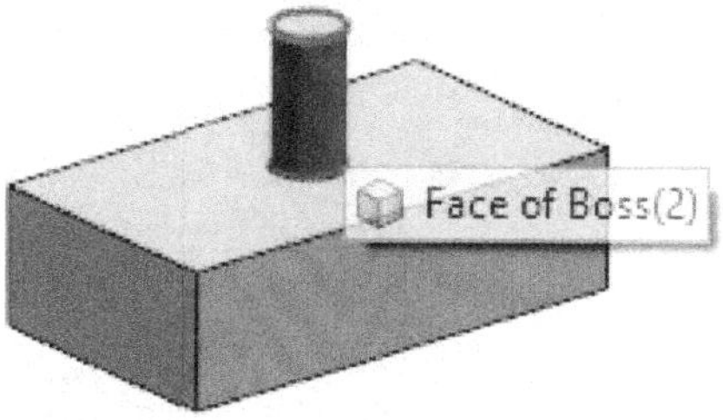

- Enter the values shown below.

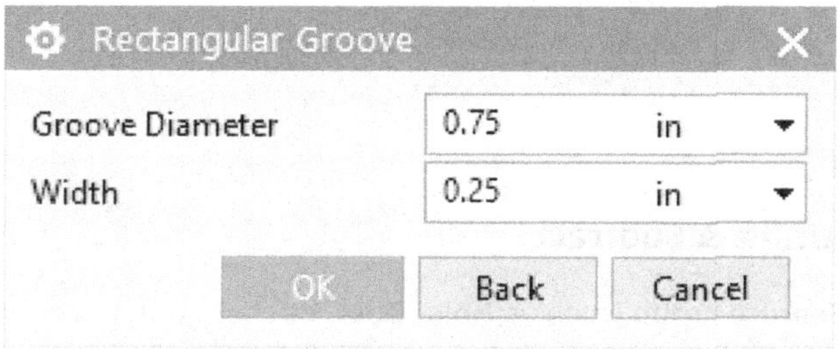

- Choose **OK**. The subtraction annulus will appear as shown below. The only meaningful dimension here is along the axis of the cylinder - the **Position Groove** dialog simply is asking you to select one of the edges on the end of the boss as the edge to which this dimension should measure. Choose the top edge, as shown below.

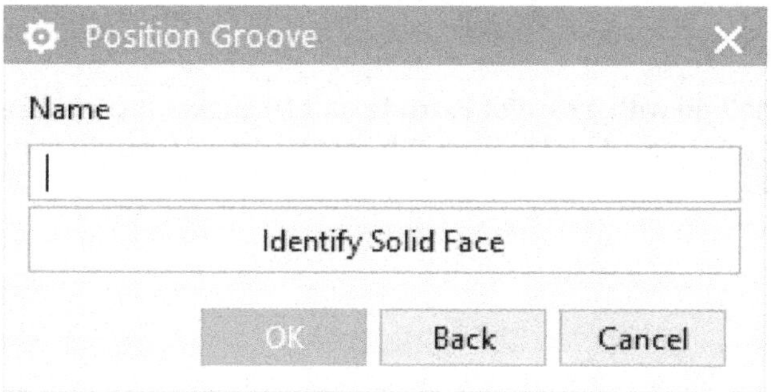

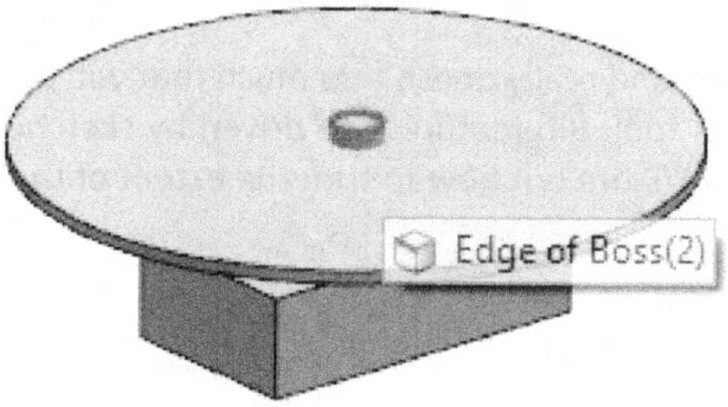

- When it prompts you for the **Tool** edge, select the top edge of the rectangular groove, as shown below.

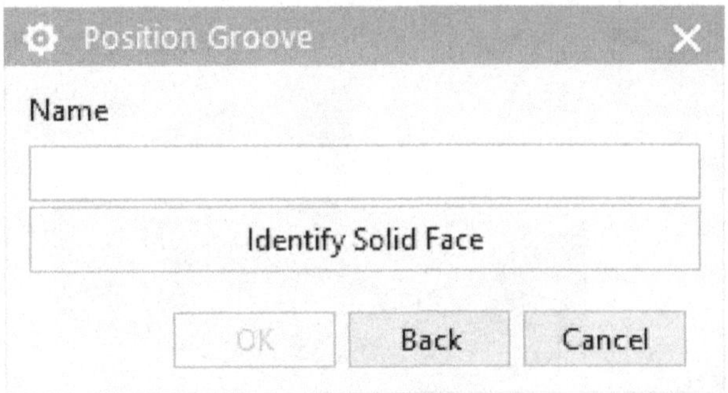

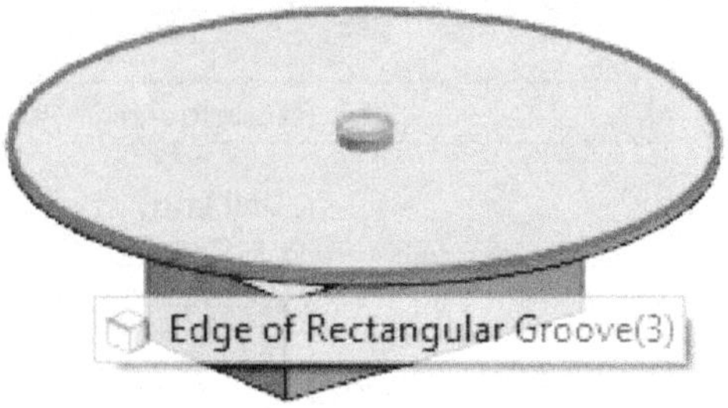

- Enter a dimension value of *0.5 in* and click **OK**. The end result is shown below!

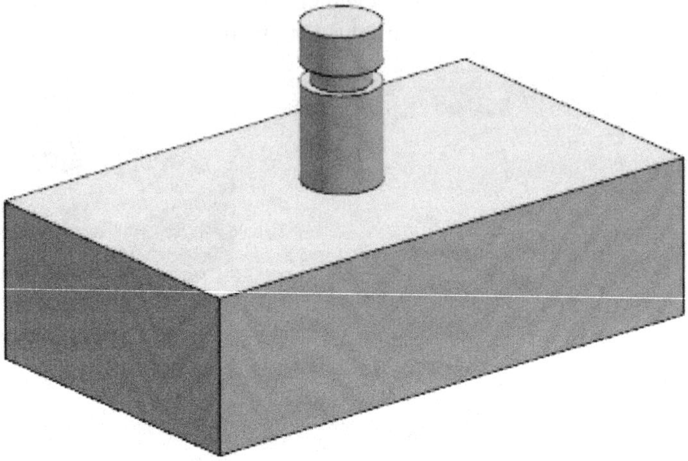

- **Save** your file!

9.7.1 Design Features vs. Unite & Subtract

Many of the tools in the **Design Feature** group (such as **Hole**, **Boss**, **Pocket**, and more) produce results that can easily be modeled using **Extrude** or **Revolve**. At this point you will have noticed that the dialogs for some of these specifically named tools are somewhat unintuitive and cumbersome, and you might well prefer to use **Extrude** or **Revolve** with an appropriate **Boolean**. It is a matter of personal preference whether you choose to use **Boss**, **Pocket**, and others, or simply **Unite** or **Subtract** features onto your solid models. You have the tools to approach your model either way!

9.8 Rib

The **Rib** command really doesn't do much that you couldn't do with a careful **Extrude**, but NX makes it very easy to create ribs with this tool. Rib features are driven by sketches, and one advantage the **Rib** tool has over **Extrude** is that NX will automatically figure out how to trim the extent of the sketch to produce the desired ribs.

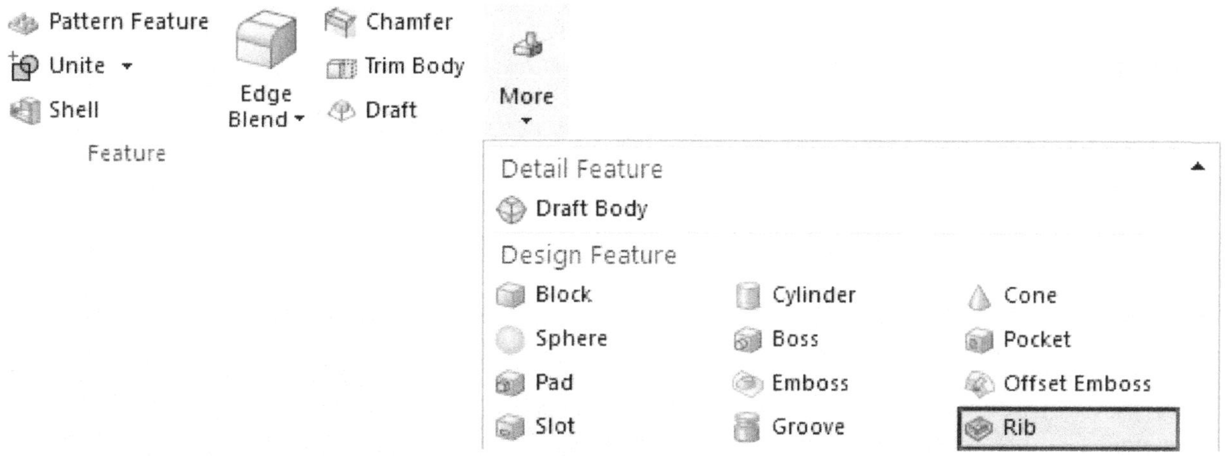

- Create a new part file (**Units**: *Millimeters*) called *"Rib.prt"* and place it in *"C:\My NX 11 Files"*.
- **Extrude** a *100 mm* x *65 mm* rectangle in the XY plane from *0 mm* to *20 mm* in the -ZC direction with a **Draft Angle** of *5 degrees*.
- Use **Edge Blend** to apply *5 mm* blends to the four outer vertical edges, as shown in the following figure.

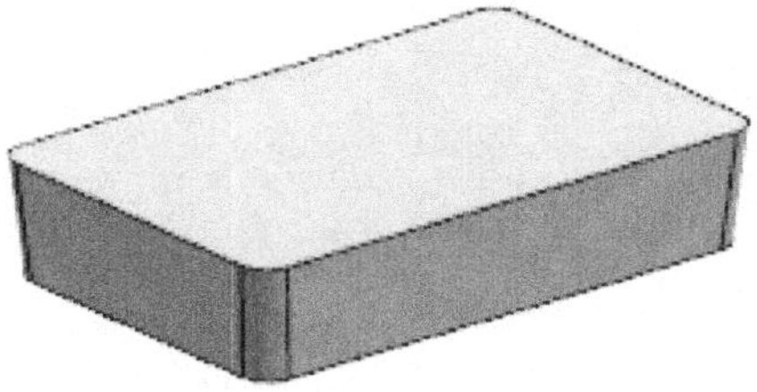

- Use **Shell**, select the top face and specify a thickness of **2mm**.

- Select the **Rib** tool and choose the body as the **Target**. Note that **Rib** will accept an internal sketch as input – click the **Sketch Section** icon and create the sketch shown below on the XY plane. Note that in the two applications of **Mirror Curve**, the *Convert Centerline to Reference* checkbox was disabled.

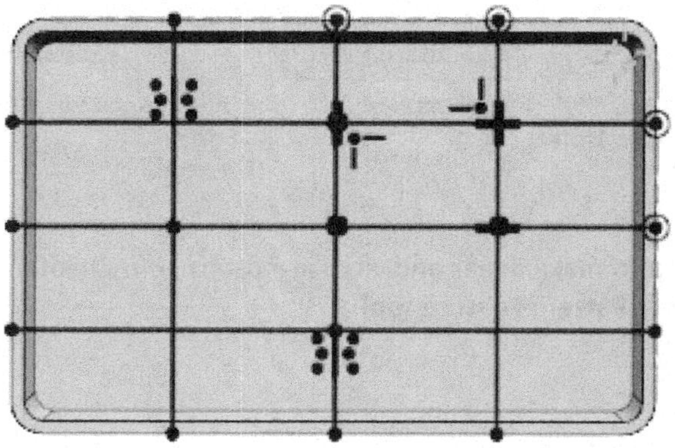

- **Finish** your sketch, and enter the parameters shown in the **Rib** dialog box below.

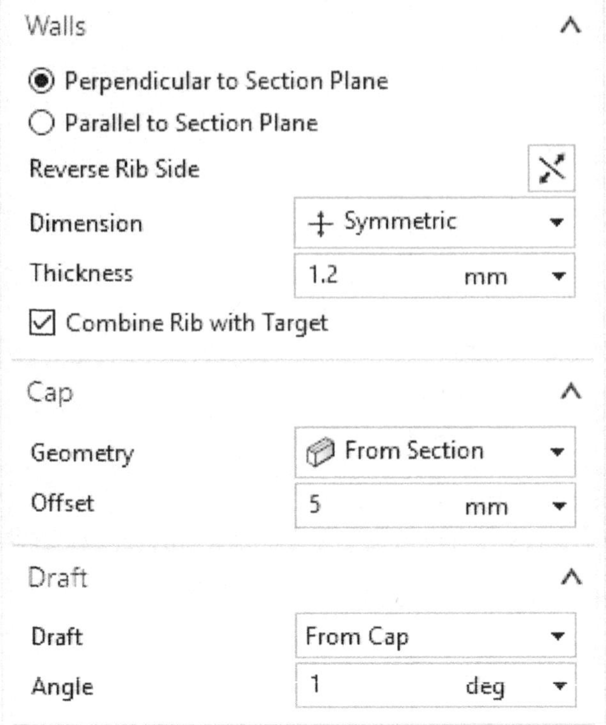

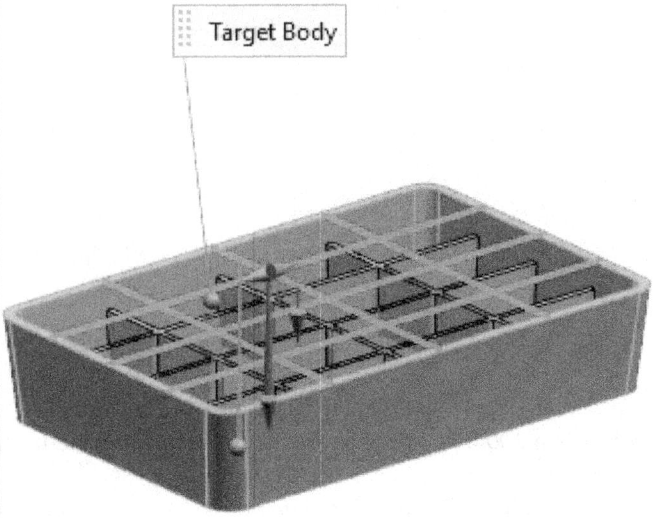

- **Save** your file!

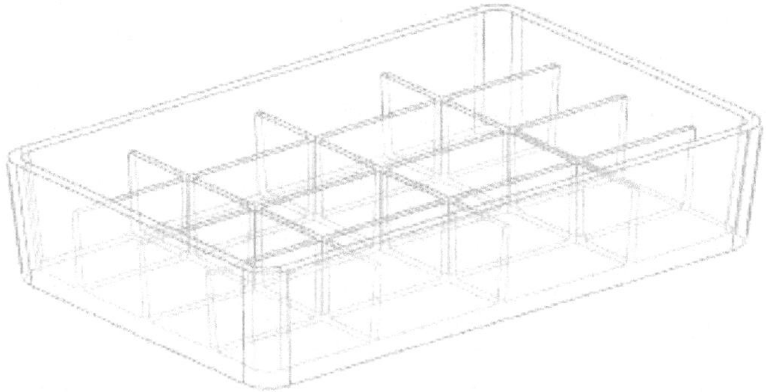

9.9 Pattern Feature

The **Pattern Feature** tool is a powerful method of creating an associative array of identical features. **Pattern Feature** is found in the **Feature** group on the **Home** tab.

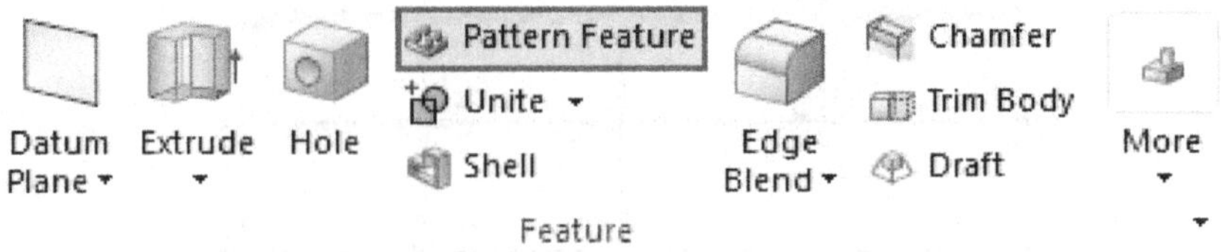

In the exercises below you will learn to make linear and circular patterns! In Chapter 17 we will delve into many of the more sophisticated options within the **Pattern Feature** tool.

9.9.1 Linear Pattern

The *Linear* option for **Layout** within the **Pattern Feature** dialog allows you to pattern features in a rectangular fashion, such as shown below.

- Create a new file (**Units**: *Inches*) called *"Pattern Feature – Linear.prt"* and place it in *"C:\My NX 11 Files"*.

- Create a **Block** with dimensions *8 in* x *6 in* x *1 in*.
- Create the initial **Boss** in the bottom left corner, with a **Height** of *1 in*, **Diameter** of *0.25 in*, *1 in* from each edge.

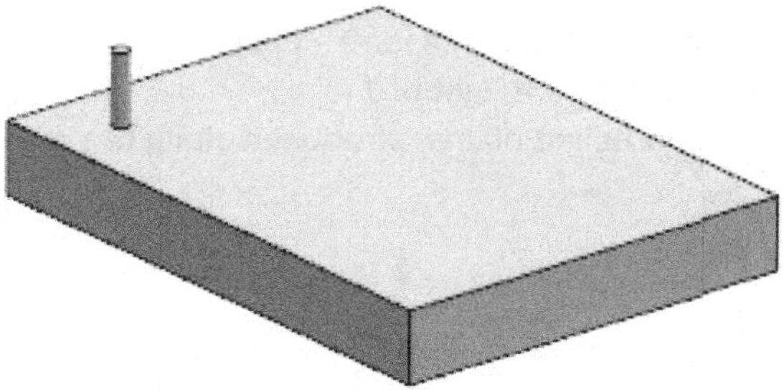

- Select the **Pattern Feature** tool, and select the boss as the **Feature to Pattern**. A yellow cube will appear. This is the feature reference point. It helps you understand what the pattern will look like before you actually have the system calculate the entire array.
- Select **Linear** from the **Layout** choice within the **Pattern Definition** menu.
- In **Direction 1**, set the **Spacing** method to *Count and Pitch*. Input a **Count** of *7* and a **Pitch Distance** of *1 in*.

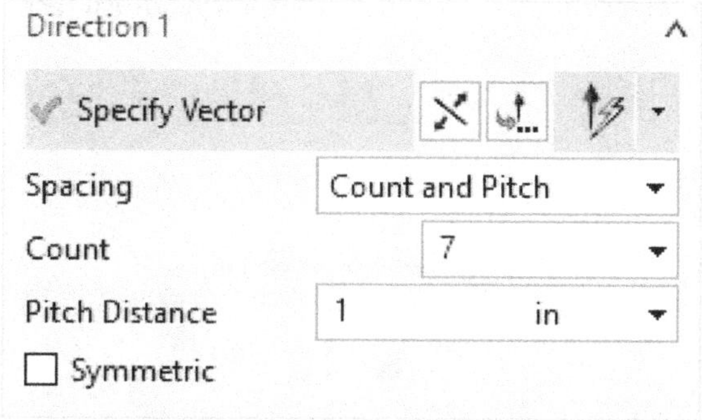

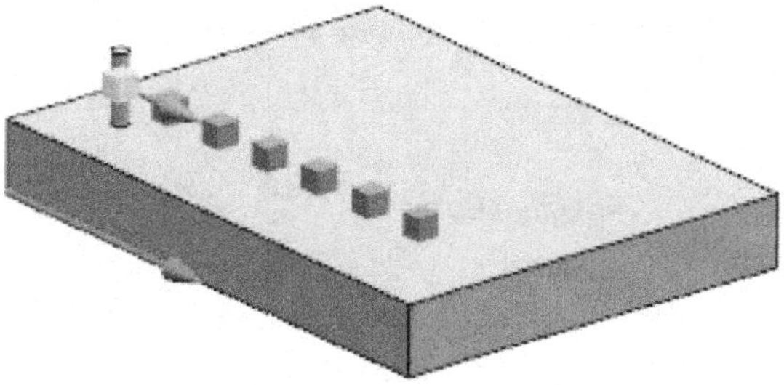

- Next, check the **Use Direction 2** checkbox. Input the parameters shown below.

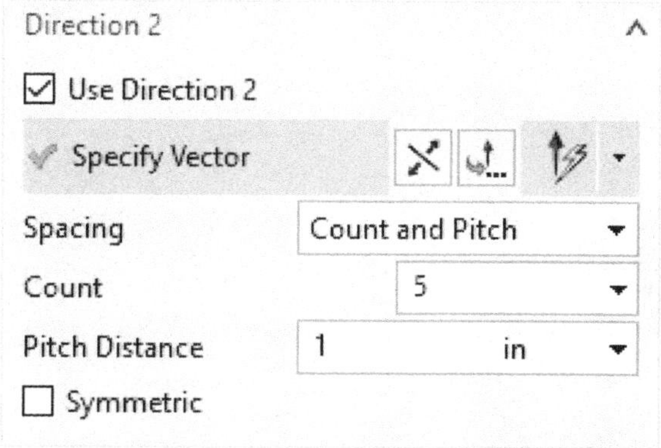

- Click **OK** to create the feature pattern. The finished product should appear as below.

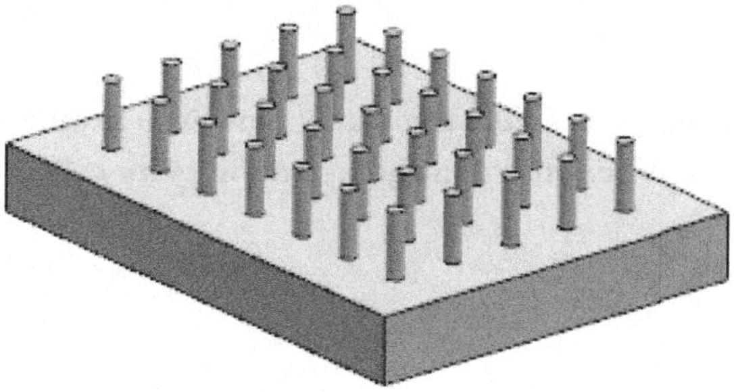

- **Save** your file!

9.9.2 Circular Pattern

You can also use the **Pattern Feature** command to create circular feature patterns.

- Create a new part file (**Units**: *Inches*) called *"Pattern Feature – Circular.prt"* and place it in *"C:\My NX 11 Files"*.
- Create a **Cylinder** with a **Diameter** of *6 in* and a **Height** of *1 in*.
- Create a **Boss** of **Diameter** of *0.25 in* and **Height** of *1 in*, positioned along the YC axis, *2 in* away from the XC axis.

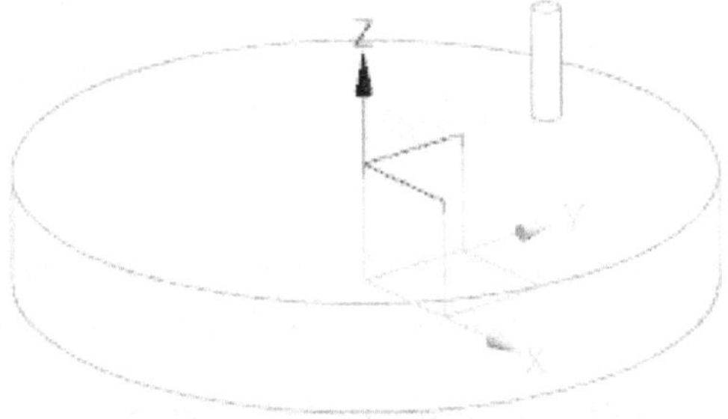

- Select the **Pattern Feature** tool. Choose the boss as the **Feature to Pattern**. Set the **Layout** to *Circular* and choose the centerline of the *Cylinder* feature as the **Rotation Axis** – to select this, you can just pick the cylindrical face. Set the **Angular Direction** parameters as shown below.

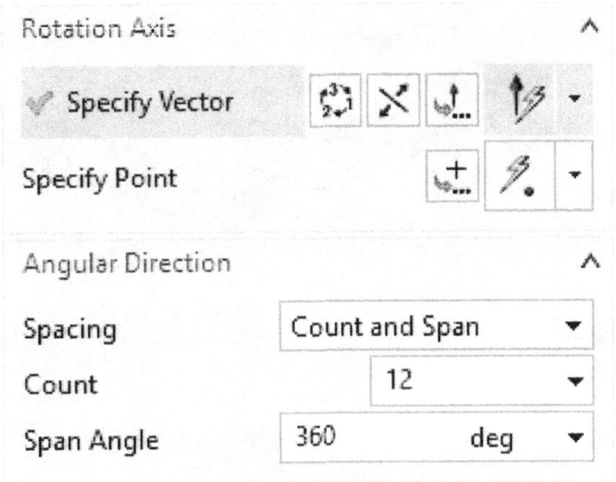

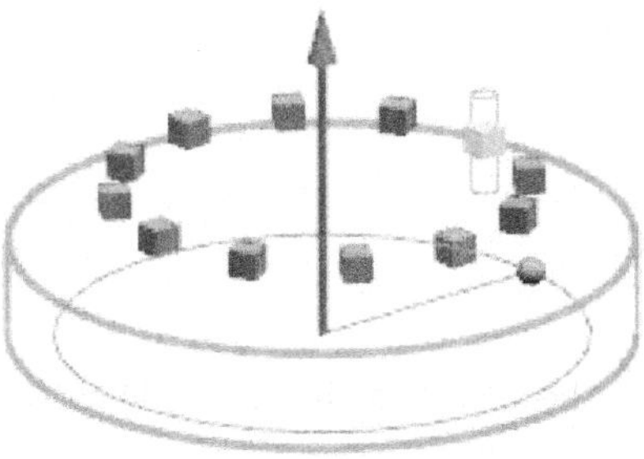

- **Save** your file!

9.10 Mirror Feature

The **Mirror Feature** command, like the **Pattern Feature** command, allows you to produce associative copies of features on a model. **Mirror Feature** is found in the **Associative Copy** group, in the **More** gallery in the **Feature** group on the **Home** tab.

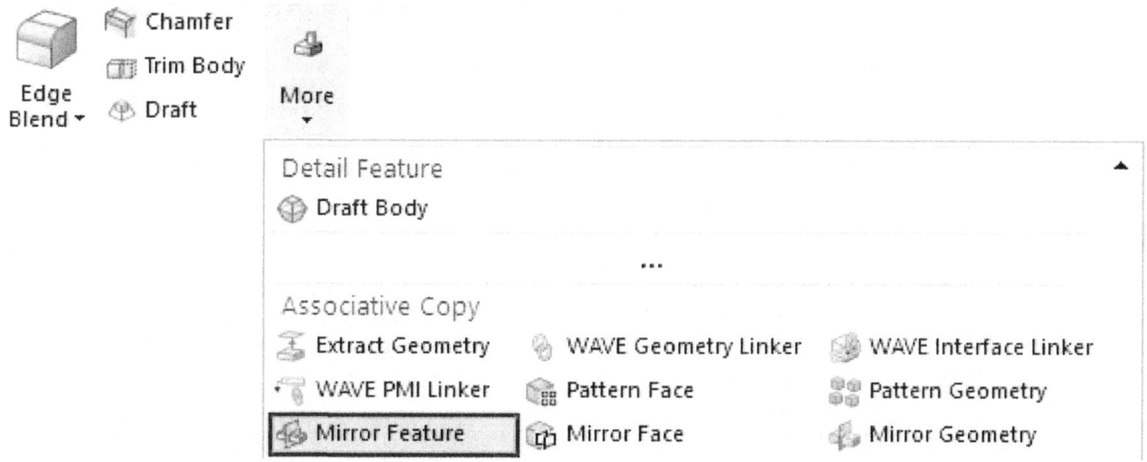

Since the features are copied about a mirror plane, certain details may become reversed. Threads in particular, you may not want to reverse. Fortunately, **Mirror Feature** has an option to maintain thread-handedness, which you will explore in the exercise below.

- Open the file *"Mirror Feature.prt"* from *"C:\My NX 11 Files\Downloaded Files"*.
- The Model History shows a non-parametric body with a threaded hole in it.

- Select the **Mirror Feature** tool and choose the threaded hole as the **Features to Mirror**. For the **Mirror Plane**, change the **Plane** drop-down to *New Plane*, and choose two mirror faces to define a *Bisector* plane as shown below.

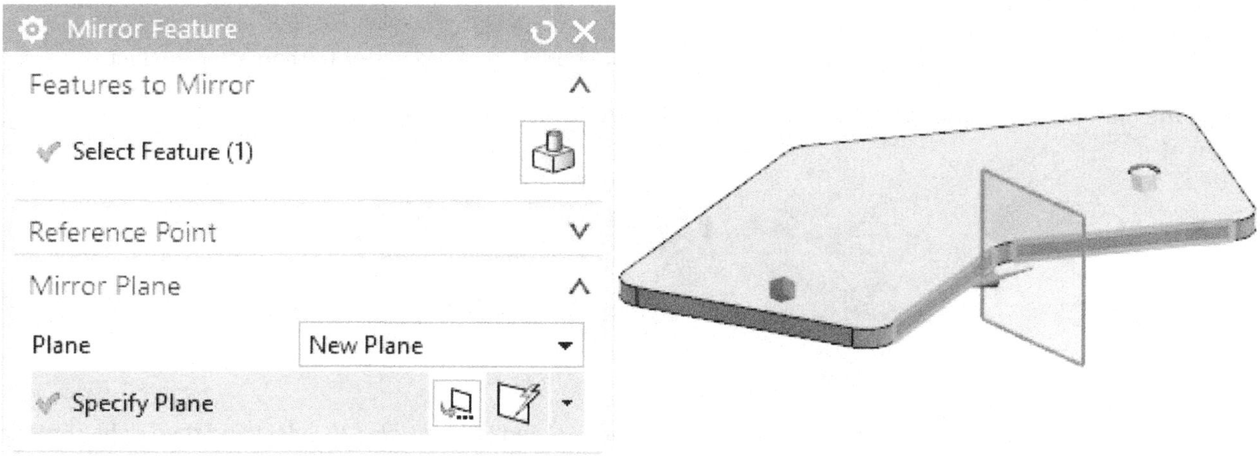

- In settings, there is a checkbox to *Maintain Thread Handedness* – make sure it's checked!

- When you produce associative copies of features containing internal sketches, such as this threaded hole, sometimes your *Mirror Feature* feature will have an alert.

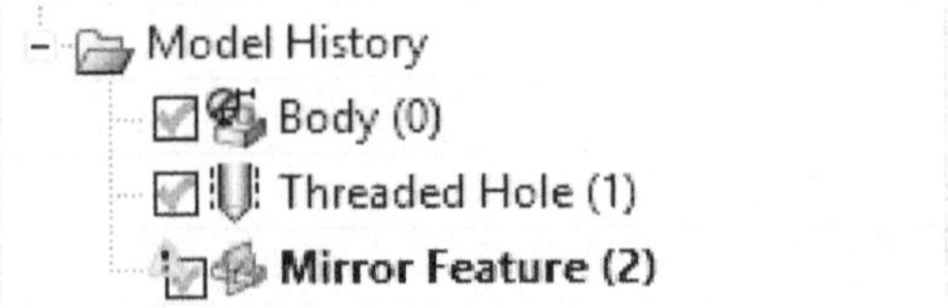

- The alert concerns references made by the internal sketch. Edit your *Mirror Feature* entity, and check the box for *Sketch External Reference* within the **Reusable References of Source Feature** section of the dialog.

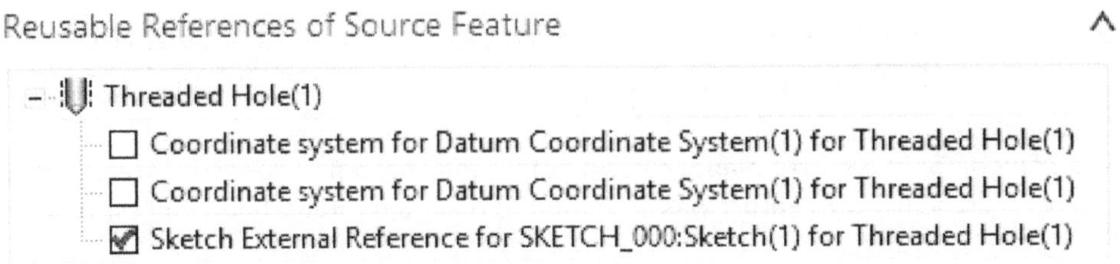

- Usually these kinds of alerts concerning internal sketches are not a big deal, but occasionally they can pose problems for maintaining full associativity of your model.

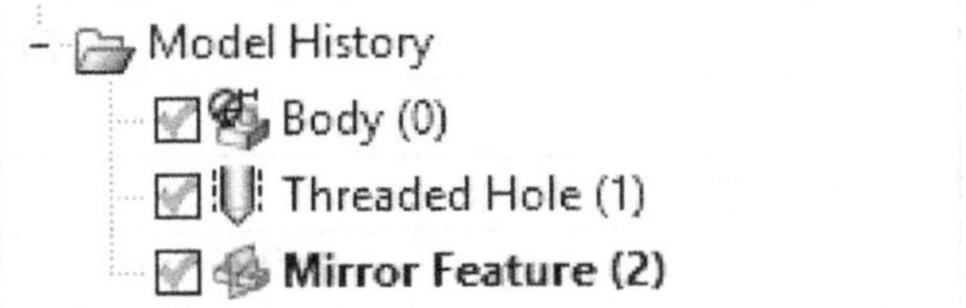

- **Save** your file!

9.11 Feature Groups

Feature Groups are useful for breaking your Model History into "chunks" that are easier to read after the fact.

- Open the file *"Feature Groups.prt"* from *"C:\My NX 11 Files\Downloaded Files"*.

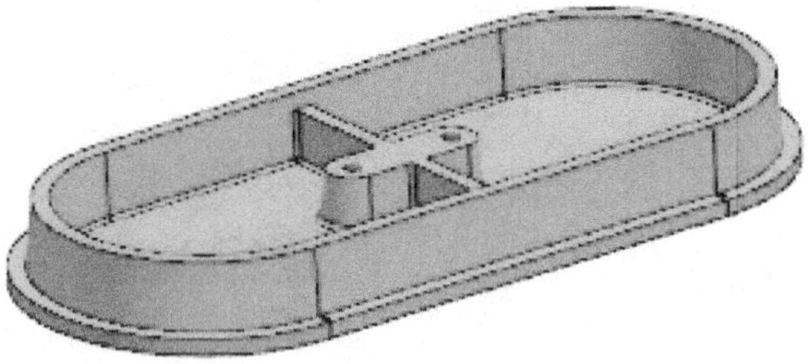

- The flange on the bottom of the body is comprised of *Extrude (1)* and *Simple Hole (2)*. Select and then right-click these features in the Model History, and choose **Feature Group**. When the dialog appears, name this group *"FLANGE"*.

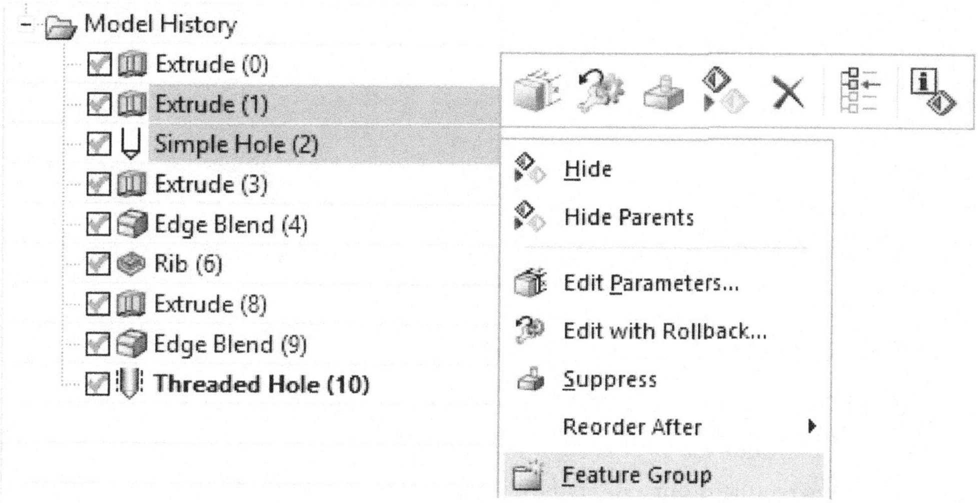

- Create another feature group consisting of the features from *Extrude (3)* to *Threaded Hole (10)*. Name this group *"CAVITY"*. As you can see, these are the features that comprise the cavity in the model. Right-click on the feature group *"CAVITY"* and select **Active**.

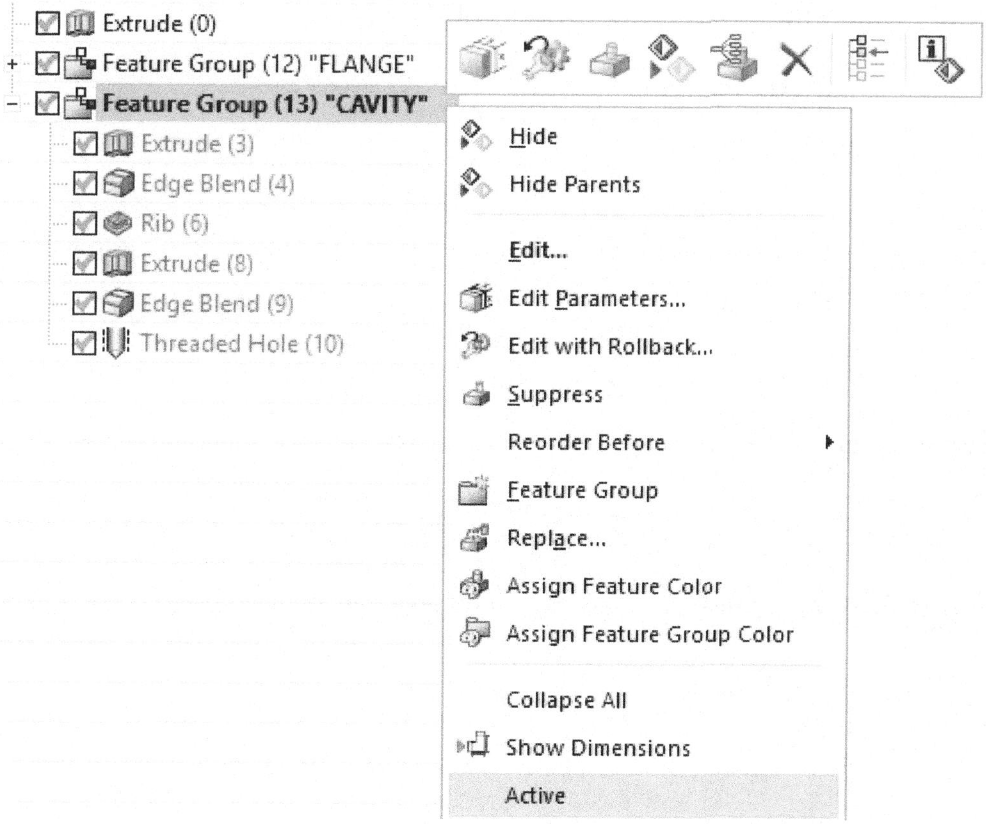

- The "active" feature group is the one in which new features are placed. In the Model History, *"CAVITY"* now appears as shown below.

- Create a **Boss** with a **Height** of *1 in*, a **Diameter** of *0.5 in*, and a **Taper** of *2 degrees*, positioned at the arc center of the rightmost interior edge on the cavity floor. Because *"CAVITY"* is the active feature group, *Boss (10)* is added to the group in the Part Navigator.

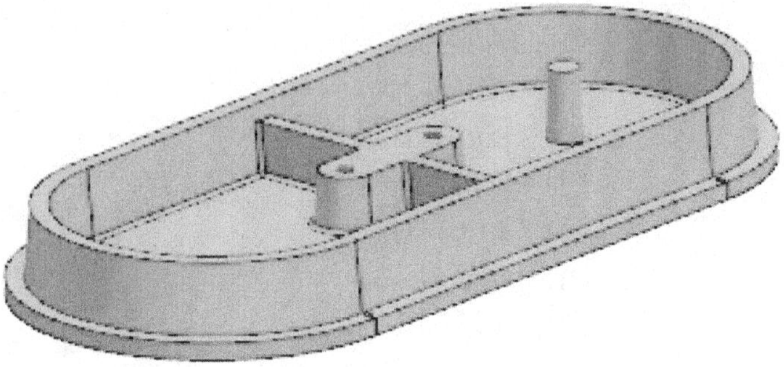

- Deactivate *"CAVITY"* by right-clicking it and unchecking **Active**.

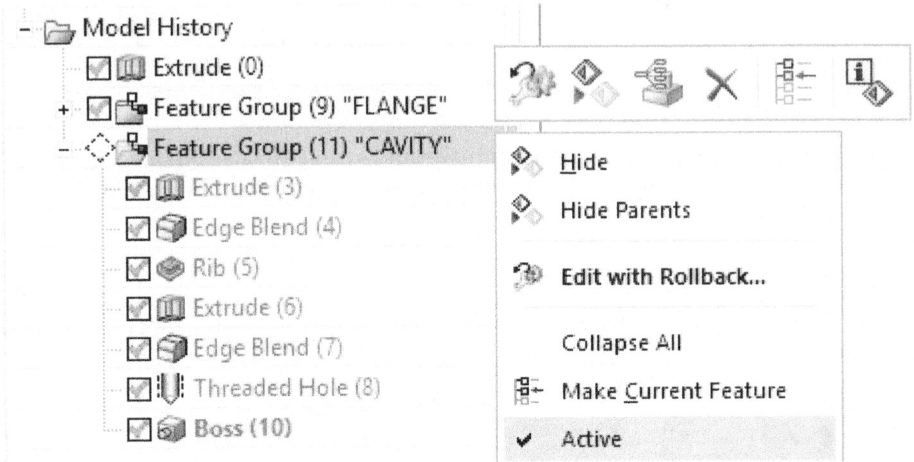

- Note that the model history is rolled to a timestamp at which the features in *"CAVITY"* are all active, but *"CAVITY"* itself is suppressed.

- Use **Make Current Feature** to roll the model history forward so that *"CAVITY"* is unsuppressed.

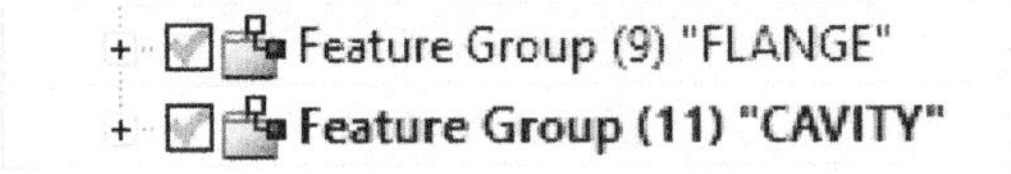

- Create a threaded hole using the **Hole** tool, positioned at the arc center of the top edge of the boss, with parameters as shown below.

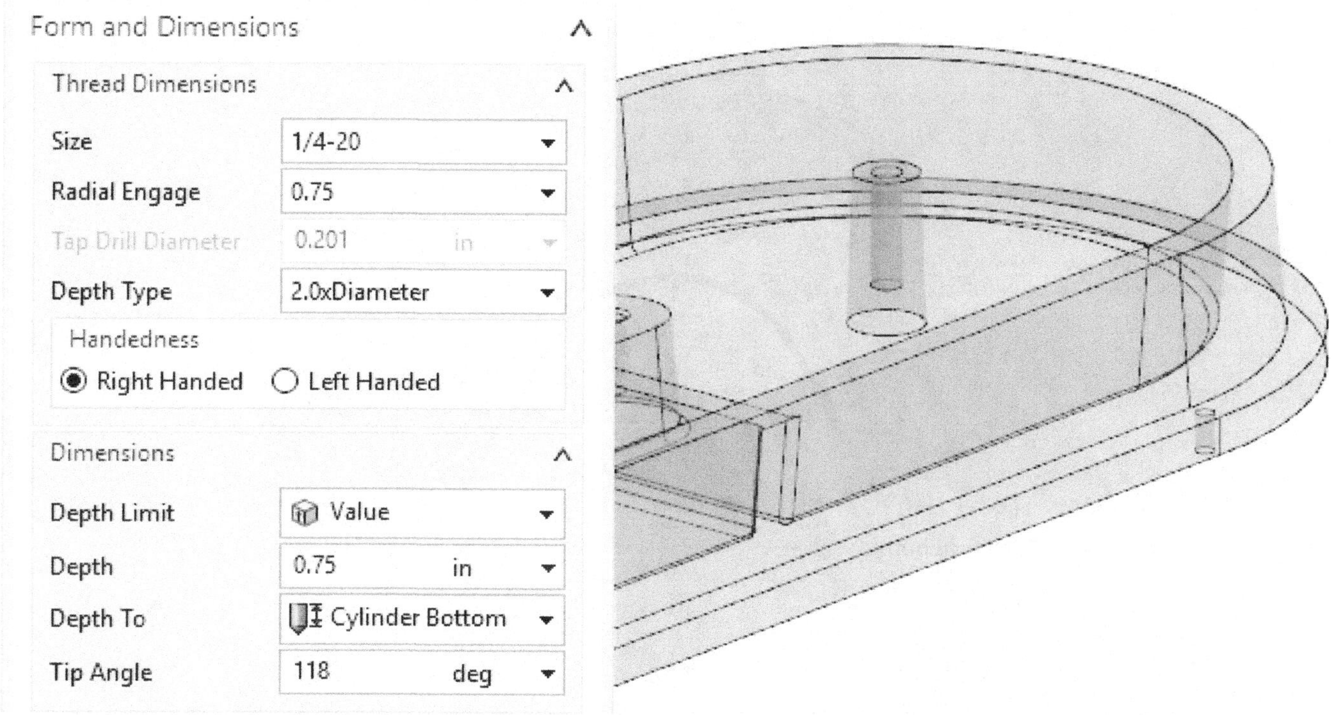

- You can also add features to feature groups with a drag-and-drop motion in the Part Navigator.

- **Save** your file!

9.12 **Project – The Pool Rack, Part II**

In this project, you will use the **Hole**, **Pocket** and **Edge Blend** features to add detail to the pool rack model that you made in Chapter 8.8. If you skipped that chapter, or otherwise don't have that file, there is a suitable replacement in the downloaded part files folder.

- Open *"C:\My NX 11 Files/Pool Rack.prt"* or *"C:\My NX 11 Files/Downloaded Files\Pool Rack.prt"*.
- Use **File / Save / Save As** to save a copy as *"Pool Rack Part II.prt"* and place it in *"C:\My NX 11 Files"*.
- Select the **Hole** tool and choose the top face of the pool rack as shown below.

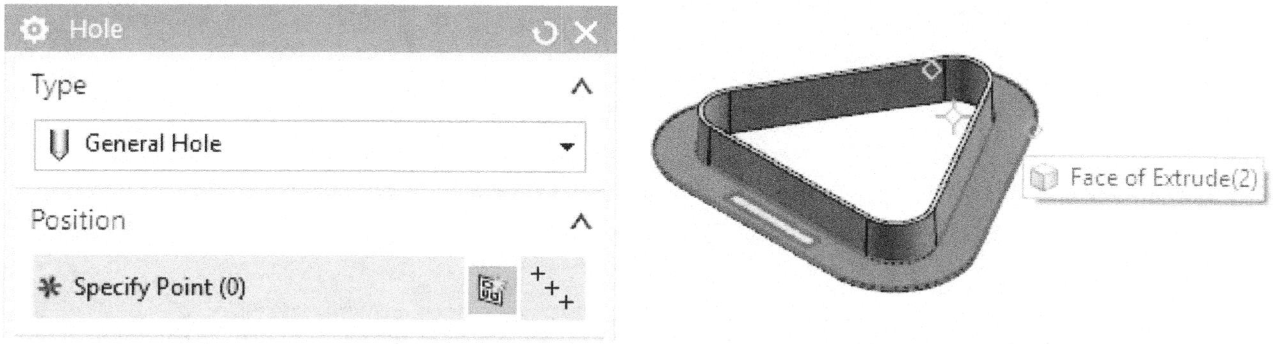

- It may be necessary to use **Reattach** to orient your sketch coordinate system as shown below. Position the hole along the sketch vertical axis, *0.5 in* from the edge of the face, as shown below.

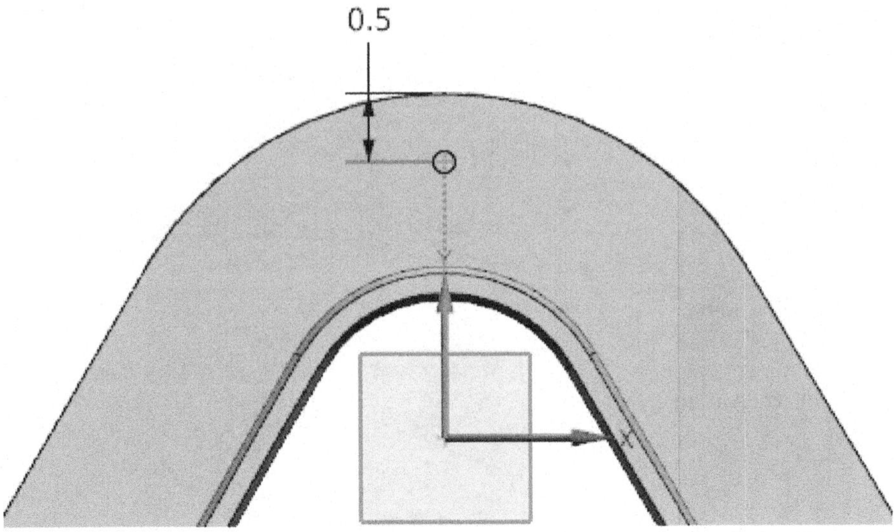

- The hole is a simple *0.25 in* thru hole.

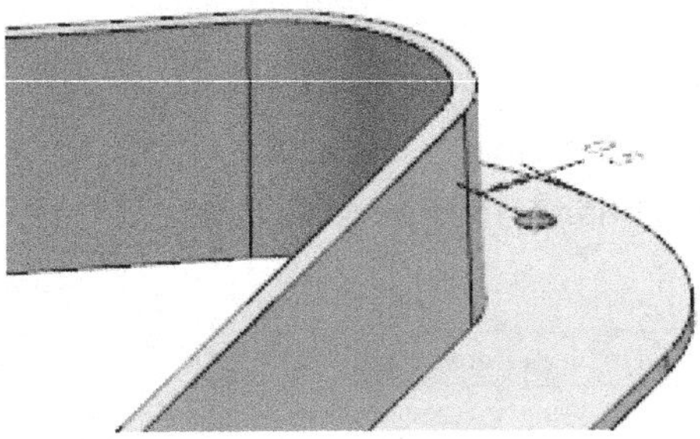

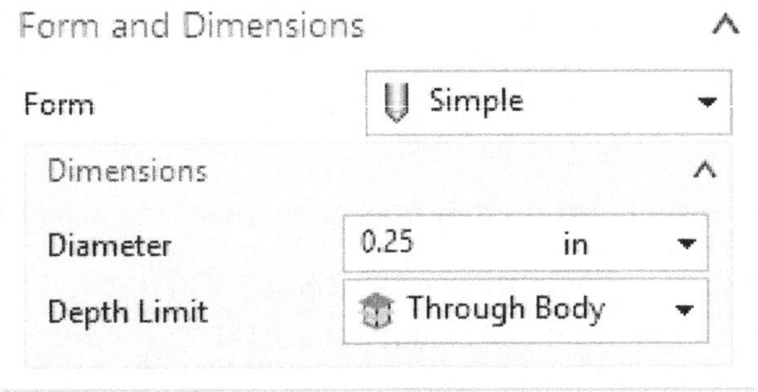

- Next, select the **Edge Blend** tool, and set your **Selection Intent** to *Tangent Curves*. Choose the edges shown below and set the radius to *0.375 in*.

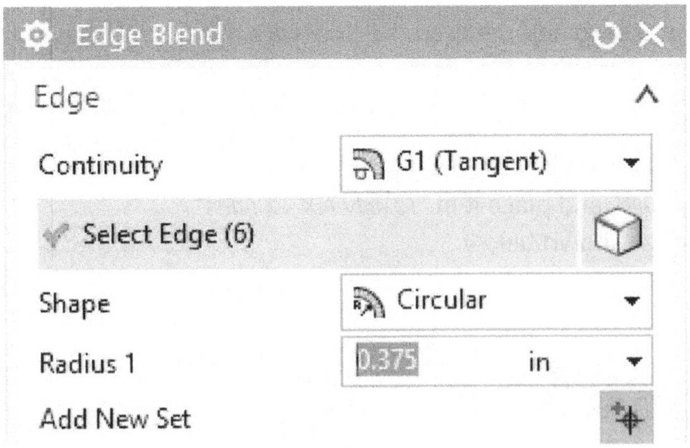

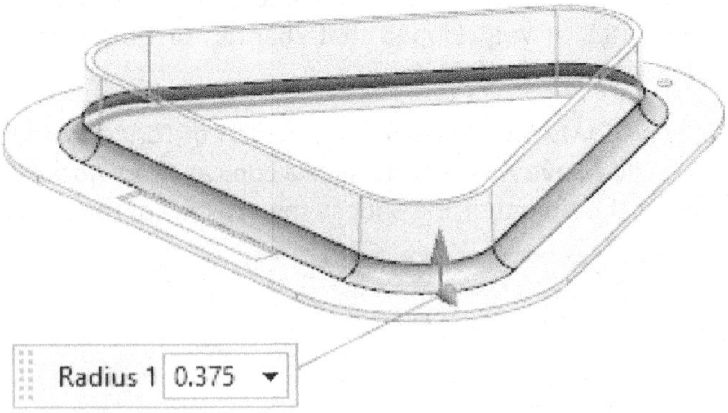

- Next, right-click on *Extrude (3)* from the Model History and select **Delete**.

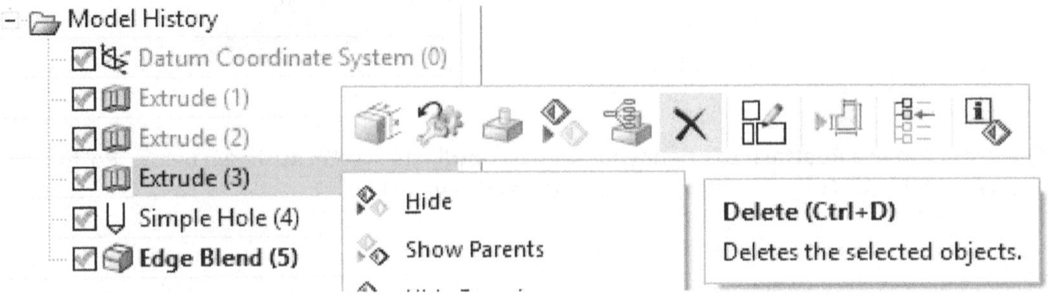

- We will rebuild the pocket using the **Pocket** tool, rather than **Extrude** with *Subtract*. To position the pocket, we will need a reference point.
- Select the **Point** tool, and set the **Type** to *Point on Face*. Choose the top face of the pool rack and position the point as shown below.

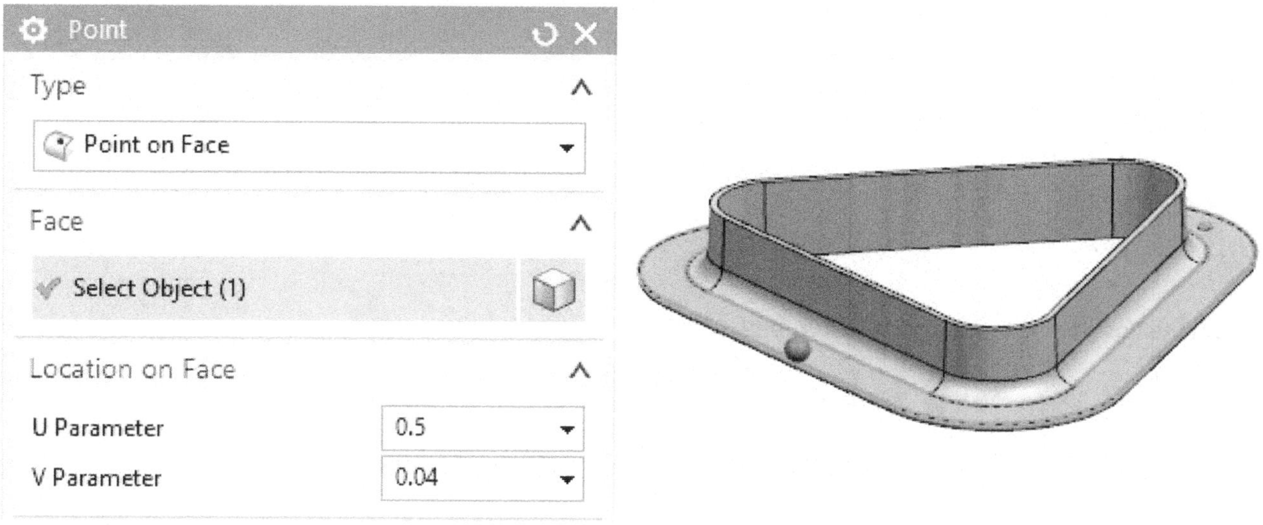

- Select the **Pocket** tool and choose **Rectangular**. When the **Rectangular Pocket** dialog appears, choose the top face of the pool rack, as shown below.

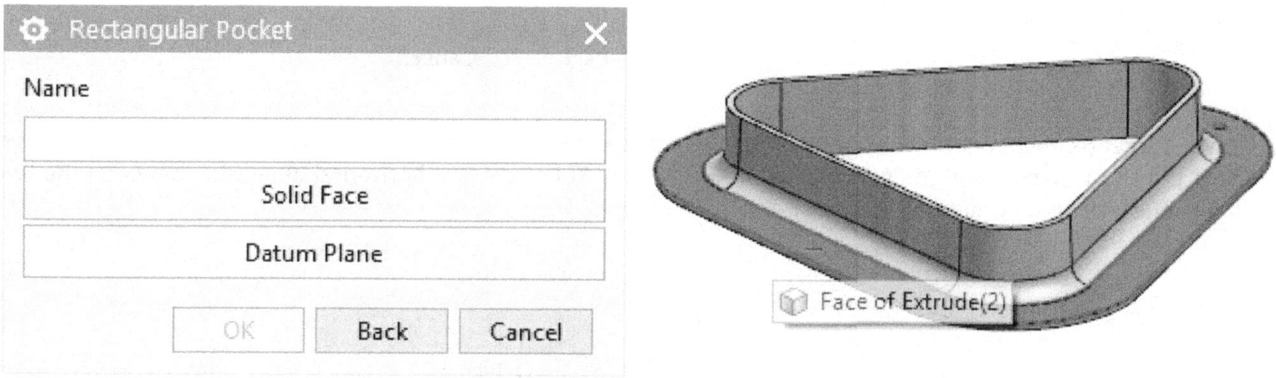

- For the **Horizontal Reference**, choose the bottom edge of the face.

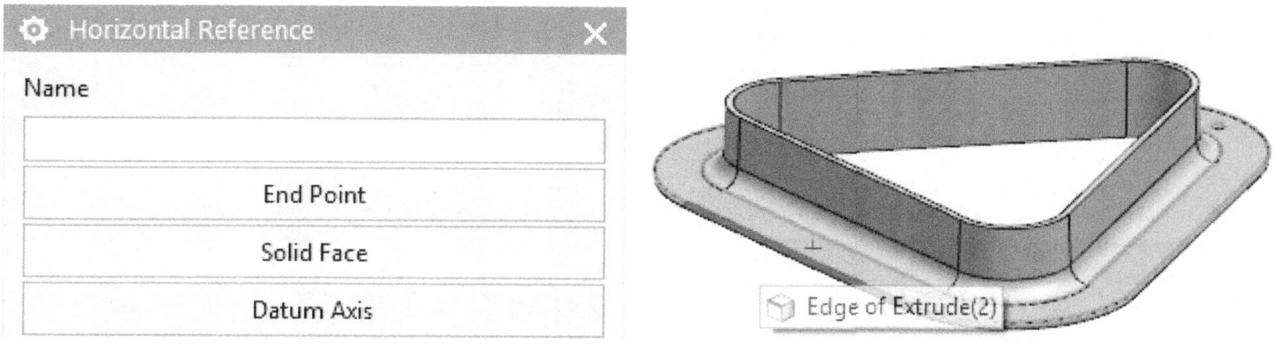

- Set the **Rectangular Pocket** dialog parameters as shown below.

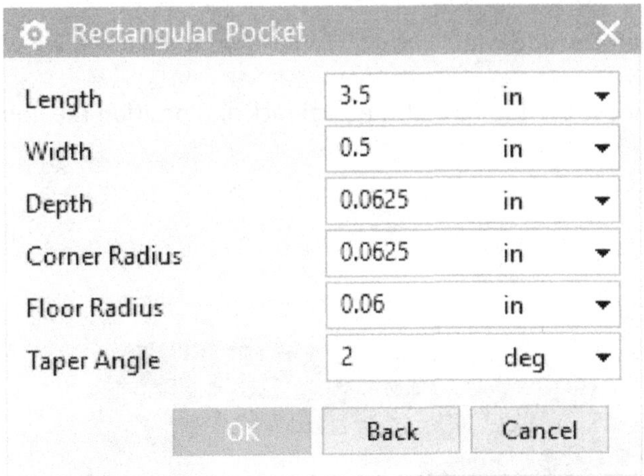

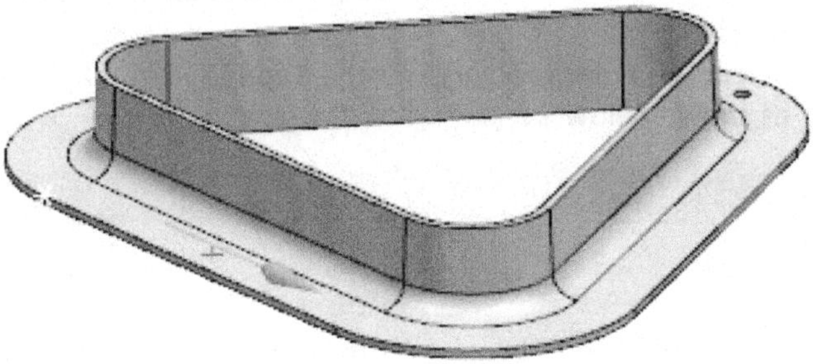

- From the **Positioning** menu, choose *Horizontal*.

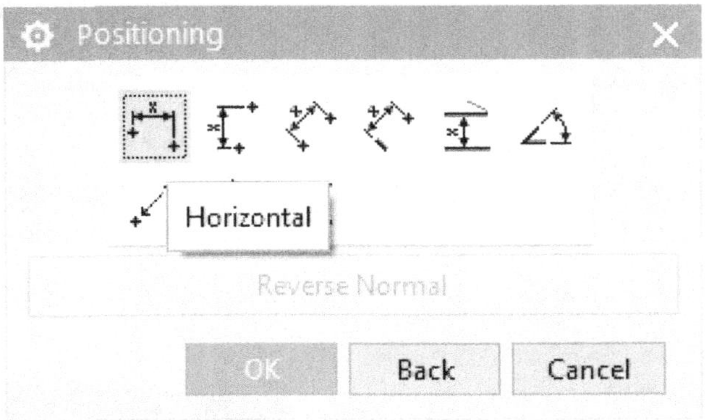

- For the **Target** (remember to read your Cue display!) choose the point on face that you constructed, and for the **Tool**, select the short vertical dotted line centered in the pocket. This creates a horizontal dimension between the two. Change the dimension value to *0 in.*

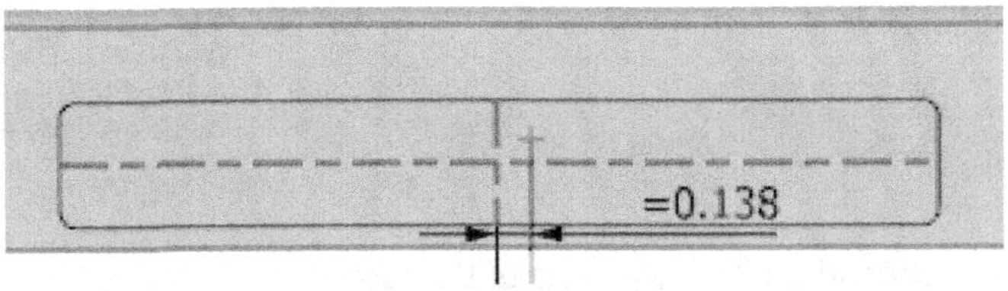

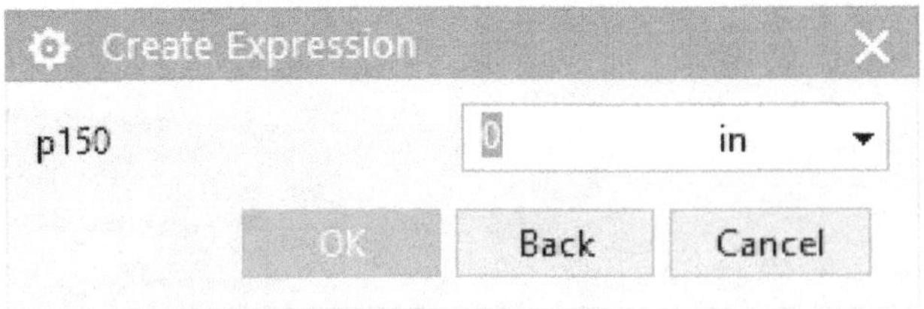

- From the **Positioning** menu, choose *Vertical*.

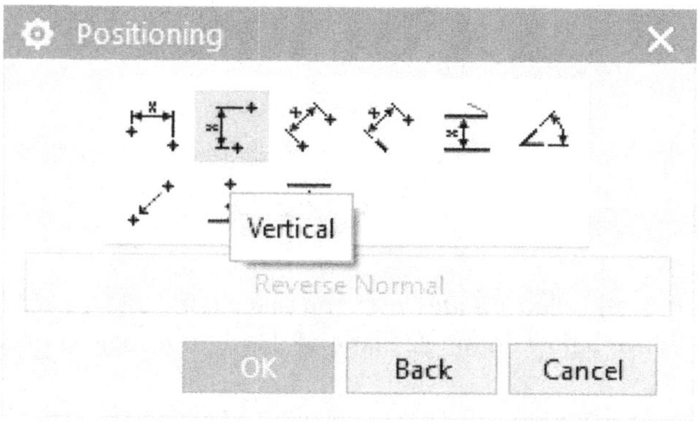

- Select the point as **Target** and the horizontal lengthwise dotted line in the pocket as the **Tool**. Enter a value of *0 in* for the dimension.

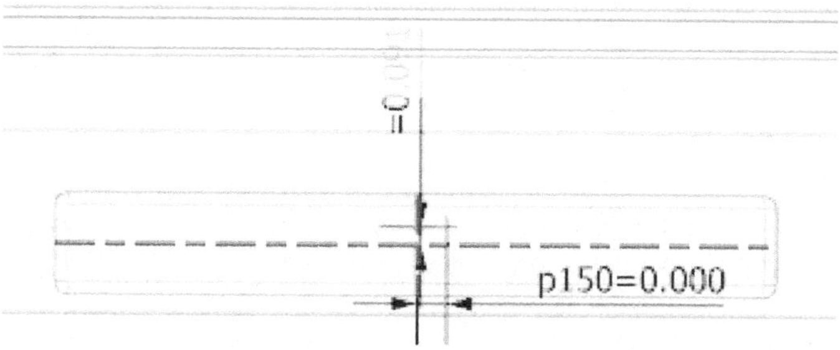

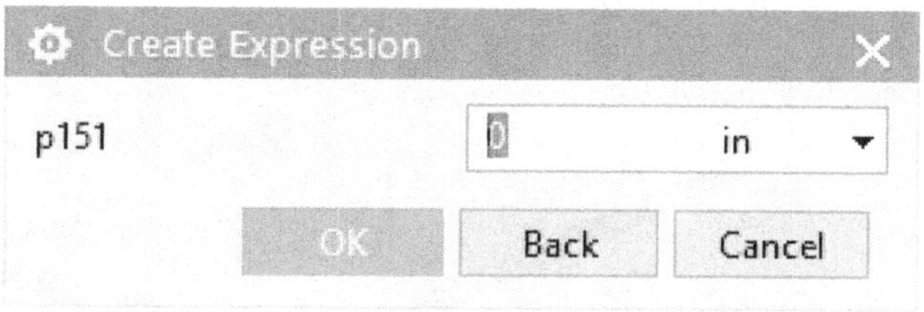

- Click **OK** through the remaining menus to center your pocket at the point you constructed!

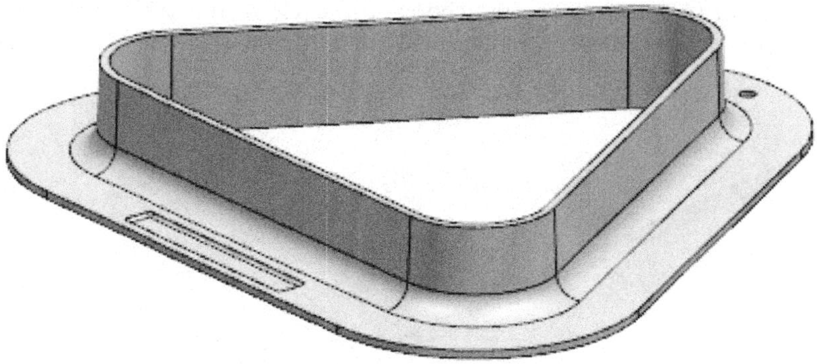

- While the **Pocket** tool leaves something to be desired where positioning is concerned, it does offer a good deal of additional control over the shape – you can specify the interior radii and the taper all in one operation rather than doing separate **Edge Blends**, etc.
- **Save** your file!

9.13 **Project – The Hinge**

In this exercise, you will practice using some of the tools and concepts covered in the last few chapters. The following figure shows the geometry of a basic hinge. Building the solid model will require knowledge of the extruded body and hole features. This model will also be used in later chapters, so please remember to save the part file as *"Hinge1.prt"*.

- Create a new part file (**Units**: *Inches*) called *"Hinge1.prt"* and place it in *"C:\My NX 11 Files"*.
- Sketch a rectangle and circle shown below on the YZ plane of the datum coordinate system.

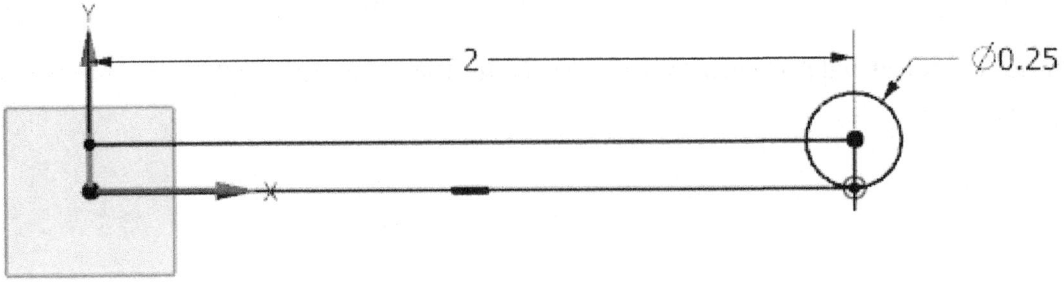

- Select the **Extrude** tool and set your **Selection Intent** to *Connected Curves*. Choose the rectangle as your **Section Curve** and enter **Start** and **End Limits** of *0 in* and *5 in*. When you finish, click **Apply**.

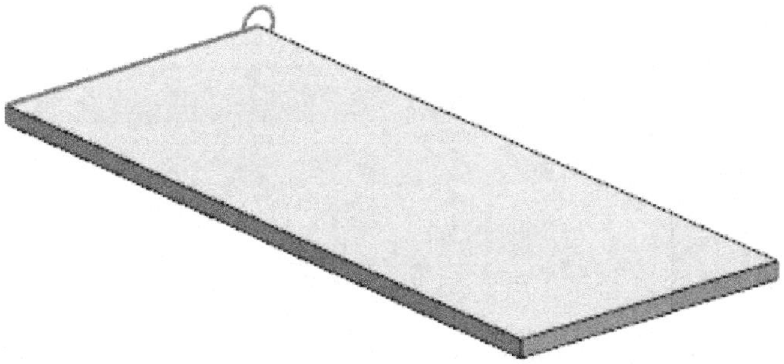

- Change your **Selection Intent** to *Single Curve*, and choose the circle as the **Section Curve** for your second **Extrude** operation. Leave the **Start** and **End Limits** set to *0 in* and *5 in*. Set the **Boolean** drop-down to *Unite*, and click **Apply**.

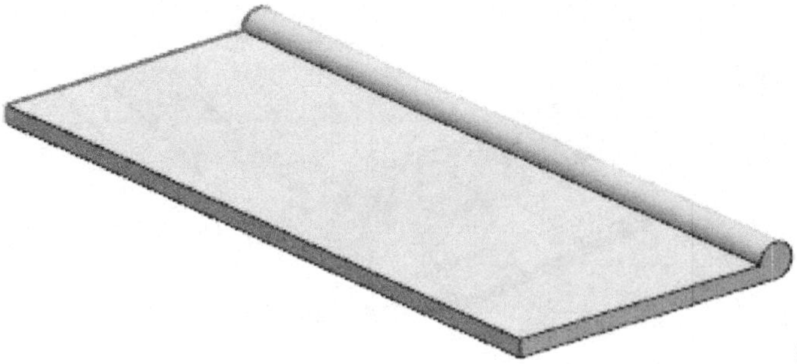

xx

- Change the **Boolean** drop-down to *Subtract*, and **Extrude** the circle from *1* to *2*. Click **Apply**.
- **Extrude** the circle from *3* to *4*, again with the **Boolean** set to *Subtract*. Your model should appear as below.

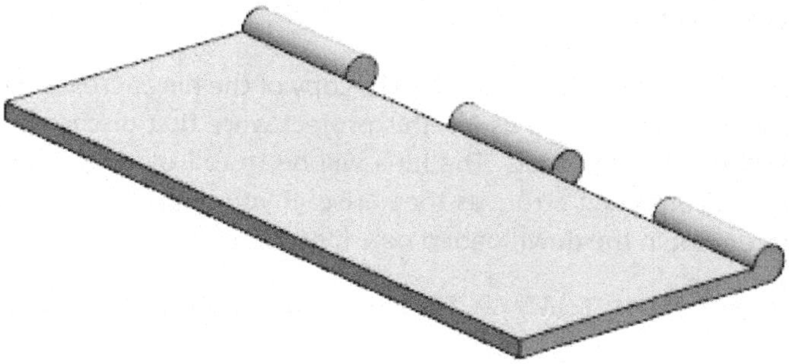

- Select the **Hole** tool and position three counterbored holes on the top face of the hinge with the parameters specified below. The middle hole is centered horizontally. Click **Apply** when you are done.

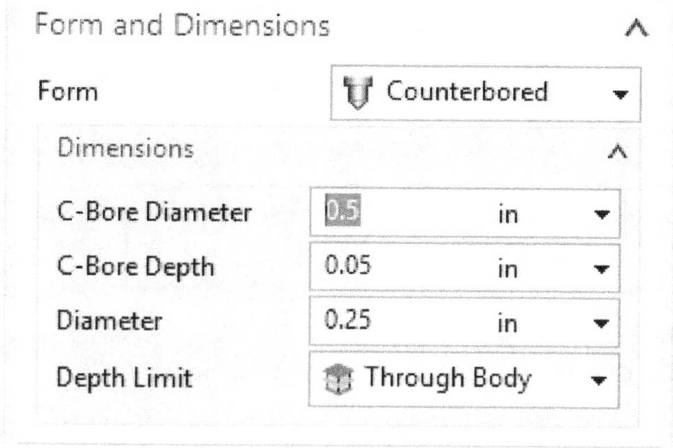

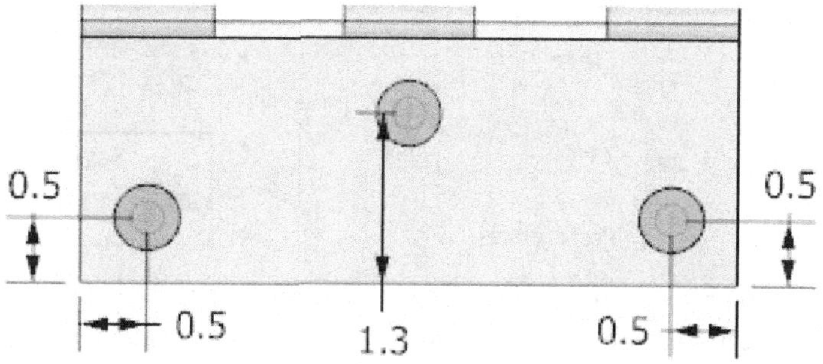

- Change the **Form** of the hole to *Simple*, and enter a **Diameter** of *0.125 in*. Leave the **Depth Limit** set to *Through Body*. As for the position, click on the arc center of the edge of the cylinder as shown below. Make sure that the **Arc Center** icon appears, so that you are guaranteed to get the arc center and not a point on the edge or the face!

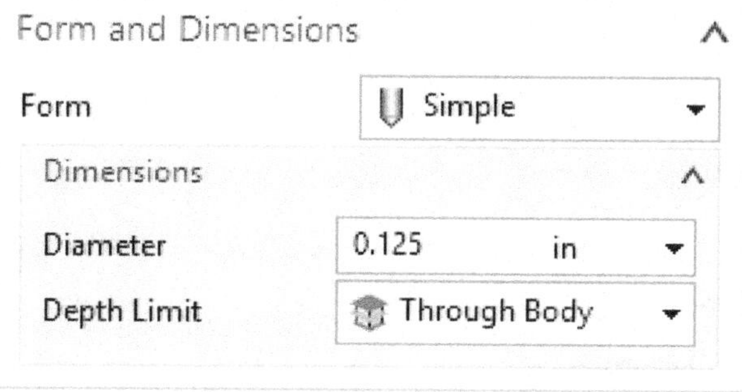

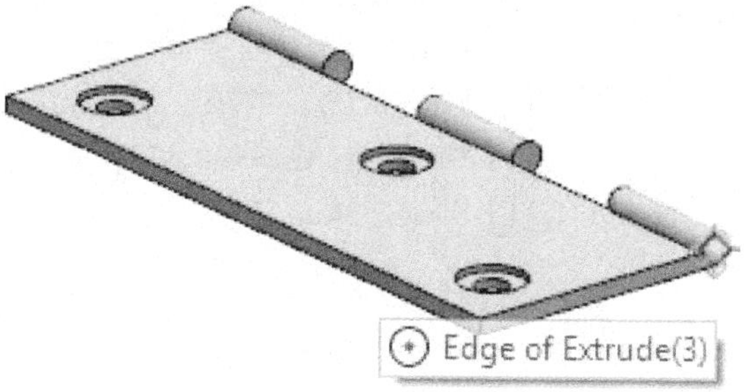

- Use **Edit Object Display** (**[Ctrl]+[J]**) to color your model yellow.

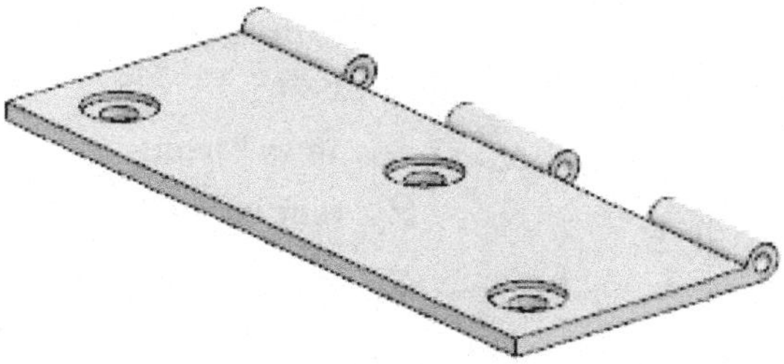

- **Save** your file!

9.14 Project – The Hinge, part II

In this exercise, you will make a complimentary hinge by saving a copy of the hinge created in the last exercise, and making edits to the Model History. The commands we will use in this project were first discussed in Chapter 4 – please refer to that chapter for further information about the tools. The idea will be to roll the clock back and then work forward one item at a time in the Model History, correcting errors as they arise. If you skipped that chapter, or otherwise don't have that file, there is a suitable replacement in the downloaded part files folder.

- Open *"C:\My NX 11 Files\Hinge1.prt"* or *"C:\My NX 11 Files\Downloaded Files\Hinge1.prt"*.
- Select **File / Save / Save As**.

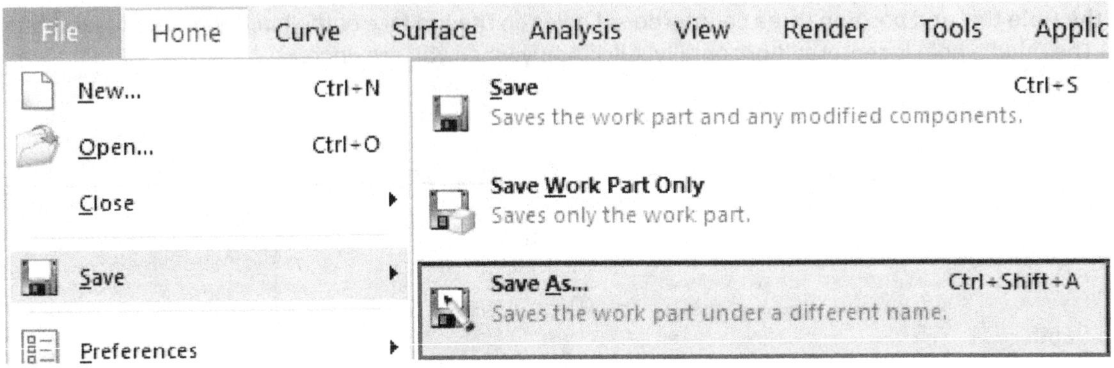

- Make sure you are saving to *"C:\My NX 11 Files"*, and name the copy *"Hinge2.prt"*.
- Right-click on *Extrude (3)* in the Model History, and select **Make Current Feature** from the pop-up menu.

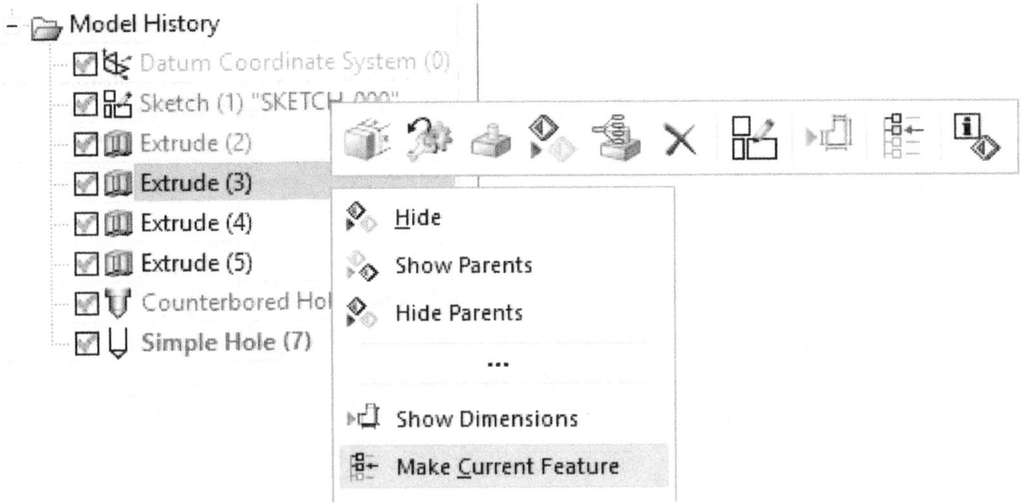

- Double-click on *Extrude (3)* to invoke **Edit Parameters**, and change the **Boolean** from *Unite* to *Subtract*.

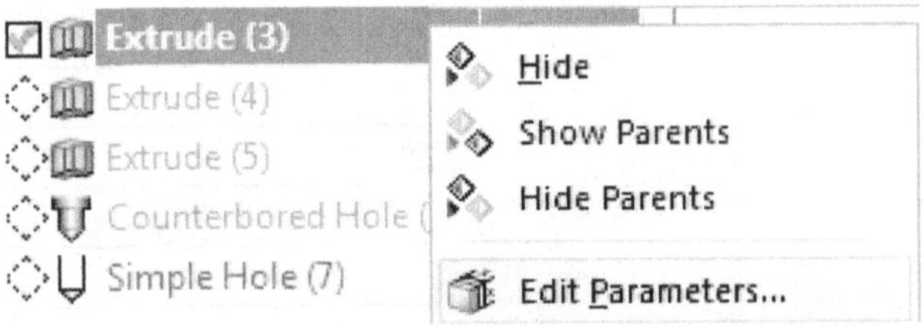

- Within the **Extrude** dialog box, change the **Boolean** from *Unite* to *Subtract*, and click **OK**. Your model should now look like this.

- At this point, we will need to revise later operations in the Model History, so right-click on *Extrude (4)* and select **Make Current Feature** (or use the keyboard shortcut **[Ctrl]+[Shift]+[→]**). This causes *Extrude (4)* to break, and you will see the following Information window.

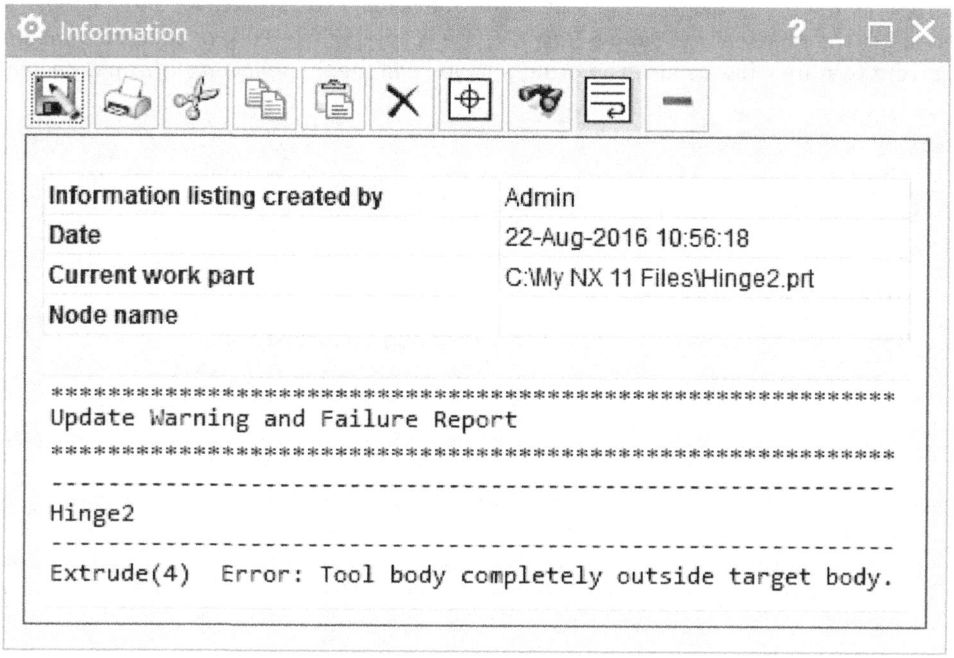

- Next, we'll see what we can do to fix the errors resulting from that change. These kinds of errors arise frequently when editing parametric models, and the ability to fix errors resulting from changes to the Model History is an extremely valuable asset to a designer, so this will be great practice! Double-click on *Extrude (4)* to make an edit, and change the **Boolean** from *Subtract* to *Unite*. This fixes the problem!.

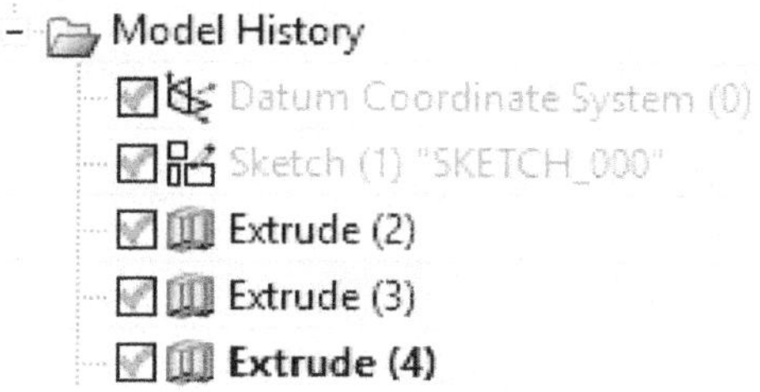

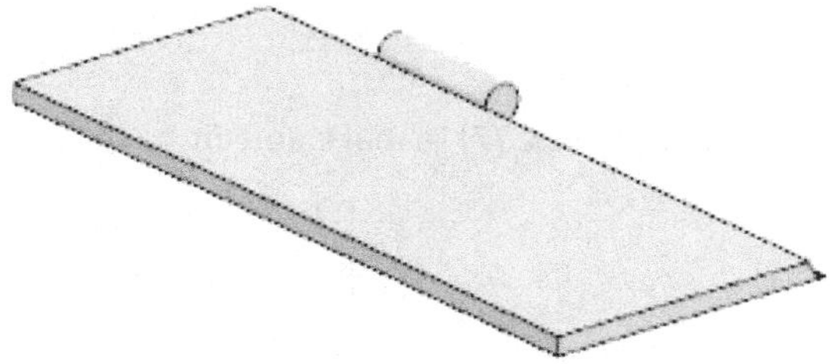

- Right-click on *Extrude (5)*, and select **Edit with Rollback**. In the **Extrude** dialog box, change the **Boolean** from *Subtract* to *Unite*. Note that upon clicking **OK**, the Model History automatically advances to *Extrude (5)*, so we didn't have to tell it to **Make Current Feature**!

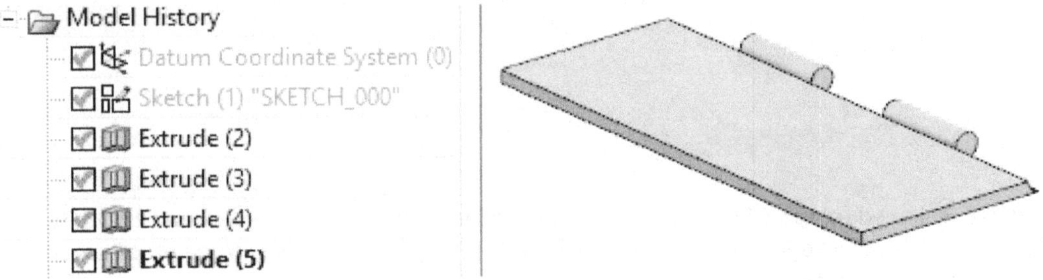

- The next feature, *Counterbored Hole (6)*, is independent of *Extrude (3)*, *Extrude (4)*, and *Extrude (5)*, and so we can skip straight ahead to the final feature, *Simple Hole (7)*. Use **[Ctrl]+[Shift]+[End]** or right-click on *Simple Hole (7)* and select **Make Current Feature**. This results in an error, and you will see the following Information window.

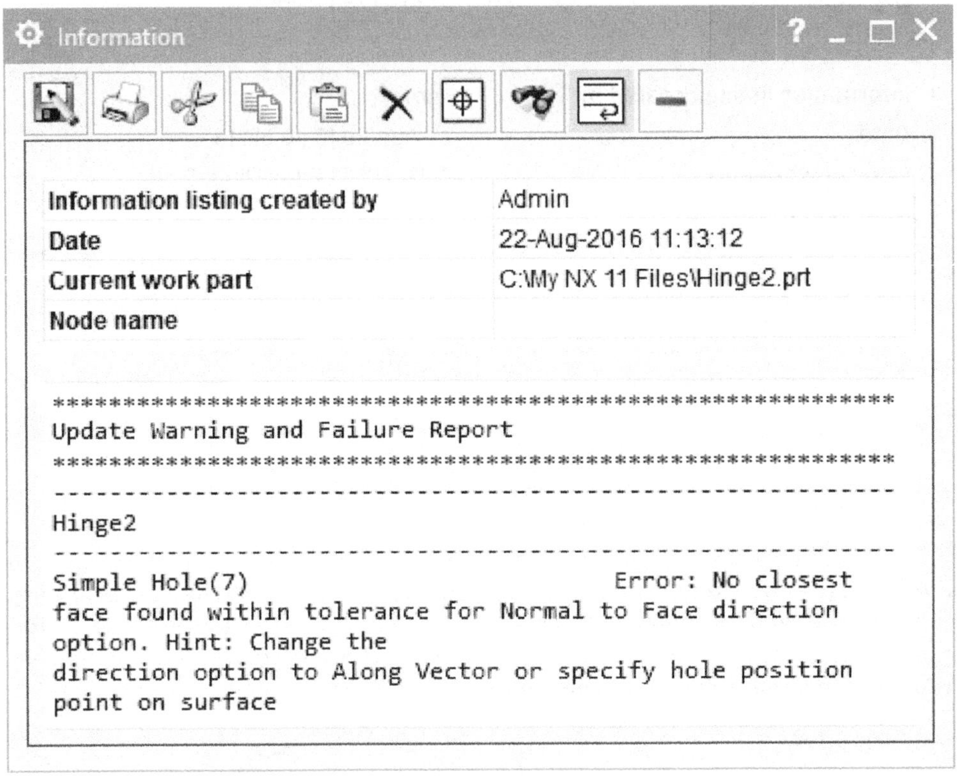

- Double-click on *Simple Hole (7)* to make an edit.

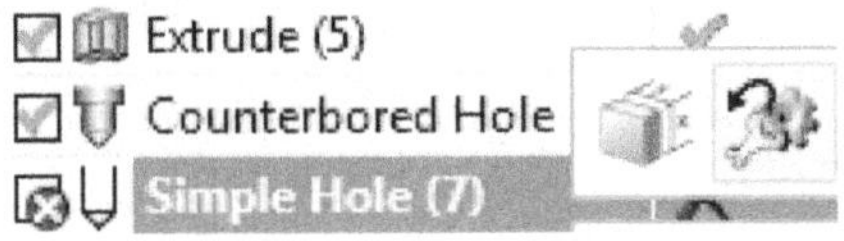

- You will need to deselect the point that previously indicated the position. You can do so by holding shift and left-clicking on it, or by using the **Deselect All** button.

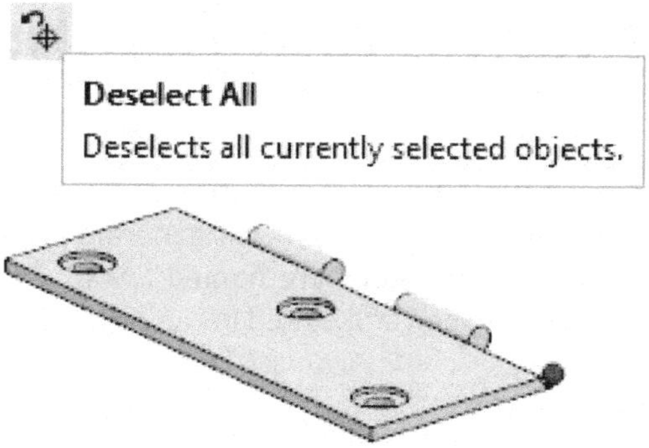

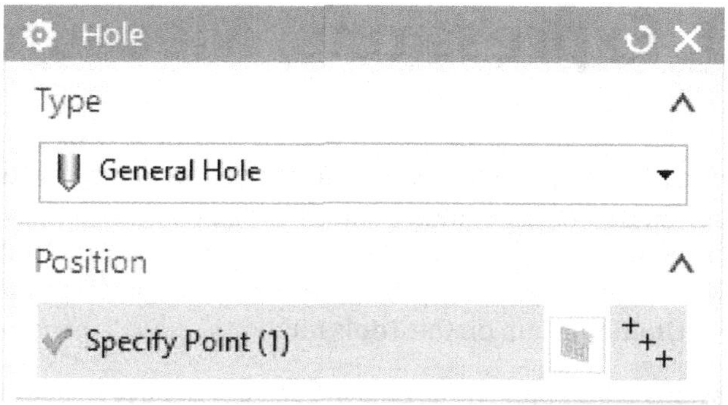

- To specify the new location for the hole, select the point indicated below. Recall that you can select the center of an arc by clicking on the arc itself, so long as the *Arc Center* snap point icon appears.

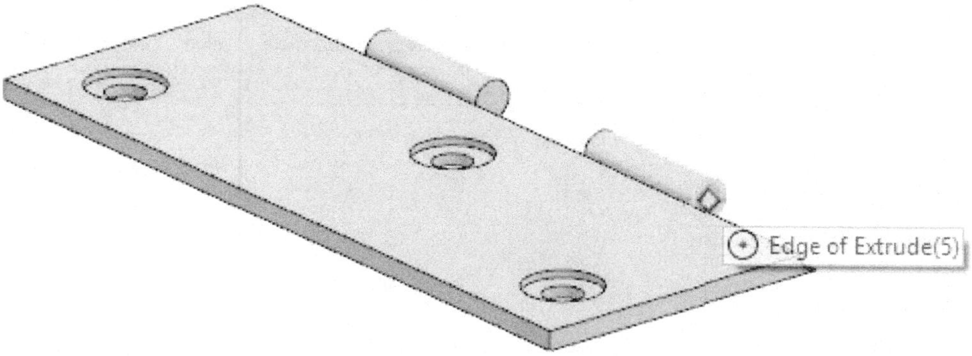

- Your Model History should now be free of errors and the complementary hinge should appear as below! Change the color!

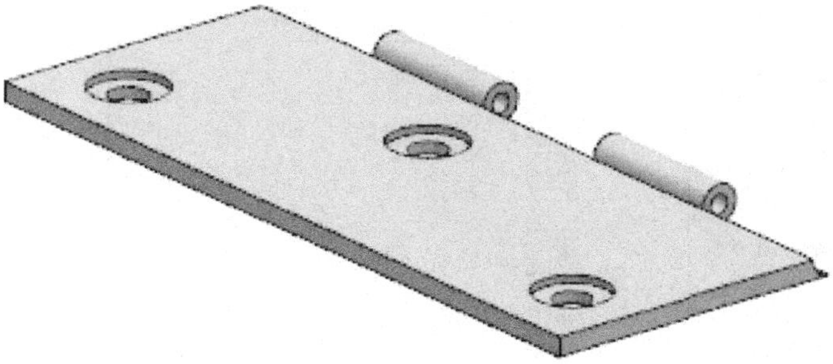

- Remember to **Save** the file!

10 Expressions, Measurements, and Attributes

In NX, *expressions* are parameters that drive the geometry in your models. Each parametric feature created has a set of expressions that define its geometry. For example, if you begin your model with an **Extrude**, NX will generate an expression *"p0"* for the **Start Limit** and a parameter *"p1"* for the **End Limit**. Feature expressions are named *"pXX"* sequentially, beginning with *p0*, then *p1*, then *p2*, etc. These expressions, and others, can be seen and edited in the **Expressions** editor, found in the **Utilities** group on the **Tools** tab.

The **Expressions** editor appears as below. Each cell behaves like a cell in an *Excel* spreadsheet. The vertical bar in the middle with the three arrows pointing left is used to minimize or expand the groups on the left.

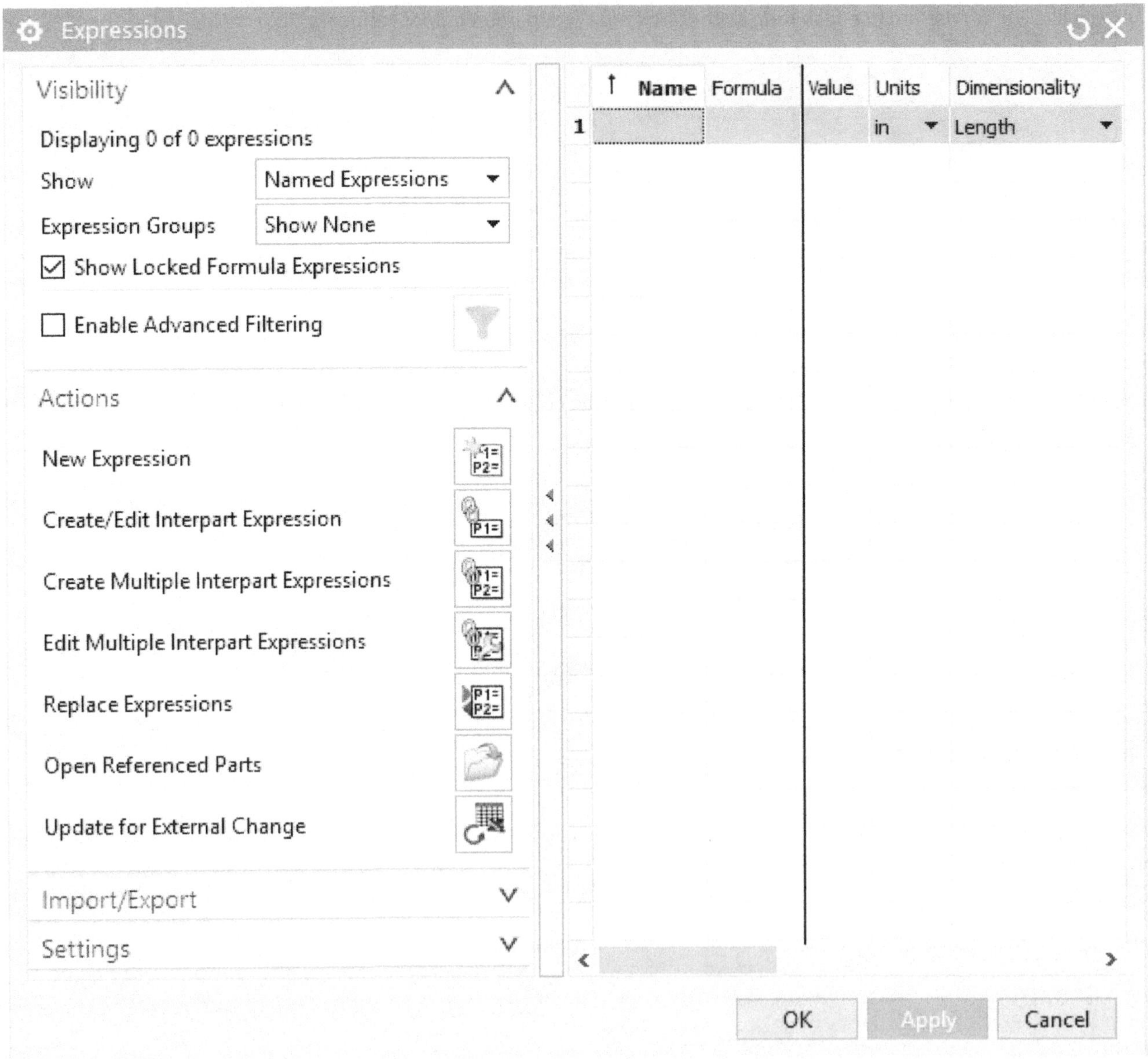

This chapter concerns the use of expressions beyond those created by NX automatically as you perform parametric operations. The **Expressions** editor allows you to create custom parameters of your own design, write formulas for expressions using a variety of mathematical and physical functions, turn measurements into expressions, share expressions between parts in an assembly and more!

10.1 User Defined Expressions

A model that has meaningful parameter names is easier to edit. For example, the parameter that controls the extrusion distance of a feature could be renamed *"extrude_dist"* instead of the standard *"pXX"* name, enabling immediate recognition of the feature to be edited.

Perhaps the simplest way to create meaningful parameter names is by entering equations into the **Expressions** menu prior to constructing a model. The parameter names can then be used during the definition of the solid feature geometry.

The following figure illustrates a model constructed with a simple extrusion and a boss feature. All dimensions of the model are to be controlled by the equations that are listed. In this project, these expressions will be defined prior to creating the solid model.

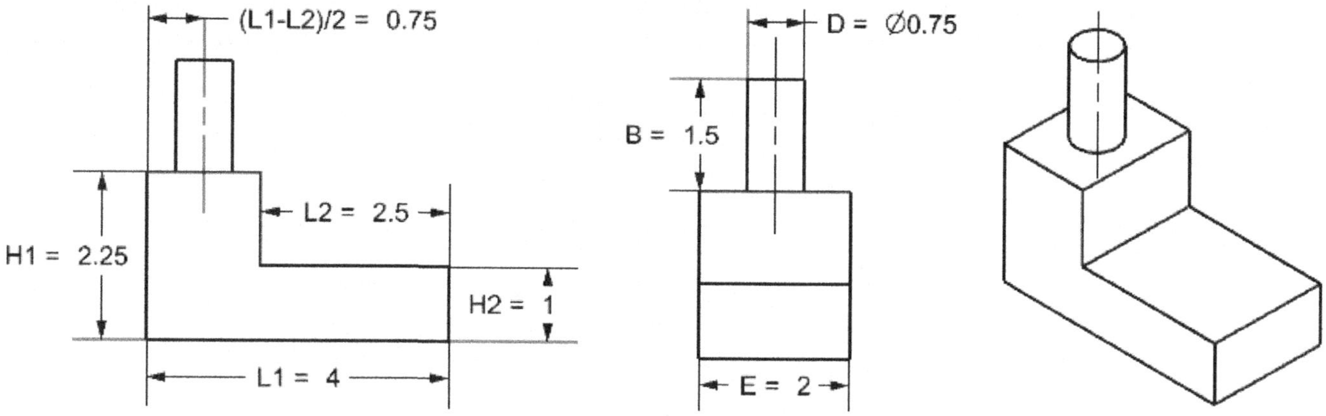

- Create a new file called *"User Defined Expressions.prt"* and place it in *"C:\My NX 11 Files"*.
- Select **Expressions** from the **Utilities** group on the **Tools** tab.

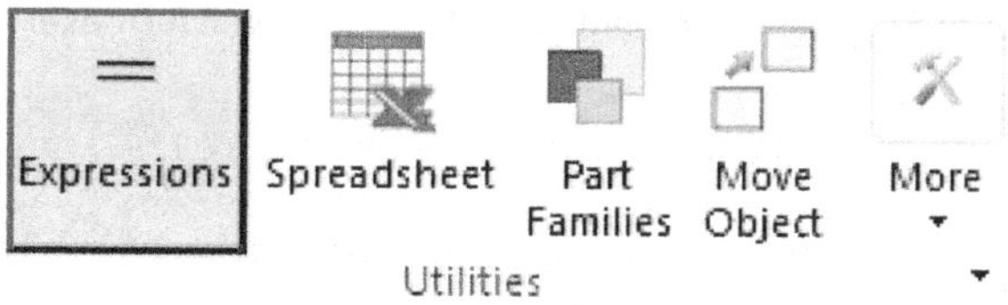

- The **Expressions** window will appear, expecting you to enter a new user-defined expression. Type *"B"* into the **Name** field as shown below, and then click in the **Formula** cell and type *"1.5"*. Click anywhere in the empty space and the other cells will populate.

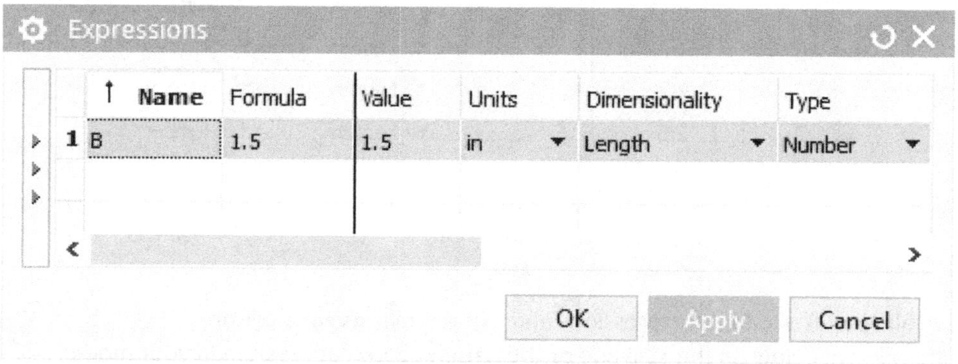

- To create another user-defined expression, use the **New Expression** button on the left.

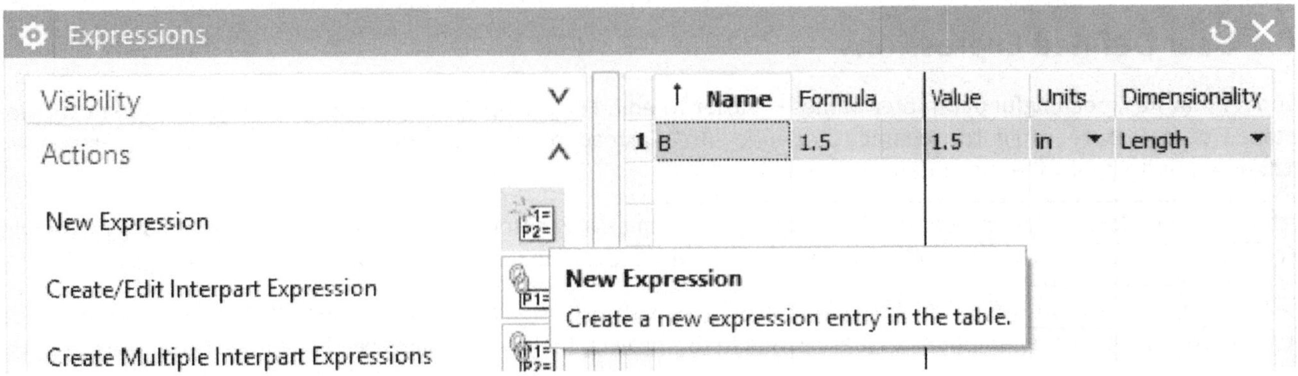

	Name	Formula	Value	Units	Dimensionality
1	B	1.5	1.5	in	Length

- This time, type *"D"* in the **Name** cell, then use the **[Tab]** key to jump to the right to the **Formula** cell, and type *"0.75"*. Use the **[Enter]** key when you are done, and note that the cursor jumps to the third line, and you can continue with another user-defined expression.

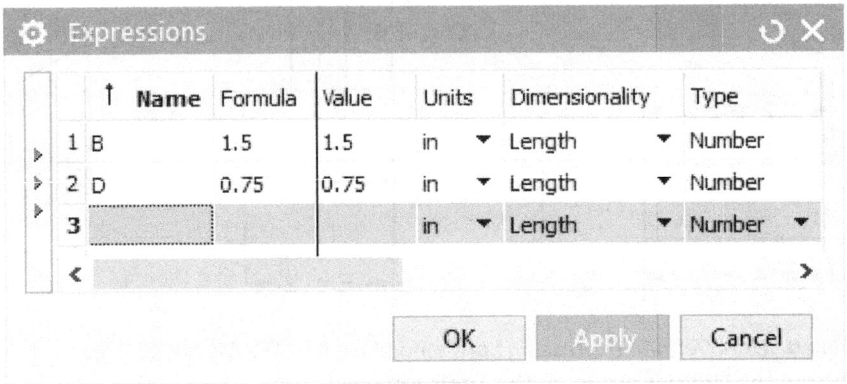

	Name	Formula	Value	Units	Dimensionality	Type
1	B	1.5	1.5	in	Length	Number
2	D	0.75	0.75	in	Length	Number
3				in	Length	Number

- Continue typing in this way to populate the **Expressions** editor with the list of user-defined expressions shown below.

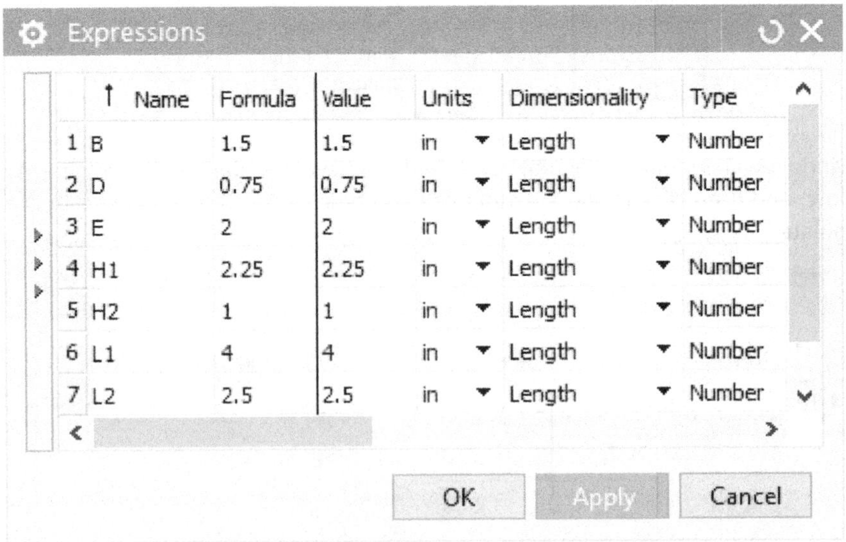

	Name	Formula	Value	Units	Dimensionality	Type
1	B	1.5	1.5	in	Length	Number
2	D	0.75	0.75	in	Length	Number
3	E	2	2	in	Length	Number
4	H1	2.25	2.25	in	Length	Number
5	H2	1	1	in	Length	Number
6	L1	4	4	in	Length	Number
7	L2	2.5	2.5	in	Length	Number

- Click **OK** to save all your user-defined expressions and exit the **Expressions** editor.
- Create the sketch shown below on the XZ plane of the datum coordinate system in your model.

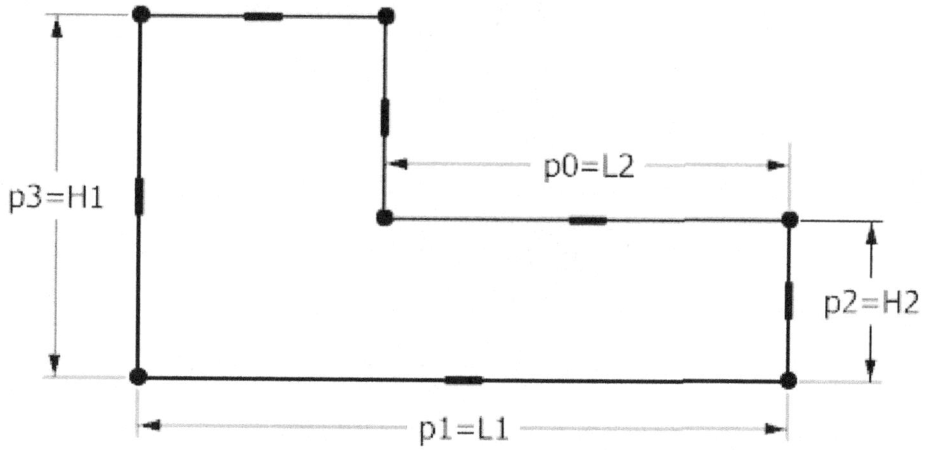

Note: *Expression names are case sensitive.*

- **Extrude** the sketch, entering a **Start Limit** of *0 in* and an **End Limit** of *E*.

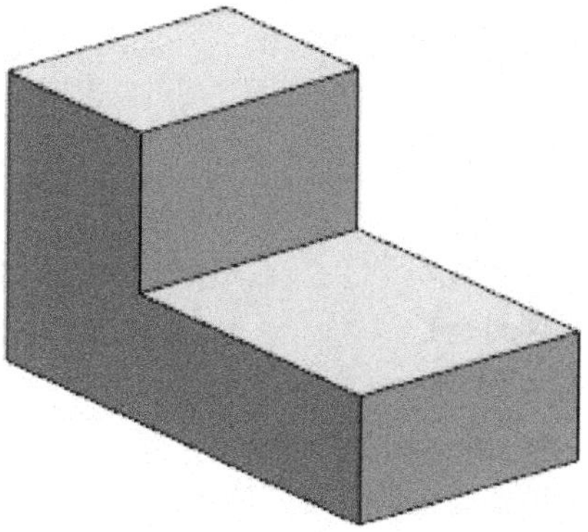

- Create the boss on the top face of the extrusion, with a **Diameter** of *D* and a **Height** of *B*.

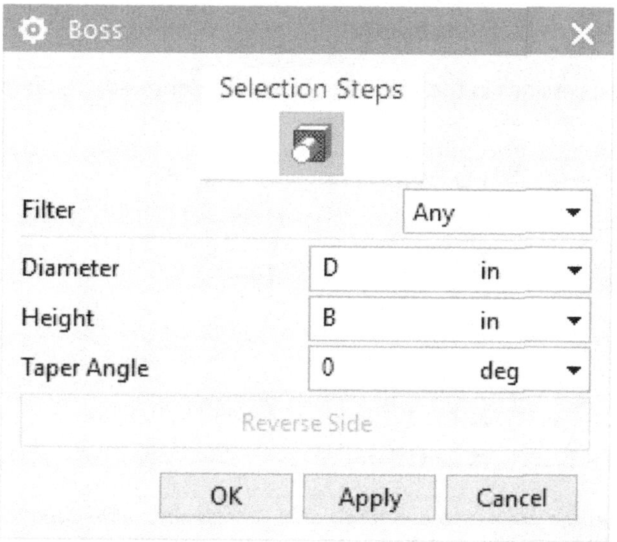

- Position the boss a distance of *E/2* from Edge 1 and *(L1-L2)/2* from Edge 2, as labeled in the figure below.

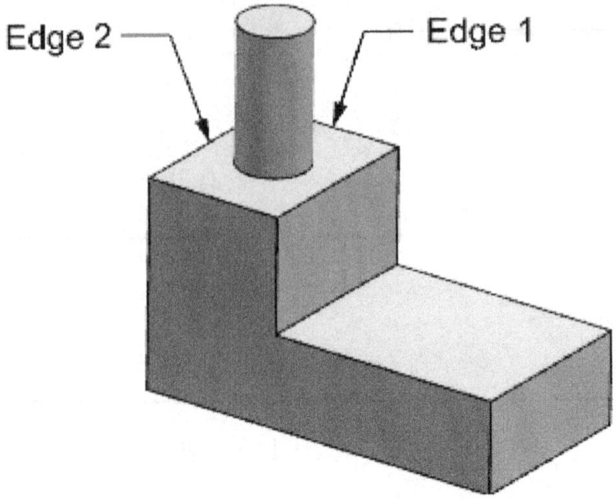

- An advantage of working with user defined expressions is the ease with which they are edited. Notice that *User Expressions* appear in the Part Navigator above the *Model History*.

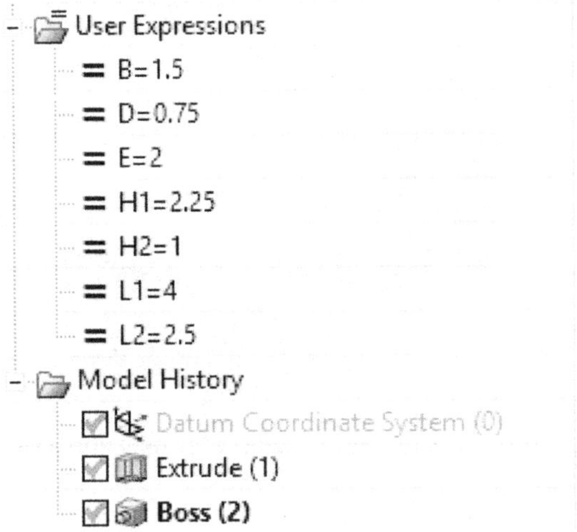

- Double-click on each of the expressions in the *User Expressions* folder to edit it. Change all the values to those shown below.

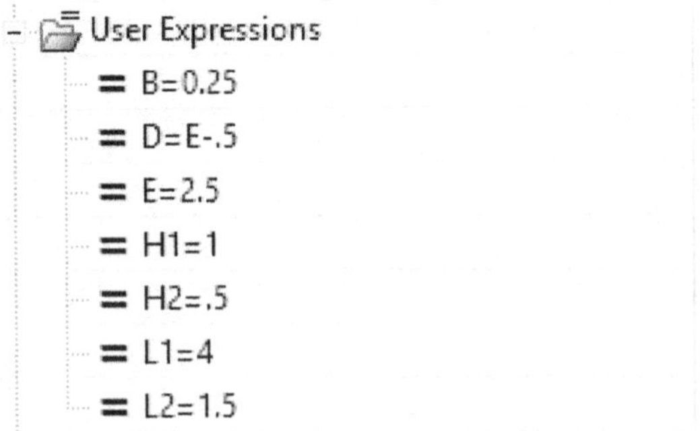

Note: The equation "d=e-0.50" *defines the expression* d *in terms of* e. *Expressions can be used to set variables equal to a function of other variables and to use mathematical operators such as sin, cos, tan, etc.*

- The following figure shows the result of these changes.

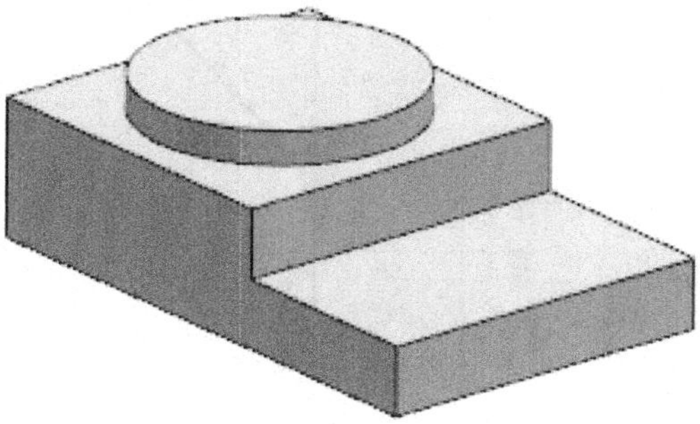

- **Save** your file!

10.1.1 Creating Expressions On-the-Fly

You have seen that expressions can be directly input during feature definition. Most of the time, you may not know all the required variables at the beginning of the design process. In this case, you can create variables on the fly as features are defined.

You will create the model shown below using named variables. This time you will define the parameters as they are needed during the design process.

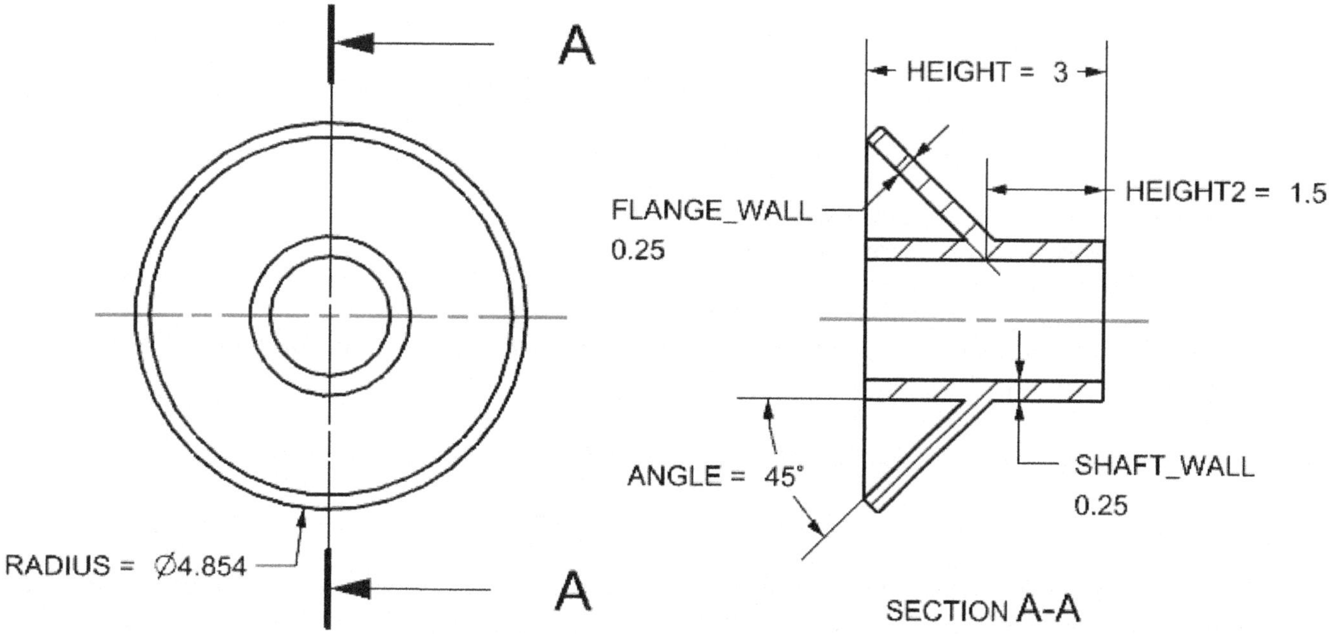

- Create a new part file (**Units**: *Inches*) named *"Expressions On-the-Fly.prt"* and place it in *"C:\My NX 11 Files"*.
- Create the sketch shown below on the YZ plane of the datum coordinate system.

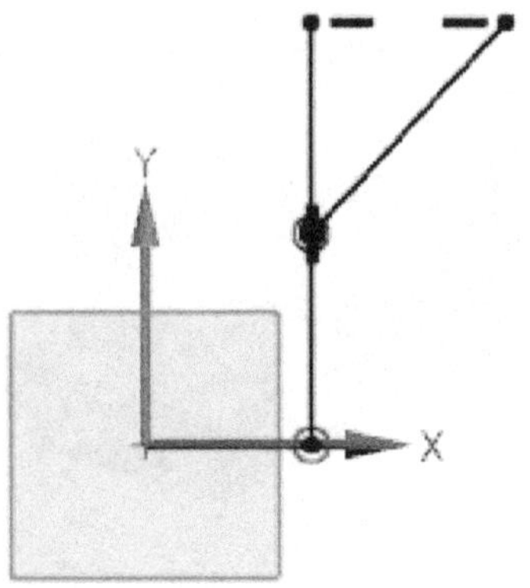

- Use **Rapid Dimension** with **Measurement Method** set to *Inferred*, and select the vertical line in the sketch. Click to place it, and when prompted for a value, instead type *"HEIGHT=3"* and press **[Enter]**. Close the **Rapid Dimension** tool.

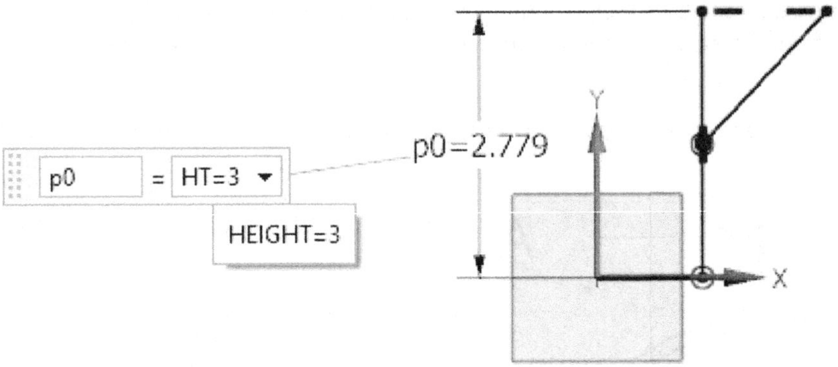

- This is a shortcut that accomplishes the following:
  - Creates a new user expression, *HEIGHT*
  - Sets *HEIGHT* equal to *3 in*
  - Defines *p0* by the formula *"p0 = HEIGHT"*
- Double-click on the dimension to edit it. Attempt to type over the value in the text entry field. You will find that you cannot overwrite the value *3.0000*.

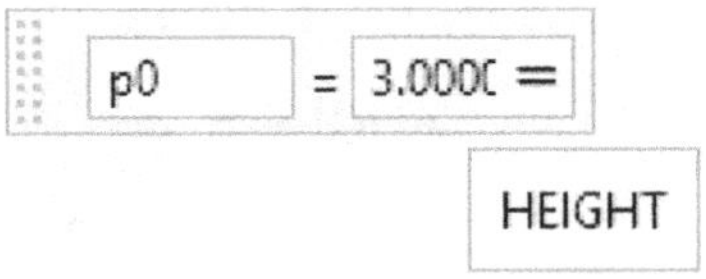

- Note the equals sign to the right of the text entry field, where there is usually a down arrow. The equals sign indicates that the value is controlled by a formula, and as such, cannot be naively edited. To modify this value, you have two choices.
  - You can click on the equals sign and select *Formula* from the drop-down menu to launch the **Expressions** editor, where you can modify the formula for *p0*.

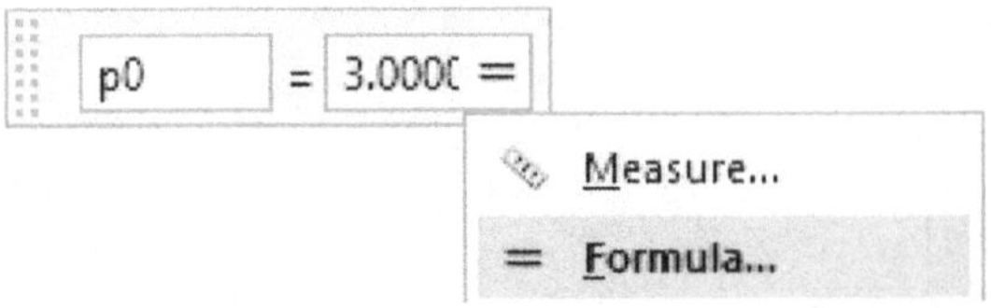

  - Your other option is to click on the equals sign and select **Make Constant** from the bottom of the drop-down menu. If you select this option, you will be able to type freely in the text entry field again.

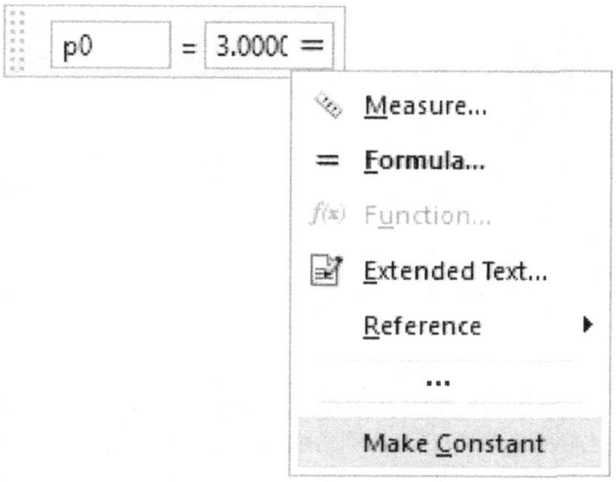

- Continue dimensioning the sketch and creating dimensions on-the-fly until your sketch appears as below, with *User Expressions* defined as below.

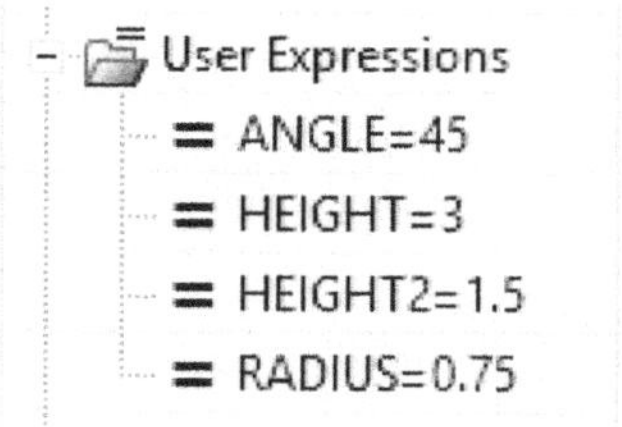

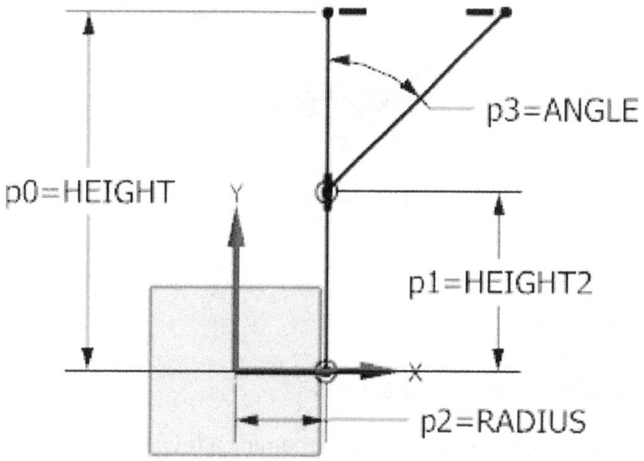

- **Finish** your sketch!
- Select **Revolve**.
- Set your **Selection Intent** drop-down to *Single Curve* so you can pick the *3 in* vertical line as the **Section**.
- To define the **Rotation Axis**, select the Z axis of the datum coordinate system.

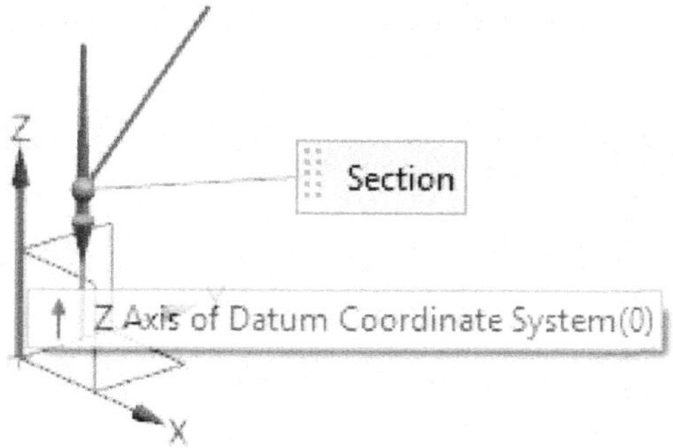

- In the **Revolve** dialog box (shown in the following figure), enter the following:
  - A **Start Angle** of *0 degrees.*
  - An **End Angle** of *360 degrees.*
  - A **Start Offset** of *0 in.*
  - **End Offset** defined by *"SHAFT_WALL = 0.25".*

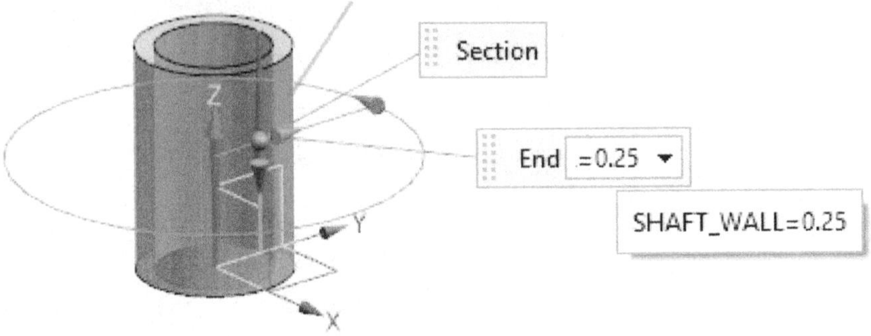

- Click **Apply** to create the hollowed cylinder, and perform another **Revolve** operation with the angled sketch line as section, and with **End Offset** = *"FLANGE_WALL=0.25"*. Set the **Boolean** to *Unite* so that the two bodies of revolution become one.

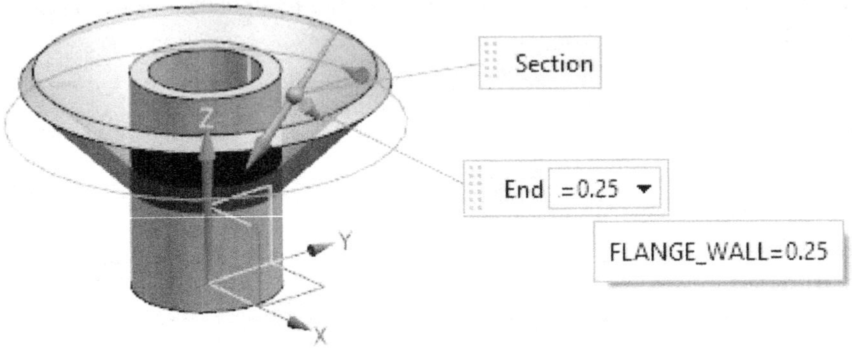

- **Save** your model!

10.1.2 Renaming User Expressions

After building a model, you may find that renaming a few variables would help to document it. Renaming variables is easily accomplished with the **Expressions** editor.

Using the part you built in the previous project, you will rename two variables that control the rotation angles of the sketch lines to *'THETA1'* and *'THETA2'*. You will then change both of their values to *180 degrees*.

- Continue working with the file called *"Expressions On-the-Fly.prt"*.
- From the **Expressions** editor, locate the two angular parameters that control the span of each **Revolve** operation. In the screenshot below, they are *p5* and *p9*.

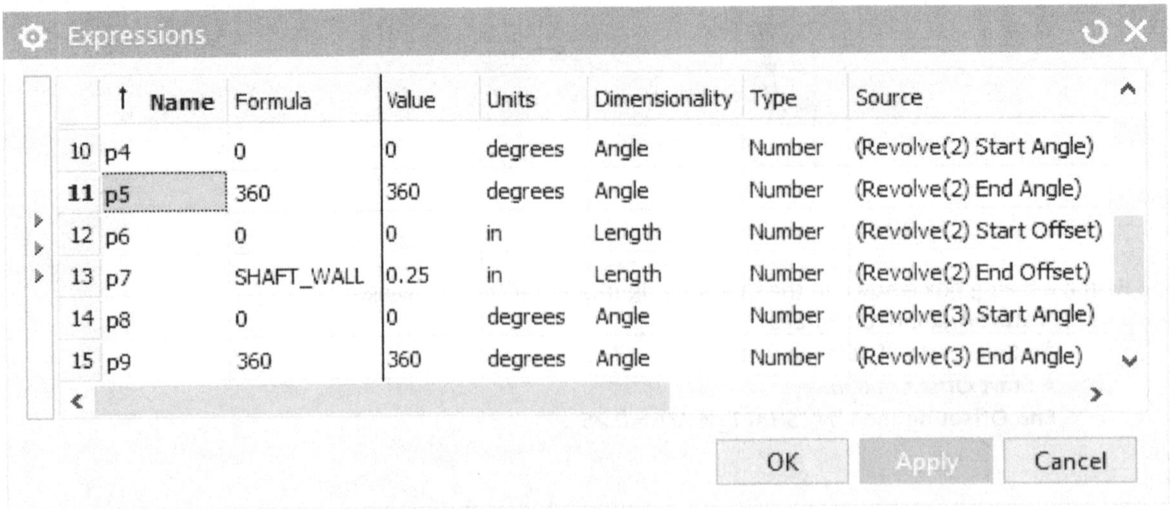

	Name	Formula	Value	Units	Dimensionality	Type	Source
10	p4	0	0	degrees	Angle	Number	(Revolve(2) Start Angle)
11	p5	360	360	degrees	Angle	Number	(Revolve(2) End Angle)
12	p6	0	0	in	Length	Number	(Revolve(2) Start Offset)
13	p7	SHAFT_WALL	0.25	in	Length	Number	(Revolve(2) End Offset)
14	p8	0	0	degrees	Angle	Number	(Revolve(3) Start Angle)
15	p9	360	360	degrees	Angle	Number	(Revolve(3) End Angle)

- To change the name of an expression, simply right-click on the value in its **Name** cell, and select **Edit**.

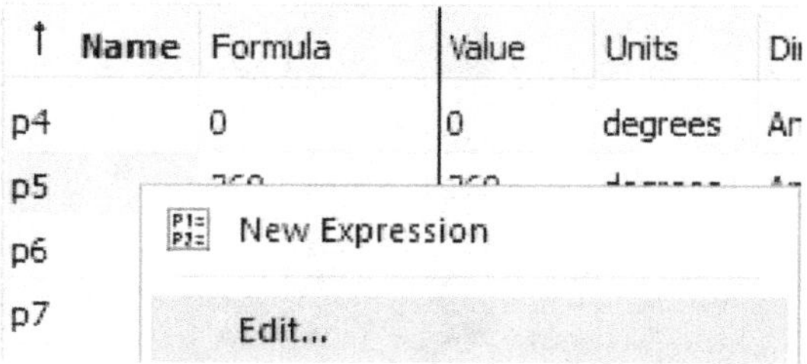

- Here you can change the name.

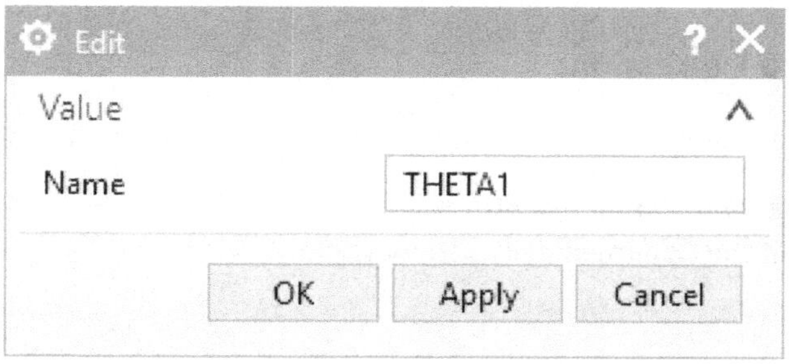

- You can also change the name by double-clicking in the **Name** cell for the expression you want to rename, and typing.

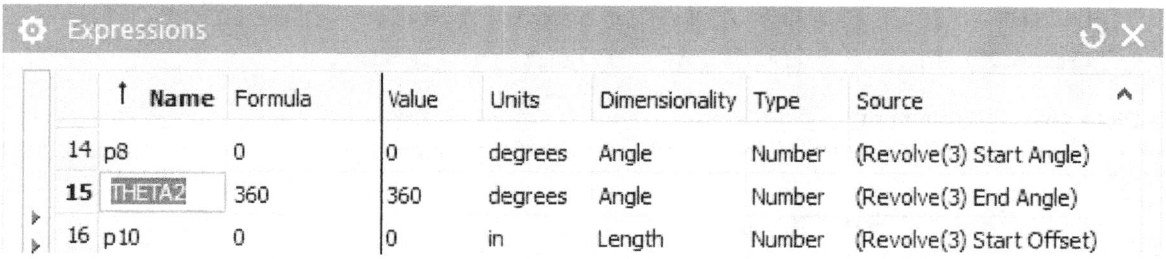

- After renaming *p5* and *p9* to *THETA1* and *THETA2*, change their formulas as below.

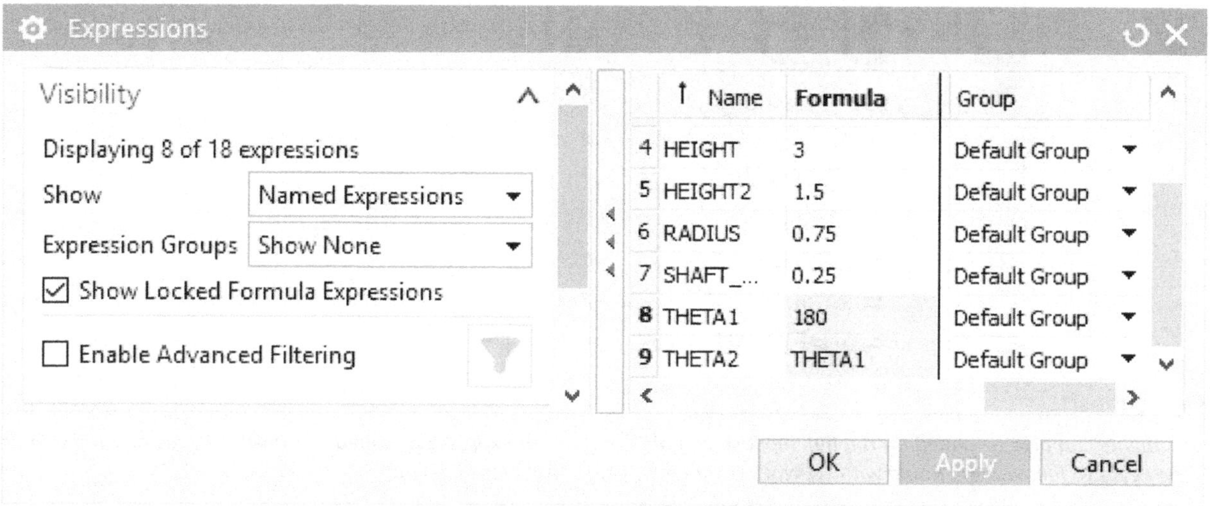

- Click **OK** to save your changes, and your part will update as below.

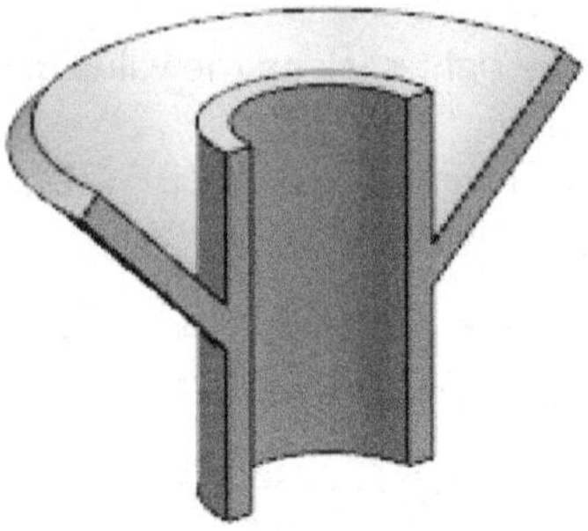

Note: You do not need to change the variable names to edit their values. You can just as easily change parameters that retain their default 'pXX' *name.*

10.1.3 Adding Comments to Expressions

NX allows you to comment on the expressions in your model. These comments are for the user's reference only.

- Continue working with *"Expressions On-the-Fly.prt"*.
- For the *HEIGHT1* variable, edit the **Comment** cell as shown below.

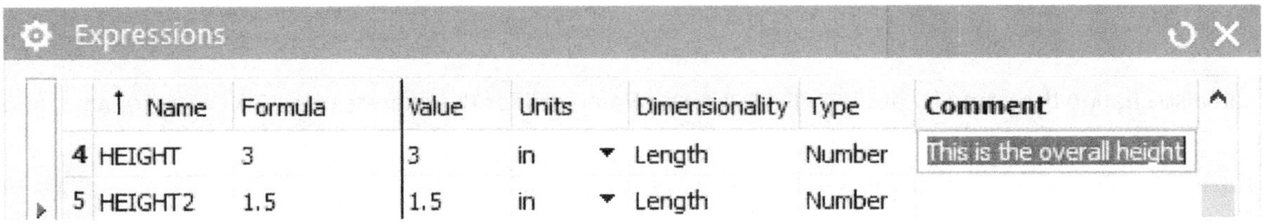

- Click **OK** to save your comment.
- Comments are visible in the Part Navigator. Widen your Part Navigator and open the User Expressions folder to see!

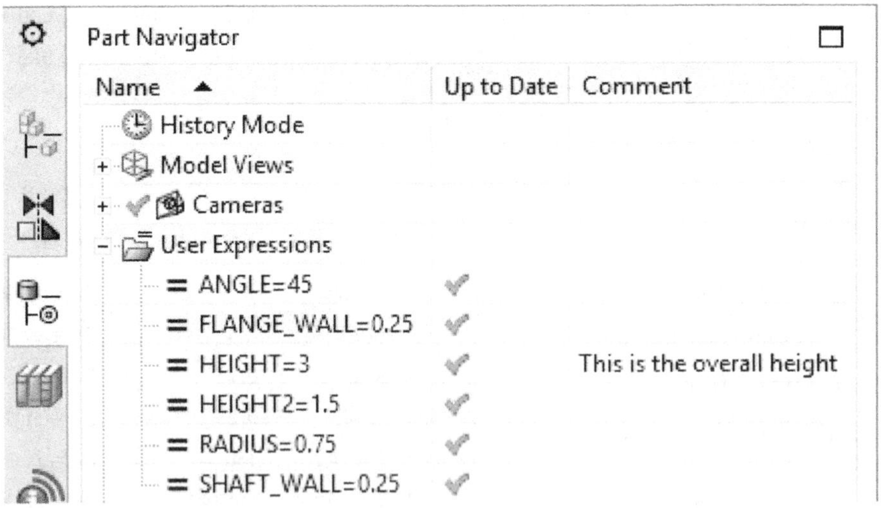

- **Save** your part!

Note: *Comments can also be added on the fly! Suppose you are creating an expression named "A" on-the-fly, as described in Chapter 10.1.1. The key is to follow the formula with* "//<your comment>", *e.g.,* "A=5//step height".

10.2 Measurements

In addition to *Feature Expressions*, and *User-Defined Expressions*, *Measurement Expressions* are another extremely important modeling tool. Measurement expressions can be created both in the **Expressions** editor, and with the help of measurement tools, found in the **Measure** group on the **Analysis** tab. We will practice both means of creating measurement expressions in the exercises below.

Despite the different names and icons for the tools in this group, most are actually variants of **Measure Distance**, so we will start with that tool!

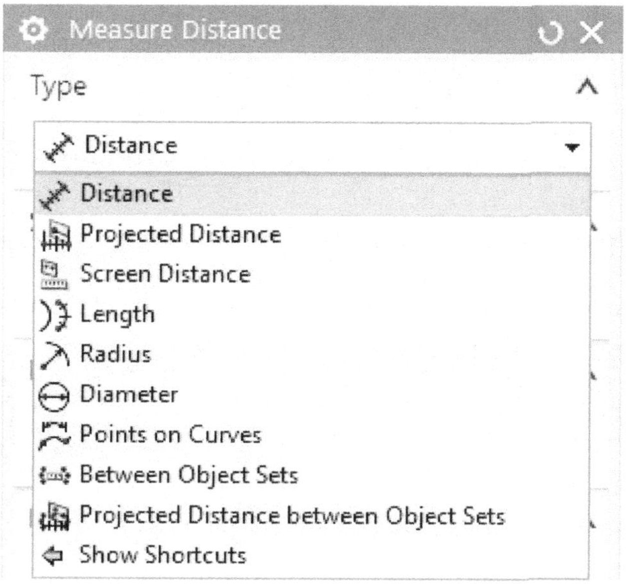

10.2.1 Distance

A distance measurement is the simplest sort of measurement available in NX – given two objects, you can find the (minimum or maximum) distance between them.

- Create a new file (**Units**: *Inches*) called *"Measurements and Attributes.prt"* and place it in *"C:\My NX 11 Files"*.
- **Extrude** a *4.5 in* x *2.5 in* rectangle from *0 in* to *0.5 in* with a **Draft Angle** of *2 degrees* in the -ZC direction.
- Apply an **Edge Blend** of radius *0.5 in* to the four drafted edges of the block.

- Apply an **Edge Blend** with a radius of *0.12 in* to the bottom edges, and then **Shell** the top face with a wall thickness of *0.06 in*.

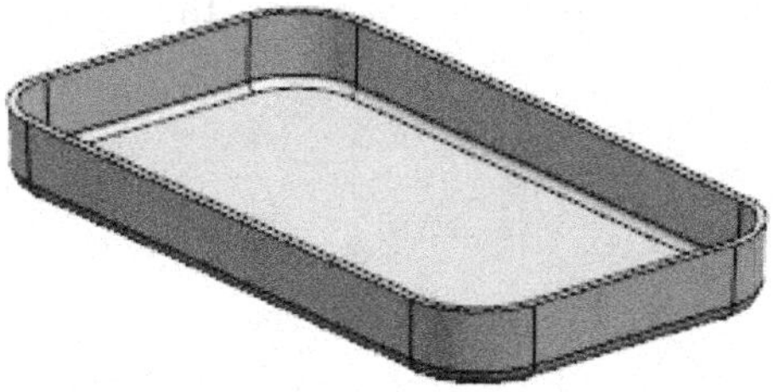

- Use the **Hole** command to place a *#5* thru hole at the arc center of each rounded corner on the interior face.

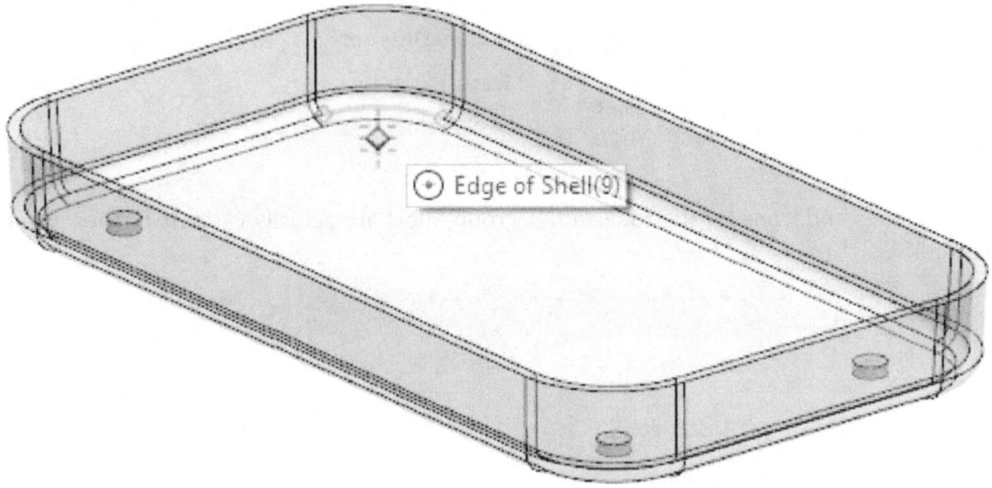

- With the holes positioned as such, the distance between the arc centers of these holes is not clear without doing some math.
- Use the **Measure Distance** command to measure between the arc centers of the holes

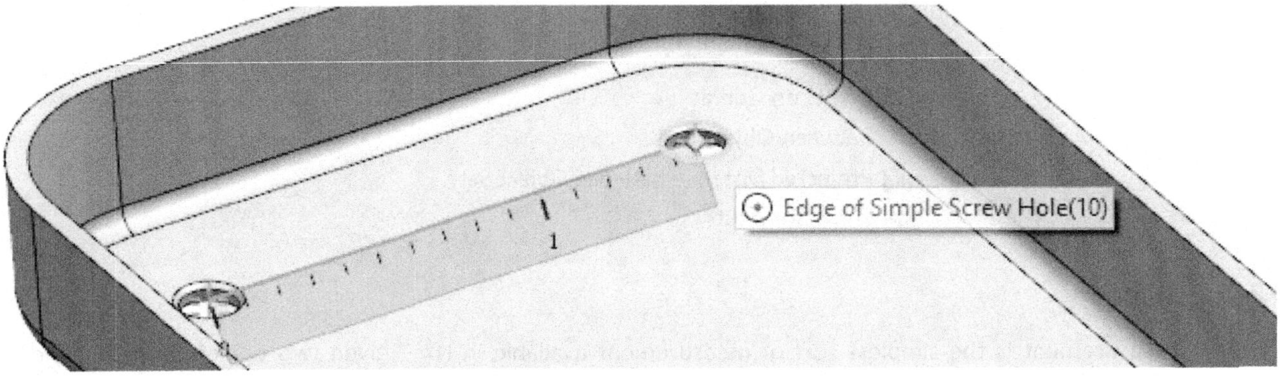

- Make sure that the **Associative** checkbox is marked. Without this, your measurement is a one-time calculation.

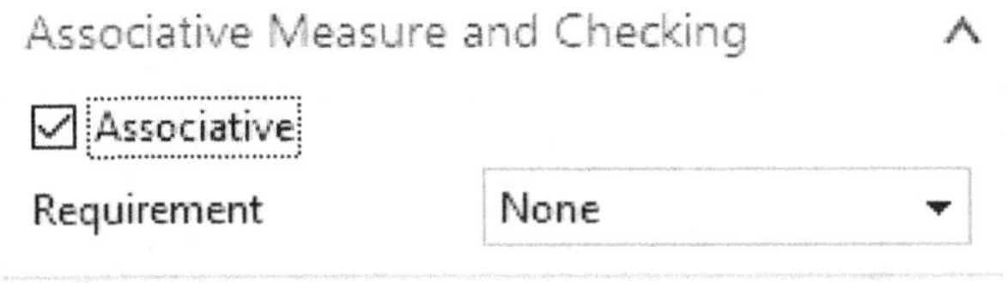

- One of the implications of the **Associative** checkbox is that your measurement is recorded in the Model History with a timestamp.

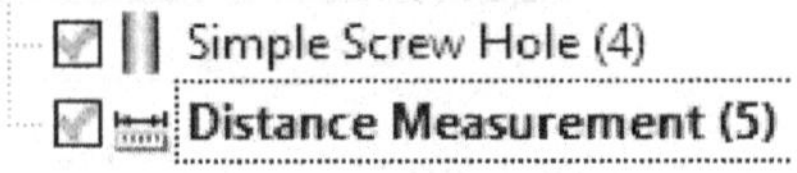

- The other implication of the **Associative** checkbox, in this case, is the creation of a measurement expression.
- Open the **Expressions** editor, and within the **Visibility** section, set the **Show** drop-down to *Measurement Expressions*.

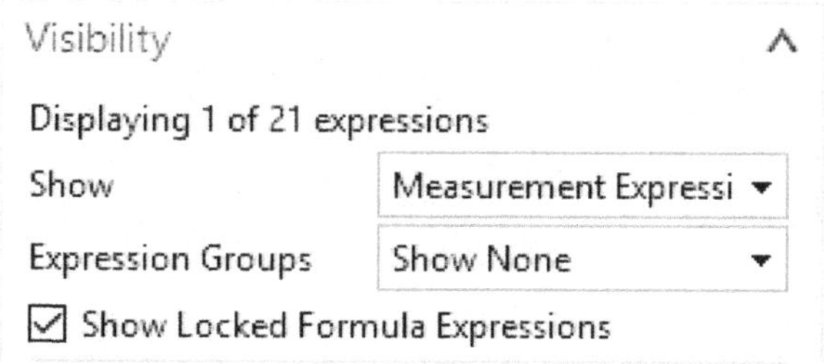

- In the spreadsheet to the right, you will then see the expression that resulted from the **Measure Distance** operation.

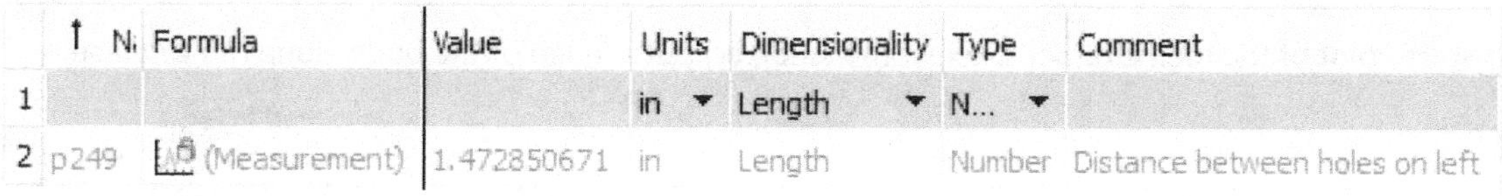

	N	Formula	Value	Units	Dimensionality	Type	Comment
1				in	Length	N...	
2	p249	(Measurement)	1.472850671	in	Length	Number	Distance between holes on left

- **Save** your part file!

10.2.2 Projected Distance

Sometimes you are not interested in the total distance between two objects, but only the distance in a specific direction. In the exercise below, you will use the *Projected Distance* measurement type to find out how far a *#5* hole in your model is from the outside edge of the part.

- Continue working with *"Measurements and Attributes.prt"*.
- You can also create associative measurement expressions from within the **Expressions** editor.
- Open the **Expressions** editor, and right-click in the **Formula** cell for the new expression being created.

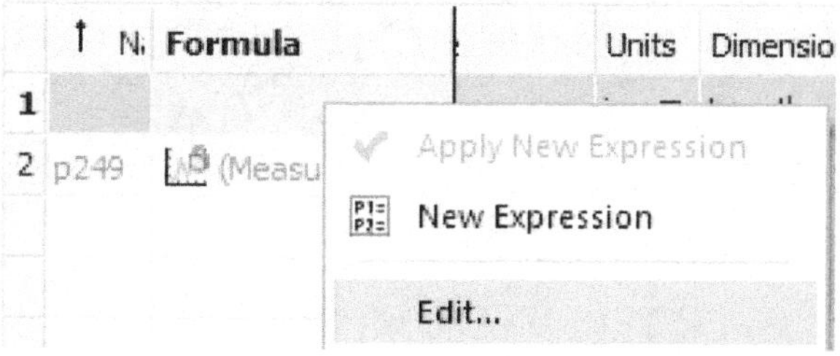

- When the **Edit** dialog appears, select **Measure Distance** from the measurements drop-down menu.

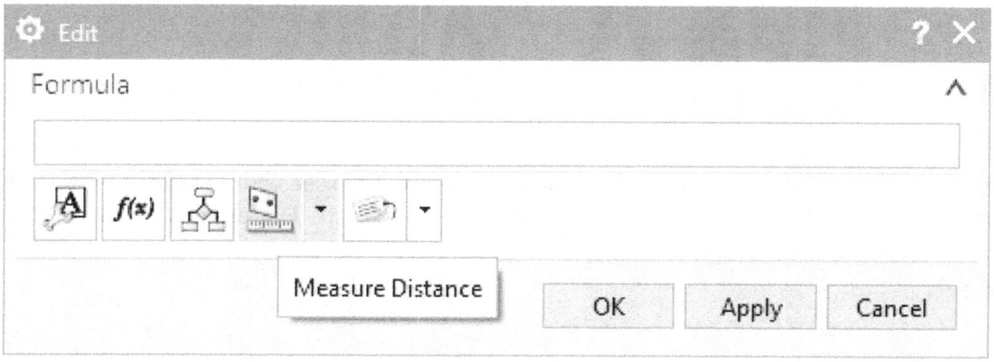

- When the **Measure Distance** dialog appears, set the **Type** to *Projected Distance*. You will be prompted for a **Vector**. Choose the X axis of the *OrientXpress* triad.

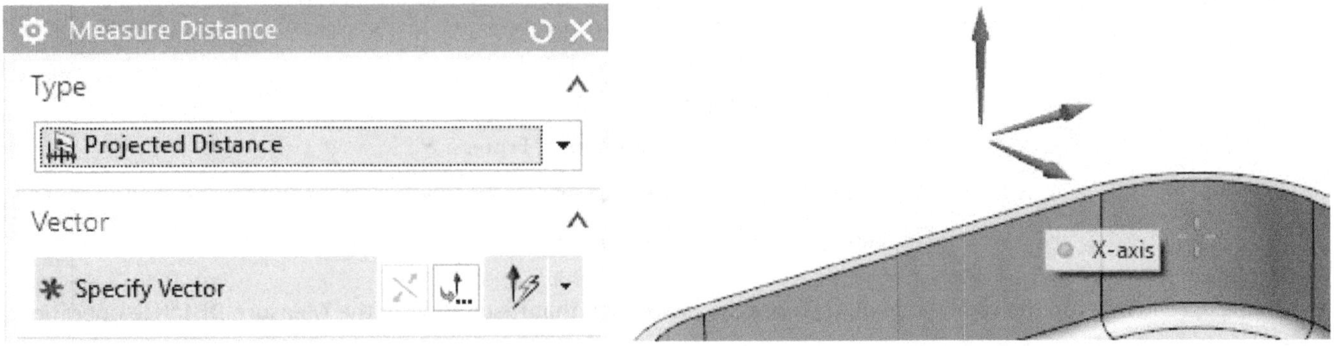

- For the **Start Point** of the measurement, select the (shorter) outer edge of the body along the top face.

- For the **End Point** of the measurement, choose the arc center of the *#5* threaded hole as shown below.

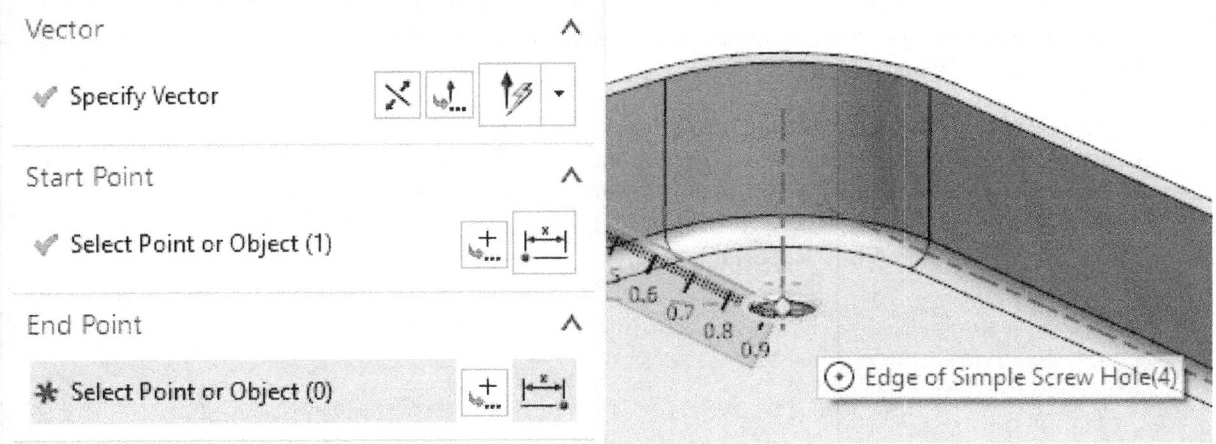

- Make sure the **Measurement Distance** is set to *Minimum*. In our last measurement, we didn't have to pay attention to this, but here one of the entities selected is an edge rather than a point.

- Again, make sure that the **Associative** checkbox is marked. You will then see your *Projected Distance Measurement* in the Model History, and the corresponding measurement expression in the **Expressions** editor.

	↑ N	Form	Value	Units	Dimei	Type	Source	Comment
1				in ▾	▾	▾		
2	p249		1.472850671	in	Le...	Nu...	(Distance Measurement(5))	Distance between holes on left
3	p252		0.5135746646	in	Le...	Nu...	(Projected Distance Measurement(6))	Distance from outside edge to hole center

- The *Projected Distance* is named as such because it projects the line segment joining the two objects orthogonally onto an axis in line with the selected vector, and calculates the distance of that projected line. If you right-click on *Projected Distance (6)* in the Part Navigator and select **Show Dimensions**, your measurement should appear in the graphics window as shown below.

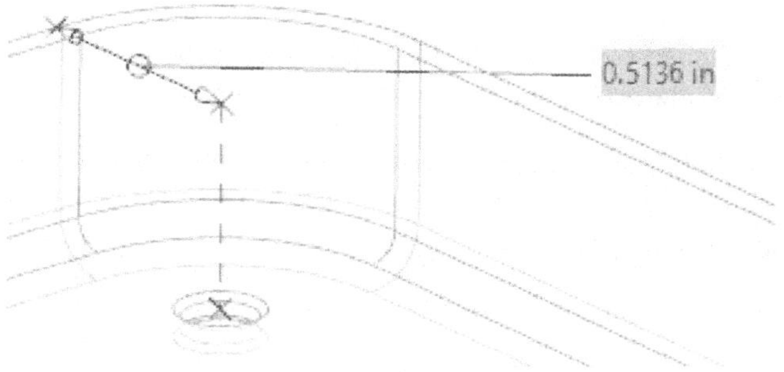

- **Refresh** to clear the measurement from the graphics window.
- **Save** your part file!

10.2.3 Measure Body

Weight and volume calculations are frequently useful in the design process. The **Measure Body** command produces an associative measurement of a body's weight, volume, and more!

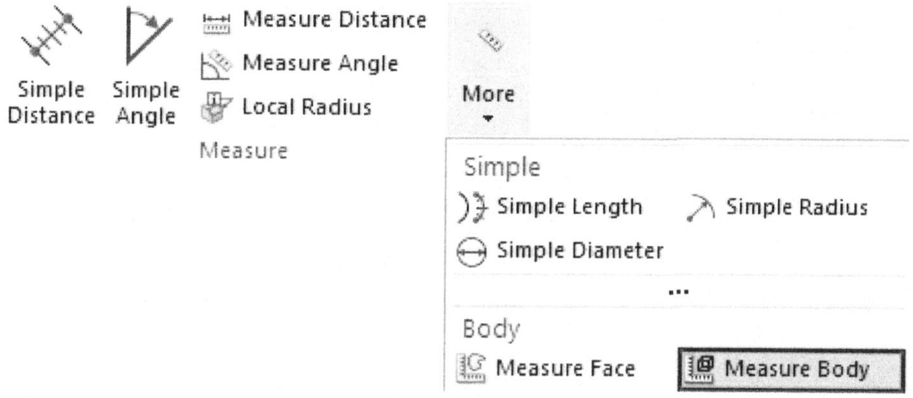

- Continue working with *Measurements and Attributes.prt.*
- Use the **Measure Body** command to create an associative measurement.

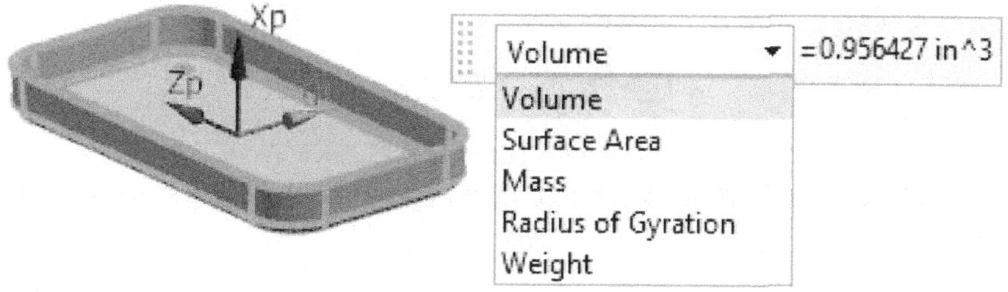

- **Measure Body** produces five measurement expressions, and an associative *Point* entity at the center of mass of the selected bodies.

4	p263	(Measurement)	32.76799912	in^2	Area	Number	(Body Measurement(8) surface_area)
5	p264	(Measurement)	0.9564270784	in^3	Volume	Number	(Body Measurement(8) volume)
6	p265	(Measurement)	0.2705732205	lbm	Mass	Number	(Body Measurement(8) mass)
7	p266	(Measurement)	0.2705732205	lbf	Force	Number	(Body Measurement(8) weight)
8	p267	(Measurement)	1.625642072	in	Length	Number	(Body Measurement(8) radius_of_gyration)

- **Save** your part!

10.3 **Attributes**

Attributes are useful bits of data attached to features, bodies, parts, components, and other entities. They facilitate the organization of assemblies and bills of materials, and the creation of drawings.

If you are using NX in a corporate environment, there may be certain number of custom-defined attributes that are essential for classifying your parts, e.g., "Official Part Number." Many of these are created and managed with custom-defined tools administered by your CAD Support Team.

In the exercises below, we will look at two part attributes of general interest – material and weight.

10.3.1 Material

The measurement expression for the weight of the part created in Chapter 10.2.3 obtained the weight as a product of the overall volume and the density of the part. NX creates solid bodies with a default density of *0.2829 lb/in³*.

A material assignment will overwrite this default density so that the value of the weight measurement expression changes. The **Assign Materials** tool is found in the **Utilities** group on the **Tools** tab.

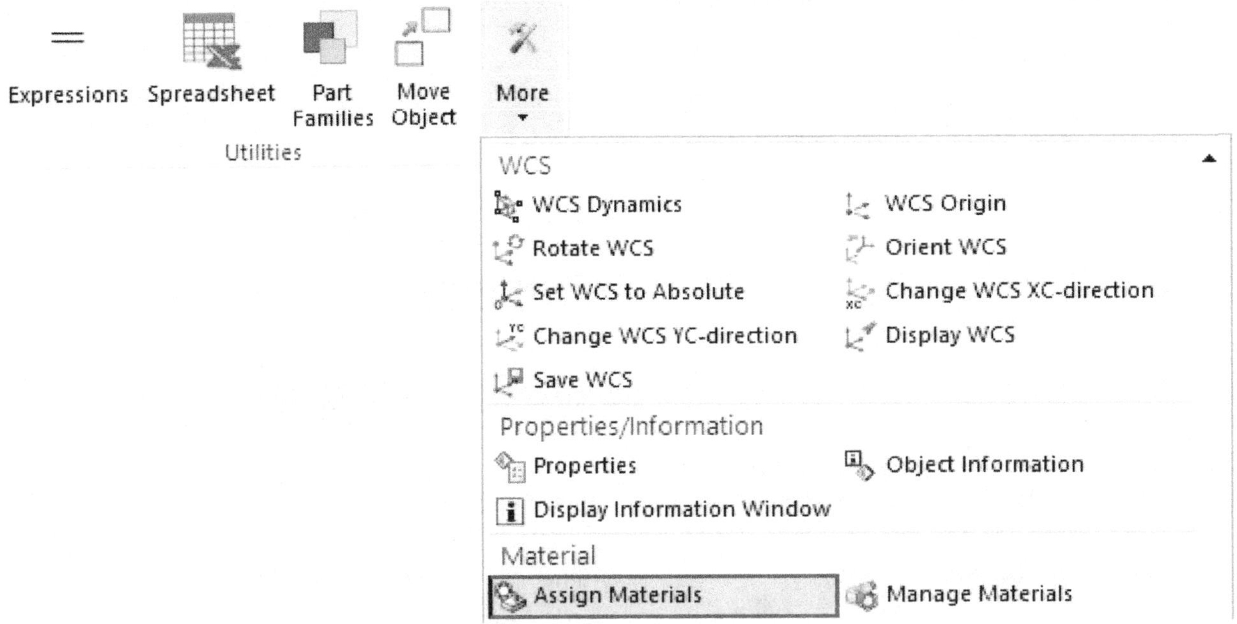

- Continue working with the part *"Measurements and Attributes.prt"*.
- Select the **Assign Materials** tool and choose the solid body of your model. Select *Polyethylene* from the list of *Library Materials*.

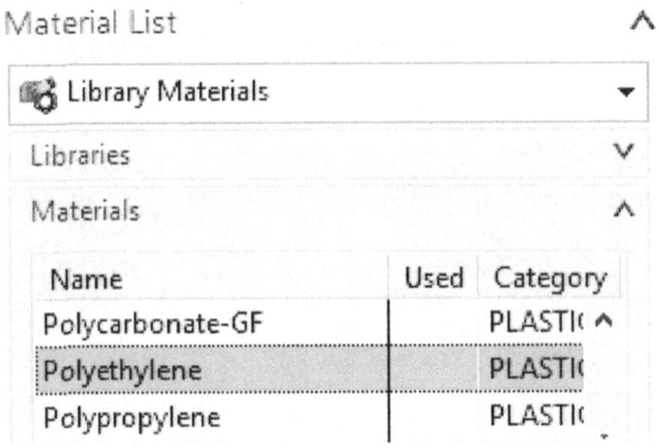

- After applying a new material, your part's measurement expressions will update immediately.

	N	Form	Value	Units	Dimension	Type	Source
1				in	Len...		
2	p263		32.76799912	in^2	Area	Nu...	(Body Measurement(8) surface_area)
3	p264		0.9564270784	in^3	Volume	Nu...	(Body Measurement(8) volume)
4	p265		0.04146368568	lbm	Mass	Nu...	(Body Measurement(8) mass)
5	p266		0.04146368568	lbf	Force	Nu...	(Body Measurement(8) weight)
6	p267		1.625642072	in	Length	Nu...	(Body Measurement(8) radius_of_g...

- You can display part attributes in the Assembly Navigator. Right-click in the empty space in your Assembly Navigator and select **Properties**. On the Assembly Navigator Properties dialog, select the **Columns** tab, and use the text entry field at the bottom of the dialog to create a column to display the *Material* attribute.

- Move it to the top of the list, following *Descriptive Part Name*.

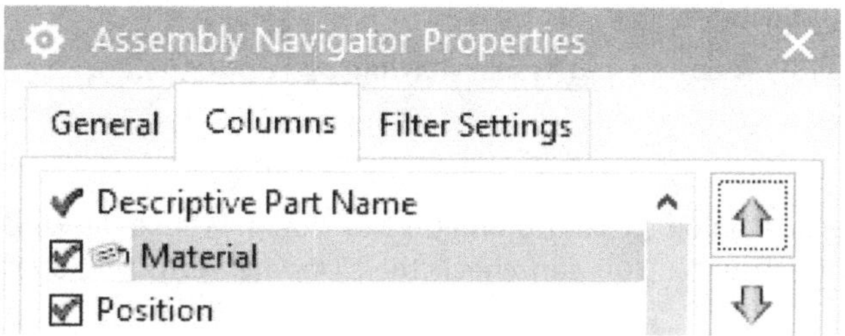

- Now the material for each component in your assembly will be displayed in the Assembly Navigator.

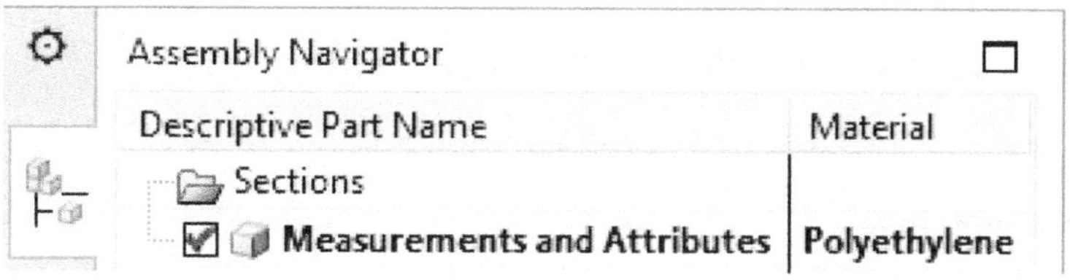

- **Save** your part!

10.3.2 Weight

The weight can be recorded with the help of a measurement expression, but there is also a part attribute called the weight that can be used in calculations and to report the current weight of the model on the drawing.

- Continue working with *Measurements and Attributes.prt*.
- Go to the **Assembly Navigator Properties dialog** again, and find *Weight (lb)*. Move it to the top, following *Material*.

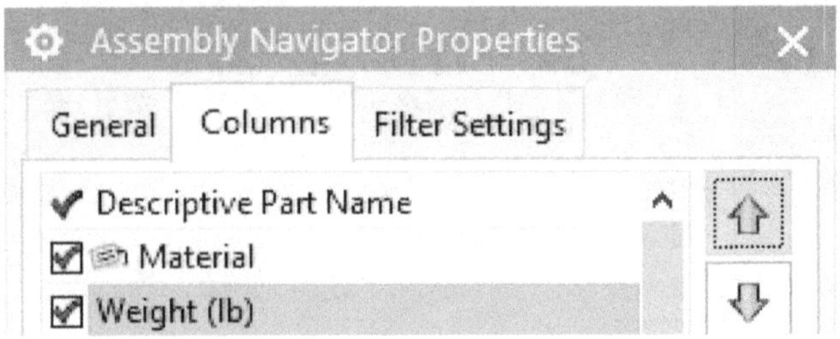

- Note that the **Weight** column is blank! The *Weight* attribute does not automatically update when changes are made to the material.

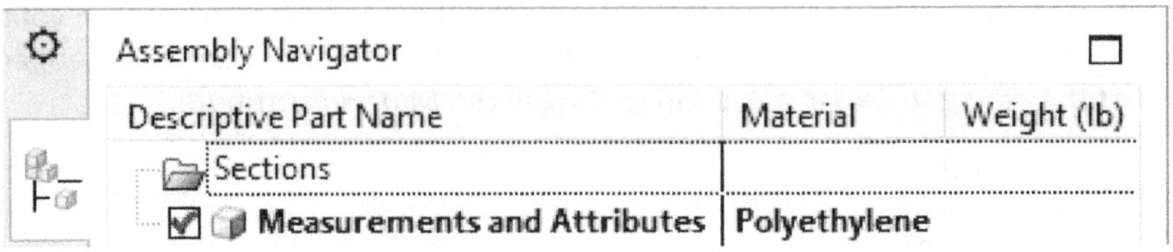

- To update the weight, you must right-click on your part in the Assembly Navigator and select **Properties**.

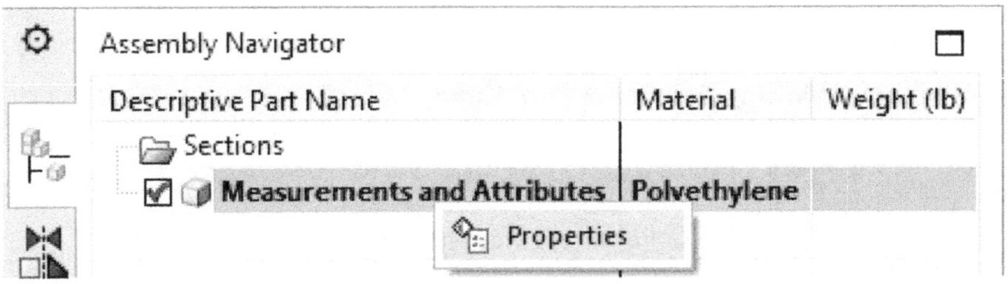

- On the **Displayed Part Properties** dialog, go to the **Weight** tab. You must use the **Update Weight Data Now** button to manually update the *Weight* attribute. You can check the *"Update Data on Save"* box to make this easier going forward.

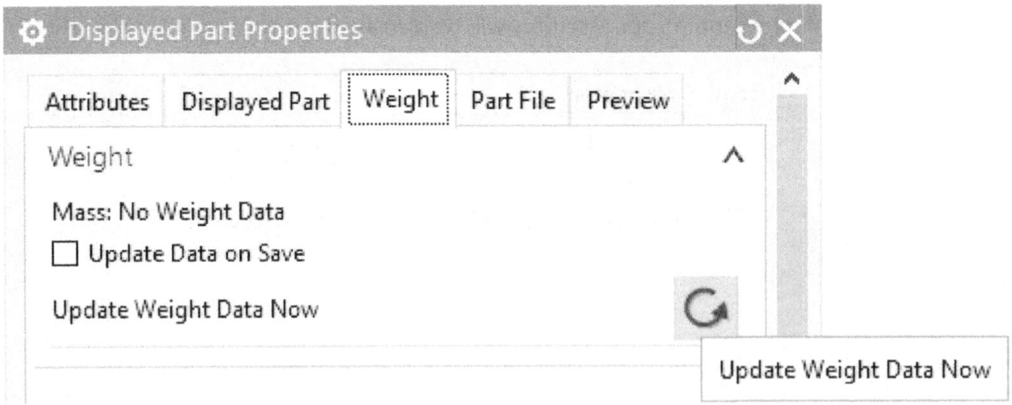

- You will now see the **Weight** column populate.

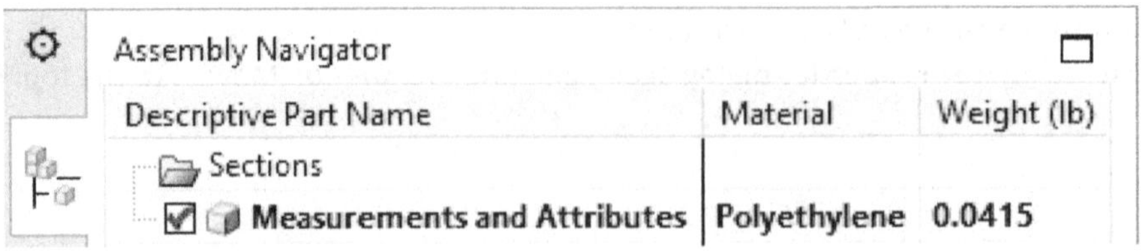

- The *Weight* attribute can also be used to create an expression. Go to the **Expressions** editor, and **Edit** the **Formula** cell for your new expression.
- From the **Edit** dialog, select **Reference Part Attribute**.

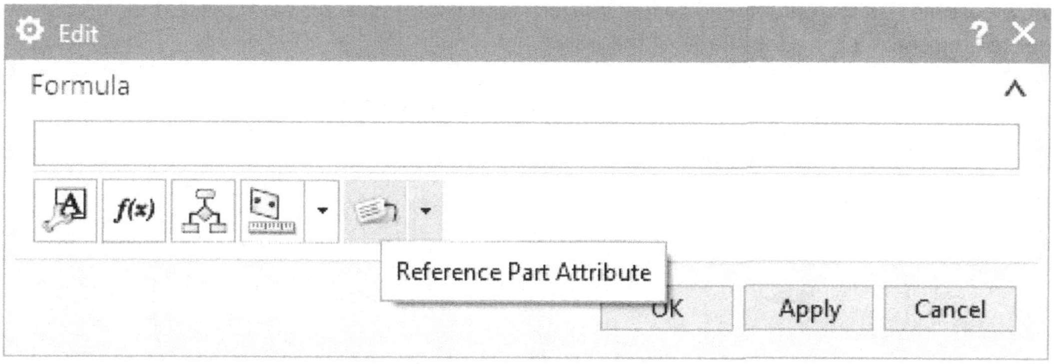

- Choose the attribute *MassPropWeight*.

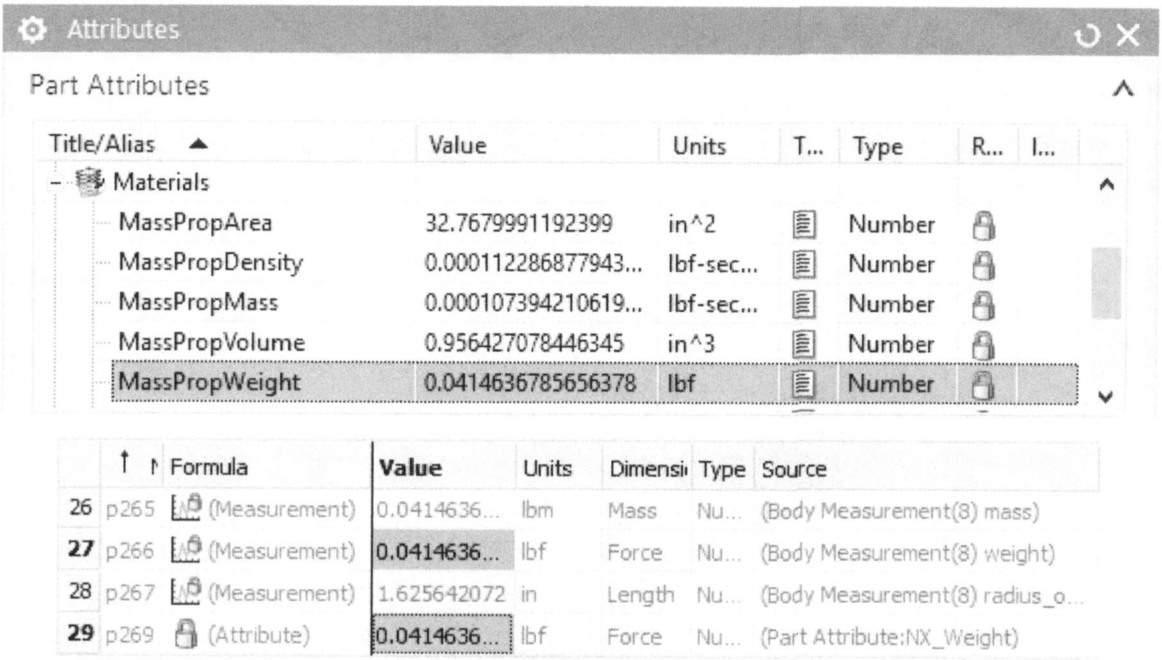

- You can use the attribute expression in formulas just as you would the measurement expression.
- **Save** your file!

10.4 **Project – Pattern Driven by Expressions**

In this exercise, you will design a plastic part for ultrasonic welding, as shown below. The "teeth" running around the top edge of the part are called "energy directors", and they are designed to facilitate the flow of heat for a clean weld.

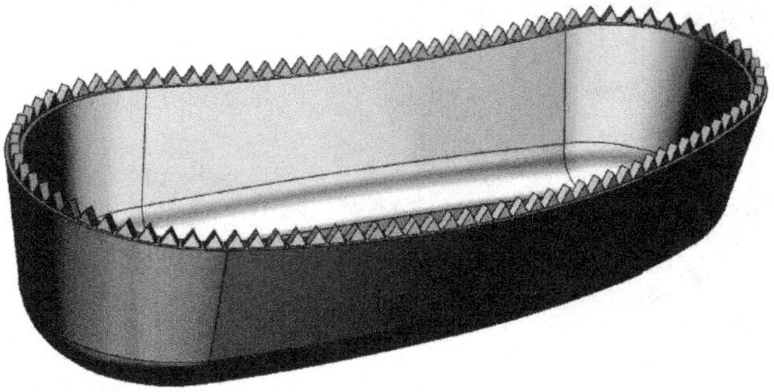

In designing this model, we will create a parameter for the wall thickness of the part and define all subsequent features in terms of that thickness, including the tooth pattern.

- Create a new part file (**Units**: *Millimeters*) called *"Pattern with Expressions.prt"* and place it in *"C:\My NX 11 Files"*.
- Create a sketch to represent the top shape of the part.

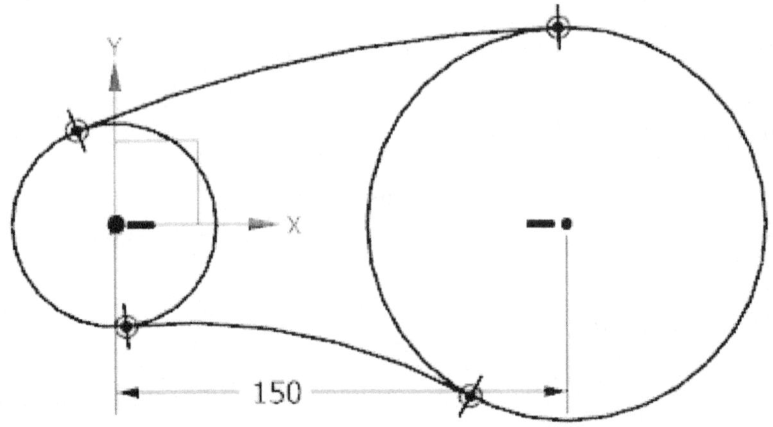

- **Extrude** this sketch from *0 mm* to *60 mm* in the -ZC direction, with a *5-degree* **Draft**. Apply a *15 mm* **Edge Blend** to the bottom edges.

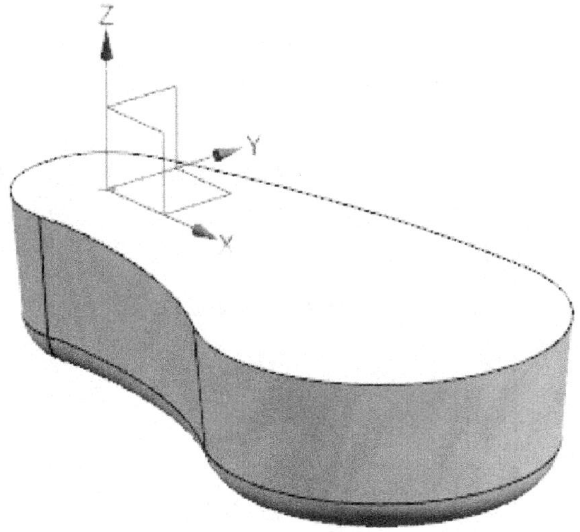

- **Shell** the top face of the body. Define the **Thickness** by typing *"THK=6//wall thickness"*. All subsequent features will be defined in terms of this wall thickness.

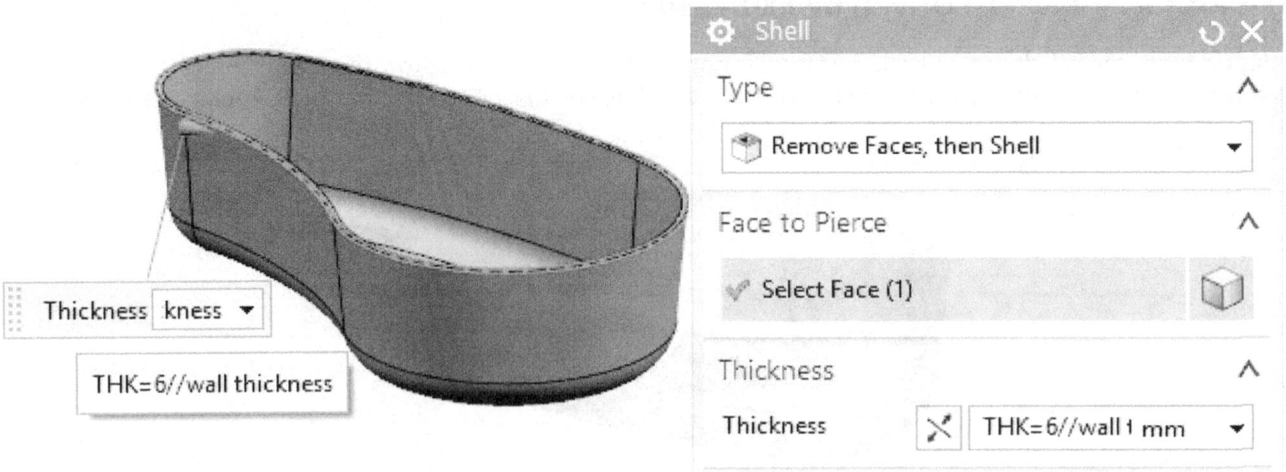

- **Extrude** the top outer edge of the body in the -ZC direction with the parameters shown below.

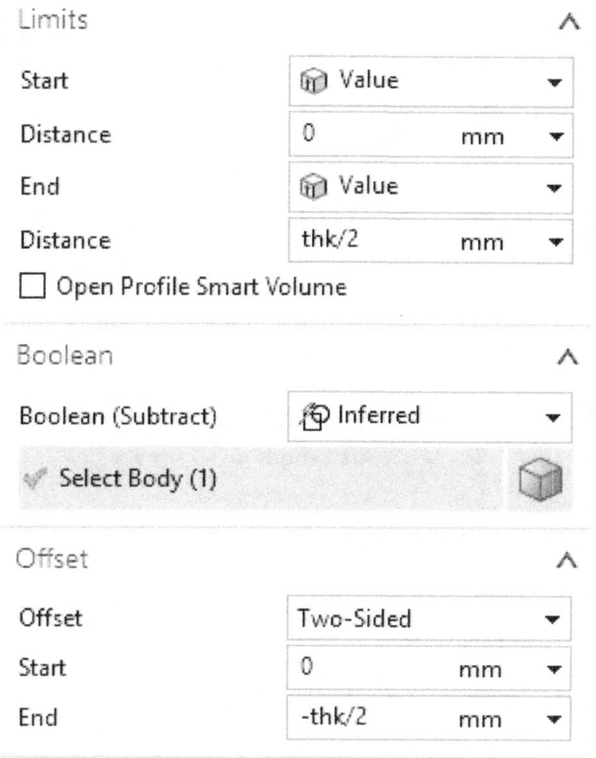

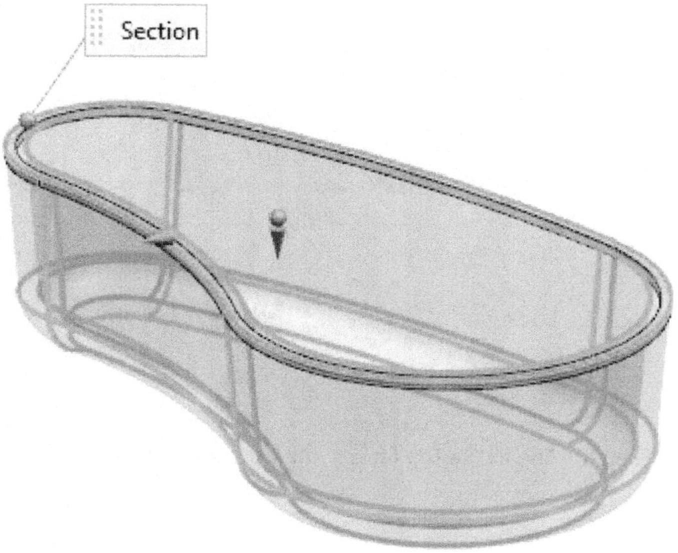

- Create a sketch on the top face of the body and use the **Offset Curve** tool to offset the inner edge by *"thk/4"* in the outward direction. This produces what is effectively a "mid-curve" on the top face.

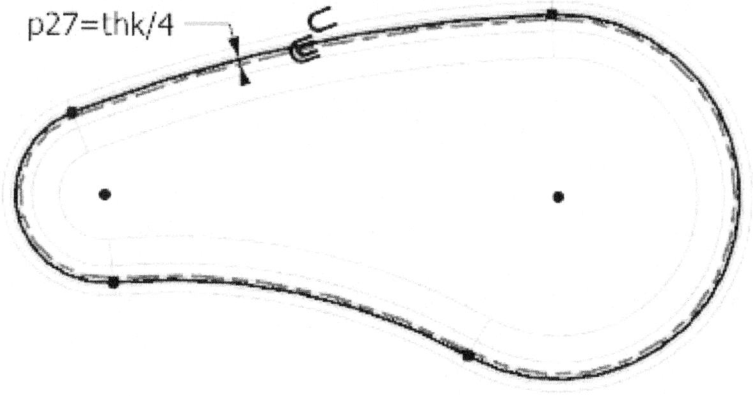

- Create a sketch of type *On Path* at the endpoint of one of the arcs from the offset curve in the previous sketch.

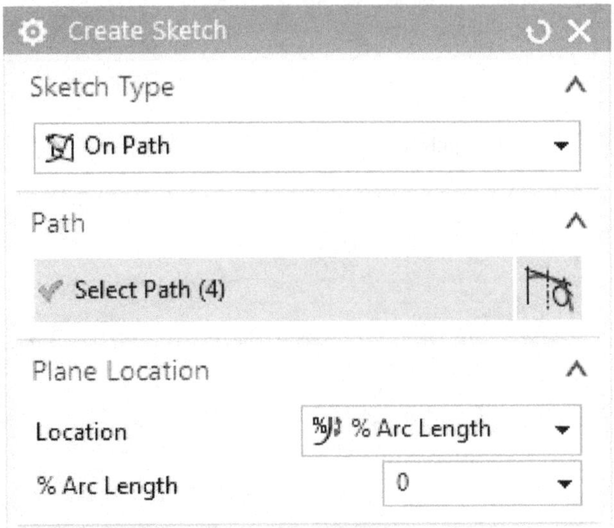

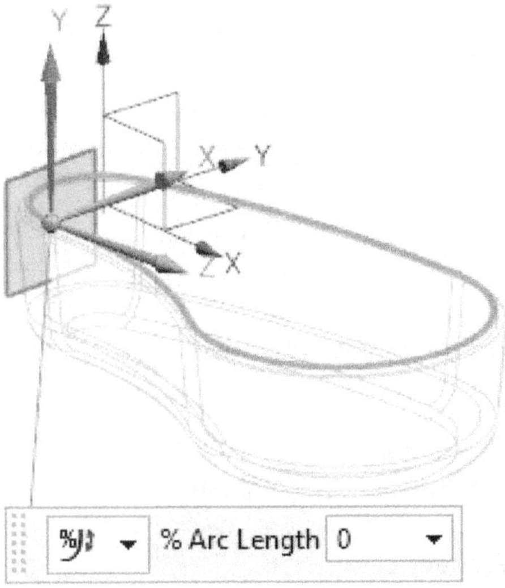

- Sketch a line in the top face, symmetric about the sketch origin, of length *"0.8*thk/2"*.

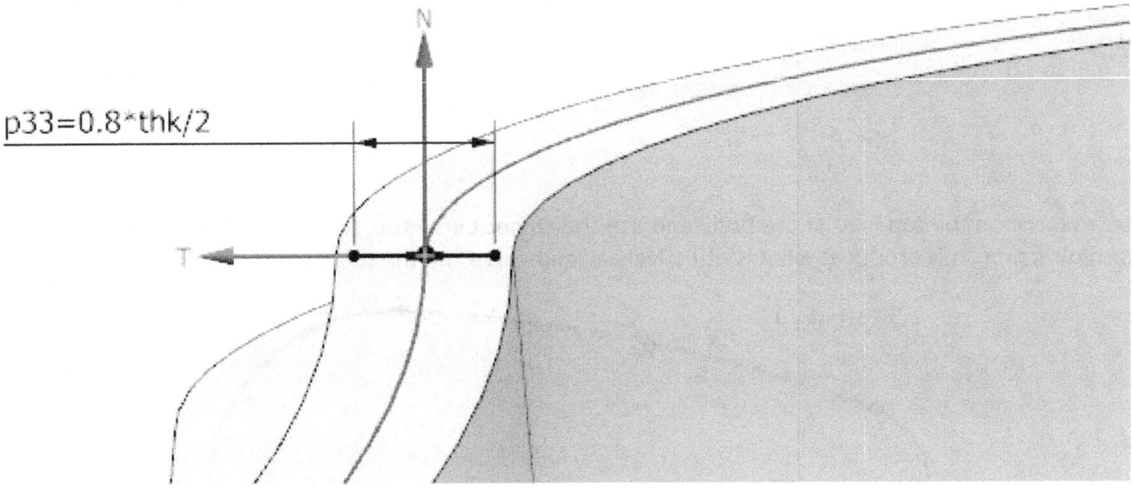

- Create another sketch *On Path*, at one of the endpoints of the line just created.

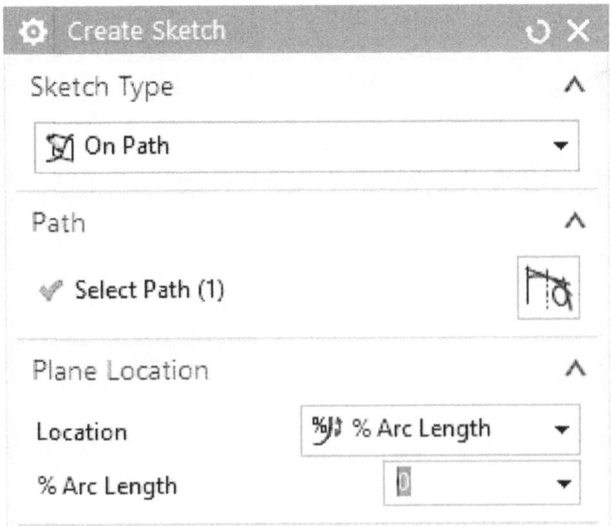

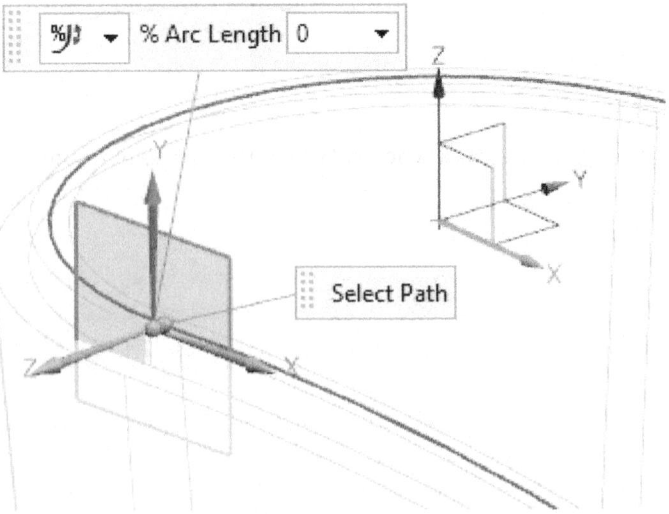

- In this sketch, use the **Polygon** tool to position a triangle sticking up and out of the body, as pictured below.

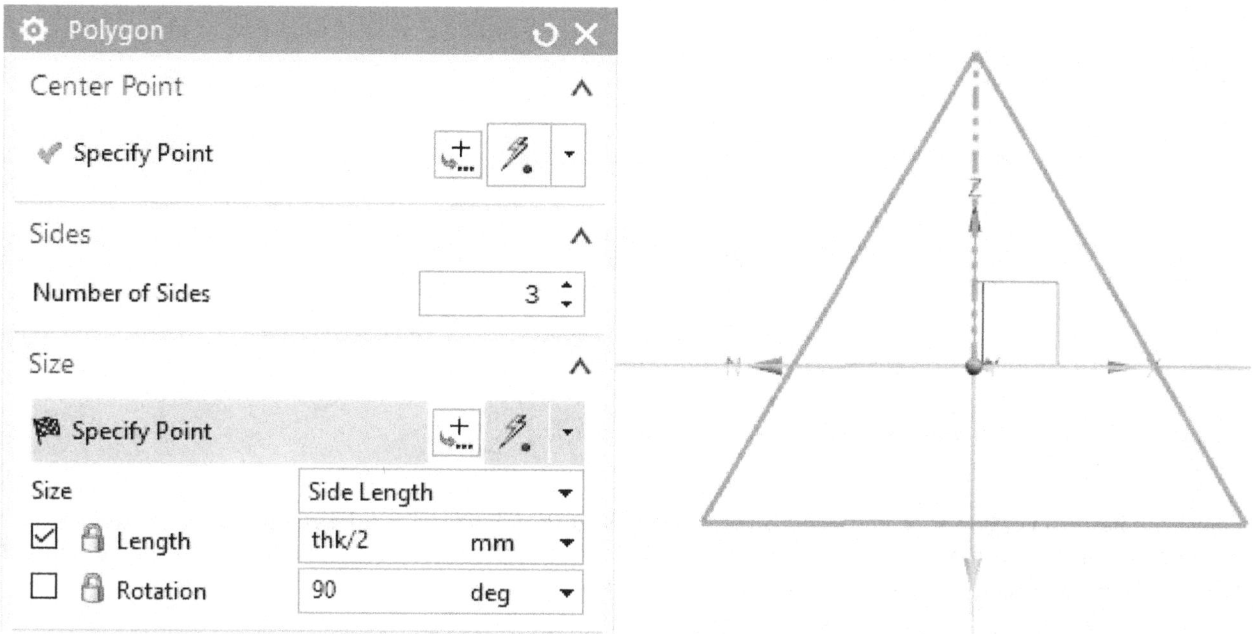

- Use **Sweep Along Guide** to sweep the triangle along the line with **Boolean** set to *Unite*.

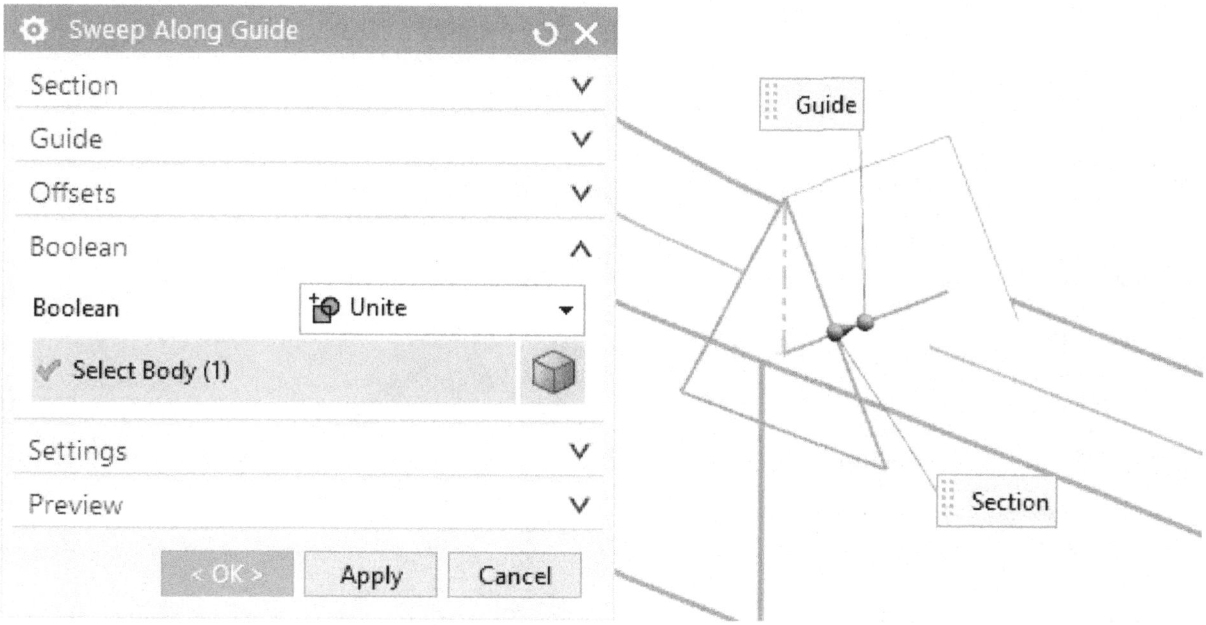

- Use **Measure Distance** to create an associative *Length* measurement for the offset curve.

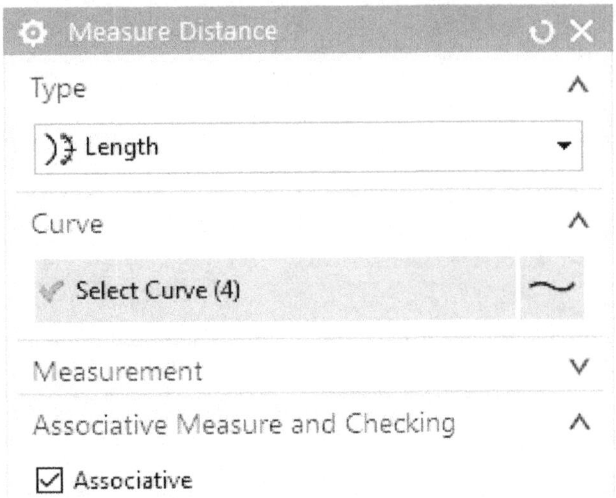

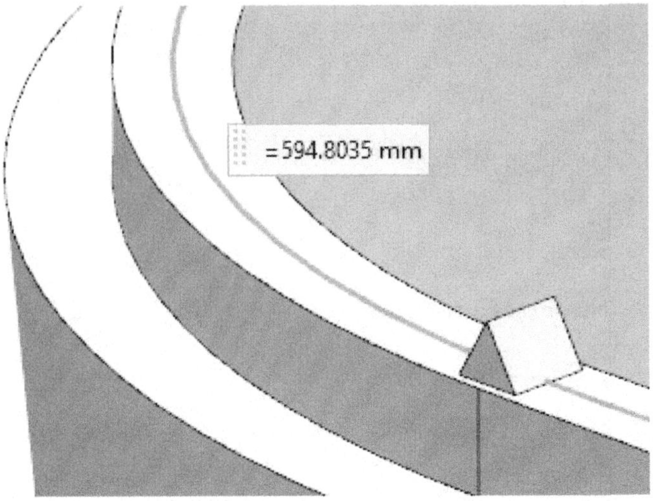

- Select the **Pattern Feature** tool, and select *Sweep* as the **Feature to Pattern**. Set the **Layout** to *Along*, and choose the offset curves as the *Path*. We would like pattern members to be uniformly spaced, and the count should depend on *"thk"*. It seems natural to choose parameters as shown below.

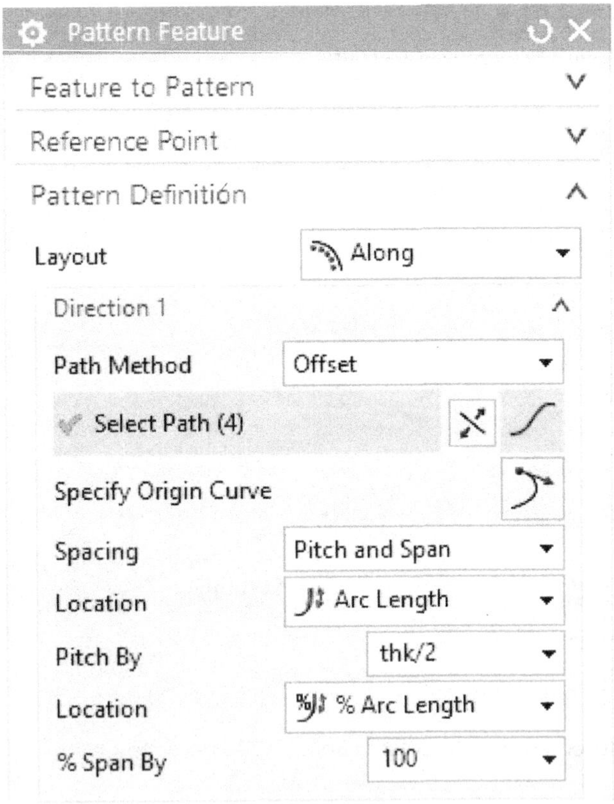

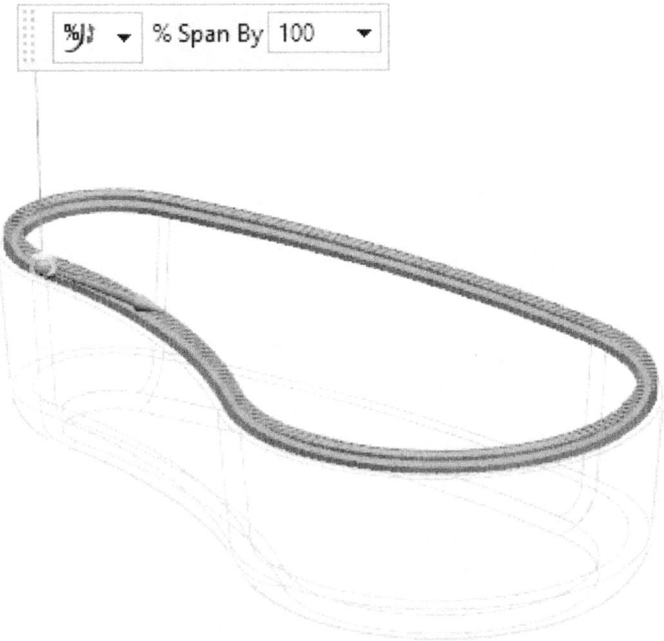

- Upon closer inspection, however, this spacing method has a problem. Namely, the arc length of the offset curve may not be evenly divisible by *"THK/2"*, and so there will be an additional gap between the last pattern member and the first.

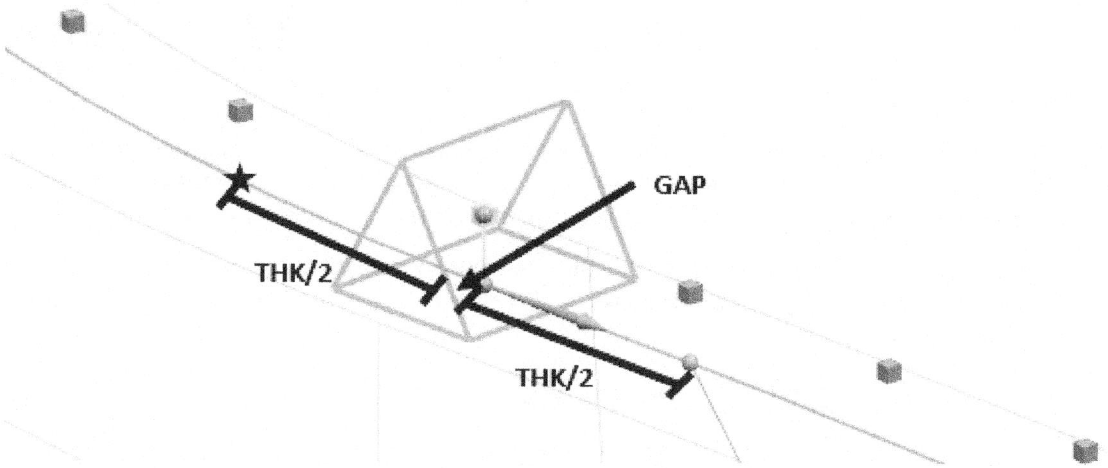

- To produce a pattern with uniform spacing of approximately *"THK/2"*, you can set the **Spacing** to *Count and Span* and enter a formula for the **Count** that makes use of the associative length measure created earlier.

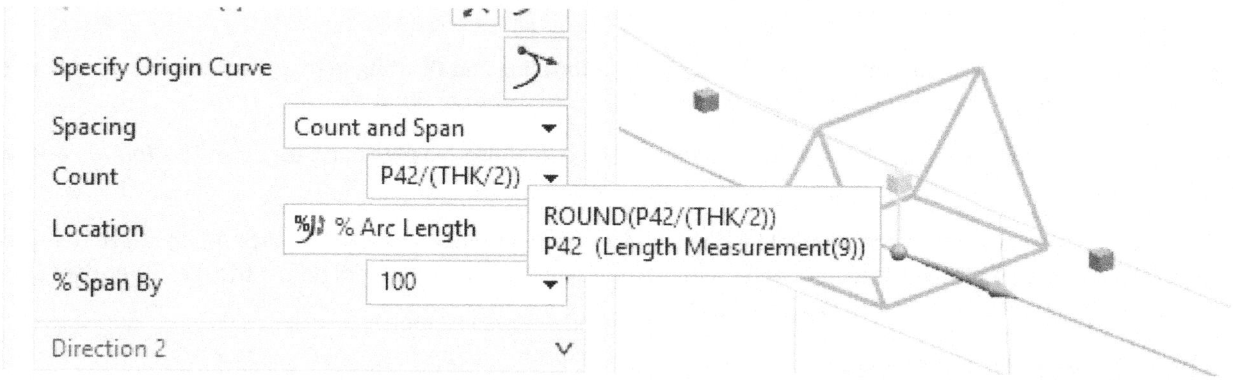

- Because your model is smart, you can change *THK* to any value between *0 mm* and *15 mm* and the features will all update nicely.

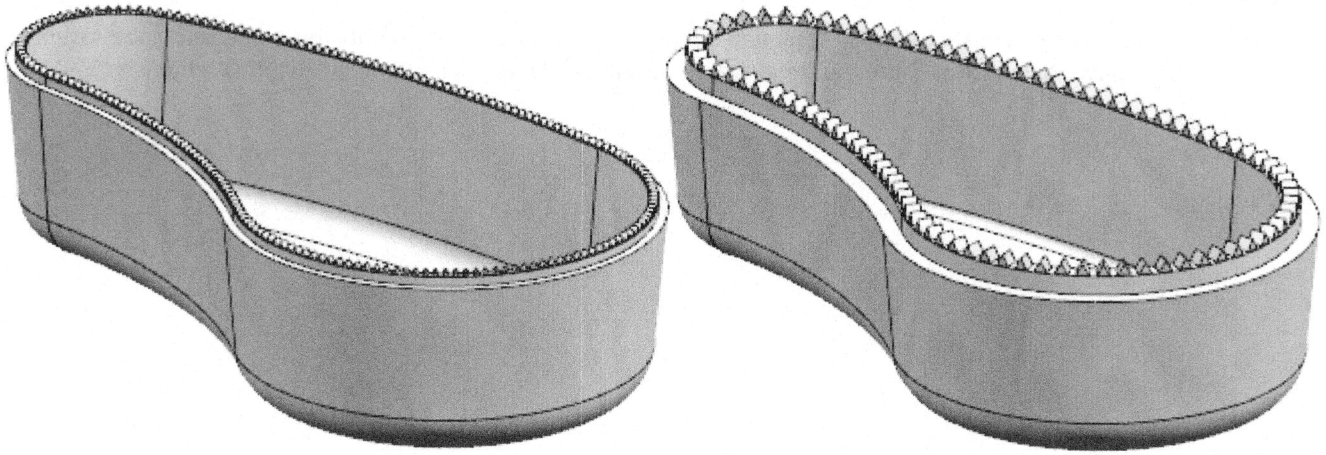

- **Save** your file!

11 Assemblies

In NX, an assembly is a *".prt"* file that manages relations between components. The key data that makes a *".prt"* file an assembly is its tree structure as viewed in the Assembly Navigator. An assembly does not actually contain the geometry within its constituent component part files, but does let you load and access the geometry in those files.

In order to manage assemblies in NX, you must first enable the **Assemblies** application. To enable or disable the **Assemblies** application while using NX, navigate to the **Applications** tab and click on the **Assemblies** icon. This icon will toggle the **Assemblies** application on or off.

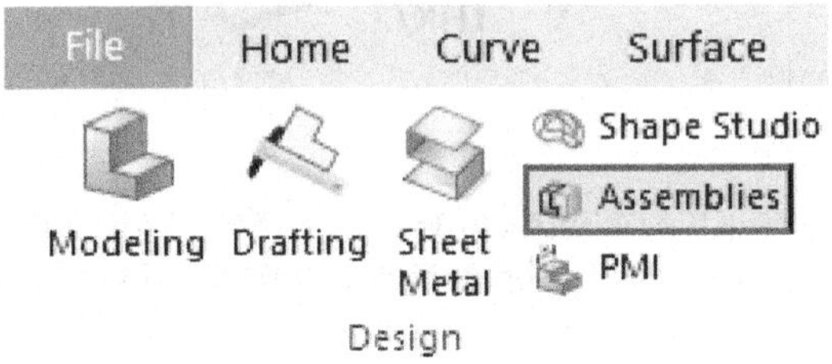

The **Assemblies** application is unlike other applications such as **Modeling** and **Drafting** in that the tools do not require their own environment – they just get added to the available tools in the current application.

There are two major ways to approach creating an assembly: by combining existing parts together (*bottom-up design*) or by creating new parts in-context (*top-down design*).

To illustrate the basic concepts in the **Assemblies** application, we will make use of the parts *"Hinge1.prt"* (created in Chapter 9.13) and *"Hinge2.prt"* (created in Chapter 9.14). If you skipped those chapters or lost the files, there are suitable replacements in *"C:\My NX 11 Files\Downloaded Files"*.

11.1 The *Assembly* Template

When you create a new file, there is an *"Assembly"* template that you can use. It is almost identical to the Modeling template.

- Create a new file. Instead of choosing the *Modeling* template, as you have done up until this point, select the *Assembly* template, as shown below. Make sure the **Units** are set to *Inches*. Name your assembly *"Hinge Assembly.prt"*.

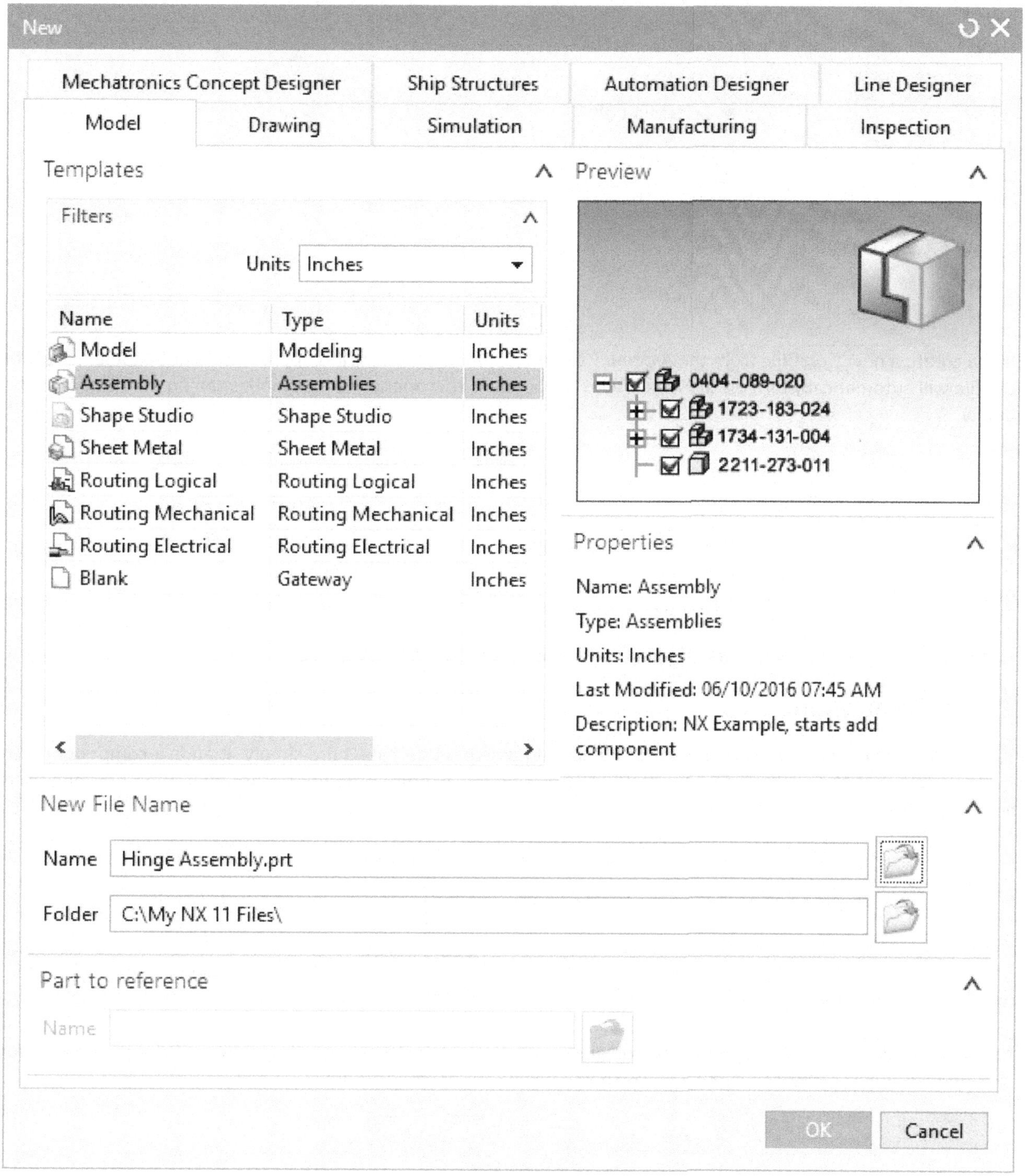

There are two main differences between the *Assembly* template and the *Modeling* template:

1. When a new file is created from the *Assembly* template, the **Add Component** dialog box will appear automatically.

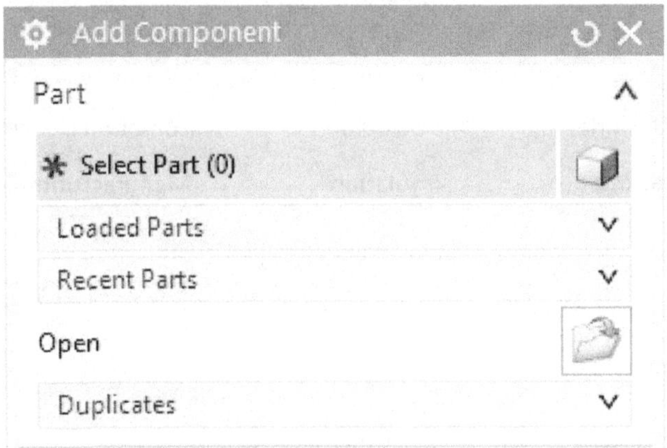

2. If you create a new *".prt"* file from the *Assembly* template, and the **Assemblies** application is not enabled, creation of the file will automatically turn on the **Assemblies** application. This opens the **Assemblies** tab on the ribbon, as shown below.

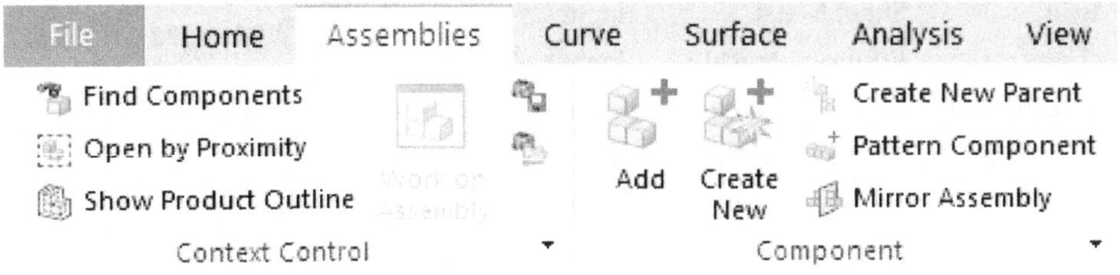

Besides those two points, the *Assembly* template and *Modeling* template are identical.

11.2 Add Component

The **Add Component** tool is the cornerstone of bottom-up assembly design. Quite simply, it adds a component into an assembly according to the positioning rules that you prescribe for it. **Add Component** is found in the **Component** group on the **Assemblies** tab.

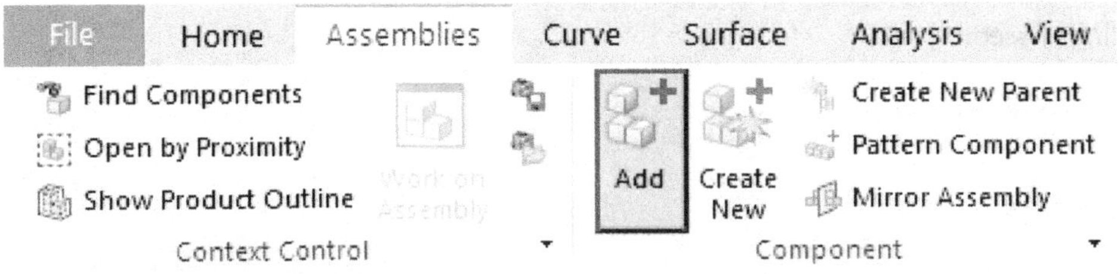

- You should still see the **Add Component** dialog box. You might see *Hinge1* and *Hinge2* in the **Recent Parts** section of the menu, but if not, click the **Open** icon, as shown below, to search for them.

- Select either *"C:\My NX 11 Files\Hinge1.prt"* or *"C:\My NX 11 Files\Downloaded Files\Hinge1.prt"*.
- Fill out the rest of the **Add Component** dialog box as shown below. You will see a preview of *Hinge1* as shown below. Note that as you add a component to an assembly, you can specify which **Reference Set** should be active (see Chapter 3.4.4). Select *Model*.

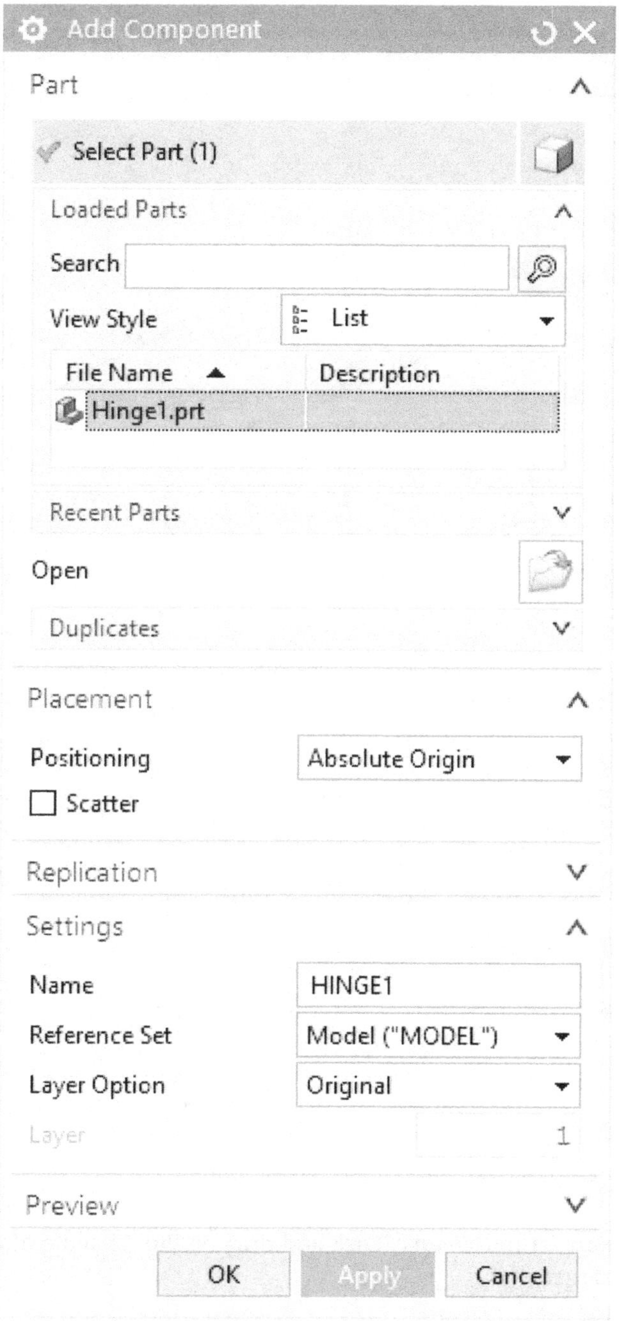

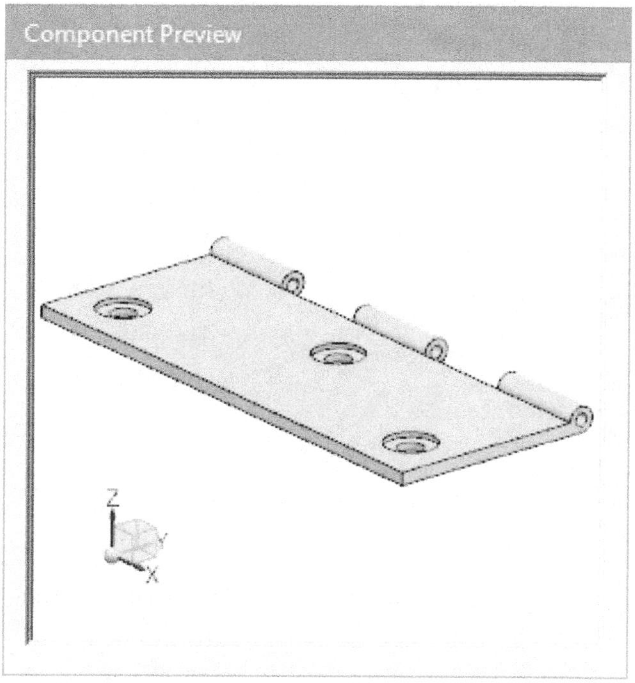

- Click **OK** to position *Hinge1* at the absolute origin of *"Hinge Assembly.prt"*.
- Next, select the **Add Component** tool from the **Component** group on the **Assemblies** tab.
- This time, select either *"C:\My NX 11 Files\Hinge2.prt"* or *"C:\My NX 11 Files\Downloaded Files\Hinge2.prt"*and set the **Positioning** to *Move*.

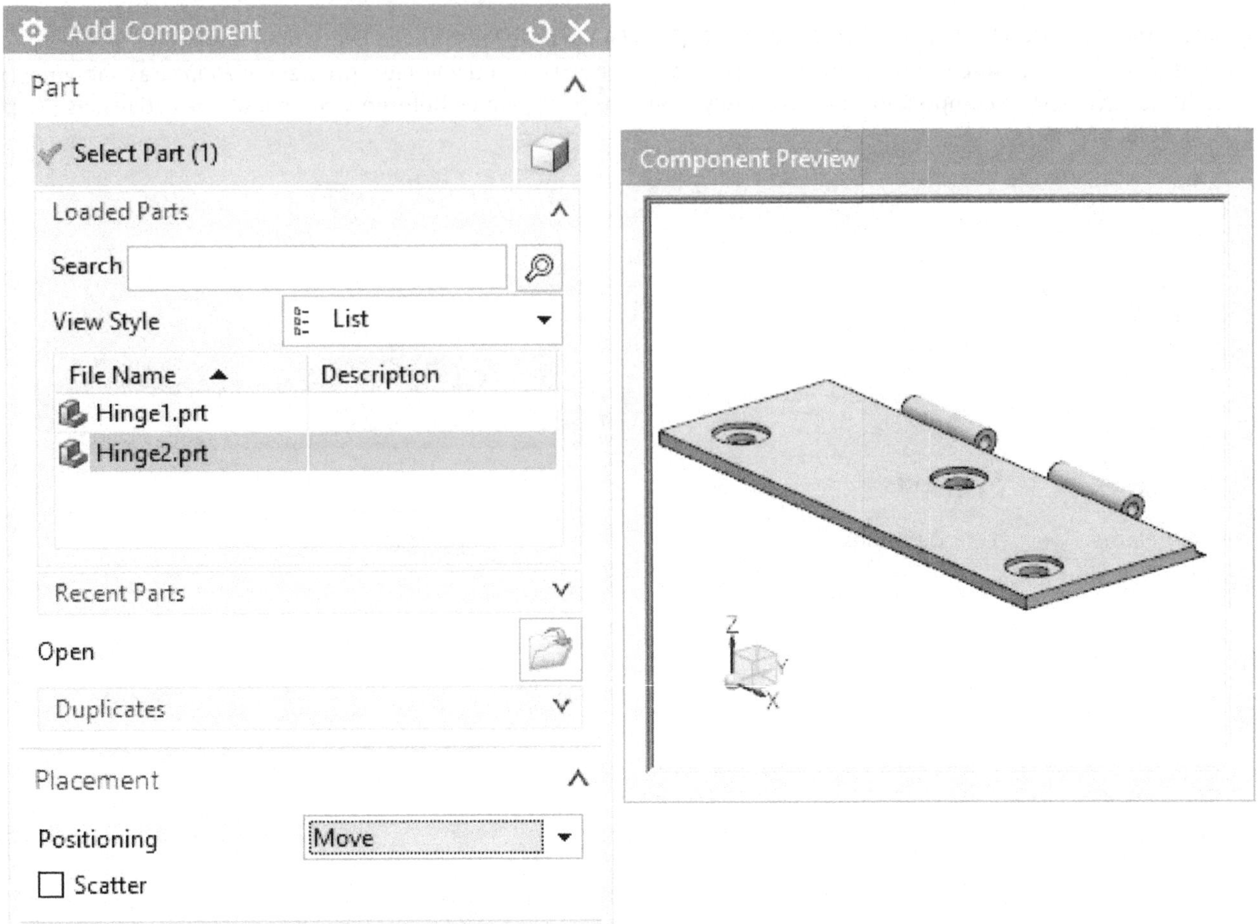

- When you click **OK**, you will be asked to specify a starting point for the motion of *Hinge2*. Click **OK** to allow it to default to the absolute origin of *Hinge Assembly*.

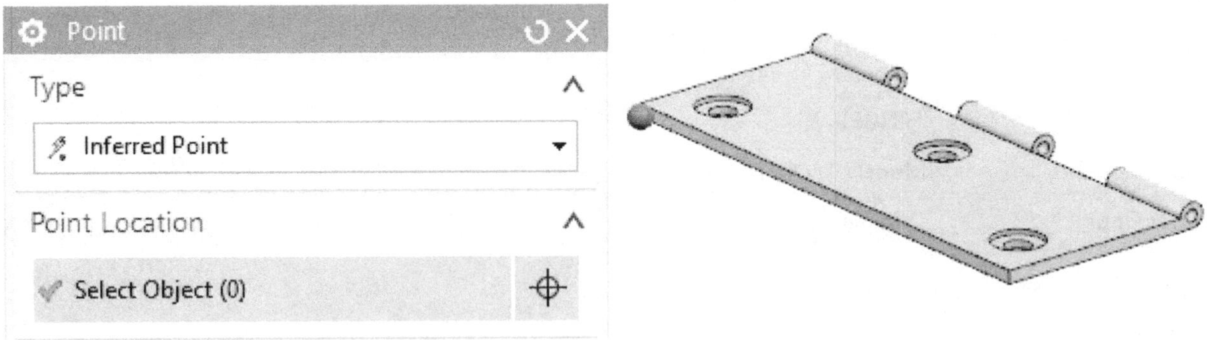

- This brings up the **Move Component** dialog with the **Motion** set to *Dynamic*. Click-and-drag on the Z handle of the dynamic coordinate system to move *Hinge2* up in the **+ZC** direction.

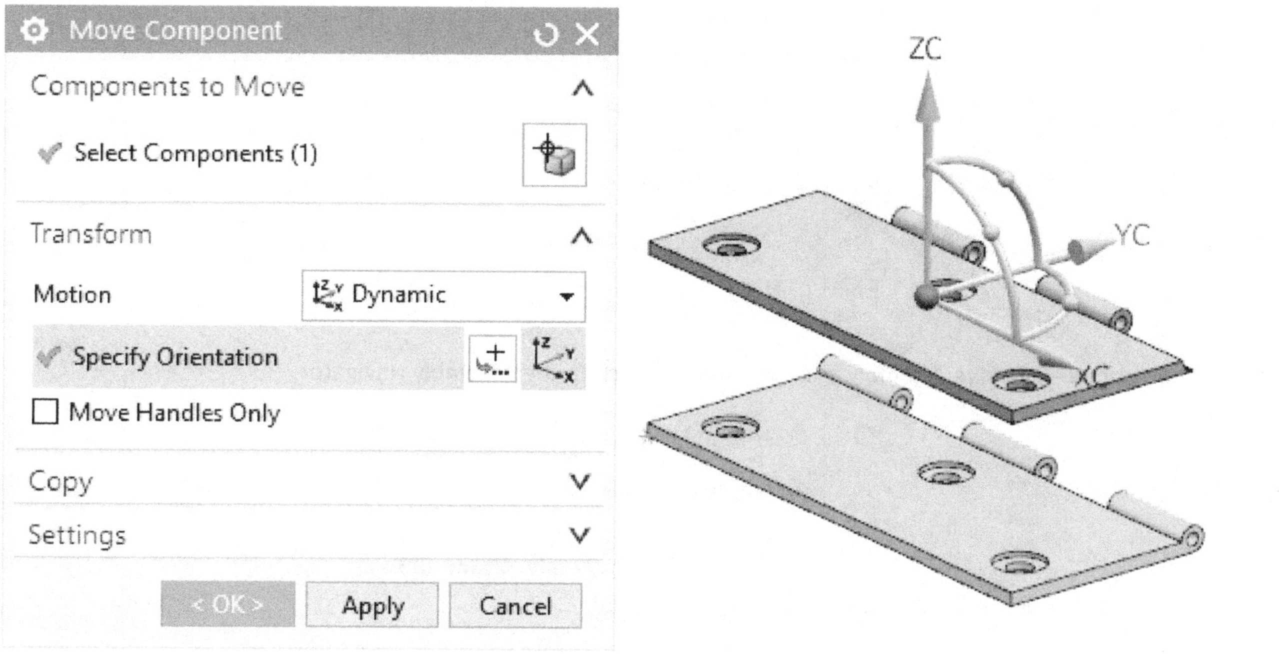

- Next, click-and-drag the rotation handle in the XZ plane of the dynamic coordinate system to rotate *Hinge2* by **180 degrees**.

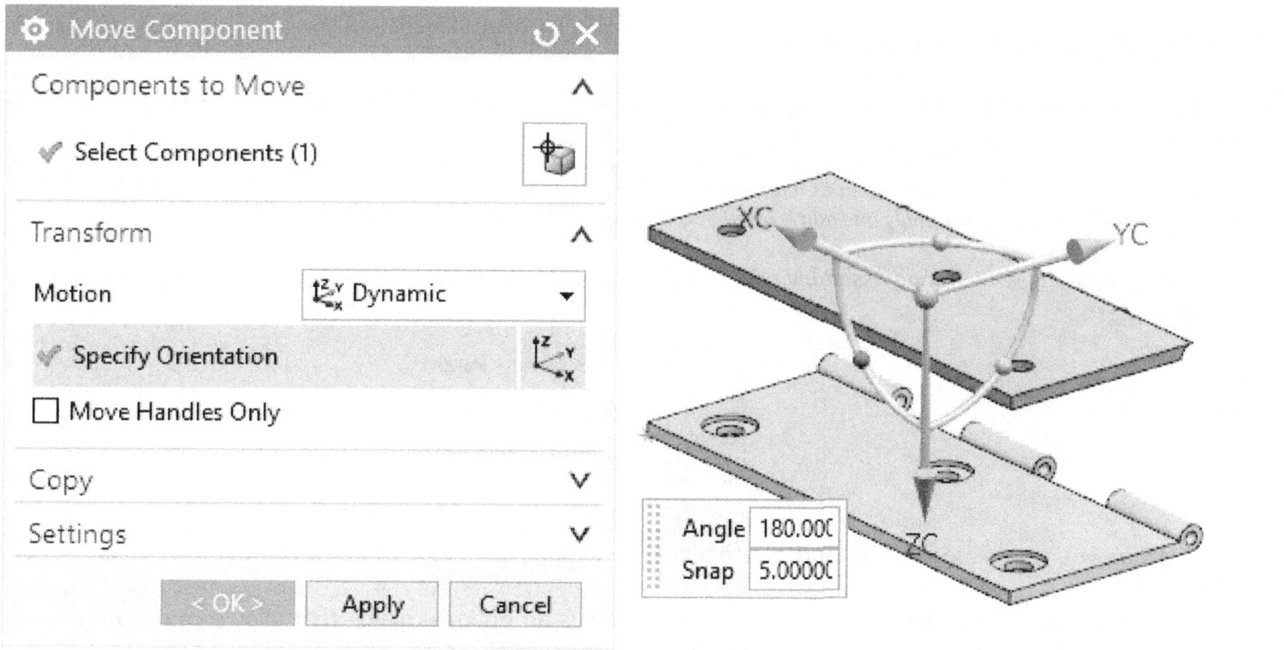

- Click **OK** to position *Hinge2*! **Save** your assembly.

11.3 The Assembly Navigator

Up until this point, we have made almost exclusive use of the Part Navigator to see the structure of our models. Now that we are working with an assembly – *Hinge Assembly*, notice that the Part Navigator is empty. The Model History for *Hinge Assembly* includes neither the model history from *Hinge1* nor *Hinge2*.

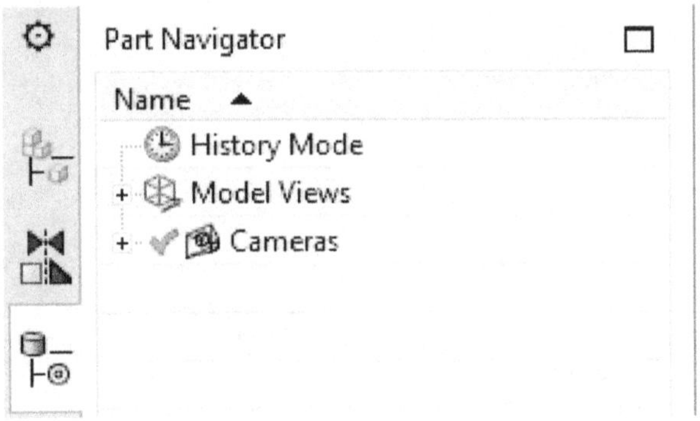

To see the structure of *Hinge Assembly* as an assembly, switch to the Assembly Navigator.

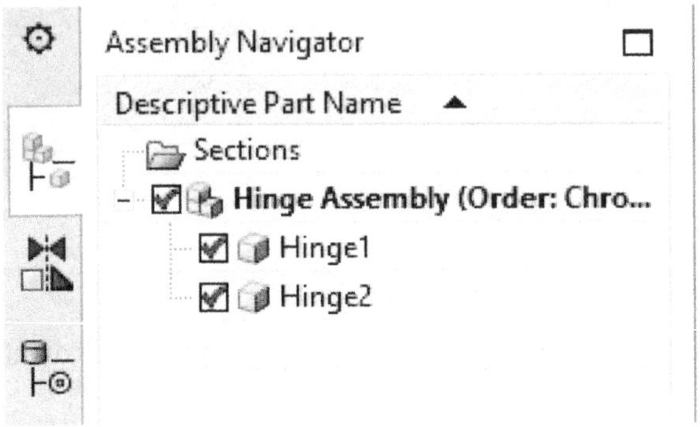

11.4 Create New Component

To perform top-down assembly design in NX, you must create parts using the **Create New Component** tool. Once you have a new empty component, you can proceed to add geometry to it.

Note that when you create a component in this manner, it inherits its units from the parent assembly. If you wish to use different units, you must create it outside the assembly and add it using bottom-up assembly design techniques.

- Continue working with *"Hinge Assembly.prt"*, and select **Create New Component** from the **Component** group on the **Assemblies** tab.

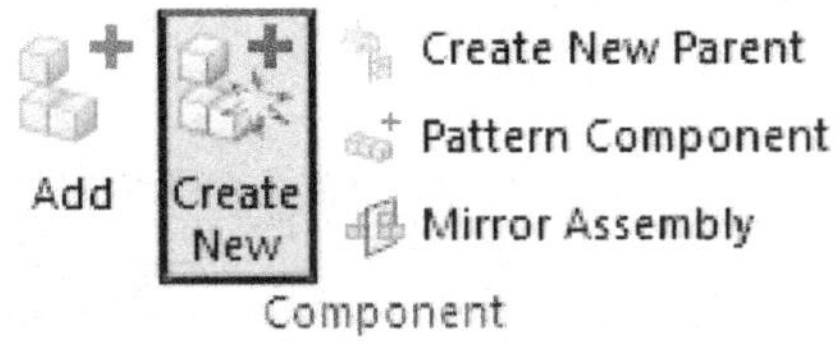

- Make sure your template is set to *Model* (rather than *Assembly*) and name this new component: *"Hinge Pin.prt"* and click **OK**.
- The **Create New Component** window then appears. This is an optional dialog that allows you to move bodies or components from the current Work Part to the new component being created. For our purposes, none of the components currently in *Hinge Assembly* need to be incorporated into *Hinge Pin*, so simply click **OK** to skip it!

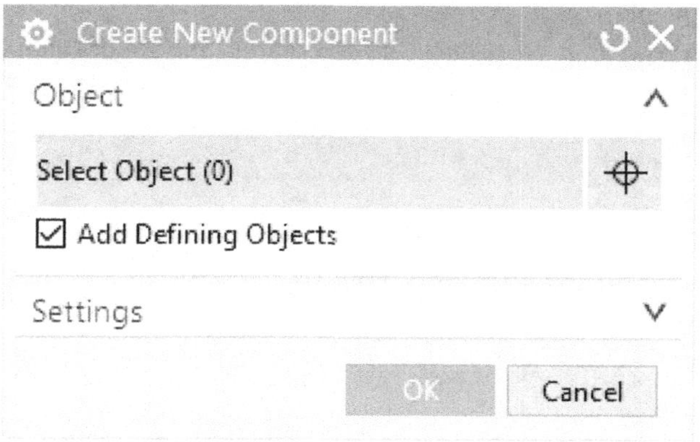

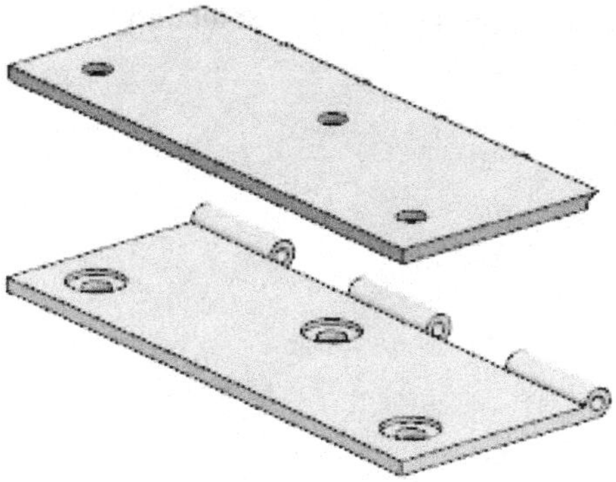

- **Right-click** on *Hinge Pin* in the **Assembly Navigator** and select **Make Work Part**.

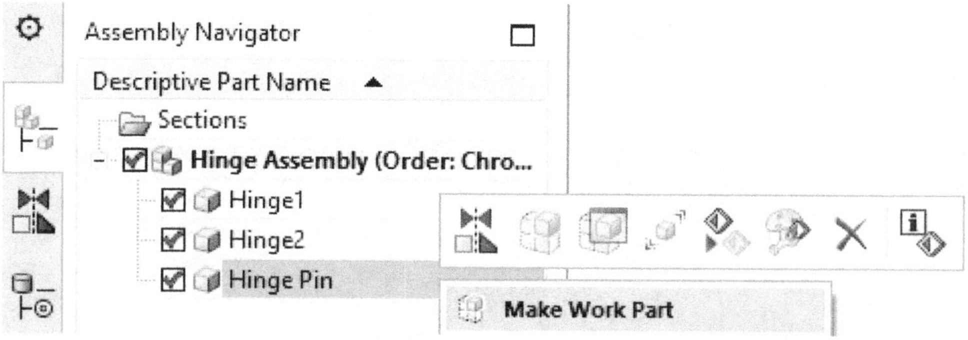

- Use the **Extrude** tool with selection drop-down menus set as shown below to extrude the inner edge of the hole in *Hinge1* shown below.

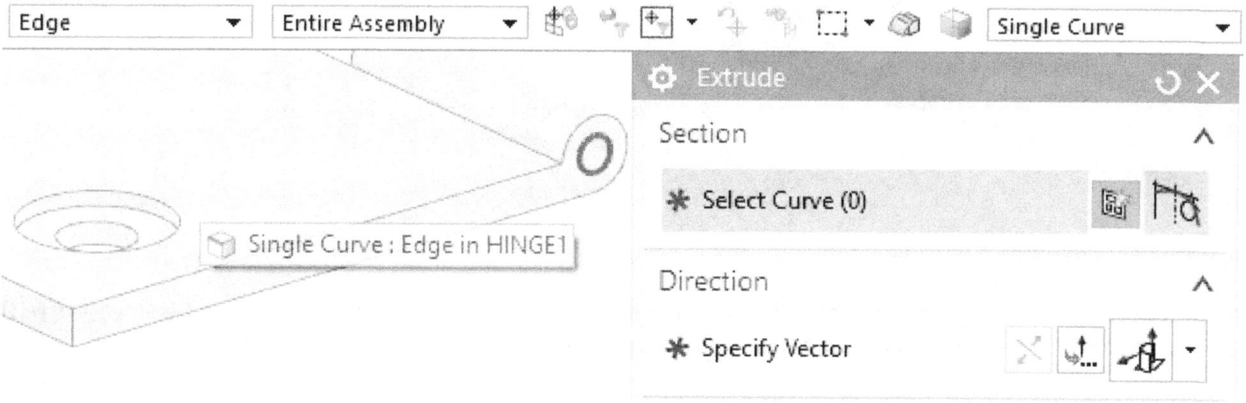

- Use **Reverse Direction** if necessary, and set the **End Limit** to *Until Extended*. Choose the opposite face of *Hinge1*, as shown below.

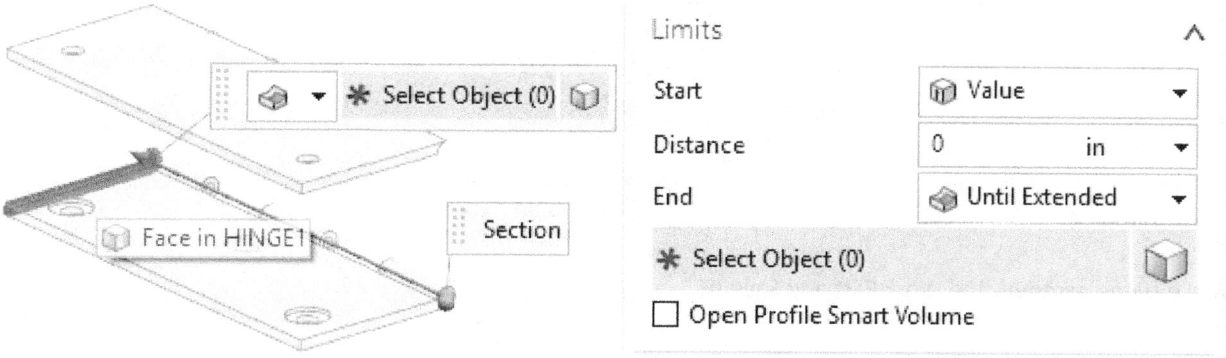

- Click **OK** to complete the cylindrical part of the hinge pin.
- Next, use the **Sphere** tool with **Selection Scope** set to *Within Work Part Only* to create a **Sphere** of **Diameter** *0.25 in*, centered as shown below.

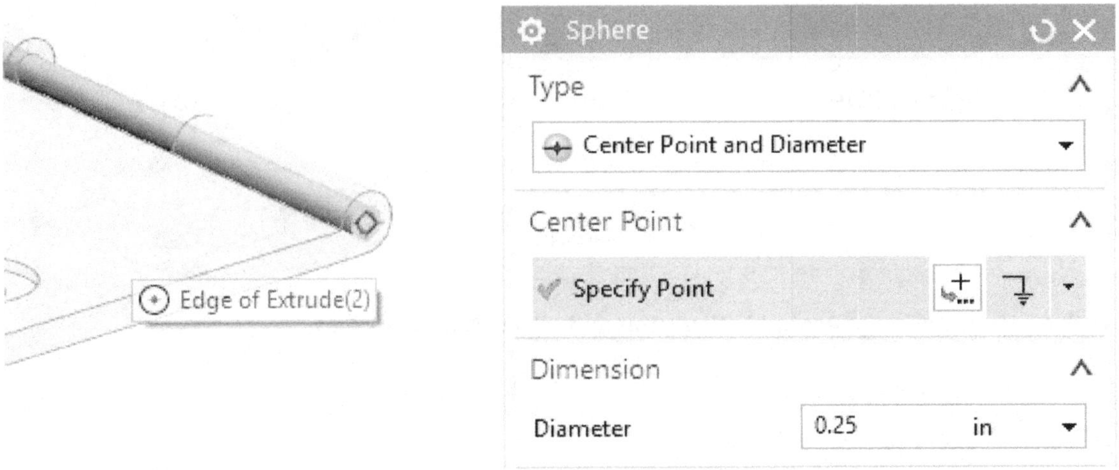

- Use **Trim Body** with the sphere as **Target** and a plane defined by the face shown below as the **Tool**. Pay close attention to your selection drop-down menus to make your selection straightforward.

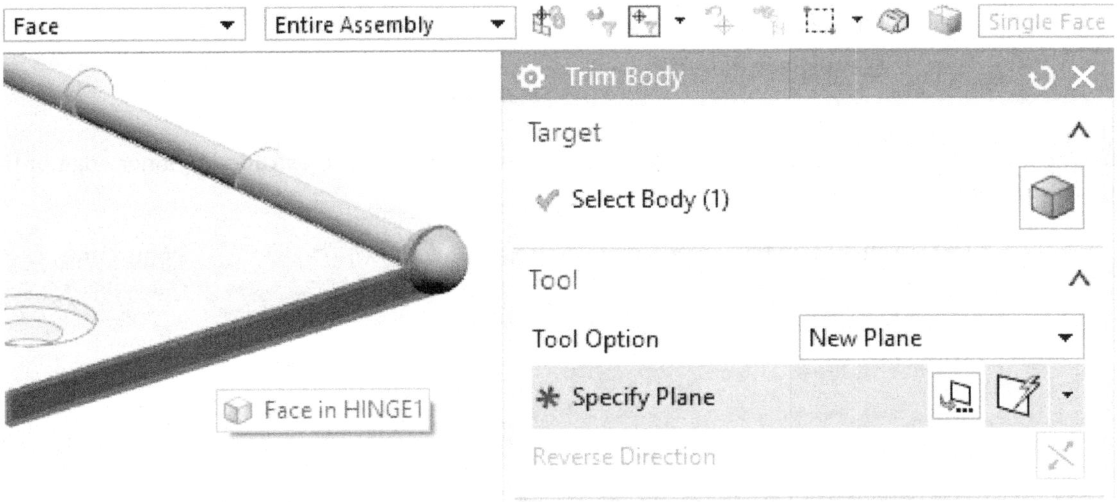

- Unite the cylinder and trimmed sphere to produce the hinge pin shown below. Use **Edit Object Display** (**[Ctrl]+[J]**) to give it a color different from *Hinge1* and *Hinge2*. **Save** your file!

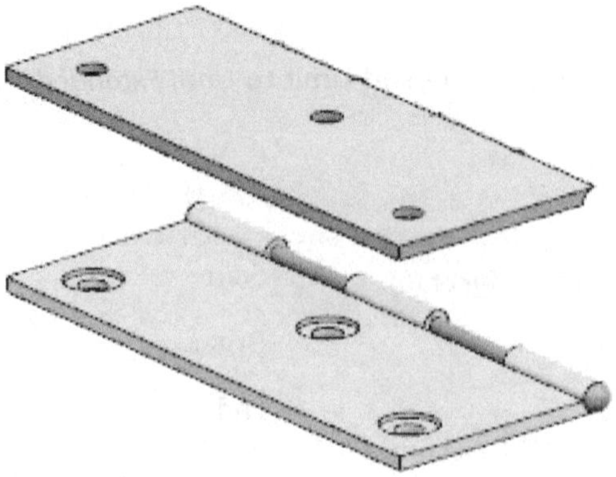

- Make *Hinge Assembly* the Work Part and **Save** it.

11.5 Assembly Constraints and Degrees of Freedom

In the course of building an assembly, you may wish to remove certain degrees of freedom from components so that they move relative to one another in a way that reflects the physical reality of the finished product. **Assembly Constraints** (sometimes called "Mating Conditions" or "Mates" in other CAD software, and even earlier versions of NX) are used to create these relationships. The **Assembly Constraints** tool is found in the **Component Position** group on the **Assemblies** tab.

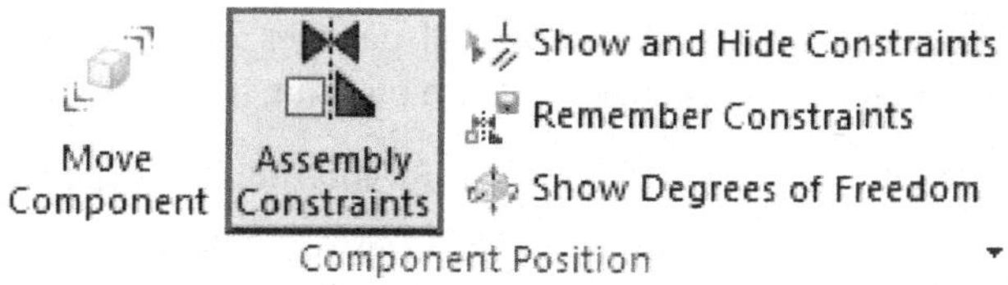

Below is a summary of the constraints available from the **Assembly Constraints** menu.

Constraint:	Icon	Applies to:	Behavior:
Touch Align	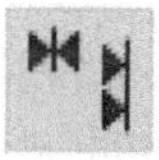	Faces, Planes, Edges, Curves, Points	Touch: normal vectors oppose Align: normal vectors agree Infer Center/Axis: cylindrical centerlines align
Concentric		Circles, Arcs, Circular Edges	Arcs become coplanar Arc centers coincident
Distance		Faces, Planes, Edges, Curves, Points	Minimum distance between (extension of, if necessary) selected entities becomes specified value, or is constrained to a range of specified values
Fix		Single Component	Removes all degrees of freedom from selected component
Parallel		Flat Faces, Planes, Edges, Lines	Selected entities become parallel
Perpendicular		Flat Faces, Planes, Edges, Lines	Selected entities become perpendicular
Align/Lock		Linear edges, Lines	Linear objects become aligned, and rotational freedom about the axis of said linear objects is removed
Fit	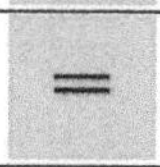	Faces or edges of equal radius	For cylindrical faces of equal radius, the centerlines align For spherical faces, the centers become constrained
Bond		Multiple Components	Selected components become constrained to move as a rigid entity
Center		Faces, Planes, Edges, Curves, Points	One component is centered between or about selected entities from one or two neighboring components
Angle		Circles, Arcs, Circular Edges	Angle between (extension of, if necessary) selected entities becomes specified value, or is constrained to a range of specified values

In many of the aforementioned constraints, "lines" include centerlines of cylindrical faces (e.g., holes, bosses). In order to make those centerlines selectable, you must usually put your cursor on the cylindrical face in question, so that the centerline appears. You can then select the centerline directly, rather than the face.

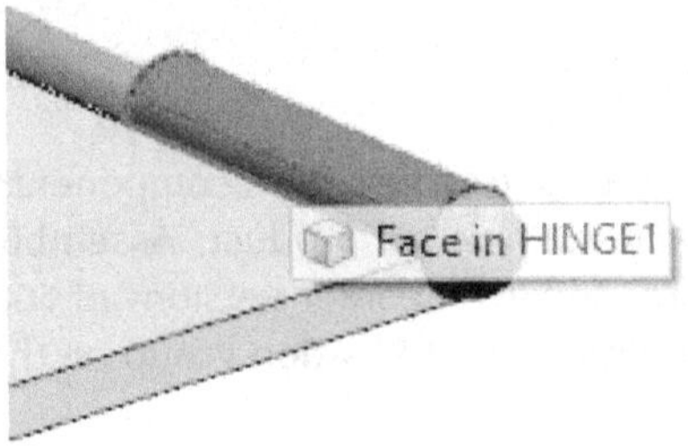

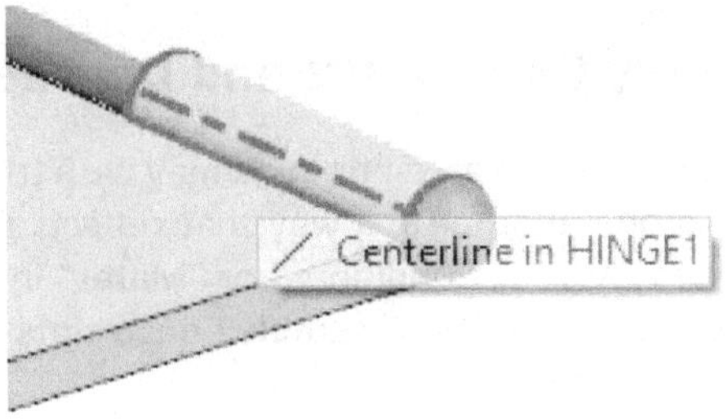

In the exercise below, we will apply constraints to remove all degrees of freedom from the components in *Hinge Assembly.prt*. We will check on the remaining degrees of freedom by using the **Show Degrees of Freedom** tool.

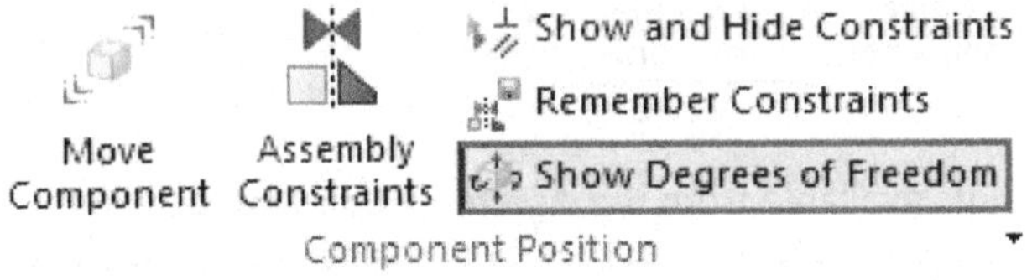

- Continue working with *Hinge Assembly.prt*, and select the **Assembly Constraints** tool.
- It is a good practice to remove all degrees of freedom from some component in your assembly before constraining other components to it. Select the **Fix** constraint and then choose *Hinge1*. You will see a small "ground" symbol appear on top of *Hinge1*, as shown below.

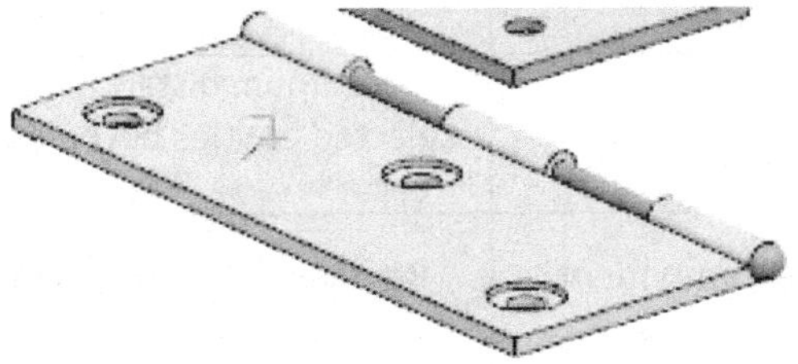

- The **Select Object** prompt on the menu will remain highlighted in yellow and with a red asterisk – the **Assembly Constraints** tool allows you to specify multiple constraints before saving all of them with either **OK** or **Apply**. We strongly encourage you to click **Apply** at this time, in case you make a mistake with one of the next constraints – the **Cancel** button will close the tool and you will lose any unsaved constraints.
- Next, set the constraint type to *Touch Align*, and from the **Orientation** drop-down menu, choose *Infer Center/Axis*.

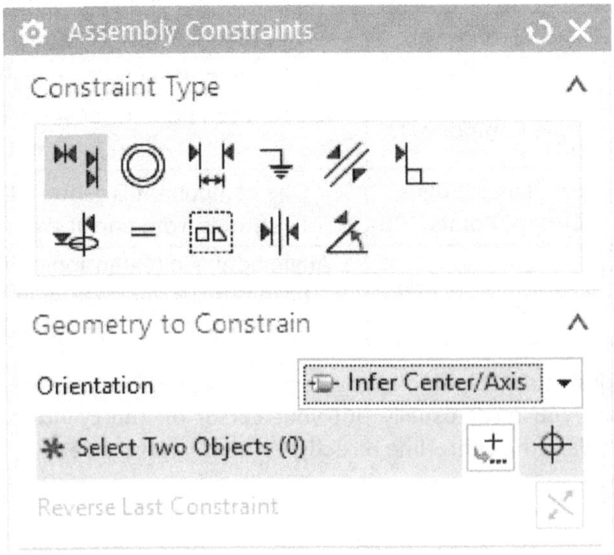

- This orientation subtype allows for a centerline-to-centerline constraint. If you are familiar with other CAD software, this is what you might expect *Concentric* to do. Note that *Concentric* in NX involves only circles and arcs! Begin by selecting the centerline of the cylindrical face in *Hinge1* shown below.

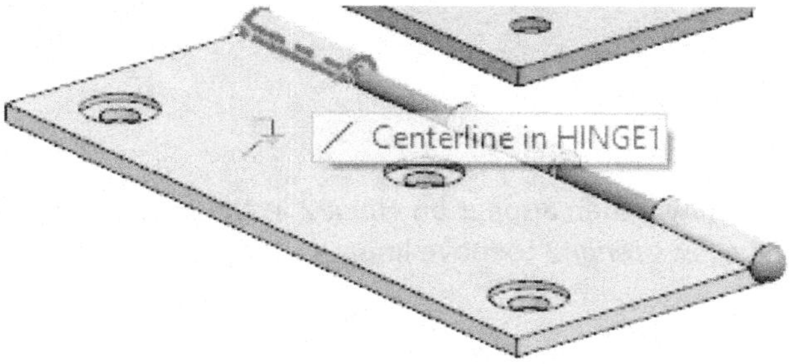

- Next, choose the centerline of a cylindrical face from *Hinge2*, as shown below.

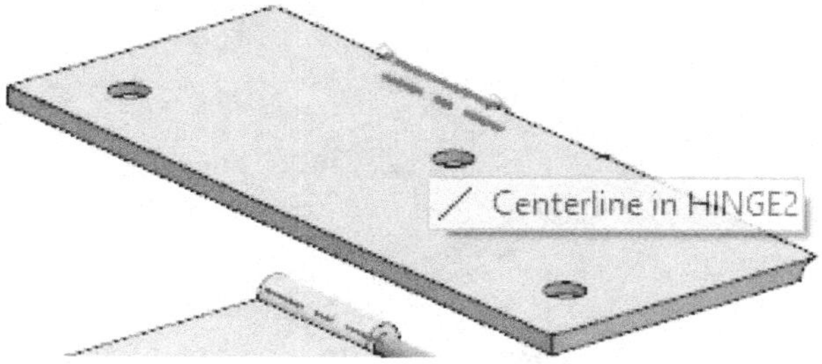

- Click **OK** to save the constraint. You will now see an alignment constraint on top of *Hinge2* in the graphics window, as shown below.

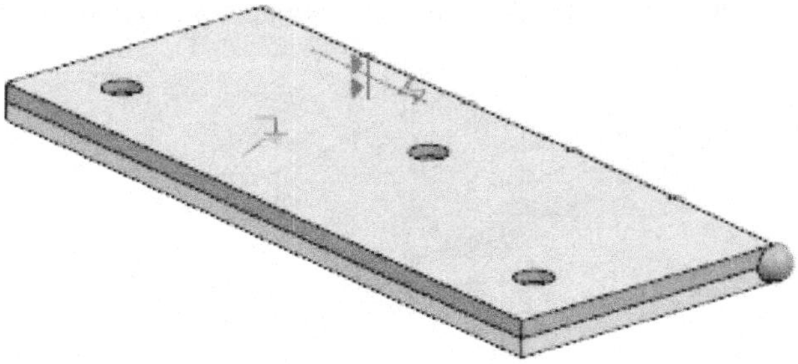

- To turn off the display of constraints in the graphics window, you can right-click on **Constraints** in the Assembly Navigator, and uncheck **Display Constraints in Graphics Window**.

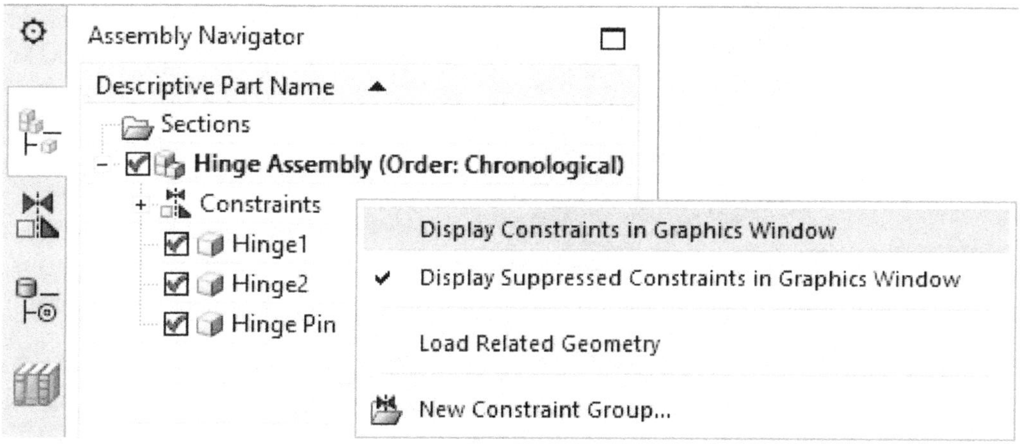

- Choose the **Show Degrees of Freedom** command from the **Component Position** group on the **Assemblies** tab. Select *Hinge2*, as shown below.

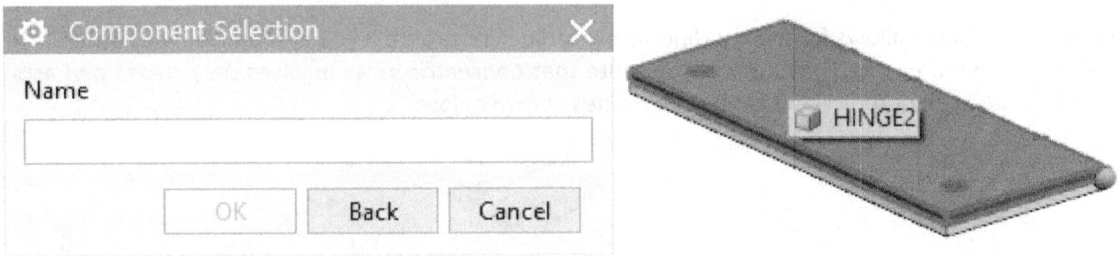

- The remaining degrees of freedom will then appear on *Hinge2* in the graphics window. Note that the *Touch Align – Infer Center/Axis* constraint still allows *Hinge2* to move linearly along, and to rotate about, the selected axis.

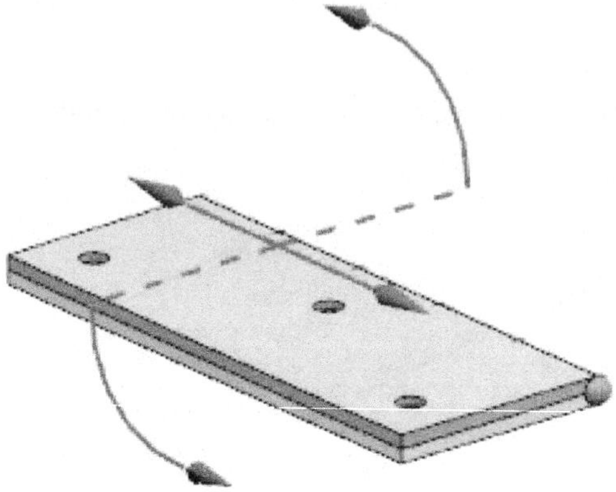

- To make the degree-of-freedom indicators disappear, you can either use **Refresh**, or **Fit**.
- Reopen the **Assembly Constraints** tool and **Reset** it so that the constraint type is set to *Touch Align*, and the **Orientation** becomes *Prefer Touch*. Click-and-hold-and-drag *Hinge2* somewhere near the edge shown below.

- Continue holding down the left mouse button and drag *Hinge2* to the left as shown below.

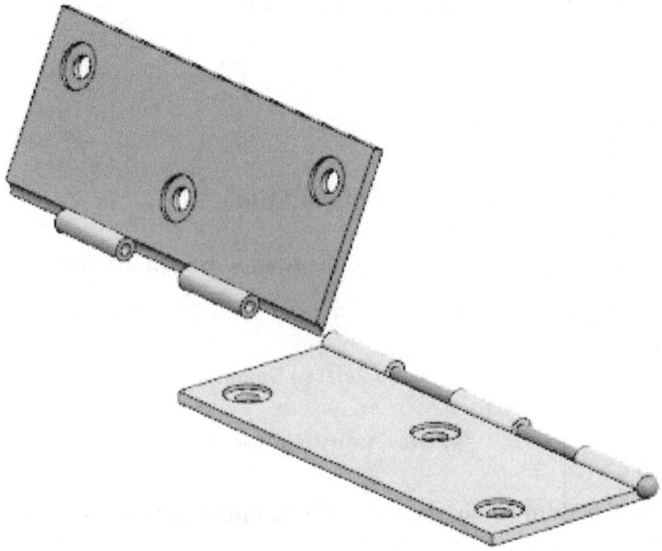

- This kind of dynamic motion is only possible when the **Assembly Constraints** tool is open. Clicking and dragging components without either the **Assembly Constraints** tool or the **Move Component** tool open will result in no action.
- Choose the inside face of the leftmost cylinder on *Hinge1*, as shown below.

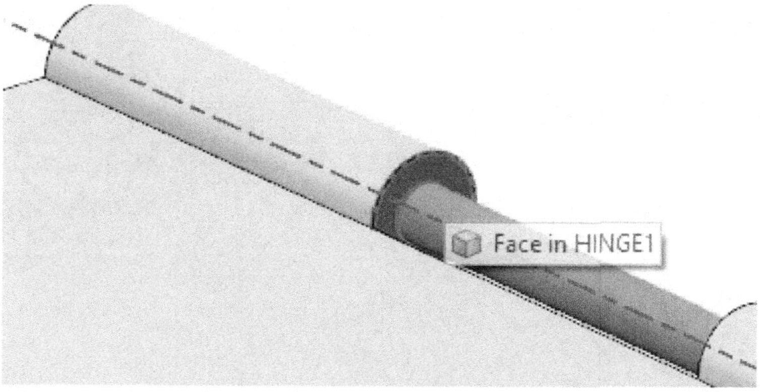

- Rotate your model **90 degrees** about **ZC** and choose the outer face of the leftmost cylinder on *Hinge2*, as shown below.

- Click **OK** to save this *Touch* constraint. Now the only motion possible for *Hinge2* is rotation about its axis.

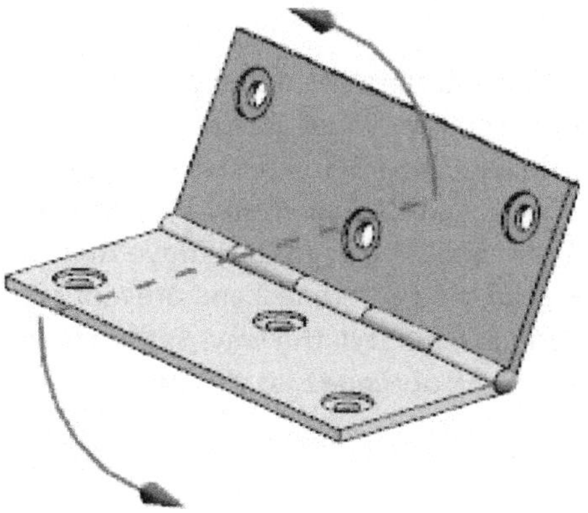

- Finally, create an assembly constraint of type *Angle*, with subtype set to *3D Angle*. Choose the inner faces of *Hinge1* and *Hinge2* as shown below.

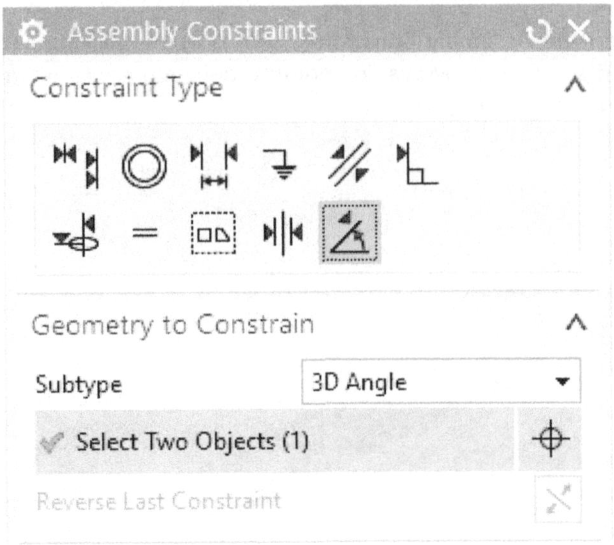

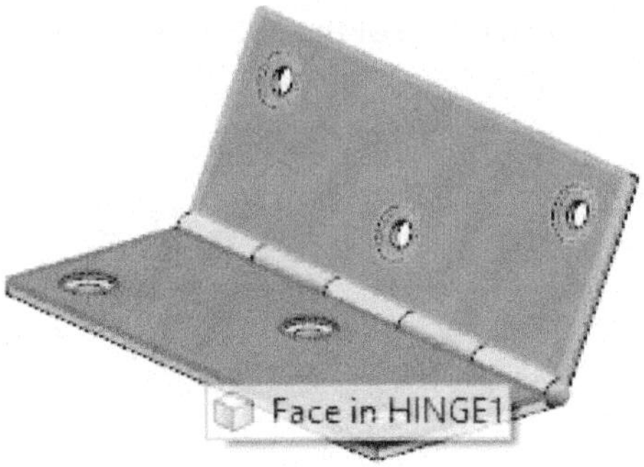

- Uncheck the box in the **Angle** section of the menu, and instead enable an **Upper Limit** and a **Lower Limit** as shown below.

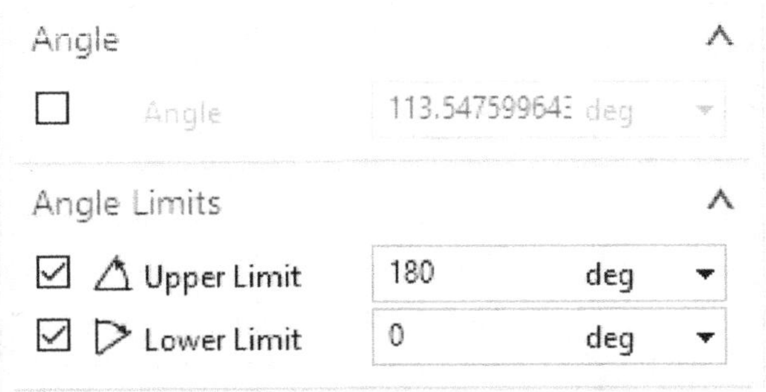

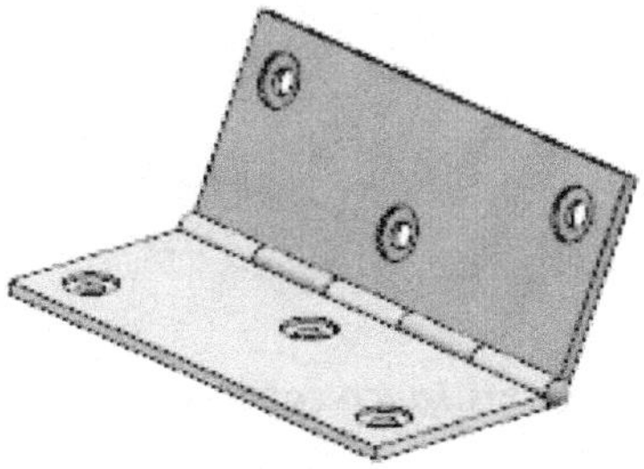

- Click **OK** to create the constraint. Note that you can still rotate about the axis of the hinge, but that the range of motion is now limited so that the bodies in the assembly cannot collide with each other.
- Next, we will constrain *Hinge Pin*. It would be natural to constrain the axis in *Hinge Pin* to the axis in either *Hinge1* or *Hinge2* using *Touch Align – Infer Center/Axis*, but this would still allow for rotational motion. Since *Hinge Pin* is a perfect body of revolution, it has no features that could be used to remove that rotational freedom. In these situations, the *Align/Lock* constraint is useful – it will align the centerlines, and prevent rotational motion.
- Open **Assembly Constraints** and drag *Hinge Pin* out of the way. Set the **Constraint Type** to *Align/Lock* and select both the centerline of *Hinge Pin* and the centerline of *Hinge1*, as shown below.

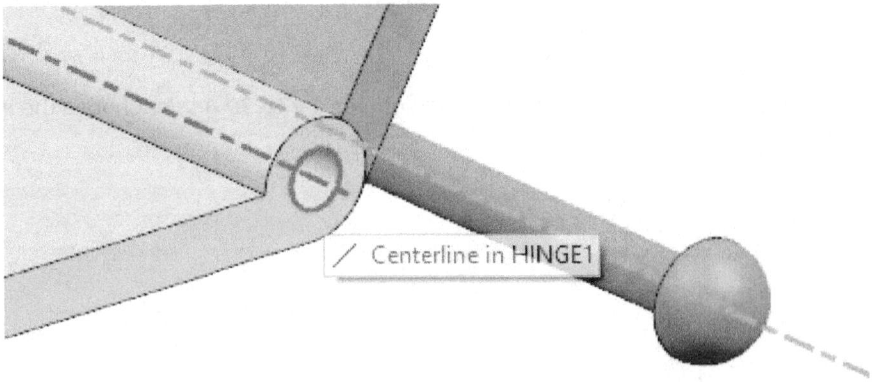

- Thanks to *Align/Lock*, the rotational motion that *Touch Align – Infer Center/Axis* would have allowed is removed, and so the only degree of freedom is linear along the axis of the hinge.

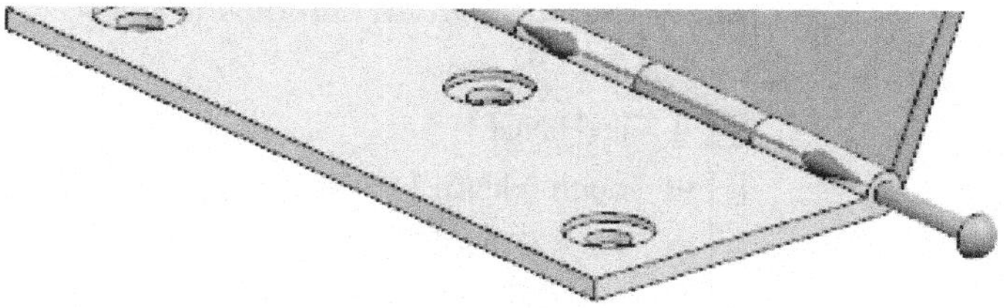

- Use a *Touch Align – Touch* constraint to position *Hinge Pin* as shown below, with no degrees of freedom remaining.

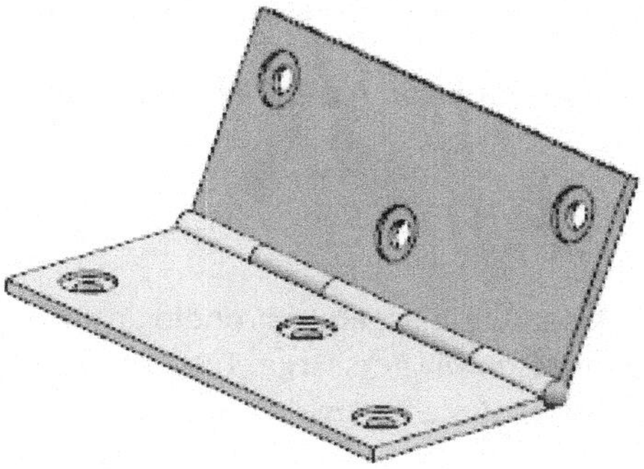

- Save your *Hinge Assembly*!

11.5.1 Modifying Assembly Constraints

You may also modify existing **Assembly Constraints** by deleting them, suppressing them, or changing their type. Components may themselves be edited, redefined, reversed, converted, deleted or suppressed.

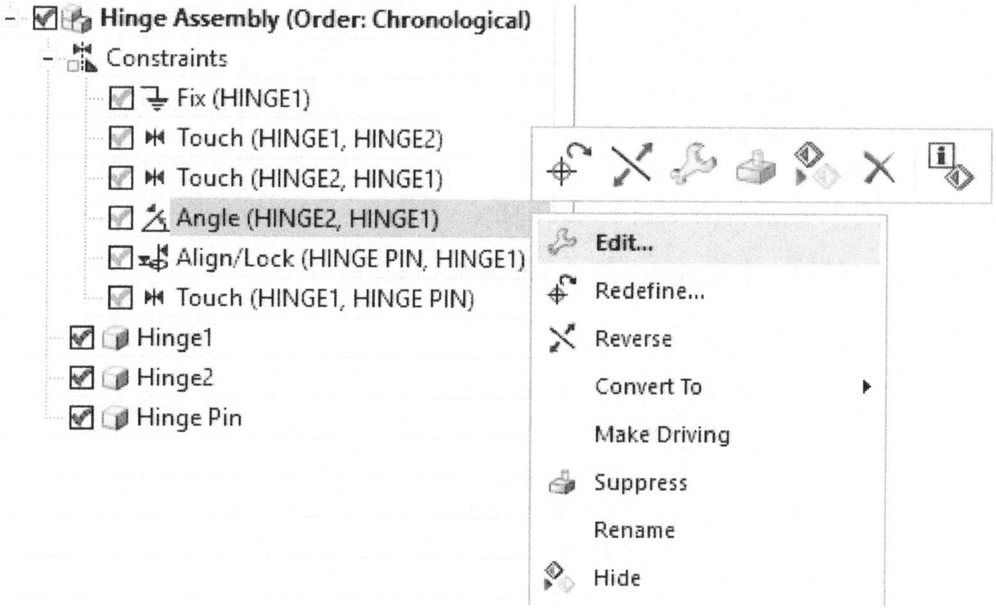

As within the Part Navigator, green check marks indicate that constraints are active. You can suppress an **Assembly Constraint** by unchecking the box next to the constraint, as shown below. The system remembers the suppressed constraint, so you can restore it by simply checking the box again.

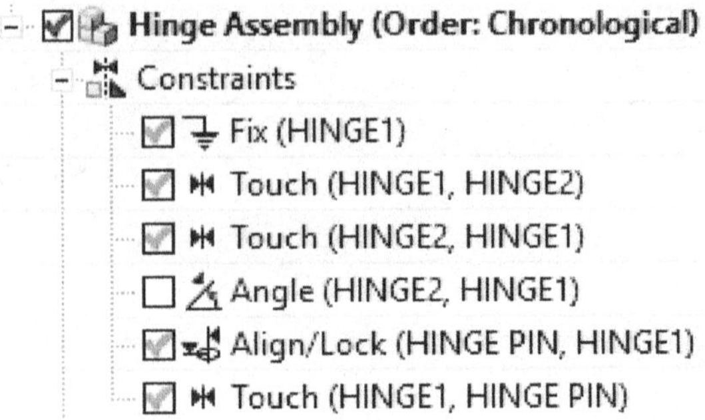

If you are certain that you want to completely remove an **Assembly Constraint**, you may delete it by **right-clicking** on the constraint and selecting **Delete**.

11.5.2 The Constraint Navigator

Assembly constraints are listed within the Assembly Navigator chronologically by default – which can make it hard to find the constraint you need to modify when your assembly gets large. The *Constraint Navigator*, which is the tab between the Assembly Navigator and Part Navigator on the resource bar, makes it easier to find the constraints you may want to modify.

One real advantage the Constraint Navigator has is the ability to group constraints. For example, you can group constraints according to the components they relate.

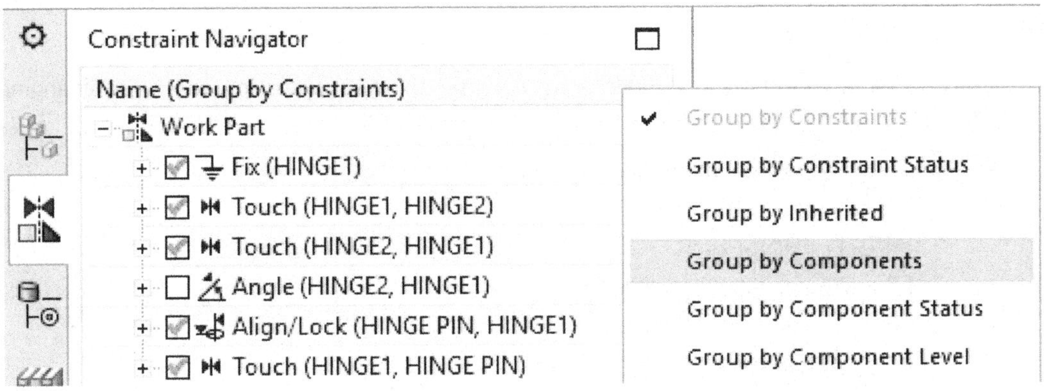

You can then expand each component to see the constraints attached to it.

11.6 Move Component

You have seen the **Move Component** command briefly in Chapter 11.2 while using **Add Component**, but it merits a discussion of its own. **Move Component** is found in the **Component Position** group, on the **Assemblies** tab.

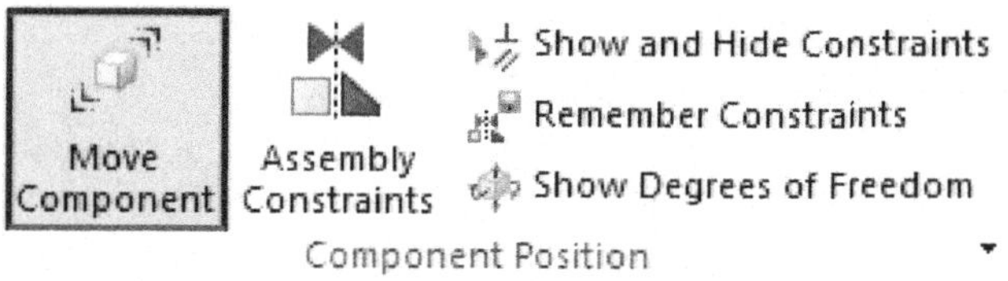

Move Component repositions components within their parent assembly, subject to any existing assembly constraints, but does not remove degrees of freedom. Once the component is positioned, it can still be repositioned as freely as the existing constraints will allow.

- Continue working with *Hinge Assembly* as the Work Part, and select the **Move Component** command.
- The default **Motion** type in **Move Component** is *Dynamic*. The motion is controlled dynamically by a yellow-and-gray coordinate system. Select the spherical handle shown below to rotate *Hinge2* about its axis.

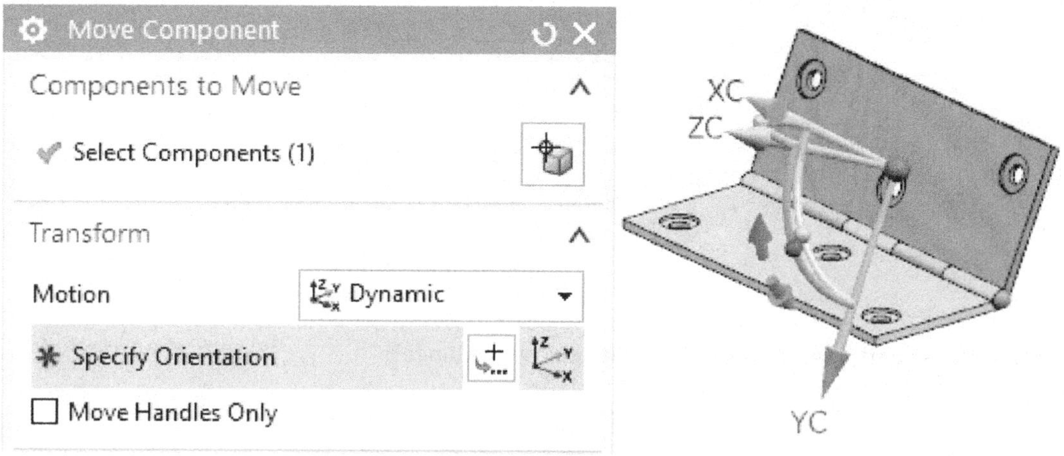

- Rotate *Hinge2* until the hinge is totally open. Remember the *Angle* constraint you imposed in the last exercise – it will keep you from opening the hinge more than **180 degrees**!

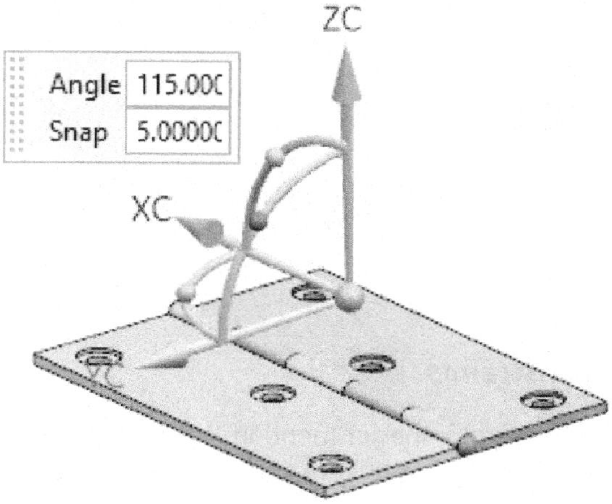

- Click **Apply** to save the motion.
- The other motion types are fairly self-explanatory. Let's get some practice with the **Motion** type set to *Angle*!
- Select *Hinge2* as the **Component to Move**, and change the **Motion** to *Angle*. To specify the axis of rotation, select the middle cylinder from *Hinge1*, as shown below.

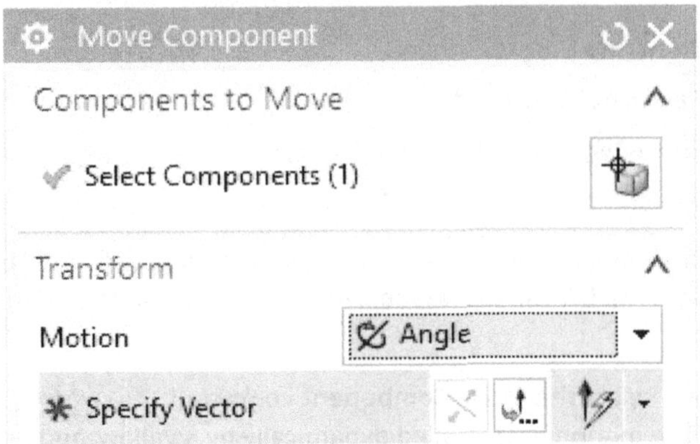

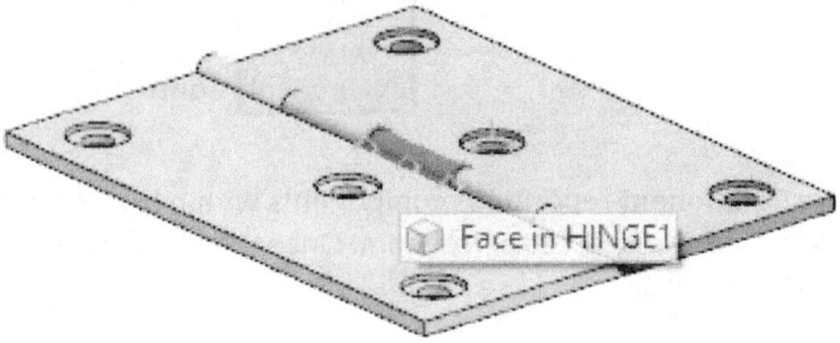

- Use the **Reverse Direction** button to change the orientation of the rotation axis. Rotation is counterclockwise about the axis when the direction vector is pointing toward you.

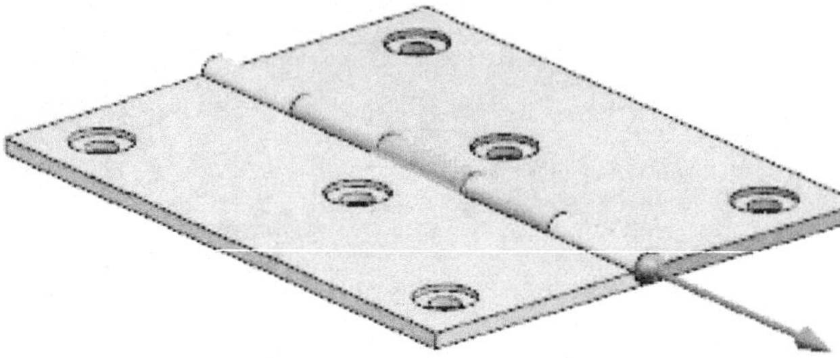

- Now you can enter a value of **45 degrees**, to partially close the hinge!

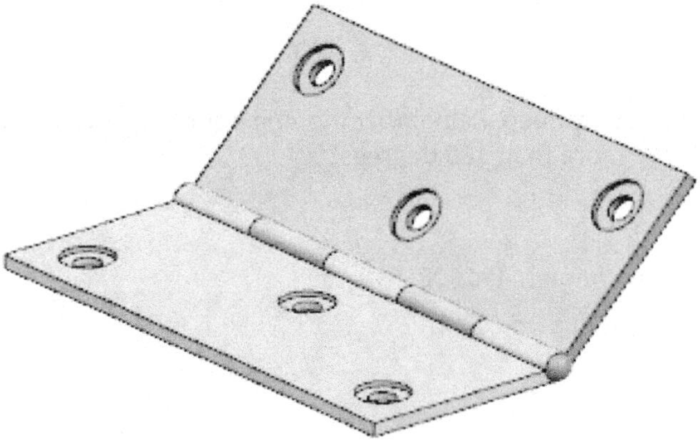

- Click **Apply** to save the motion.

11.6.1 Move Component – By Constraints

The *By Constraints* movement method deserves special mention. When you move a component *By Constraints*, the resulting motion positions the component as though it were subject to the specified assembly constraint, but does not actually remove degrees of freedom.

- In the **Move Component** dialog, set the **Motion** to *By Constraints*. Note that it is not necessary to specify any **Components to Move** prior to selecting this motion type.

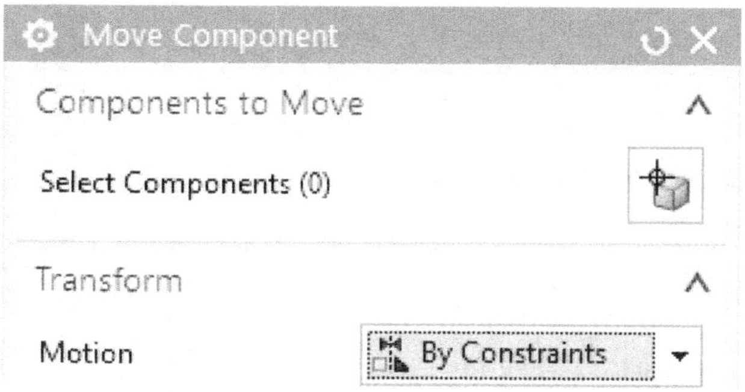

- When performing **Move Component** – *By Constraints*, if no component is selected, your first click must come from the component that should move. Make sure the **Constraint Type** is set to *Touch Align – Prefer Touch*, and first click on the inside face of *Hinge2*. The second face to be selected is the inside face of *Hinge1*, as shown below.

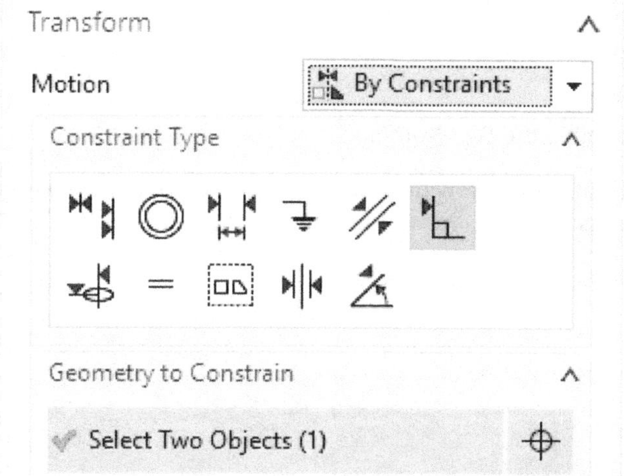

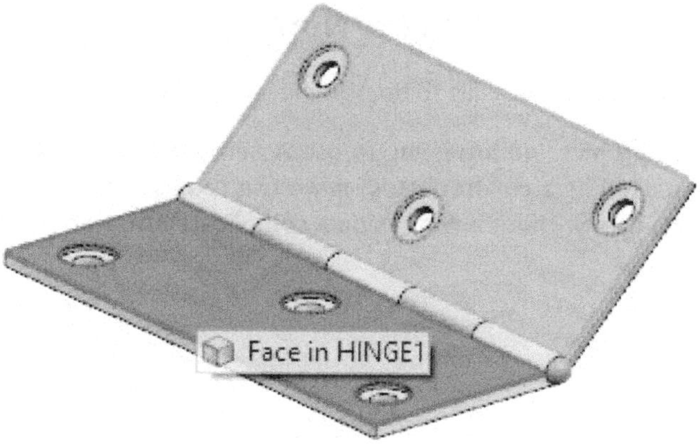

- The resulting motion temporarily places a *Perpendicular* constraint in the list of constraints in the Assembly Navigator.

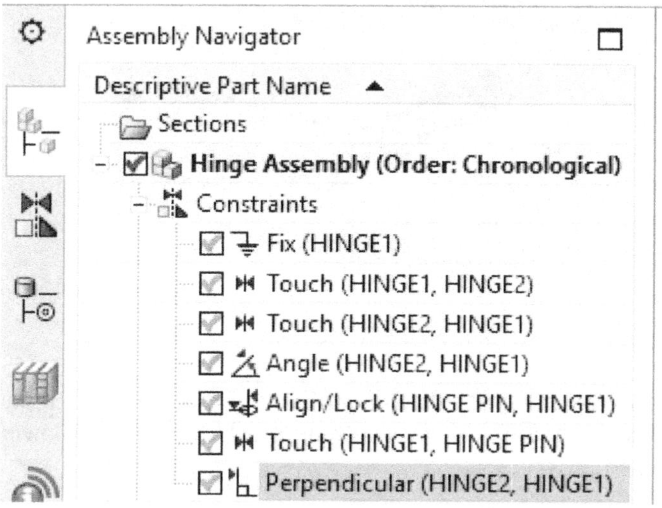

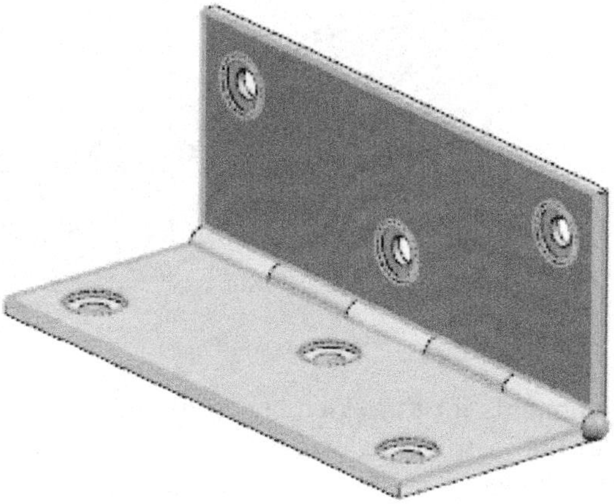

- As soon as you click **OK** or **Apply**, the *Perpendicular* constraint vanishes. If you apply a motion by multiple constraints, you will see them added to the list here before vanishing when you click **OK** to save the motion.

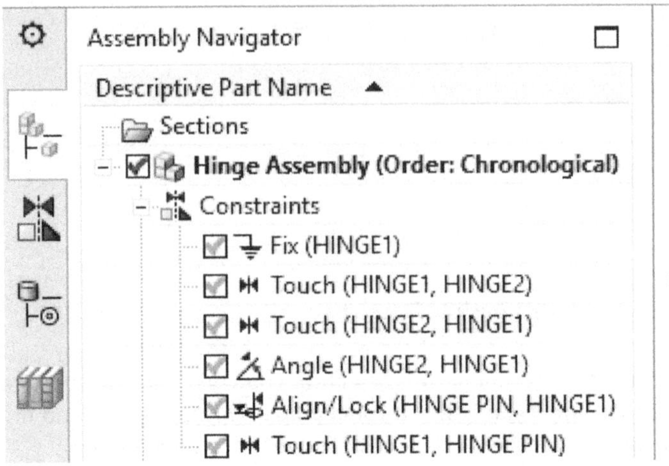

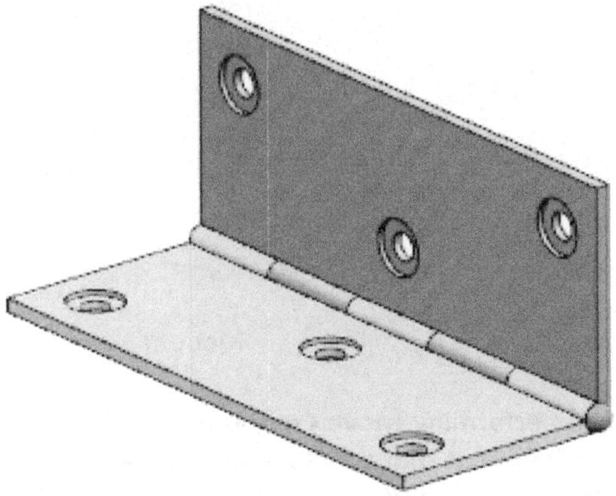

- Save your *Hinge Assembly*!

11.7 **Project – The Do-Nothing**

Now you have the skills to do nothing! No really; this next toy is called the *"Do-Nothing."*.

In this project, we are going to use **Assembly Constraints** to create the following assembly. This is a small mechanism, often used as a children's toy. If we use the assembly constraints properly, we will be able to show the motion of the mechanism without having to do a complicated motion simulation.

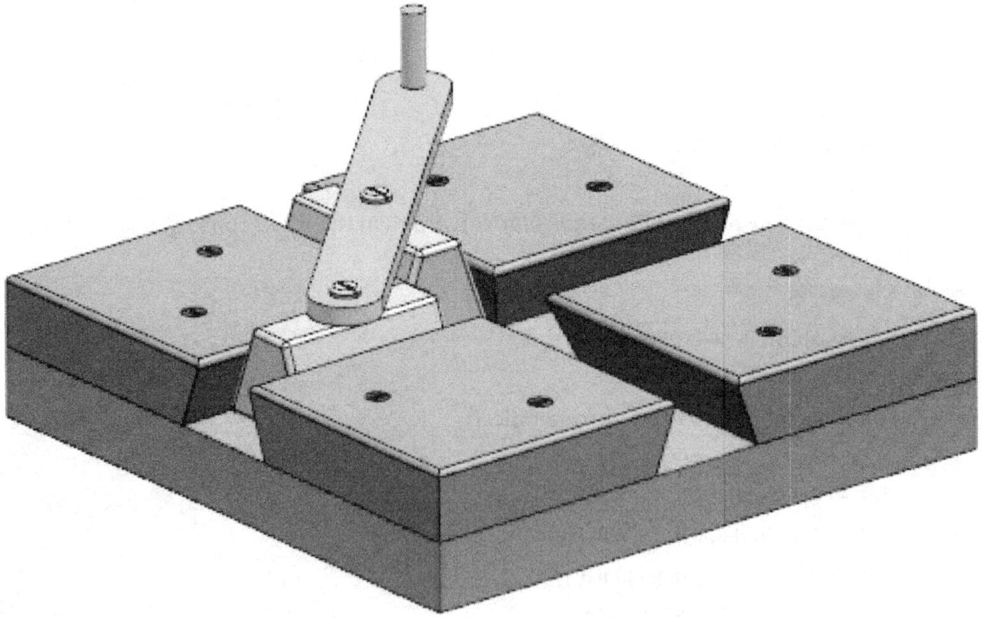

The components for this assembly are found in their own folder called *"Do-Nothing"* within the downloaded part files folder. In the instructions below, you should be able to find the named files within *"C:\My NX 11 Files\Downloaded Files\Do-Nothing"*.

- Create a new part file called *"Do-Nothing.prt"* using the *Assembly* template, and place it in *"C:\My NX 11 Files"*.
- When the **Add Component** dialog appears, find *"Base.prt"*, and set the **Positioning** to *Absolute Origin*. Click **OK** to add *Base* to your assembly.
- As we begin adding additional components to the assembly and positioning them with constraints, we will want them to become fully constrained, so at this time, use **Assembly Constraints** to apply a *Fix* constraint to *Base*.

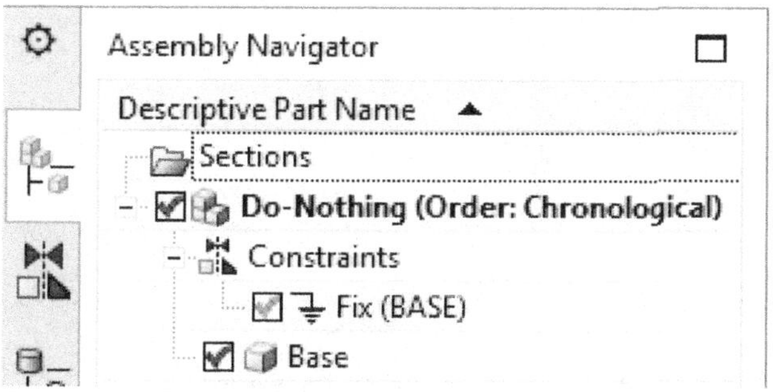

- Now select **Add Component** and select *"Body Part.prt"*. Set **Positioning** to *Move*. Click **OK**.
- The **Point** dialog will appear. This is to give the component an initial position, from which it can be moved. Simply click **OK** and the component will be positioned at the absolute origin of the work part.

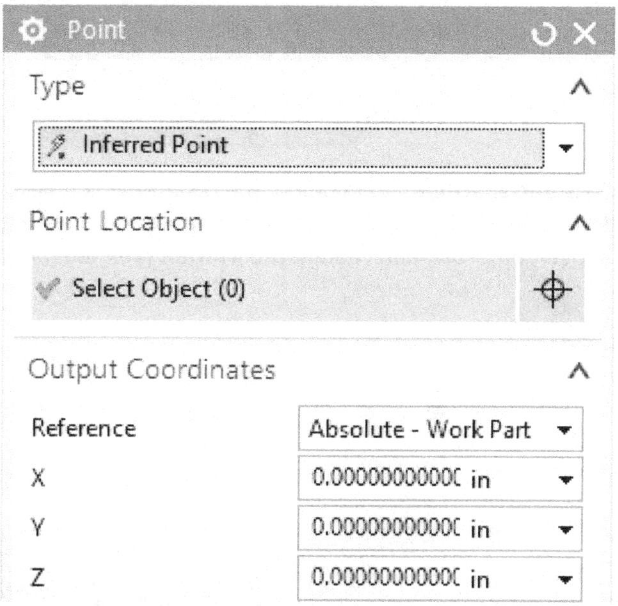

- The **Move Component** dialog will appear, with the default **Motion** of *Dynamic*. Click-and-hold-and-drag the **ZC** arrow of the dynamic coordinate system to a distance of *0.75 in* (this is the overall height of the block in *"Base.prt"*). Click **OK** to finish the move!

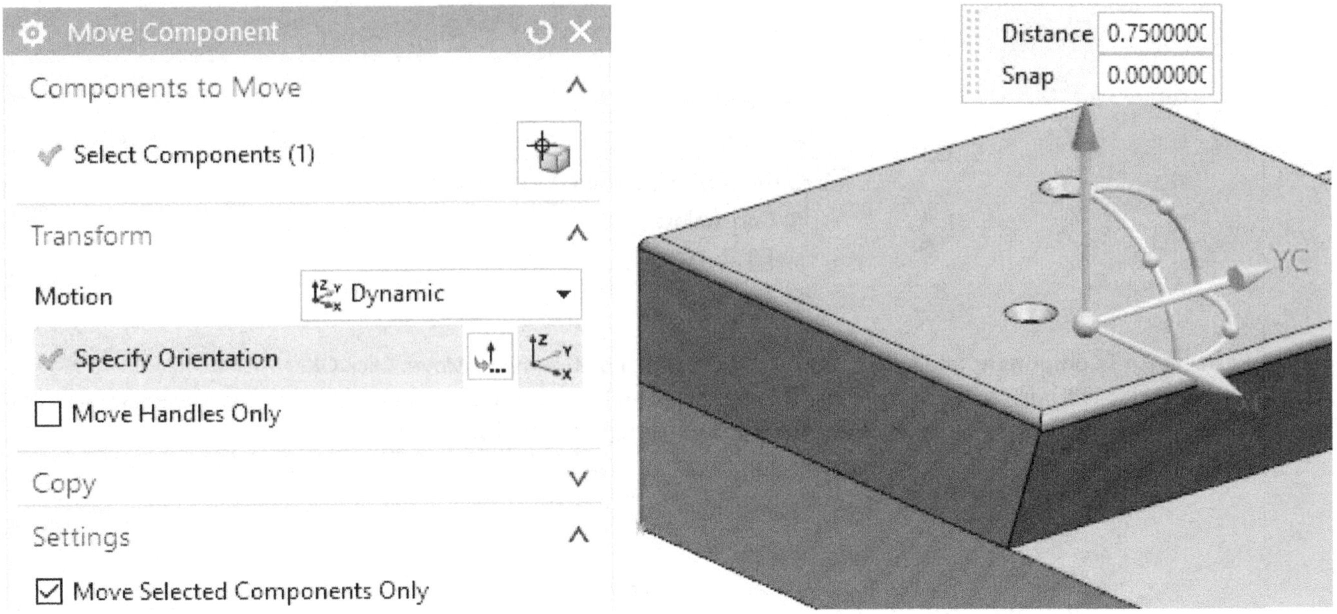

- Next you will add a copy of *Wood Screw* to the assembly, positioned in one of the countersunk holes of *Body Part*. Use **Add Component** with **Positioning** again set to *Move* to bring a copy of *Wood Screw* into the assembly.
- Use *Dynamic* motion to reposition the screw above *Body Part*, and change the **Motion** type to *Point to Point*.

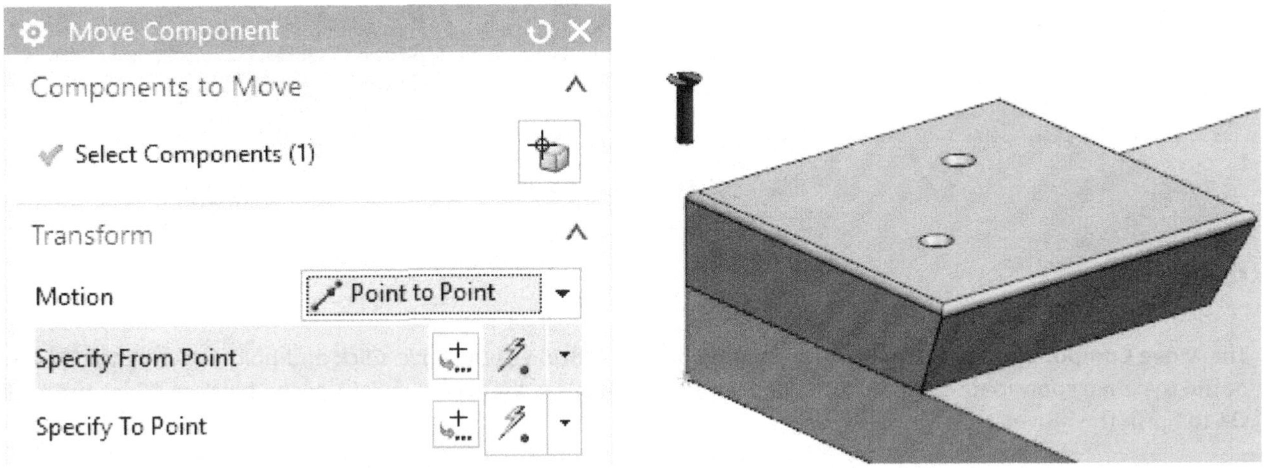

- For the *From Point* (the starting point for the motion), select the arc center of the top edge of *Wood Screw*.

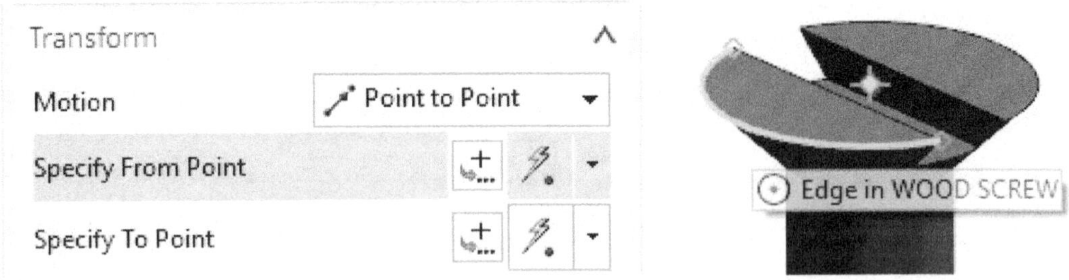

- For the *To Point* (the end point for the motion), select the arc center of the top edge of the countersunk hole. Click **OK** to finish the motion!

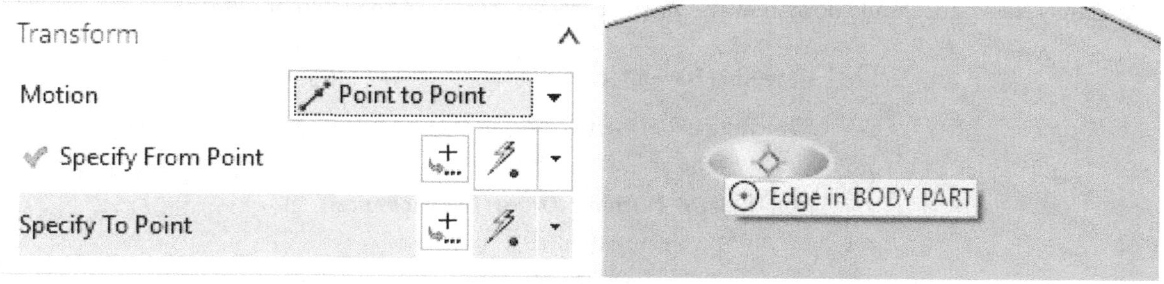

- Use **Add Component** again to add and then position a second copy of *Wood Screw* in the other hole.

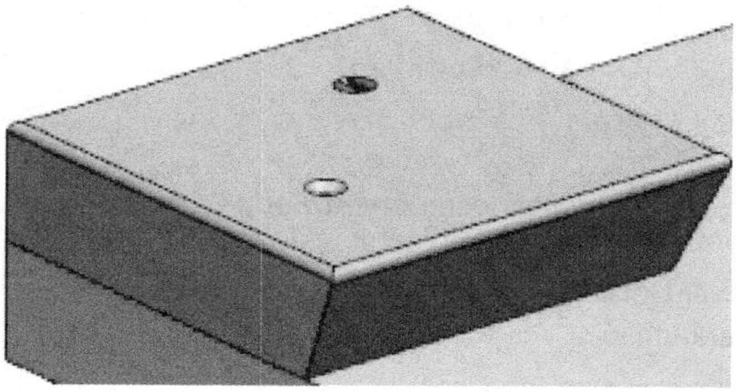

- **Save** your assembly.

11.8 Subassemblies

When building an assembly of any complexity at all, you will more than likely find yourself wanting to group components together into subassemblies. Creating a subassembly in-context is no different than creating a component in-context, and adding an assembly to another assembly is no different than adding any other component to an assembly.

Degrees of freedom for components within subassemblies exhibit different behavior, however. An important point about **Assembly Constraints** is that they only affect degrees of freedom within the immediate parent assembly. This means that if you move components into a subassembly, any constraints involving those components will break! This is why we positioned *Body Part* and *Wood Screw* within *Do-Nothing* using **Move Component** in the previous exercise. In the next exercise, you will create a subassembly called *Body Part Assembly*, and bring both *Body Part* and *Wood Screw* into it.

- Continue working with *"Do-Nothing.prt"*.
- Use **Assembly Constraints** with **Constraint Type** set to *Touch Align* and **Orientation** set to *Align* to add a constraint between the aligned faces of *Body Part* and *Base*, as shown below.

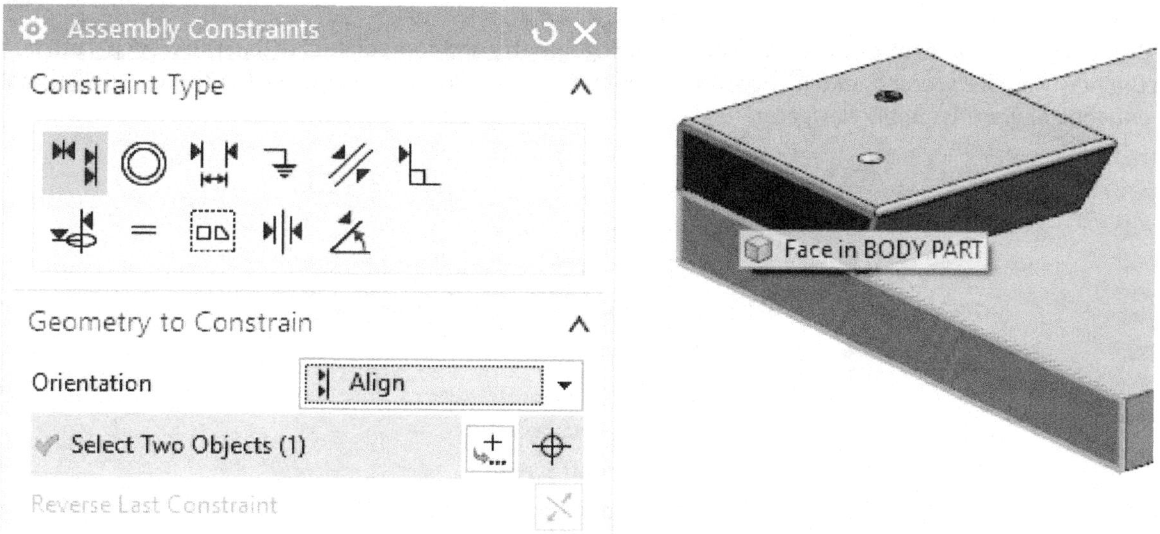

- Your Assembly Navigator should now show the *Align* constraint between *Body Part* and *Base*.

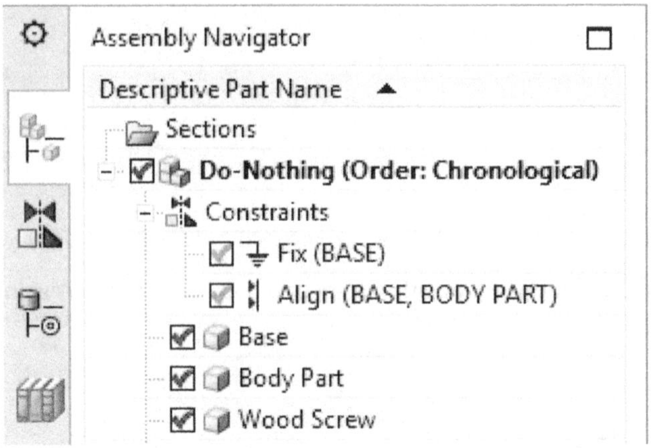

- Select **Create New Component**, name the new component *Body Part Assembly* and place it in *"C:\My NX 11 Files"*. Click **OK**.
- When the **Create New Component** dialog appears, select *Body Part* and *Wood Screw*. Make sure that the **Delete Original Objects** checkbox is marked!

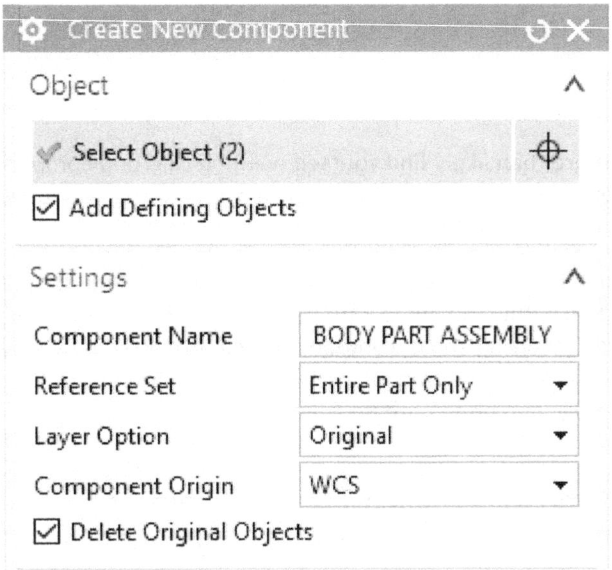

- The **Create New Component** dialog allows you to select geometry and components <u>within the work part</u> to be brought into the newly created component. If you select components, your new component will become a subassembly! Click **OK**, and look at your Assembly Navigator!

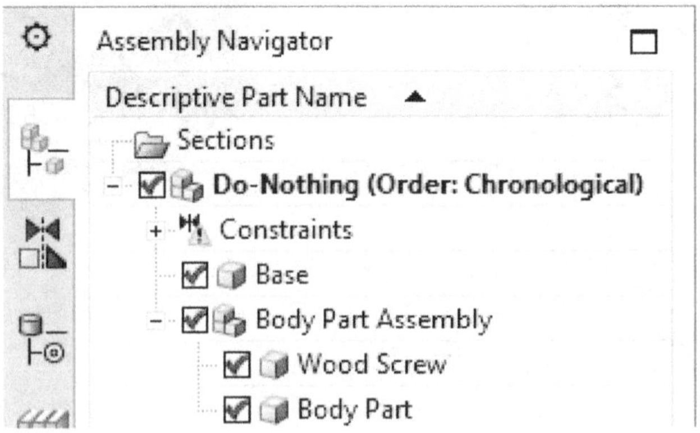

- The most obvious change is that the assembly tree has now gained a new level, and *Wood Screw* and *Body Part* are now components within *Body Part Assembly*, rather than *Do-Nothing*. The subtler change is to the assembly constraints. The *Align* constraint between *Base* and *Body Part* has lost its reference to *Body Part*!

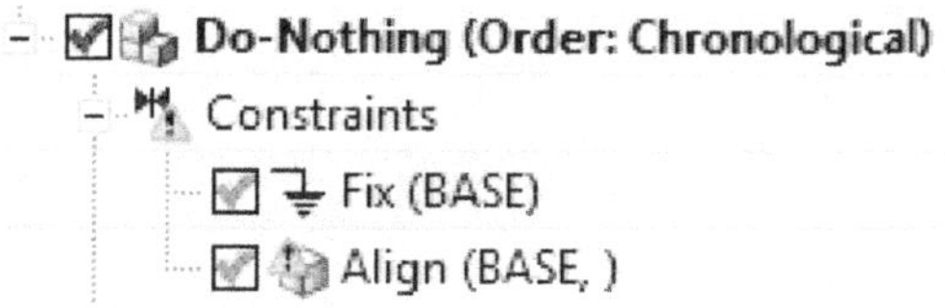

- **Assembly Constraints** can only make reference to top-level components within the Work Part, and not to components within subassemblies. Let's repair the broken constraint! Right-click on *Align*, and choose **Redefine**.

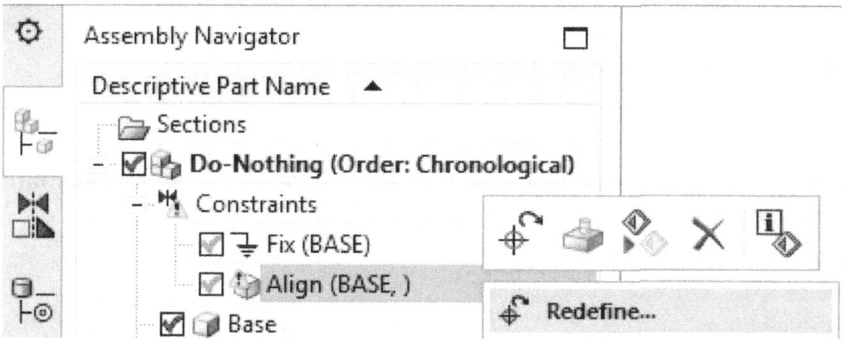

- The constraint still makes use of the face of *Base*, so all you have to do is reselect the face from *Body Part*.

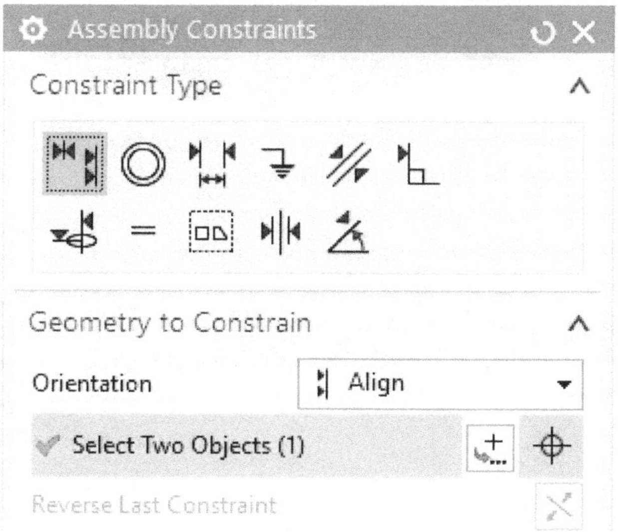

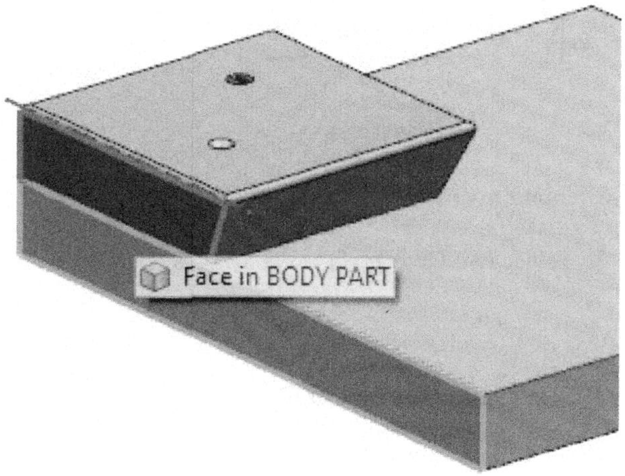

- This repairs the constraint.

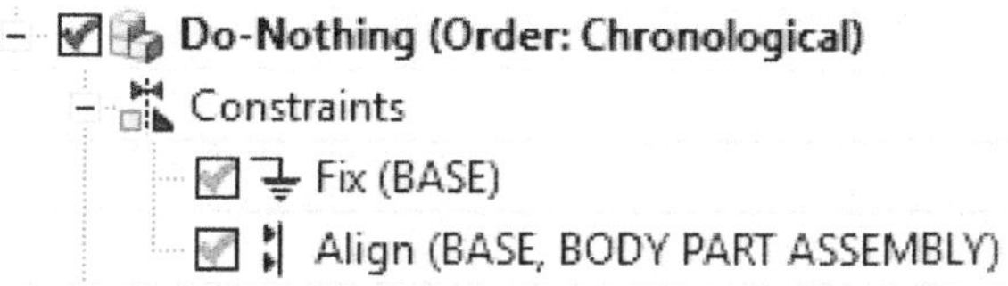

- Next, rotate the model and apply a similar *Align* constraint between the other outside faces of *Body Part* and *Base*, respectively.

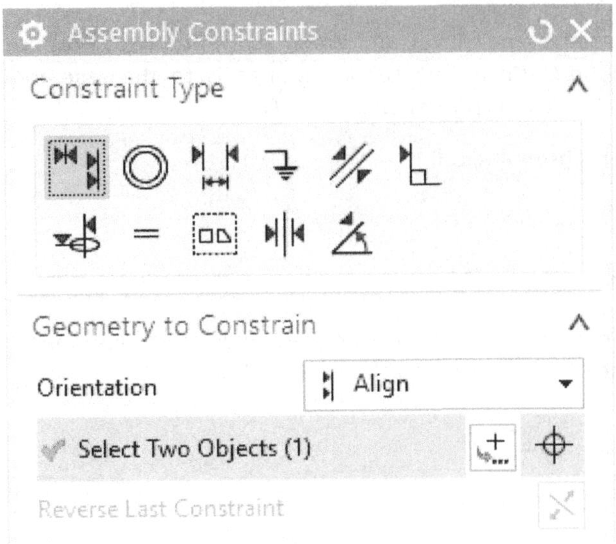

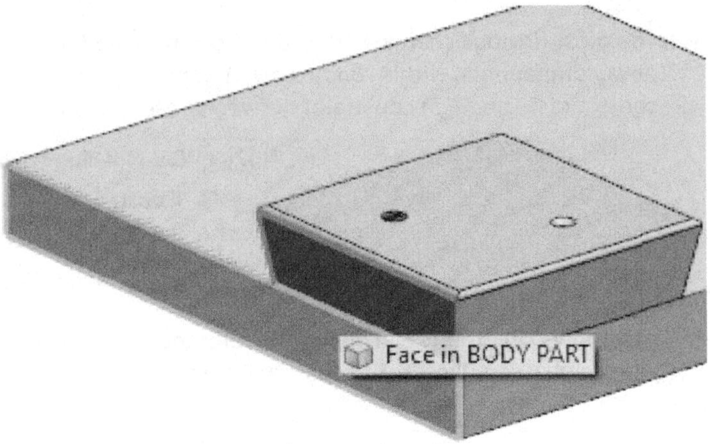

- Drag *Body Part Assembly* upward so that you can see the bottom face. Change the **Orientation** to *Touch*, and then choose the bottom face of *Body Part.*

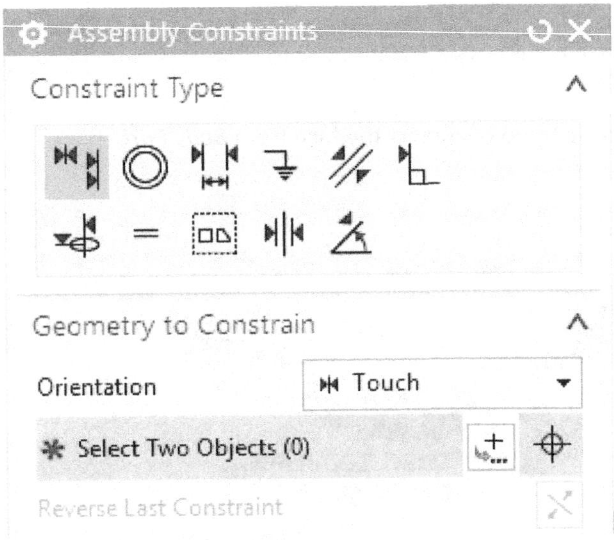

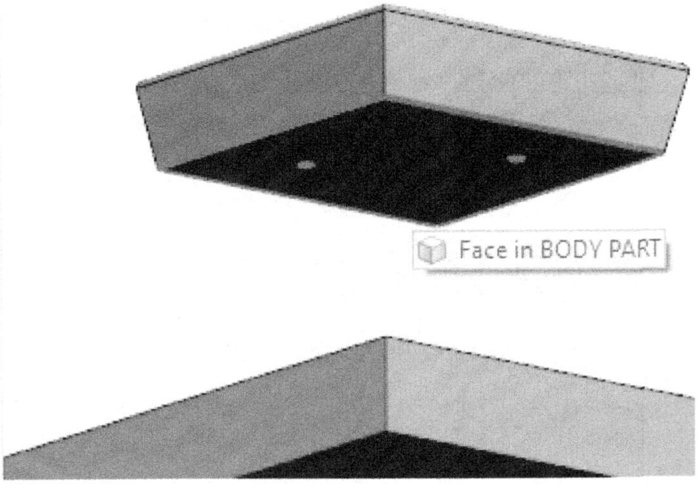

- Now select the top face of *Base*.

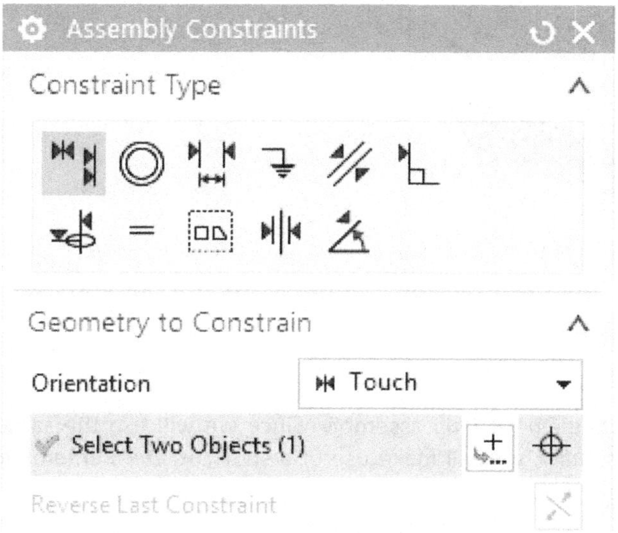

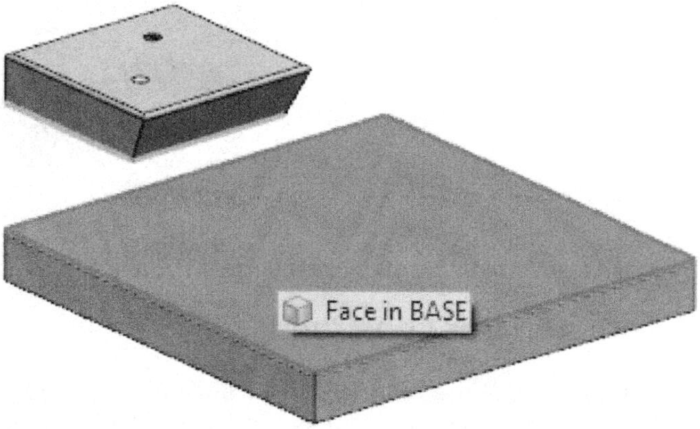

- Click **OK** to save all your constraints. The end result of these constraints is that *Body Part Assembly* is fully constrained. To confirm this, you can use **Show Degrees of Freedom** and select *Body Part Assembly* from the Assembly Navigator. The Status Display at the bottom center of the Graphics Window should appear as shown below.

There are no degrees of freedom.

- You have to be careful to distinguish between *Body Part Assembly* and any of its components when using **Show Degrees of Freedom**. If you select *Body Part* and then click **Show Degrees of Freedom**, NX will tell you that all degrees of freedom remain!

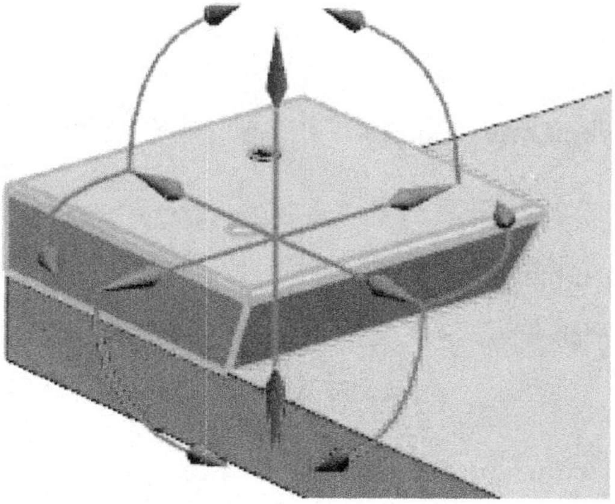

There are 3 rotational and 3 translational degrees of freedom.

- To make the degree-of-freedom indicators disappear, you can either use **Refresh**, or **Fit**.
- Next, we will add the second *Wood Screw* to *Body Part Assembly*. Make *Body Part Assembly* the Work Part. **Add Component** will add the specified component to the Work Part, so this is really necessary!
- Select the **Add Component** command. You can click on the copy of *Wood Screw* already in your assembly. Set **Positioning** to *Move*, and use the same technique as before to position this copy of *Wood Screw* in the other countersunk hole.

- Make *Do-Nothing* the Work Part again and **Save**!

11.9 Remember Constraints

We will now add three more copies of *Body Part Assembly* to the *Do-Nothing* assembly. Since we will use the same constraints to position the additional instances of *Body Part Assembly*, we will make use of a shortcut: the **Remember Assembly Constraints** tool.

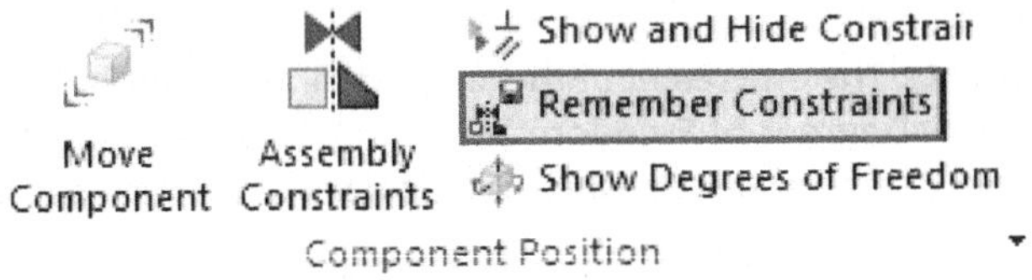

- Continue working with *"Do-Nothing.prt"*, and select the **Remember Constraints** tool.
- From the Assembly Navigator, select *Body Part Assembly*.

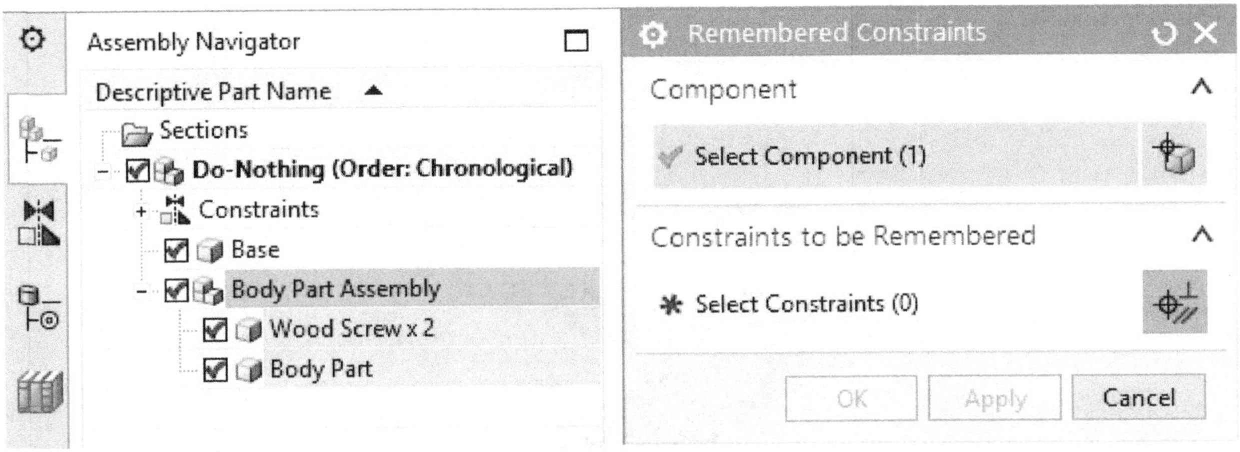

- For the **Constraints to be Remembered**, select the constraints involving *Body Part Assembly*, as shown below.

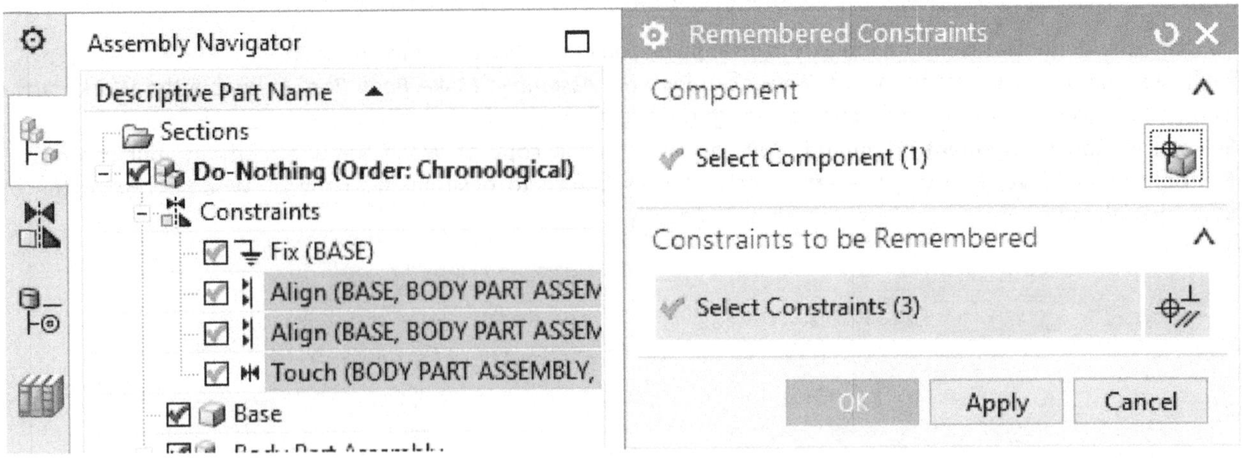

- Click **OK**. Now, when you add additional copies of *Body Part Assembly* to *Do-Nothing*, you will be asked to pick faces for each of the remembered constraints.
- Use **Add Component** and select *Body Part Assembly* from the Assembly Navigator. Set **Positioning** to *By Constraints*. When you click **OK**, you will meet the **Redefine Constraints** menu.

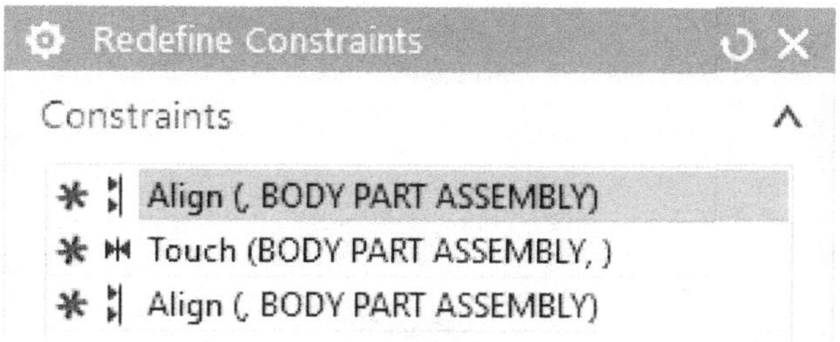

- Each constraint in the list comes with one face from *Body Part Assembly*, as indicated in the Graphics Window. One by one, you will select the correct face for the constraint so that the new instance of *Body Part Assembly* comes in as desired. For the first *Align* constraint, select the face shown below.

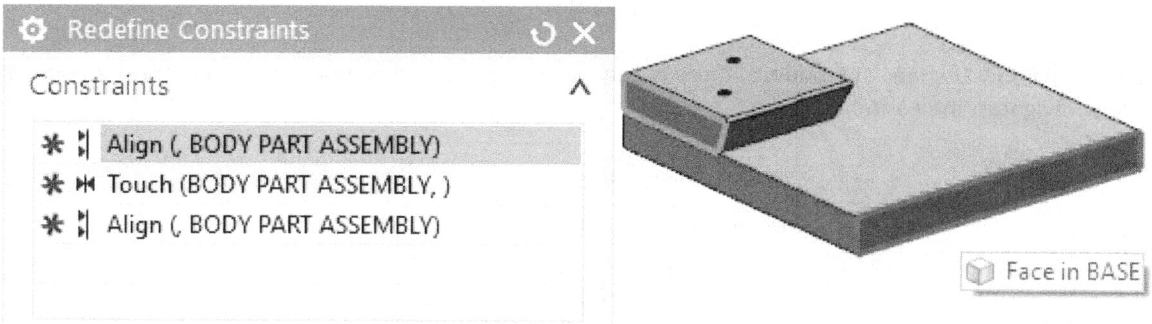

- For the *Touch* constraint, choose the top face of *Base*.

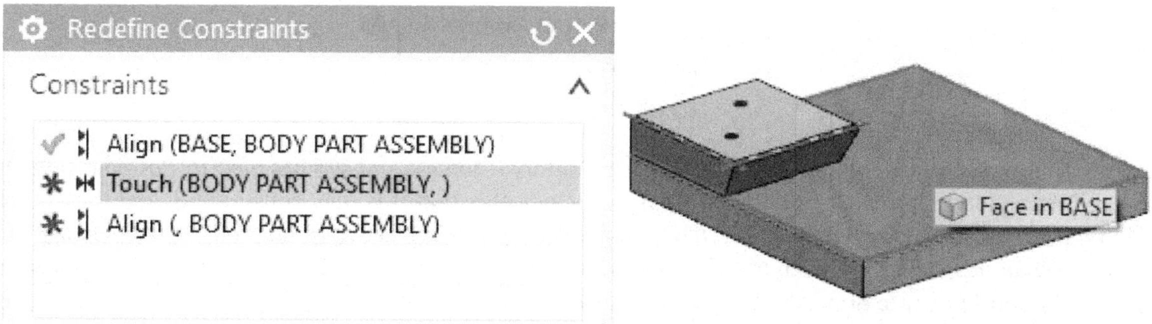

- For the second *Align*, choose the face that neighbors both the top and the previous face selected.

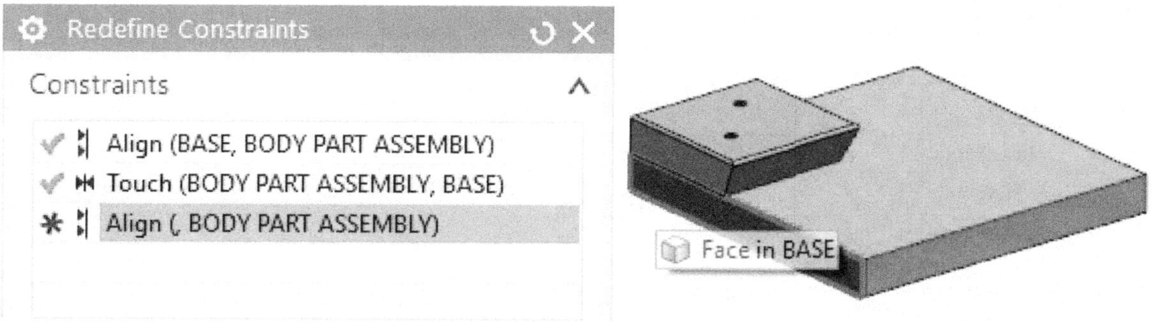

- Once all three constraints have been resolved, click **OK** and the second instance of *Body Part Assembly* will be added to *Do-Nothing*!

- Repeat the steps above for each of the other corners.

- In addition to **Show Degrees of Freedom**, there is another way to see if a component is fully constrained, within the Assembly Navigator: the *Position* column.

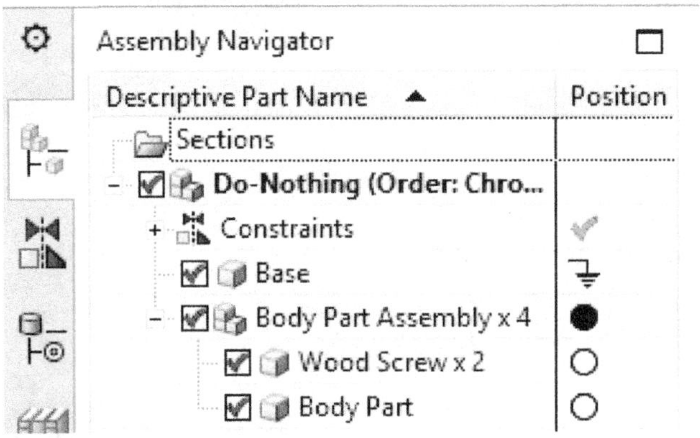

- The symbols in the position column are fairly self-explanatory – the ground symbol next to *Base* indicates that it is fixed, the fully shaded circle next to *Body Part Assembly* indicates that it is fully constrained, and the empty circles next to *Wood Screw* and *Body Part* indicate that they are totally unconstrained within *Body Part Assembly.*
- The *Position* column will make it much easier to see which components still require constraints, going forward.
- Next, use **Add Component** to bring a copy of *Slider* into your assembly. Set **Positioning** to *By Constraints* and click **OK**.
- On the **Assembly Constraints** dialog, set the **Constraint Type** to *Touch Align*, and the **Orientation** to *Touch*. From the **Component Preview** window, select the bottom face of *Slider*, as shown below.

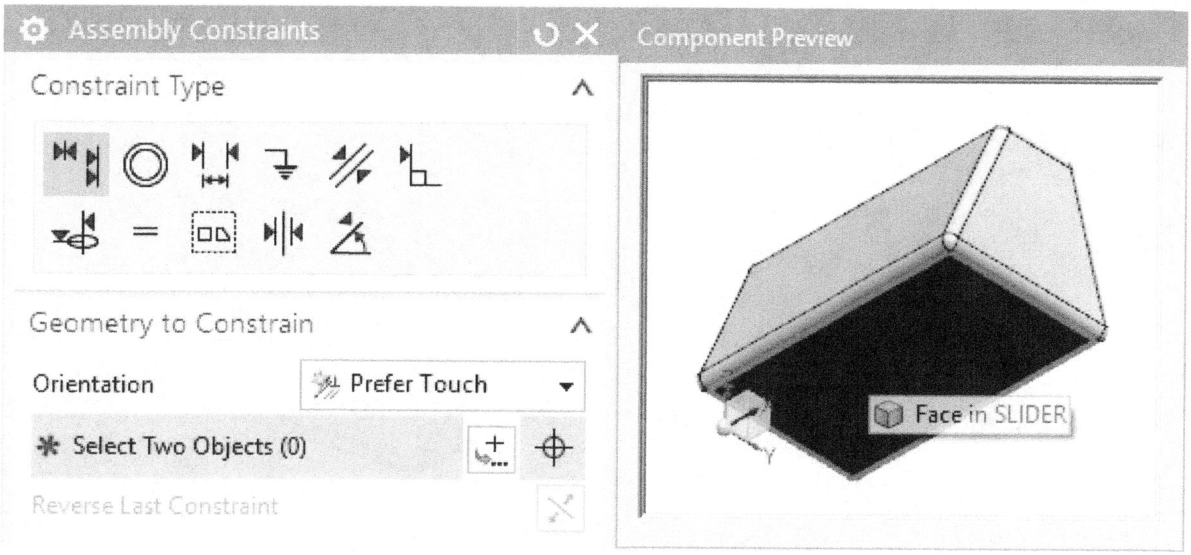

- Select the top face of *Base* and click Apply to save the *Touch* constraint.
- Next, set the **Constraint Type** to *Center*, and set the **Subtype** to *2 to 2*. You will now select two lines from *Slider* that should be centered between two faces on *Base*.
- First, select the centerline of the blend shown below.

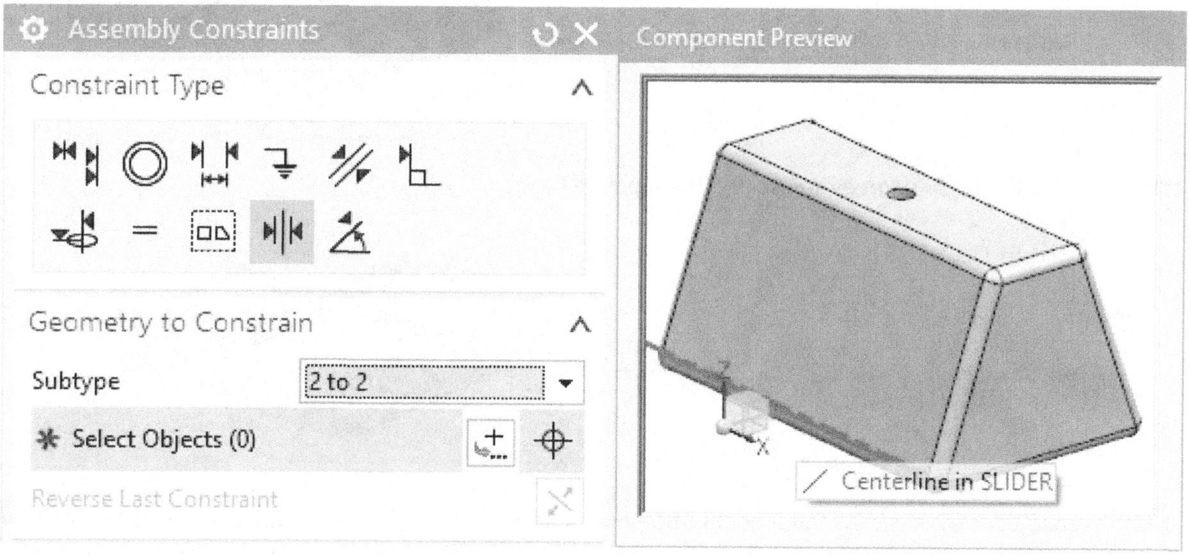

- Rotate *Slider* and select the centerline of the opposite blend.

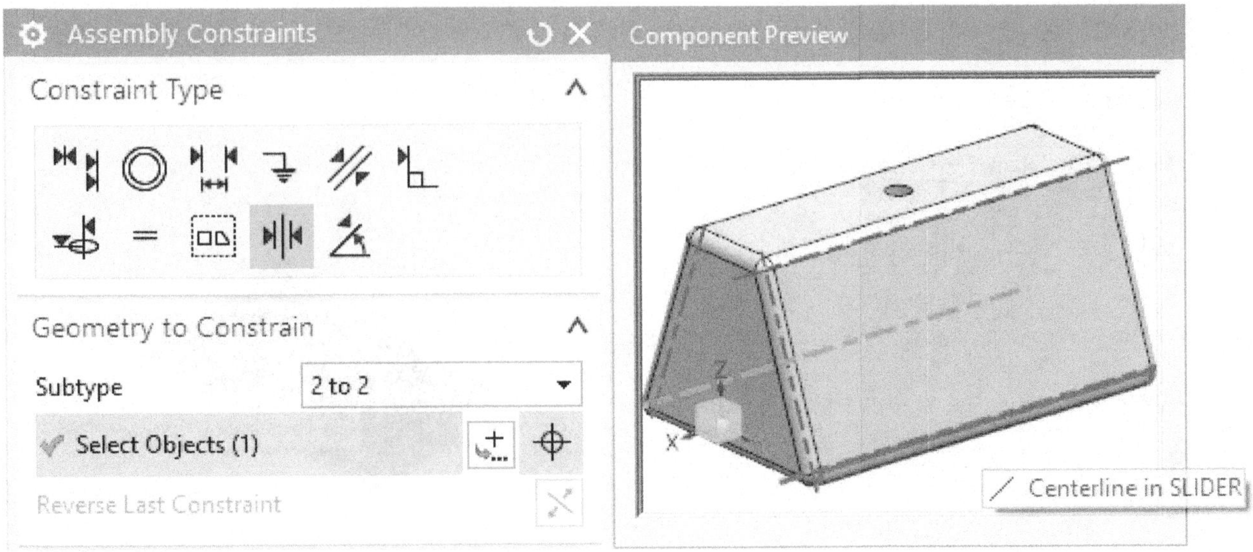

- Next, choose the side face of *Base* shown below.

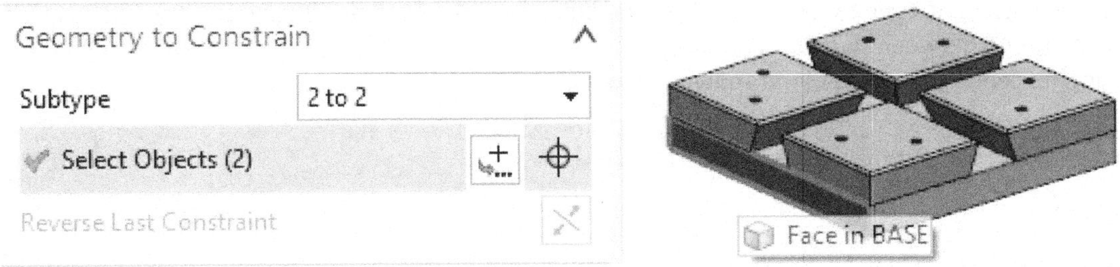

- Rotate the assembly by *90 degrees* and select the opposite face.

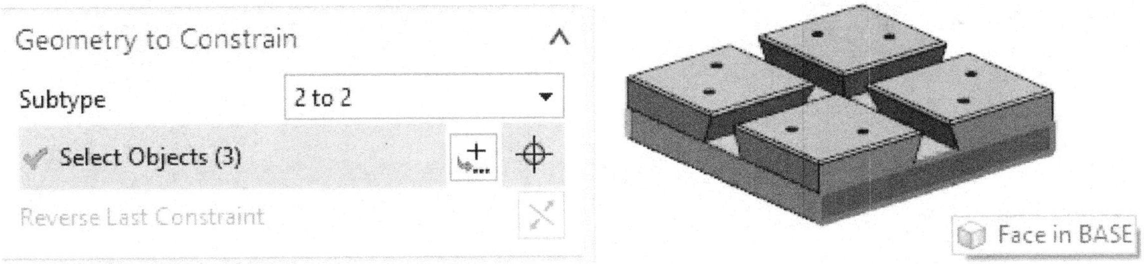

- Click **OK** to save your constraints. *Slider* should be positioned evenly between the different instances of *Body Part Assembly*, as shown below! Since the slider will move when the crank is added to the assembly, we'll leave it only partially constrained, as indicated by the partially shaded circle.

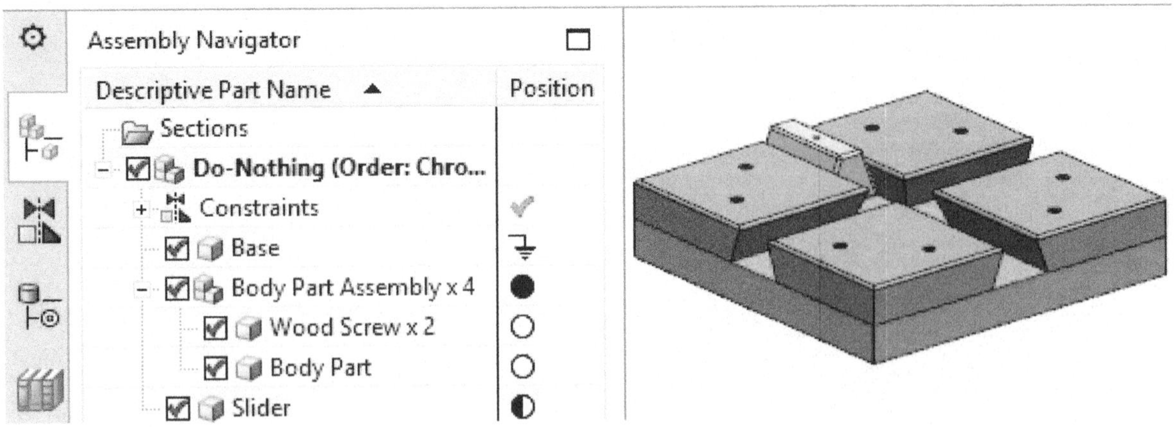

- Use the same combination of constraints to position a second copy of *Slider* between the other pair of faces on *Base*.

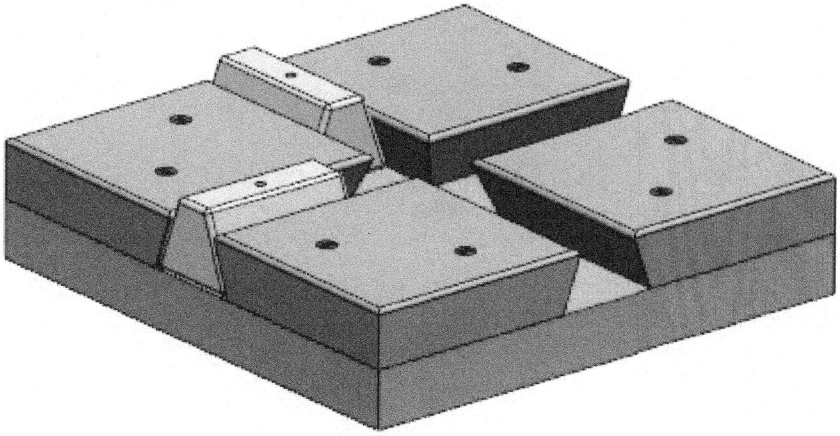

- **Save** your assembly!

11.10 **Align/Lock vs. Concentric vs. Touch Align – Infer Center/Axis**

We mentioned in Chapter 11.5 that the *Concentric* constraint does not behave as you might expect it to, if you are familiar with concentricity either from other CAD software, or the GD&T characteristic. In NX, there are three assembly constraints that can be used to align holes and other cylindrical geometry: *Align/Lock*, *Concentric*, and *Touch Align – Infer Center/Axis*. You have used both *Align/Lock*, and *Touch Align – Infer Center/Axis* in Chapter 11.5, so only *Concentric* is new in the exercise below, but we believe that it is instructive to compare the three side-by-side.

As we wrap up the *Do-Nothing* assembly, you will use each of these three constraints to position the crank and its screws. Each has its own advantages and disadvantages, and with a thorough grasp of them all, you can choose the best one for the task at hand.

- Continue working with *Do-Nothing*, and use **Add Component** to bring a copy of *Crank* into the assembly with **Positioning** set to *By Constraints*.
- Select the *Concentric* constraint. Choose the bottom edge of one of the two holes on *Crank*, as shown below.

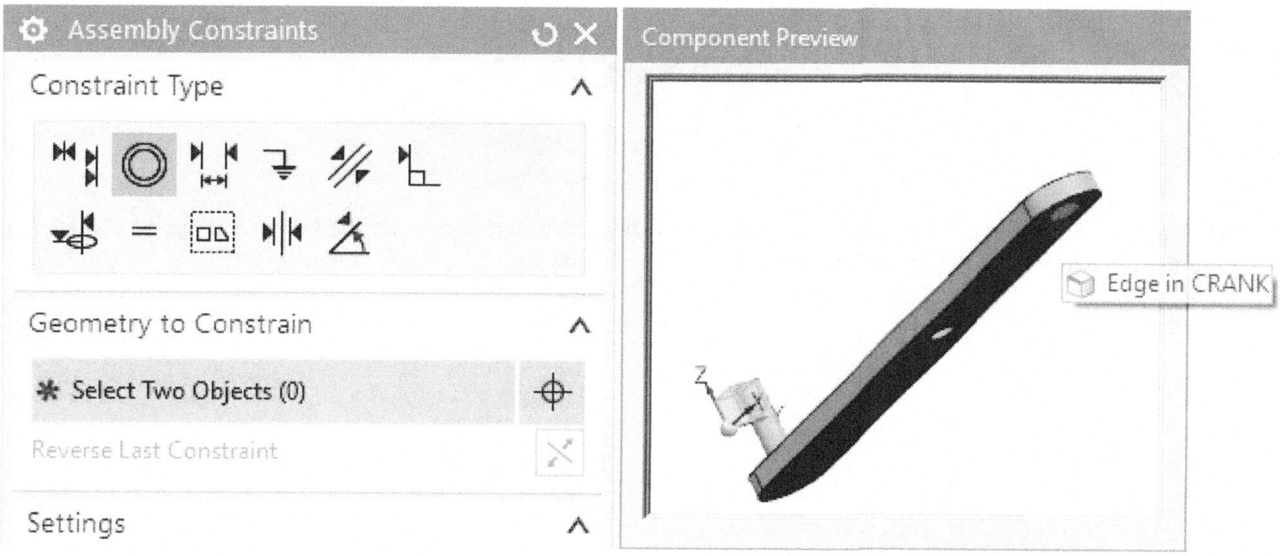

- For the second selection, choose the edge of the hole on the top face of one of the sliders, as shown below.

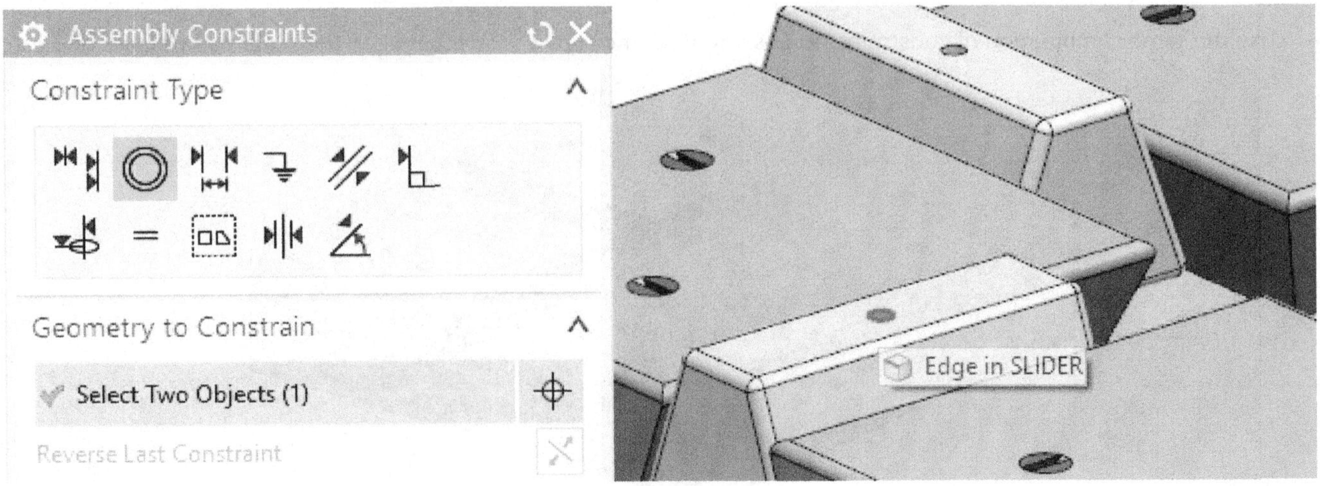

- The *Concentric* constraint forces the two circles to become coplanar, and then forces their arc centers to coincide. Click **OK** to save the constraint, and use **Show Degrees of Freedom** to see how *Crank* can move.

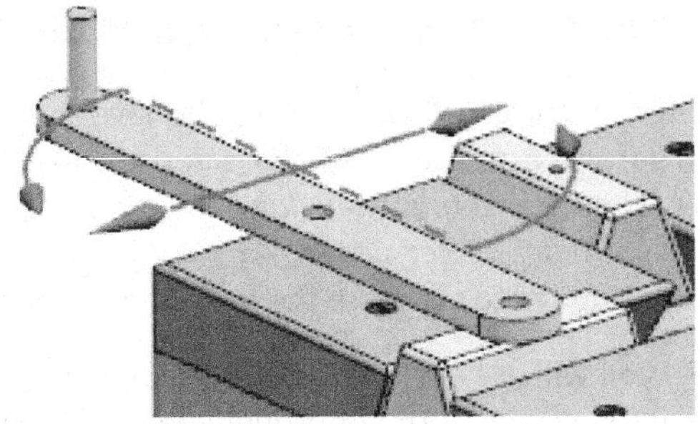

- Use a second *Concentric* constraint to align the other hole on *Crank* with the hole on the other *Slider*. After this, the only degree of freedom remaining is rotation about the centerline of the middle hole on *Crank*.

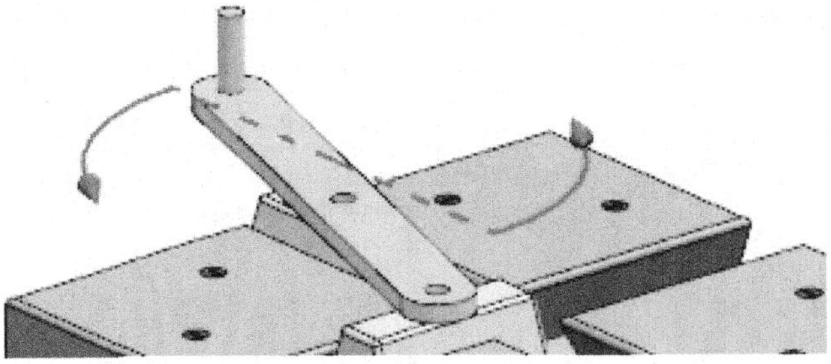

- Next, use **Add Component** *By Constraints* to bring in another *Crank Screw*, and set the **Constraint Type** to *Touch Align – Infer Center/Axis*. Select the centerline in *Crank Screw* shown below.

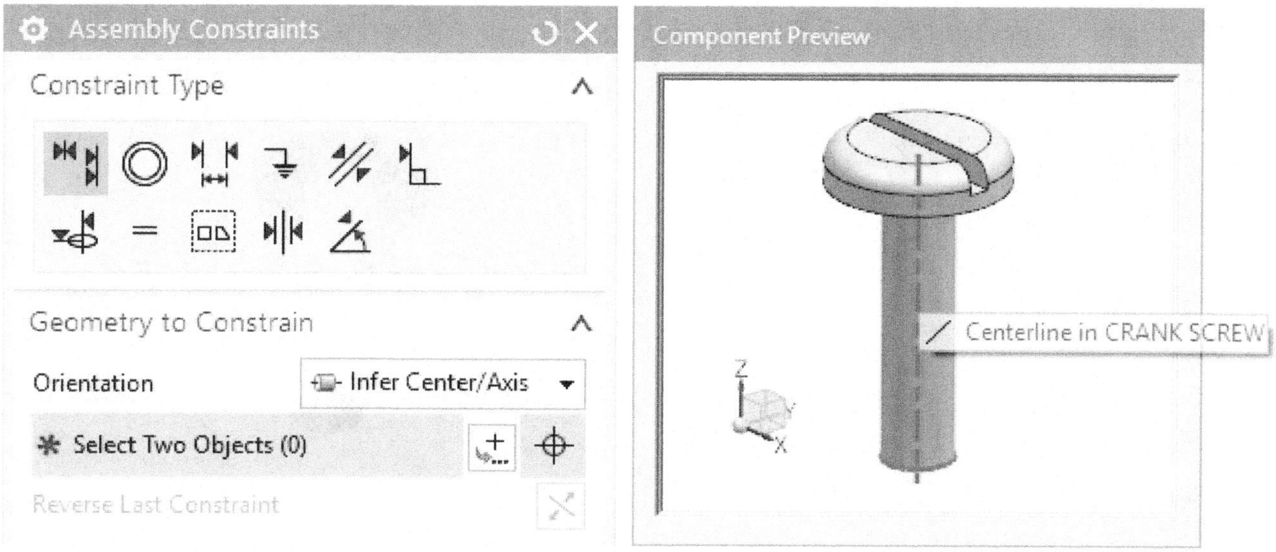

- Choose the centerline of the hole in the other copy of *Slider* – be sure that you are selecting the centerline from *Slider* and not from *Crank*!

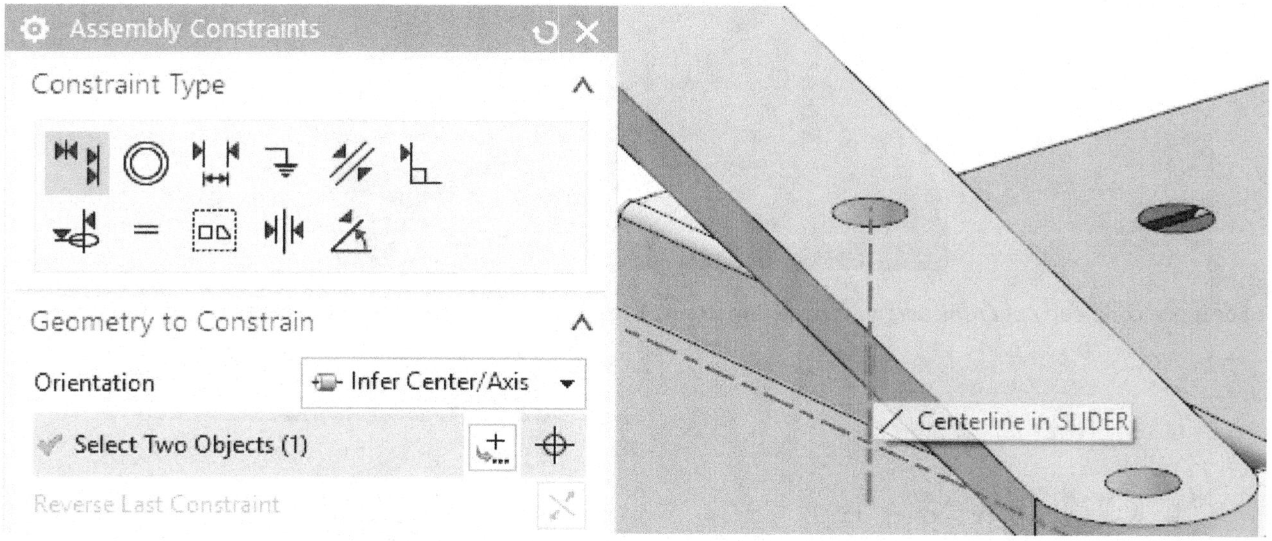

- This aligns the screw with the hole, and now it remains to press the screw into the hole. Set the **Constraint Type** to *Touch Align – Touch*, and choose the face of the *Crank Screw* shown below.

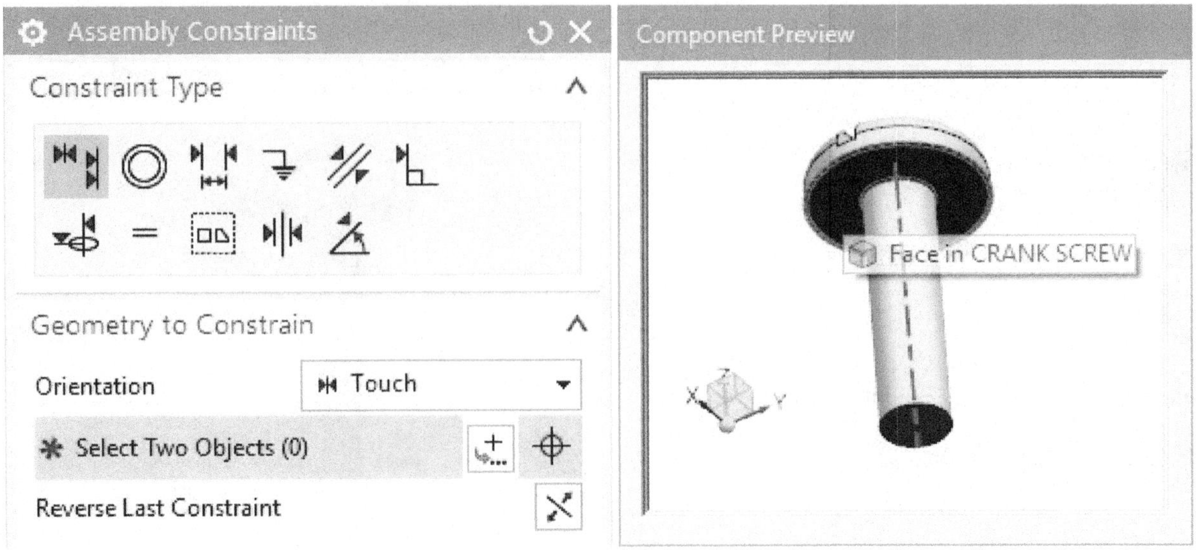

- Select the top face of *Crank* and click **OK** to save the constraints. The first copy of *Crank Screw* has rotational freedom remaining, as well as lateral motion in the same direction as the slider to which it is constrained.

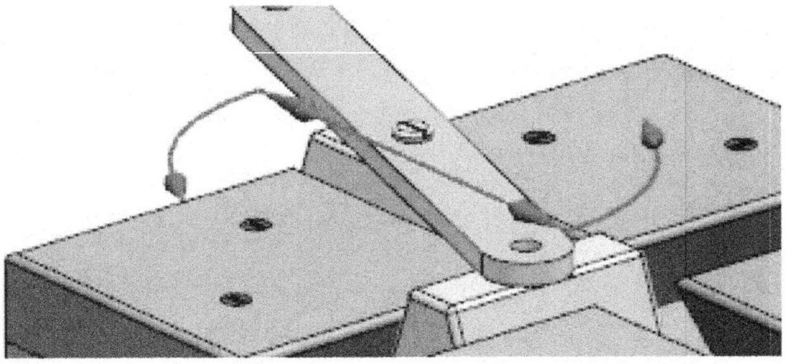

- For the second copy of *Crank Screw*, choose **Constraint Type** *Align/Lock* and choose the centerline of the screw.

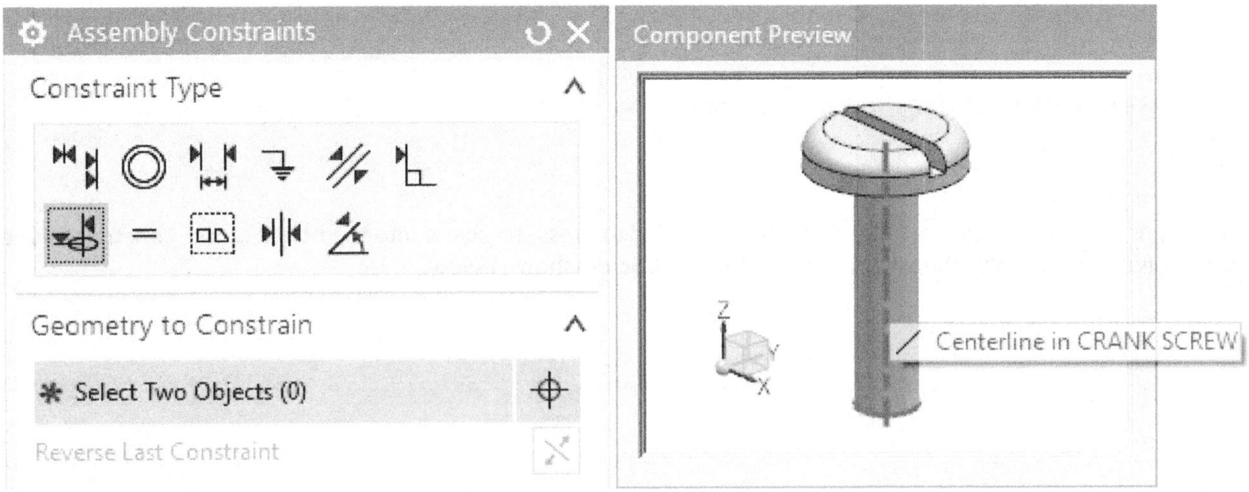

- Select the centerline of the hole from the other copy of *Slider* (again taking care not to select the centerline of the hole in *Crank*).

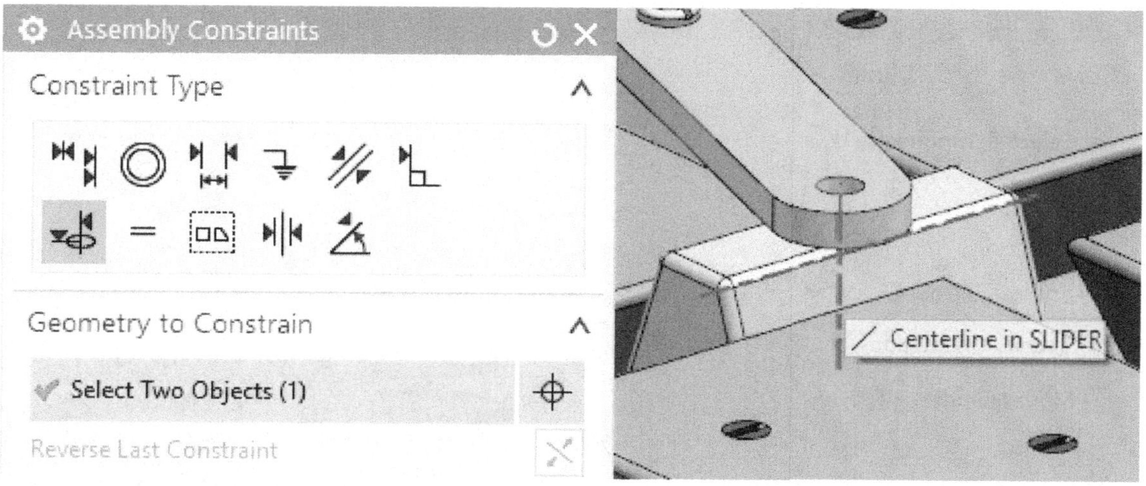

- Apply a *Touch* constraint between *Crank Screw* and the top face of *Slider.* The *Align/Lock* constraint removes the rotational freedom from the screw, so that the only degree of freedom it has is the same as that of the *Slider* it is constrained to.

- Use **Move Component** to see your assembly *do nothing*!

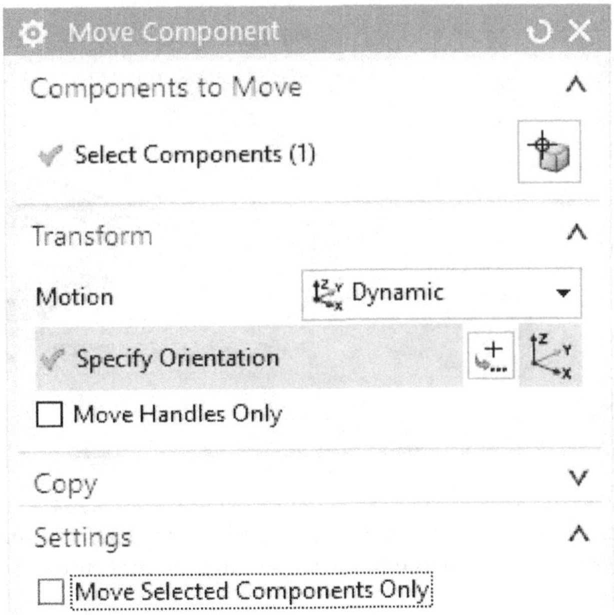

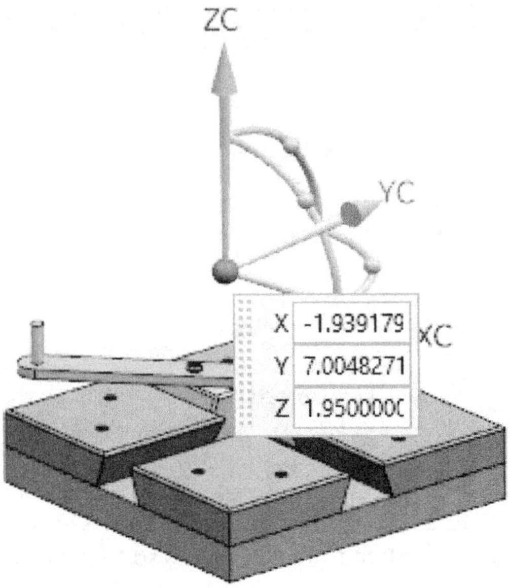

- Note that your *Crank Screw* with only the *Touch Align – Infer Center/Axis* constraint has rotated!

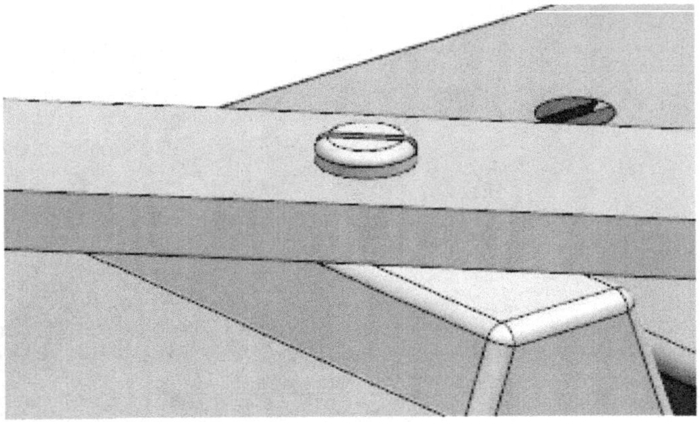

- Use a *Parallel* constraint to align the faces of the slots on the *Crank Screws*. This will keep the screw from rotating regardless of how you move the crank!

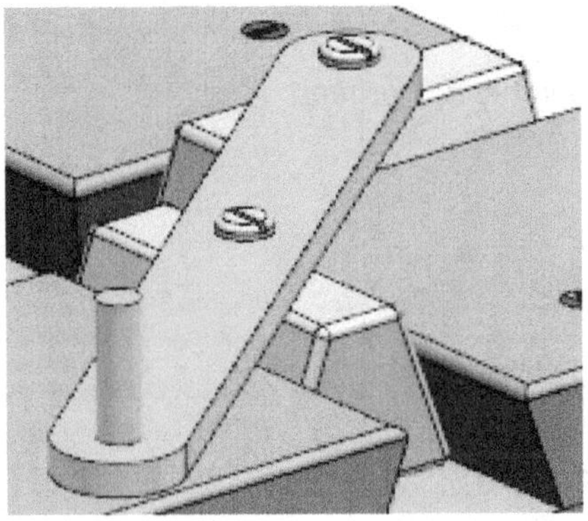

- **Save** your assembly!

11.11 Constraining a Cylinder to a Slot

Given the special attention we gave to making cylinders coaxial, we would be remiss not to discuss what to do when one of the faces you need to constrain is not cylindrical, but slot-shaped instead. There are several approaches to constraining a cylinder to a slot in NX. In the exercises below, we outline some methods for doing so, depending on your design intent. Although we use a cylinder that fits into each slot in the examples below, the same principle applies to a hole that should line up with a slot.

- Open *"Slot Assembly.prt"* from *"C:\My NX 11 Files\Downloaded Files\Slot Assembly"*.
- This assembly has five components – one copy of *"Slot.prt"*, which is a plate with different types of slots on it, and four copies of *"Cylinder.prt"*, each of which is aligned with the bottom face of the body in *Slot*.

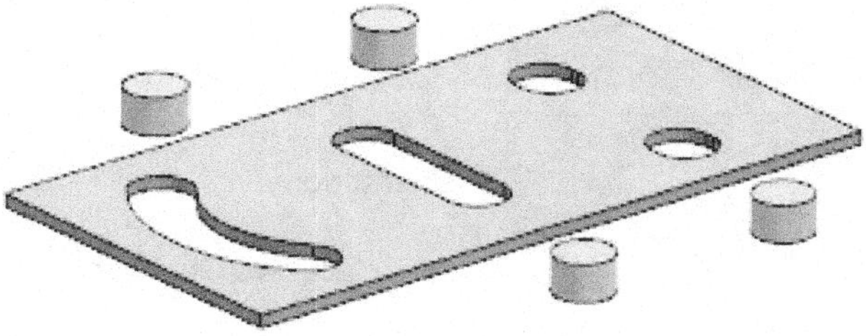

11.11.1 Fixed Cylinder Centered in Slot

Our first exercise concerns the case in which you might use a slot instead of a hole. You would take this approach if designing for ease of assembly – where the slot allows for variance in manufacturing, but a hole would not.

- Select the **Assembly Constraints** tool and set the **Constraint Type** to *Center*, with **Subtype** of *1 to 2*. One way to center the cylinder in the slot is to center its centerline between the centerlines of the two cylindrical faces of the slot.

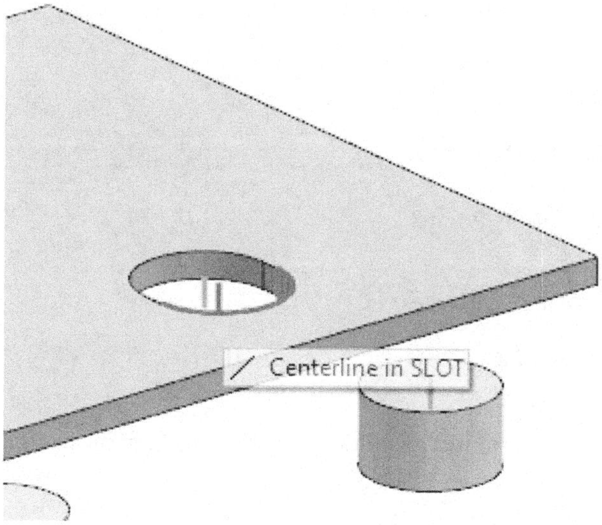

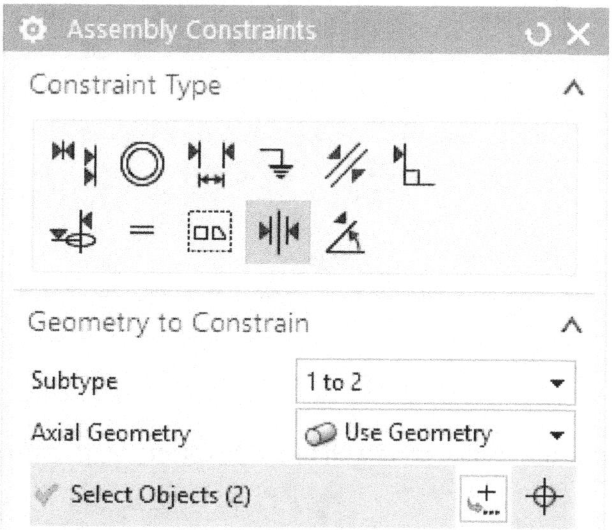

- In conjunction with the *Align* constraint already present, this allows only for rotational motion about the axis of the cylinder.

- Another way to center the cylinder in the slot is to constrain the center point of the slot to the centerline of the cylinder. When the **Constraint Type** is *Touch / Align*, you can pick the centerline of the cylinder and then use the **Point Constructor** to build the center point of the slot on-the-fly.

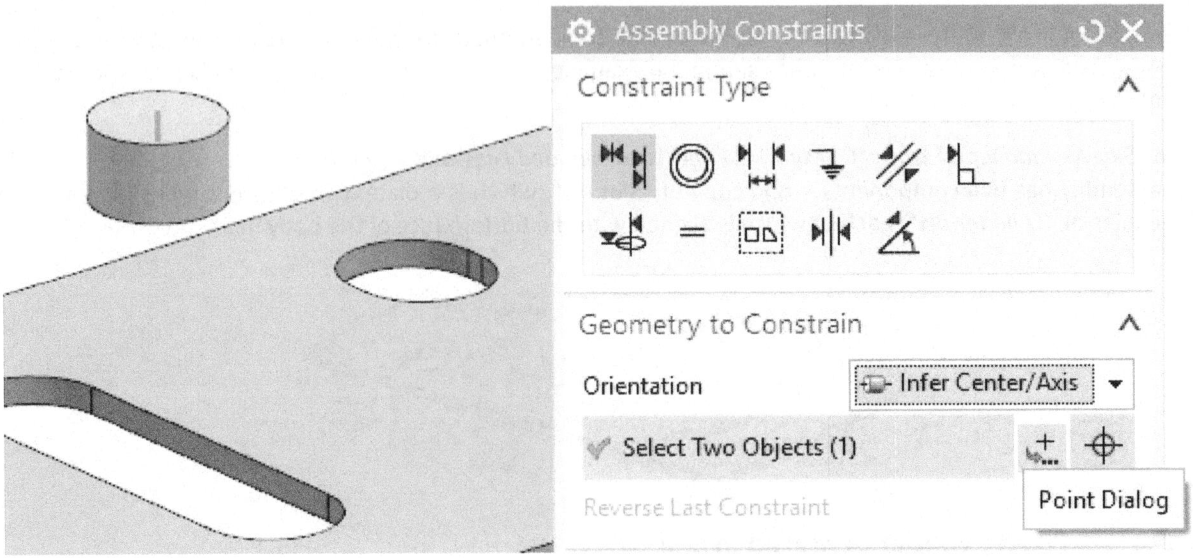

- The easiest way to model the center point of the slot is with **Point Type** *Between Two Points*.

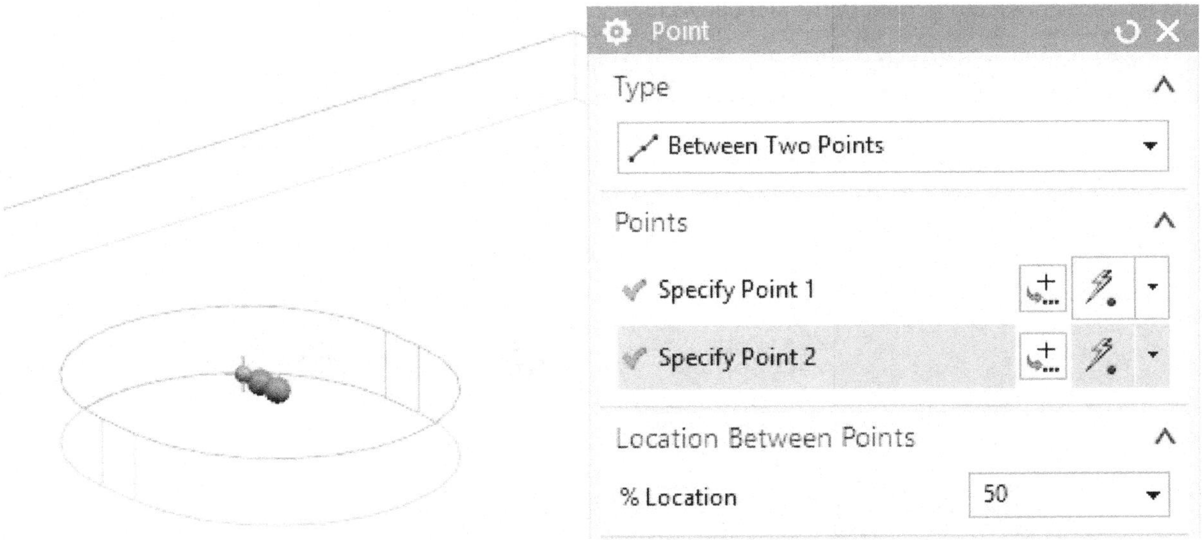

- Again, in conjunction with the *Align* constraint, this allows only for rotational motion about the cylindrical axis.

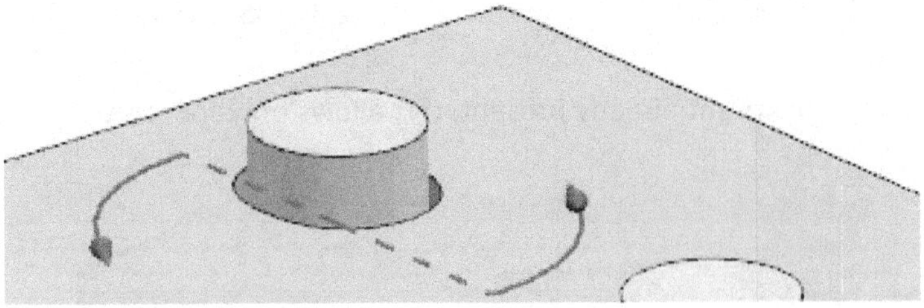

11.11.2 Cylinder Free to Move Within Slot

When the positioning of the cylinder within the slot should be adjustable, you should use a different combination of constraints.

- For a straight slot, you can use the *Center* constraint with **Subtype** *1 to 2* to center the centerline of the cylinder between the parallel faces of the slot.

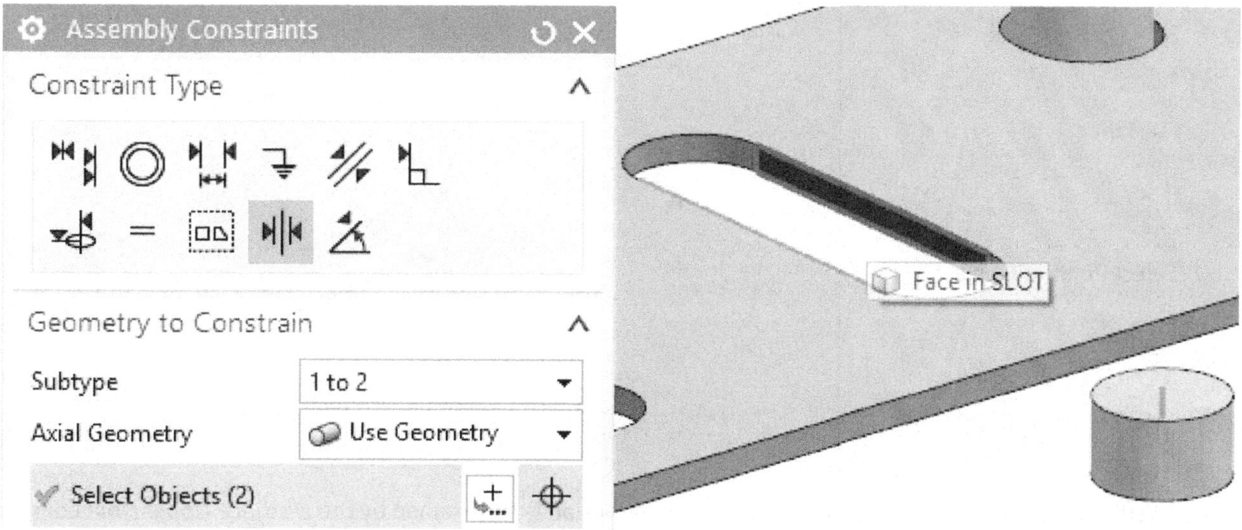

- To allow the cylinder to move within the slot but not outside of it, you will apply a *Distance Upper Limit* constraint between the centerline of the cylinder and the midpoint of the slot[4]. Set the **Constraint Type** to *Distance*, and push the **Point Dialog** button. Set the **Point Type** to **Between Two Points** and construct the midpoint of the slot, as in the previous exercise. Then choose the centerline of the cylinder.

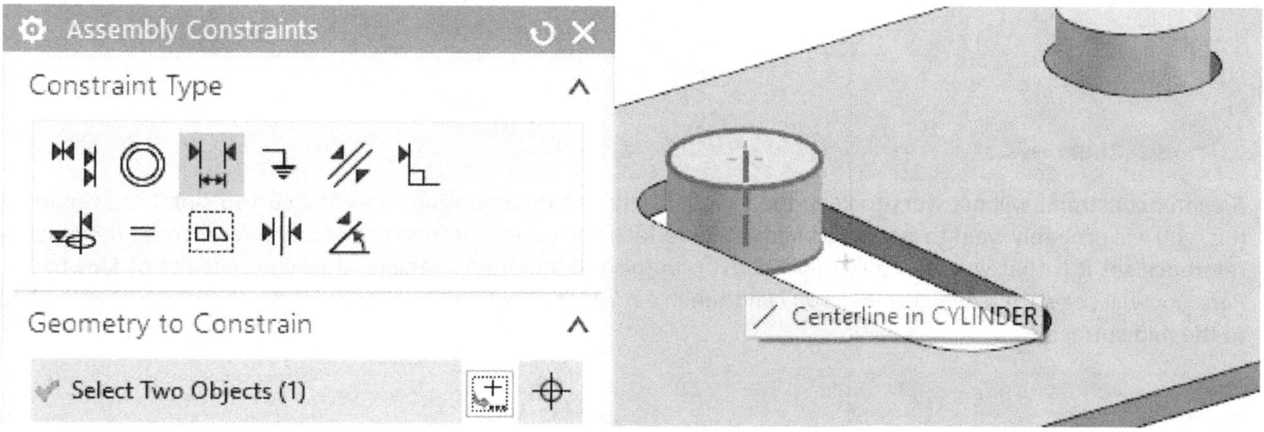

- Uncheck the **Distance** checkmark. Instead, check the **Upper Limit** checkmark, and create a measurement on-the-fly.

- The measurement should be the distance from the center of the slot, to the centerline of one of the cylindrical faces.

[4] You may wonder whether it is possible to measure to the centerlines of one of the cylindrical faces of the slot, and then use an **Upper Limit** (distance between opposite centerlines) and a **Lower Limit** of zero. This does not work because the **Distance** constraint is *unoriented* (it doesn't take direction into account).

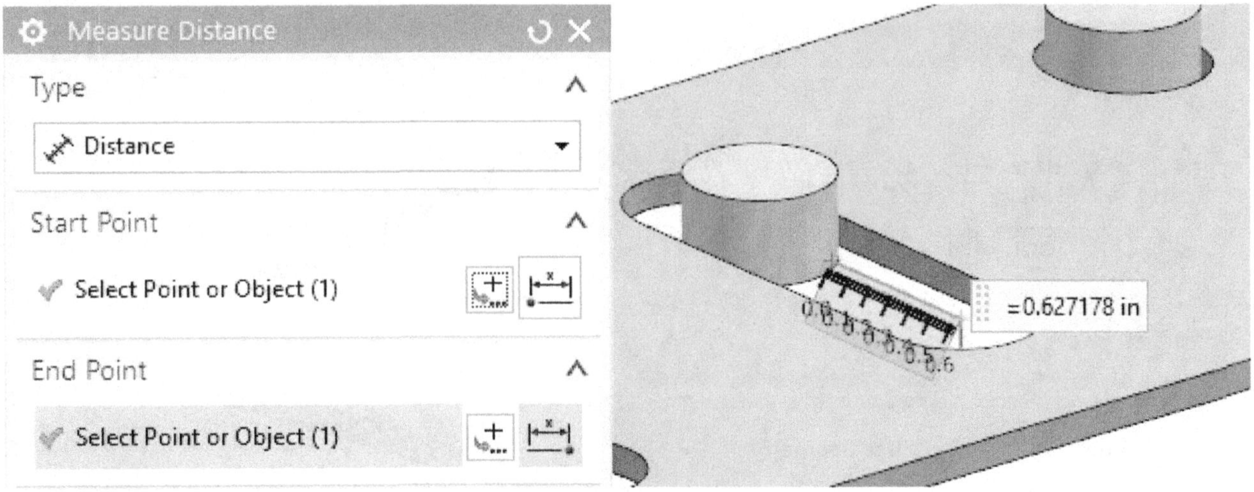

- There is still a linear degree of freedom, but the range of motion is constrained by the *Distance Upper Limit* constraint!

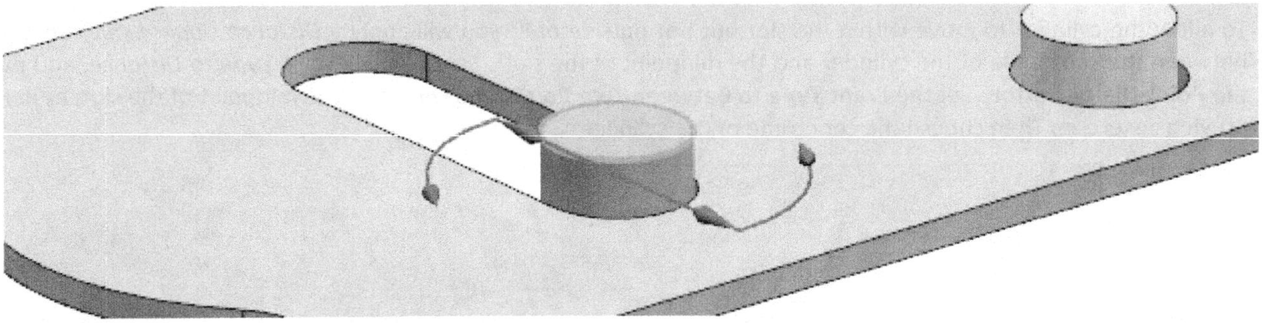

- A *Center* constraint will not work to keep the cylinder between the opposing faces of a curved slot. In a scenario like this, you will probably want to manually build a mid-surface and exclude it from the *Model* reference set (or whichever reference set it is that you intend to make active). In this model, upon changing the reference set of **Slot** to *Entire Part*, you will see such a mid-surface. You can then use a *Touch* constraint to constrain the centerline of the cylinder to the mid-surface.

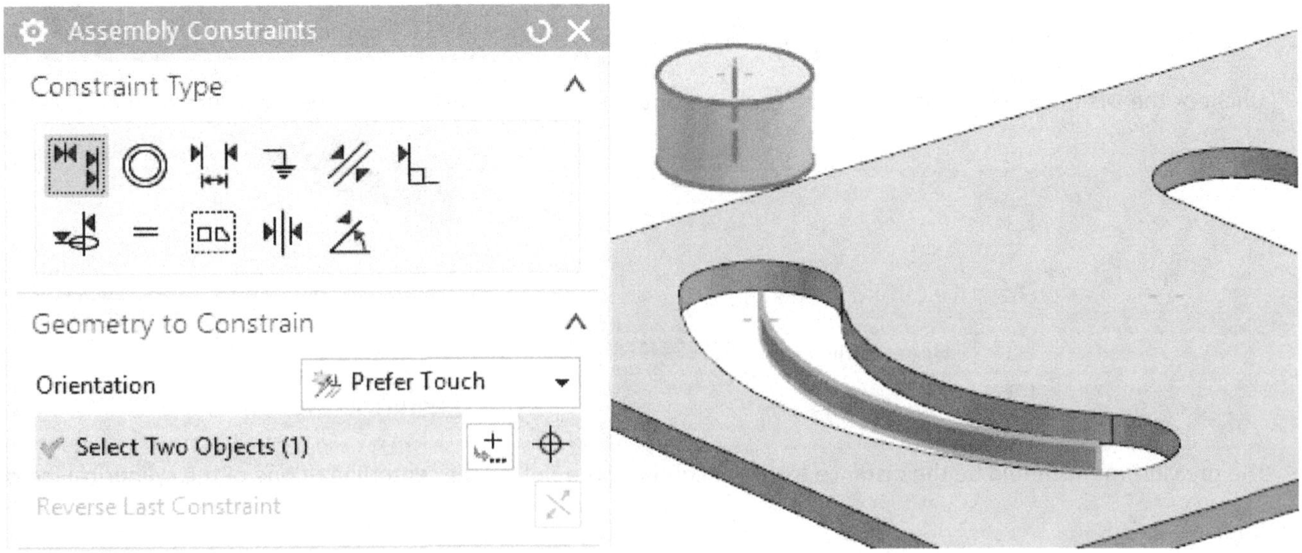

- To limit the range of motion, use another *Distance Upper Limit* constraint, where quantity to be limited is the distance from the centerline of the cylinder to the midpoint of the midsurface.

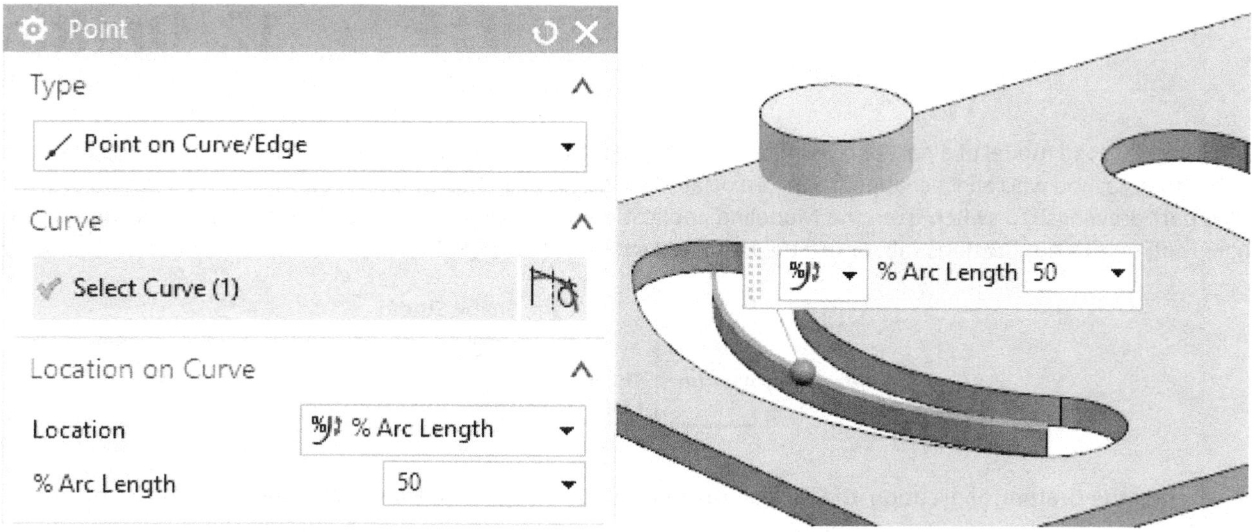

- The limiting value should be the distance from the midpoint of the mid-surface to the end.

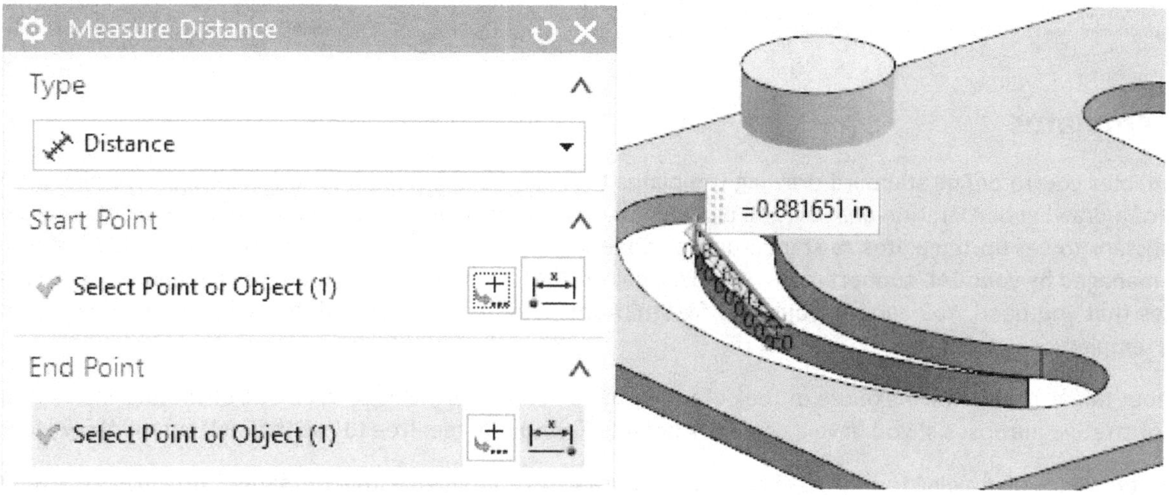

- In this example, the measurement ends up being about *0.88 in*.

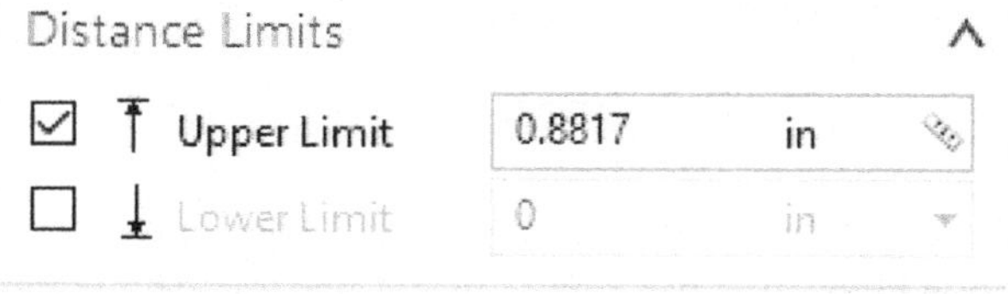

- You might expect the *Touch* and *Distance Upper Limit* constraints to break upon changing the reference set back to *Model*, but they don't. **Assembly Constraints** can make use of entities excluded from the active reference set.
- Again, there is a single degree of translational freedom, but the range of motion is limited.

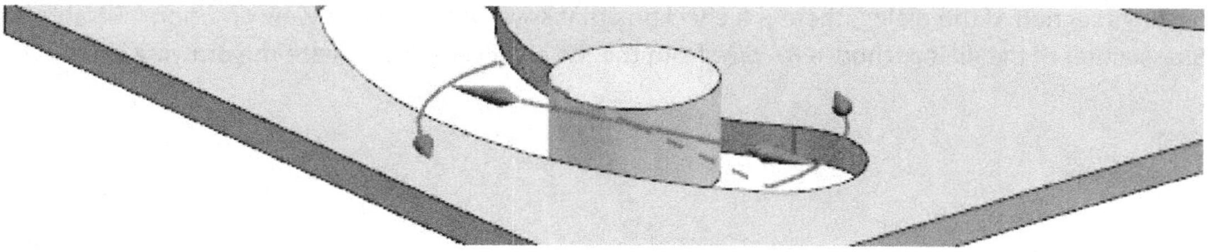

- **Save** your file!

12 Drafting

Once you have a solid model of a part or assembly, you can use it to create a manufacturing drawing. Drawings are defined using the 3D data, and will reflect changes to the associated solid model. In this chapter, you will learn the basics of creating detailed part drawings. To switch from the Modeling application to the Drafting application, find the **Drafting** icon in the **Design** group on the **Applications** tab, or use the keyboard shortcut **[Ctrl]+[Shift]+[D]**.

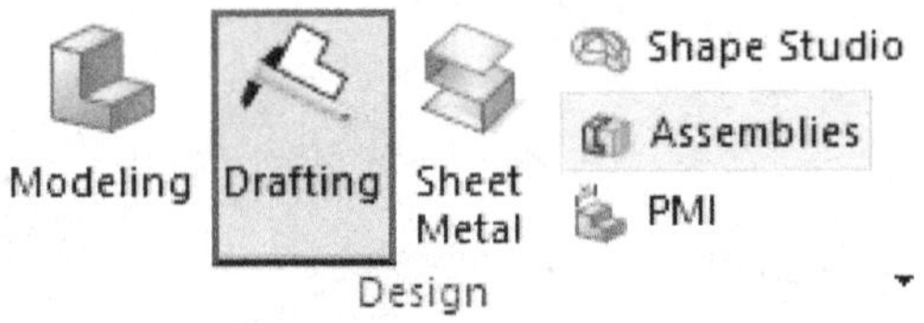

Upon entering the Drafting application, the tabs on the ribbon will change. The **Home** tab appears as shown below.

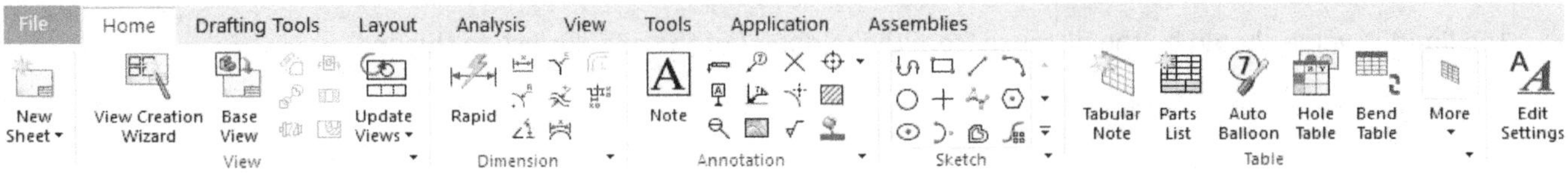

12.1 Templates

NX 11 enables you to define standard drawing templates that can be imported into any drawing. Typically, a template consists of a drawing border, title and revision blocks, and other information that should appear on all of your drawings. It is necessary to set up templates to standardize engineering drawings across a company, but this is a rather painful process managed by your CAD support team. As a designer or drafter, you will more than likely already have access to the templates that you need. You can ask your CAD System Administrator for information about your company's standard drawing templates.

Throughout this book, we will make use of the default Siemens templates included in an out-of-the-box installation of NX for demonstrative purposes. If you have a company-provided template, feel free to use that and try to follow along.

There are two kinds of drawing templates – templates for drawings contained within the same *".prt"* file as the model, and templates for drawings that make reference to the *".prt"* file of the model. The latter type of template is actually an assembly which contains the model as a component (the drawing is contained in the assembly file). We will practice both approaches below.

12.1.1 Drawing in Same File

In the approach below, we will make use of the part file *"The Pool Rack Part II.prt"* that you created in Chapter 9.12. If you skipped that chapter, or otherwise don't have that file, there is a suitable replacement in the downloaded part files folder.

- Open *"C:\My NX 11 Files\The Pool Rack Part II.prt"* or *"C:\My NX 11 Files\Downloaded Files\Pool Rack Part II.prt"*.
- When the part file opens, you will be in the Modeling application. From the **Application** tab, select the **Drafting** icon.
- In the Drafting environment, you will be prompted to create a drawing sheet by the **Sheet** dialog that appears.
- In the **Settings** section of the dialog, there is a checkbox that says *"Always Start View Creation"*. Disable it.
- In the **Size** section of the dialog, choose *B – Size* from the list and click **OK** to create the drawing sheet.

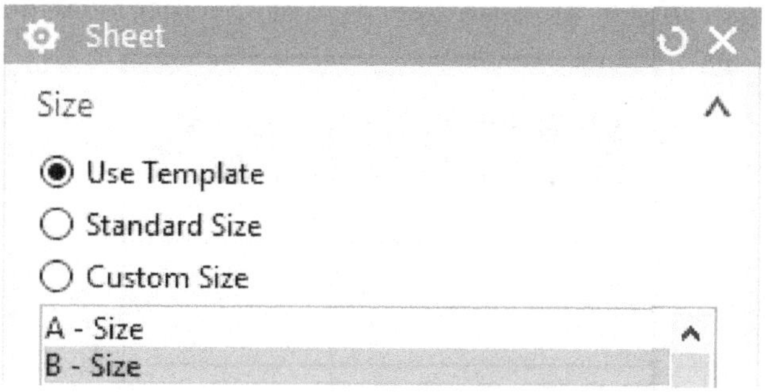

- The template you have chosen is one of the default templates provided by Siemens, and it should appear as below.

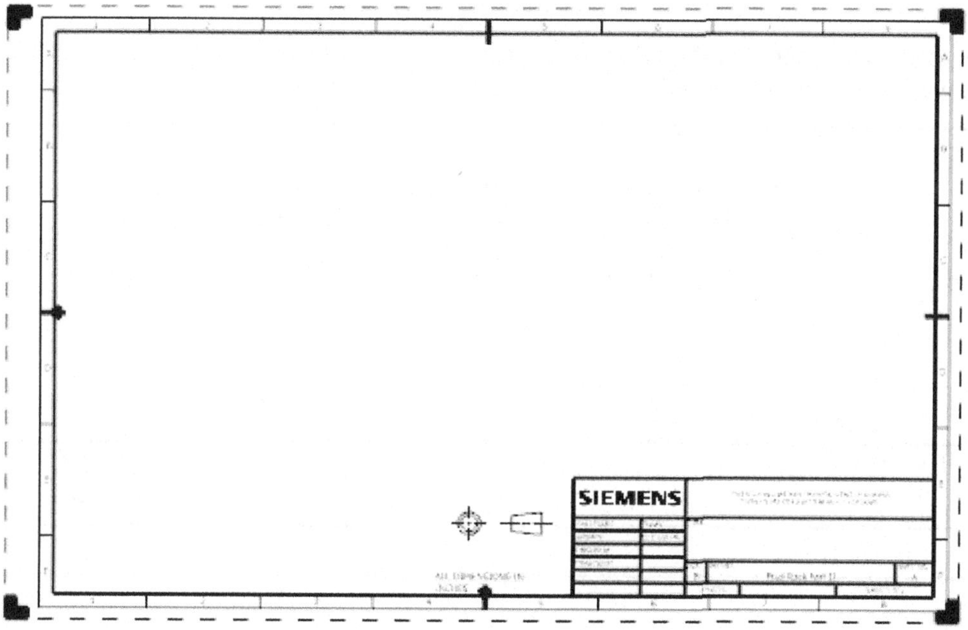

- With the sheet created, the **Populate Title Block** dialog should appear. Enter *"TODAY"* for **First Issued**, and *"BEST STUDENT EVER"* for **Drawn By**. **Close** the **Populate Title Block** dialog.

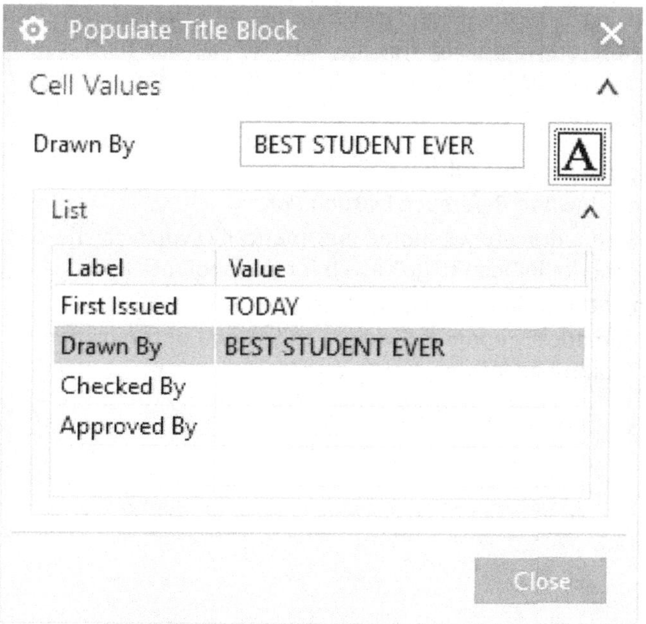

- This fills in some of the cells in the title block in the lower right hand corner of the drawing.

SIEMENS

FIRST ISSUED	TODAY
DRAWN BY	BEST STUDENT EVER
CHECKED BY	
APPROVED BY	

- Note the values already entered in the other cells – some of these, such as the *Size*, *Scale*, and *DWG No.* are pre-populated by way of certain part attributes.

THIS DRAWING HAS BEEN PRODUCED USING AN EXAMPLE TEMPLATE PROVIDED BY SIEMENS PLM SOFTWARE

TITLE			
SIZE B	DWG NO. Pool Rack Part II		SHEET REV A
SCALE 1:1		SHEET 1 OF 1	

- **Save** your file!

12.1.2 Drawing in Separate File

In this exercise, we'll create a drawing using the second kind of template – the template that makes reference to an existing part. You won't practice making the drawing using this file, but the exercise here is included so that you know how to create a drawing in this way.

- Select **File / New**.
- On the **New** dialog, switch to the **Drawing** tab.
- Change the **Relationship** drop-down to *Reference Existing Part*.
- At this point, the Part to create a drawing of should automatically switch to the current Displayed Part (*"Pool Rack Part II.prt"*). If it doesn't, use the folder icon to go search for the pool rack.
- From the **Templates** section of the dialog, choose *B – Size*.
- The **Name** of the file will automatically append *"_dwg1"* to the end of the filename – if you want some other suffix, feel free to type it, but we recommend leaving it as *"_dwg1"*.

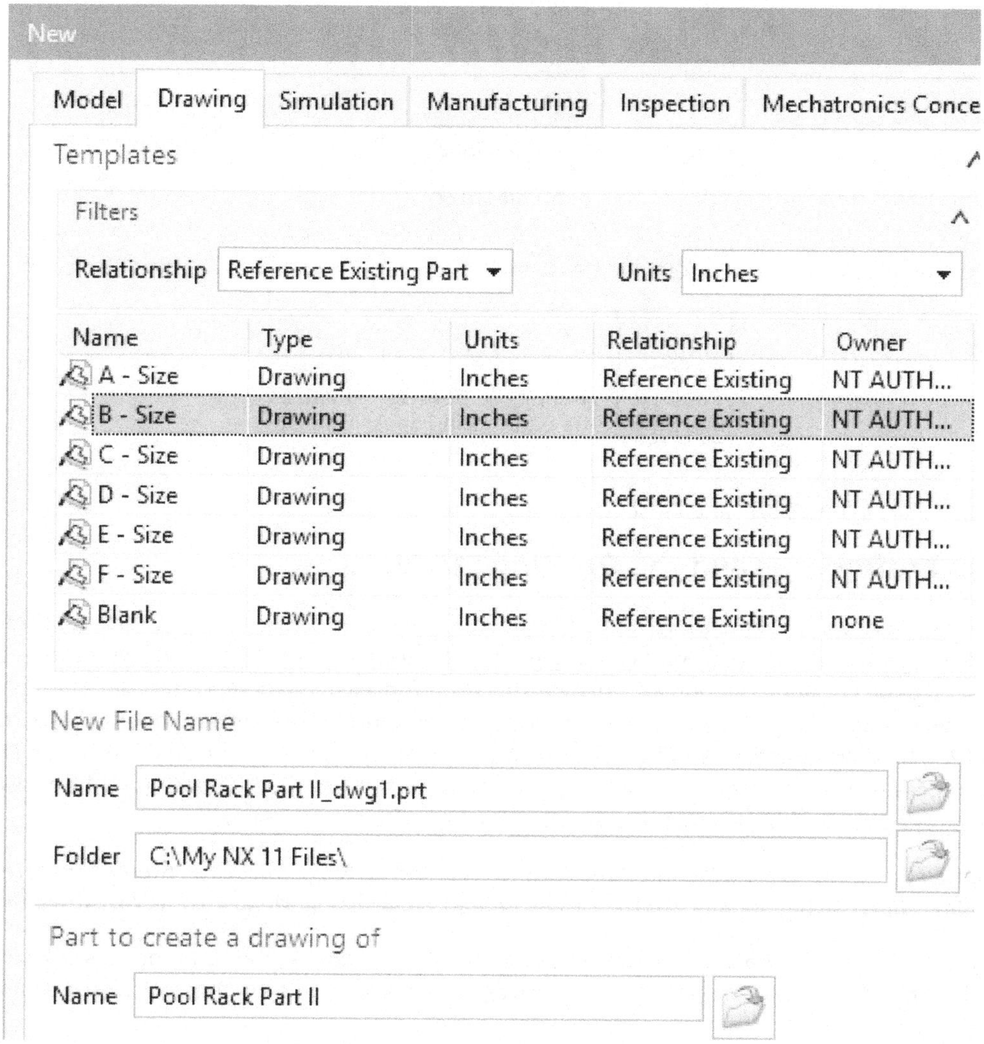

- When the new file is created, you will enter the Drafting application, but the neither the **Sheet** dialog nor the **Populate Title Block** dialog will appear. In the Assembly Navigator, you can verify that the drawing is actually an assembly which has one component – the part being drawn.

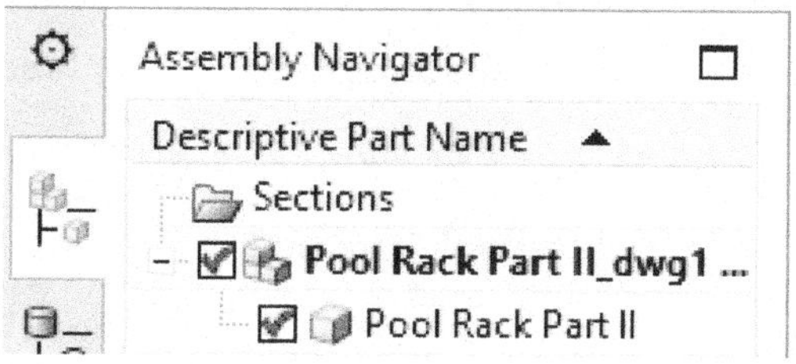

- The **Sheet** dialog is the result of the **New Sheet** command, which from here can be used to add additional sheets to the drawing. **New Sheet** is found in the **Sheet** drop-down on the far side of the **Home** tab.

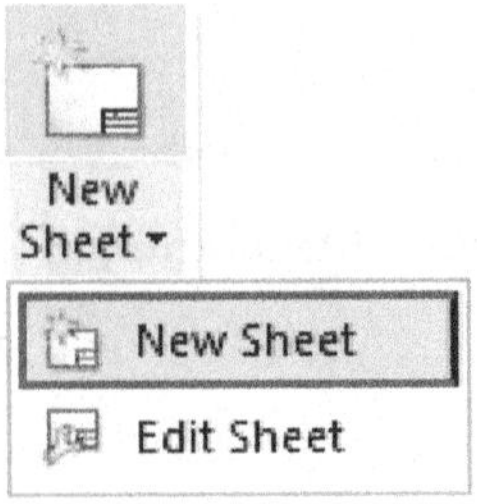

- The **Populate Title Block** command is found in the **Drawing Format** group on the **Drafting Tools** tab.

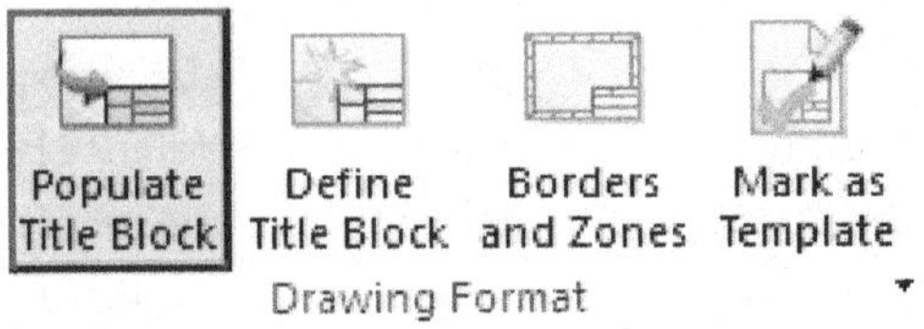

- **Save** *"Pool Rack Part II_dwg1.prt"* and **Close** it.

12.2 Drafting Preferences and the Drafting Standard

Before you begin placing views and dimensions and other drafting essentials, you may want to change some of the defaults that determine how your dimensions, views, etc. appear on the drawing sheet. Once drafting entities are placed, their display settings need to be individually modified. The **Drafting Preferences** menu determines these defaults. Much like the **Modeling Preferences** menu, any changes made to **Drafting Preferences** are only temporary – in fact, even more temporary than **Modeling Preferences**, which persist for the duration of your session (see Chapter 2.9.1). Changes made to **Drafting Preferences** only affect the Displayed Part, so if you need to make many drawings with consistent formatting, it is crucial that you set your **Drafting Preferences** and **Drafting Standard** beforehand.

Drafting Preferences are found in the second column of the **File** drop-down as well as at **Menu / Preferences / Drafting**, and the **Drafting Standard** can be modified for the Displayed Part by selecting **Menu / Tools / Drafting Standard**.

The better place to set up your **Drafting Preferences** is actually in **Customer Defaults**, because **Drafting Preferences** themselves are determined on a part-by-part basis. In **Customer Defaults**, you have the option to customize the **Drafting Standard**.

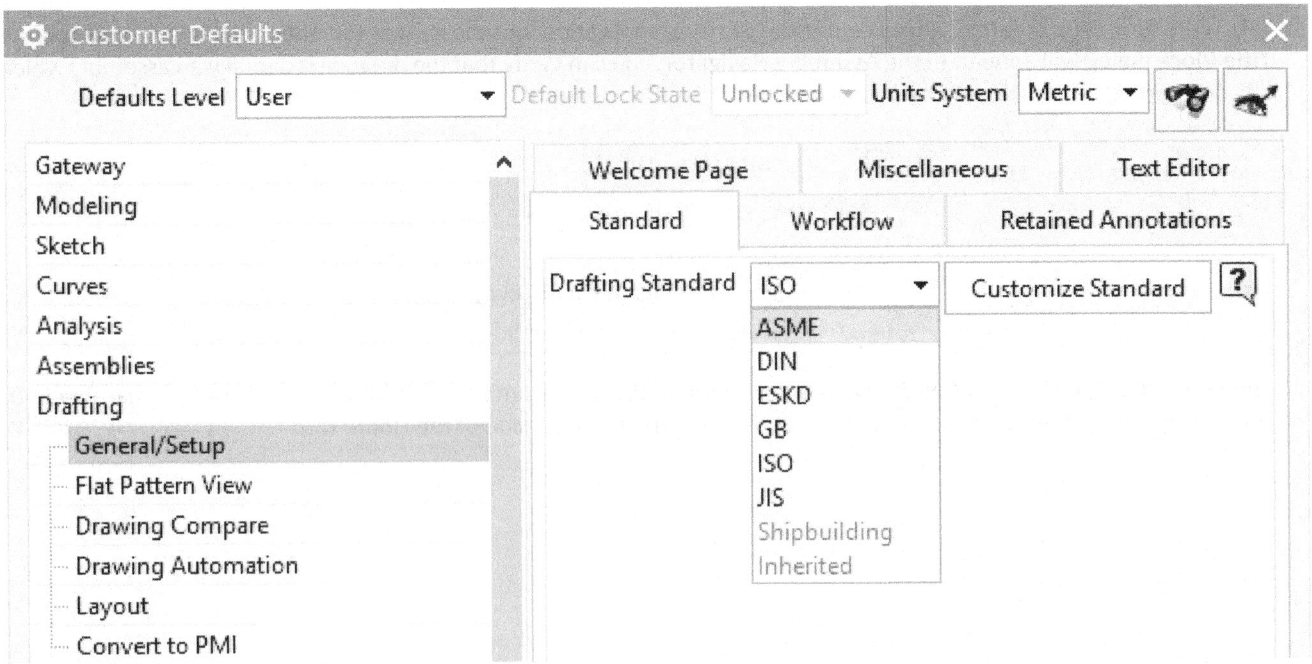

To modify the **Drafting Standard**, select a starting standard (ASME, ISO, etc.), and push the **Customize Standard** button. This will bring you to a dialog that allows you to further specify the standard. For instance, if you wish to use an *ASME* standard, which version of Y14.5 should your drawing objects follow? The dialog below shows what the **Customize Standard** options are for the *ASME* choice.

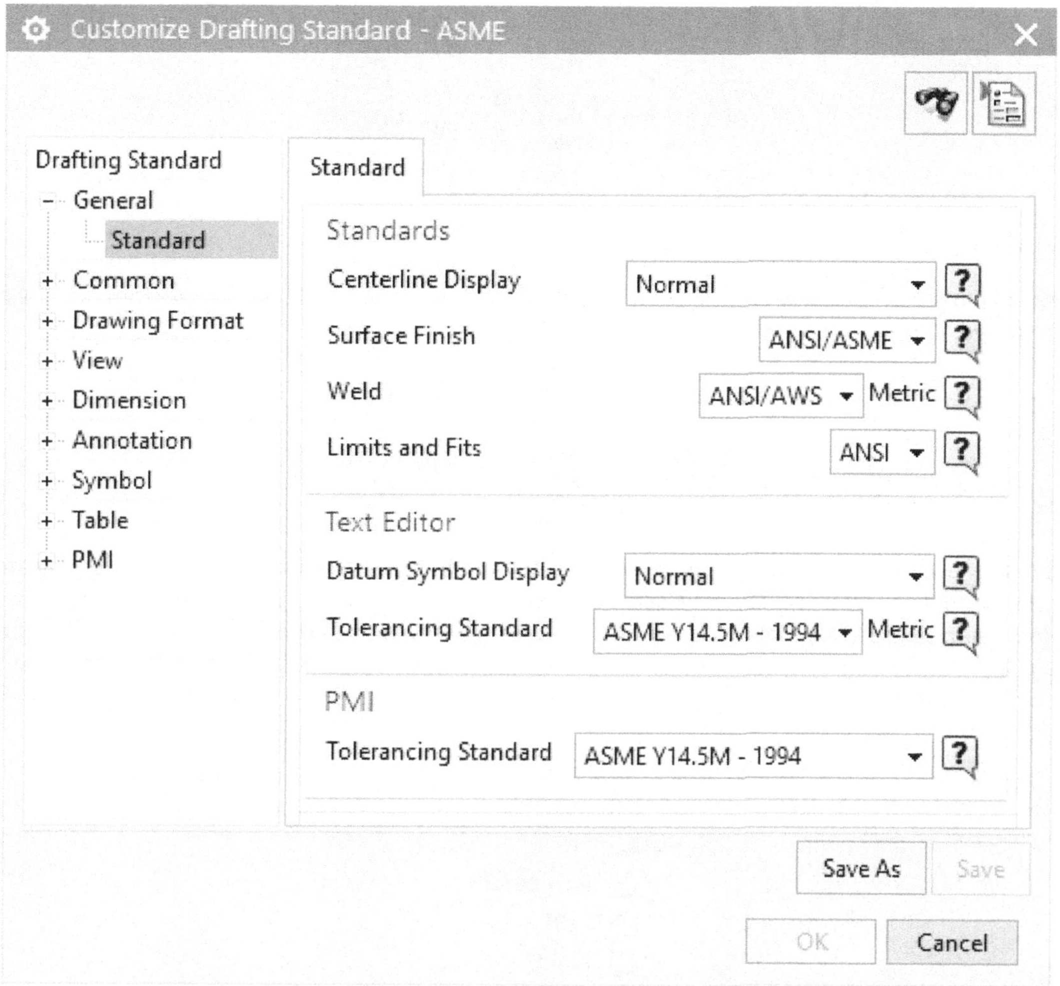

Again, in many corporate environments, your CAD support team will already have done a lot of the hard work of sorting out which settings are most beneficial for your company's drafting standards.

12.2.1 Edit Settings

The tool that allows you to modify the appearance and properties of drafting entities is **Edit Settings**. It is found on the far-right side of the **Home** tab.

You will see different options for different selected drafting entities – this tool basically customizes the same options that the **Drafting Standard** and **Drafting Preferences** control. You will make use of it once you have placed some objects on the drawing.

12.3 **Placing Views on the Drawing**

Once you have created a drawing sheet and set **Drafting Preferences** and the **Drafting Standard** to your liking, the next step is to place views of your model on the drawing. The tools in the **View** group found on the **Home** tab create different types of views on the drawing – we begin with an overview of the tools in this group.

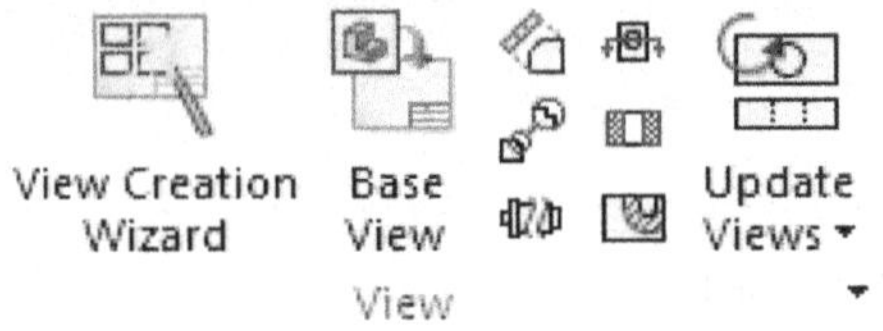

12.3.1 Base View

The **Base View** command produces a primary view of your model upon which derived views (projected views, section views, etc.) are based.

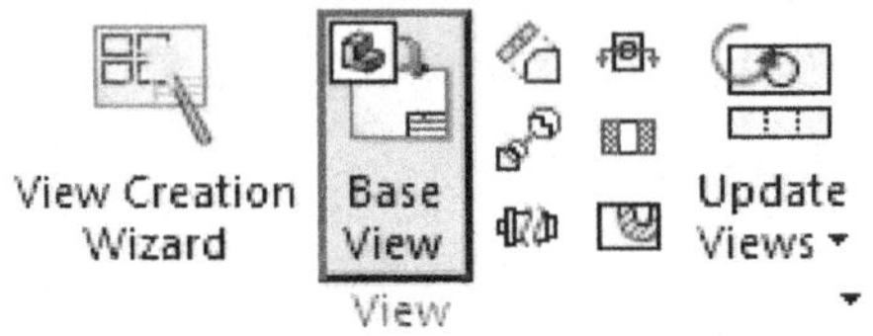

Base View places a view on the drawing which comes from one of the Model Views listed in the Part Navigator while you are in the Modeling application. Once you have selected a Model View, it is possible to modify the orientation using the **Orient View** tool on the **Base View** dialog, but we will not practice this in the exercises below. Try it for yourself with a drawing of your own!

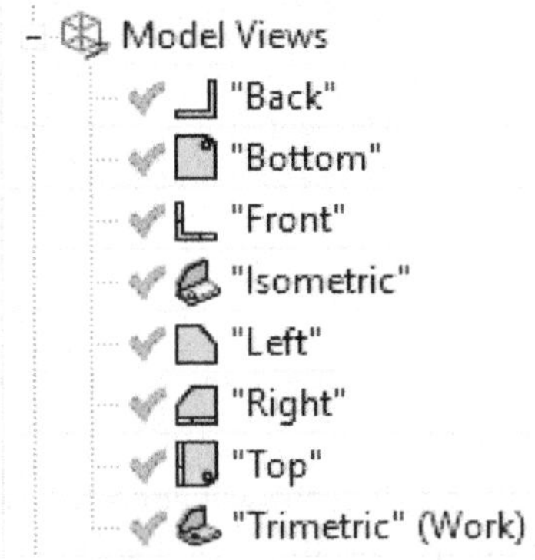

- Continue working with *"Pool Rack Part II.prt"*.
- Select the **Base View** tool and set the **Model View to Use** to *Right*. Set the **Scale** to *1:4* and click in the Graphics Window to position the view in the middle of the drawing sheet.

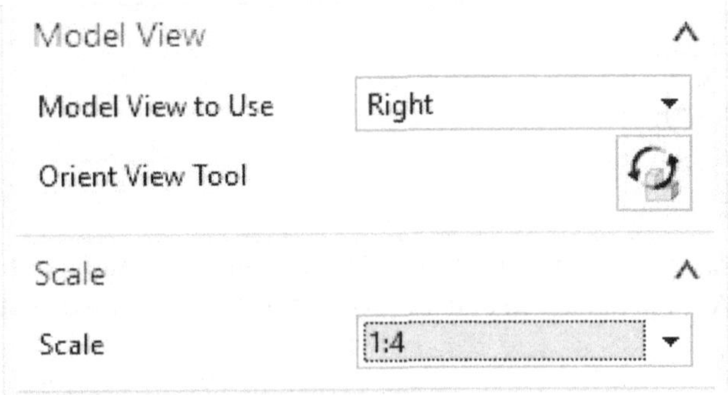

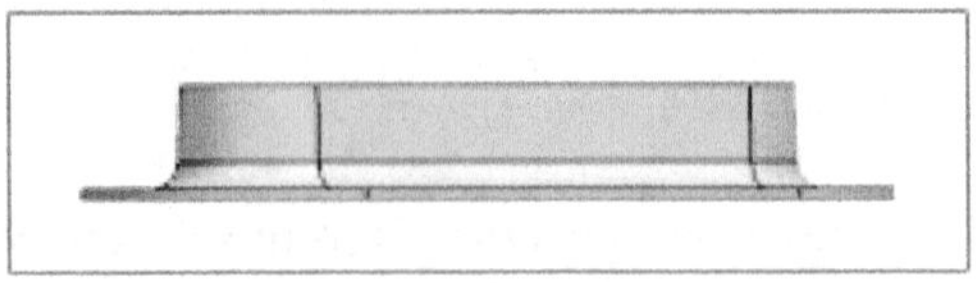

- After your base view is placed, NX will immediately switch to the **Projected View** tool, which positions an orthographic projection of the base view in its own view. Move your cursor directly above the base view and click to position a projected top view as shown below.

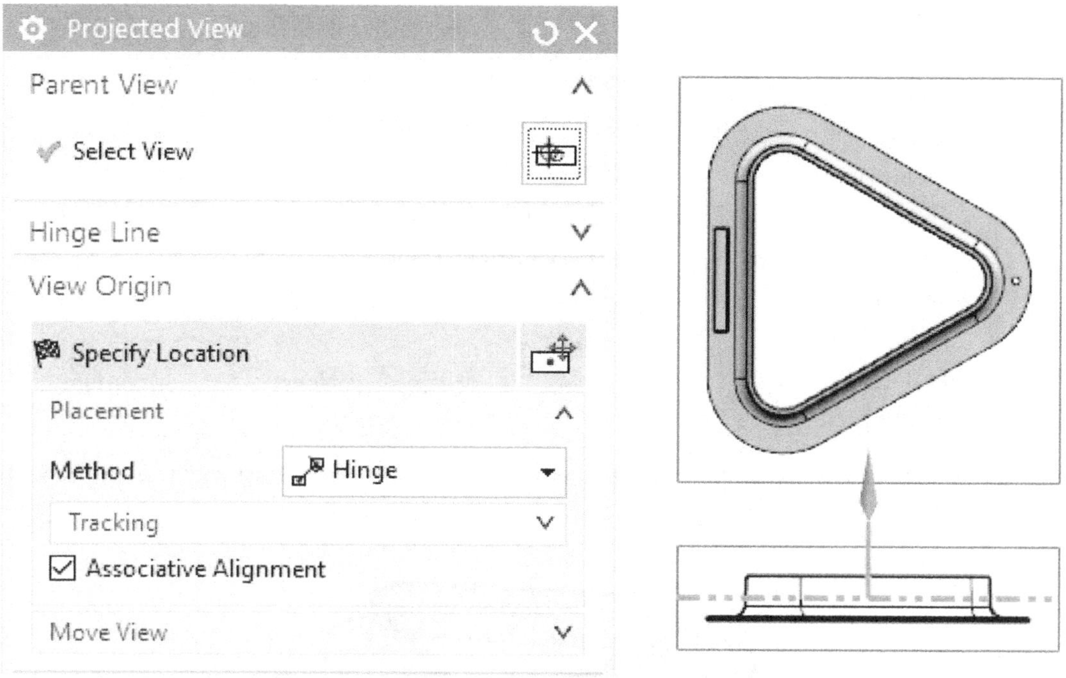

- The **Projected View** tool will remain open, asking you to specify additional projected views – close it.
- Select the **Base View** tool again and set the parameters shown below. Click to the right side of the drawing sheet to position a trimetric view of the pool rack.

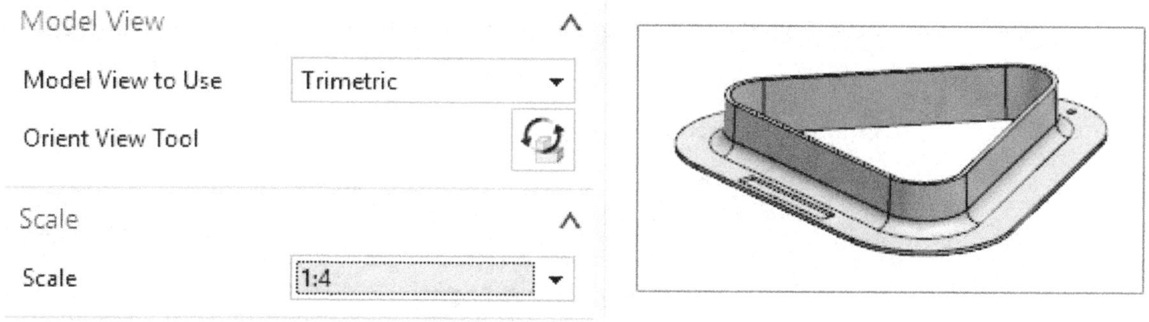

- When the **Projected View** dialog appears, close it.
- An important point about the borders around the views that you have placed is that they do not appear when the drawing is printed or exported to pdf. They are merely visible to help you select the drawing views – views, like datum planes, must be selected from their visible boundaries. Click on the boundary of the trimetric view to select it.

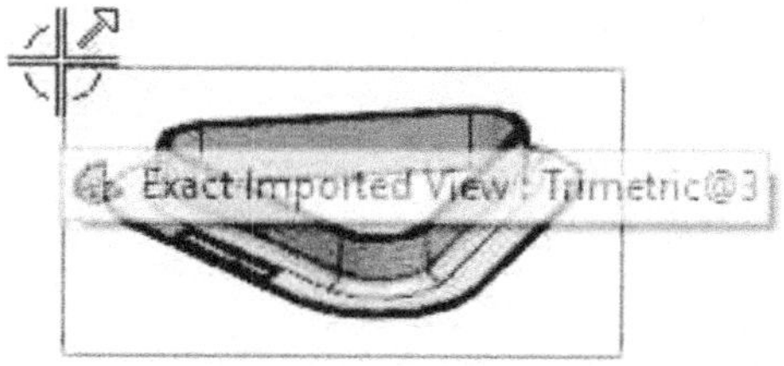

- With the trimetric view selected, push the **Edit Settings** button. In the **Settings** dialog, go to the *Shading* settings and change the **Rendering Style** to *Fully Shaded*. Click **OK** to save your changes.

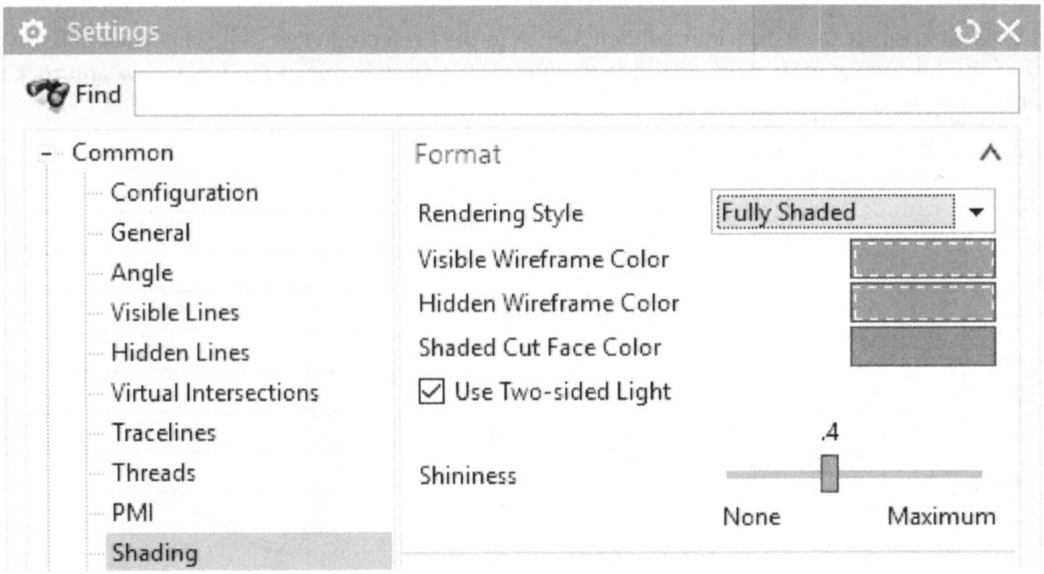

- Your trimetric view should now appear fully shaded on the drawing sheet, in contrast with the wireframe top and right views.

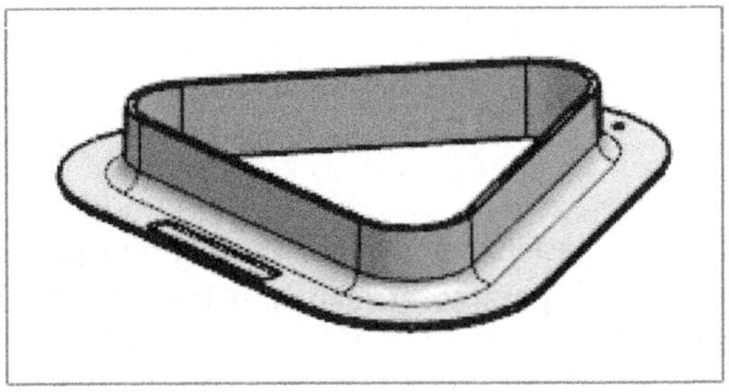

- You can also select drawing views from the Part Navigator. Select the right view, and then hold **[Ctrl]** and select the projected view.

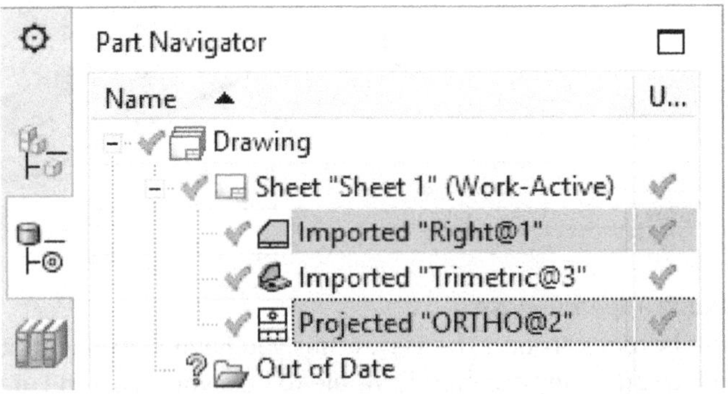

- With the top and right views selected, push the **Edit Settings** button. In the **Settings** dialog, find the *Smooth Edges* settings, and uncheck the *"Show Smooth Edges"* checkbox.

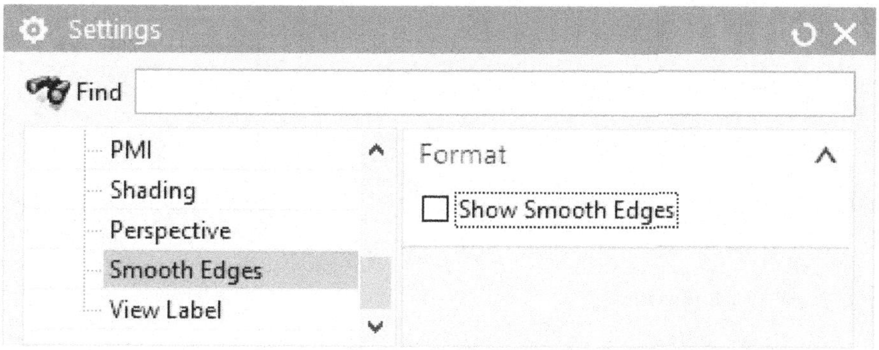

- The result is that edges between tangent faces will no longer appear on the top and right views on the drawing!

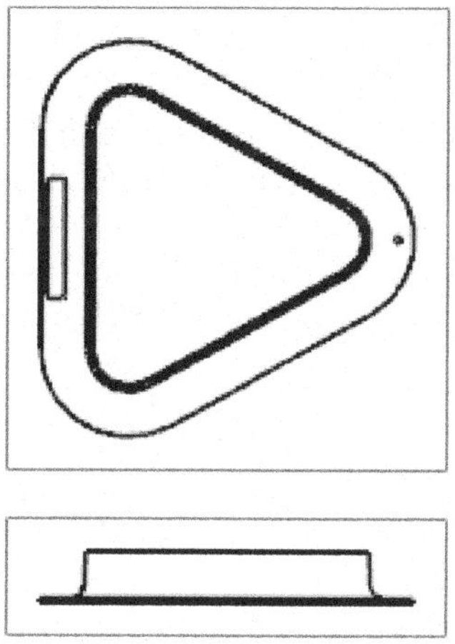

12.3.2 Projected View

You have already seen that the **Projected View** tool activates immediately following the use of the **Base View** tool. If you wish to use the tool when there are multiple base views already on the drawing sheet, you will be prompted to select one.

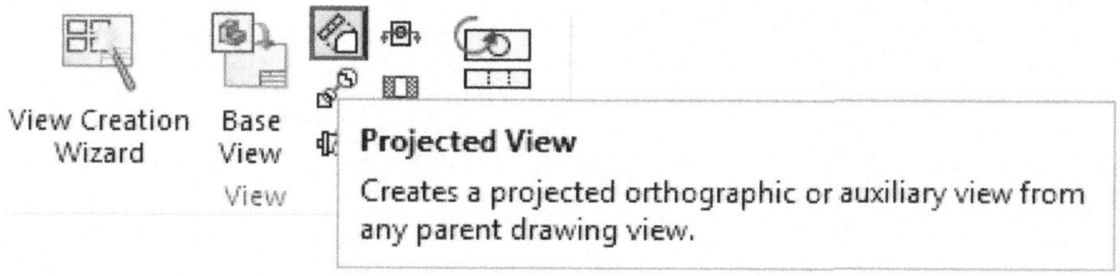

- Select the **Projected View** tool. By default, it will begin generating a preview of an orthographic projection to the last base view placed – the trimetric view in this case. On the **Projected View** dialog, click on *Select View* under **Parent View**, and choose the right view.

- Upon selecting the right view, the preview will change to an orthographic projection of the right view but it won't track your cursor – you must click on *Specify Location* under **View Origin**. Set the projected view below the right view – the result will be a projected bottom view!

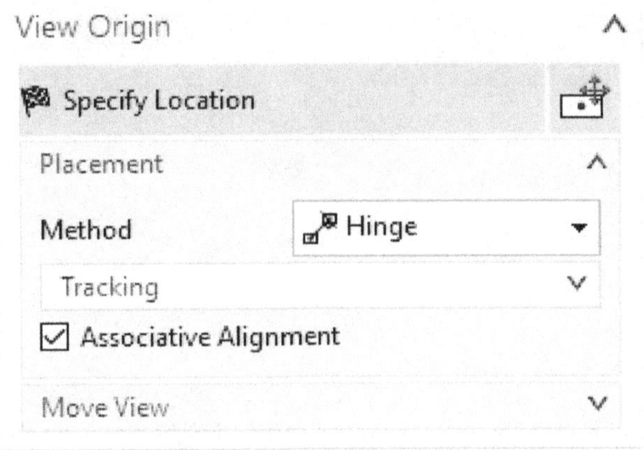

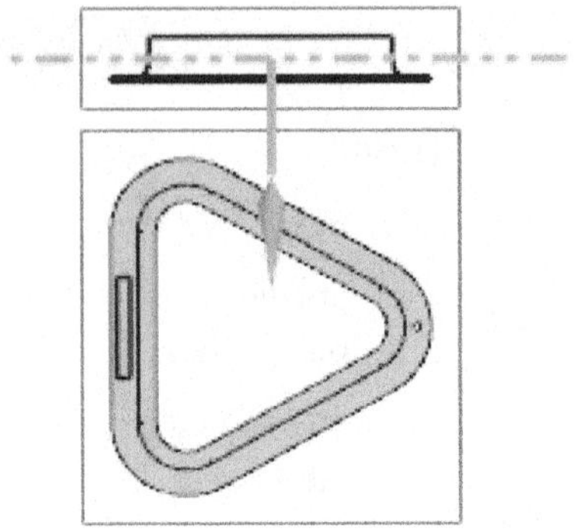

12.3.3 Layer Visible in View

The objects visible in a drawing view are by default, the same as those that are visible throughout the entire part, whether in the Modeling or Drafting application. In other words, they are only those that are on visible layers. Each view actually has its own layer visibility settings, which by default are the same as the global layer visibility settings determined by the **Layer Settings** command (see Chapter 3.5.1 for a refresher). The command that controls each view's individual layer visibility settings is called **Layer Visible in View**, and it is found in the **More** gallery of the **Visibility** group on the **View** tab.

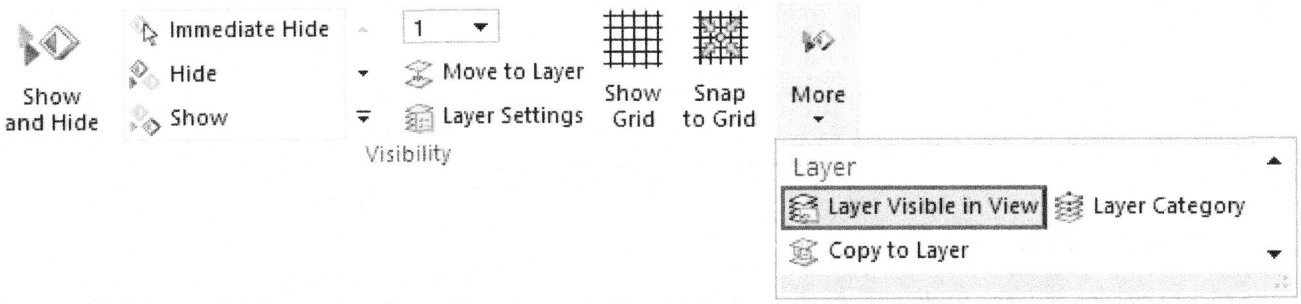

- Use the **Show and Hide** command (**[Ctrl]+[W]**) to **Show** *All*.

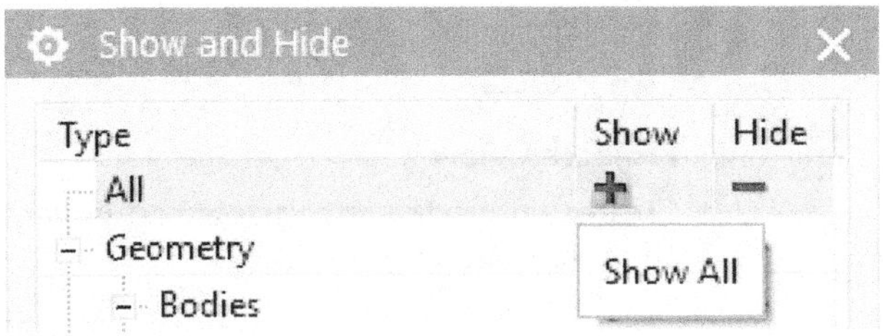

- This renders two previously-invisible objects from the Modeling environment visible in each drawing view – the datum coordinate system, and the point entity used to position the pocket.

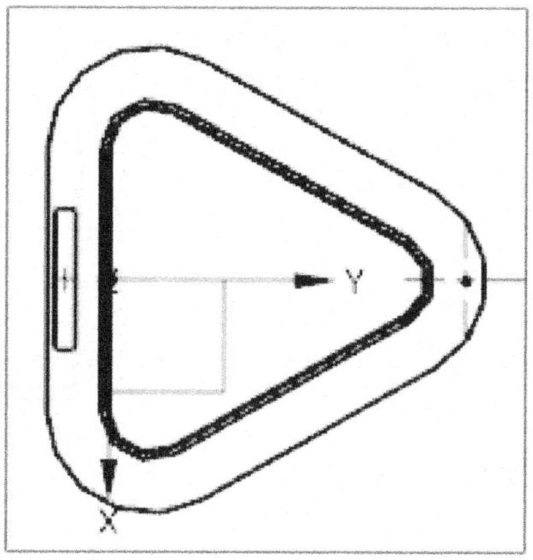

- Switch to the Modeling application and select the **Move to Layer** command (see Chapter 3.5.3 for a refresher). When the **Class Selection** dialog appears, choose the *Point* feature and click **OK**.

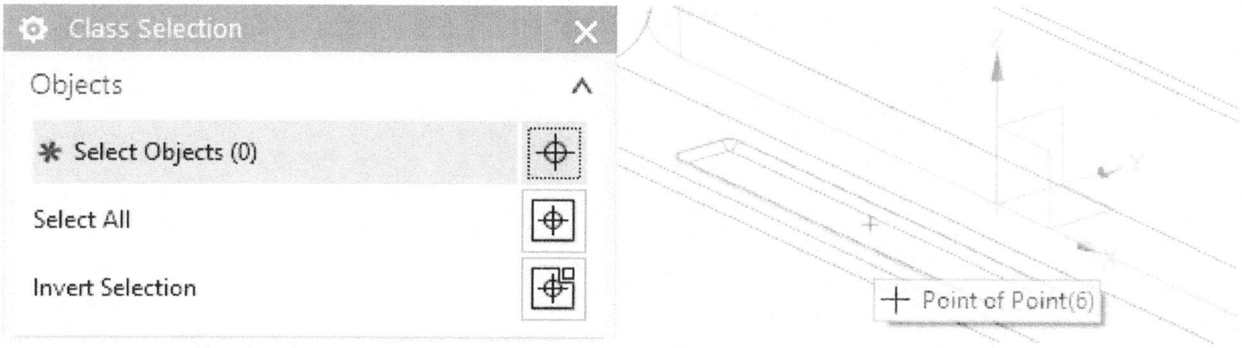

- When the **Layer Move** dialog appears, enter *61* as the **Destination Layer** and click **OK**. This moves the *Point* feature to layer *61*, where the datum coordinate system already resides.

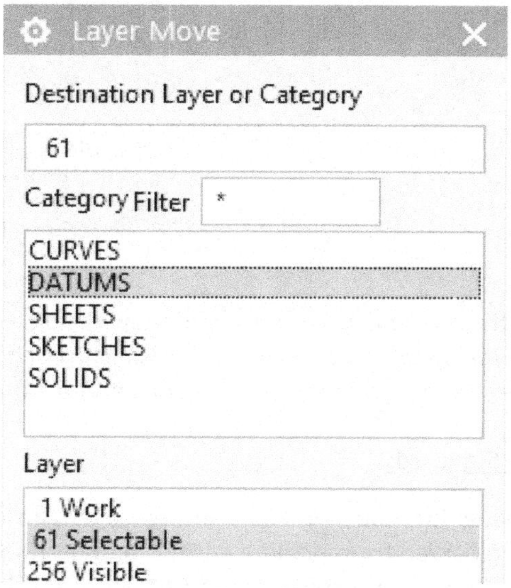

- Return to the Drafting application and select the **Layer Visible in View** command. Click on the top view. The **Layer Visible in View** dialog is slightly strange – instead of requiring you to click **OK**, it will immediately bring up another dialog called **Layer Visible in View**.

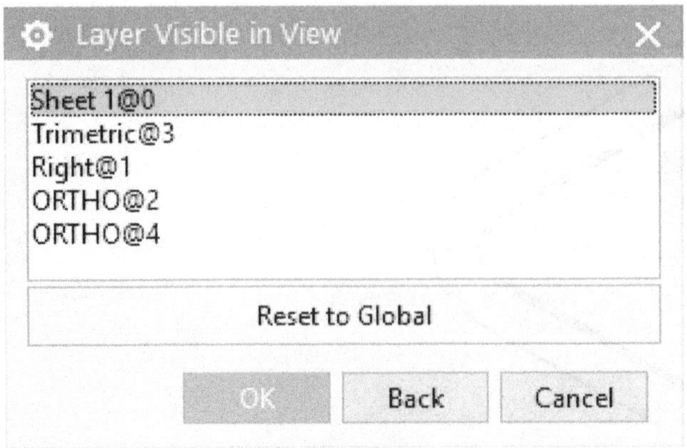

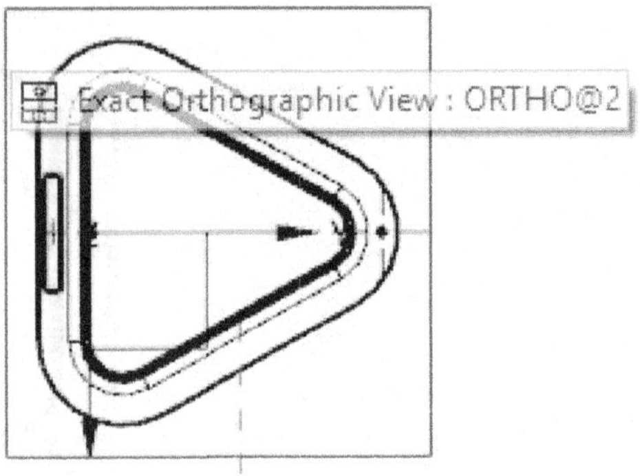

- In this dialog, click on *"ALL"* in the **Category** section, and then hold **[Ctrl]** and click on *1 Visible* in the **Layers** section to deselect it. Push the **Invisible** button, and click **OK**. The result is that only layer *1* will be visible in the top view.

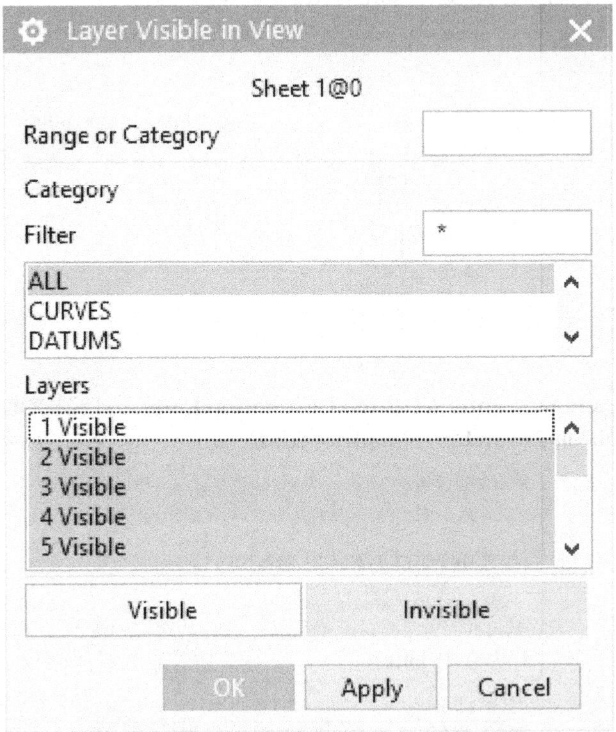

- At this point you will be looking at the previous **Layer Visible in View** dialog – go ahead and click on another view on the drawing and repeat the above process to render only layer 1 visible in each view. Take care not to change the layer visibility settings for the whole sheet – notice that the sheet is considered a drawing view by **Layer Visible in View**!

12.3.3.1 Drawing and Modeling Layer Standards

Most companies have standard procedures for modeling and drafting. Layer conventions are an important part of these standards. Usually, a range of layers is reserved for modeling entities and another range for drafting entities. It is good practice to organize your work in this fashion. Please contact your CAD System Administrator to obtain more information on how this should be done at your site.

12.3.4 Update Views

If you make changes in the Modeling application, or to the visibility of layers in views on your drawing with **Layer Visible in View**, your drawing will become out-of-date. Drawing views will show the latest version of the model, but only once they are updated.

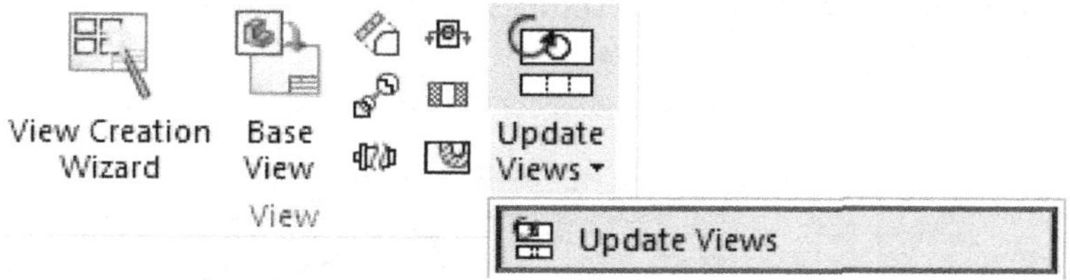

- In the Part Navigator, note that the **Up to Date** column shows that all views and the drawing sheet itself have become out-of-date. Rather than using the **Update Views** command in its location on the ribbon shown above, it is simplest to right-click on *Drawing* in the Part Navigator, and choose **Update** from the pop-up menu.

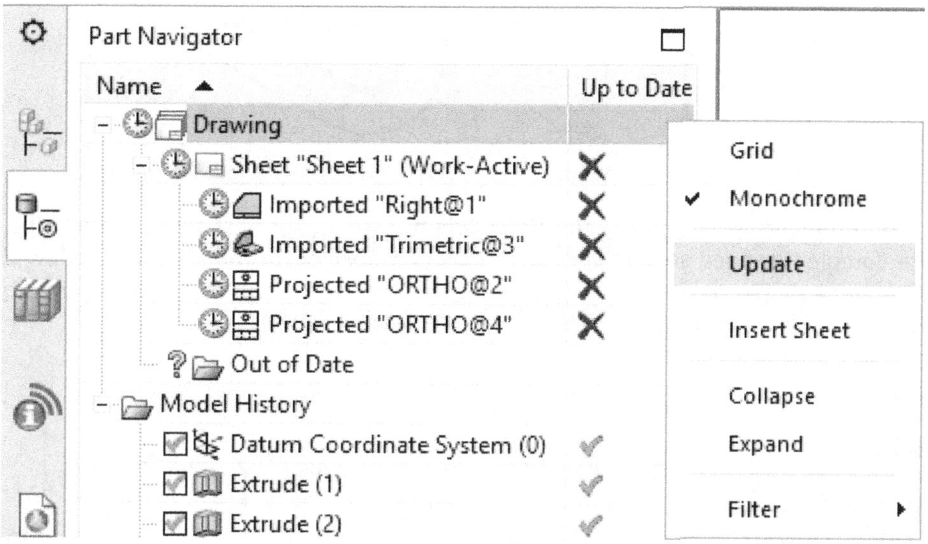

- This brings all views, and the drawing sheet up to date.

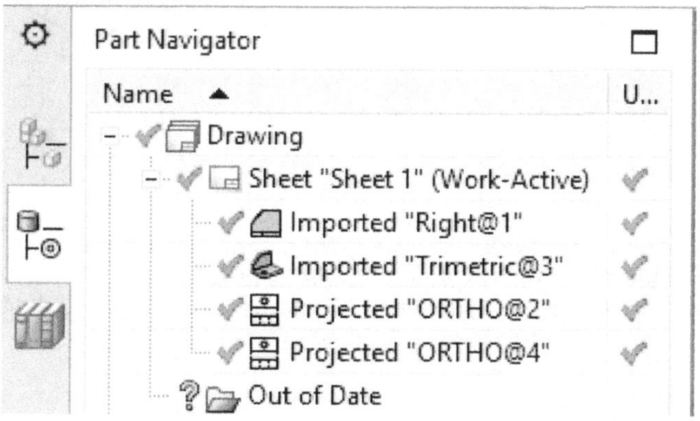

12.3.5 Section View

NX offers a variety of section views in the Drafting application, the most basic of which shows the effect of cutting your model with a section line.

- Select the **Section View** tool. With the **Section Line** settings as shown below, choose the arc center shown below to determine the location of the **Section Line Segments**.

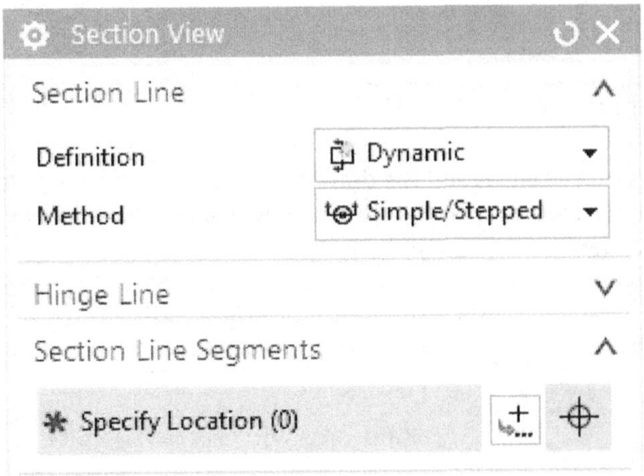

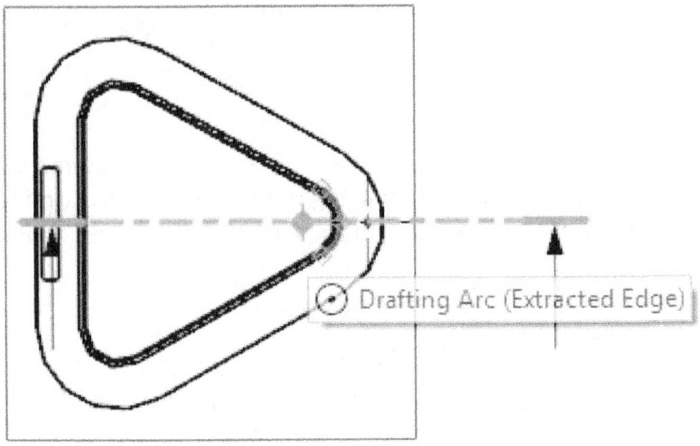

- Your selection will automatically advance to the **View Origin** so that your next click positions the section view. You can create more complex stepped section lines by clicking on *Specify Location* within **Section Line Segments** and continuing to click, but a simple section line is good enough for us. Move your cursor below the top view and click to place the section view.

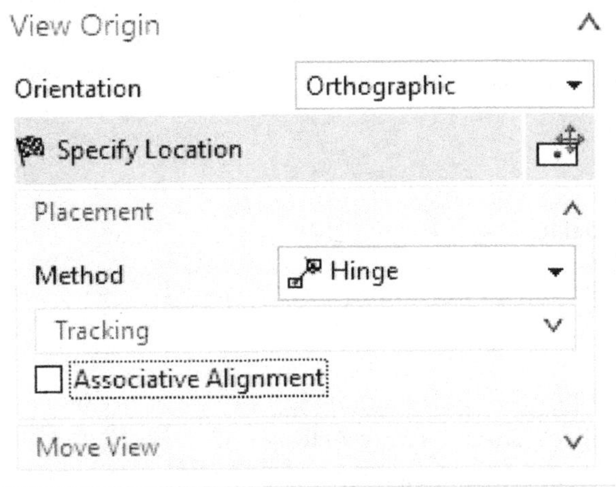

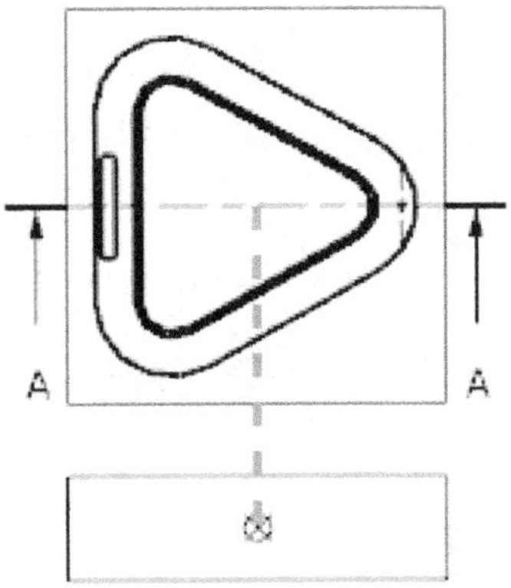

- Although the orientation of the section view is an orthographic projection indicated by the labeled arrows, the view itself need not remain below the parent view. Uncheck *Associative Alignment*, and drag the section view over to the right, as shown below.

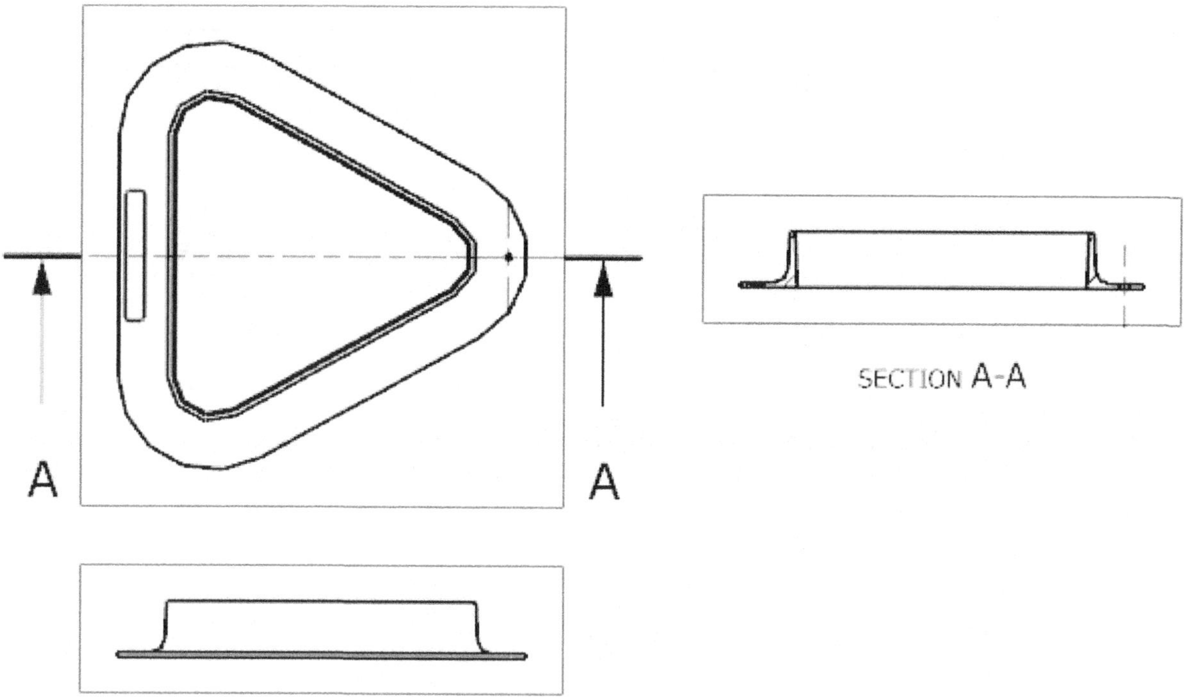

12.3.6 Detail View

Detail views are used to magnify regions on a drawing view that show features of a dramatically different scale than the rest of the part. The regions displayed by the **Detail View** tool are typically circular or square.

- Select the **Detail View** command and specify a **Center Point** approximately as shown below. For the **Boundary Point**, move your cursor until the circular preview is similar to that shown below. If you are unhappy with the position of either the **Center Point** or the **Boundary Point**, you can drag them around dynamically – the golden dots are dynamic handles.

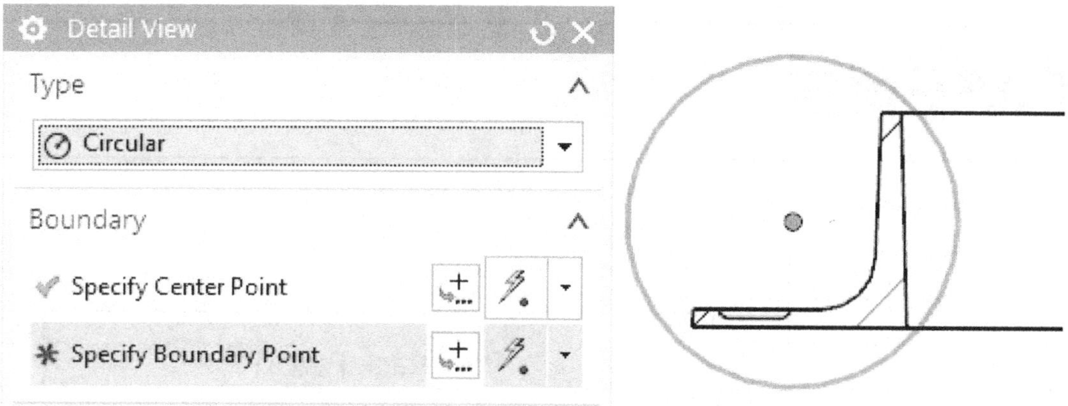

- Set the **Scale** to *1:1*, and specify a **Label on Parent** of *Label*. Click to place the detail view.

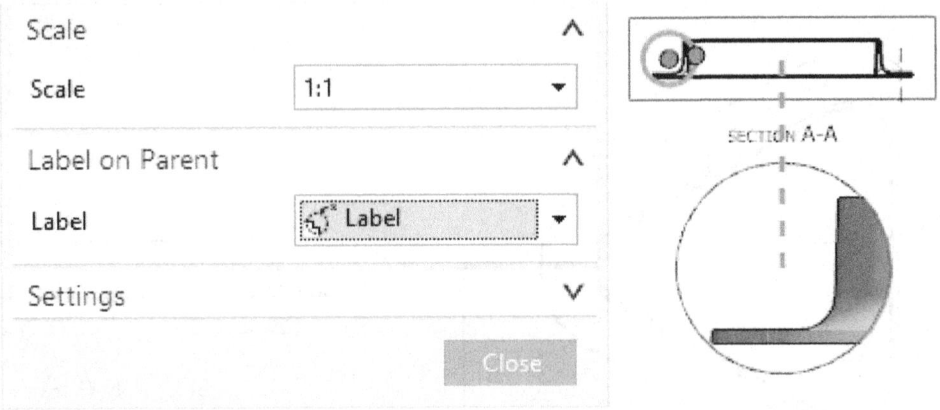

- The result is a view that cleanly illustrates the pocket. The label position can be dynamically modified by dragging it. If you wish to change the point of contact the leader arrow makes with the detail view boundary, hold **[Shift]** while you drag the label to move it dynamically.

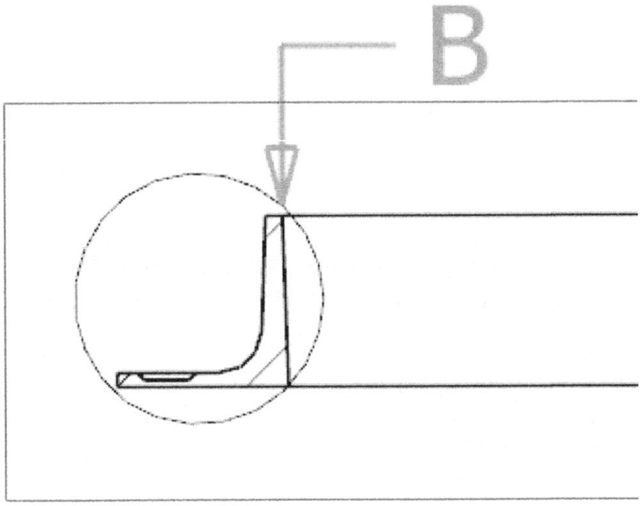

12.4 The Title Block

The *Siemens* template used to create our drawing sheet comes equipped with a title block in the lower right corner. The title block itself is nothing more than a tabular note, designated as the title block. Any modification that you can make to a tabular note, you can also make to the title block. Since it is flagged as a special table, it comes with its own special set of tools that enable you to populate its cells easily. Some of the cells in the title block are linked to part attributes, and you can link blank cells to existing part attributes and expressions to your heart's content.

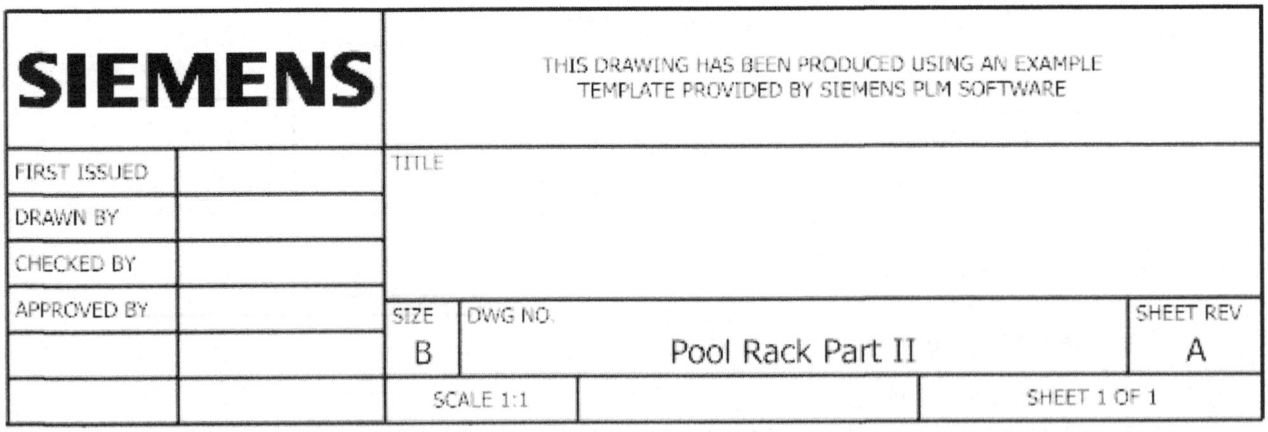

- Switch to the Modeling application and use the **Assign Materials** command to specify a material of *ABS* for the solid body of the pool rack.

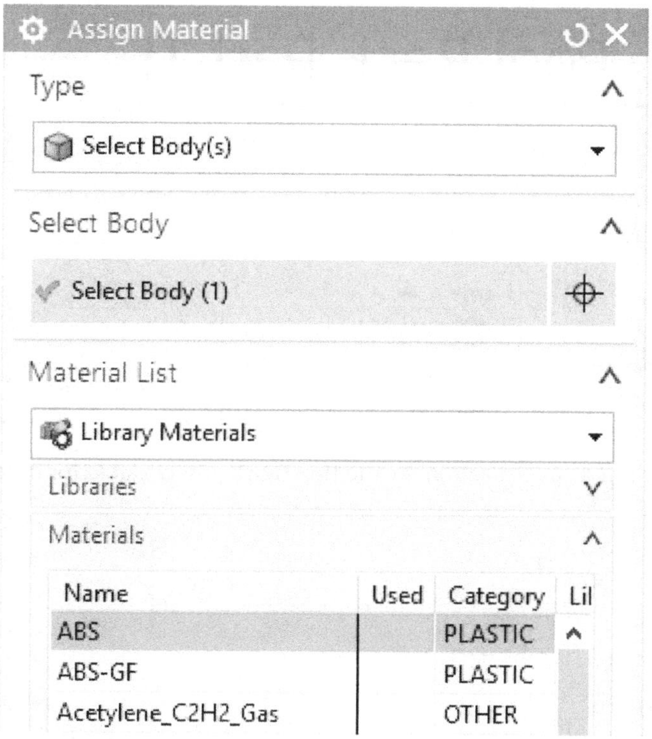

- Return to the drafting application, right-click on the drawing sheet from the Part Navigator, and select **Edit Sheet**.

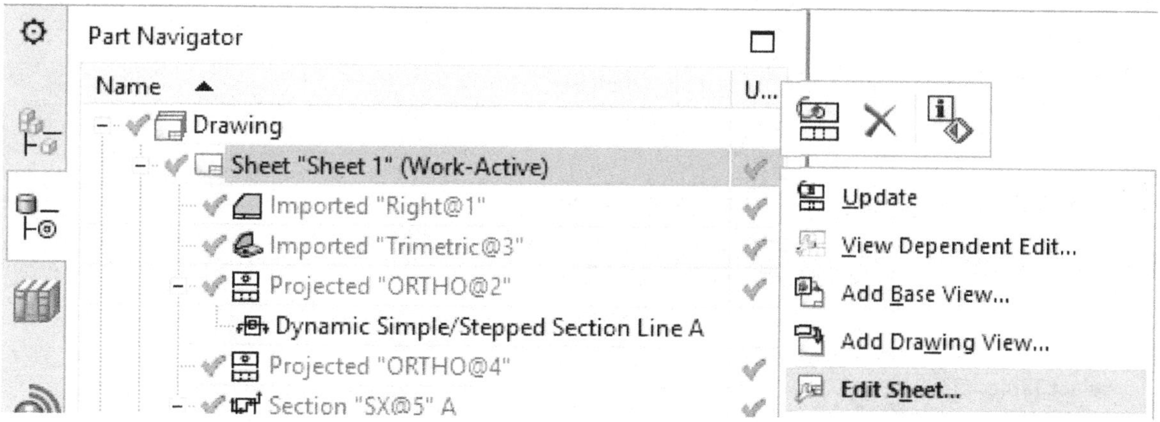

- The scale of the drawing sheet is an attribute that is controlled here in the **Sheet** dialog – set the **Scale** to *1:4* to match the scale used for the base views on the drawing sheet.

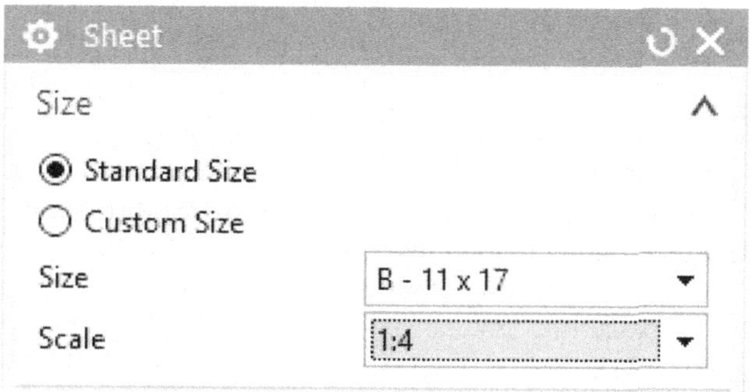

- After modifying the scale, the corresponding cell in the title block will update.

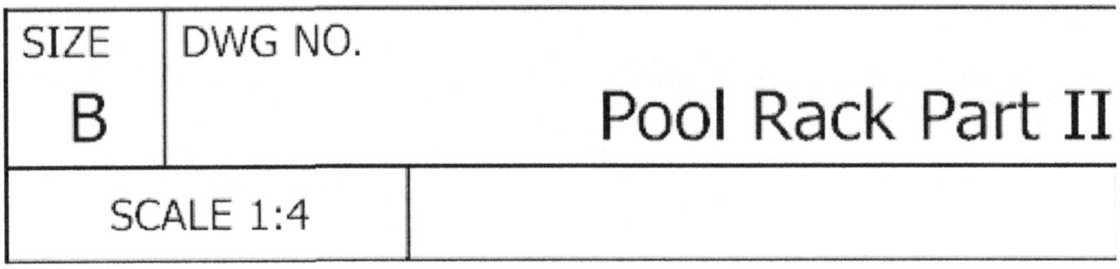

- Change layer *256* from *Visible Only* to *Selectable* in the **Layer Settings** menu.

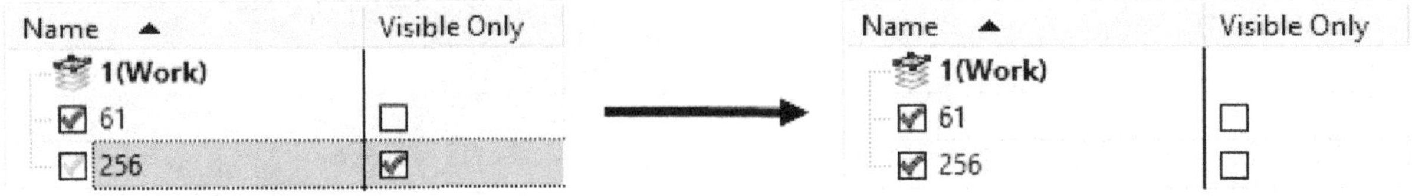

- Right-click on the title block and select **Edit Definition**. This is a special action for the title block not available for other tabular notes.

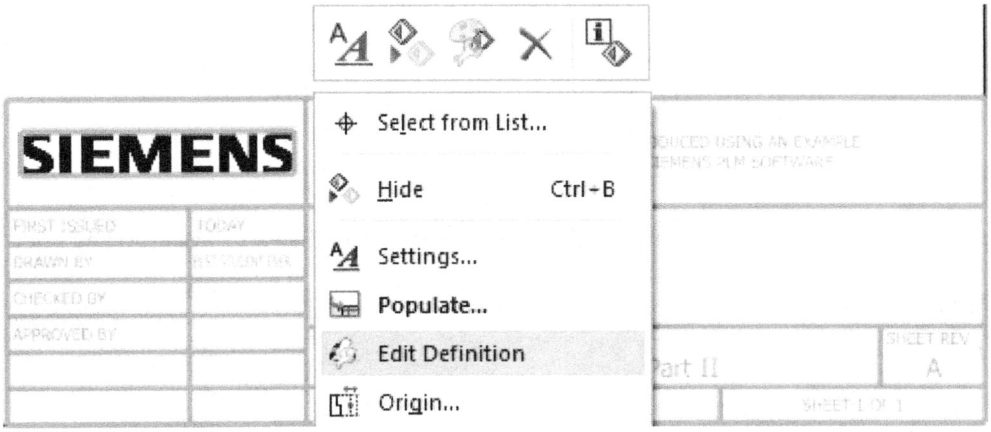

- In the **Define Title Block** dialog, push the **Edit Table** button.

- This brings you into the *"Edit Title Block Table"* environment – it is similar to the Sketch Task Environment in the sense that the ribbon has changed to show only tools pertinent to editing the title block, but you are still in the Drafting application. Like the Sketch Task Environment, you must use either **Exit Edit Table** or **Finish Edit Table** to return to the broader Drafting application and continue making other drawing objects.

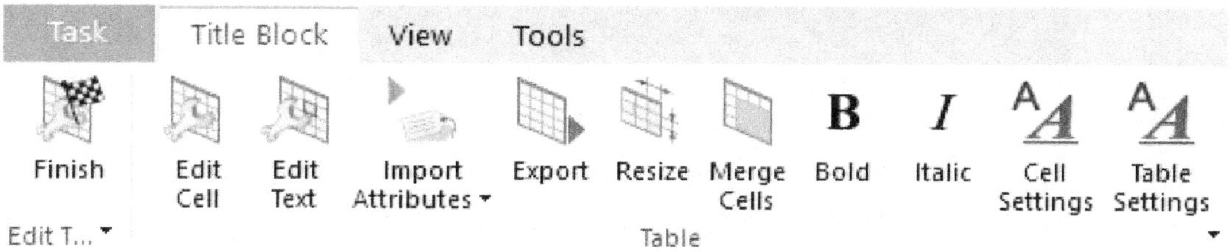

- Right-click the cell shown below and select **Edit Cell**. Type *"MATERIAL"* in the text entry field and press **[Enter]** to save the change.

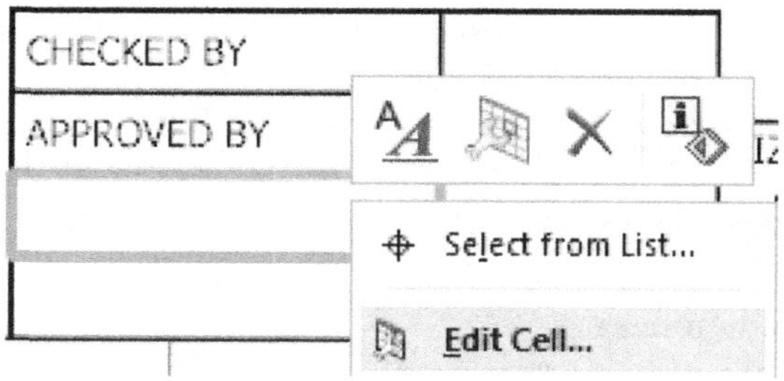

- Right-click the cell shown below, and select **Edit Text**.

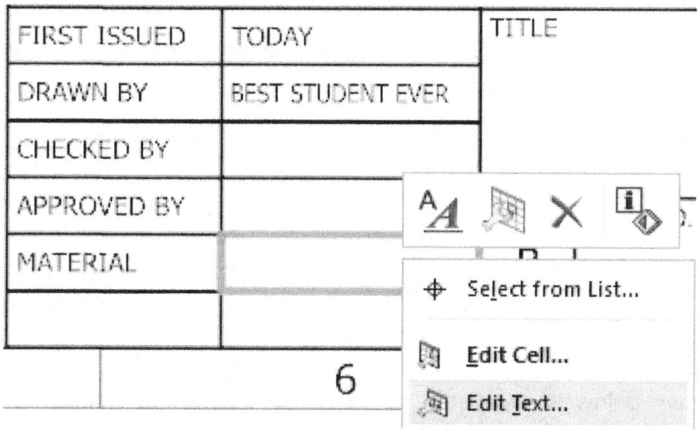

- This brings up a more involved **Text** dialog with an extended text entry field. In the **Symbols** section of the dialog, change the **Category** to *Relationships*, and push the **Insert Part Attribute button**.

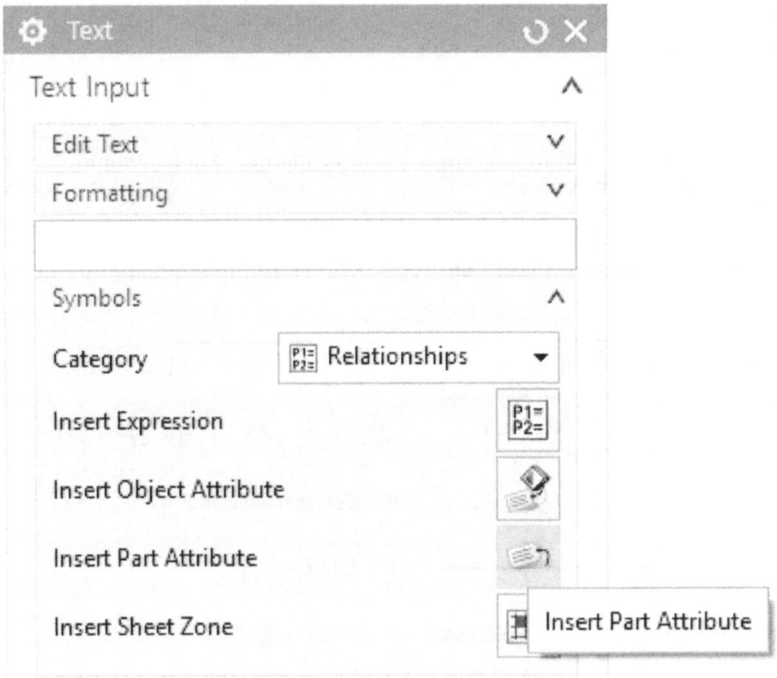

- This brings you to the **Attributes** dialog, where you can easily find the *"Material"* attribute. Select it and click **OK**. Click **OK** when you return to the **Text** dialog as well.

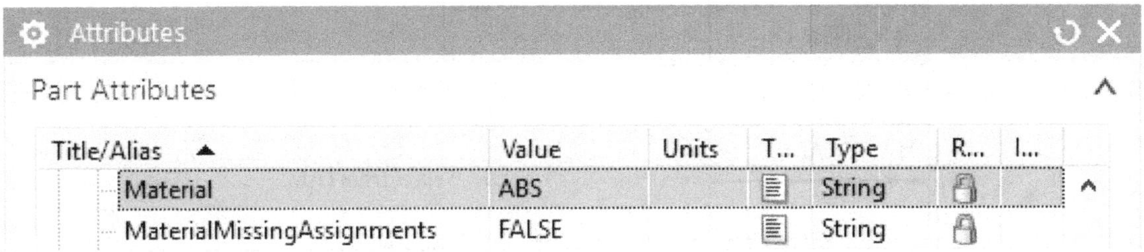

- The value in the cell shown below is associative in the sense that if you change the part material using **Assign Materials**, the value in the cell will update.

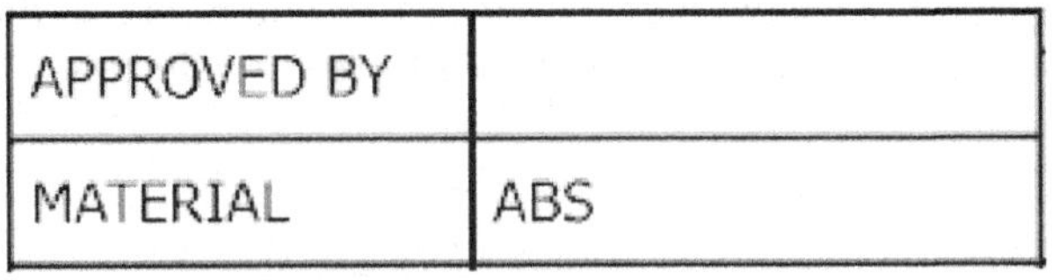

- Enter *"WEIGHT"* for the cell below *"MATERIAL"* and then right-click the cell to its right and select **Edit Text**.

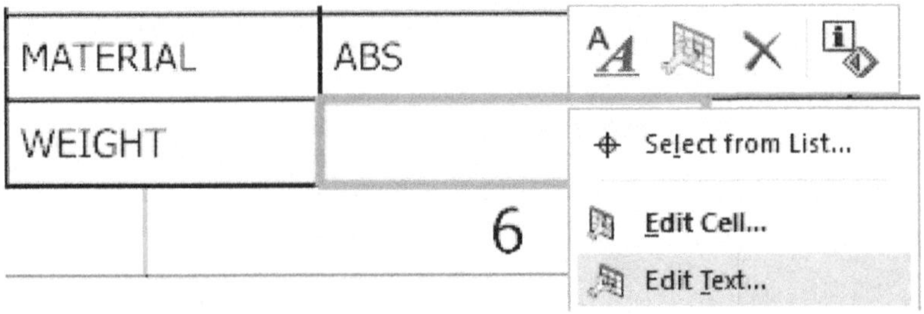

- The weight is another part attribute of interest – find it in the same way you found the material. Notice that both the material and the weight have special symbols you can type directly into the text entry field. Add *" LBS"* to the text entry field following the weight.

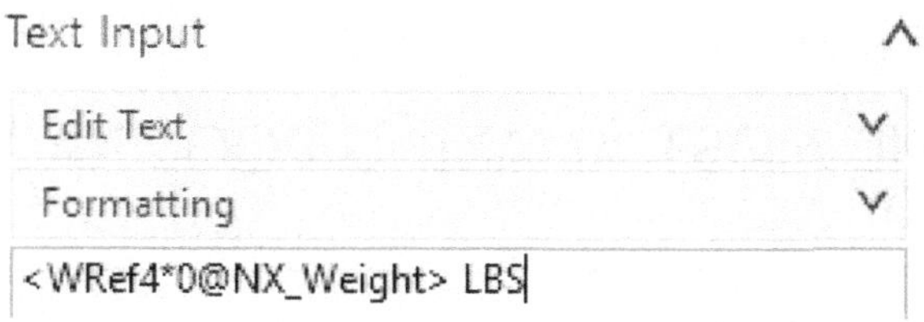

- The weight displays 6 decimal places by default. There are several ways to display a different number of decimal places – for instance, you could use a measurement expression instead of the attribute!

MATERIAL	ABS
WEIGHT	0.511238 LBS

- Change the value in the cell underneath *"TITLE"* to *"THE MOST AMAZING POOL RACK"*. Click on the cell to select it when you are done.

- Push the **Cell Settings** button.

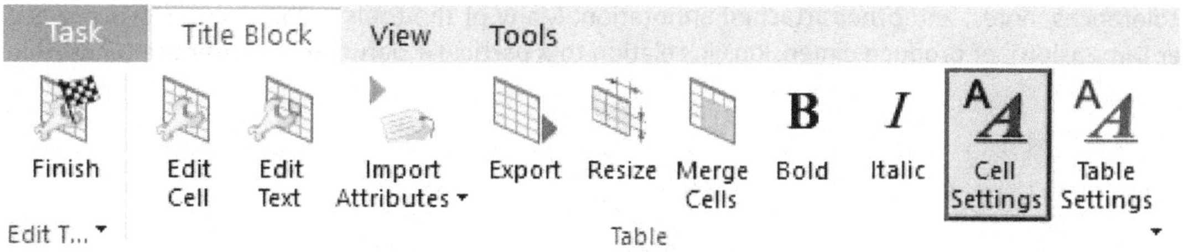

- In the **Settings** dialog, navigate to *Cell* settings, and change the **Text Alignment** to *Middle Center*.

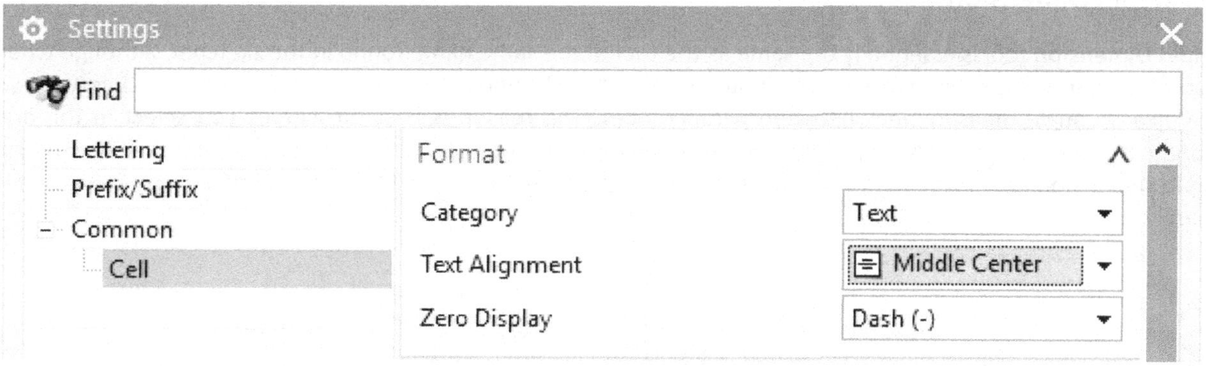

- Now the text becomes centered in the cell.

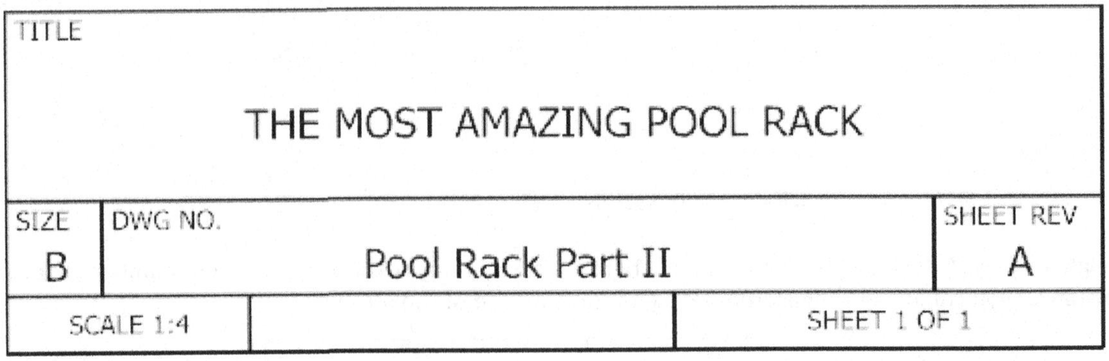

- **Finish Edit Table** to return to the Drafting environment.

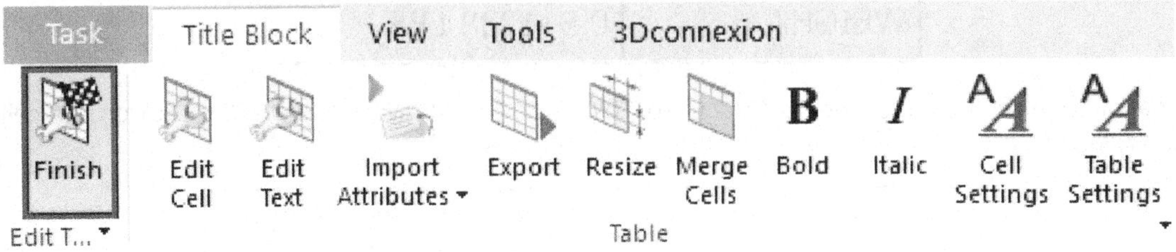

- Click **OK** on the **Define Title Block** dialog to save your changes.
- Make layer *256 Visible Only* again.
- **Save** your file!

12.5 Dimensions

The tools in the **Dimension** group on the **Home** tab are used to call out dimensions of interest on the drawing, with appropriate tolerances, notes, and other attached annotation. Many of the tools in this group call out specific features, (e.g., **Chamfer Dimension**), or produce dimensions in relation to a particular datum (e.g., **Ordinate Dimension**).

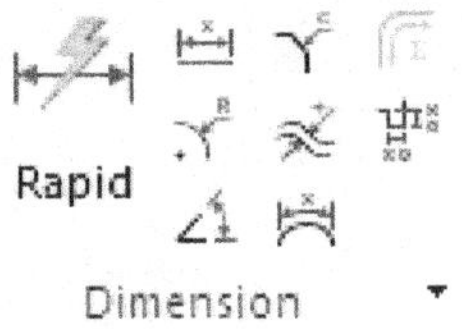

12.5.1 Rapid Dimension

The **Rapid Dimension** tool is essentially the same as the tool of the same name found in the sketcher, although when the **Measurement Method** is set to *Inferred*, it isn't quite as smart as the one in the sketcher. Sometimes the inference will be strange, like an angle measurement between parallel faces. The reason is that the curves you select in the drafting application are often projected faces, and the inference is based on the nature of the selected object. It doesn't render the **Rapid Dimension** tool unusable, you will just have to set the **Measurement Method** explicitly more often than in the sketcher.

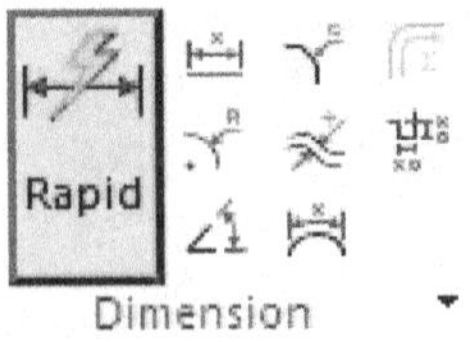

As you place dimensions using the **Rapid Dimension** tool, you will encounter a very handy dialog that appears if you pause, called the *Dimension Scene Dialog*. This menu offers shortcuts to a variety of dimension settings that you would otherwise need to manipulate with the **Edit Settings** tool, and also lets you add appended text.

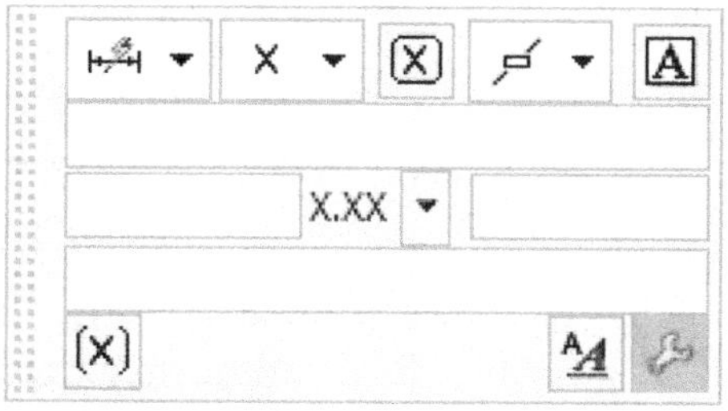

- Select the **Rapid Dimension** tool and choose the point from the right view shown below as the **First Object**, and the endpoint of the bottom edge as the **Second Object**.

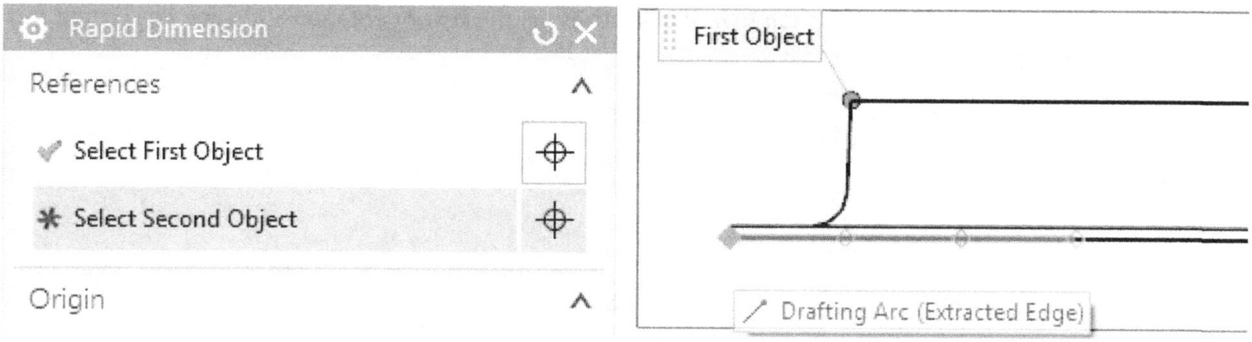

- Moving your cursor away up and to the left results in a *Point to Point* measurement. Do not click to place the dimension.

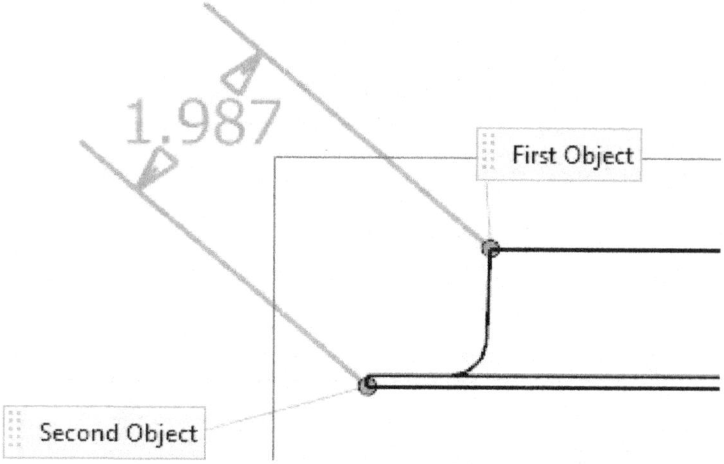

- Pause for a moment and the *Dimension Scene Dialog* will appear. The top-left drop-down in this menu is the **Measurement Method** from the **Rapid Dimension** dialog. Change it to *Vertical*, as shown below.

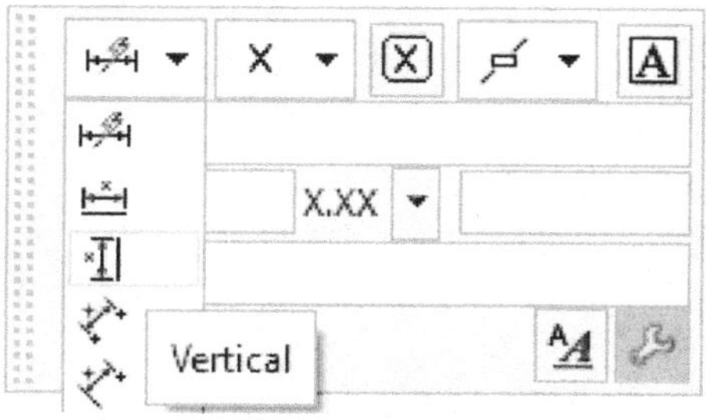

- This forces the dimension to measure the vertical distance between the points – a more sensible measurement on this part to be sure. Again, do not click to place the dimension yet.

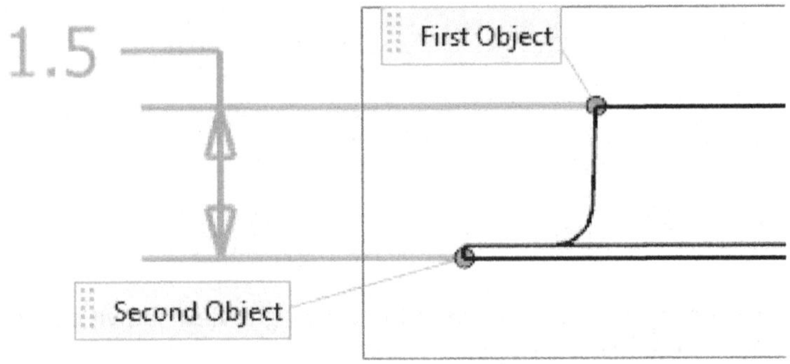

- The rightmost drop-down menu on the top row of icons controls the text orientation – change the setting to *Text Perpendicular to Dimension Line*.

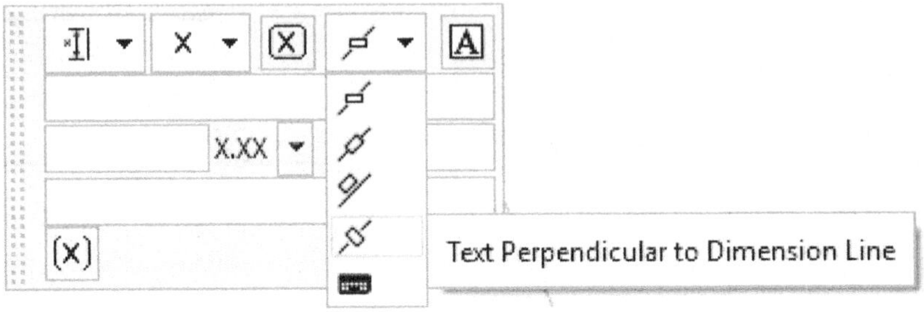

- This eliminates the *stub*, so that the text sits directly above the dimension line.

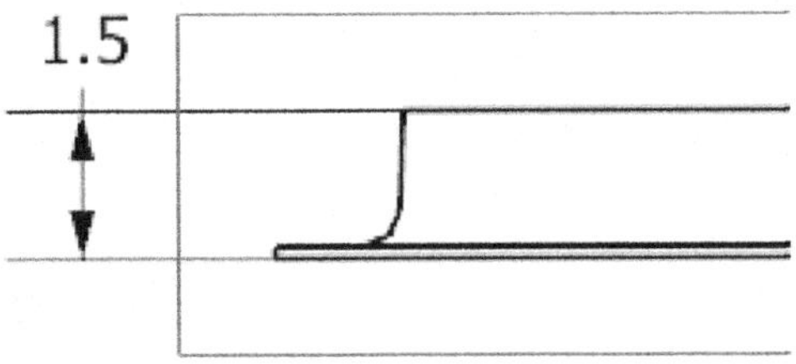

- Add two more dimensions to the right view, as shown below.

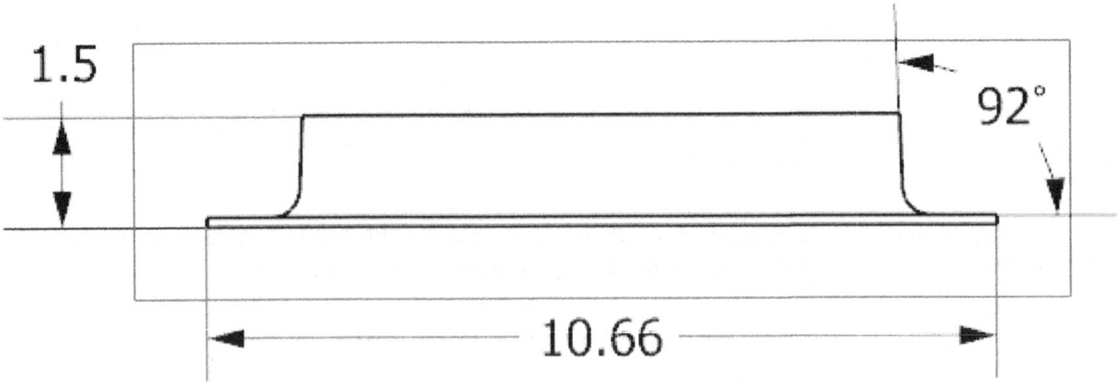

- Add the dimensions shown below to the pocket.

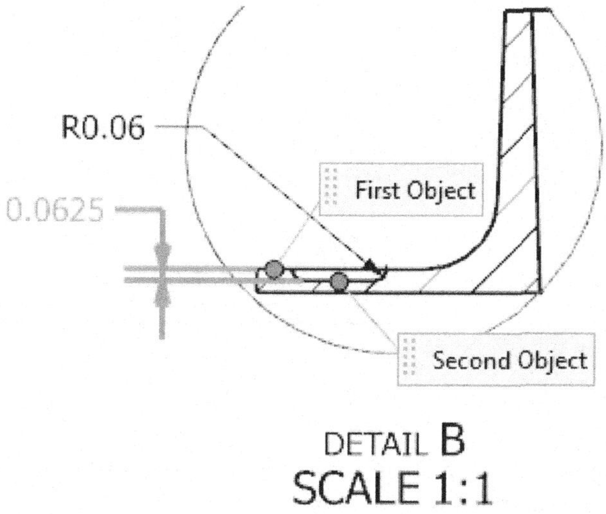

- For the depth of the pocket, there is a useful setting within the *Dimension Scene Dialog* that allows you to display more decimal places – change this drop-down to show *4* decimal places.

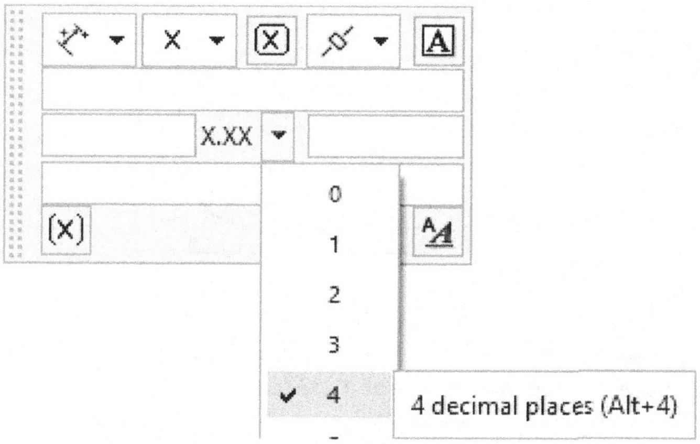

- A useful property of drafting dimensions is associative alignment. Drag the radial dimension over the vertical dimension as shown below, so that a dotted vertical line appears, and then drag the dimension straight up and release.

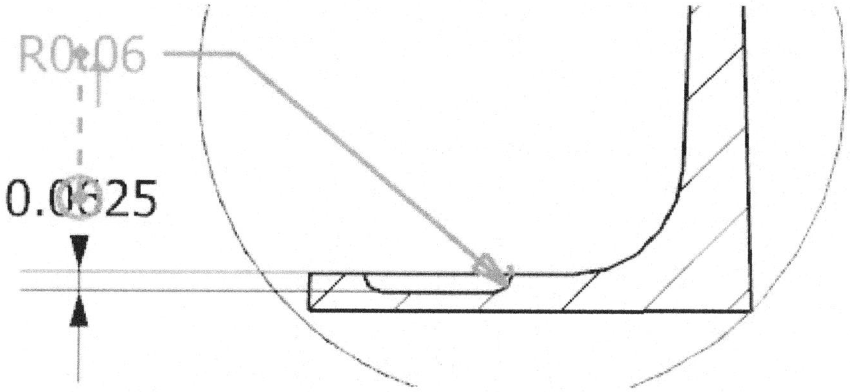

- Now, if you move the vertical dimension to the left, the radial dimension will follow – notice the asterisk at the top right corner of the "R".

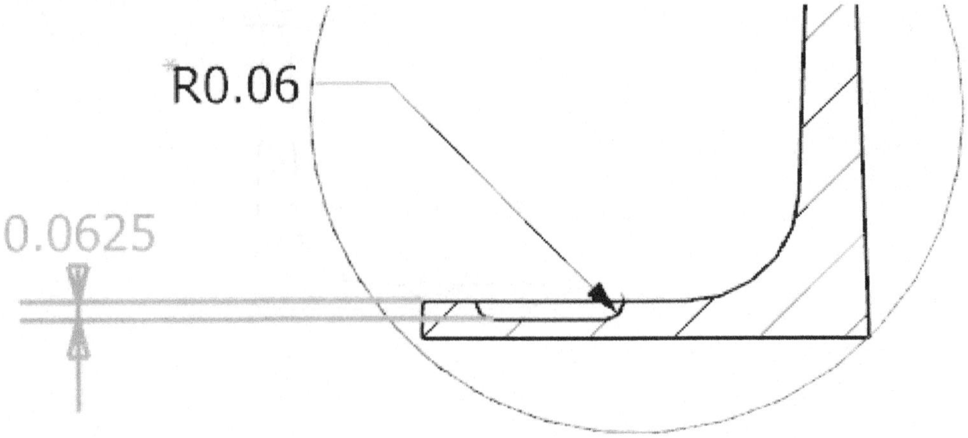

- Add a radial dimension to call out the edge blend, and use the same technique to guarantee vertical associative alignment with the radial dimension beneath.

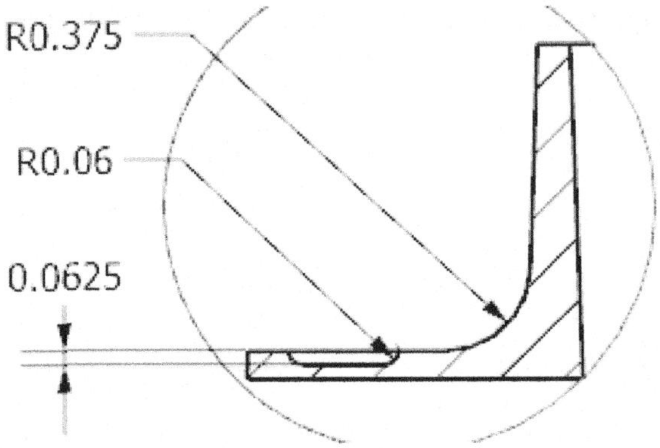

- Turn your attention to the top view – our next aim is to dimension the length and width of the pocket, but the feature is too small on the top view. Create a new detail view at a *1:2* scale.

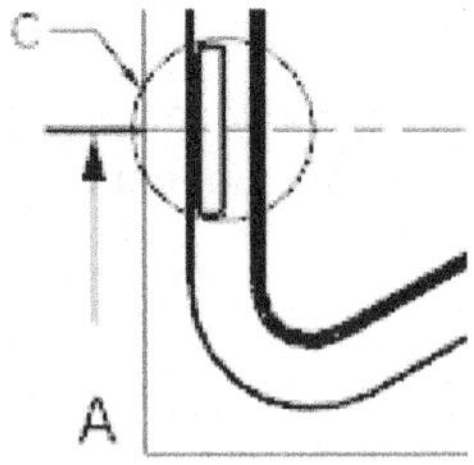

- Add the dimensions shown below to the detail view instead.

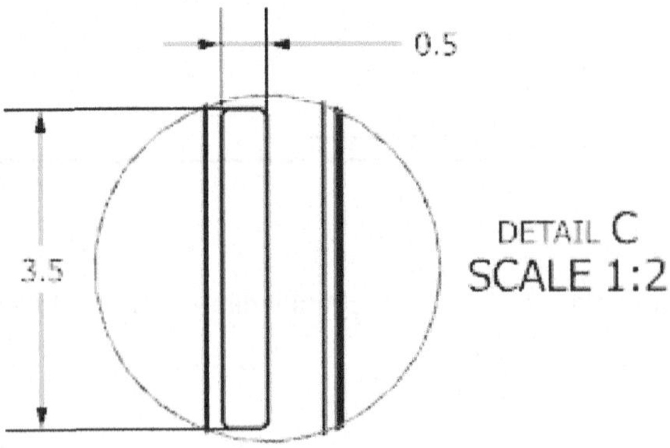

12.5.2 Hole Callout

One very important kind of dimension that the **Rapid Dimension** tool is unable to produce is a hole callout. To produce hole callouts, you must use the **Radial Dimension** tool.

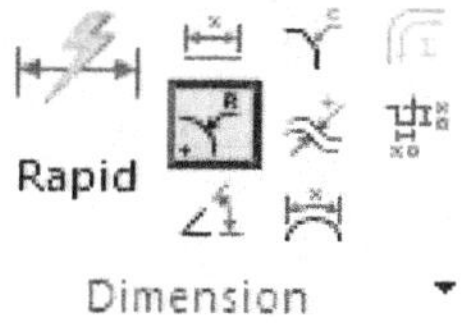

The key setting within the **Radial Dimension** dialog that enables hole callouts is a special **Measurement Method**.

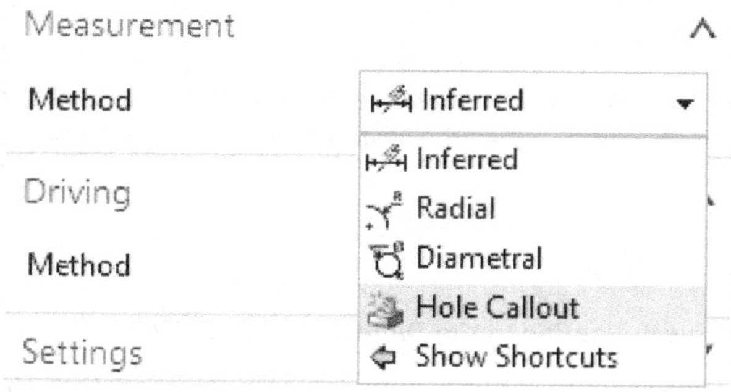

- Our intention is to call out the simple thru hole visible from the top view, but it is far too small in this view.

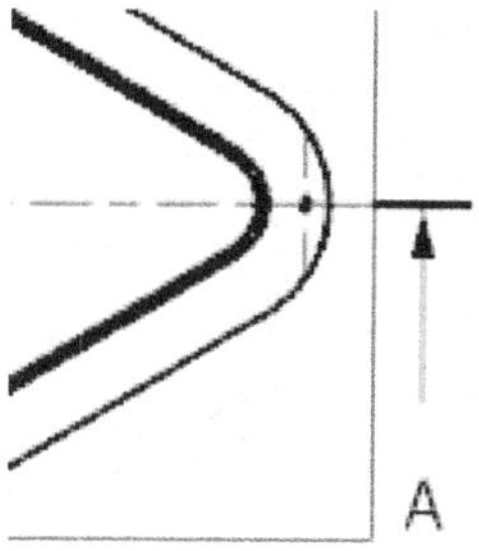

- Add a **Detail View** just big enough to show the hole and the edges of the body nearest it. Set the Scale to *1:1*.

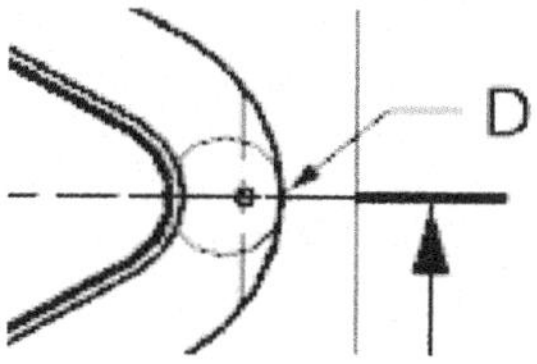

- Select the **Radial Dimension** tool and set the **Measurement Method** to *Hole Callout*. Choose the hole and place the dimension as shown below. Note that it calls out the depth of the hole – in this case, *"thru"*.

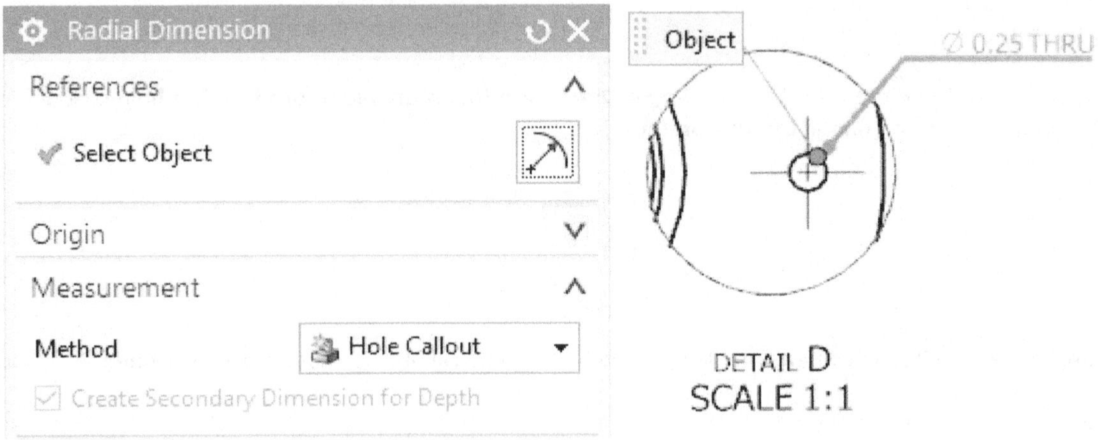

- For other kinds of holes (counterbored holes, threaded holes), this tool will report on the features of interest (counterbore depth & diameter, thread size). If your hole is a member of a feature pattern, it will also include the pattern count!

12.5.3 Appended Text

The appended text entry fields are a useful option in the *Dimension Scene Dialog*.

- On the bottom view, place two radial dimensions as shown below. Since there are three corners of the pool rack, and each corner has the same radius, these dimensions can include a note for *typical* (*"TYP"*) to indicate that the dimensions are the same in the other corners.

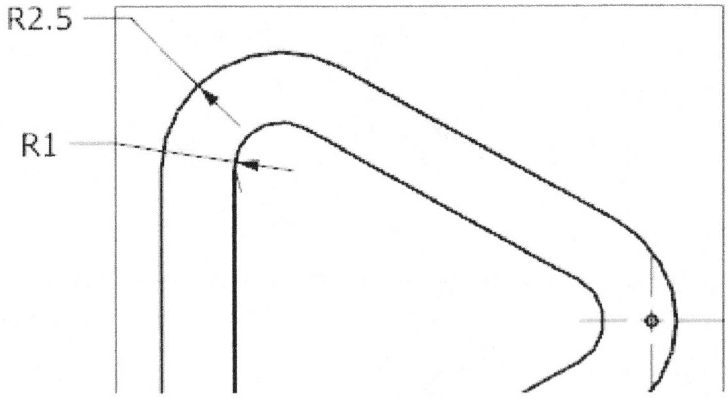

- Edit each dimension, and enter *"TYP"* in the appended text entry field in the *Dimension Scene Dialog*, as shown below. There are four text entry fields – above, before, after, and below.

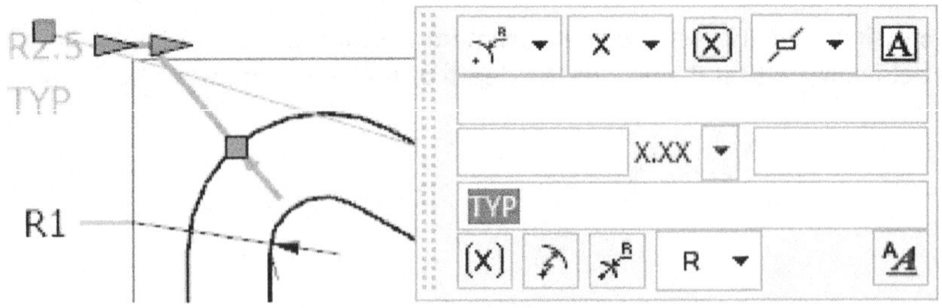

12.5.4 Tolerances

Another very useful tool within the *Dimension Scene Dialog* is the **Tolerance** drop-down. This drop-down is second to the left on the top row of the dialog.

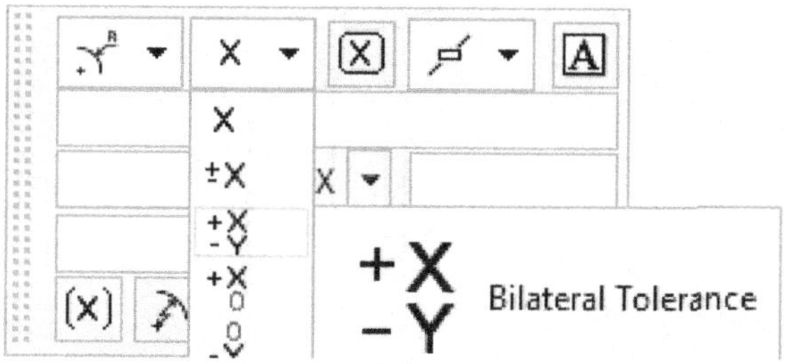

- Edit the horizontal dimension on the detail view showing the pocket. Change the **Tolerance** drop-down to *Bilateral Tolerance*. In the text entry fields that appear, enter the tolerances shown below, and finish editing the dimension.

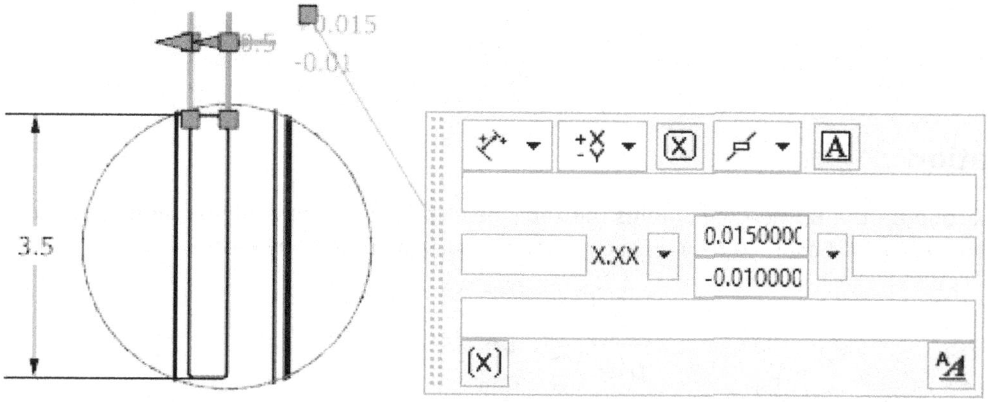

- Tolerances can also be added using the **Edit Settings** tool. A useful operation within the **Edit Settings** tool is the **Inherit** button. Select the vertical dimension on this view and then select the **Edit Settings** tool. In the **Inherit** section of the dialog, change **Settings Source** to *Selected Object*, and choose the horizontal dimension with the tolerance.

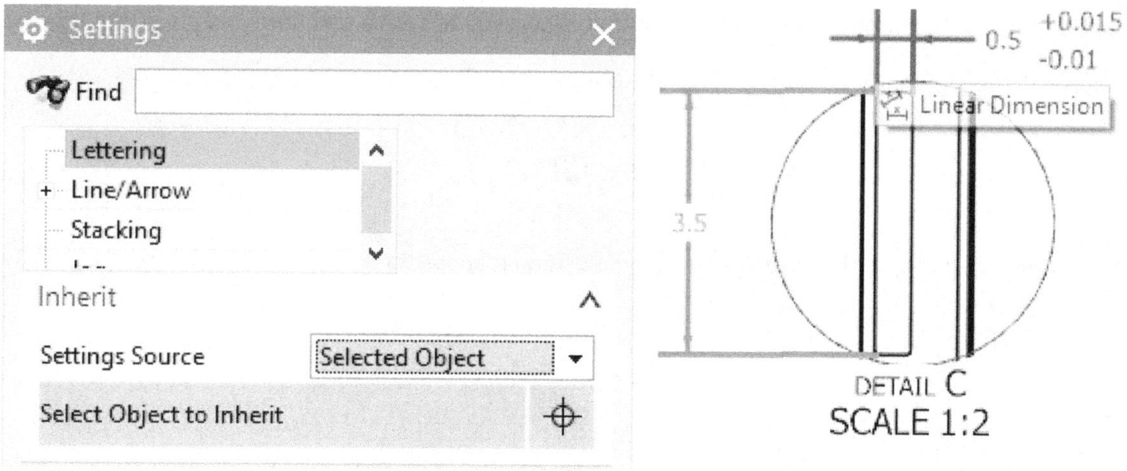

- This adds the same tolerance to the vertical dimension, and will copy any other modifications made to the dimension settings. Inheritance is an extremely powerful method for formatting your drawing dimensions and annotation, you can set a single dimension or note the way you want it, and then use **Edit Settings** to inherit all the changes to many other dimensions and notes. Quite the time-saver!

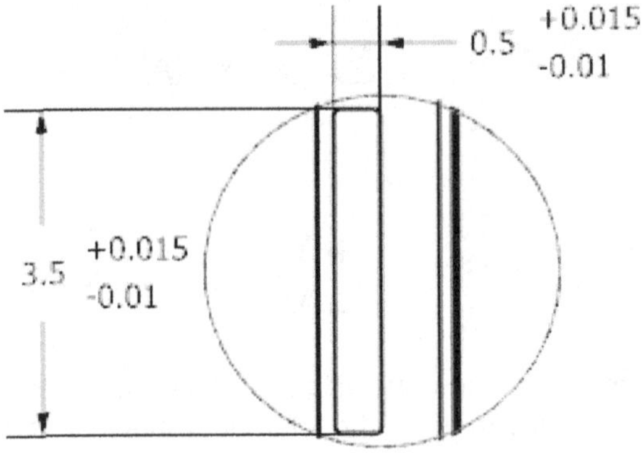

DETAIL C
SCALE 1:2

12.6 **Annotation**

The **Annotation** group on the **Home** tab contains the tools necessary for adding additional markup to your drawings – symbols like centerlines, GD&T characteristics, callout balloons, weld symbols, and more!

12.6.1 Note

The **Note** tool is the basic tool in the **Annotation** group that allows you to type a text note and place it on the drawing.

- Select the **Note** tool and type the note shown below. Position it above the title block.

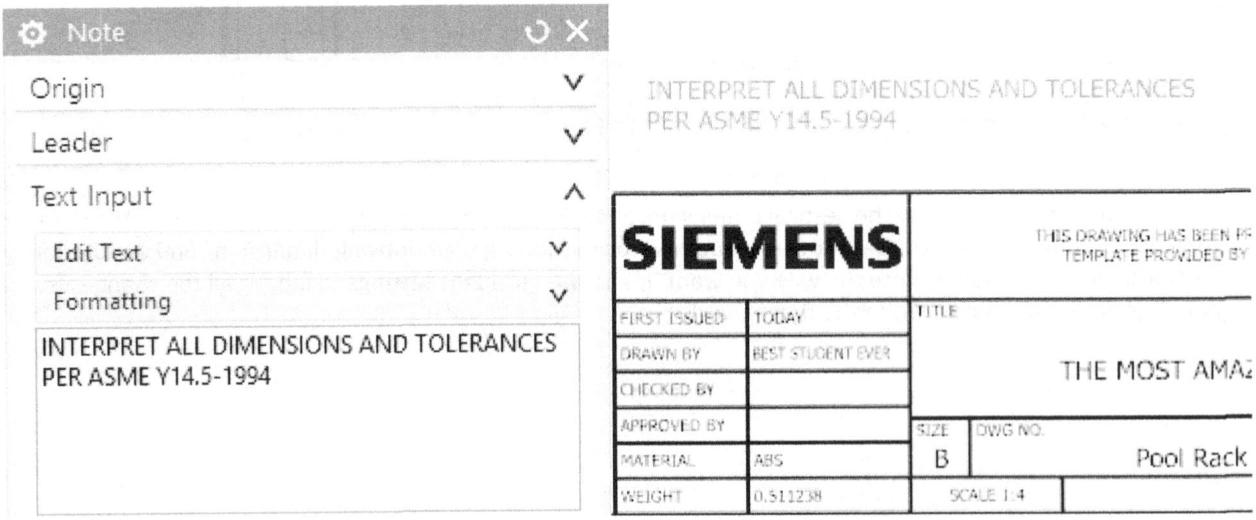

12.6.2 Datum Feature Symbol

If you wish to add GD&T markup to your drawing, you will first need to be able to specify datums – enter the **Datum Feature Symbol**.

- Move to the right view, and select the **Datum Feature Symbol** tool. Set the **Datum Identifier Letter** to *A*, and click-and-hold-and-drag on the bottom face of the model, as shown below. Move the datum symbol to the left, keeping it horizontally aligned with the horizontal dimension. Continue holding down the left mouse button until you are satisfied with the position of the symbol, then release and click to place it.

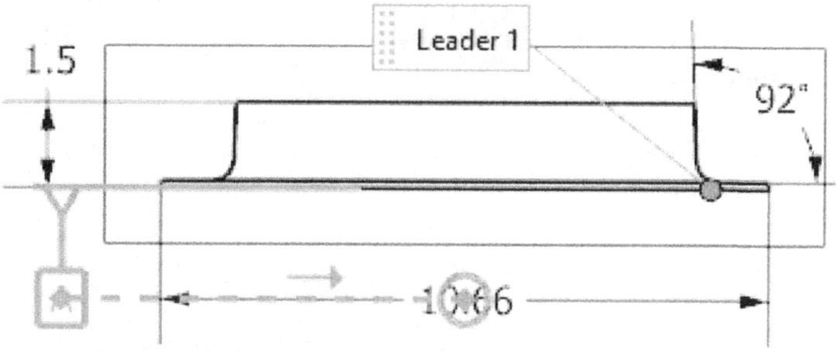

- If you wish to adjust the position of the symbol, hold **[Shift]** while you drag it around and the datum symbol will move along the extension line.

12.6.3 Feature Control Frame

Feature Control Frame is the tool that allows you to add specific GD&T characteristics to features and faces, with options for any kind of frame you might wish to depict.

- Select the **Feature Control Frame** tool. Within the **Frame** section of the dialog, change the **Tolerance** value to *0.015*.

- Also within the **Frame** section of the dialog is the **Primary Datum Reference** – change the drop-down here to *"A"*.

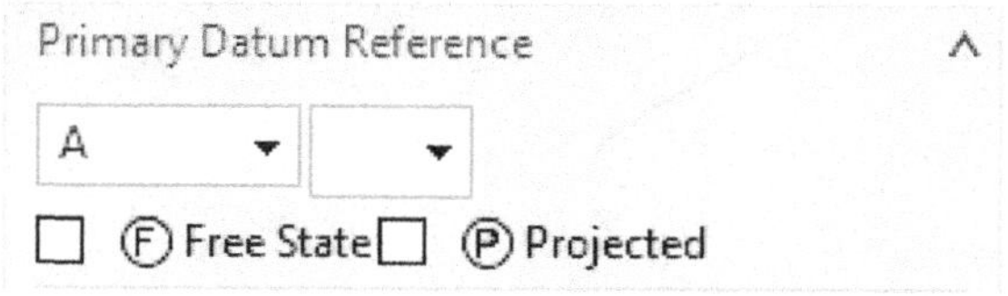

- Change the **Characteristic** to *Parallelism*, then click-and-hold-and-drag along the top face of the model as shown below. When you are content with the attachment point of the leader line, release the left mouse button, then drag the frame to the desired position and click to place it.

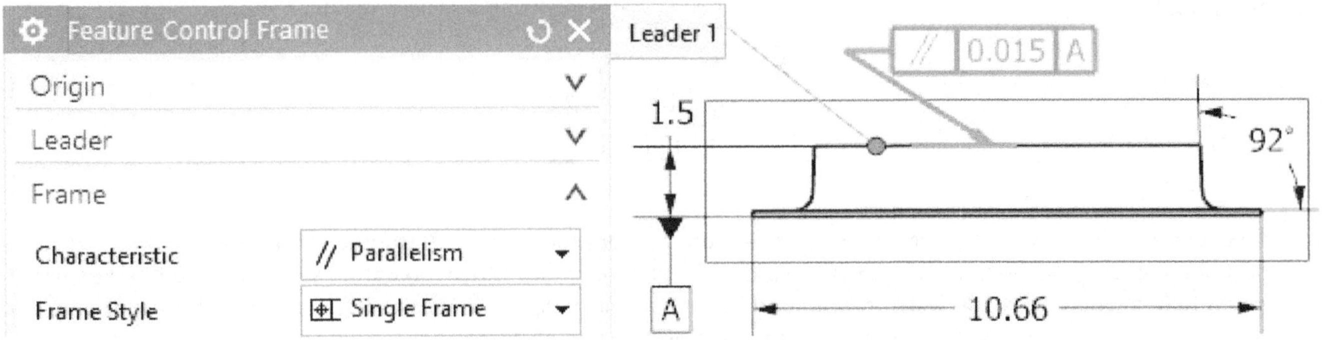

- Much like the datum feature symbol, the frame itself can be moved dynamically just by dragging on it, but if you want the arrow and leader to move as well, you must hold **[Shift]**. This is typical behavior for any such symbols in the **Annotation** group.

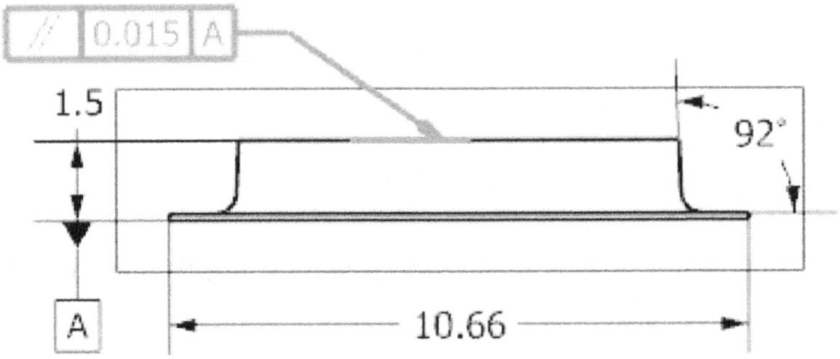

12.6.4 Center Marks and Centerlines

The tools in the **Centerline** drop-down in the **Annotation** group all add center marks or centerlines of various types – bolt circle centerlines, 2D and 3D centerlines, and more.

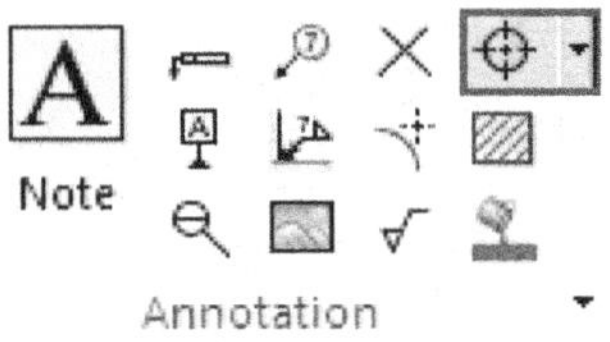

- At the time of view creation, there is an option (enabled by default in an out-of-the-box installation of NX) to create center marks on drawing views. For instance, there is a center mark indicating the position of the hole on the bottom face on the bottom view. **Delete** this center mark.

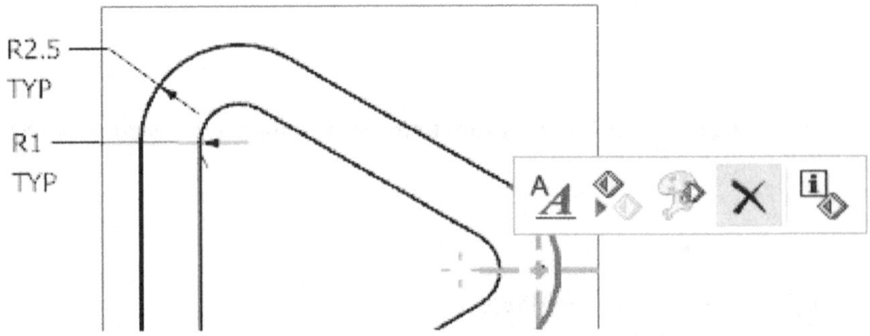

- Select the **2D Centerline** tool from the **Centerline** drop-down.

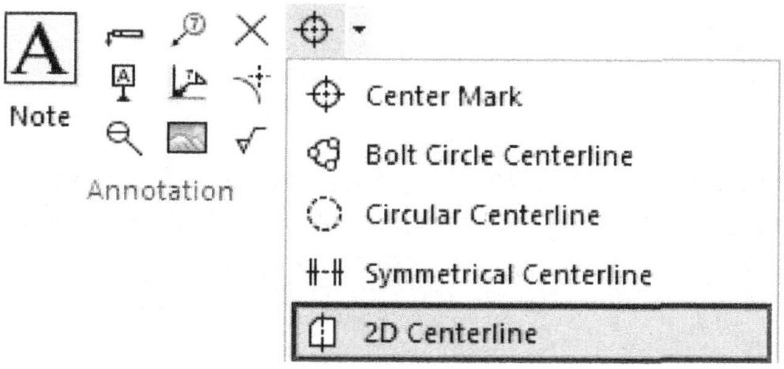

- This tool requires two inputs when the **Type** is set to *From Curves* – choose the curves shown below, and a horizontal centerline will appear. Click **OK** to create it. This centerline indicates a plane of symmetry on the model.

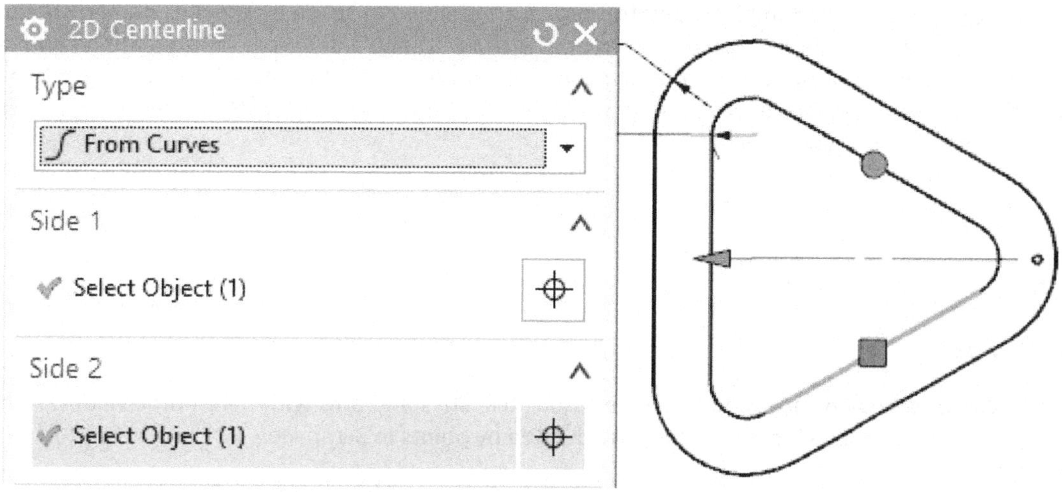

- The golden arrows indicate handles to control the length of the centerline. Edit the center mark indicating the position of the hole on the detail view created from the top view of the model. You can dynamically resize the center mark with the handle provided.

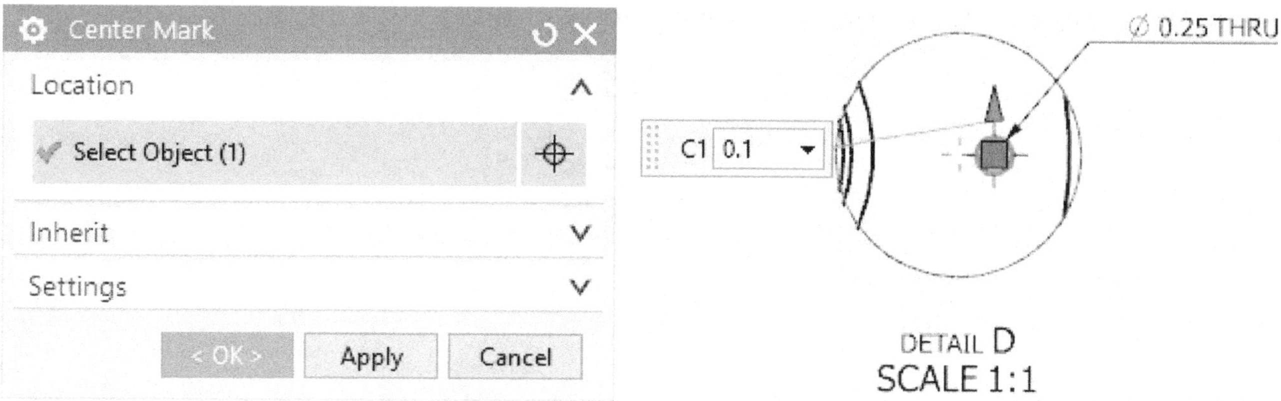

- **Save** your file!
-

13 Curves

In NX, the definition of *"curve"* includes lines, arcs, circles, ellipses, and splines, and a few other types of specially defined curves. Sketch curves are a special class of curves, but they are not the focus of this chapter. The tools discussed in this chapter are mostly found on the **Curve** tab, and they can produce curves that are twisted in 3D space.

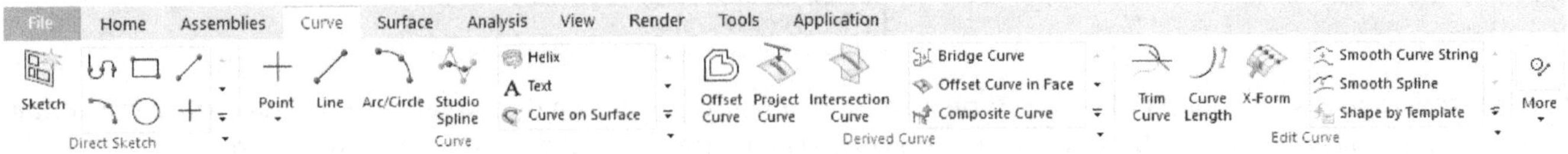

Curves can either be parametric or 'dumb'. Parametric curves are defined in space by specific, numeric parameters, and they have a timestamp that earns them a place in the Model History. 'Dumb' curves are non-quantified entities which do not have parameters defining their position. A 'dumb' curve will not be listed in the Model History. Virtually every tool on the **Curve** tab will have an *Associative* checkbox – keep an eye out for it in every dialog. When this checkbox is unmarked, the output of your operation is a dumb curve.

13.1 Lines & Arcs

The **Line** and **Arc/Circle** commands are used to create associative lines and arcs with parametric values. Although these curves are planar, they are <u>not</u> sketch curves – they are defined by points in 3D space.

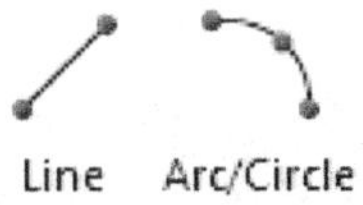

If your **Direct Sketch** group is visible, be careful not to confuse the parametric 3D curve tools with the sketch tools on the **Curve** tab – they are right next to each other and their icons are very similar!

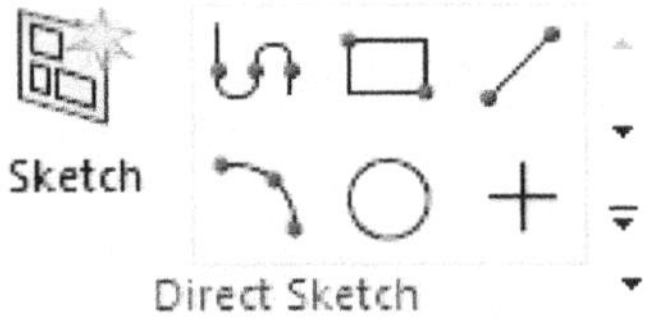

<u>Note:</u> *Remember, to avoid this confusion, you can use the* Learning NX 11 *role created in Chapter 2.9, which replaced the* ***Direct Sketch*** *group with the* ***Sketch in Task Environment*** *tool.*

13.1.1 Line

The **Line** tool connects any two points in 3d space.

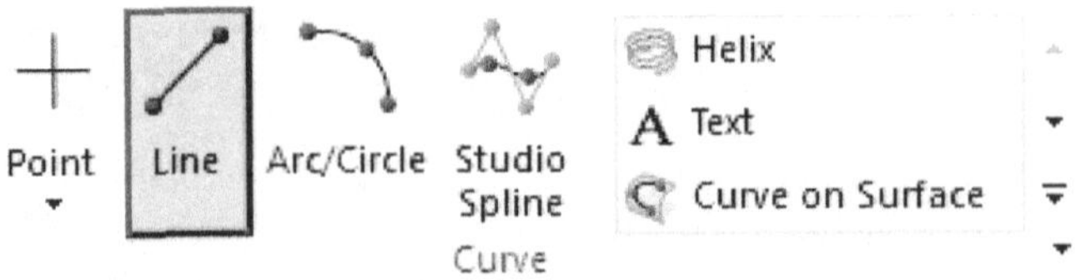

- Create a new file (**Units:** *Inches*) called *"3D Lines and Arcs.prt"* and place it in *"C:\My NX 11 Files"*.
- Select the **Line** tool from the **Curve** tab.
- For the **Start Point**, select the point of the datum coordinate system already in your file.

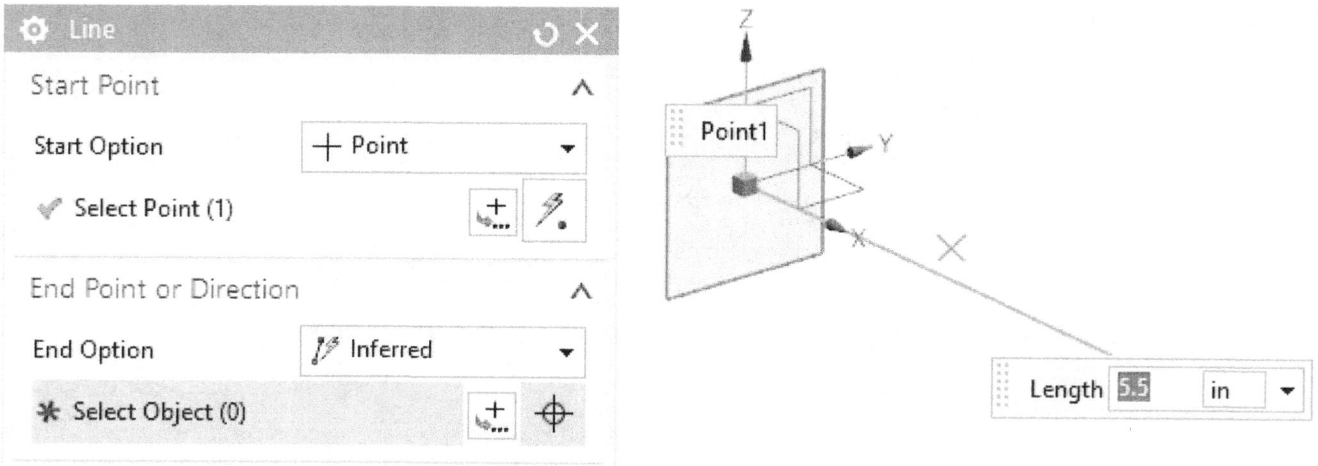

- The line tool has a variety of **Start** and **End Options** that give it quite a bit of utility, and the *Inferred* option does a good job of covering the others. For instance, notice the "X" hovering above the preview of the line under construction.
- Upon clicking, the **End Option** becomes *Along XC* – the **X** in the preview above was the *Inferred* **End Option** aligning the line with the X axis of the WCS.

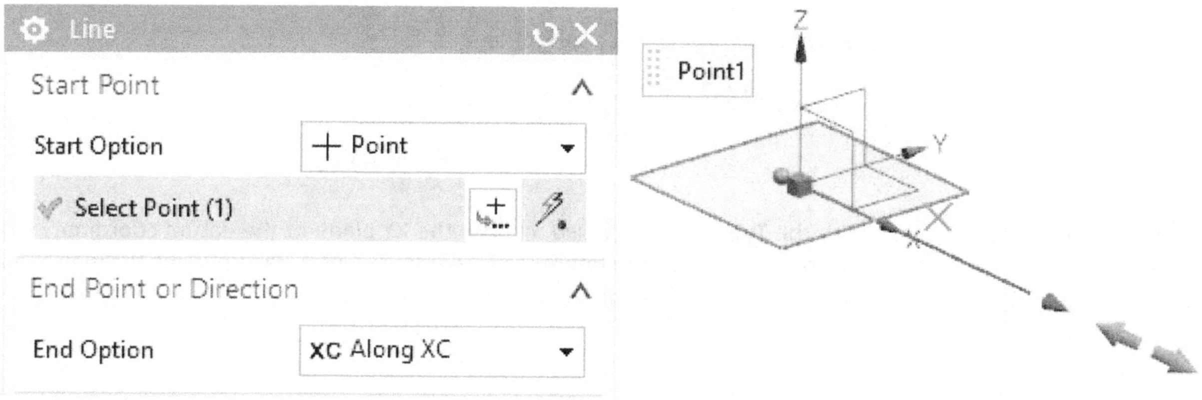

- Change the **End Limit** to *5 in*.
- Within settings, check the *Associative* box. Without this, your line will be dumb!
- Click **OK** to create your line.
- **Save** your file!

13.1.2 Arc/Circle

The **Arc/Circle** tool is, unfortunately, not quite as easy to control as the **Line** tool since – although it is not constrained to any particular plane – it sometimes gets stuck on familiar planes. In the exercise below, you will practice constraining the arc to a specific plane.

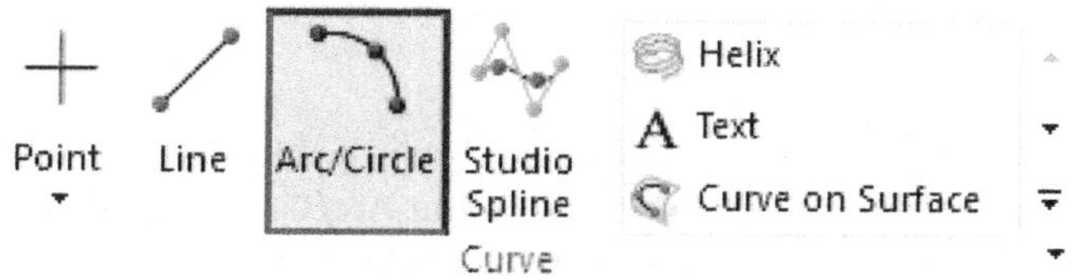

- Continue working with *"3D Lines and Arcs.prt"*.
- Select the **Arc/Circle** tool from the **Curve** tab and set the **Type** to *3 Point Arc*.
- For the **Start Point**, select the endpoint of the line along the X axis.
- For the **End Point**, set the **End Option** to *Tangent* and select the line.

- Set the **Mid Option** to *Diameter* and enter a diameter of *5 in*.

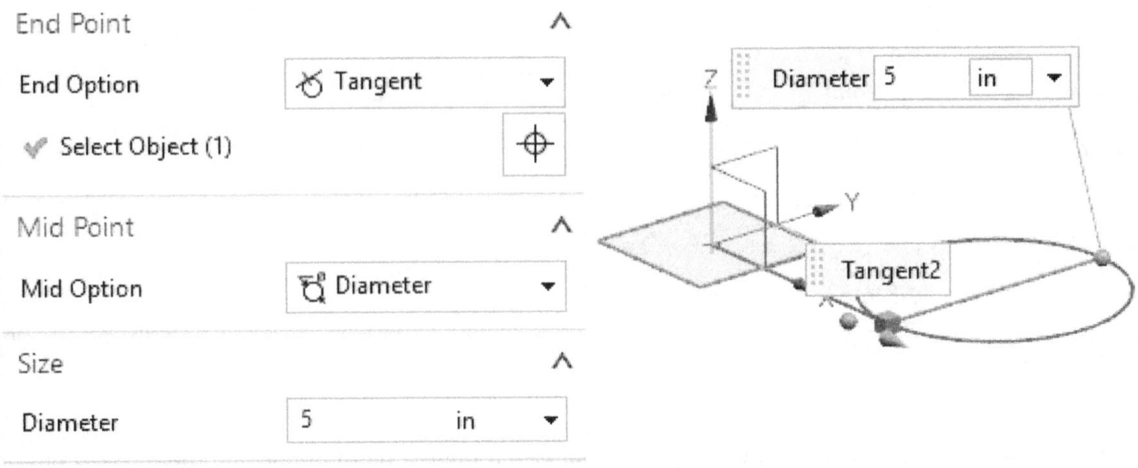

- In the **Support Plane** section of the dialog, set the **Plane Options** drop-down to *Select Plane*, and then click the **Plane Dialog** button to build a datum plane in-context and on-the-fly.

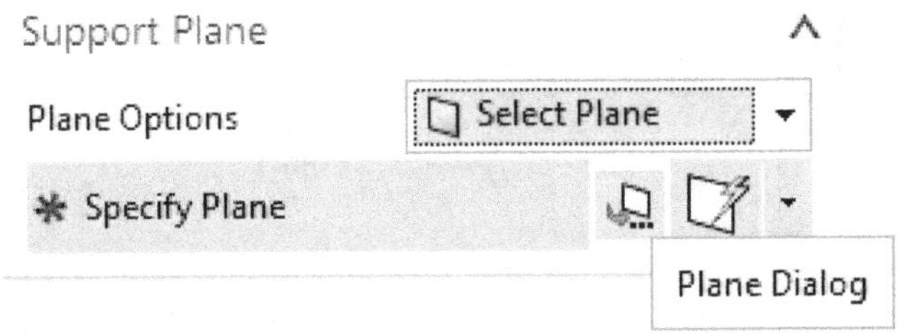

- When the **Plane** dialog appears, with the **Type** set to *Inferred*, choose the XY plane of the datum coordinate system, and then the X axis of the datum coordinate system. The inference is that you want a plane passing through the X axis and at an angle to the XY plane. Enter a value of *135 degrees*, as shown below.

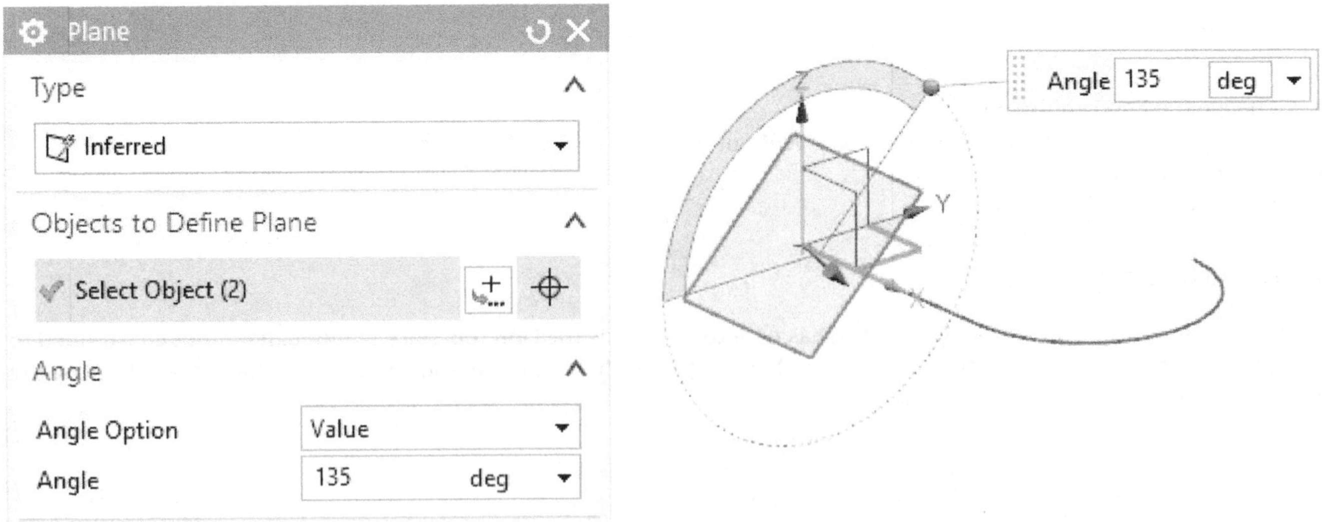

- After clicking **OK** to set the datum plane, you will return to the **Arc/Circle** dialog box. Change the **End Limit** to *180 degrees*, as shown below. If necessary, use the **Complement Arc** button.

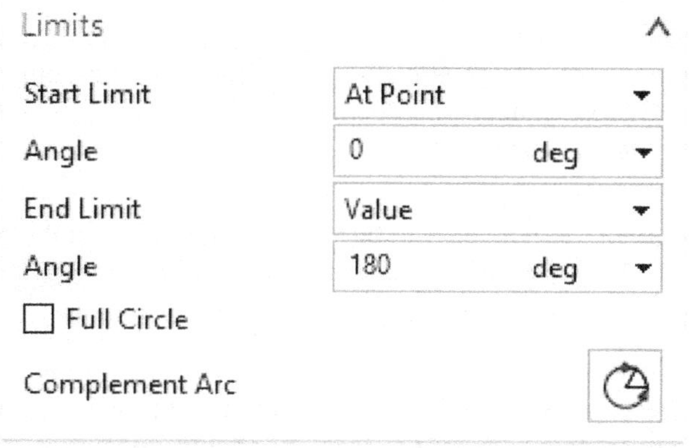

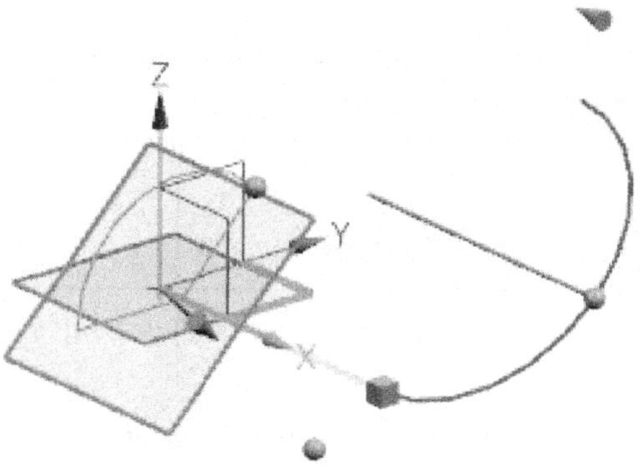

- As with the **Line** tool, be sure to check the *Associative* box within **Settings**.
- Click **OK** to create your arc!
- For extra credit, create an arc spanning *180* degrees, starting from the endpoint of the last arc, tangent to it, lying in a plane parallel to the XZ plane, and with a **Diameter** of *2.5 in*.

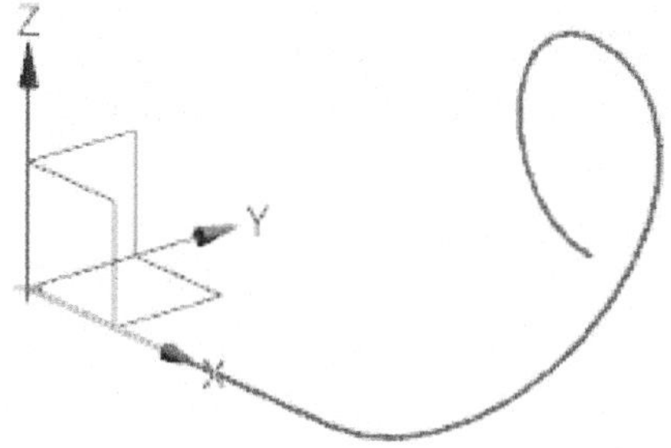

- For even more extra credit, create another **Line** of total length *5 in*, tangent to the endpoint of the last arc!

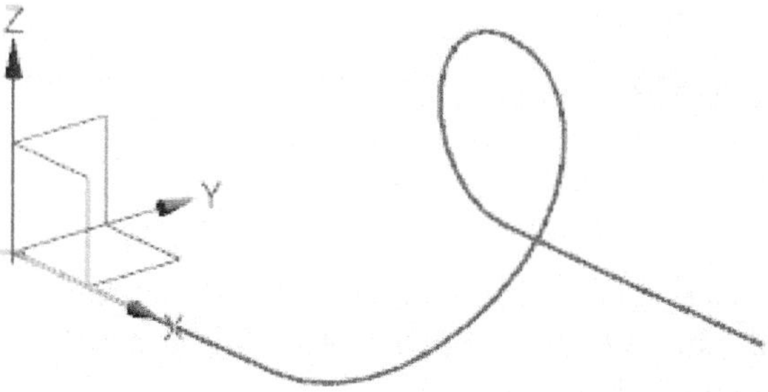

- The front and right views of your curves will appear as below.

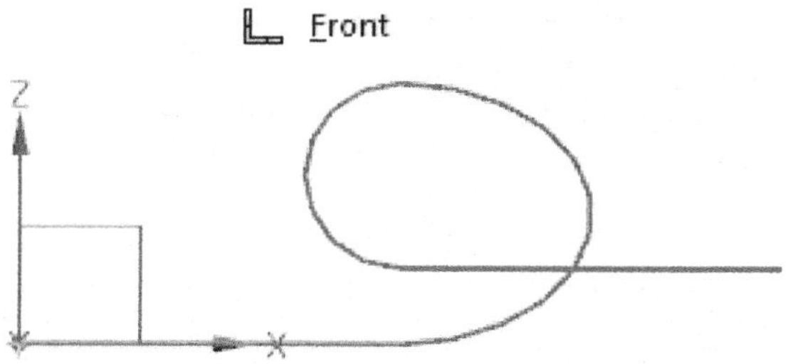

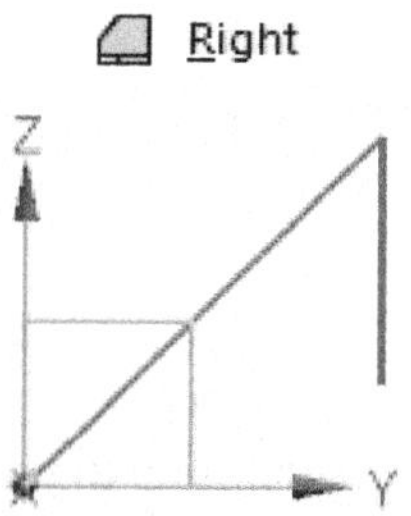

- **Save** your file!

13.2 Splines

The majority of the curves that we have encountered so far in NX – lines, arcs, occasional ellipses and conics – are all defined by polynomial equations of degree 1 or 2. To model more challenging geometry, it is necessary to make use of splines – curves defined *locally* by polynomial equations – of degree three or higher.

Line	Circle	Cubic Curve	Degree n polynomial equation
$y = mx + b$	$x^2 + y^2 = 1$	$y = 2x^3 + 5x^2 - 1$	$y = a_n x^n + a_{n-1} x^{n-1} + \cdots + a_1 x + a_0$

Splines allow you to build surfaces that are more aesthetically pleasing than those that you can build using only lines, arcs, and conics – and so they are invaluable when designing anything cosmetic, from handheld consumer gadgets to automotive bodies. They are also critical for the design of highly-engineered components, such as airfoils.

There are two tools, both named **Studio Spline**, for creating arbitrary freeform splines – one in the sketch environment, and one on the **Curve** tab. We will begin with splines in the sketch environment, as they are slightly easier to control.

13.2.1 Studio Spline (Sketch) – By Points

The default spline **Type** in NX 11 is *Through Points*. A spline constructed in this way is required to pass through all of the selected points, and it interpolates between them.

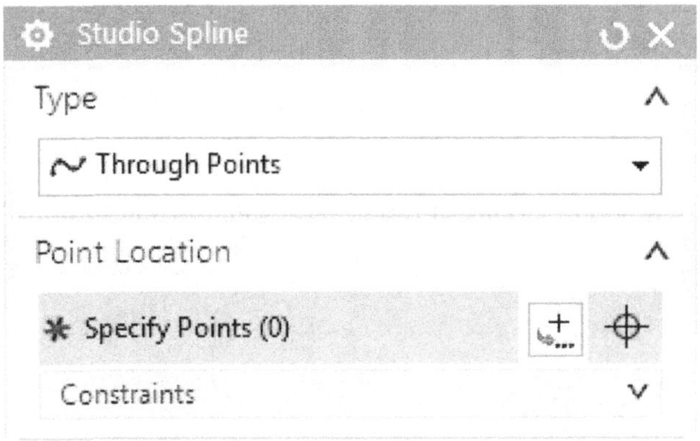

- Create a new file (**Units**: *Inches*) called *"Splines.prt"*, and place it in your folder *"C:\My NX 11 Files"*.
- Begin with a **Sketch in Task Environment** on the X-Y plane. Name it *"THROUGH POINTS"*.
- Select the **Studio Spline** tool from the ribbon.

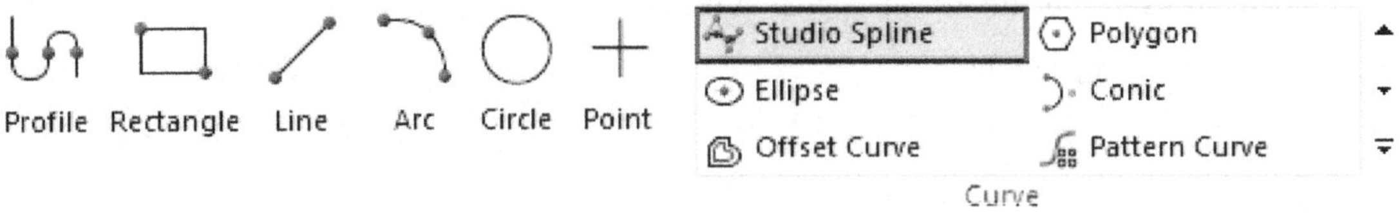

- Studio Splines are cubic (degree 3) by default. Cubic splines are a very flexible class of objects, which can be quickly and efficiently manipulated to model challenging geometry, and they are smooth enough for nearly all applications. NX 11 supports splines up to degree 24.

- Try the following exercise: to the best of your ability, select points on a vertical line, starting from the top and working your way down. You should not cheat and use constraints to get these points exactly on a line, just do your best to align them vertically by hand. Place twenty defining points for your spline in this way.

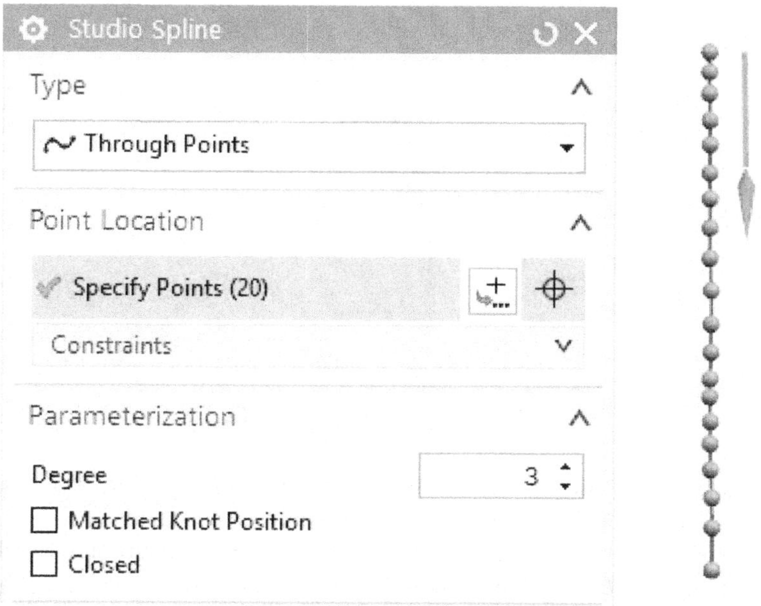

- Now increase the degree. Keep doing so, and eventually you will see wild swings in one direction or another on the spline.

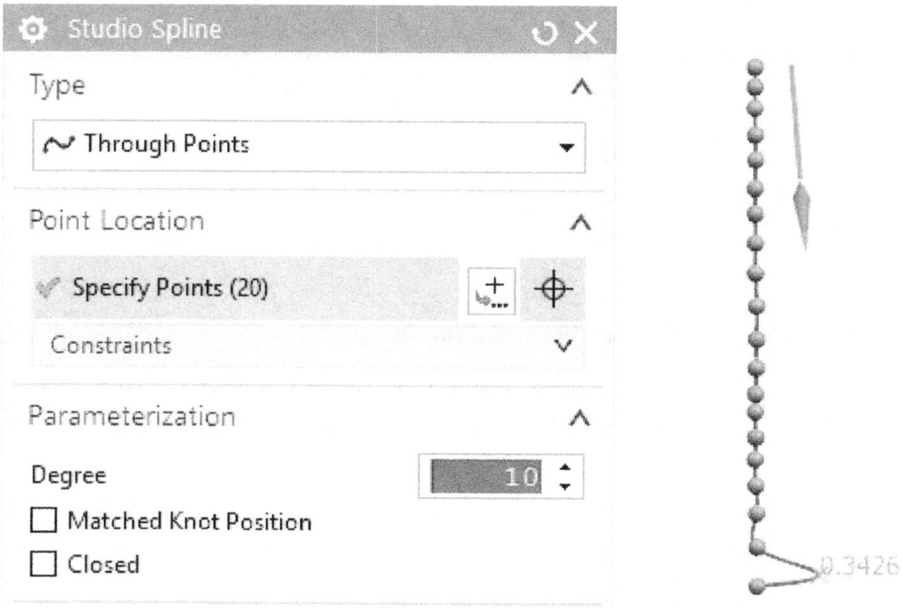

- In general, as the degree of a *Through Points* spline increases, your control over the shape of the spline decreases. With the degree of your spline set high, try moving one of the defining points (click and hold and drag it).

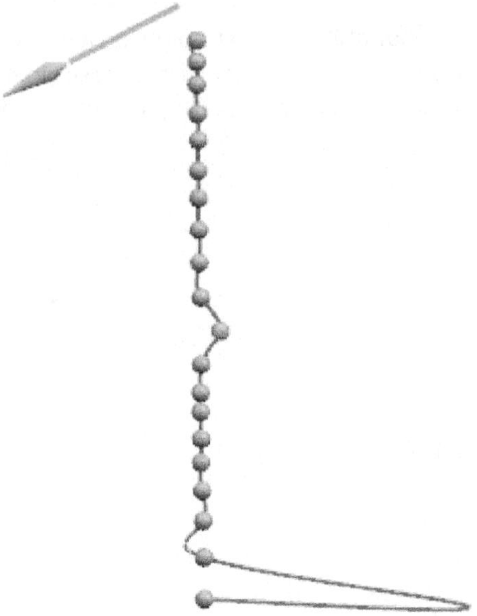

- Notice that a small change in the position of any one point can have *global* consequences for the shape of the entire spline – in the image above, the last few sections of the spline reversed concavity!
- Click **Apply** to finish your spline, keeping the **Studio Spline** tool open. **Reset** the tool to the default input parameters.
- For your next spline, make sure the **Degree** is *3* and choose four defining points as shown below.

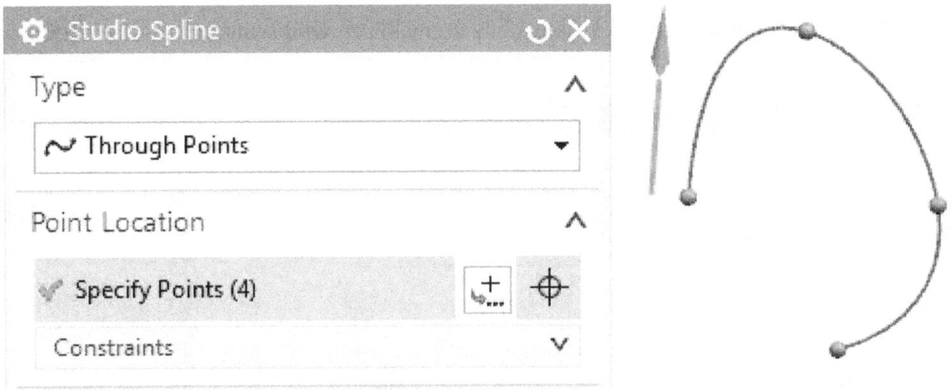

- Mark the **Closed** checkbox and the spline will become a closed loop.

- Click **OK** to create the spline. The defining points of your spline are selectable and can be manipulated with click-and-drag motion like most sketch entities.

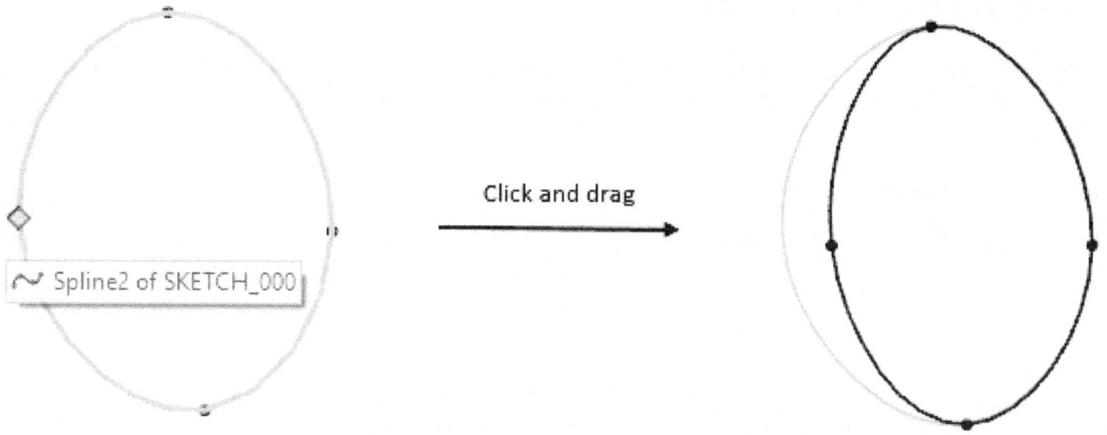

- Defining points of splines are selectable if the *Defining Point* snap point icon is enabled – if you are unable to select and drag your spline's defining points as shown above, check your snap points!

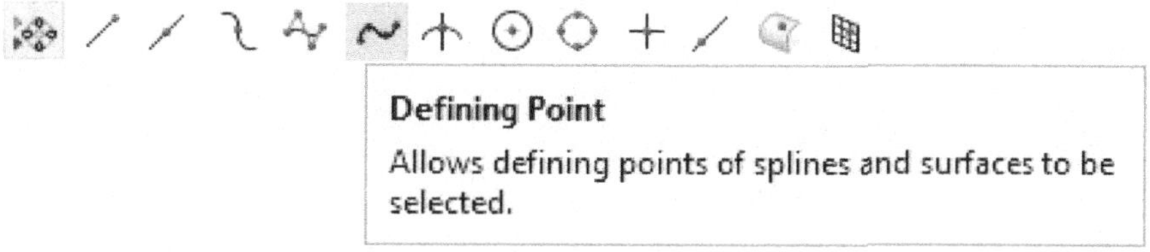

- Finish your sketch and create a new sketch called *"BY POLES"*.
- Select the **Studio Spline** too, set the **Degree** to *3*, and choose three defining points as shown below.

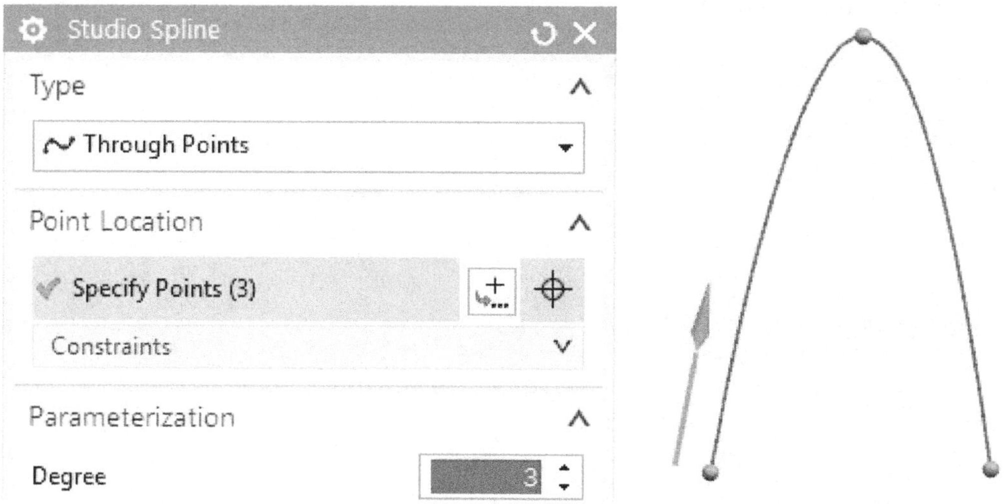

- Click **OK**. You will encounter the following message.

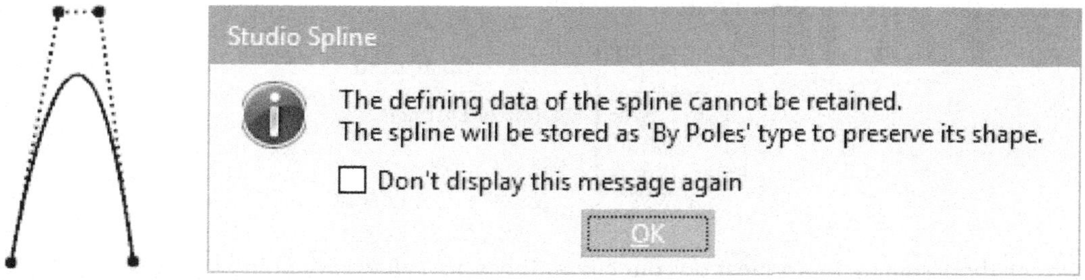

- The lesson here is that <u>the number of defining points must be greater than the degree</u>, or else the spline will be saved using the *By Poles* method. What are poles? Read on!

13.2.2 Studio Spline (Sketch) – By Poles

In defining a **Studio Spline** *By Poles*, the defining points need not lie on the curve – in contrast with the *Through Points* method of constructing a spline, which interpolates between the defining points, the *By Poles* method approximates the specified points.

- Continue working with the last spline constructed.

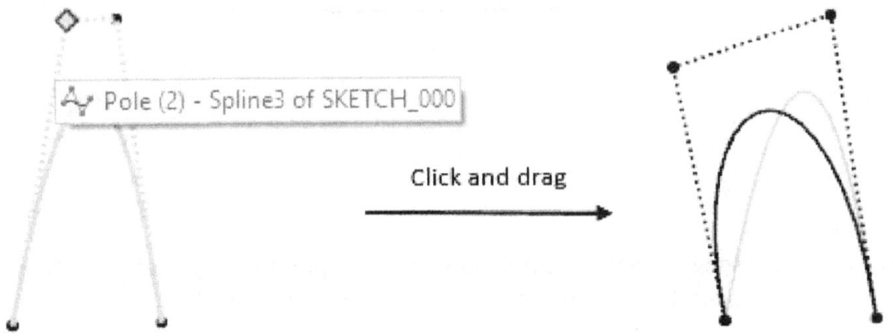

- Make sure that your *Poles* snap point icon is enabled.

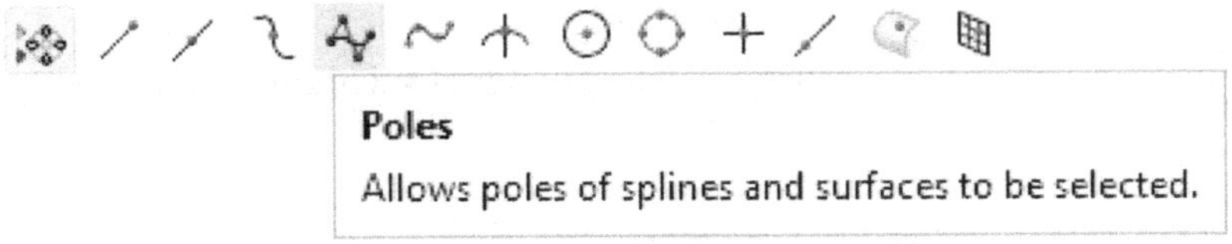

- Select the **Studio Spline** tool again, **Reset** it, and set the **Type** to *By Poles*.
- Specify seven poles as shown below.

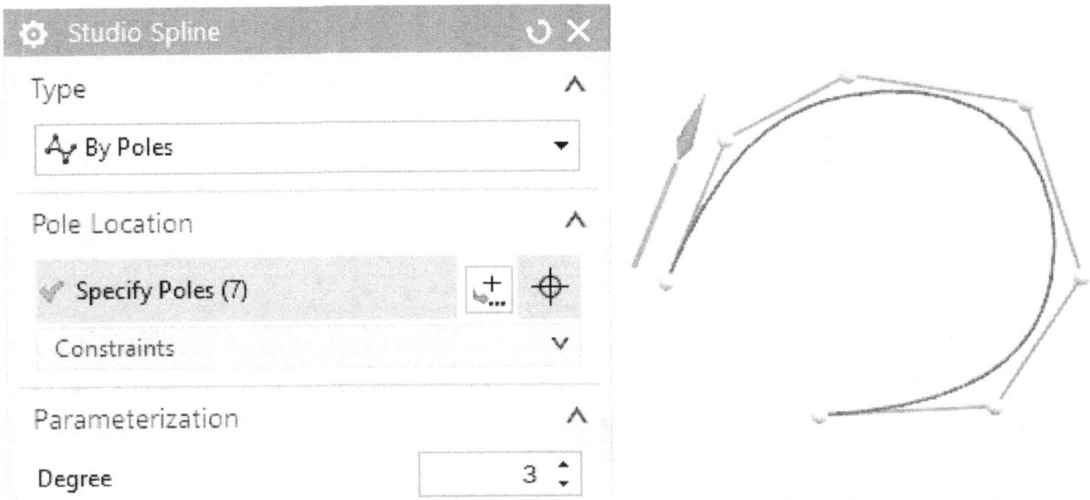

- From the **Analysis** tab, select the **Show Knots** command.

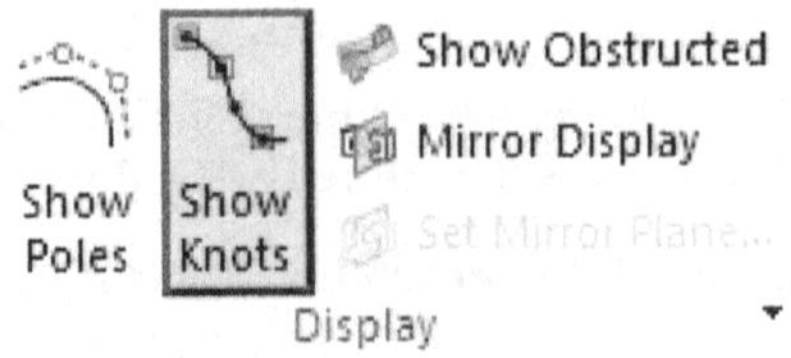

- You will then see three square dots appear on your curve. These are called knot points. The portion of a spline between two knot points is called a segment. Roughly speaking, on each segment, a spline has a different description in terms of polynomial equations. The spline below has four segments, labeled clockwise.

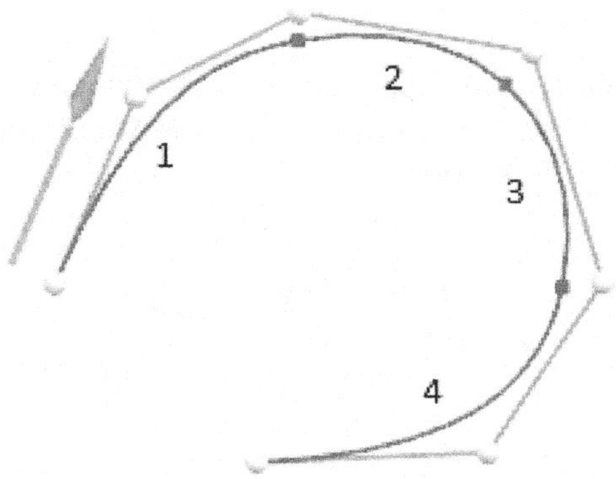

- You saw earlier that a small change to the defining points of a *Through Points* spline, especially when the degree is high, can have drastic consequences for the global shape of the spline.
- *By Poles* splines, in contrast, offer better local control over the shape of the spline.
- Click-and-hold-and-drag the third pole to the location shown below.

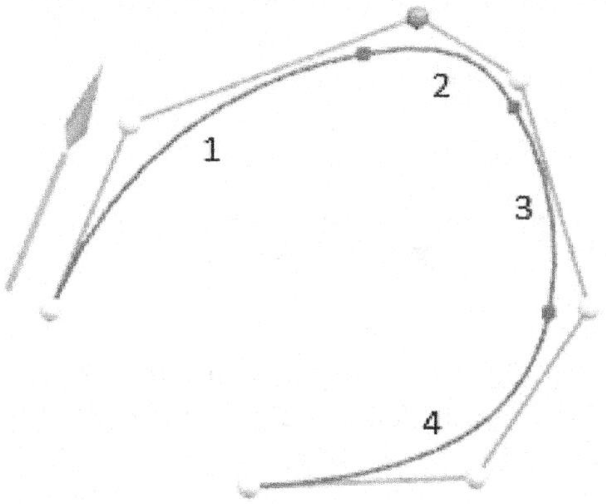

- Notice that segments *1*, *2*, and *3* have all changed, but that segment *4* is unchanged.
- Modifications to a single pole only affect nearby segments of the spline.
- Next, check the box within the **Parameterization** section of the dialog that says *Single Segment*.

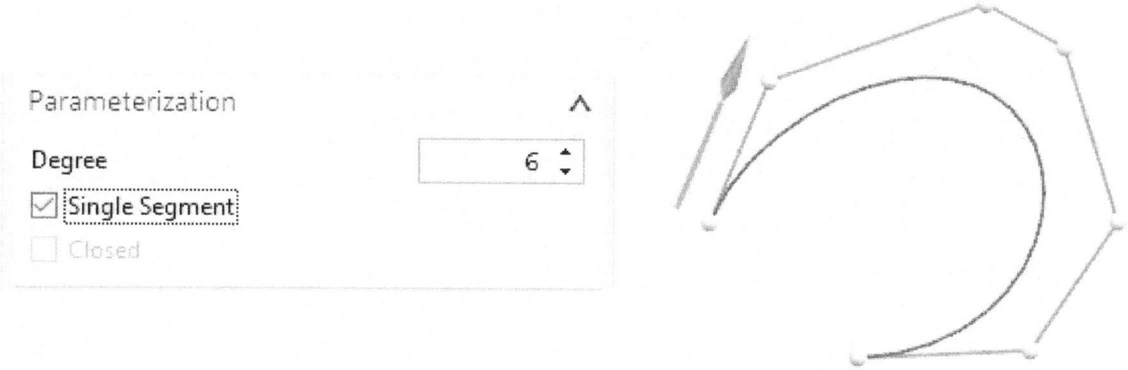

- Note that the degree increases to six – one less than the number of poles.
- Splines having only a single segment necessarily have degree equal to one less than the number of poles. Such splines are called *Bézier curves*.
- Click **OK** to create your spline!
- **Finish** your sketch, and **Save** your file!

13.2.3 Studio Spline (3D)

When you need a freeform curve that is twisted in 3D space and not constrained to a plane, you will want to use the **Studio Spline** tool found on the **Curve** tab. It has a nearly identical dialog to the **Studio Spline** tool found in the sketcher, but the defining points can be positioned anywhere in 3D space.

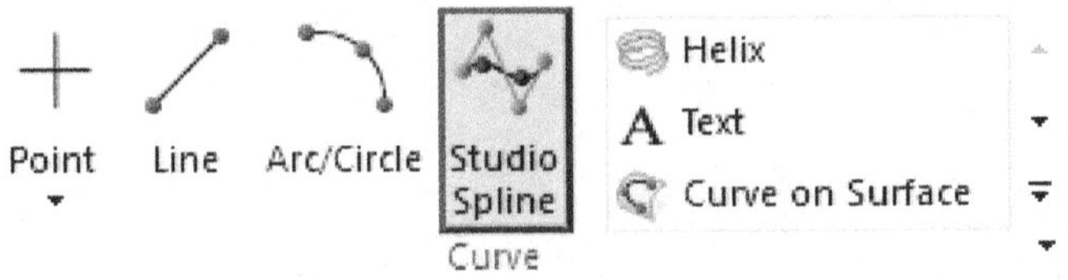

The key controls that the 3D **Studio Spline** tool offers for controlling the position of the defining points are found in the **Drawing Plane** and **Movement** sections of the dialog.

Let's get some practice positioning defining points and moving them in 3D space.

- Continue working with *"Splines.prt"* and select the **Studio Spline** tool from the **Curve** tab. The default setting for the **Drawing Plane** is the *View Plane*, which is the XY plane of the *View Coordinate System* as described in Chapter 3.1.6.

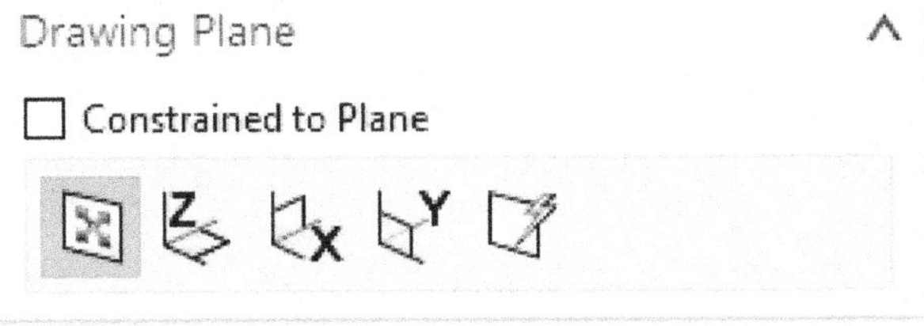

- Orient the view to a *Top* view – this will align the *View* plane with the XY plane of the Absolute Coordinate System in the part. Create a spline *Through Points* with six defining points and check the *Closed* box. Also, note that the default degree for a 3D **Studio Spline** is *5*.

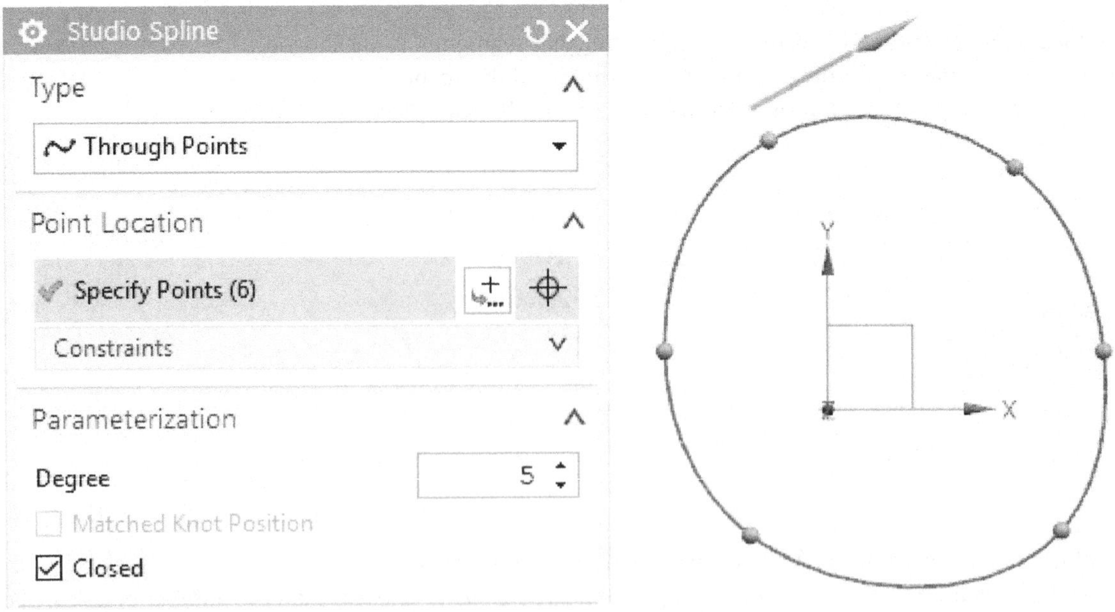

- Your spline should appear as shown below.

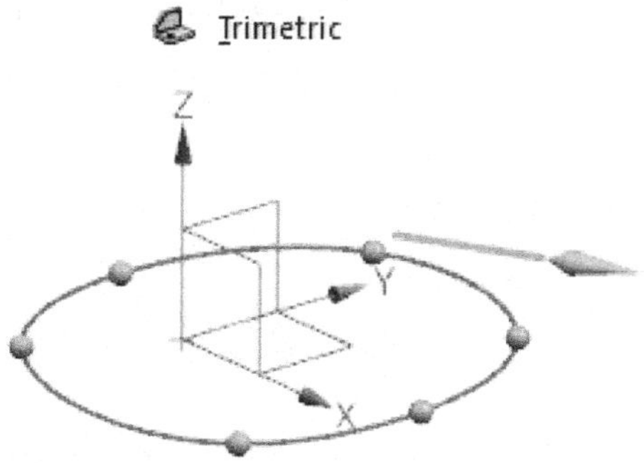

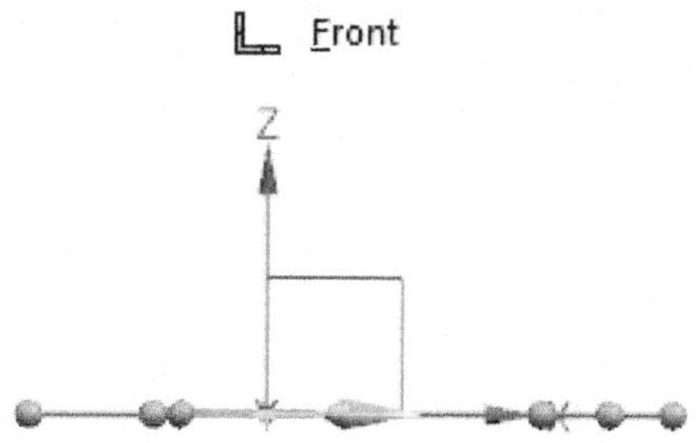

- In the **Movement** section of the dialog, the radio button is by default set to *View*. This will restrict the motion of the defining points to the *View* plane.

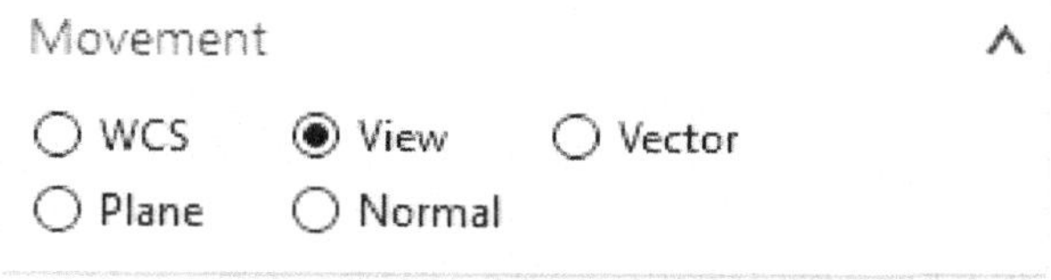

- Orient the view to a *Front* view and select the right-most defining point.

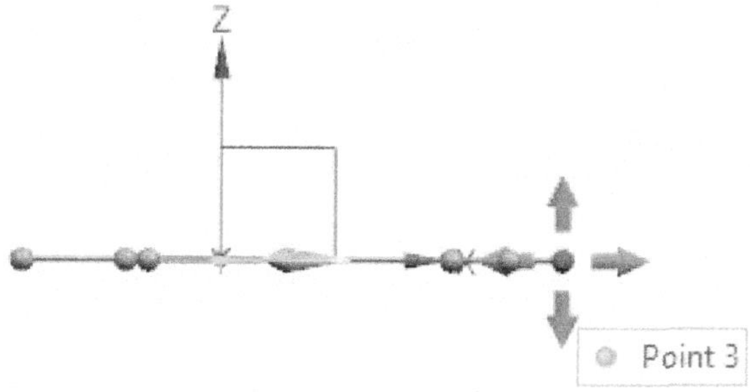

- Change the **Movement** radio button to *Vector*, and select the Z axis of the *OrientXpress*.

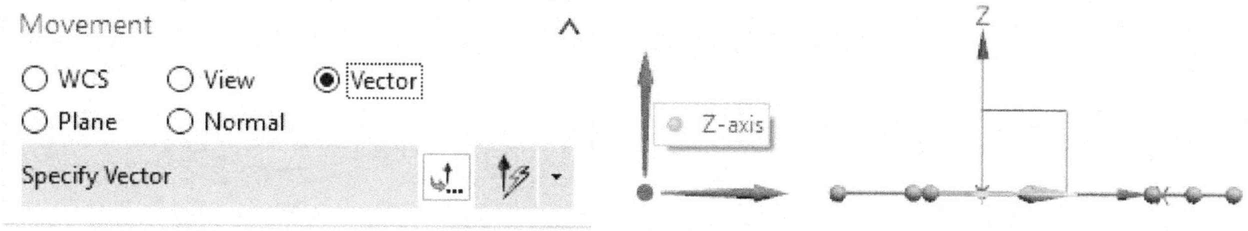

- Now when you drag the selected pole, it will only move in the Z direction of the WCS.

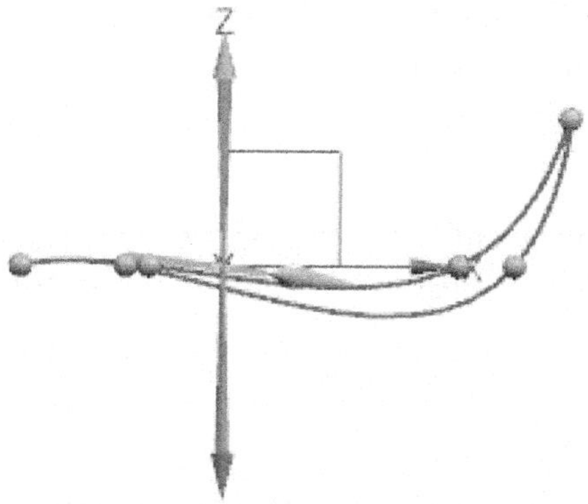

- Set the **Movement** radio button back to *View*, and orient the view to *Trimetric*. Choose the defining point shown below and drag it up and to the left slightly.

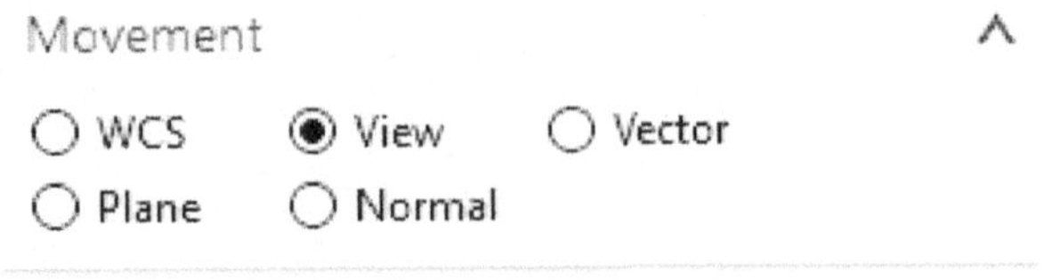

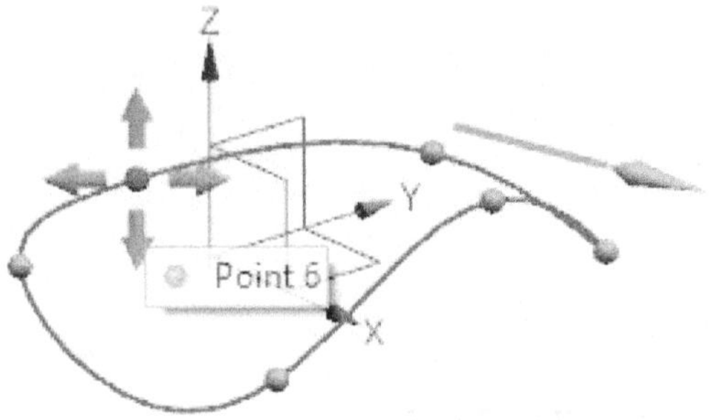

- This deforms the spline into a saddle shape.

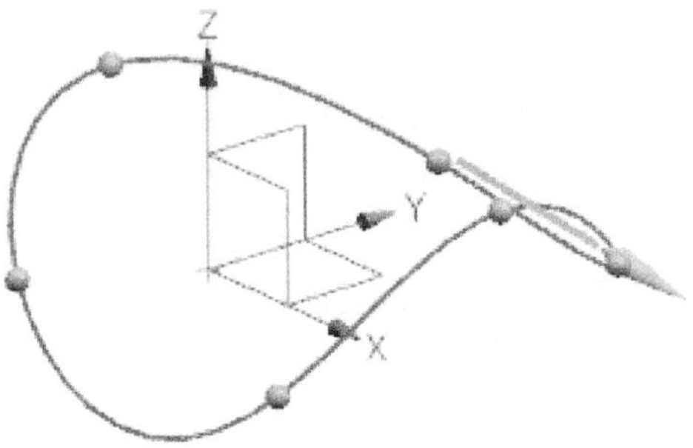

- Your spline should appear as shown below.

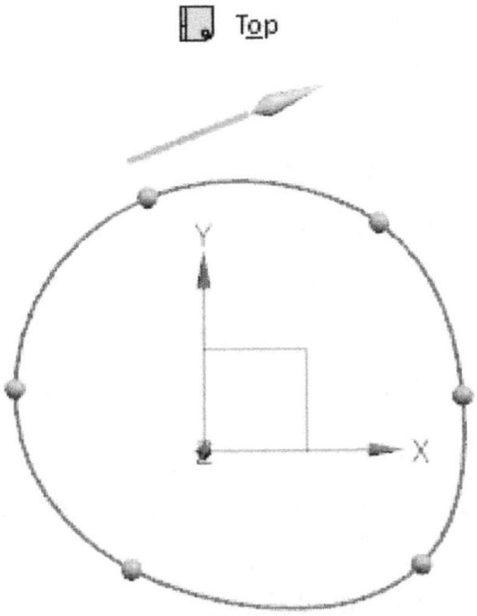

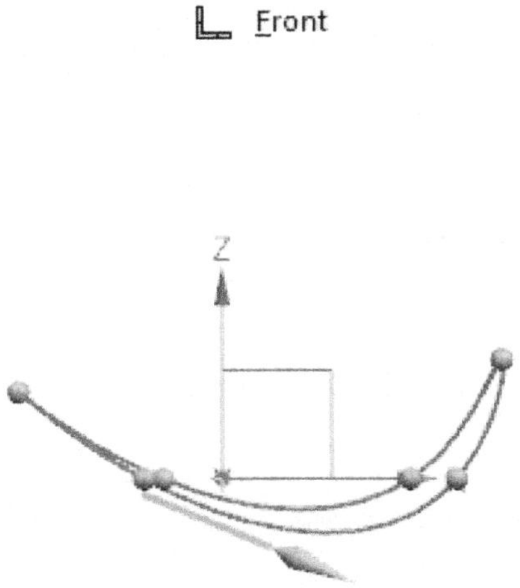

- Experiment with the other **Movement** methods!
- Click **OK** to create your spline, and then **Save** your file!

13.2.4 Specify Constraint

So far, we have seen the effects of modifying either the degree, or the position of defining points for splines. There is additional control over the shape in the form of tangency and curvature constraints at defining points.

At any defining point that a spline passes through, you can use the **Specify Constraint** command to constrain it associatively up to the second order.

- Create another sketch and name it *"SPECIFY CONSTRAINT"*.
- Use the **Studio Spline** tool to create a *Through Points* spline with three defining points, and then right-click on the second defining point and choose **Specify Constraint**.

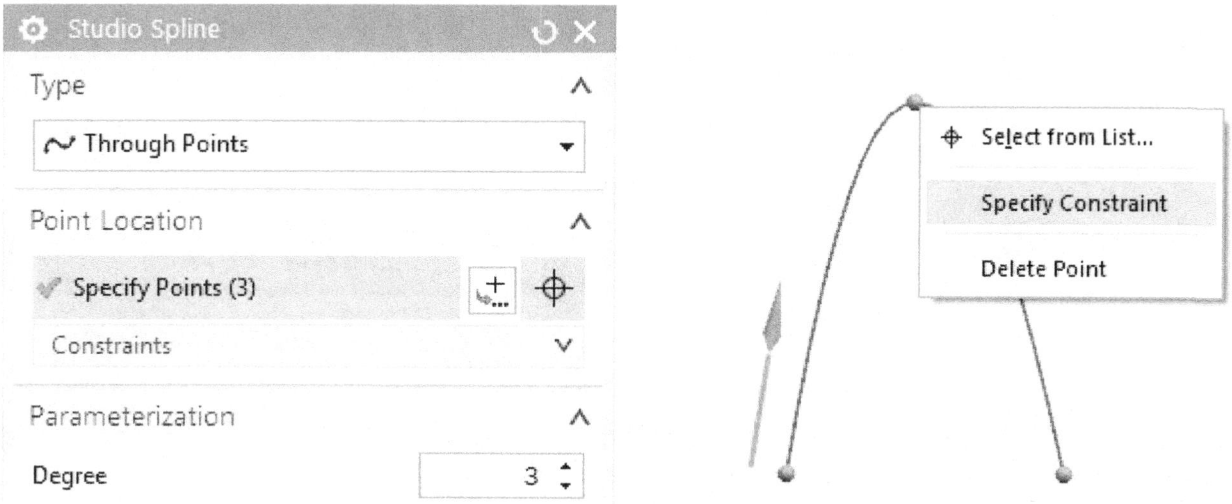

- In the **Constraints** section of the dialog, set the **Continuity Type** to *G2 (Curvature)*. This activates two controls for the shape of the spline near the defining point – the magnitude of the tangent vector, and the radius of curvature at the defining point. These controls come with dynamic handles in the graphics window – for the tangent vector, the magnitude is adjusted with the attached arrow.

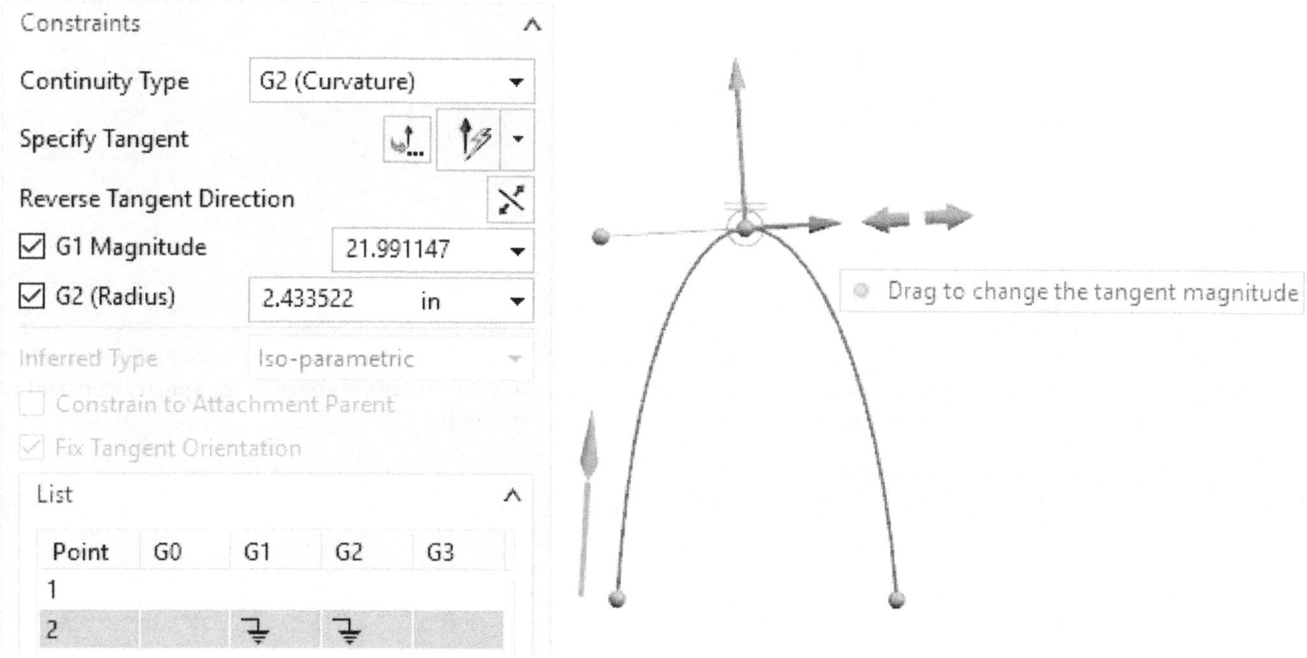

- The direction of the tangent vector is controlled with the attached rotation handle.

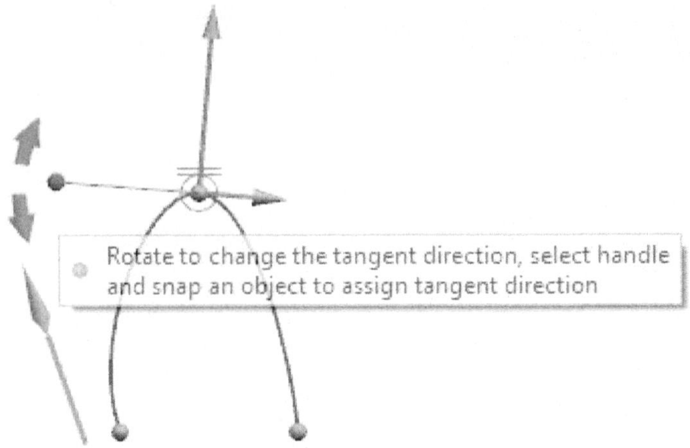

- The radius of curvature is controlled by the normal vector.

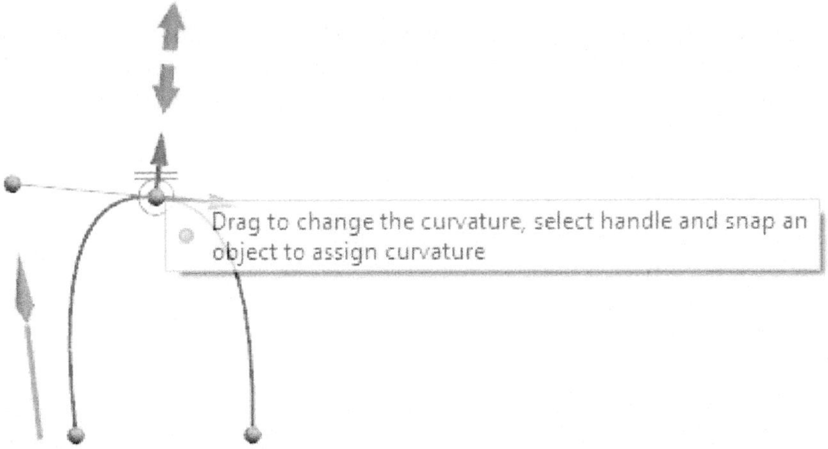

- Click **Apply** to create your spline.
- Set the **Type** to *By Poles* and make a spline with six poles as shown below. Right-click on the leftmost pole and select **Specify Constraint**. The poles at the ends of a *By Poles* spline are the only control points eligible for **Specify Constraint**, since they are the only defining poles that the spline actually passes through.

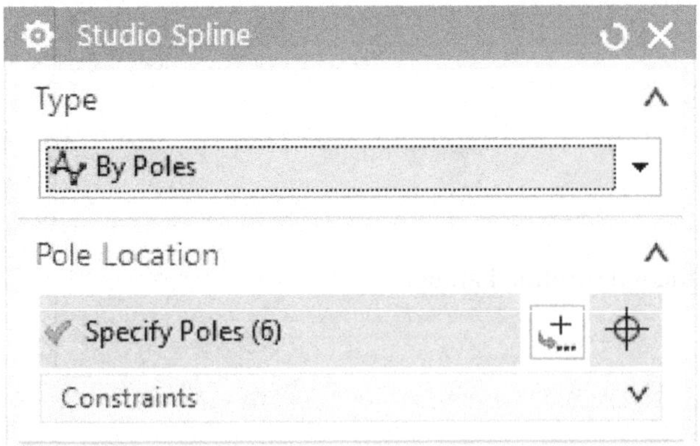

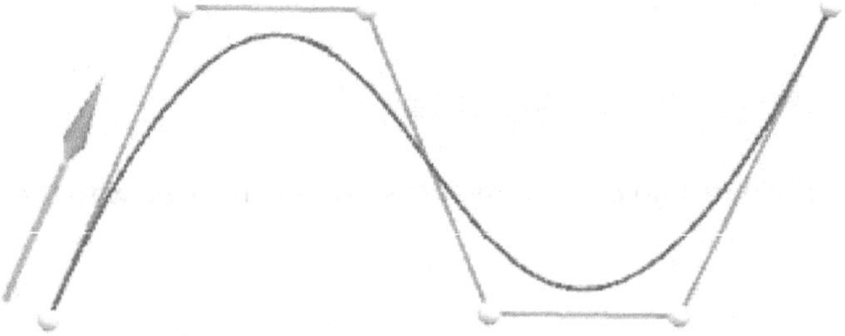

- Drag the rotation handle to change the direction of the tangent vector. Observe that the second pole always stays aligned with the tangent vector. In fact, the line joining the first and second pole is always tangent to the spline at the first pole.

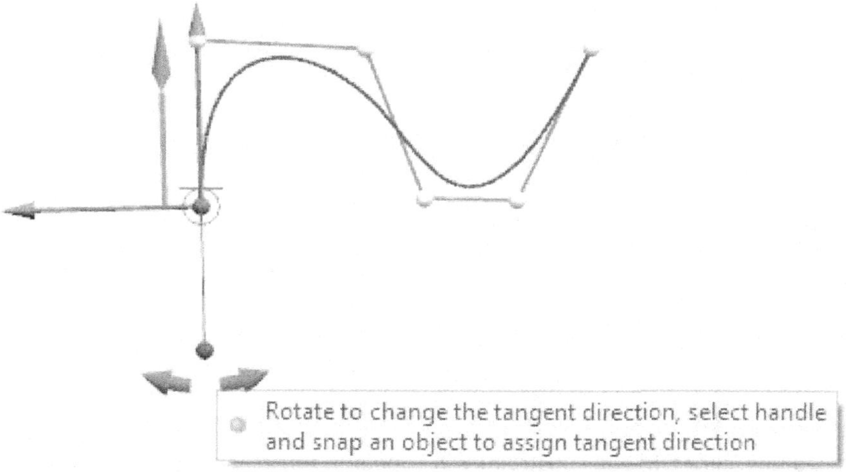

- Next, drag the tangent vector. Note that the second pole always stays attached to the arrow. The distance between the first and second poles directly correlates to the magnitude of the tangent vector to the spline at the first pole.

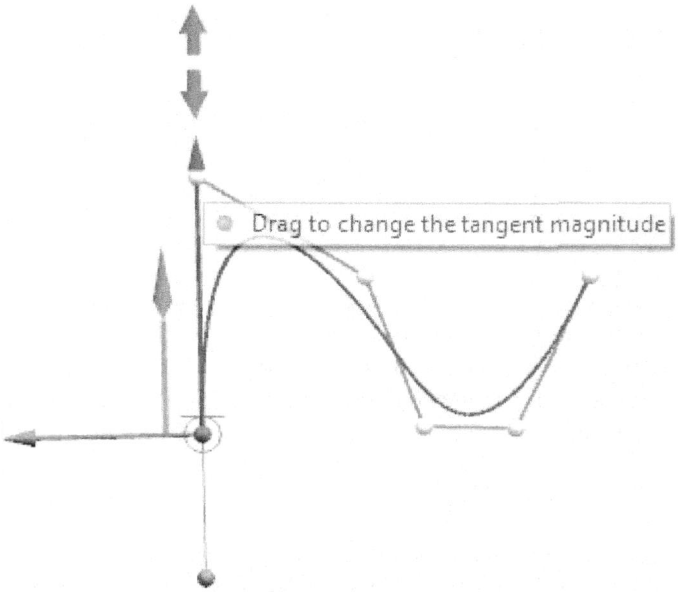

- So far, in manipulating the tangent vector, only the first and second poles have changed position.
- Next, drag the normal vector to change the curvature of the spline at the first pole. Note that the first and second poles remain fixed but that the third pole moves. The position of the third pole, for reasons that are not obvious, controls the curvature of the spline at the first pole.

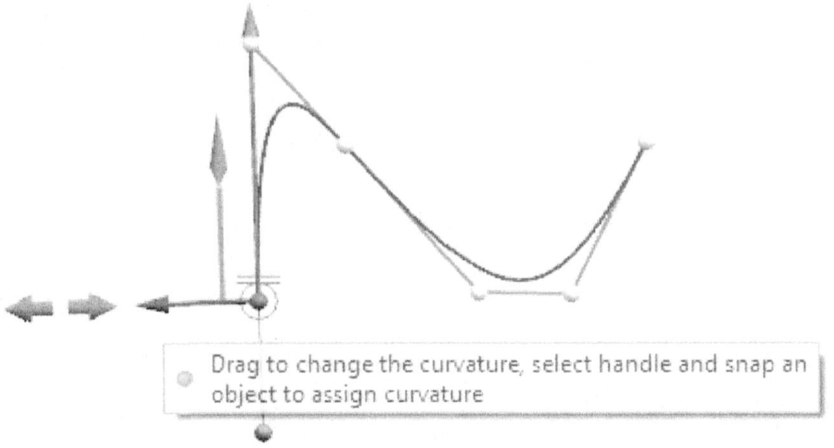

- When you are done editing, click **OK**. The constraint will appear in the graphics window at the end of the spline as an arrow with two lines on each side – the two lines indicate associative control of the shape up to the curvature.

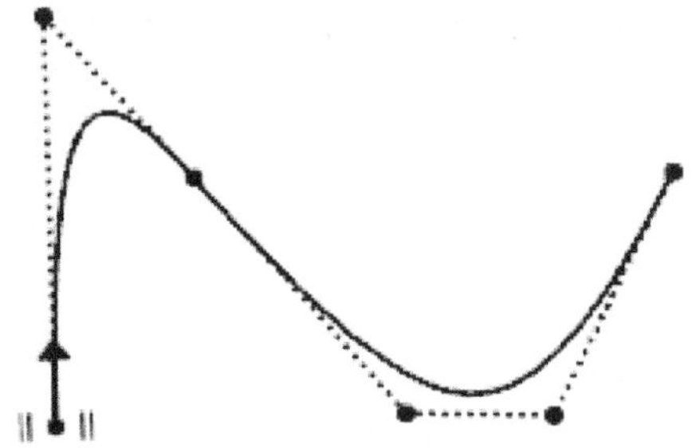

13.3 Geometric Continuity

Often times the curves required in the design of a product are composed of lines, arcs, and splines that meet each other "smoothly". The degree of smoothness is formalized by the notion of *geometric continuity*, and in NX there are four available degrees of geometric continuity for splines meeting other curves – *G0, G1, G2, G3.* In the exercises below, we will illustrate the meaning of these terms along with the help of some tools from the **Analysis** group.

13.3.1 Curvature Combs

To fully understand geometric continuity, it will be helpful to display the curvature combs of different curves as we build them. The curvature comb is a graph of the curvature along the length of the curve. There are two tools that enable you to see the curvature comb for a curve in NX: **Show Combs**, and **Curve Analysis**. **Show Combs** toggles the display of the curvature comb on or off, and **Curve Analysis** lets you modify the of the comb itself. Both of these tools are found in the **Curve Shape** group on the **Analysis** tab.

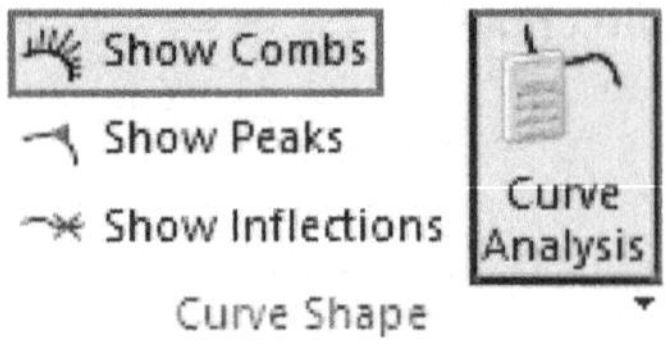

Below, the curvature comb of the *By Poles* spline you constructed in Chapter 13.2.2 is displayed. For planar curves, the curvature is inversely related to the radius, and so the curvature comb is particularly useful for understanding how the radius of a planar curve changes – when the curvature is large, the radius is small, and vice versa.

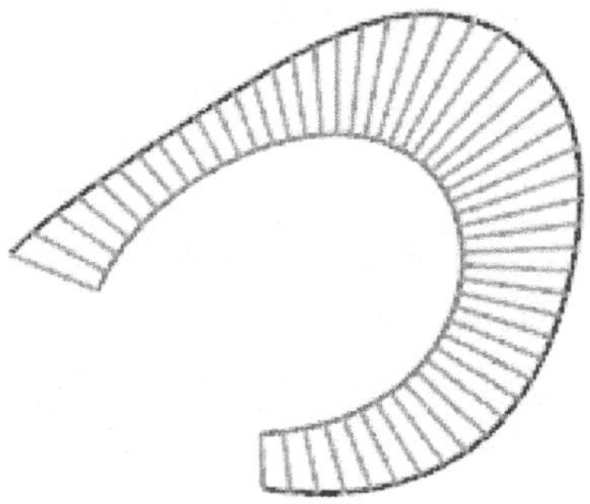

- Open *"Continuity.prt"* from *"C:\My NX 11 Files\Downloaded Files"*.
- The Model History for this part should show a number of suppressed sketches and curves, and one unsuppressed sketch called *"CURVE ANALYSIS"*. The curve in this sketch is a cubic spline.
- Select the **Curve Analysis** tool, and choose the spline from the sketch *"CURVE ANALYSIS"*. In the **Analysis Display** section of the dialog, make sure the *Show Comb* checkbox is marked. The *Suggest Scale Factor* checkbox will adjust the **Needle Scale** according to your current view of the selected curve. Slide the **Needle Scale** to *25*, and set the **Number of Needles** to *25*.

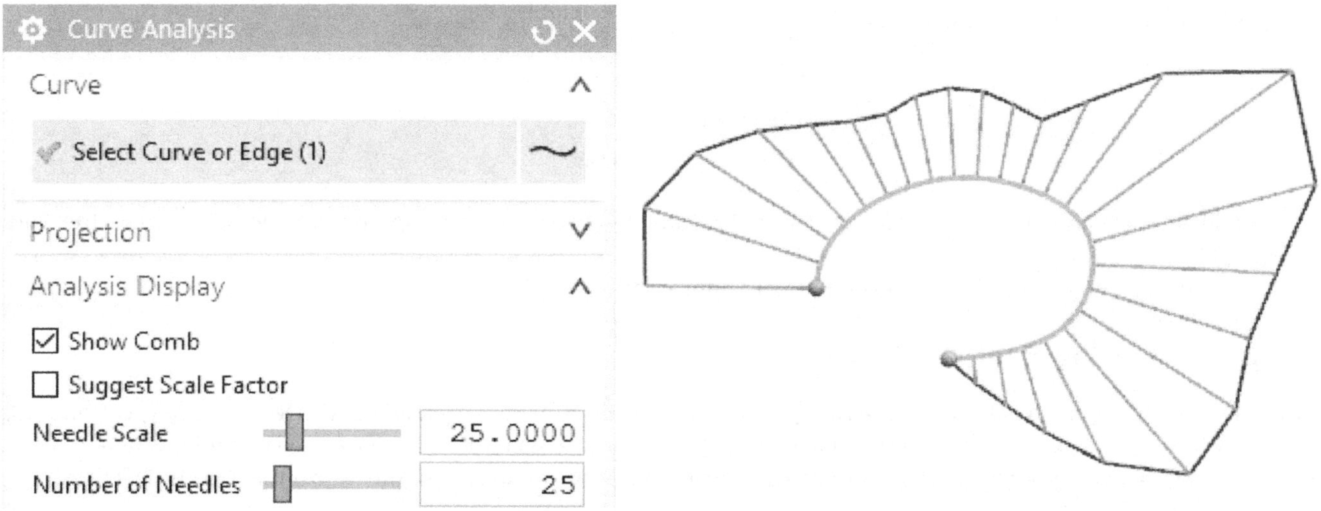

- The curvature comb so far appears jagged – each needle is the result of a curvature calculation made at a point on the selected curve, and so more needles means the curvature comb will smooth out. Increase the **Number of Needles** to *100* and click **OK** to save the curvature comb.

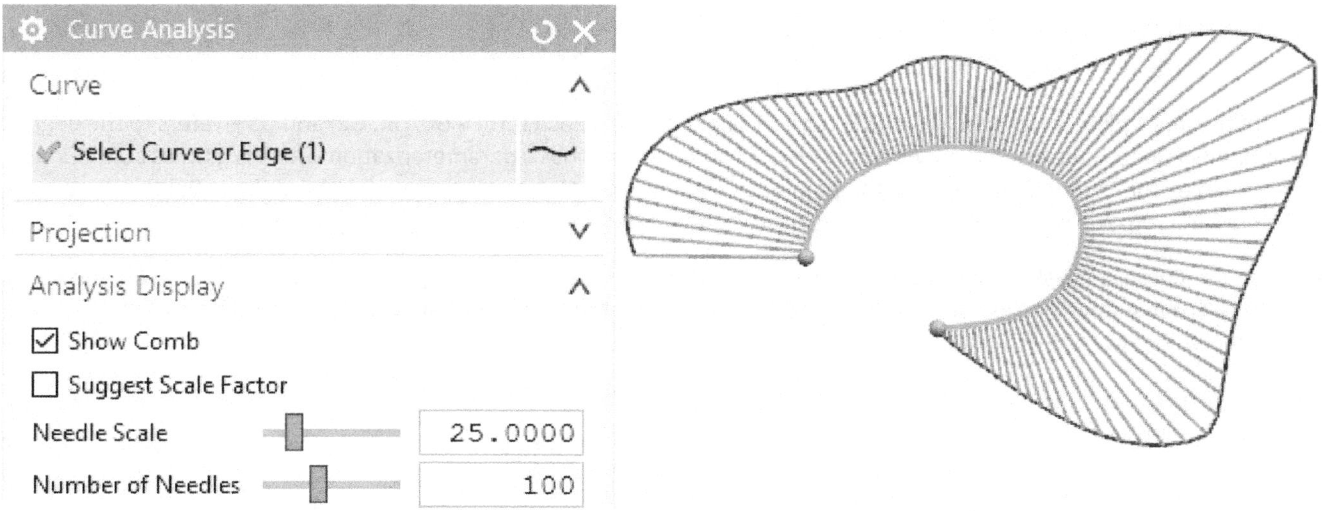

- The remaining "kinks" in the curvature comb only occur over knots in the spline – use the **Show Knots** command to display the knots to see for yourself!

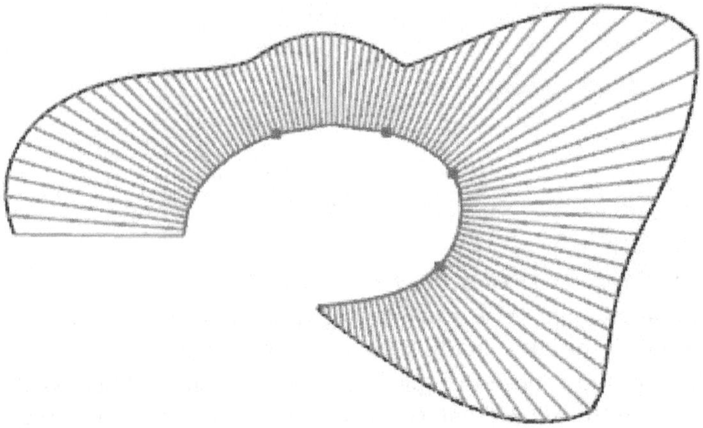

- The output of **Curve Analysis** -the curvature comb – is an *Analysis Object*, found in the *Analysis* folder in the Part Navigator. You can show or hide the comb by checking or unchecking the checkbox to the left, or with the **Show Combs** command from the **Analysis** tab.

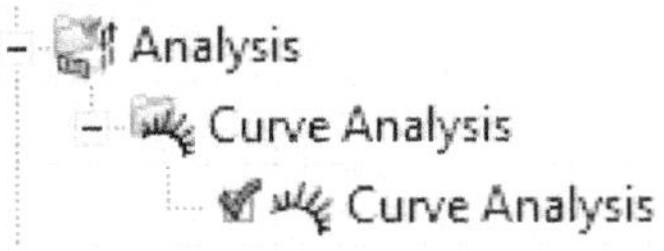

- To edit an existing curvature comb, you can either double-click on it in the graphics window, or from the Part Navigator.
- Finish the *"CURVE ANALYSIS"* sketch and **Hide** or **Suppress** it!

13.3.2 The Meaning of G0, G1, G2, G3

Now that you are equipped with the **Curve Analysis** tool, we can illustrate the meaning of the terms *G0*, *G1*, *G2*, and *G3*. In the examples below, you will see splines joining otherwise identical pairs of line segments. The middle curve in each sketch is a degree five *Through Points* spline with defining points at the endpoints of the horizontal and vertical lines.

- **Unsuppress** each of the sketches, *"G0"*, *"G1"*, *"G2"*, and *"G3"*.

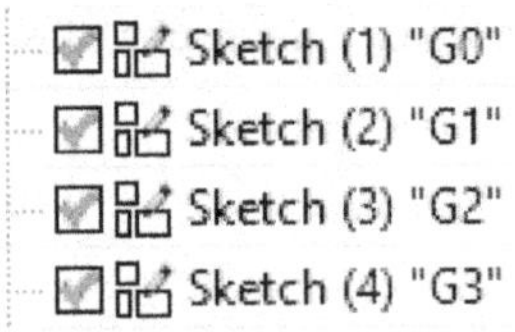

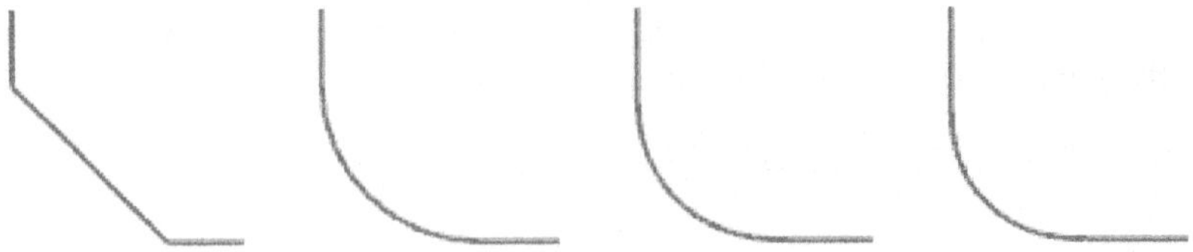

Think of the line–spline–line chain in each sketch as a single "curve". Each term – *G0*, *G1*, *G2*, and *G3* – refers to the degree of continuity (smoothness) of this curve. A curve is *Gn-continuous* if it has a parameterization which is n times continuously differentiable. The term *G0* means that the parameterization is continuous but might not be differentiable. The curves look very similar to the naked eye, but their curvature combs exhibit very different qualitative behavior.

- Use the **Curve Analysis** tool to display the curvature combs of the curves in each sketch, *"G0"*, *"G1"*, *"G2"*, and *"G3"*. In the **Curve Analysis** dialog, set the **Needle Scale** to *0.25*, and the **Number of Needles** to *25*. The combs should appear as shown below.

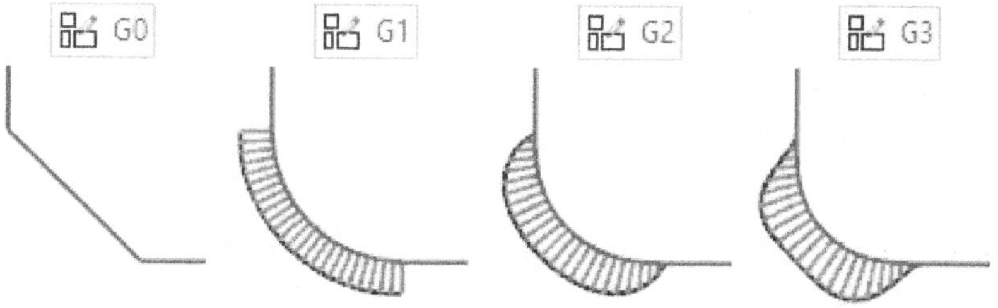

13.3.2.1 G0 – Positional Continuity

The term *G0* is called *"positional continuity"*. It means simply that the curve is continuous. For the line – spline – line chain in the sketch *"G0"*, it means that the endpoints of the spline are touching the endpoints of the lines! A line has no curvature, so the curvature comb is moot here.

13.3.2.2 G1 – Tangent Continuity

The term *G1* is called *"tangent continuity"*. It means that the curve is tangent-continuous. For the line–spline–line chain in the sketch *"G1"*, it means that the spline meets the lines with tangency!

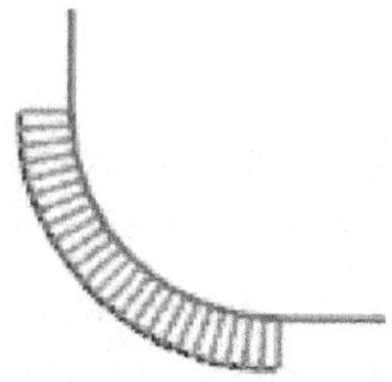

Note that the curvature comb for the line–spline–line chain in the sketch *"G1"* is discontinuous – the curvature jumps from zero to nonzero at the endpoints of the spline. The curvature also appears to be constant. Since the curvature of a planar curve is inverse to the radius, this suggests that the radius of this spline is constant – in other words, the spline is an arc! This is not quite true, but it is very nearly true.

- Hide the curvature comb for the sketch *"G1"*, suppress the other sketches, and **Unsuppress** the sketch *"G1 – Arc"*. This sketch is the same as the others, but with an arc joining the line segments instead of a degree *5* spline.

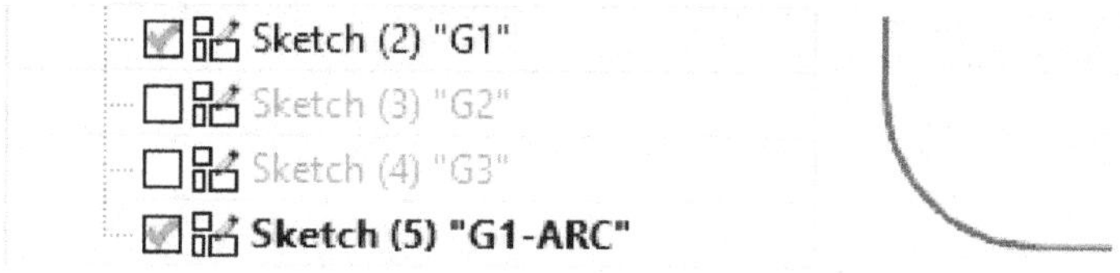

- Select the **Deviation Gauge** tool found in the **Relation** group on the **Analysis** tab.

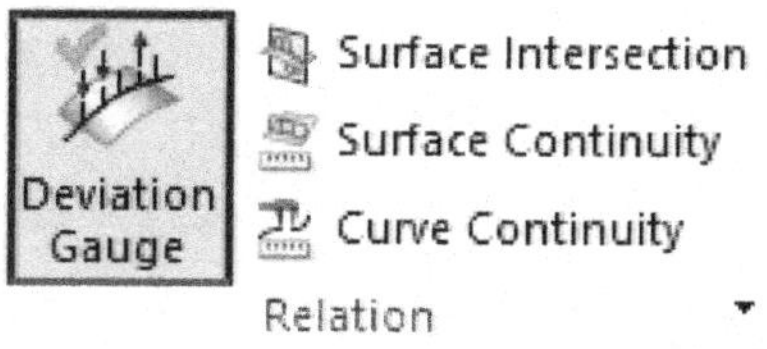

- Select the arc and the spline as the **Objects to Compare**, and within the **Plot** section of the dialog, check the *Suggest Scale Factor* to resize the needles. As you can see, the spline is awfully close to being identical to the arc, but there is a small amount of deviation.

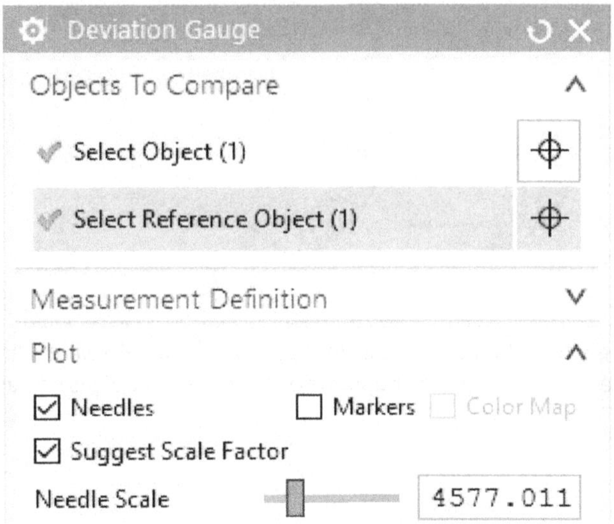

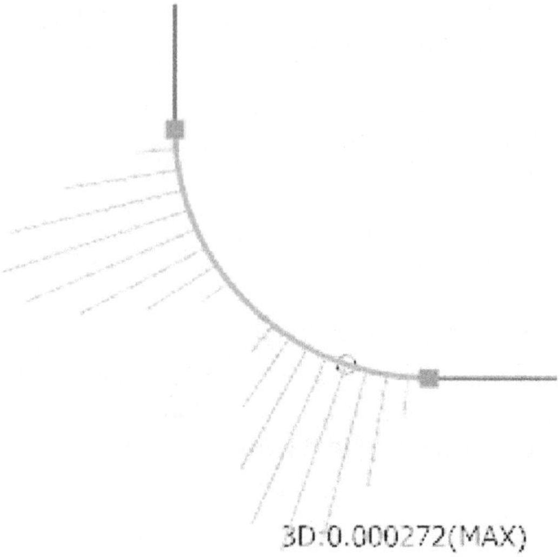

- Click **OK** to create the *Deviation Gauge* analysis object.

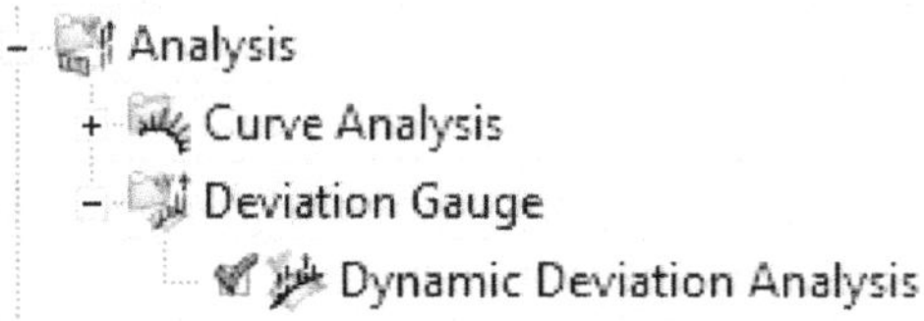

13.3.2.3 G2 – Curvature Continuity

The term *G2* is called *"curvature continuity"*. It means that the curve is curvature-continuous. For the line–spline–line chain in the sketch *"G2"*, it means that the curvature is continuous along the length of the whole chain.

The curvature comb illustrates the continuity of the curvature clearly – in contrast with the *G1* spline, the curvature of the spline in the sketch *"G2"* decreases to zero at the endpoint of the spline where it meets the line.

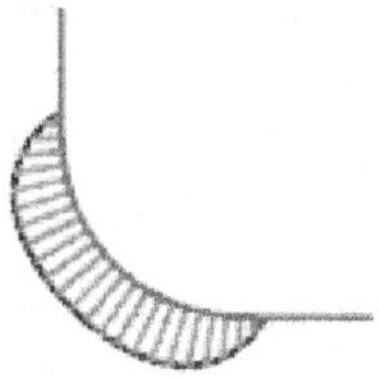

- Edit the curvature comb for the curves in the sketch *"G2"* – change the **Number of Needles** to *100*. Zoom in near the endpoint of the spline to convince yourself that the curvature goes to zero continuously!

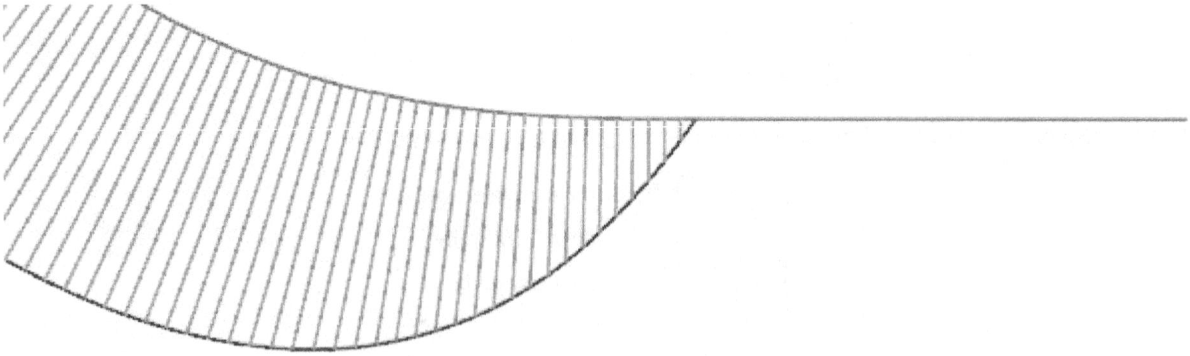

13.3.2.4 <u>G3 – Flow Continuity</u>

The term *G3* is called *"flow continuity"*. It means that the rate of change of the curvature is continuous along the length of the curve. For the line–spline–line chain in the sketch *"G3"*, it means that the curvature is tangent-continuous along the length of the whole chain.

The curvature comb illustrates the continuity of the curvature clearly – as with the *G2* spline, the curvature of the spline in the sketch *"G3"* decreases to zero at the endpoint of the spline where it meets the line. But it does so smoothly, so that the curvature comb is tangent-continuous.

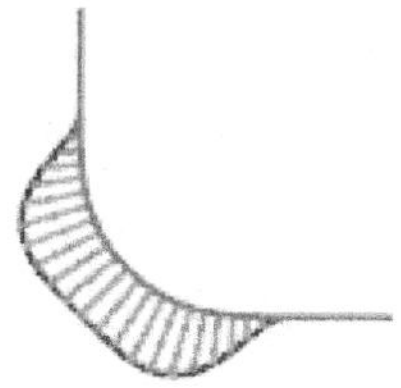

- Edit the curvature comb for the curves in the sketch *"G3"* – change the **Number of Needles** to *100*. Zoom in near the endpoint of the spline to convince yourself that the curvature goes to zero with tangency!

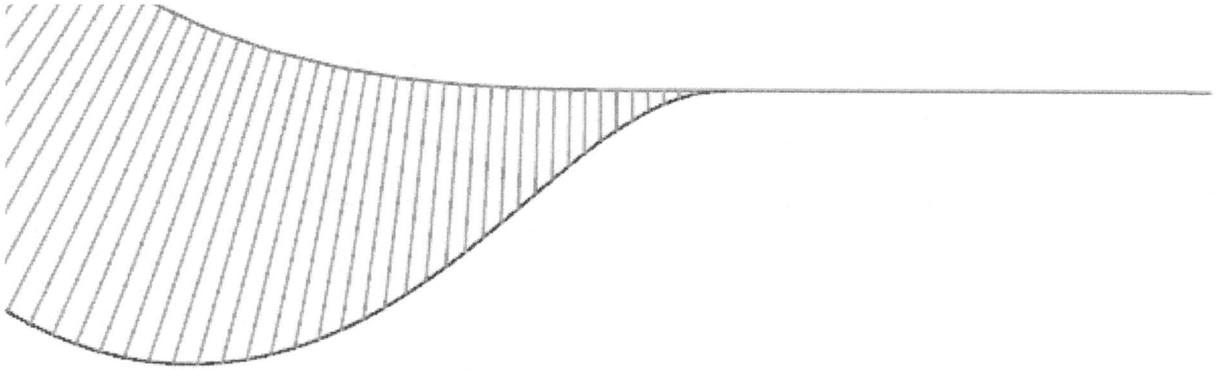

13.3.3 Studio Spline (Sketch) – G2 Associative Constraint

Tangency is a "first order" control, and curvature is a "second order" control. The constraints that you apply using **Specify Constraint** are associative up through the second order. It is possible to apply higher order constraints at the endpoints of a spline, though third-order constraints are not associative. We explore the difference in the exercises below.

- **Unsuppress** and then **Open** the sketch *"G2 Associative"*. **Hide** all other sketches and analysis objects.
- Inside, you will find two identical pairs, each consisting of a line and an arc, as shown below. You will build a spline that joins the line/arc pair and apply the *G2 Associative* constraint at each endpoint – first for a *Through Points* spline, and then for a *By Poles* spline.

- Select the **Studio Spline** tool, set the **Type** to *Through Points*, and the **Degree** to *3*. For the first defining point, choose the endpoint of the line. You will see a pop-up toolbar appear – push the **Infer G2** button.

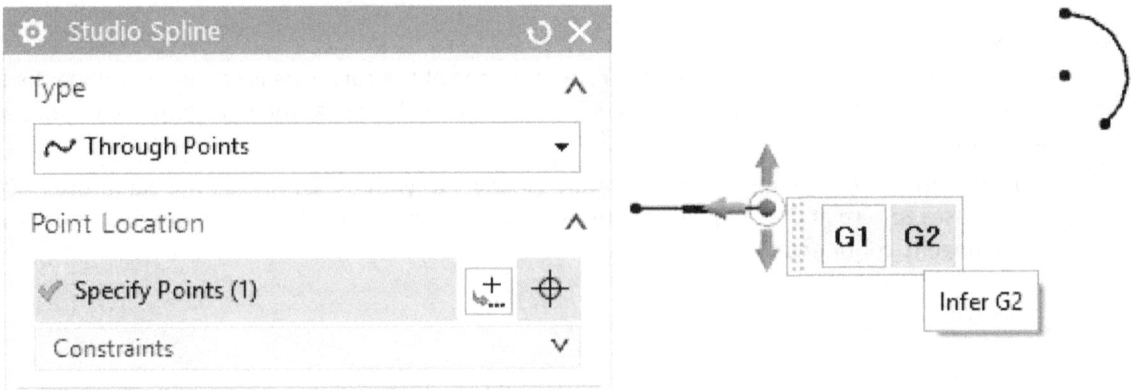

- For the second defining point, choose the endpoint of the arc, and again push the **Infer G2** button on the pop-up toolbar.

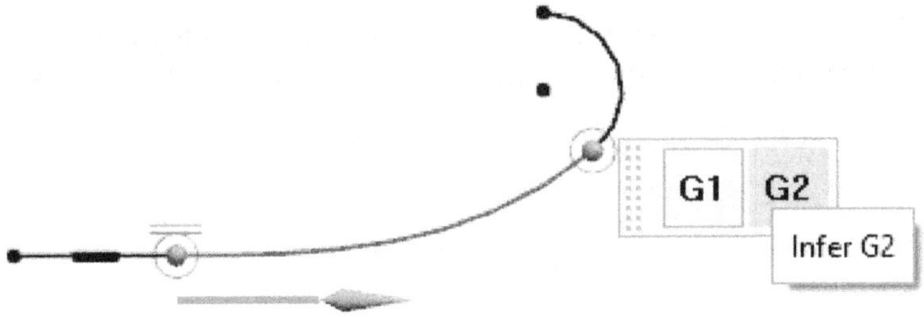

- Click **OK** to create your spline, and you will see two *G2 Associative* constraints appear.

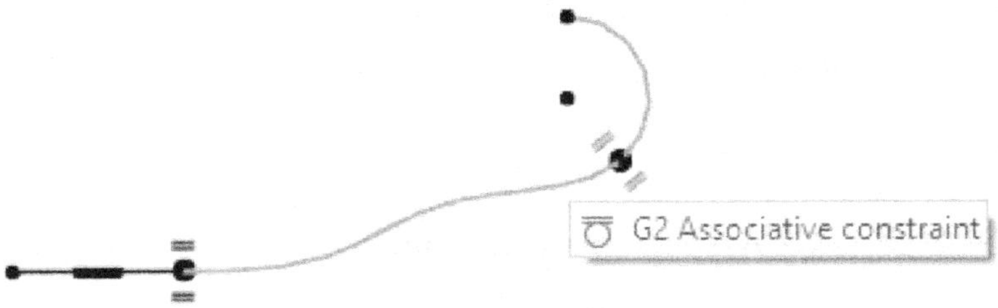

- Drag the arc center down and to the left so that the inflection in the spline reverses.

- The *Associative G2 Constraint* is also valid for *By Poles* splines.
- Create another horizontal line and arc, and then create another spline with **Type** set to *By Poles* and **Degree** set to *3*. Place six poles as shown below, taking care to position the first and last poles at the endpoints of the line and arc, respectively.

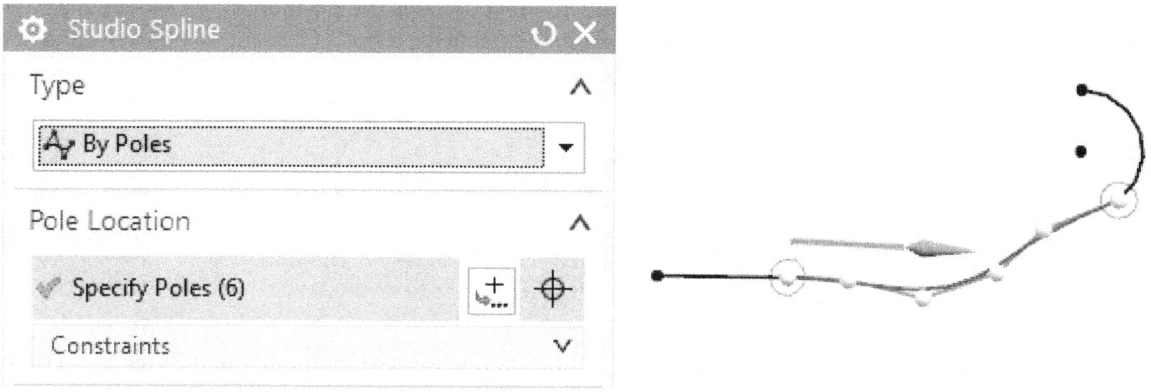

- Right-click on the first pole and select **Infer G2**.

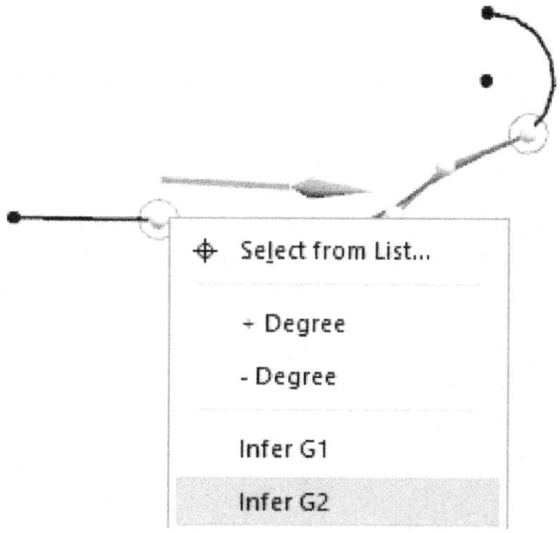

- Notice that the first, second, and third poles are constrained to be horizontally aligned – this is typical when you ask a line for a certain degree of continuity with a neighboring line – poles will be constrained to that line.

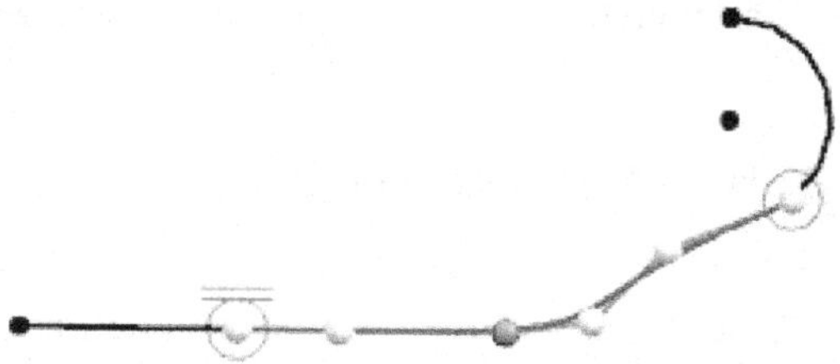

- Right-click on the last pole and select **Infer G2**.

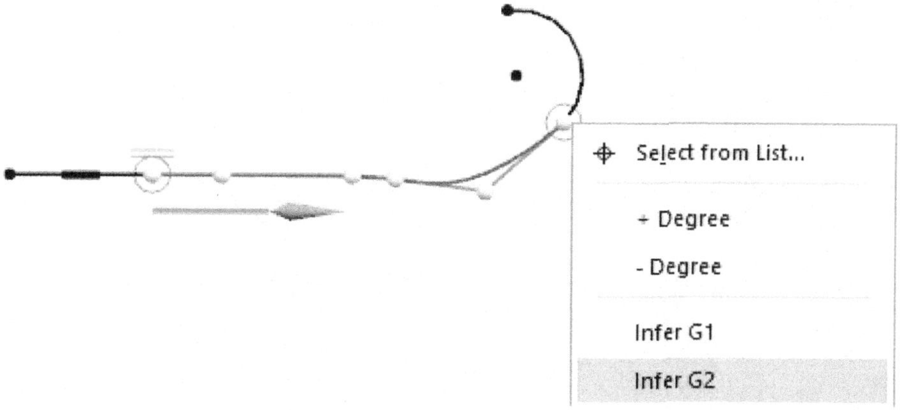

- Note how far the fourth pole jumped! Again, the *G2* constraint constrains the position of the second pole away from the endpoint, but when the endpoint is attached to an arc, the behavior is not as cut-and-dry as that for a line.

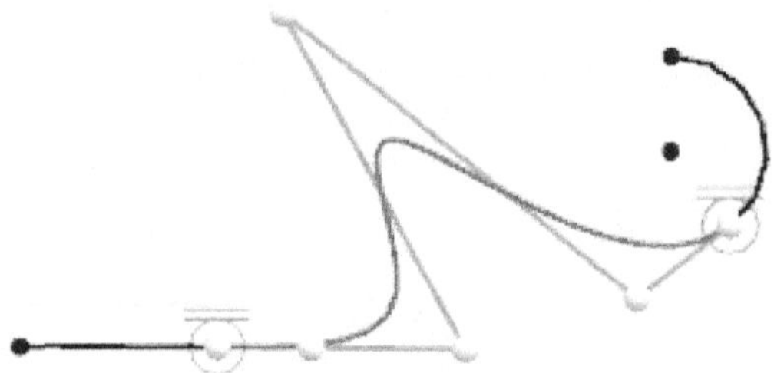

- Drag the fifth pole closer to the sixth and the fourth will move in toward both of them and the spline will smooth out.

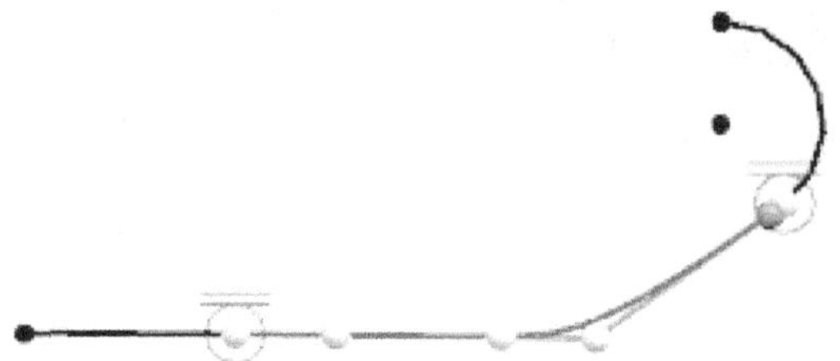

- Now try dragging the fourth pole. Notice that its range of motion is restricted to a line parallel to the tangent line to the spline at the sixth pole! There is a lesson here: the positions of a certain number of poles beyond the endpoints have something to do with the degree of continuity a spline has with a neighboring curve at that endpoint.

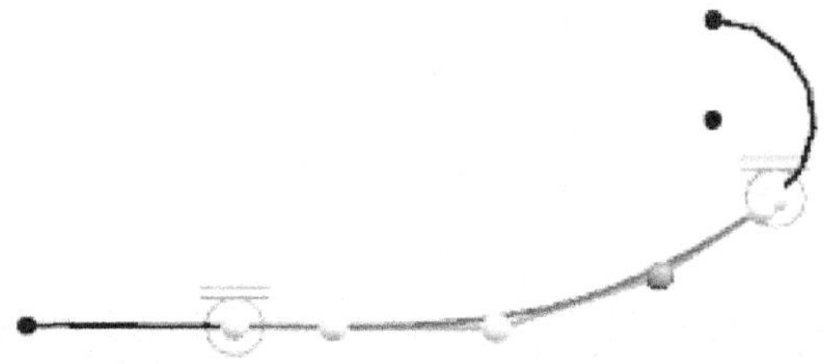

- In the same way that the *G2* constraint was associative for a *Through Points* spline, it will be associative for a *By Poles* spline. Click **OK** to finish your spline and then move the arc at the end – you will see that the three rightmost poles remain constrained!

13.3.4 Studio Spline (Sketch) – G3 Non-Associative Constraint

Now we will explore the *G3* sketch constraint and its lack of associativity.

- Unsuppress and open the sketch *"G3 NON-ASSOCIATIVE"*. It contains another line/arc pair, identical to those in the sketch *"G2 ASSOCIATIVE"*.
- Create another spline – this time, set the **Degree** to *4*, and specify seven poles, as shown below. Right-click on the first pole and select **Infer G3**.

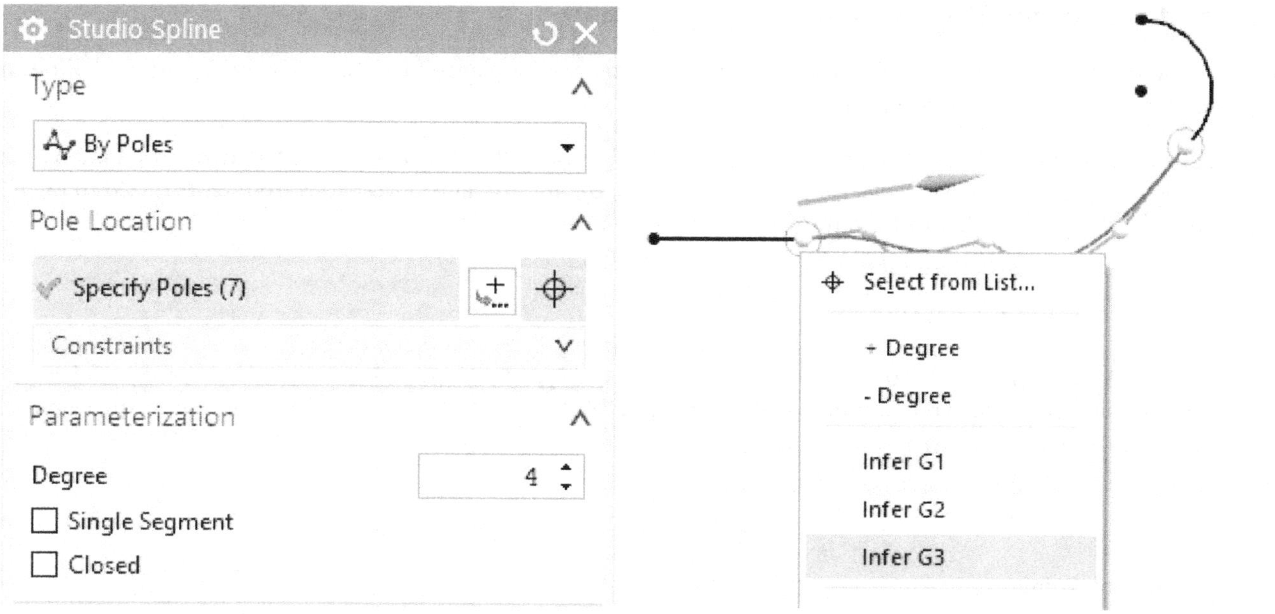

- This snaps the four leftmost poles to the horizontal line. Right-click on the last pole and select **Infer G2**. Adjust the poles that your spline appears as shown below.

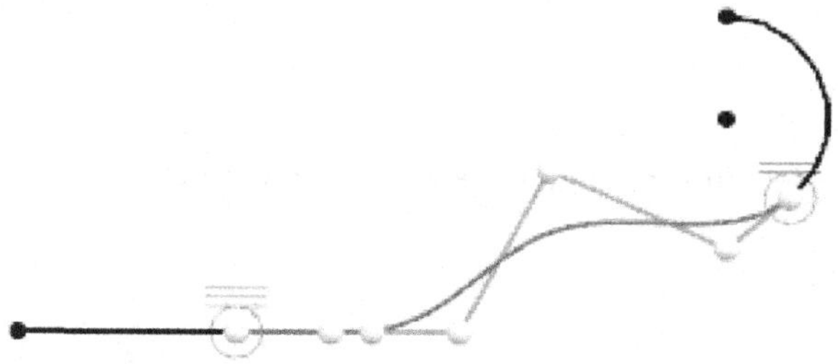

- Upon clicking **OK**, you will find the *G2 Associative constraint* on the right side...

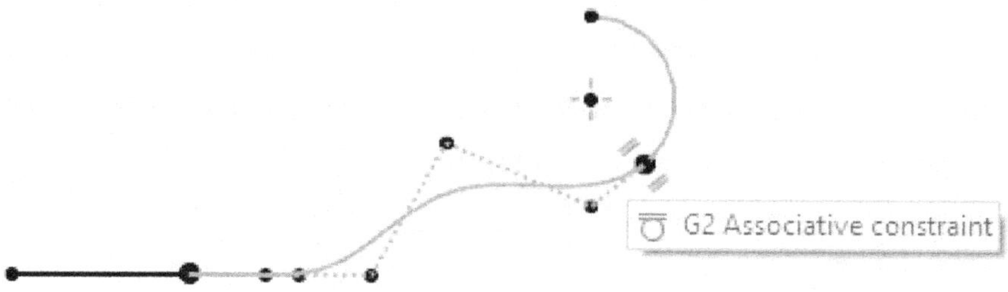

- ... but there is no *G3* constraint on the left, only a *Coincident* constraint between the first pole and the endpoint of the line. The *G3* condition which constrained Pole 2, Pole 3, and Pole 4 to the horizontal line no longer keeps them there – and you can drag them freely. Move Pole 2 upwards so that the spline appears as shown below.

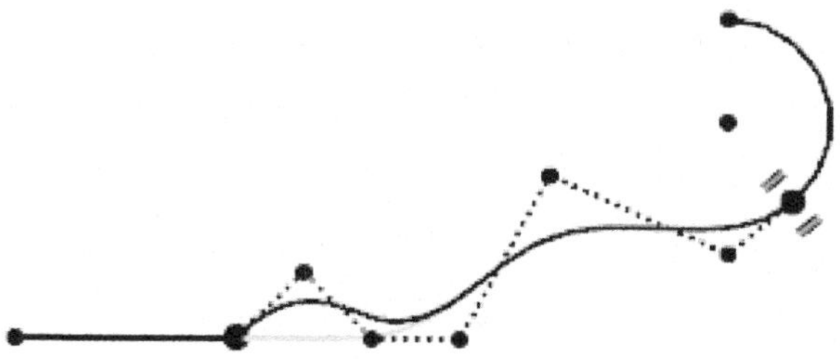

13.3.5 Studio Spline (3D) – G3 Associative Constraint

In light of the limitations of the sketch **Studio Spline** tool, you will be pleased to learn that the 3D **Studio Spline** tool is capable of creating *G3 Associative* constraints.

- **Suppress** all other features in the Model History of *"Continuity.prt"*, and **Unsuppress** the two lines named *"START"* and *"END"*. The line *"START"* goes from *(0, 0, 0)* to *(1, 0, 0)*, and the line *"END"* goes from *(2, 0.5, 0.5)* to *(3, 1, 0.5)*. The lines will appear as below from the *Trimetric* view.

- Select the **Studio Spline** tool from the **Curve** tab, and set the **Type** to *Through Points* and the **Degree** to *5*. For the first defining point, select the endpoint of *"START"* closest to *"END"*, and push the **Infer G3** button on the pop-up toolbar.

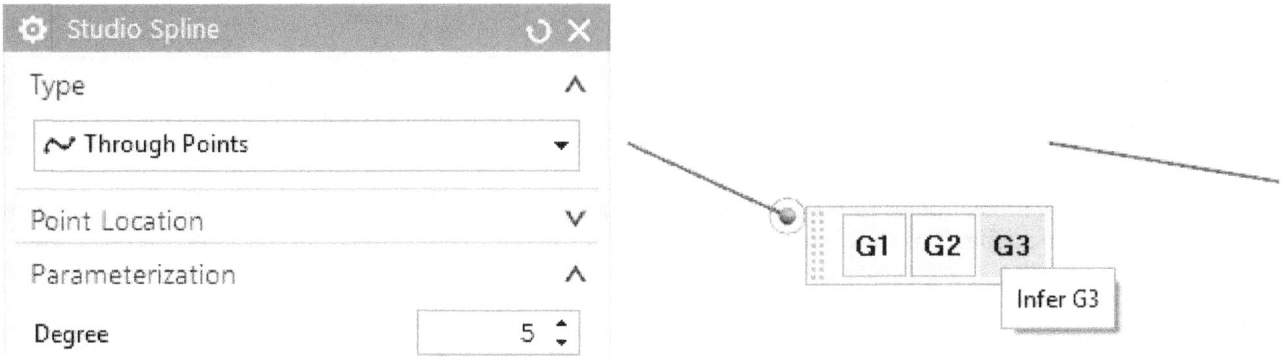

- For the second defining point, select the endpoint of *"END"* nearest *"START"* and again push the **Infer G3** button from the pop-up toolbar.

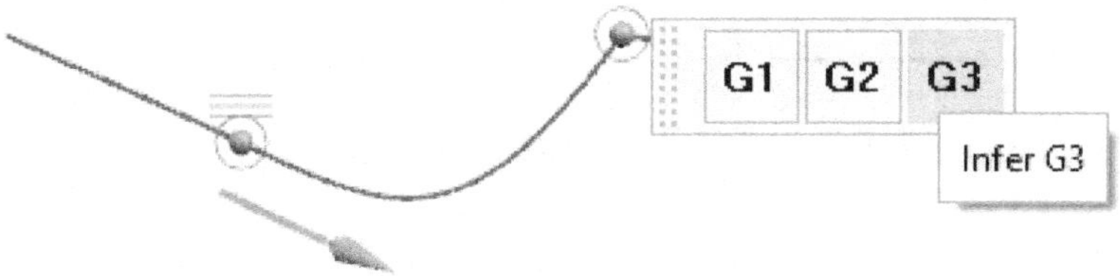

- Click **OK** to create the spline and use **Curve Analysis** to display the curvature comb as shown below.

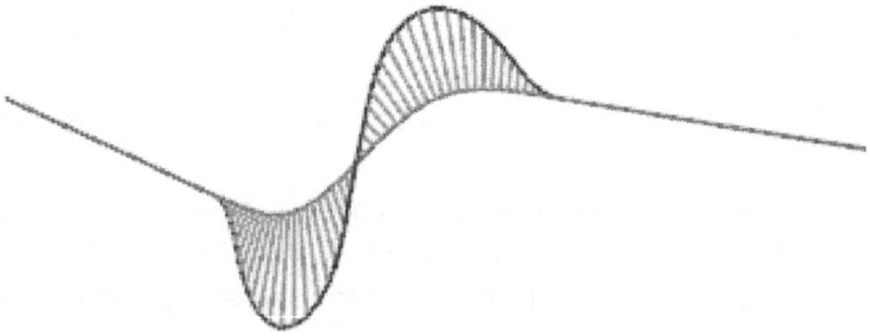

- Double-click on the line *"END"* to modify its **End Point** – change the Z coordinate from *0.5* to *1*.

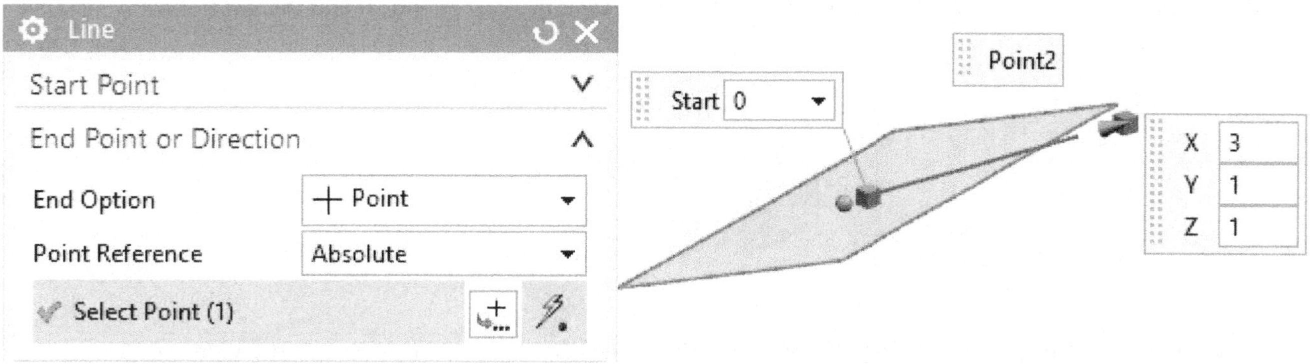

- The *G3* constraint applied to the end of the spline is associative, and so the spline maintains *G3* continuity with the line *"END"* in its new position.

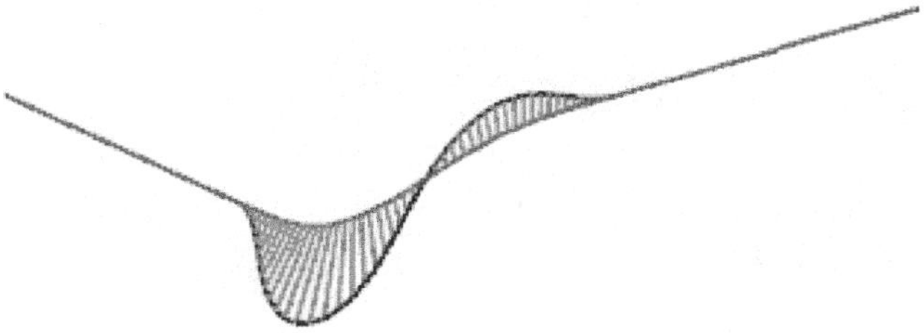

- **Save** your file!

13.4 **Bridge Curve**

Given a starting object and an ending object, the **Bridge Curve** tool produces a high-quality spline that joins those objects with a specified degree of continuity, and a modest amount of overall shape control. Bridge curves are easy splines to manipulate, in comparison with their freeform counterparts produced by the **Studio Spline** command. **Bridge Curve** is an excellent choice when high quality curves are needed for constructing the skeleton of a surface, modeling routed cables, and a variety of other applications. The **Bridge Curve** command is found in the **Derived Curve** group on the **Curve** tab.

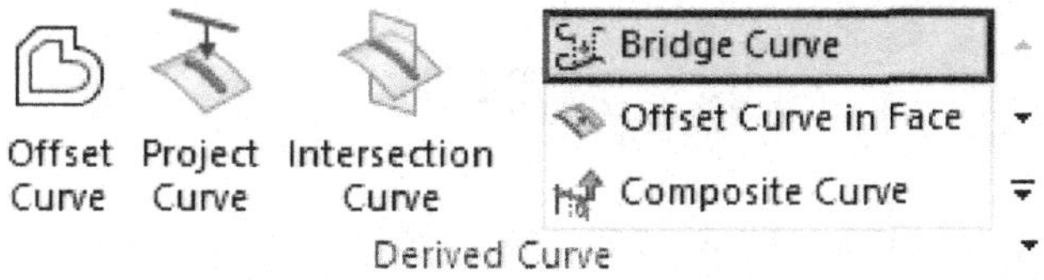

- Create a new file (**Units**: *Inches*) called *"Bridge Curves.prt"* and place it in *"C:\My NX 11 Files"*.
- Create the following sketch on the XZ plane of the datum coordinate system. (Unpictured: four perpendicular constraints in the corners of the rectangle, one midpoint constraint between the right arc center and the right edge of the rectangle.)

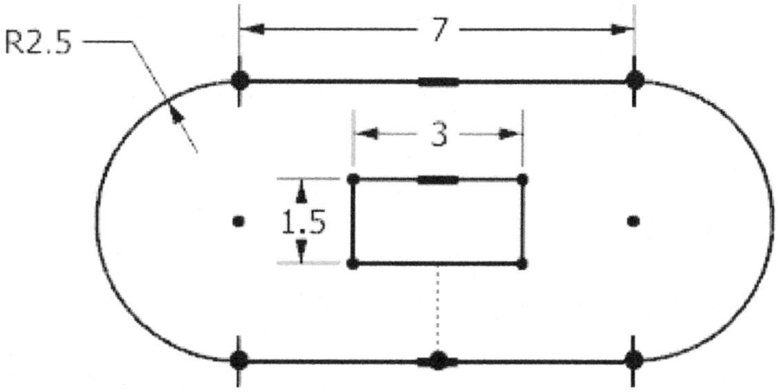

- **Extrude** the center rectangle in the -YC direction with a **Start Distance** of *1 in* and an **End Distance** of *4 in*.

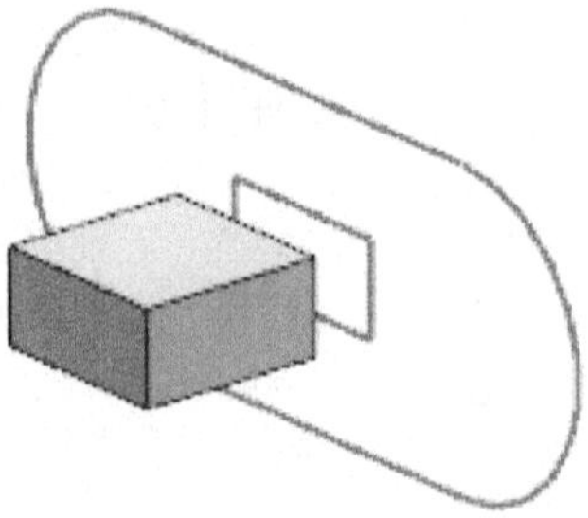

- **Extrude** the slot-shaped section in the +YC direction with a **Start Distance** of *4 in* and an **End Distance** of *8 in*.

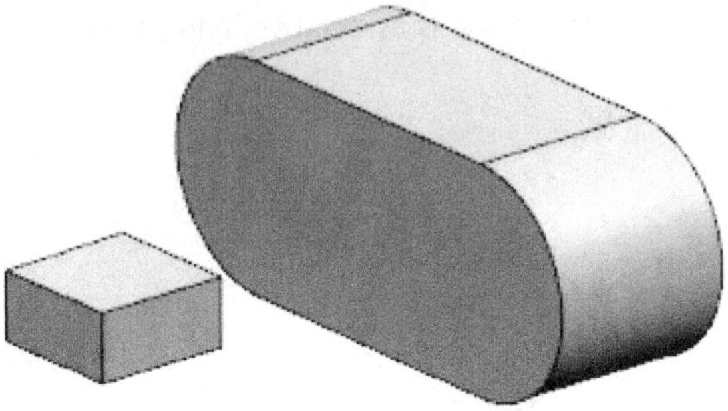

- Select the **Bridge Curve** tool from the **Curve** tab.
- For the **Start Object**, select the edge of the slot-shaped body as shown below. Be sure to put your cursor somewhere on the edge near the endpoint where you want the **Bridge Curve** to originate! If your selection does not match what you see below, use the **Reverse Direction** button!

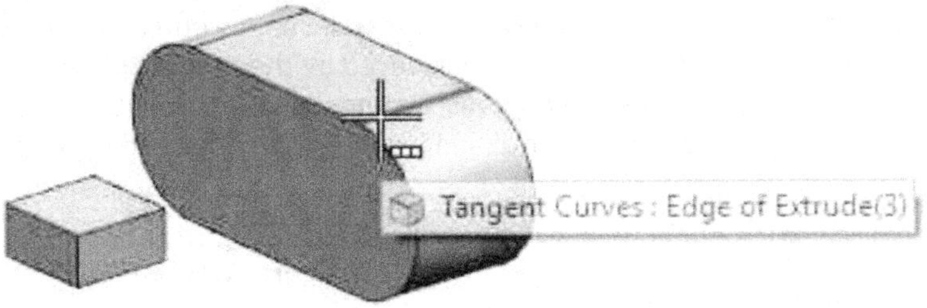

- Note that the dialog will allow you to continue adding to your selection, and that you must manually advance the menu to the next required selection. This is a good opportunity to practice the middle-mouse technique you learned in Chapter 2.7.12!

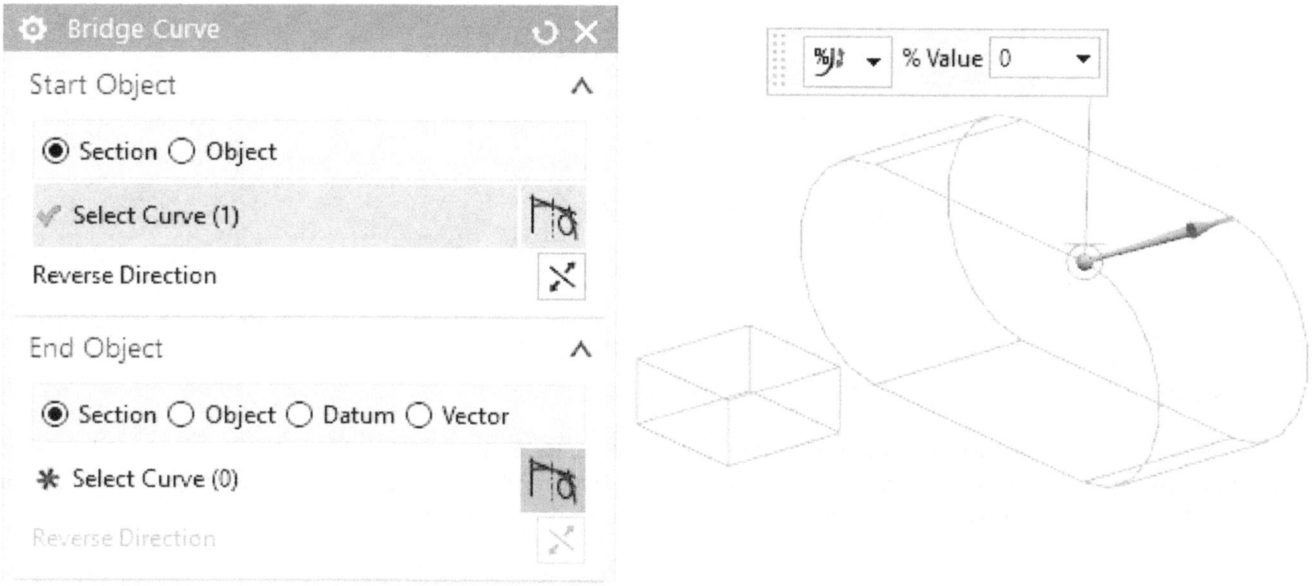

- Select the edge of the block shown below, and open the **Connectivity** section of the **Bridge Curve** menu. On the **Start** and **End** tabs, make sure **Continuity** is set to *G1 (Tangent)*.

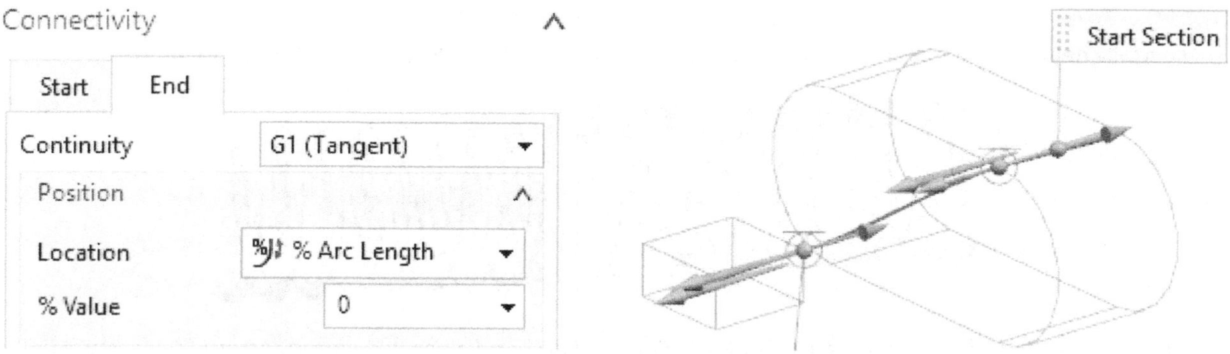

- Next, open the **Shape Control** section of the menu. You can adjust the magnitude of the tangent vectors to the bridge curve at its start object and end object either by dragging the sliders, or by dragging the arrows, as shown below.

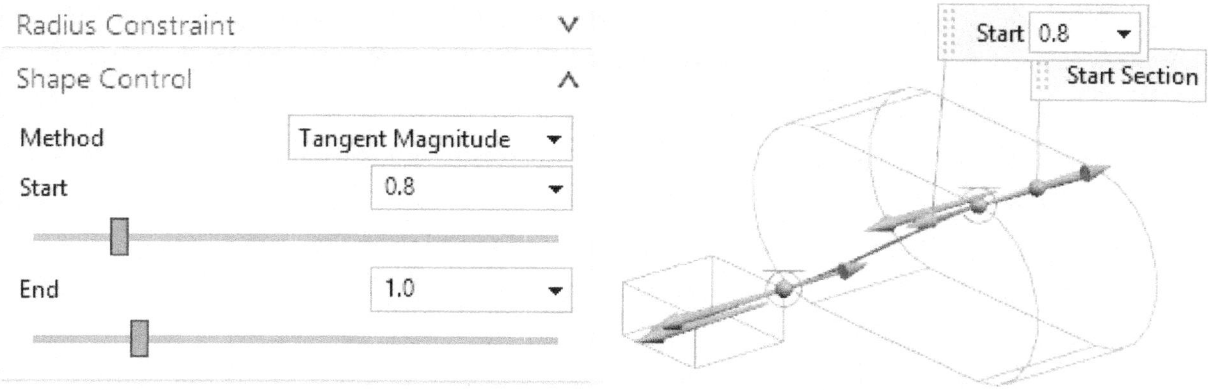

- Set both the **Start** and **End** tangent magnitudes to *1*, and click **Apply** to create your **Bridge Curve**!
- Repeat this process three more times to create the **Bridge Curves** shown below. Use the middle-mouse button to advance the menu and click **Apply** each time for practice!

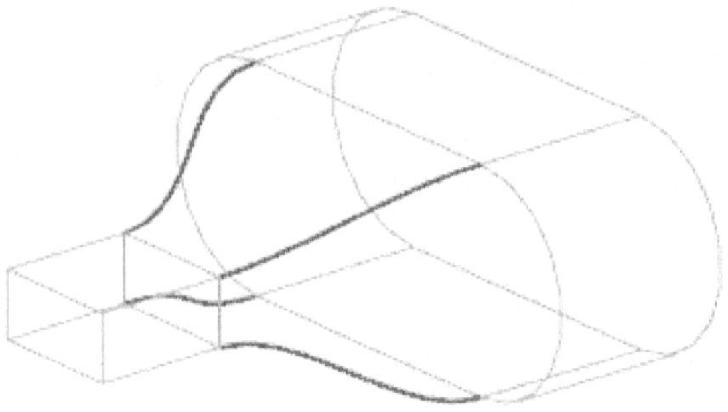

- **Save** your file!

13.5 Helix

A helix is a very useful shape for use as a guide string. The parametric **Helix** command makes it easy to create helices that coil about an axis, or even about a guide curve! It also allows you to vary the basic parameters according to a law, if you need the radius to vary. The **Helix** command is found in the **Curve** group on the **Curve** tab.

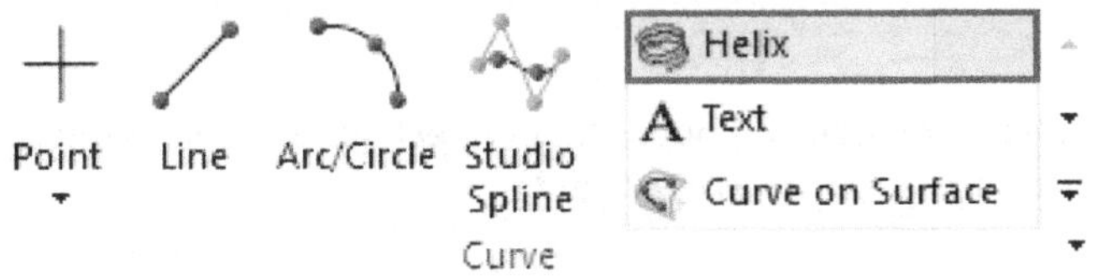

This is especially helpful for designing springs, heat exchange coils, and a variety of other routed shapes.

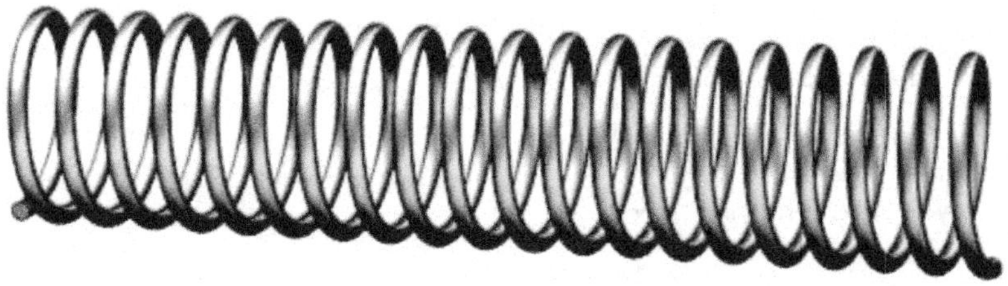

- Create a new file (**Units**: *Inches*) called *"Heat Exchanger Coil.prt"* and place it in your directory *"C:\My NX 11 Files"*.
- Use the **Helix** tool to produce a curve with the parameters shown below.

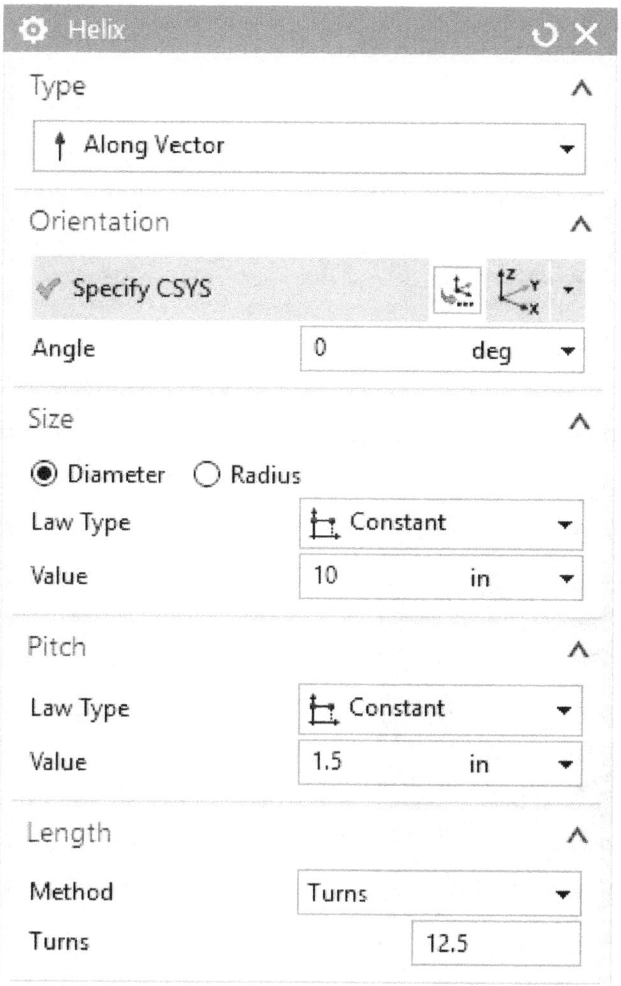

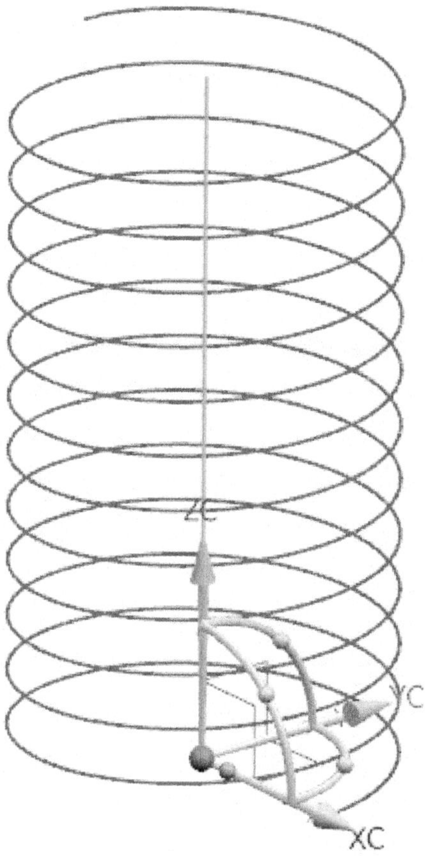

- Create two horizontal lines using the **Line** tool.
    - One connecting the points *(0, -13, 18)* and *(0, -10, 18)*.
    - The other connecting the points *(0, -13, 0.75)* and *(0, -10, 0.75)*.

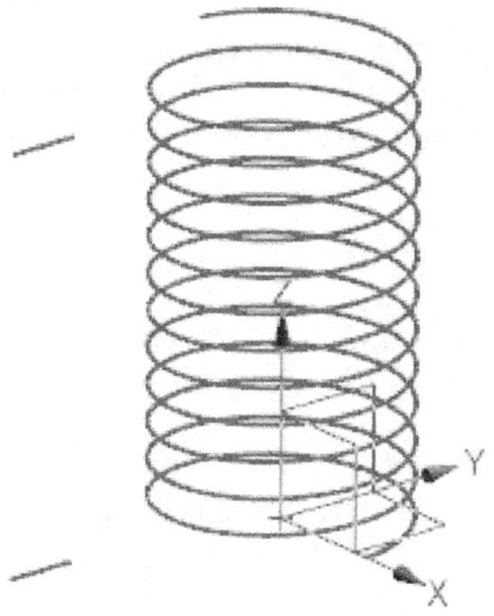

- Connect the two lines to the helix using the **Bridge Curve** command. For each, set **Continuity** to *G1* and **Tangent Magnitude** to *1* at both ends.

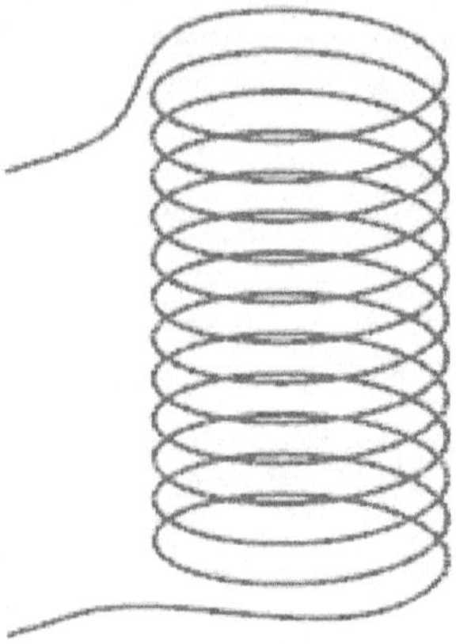

- Use the **Tube** command with **Outer Diameter** *1 in* and **Inner Diameter** *0.75 in*. Set the **Output** to *Single Segment*.

- **Save** your file!

13.6 **Derived Curves**

In the next few exercises you will learn how to combine curves and other entities to produce derived curves. This is a very common and expedient way to produce 3D curves. The majority of these curves end up parameterized as cubic splines. The tools discussed in this section are found in the **Derived Curve** group on the **Curve** tab.

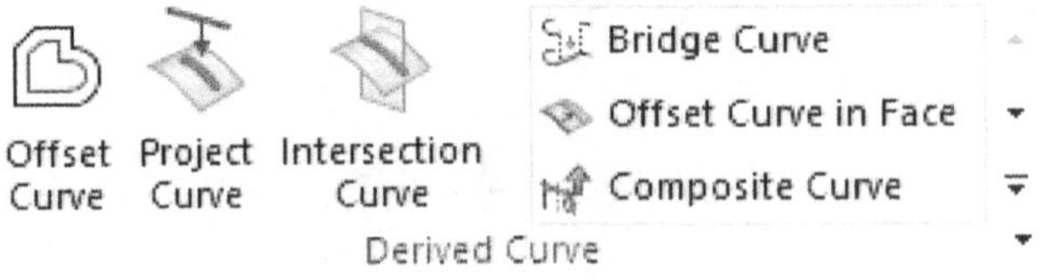

- Create a new file (**Units**: *Inches*) called *"Derived Curves.prt"* and put it in *"C:\My NX 11 Files"*.
- Place a *5 in* x *5 in* **Cylinder** at the origin in your part. **Shell** the top and bottom faces with a wall thickness of *0.5 in*.

- **Save** your part file for the exercises below!

13.6.1 Intersection Curve

Intersecting two sets faces and/or planes is a great way to produce 3D curves that would otherwise be challenging to model. The **Intersection Curve** tool finds the "edges" between two such sets of faces.

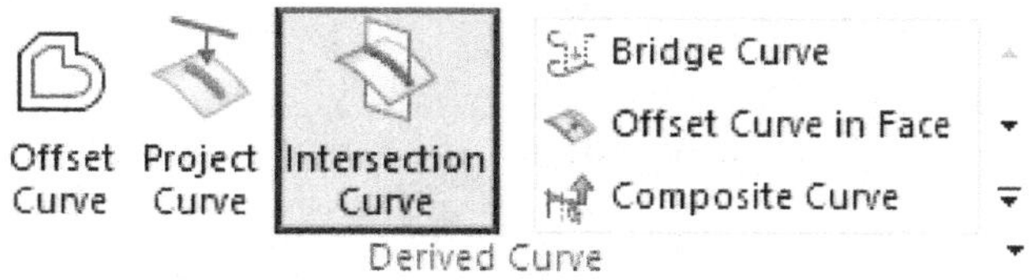

- Continue working with *"Derived Curves.prt"*.
- Create the sketch shown below on the YZ plane of the datum coordinate system in the model. Note that the arc center of the arc in the sketch is made coincident with the arc center of the bottom edge of the cylinder.

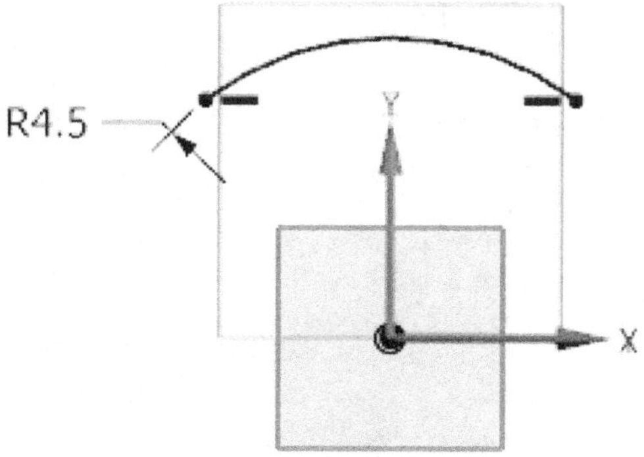

- **Extrude** your sketch symmetrically *5 in* in each direction. Select the **Intersection Curve** tool and choose the outer face of the cylinder as **Set 1**. Choose the face of the extruded body as **Set 2**.

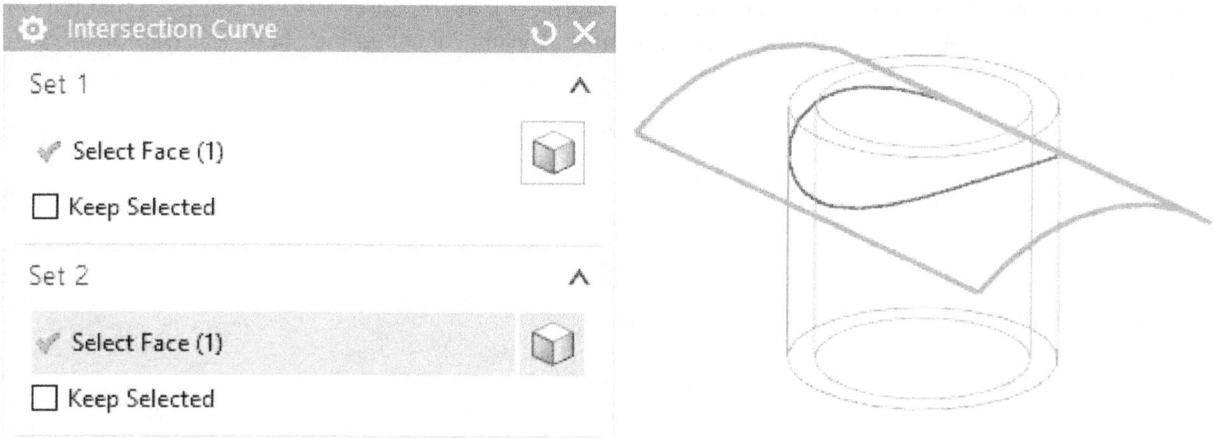

- **Save** your file!

13.6.2 Offset Curve

The **Offset Curve** tool on the **Curve** tab is similar to the **Offset Curve** tool in the sketcher discussed in Chapter 6.9.1. It has a few additional modes available for curves that are twisted in 3D space however.

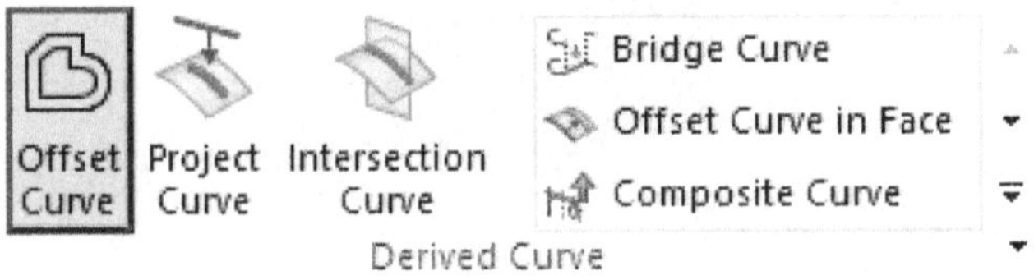

- Continue working with *"Derived Curves.prt"*.
- **Hide** the *Intersection Curve* feature and extruded body.
- Use the **Offset Curve** tool and choose the outermost edge of the top face of the cylinder. Enter an **Offset Distance** of *0.25 in*, and make sure the *Associative* checkbox is marked, as shown below. Click **Apply** to create the **Offset Curve** feature.

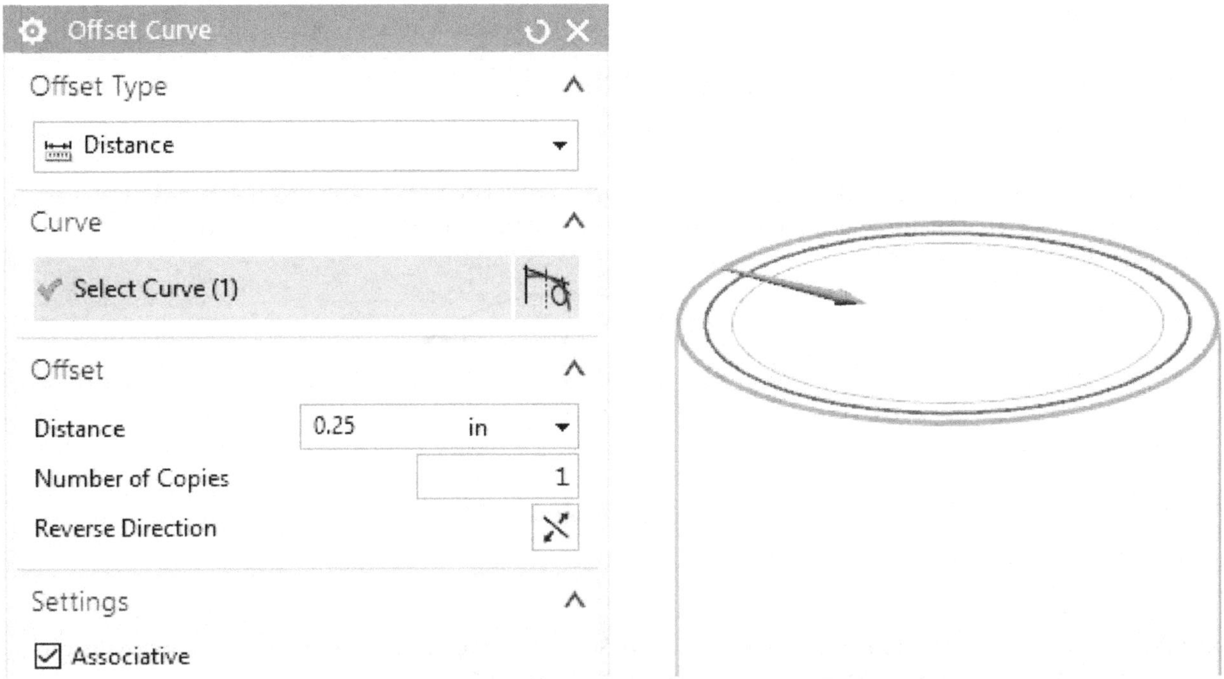

- Show the **Intersection Curve** feature previously hidden and select it as the **Curve**. NX will warn you that the curve is not planar. Set the **Offset Type** to *3D Axial* and it will infer a cylindrical centerline about which to offset. Enter an **Offset Distance** of *0.25 in*.

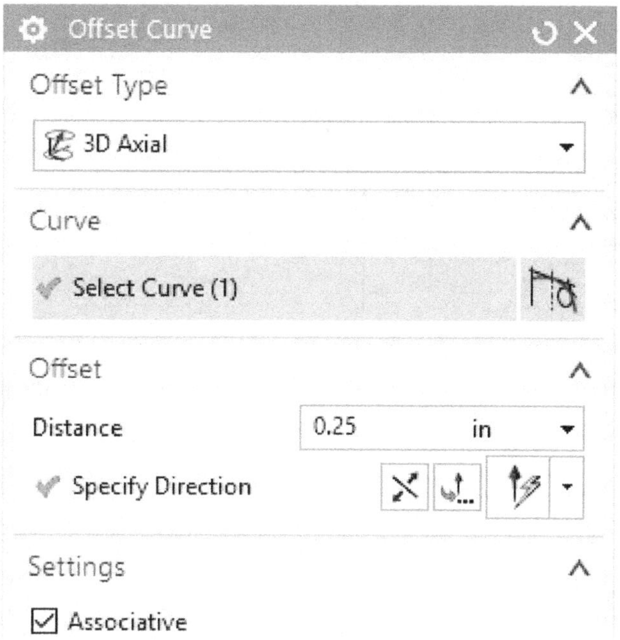

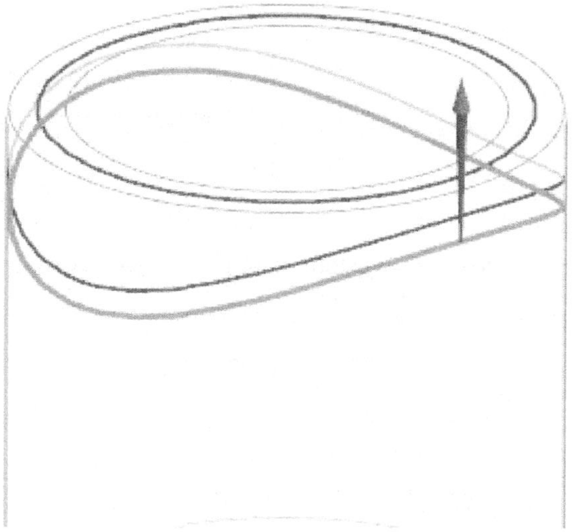

- Click **OK** to create your **Offset Curve** feature.
- **Save** your file!

13.6.3 Project Curve

Curve projection is a great way to get an image of some curves on a surface, whether faithful or deformed in a natural way. The **Project Curve** command found on the **Curve** tab differs from the **Project Curve** command in the sketcher insofar as it allows curves to be projected onto curved surfaces.

- Continue working with *"Derived Curves.prt"*.
- Create a datum plane tangent to the outside face of the cylinder and parallel to the XZ plane.

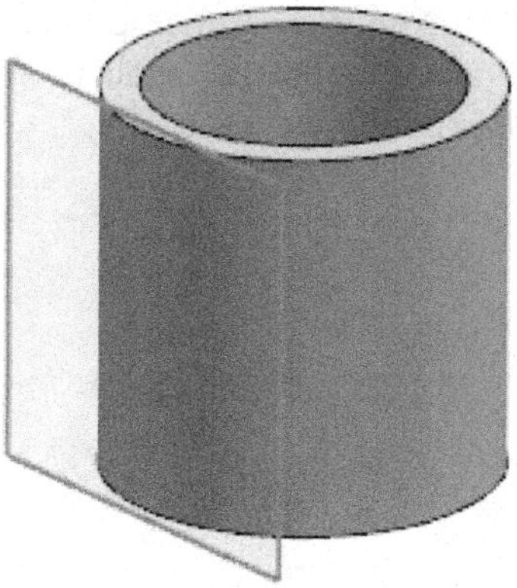

- On that datum plane, create the following sketch, centered within the face of the cylinder.

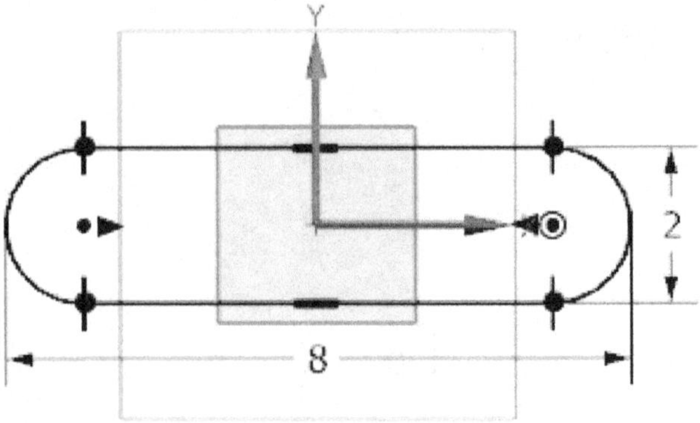

- Use the **Project Curve** command to map your sketch onto the face of the cylinder with the **Direction** set to *Along Face Normal*, as shown below.

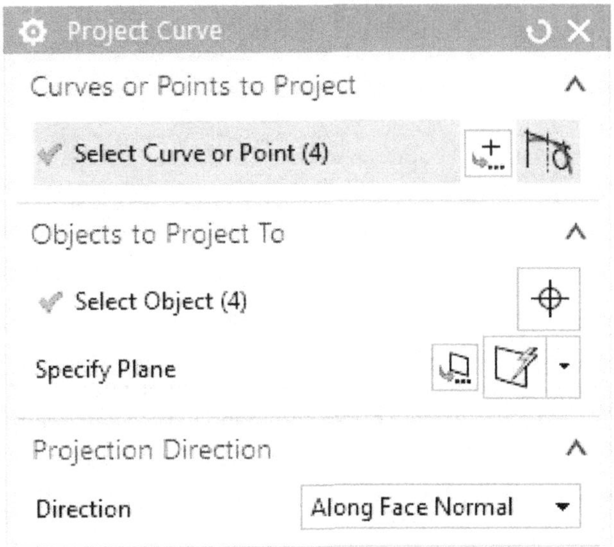

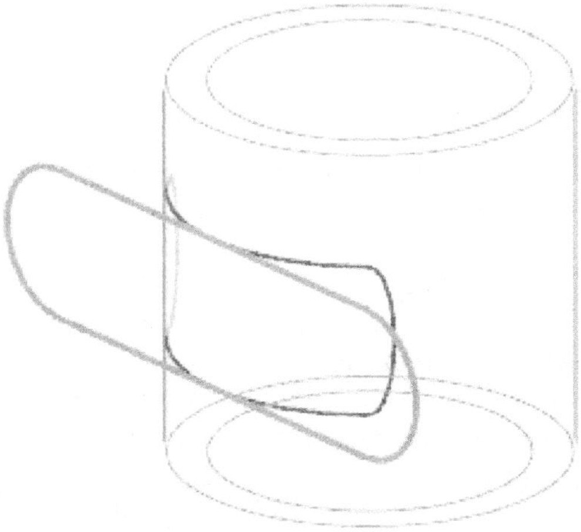

- **Save** your file!

13.6.4 Wrap/Unwrap Curve

A *developable surface* is a curved surface that can be flattened onto a plane without distortion. The most basic example of a curved developable surface is the face of a cylinder. For developable surfaces, there is another way to create curves based on existing curves – namely, to wrap them. The **Wrap/Unwrap Curve** command does exactly this – takes a curve and wraps it about a developable surface. The transformation is guided by a plane tangent to the surface in question – the plane indicates how the surface should be wrapped or unwrapped.

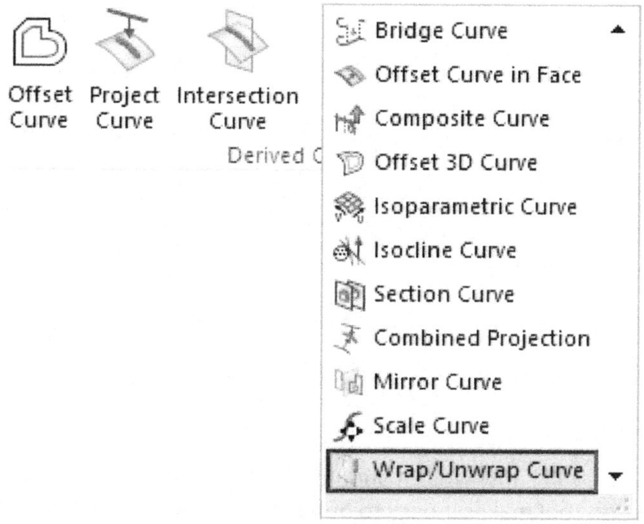

- Select the **Wrap/Unwrap Curve** command. The **Curve** in question is the sketch, the **Face** in question is the face of the cylinder, and the **Plane** in question is the datum plane.

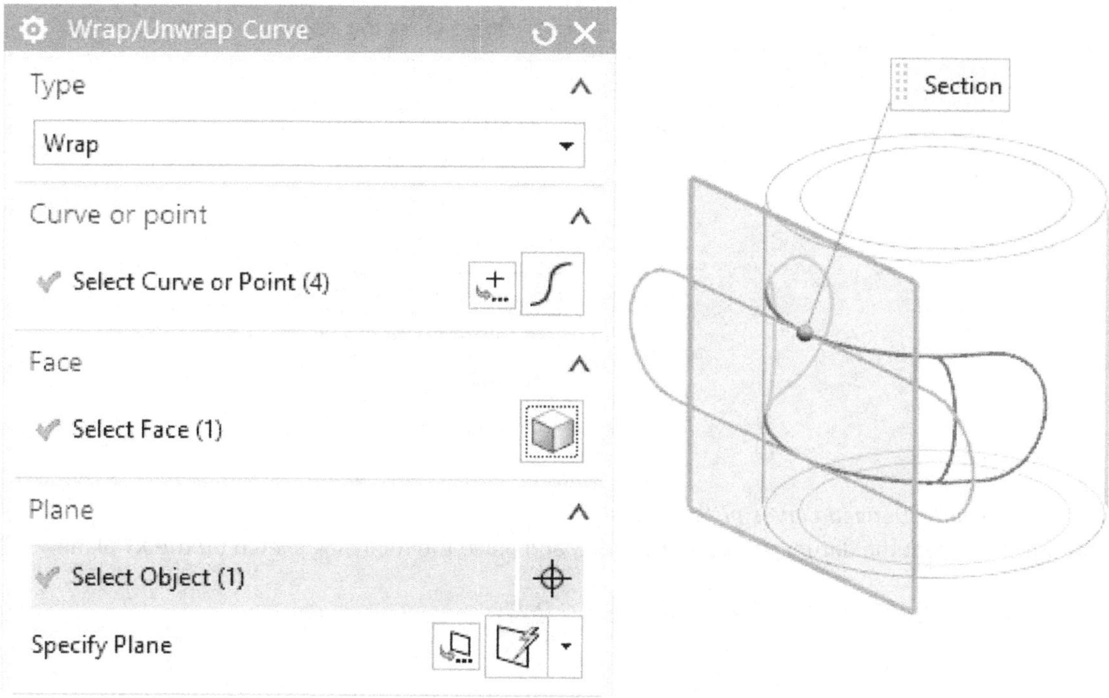

- The **Wrap/Unwrap Curve** is (well, to within a tolerance) an arc length preserving operation, unlike **Project Curve**. Imagine rolling the cylinder across the plane containing the sketch – the "wrapped" curve would be the imprint of the sketch on the cylindrical face.

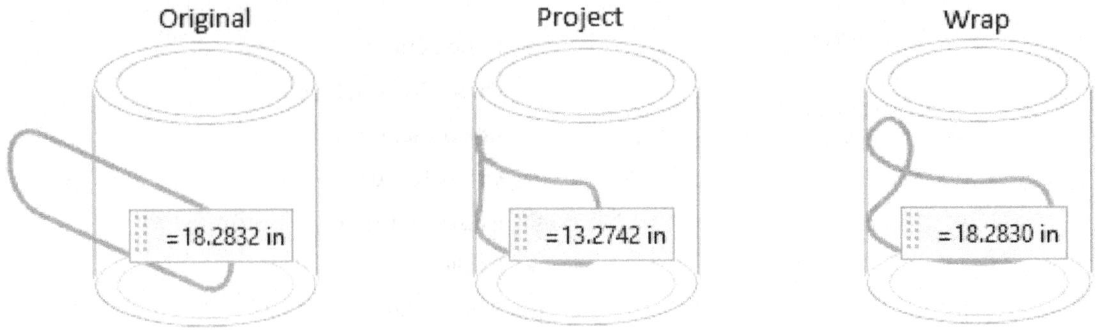

- **Save** your file!

13.6.5 Combined Projection

Oftentimes in designing a curve, you will know what it should look like from two different orthographic views. When this is the case, the **Combined Projection** tool makes it easy to build that 3D curve from two of its "shadows" on perpendicular planes.

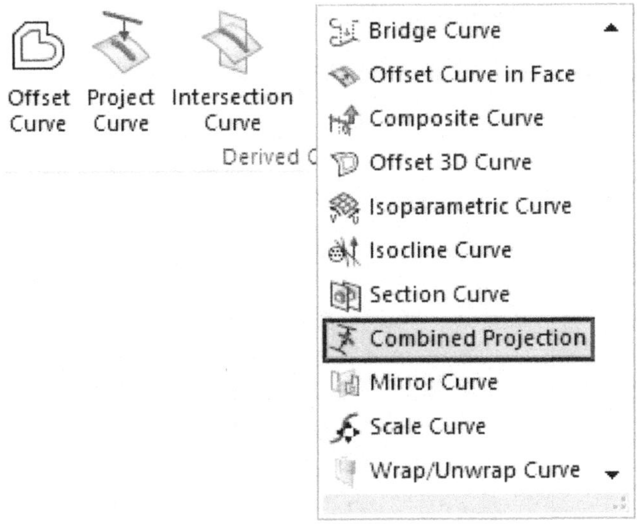

- Continue working with *"Derived Curves.prt"*.
- Hide all geometry except the datum coordinate system, and place the following sketch on the XY plane

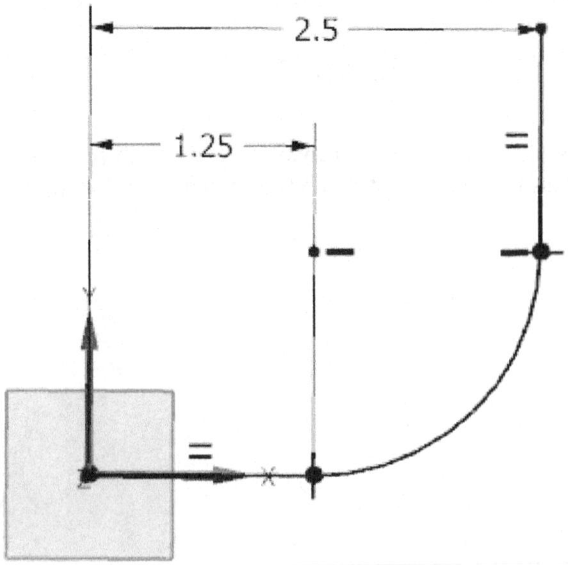

- Create the sketch shown below on the XZ plane.

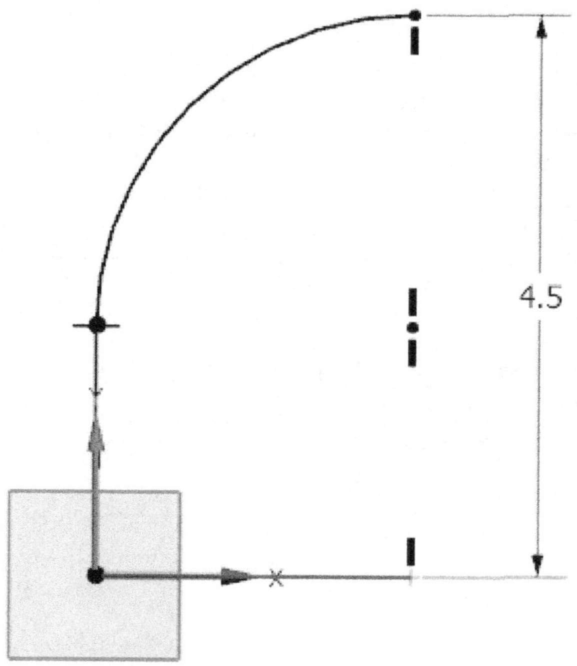

- Select each of the two preceding sketches as the inputs for the **Combined Projection** tool.

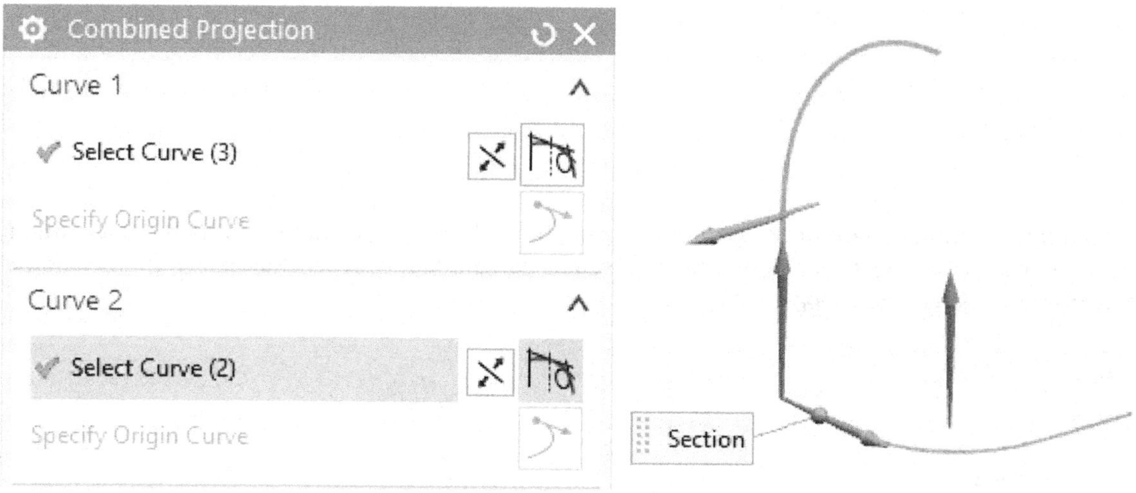

- This results in a spline which curves along both arcs in a nontrivial way – this would be difficult to model as a freeform spline!

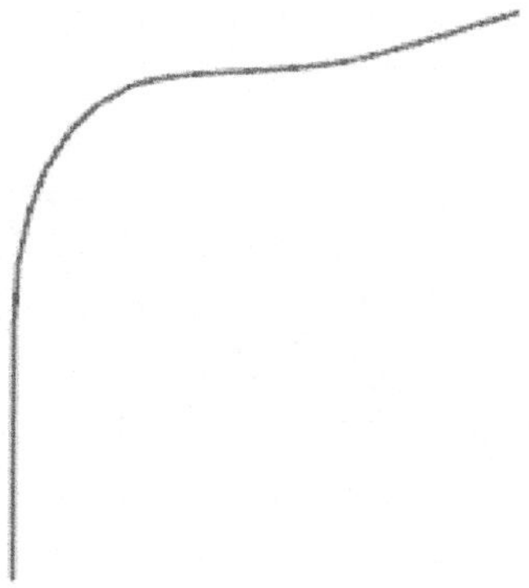

- **Save** your file!

13.7 Text

The **Text** tool allows you to wrap text curves onto the faces of a body. For example, when creating a casting model with a brand name etched onto one side, text curves can be used to emboss letters of the brand name onto the part. It can also be used to etch the material or part number on a prototype by linking to the appropriate attributes.

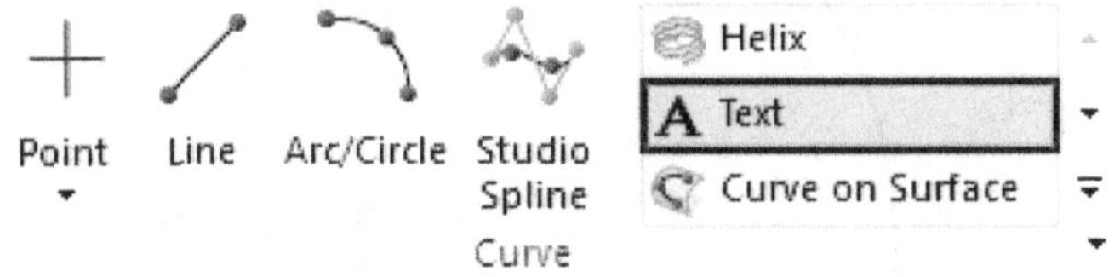

- Create a new file (**Units**: *Inches*) called *"Text.prt"* and place it in *"C:\My NX 11 Files"*.
- Place a *5 in* x *5 in* **Cylinder** at the origin in your part.
- Create a circle on the surface of the cylinder, offset *1 in* from the top edge.

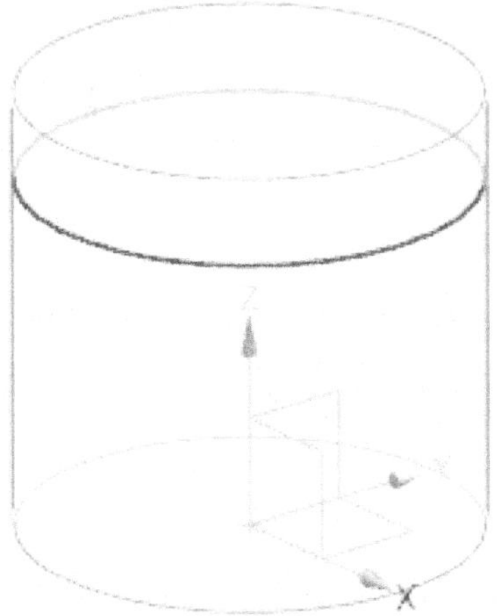

- Select the **Text** tool and set the **Type** to *On Curve*. Choose the circle you just created, and set the **Orientation Method** to *Vector*. Choose +ZC as the vector so that the text stands upright. In the screenshot below, the **% Parameter** from the **Text Frame** section of the dialog is set to *80*.

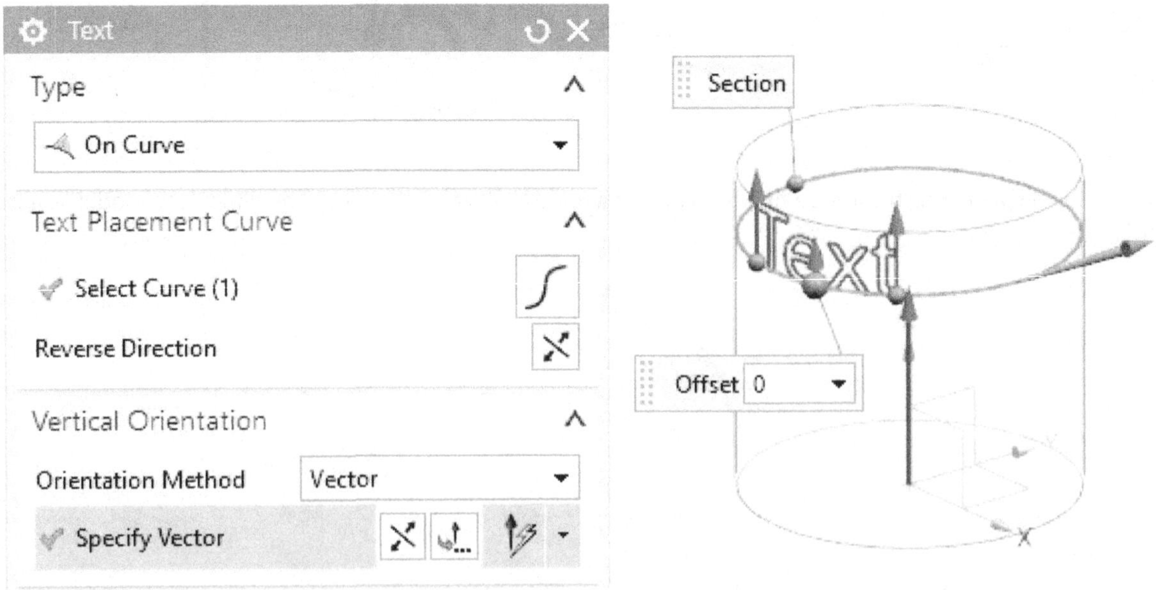

- Click on one of the vertical golden arrows and a dialog will appear that lets you modify the size of the text. Set the **Height** to *0.75* and the **W Scale** to *100*.

- Within the **Text Properties** section of the dialog, change the text to say *"PART # 1"*.

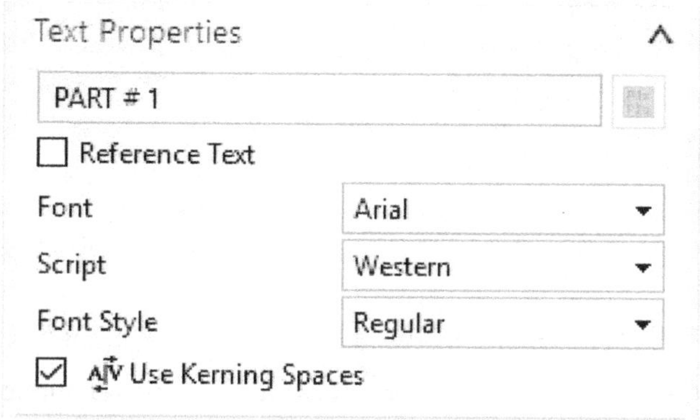

- Click **OK** to create your text, and **Save** your file!

13.8 **Surgery on Curves**

In this section, we will review a number of tools that are used for performing various types of surgery on curves – merging them together, breaking them apart, and trimming them.

13.8.1 Join Curves

The **Join Curves** command allows you to select distinct curves and connect them to provide one continuous spline. Like many tools on the **Curve** tab, it can be produce associative or dumb curves, and provides options for how the originals should be handled. **Join Curves** is found the **More** gallery on the far-right side of the **Curve** tab.

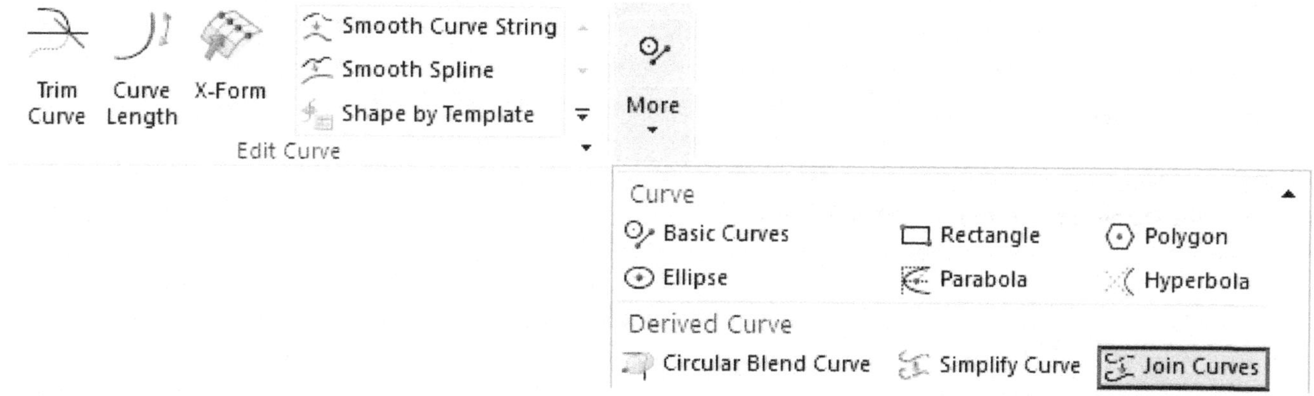

Below are shown three curves in 3D space. One curve was created on the XY plane and two curves (one line and one arc) were created in the YZ plane.

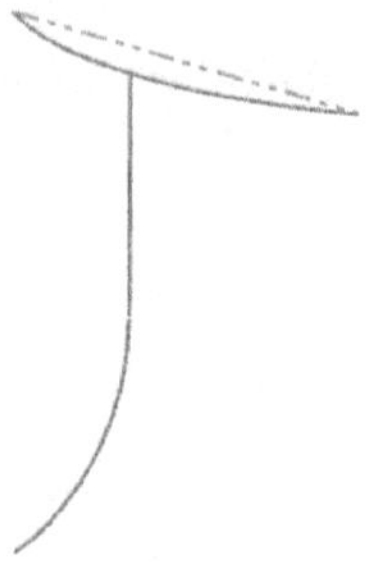

The following figures illustrate the difference between swept features created with and without the **Join Curves** command. Notice the edge that divides the swept feature in the first model into two separate faces. The second feature does not have this edge as a result of the **Join Curves** command used to merge the line and arc of the guide curve.

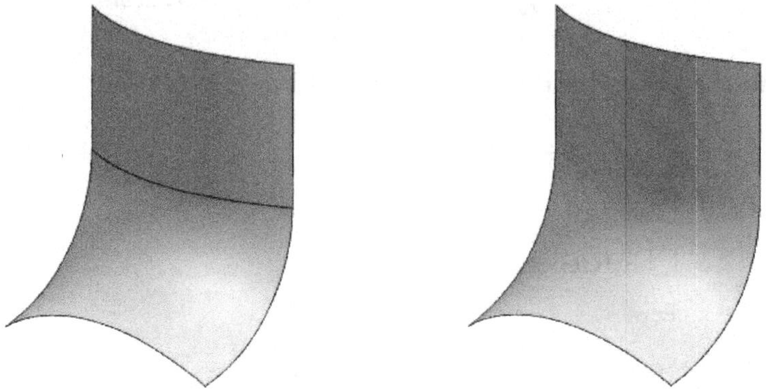

In the exercises below you will use the **Join Curves** command to produce a spline from lines and arcs.

- Open the file *"3D Lines and Arcs.prt"* that you created in Chapter 13.1. If you skipped that exercise, you can use the part file found at *"C:\My NX Files\Downloaded Files\3D Lines and Arcs.prt"*.
- Use **File / Save / Save As** to create a copy called *"Surgery on Curves.prt"*.
- Select the **Join Curves** tool and choose the lines and arcs in the file. Check the **Associative** checkbox and set **Input Curves** to *Keep*.

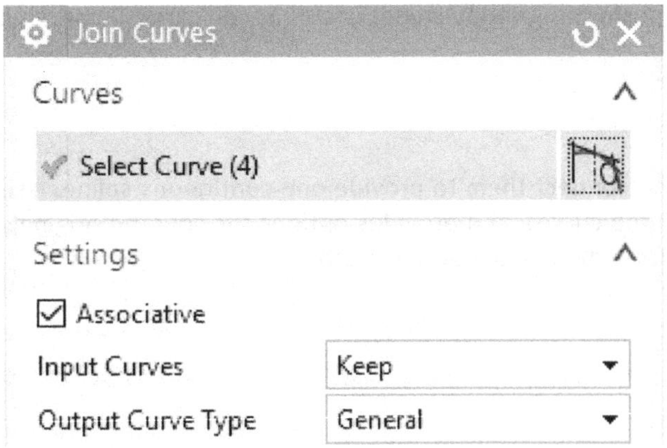

- Hide your lines and arcs.
- Select the **Detail Filtering** tool and filter for *Splines*.

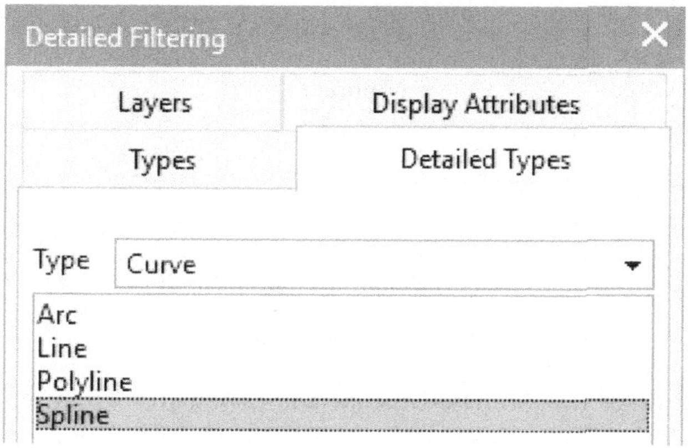

- The curve produced by **Join Curves** is in fact a spline!

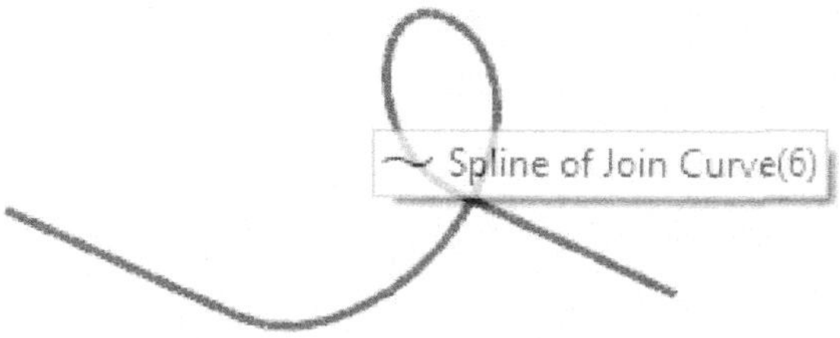

- **Save** your file!

13.8.2 Simplify Curve

In some sense, the inverse operation to **Join Curves** is **Simplify Curve**. This takes a spline as input, and breaks it down into a series of lines and arcs that approximate that spline.

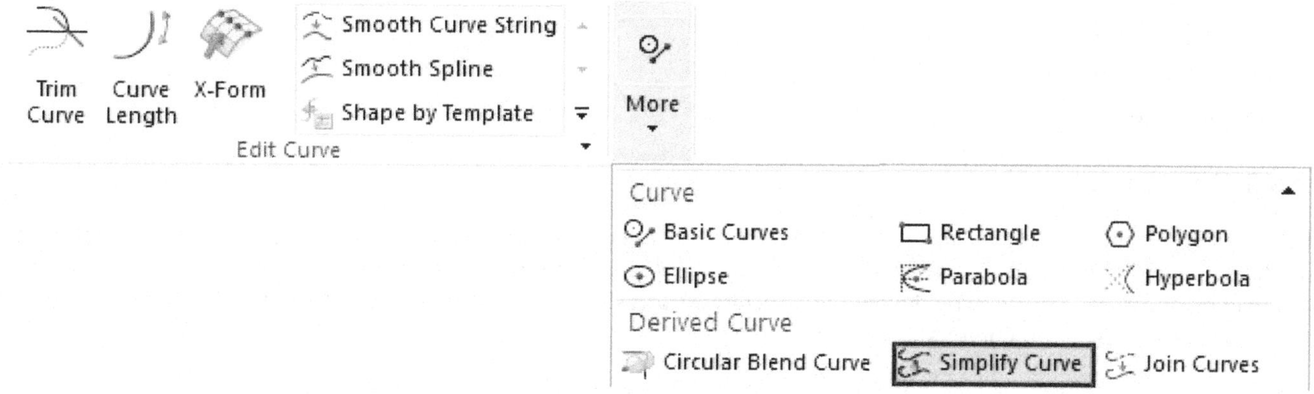

Simplify Curve is occasionally a useful tool when exporting a DXF file to send to a manufacturer – some machine tools are only capable of processing lines and arcs, and may reject DXFs that contain splines.

- Continue working with *"Surgery on Curves.prt"*.
- Select the **Simplify Curve** command.
- On the **Simplify Curve** dialog, select **Hide**. This means that the input spline will be hidden.

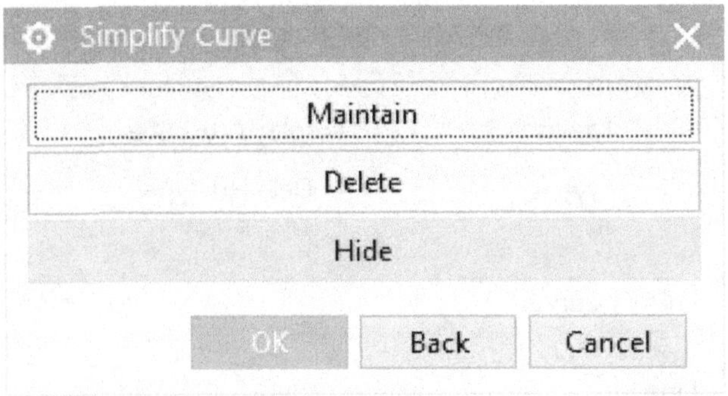

- The output of **Simplify Curve** will be a set of non-associative lines and arcs – they don't appear in the Model History.

- You might find that the way the spline breaks down into lines and arcs does not match the original input to **Join Curves**.

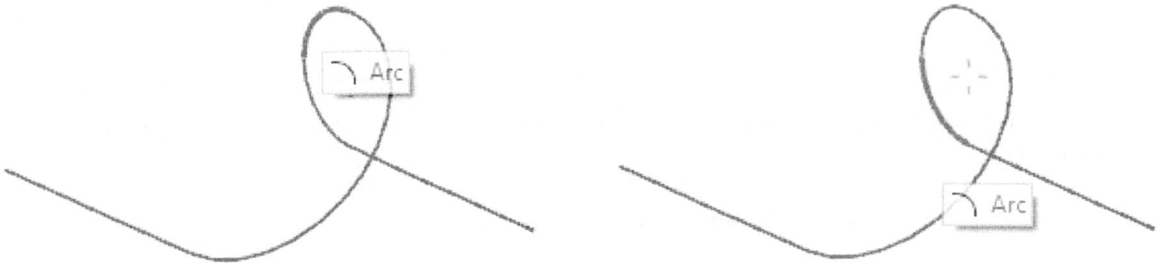

- **Save** your file!

13.8.3 Trim Curve

The **Trim Curve** command is similar to the sketch command **Quick Trim** covered in Chapter 6.9.6, in the sense that it lets you delete a portion of a curve. It is slightly clumsier of a tool, and there is an artifact of the deleted curve that remains, but it can be necessary for trimming down non-associative curves that have no parameters to edit. **Trim Curve** is found in the **Edit Curve** group on the **Curve** tab.

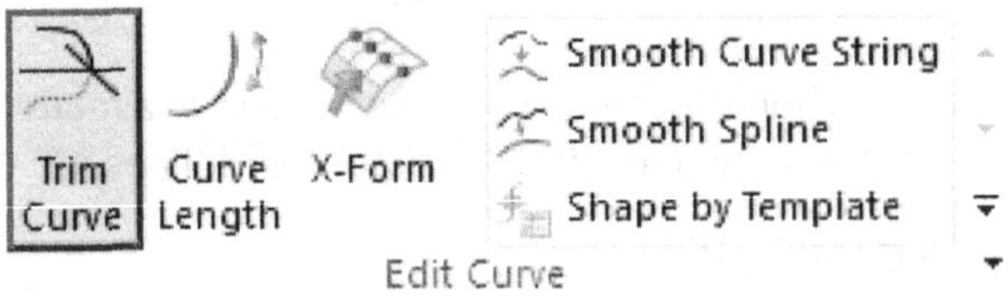

- Continue working with *"Surgery on Curves.prt"*.
- Position a datum plane *25%* of the way on the first line that resulted from **Simplify Curve**.

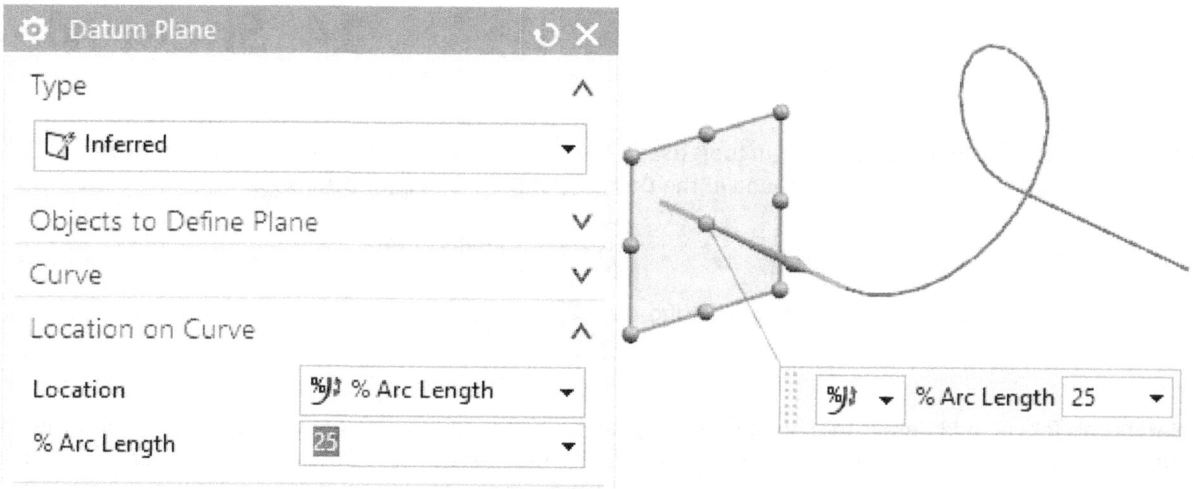

- Open the **Trim Curve** command and select the line. The tool is sensitive to which side of the line you click on. The **End to Trim** (whether that be **Start**, or **End**) is indicated in the graphics window with a small orange circle around the endpoint of the curve to be trimmed. In the image below, the arrow shows the position of this orange circle.

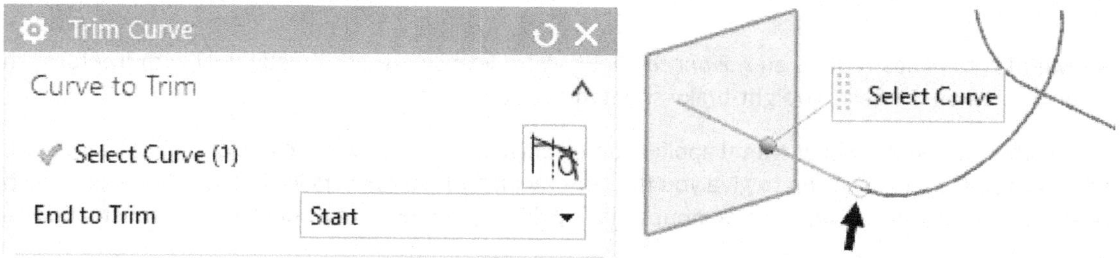

- Select the datum plane as **Bounding Object 1**. The tool will allow for a second, optional bounding object. Click **OK**.

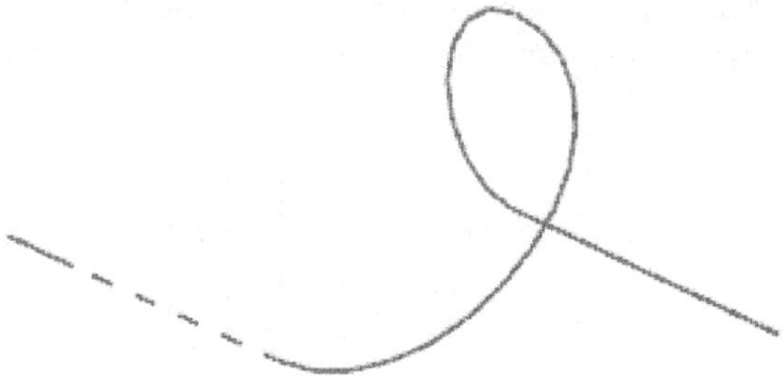

- **Save** your file!

14 Sheet Metal

The **NX Sheet Metal** application includes design tools useful for the creation of straight-brake sheet metal parts. To launch the **Sheet Metal** application, select its icon found in the **Design** group on the **Applications** tab.

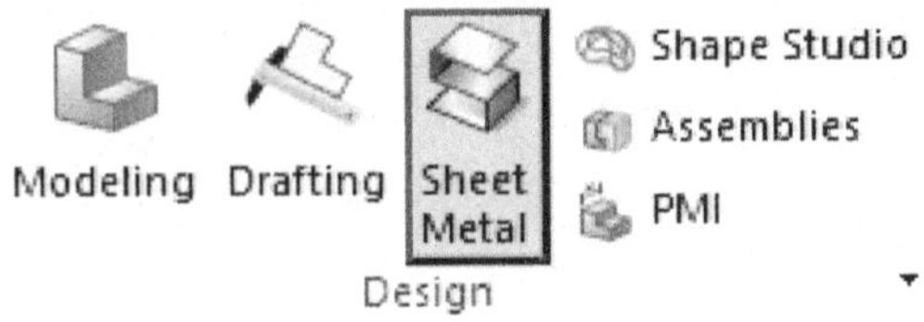

Sheet Metal tools let you add details that are parametric, can be added to any appropriate solid model, and will account for distortion due to forming. You will learn how to create sheet metal parts from scratch, and by converting existing solid models.

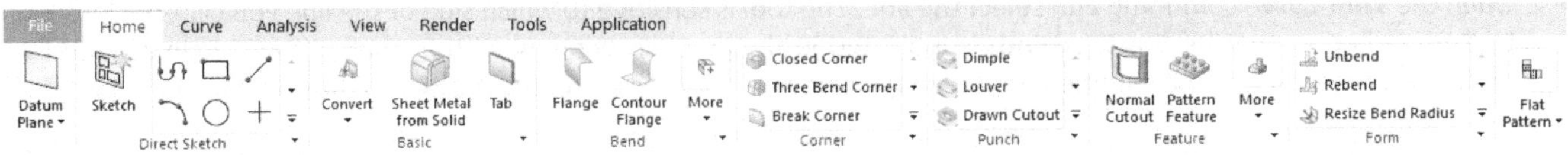

Certain **Sheet Metal** commands require an *Advanced Sheet Metal* license – these mostly deal with freeform shapes. We will focus our discussion on the basic straight-brake sheet metal operations.

We will not cover the tools in the **Sheet Metal** application exhaustively in the exercises below – instead, we will give you a sample of a few tools from each group to give you the general idea of how others in that same category might behave. The tools are all straightforward, so we have no doubt that by this point you will be able to master them all with practice!

14.1 Tab

The **Tab** command is to the **Sheet Metal** application as **Extrude** is to the **Modeling** application. It creates a uniform thickness solid body from a closed planar section string. **Tab** is one of two tools that can create a sheet metal part from a sketch.

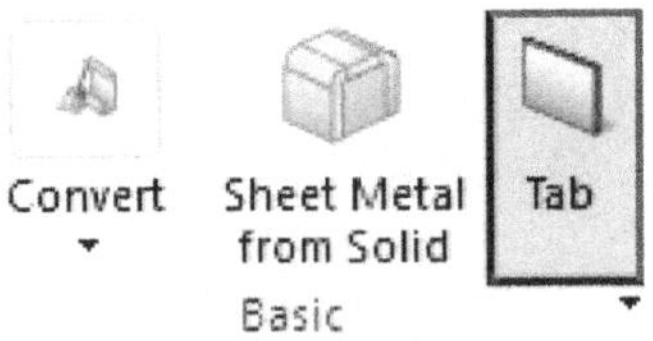

Tab can also be used to add material to an existing sheet metal part, once you already have one – the key is to set the **Type** to *Secondary*. We leave it to you to experiment with this option.

- Create a new file (**Units**: *Inches*) called *"Tab, Flange, Corner, Bend.prt"*.
- Select the **Tab** command. Note that since it has the **Sketch Section** icon, when you click on the XY plane of the datum coordinate system, you will be thrown into the sketcher. Sketch a *6 in* x *6 in* rectangle. The thickness added to your tab is controlled by a global parameter named *Sheet_Metal_Material_Thickness*. If you wish to create a tab of some other thickness, you can click the **"="** icon and select *Use Local Value*.

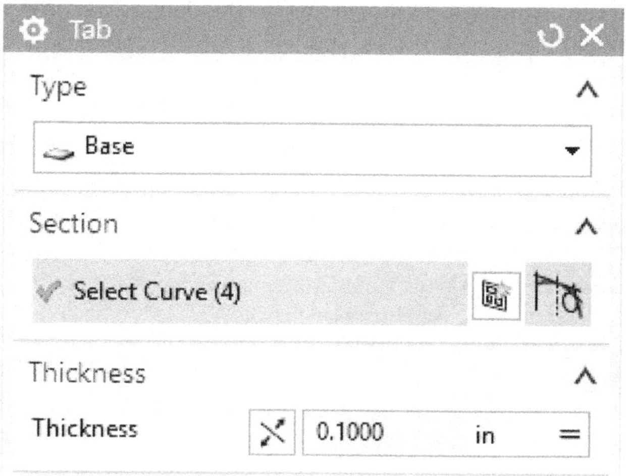

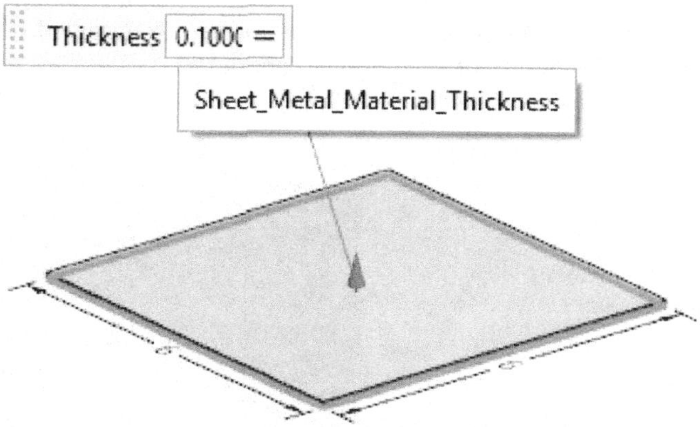

- **Save** your part file!

14.2 Flange

The **Flange** tool is the next basic operation to know in the **Sheet Metal** application. It adds a bend along an edge of an existing piece of sheet metal, and adds material to the part along that bend. This is basically what all of the tools with "flange" in their name do – turn an edge or set of edges into bends, and add material to a sheet metal part.

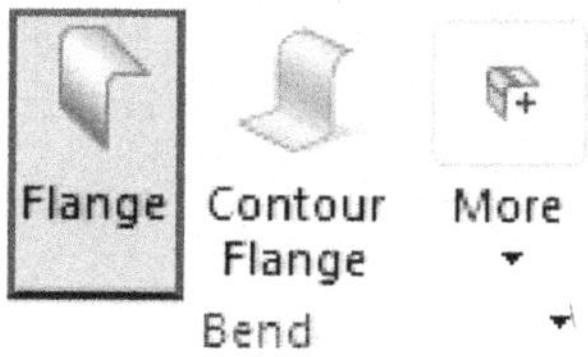

Following the exercise below, there are screenshots detailing a number of the important options in the flange dialog.

- Continue working with *"Tab, Flange, Corner.prt"*.
- Select the **Flange** tool and select the edge shown below. Set the **Width** to *Full*, the **Length** to 1 in, the **Length Reference** to *Outside*, and the **Inset** to *Bend Outside*.

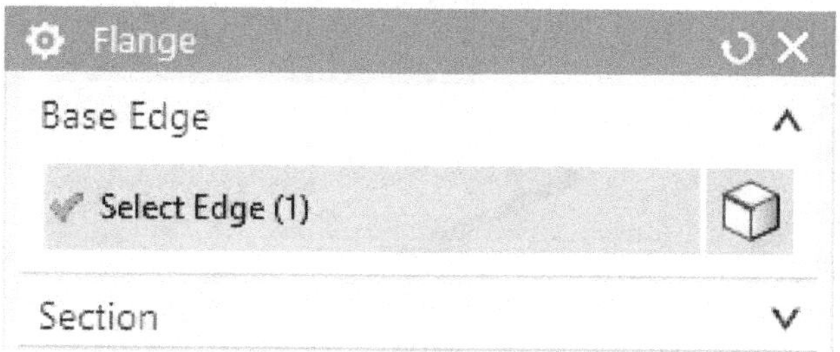

- Create a second flange with the same options on the edge shown below.

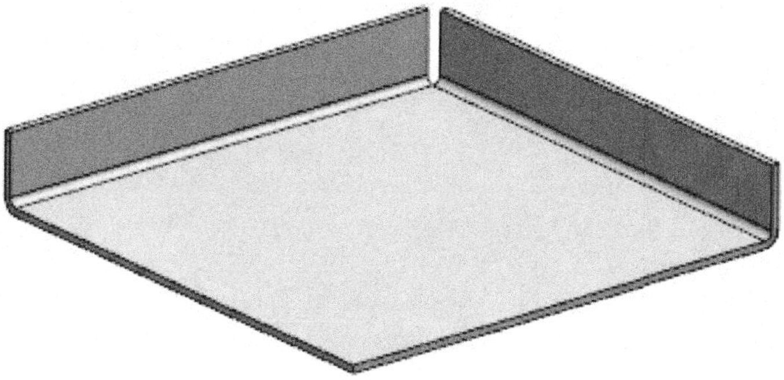

- Create a third flange with the same options from the edge of the last flange, as shown below. The flanges will now be colliding, and obviously cannot be formed as modeled below. In the next exercise, you will treat the corner so that these flanges meet each other properly.

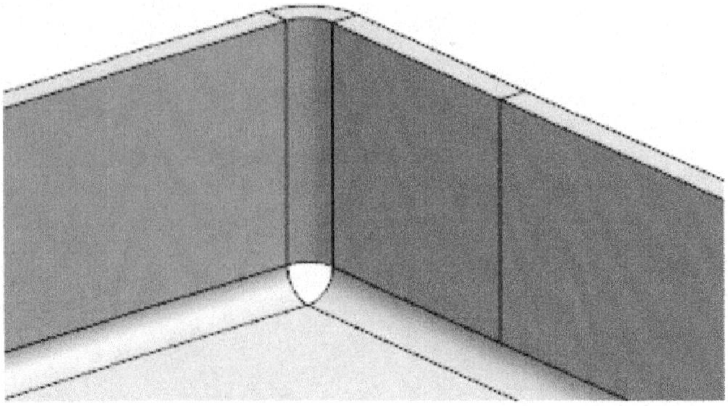

- **Save** your file!

14.2.1 Width Options

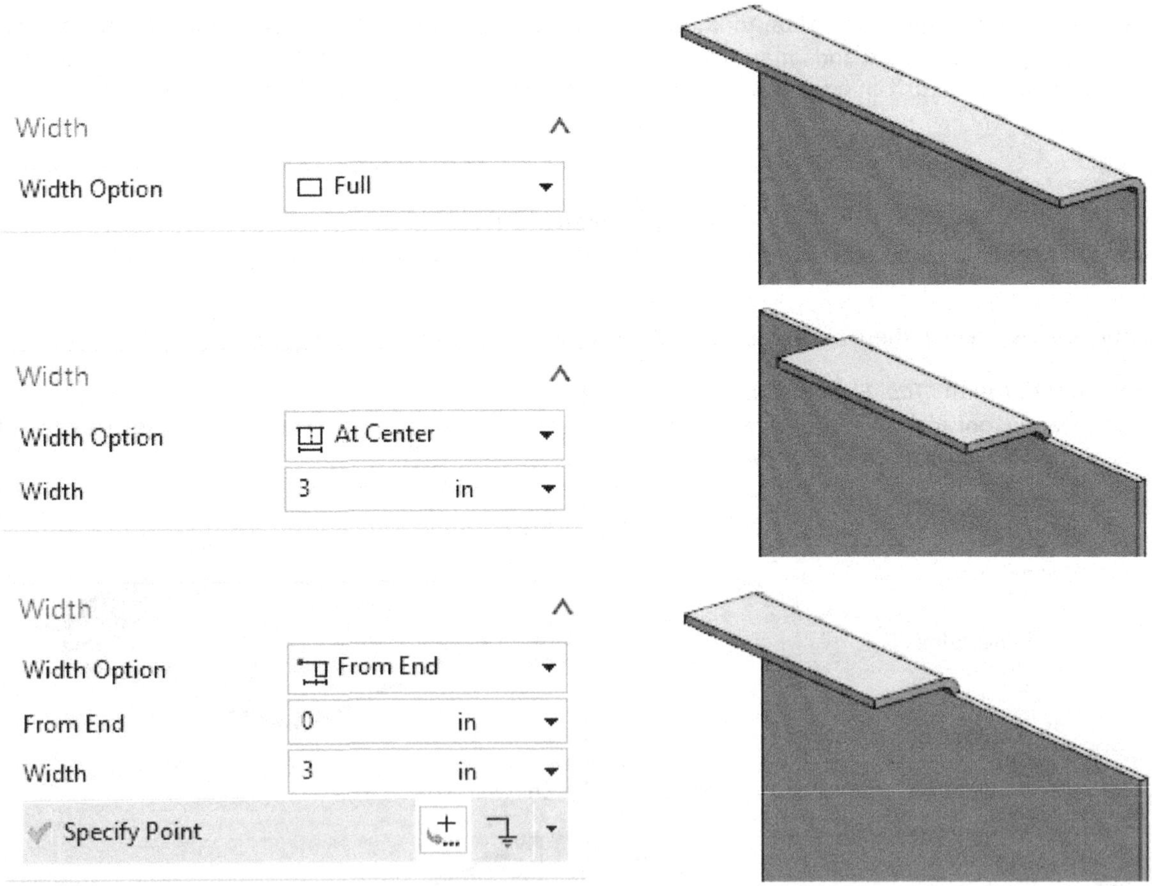

14.2.2 Length Reference

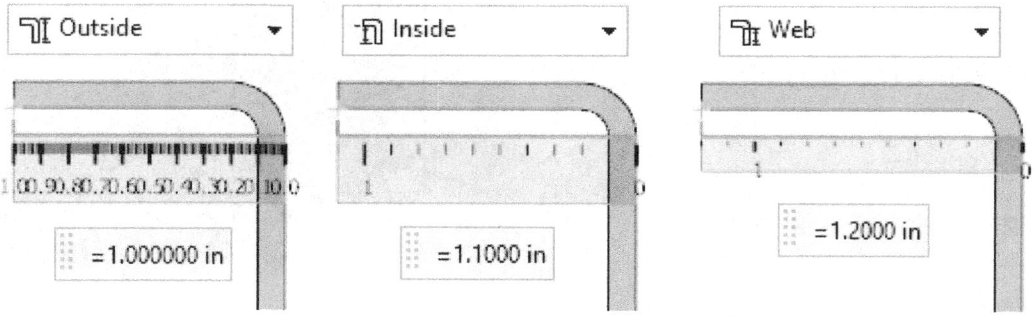

14.2.3 Inset

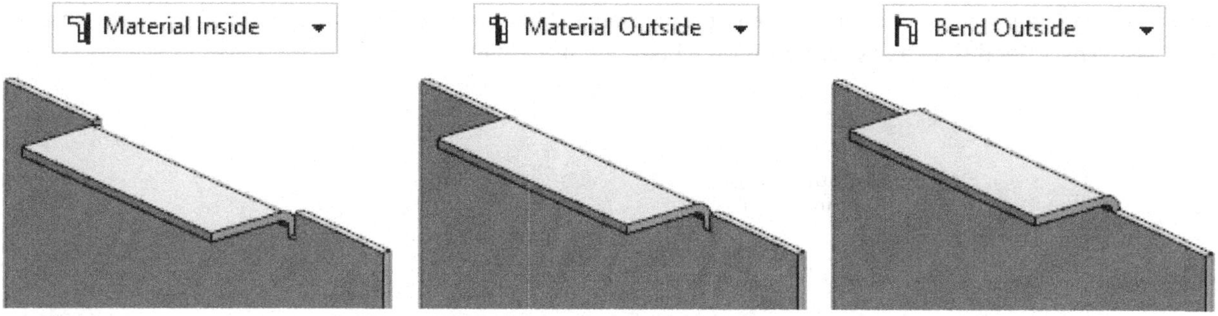

14.2.4 Relief

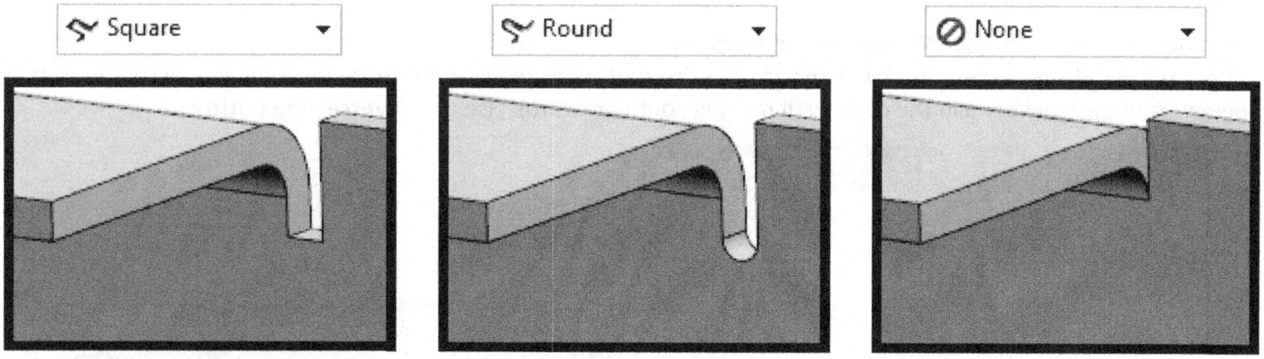

14.3 Three Bend Corner

The **Three Bend Corner** tool is one of several tools in the **Corner** group that modify bends near corners. These tools, generally speaking, allow you to provide relief to your specifications so that the part doesn't rip.

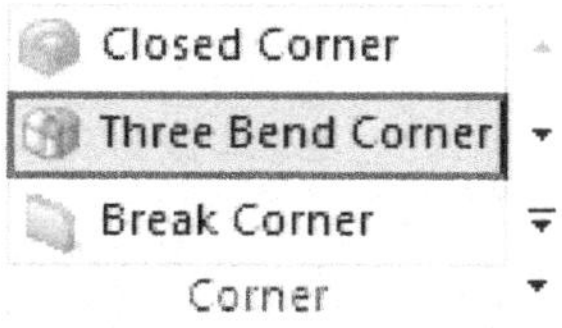

- Continue working with *"Tab, Flange, Corner.prt"*.
- Select the **Three Bend Corner** tool, and choose the bends shown below. Note that it breaks material off the two overlapping flanges so that they don't collide with each other. Click **OK**.

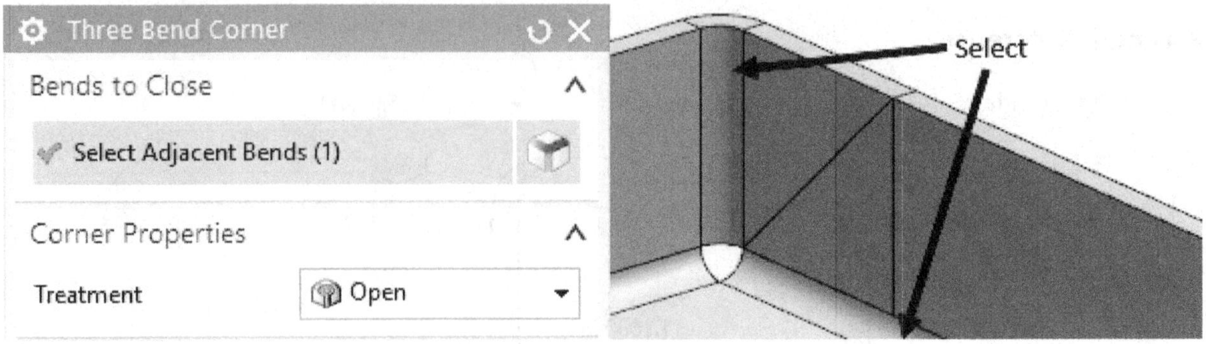

- **Save** your file!

14.4 Bend

The **Bend** tool is unlike the other tools in the **Bend** group (flanges) in that it does not add material to a sheet metal part. Instead, you specify a **Bend Line**, and your part is then bent about that line.

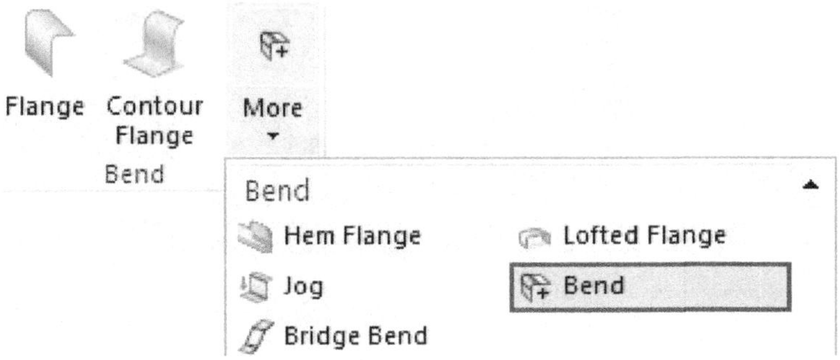

- Continue working with *"Tab, Flange, Corner, Bend.prt"*.
- Select the **Bend** tool. Select the inner top face of the part and sketch an arbitrary diagonal line across it as shown below. Your part will bend in the corner – if it comes out backwards, use the **Reverse Side** button.

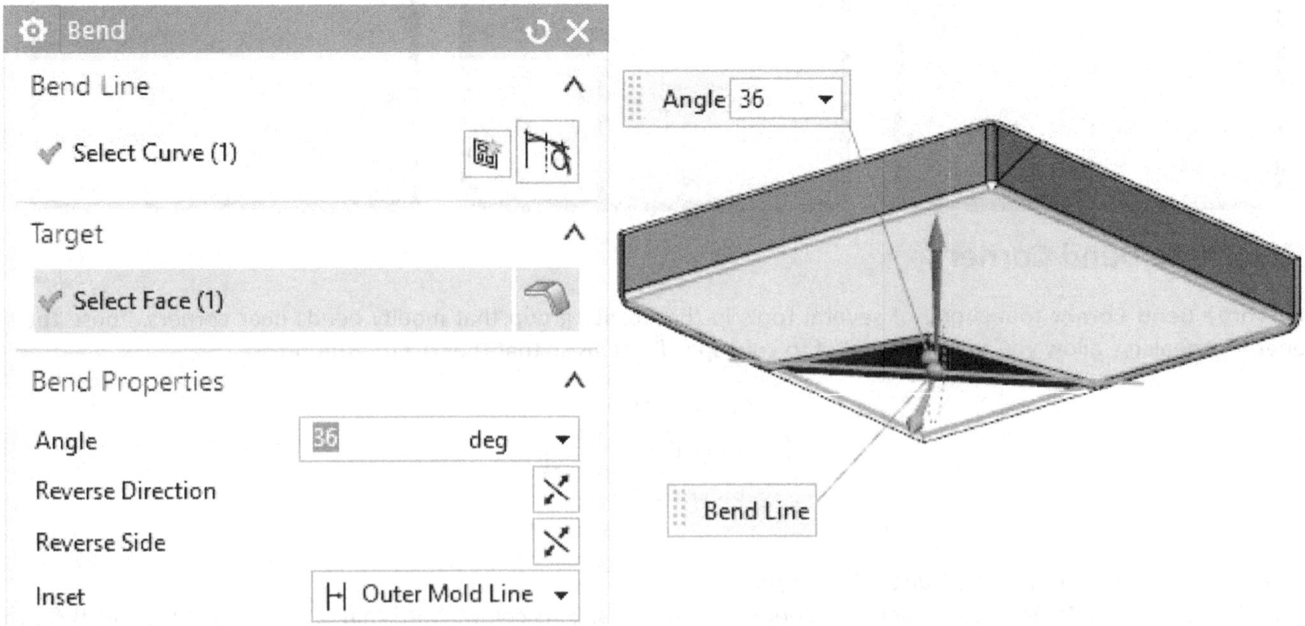

- **Save** your file!

14.5 Contour Flange

Besides the **Tab** tool, the **Contour Flange** tool is the only tool in the **Sheet Metal** application that will create a sheet metal part starting from a sketch.

It is also incredibly useful in its *Secondary* mode for quickly generating a series of flanges based on a sketch, and can save you a lot of work if you use it properly!

14.5.1 Base

In *Base* mode, the **Contour Flange** tool takes as input an open sketch section, and then extrudes it with an offset equal to *Sheet_Metal_Material_Thickness* to produce a sheet metal body that has a silhouette based on your sketch.

- Create a new part file (**Units**: *Inches*) called *"Conduit Strap.prt"*.
- Select the **Contour Flange** tool, and then select the YZ plane.
- Create the sketch shown below on the YZ plane.

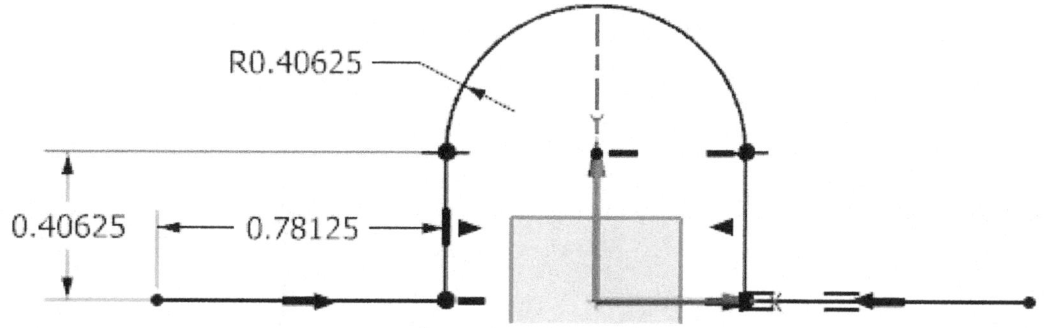

- Set the **Width Option** to *Symmetric* and provide a value of *0.75 in*. Make sure that the radius of the inner face of the clamp is *0.40625 in* – you might need the **Reverse Direction** button.

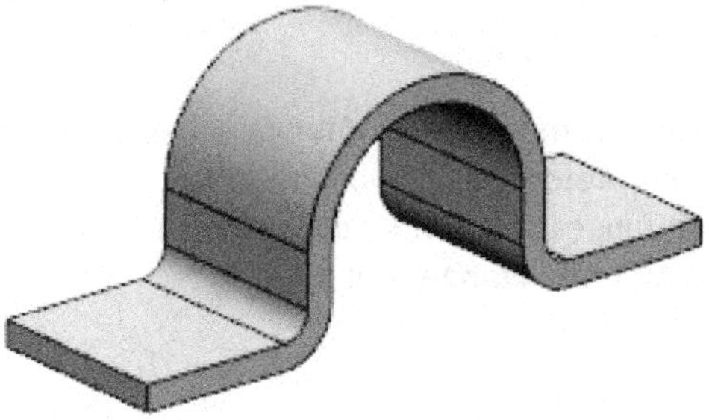

- **Save** your file!

14.5.2 Secondary

The **Contour Flange** tool, when used in *Secondary* mode, behaves like a sweep with an offset. You sketch an open section curve attached to an edge of a sheet metal body, and it is swept along that edge and any selected neighbors with an offset. Sharp corners in the sketch are converted into rounded bends of the sheet metal bend radius.

- Create a new file (**Units**: *Inches*) called *"Louvered Vent.prt"*.
- Create a **Tab** from a rectangle with dimensions *11 in* x *14 in*.

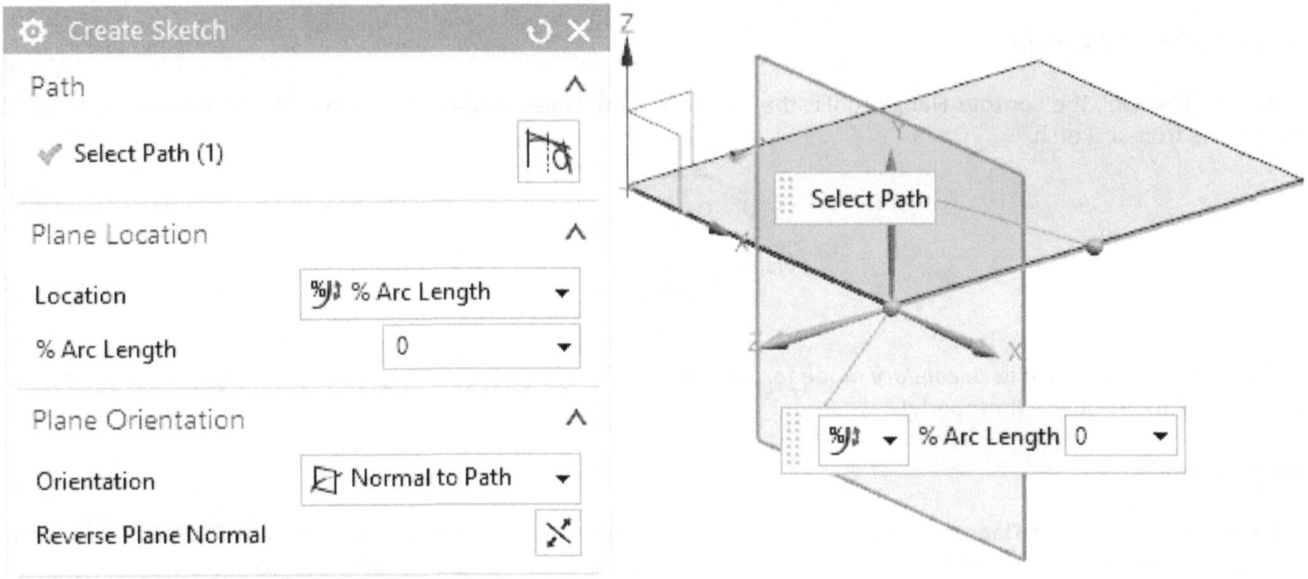

- Create the sketch shown below. An important point about sketches for the **Contour Flange** tool is that they must not cross the sketch horizontal axis.

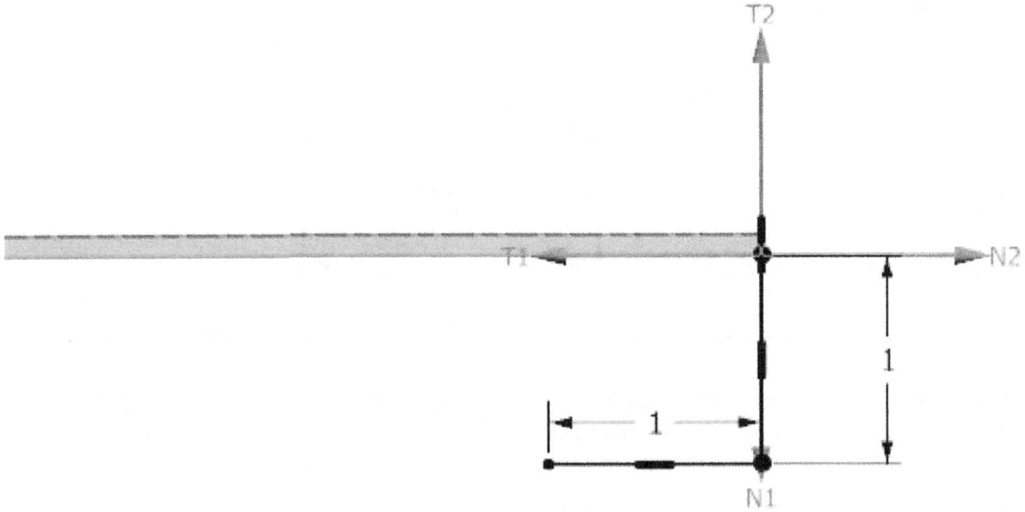

- When you finish the sketch and return to the **Contour Flange** dialog, you will have various choices for the **Width Option**. *Finite* allows you to sweep your sketch along a portion of the edge from the starting point. This is a great way to quickly model a complex series of flanges! The only downside is that it doesn't give you control over the **Length Reference** and **Inset** in the same way that the **Flange** tool does.

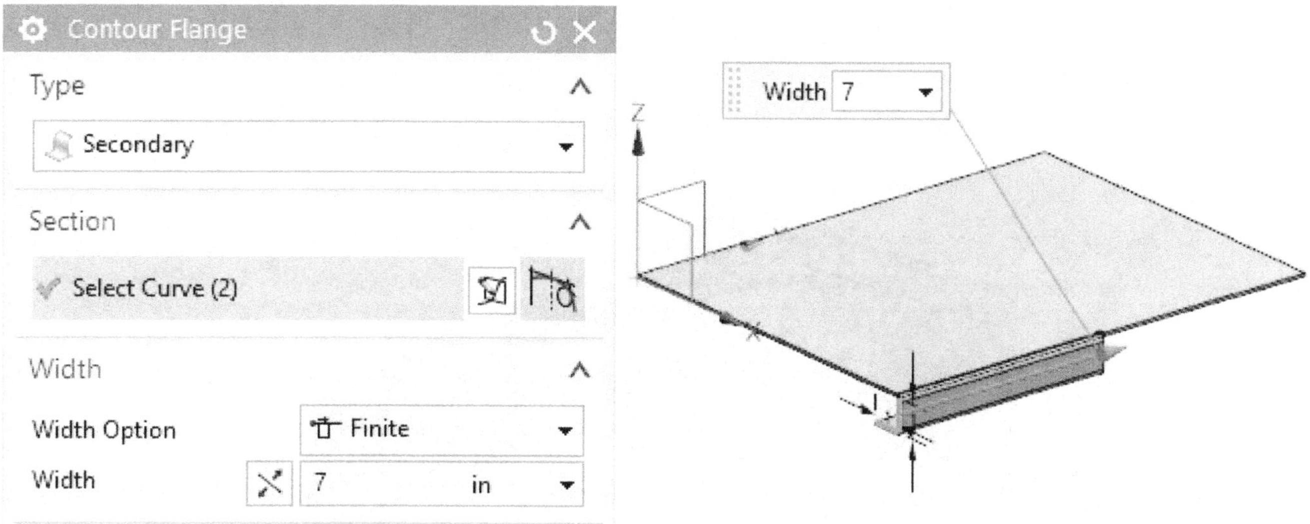

- The really powerful **Width Option** is *Chain*. Set the **Width Option** to *Chain* and select the three other edges on the bottom face of the model, and the tool will sweep the sketch along all four edges. At the corners, it will automatically break the flanges so that they don't collide with each other!

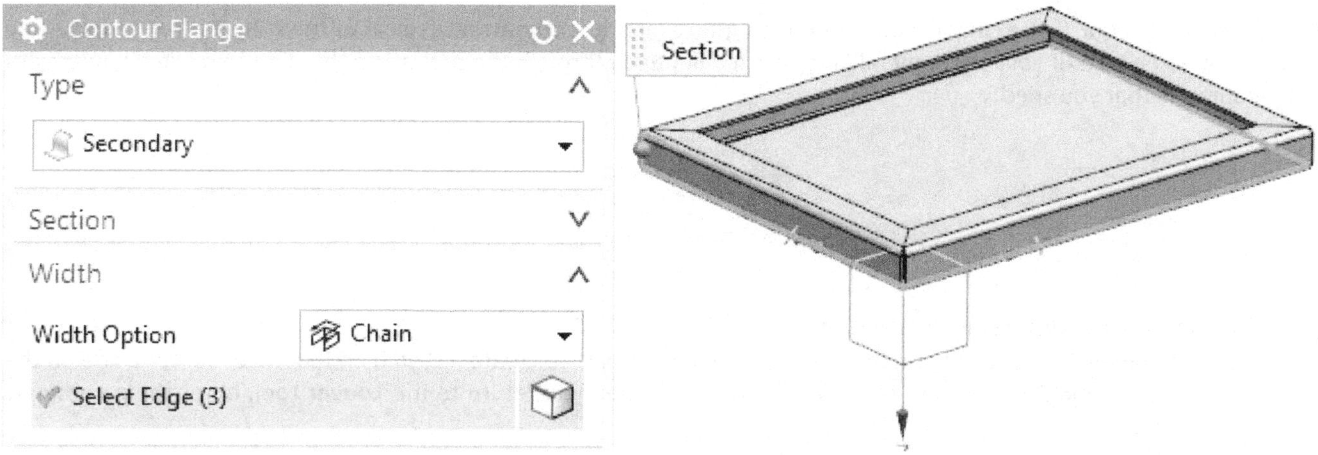

- Within the **Contour Flange** dialog, you can also specify a **Corner** treatment. Set it to *Open*.

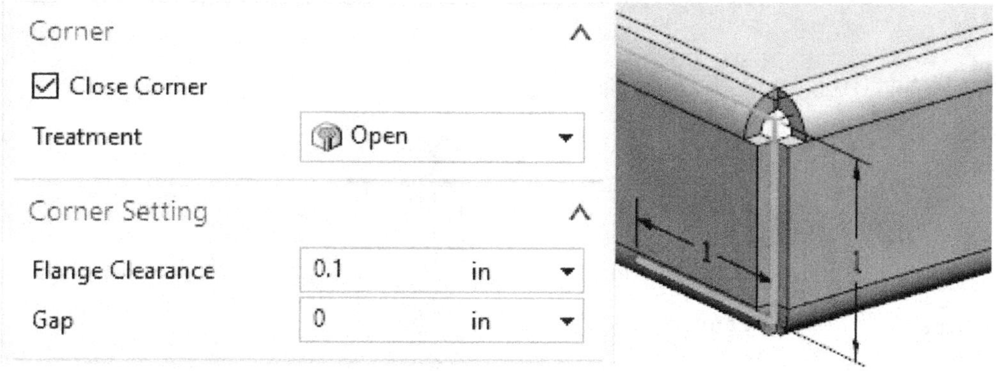

- You could have built this same shape with eight applications of the **Flange** tool, four applications of **Three Bend Corner**, and four applications of **Closed Corner**. **Contour Flange** can be quite the timesaver!

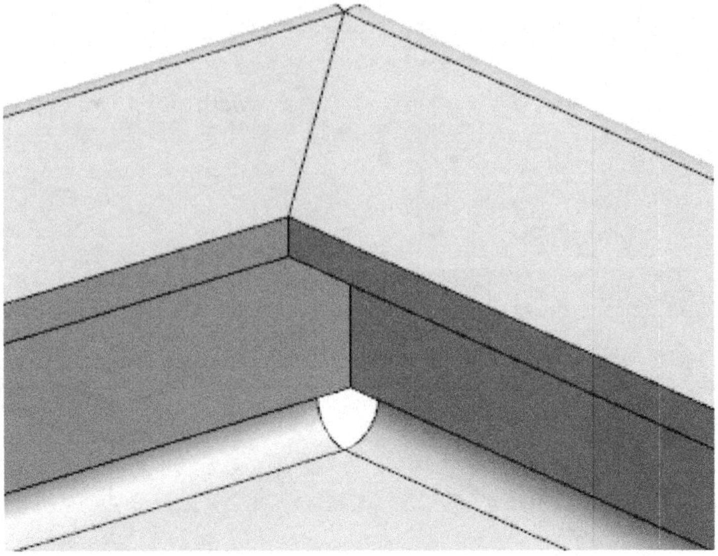

- **Save** your file!

14.6 **Louver**

The **Louver** tool is our first example from the **Punch** group, and it is somewhat typical of the tools in that group – you sketch a line on a face of your sheet metal part, and it will punch through the part along that line to make a louver with the parameters that you specify.

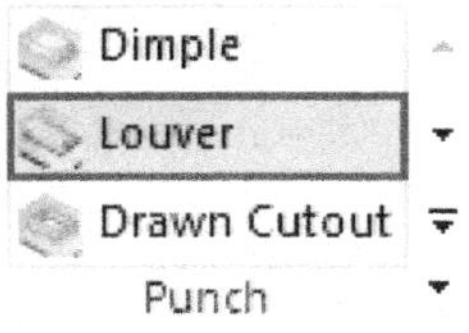

- Continue working with *"Louvered Vent.prt"*.
- Select the **Louver** tool, and click on the bottom face of your part to create a sketch on that face. Center a line on the face, *8 in* wide, and *2 in* from the top edge of the part. When you return to the **Louver** tool, enter the parameters shown below.

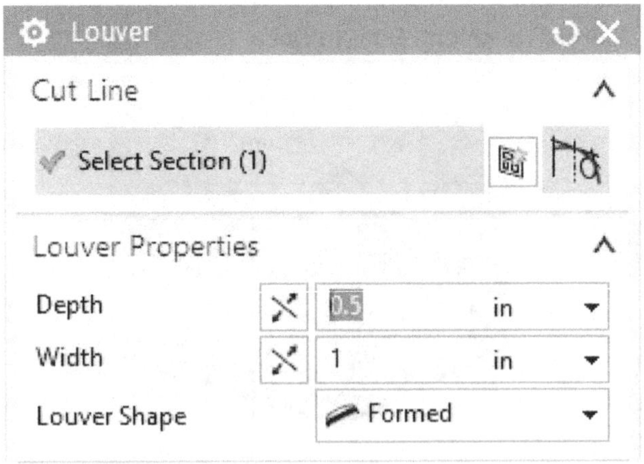

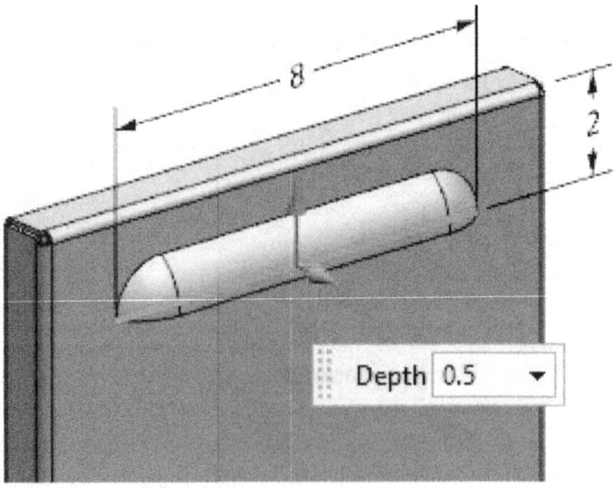

- Use **Pattern Feature** to create three more copies, spaced with a **Pitch Distance** of *2 in*.

- **Save** your file!

14.7 **Sheet Metal Preferences**

We have made mention of a global parameter called *"Sheet_Metal_Material_Thickness"* several times now, and in the exercise below, you will learn to modify this parameter, and others that are specific to sheet metal parts. You can do so from the **Expressions** editor, but these parameters naturally live in the **Sheet Metal Preferences** menu.

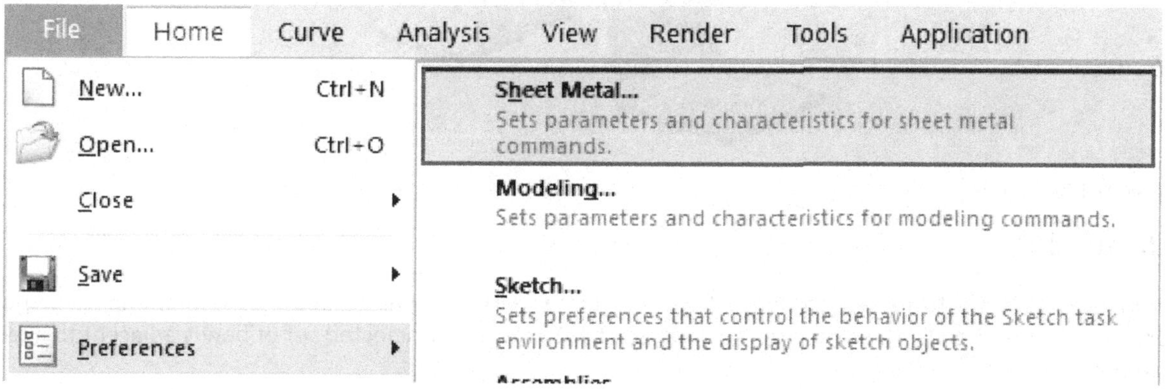

- Continue working with your part *"Conduit Strap.prt"*.
- Open the **Sheet Metal Preferences** menu. In the **Global Parameters** section, change the value of **Material Thickness** and **Bend Radius** as shown below. Click **OK** to save your changes.

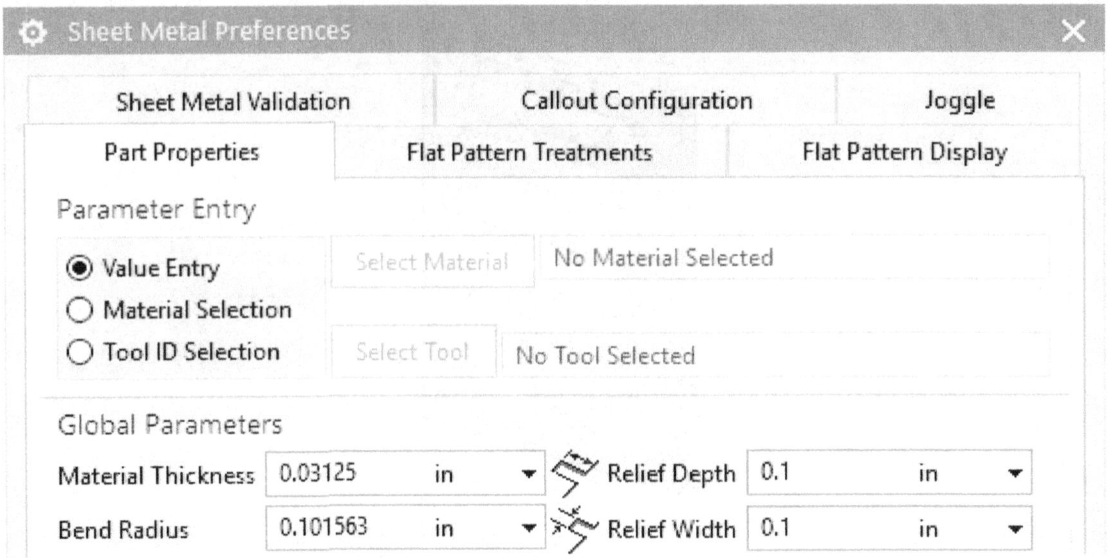

- This has immediate consequences for the shape of the strap.

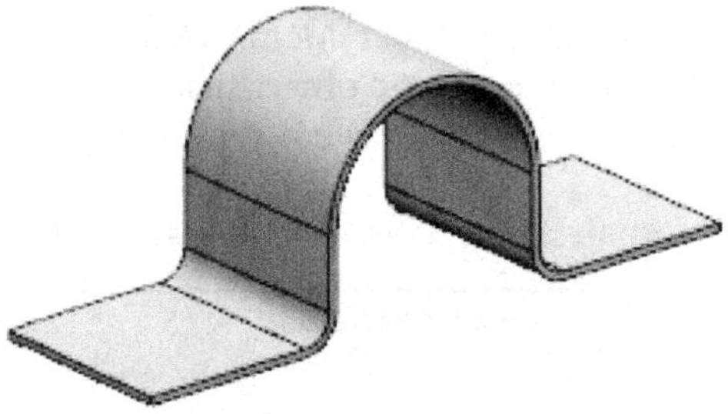

- **Save** your file!

14.8 **Unbend**

It is sometimes necessary to flatten portions of a sheet metal part in order to apply certain stamping or punching operations across those bends. The **Unbend** tool does exactly that – takes a selected set of bends adjacent to a face, and unforms them.

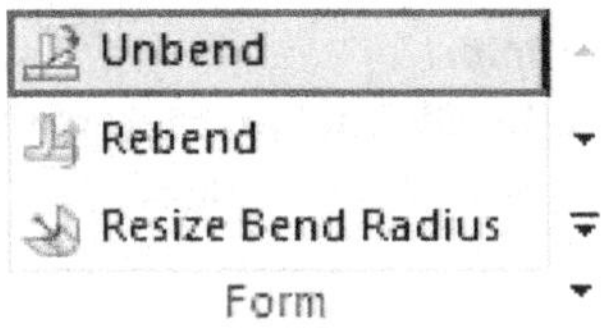

Unfortunately, there is no "Unbend All Bends" command in NX.

- Continue working with *"Conduit Strap.prt"*.
- Select the **Unbend** tool, and choose the leftmost flat face as the **Stationary Face**, and the adjacent bend as the **Bend**.

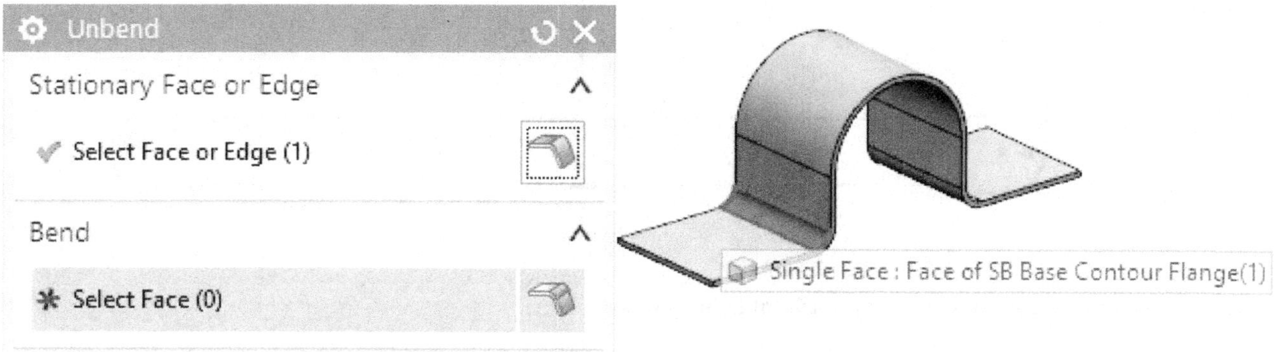

- **Unbend** also recognizes the large circular section as a bend.

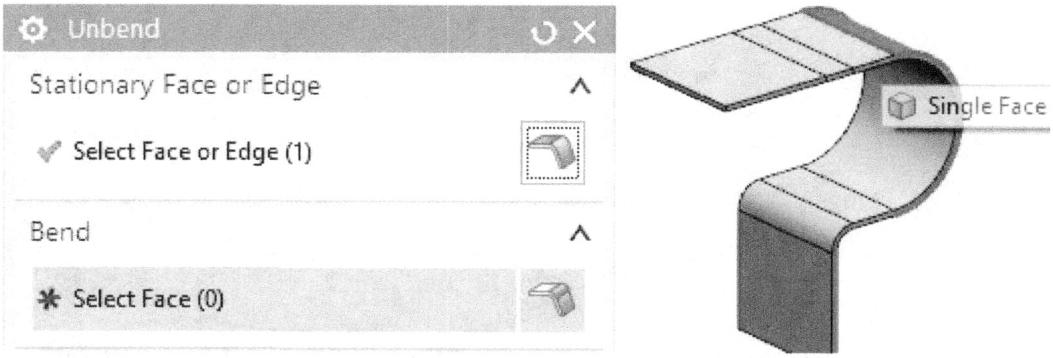

- When you are done, your part will be totally flattened.

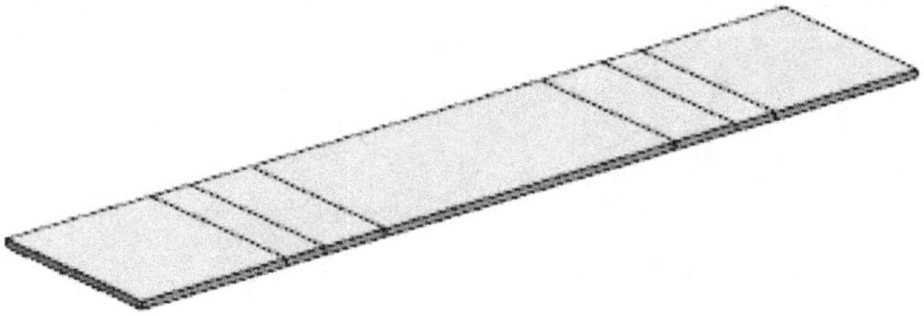

- **Save** your file!

14.9 Dimple

The **Dimple** tool is our second example from the **Punch** group. Like the **Louver** tool, it takes a sketch as input, although it needs a closed contour rather than a line to determine the shape of the dimple.

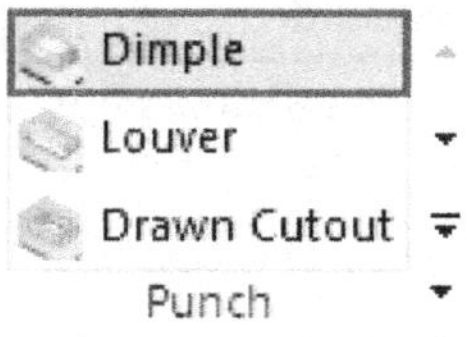

Dimple and **Bead** features can be placed on flattened bends and then reformed.

- Continue working with *"Conduit Strap.prt"*.
- Select the **Dimple** tool, and choose the middle face of your flattened body to sketch on. Create the sketch below. You will need to use **Reattach** to center the **Sketch Origin** at the middle of the face, and you will need the **Make Symmetric** command several times.

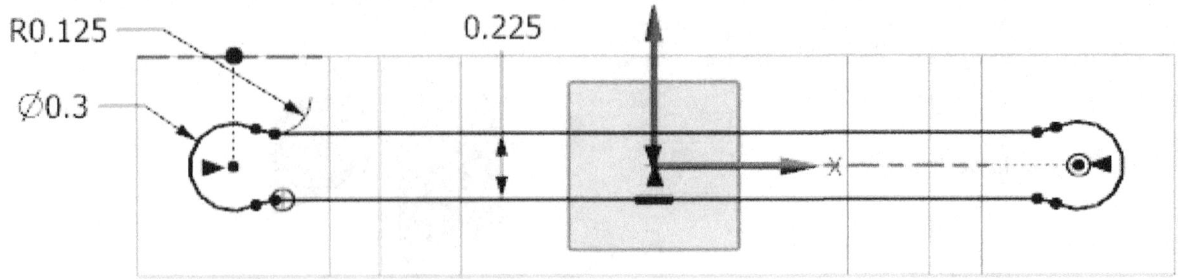

- Enter the following parameters into the **Dimple** dialog box.

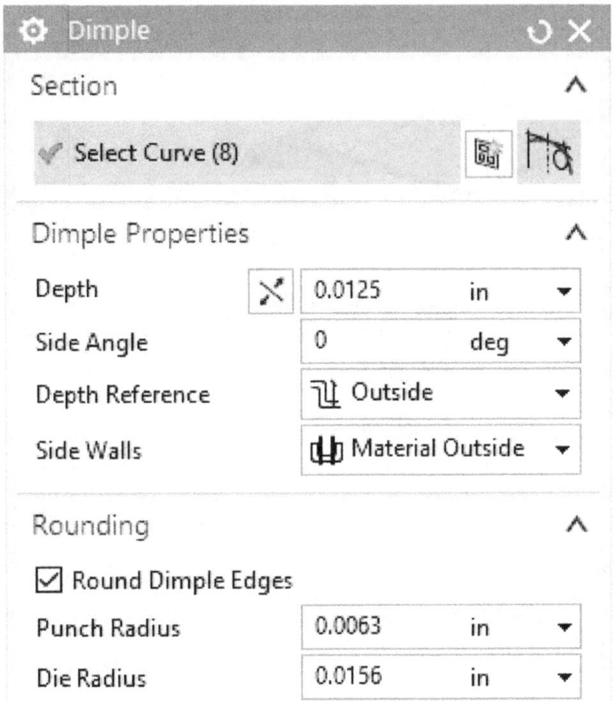

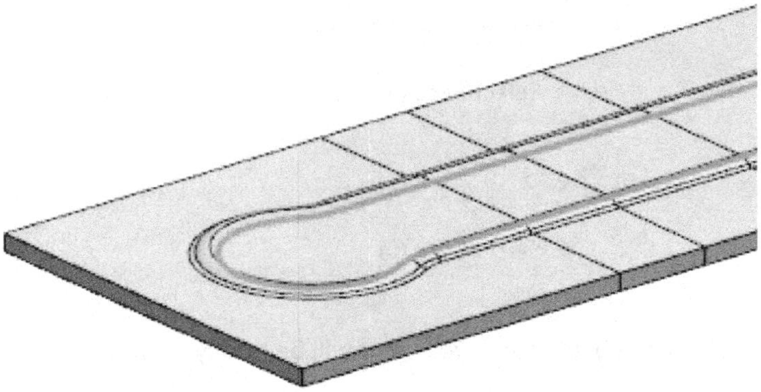

- **Save** your file!

14.10 Rebend

When you have used **Unbend** to flatten a sheet metal part, **Rebend** will reshape your part.

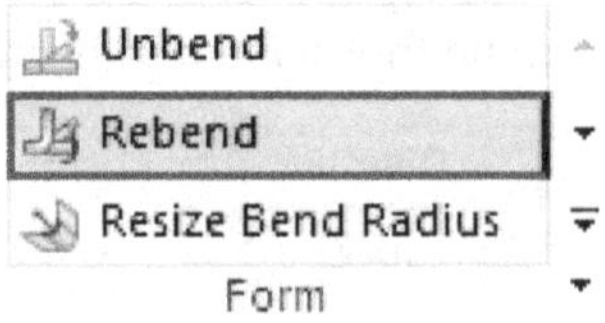

- Continue working with *"Conduit Strap.prt"*.
- Open the **Rebend** tool, and select each unbent bend one-by-one.

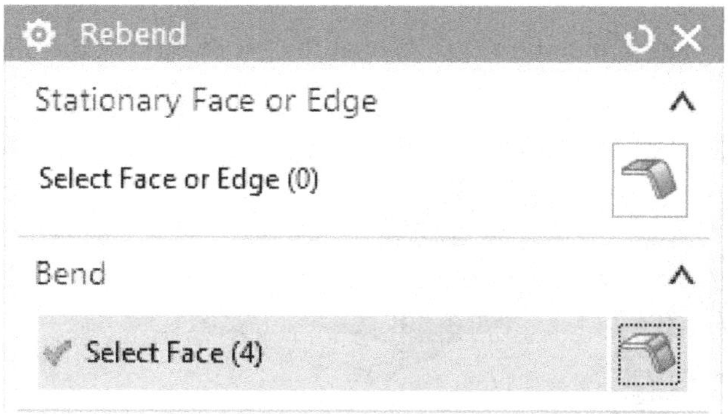

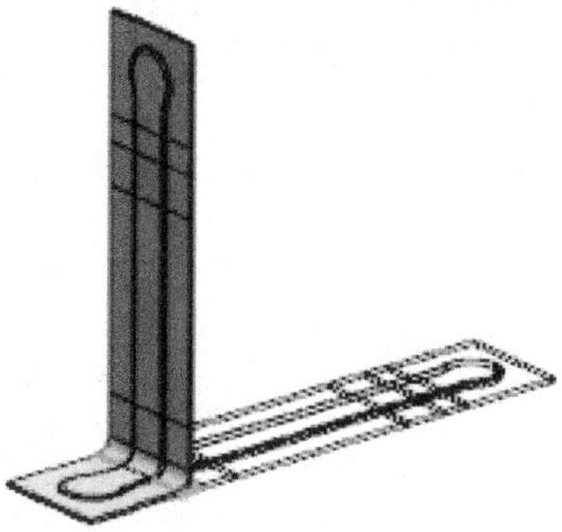

- When you are done, your part will be in its original shape, now with the dimple laid out over the bends.

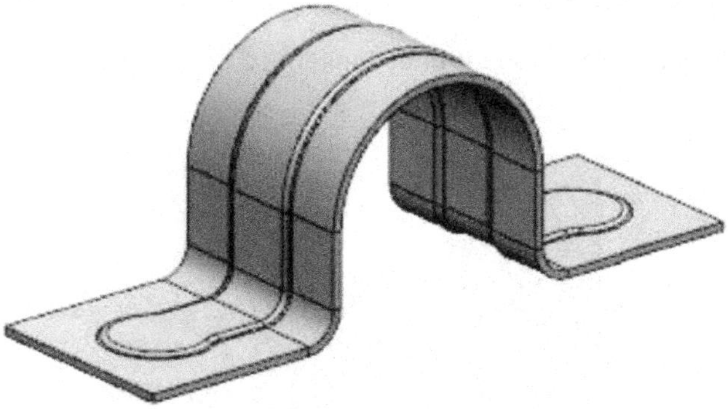

- **Save** your file!

14.11 The Feature Group

An important point about tools in the **Sheet Metal** application is that they only work on parts that NX recognizes as sheet metal – either built directly in the **Sheet Metal** application, or converted from an appropriate solid model. Generally speaking, when you apply operations from the **Modeling** application, those tools stand a good chance of making your part unusable in the **Sheet Metal** application.

It is often necessary to apply **Modeling** features to your sheet metal parts (e.g., holes), and so a small group of tools from the **Modeling** application have been made available within the **Sheet Metal** application, in the **Feature** group. You have already used Pattern Feature to copy the louver above, and in the exercise below, you will use some of these other tools to add finishing details to your part *"Conduit Strap.prt"*.

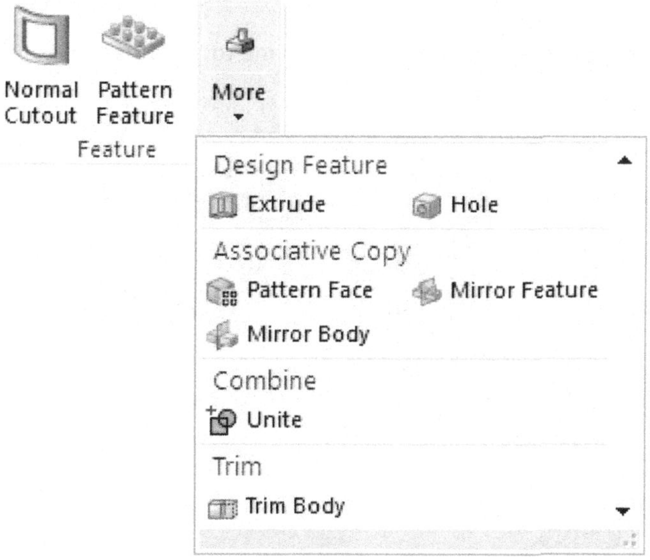

- Continue working with *"Conduit Strap.prt"*.
- Use the **Hole** tool to position two simple thru holes of diameter *0.1875 in* at the arc centers of the dimple.

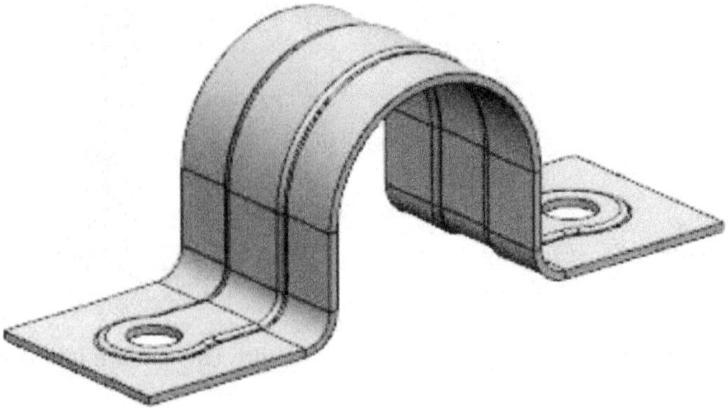

- **Extrude** an arc of radius *0.975 in* to create a sheet body tangent to the edge of the sheet metal part, and centered on the face, as shown below.

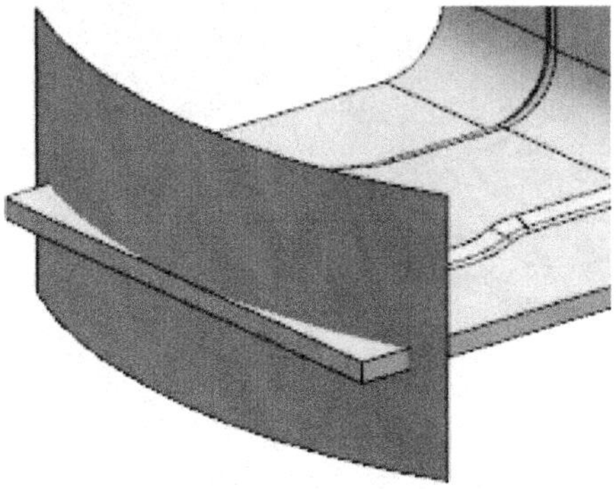

- Use **Trim Body** to cut your sheet metal part with this sheet body, and then **Hide** the sheet body.

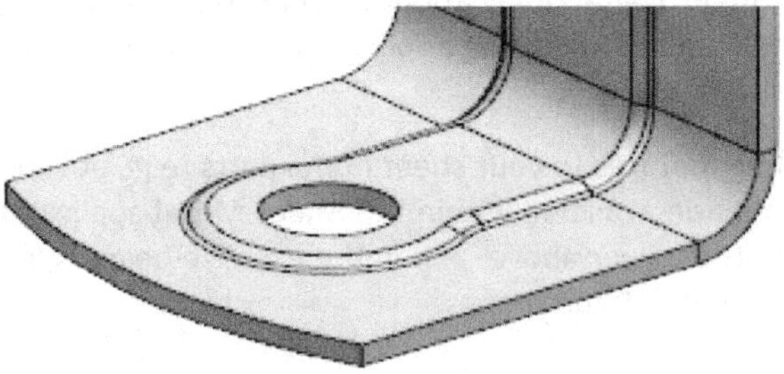

- Use **Mirror Feature** to mirror the curved face about the XZ plane to the other side of the sheet metal part.

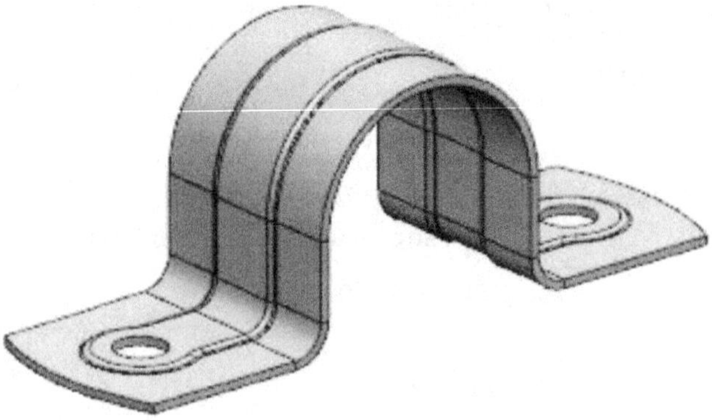

- **Save** your file!

14.12 **Normal Cutout**

Now that you have seen the **Unbend** and **Rebend** commands, you can imagine the following workflow: you model a sheet metal part, unbend it into its flattened state, use **Extrude** with **Boolean** set to *Subtract* to cut a shape out of the part in its flattened state, and then use **Rebend** to put it back into its formed state.

The **Normal Cutout** tool makes it easy to cut through a sheet metal part along a curve, without flattening the part first. The main challenge is to model the curve that you wish to cut along accurately

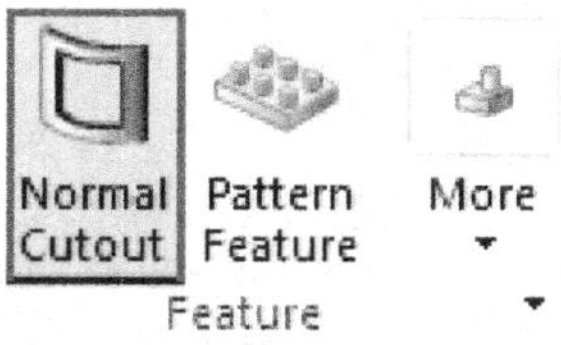

Normal Cutout allows you to cut a sheet metal part using a sketch, or using 3D curves. If you use 3D curves, they should be attached to the faces of your part exactly where you want to cut.

- Create a new part file (**Units**: *Inches*) called *"Normal Cutout.prt"*.
- Create the sketch shown below on the YZ plane of the datum coordinate system, and use the **Contour Flange** tool to create a part from that sketch with **Width** set to *5 in*. The arc has a **Perimeter Dimension** (unpictured) of *3 in*.

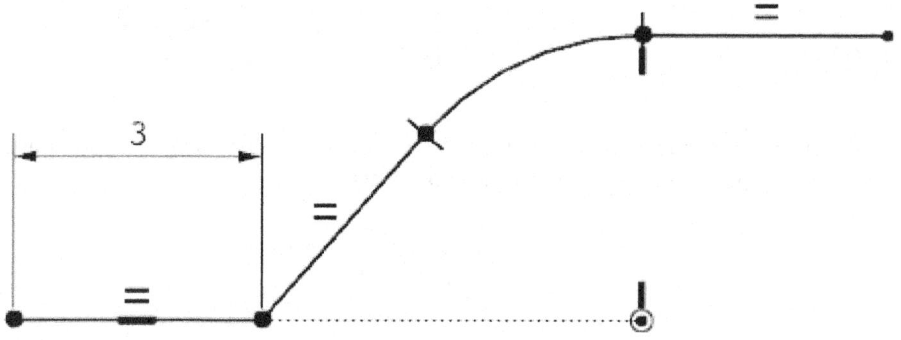

- Create the sketch shown below.

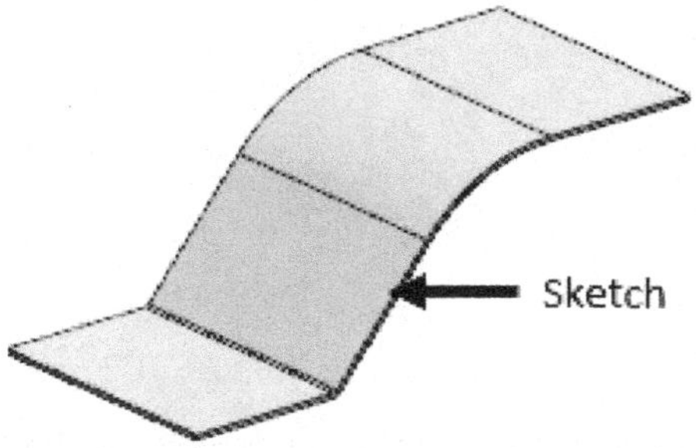

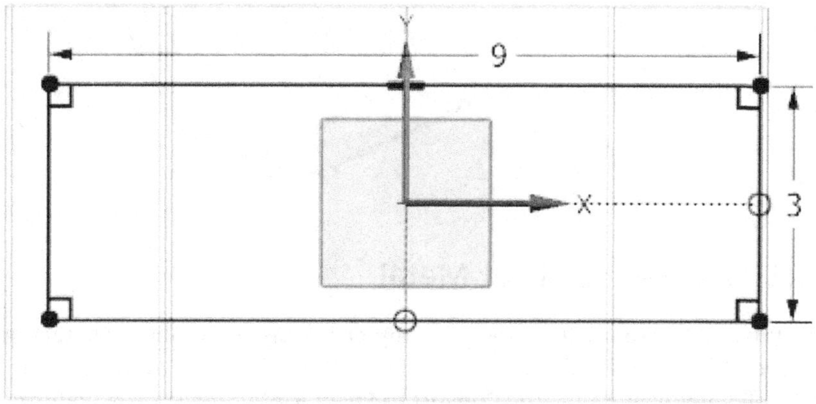

- Use **Wrap/Unwrap Curve** to wrap your sketch on the faces shown below. For the **Plane** object in the **Wrap/Unwrap Curve** dialog, select the face you sketched on.

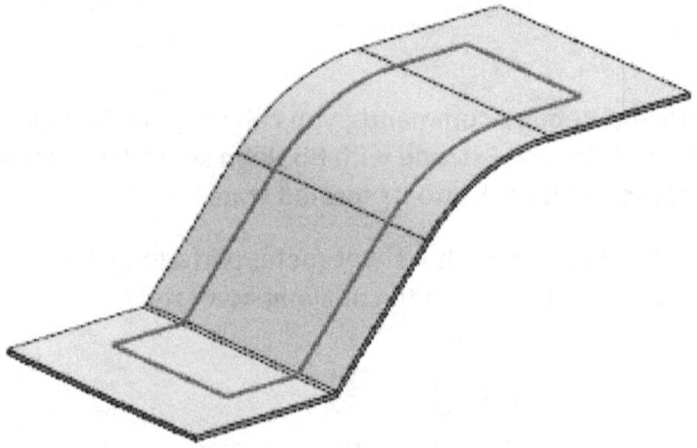

- Use the **Normal Cutout** tool with **Type** set to *3D Curves* and select the wrapped curves.

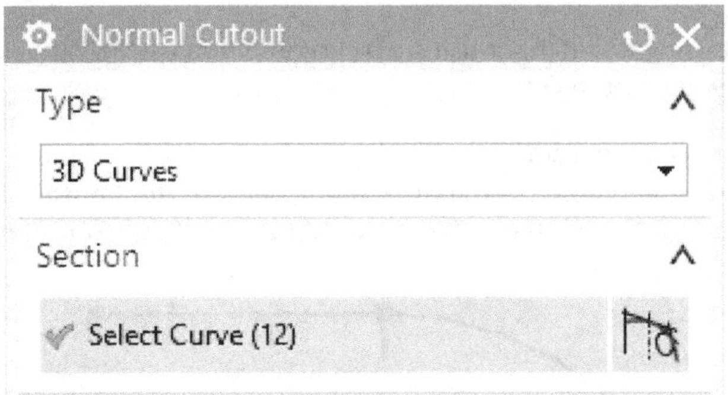

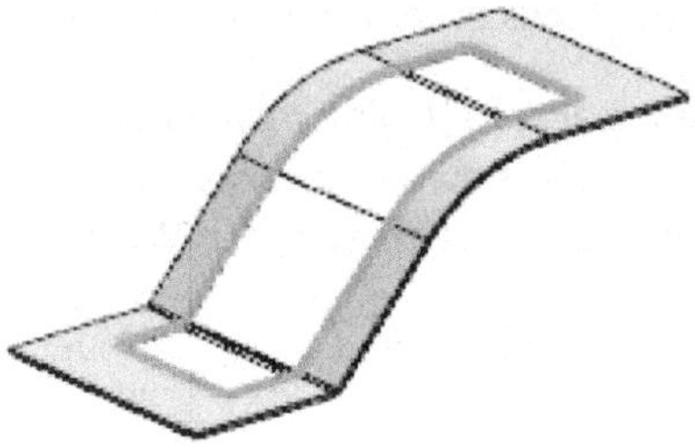

- This effortlessly cuts out your shape with the cut direction normal to the faces of your part. Much easier than unbending and rebending, especially when your part is complicated!

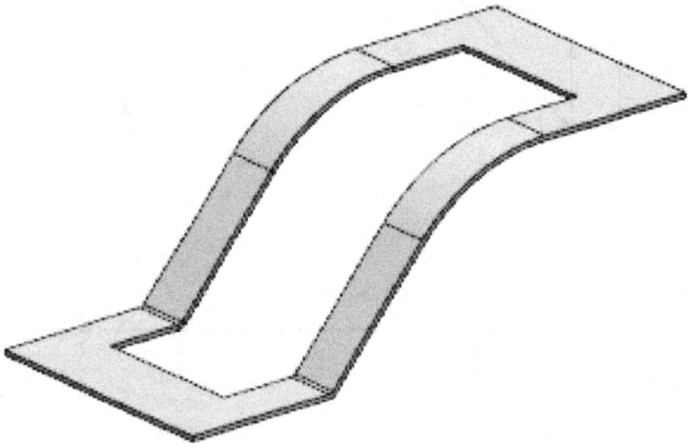

- **Save** your file!

14.13 Convert to Sheet Metal

The **Tab** and **Contour Flange** tools are the only tools in the **Sheet Metal** application that are capable of building a sheet metal part "from scratch" – i.e., from a sketch. The other tools available for creating sheet metal parts require a previously-built solid model as input. The **Convert to Sheet Metal** tool takes a uniform thickness solid body and turns it into a sheet metal model, transforming sharp edges into bends with *Sheet_Metal_Bend_Radius* as the radius.

Sometimes your uniform thickness solid body will have certain edges that are impossible for a piece of sheet metal to have – in these cases you will need to rip the metal along those edges, so that it can be folded into shape from an appropriate cutout.

- Create a new part file (**Units**: *Inches*) called *"Convert to Sheet Metal.prt"*.
- Create a **Block** with dimensions *2 in* x *2 in* x *1 in*.
- Use the **Shell** tool to pierce the top face of the block with thickness set to *0.1 in*.
- Enter the **Sheet Metal** application and select the **Convert to Sheet Metal** tool.
- For the **Base Face**, select the bottom interior face of the shelled body. The four edges connecting that face to the top of the body need to be ripped. Note that when you click a single edge, it will select its partner on the other side of the body.

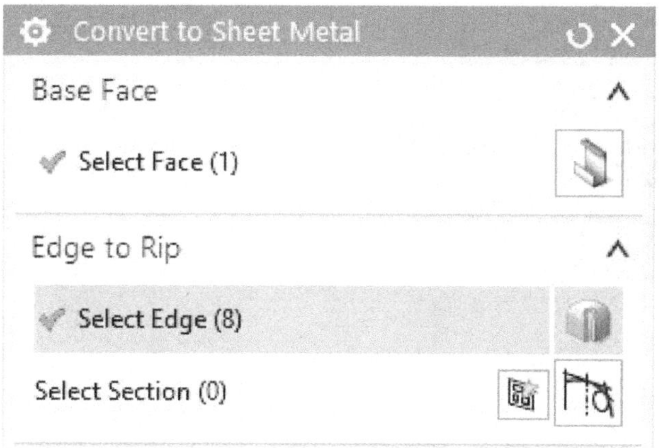

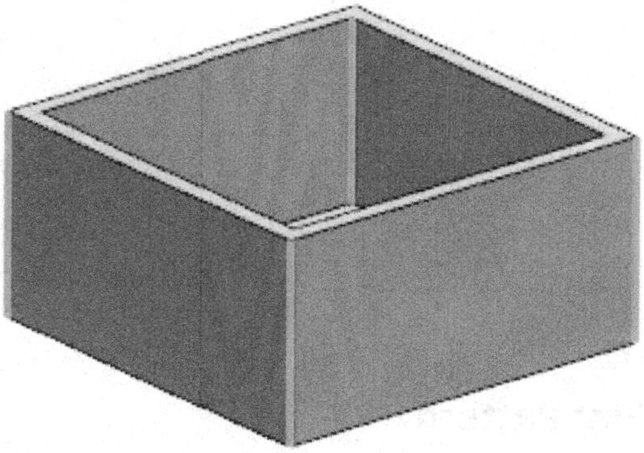

- Upon performing the conversion, the edges are ripped and the sharp edges at the bottom are converted into bends. Also, notice the corner treatment.

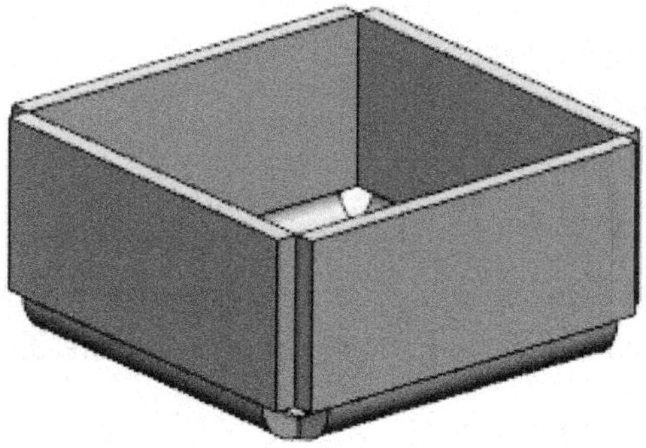

- **Save** your file!

14.14 Solid Punch

A **Solid Punch** can be used when you want to inherit a shape from a solid tool body.

- Continue working with *"Convert to Sheet Metal.prt"*.
- Enter the **Modeling** application and create a **Sphere** with a **Diameter** of *1 in*, centered at the point with coordinates *(1, 1, -0.25)*.
- Re-enter the **Sheet Metal** application and select the **Solid Punch** tool.
- Choose the bottom face of the sheet metal part as the **Target** (top in the image below), and the sphere as the **Tool**.

- The **Tool** body is deleted and its imprint left on your sheet metal part.

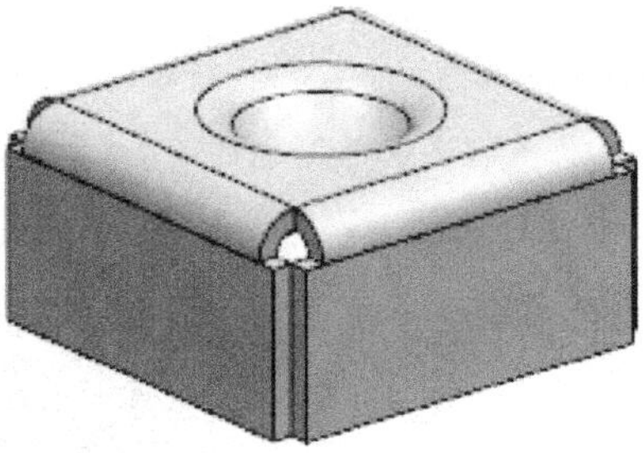

- **Save** your file!

14.15 Flat Pattern

The **Flat Pattern** tool creates a model view of a sheet metal part in its totally flattened state, for inclusion on a manufacturing drawing.

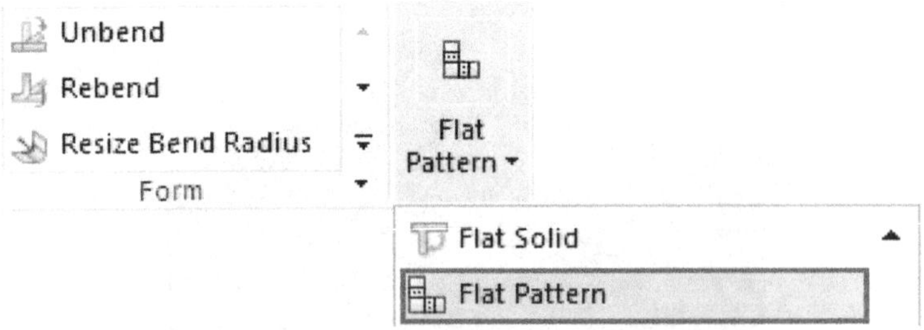

The **Flat Pattern** tool will typically ignore features from the **Punch** group in considering what to display in the model view.

- Continue working with *"Convert to Sheet Metal.prt."*.
- Select the **Flat Pattern** tool, and choose the top face of your part as the **Upward Face**.

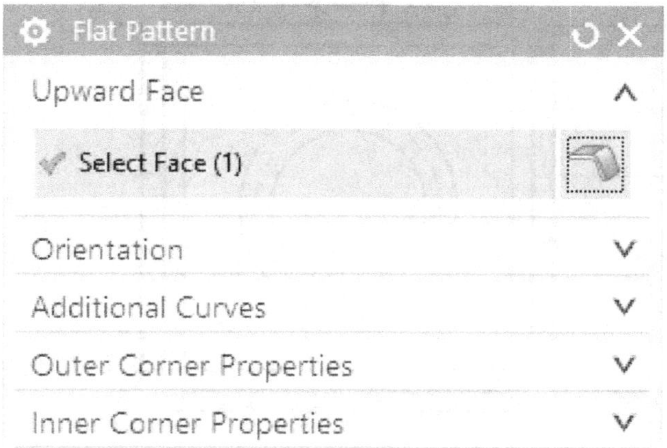

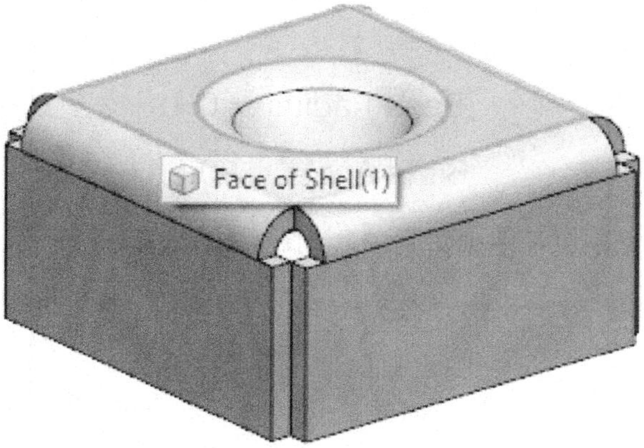

- The output of **Flat Pattern** is a model view that appears in the part navigator – *"FLAT-PATTERN#1"*.

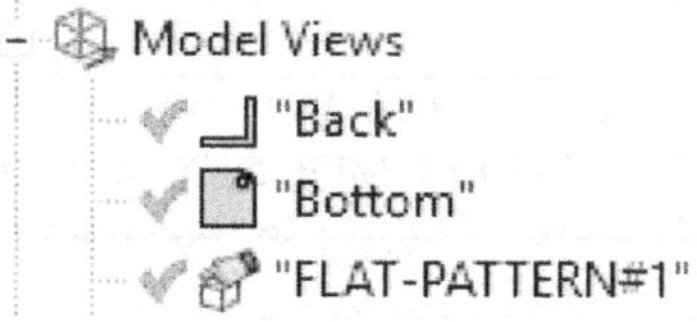

- Enter the **Drafting** application. Create a *B*-sized sheet with the *Siemens* template.
- Use the **Base View** command to place *FLAT-PATTERN#1* at a **1:1** scale on the drawing sheet.

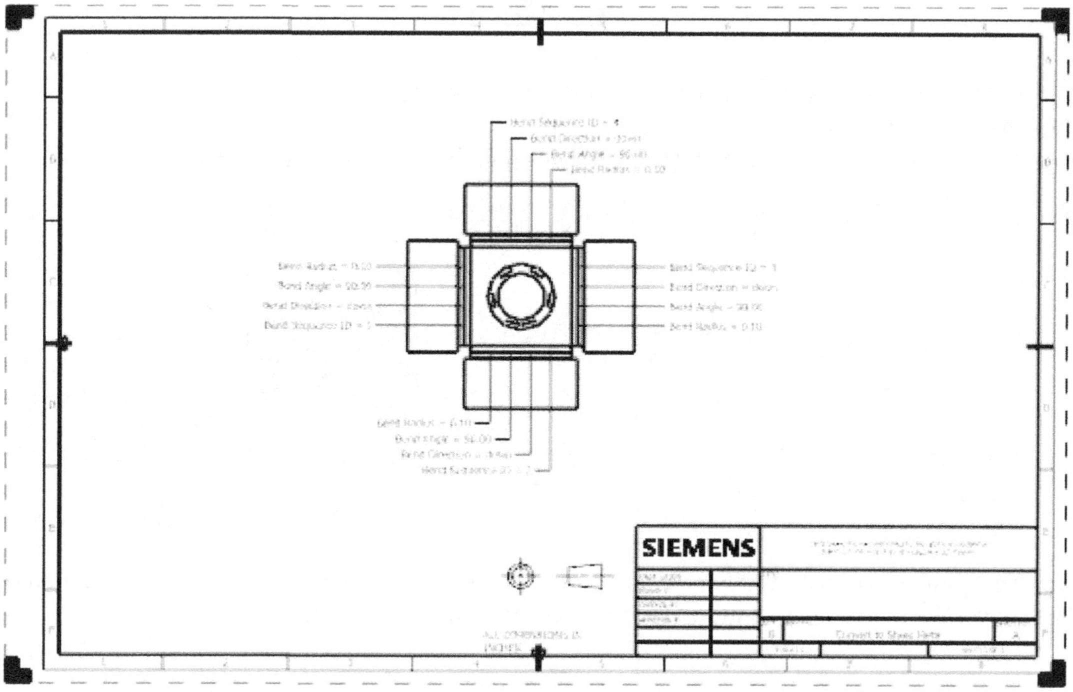

- The model view *FLAT-PATTERN#1* includes dimensions and annotation for each bend in the sheet metal part.

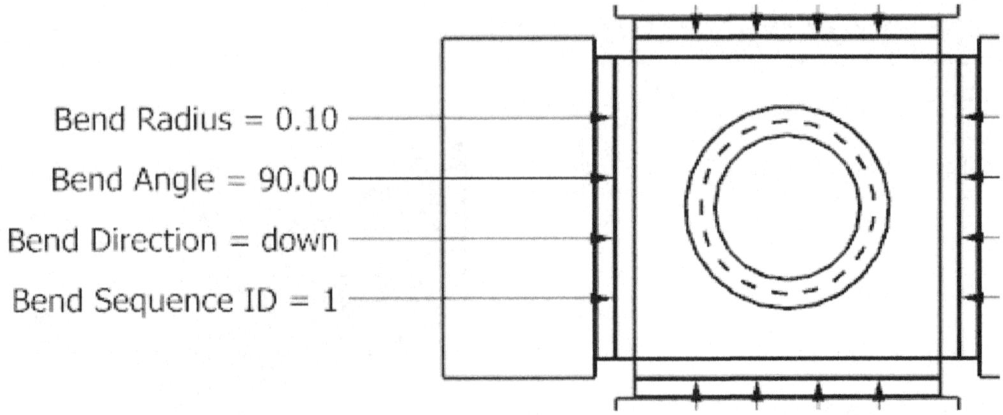

- **Save** your file!

14.16 **Sheet Metal from Solid**

Sheet Metal from Solid is used to create a sheet metal feature that "wraps" around existing geometry. This is a really good technique especially when you have a part that needs a sheet metal inlay or something to that effect. Pockets and bosses and other features added onto your part will result in normal cutouts in the sheet metal part.

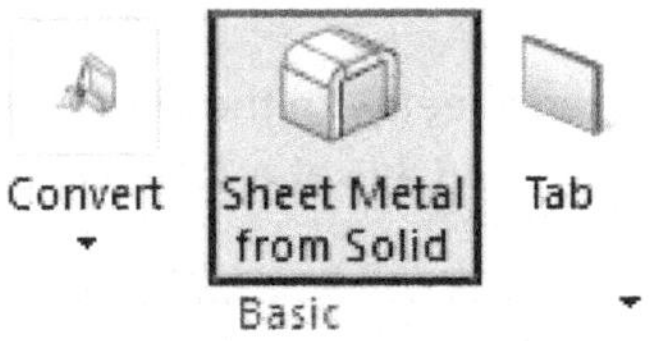

- Open the file *"Sheet Metal from Solid.prt"* from your downloaded part files folder.
- Select **Sheet Metal from Solid**.
- Choose the five faces shown below as **Web Faces**. NX will infer which **Bend Edges** as long as it is able.

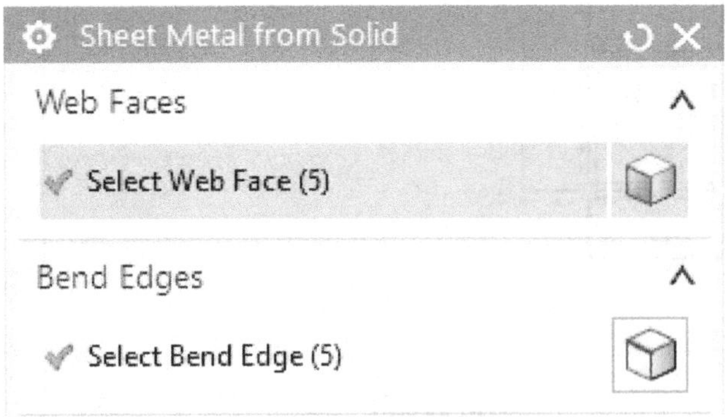

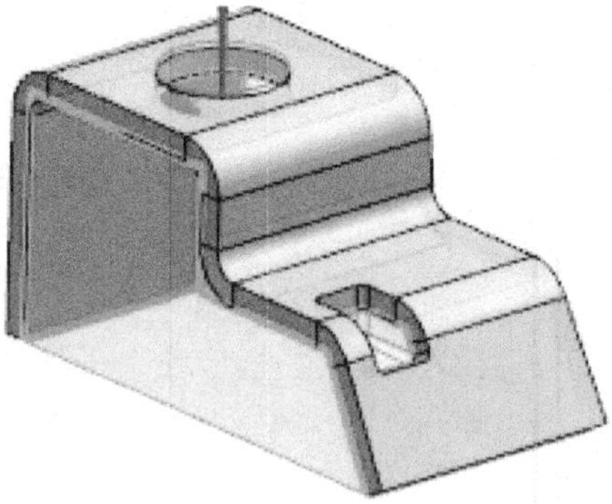

- Once you wish to add a face that neighbors two or more already-selected faces, you will have to manually specify the **Bend Edge**. Select the face/edge pair shown below, and then select their mirror image.

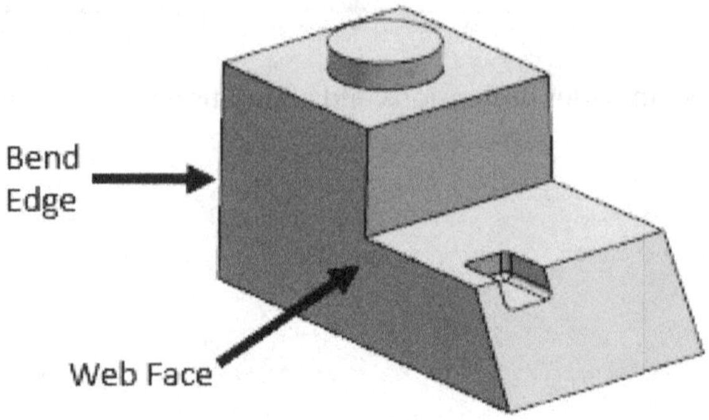

- This is what your sheet metal part looks like!

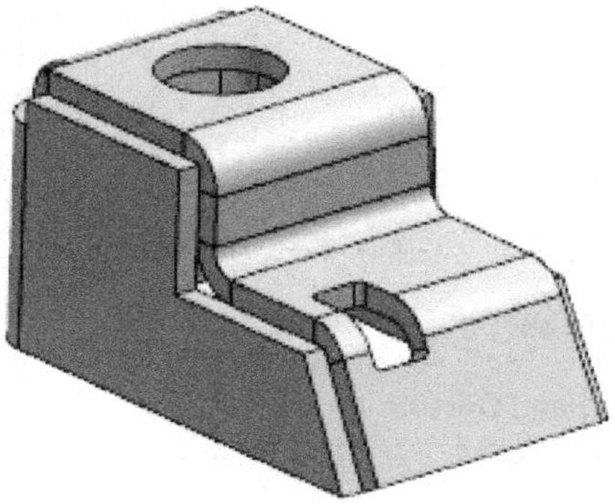

- **Save** your file!

14.17 Flat Solid

The **Flat Solid** tool creates a totally flattened version of your sheet metal part that remains current as you apply additional operations from the Sheet Metal application. It also produces a reference set named *"FLAT SOLID"* that is particularly useful with **Assembly Arrangements** (see Chapter 18.7) for producing drawing views of sheet metal parts with attached components in both their formed and flattened states.

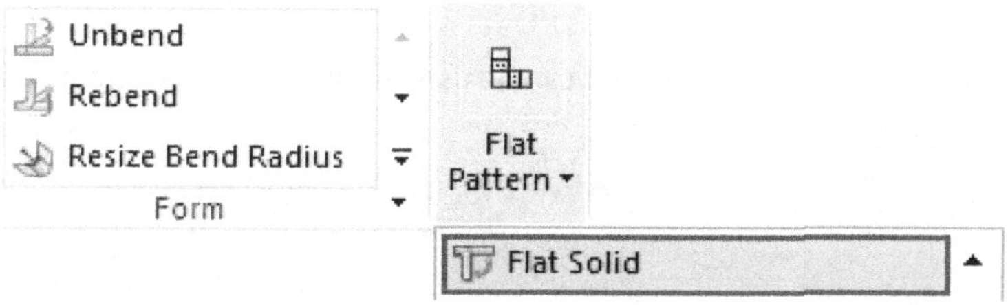

- Continue working with *"Sheet Metal from Solid.prt"*.
- Select the **Flat Solid** tool, and choose the top face as shown below.

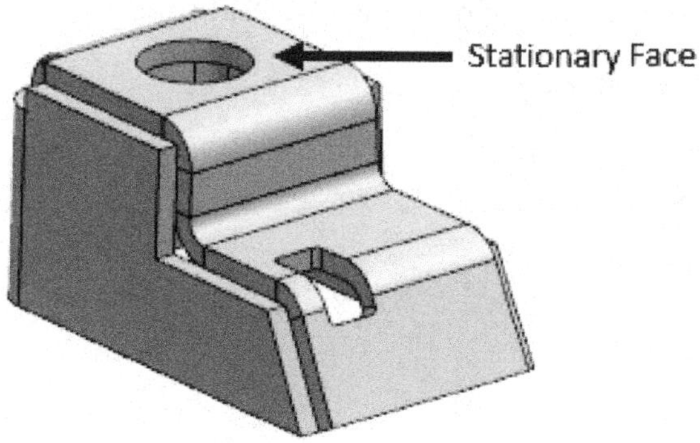

- The totally flattened body then appears in the graphics window, with the **Stationary Face** in the same position as on the formed body.

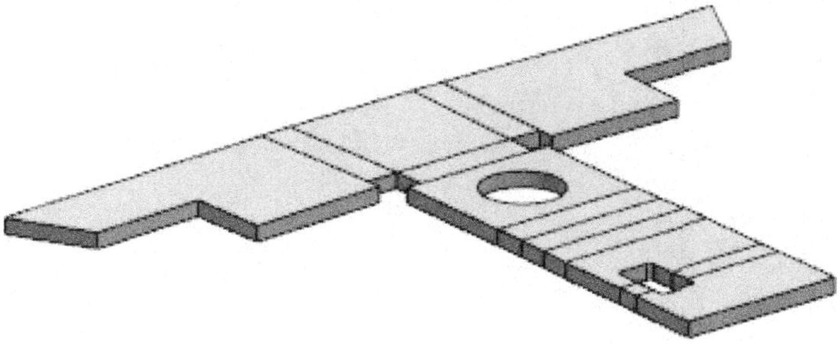

- Use the **Flange** tool to add a flange as shown below.

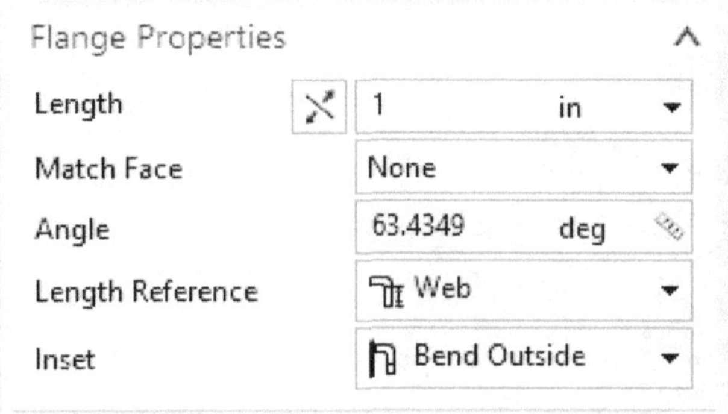

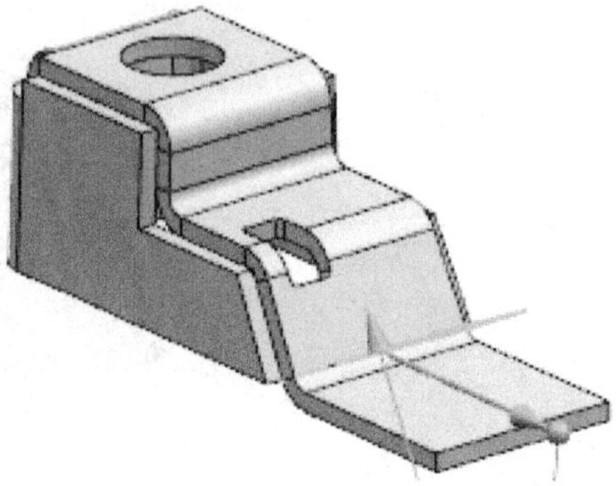

- A remarkable aspect of the **Flat Solid** feature is that it automatically jumps to the bottom of the Model History, so that it stays current with your sheet metal part as parametric features are added!

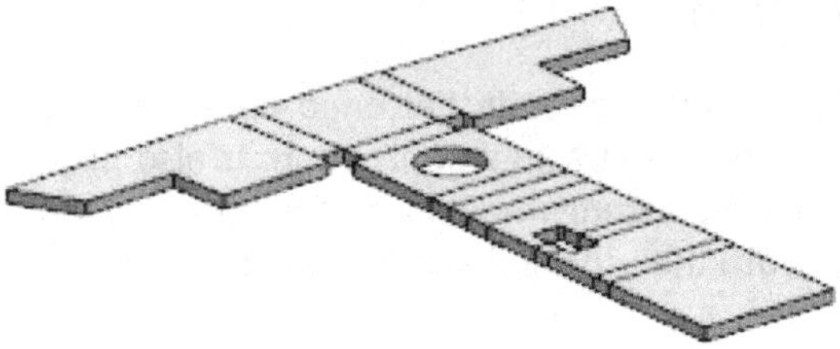

- The **Flat Solid** tool also produces a reference set that can be used to display your sheet metal part in its totally flattened state within an assembly.

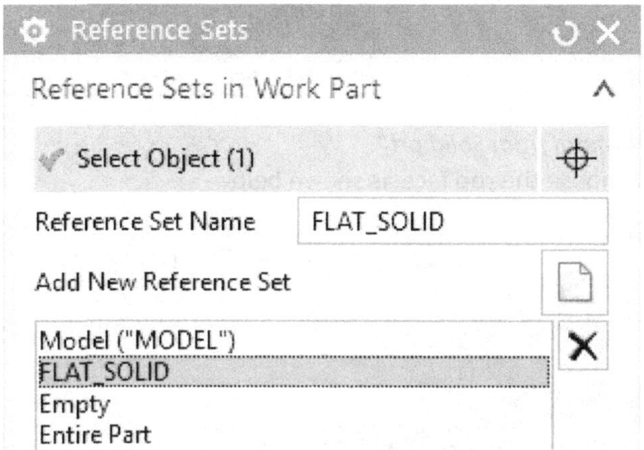

- **Save** your file!

15 Advanced Sweeps and Blends

In Chapter 7 you learned about several basic modeling tools that create solids from sketches – among them, **Extrude**, **Revolve**, **Tube**, and **Sweep Along Guide**. All four of these solid modeling tools are examples of *sweeps*. In this chapter, you will learn about several more advanced sweep tools. The **Sweep** group is pictured below.

Blends are closely related to sweeps, in that they create a surface by sweeping some section along edges of a sheet or solid body. You have already used the **Edge Blend** and **Face Blend** tools in Chapter 9.3, and in this chapter, you will learn how to create blends with more control over the shape of the cross section. The **Blend** gallery is found in the **Surface** group on the **Surface** tab.

In the text, you will learn about options within the **Edge Blend** and **Face Blend** tools that should prepare you to use the other advanced blends if necessary. However, much of the time, if the surface is complex enough, you will want to build it by hand using a sweep tool or a surfacing tool – for example, the **Swept** command, which you will learn about first.

15.1 Swept

So far you have used the **Sweep Along Guide** and **Tube** commands to create simple swept features. The **Swept** tool is far more sophisticated and can combine many sections with up to three guides to produce a high-quality surface or solid body. The **Swept** command is found in several places – the most direct is the **Surface** group on the **Home** tab.

It is worth mentioning a few points regarding the definition of **Guide** and **Section Strings** when defining **Swept** features:

- The orientation of each section must be consistent; otherwise the resultant surface feature will be twisted.
- Sections may be composed of one or many curve segments.
- Sections are not required to have a corresponding number of curve segments, but if so, the resultant surface shape will often be smoother.

These comments are applicable to most surfaces created in NX. If the sections are closed loops, the resultant geometry will be a solid feature. If the sections are open loops, the resultant geometry will be a sheet body feature.

The **Swept** tool is arguably the best and most flexible sweep tool – the others have their own subtleties that occasionally make them the correct choice, but generally speaking, the **Swept** tool is <u>the</u> tool to begin with when creating an arbitrary sweep.

15.1.1 Swept Surfaces with Multiple Sections and Multiple Guides

One advantage the **Swept** tool has over other sweep tools is that it offers you the ability to create swept bodies with multiple sections and multiple guides. In this exercise, you will learn to add multiple sections and multiple guides.

- Open the part file *"C:\My NX 11 Files\Bridge Curves.prt"* that you created for the exercise in Chapter 13.4. If you skipped that exercise, you can open *"C:\My NX 11 Files\Downloaded Files\Bridge Curves.prt"*.
- Use **File / Save / Save As** to save a new copy called and *"Swept – With Multiple Sections and Multiple Guides.prt"*.
- Select the **Swept** tool.
- Select the edge shown below as the first **Section**.

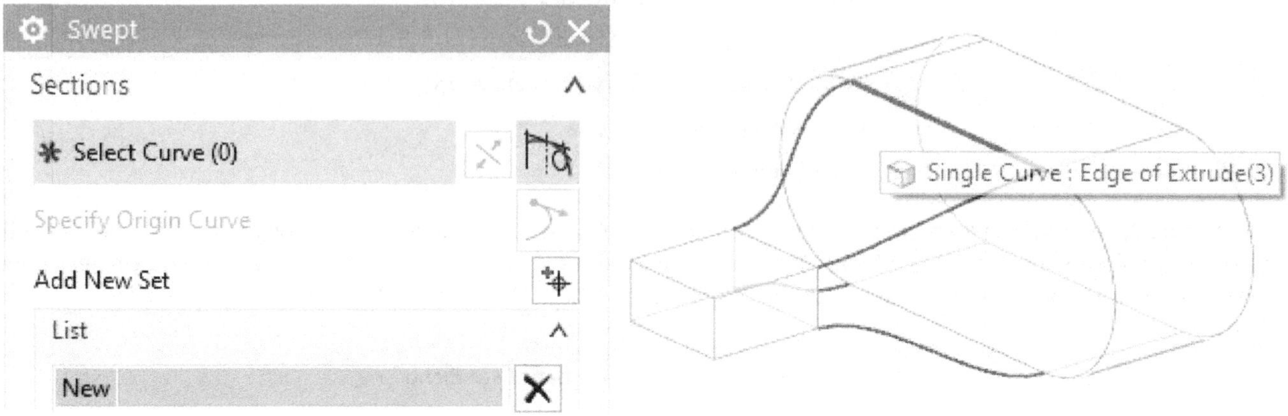

- Select **Add New Set**. This creates a line item for a new **Section** in the **List**. (Note: the middle-mouse button MB2 will also act as **Add New Set** here!) Select the edge of the rounded side of the second body connecting the same bridge curves as the **Second Section**.

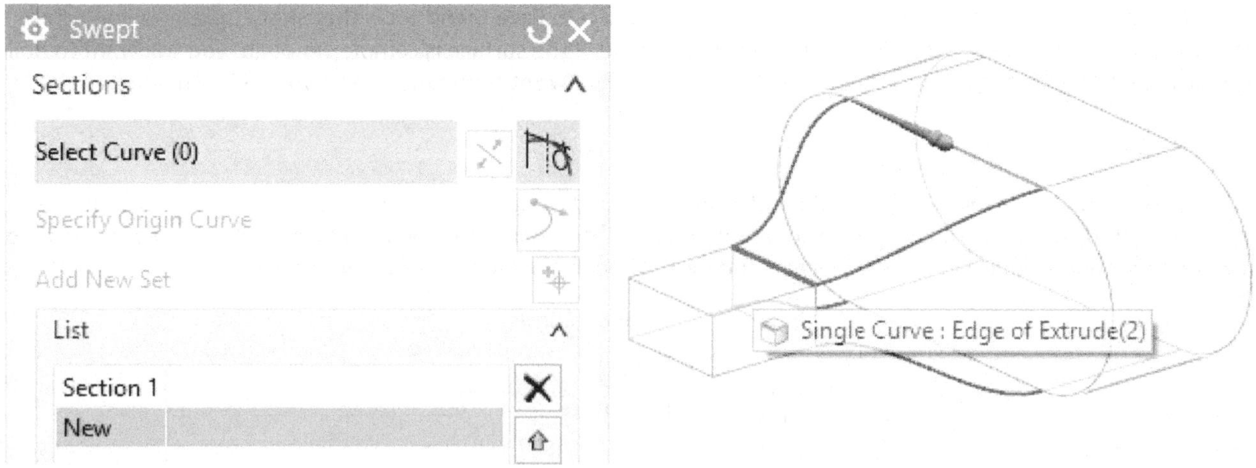

- Now advance the menu to the **Guides** section. Set your **Selection Intent** to *Single Curve* and choose the *Bridge Curve* shown below.

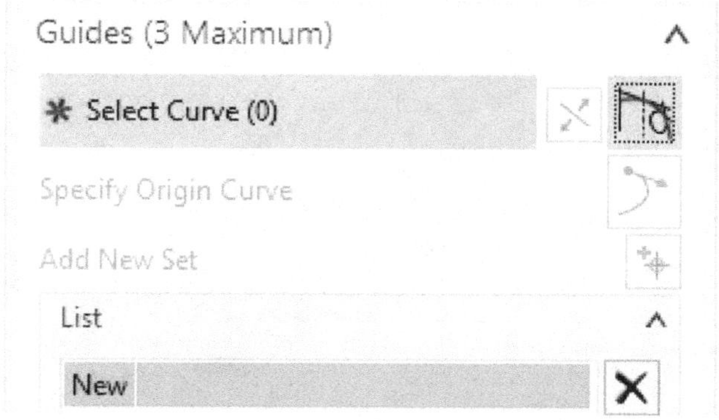

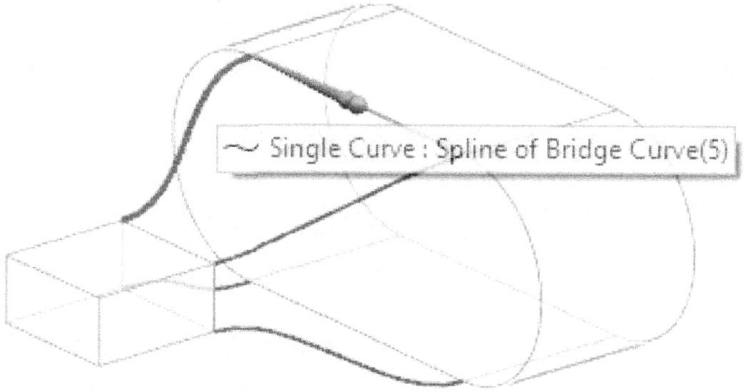

- Select **Add New Set** (or middle-click) and choose the other *Bridge Curve* as shown below.

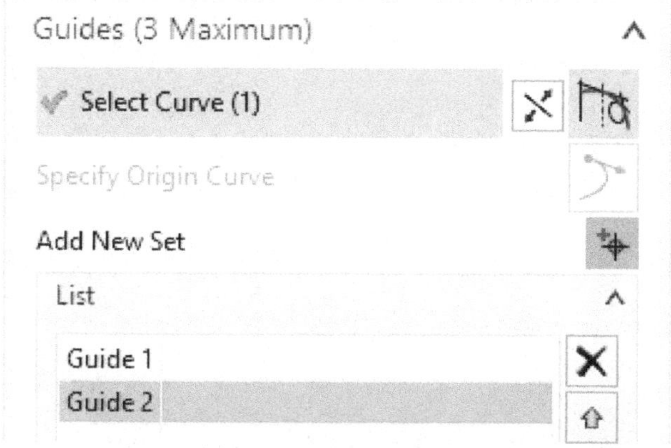

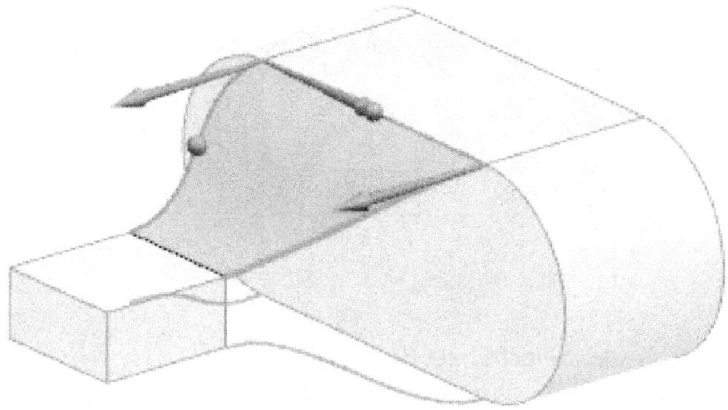

- In the **Section Options** section of the dialog, make sure the *Preserve Shape* checkbox is marked.
- Click **Apply** to produce your **Swept** surface!
- Repeat this process three more times to fill in the other missing faces.

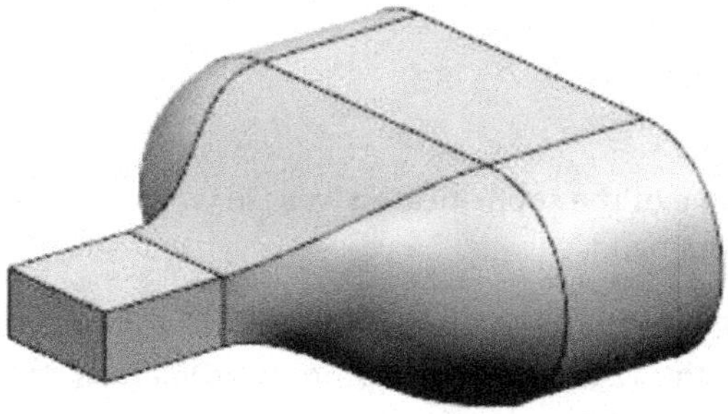

- **Hide** all sheet bodies and curves in your model, and use the **Unsew** command to remove the faces of the solid bodies that are facing each other.

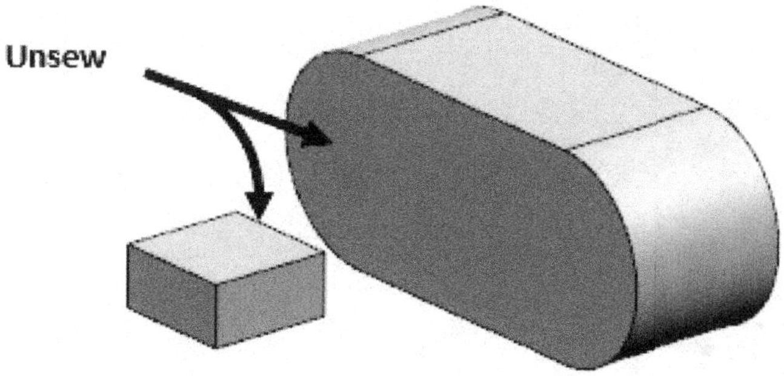

- **Hide** the larger sheet bodies so that only the unsewn faces remain. Use **Delete Body** to associatively delete them!

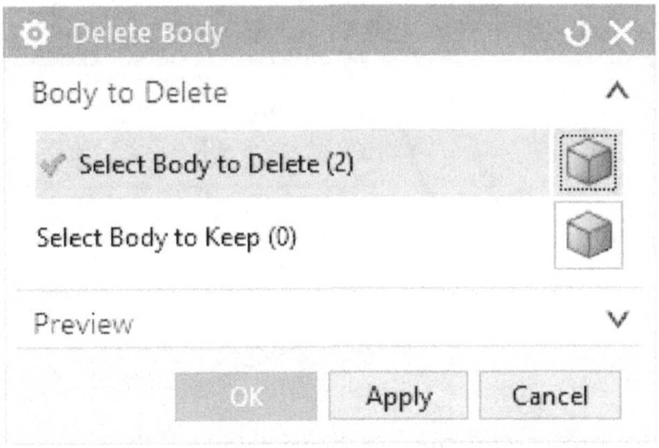

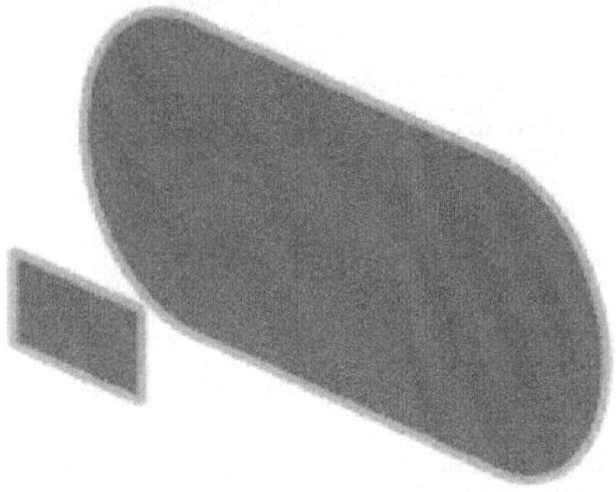

- **Show** all sheet bodies again and **Sew** them together.

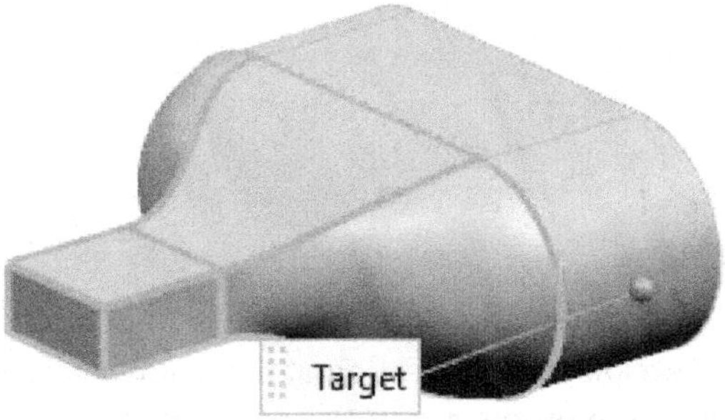

- This produces a solid body. **Save** your file!

15.1.2 Project – Twisted Wires

In this exercise, you will learn about some of the sophisticated ways in which you can modify a section string as it is swept along a guide string using the **Swept** command.

- Create a new part file (**Units**: *Millimeters*) called *"Twisted Wires.prt"* and place it in *"C:\My NX 11 Files"*.
- Create the sketch shown below in the YZ plane of the datum coordinate system. Use the **Perimeter Dimension** tool to constrain the total arc length to *250 mm*.

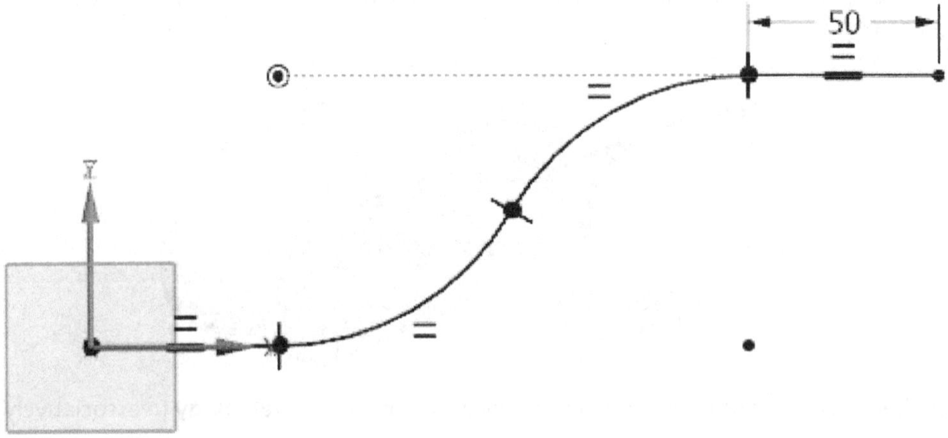

- Create a sketch with **Type** set to *On Path* at the end of the first sketch you created.

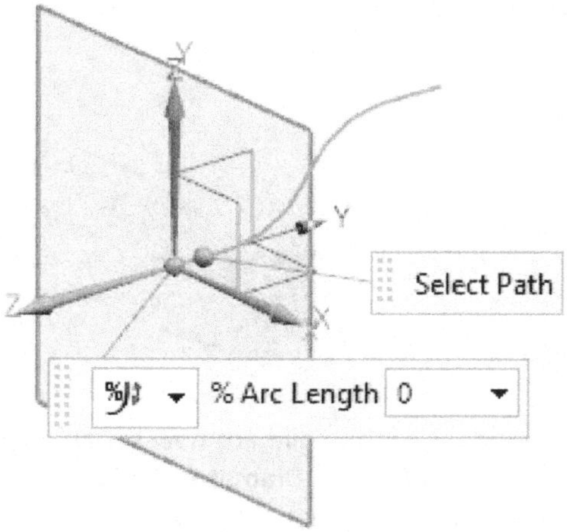

- In the sketch on path, use the **Polygon** tool create an equilateral triangle centered at the sketch origin, with a **Side Length** of *3*. The **Rotation Angle** is irrelevant.

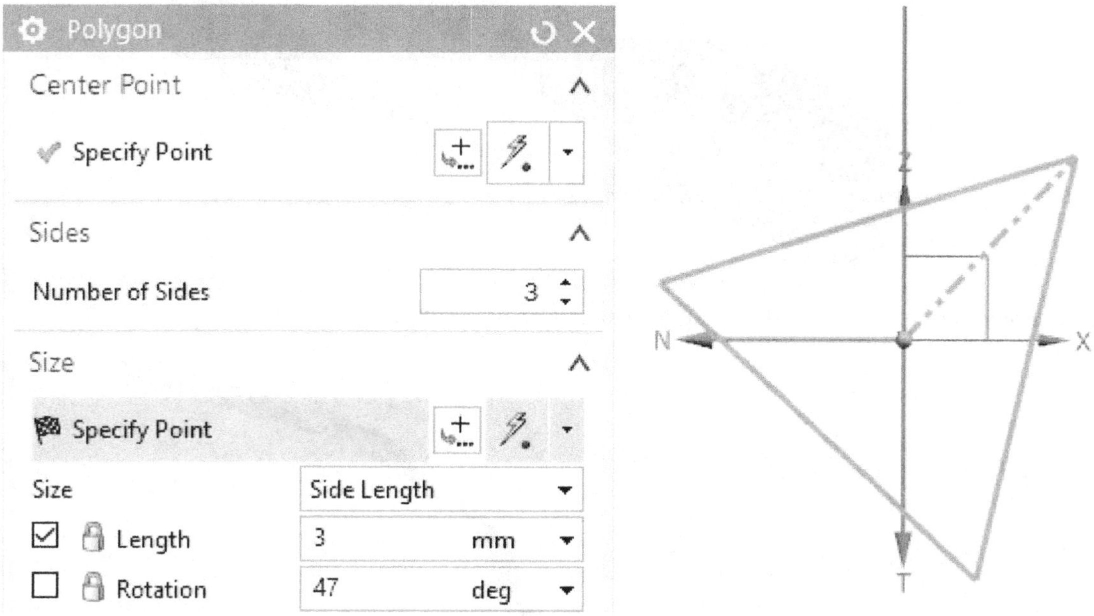

- Use the **Swept** tool and select the triangle as **Section** and the first sketch as **Guide**. The preview will shed some light on the meaning of the *Preserve Shape* checkbox – in this case, one consequence of *Preserve Shape* is that vertices of the triangle become edges of the *Swept* feature. This is certainly our intent, so make sure to check the *Preserve Shape* box!

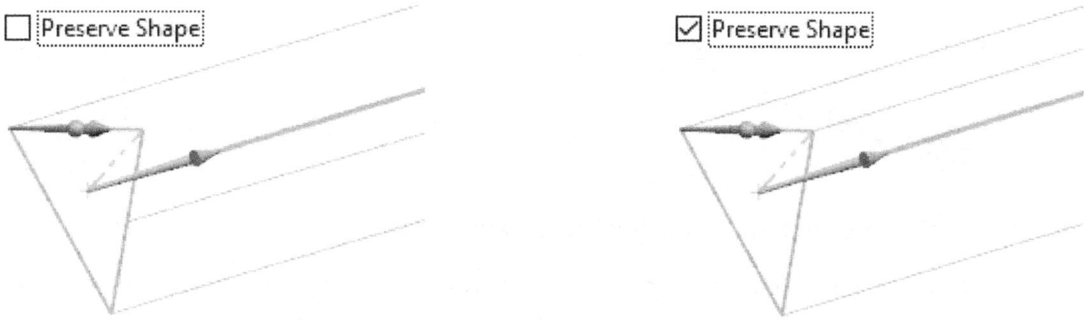

- Within the **Section Options** section of the dialog, there is a subsection called **Orientation Method**. Set the **Orientation Method** parameters as shown below. Essentially, we are asking the triangle to rotate a full *360 degrees* seven times in its plane as it traverses the guide curve.

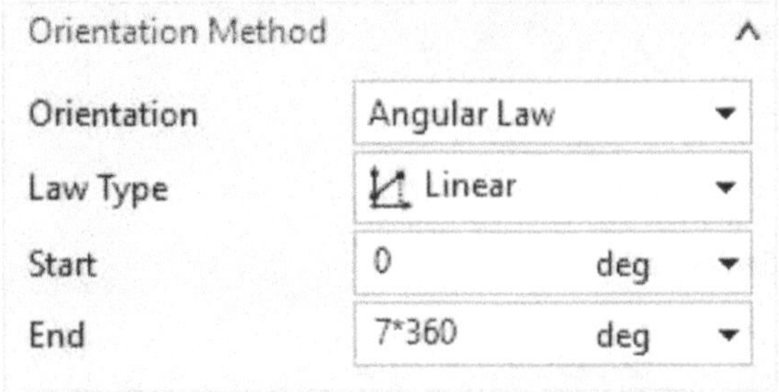

- Click **OK** to create the swept solid. The edges of this solid are now suitable guide curves for the wires. Use the **Tube** command three times to create three wires of **Outer Diameter** *2.75 mm*. Be sure to set **Output** to *Single Segment* within the **Tube** dialog. Color your three wires red, green, and black.

- Use **Delete Body** to remove the twisted triangular swept body, and **Save** your file!

15.1.3 Swept Bodies with Constant Cross-Sectional Area

In this exercise, you will explore further options for controlling the cross-section of a **Swept** feature along the guide curves.

- Open *"C:\My NX 11 Files\Twisted Wires.prt"*. If you skipped the last exercise, you can open *"C:\My NX 11 Files\Downloaded Files\Twisted Wires.prt"*.
- Select **File / Save / Save As** to create a copy named *"Swept – Constant Cross-Sectional Area.prt"* in *"C:\My NX 11 Files"*.
- **Delete** the three tubes, the **Swept** feature, and the sketch of the triangle.
- Create the sketch shown below at the far end of the guide curve, away from the datum coordinate system. There is a **Perimeter Dimension** of *200 mm* on the whole sketch.

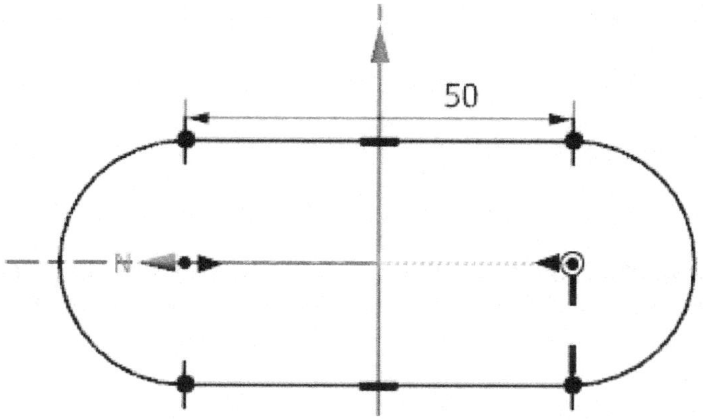

- Create the sketch shown below at the start of the guide curve (on top of the datum coordinate system). Start with the arc and *45-degree* reference line, and then use **Mirror Curve** three times to create the whole sketch.

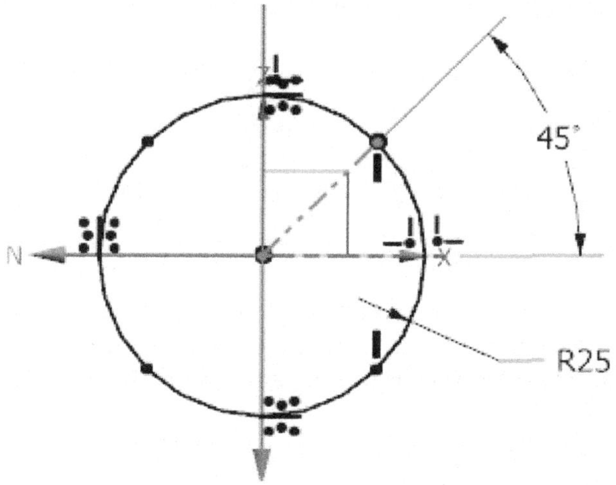

- The reason we subdivided the circular section in the last step is so that the sections for the **Swept** feature can be oriented consistently. Click near the endpoints attached to the orientation arrows shown below to ensure that the sections are aligned.

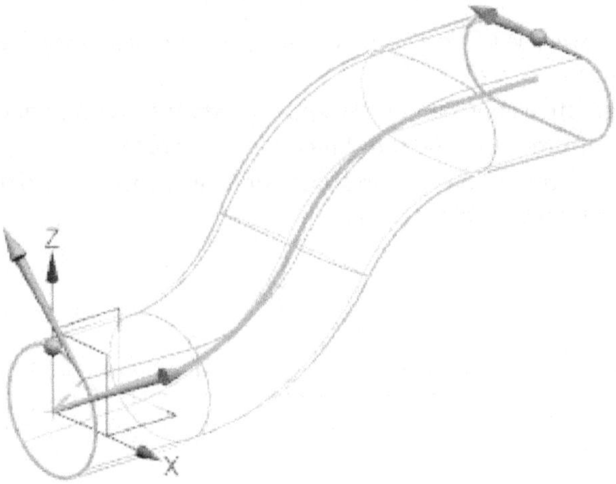

- Within the **Section Options** section of the dialog, there is a subsection called **Scaling Method**. This allows you to define a law that controls the cross-sectional area of the section along the guide. Set the parameters as shown below.

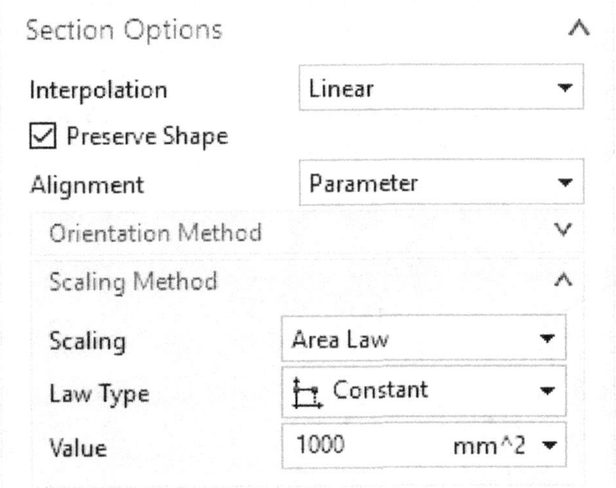

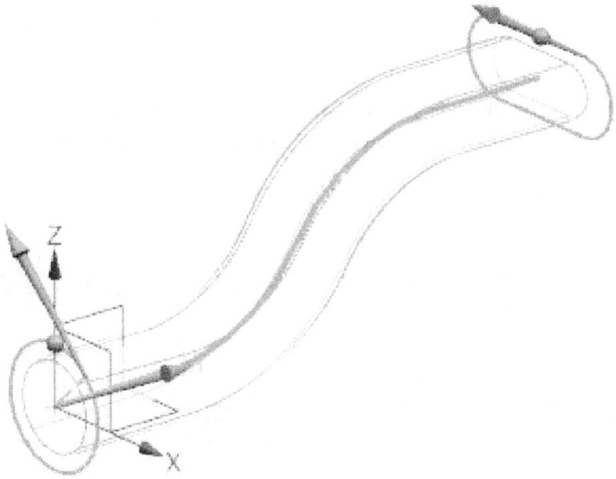

- After creating your solid body, **Shell** the opposite ends of the **Swept** feature to add *5 mm* thickness to the outside. You will probably need the **Reverse Direction** button. This ensures that the interior of your pipe has constant cross-sectional area.

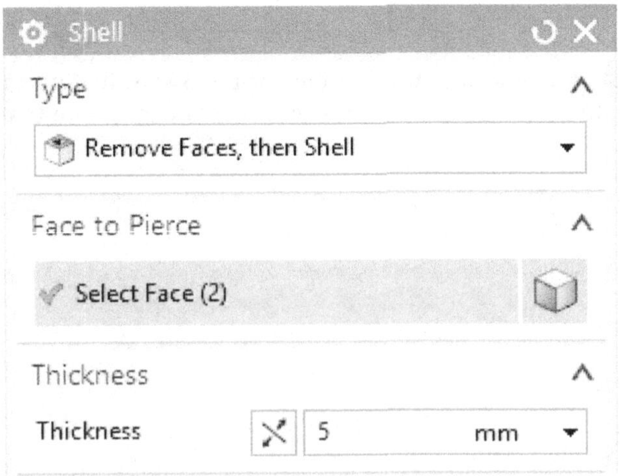

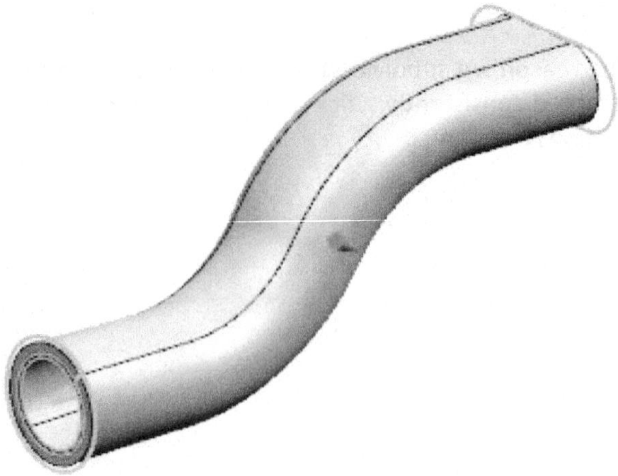

- **Save** your file!

15.2 **Variational Sweep**

Another example of a sweep tool that offers additional control over the shape of the section string is the **Variational Sweep**. This sweep tool is more sophisticated than **Sweep Along Guide**, and allows a different sort of control over the shape of the section string than **Swept**. **Variational Sweep** is found in the **Surface** group on the **Home** tab.

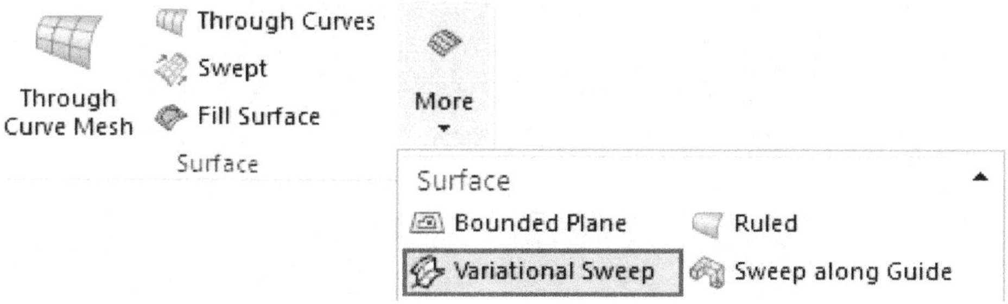

The basic input for **Variational Sweep** is a sketch on a path. It allows you to select control points for additional sections on that path and modify the sketch parameters for each copy of the sketch at each control point.

- Create a new file (**Units**: *Inches*) called *"Variational Sweep.prt"* and place it in *"C:\My NX 11 Files"*.
- Create the sketch shown below in the XY plane.

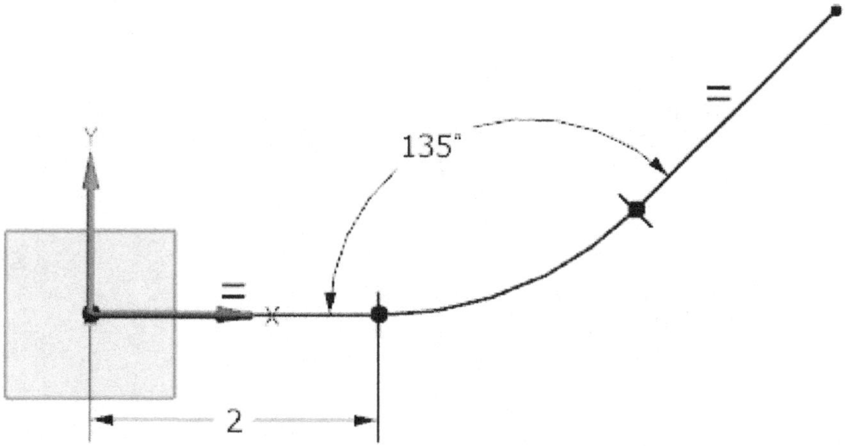

- Create the sketch shown below with **Type** set to *On Path*, attached to the first sketch at *0%* **Arc Length**.

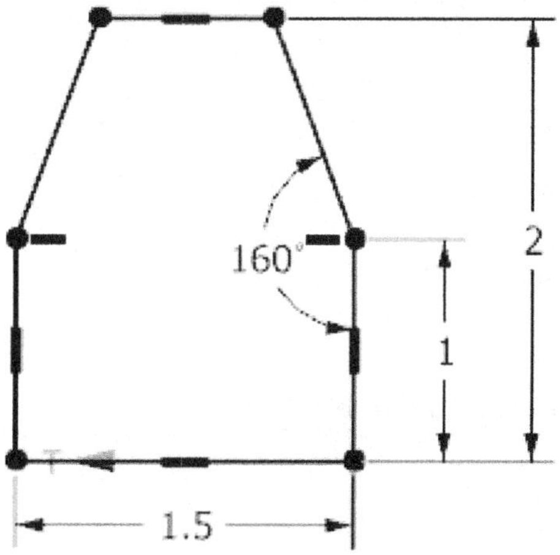

- Use the **Variational Sweep** tool and select the second sketch.
- Within **Secondary Sections**, use the **Add New Set** button twice to add two intermediate sections, *Section1* and *Section2*, between the *Start Section* and *End Section*. Position them at *33%* and *66%*, respectively.

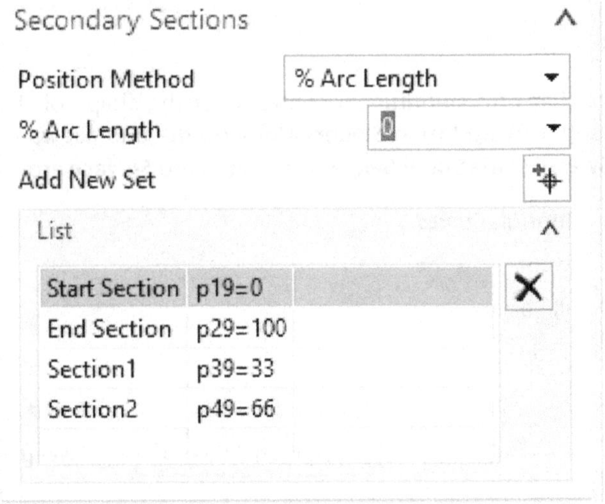

- To modify the parameters for a section, select it from the **List**, and double-click on the dimension you want to modify. The parameters to be modified in each section are the two vertical dimensions. Parameters given below.

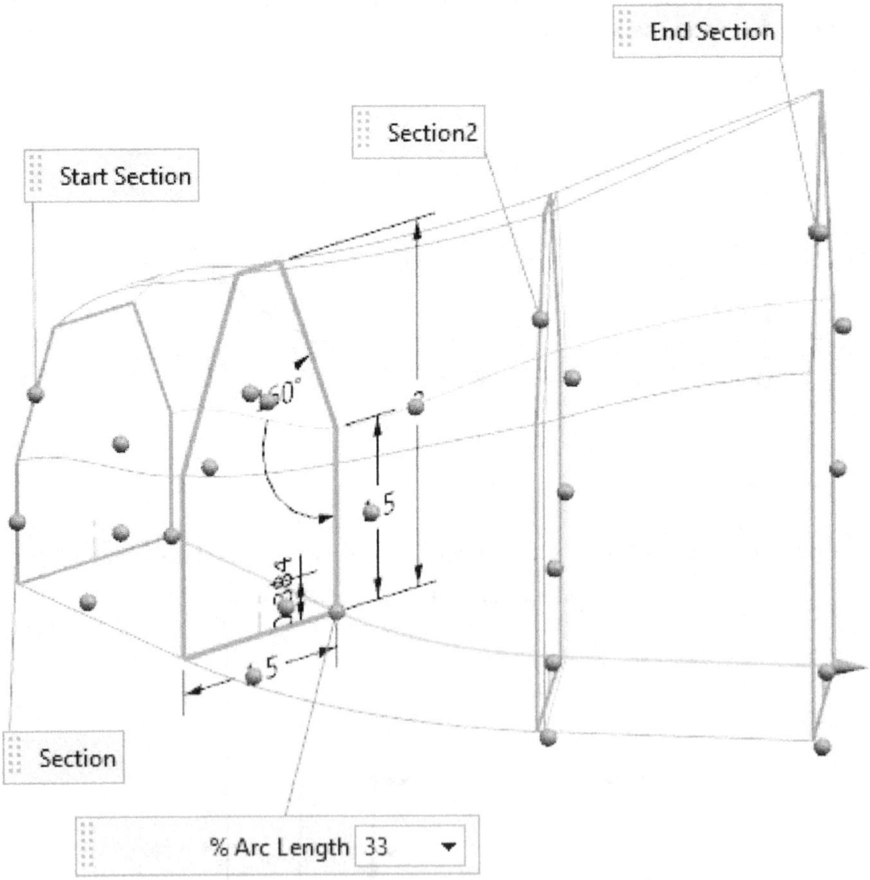

- *Start Section*: *1 in, 2 in* (unmodified from original sketch)
- *Section 1*: *1.5 in, 3 in*
- *Section 2*: *2.5 in, 4 in*
- *End Section*: *3 in, 5 in*
- You will have created a swept solid like the one shown below!

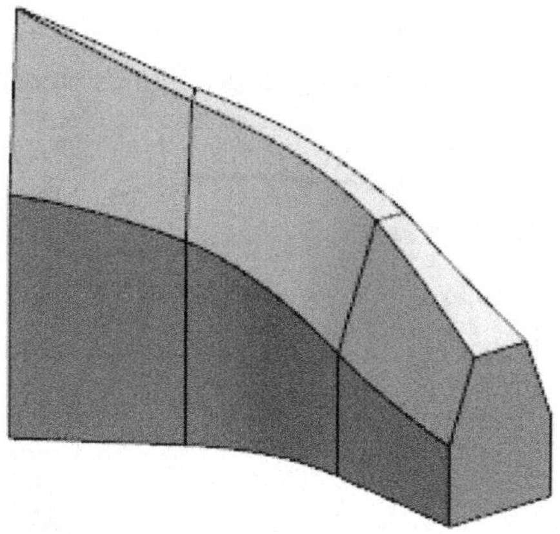

- **Save** your file!

15.3 Edge Blend Revisited

In Chapter 9.3.1, you practiced using the **Edge Blend** tool in its simplest mode – to create a fillet between two faces on a solid, whose cross section is an arc of constant radius. In the exercises below, you will see how to create fillets with more interesting cross sections using the **Edge Blend** tool.

15.3.1 Variable Radius Edge Blend

The **Edge Blend** function can also be used to vary the radius of the blend as it traverses the distance of the edge. Use the geometry from the previous project.

- Create a new file (**Units**: *Inches*) named *"Edge Blend Revisited.prt"*.
- Create a **Block** with dimensions *6 in* x *6 in* x *1 in*.
- Select the **Edge Blend** tool.
- Select the edge shown below.
- Within the **Variable Radius** section of the menu, use the **Point** constructor icon to add five points to the list, spaced evenly. At each point, specify the radius shown below.

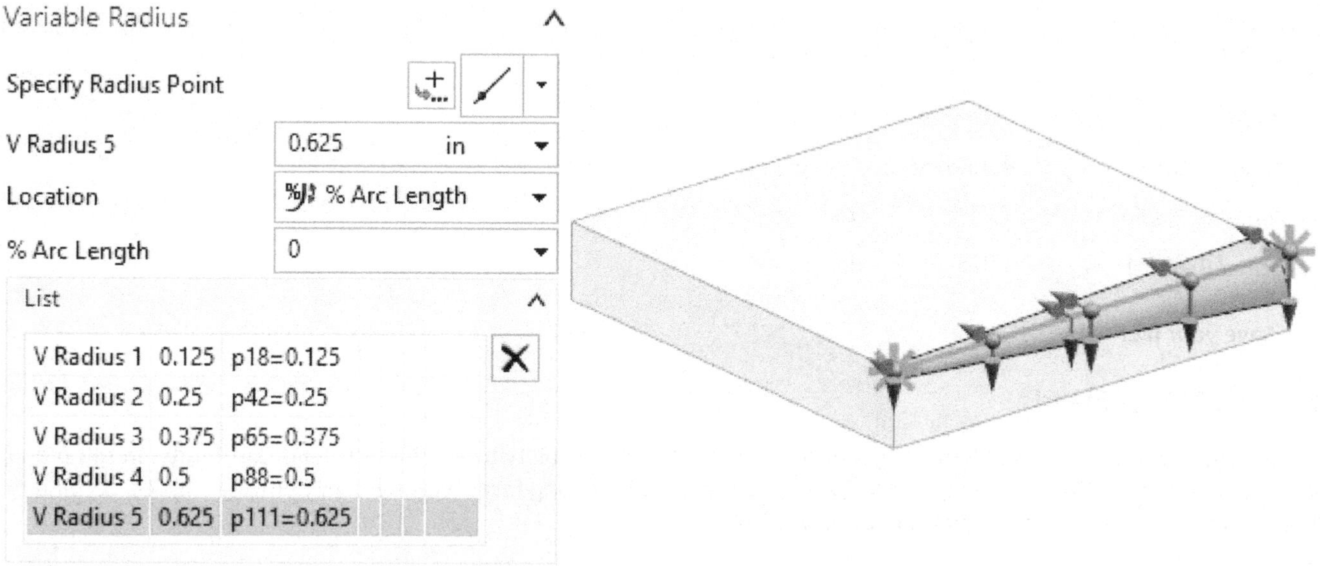

- **Save** your file!

15.3.2 Conic Cross Sections

Besides blends whose cross sections are arcs, you can also produce blends whose cross sections are conics.

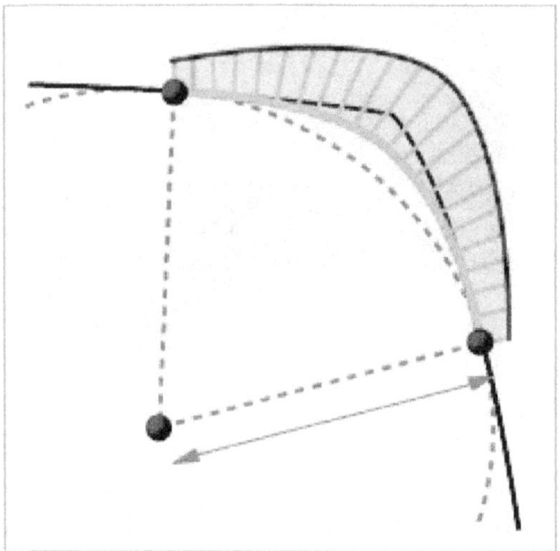

Conical fillets are sometimes considered to be more aesthetically pleasing than circular fillets. The conic edge blend has two basic inputs – the boundary radius and center radius. The boundary radius indicates how far the edges of the blend face are from the input edge, and the center radius indicates the radius of an arc tangent to the conic at the center. The center radius must be a nonnegative number.

- Continue working with *"Edge Blend Revisited.prt"*.
- Rotate the model *180 degrees* about ZC.
- Select the **Edge Blend** tool and set the parameters as shown below.

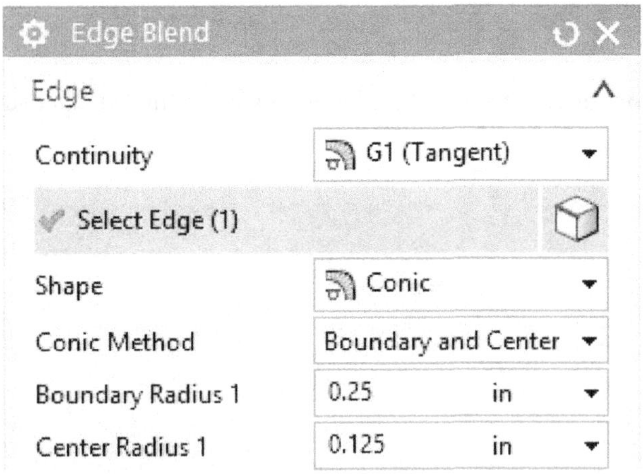

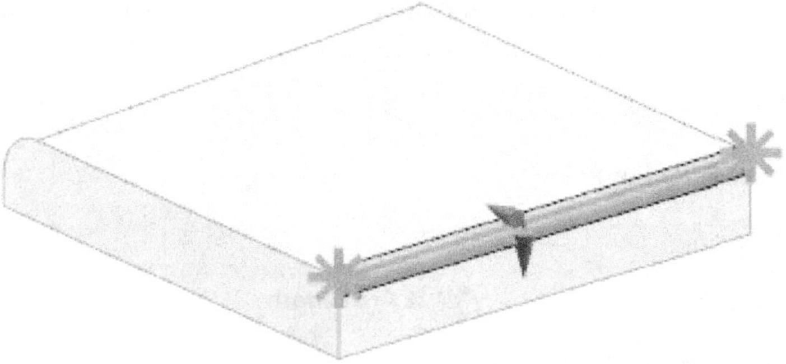

- **Save** your file!

15.3.3 G2 Edge Blends

A new option within the **Edge Blend** tool, introduced in NX 10, is the ability to modify the blend **Continuity**. In this menu, you can choose between producing a blend which meets its neighboring faces with *G1* continuity, or with *G2* continuity. The curvature combs of these blends are displayed below.

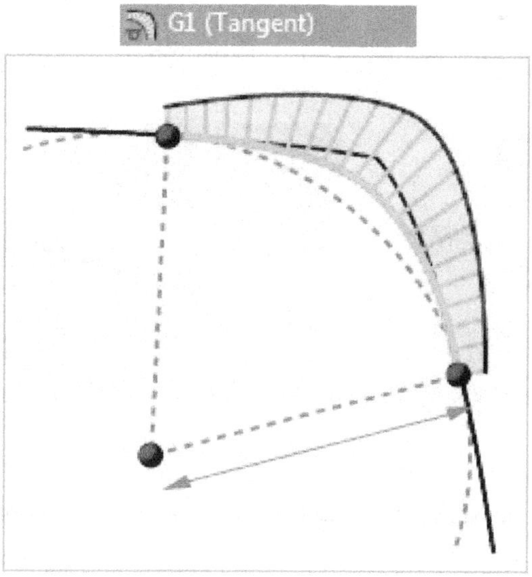

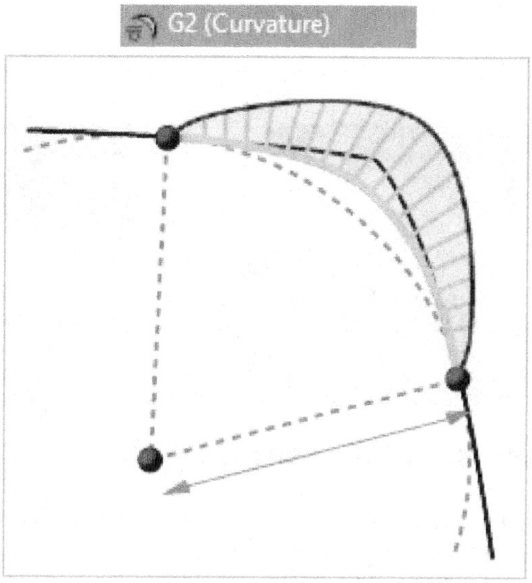

The cross section of a *G2* blend is a curvature continuous spline (see Chapter 13.3.2). Like the conic edge blend, it has two controls, a **Radius** and a **Rho**. The radius is like the **Boundary Radius** for a conical edge blend, and the rho is similar to the **Center Radius** insomuch as it controls the radius of an arc tangent to the spline at its center. The rho parameter, however, varies between *0.01* and *0.99*.

- Continue working with *"Edge Blend Revisited.prt"*.
- Flip the model **180 degrees** about the XC axis.
- Select the **Edge Blend** tool and set the **Continuity** to *G2 (Curvature)*.
- Select the edge shown below and enter the parameter values shown in the dialog.

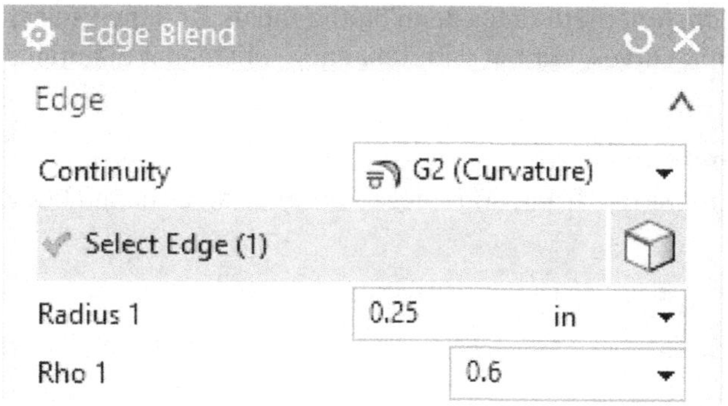

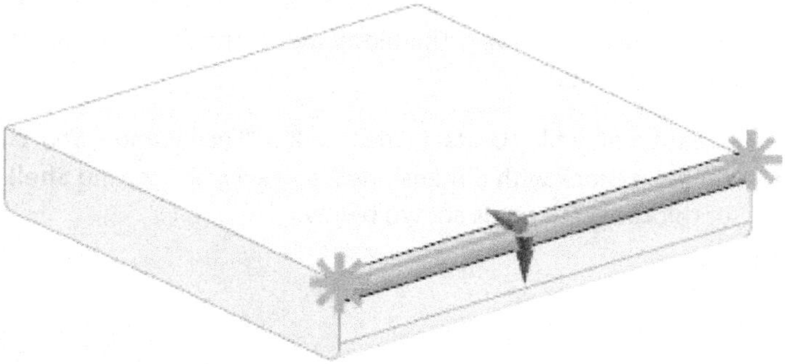

- **Save** your file!

15.3.4 G1 – Circular vs. G1 – Conic vs. G2

Now that you know about the different shapes the cross section of an **Edge Blend** can take on, you might wonder why you would choose a conic or *G2* edge blend over a simple *G1* circular edge blend. The main reason to use **Edge Blend** with these other cross-sectional shapes is aesthetic – in the image below, note that the circular edge blend has a visible "edge" where the arc meets the blend. The conical edge blend also has an edge, although it is not quite as noticeable, and the light gradient transitions more smoothly. The *G2* edge blend has no visible "edge" and the light gradient is even smoother.

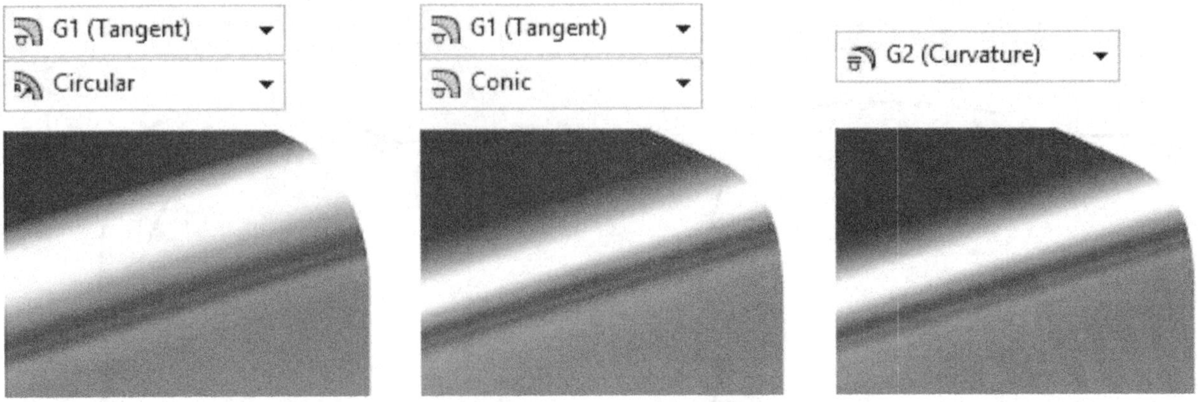

15.4 **Face Blend Revisited**

In Chapter 9.3.2, you learned the basic use of the **Face Blend** command. Now you will learn some additional options that help to vary the shape of the cross section.

There are two main types of face blends: *Rolling Ball* and *Swept Disc*. *Rolling Ball* is the simpler and more commonly used face blend. *Swept Disc* is a blend whose cross-section is normal to a spine curve. The exercises below discuss the *Rolling Ball* type face blend. *Swept Disc* is not discussed in this book.

There are also three different options to define radii: *Constant*, *Variable* and *Limit Curves*. In the exercises below, we explore two approaches to creating a blend by *Limit Curves*.

15.4.1 Sharp Limit Curves

One advantage the **Face Blend** tool has over the **Edge Blend** tool is the ability to create blends that do not meet their neighboring faces with tangency. To do so, you must set use the *Constant* method for defining the radius, and then select certain *Sharp Limit Curves*. The blend meets the faces along the limit curves, but not with tangency – instead it maintains the radius specified.

- Create a new file (**Units**: *Inches*) called *"Face Blend – Sharp Limit Curves.prt"*.
- Create a **Block** with dimensions *1.5 in* x *4 in* x *1 in*, and **Shell** out the top, front, and sides to produce an L-shaped body of thickness *0.1 in*, as shown below.

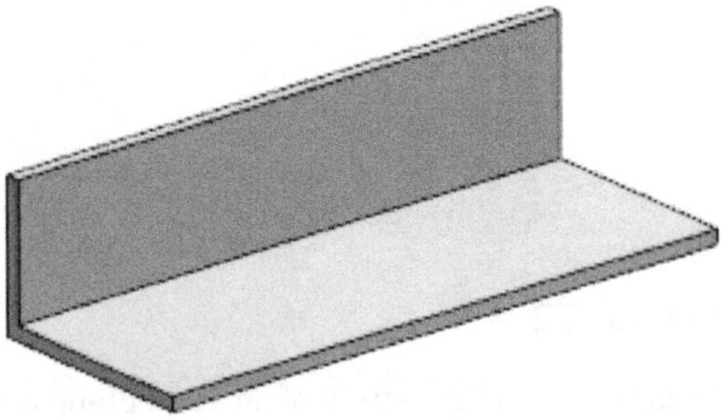

- If you attempt to apply an **Edge Blend** to the interior edge of this body with radius too large, it cannot possibly be tangent to the neighboring faces, and so the blend will fail.

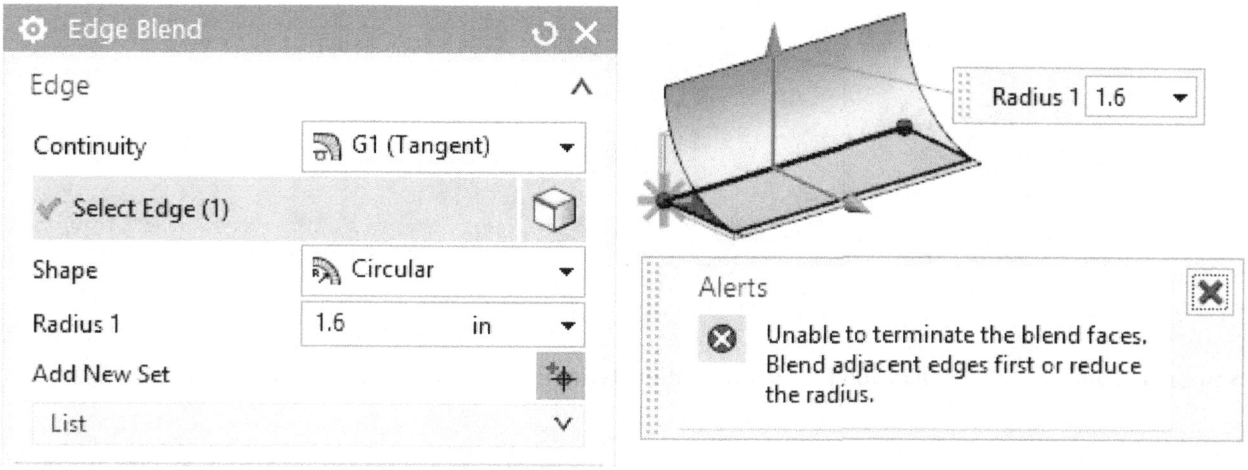

- Use the **Face Blend** with **Type** set to *Two-face* and choose the interior faces in the order shown below. If you attempt to enter a value of *1.6 in* for the **Radius** at this point, you will again meet an error message. In the **Width Limits** section of the menu, set **Location** to *Face 1* and choose the top edge of the interior faces as shown below.

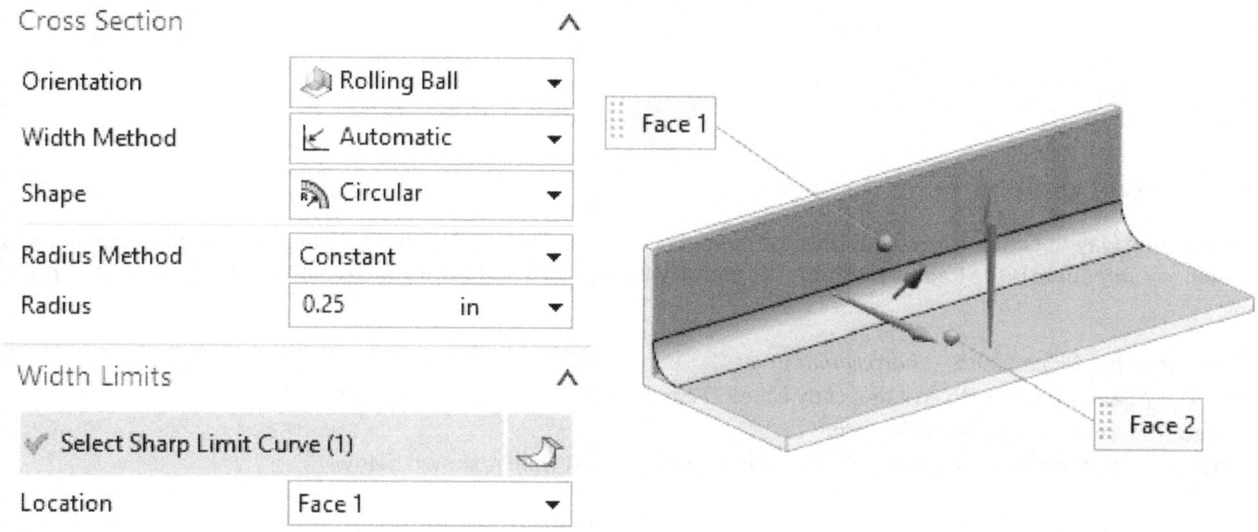

- Change the **Location** drop-down to *Face 2*, and select the edge of the second face as shown below.

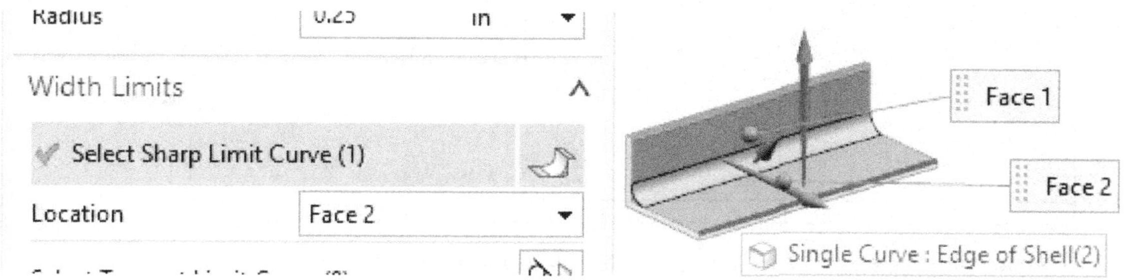

- Now you can change the value for the radius to *1.6*, the value that failed for **Edge Blend**, and the tool is happy to accommodate!

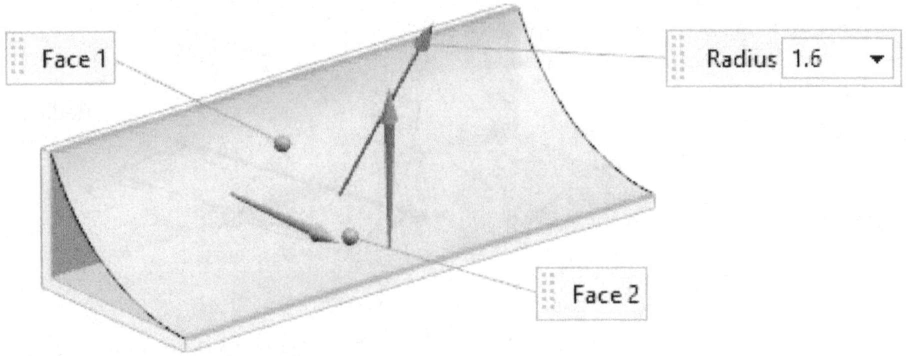

- You can even set a radius well beyond the width of the interior face. In the shot below, the radius is *2.5 in*!

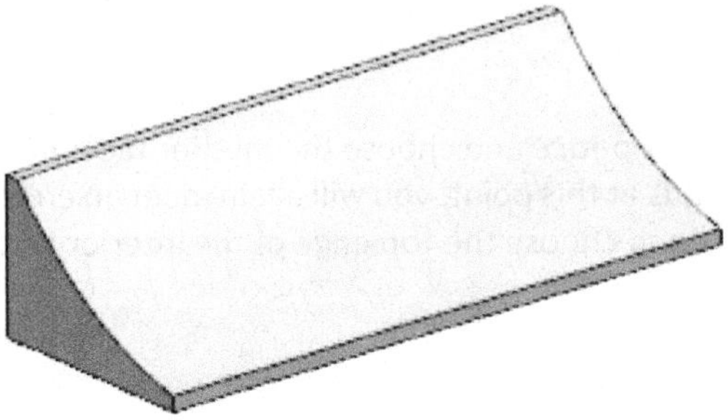

- **Save** your file!

15.4.2 Tangent Limit Curves

One interesting way to control the radius of the cross-sectional arc of a **Face Blend** is to make selections for *Tangent Limit Curves* when *Limit Curve* is the **Radius Method**. The blend will meet the neighboring faces with tangency along the specified curves.

- Open your file *"Face Blend – Sharp Limit Curves"*.
- Use **File / Save / Save As** to create a copy called *"Face Blend – Tangent Limit Curves.prt"*.
- **Delete** the existing *Face Blend* feature.
- **Extrude** with **Boolean** set to *Subtract* from the interior vertical wall as shown below.

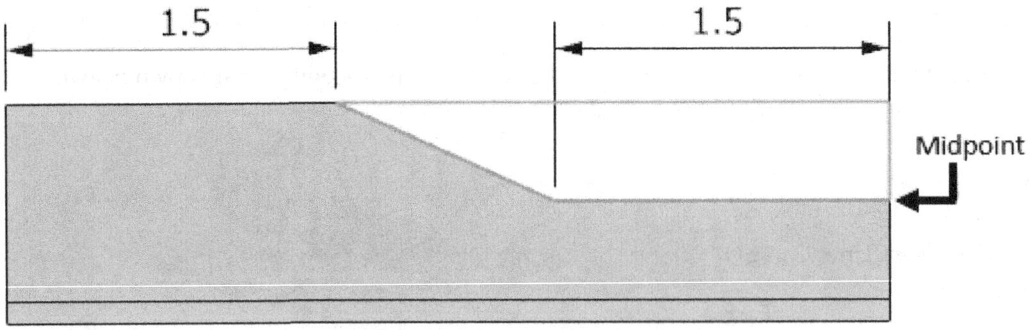

- Select **Face Blend** with **Type** set to *Two-face*, and select **Face 1** and **Face 2** as shown below.

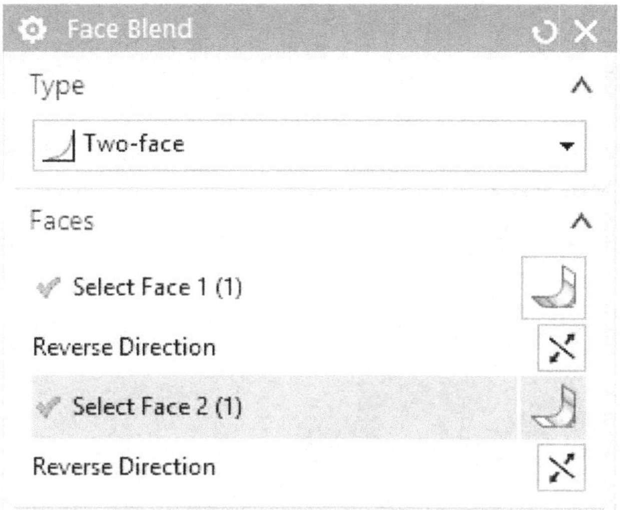

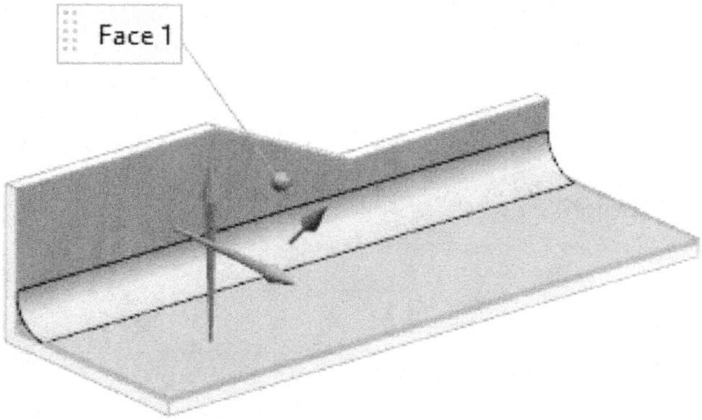

- Within the **Cross Section** part of the dialog, set the **Radius Method** drop-down to *Limit Curve*. This will make the *Tangent Limit Curve* input within the **Width Limits** section of the dialog a required input. Choose the three edges at the top of Face 1 as the *Tangent Limit Curves*.

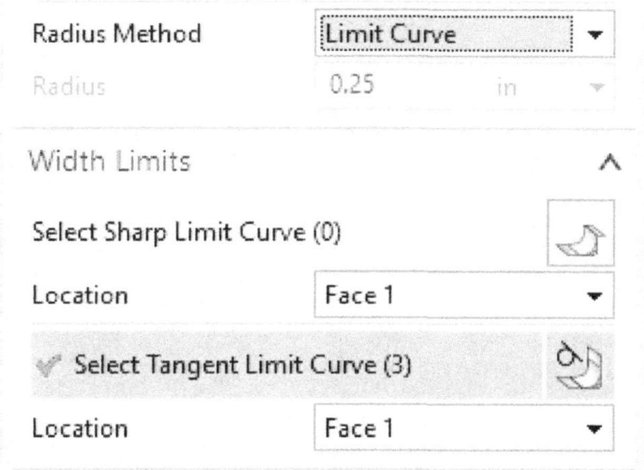

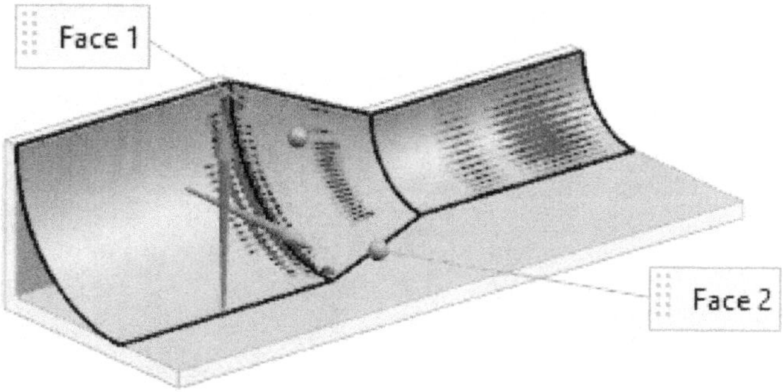

- Although the preview appears to shown an error, NX is able to resolve any issues and the **Face Blend** feature is created without issue!

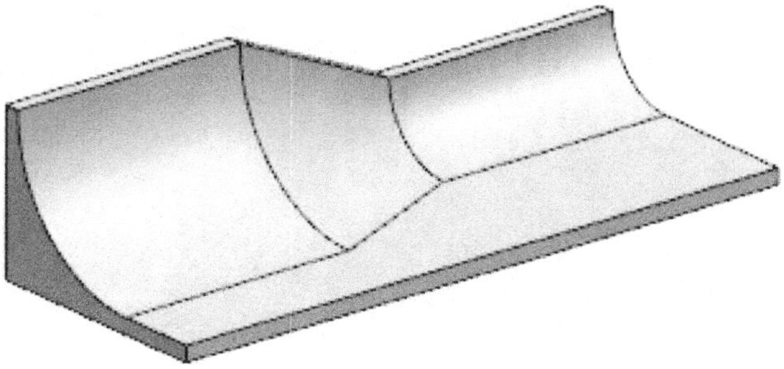

- **Save** your file!

You might wonder whether the same feature can be created using **Edge Blend** with **Variable Radius** points (see Chapter 15.3.1) rather than **Face Blend**. In fact, it cannot be modeled with a single *Edge Blend* feature using **Variable Radius** points. The reason is that the tool creates faces whose edges are cubic splines.

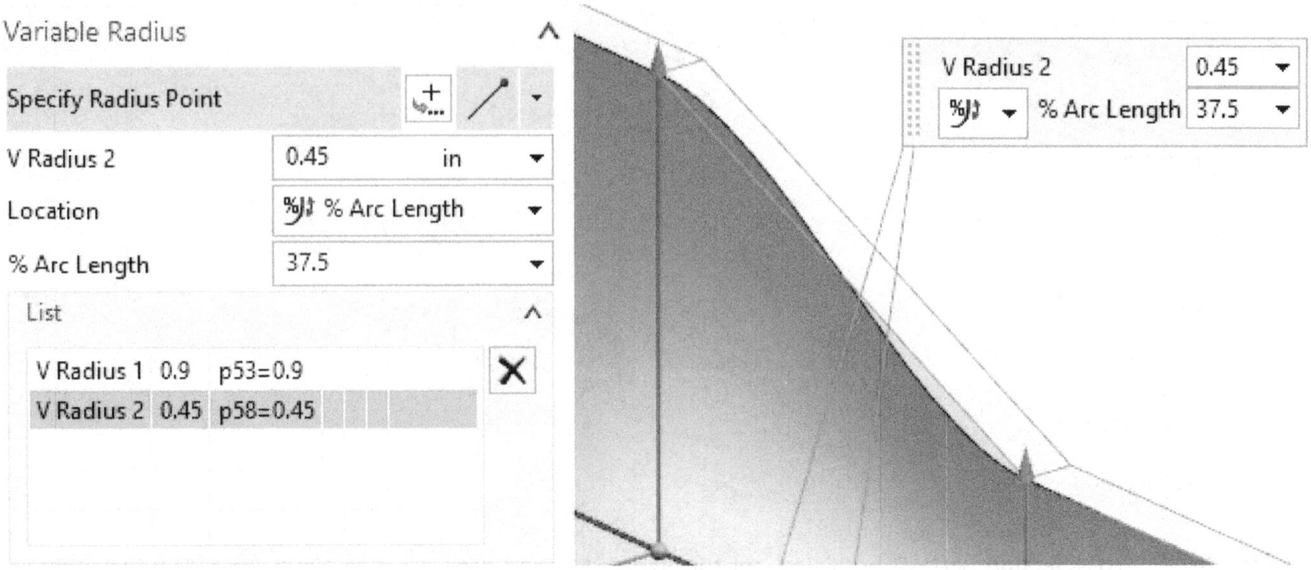

As a bonus exercise, try to create the geometry of the **Face Blend** above using the **Edge Blend** tool. Hint: it will require three blends, and the use of the **Stop Short of Corner** option in the **Edge Blend** dialog. See the part files you downloaded from www.designviz.com/goodies for the answer!

16 Synchronous Modeling

If there is one good argument against parametric modeling in general, it is that when used, models get too complicated. It becomes difficult to make edits because one must understand everything that was done before in that model.

NX has a fantastic suite of tools suitable for modifying both parametric and non-parametric models that do not require the designer to understand the complete model history – these tools are found in the **Synchronous Modeling** group on the **Home** tab.

Using **Synchronous Modeling** tools on an already-parametric model tends to obfuscate the original design intent, because Synchronous Modeling tools add parameters to the model that may override the original parameters. The model will no longer work in the way that was originally intended to, and in some sense the value of its parameters are diminished.

However, there are enormous advantages to using **Synchronous Modeling** tools, even on parametric models. Since changes do not require a thorough understanding of the Model History, they can be made much more quickly. At the end of the day, your goal as a designer is to get the best product to market as quickly as possible, not to produce a "perfect parametric model", and the **Synchronous Modeling** suite excels at rapid direct modeling.

Synchronous Modeling tools can be grouped into three classes, based on the nature of the parameter or modification they make to a model.

- <u>Differential Parameters:</u> Some **Synchronous Modeling** tools add a parameter that is essentially a displacement – faces are transformed by way of a parameter that measures the difference between the starting position and the ending position. These "differential" parameters are sensitive to the model history. Tools that add differential parameters include **Move Face**, **Pull Face**, **Offset Region**, **Copy Face**, **Pattern Face**, **Move Edge**, **Offset Edge**.
- <u>Absolute Parameters:</u> Some **Synchronous Modeling** tools measure an existing quantity – e.g., a distance, or a radius – and turn it into a parameter that can be varied. These "absolute" parameters are not sensitive to the model history in the same way that differential parameters are. You can think of these tools as "reparametrizing features" that may or may not be actual features in the Model History. Tools that add absolute parameters include **Resize Face**, **Resize Blend**, **Linear Dimension**, **Angular Dimension**, and **Radial Dimension**.
- <u>Geometric Operations:</u> Some **Synchronous Modeling** tools modify faces of a model in purely by imposing geometric relations, without adding a parameter to the model. Tools in this class include **Delete Face**, **Make Tangent**, **Make Parallel**, **Make Symmetric**, and other tools from the **Relate** group.

Basic behaviors of synchronous modeling tools: trimming & extension of motion faces, and neighbors. Incident blends to motion faces are "special" neighbors in that they often remain incident to the motion face without modifying their dimensions.

16.1 Move Face

The **Move Face** command is in some sense the fundamental **Synchronous Modeling** operation. You select a face or chain of faces and specify a transformation. Adjacent edges/faces will adapt to accommodate the change. A good grasp of this tool will equip you with the insight to predict the behavior of other **Synchronous Modeling** tools.

Move Face is our first example of a **Synchronous Modeling** tool that adds a differential parameter to a model.

- Create a new file (**Units**: *Millimeters*) called *"Synchronous Modeling.prt"*.
- Create a **Block** with dimensions *200 mm* x *100 mm* x *50 mm*.
- Use the **Hole** tool to place a simple thru hole of diameter *30 mm* on the top face of the block, centered and at a distance of *75 mm* from one of the *100 mm* edges.
- Apply an **Edge Blend** of radius *5 mm* to the four vertical edges and the top edge of the hole.

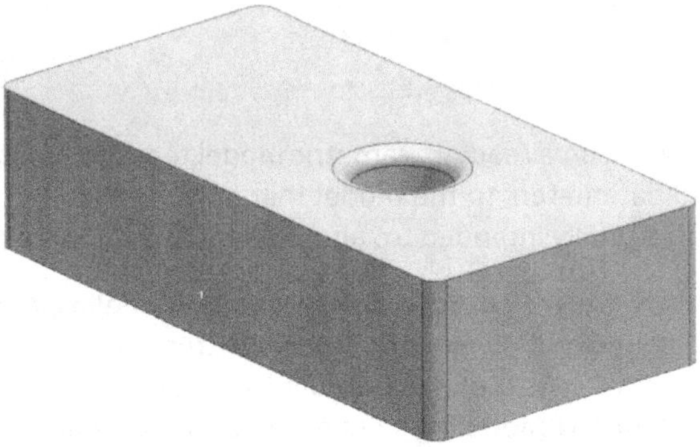

- Select the **Move Face** tool and set the **Motion** to **Distance**. Select the face shown below, and set the **Distance** to **50 mm**. Notice that the blends attached to the motion face follow it.

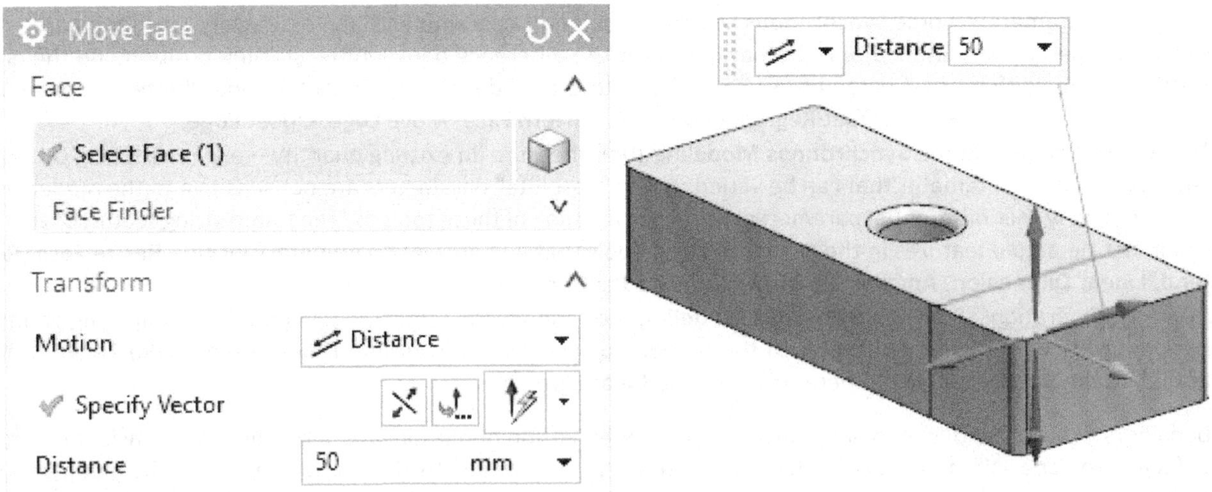

IMPORTANT: *The hole was placed by specifying a dimension of* 75 mm *from the edge of the block, but after applying* ***Move Face****, that dimension does not update based on the change to the face of the block! The hole remains positioned* 75 mm *from the original position of the edge of the block.*

- If you wish for the hole to move along with the face, you must select it too. Choose the cylindrical face of the hole as shown below. Click **OK** to complete the motion.

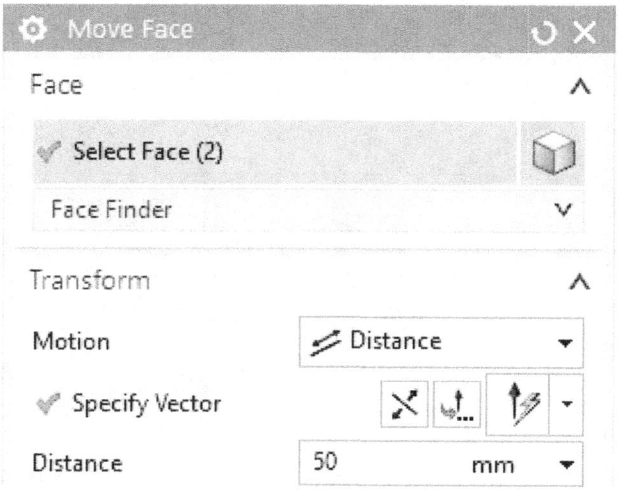

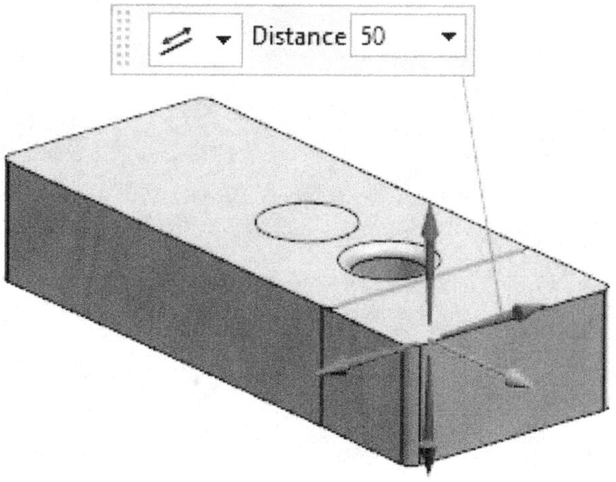

- **Save** your file!

16.2 **Copy Face and Paste Face**

Copy Face allows for many of the same types of motions as **Move Face**, but instead of moving the face itself, a copy is moved. The copy operation alone results in the creation of a sheet body, and in order to ensure that the copied face ends up being a part of the same body, you must also paste it.

The **Paste Face** tool takes a sheet body (not necessarily one that resulted from **Copy Face**!), extends or trims as necessary to find proper intersections with the body in question, and then adds or removes material from the body as necessary to "paste" the selected faces.

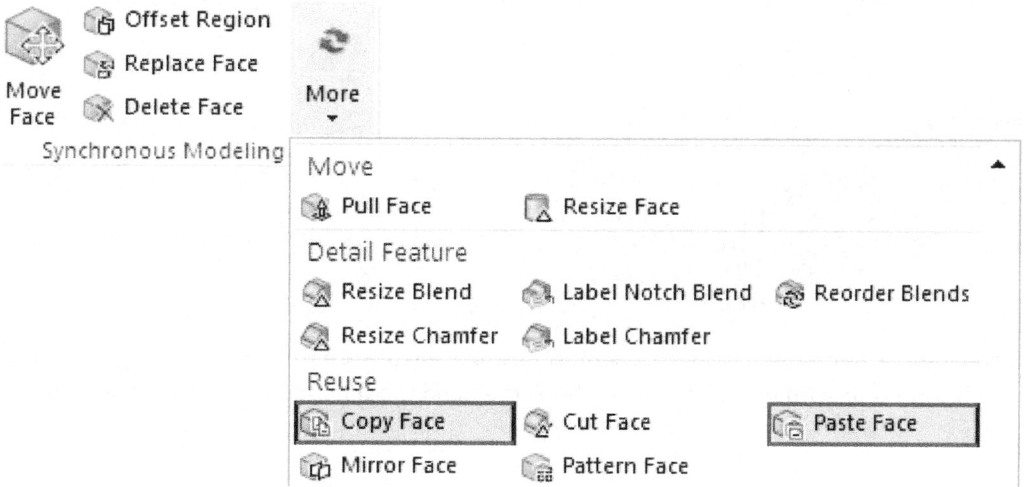

Copy Face actually has the functionality of **Paste Face** built into it for convenience, as you will see below.

- Continue working with *"Synchronous Modeling.prt"*.
- Select the **Copy Face** tool and then choose the cylindrical face of the *Simple Hole* feature. Set the **Motion** to *Distance* and move the copy *100 mm* as shown below.

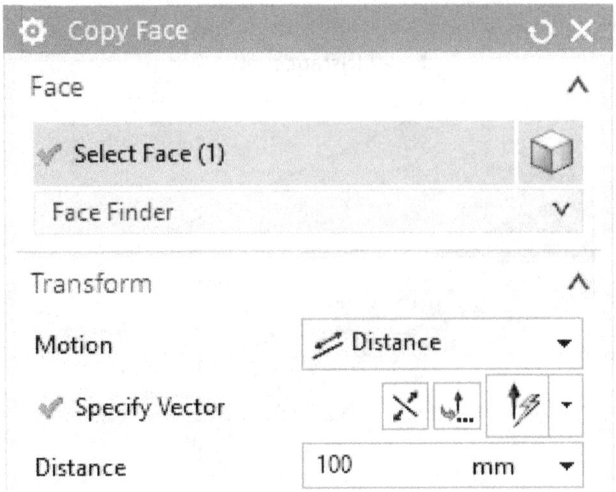

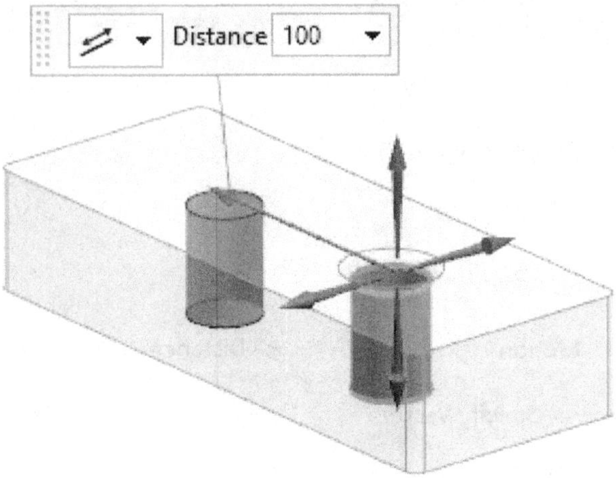

- Make sure the *Paste Copied Faces* box is unchecked.

- Click **OK**. The output of **Copy Face** is then a sheet body that happens to be submerged in the solid body. Note that the incident blend is not automatically included – with **Copy Face** you must explicitly say exactly which faces are to be copied and moved.

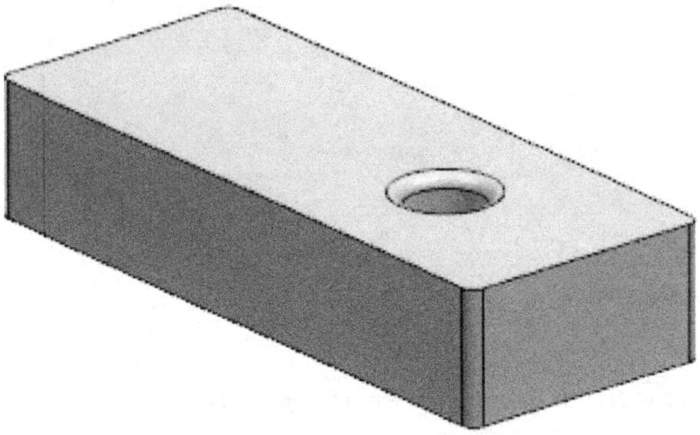

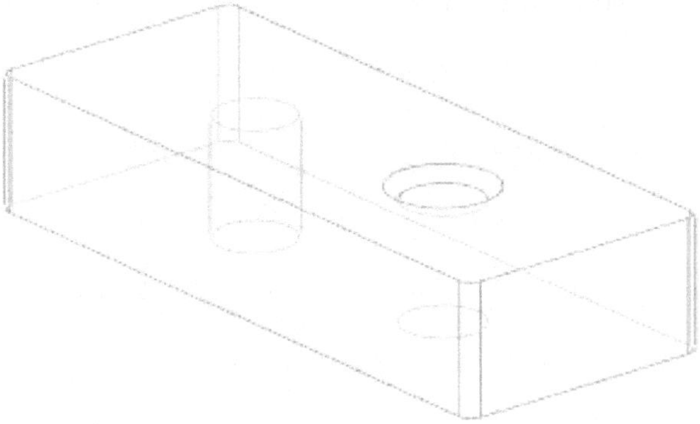

- In order to generate faces on the solid body, we must now use **Paste Face**.
- Select the **Paste Face** tool, with the solid body as the **Target**, and the sheet body of **Copy Face** as the **Tool**.

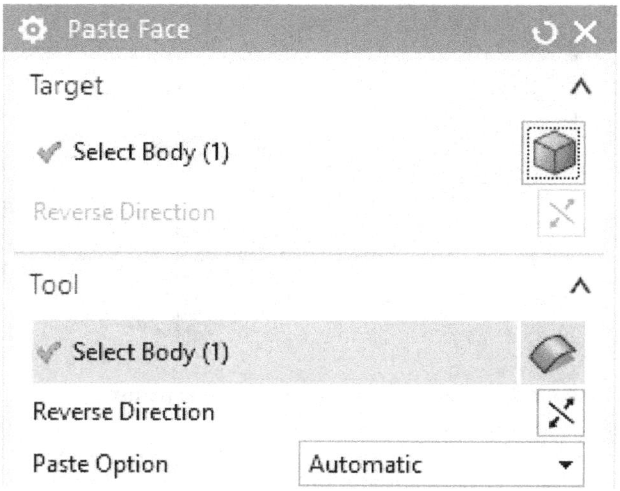

- Note that the cylindrical face is extended until a proper intersection is found with the faces of the block, at which point said faces are trimmed to the boundary of that intersection. Click **OK** to create a new hole!

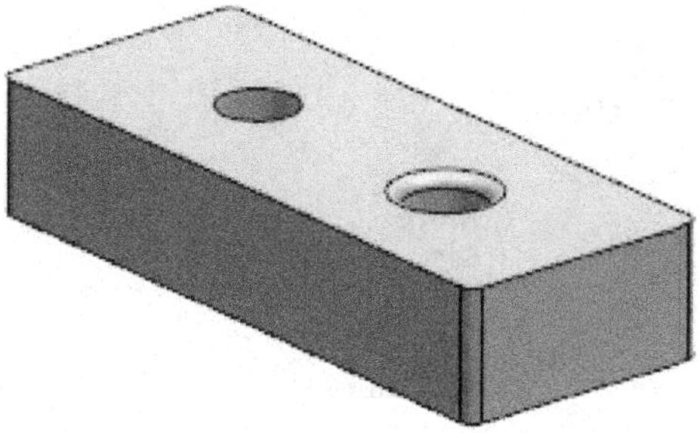

- **Save** your file!

16.3 **Offset Region**

The **Offset Region** command is yet another tool that produces a differential parameter. Unlike **Move Face**, there is only one type of motion allowed by **Offset Region** – as the name suggests, the chosen faces are moved along normal vectors to each point of those faces. As you might expect by now, the motion face and neighbors extend or trim to accommodate the change to the selected face(s).

- Continue working with *"Synchronous Modeling.prt"*.
- Select the **Offset Region** tool and choose the cylindrical face of the *Simple Hole* feature. Make sure the direction is pointing outward and enter a value of *10 mm*.

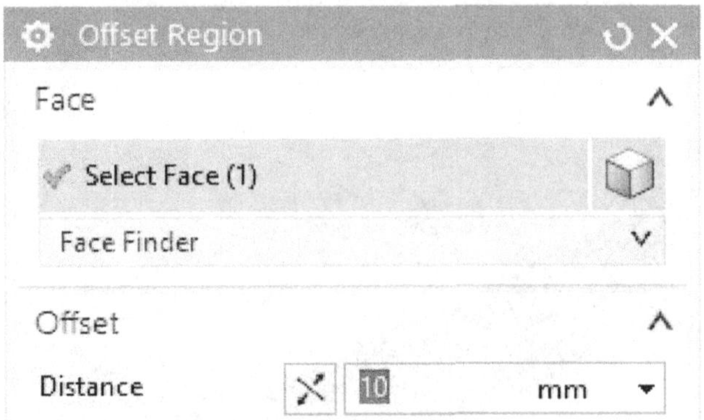

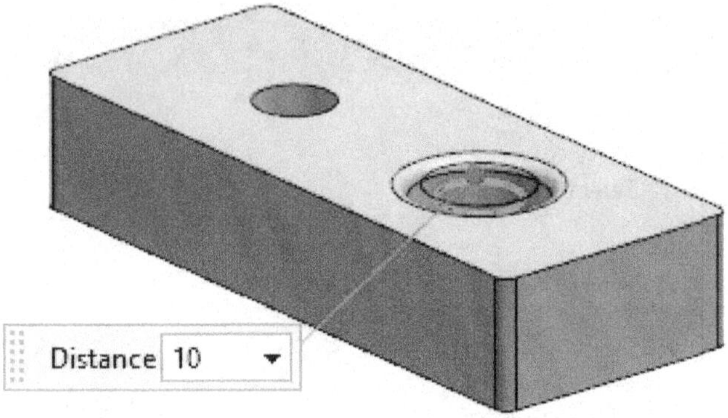

- As with **Move Face**, the incident blend follows the offset face, and the neighboring planar face is trimmed.
- **Save** your file!

16.4 **Resize Face**

The **Resize Face** command is used to change the diameter of cylindrical or spherical faces. The position of the faces remains unaffected when the **Resize Face** feature is created. If a cylindrical face with a draft is selected, **Resize Face** will instead modify the draft angle.

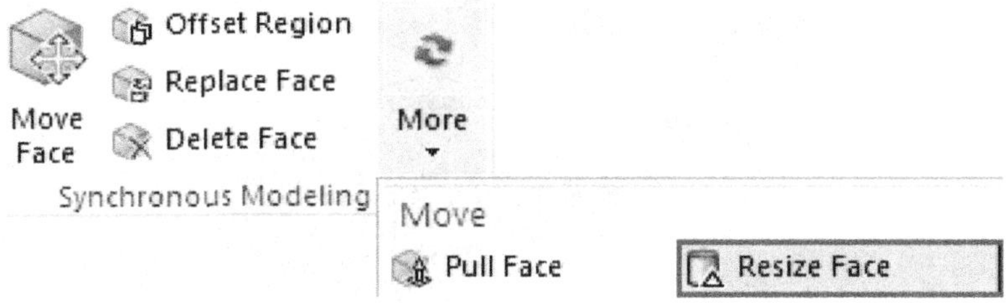

Resize Face is our first example of a **Synchronous Modeling** tool that adds an absolute parameter. In the exercise below, you will see firsthand how this is different from a displacement parameter.

- Continue working with *"Synchronous Modeling.prt"*.
- Open the **Resize Face** tool and select the interior of the copied hole, as shown below.
- **Resize Face** measures the hole and reparametrizes it so that the **Diameter** can be varied.

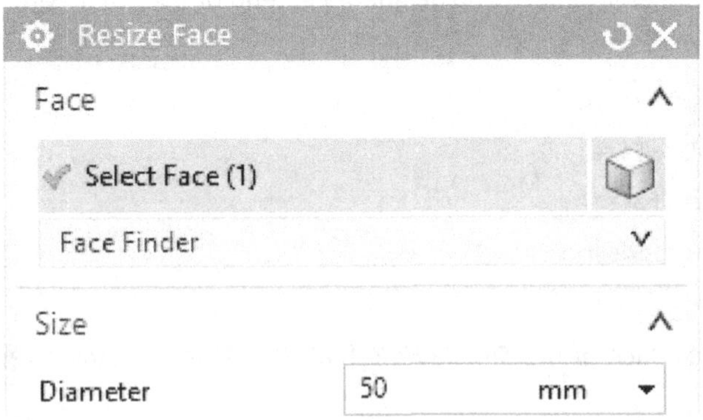

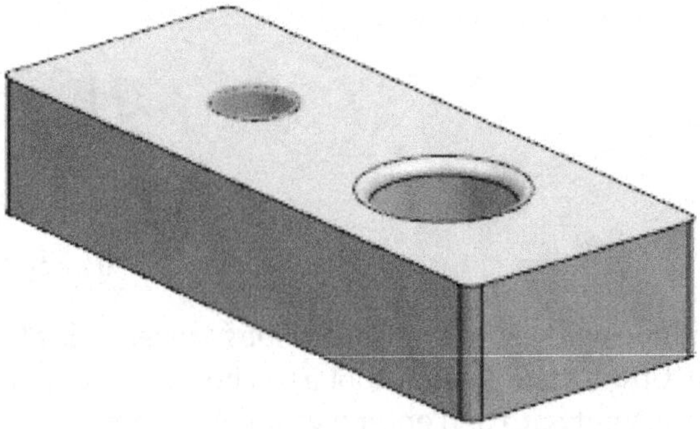

- Enter a **Diameter** of *50 mm* and click **OK**. The result is below!

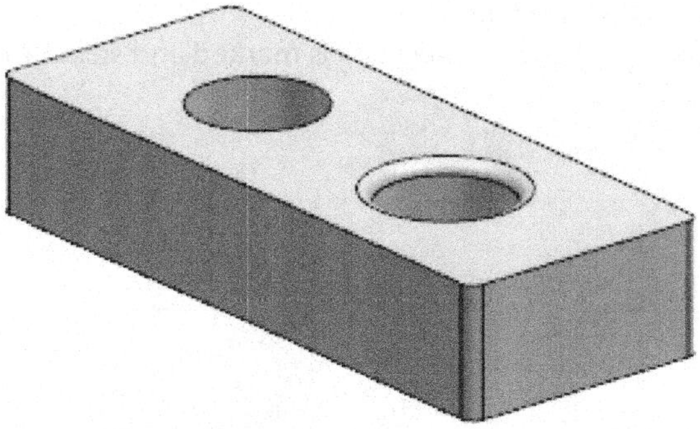

- Absolute parameters that arise from measurements in this way are not affected by the Model History in the same way that displacement parameters are. To illustrate this, edit your *Simple Hole* feature – change the **Diameter** to *10 mm*.

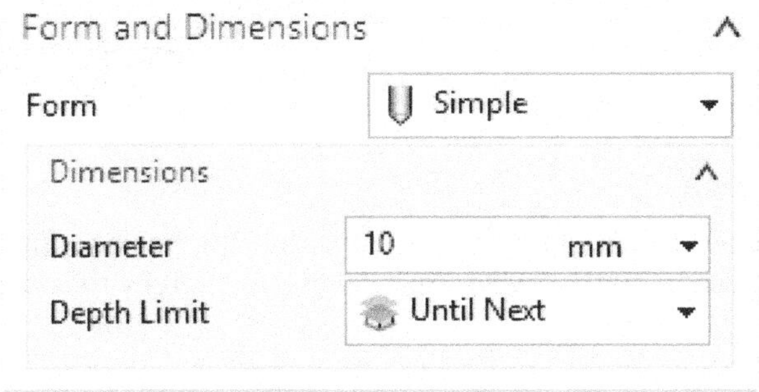

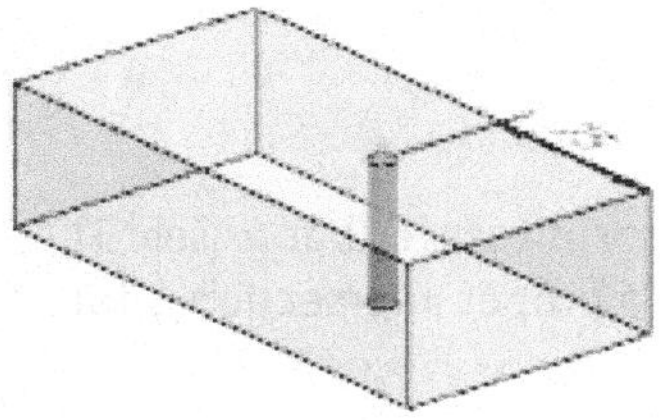

- The *Offset Region* feature is affected – it adds *20* to the diameter of the hole, in this case making the total diameter *30 mm*. The diameter of the hole modified with **Resize Face** is not changed, since the *50 mm* is an absolute parameter.

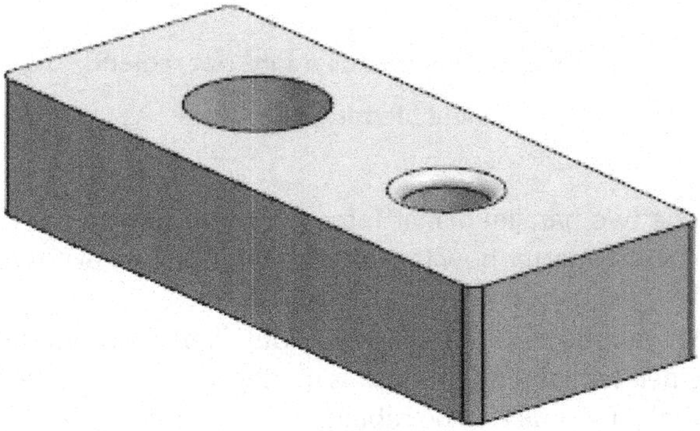

- **Save** your file!

16.5 Delete Face

The **Delete Face** command deletes a face or set of faces on a body and extends or contracts the neighboring faces to adapt to the change. Not all faces can be deleted with **Delete Face** - if the adjacent faces cannot close the void left by the face you intend to delete, you will be faced with an error message. This limitation can sometimes be circumvented by changing the **Cap Face** settings, but first you must thoroughly understand the geometry and the expected behavior.

Delete Face is our first example of a command that works purely by imposing geometric relations between faces on a model.

- Continue working with *"Synchronous Modeling.prt"*.
- Select the **Delete Face** tool. Make sure the *Heal* checkbox is marked and select the face shown below.

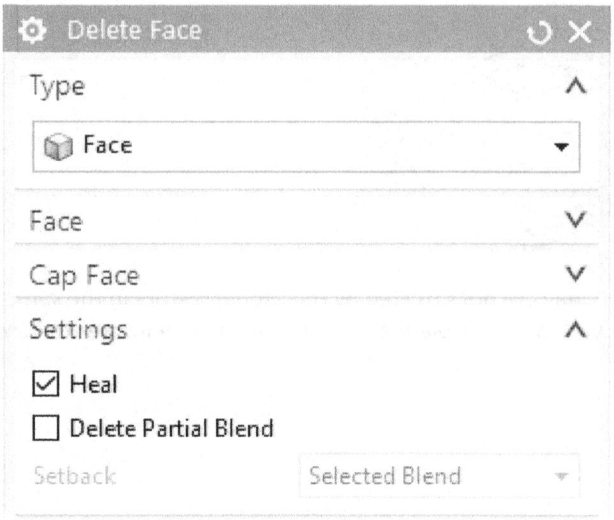

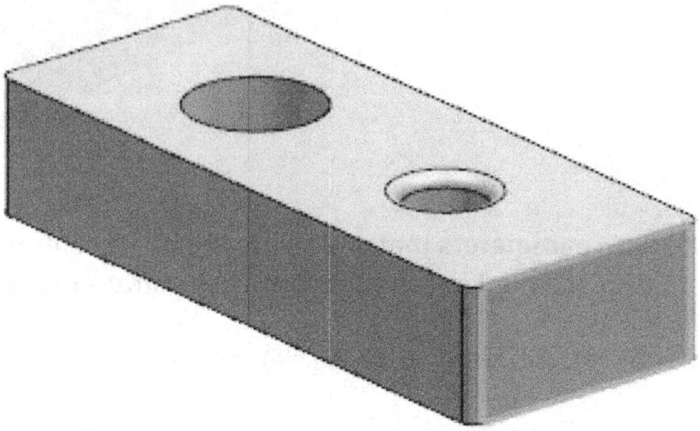

- Note the error message that arises. The *Heal* checkbox being marked means that the neighboring faces will try to extend until a proper intersection is found, at which point they will trim to fill the void left by the deleted face.

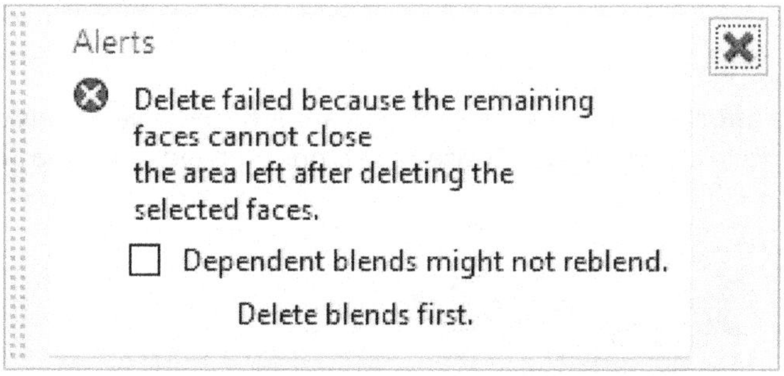

- In this case the neighbors include two parallel planar faces perpendicular to the face to be deleted, and two blends that meet the face with tangency. No matter how far the parallel faces are extended, they will never intersect, so NX is unable to "heal" the void left by deleting the face.
- Uncheck the *Heal* checkbox. When you use **Delete Face** without *Heal* checked, the output of the operation will be a sheet body – the face is simply deleted, and no attempt is made to fill the void left by deleting that face. This can be a convenient way to prepare a solid for a face to be rebuilt or replaced by something geometrically more complex.
- Note that **Delete Face** without *Heal* is the same as **Unsew** followed by **Delete Body** on the unsewn face.

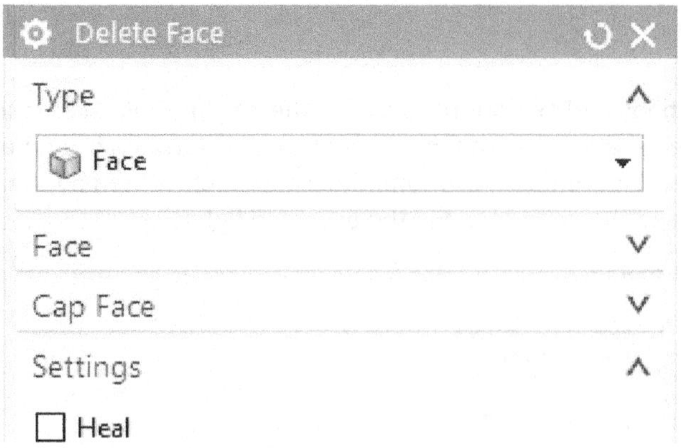

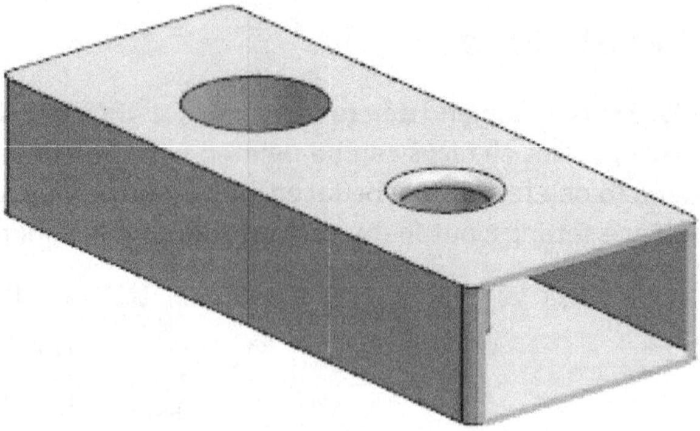

- Deselect the end face of the solid body, and instead select the blend on the top edge of the *Simple Hole* feature. In the case of the blend, as the cylindrical face of the hole and the planar face atop the body are extended, they intersect properly along a circle, so NX is able to heal the void left by deleting the blend. Click **Apply** to delete the blend.

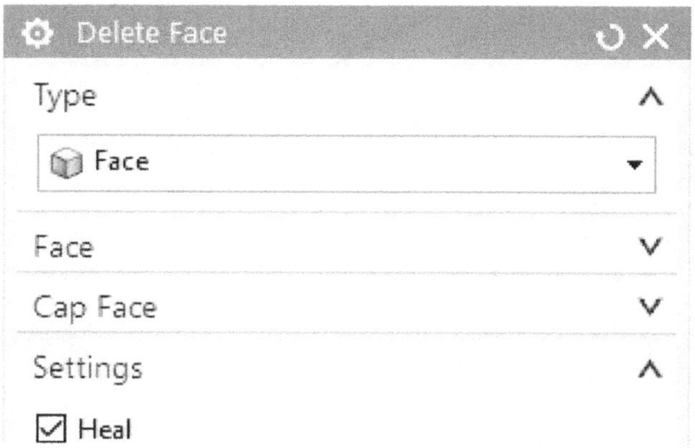

- Next choose the face of the copied hole. **Delete Face** has no problem healing the void in this case, since the planar faces can be extended over the hole without the need to find an intersection with some neighbor.

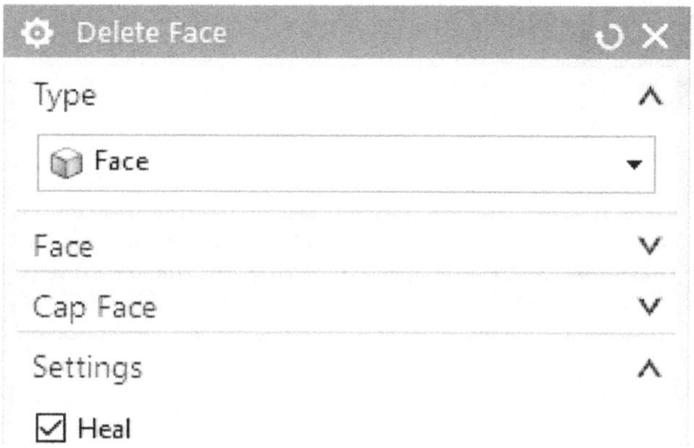

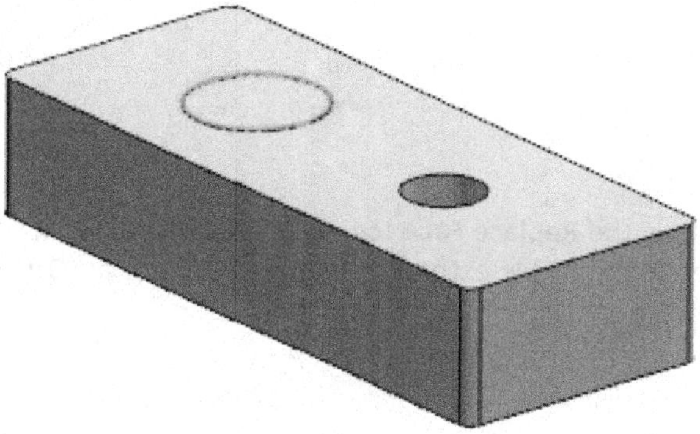

- **Save** your file!

16.6 Replace Face

The **Replace Face** tool requires two inputs. The face to replace should be attached to a solid body, while the replacement face need not be part of a solid. **Replace Face** will delete the face to be replaced, and will extend or trim its neighbors in order to find a proper intersection with the replacement face. When it finds such an intersection, a copy of the replacement face is trimmed to within the intersection, and becomes the new face.

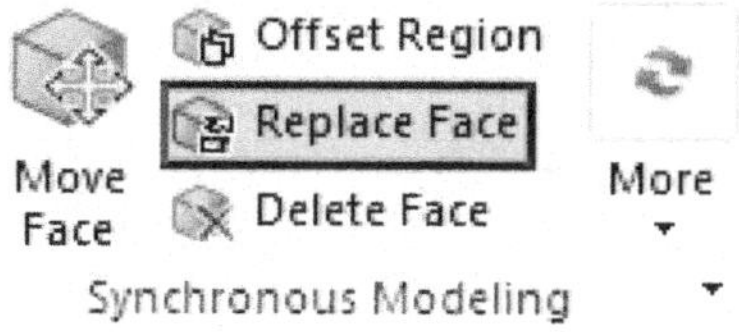

Replace Face is a fairly loaded tool and quite powerful for modeling parts that fit each other with the right amount of clearance.

- Continue working with *"Synchronous Modeling.prt"*.
- Sketch an arc on the face of the block as shown below.

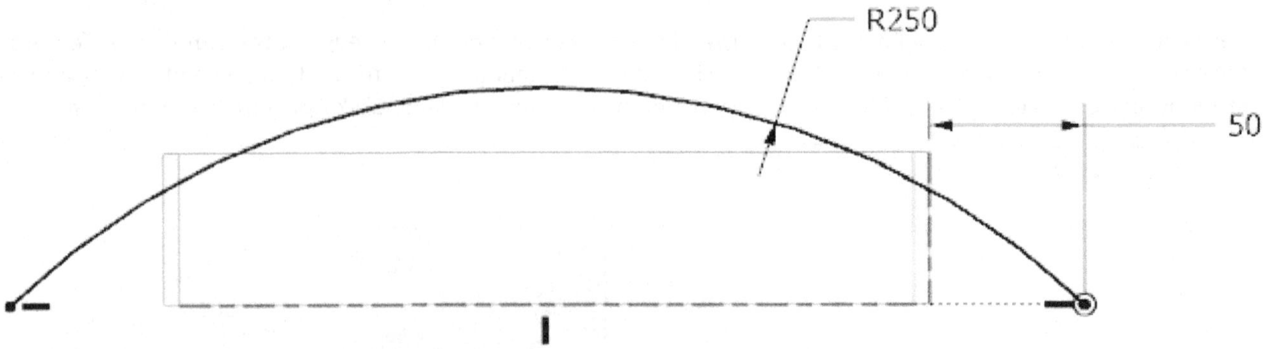

- **Extrude** the arc so that it spans across the block.

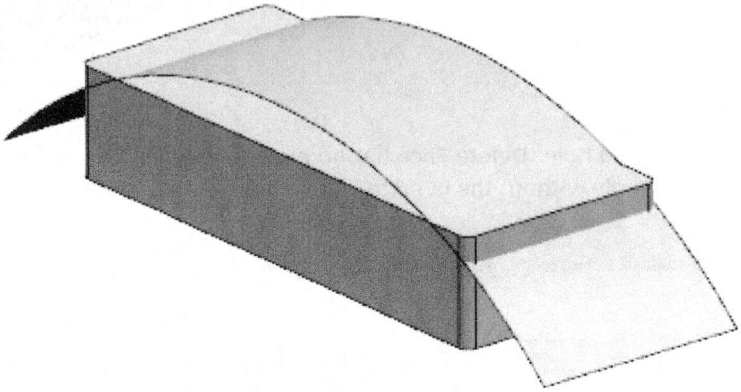

- Select the **Replace Face** tool. For the **Original Face**, select the top face of the block, and for the **Replacement Face**, select the newly extruded sheet body. Enter an **Offset Distance** of *10 mm* and make sure the direction is pointing inward.

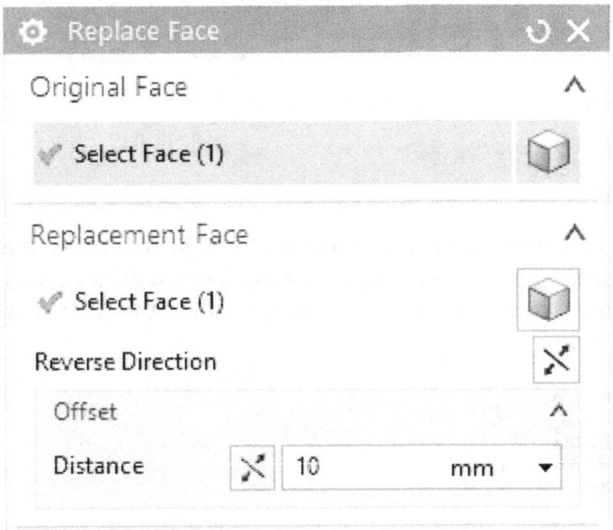

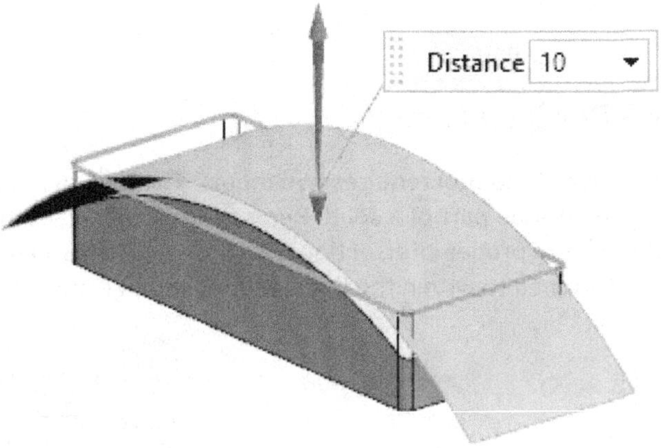

- Here **Replace Face** is exhibiting the same behavior as three other operations – **Move Face**, followed by **Trim Body**, followed by **Offset Region**. The offset functionality allows you to quickly generate exactly the appropriate amount of clearance when designing parts with a specific fit.

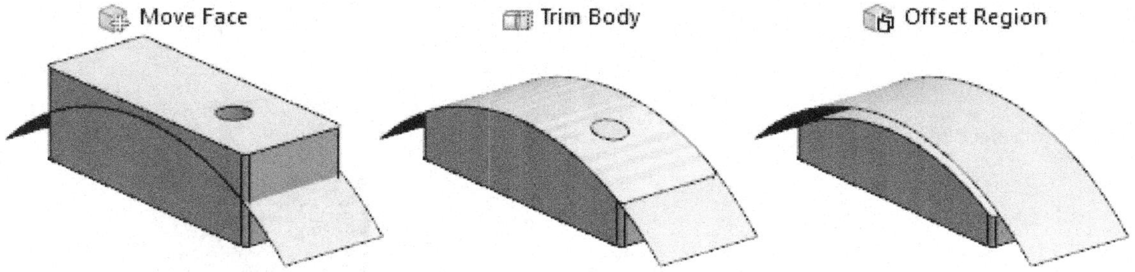

- Use the **Delete Body** command to associatively delete the curved sheet body.

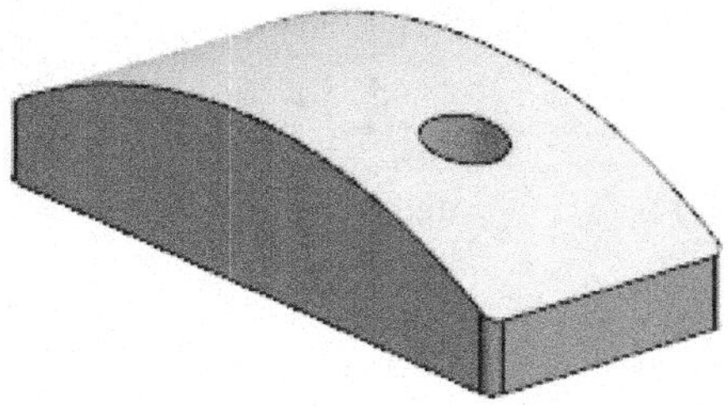

- **Save** your part!

16.7 Move Edge

The **Move Edge** command acts in much the same way that **Move Face** does, except that it applies to edges rather than faces. The same types of motion are available for both **Move Face** and **Move Edge**, but the effects of those transformations are quite different.

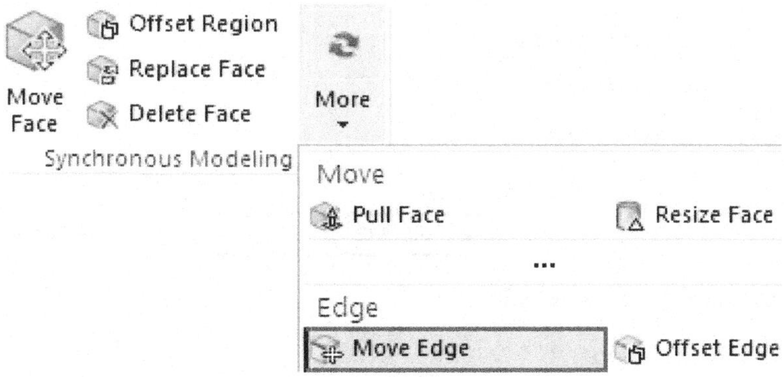

- Create a new part file (**Units**: *Millimeters*) called *"Move Edge.prt"*.
- If your model does not already contain a **Datum CSYS**, place one at the origin. On the X-Y plane, place a sketch of a hexagon with an **Inscribed Radius** of *100 mm* and **Extrude** it *300 mm* in the +Z direction to create a hexagonal solid.
- Select the **Move Edge** tool and select the top six edges of the hexagonal body, as shown below.

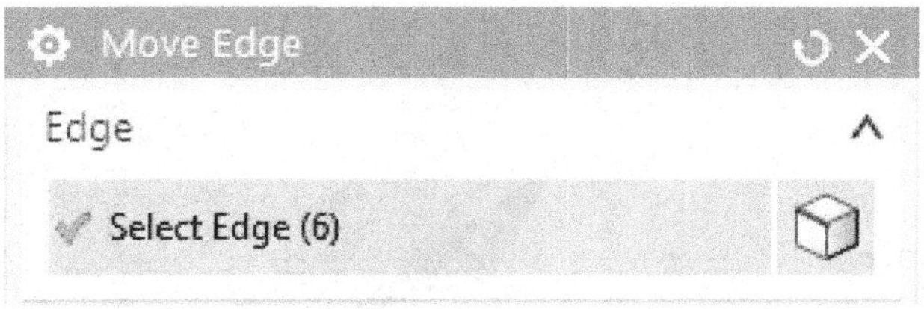

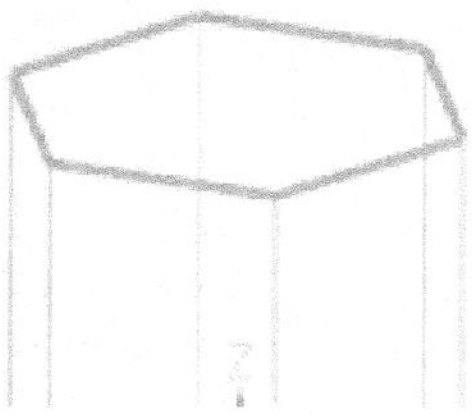

- Set the **Motion** to **Angle**. Select the **Z axis** of the datum CSYS as the axis of rotation, as shown below. Set the **Angle** to **60 deg**.

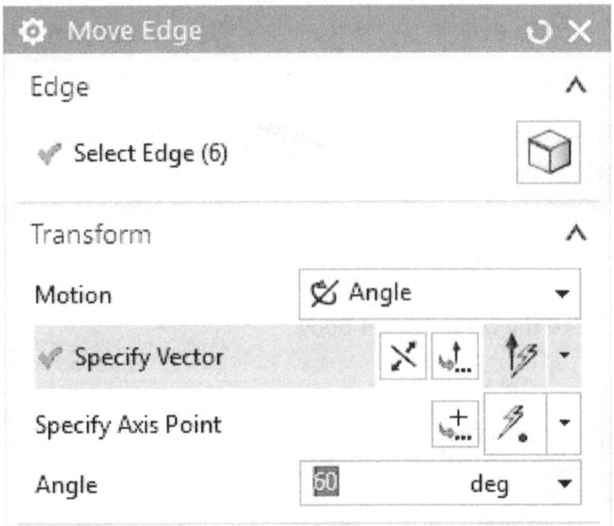

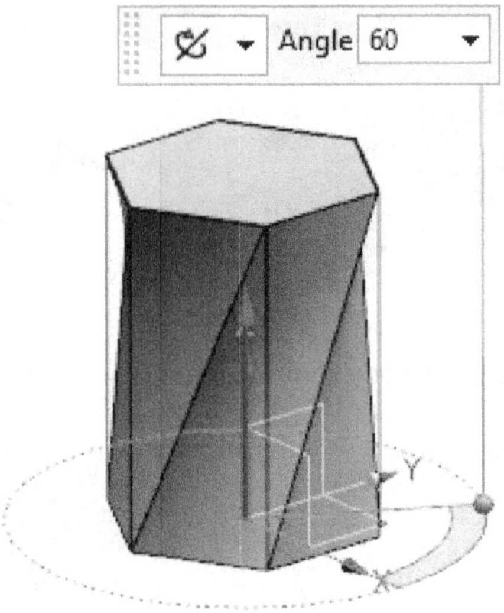

- **Save** your file!

Note that the **Move Edge** command results in the neighboring faces becoming curved. This is something to be aware of when using **Move Edge** vs. other *Synchronous Modeling* tools. It is easy to accidentally produce geometry that is difficult to manufacture!

16.8 Pull Face

The **Pull Face** command is similar to the **Move Face** command, but if it is unable to extend neighboring faces to accommodate the change, it will create new ones to preserve the shape of the motion face. It's very easy to use!

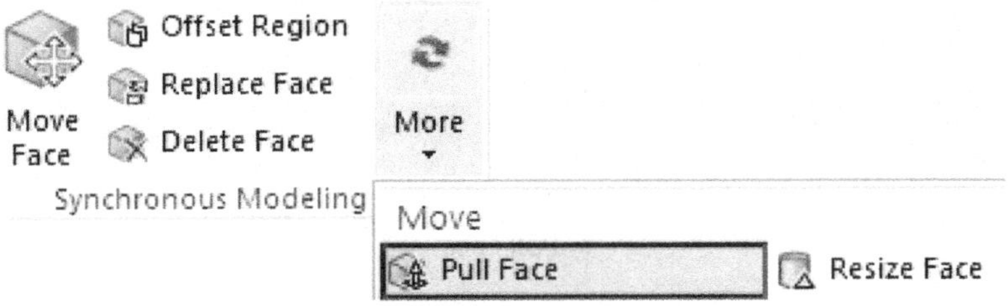

Pull Face can also be used to "extrude" a sheet body.

- Continue working with *"Move Edge.prt"*.
- Select the **Pull Face** tool. Select the face shown below and enter a value of **100 mm**.

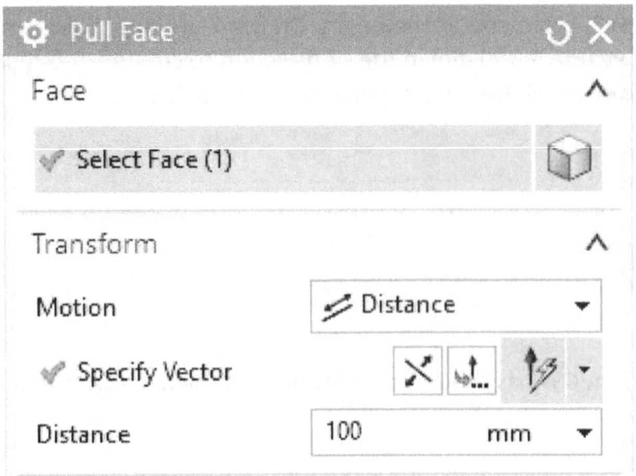

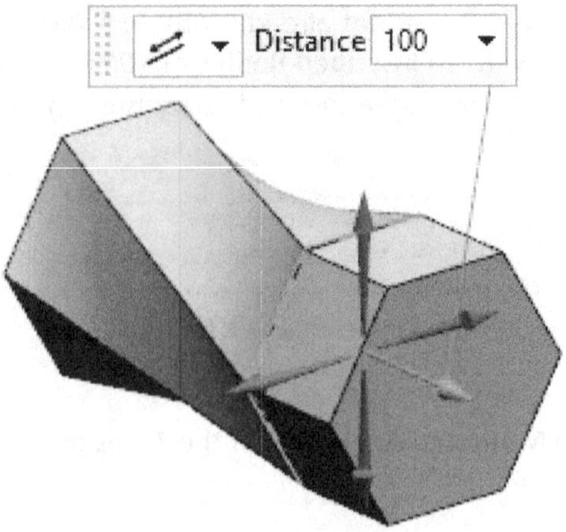

- Note that the selected face retains its shape, as do the neighbors. In order to accommodate the change, new faces are introduced adjacent to the transformed face.
- For comparison, select **Move Face** and apply it to the opposite end of the body. This operation results in the surface area of the motion face growing, and extension of the neighboring faces to maintain proper intersections with the motion face.

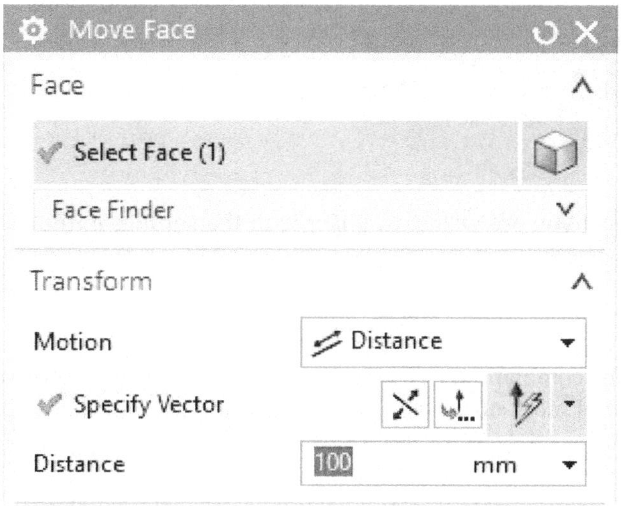

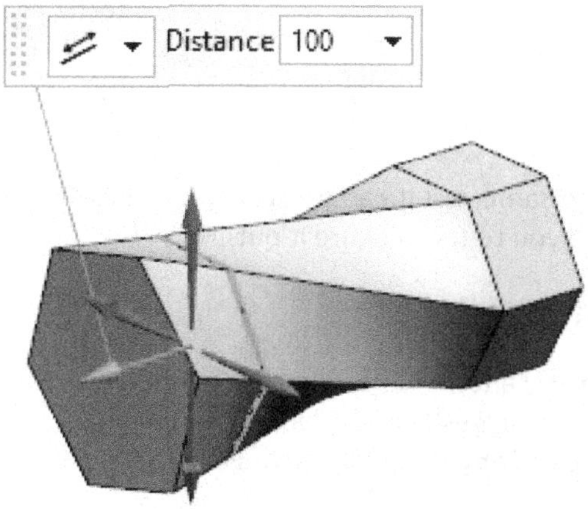

- **Move Face** and **Pull Face** are both incredibly useful – it is important to know the difference so that you can decide which is best for your model!

Another great elementary example that illustrates a fundamental difference between **Pull Face** and **Move Face** is this block with top face subdivided into two faces.

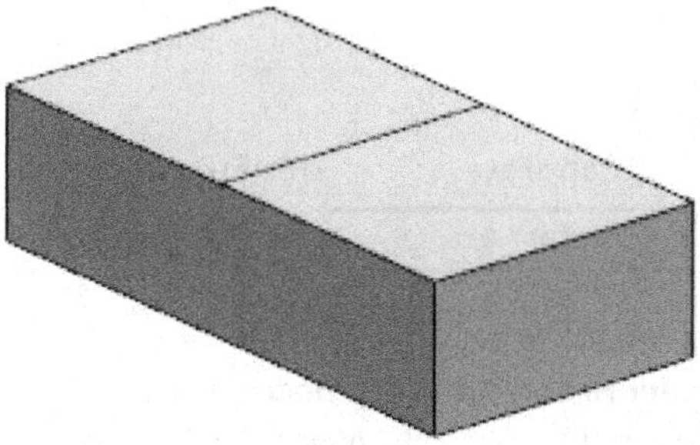

Move Face, with its default settings, is unable to move one of the two faces, since it seeks to maintain proper intersections with the neighbors.

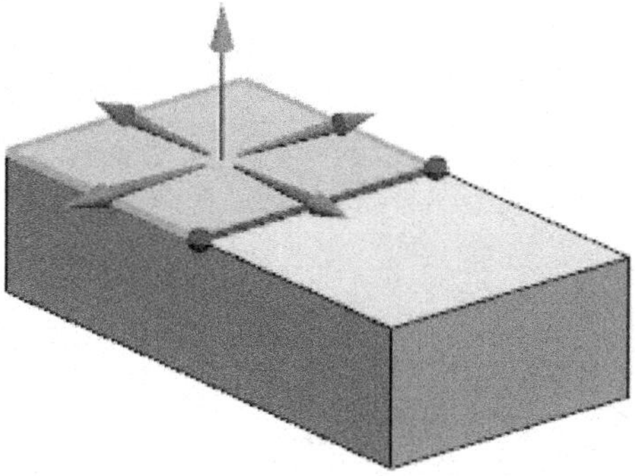

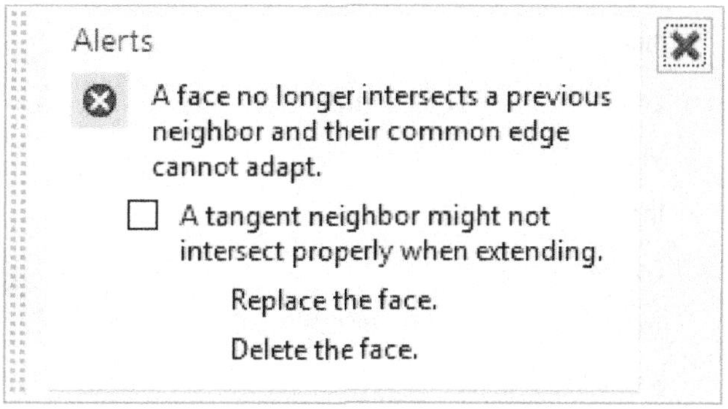

Pull Face, however, has no qualms about introducing new neighbors to get the job done.

In fact, the same result can be achieved with **Move Face**, if you are willing to tinker with the settings in the dialog. We encourage you to try to figure it out!

16.9 **Mirror Face**

Mirror Face is another example of a **Synchronous Modeling** command that works purely by imposing geometric relations. As the name suggests, it produces a mirror image of a set of faces on a body, and combines them with that body by the same logic as **Copy Face** and **Paste Face**.

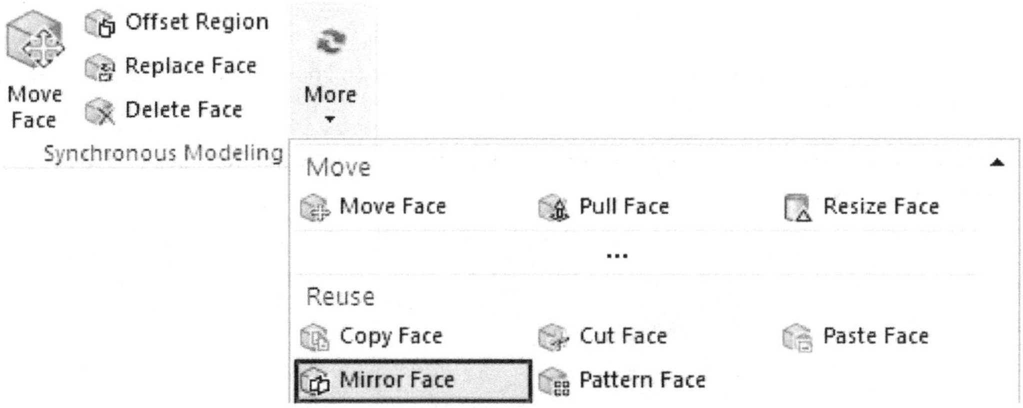

- Continue working with *"Synchronous Modeling.prt"*.
- Until now, we have retained the Model History, it is not necessary to do so in order to use **Synchronous Modeling** tools. Let's get rid of the Model History right now, with **Menu / Edit / Feature / Remove Parameters** (this command is buried rather deep in the menu).

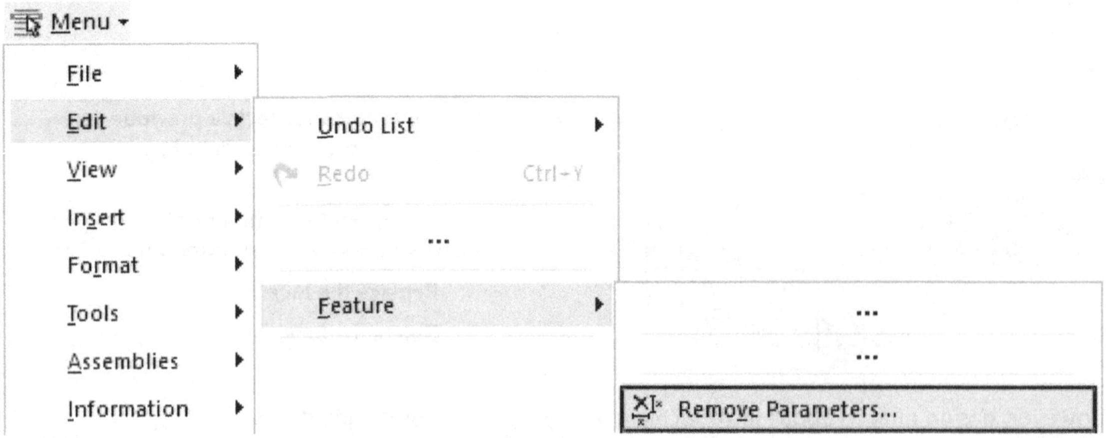

- Select the body and click **OK**. NX will ask if you really want to do this. Tell it **Yes**!

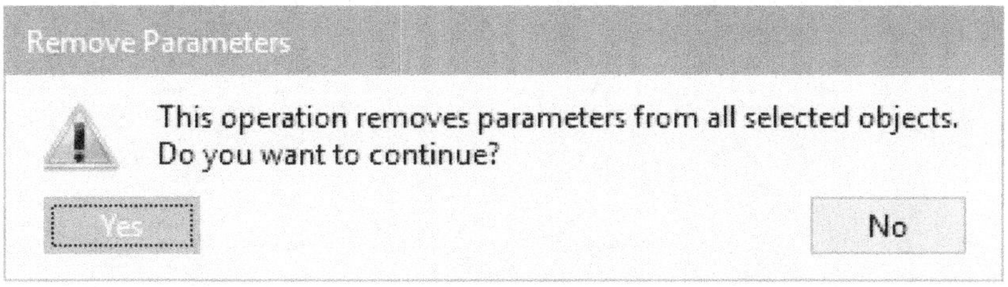

- Not only does **Remove Parameters** throw away all the expression driving your model, it also discards the Model History – the body you selected simply becomes *Body (0)* in the Model History.

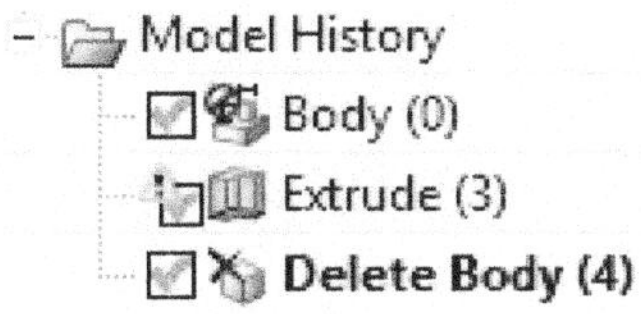

- There is no longer any dependency between the body and *Extrude (3)* (the sheet body used in the **Replace Face** operation), so you can safely delete both *Extrude (3)* and *Delete Body (4)*.
- Now that we've converted our model into a dumb solid, let's use **Mirror Face** to illustrate how well **Synchronous Modeling** tools work, even on non-parametric models!
- Open the **Mirror Face** tool and select the face of the thru hole.

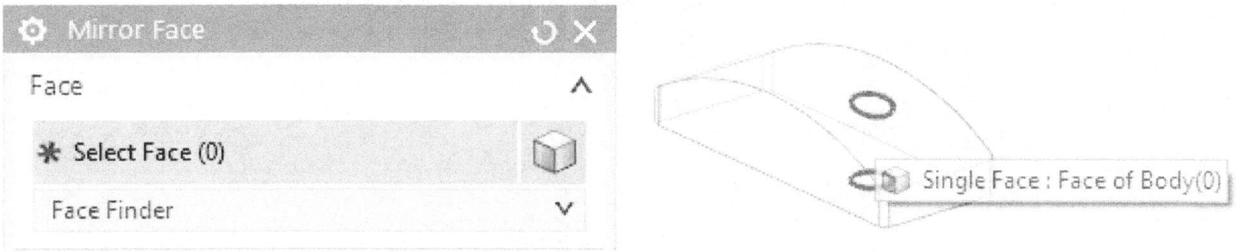

- Within the **Mirror Plane** section of the dialog, set the **Plane** drop-down to *New Plane*, and set the **Plane Type** drop-down to *Bisector*.

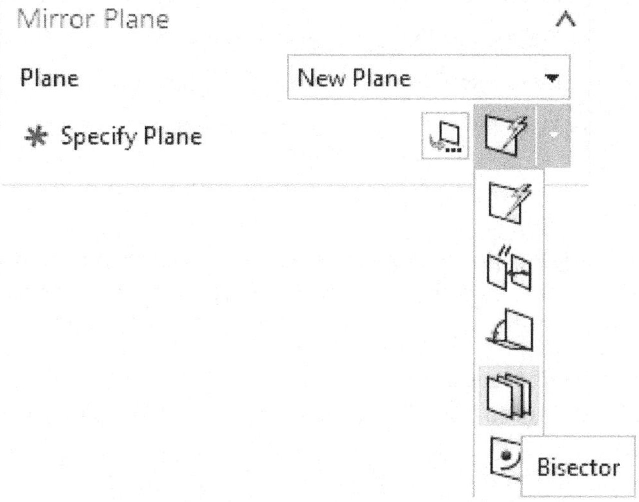

- Choose the pair of opposite faces of the model to specify a midplane, and click **OK** to mirror the hole.

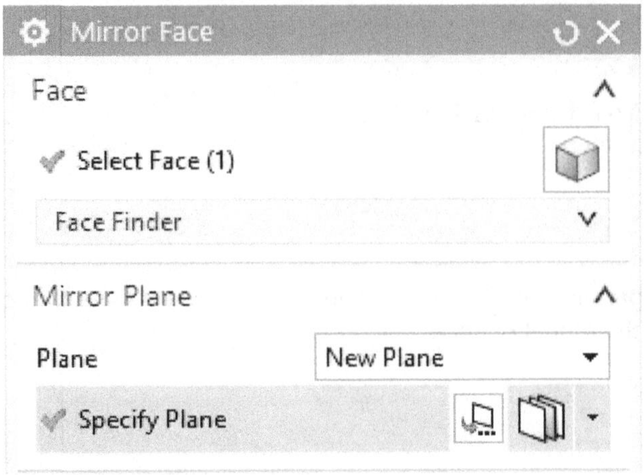

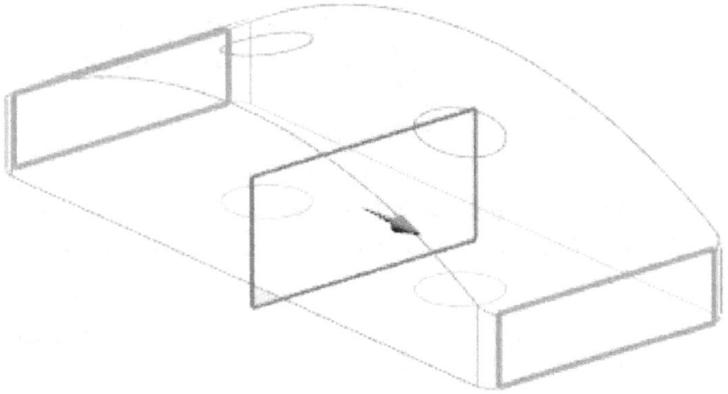

- **Save** your file!

16.10 The Relate Group

The **Relate** group within the **Synchronous Modeling** group contains a suite of tools that can be used to "reparametrize" non-parametric bodies, where there are none. The tools in the **Relate** group break down into two categories – *dimensions*, and *constraints*. We choose this terminology because of the analogy with the corresponding relations available in the sketcher. These tools are in some sense the three-dimensional versions of those sketch tools.

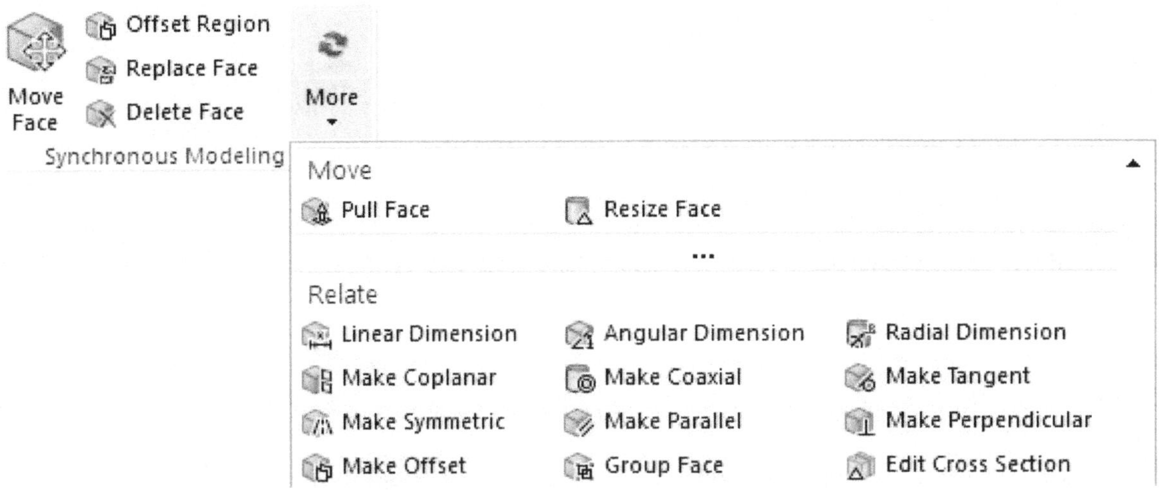

16.10.1 Dimensions

The dimension tools within the **Relate** group provide further examples of tools that add absolute parameters to your model. They can also be used on non-parametric, as well as parametric models, to override existing dimensions, if the design intent is buried in a complicated model history.

- Continue working with *"Synchronous Modeling.prt"*.
- **Extrude** the bottom edge of the mirrored thru hole from *0 mm* to *100 mm*, with **Boolean** set to *Unite*.

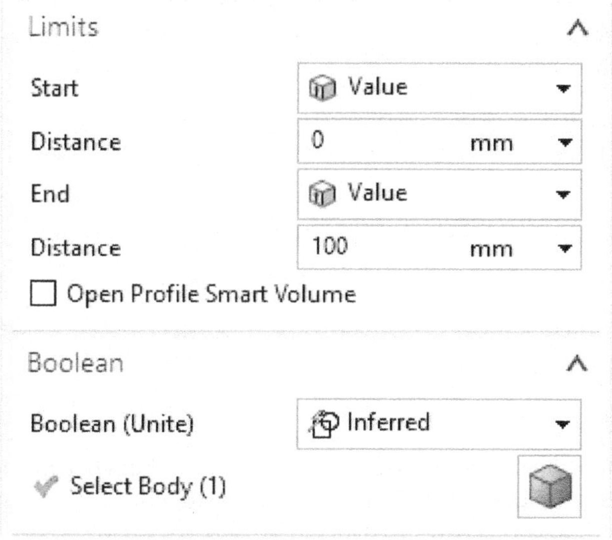

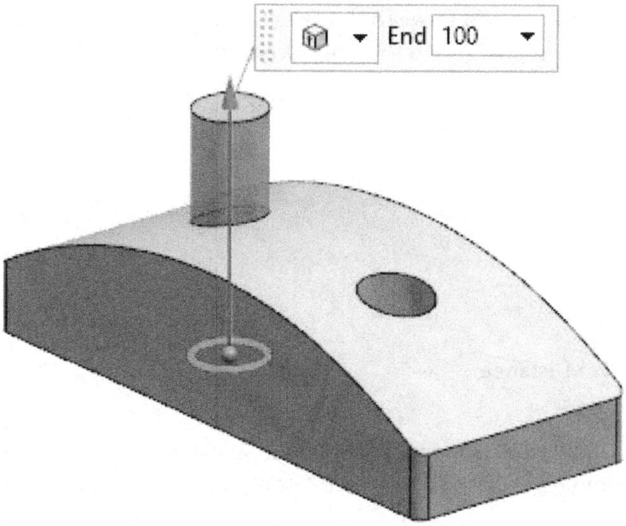

- Apply a *5 mm* **Edge Blend** to the top edge of the hole, and the top face of the cylindrical extrusion.

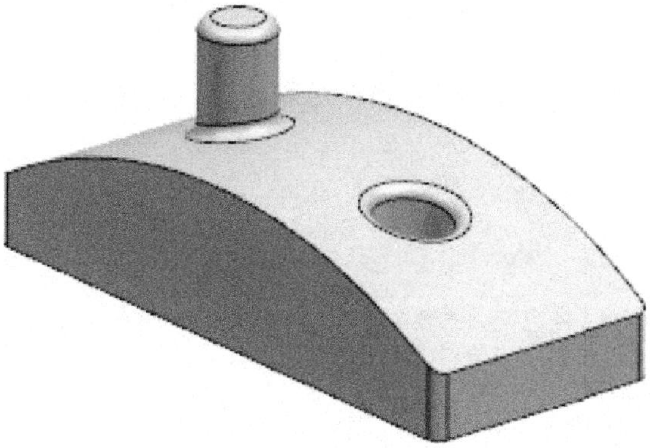

- Select the **Linear Dimension** tool from the **Relate** group. For the **Origin**, choose the bottom right edge of the body, as shown below, and for the **Measurement Object**, choose the centerline of the thru hole.

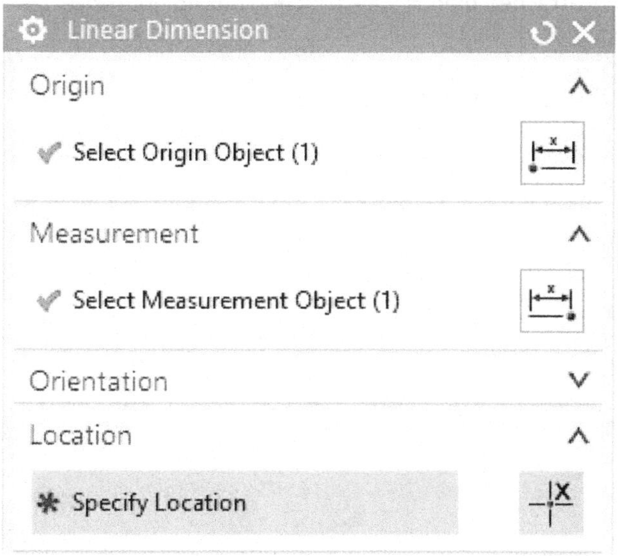

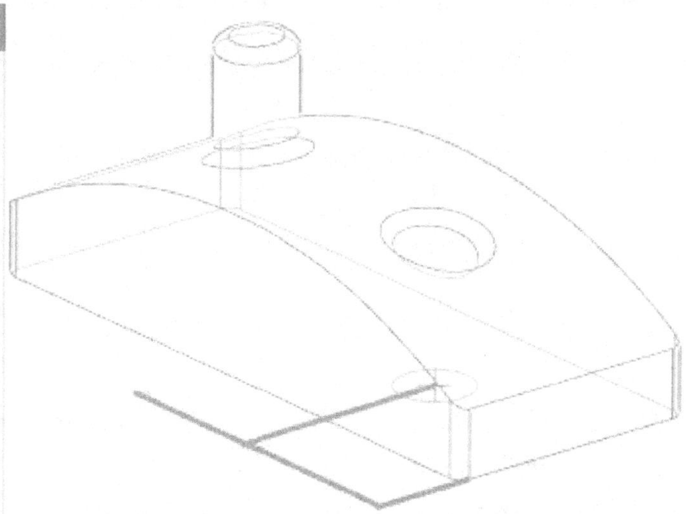

- The **Face to Move** section of the **Linear Dimension** dialog is critical for establishing the design intent of your dimension – you can add faces to the selection that NX has inferred here to ensure that other faces of the model maintain their relative position to the **Measurement Object**. Enter a **Distance** of *50 mm*.

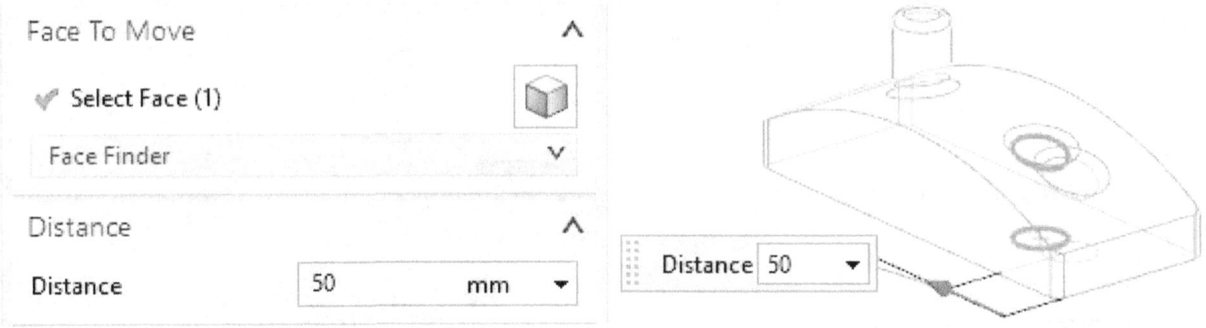

- The **Radial Dimension** command works in very much the same way as the **Resize Face** command, although it will not let you choose conical faces – only purely cylindrical faces. Set the **Diameter** of the thru hole to *10 mm*.

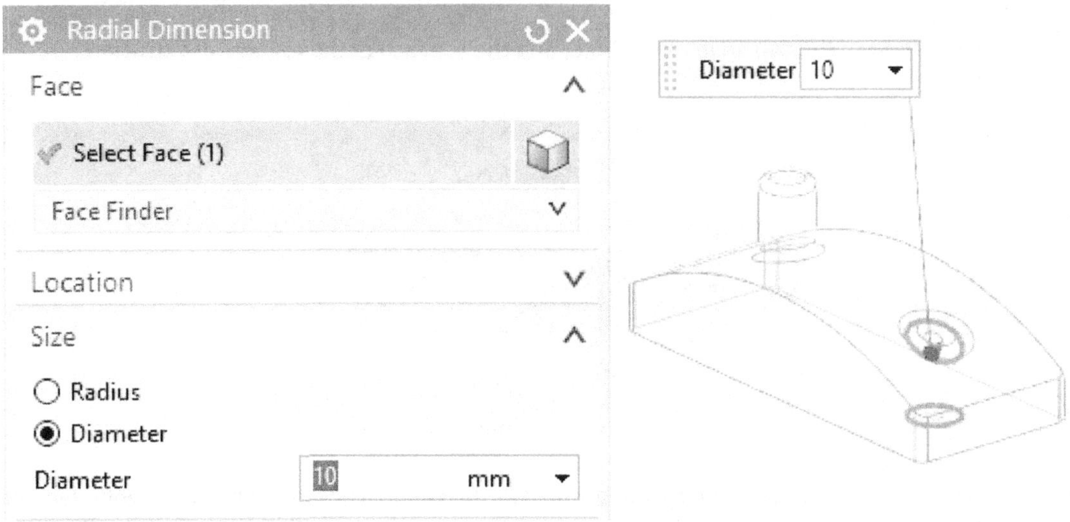

- Use **Radial Dimension** again to resize the boss. Notice that the attached blends stay attached but do not resize.

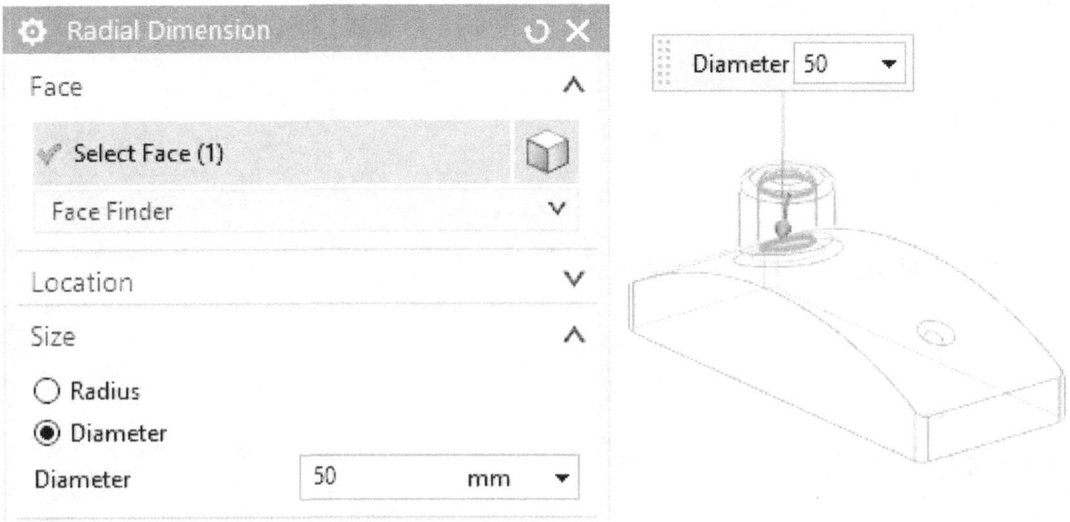

- Use **Radial Dimension** once more to resize the blend on the corner of the model.

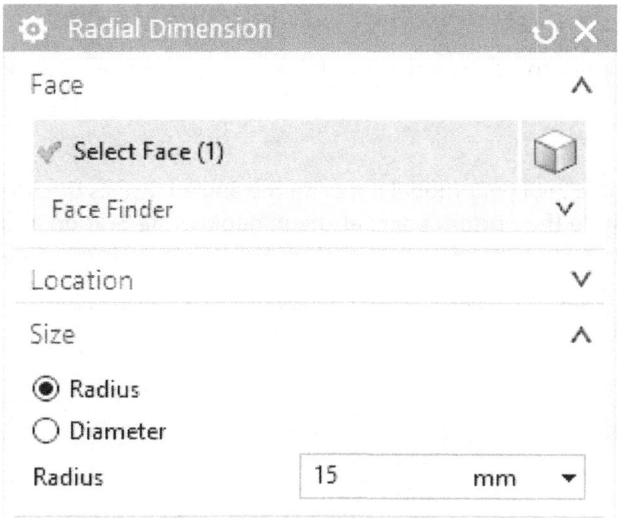

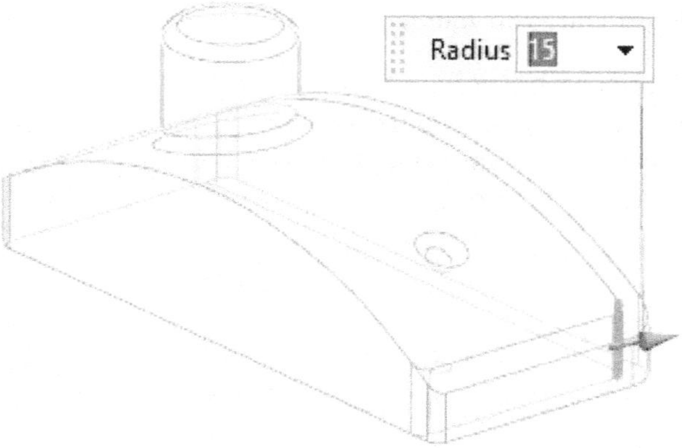

- The **Angular Dimension** tool (much like its counterpart in the sketcher) is straightforward. Choose the bottom face of the model as the **Origin**, and the face shown below as the **Measurement Object**, and change the angle to *45 degrees*.

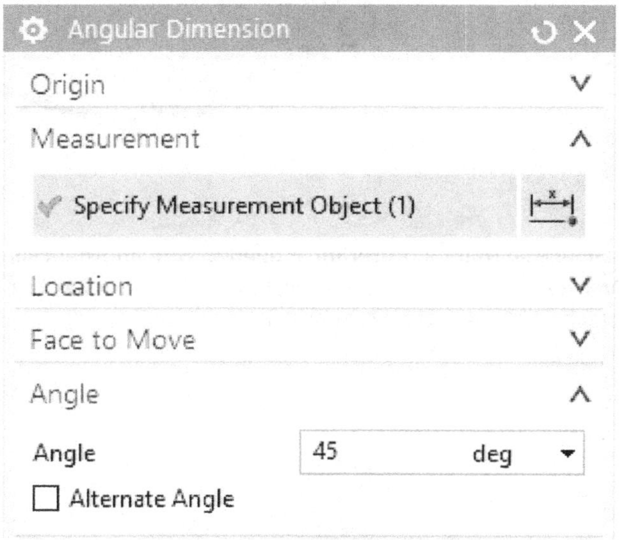

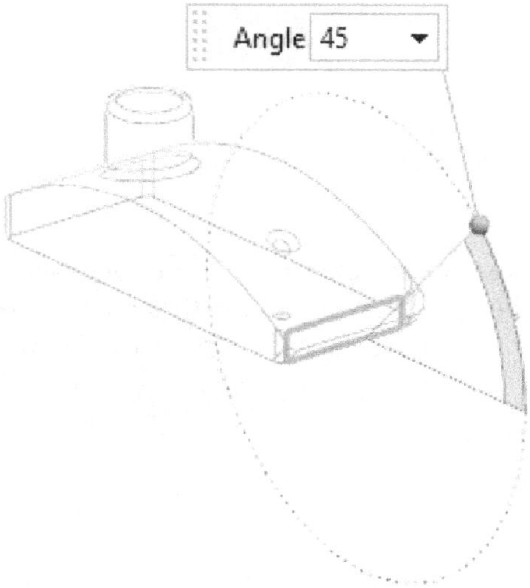

- Your model will now appear as shown below!

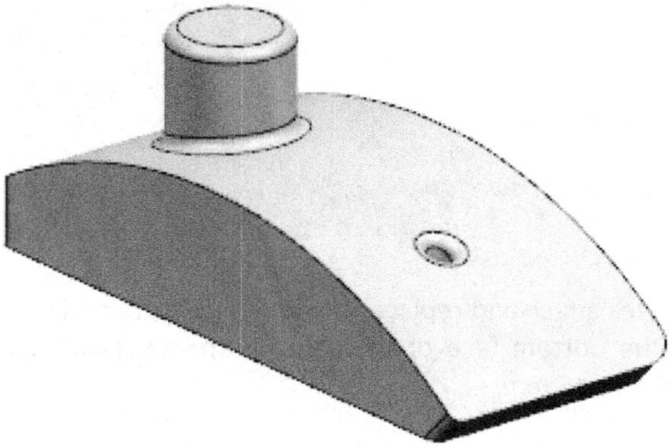

16.10.2 Constraints

The constraint tools within the **Relate** group add no parameters to your model, but instead impose purely geometric relations to faces on your model.

- Continue working with *"Synchronous Modeling.prt"*.
- Select the **Make Symmetric** command, and choose the flat face of your model opposite the angled face as the **Motion Face**. For the **Symmetry Plane**, choose a midplane normal to the bottom edge, at the midpoint. The **Stationary Face** is the angled face.

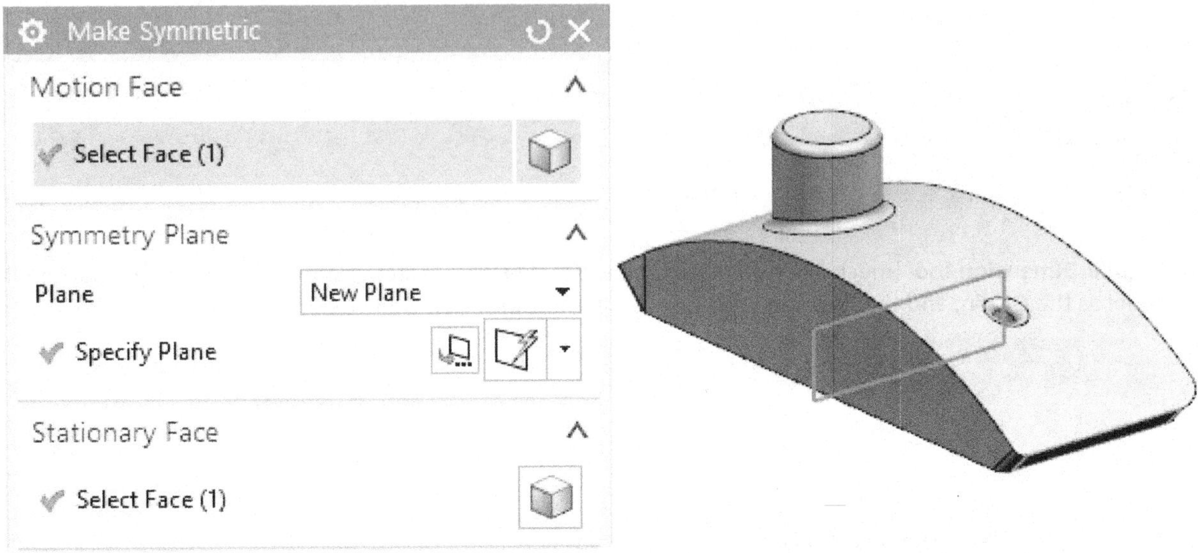

- The **Make Coaxial** command relocates one cylindrical face so that its centerline is aligned with another cylindrical axis. Select the cylindrical face of the thru hole as the **Motion Face**, and the cylindrical face of the boss as the **Stationary Face**. Note that it is not necessary to include the attached blend in the **Motion Group**.

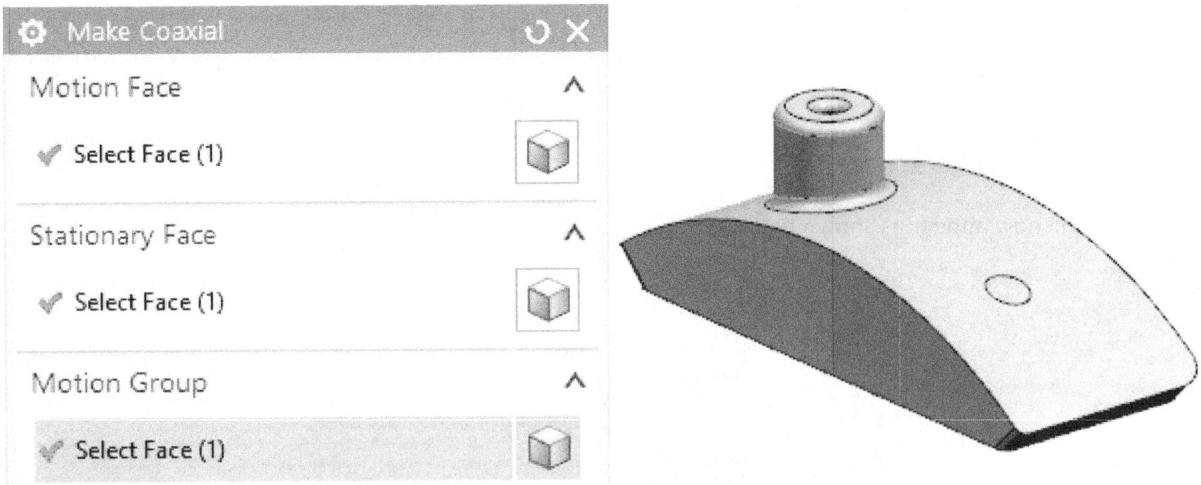

- The **Make Offset** command deletes a face and replaces it with a face obtained by offsetting another. Select the **Make Offset** command and choose the bottom face of the solid as the **Motion Face**, and the curved top face as the **Stationary Face**. Enter a **Distance** of *15 mm*.

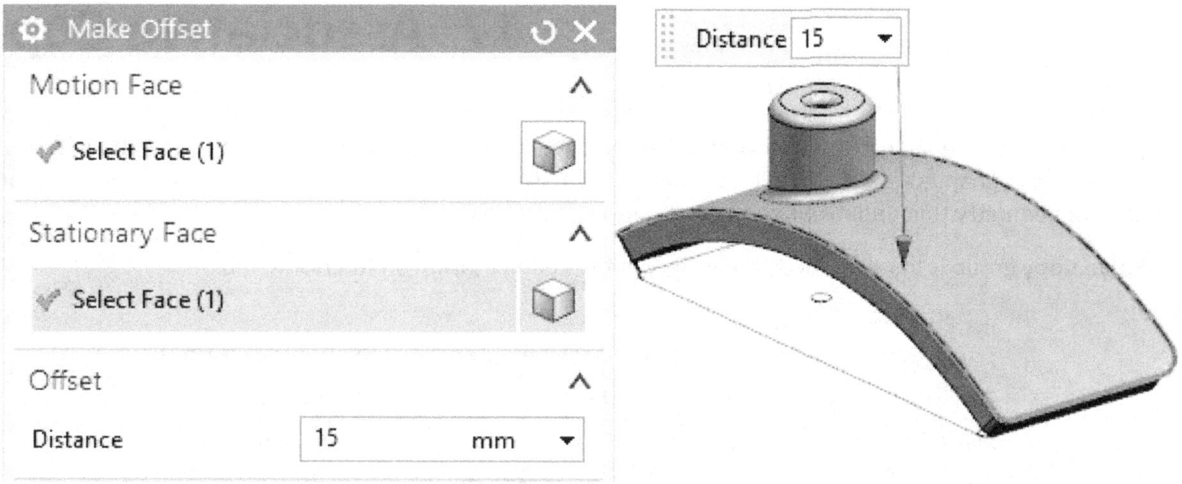

- Hopefully these examples have given you a taste of what to expect from tools in the **Relate** group. We strongly encourage you to try others.
- **Save** your file!

17 Associative Copies

In this chapter, we delve deeper into the **Associative Copy** group. You have already worked with the **Pattern Feature** and **Mirror Feature** tools in Chapters 9.9 and 9.10 respectively. The **Associative Copy** group has a variety of tools that produce "smart" copies of geometry that update when the original geometry changes.

The **Associative Copy** group is found in the **More** gallery in the Feature group on the **Home** tab.

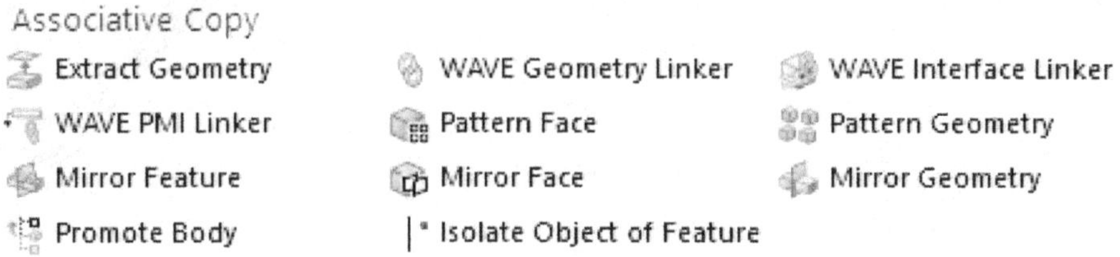

Of these tools, patterning tools have by far the most depth, and so we will focus our discussion on various options in the patterning tools dialogs. Notice that there are three tools named "Pattern" – namely, **Pattern Face**, **Pattern Feature**, and **Pattern Geometry**. Also, notice that there are three tools named "Mirror" – **Mirror Face**, **Mirror Feature**, and **Mirror Geometry**. We will begin with an exercise showing the difference between these three tools, and the advantages and disadvantages of each.

17.1 Pattern Face vs. Feature vs. Geometry

These three tools might sound similar, but there are subtle differences!

- Create a new file (**Units**: *Inches*) called *"Pattern Face vs Feature vs Geometry.prt"*.
- Create a **Block** with dimensions *4 in* x *2 in* x *0.5 in*.
- Place a counterbored hole *0.667 in* from the left edge, and centered on the face.
    - **Counterbore Diameter**: *1 in*
    - **Counterbore Depth**: *0.25 in*
    - **Thru Diameter**: *0.5 in*

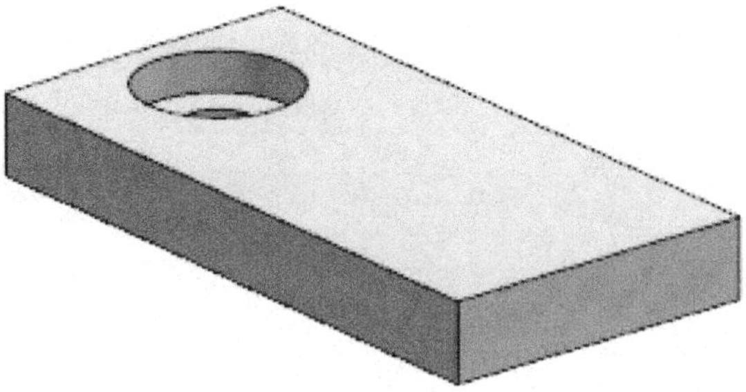

- Select the **Pattern Geometry** tool. Note that **Pattern Geometry** is unable to select the *Counterbored Hole* feature – when you put your cursor over the hole, the entire body highlights. **Pattern Geometry** is only able to select entire bodies, and not features on bodies.

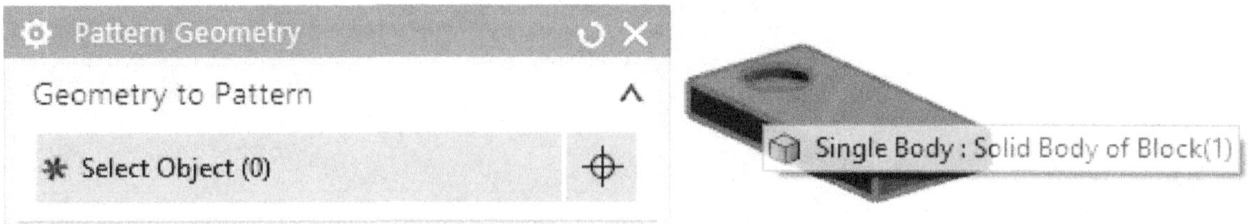

- Set the **Pattern Layout** to *Linear* and make a copy of the body *2.5 in* away in the YC direction.

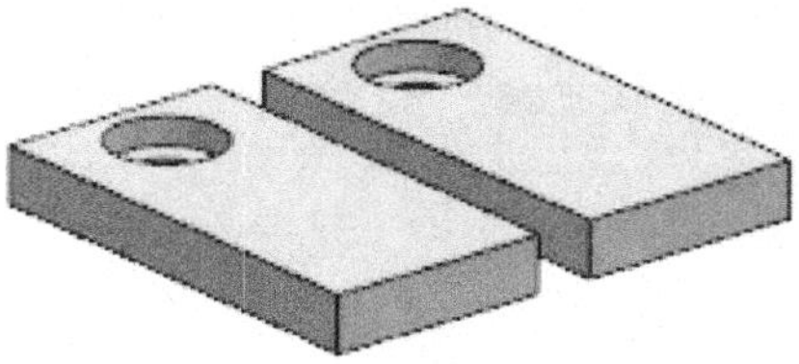

- Next, select the **Pattern Feature** tool.
- **Pattern Feature** is of course able to distinguish the *Counterbored Hole* feature from the *Block*.

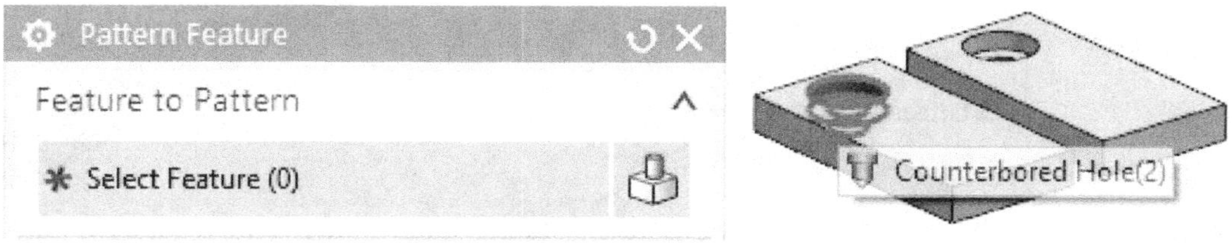

- However, on the body created by **Pattern Geometry**, the **Pattern Feature** tool is unable to select the hole. The hole on the copy is not – strictly speaking – a feature, and so it cannot be selected. When you put your cursor over any point on the copied body, the entire body highlights, since it is regarded as a feature – namely, *Pattern Geometry (3)*.

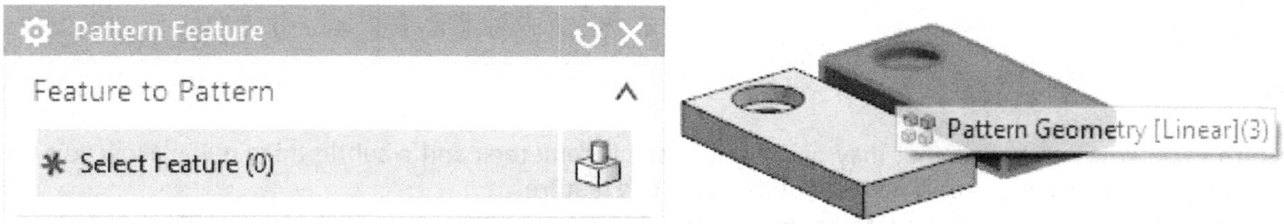

- Set the **Pattern Layout** to *Linear* and produce a copy of the *Counterbored Hole 1.333 in* away from the original.

- Next, select the **Pattern Face** tool.
- When you put your cursor over the *Counterbored Hole* feature, individual faces of it will highlight.

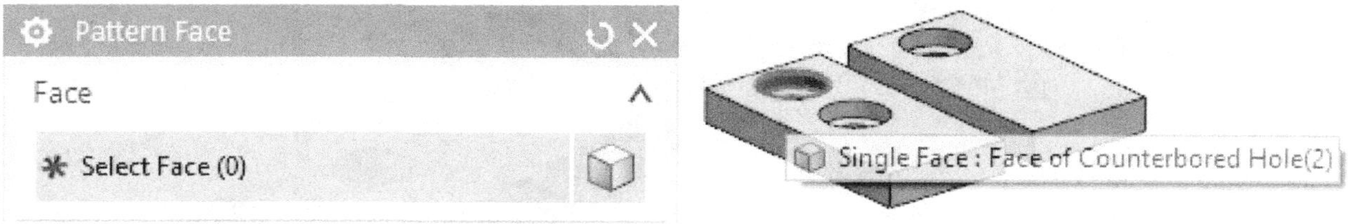

- Since **Pattern Face** is a **Synchronous Modeling** tool, it doesn't care whether the faces to be patterned belong to a feature or not!

- Set the **Pattern Layout** to *Linear* and copy the three faces of the hole on the second body as shown below.

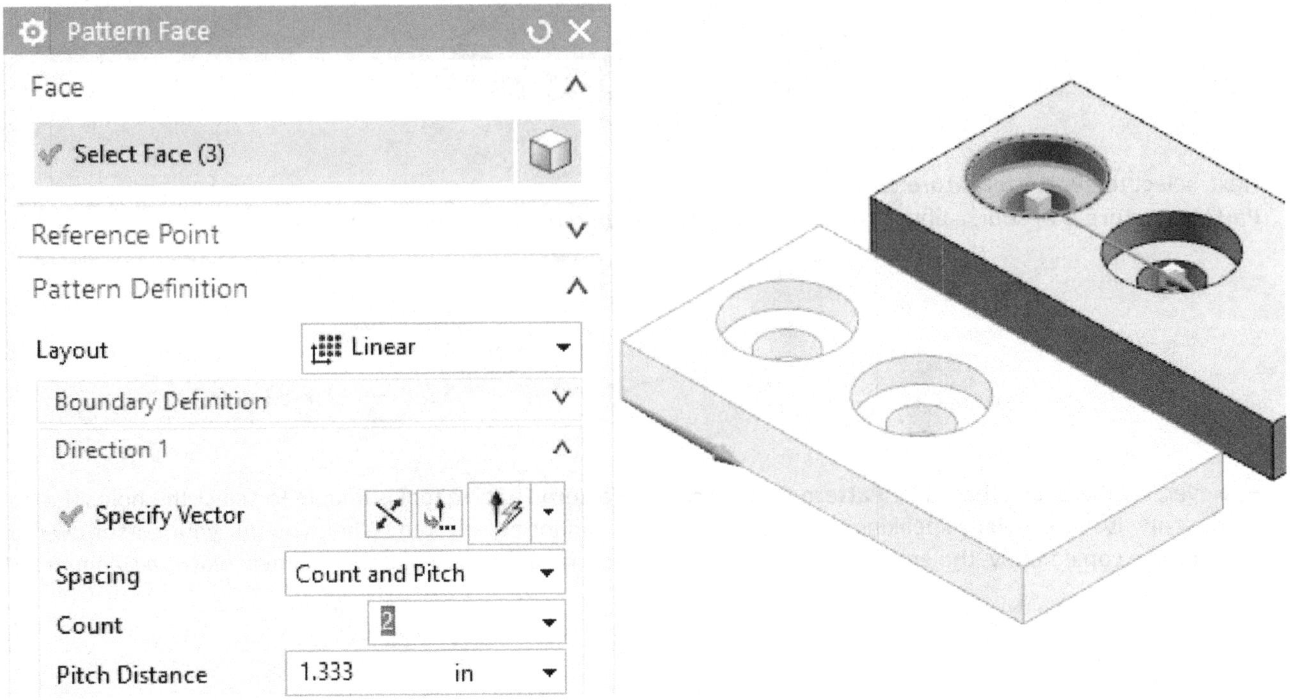

- **Pattern Face** works on bodies that may or may not have parameters, and a subtle thing it will allow you to do on parametric models is pattern only a subset of the faces in a feature.
- Select only the thru hole of the *Counterbored Hole* feature and enter the parameters shown below. Even though the thru face of the hole does not intersect the top face of the block, since **Pattern Face** is a **Synchronous Modeling** tool, it is able to produce a copy with the face of the thru hole extended!

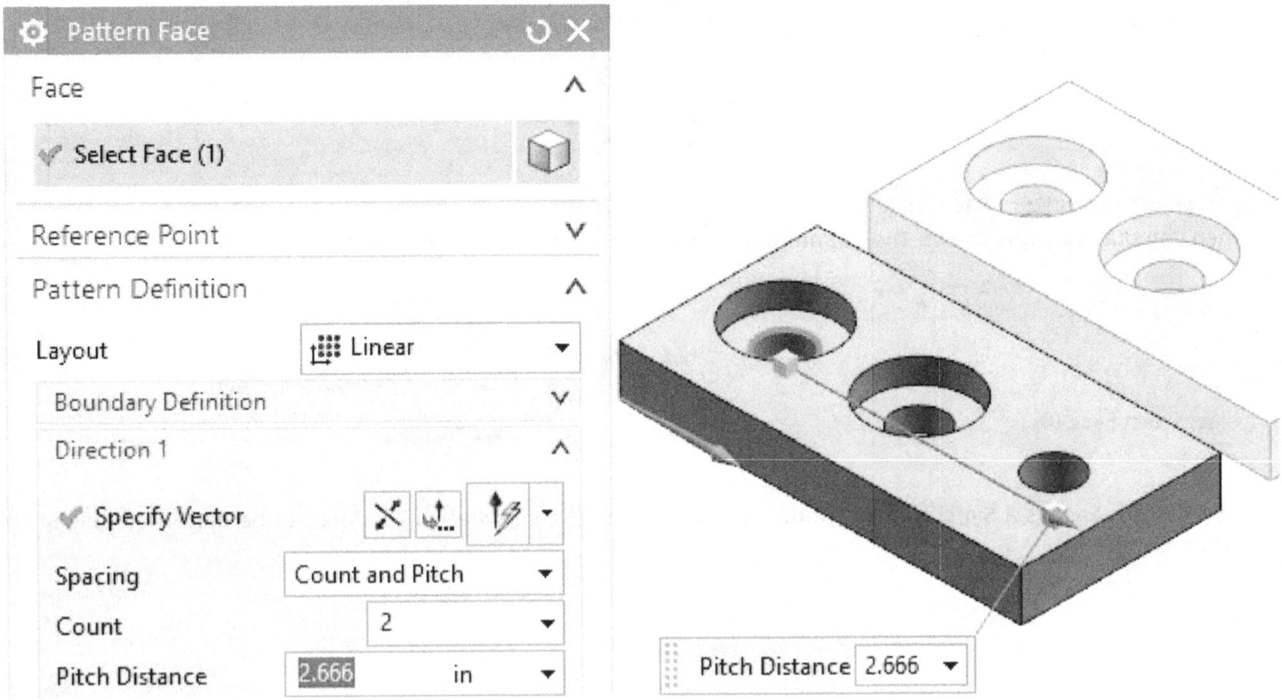

- **Save** your file!

17.2 Pattern Boundary Definition

The **Boundary Definition** option in each of the patterning tools that allows you to contain the pattern members by a set of curves, with a face. In this project, you will explore the use of a pattern boundary to keep pattern members from entering certain regions on a plate.

- Create a new file (**Units**: *Inches*) called *"Pattern Boundary Definition.prt"*.
- First create a **Block** with dimensions *10 in* x *6 in* x *0.25 in*.
- Next create the following sketch on the top face of the block. Note that all vertical lines have equal length.

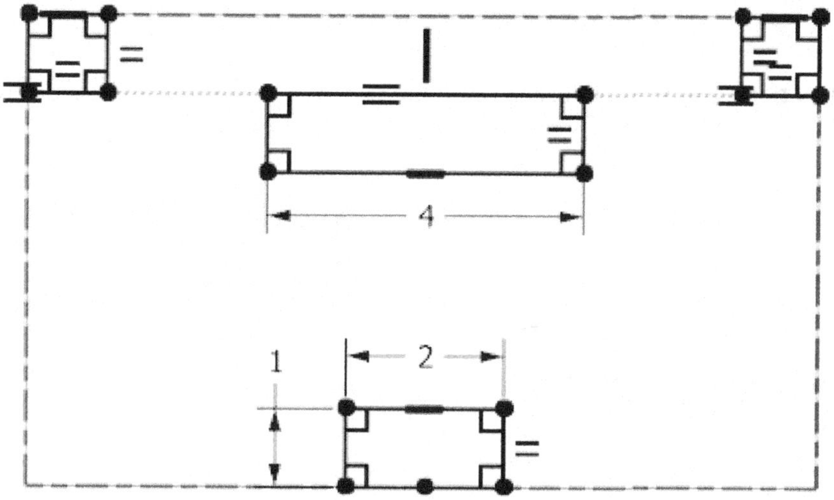

- Next create a *0.25 in* diameter simple thru **Hole** in the center of the block as shown.

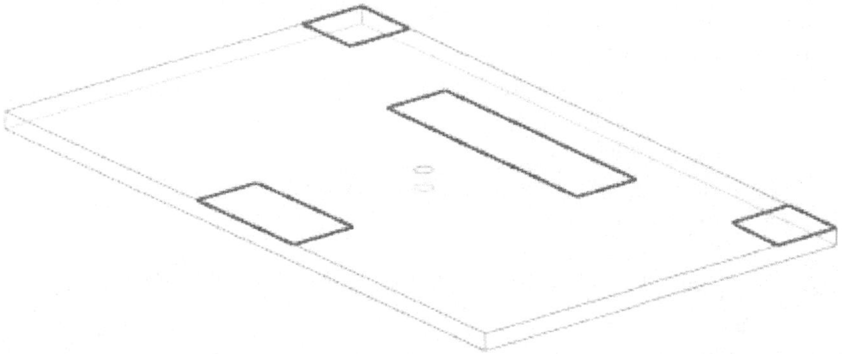

- Select **Pattern Feature** and set the **Pattern Type** to *Linear*.
- Pattern the hole in **Direction 1** with the parameters shown below.

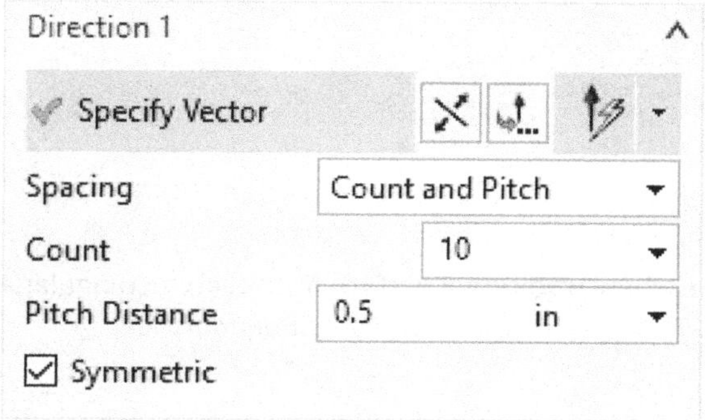

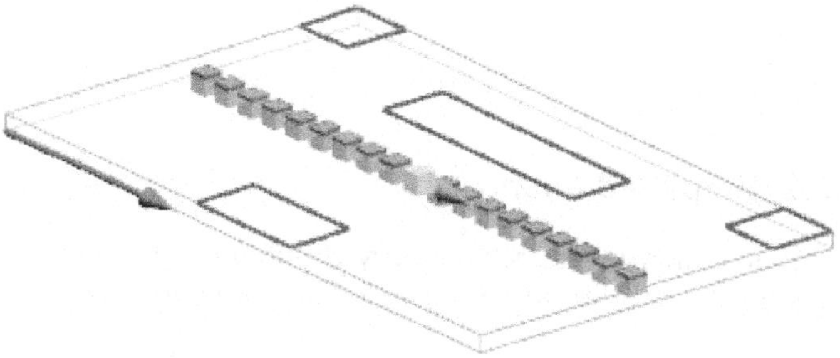

- Next check the **Use Direction 2** checkbox and set the **Direction 2** parameters as shown below.

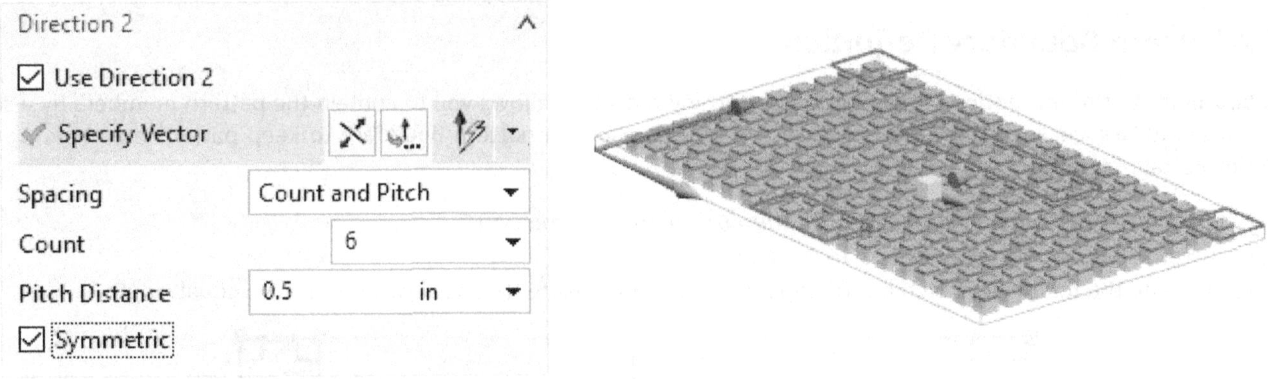

- Change the **Boundary Definition** to **Exclude**, and select the sketch as shown in the figure below.

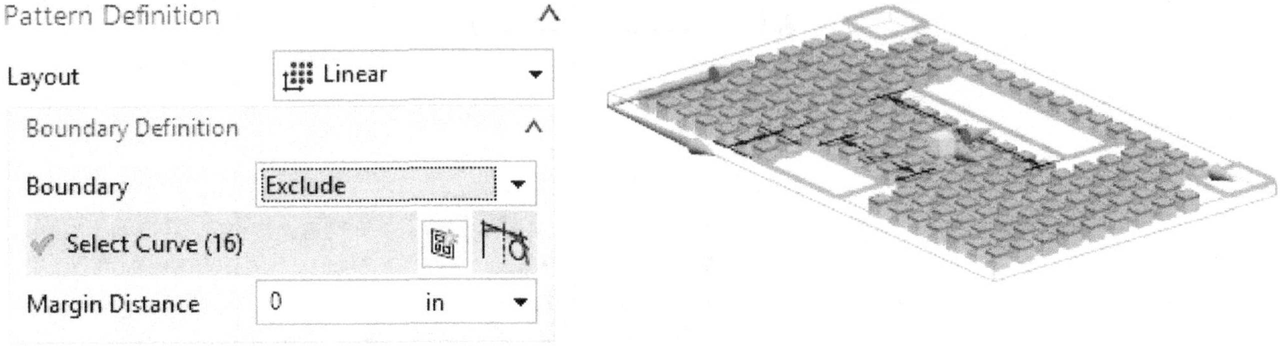

- Notice the instances do not enter the sketched areas.
- If more space is needed from the sketch lines the **Margin Distance** can be used to put an offset barrier around the sketch lines that the pattern instances cannot pass.
- Compare your pattern to the following figure.

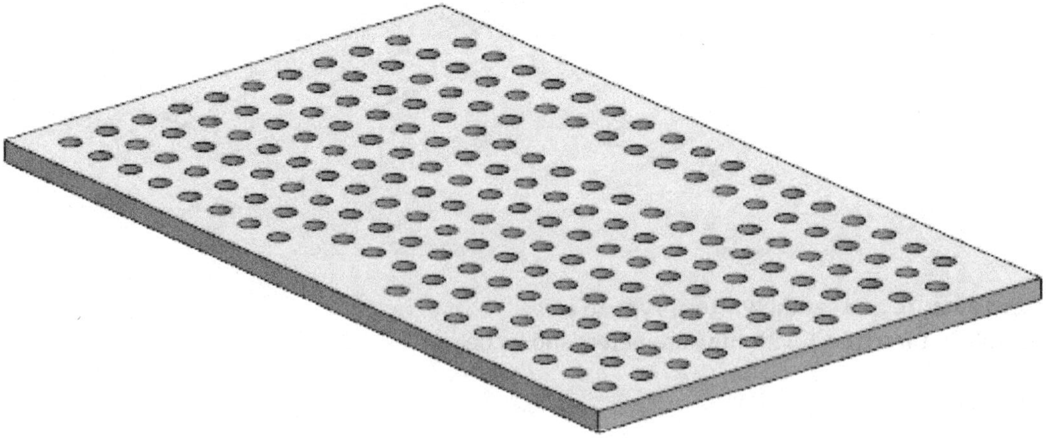

- **Save** your file!

17.2.1 Simplified Boundary Fill

There is an option within the **Boundary Definition** that allows you to eschew traditional pattern forms like rectangular and circular arrays, and instead fill a given region with pattern members. This option is the *Simplified Boundary Fill*.

- Continue working with *"Pattern Boundary Definition.prt"*.
- Use **File / Save / Save As** to create a copy called *"Pattern Simplified Boundary Fill.prt"*.
- **Delete** the existing *Pattern Feature* feature, and select the **Pattern Feature** tool again. Pick the hole.

- Select *Face* from the **Boundary** drop-down. Select the top face as the **Boundary Face**.
- Check the **Simplified Boundary Fill** checkbox. Notice that the **Direction 1** and **Direction 2** sections of the dialog have disappeared, and in their place, you now have a section called **Simplified Layout**. Set the parameters for **Simplified Layout** as shown below.

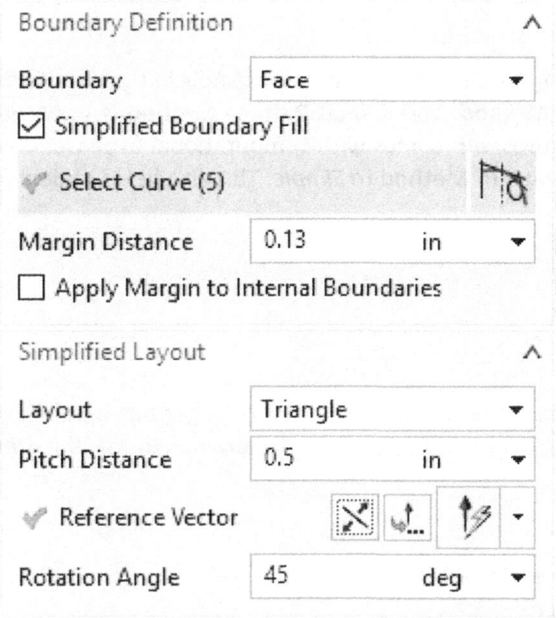

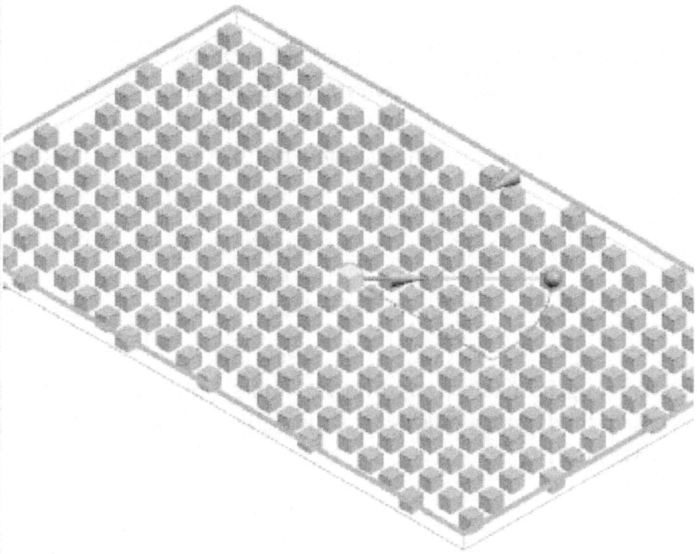

- This allows a quick way to create staggered patterns easily and quickly onto faces without the need to specify an axis or symmetry.

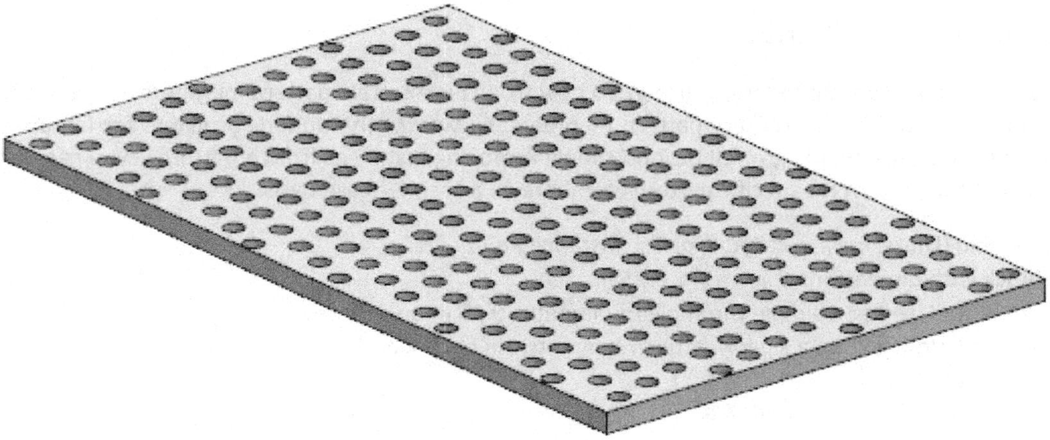

- **Save** your file!

17.2.2 Pattern Method

You might have noticed that the *Pattern Feature* entity created in the Model History for the last pattern came with an alert. In the Part Navigator, you may see something like this.

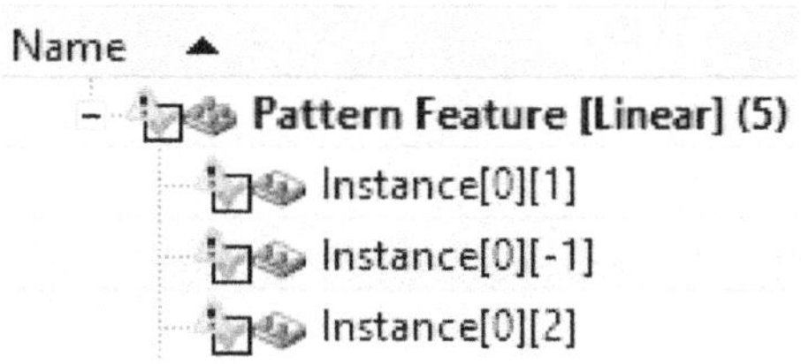

If you go to edit *Pattern Feature* from the Part Navigator, you may see an alert like this.

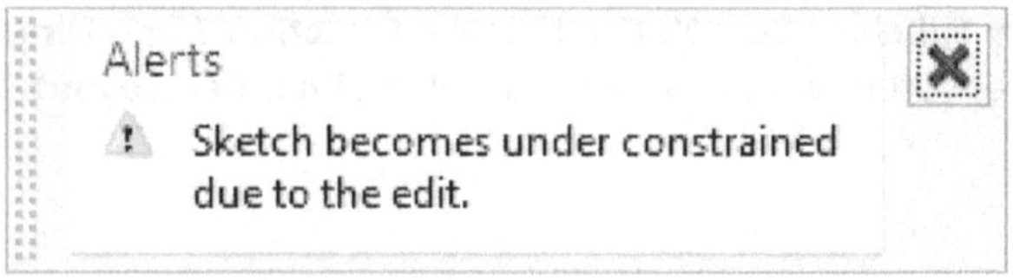

Sometimes **Pattern Feature** is finicky about patterning features with internal sketches, such as the hole in the last exercise. In those situations, it can be advantageous to change the **Pattern Method**. The default **Pattern Method** is *Variational*, which allows for some more robust patterns involving multiple features, but comes with computational expense. If your pattern is simple, then to reduce loading times you can change the **Pattern Method** to *Simple*. This also helps to deal with finicky internal sketches!

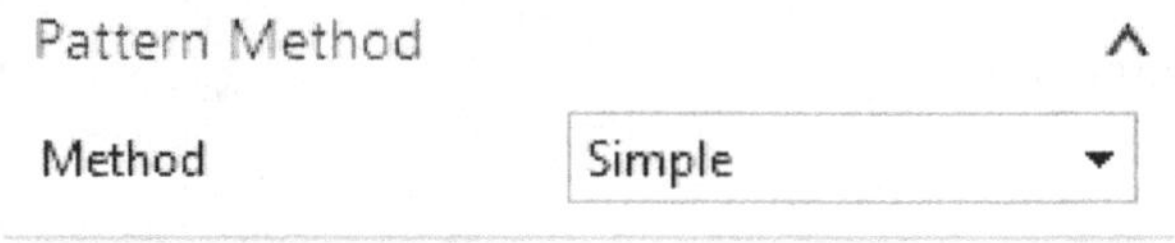

After changing the pattern method to *Simple*, you will see that the *Pattern Feature* entity no longer has an alert. The pattern also loads more quickly. If your pattern involves patterning simple features, it's a no-brainer to use the *Simple* **Pattern Method**!

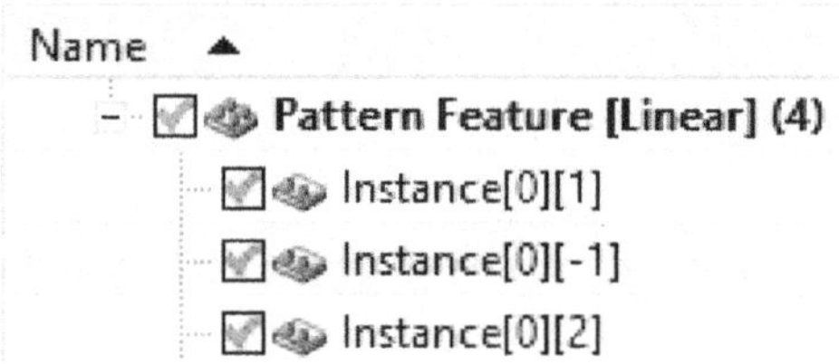

17.3 Pattern Instance Points

With the option to set a **Pattern Boundary**, you learned how to exclude certain pattern members from the pattern. In fact, you have the ability to do this for arbitrary pattern members, by way of the options available for **Instance Points**. The options available for instance points go well beyond simply removing instances from a pattern – you can also modify the basic parameters of the pattern for specific pattern members on a case-by-case basis.

- Create a new file (**Units**: *Inches*) called *"Pattern Instance Points.prt"*.
- Create a **Block** with dimensions *8 in* x *8 in* x *0.25 in*.
- Using the **Hole** tool, center a simple thru hole of diameter *0.25 in* on its face.
- Use **Pattern Feature** to set up a *Linear* pattern, symmetric in both X and Y.
    - **Pitch**: *1 in* (both X & Y)
    - **Count**: *4* (both X & Y)

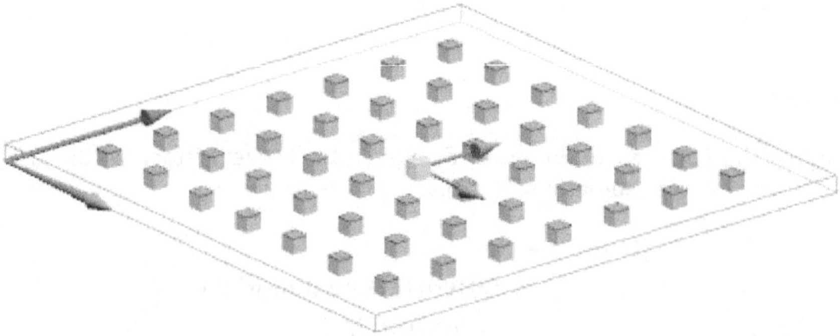

- Go to the **Instance Points** section of the **Pattern Feature** dialog, and click on *"Select Instance Point (0)"*.

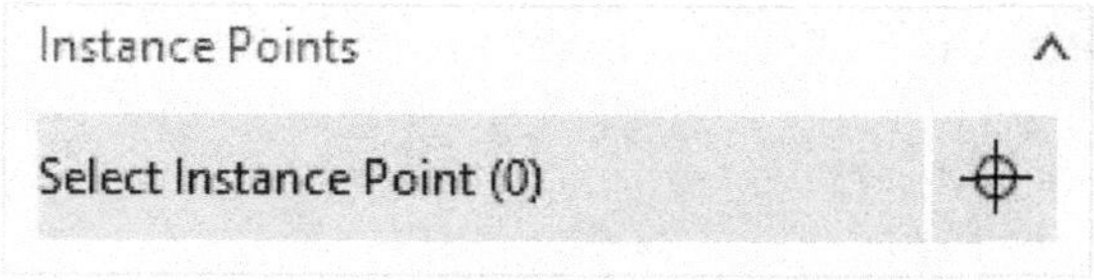

- Notice now that if you put your cursor on one of the golden preview cubes, it highlights.

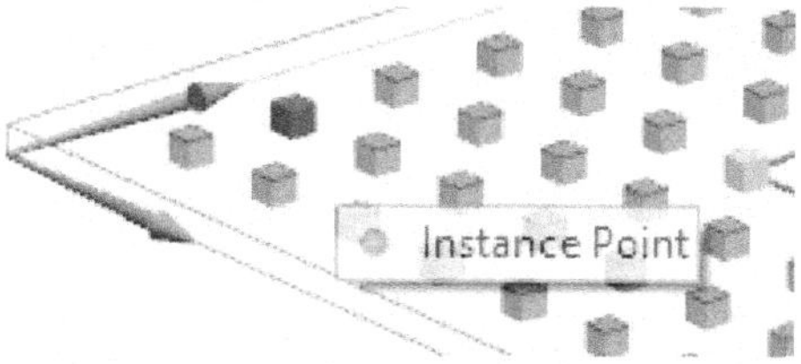

- Right-click on an instance point and examine your options. Choose **Delete**.

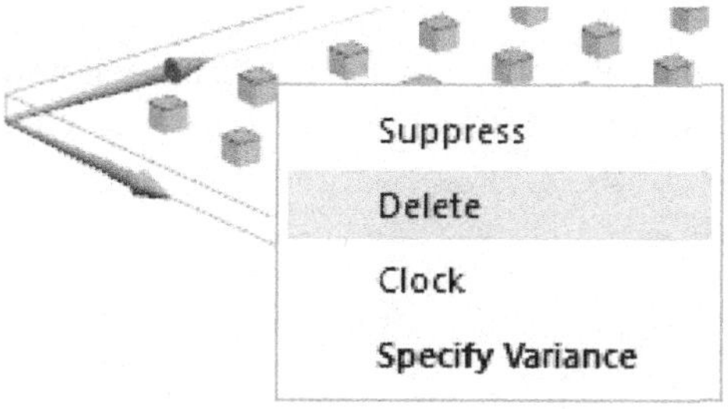

- Once deleted, the instance point becomes a red sphere. It can be undeleted. Next, click on three other instance points and then right-click on them and select **Clock**.

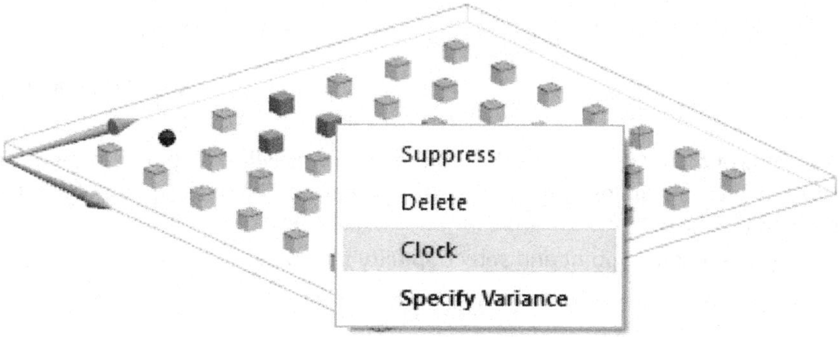

- To clock is to displace the instance points by a certain amount in either of the pattern's directions.

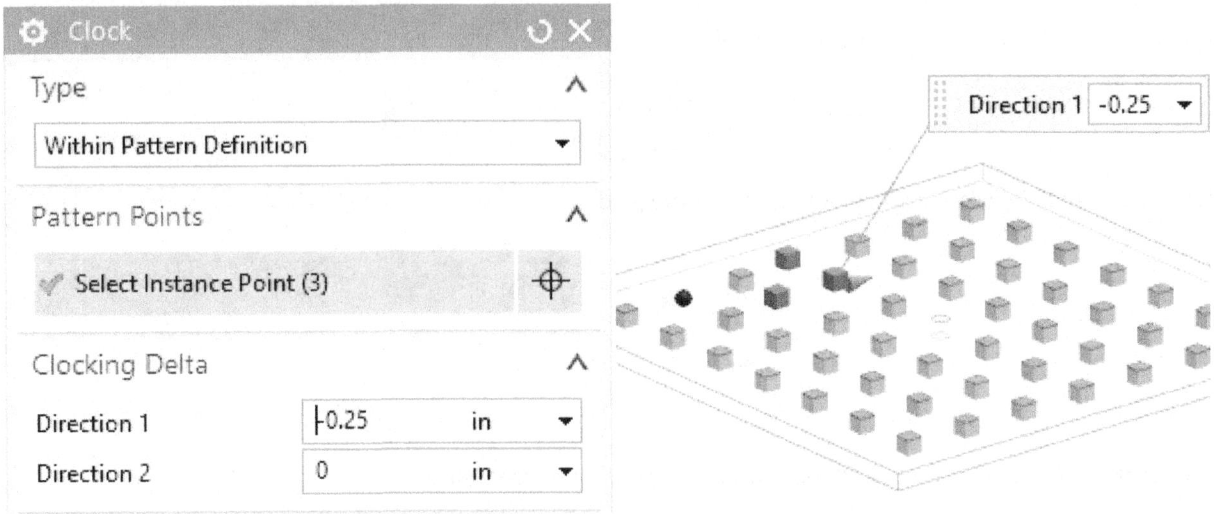

- Once clocked, the dynamic preview will show the instance points in their new locations. They can be unclocked, and you can modify the clock parameters. Select another group of instance points, right-click on any one of them, and select **Suppress**.

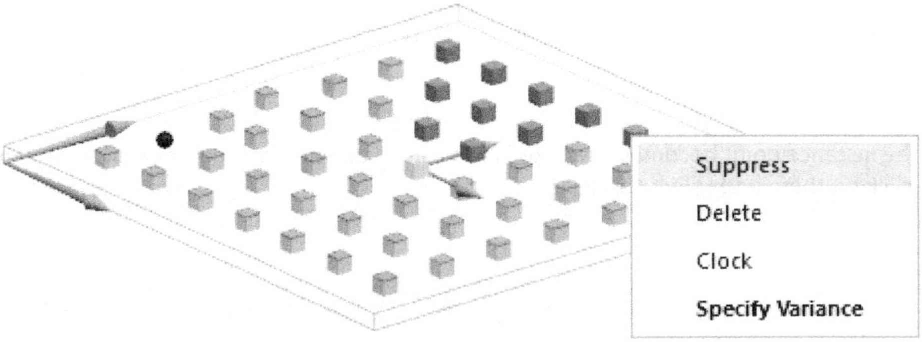

- Suppression is slightly different from deletion insofar as the instances will still appear in the Model History and can be unsuppressed. In order to undelete an instance after pattern creation, the pattern must be edited so that the instance point can be selected and undeleted.
- Finally, right-click on any one instance point and select **Specify Variance**.

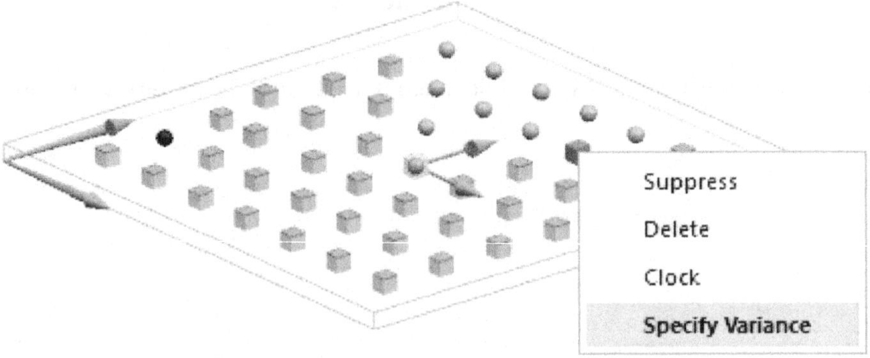

- **Specify Variance** is the juiciest option – it allows you to modify the underlying parameters of the feature driving this particular instance. Select the hole *Diameter* from the **Parameters** list and use the **Add New Set** button (or double-click on it) to make it available in **Values**.

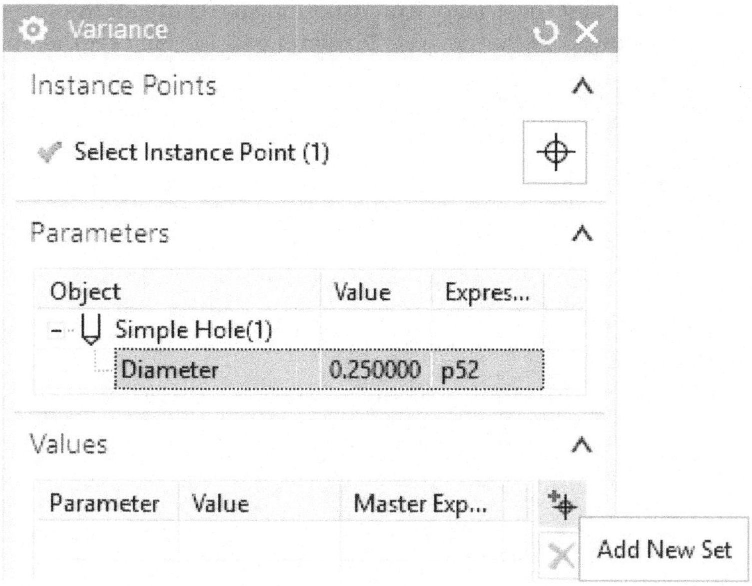

- Change the value to *0.5 in* and click **OK**. Instance points for which the parameters have been varied appear as green arrows in the preview.

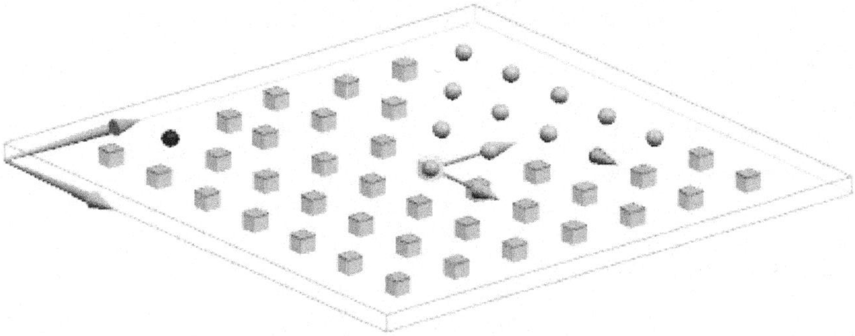

- Click **OK** to create your pattern.

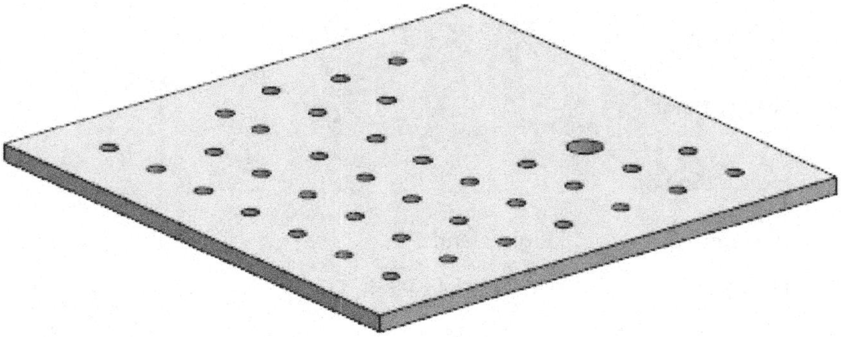

- **Save** your file!

17.4 Pattern Increment

The preceding exercise on instance points was a warm-up for some of the more robust means of varying the parameters of a patterned feature. The next level of sophistication offered in patterning tools tool is the ability to systematically increment either the pattern parameters (e.g., the **Pitch**) or the feature parameters, as the index of the instance increases.

- Create a new file (**Units**: *Inches*) called *"Pattern Increment.prt"*.
- Create a **Cylinder** with a **Diameter** of *1 in* and a **Height** of *6.5 in*.

- Select the **Sphere** tool and use the **Point** constructor icon to position the center of the sphere *0.25 in* from the bottom edge of the cylinder, as shown below. Give the sphere a **Diameter** of *0.1 in*, and set the **Boolean** to *Unite*.

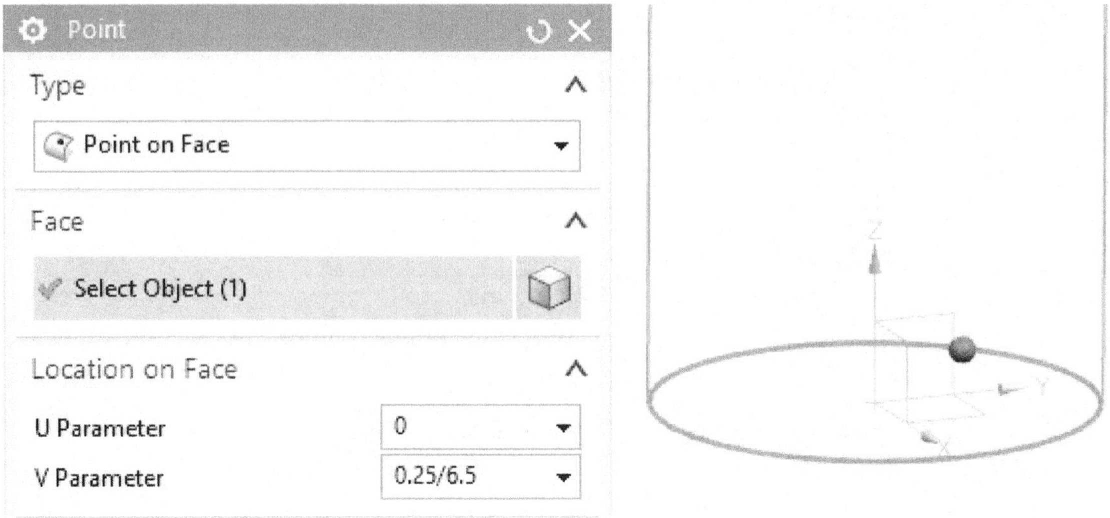

- Select the **Pattern Feature** tool and set the **Type** to *Helix*. Set the pattern parameters as shown below.

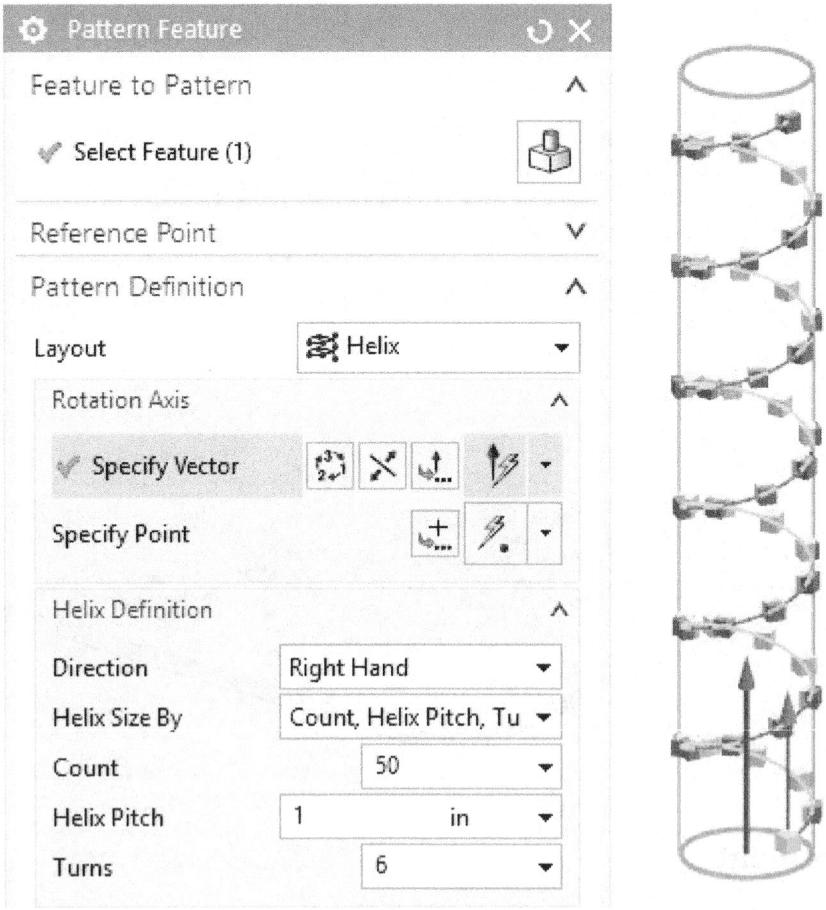

- Open the **Pattern Increment** section of the dialog, and use the icon to bring up the **Pattern Increment** dialog.

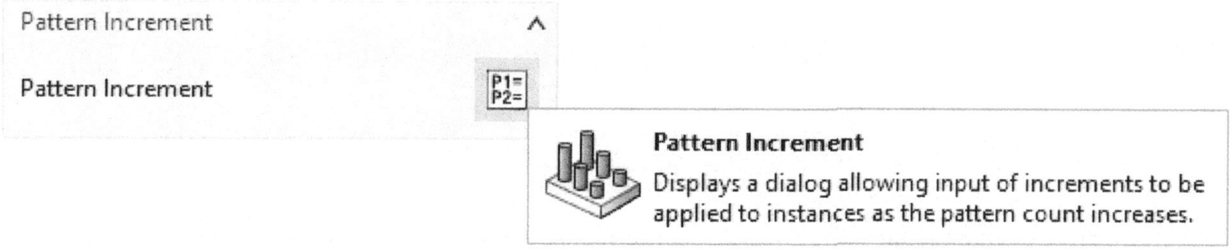

- Notice that the **Pattern Increment** dialog is very similar to the **Specify Variance** dialog for **Instance Points**.
- Set an **Increment** of *0.01 in* for the **Diameter** of the *Sphere*.

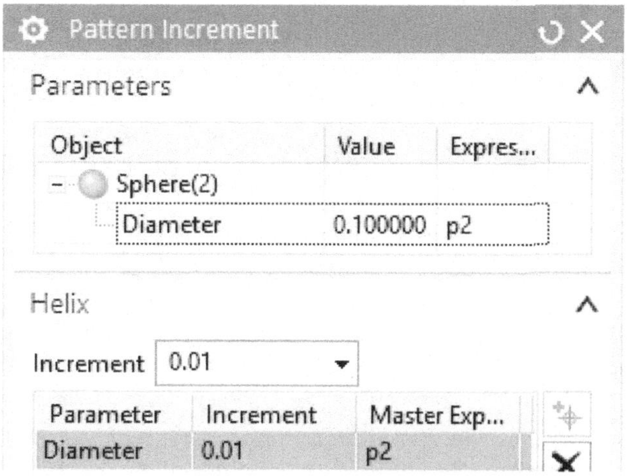

- The result will be a shape that resembles a cudgel as shown below.

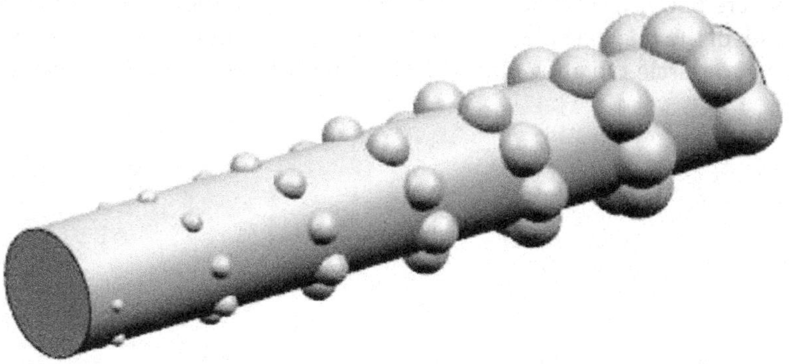

- **Save** your file!

17.5 Patterns Using a Spreadsheet

The **Instance Points** and **Pattern Increment** options are great tools for varying your pattern parameters, but they both pale in comparison to the power of the spreadsheet as a tool for controlling every detail of your pattern. In this exercise, you will learn to leverage the *Use Spreadsheet* functionality within Pattern Feature to create a showerhead with exact spacing in the angular direction on each concentric circle in a circular pattern!

- Create a new file (**Units**: *Millimeters*) called *"Pattern With Spreadsheet.prt"*.
- Create a **Cylinder** with a **Diameter** of *200 mm* and a **Height** of *10 mm*.
- Using the **Hole** tool, position a *6 mm* thru hole at the center of the cylinder.

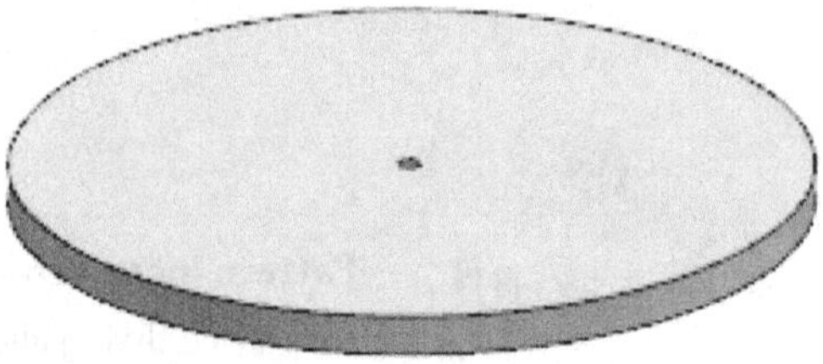

- Use **Pattern Feature** with **Layout** set to *Circular* to pattern the hole with the parameters shown below.

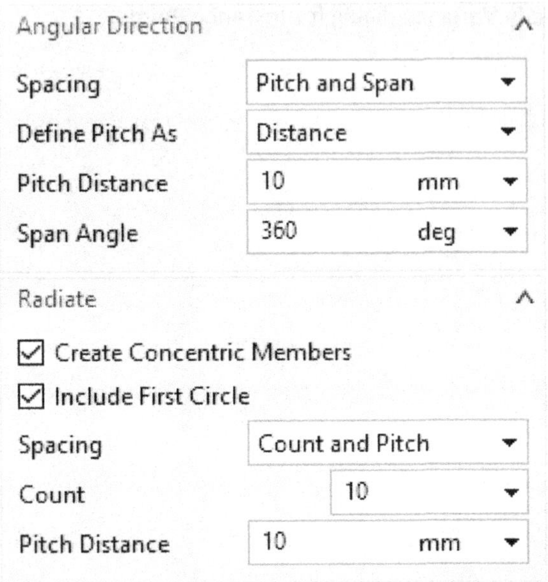

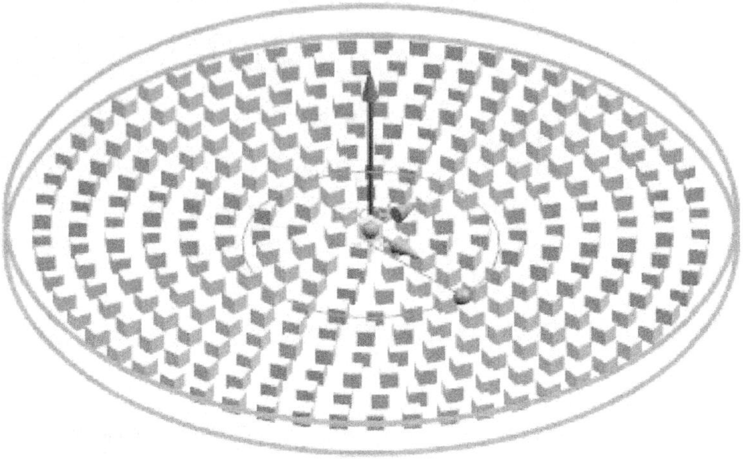

- After generating the feature pattern, create a sketch on the top face of the cylinder consisting of nine arcs. Each arc should have the arc center of the original hole as its center point, and should have at its endpoints the first and second instances on each concentric circle radiating outward. Note that each concentric circle has its first pattern instance along the XC axis.

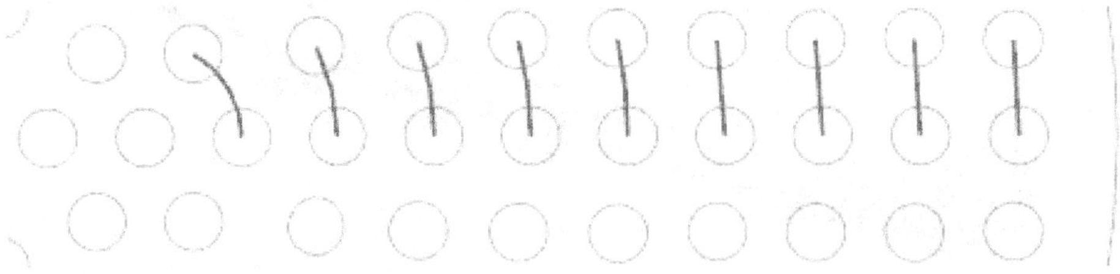

- Use **Measure Length** to create nine associative length measurements, for each of the arcs just created.
- In the **Expressions** editor, you can see that the lengths of these arcs are all close to *10 mm*, but none match exactly, despite the fact that your pattern asked for an angular pitch of *10 mm*.

Expressions

	↑ Name	Formula	Value	Units	Dimensi	Type	Source
1				mm	Le...	Nu...	
2	p75	(Measurement)	10.47197551	mm	Length	Number	(Length Measurement(4))
3	p77	(Measurement)	9.666438934	mm	Length	Number	(Length Measurement(5))
4	p79	(Measurement)	9.920818906	mm	Length	Number	(Length Measurement(6))
5	p81	(Measurement)	10.05309649	mm	Length	Number	(Length Measurement(7))
6	p83	(Measurement)	10.13416985	mm	Length	Number	(Length Measurement(8))
7	p85	(Measurement)	9.920818906	mm	Length	Number	(Length Measurement(9))
8	p87	(Measurement)	9.995976625	mm	Length	Number	(Length Measurement(10))
9	p89	(Measurement)	10.05309649	mm	Length	Number	(Length Measurement(11))
10	p91	(Measurement)	9.920818906	mm	Length	Number	(Length Measurement(12))

- The problem is that the circumference of each concentric circle in the pattern may not be evenly divisible by *10 mm.*
- To come up with a pattern that has a true angular pitch of *10 mm*, we will modify the radius of each concentric circle in the pattern. This cannot be done with **Pattern Increment**, and while possible with **Instance Points**, would be terribly inconvenient. In this example, the spreadsheet really shines.
- Edit your **Pattern Feature** entity. In the dialog, below Instance Points, there is a checkbox that says *Use Spreadsheet*, along with a **Display Spreadsheet for Pattern** button. Click either.

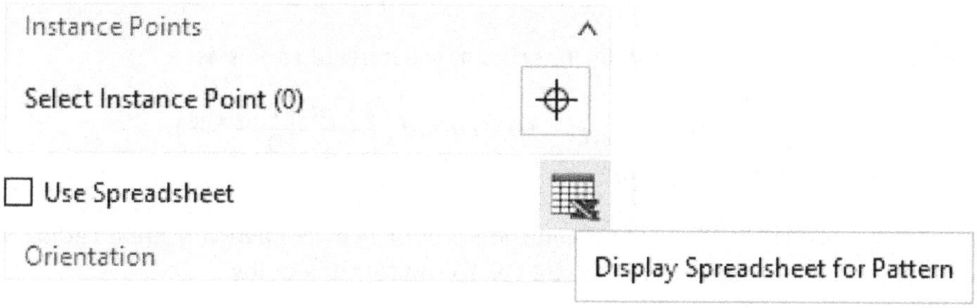

- This brings up an *Excel* window as shown below[5]. Columns **A** and **B** indicate the "coordinates" of each instance point, measured in degrees and millimeters respectively, and column **C** shows the feature parameter – in this case, the **Diameter** for the hole.

	A	B	C
1	*Angular Direction Offset*	*Radiate Offset*	*SIMPLE HOLE(1):Diameter Default Value: 6*
2	0	10	
3	60	10	
4	120	10	
5	180	10	
6	240	10	
7	300	10	
8	0	20	
9	27.69230769	20	
10	55.38461538	20	

[5] You must have *Microsoft Excel* installed to use this functionality.

- You can change any of the values in columns **A**, **B**, and **C**, to vary the position or diameter of each hole in the pattern!
- Copy and paste the values in column **B** into column **D**.

	A	B	C	D
1	*Angular Direction Offset*	*Radiate Offset*	*SIMPLE HOLE(1):Diameter Default Value: 6*	
2	0	10		10
3	60	10		10
4	120	10		10
5	180	10		10
6	240	10		10
7	300	10		10
8	0	20		20
9	27.69230769	20		20
10	55.38461538	20		20

- The ideal circumference and radius are determined by the equation below.

$$circumference_{ideal} = 2 * pi * r_{ideal} = 10 * count$$

- The count (# of instances on a given circle) is determined by the equation below.

$$count = round\left(\frac{2 * pi * r_{current}}{10}\right)$$

- This means that we can solve for the ideal radius in terms of the current radius as:

$$r_{ideal} = \frac{10 * count}{2 * pi} = \frac{10 * round\left(\frac{2 * pi * r_{current}}{10}\right)}{2 * pi}$$

- Column **D** represents the current radius values, and we would like to have the ideal radius values in column **B**. Consequently, we should change the value for the **B2** cell to the formula below.

```
=10*round(2*pi()*D2/10,0)/(2*pi())
```

- Copy and paste the formula from **B2** into all the subsequent cells in the **B** column. You can do this in several ways.
    - Dragging the square in the bottom right corner of the cell down the **B** column.

Angular Direction Offset	*Radiate Offset*
0	9.549296586
60	10
120	10
180	10
240	10
300	10
0	20
27.69230769	20
55.38461538	20
83.07692308	20

- Use **[Ctrl]+[C]** to copy the cell, then hold **[Ctrl]+[Shift]** and press the down key **[↓]** to select the rest of the column, and then **[Ctrl]+[V]** to paste.

B2 =10*ROUND(2*PI()*D2/10,0)/(2*PI())

	A	B	C	D
1	*Angular Direction Offset*	*Radiate Offset*	*SIMPLE HOLE(1):Diameter Default Value: 6*	
2	0	9.549296586		10
3	60	9.549296586		10
4	120	9.549296586		10
5	180	9.549296586		10
6	240	9.549296586		10
7	300	9.549296586		10
8	0	20.6901426		20
9	27.69230769	20.6901426		20
10	55.38461538	20.6901426		20

- **Exit** Excel. It will prompt you to save the data. Click **OK**.

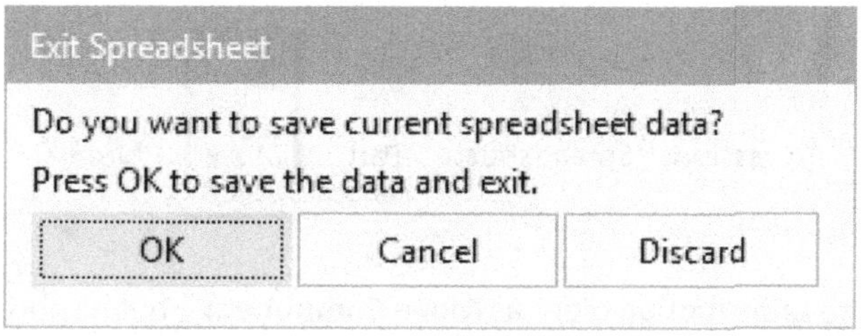

- Click **OK** to regenerate the feature pattern.

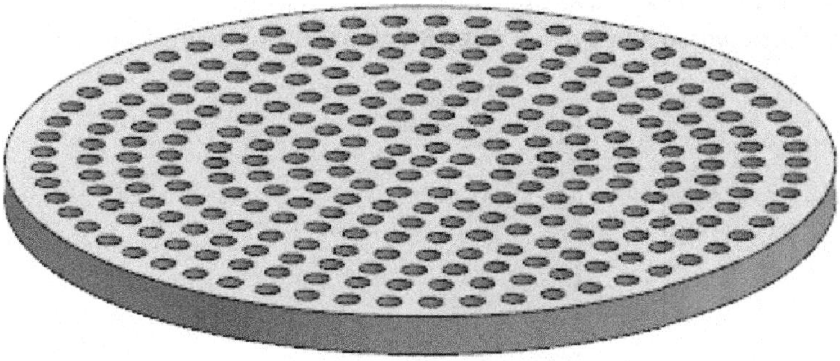

- Upon inspecting the measurement expressions, all angular pitches are now truly **10 mm**.

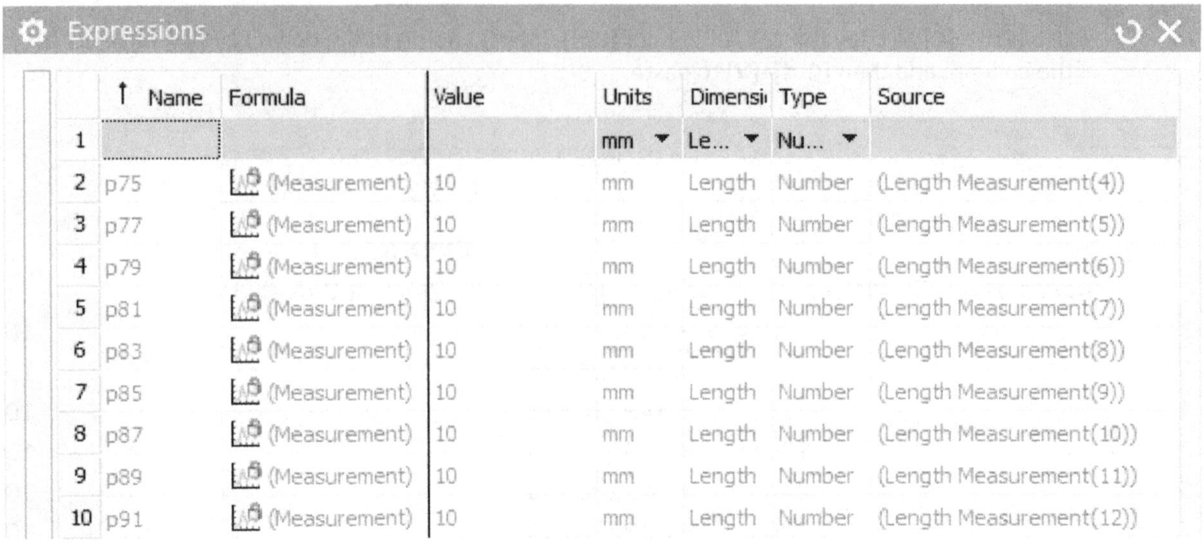

	Name	Formula	Value	Units	Dimensi	Type	Source
1				mm	Le...	Nu...	
2	p75	(Measurement)	10	mm	Length	Number	(Length Measurement(4))
3	p77	(Measurement)	10	mm	Length	Number	(Length Measurement(5))
4	p79	(Measurement)	10	mm	Length	Number	(Length Measurement(6))
5	p81	(Measurement)	10	mm	Length	Number	(Length Measurement(7))
6	p83	(Measurement)	10	mm	Length	Number	(Length Measurement(8))
7	p85	(Measurement)	10	mm	Length	Number	(Length Measurement(9))
8	p87	(Measurement)	10	mm	Length	Number	(Length Measurement(10))
9	p89	(Measurement)	10	mm	Length	Number	(Length Measurement(11))
10	p91	(Measurement)	10	mm	Length	Number	(Length Measurement(12))

- **Save** your file!

17.6 **Move Object**

The **Move Object** command is not a member of the **Associative Copy** group, but has options within its dialog that merit discussion in the context of associative copies. The **Move Object** command is found in the **Utilities** group on the **Tools** tab.

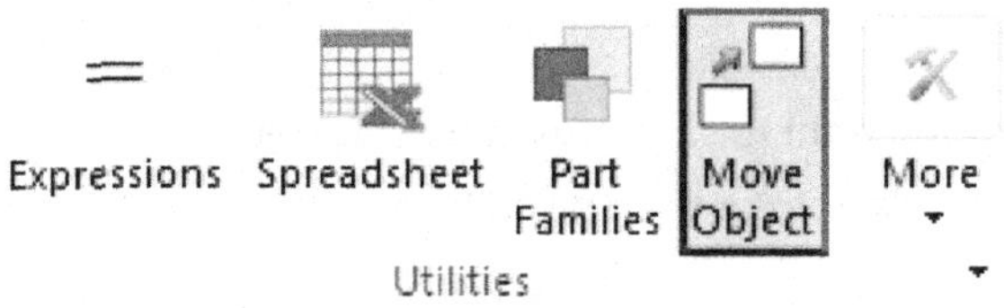

Move Object allows for many of the same motion types as **Move Component** – but its purpose is to move geometry around within a part file, rather than components within an assembly.

- Create a new file (**Units**: *Inches*) called *"Move Object.prt"*.
- Create a *2 in* x *1 in* x *2 in* **Block** with a **2 in** *Circular* **Edge Blend** as shown below.

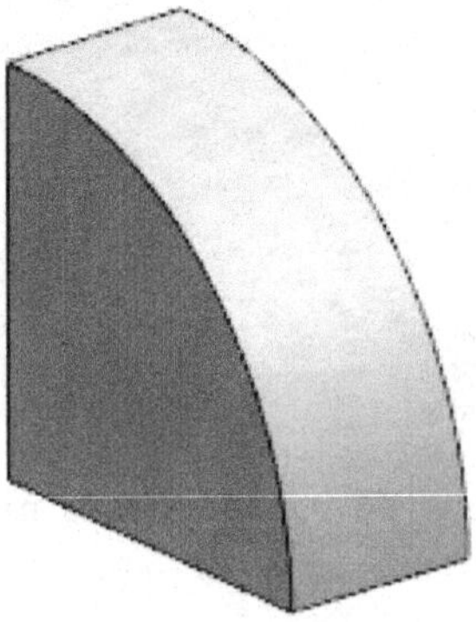

- **Move Object** gives you the option to either move the original object, or a non-associative copy. Select the **Move Object** tool, choose the body, and drag the dynamic coordinate system 2.5 in in the YC direction.
- In the **Result** section of the menu, change the radio button to *Copy Original*. You will then be asked to provide a **Number of Non-associative Copies** – set it to *2*. In this way, you can create arrays of non-associative copies of the object in question.

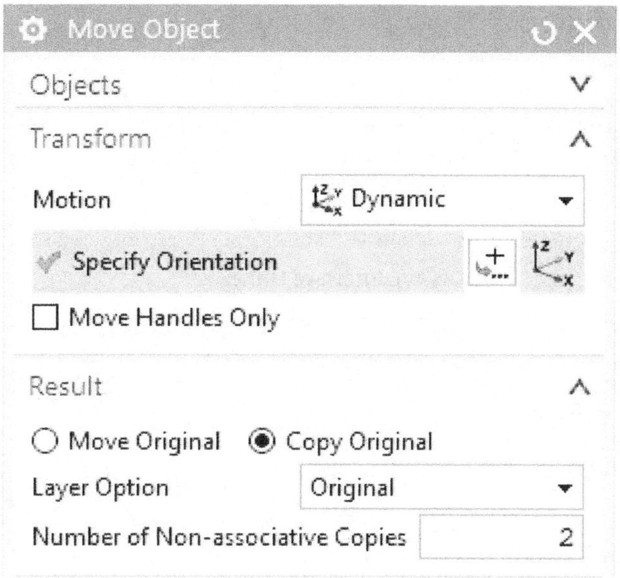

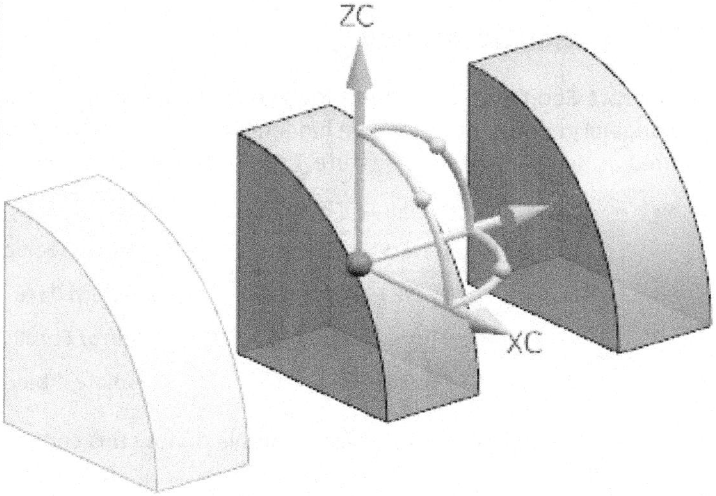

- The non-associative copies created appear in the Model History as *"Body"* – they are dumb solids.

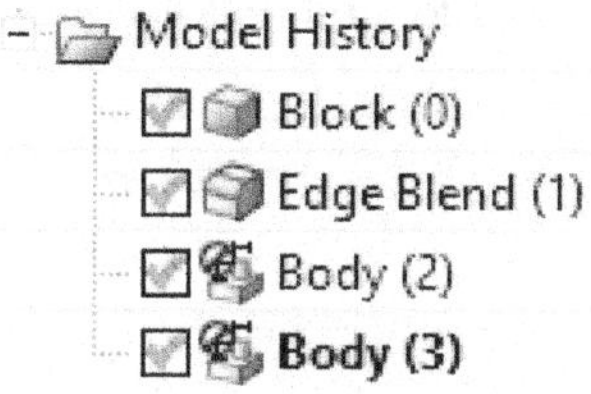

- You can also use **Move Object** to associatively move the original. Use the tool again to move the original body *2.5 in* to the left, with the **Result** set to *Move Original* and the *Associative* checkbox marked.

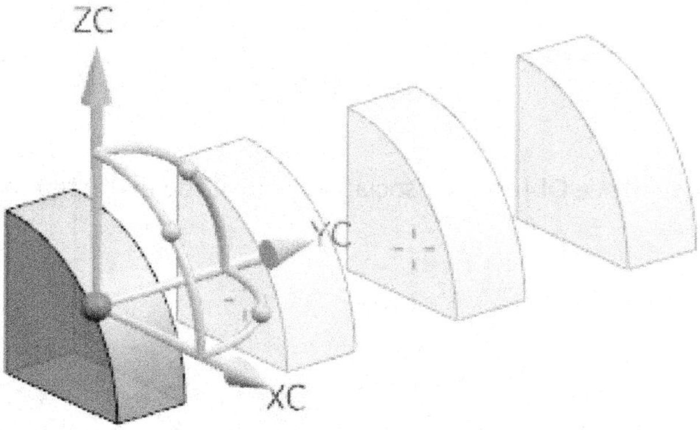

- Note: NX will not allow you to check both *Move Parents* and *Associative*. When *Associative* is checked, **Move Object** appears as a feature in the Model History. When *Move Parents* is checked, the operation is not recorded in the Model History, but all relevant bodies are moved in relation to the absolute coordinate system.

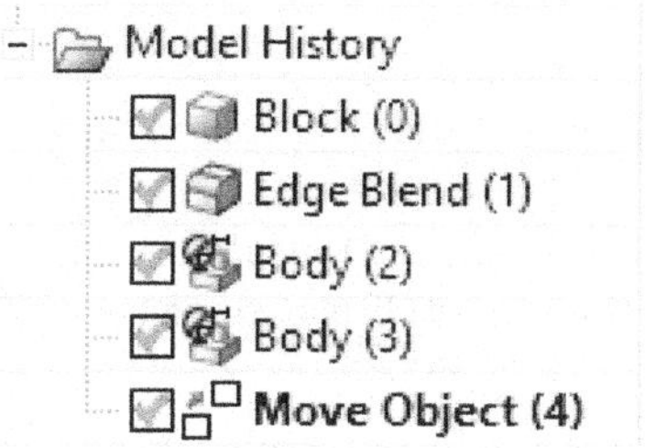

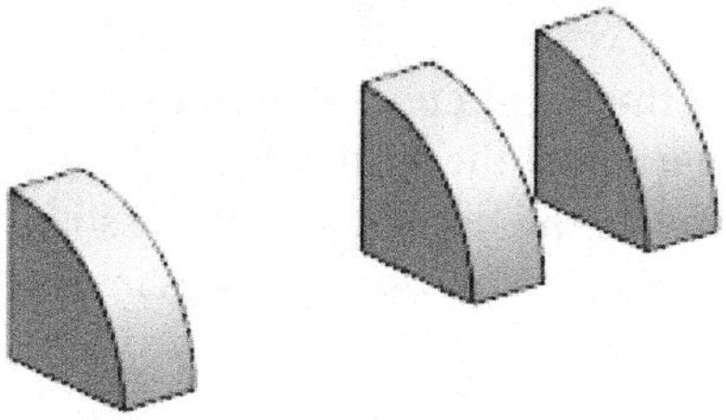

- **Save** your file!

17.7 Extract Geometry

The **Extract Geometry** command allows you to make copies of existing geometry, either associatively or non-associatively. The original geometry can then be hidden leaving you with only the face or region you desire. New geometry can then be created using your extracted feature, ensuring a perfect match.

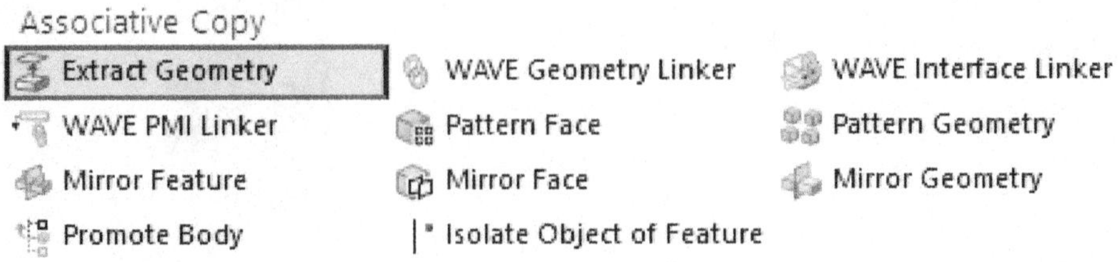

The following exercise is a very basic example of how this command is used.

- Continue working with *"Move Object.prt"*.
- Use **File / Save / Save As** to create a copy called *"Extract Geometry.prt"*.
- Select **Extract Geometry**, and set the **Type** to **Face** as shown in the following figure. Select the curved surface and the back vertical surface of your associatively-moved body as the faces to extract. In the **Settings** section of the dialog, make sure that the *Associative* checkbox is checked, and that *Fix at Current Timestamp* is unchecked.

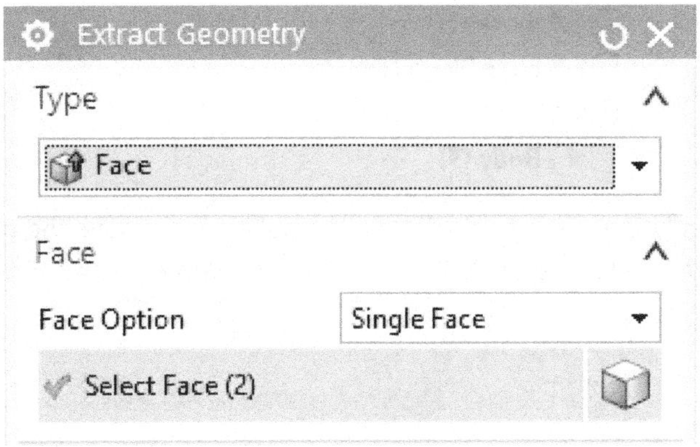

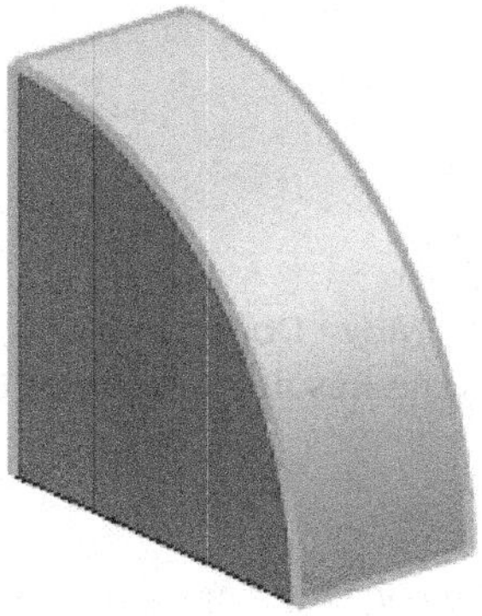

- Use **Move Object** to associatively move the extracted faces *2.5 in* along the YC axis.

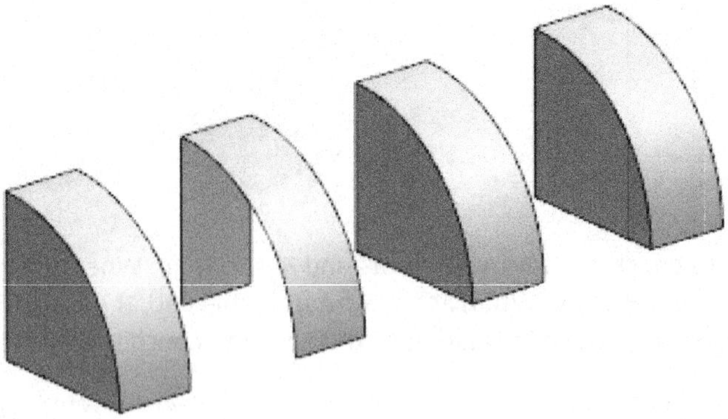

- The effect of leaving the *Fix at Current Timestamp* box unchecked is that the copied faces will remain associatively linked to the latest version of the solid body.
- Use **Delete Face** to delete the bottom face of the solid. Since *Fix at Current Timestamp* was unchecked for the **Extract Geometry** operation, the *Delete Face* feature is inserted into the Model History prior to the *Extracted Face* features.

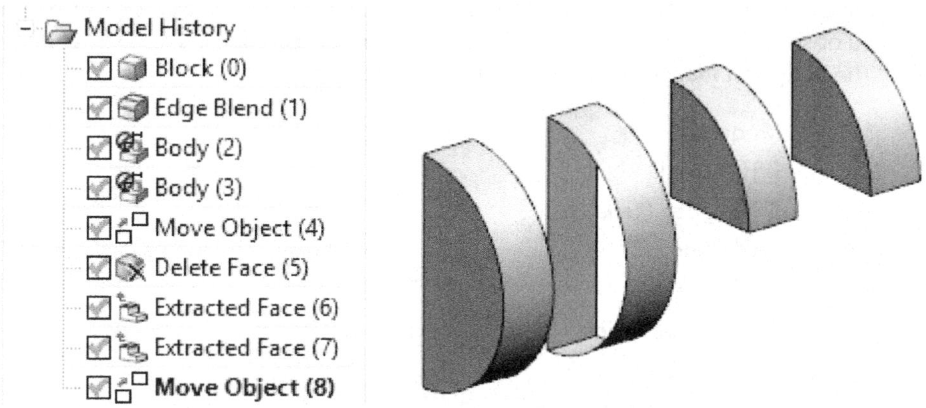

- Next, we will see what the effect of the *Fix at Current Timestamp* checkbox is when it is active.
- Double-click each of *Extracted Face (6)* and *Extracted Face (7)* to edit them. In the **Extract Face** dialog, go into **Settings**, and check the *Fix at Current Timestamp* checkbox.

- After modifying the extracted faces, use **Delete Body** to delete the original body.

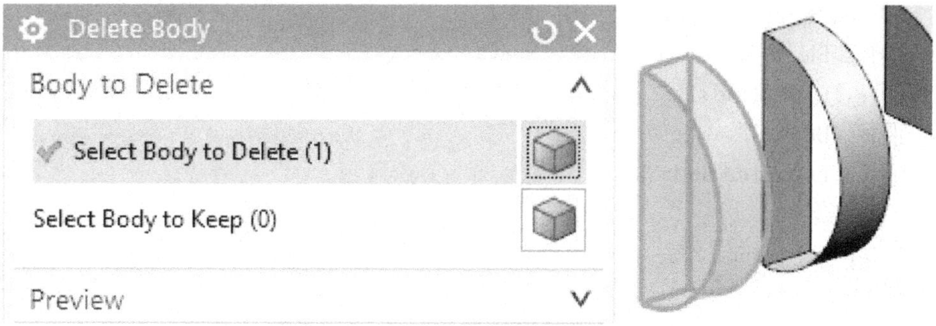

- The extracted faces remain, since each **Extracted Face** feature refers to the body prior to when it was deleted using **Delete Body**.

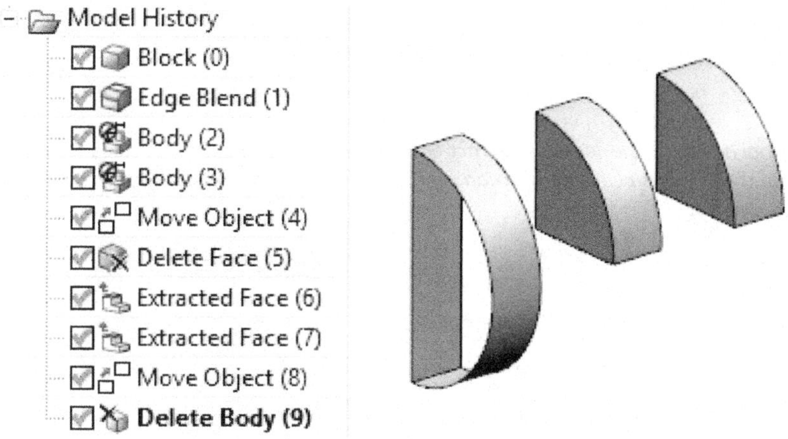

- Sometimes you can use a shortcut for **Extract Geometry** from within the Part Navigator. For features that are a part of some existing solid or sheet body within your part file, you can right-click and select *Extract Body Here* to create an associative copy of the body as it looked at the time of that timestamp.

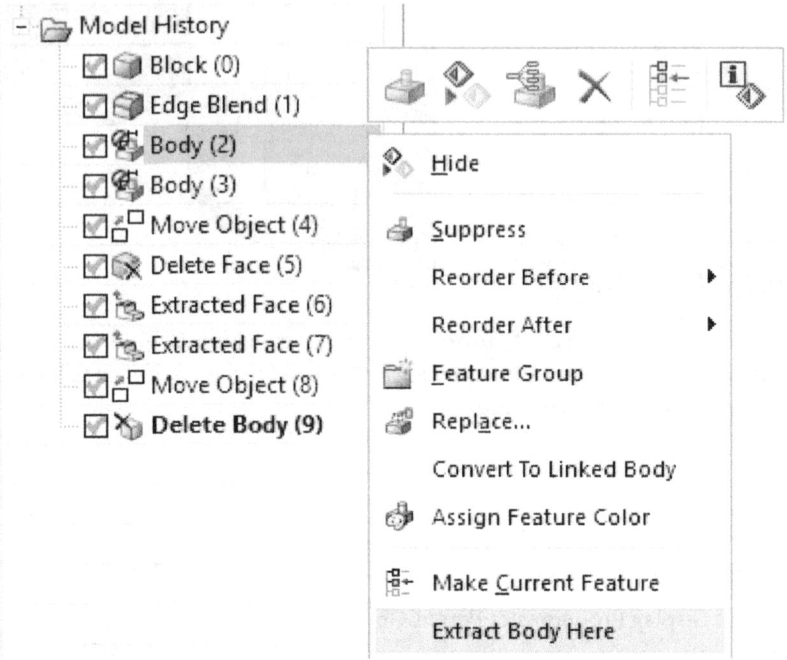

- Double-clicking on the *Extracted Body* feature shows that in **Settings**, both the *Associative* and *Fix at Current Timestamp* checkboxes are automatically marked when you extract in this way.

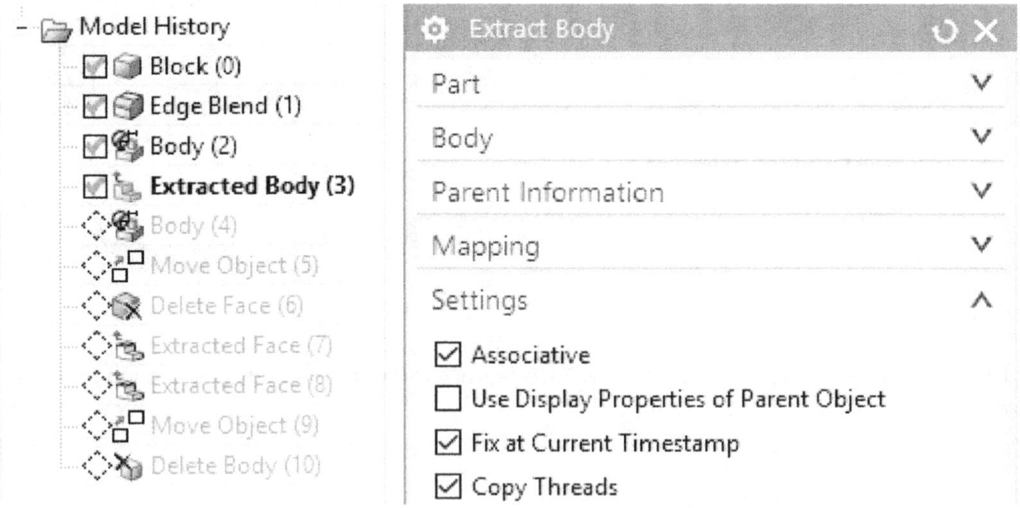

- **Save** your file!

*Note: The **Composite Curve** tool is exactly the same as **Extract Geometry** with **Type** set to* Composite Curve*. **Composite Curve** is found in the **Derived Curve** group on the **Curve** tab. In fact, you can switch to the **Composite Curve** tool from **Extract Geometry** by way of the **Dialog Options** for **Extract Geometry**.*

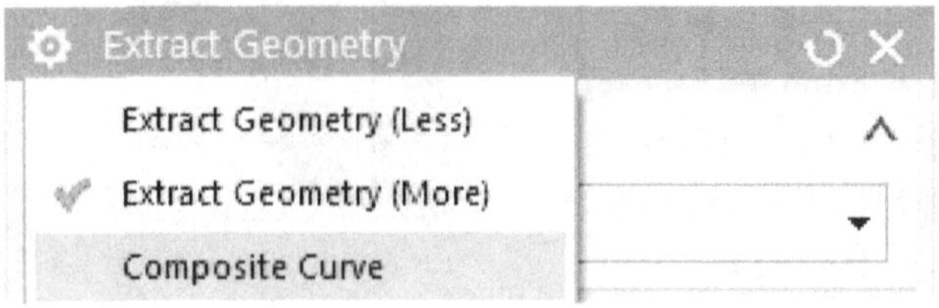

17.8 The WAVE Geometry Linker

We conclude this chapter with a lesson on the tool that is essential for creating "smart" assemblies – the **WAVE Geometry Linker**. Functionally, the **WAVE Geometry Linker** is the same as **Extract Geometry**, but with one key difference – the **WAVE Geometry Linker** can only make copies of geometry from other parts within an assembly.

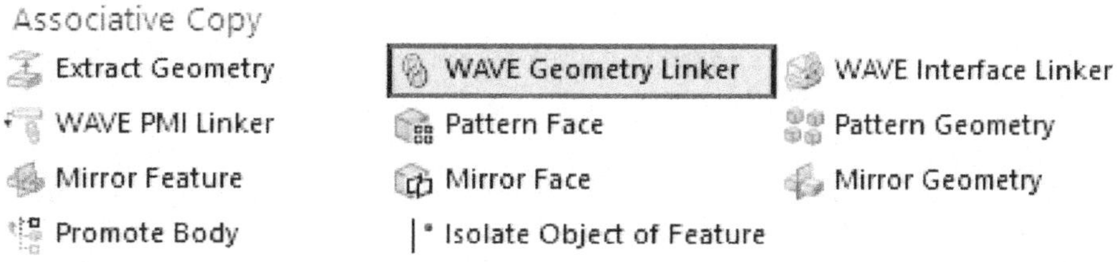

Since the **WAVE Geometry Linker** is only used in the context of an assembly, it is also in the **General** group on the **Assemblies** tab.

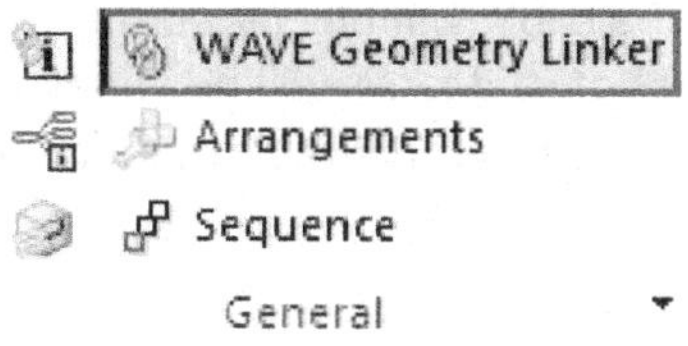

Recall the construction of the *Hinge Pin* in the *Hinge Assembly* in Chapter 11.4. The main cylinder was constructed with **Extrude**, and made use of a circular edge from *Hinge1*. If for some reason that circular arc on *Hinge1* were to change diameter or otherwise change shape, the corresponding cylinder from *Hinge Pin* would not follow suit. (Try it yourself, **Open** *Hinge Assembly*, make *Hinge1* the work part, and use **Offset Region** to resize the hole – *Hinge Pin* will not follow suit!)

The reason for this is that, unless an associative copy of geometry from another part file in an assembly is explicitly made within the work part, any reference to geometry outside of the work part is "dumb" and non-associative. Use of the **WAVE Geometry Linker** is the only way to guarantee fully parametric and associative assemblies with proper parent-child dependencies between components.

- Create a new file (**Units**: *Inches*) called *"WAVE Assembly.prt"*.
- Use **Create New Component** to add a component in-context called *"WAVE Tray.prt"*.
- Make *WAVE Tray* the work part and build it as shown below. It consists of:
    - **Block**, *8.5 in* x *11 in* x *1 in*
    - **Shell** top face, *0.25 in*
    - **Hole** x *2*, *0.375 in* **Diameter**, *2 in* from vertical edge, centered on face

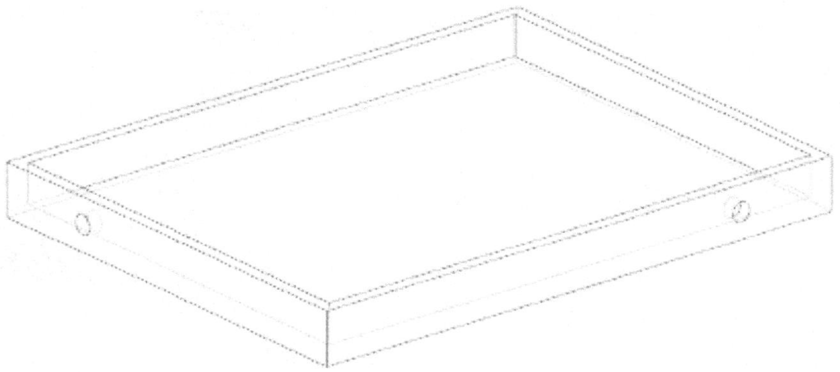

- After building *WAVE Tray* in-context, make *WAVE Assembly* the work part and use **Create New Component** again to create a second new file in-context named *"WAVE Cable.prt"*.
- Make *WAVE Cable* the work part and select the **WAVE Geometry Linker**.
- Set the **Type** to *Face*, and **Face Option** to *Single Face*. Choose the faces of the two simple holes in *WAVE Tray*.

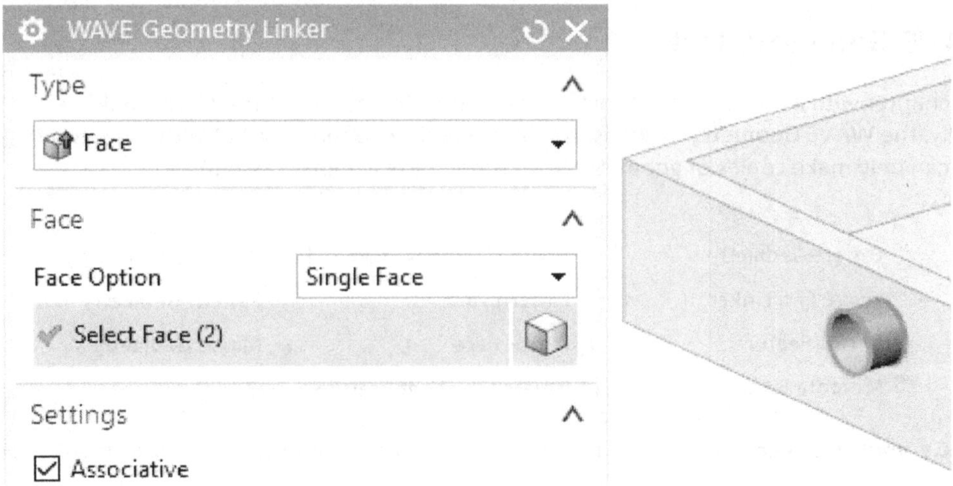

- The idea of this construction here is that the cable will be routed through the tray, and so the extracted faces should guide the cable. Specifically, the centerlines of said faces will be joined using **Bridge Curve**.
- First, we must extract the centerlines using **Extract Virtual Curve**, found in the **More** gallery on the far-right end of the **Curve** tab.

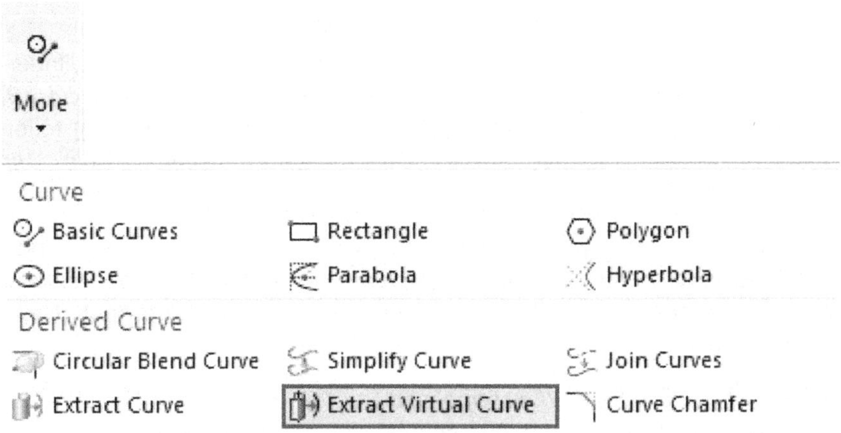

- Set the **Type** to *Rotation Axis* and choose each face. Make sure the *Associative* checkbox is marked.

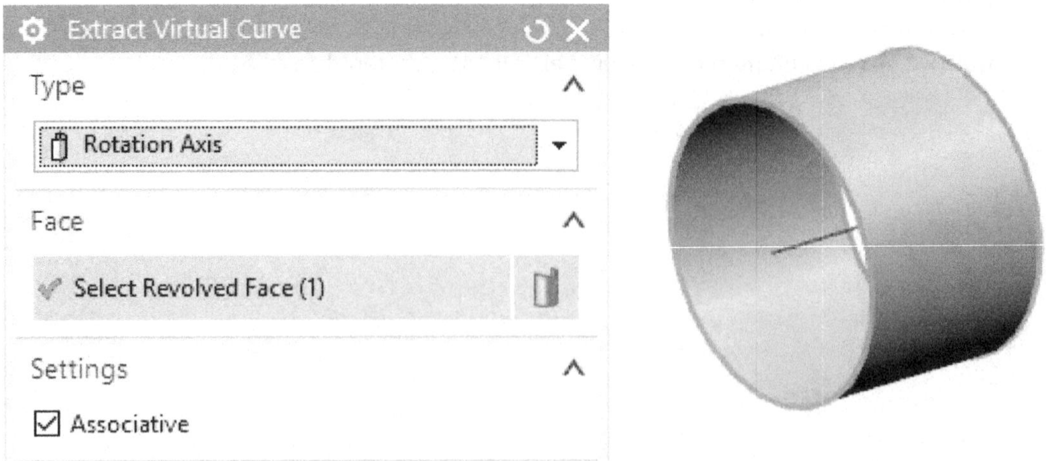

- One could create these centerlines by hand using the **Line** tool, but the **Extract Virtual Curve** tool makes it even easier!
- Use **Delete Body** to associatively delete the sheet bodies of the two *WAVE*-linked faces.

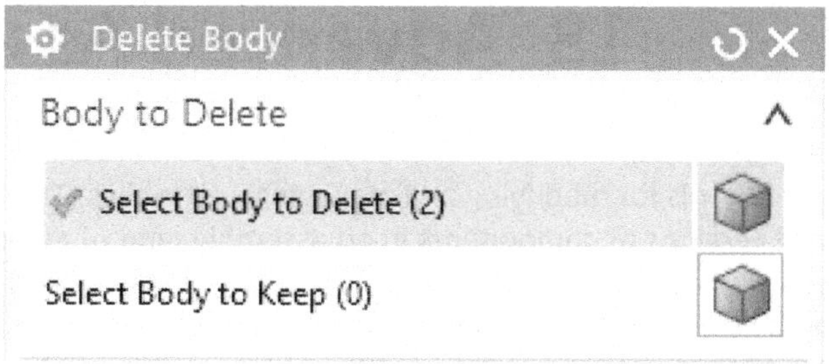

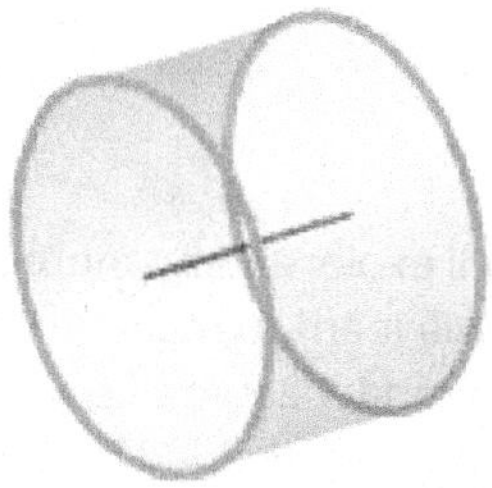

- Use the **Bridge Curve** tool with **Continuity** set to *G1* at each end to make a curve that joins the two centerlines. Adjust the **Tangent Magnitudes** to your heart's content.

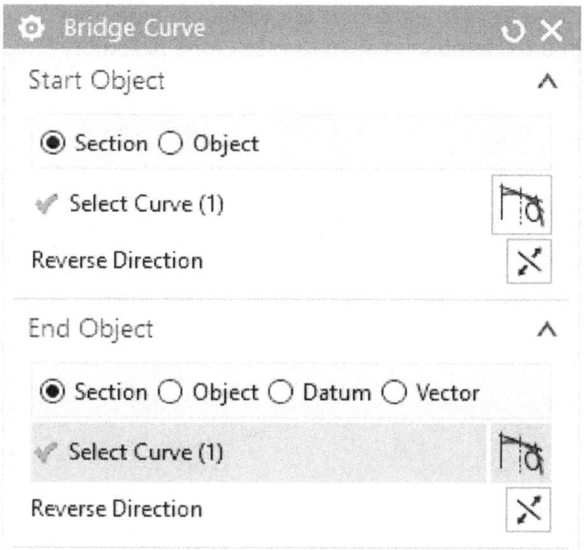

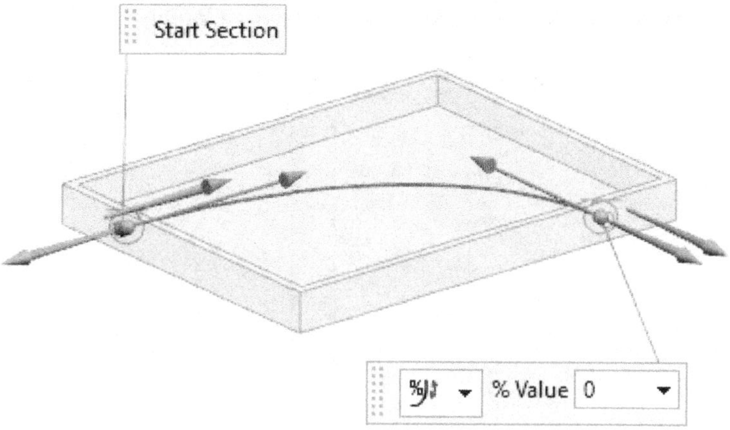

- Use the **Tube** command with **Inner Diameter** *0.25 in*, and **Outer Diameter** *0.375 in* to model the cable.

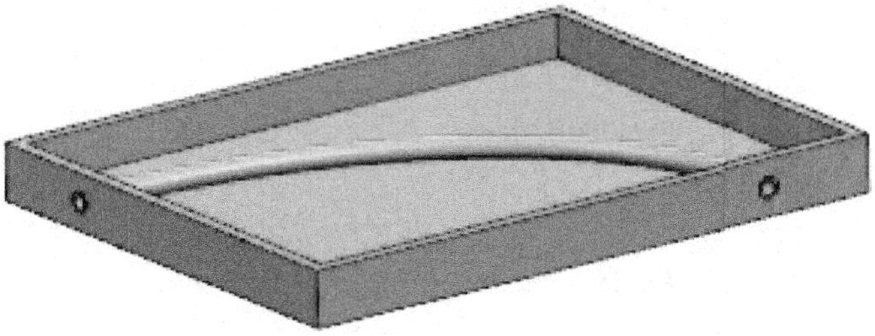

- Now for the magic of the **WAVE Geometry Linker**... make *WAVE Tray* the work part.
- Use **Move Face** to scoot one of the holes *3 in* to the side.
- Marvel at your smart, parametric assembly! The cable follows suit!

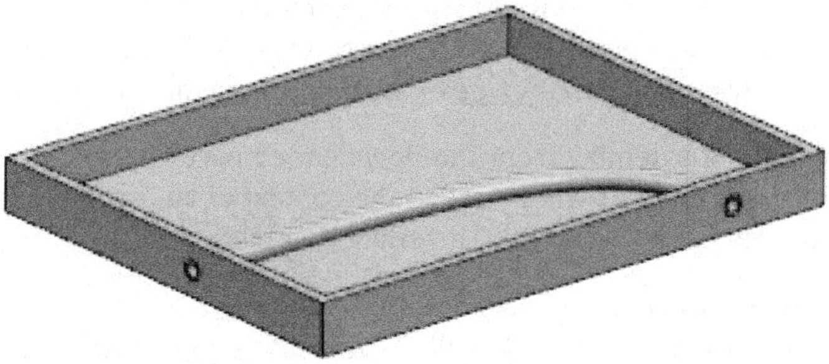

- **Save** your assembly.

18 Advanced Assemblies

In this chapter, we will explore some more advanced tools for modifying and managing assemblies. There are a few themes in this chapter. You will learn to manage different *versions* of components in an assembly, and of assemblies themselves. You will also learn to model assemblies in different states of interest and to display those states on the drawing. Finally, we will share the techniques needed for building fully associative, parametric assemblies with interpart data following our lesson on the **WAVE Geometry Linker** in Chapter 17.8.

18.1 Clone Assemblies

The **Save As** command, when applied to an assembly, does not allow you to copy each part file within an assembly. The operation in NX that best describes a *"Save As"* for each component in an assembly is called **Create Clone Assembly**. The cloning group is found at **Menu / Assemblies / Cloning**.

You may wonder why cloning is necessary, as opposed to simply copy-and-pasting all the files in an assembly from one folder to another within Windows and then working with the copy. One advantage cloning offers is that *WAVE* links are cloned and point to the correct part files, which is not true of assemblies that are copy-pasted within Windows.

In the exercise below you will clone an assembly from the downloaded part files folder into your folder *"C:\My NX 11 Files"*. In light of the issues surrounding **Save As** when used in the context of an assembly (as described in Chapter 2.3.8), assembly cloning offers a real advantage for "versioning" an entire assembly.

- Within *"C:\My NX 11 Files"*, create a new folder called *"Table Assembly"*.
- Your downloaded part files folder contains a folder called *"Table Assembly"* with two folders inside – *"Incomplete"*, and *"Finished"*. Copy and paste the contents of the *"Incomplete"* folder into *"C:\My NX 11 Files\Table Assembly"*.
- Within *"C:\My NX 11 Files\Table Assembly"*, create a new folder called *"Finished"*.

- Select the **Create Clone Assembly** tool.
- On the **Naming** tab, specify a **Default Output Directory** of *"C:\My NX 11 Files\Table Assembly\Finished"*. Click the **Define Naming Rule** button.

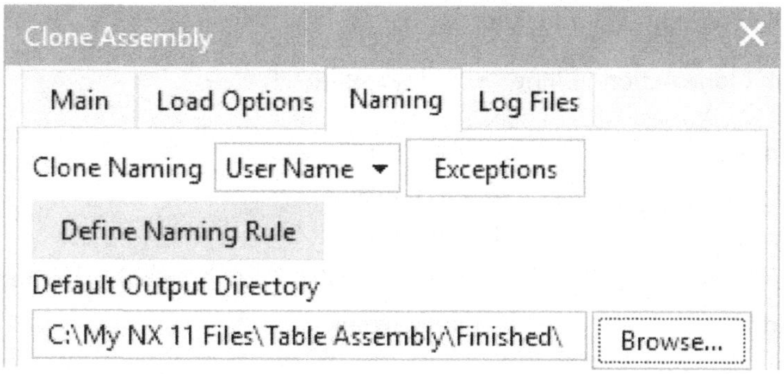

- On the **Naming Rule** dialog, set the radio button to **Add Suffix**. For the suffix to add, specify the string *"_FINISHED"*, and click **OK** to return to the **Clone Assembly** dialog.

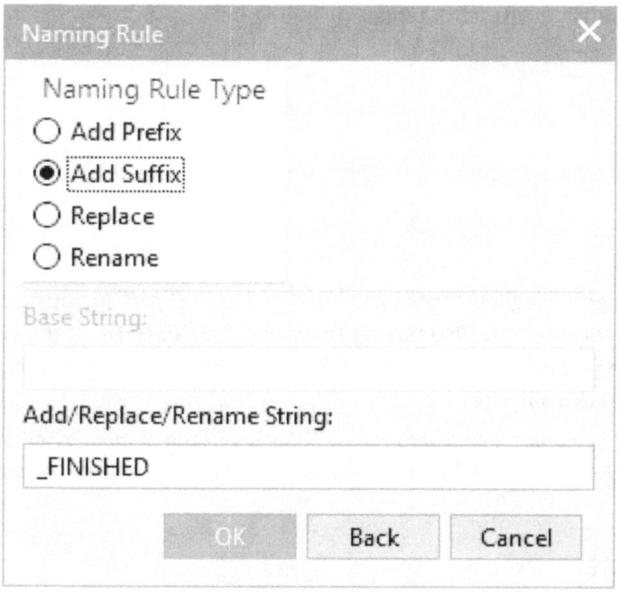

- Switch back to the **Main** tab and push the **Add Assembly** button.

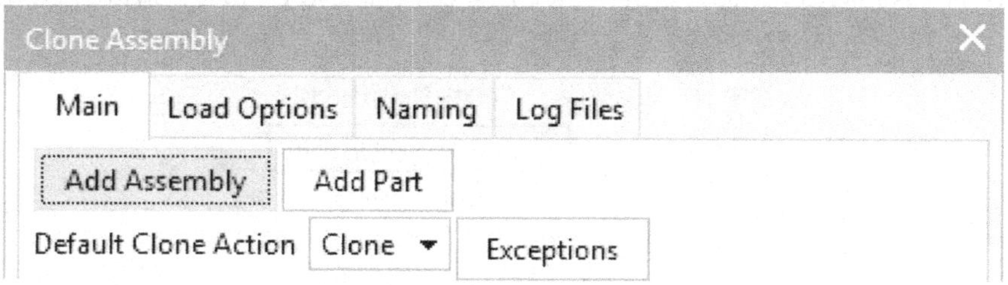

- Select the assembly found at *"C:\My NX 11 Files\Downloaded Files\Table Assembly\Finished\Table Assembly.prt"*.

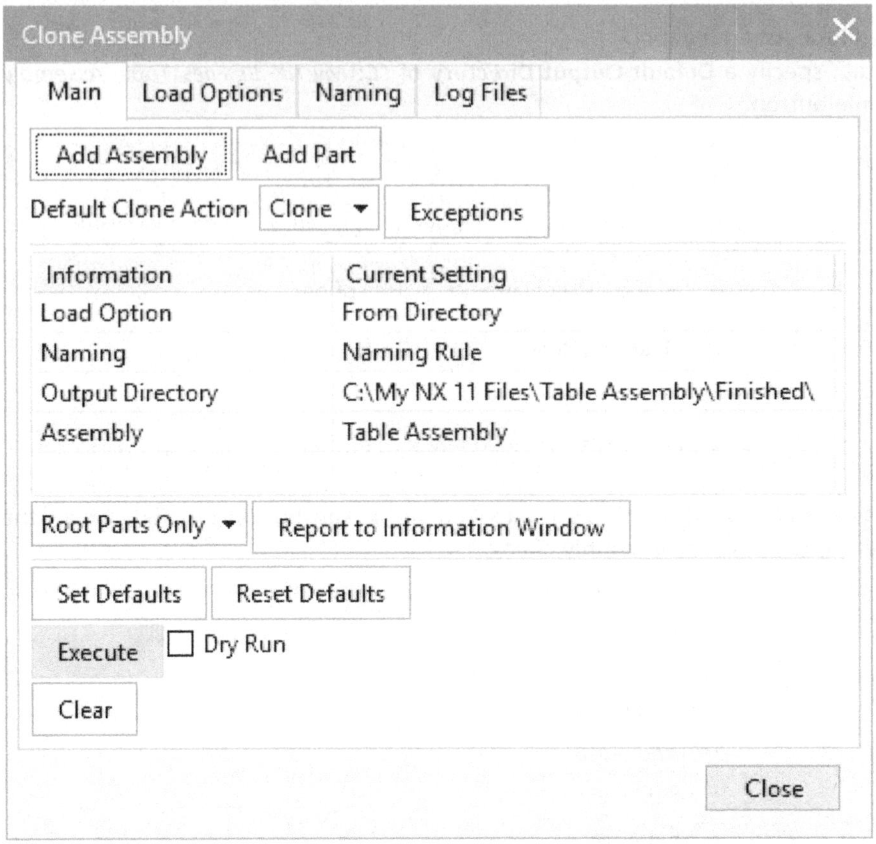

- **Execute** produces a clone assembly called *"Table Assembly_FINISHED.prt"* in your directory *"C:\My NX 11 Files\Table Assembly\Finished"*, with all components also having the string *"_FINISHED"* at the end of their name.
- Thanks to the names not being identical, you can keep the assembly *"Table Assembly_FINISHED"* open while you work on the exercises below, to check your work!

Tip: To create a clone assembly with all names identical to the components of the original assembly, use a suffix consisting of a blank space – just " ". Part file names in NX do not allow trailing whitespace, so this trick will let you clone an assembly to a new directory without modifying the names!

18.2 **Pattern Component**

Component patterns are a great way to bring copies of a component into an assembly in a controlled way. So long as the component being patterned is fully constrained, the new instances will also be fully constrained, thanks to the pattern parameters.

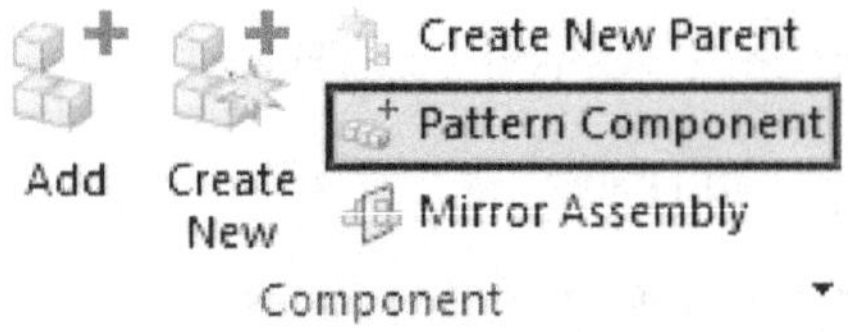

In the exercises below, you will see how to use **Pattern Component** for positioning components using parameters, and by referencing existing patterns.

- Open the file *"Table Assembly – Incomplete.prt"* from *"C:\My NX 11 Files\Table Assembly"*.

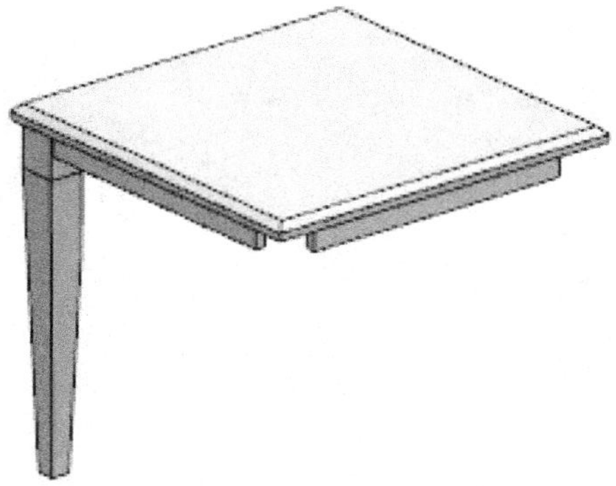

- Your assembly tree should appear as shown below. Note that there is already a circular component pattern in the assembly!

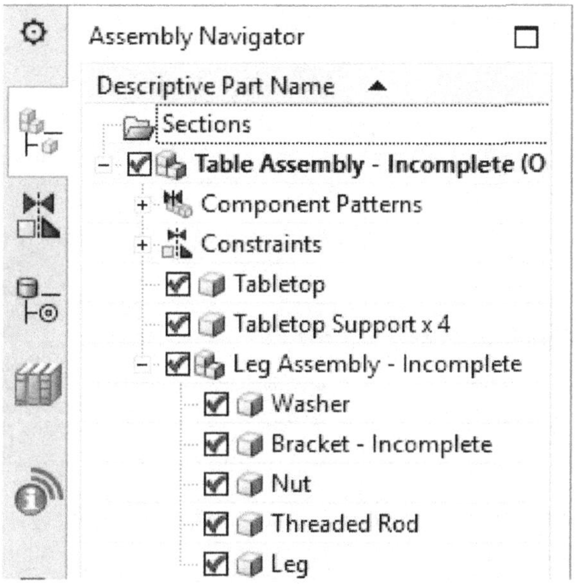

- Select the **Pattern Component** tool. Select *"Leg Assembly – Incomplete"*, as shown below. If you are selecting from the Graphics Window, make sure that you see *"Leg Assembly – Incomplete"* highlight, and not one of the components within the subassembly. If necessary, you can pick *"Leg Assembly – Incomplete"* from the Assembly Navigator.

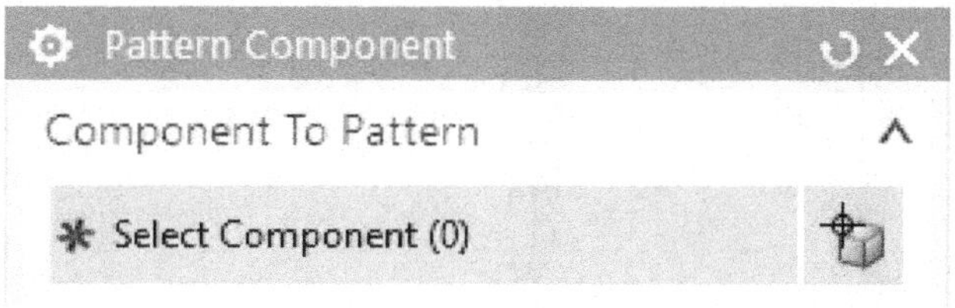

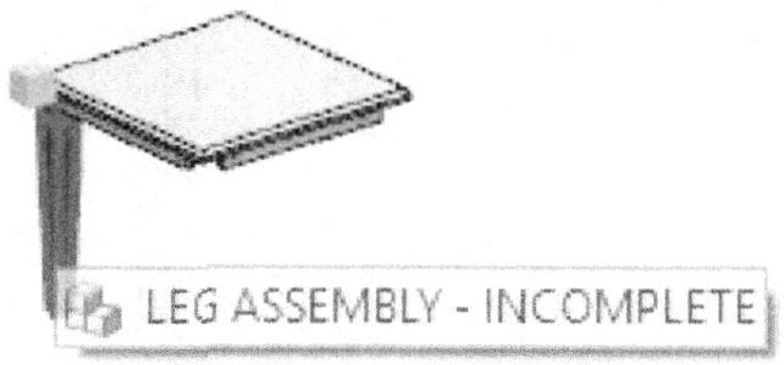

- For the **Rotation Axis**, select the top face of the tabletop. It will infer a vector normal to that face, and positioned at the center of the face. Pretty convenient!

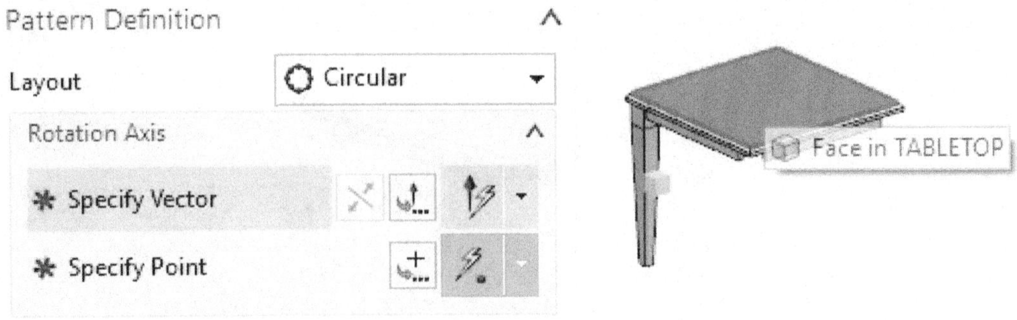

- Set **Spacing** to *Count and Span*, and set the **Count** and **Span Angle** to *4* and *360 degrees*, respectively.

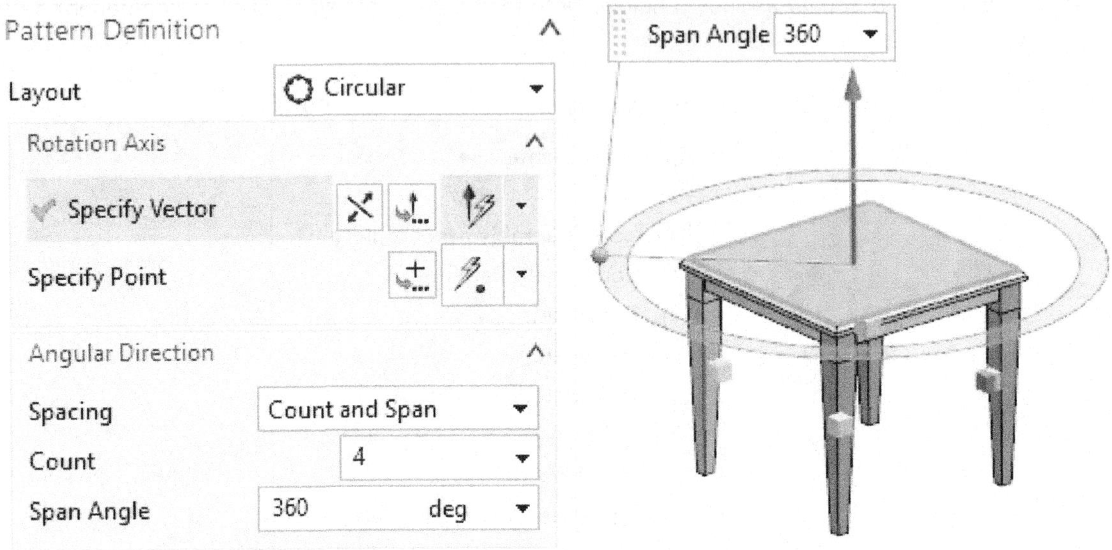

- This positions four copies of *"Leg Assembly – Incomplete"* at the four corners

18.2.1 Pattern By Reference

Patterns created by reference to another pattern follow the parent pattern associatively. This is a great way to position fasteners and related parts, but requires some planning in terms of how you build your models. In this exercise, you will modify the bracket that holds the leg to the supports so that you can create a component pattern by reference. In doing so, you will learn some of the behavior of **Save As** in the context in an assembly.

- Continue working with *"Table Assembly – Incomplete"*.
- Within *"Leg Assembly – Incomplete"*, find *"Bracket – Incomplete"*, right-click, and select **Make Displayed Part**.

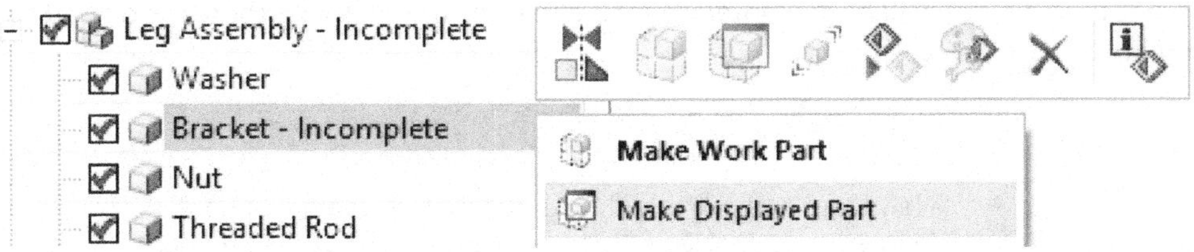

- For each of the two final *Simple Screw Hole* features in the part navigator, make the sketch external.

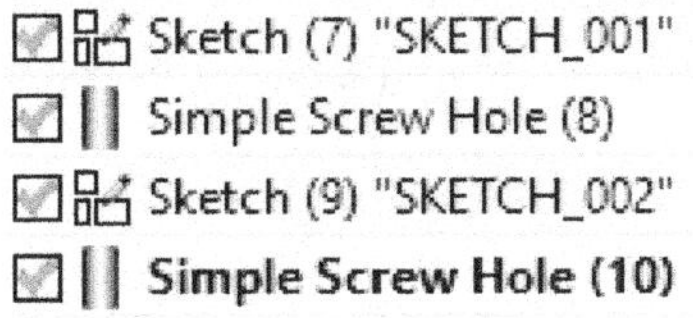

- You will now be able to see the points that position the holes on the flanges of the bracket.

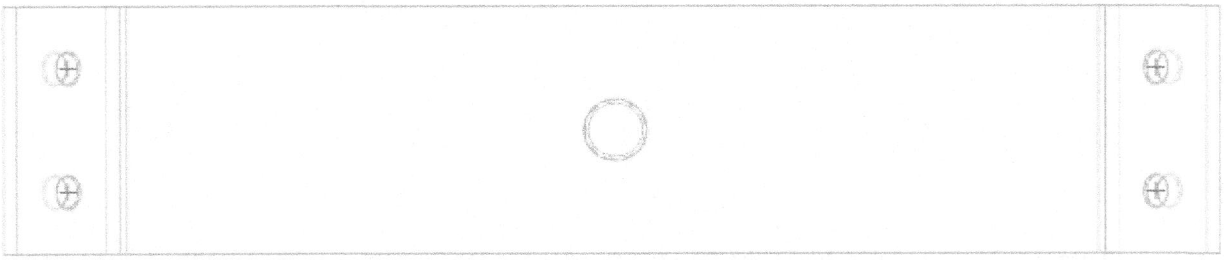

- **Delete** *Simple Screw Hole (8)*. Double-click *Simple Screw Hole (10)* to edit it.
- You must deselect the points of the sketch using **Deselect All**.

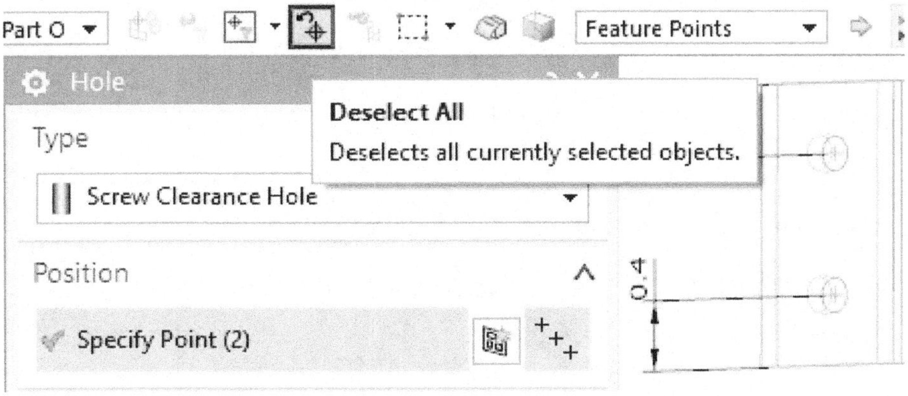

- Change your **Selection Intent** drop-down to *Single Point* and choose only the top point in the sketch.

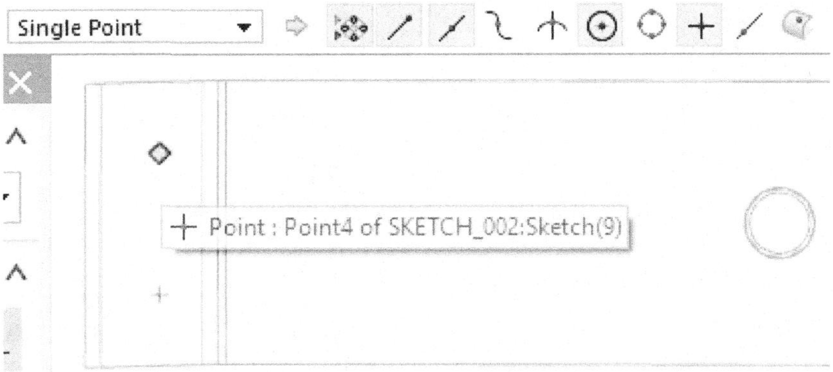

- Click **OK** to accept the edit. You will be prompted to map curves of the section – there is no need to do this, just click **No** when the **Map Parents** dialog appears.

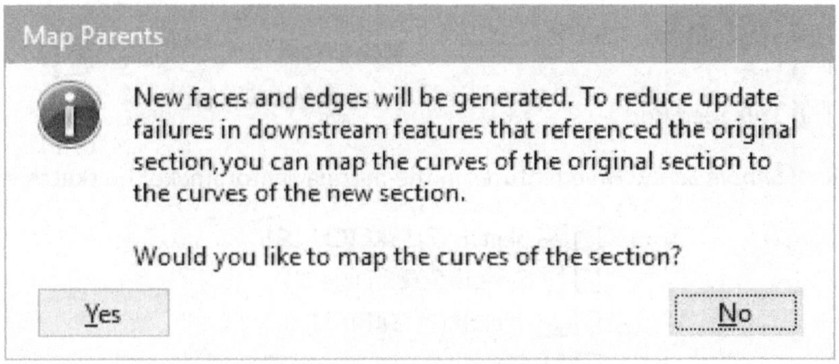

- At this point you should have only one simple screw hole on the flanges, as shown below.

- Select the **Pattern Feature** tool and choose the *Simple Screw Hole* as the feature to pattern. Set the **Layout** drop-down to *General*. For the **From** point, select the point at the center of the hole, and for the **To** points, select the other three sketch points. Set the **Pattern Method** to *Simple*. You will need to check the *Follow Face* box in the **Orientation** section of the dialog, and select the tangent faces of this side of the bracket.

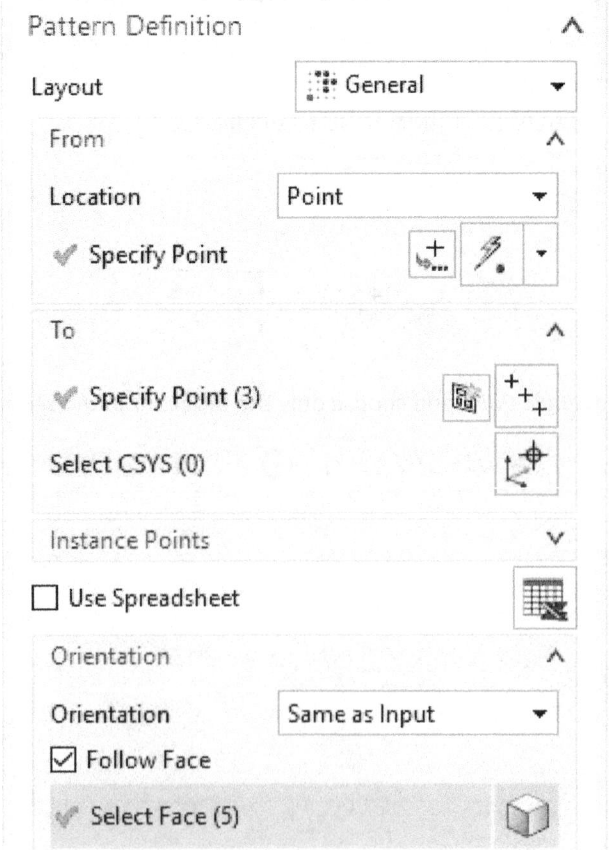

- This rebuilds the holes in a way that will allow for easy component patterning.

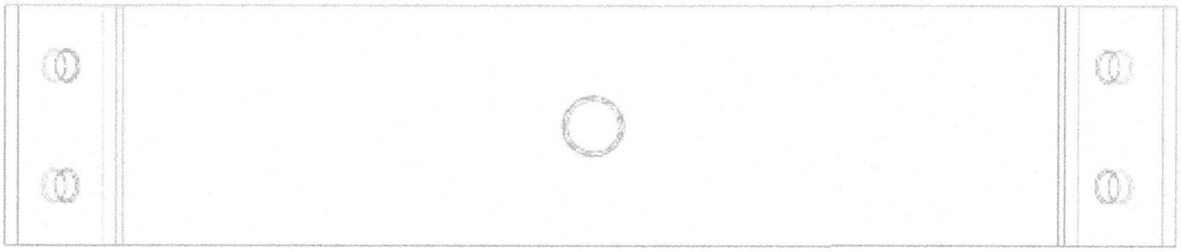

- Use **File / Save / Save As** to save your modified version of the bracket as *"Bracket.prt"*.
- Upon clicking **OK**, the **Save As** dialog will reopen, asking you to name a copy of *"Leg Assembly – Incomplete"*. Type in *"Leg Assembly.prt"*.
- Once more, the **Save As** dialog will appear, this time wanting a name for a copy of *"Table Assembly – Incomplete"*. Type in *"Table Assembly.prt"*.
- Click **OK** on the **Save As** dialog that asks if you want the **Save As** to continue.

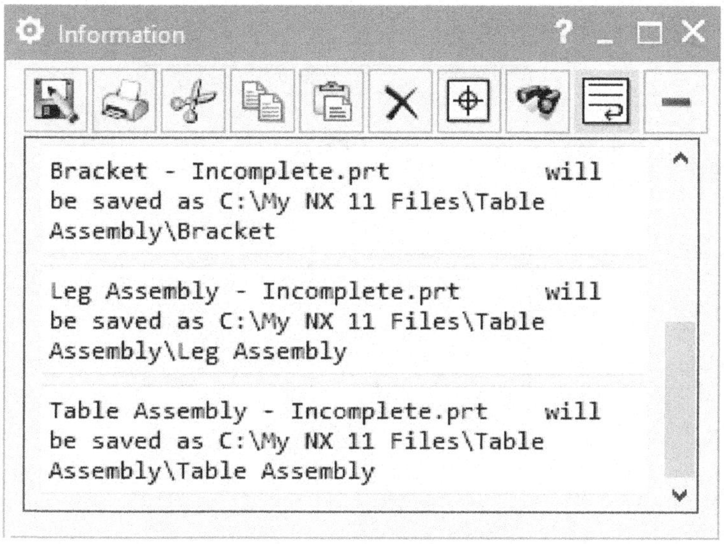

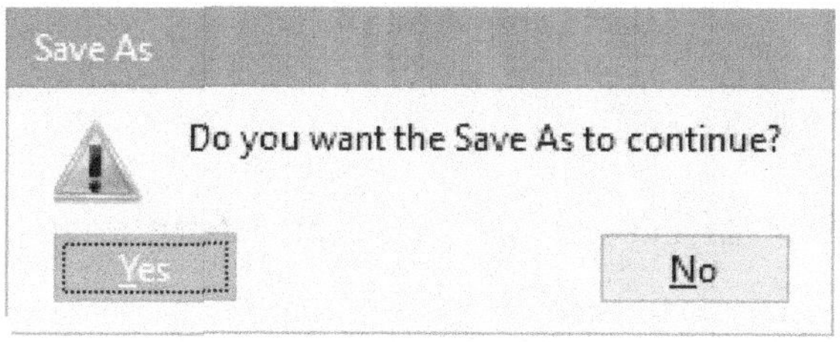

- In *"Table Assembly"*, the Assembly Navigator now should appear as shown below. All four instances of *"Leg Assembly"* were renamed.

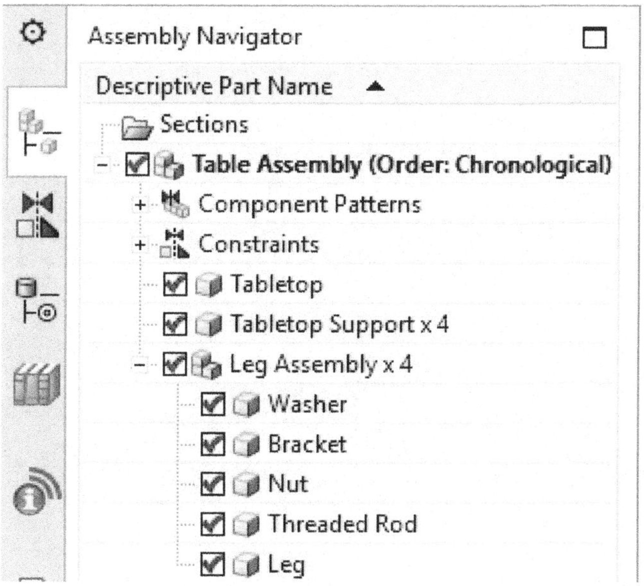

- Make *"Leg Assembly"* the Displayed Part.
- Use **Add Component** to add a copy of *"Screw.prt"* (found in *"C:\My NX 11 Files\Table Assembly"*) to *"Leg Assembly"*, and use a *Concentric* constraint to position it in the first screw hole on the bracket, as shown below.

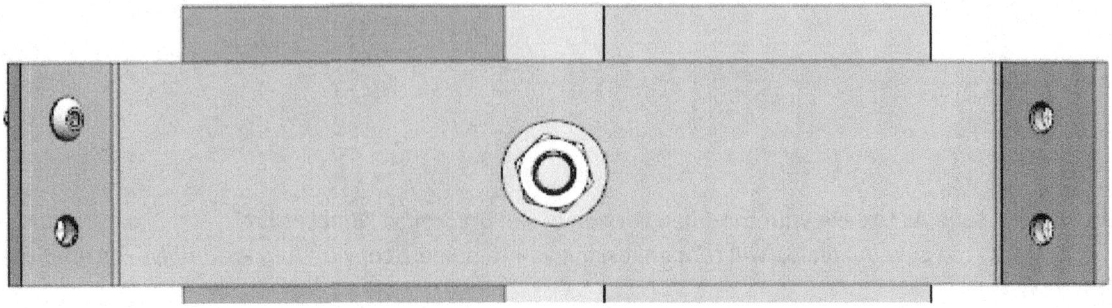

- Select the **Pattern Component** tool, and set the **Layout** to *Reference*. It will detect and automatically select the existing **Pattern Feature** entity since the screw is constrained to an edge of that feature. It then automatically places three more screws in the other holes. Click **OK** to generate the component pattern!

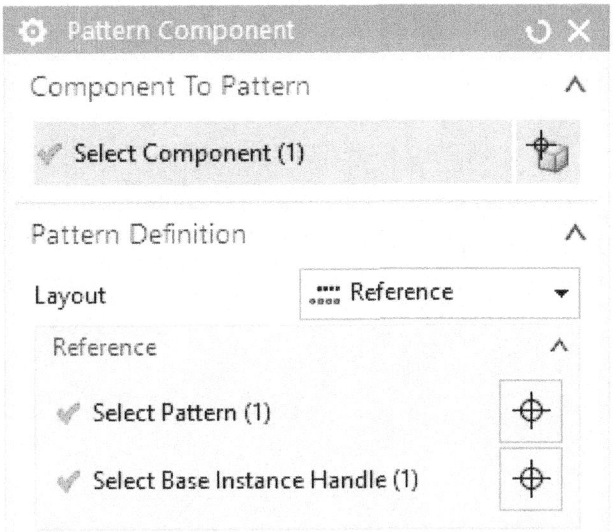

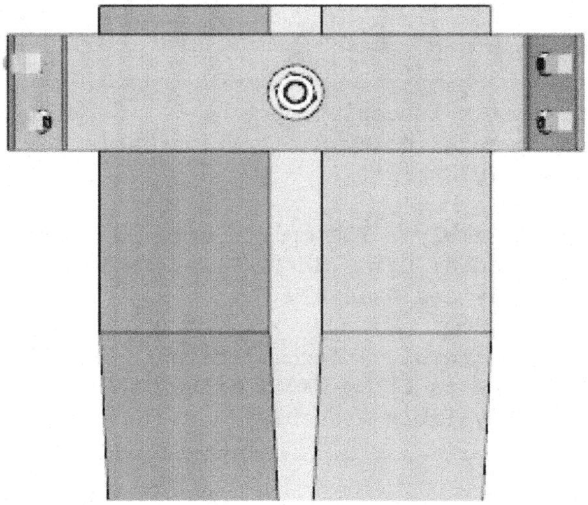

- Upon closer inspection, you will find that the patterned screws don't quite align flush with the face of the bracket.

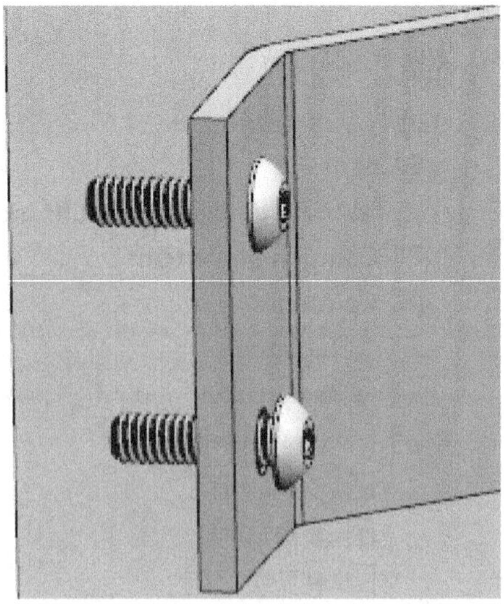

- The reason for this behavior is a rather subtle aspect of **Pattern Feature** called the **Reference Point**. Make *"Bracket"* the Work Part and edit the **Pattern Feature** entity. Notice that the **Reference Point** is at the center of mass of the screw hole.

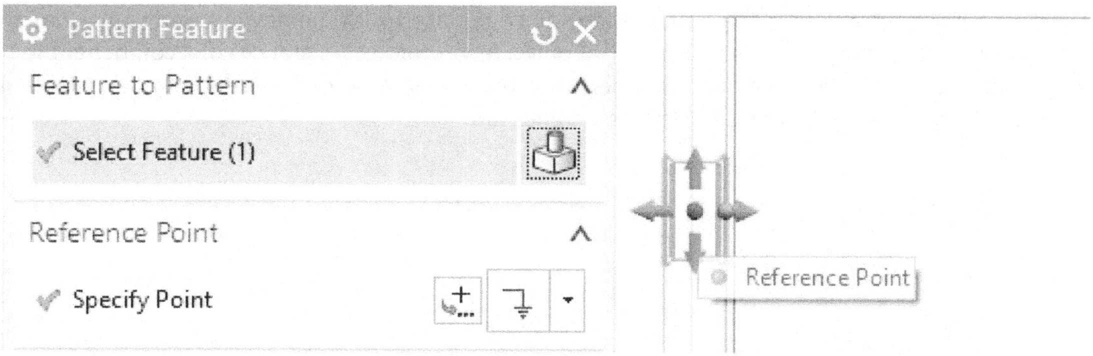

- Drag the reference point until it coincides with the origin point of the general pattern.

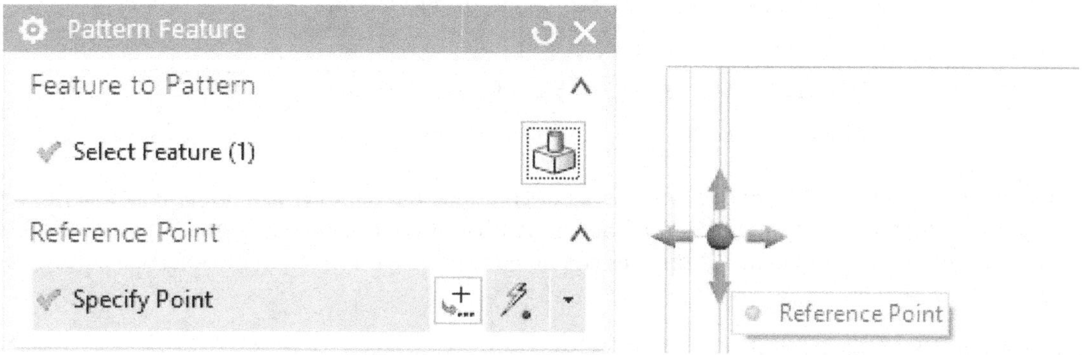

- The precise positioning of pattern members depends on the relationship between the *Pattern Origin* and the **Reference Point**. Practically speaking it often has no effect on whether the pattern will be created or not, but in this case, it impacted the reference pattern. After making the **Reference Point** and *Pattern Origin* coincide for the **Pattern Feature** entity, click **OK** and your screws will reposition.

- **Save** your *Table Assembly*!

18.3 Make Unique

So far in this chapter, we have seen that "**Save As**" for assemblies is a bit more complex than it is for parts. Recall from Chapter 2.3.8, that when you perform a **Save As** on a component within an assembly, all instances of that part in all assemblies that you have open in your current NX session will be replaced with the newly saved copy. Sometimes this is good, but there are also many situations in which you want only a few – not ALL – instances of a component to be replaced with some new version of a part. The command that performs this special kind of "Save As" is called **Make Unique**. It is found in the **More** gallery on the far-right side of the Assemblies tab.

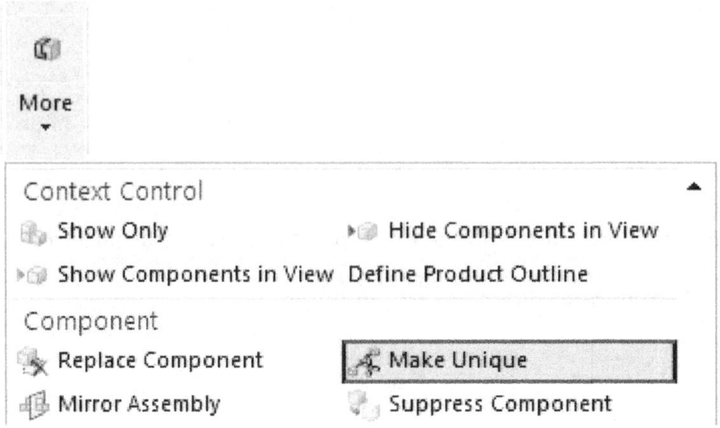

You will find this tool extremely useful when, in the course of creating an assembly, you find it necessary to modify some number of components that you first thought would be all identical. In our example, suppose that instead of our table assembly having four identical legs, that two of the legs should have casters on them. In the exercise below, you will make this modification to the assembly.

- Continue working with *"Table Assembly"*.
- Although **Make Unique** is found on the ribbon, it is much more convenient to access with a right-click in practice. Unpack and select the two front instances of *"Leg Assembly"* (either from the Assembly Navigator or the Graphics Window) and then right-click on one and select **Make Unique**.

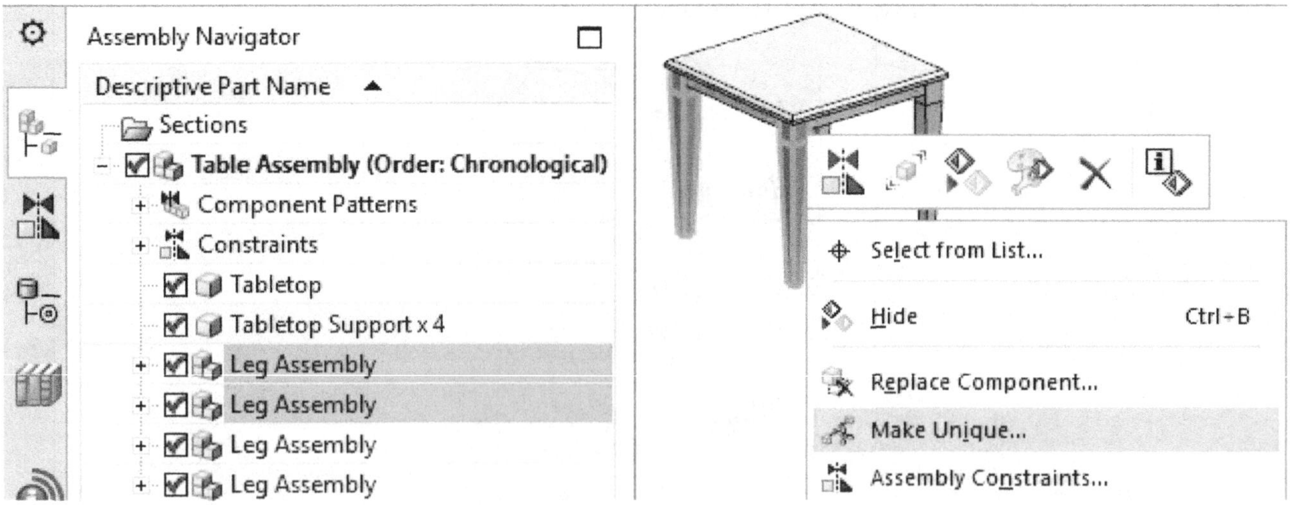

- On the **Make Unique** dialog, click the **Name Unique Parts** button, and name the new copies *"Leg Assembly – Short – Left.prt"*. Click **OK** to run the **Make Unique** tool.

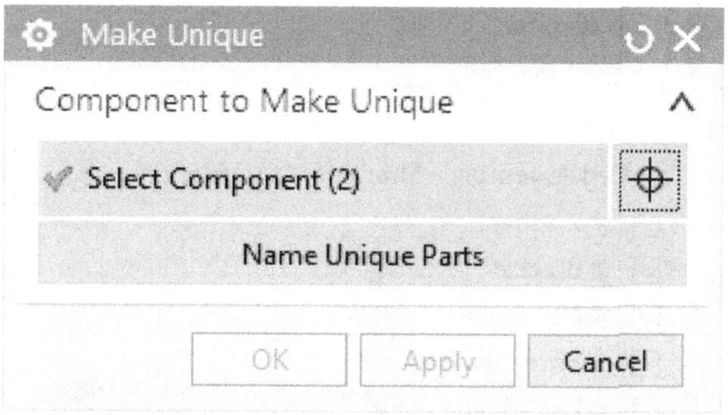

- Your Assembly Navigator will now appear as below.

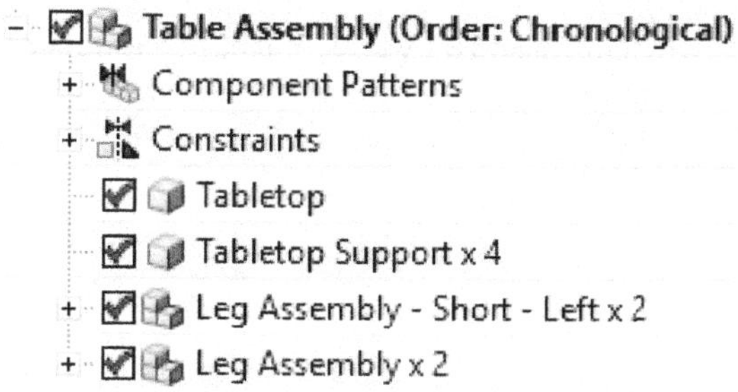

- Make *"Leg Assembly – Short – Left"* the Displayed Part, and use **Make Unique** to rename a copy of *"Leg"* to *"Leg – Short – Left"*, so that your assembly tree appears as shown below.

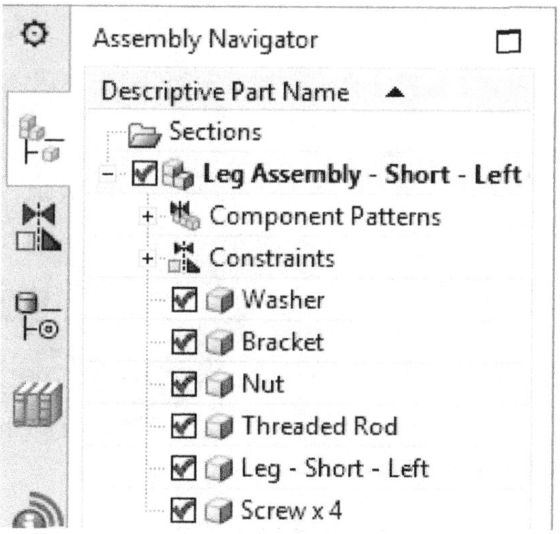

- Use **Add Component** to bring in a copy of *"Caster Assembly – Incomplete"* (found in *"C:\My NX 11 Files\Table Assembly"*).

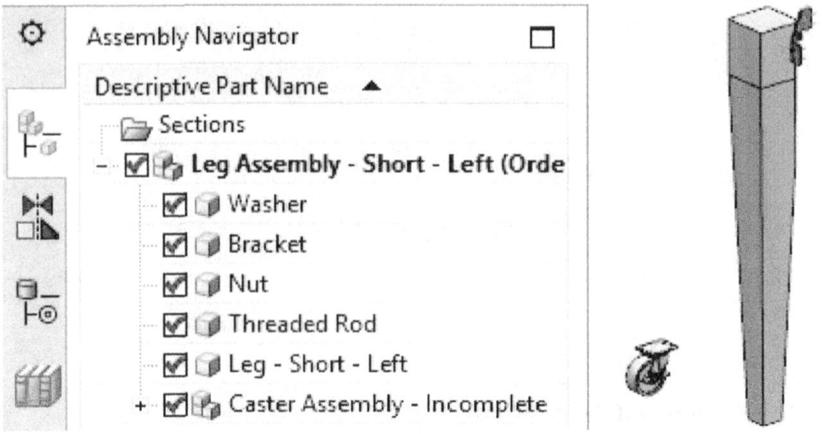

- We want to shorten the leg, but we need to know by how much. Use **Measure Distance** to make a *Projected Distance* measurement as shown below to find the overall height of the caster.

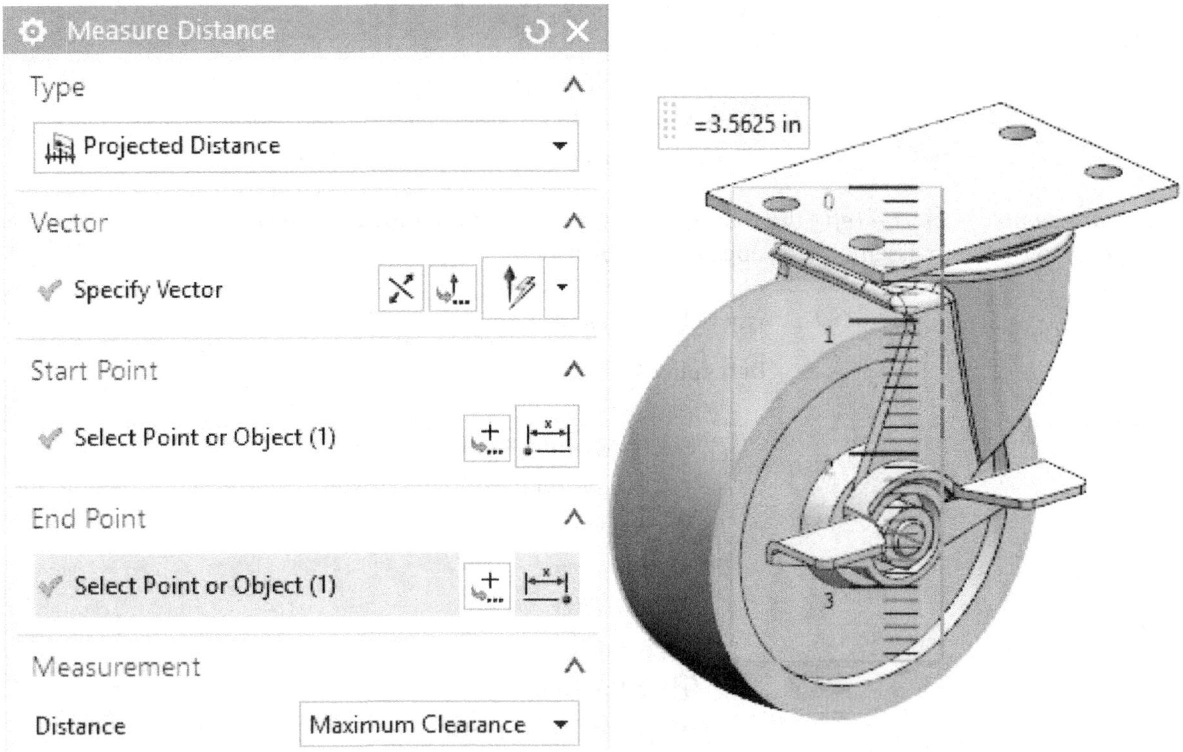

- Make *"Leg – Short – Left"* the Work Part, and use **Move Face** to shorten the leg by *3.5625 in*.

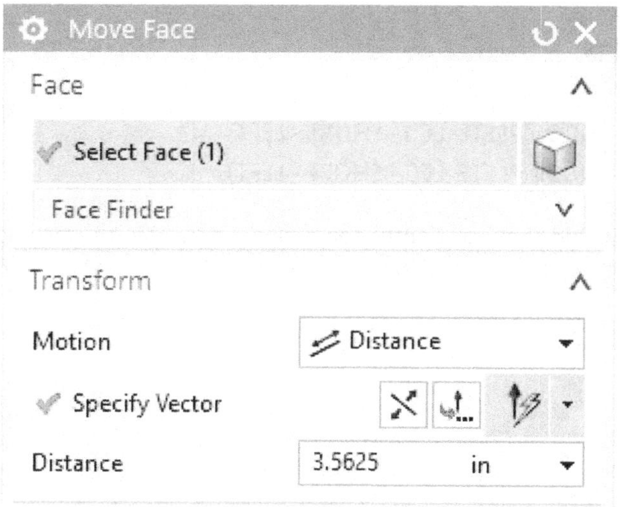

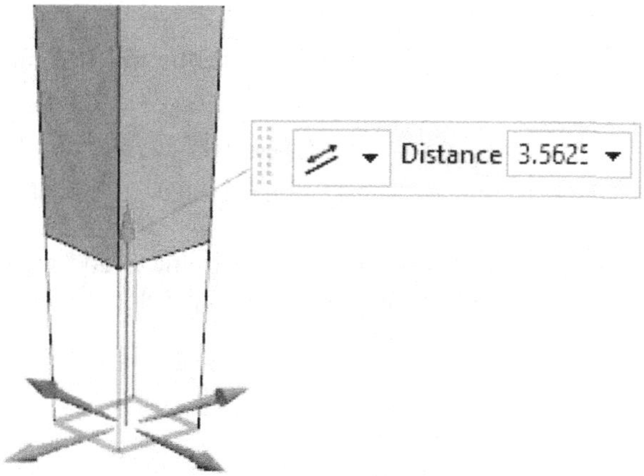

- Use **Assembly Constraints** – *Fix* to keep the leg from moving.
- Next, use **Assembly Constraints** – *Center*, with **Subtype** *2-to-2*, to center the edges of the caster plate between the edges of the shortened leg. The caster plate is actually too big to fit the bottom of the leg!

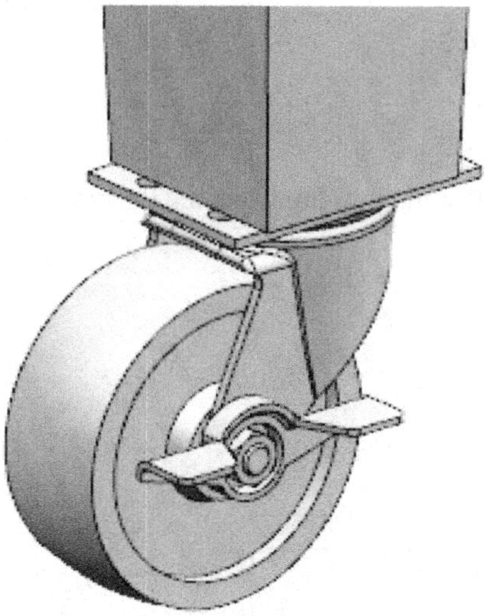

- Make *"Leg – Short – Left"* the Work Part and use **Delete Face** to delete the outer drafted faces so that they become rectangular. This will clear enough space for the caster to fit to the bottom of the leg.

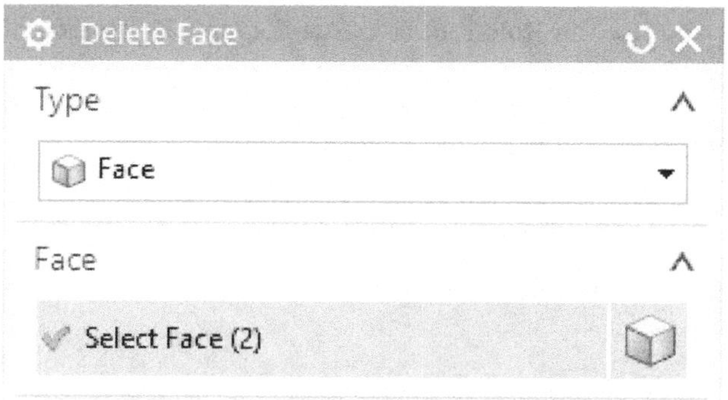

- **Delete Face** screws up your *Center* constraints because the edges referenced no longer exist!

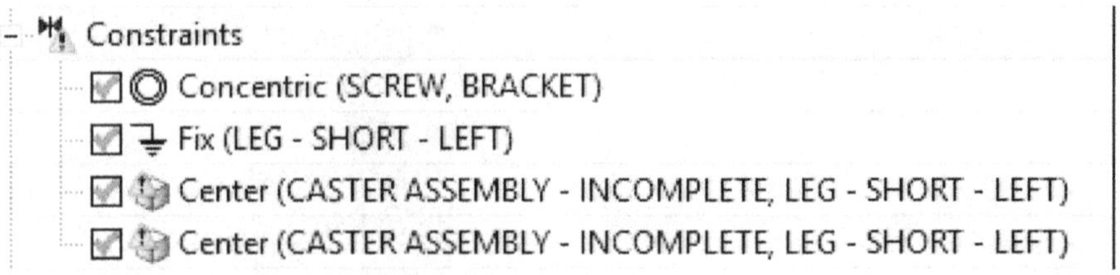

- Double-click a broken *Center* constraint to edit it. You will find that the *Center* constraint is only able to track down three of the edges involved originally. Select the missing edge, as shown below.

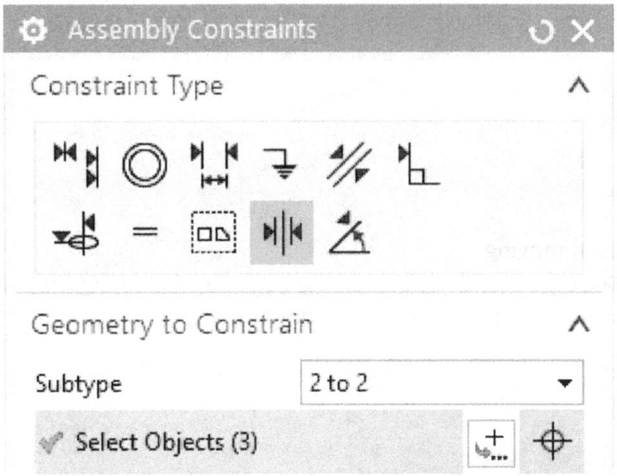

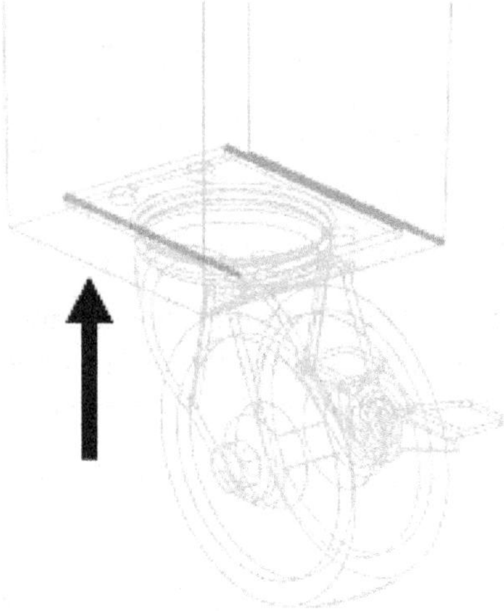

- Do the same for the other *Center* constraint. Now the leg is big enough to fit the caster.

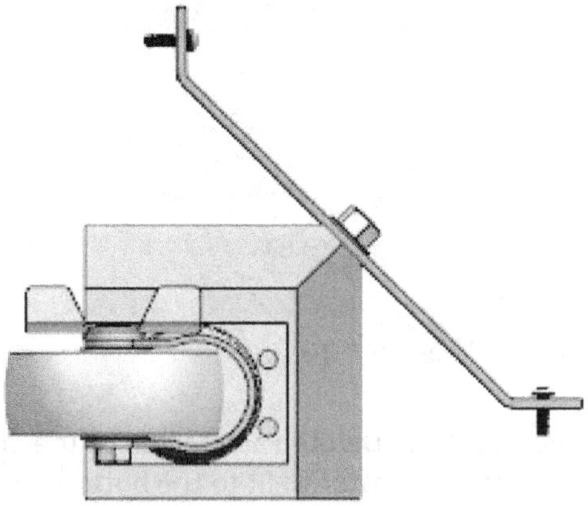

- Make *"Leg – Short – Left"* the Work Part and use the **Hole** tool to create four *#8-32 Threaded Holes 3 in* deep, and aligned with the holes in the caster plate.

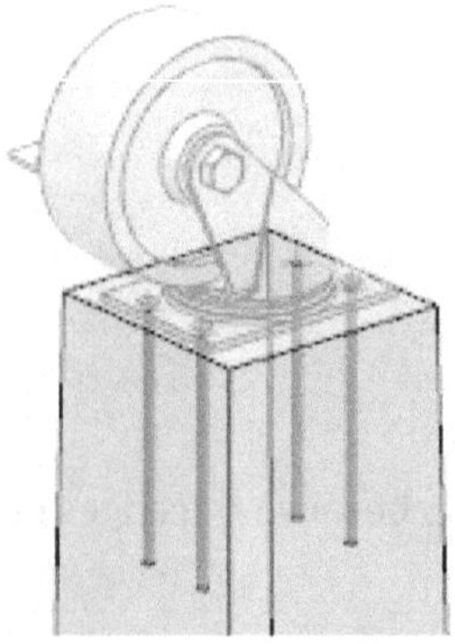

- Next you will modify the caster mount plate so that the holes are created as a pattern, and you can easily position screws using **Pattern Component**. First, use **Make Unique** to rename *"Caster Assembly – Incomplete"* and *"Caster Plate – Incomplete"* as shown below.

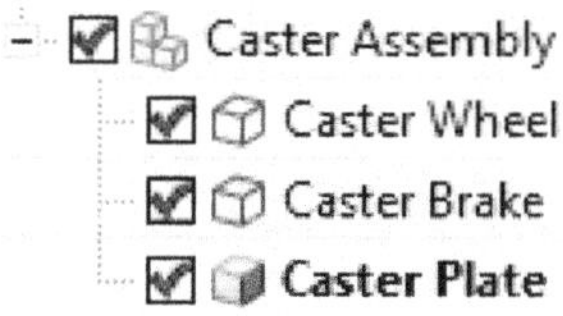

- Make *"Caster Plate"* the Work Part.
- Create a sketch consisting of four points, at the centers of each hole.
- Use **Delete Face** to delete three of the holes.

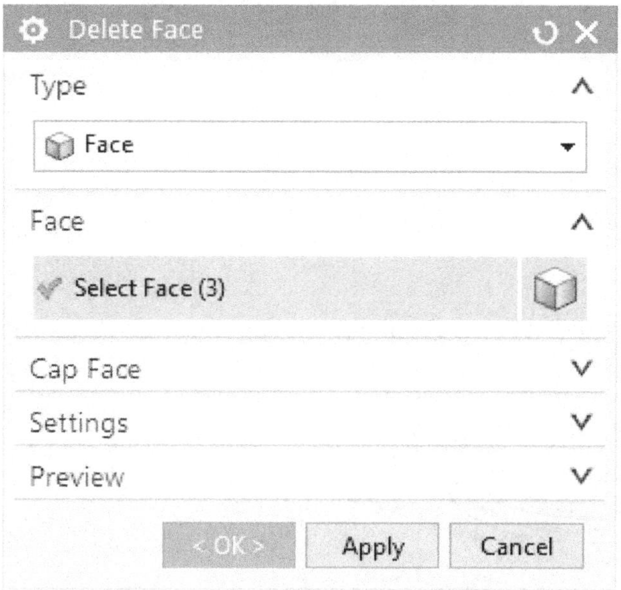

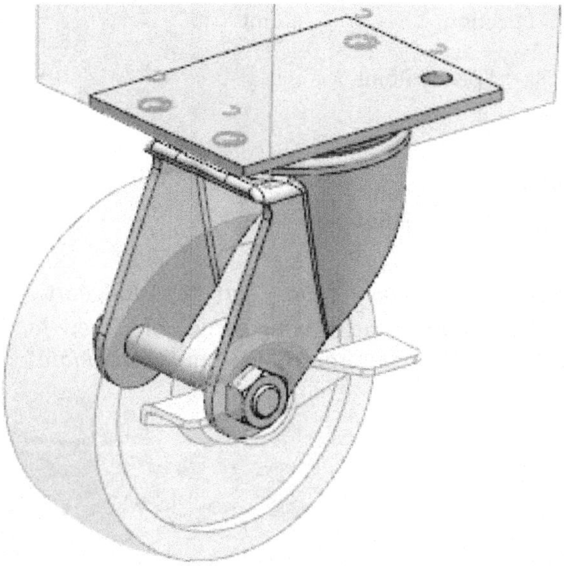

- Use **Pattern Face** with **Layout** set to *General* to copy the last hole back to the other three locations.

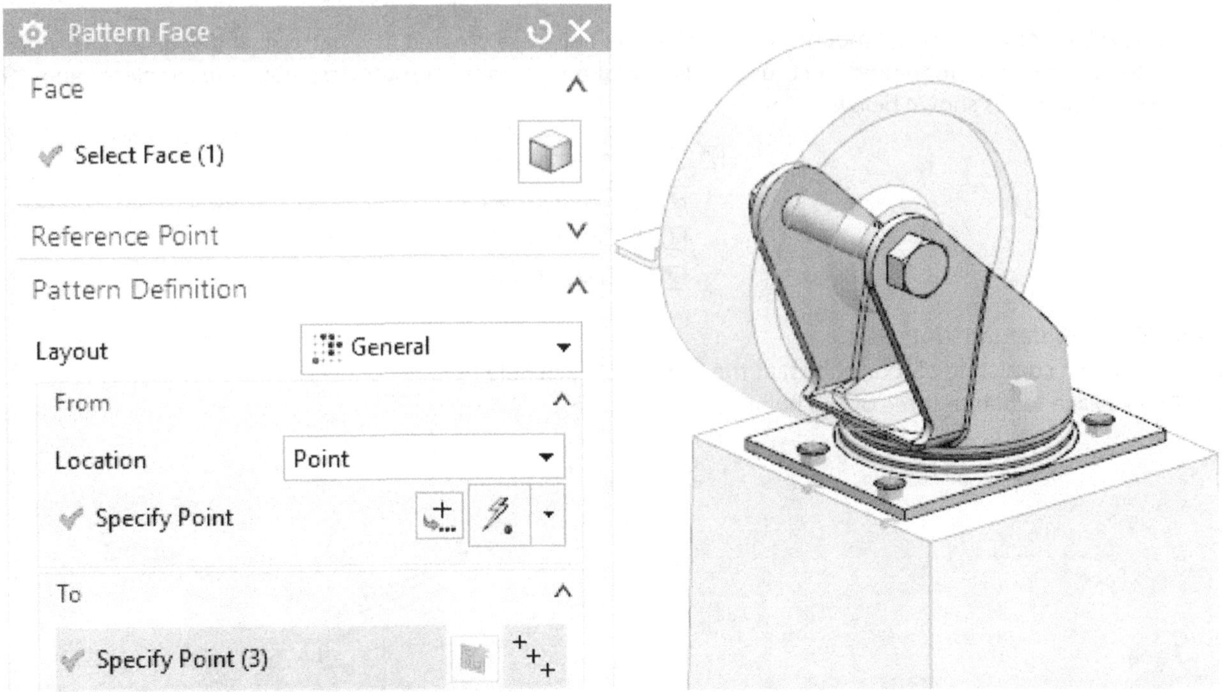

- Make *"Leg Assembly – Short – Left"* the Work Part.
- Use **Add Component** to bring a copy of *"Screw for Caster"* into the assembly, and use a *Concentric* constraint to position it in the patterned hole from *"Caster Plate"*.

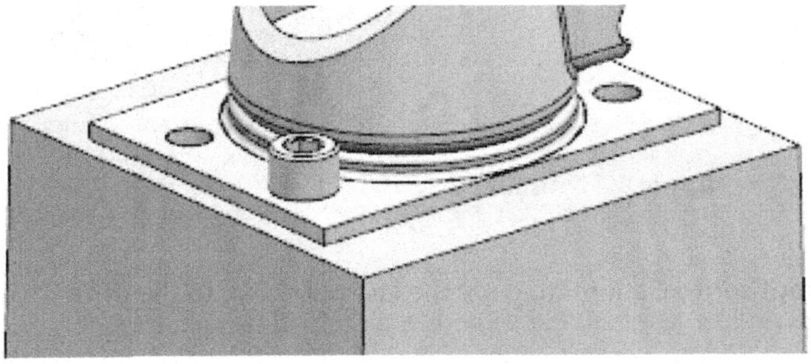

- When prompted for a *Base Instance Handle*, choose the *Instance [0,0] Handle*, as shown below.

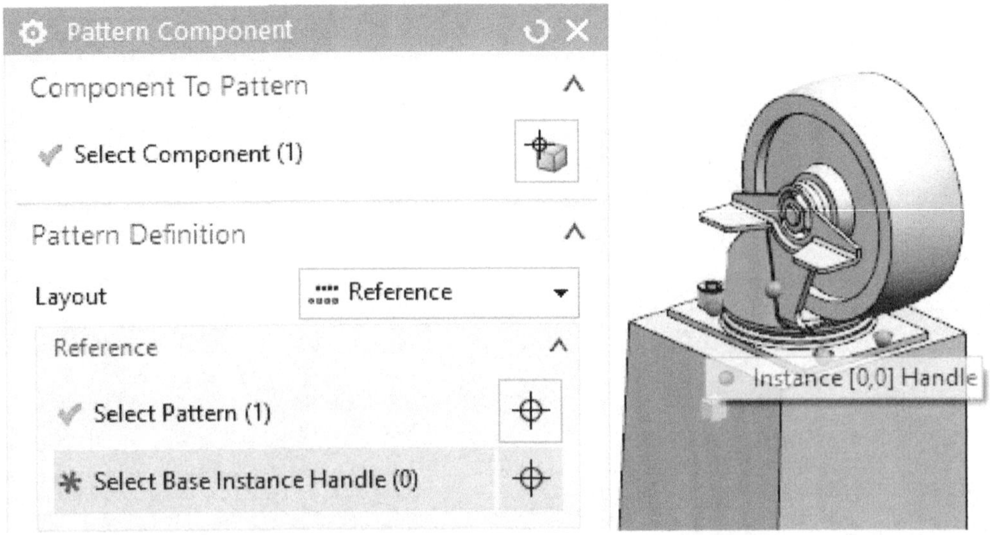

- It is sometimes necessary to input the *Base Instance Handle* as we just did, although often NX is able to infer it.

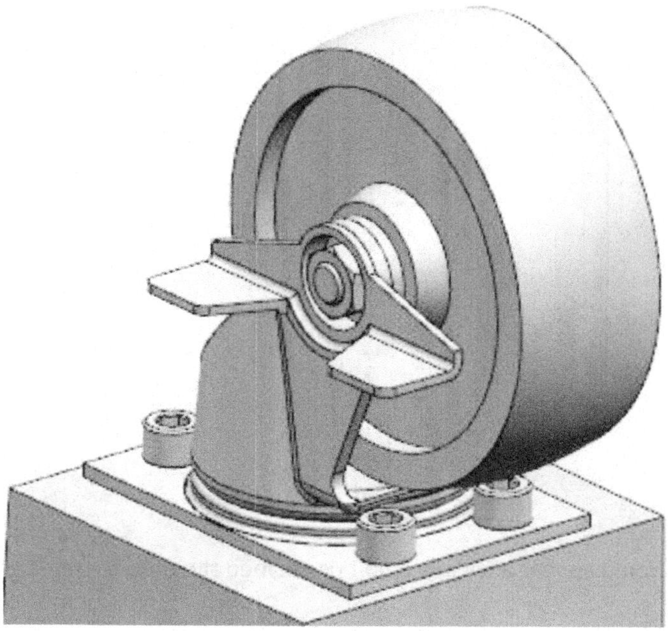

- Return to the *"Table Assembly"*. Uh-oh! One of the legs is sideways.

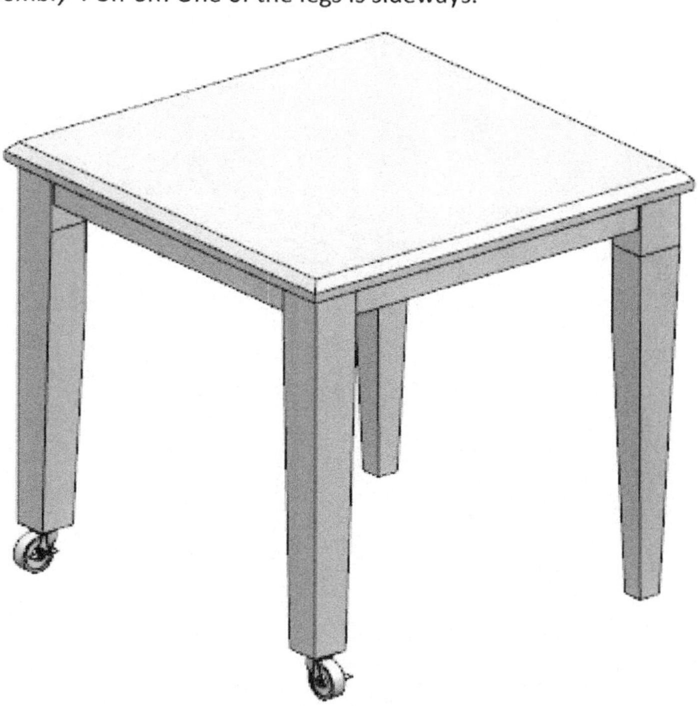

- Use **Make Unique** to rename the second copy of the leg assembly with the caster as *"Leg Assembly – Short – Right"*. Perform the necessary modifications to *"Leg Assembly – Short – Right"* so that the caster is pointing the right direction. (You can check your work against the clone assembly *"Table Assembly_FINISHED"* that you created in Chapter 18.1).

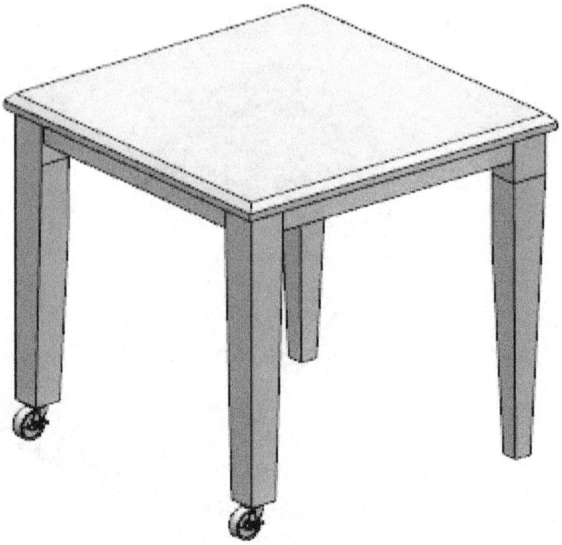

- Your Assembly Navigator should appear as below if you performed the modifications correctly!

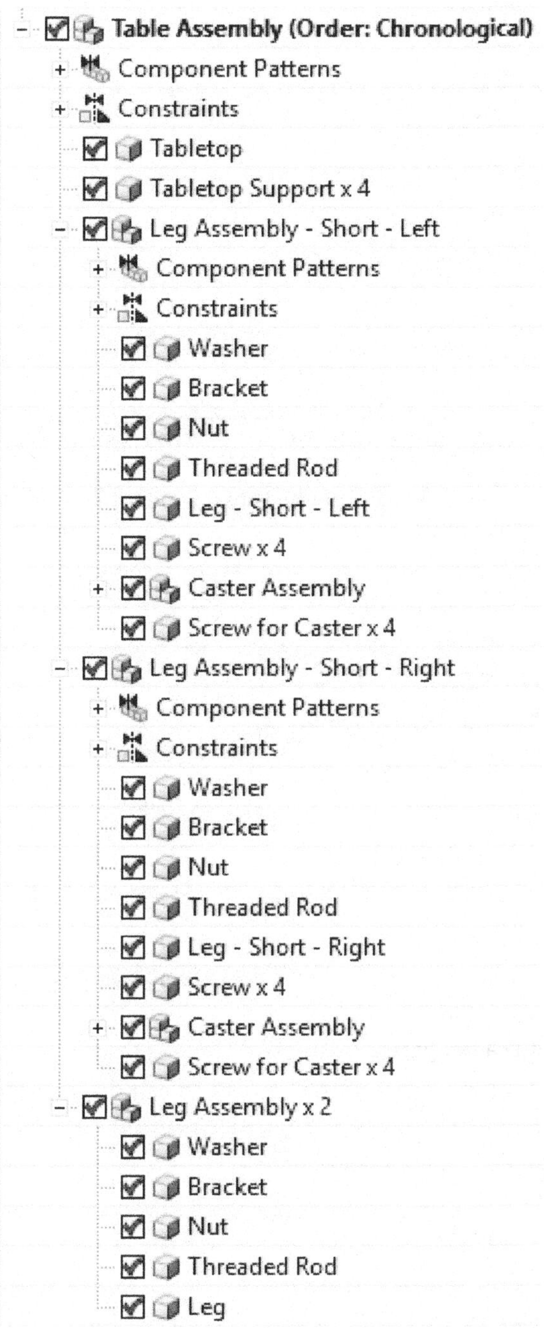

- **Save** your assembly!

18.4 **Replace Component**

Once you begin handling multiple versions of the same part, or parts that were made by modifying a part and then doing a **Save As** or a **Make Unique**, you may want to interchange those parts in an assembly. **Replace Component** allows you to replace any one component in an assembly with any *".prt"* file, but is ideally suited for replacing one version of a part with another.

Replace Component is found in the **More** gallery on the far-right side of the Assemblies tab, but much like **Make Unique**, you will find it easier to access by right-clicking from the Assembly Navigator.

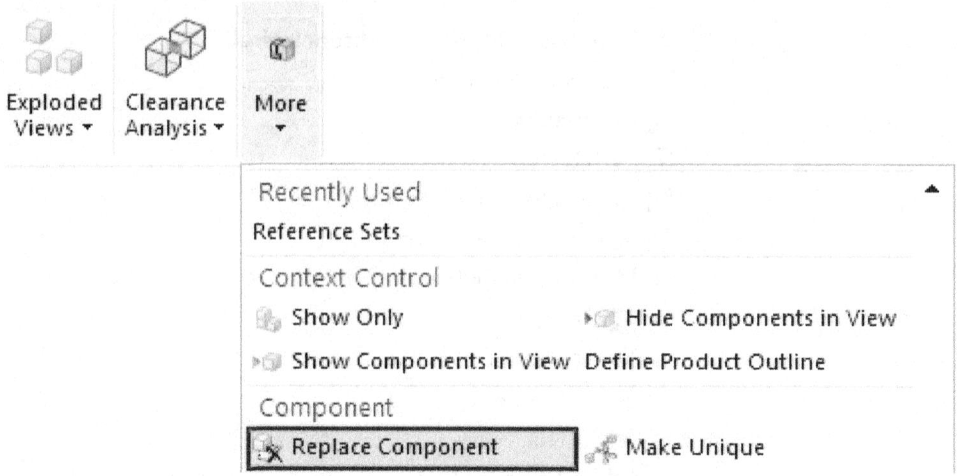

Replace Component is very easy to use – the exercise below is really meant to make sure you know where to find it!

- Continue working with *"Table Assembly"*.

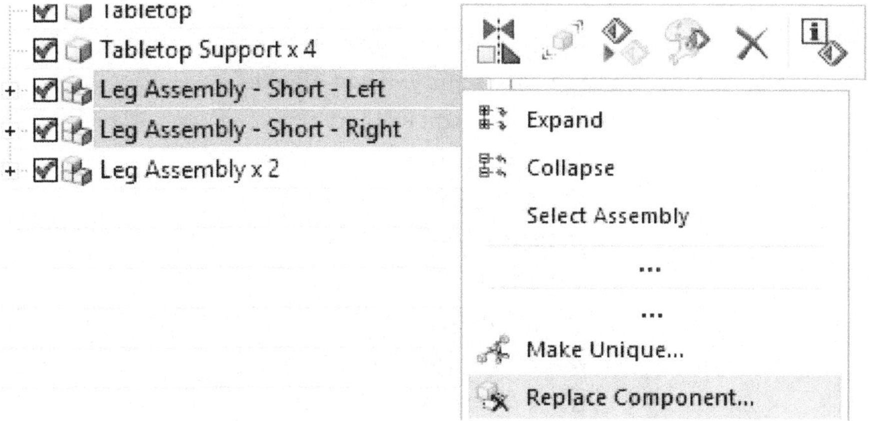

- Find *"Leg Assembly"* in the list of **Loaded Parts**.

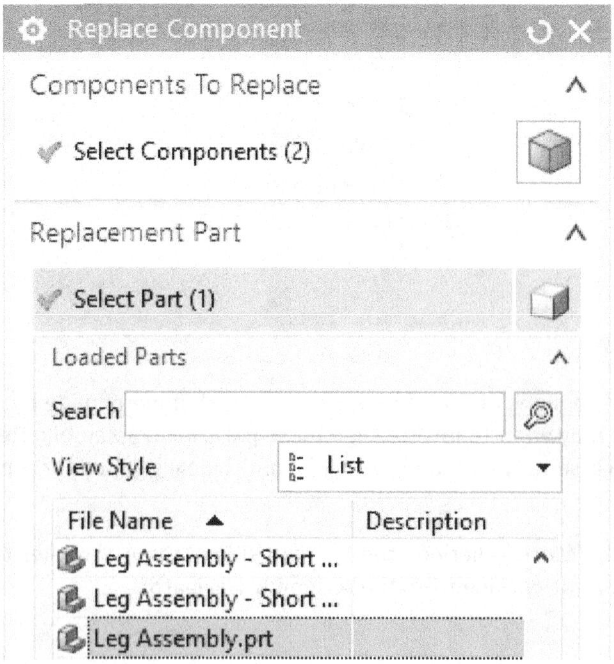

- The result is an earlier version of *"Table Assembly"*.

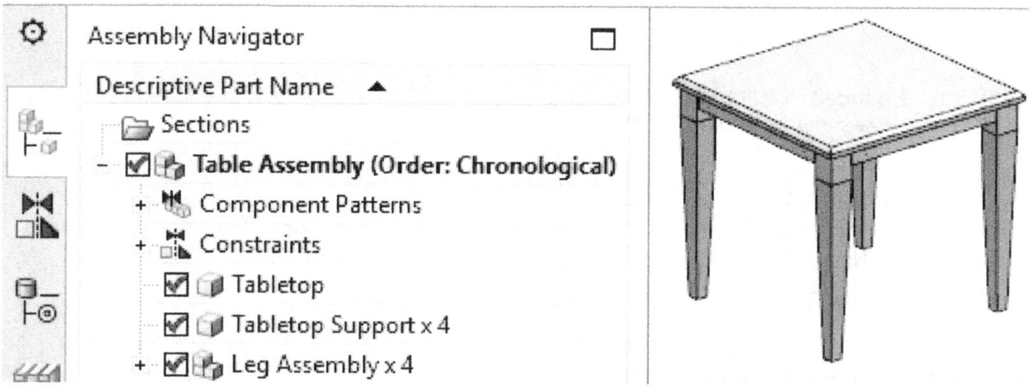

- For practice, use **Replace Component** twice more to bring back the legs with casters!

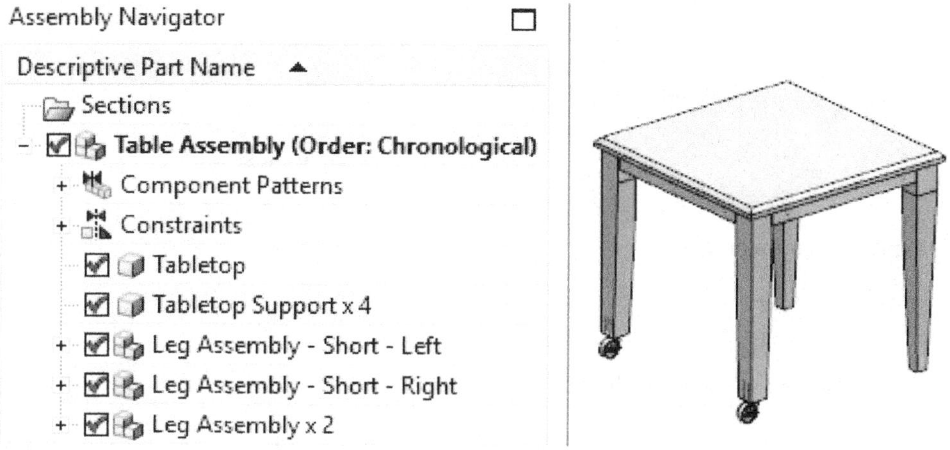

- **Save** your assembly!

18.5 Exploded Views

Exploded views allow you to "explode" parts away from their assembly location to better show how parts should be put together in an assembly. These are helpful for assembly schematics and instructions. Tools for managing exploded views in an assembly are found in the **Exploded Views** gallery, on the far-right side of the **Assemblies** tab.

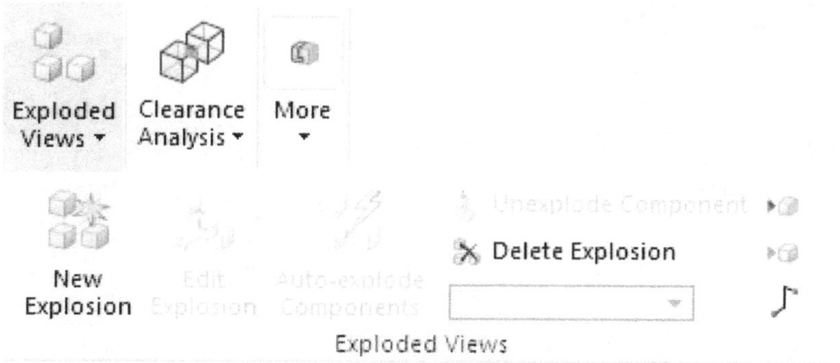

You can create and manage one or several exploded views of an assembly. They are stored as a "snapshot" of the assembly that includes the exploded positions of the components. Exploded views can even be used for display in assembly drawings.

In this project, you will create an exploded view of your table assembly and place it on a drawing.

- Continue working with *"Table Assembly.prt"*.

- Select the **New Explosion** command.

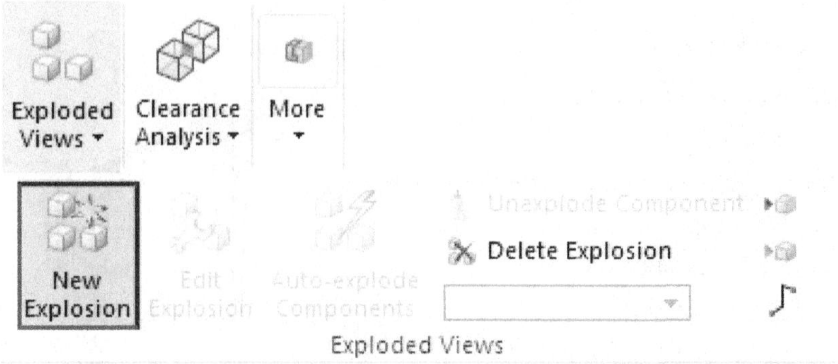

- Choose **OK** to accept the default name *"Explosion 1"*.

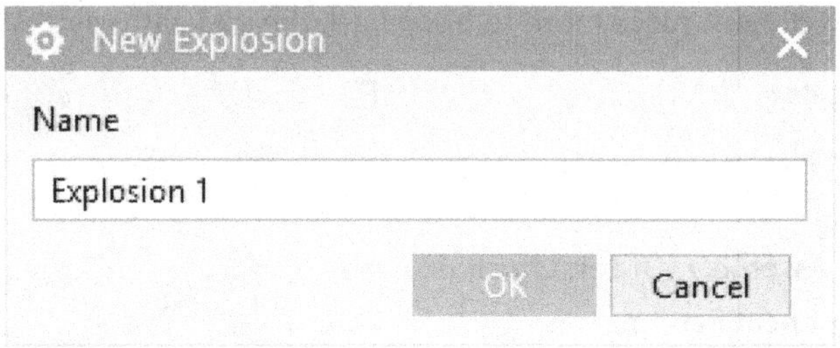

- Select the **Edit Explosion** command.

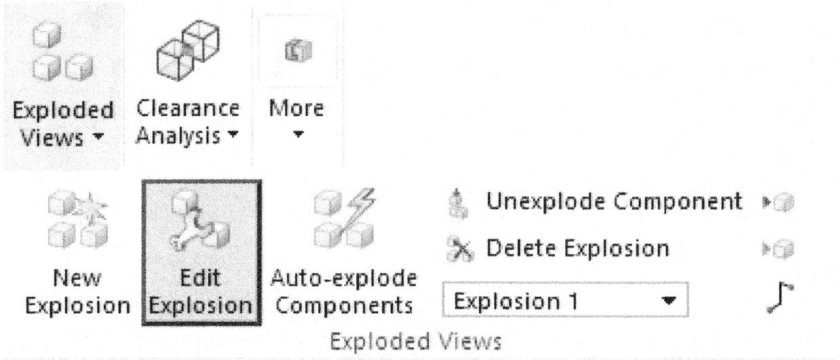

- Select the *"Tabletop"* part.

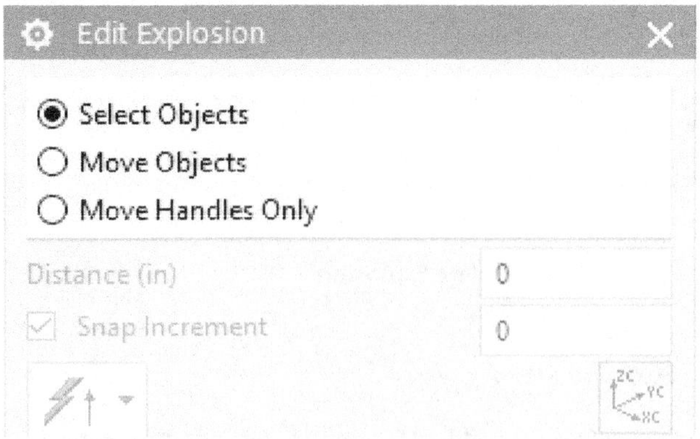

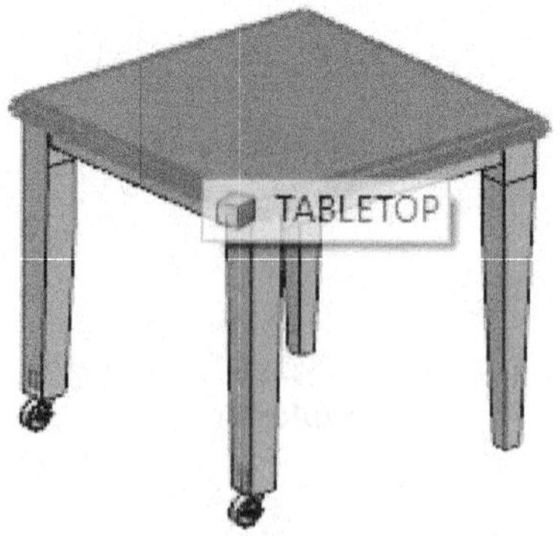

- Toggle the radio button to *Move Objects*. Drag the ZC handle of the dynamic csys to move the tabletop.

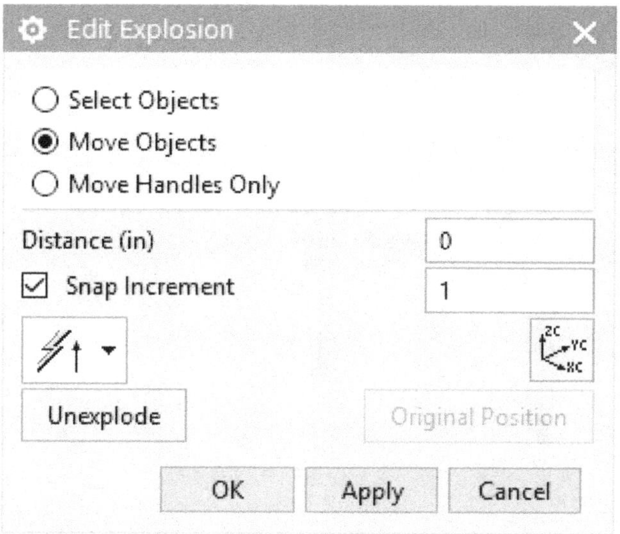

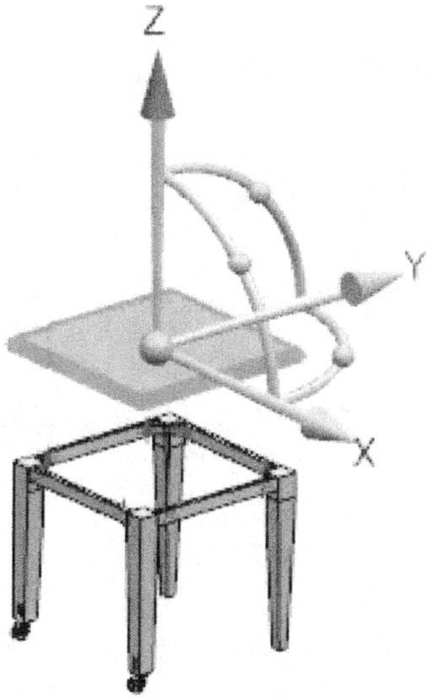

- Toggle the radio button back to *Select Objects*.
- Use **Deselect All**, and then select the left and right leg assemblies with casters, along with the *"Tabletop Support"* between them. Select the YC handle of the dynamic CSYS and enter a **Distance** of *-10 in* to move them.

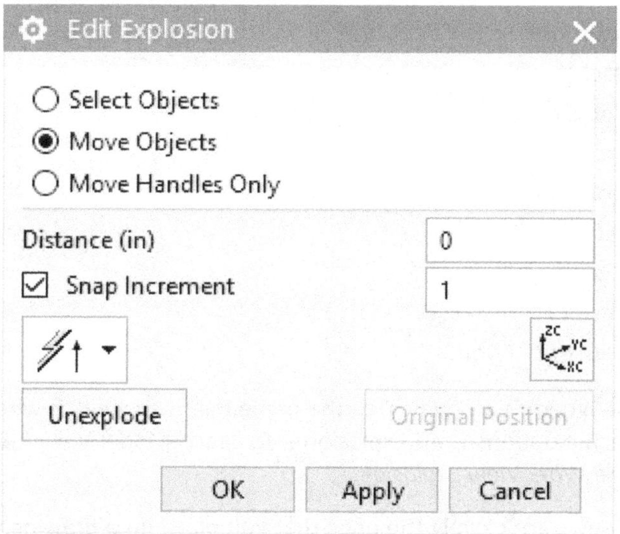

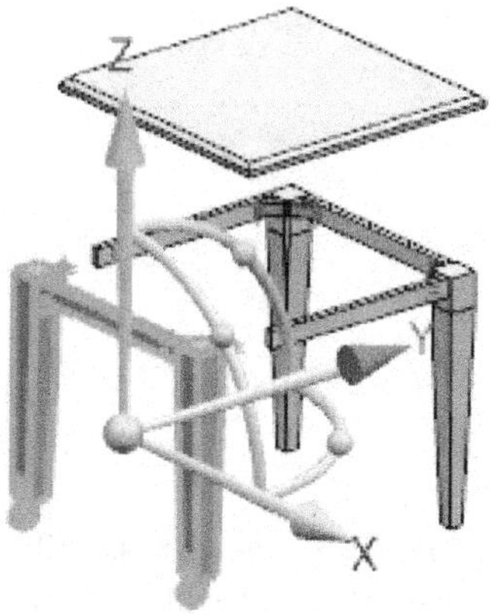

- Continue moving leg assembly pairs with the tabletop support between them *10 in* outward until your assembly is exploded as shown below.

- Choose **OK** to save your edits.

18.5.1 User-Defined Model Views

Now is a good time to discuss the use of user-defined *Model Views*. The Model Views group found in the Part Navigator includes standard orthographic views of models, such as *Top*, *Front*, and angled views such as *Isometric* and *Trimetric*.

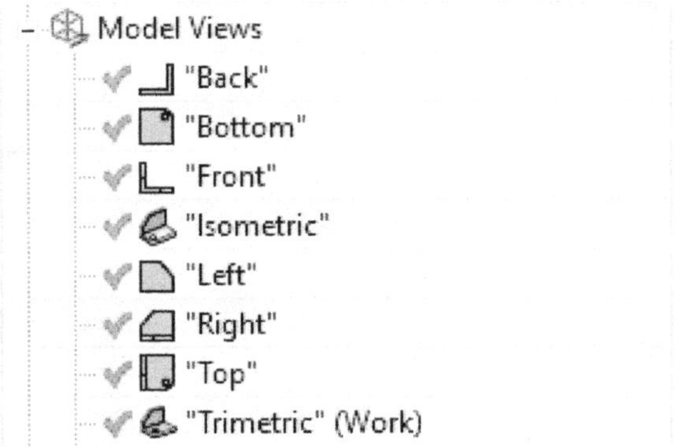

Double-clicking any of these views makes it into the Work View – which, as the name might suggest, is where your view-specific choices take effect. One of those view-specific options is an explosion – for each Model View here, any of the explosions created in your assembly can be made the *Work View Explosion*.

The main reason this matters is that these Model Views are exactly the ones that you place on a drawing with the **Base View** command. If you want to be able to place standard views on the drawing alongside exploded views, you will need to create a user-defined Model View for use on the drawing. This is exactly what you will do in the exercise below.

- From the Part Navigator, right-click on *Model Views* and select **Add View**. This saves the current camera angle as well as the current *Work View Explosion* (indicated by the drop-down menu in the **Exploded Views** group)

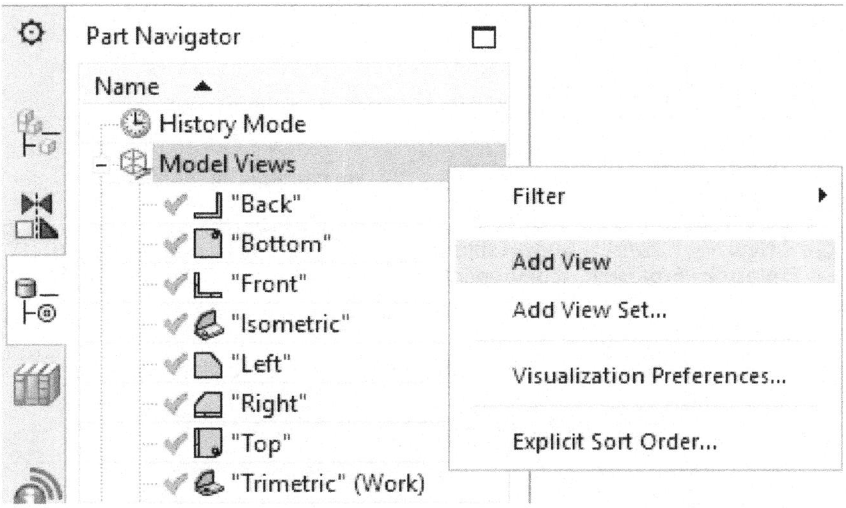

- Name your newly created Model View something fancy, like *"Explosion 1"*.

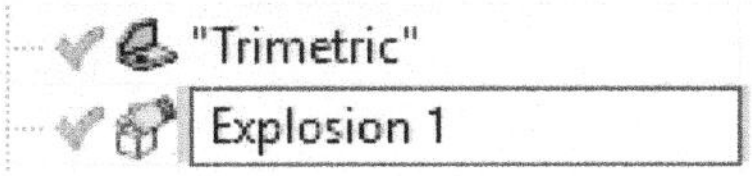

- This is how your *Model Views* group will appear now.

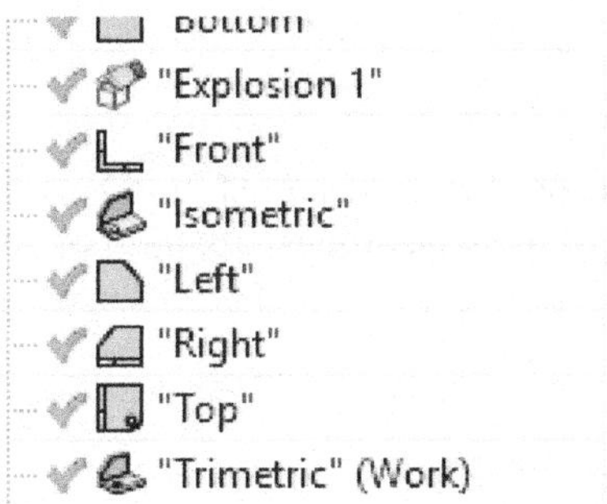

- Since the *Trimetric* view is the *Work View*, and we might want to include that on the drawing, set the **Work View Explosion** drop-down in the **Exploded Views** group to *(No Explosion)*.

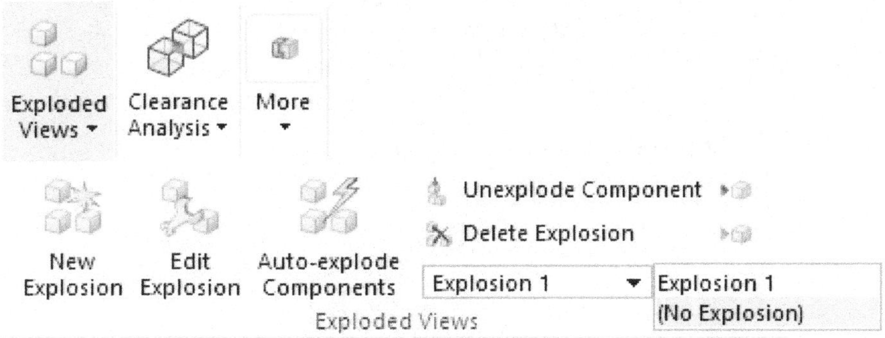

- **Save** your assembly!

18.5.2 Tracelines

You may want to add tracelines to an exploded view to show how fasteners, holes, and other entities align with each other. The **Tracelines** tool is found in the bottom right corner of the **Exploded Views** group.

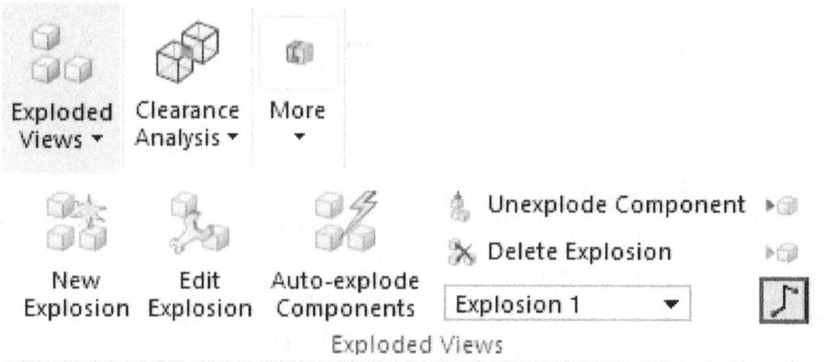

- Continue working with *"Table Assembly.prt"*, and create a new explosion called *Explosion 2*.

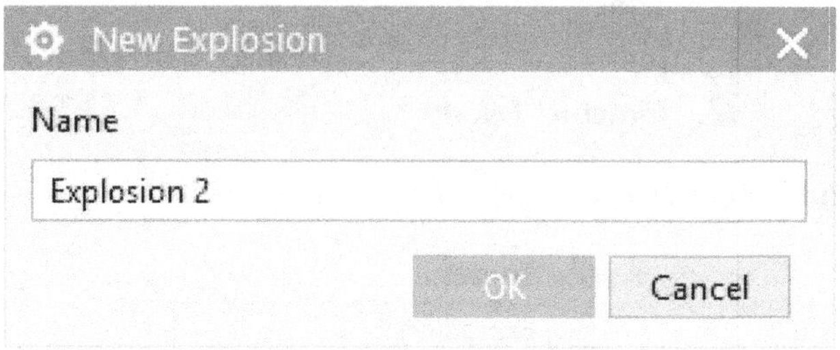

- Add a new model view, also called *Explosion 2* and make it the Work View.

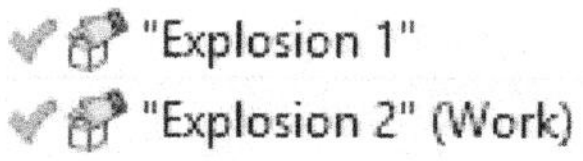

- Next, use the **Hide Components in View** command. This tool found both here in the **Exploded Views** group, as well as in the **More** gallery on the far-right side of the **Assemblies** tab. It is not specific to exploded views.

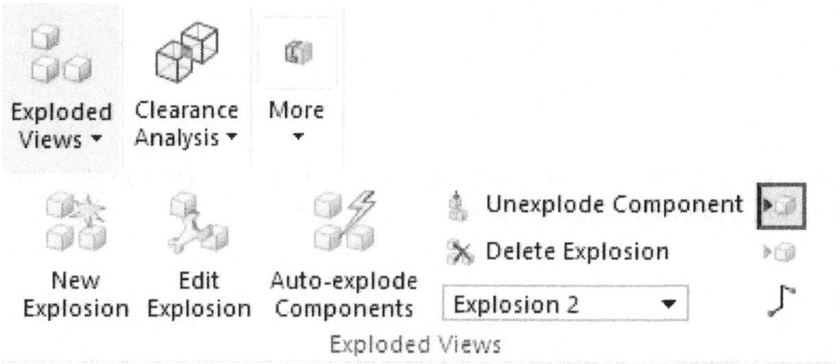

- Select all components except those shown below, and click **OK**.

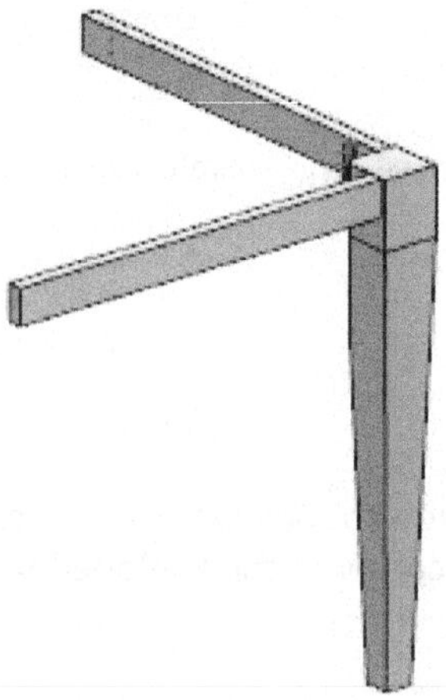

- **Edit Explosion** to explode the hardware in the corners, as shown below.

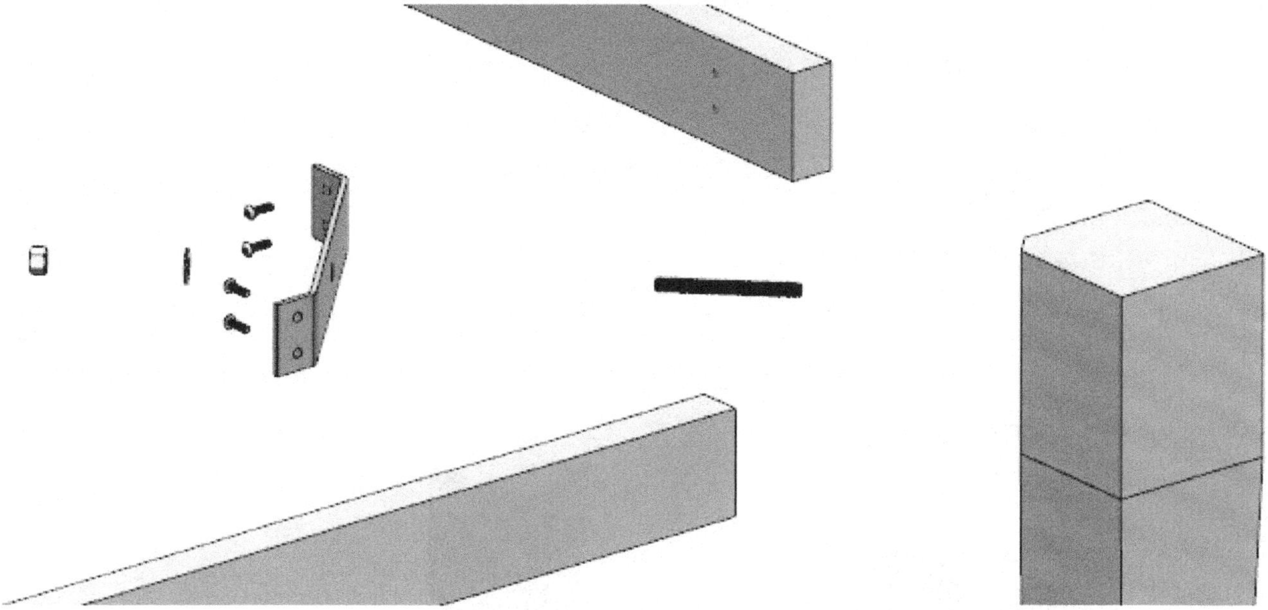

- Next, we will need some tracelines in this exploded view. Select the **Tracelines** command.
- Select the arc center of the edge of the nut, as shown below.

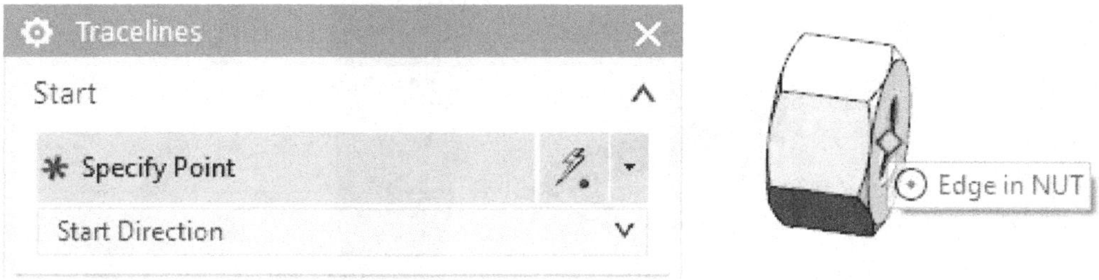

- For the **End Object**, select the arc center of the edge of the threaded rod facing the nut, as shown below.

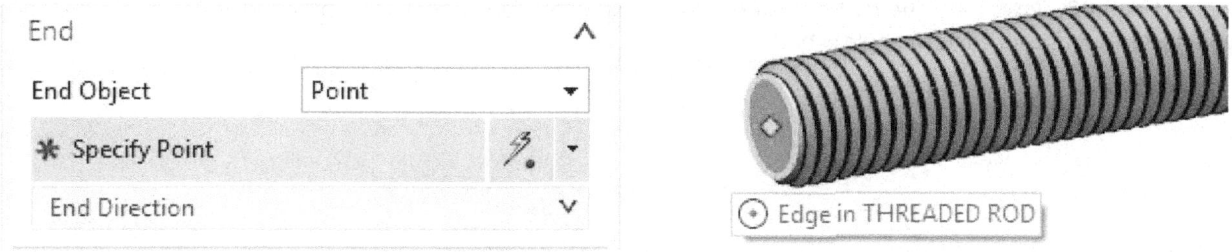

- This produces a traceline as shown below.

- Use the **Tracelines** tool five more times to place the tracelines shown below.

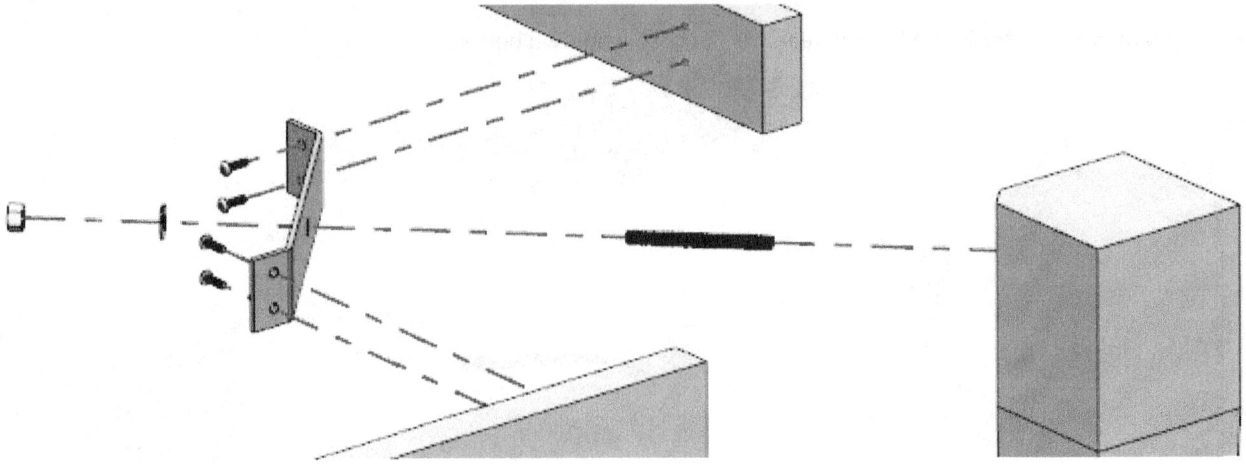

- Orient the view as shown below.

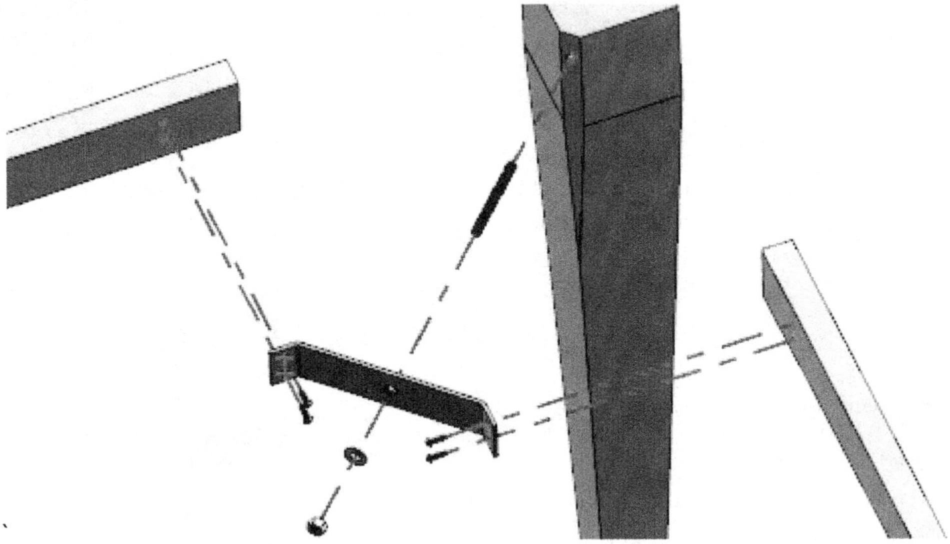

- Right-click on *Explosion 2* in the Part Navigator and select **Save**. This is how you update user-defined *Model Views* if you decide a different camera angle is better.

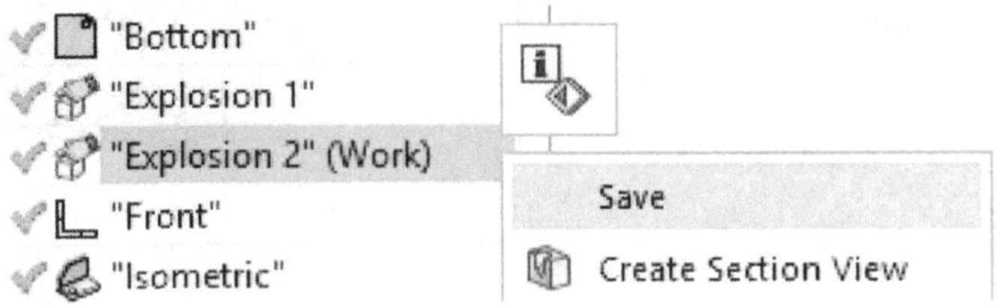

- **Save** your assembly!

18.5.3 Exploded Views on the Drawing

Now that you have created several explosions and custom views, let's place them on the assembly drawing.

If your company requires that you create drawings as separate files from your assembly files (as you practiced in Chapter 12.1.2), you need to be aware that exploded views created in the assembly will not be visible in the drawing – exploded views must be created at the Displayed Part level in order to appear on a drawing.

- Continue working with *"Table Assembly.prt"*, and switch to the **Drafting** application.
- Create a *B*-sized sheet based on the *Siemens* template.

- Use the **Base View** command to place the Model View *Explosion 1* at a *1:16* scale.

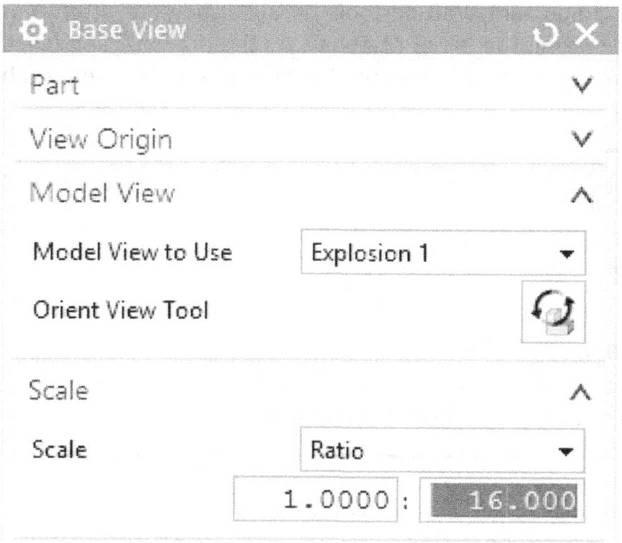

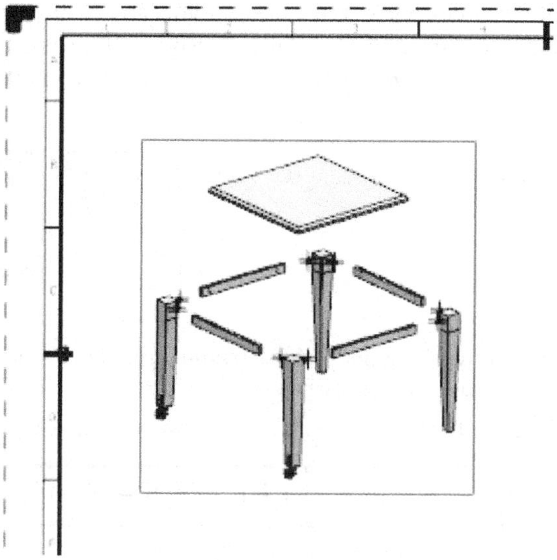

Important side note: when the drawing is a separate file from the assembly, you must create the exploded view at the drawing level, and you must be sure to select the correct part in the ***Part*** *section of the* ***Base View*** *dialog!*

- Create a second drawing sheet with the same *B*-sized *Siemens* template, and use the **Base View** tool to place *Explosion 2* on the sheet at a *1:4* scale. Note that the tracelines appear on the drawing.

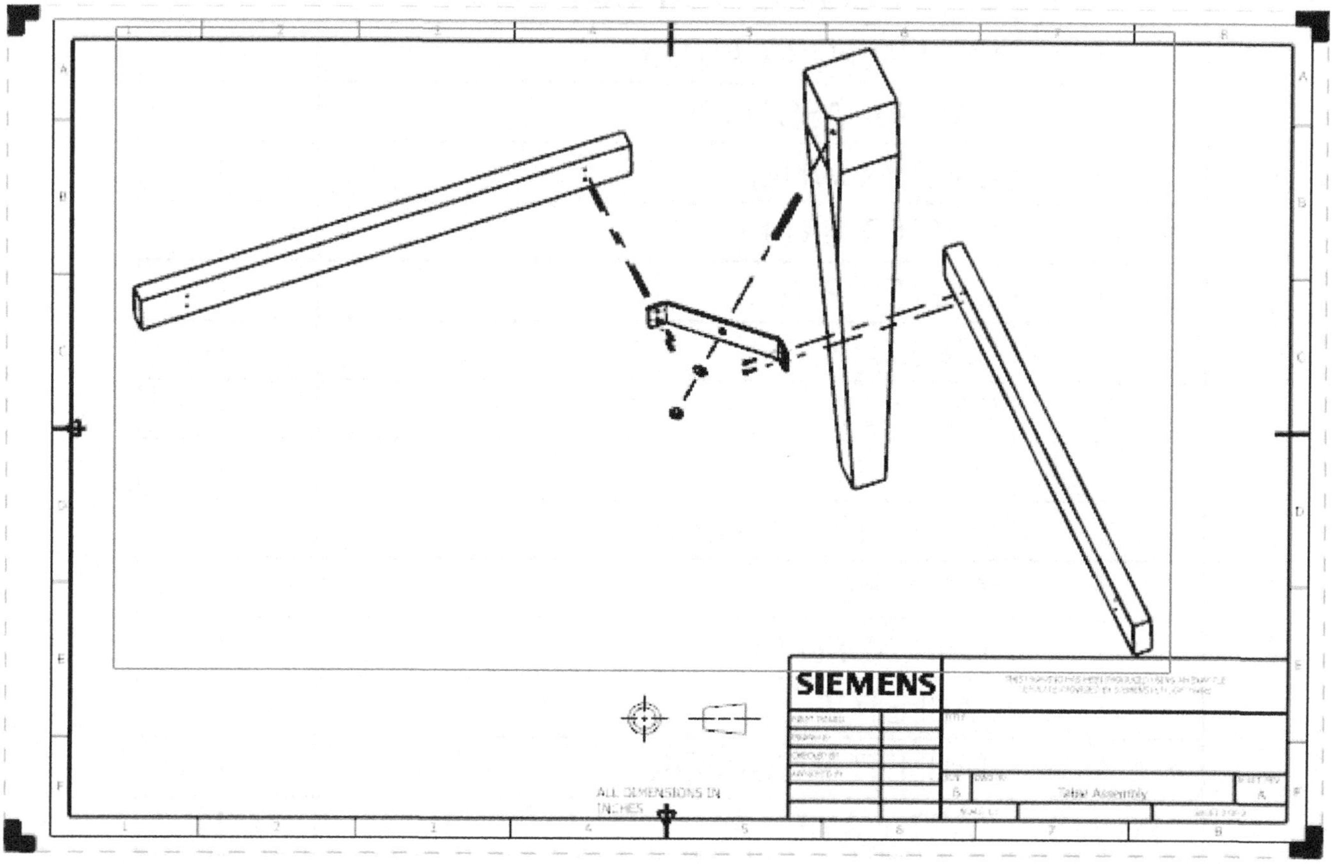

- **Save** your assembly.

18.5.4 Parts List & Auto Balloon

In Chapter 12 we neglected to teach you about two very important drafting tools because they are often applied to exploded view assembly drawings. These are the tools for making the Bill of Materials on the drawing, and labeling the components in a view, namely **Parts List**, and **Auto Balloon**. They are found in the **Table** group on the **Home** tab in the **Drafting** application.

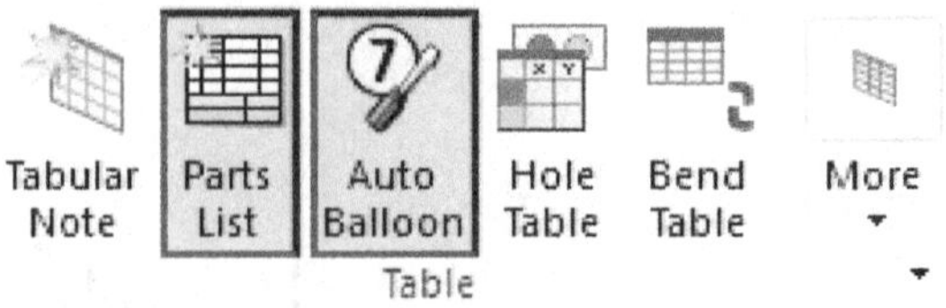

- Continue working on *"Table Assembly.prt"* in the **Drafting** application.
- Make *Sheet 1* the *Work Sheet*.
- Select the **Parts List** tool. It's very easy to use, you just click to place the bill of materials.

18	LEG ASSEMBLY	2
17	LEG	2
16	LEG ASSEMBLY - SHORT - RIGHT	1
15	LEG - SHORT - RIGHT	1
14	LEG ASSEMBLY - SHORT - LEFT	1
13	SCREW FOR CASTER	8
12	CASTER ASSEMBLY	2
11	CASTER WHEEL	2
10	CASTER BRAKE	2
9	CASTER PLATE	2
8	SCREW	16
7	BRACKET	4
6	NUT	4
5	LEG - SHORT - LEFT	1
4	WASHER	4
3	THREADED ROD	4
2	TABLETOP	1
1	TABLETOP SUPPORT	4
PC NO	PART NAME	QTY

- The parts list comes out with components at all levels in the assembly tree showing. Suppose that you only want to see the top-level components.

- You can edit the **Parts List** (and other tabular notes) by putting your cursor in the top left corner and waiting until a plus sign in a box appears. When it does, you can right-click to choose editing tools.

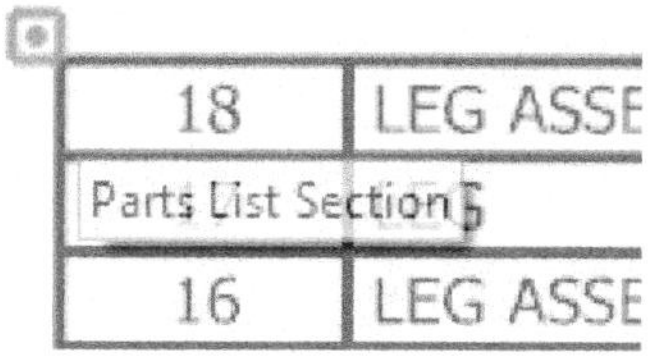

- One such tool is called **Edit Levels**.

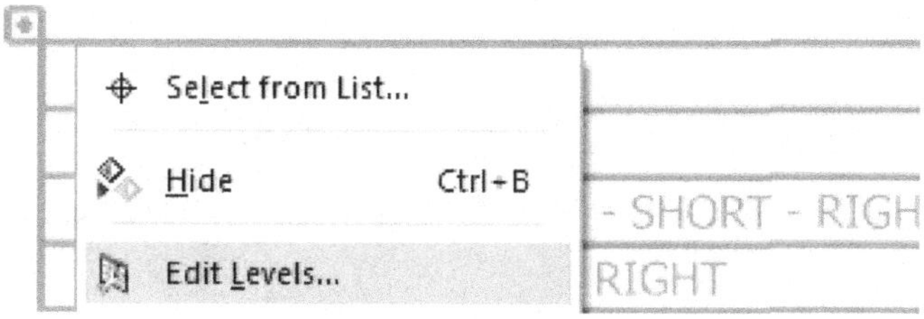

- In the **Edit Levels** dialog, push the **Top Level Only** button.

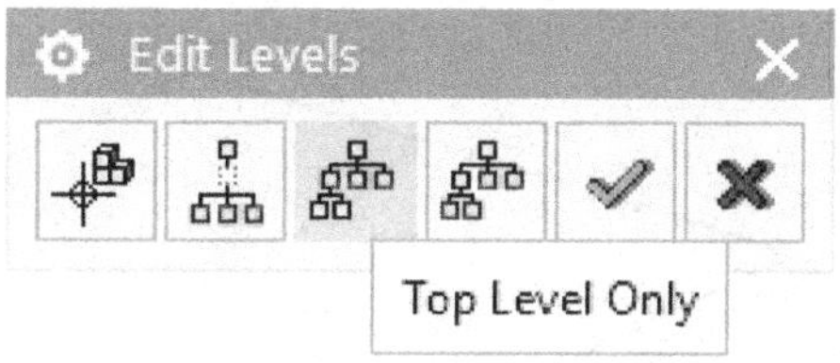

- This makes the parts list substantially shorter.

5	LEG ASSEMBLY	2
4	LEG ASSEMBLY - SHORT - RIGHT	1
3	LEG ASSEMBLY - SHORT - LEFT	1
2	TABLETOP	1
1	TABLETOP SUPPORT	4
PC NO	PART NAME	QTY

- To label your drawing view, select the **Auto Balloon** tool. It will prompt you to choose a drawing view if you placed multiple views on the sheet. If necessary, pick *Explosion 1*.
- The **Parts List Auto Balloon** dialog wants you to choose a parts list. Pretty straightforward.

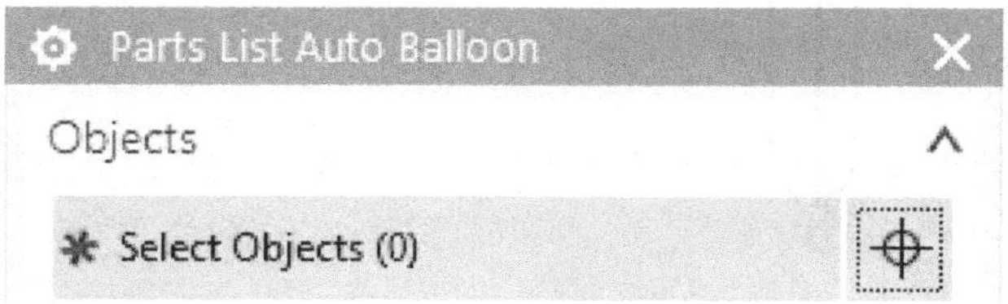

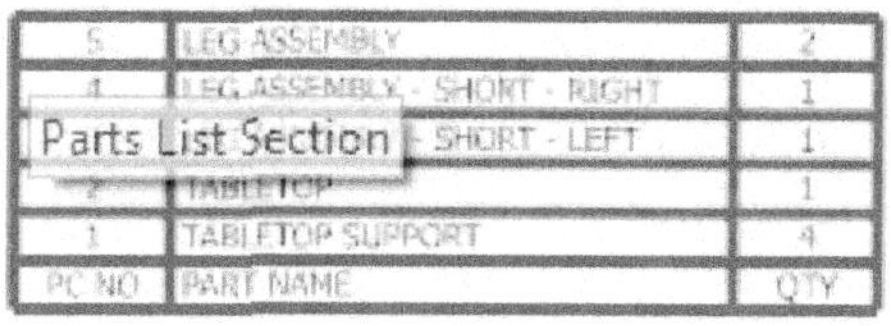

- The balloons may appear pointing at strange locations. If so, you can double-click on them to edit them, and reposition the leader arrow by clicking on another control point. In the screenshot below, the leftmost corner of the tabletop is the new position for the leader arrow.

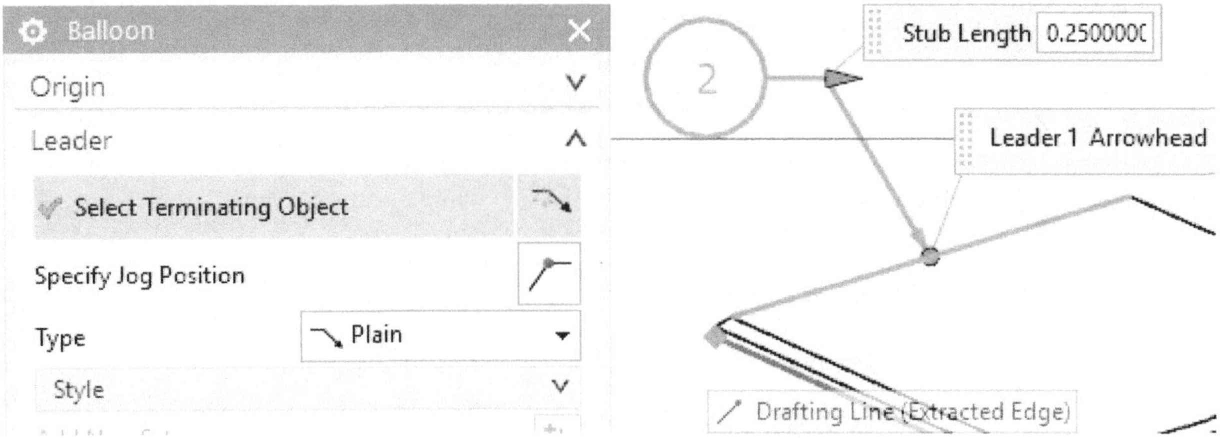

- You can make the balloon leader arrows all point nicely, and then line them up on one side of the view.

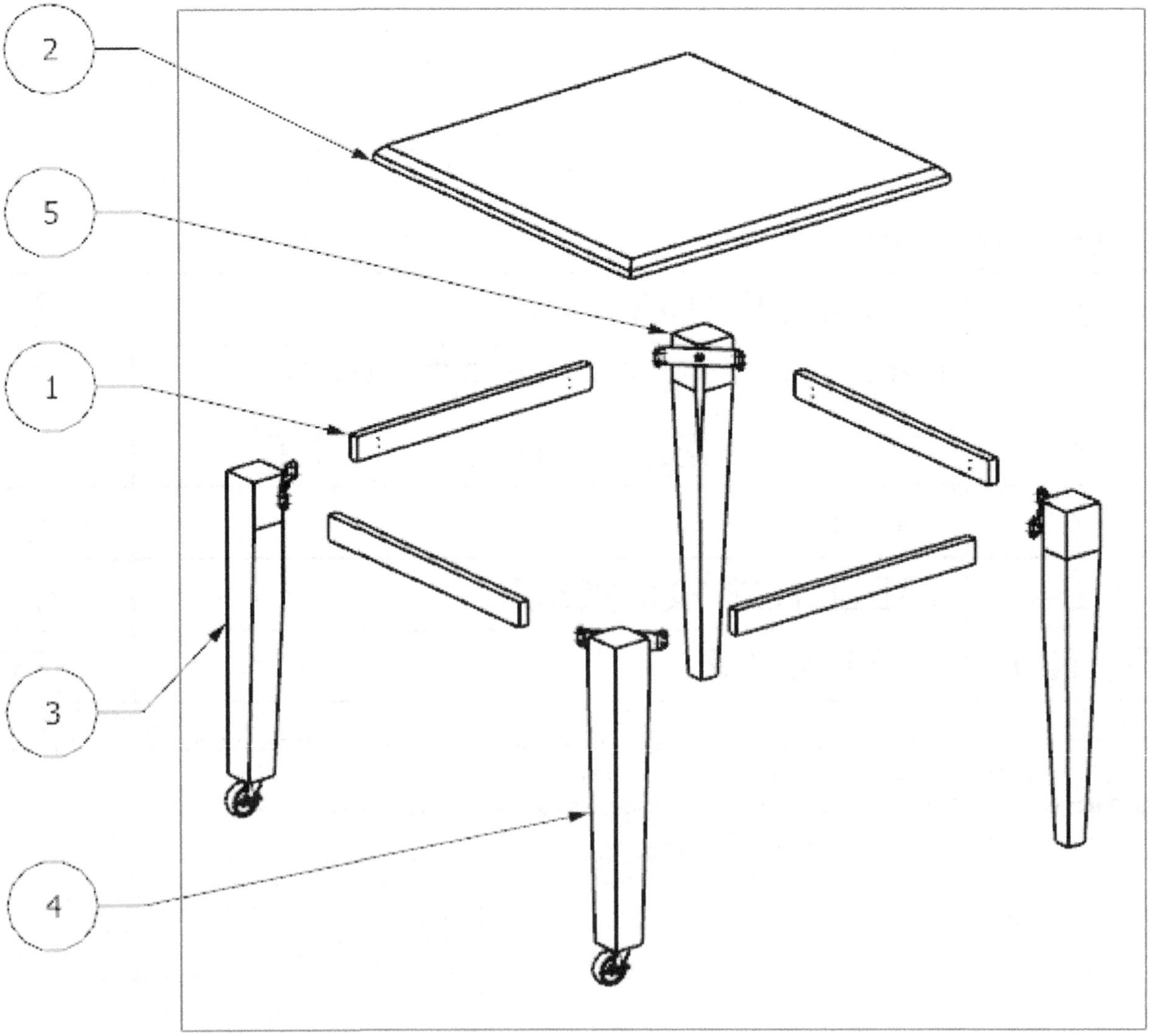

- Good work!

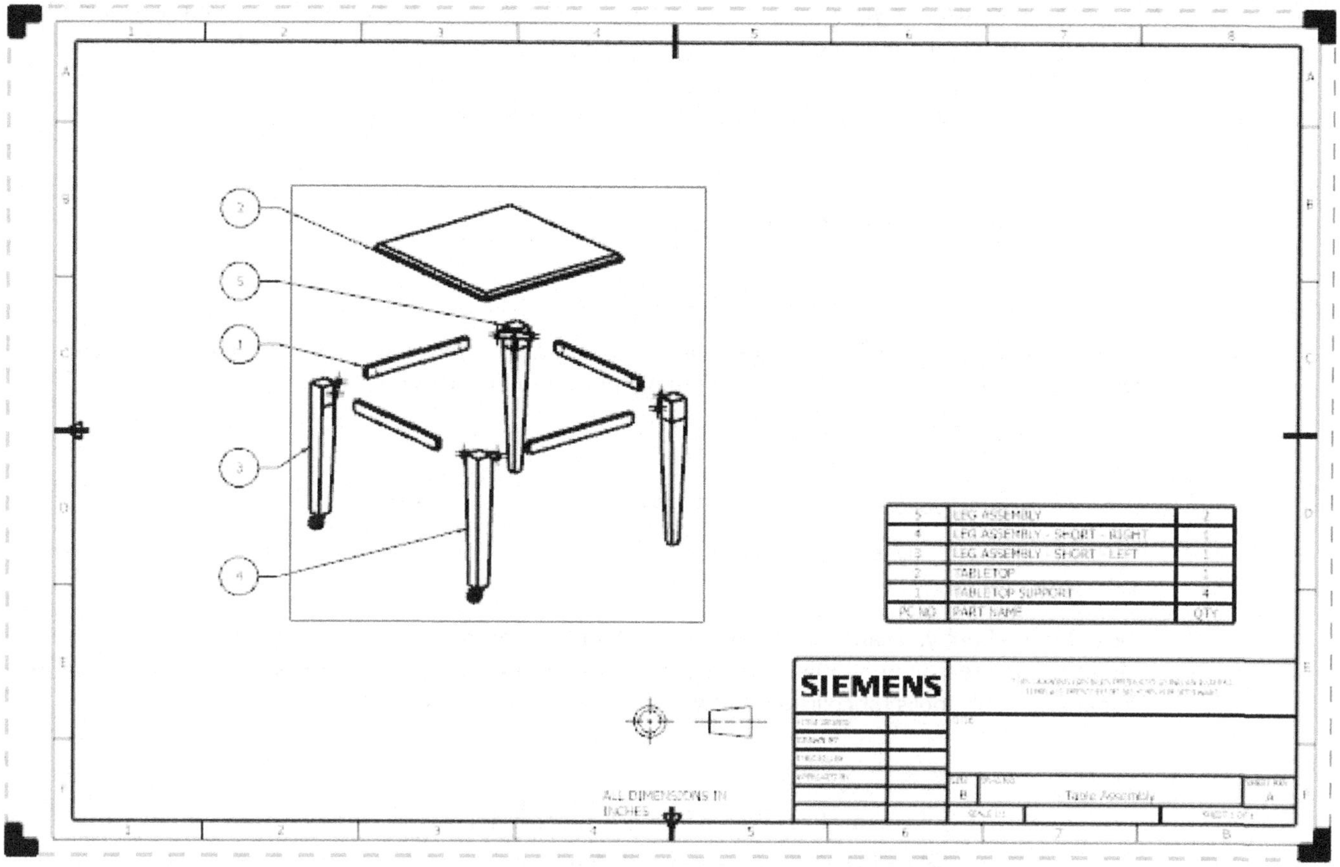

- **Save** your assembly!

18.6 Clearance Analysis

Clearance Analysis is an incredibly useful tool for finding collisions between bodies in your assemblies. It can be used in a more sophisticated manner to detect violations of clearance zones, but in the exercise below, you will use it to check for interferences. The **Clearance Analysis** group is found on the far-right side of the **Assemblies** tab, next to **Exploded Views**.

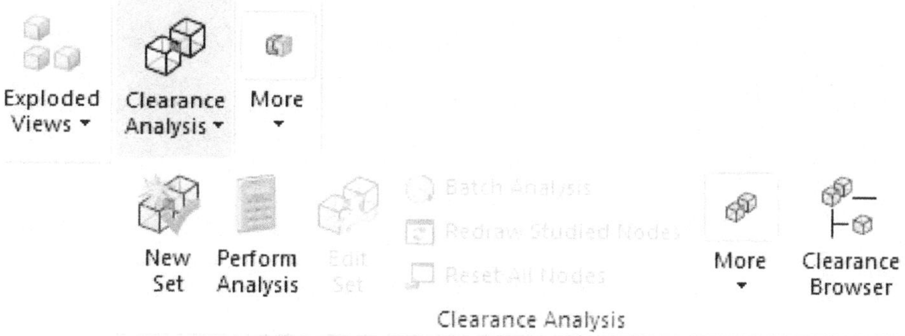

- Continue working with *"Table Assembly"*, or open up *"Table Assembly_FINISHED"* from the cloning operation in Chapter 18.1.
- From the **Clearance Analysis** group, select the **New Set** tool.

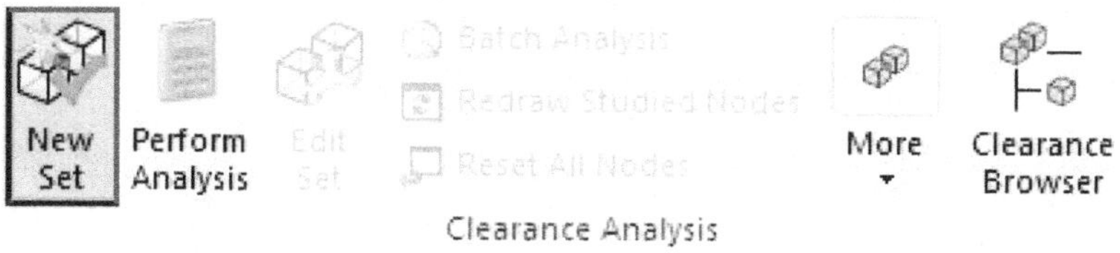

- Set the options in the **Clearance Analysis** dialog as shown below. Several of these settings deserve some explanation.
  - **Clearance Between**: *Components* – we don't care about collisions within part files for this analysis.

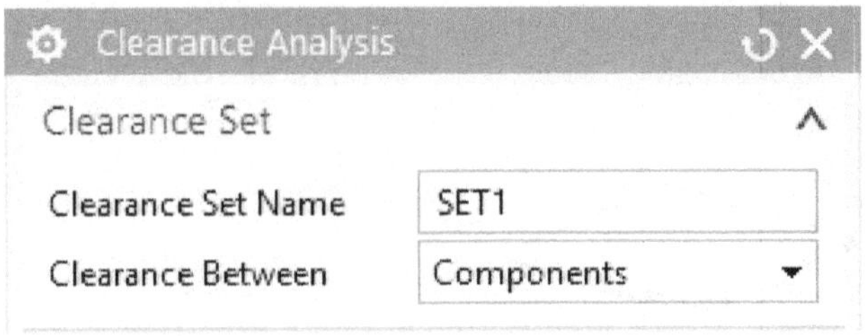

  - **Objects to Analyze**: *All Objects* – this is OK because the assembly is small but for larger assemblies you will want to target "hotspots" where you think there may be collisions.

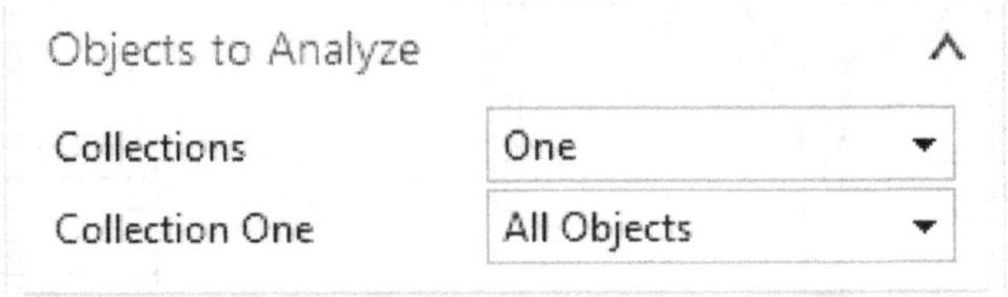

  - **Exceptions** – here you can individually pick out known clearance violations that you happen to not care about for one reason or another. In this section, check *"Exclude Pairs from Same Subassembly"* so that our analysis only looks for collisions between top-level components.

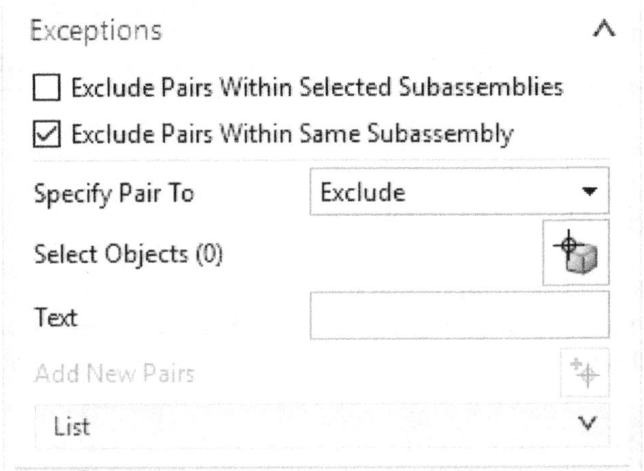

- **Clearance Zones** – this is where you specify what the default amount of clearance needed is, and specify other clearance zones with different clearance requirements. Since we are checking only for collisions, our **Default Clearance Zone** is zero inches, and we don't need to modify any part of this section!

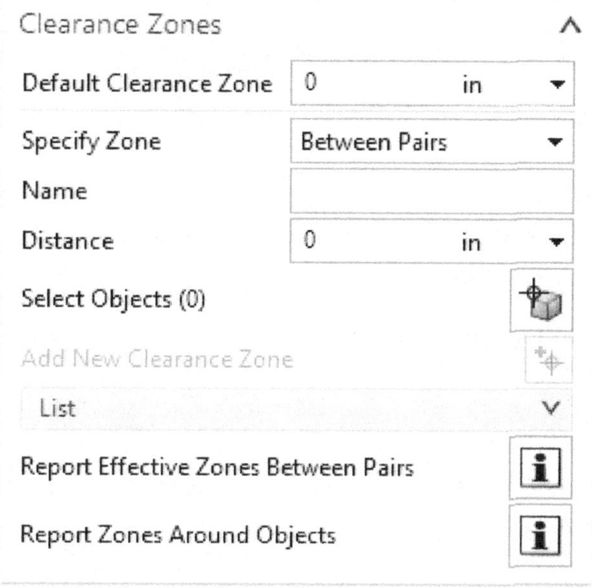

- **Interference Geometry** – the *"Save Interference Geometry"* checkbox will create bodies that represent the collisions found. These bodies are smart, insofar as they will recalculate each time you perform the analysis.

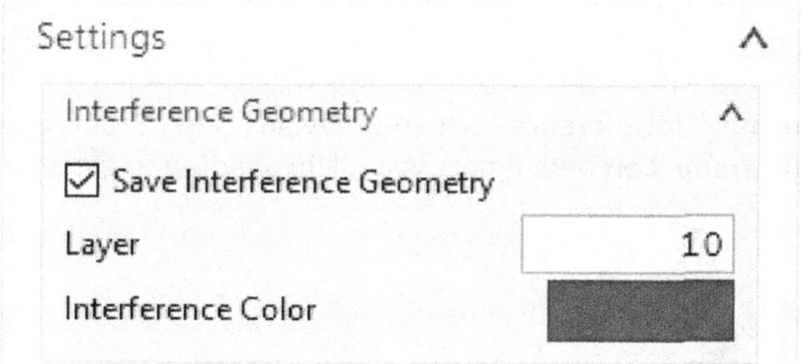

- *Perform Analysis* – if this checkbox is marked, the tool will run the analysis when you click **OK**, otherwise, you need to use the **Perform Analysis** button found in the **Clearance Analysis** Group after defining the **Clearance Set**!

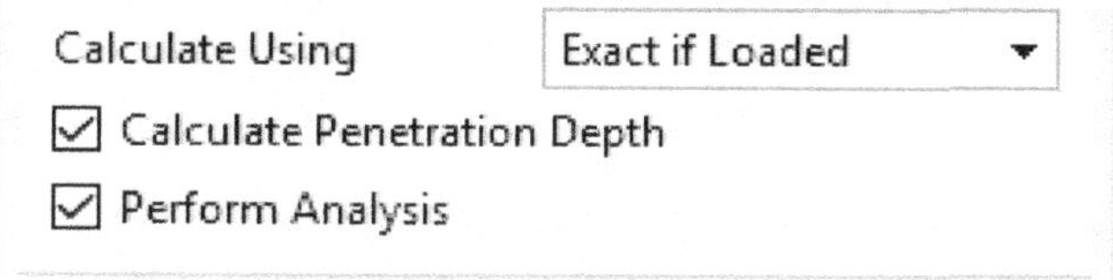

- The clearance set will be generated and the analysis performed. The results of the analysis then appear in the **Clearance Browser**.

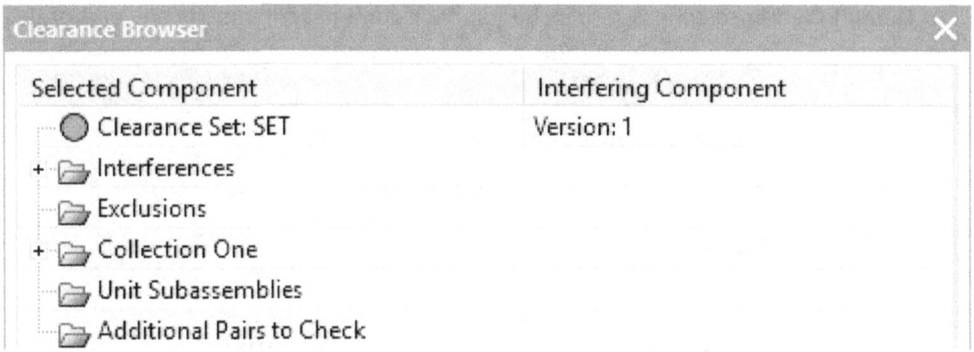

- Each interference has an icon to the left – the "touching boxes" icon indicates that faces are touching. These are usually not problematic collisions.

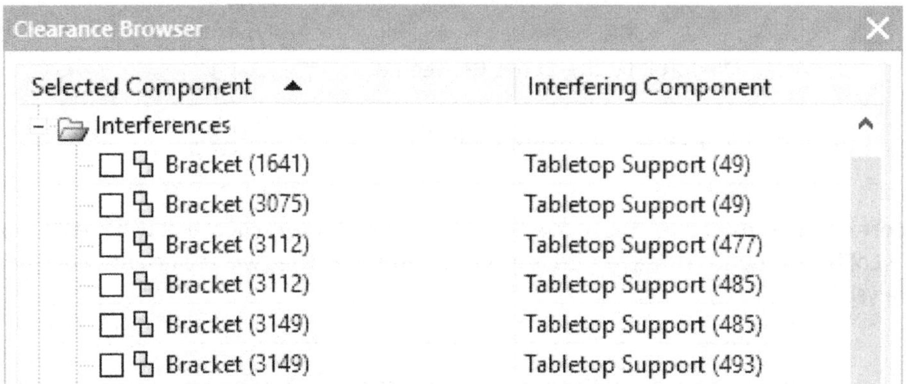

- Right-click on one of the "touching" interferences between instances of *"Bracket"* and *"Tabletop Support"*. You can right-click again and select **Wireframe Left** and, if necessary, **Flip Shading** to get a better view of the interference.

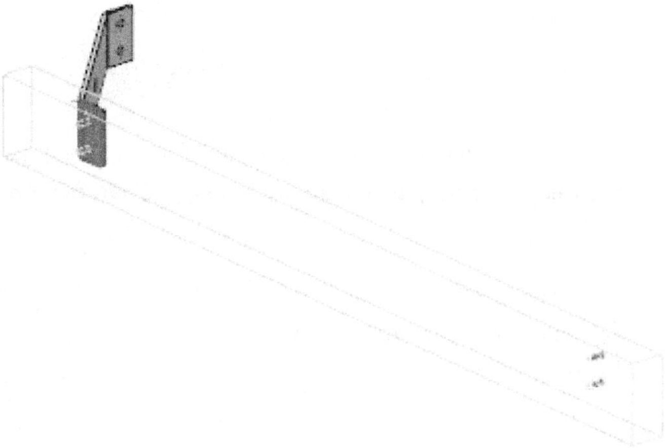

- Faces touching really isn't a big deal, so right-click again and select **Ignore**.

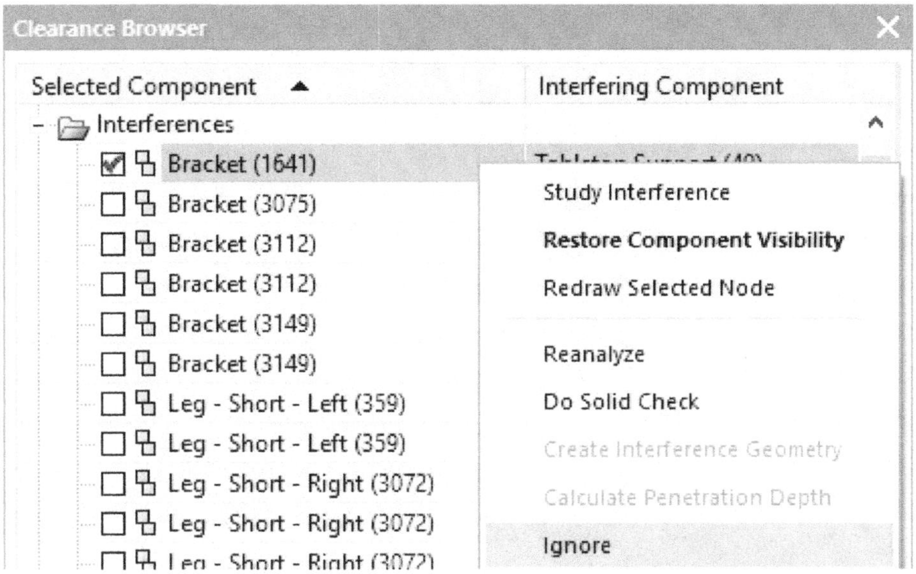

- This removes the interference from the list.
- Select all "touching" interferences in your assembly, right-click, and select **Ignore**.

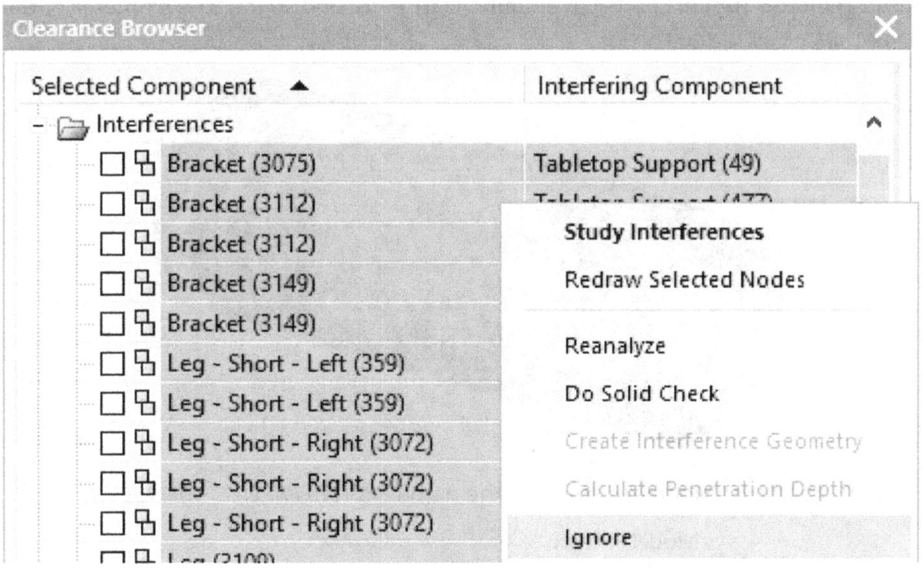

- This leaves only "hard" collisions in your **Clearance Browser** – the icon shows overlapping boxes.
- Right-click on one of the "hard" interferences and select **Study Interference**.

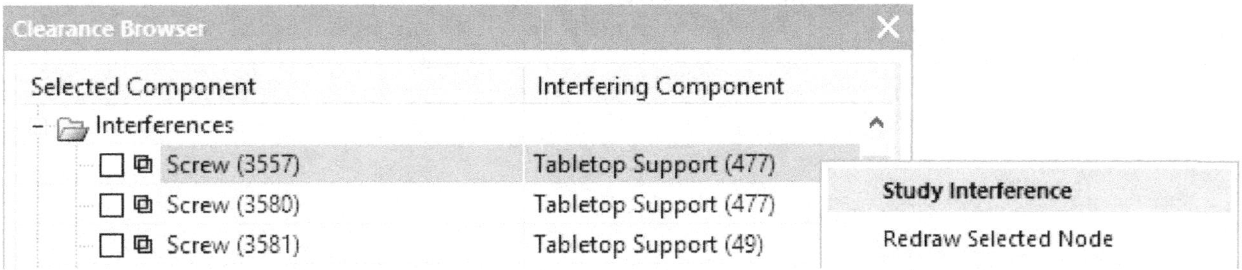

- Use **Wireframe Left** and **Flip Shading** as necessary, and zoom in to get a better view. The planes and measurement that appear are the result of the *"Calculate Penetration Depth"* checkbox in the **Clearance Analysis** dialog.

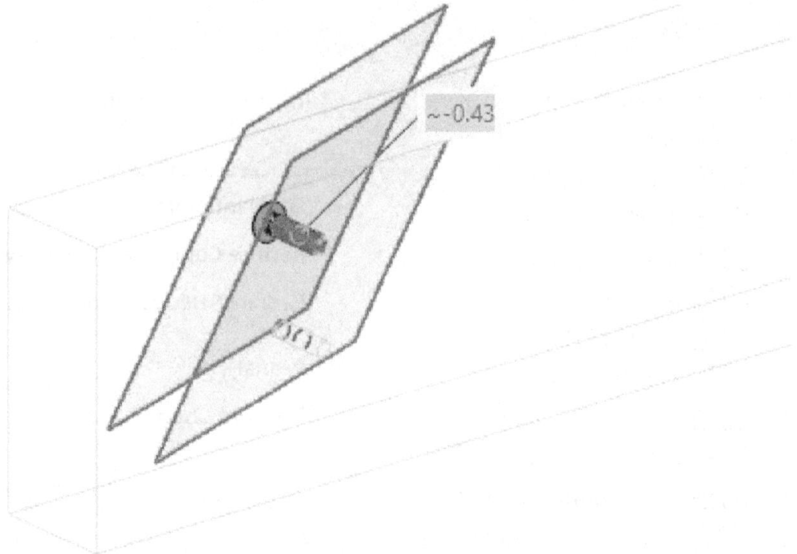

- Close your **Clearance Browser**, and use **[Ctrl]+[W]** to hide all components. The only remaining bodies visible in the graphics window will be the interference geometry created on layer *10*. Looking closely at each such body, you can surmise that all interferences are simply screw threads colliding with threaded holes – not an issue! The **Threaded Hole** features in the **Tabletop Support** parts have symbolic threads, and so the true diameter of the hole in the model is equal to the minor diameter of the thread.

- These bodies are collected in the Model History with the name *"INTERFERENCE_GEOMETRY"*.

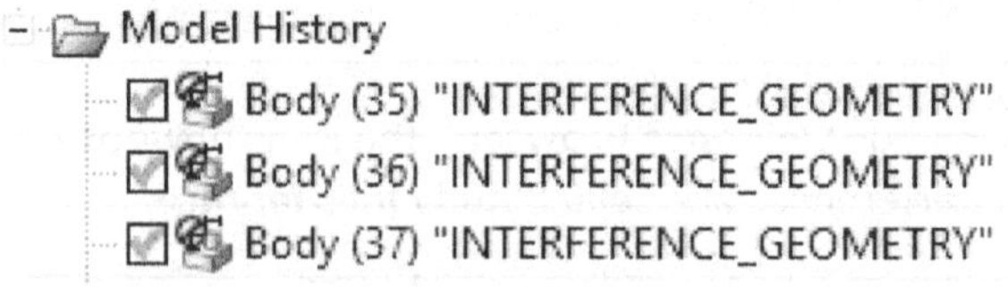

- Reopen the **Clearance Browser**.

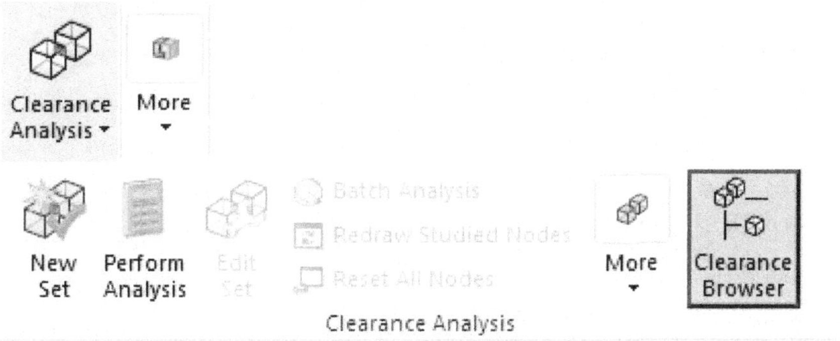

- Remember when you grabbed all the "touching" interferences, right-clicked, and selected **Ignore**? Well, in the **Clearance Browser**, that moved those interferences into the *Exclusions* folder. You can also move interferences to the *Exclusions* folder by selecting them, and then doing a drag-and-drop.

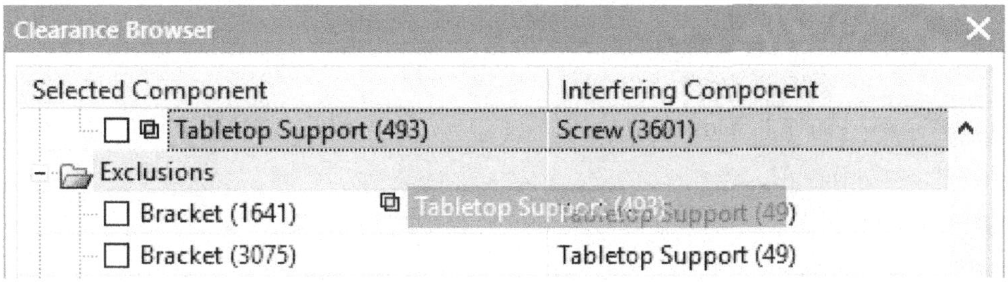

- The interference geometry is deleted when you choose to ignore collisions. Your *Model History* should now be clear.

- For extra credit, try using the **Clearance Analysis** tool on the version of *"Leg Assembly – Short – Right"* that you built, to see if you did it right!
- **Save** your assembly.

18.7 Arrangements

Assemblies are rarely static – more often than not, when you build an assembly, you want to see its components in different configurations. The tool in NX that allows you to model your assembly in different states of interest is called **Assembly Arrangements**. The **Assembly Arrangements** tool is found in the **General** group on the **Assemblies** tab.

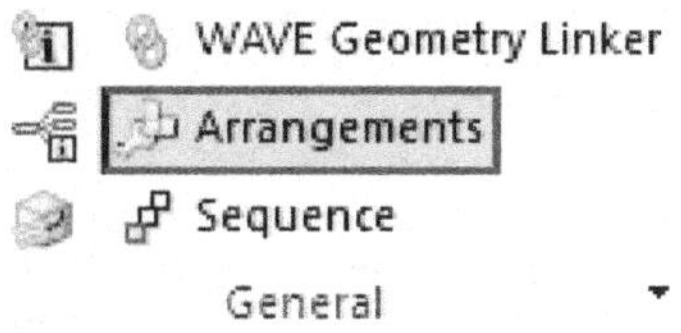

In the exercise below, you will see the use of the **Arrangements** tool for modeling the Hinge assembly created in Chapter 9.13 in four states of interest – *Fully Open*, *Fully Closed*, *Partially Open*, and *Free*.

- Open the file *"Hinge Assembly.prt"* that you created in Chapter 11 from *"C:\My NX 11 Files"*. If you skipped that exercise, open *"C:\My NX 11 Files\Downloaded Files\Hinge Assembly.prt"*.
- Select the **Arrangements** tool.
- Use the **New Arrangement** button to create two new arrangements – *"Fully Open"* and *"Fully Closed"*.

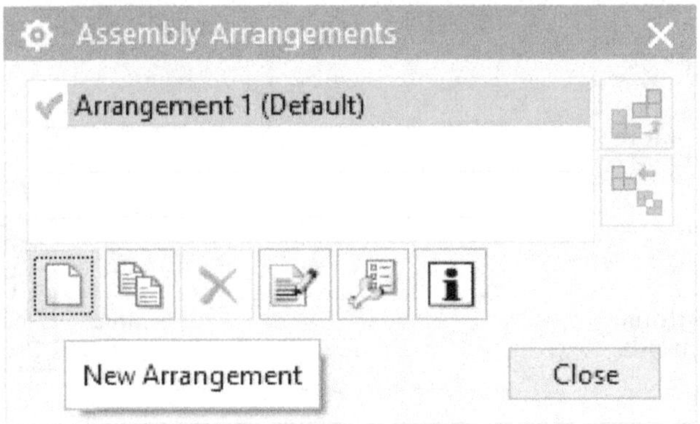

- Rename the default arrangement to *"Free"*.

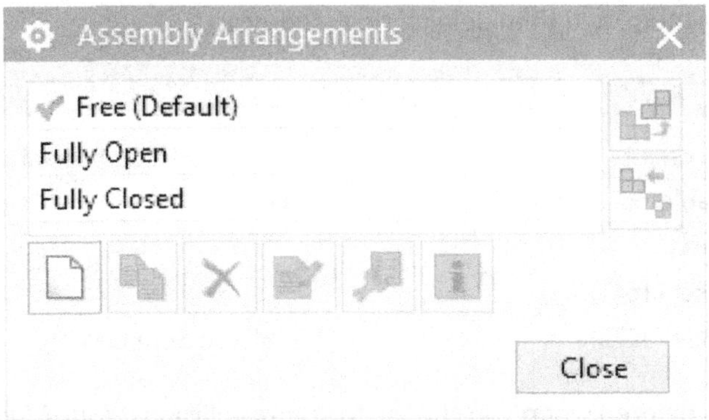

- We will want a new constraint in order to make use of these arrangements. Add a *3D Angle* constraint between the edges of *Hinge2* and *Hinge1*, and enable the **Angle** parameter, but not the **Angle Limits**. The value is irrelevant. For mysterious reasons, this will work more consistently if your constraint involves the edges rather than the faces.

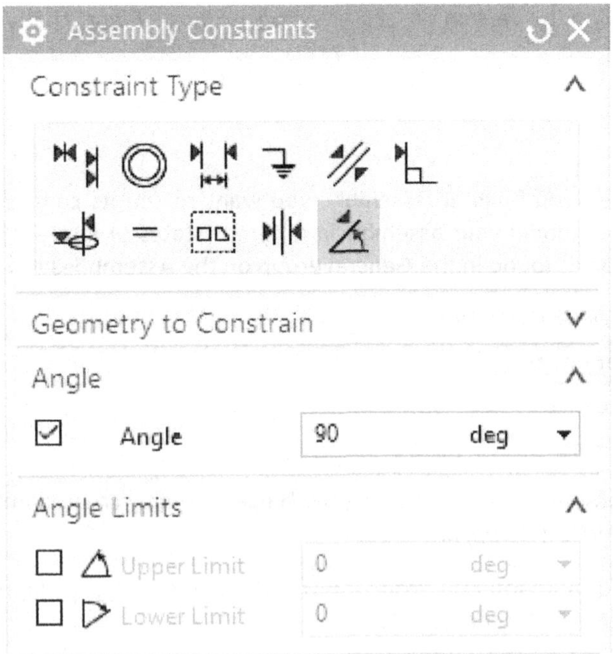

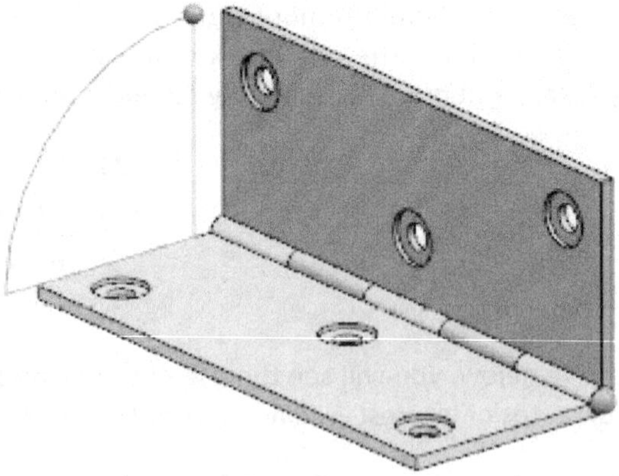

- Right-click on your newly-created *Angle* constraint, and from the bottom of the drop-down menu that appears, select **Edit in Arrangements**. Note the new icons in the **Position** column – these now indicate degrees of freedom that are specific to the active arrangement.

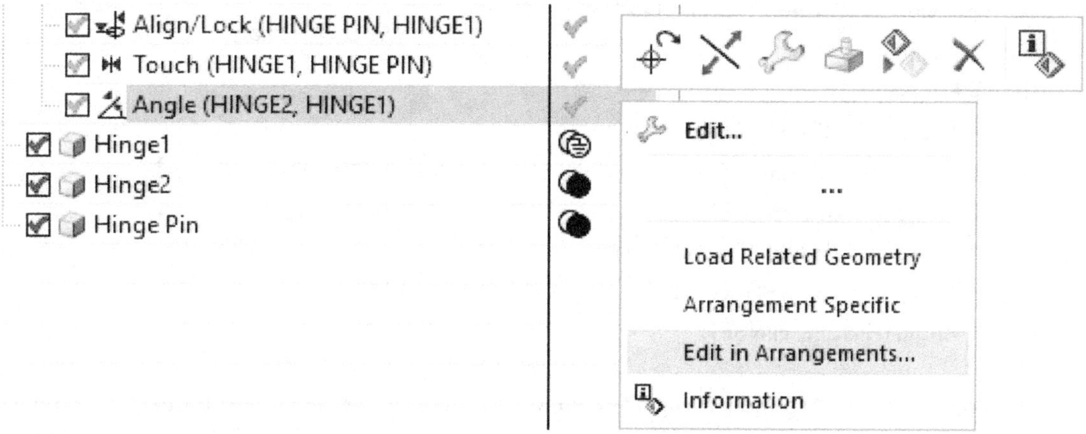

- For the *Fully Closed* and *Fully Open* arrangements, choose the **Specific** radio button. Change the value of the **Angle** to *0 degrees* and *180 degrees* each, respectively. Uncheck the red checkmark in the box next to *Shared State* – this will suppress the constraint in the *Free* arrangement (and any others that you might want to include in the *Shared State* group).

- Since *Free* is the active arrangement, when you finish editing the *Angle* constraint, you will find that the *Angle* is suppressed. You can now move *Hinge2* freely (subject to the *Angle Limit* constraints from the other *Angle* constraint).

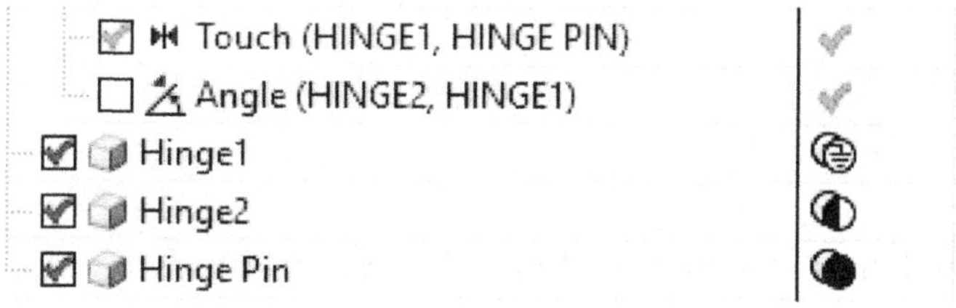

- To change the active arrangement, simply right-click *Hinge Assembly* in the Assembly Navigator, and choose another arrangement.

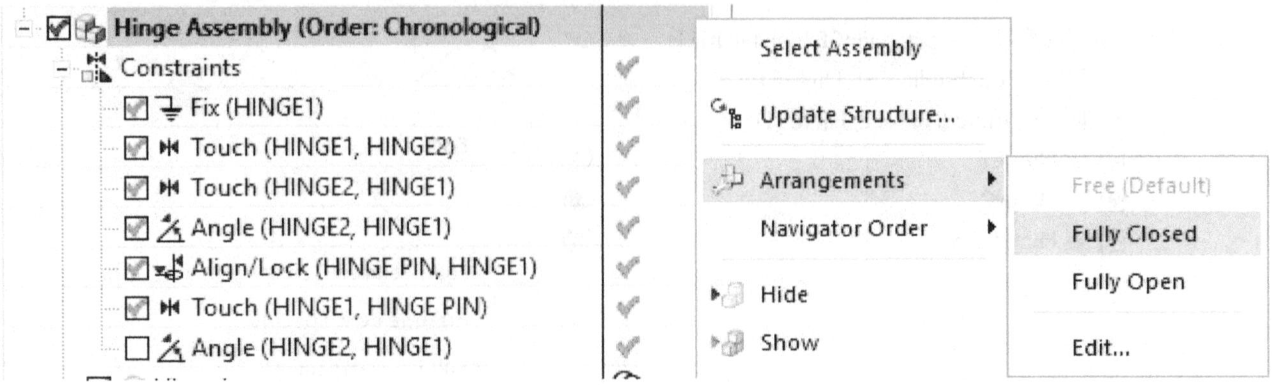

- You can also configure your Assembly Navigator columns to show the **Arrangement** for each assembly in the tree.

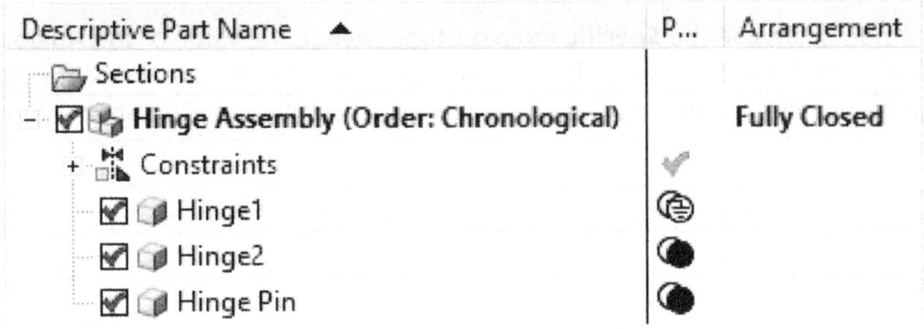

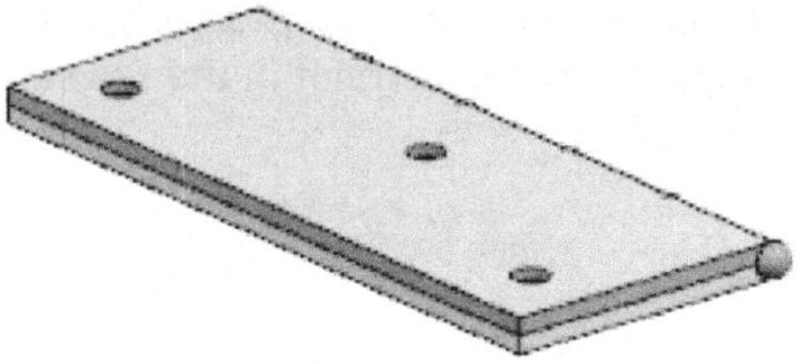

- **Arrangements** are useful for toggling between different states of interest in the Graphics Window, but their real value comes from their ability to communicate those different states of interest on an assembly drawing.
- Switch to the **Drafting** application, and create a *B*-sized sheet based on the *Siemens* template.
- Select the **Base View** tool.
- Note that there is a new section within the **Base View** tool that allows you to choose an **Arrangement** for each view.

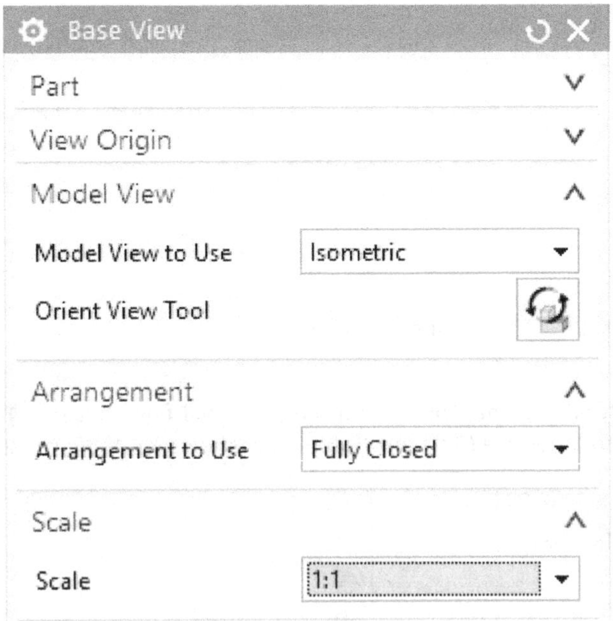

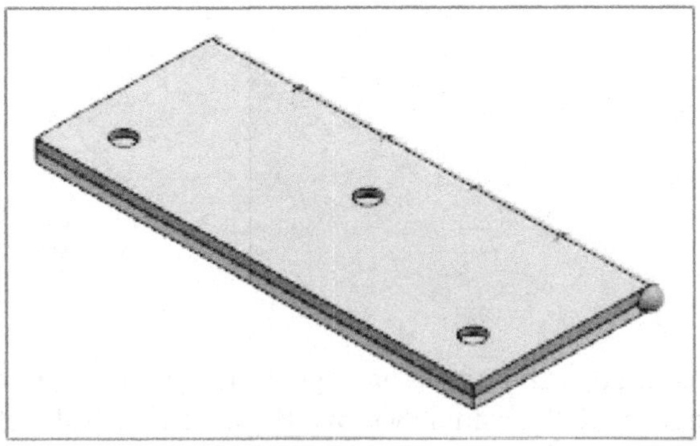

- Use the **Base View** command three times to place isometric views of all three arrangements on your drawing sheet.

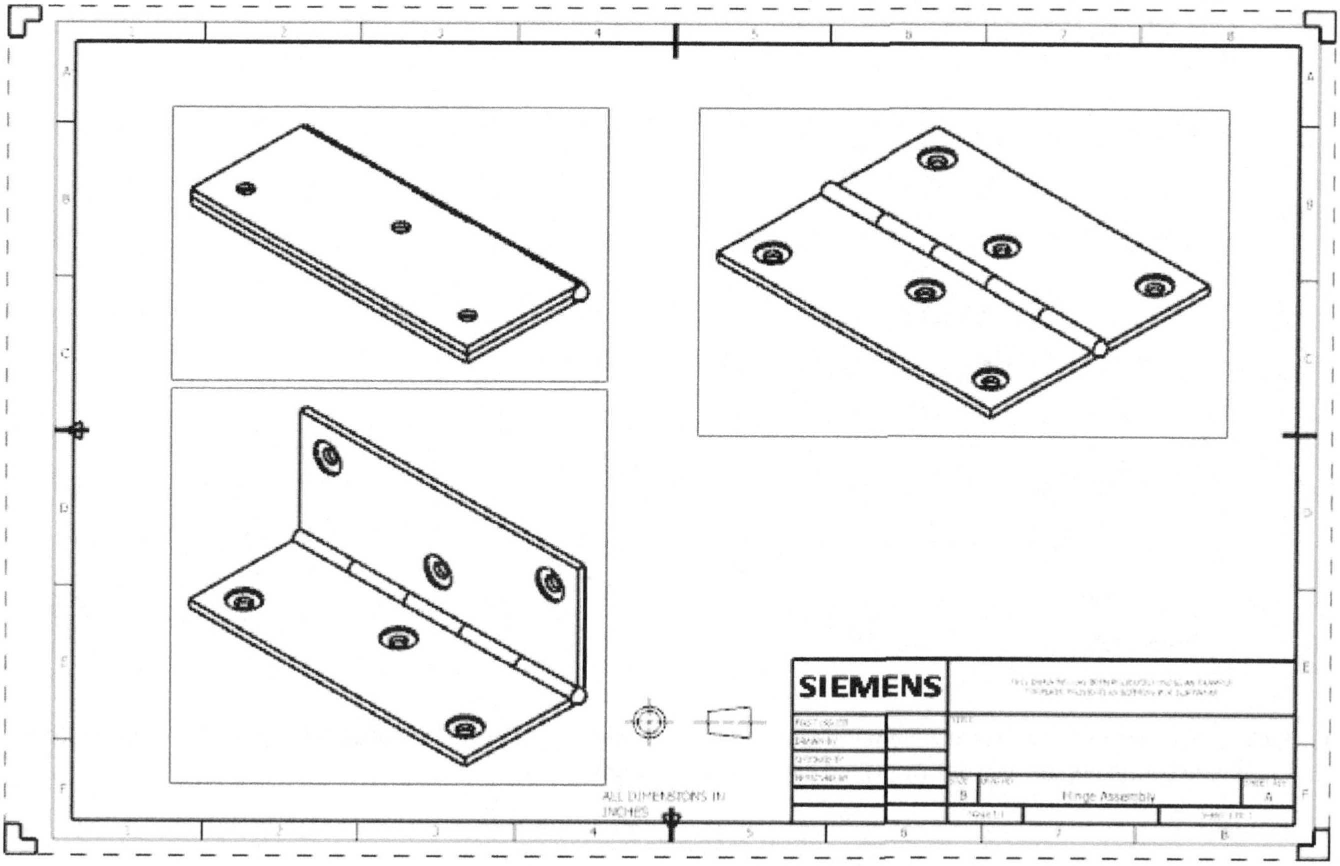

- **Save** your assembly!

18.8 **Interpart Expressions**

You learned how to share geometry associatively between components in an assembly with the **WAVE Geometry Linker** in Chapter 17.8. Sometimes it is not geometry that needs to be shared in an assembly, but parameters within the components. In this exercise, you will learn to use *Interpart Expressions* to drive a model of a DVD spindle.

- Create a new part file (**Units**: *Millimeters*) called *"DVD.prt"*.
- In the **Expressions** editor, create the following parameters.

Expressions

	↑ Name	Formula	Value	Units	Dimensional	Type
1	DVD_DIAM	120	120	mm	Length	Number
2	DVD_HOLE_DIAM	15	15	mm	Length	Number
3	DVD_THK	1.2	1.2	mm	Length	Number

- Model the DVD as a **Cylinder** with a **Simple Hole** at the center.

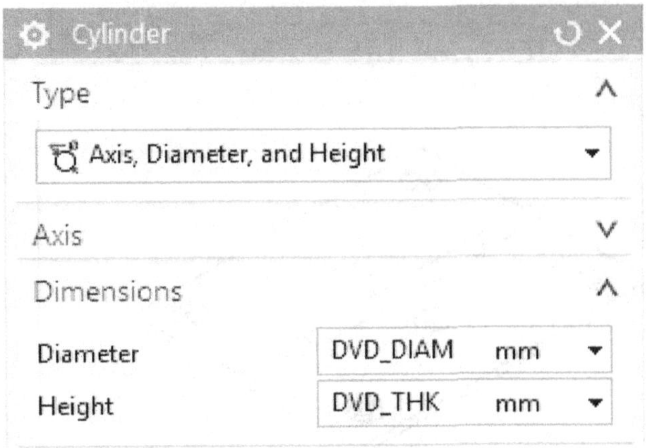

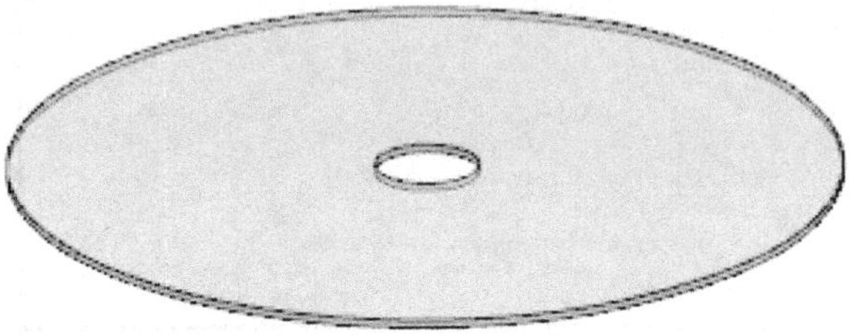

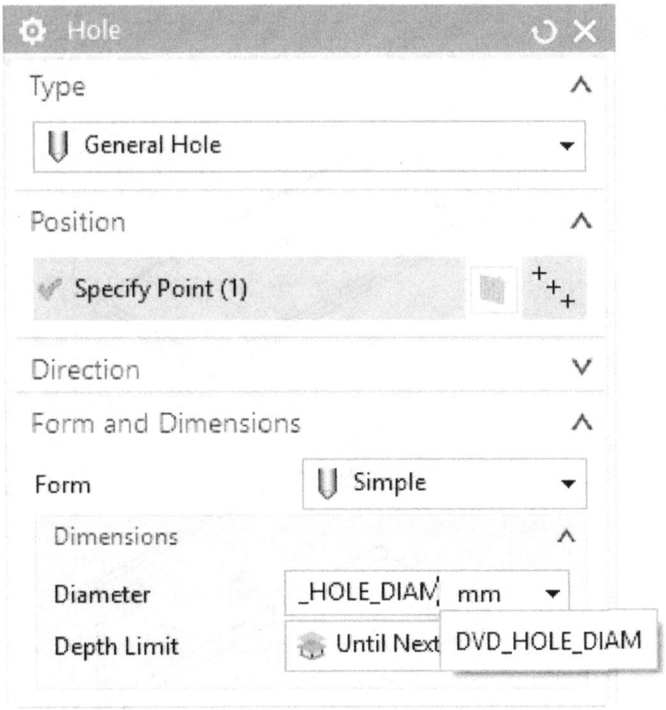

- Select the **Create New Parent** command from the **Component** group on the **Assemblies** tab. This will create an assembly containing the current Displayed Part as a component.

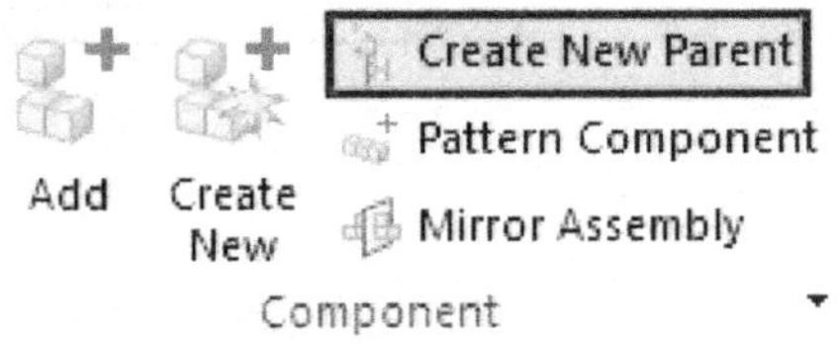

- Use **Create New Parent** followed by **Create New Component** to build your assembly as shown below.

- With *DVD Spindle Assembly* as the Work Part, enter the **Expressions** editor. In the **Actions** group, find the **Create/Edit Interpart Expression** button and click on it.

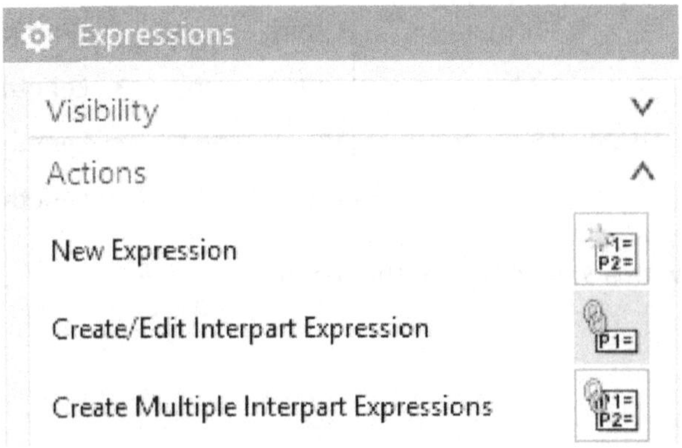

- Use the **Create/Edit Interpart Expression** tool to find *DVD_THK* from *"DVD.prt"*, and then add a new user-defined expression named *"N"* as shown below.

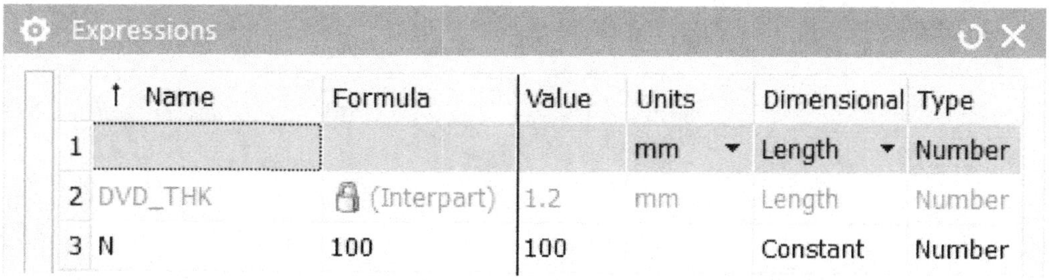

- Make *DVD Spindle* the Work Part.

- Enter the **Expressions** editor and select **Create/Edit Interpart Expression**.
- From *DVD Spindle Assembly*, you will want to borrow *N*.

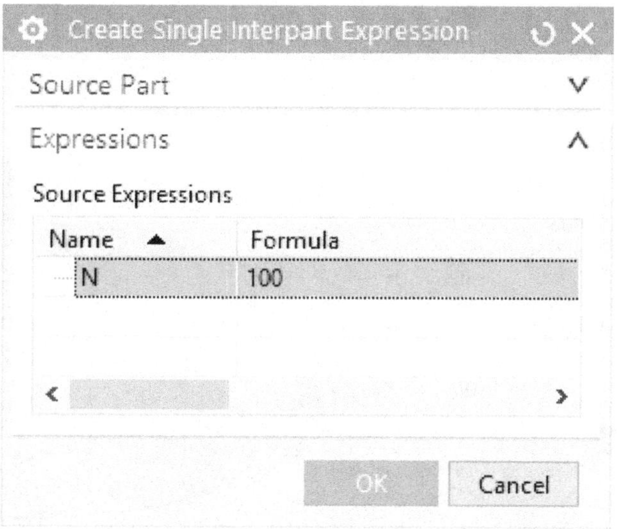

- Use **Create/Edit Interpart Expression** one more time to bring *DVD_THK* into *"DVD Spindle.prt"*.

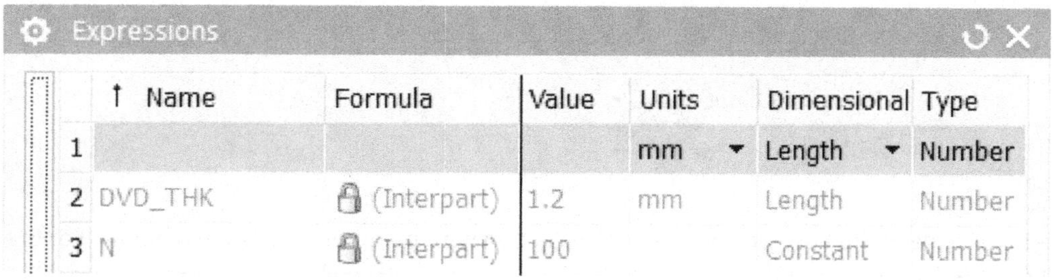

- Click **OK** To save your expressions.
- Next, select the **WAVE Geometry Linker** and set the **Type** to *Face*. Choose the bottom face of the DVD.

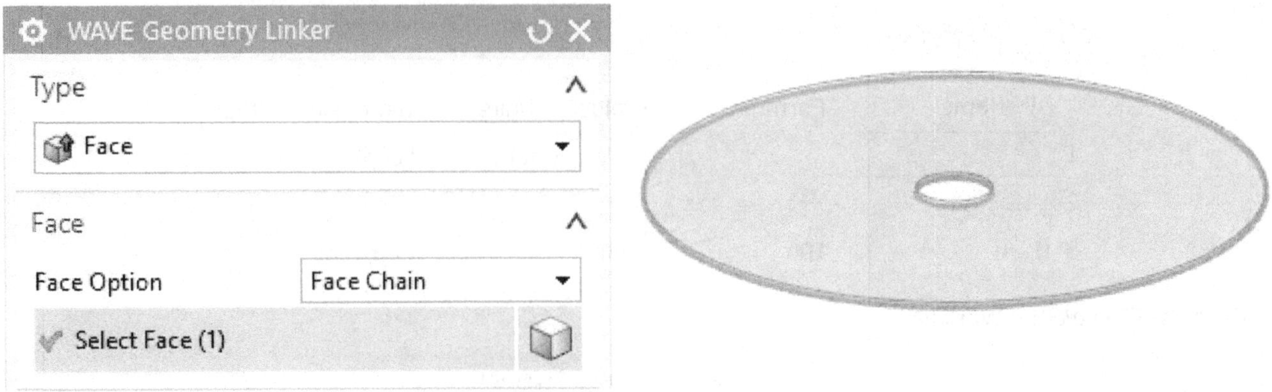

- Hide *"DVD.prt"* and **Extrude** the outer edge of the linked face in the -ZC direction with parameters as shown below.

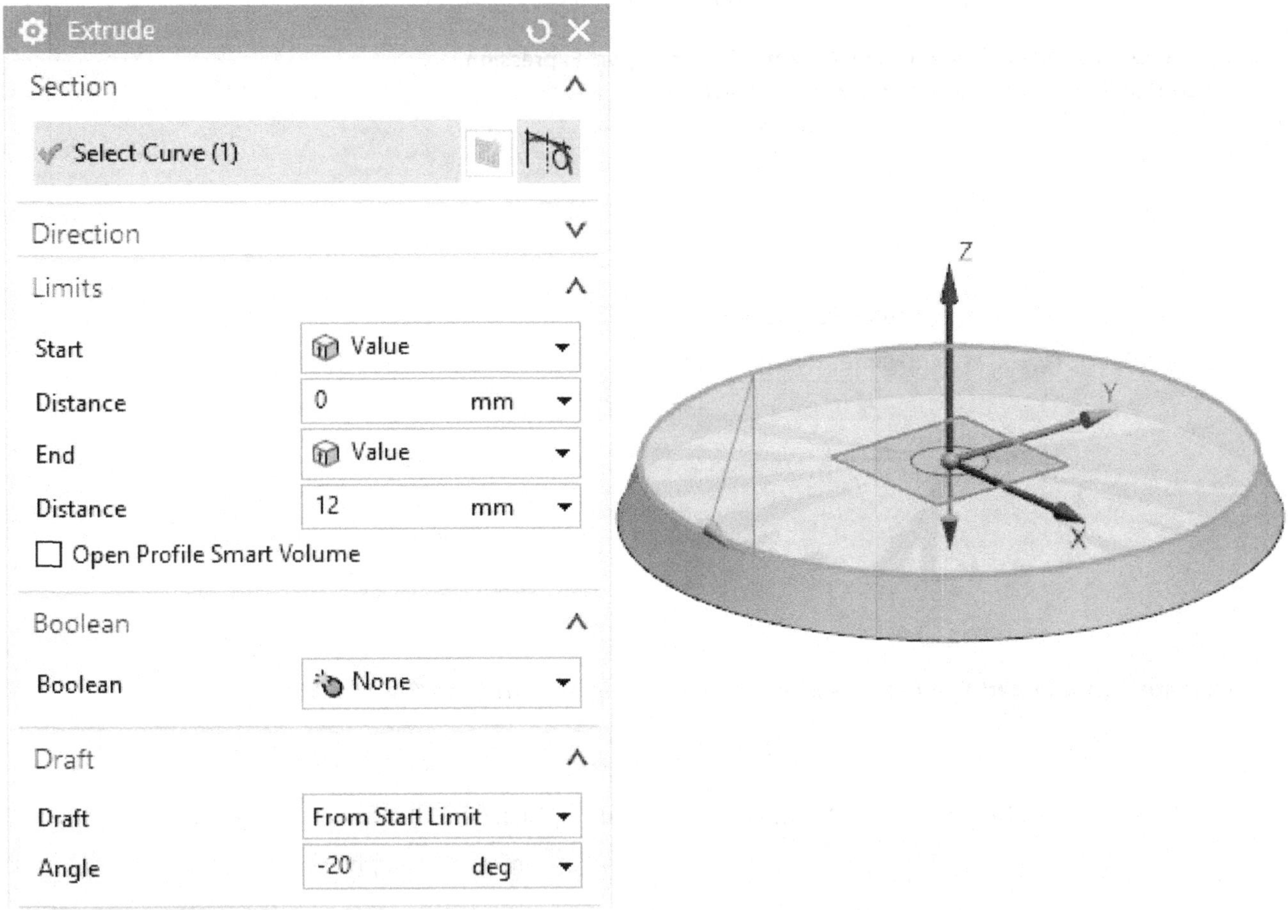

- **Extrude** the inner edge of the linked face in the +ZC direction with parameters as shown below. For the **End Distance**, enter *"N*DVD_THK"*.

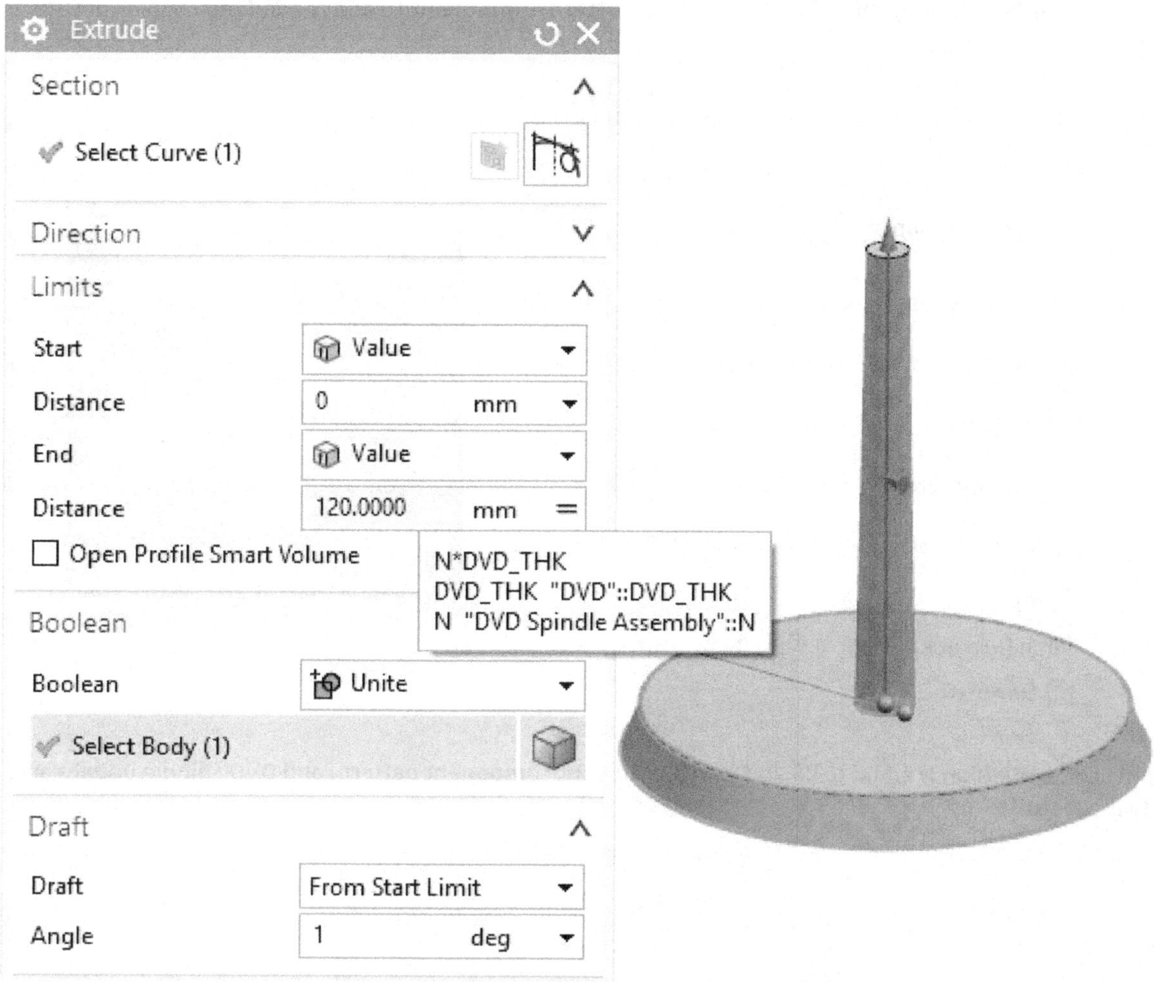

- Use **Delete Body** to delete the *WAVE* linked sheet body associatively, and cap off the top of the spindle with a **Sphere** of diameter equal to the diameter of the edge there (make a measurement on-the-fly!).

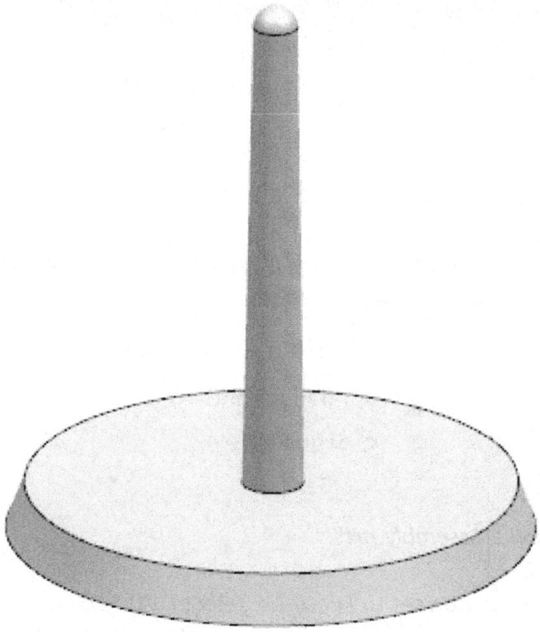

- Make *DVD Spindle Assembly* the Work Part and use **Pattern Component** to add *N* additional copies of DVD as shown below.

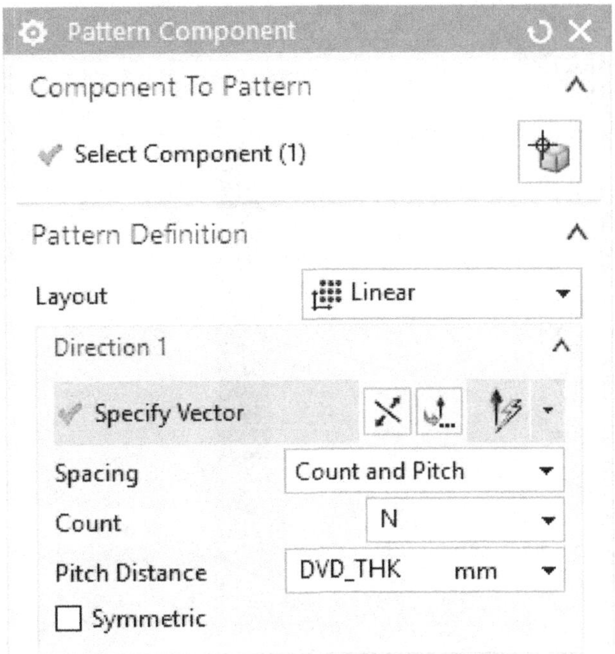

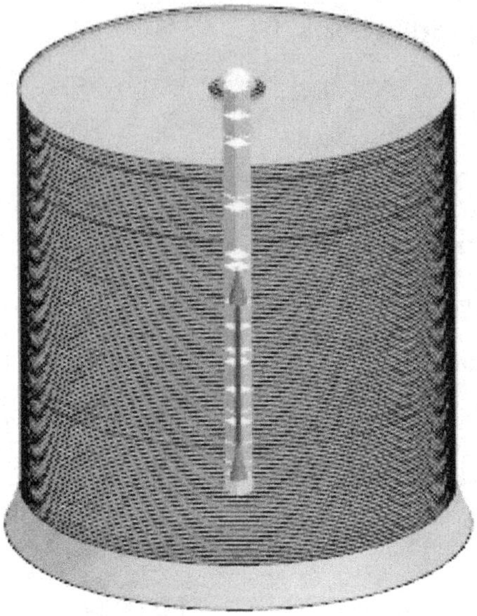

- Set the user expression *N* equal to *25* and watch both your component pattern, and DVD spindle update accordingly. Pretty great stuff!

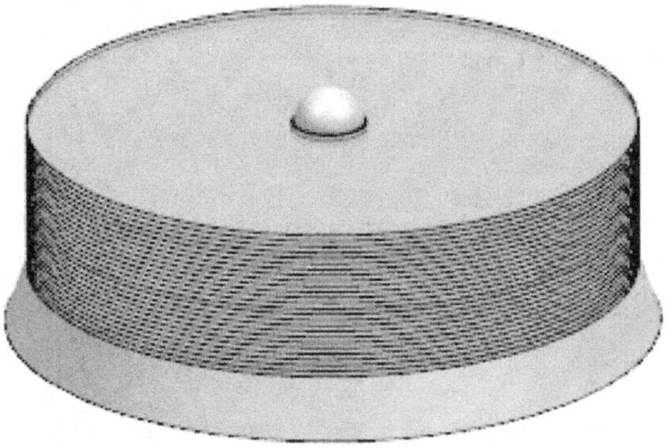

- **Save** your assembly!

18.9 Interpart Link Browser

The **Interpart Link Browser** is a great tool for keeping track of *WAVE* links and interpart expressions in an assembly. The **Interpart Link Browser** tool is found next to the **WAVE Geometry Linker** in the **General** group on the **Assemblies** tab.

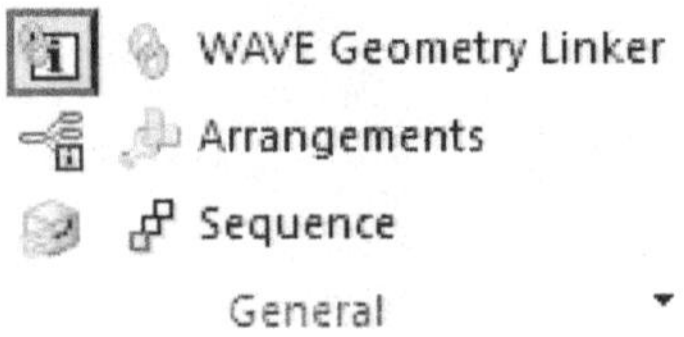

- Continue working with *"DVD Spindle Assembly.prt"*.
- Open the **Interpart Link Browser**.
- With the **Links to Investigate** radio button set to *Features*, select *DVD Spindle* from the list of **Parts**.
- The list of **Interpart Links** is short – you can easily find the *Linked Face* entity. Select it.

- Use the **Break Link** button on the bottom right of the menu. This is an important step to take when your parts are finalized and released.

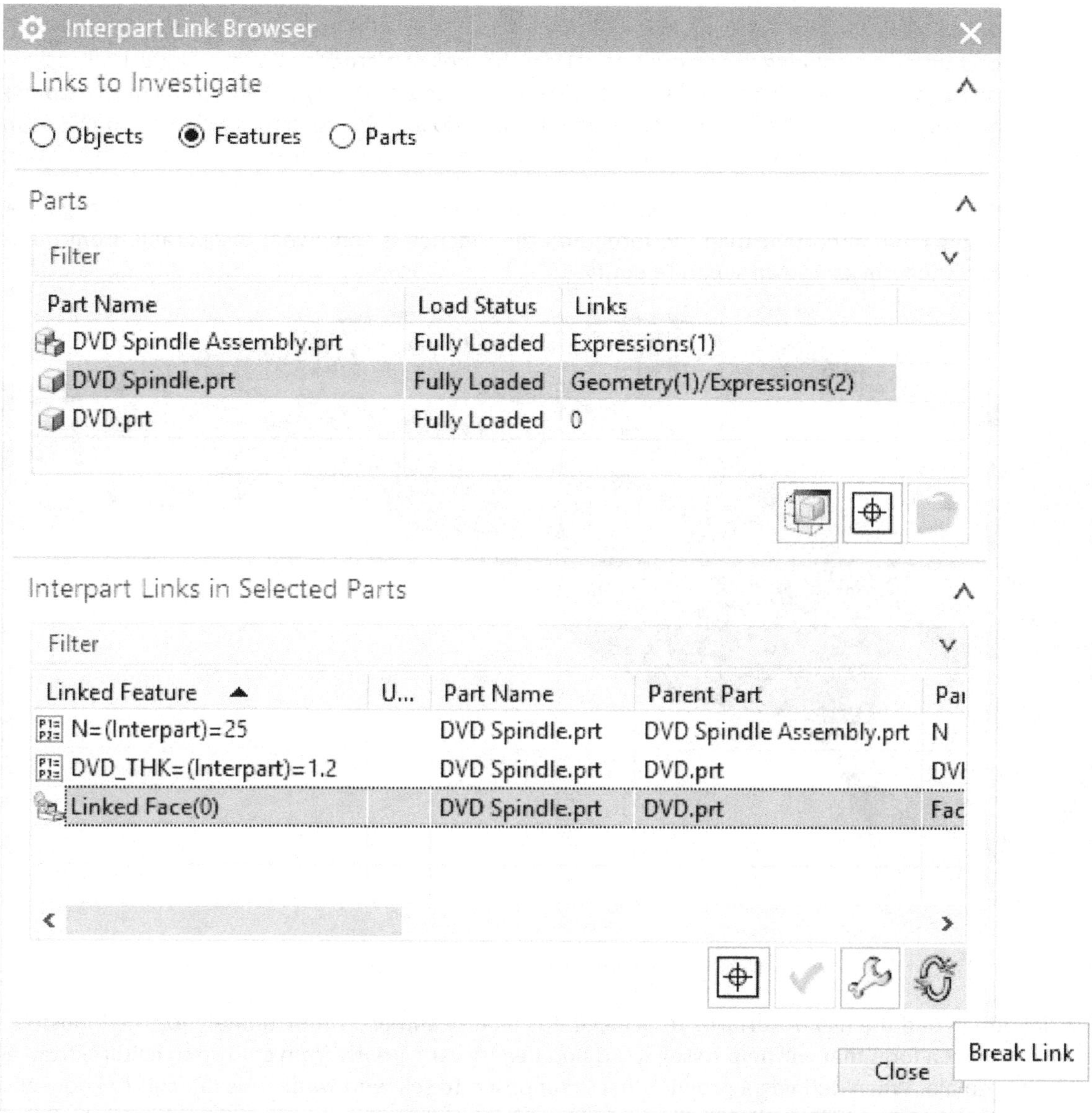

- You can also break the link by making *DVD Spindle* the Work Part, editing *Linked Face (0)*, and unchecking the *Associative* checkbox.
- The **WAVE Geometry Linker** is a very powerful tool for creating "smart" assemblies, but when your assembly approaches its final design, you should break WAVE links so that the released, production version of those parts cannot be accidentally modified by making changes to other parts.
- **Save** your assembly!

19 Freeform Surfaces

It's amazing how good surfacing, coloring and texture can affect the look of a product. As many industrial designers know, a product's form can attract and delight the end user. The form can capture the eye of a prospective customer, beacon them over to the shelf, or click on the link and input their credit card info then proceed to checkout. Product form can give a product owner a sense of pride as they are seen using the product by friends, family members or even total strangers.

Form gives the user something that somehow appeals to an underlying sense of beauty and value. Although the function of a product is arguably more important than the form, and the function is somewhat inseparable from the form, it is instructive to consider the form as its own separate entity.

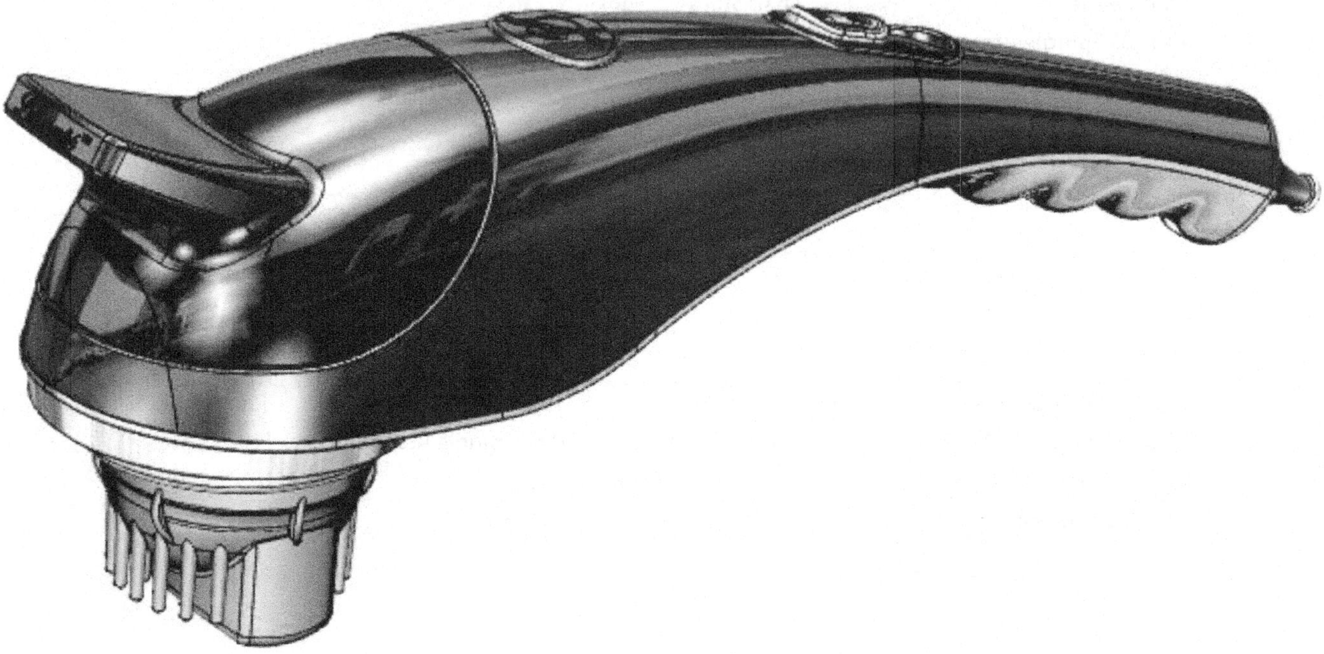

Massager unit designed by Design Visionaries

Great industrial design and the use of surfaces to achieve it is beauty and art. Unfortunately for the industrial designer trying to give a product a form that will help to sell it, art and beauty vary greatly from culture to culture, from age to age and from male to female. When defining a product that is supposed to sell worldwide, it is difficult to know what will be appealing to such a varied user community. That said, industrial designers who are trying to suit the tastes of a huge and varied audience benefit by considering what is universal.

The **Surface** tab contains a variety of tools that are used to construct organically sculpted geometry. The surfacing tools available in NX allow you to build and edit any type of freeform surface, whether swept, lofted, blended, or offset.

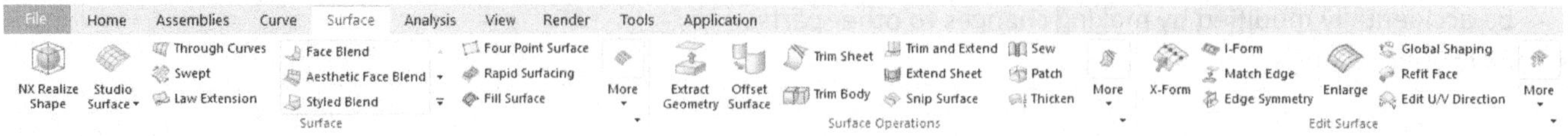

The figure below illustrates two typical examples of surfaces – the left is an example of a *mesh surface*, and the right is an example of a *swept surface*.

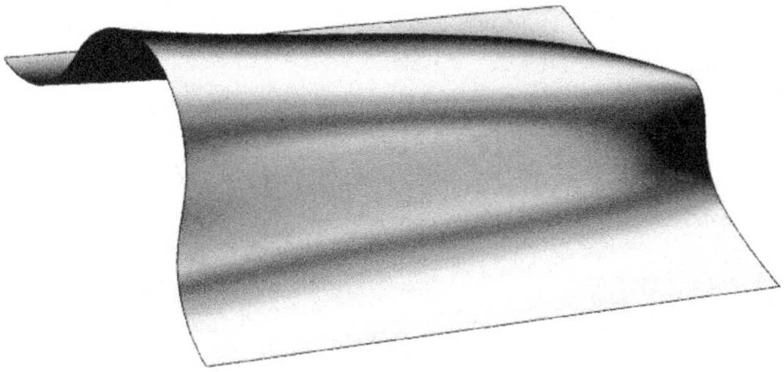

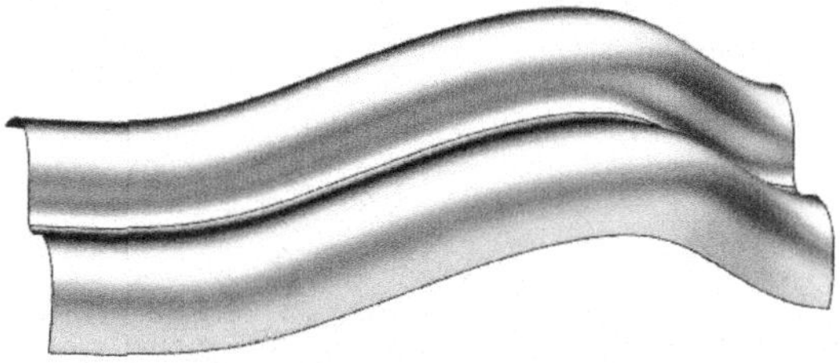

By their nature, surfaces are highly dependent upon other geometry. In the most rigorous situations, surfaces are created using fully parametric sketch curves, offset curves, or projected curves. By using these intelligent entities, you can create the most difficult of geometries.

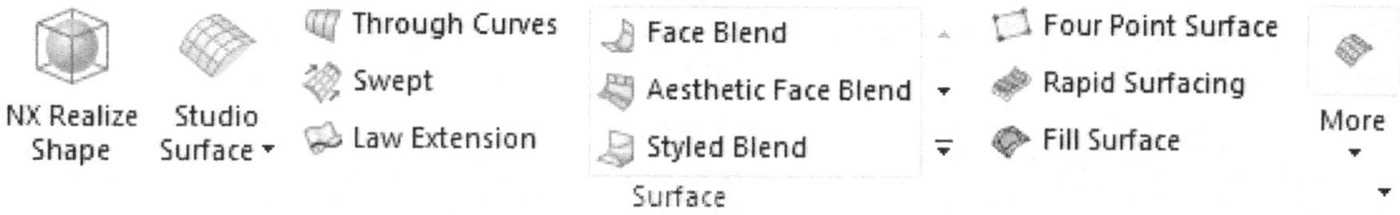

In addition to the **Surface** tab, a limited set of surfacing tools can be found in the **Surface** group, on the **Home** tab.

19.1 **Bridge**

We begin our discussion of lofted surfaces with the 2D analogue of the **Bridge Curve** command. The **Bridge Surface** command produces a high-quality surface that transitions between two specified faces, offers modest shape control, and continuity at the ends up to *G2*.

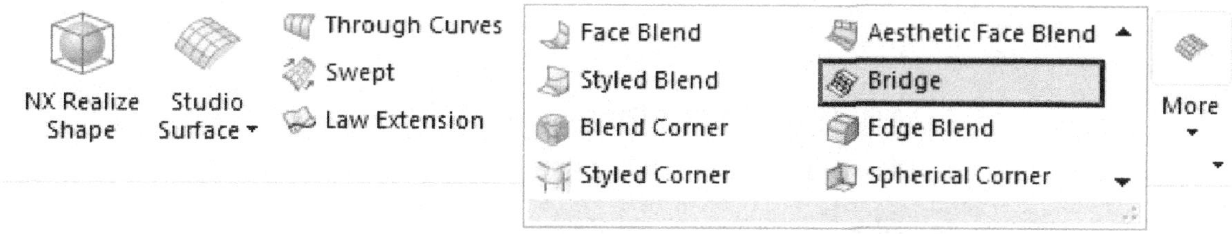

- Create a new file (**Units**: *Inches*) called *"Bridge Surface.prt"*.
- Create the sketch shown below in the YZ plane.

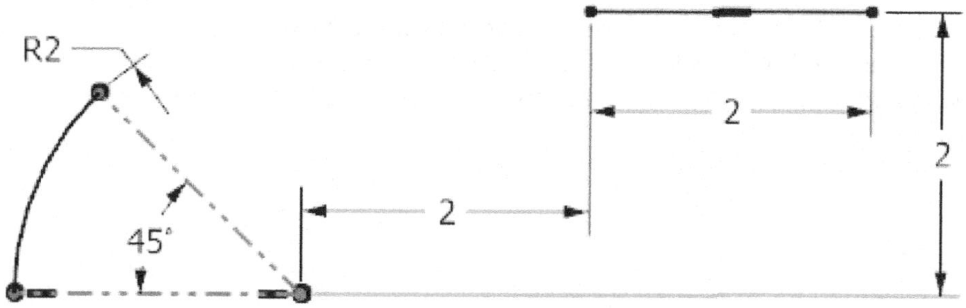

- **Extrude** the arc with a *Symmetric Value* of *2 in* to create a cylindrical surface.
- **Extrude** the line with a *Symmetric Value* of 1 *in* to create a planar surface.

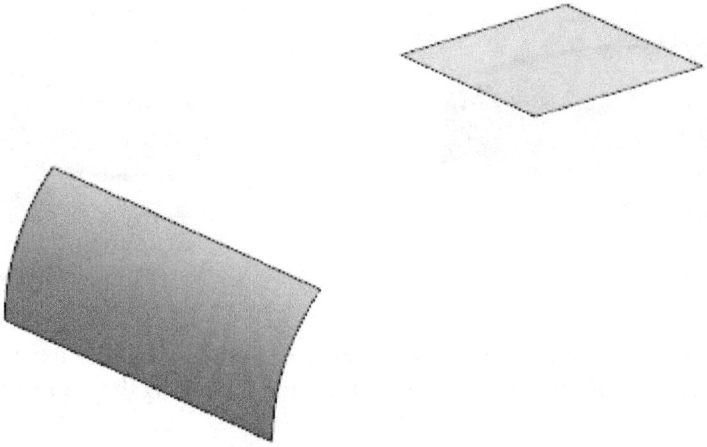

- Select the edge of the planar surface as *Edge 1* and the edge of the cylindrical surface as *Edge 2*. This builds a lofted surface joining those edges.

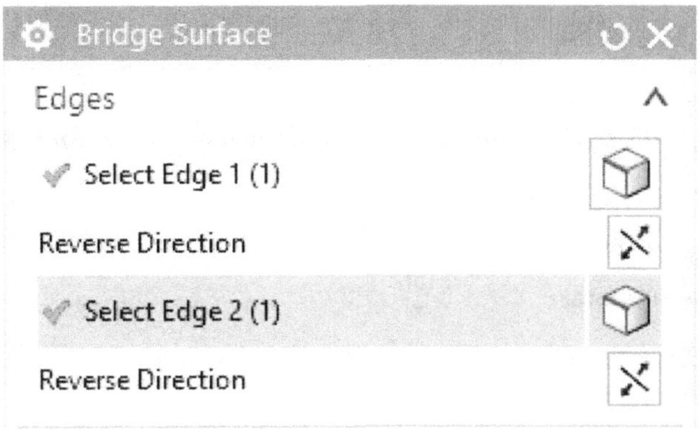

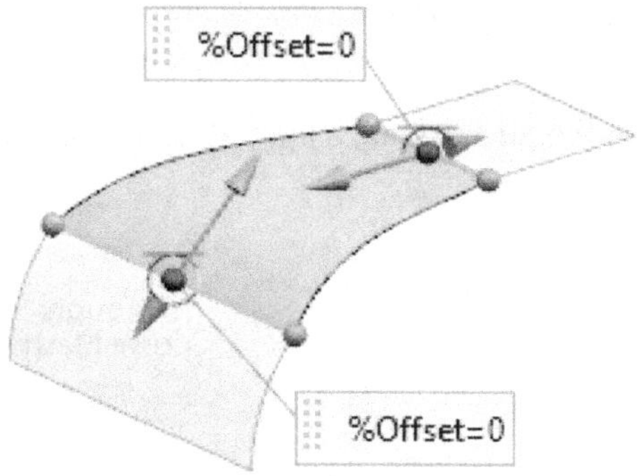

- As with the **Bridge Curve** tool, you can adjust the degree of continuity the bridge surface where it meets its neighbors. Leave the **Continuity** set to *G1 (Tangent)* at both **Edge 1** and **Edge 2**.

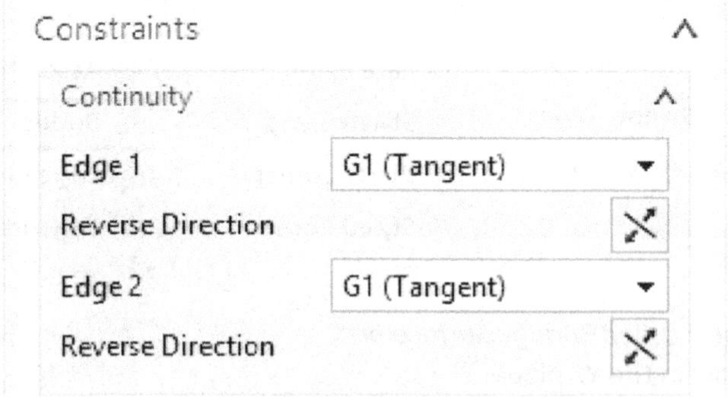

- The **Edge Limit** section of the dialog allows you to control how much of the edges are used in building the lofted surface. The **%Start** and **%End** sliders allow you to offset the corners of the surface from the corners of the original surfaces. The *End to End* checkbox guarantees that the corners of your bridge surface coincide with the endpoints of the parent edges.

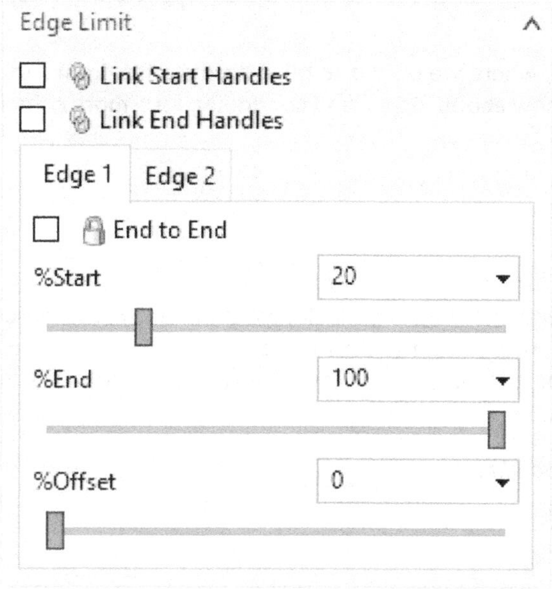

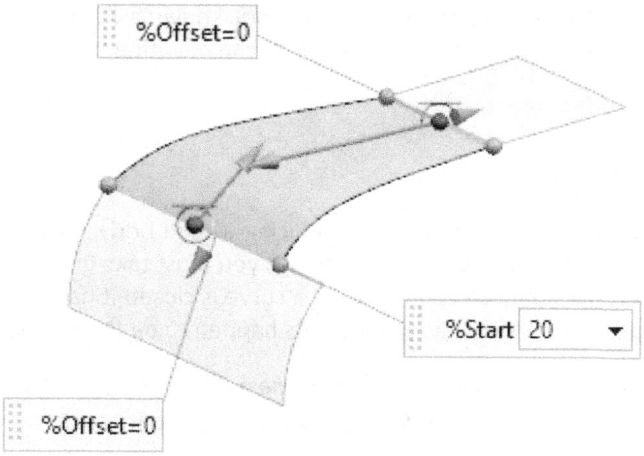

- The **Flow Direction** is the direction that connects *Edge 1* and *Edge 2*. You can control the tangency of the edges in the **Flow Direction** with the **Edge 1 and 2** drop-down. *Isoparametric* will result in the "flow" edges being tangent to the edges of the parent surfaces.

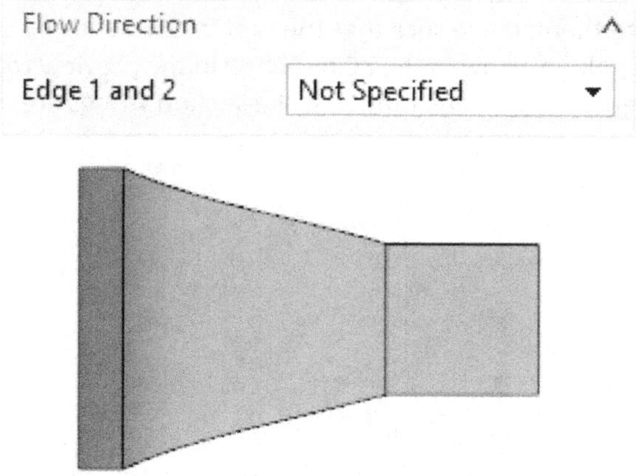

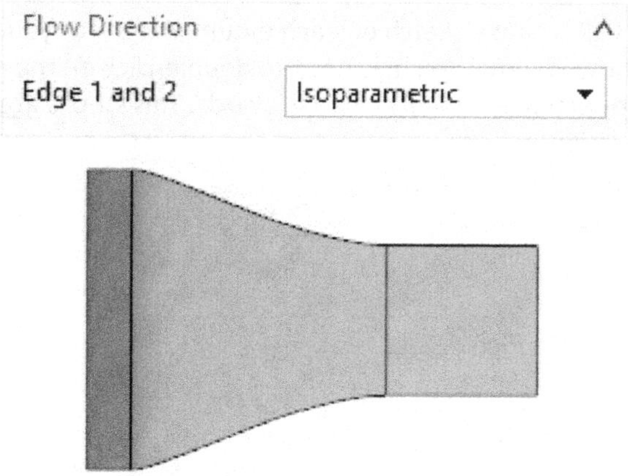

- Finally, just like a **Bridge Curve**, you can adjust the magnitude of tangent vectors to the bridge surface at its start and end.

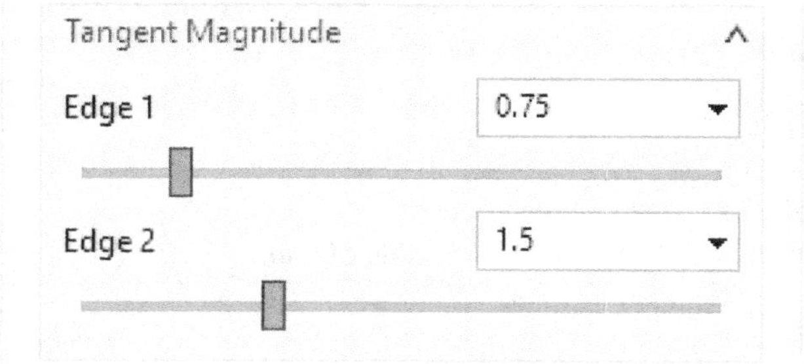

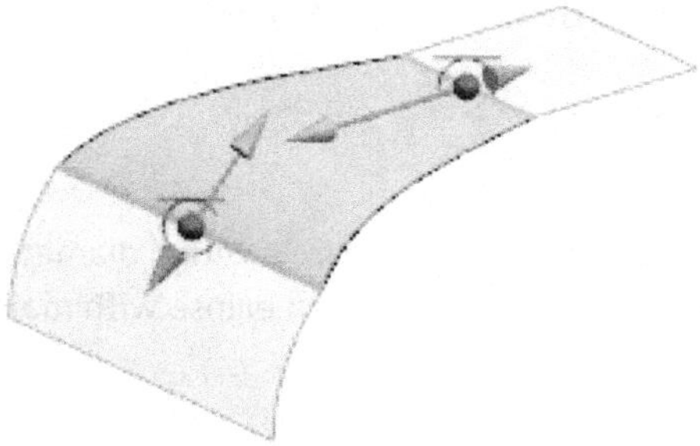

- **Save** your file!

19.2 Through Curves

We first encountered the **Through Curves** command in Chapter 7.7, where we used it to build a simple freeform surface. We would be remiss not to revisit the tool, knowing what we now know about curves, and continuity, for a more complete discussion of the control you have over the resulting body.

19.2.1 Solid

Through Curves is capable of producing a solid body, when the section strings are all closed. When using the **Through Curves** tool with *closed* section strings, you must take great care to ensure that the start points of those closed curves are aligned. This may be confusing – if a curve is closed it has no start or end, right? In NX, even closed curves have start and end points – the start and end points happen to be the same point.

- Create a new part file (**Units**: *Inches*) called *"Through Curves – Solid.prt"*.
- Create three parallel datum planes, two inches apart.

- You will create a sketch on each datum plane – as you create them, make sure that the sketch CSYS ends up oriented as shown below. You may need to double-click on the small golden arrow in the ZC direction in the preview to reverse the direction if it comes out backwards. This step is absolutely critical to ensuring that the section strings are oriented consistently.

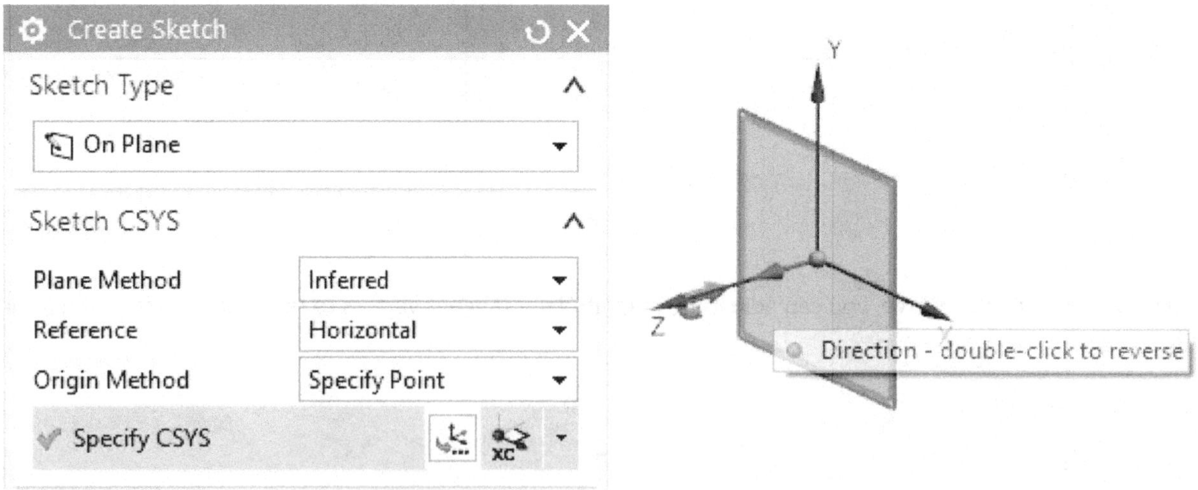

- On the first plane, sketch a circle with a diameter of *2 in*.
- On the second plane, sketch an ellipse with major radius of *2 in* and minor radius of *1 in*.

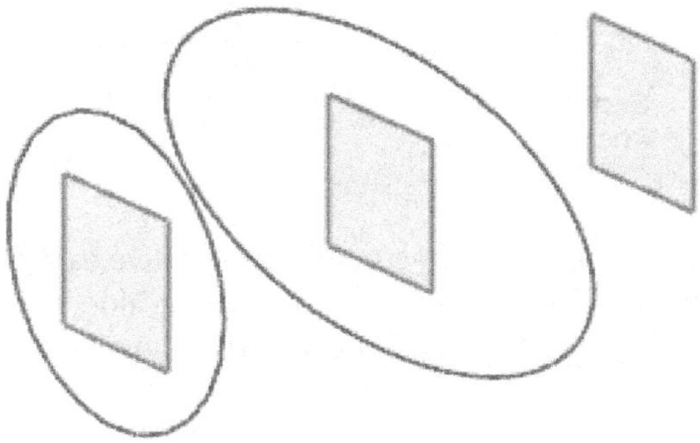

- On the third sketch plane, use the **Studio Spline** tool with type set to *By Poles* to sketch the closed cubic spline shown below. The positioning of the first pole is critical – this will ensure that the start points of your curves are consistently positioned.

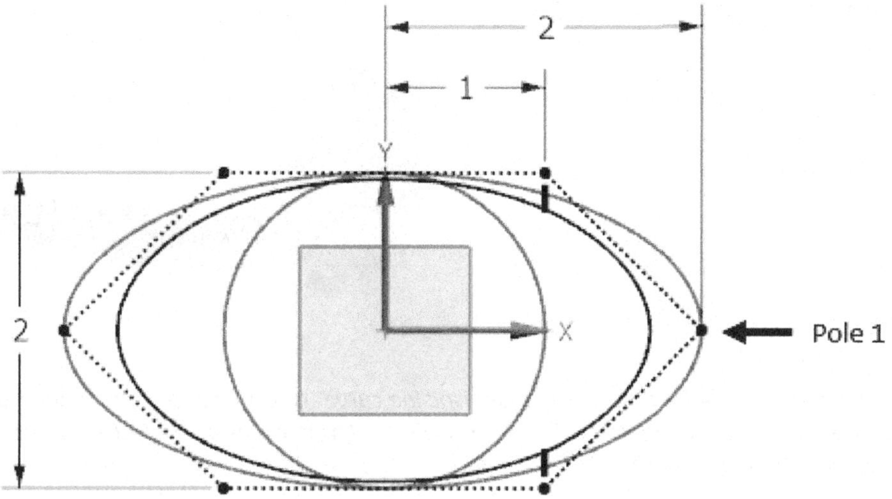

- Select these sketches, in order, as *Section 1*, *Section 2*, and *Section 3*, respectively.

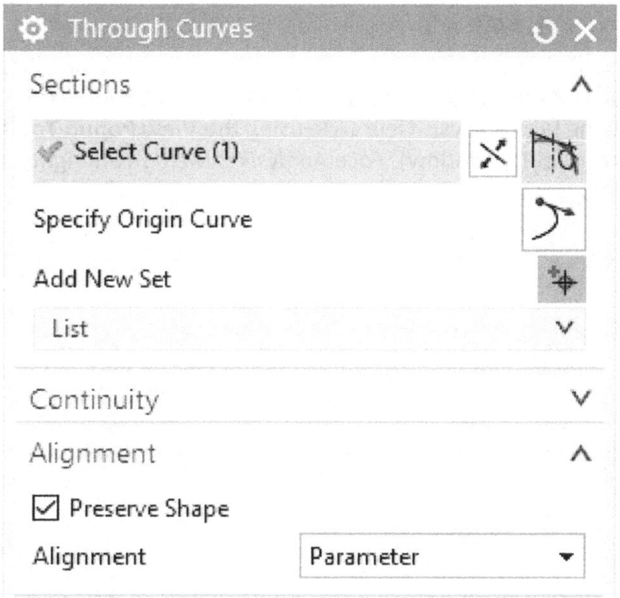

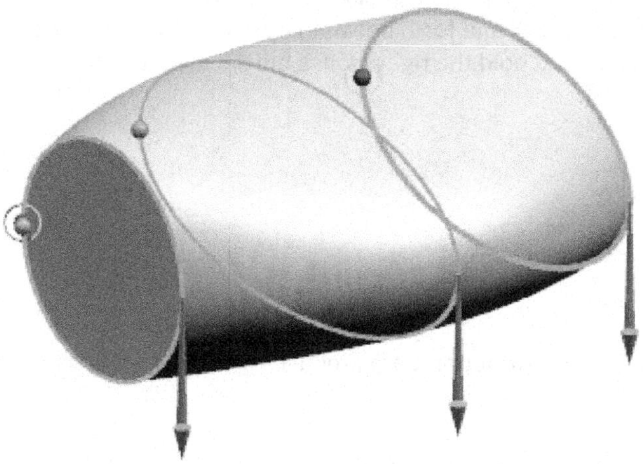

- Each of these curves has its start point positioned along the positive sketch horizontal axis. To verify that the "flow curve" starting at the start point of the circle will actually pass through the start point of the ellipse and of the spline, we will use the **Draft Analysis** command, found in the **Face Shape** group on the **Analysis** tab.

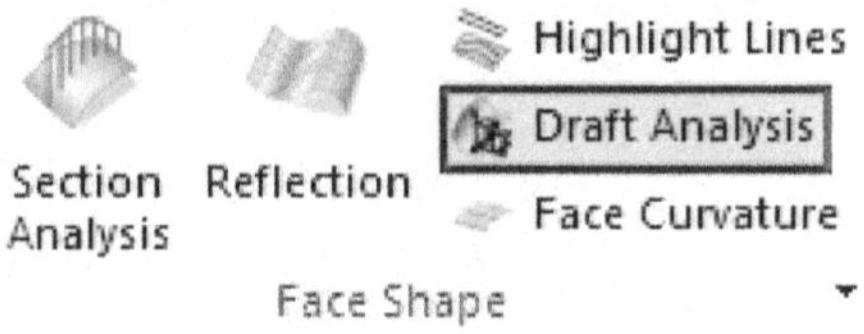

- The **Draft Analysis** tool will display the draft on a selected surface relative to the **Draw Direction** with a color map. Set the **Draw Direction** radio button to *Vector* and choose ZC. The parting "line" is the curve on the selected face defined by having a draft angle of zero in relation to the +ZC direction. The *Show Parting Line* checkbox determines whether this curve will be displayed.

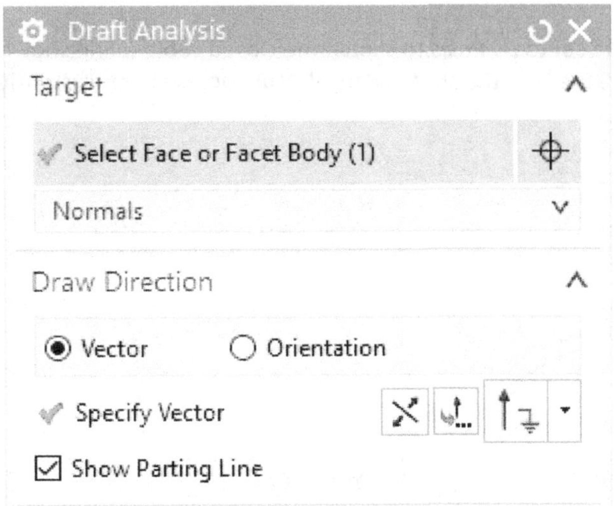

- The parting line a special example of a curve called an *isocline curve*. Isocline curves are defined by having a constant draft angle in relation to the draw direction. You can have the **Draft Analysis** tool produce as its output either an *Analysis Object*, *Isocline Curves*, or *Both*.

- The color map will remain visible on your model so long as you are using the *Face Analysis* rendering style. You can switch back and forth between a standard shaded view and the face analysis view style using the **View Popup Toolbar** (click and hold the right mouse button in empty space in the graphics window). *Face Analysis* is the bottom right icon.

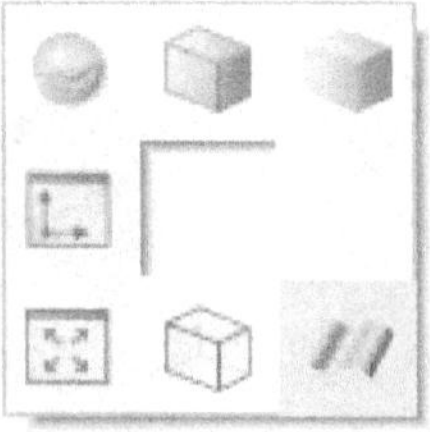

- The **Output** option *Both* produces an Analysis Object called *Draft Analysis*, and a dumb, non-parametric spline.

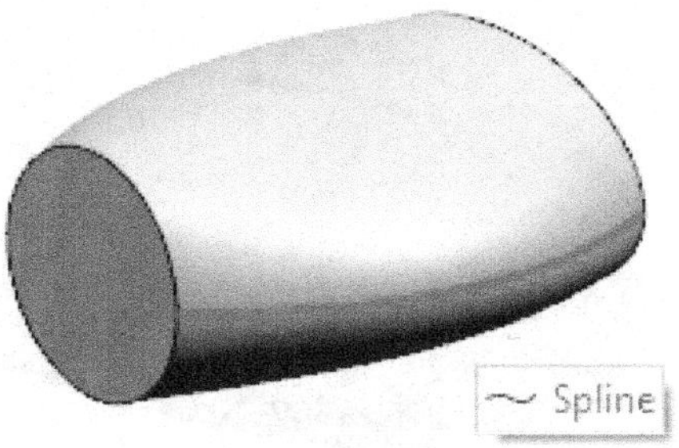

- You can also produce parametric isocline curves using the **Isocline Curve** tool, found in the **Derived Curve** group on the **Curve** tab.

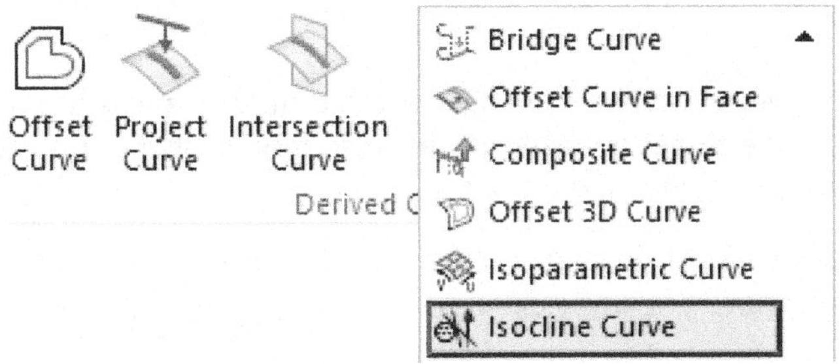

- The basic inputs for the **Isocline Curve** tool are similar to those in the **Draft Analysis** tool. Set the **Reference Direction** to +ZC. In the **Isocline Angle** section of the dialog, you can specify which constant angles you would like to use to define isocline curves – set the radio button to *Single* and specify an **Angle** of *0 degrees*. In the **Settings** section of the dialog, check the *Associative* checkbox.

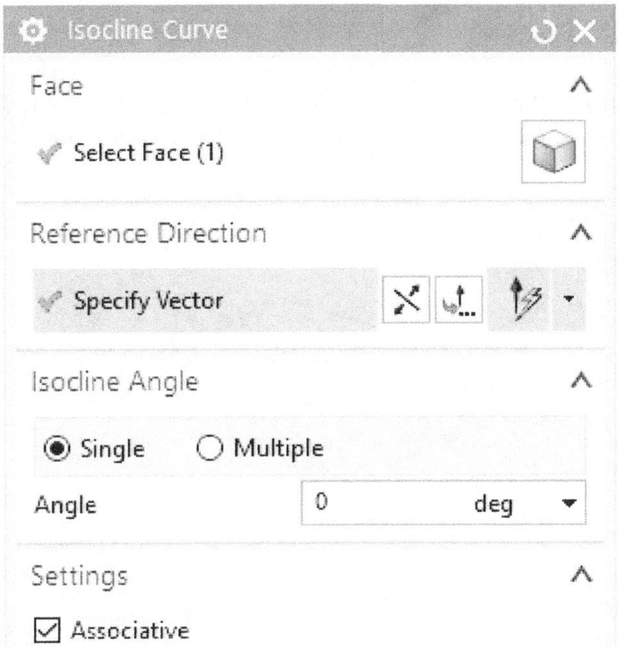

- The output of the **Isocline Curve** tool is exactly the same as the parting line, except that it is a fully associative and parametric feature in the Model History.

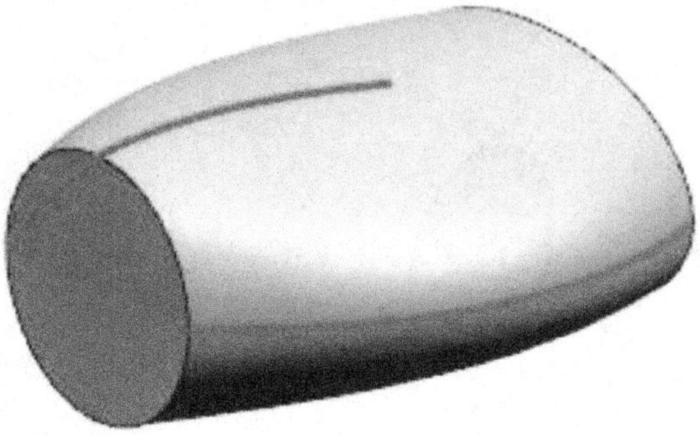

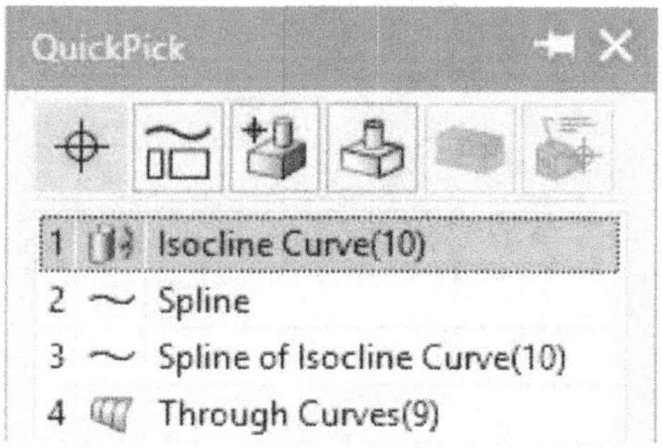

- **Save** your file!

19.2.2 Continuity

In this exercise, you will use the **Through Curves** tool to build a lofted surface joining the surfaces from the part created in Chapter 19.1. Besides the shape of the curves themselves, the main form of shape control **Through Curves** offers for the resultant surface is in the **Continuity** section of the menu[6].

- Open the part file *"Bridge Surface.prt"* from *"C:\My NX 11 Files"*, or from your downloaded part files folder.
- **Suppress** the *Bridge Surface* feature.

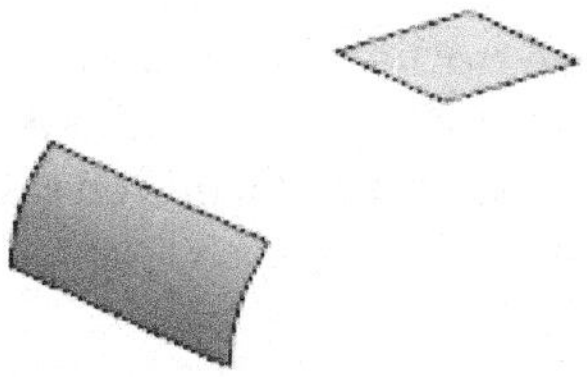

- Select the edges of the planar and cylindrical surfaces as *Section 1* and *Section 2*. Notice that when you have only two section strings, **Through Curves** will automatically result in a ruled surface, it simply draws lines between the points of the two sections.

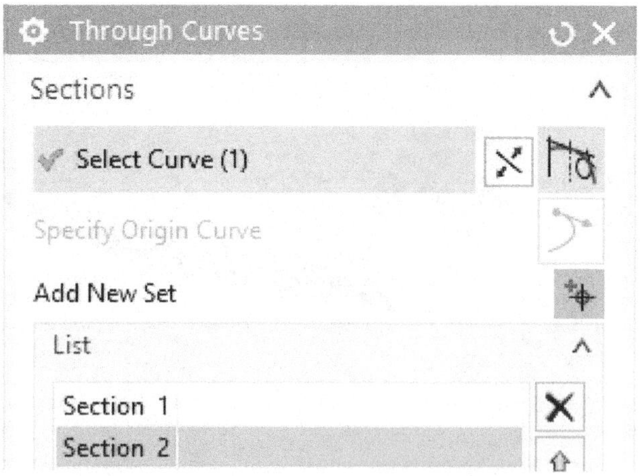

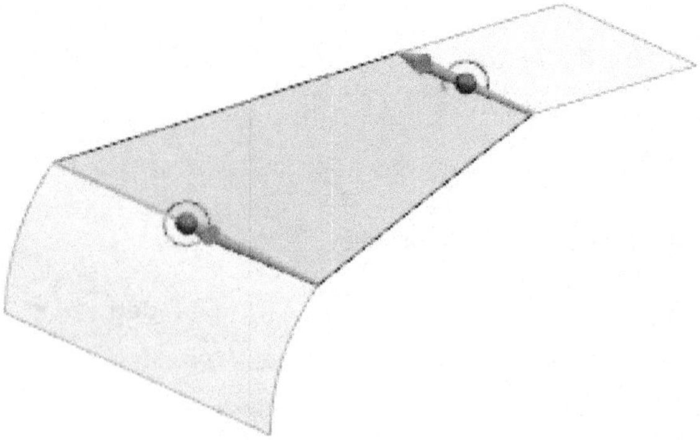

- To get a surface that smoothly transitions from one surface to the other, change the **Continuity** to *G1 (Tangent)* for the **First Section** and select the adjacent surface.

---

[6] Another important option for controlling the shape of a **Through Curves** surface is the **Lofting** degree, which is not covered in this text. It is analogous to the degree of a **Studio Spline**, as described in Chapter 13.2.

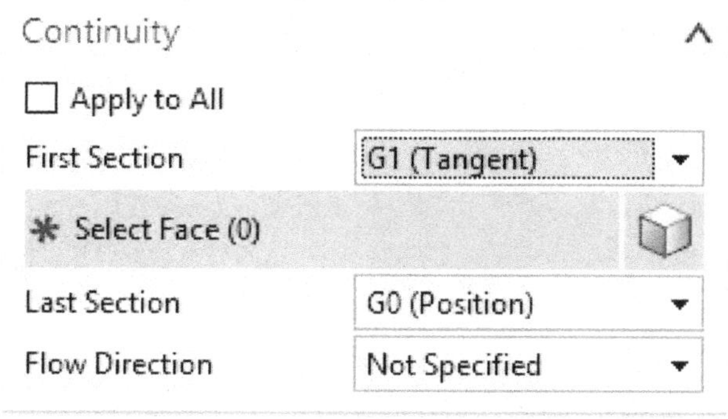

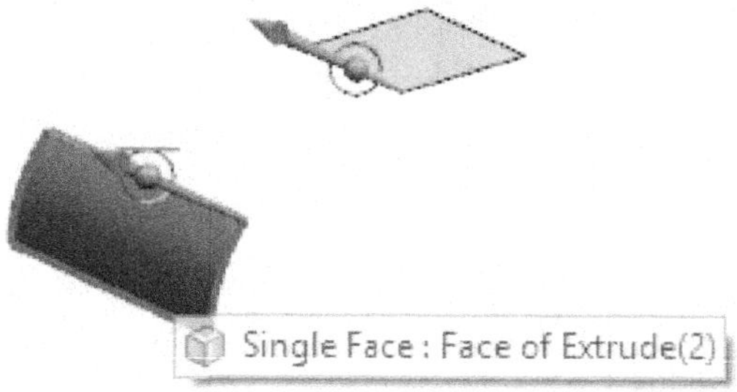

- Do the same for the **Last Section**. Set the **Flow Direction** to *Isoparametric*.

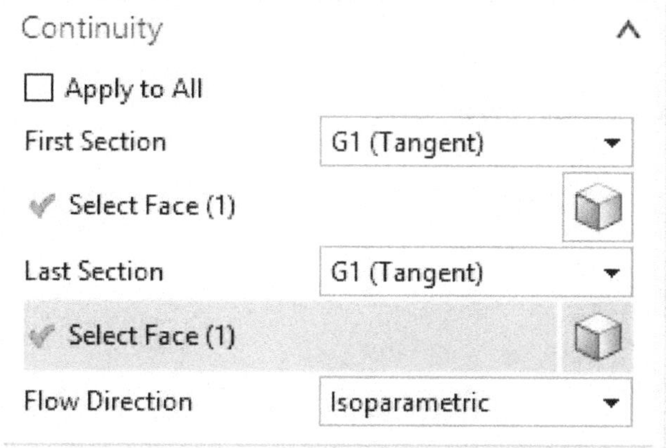

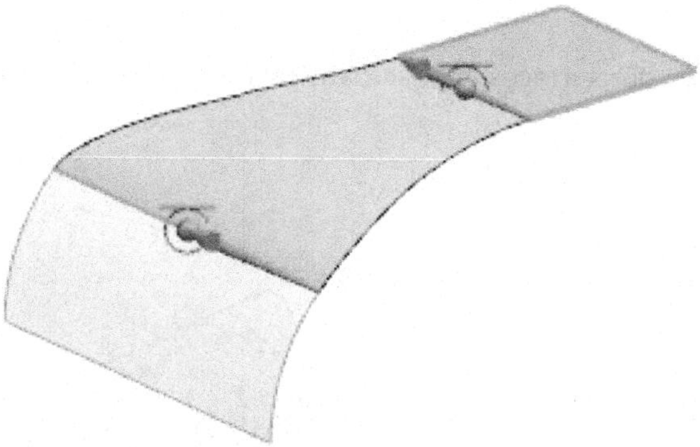

- Your surface looks awfully similar to the bridge surface built previously – if you **Unsuppress** the *Bridge Surface* feature, you will see that their shape is similar but the edges obviously don't align.

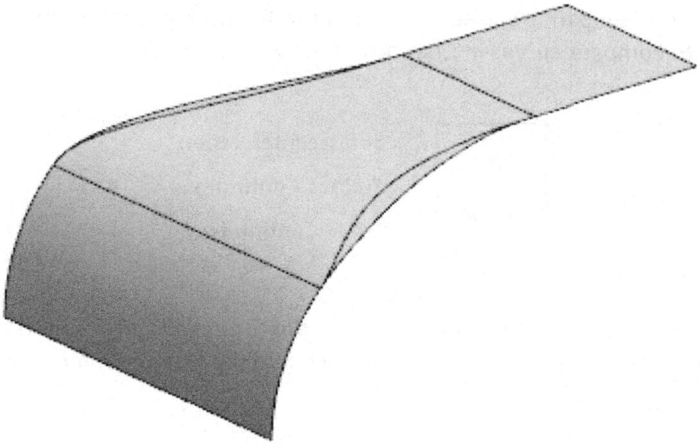

- Modify the bridge surface so that the **Tangent Magnitude** is *1* at each end.

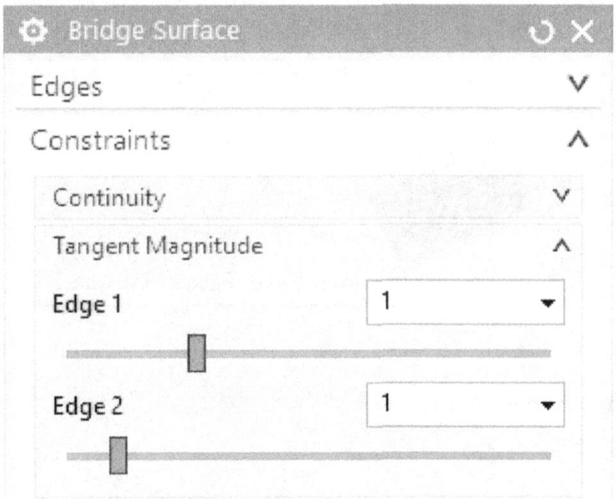

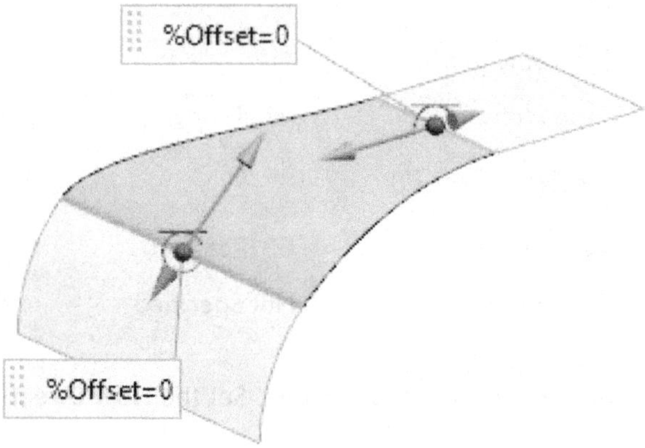

- Now your surfaces appear identical to the naked eye! But how can you be sure?

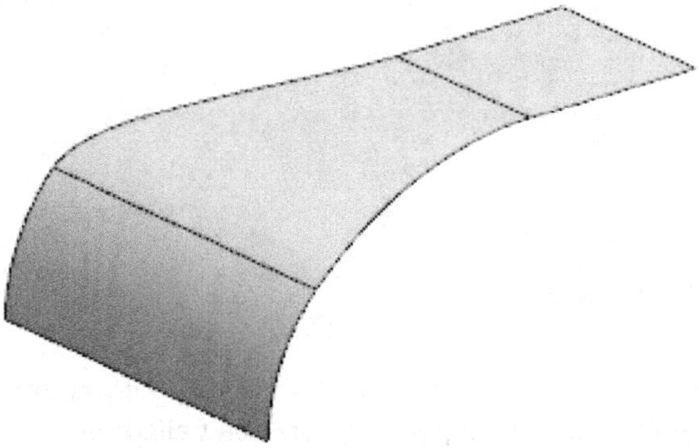

- There is a very useful tool for comparing surfaces found in the **Relation** group on the **Analysis** tab, called **Deviation Gauge**. You used it once to compare curves in Chapter 13.3.2.2, but it really excels at identifying and distinguishing surfaces.

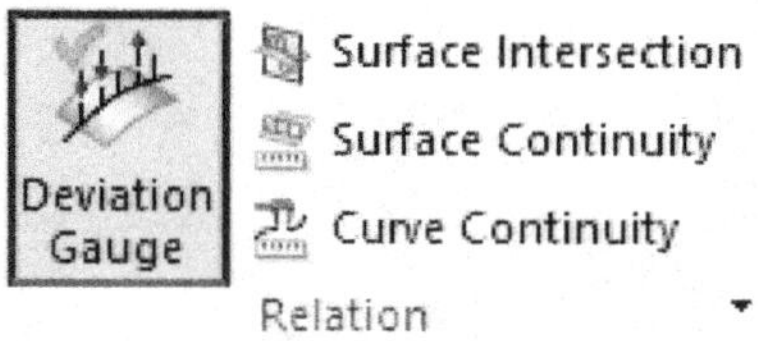

- For the **Objects To Compare**, set the **Selection Intent** drop-down to *Single Face* (for both the **Object** and the **Reference Object**), and choose the face of *Through Curves* as the **Object**, and the face of *Bridge Surface* as the **Reference Object**.

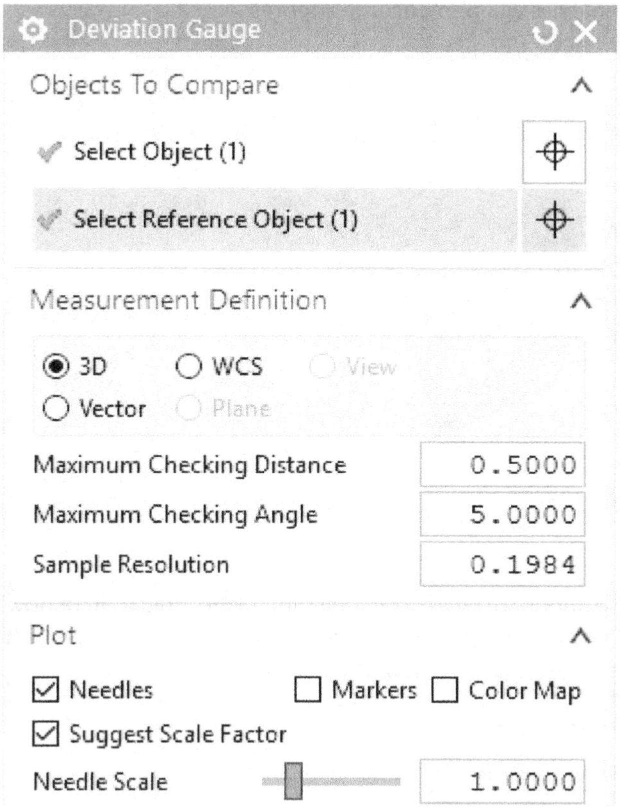

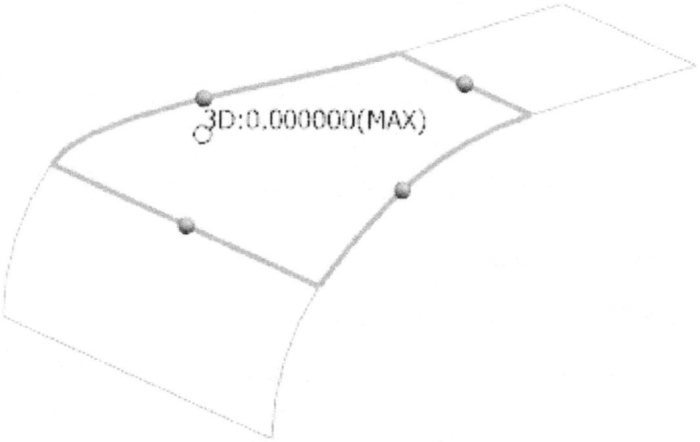

- The **Deviation Gauge** tool takes a number of samples at specified intervals on the surfaces and measures the total deviation between the surfaces at those points. In this case, it finds a perfect match everywhere! The results of your **Deviation Analysis** are stored as an analysis object, and placed in the *Analysis* folder in the Part Navigator. *Analysis* objects can be suppressed, deleted, hidden, shown, etcetera! *Analysis* objects dynamically update as changes are made to the objects that they measure or compare, and they can be extremely helpful in seeing the effects of each change you make to a surface as you build it.

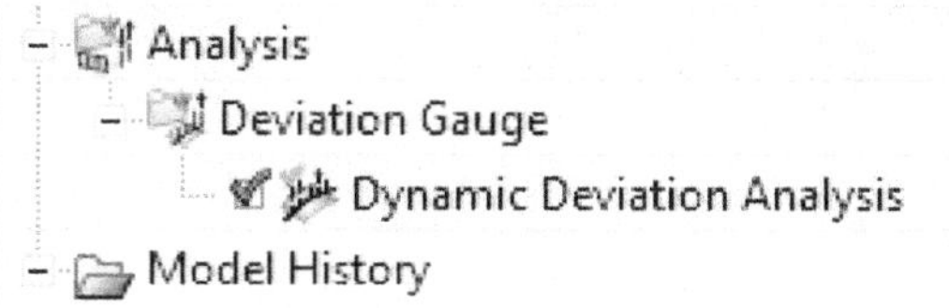

- Modify the *Through Curves* feature so that the **Continuity** at the **First Section** and **Last Section** is *G2 (Curvature)*. Your dynamic deviation gauge will update – as you edit – to show the total deviation at the sample points with a needle plot.

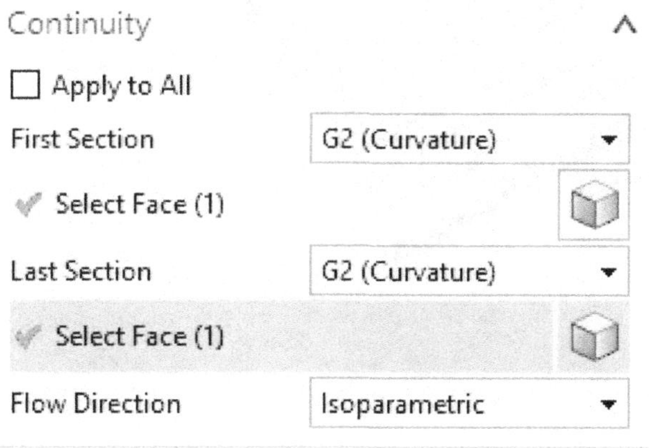

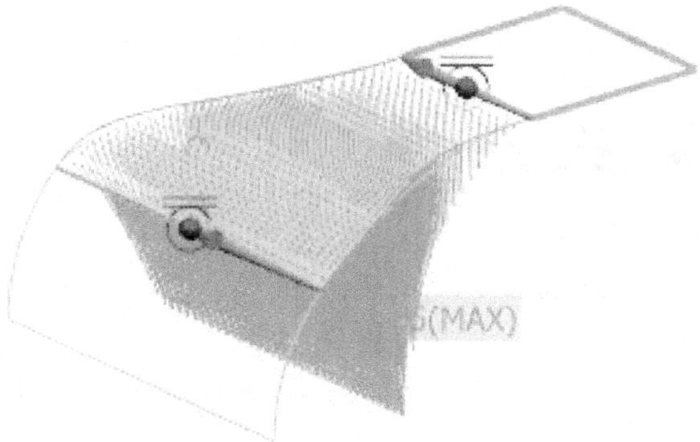

- If you suppress your **Bridge Surface**, the **Dynamic Deviation Analysis** will also become suppressed.

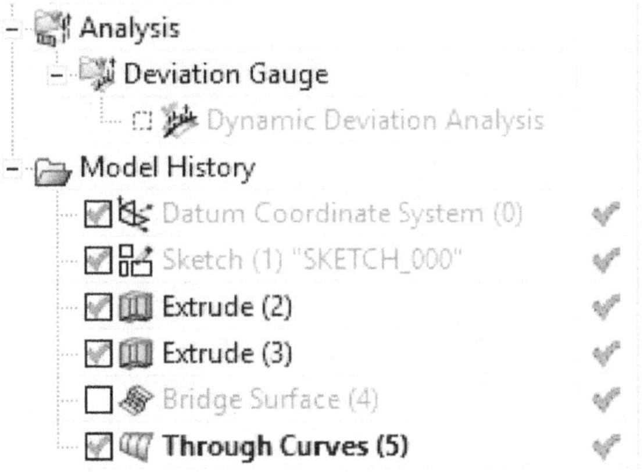

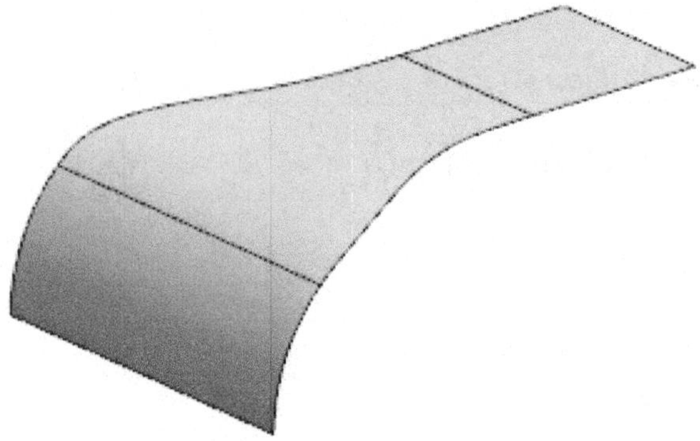

- **Save** your model!

19.3 **Reflection Analysis**

The **Reflection Analysis** tool is extremely useful when developing cosmetic surfaces, as it gives you some indication of how light will reflect off surfaces. Its main use is in identifying visible smooth edges, where two faces meet each other with tangency, but a discontinuity or kink in the reflection shows a sharp transition between those faces.

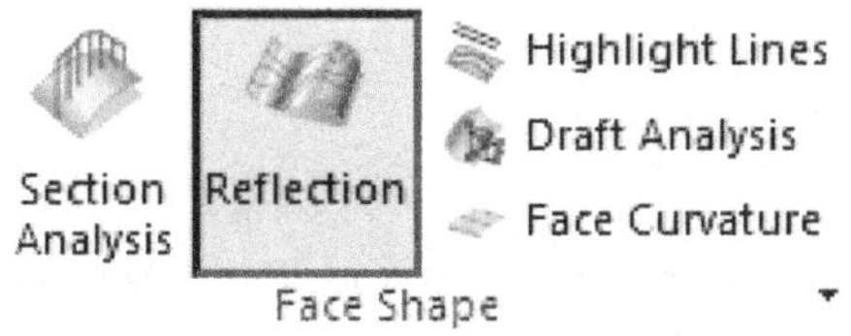

The reflection tool can take any number of faces as input, and allows you to select a variety of "images" whose reflection you would like to see on those faces. A common use of the tool is to see "zebra stripes" which make it easy to identify discontinuities and kinks in the reflections along edges. The rendering of the reflection is only visible in the *Face Analysis* rendering style.

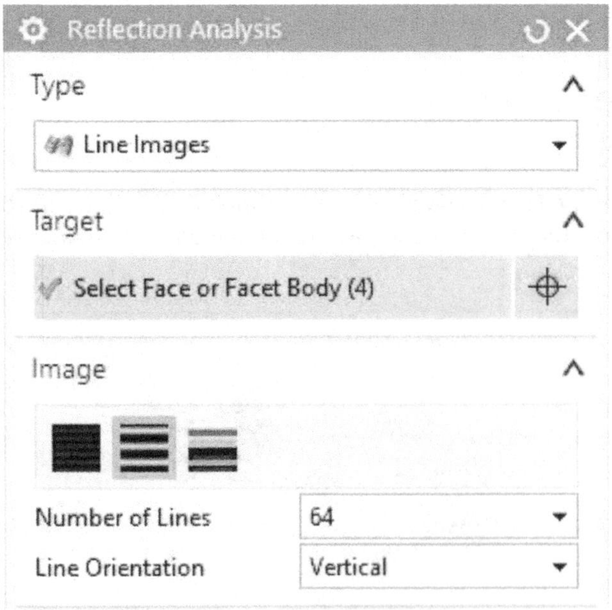

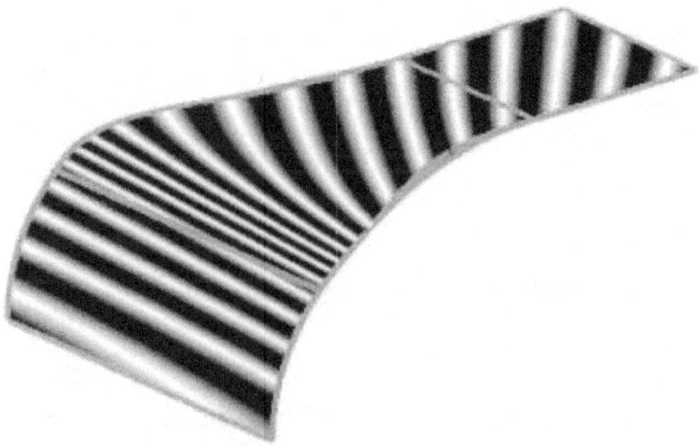

One very important special case where the **Reflection Analysis** tool shows smooth edges clearly, is that of a fillet with circular cross section, or more generally any surface meeting a planar surface with *G1* continuity. Rotating the model slowly will animate the kinks in the reflection, and show a "sharp" smooth edge.

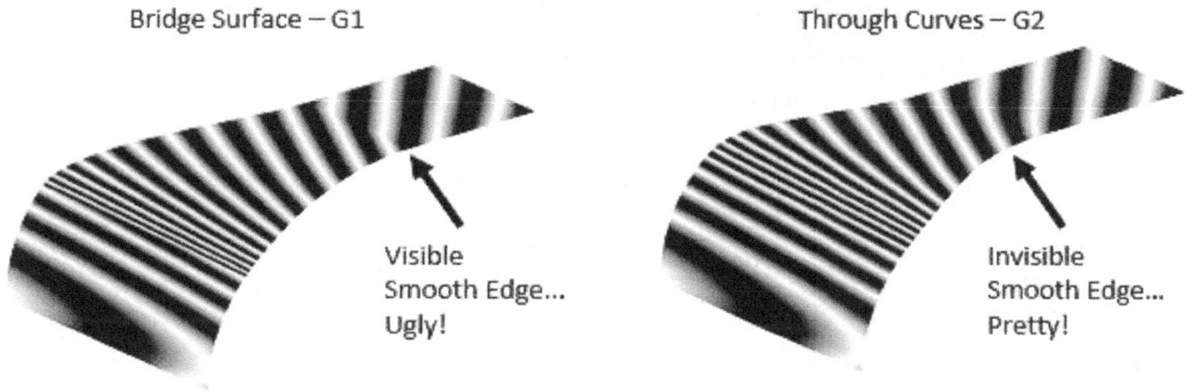

Most designers agree that *G2* is the minimum degree of continuity needed to ensure invisible smooth edges between cosmetic surfaces, although sometimes higher degrees of continuity, such as *G3* and *G4* are used to produce Class A surfaces. Class A means "surfaces that are immediately visible". This is of great concern when designing highly reflective surfaces that need to appeal to consumers, such as automotive bodies.

19.4 **Through Curve Mesh**

The **Through Curve Mesh** tool creates a lofted surface that interpolates between disconnected section strings, but also allows you to specify guide strings running cross to the section strings. As per the usual, using closed section string entities allows you to create a solid body; using open section strings will result in a surface body.

An important point about lofted mesh surfaces in NX is that they are rectangular in nature – when building a mesh surface, you should be able to identify four "sides" that meet in "corners".

The main difference between **Through Curves** and **Through Curve Mesh** is that **Through Curves** only requires section curves, and **Through Curve Mesh** requires both section curves and guide curves. The minimum number of required input curves for **Through Curves** is two – you can think of these as the top and bottom edges of the "rectangular" surface. The minimum number of required input curves for **Through Curve Mesh** is four – you must specify all four "sides" of the "rectangle".

Because of the control this tool affords you in both "directions", **Through Curve Mesh** is probably the most important tool for creating freeform rectangular surfaces in NX. It is placed – rather unintuitively – in a drop-down menu in the **Surface** group on the **Surface** tab.

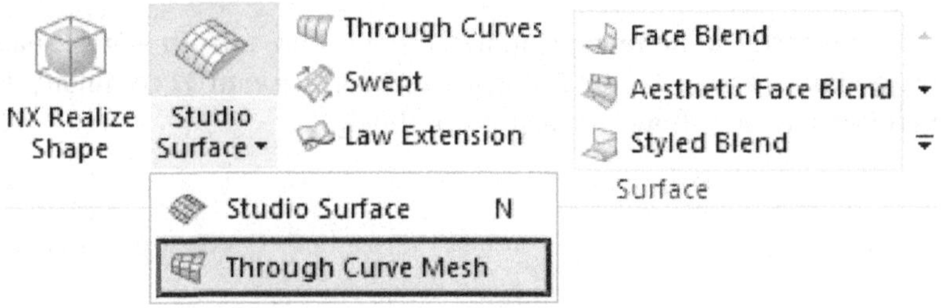

But it is featured much more prominently in the **Surface** group on the **Home** tab.

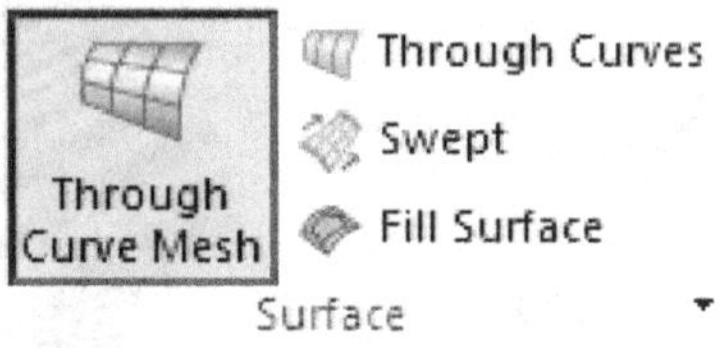

19.4.1 Sheet

In this exercise, you will build a lofted surface using **Through Curve Mesh** from a "skeleton" of previously-built curves.

- Open the file *"Through Curve Mesh – Sheet.prt"* from your downloaded part files folder.
- This file has three arcs which are connected by splines at their endpoints. We will construct a lofted surface from these curves, with the three arcs as "sections" (primary curves) and the two splines as "guides" (cross curves).

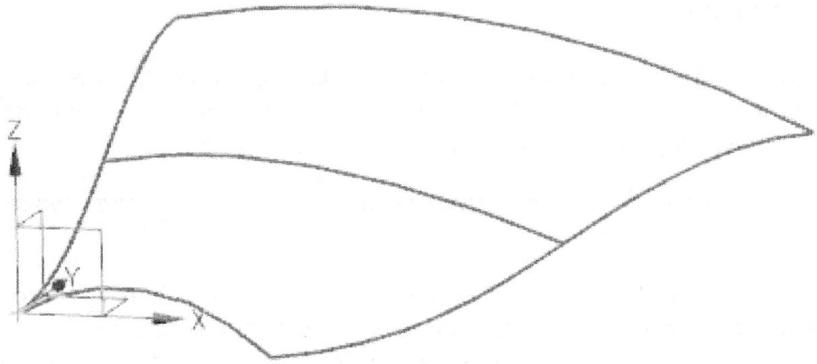

- Open the **Through Curve Mesh** tool.
- From smallest to largest, select *Primary Curve 1*, *Primary Curve 2*, and *Primary Curve 3* as shown below. Remember to use the **Add New Set** button between selections.

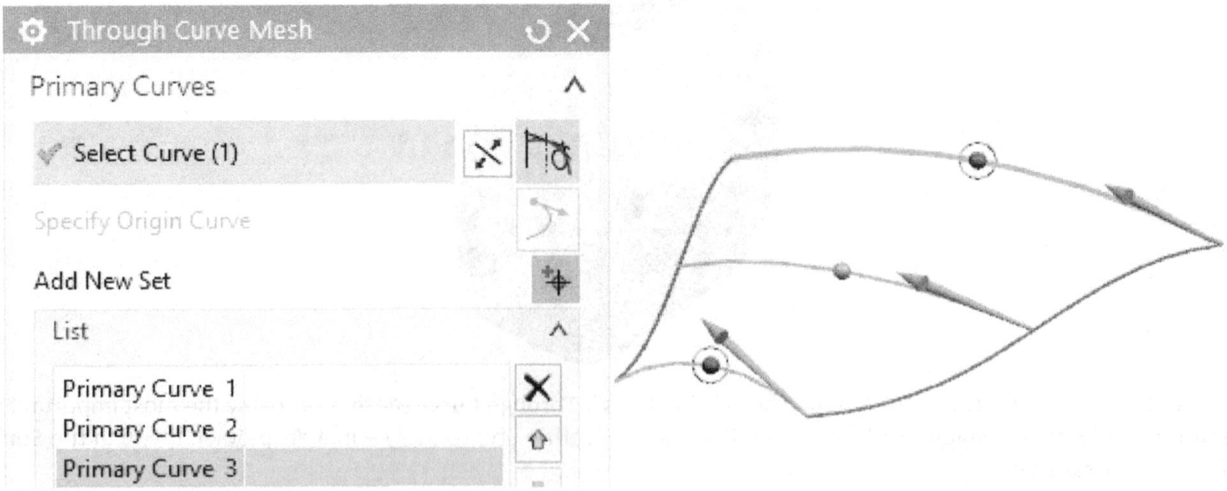

- Select the splines as the cross curves as shown below. By the way, you can use the middle mouse button to push the **Add New Set** button.

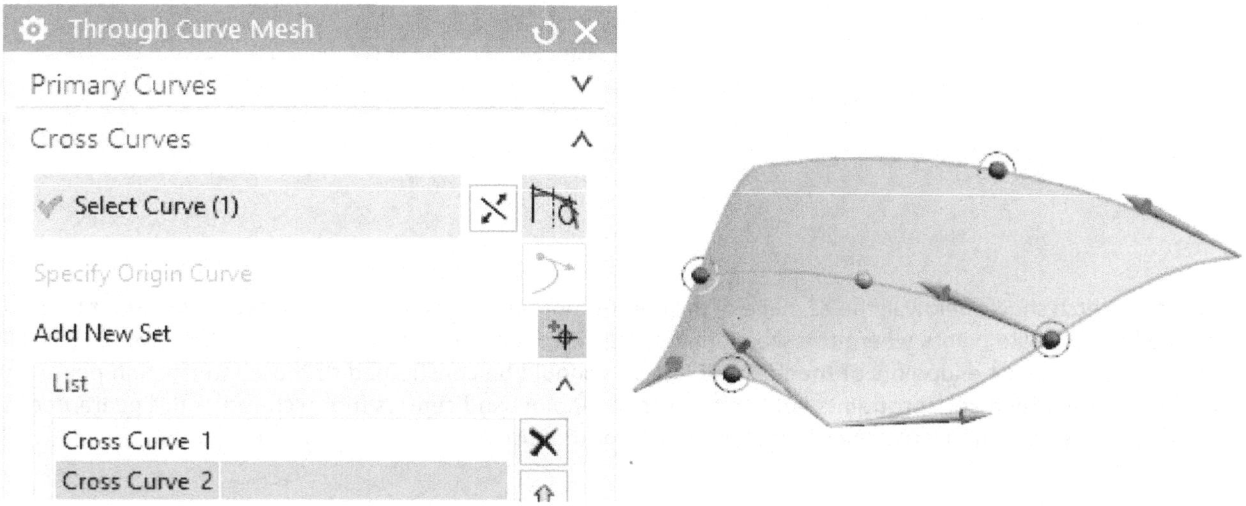

- **Save** your part!

19.4.2 Solid

In this exercise, you will use **Through Curve Mesh** to produce a solid body. In order to obtain a solid, you must use closed section (primary) curves, and the first and last guide (cross) curve must be the same.

- Create a new file (**Units**: *Inches*) called *"Through Curve Mesh – Solid.prt"*.
- Create the sketch shown below on the YZ plane, by creating a rectangle centered at the sketch origin, adding dimensions, using the **Fillet** tool to round the corners, and then adding appropriate dimensions and constraints.

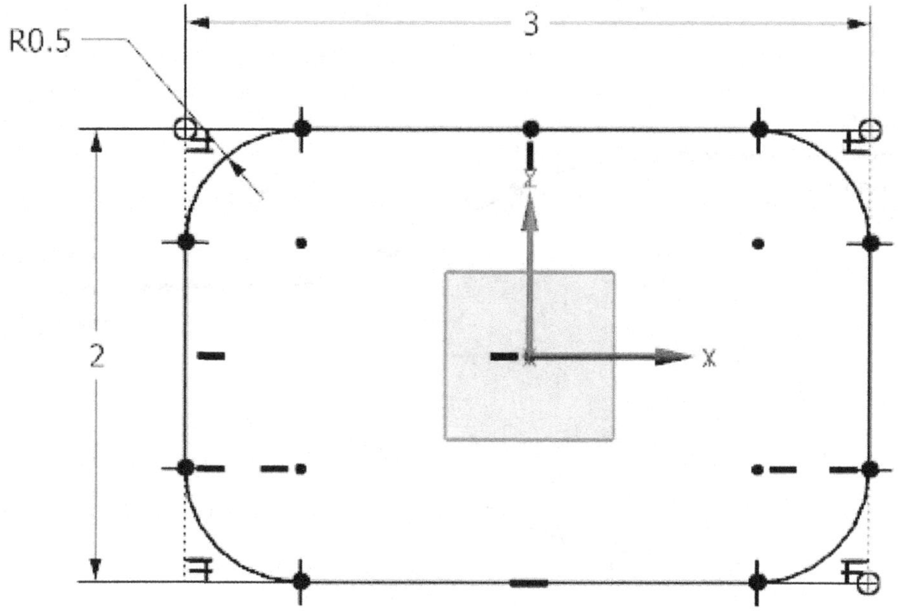

- Create a new datum plane offset by *3 in* in the +XC direction.
- Sketch a circle with a **Diameter** of *1 in* on the offset datum plane. Center it in the sketch plane.

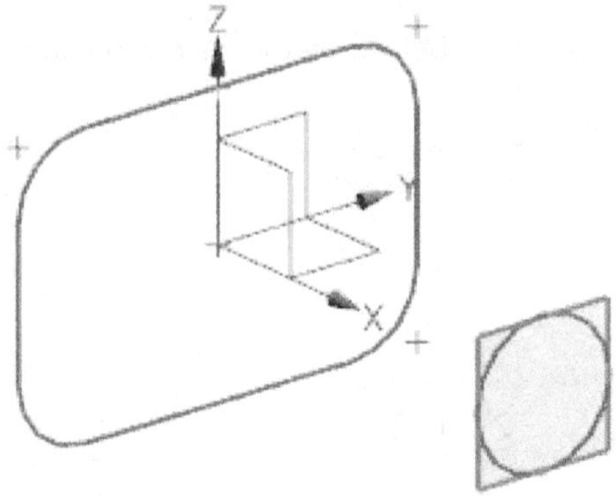

- Create the sketch shown below in the XZ plane of the datum coordinate system. Use the **Intersection Point** tool within the sketcher to create points where the sketch plane intersects the line and circle from the first two sketches. The leftmost and rightmost endpoints of the first and last arcs should be constrained to those intersection points. Also, note the vertical alignment constraints between the arc endpoints and centers (not pictured) – this guarantees that the arcs are tangent to horizontal lines at their ends. Mirror the top to produce the bottom.

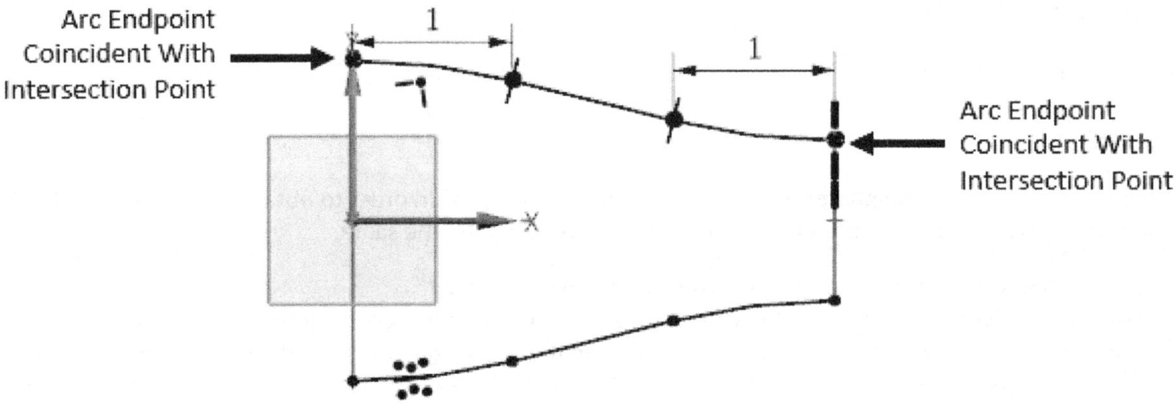

- Use the same technique to create the sketch shown below in the XY plane of the datum coordinate system.

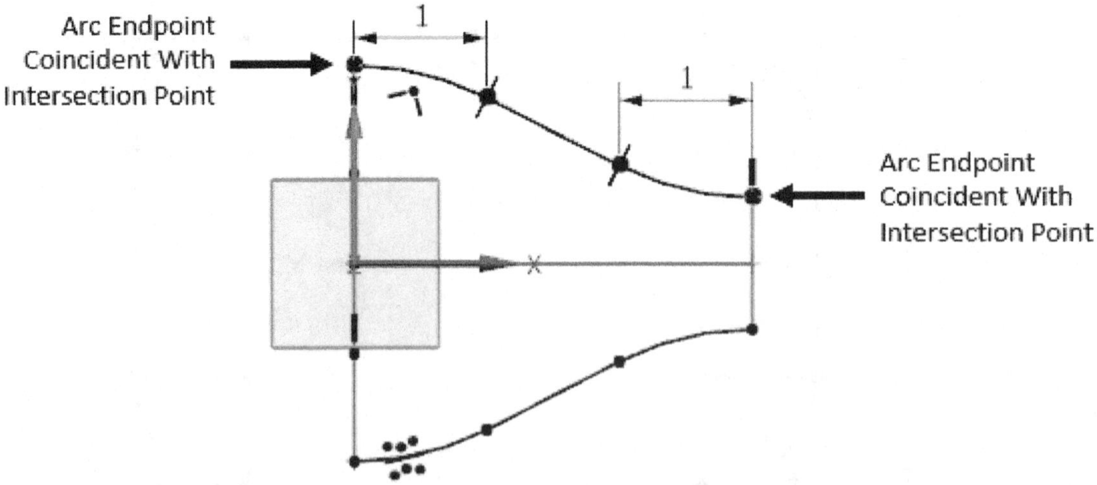

- Select the **Through Curve Mesh** tool, and choose the first two sketches as the **Primary Curves**. Note that the orientation arrows indicate the start point of each section.
- Before you select the **Cross Curves**, set your **Selection Intent** drop-down to *Tangent Curves* to avoid selecting both halves of each sketch. For *Cross Curve 1*, select the sketch curve in the three o'clock position (the one that originates at the start point of the circle).

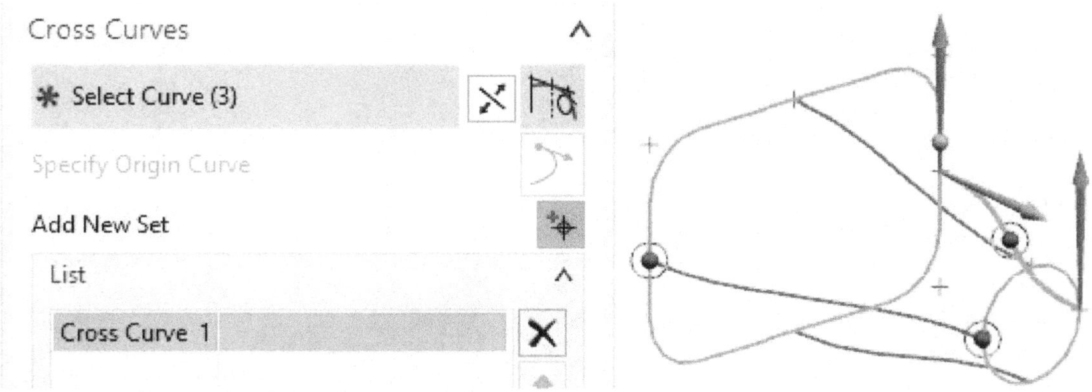

- Select the top sketch curve as the second cross curve. The preview will now be a weird surface – certainly not what you're trying to build. This is because *Cross Curve 1* does not connect the start points of *Primary Curve 1* and *Primary Curve 2*.

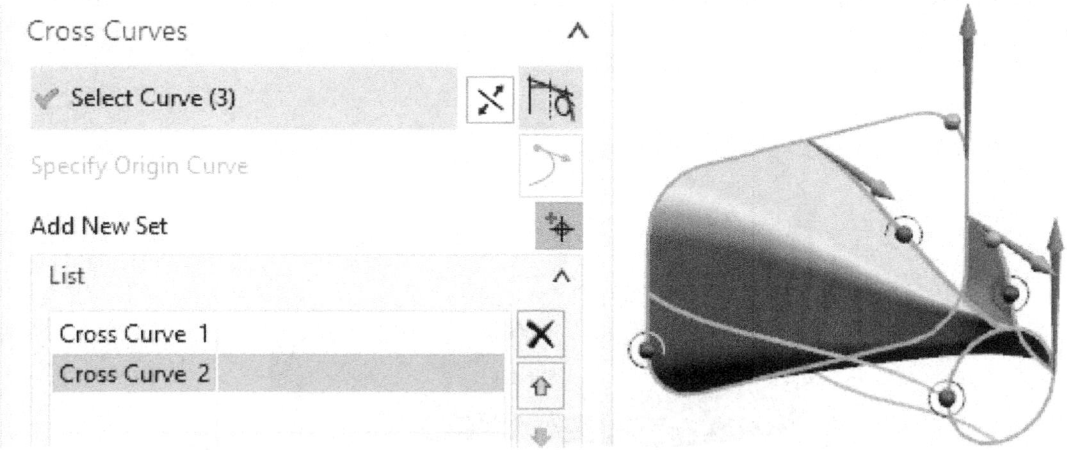

- Continuing counterclockwise, select the sketch curves in the 9 o'clock position as *Cross Curve 3*. Now your preview will improve. Unlike **Through Curves**, where misaligned section start points cause the tool to produce solid bodies with terrible curvature properties and self-intersections, **Through Curve Mesh** is often more forgiving.

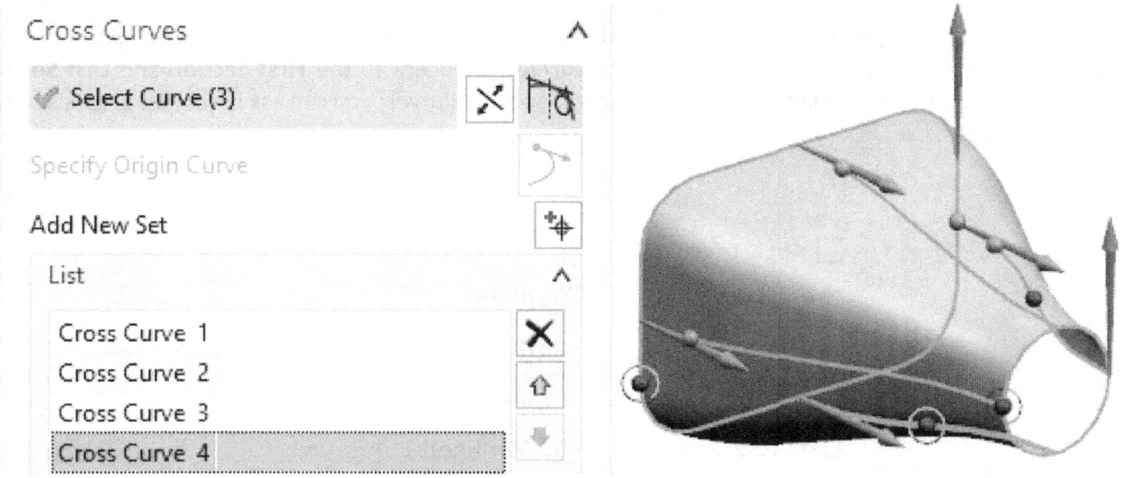

- To complete your selection, add *Cross Curve 5* and choose the same curves you chose for *Cross Curve 1*. Click **OK** to create the solid body.

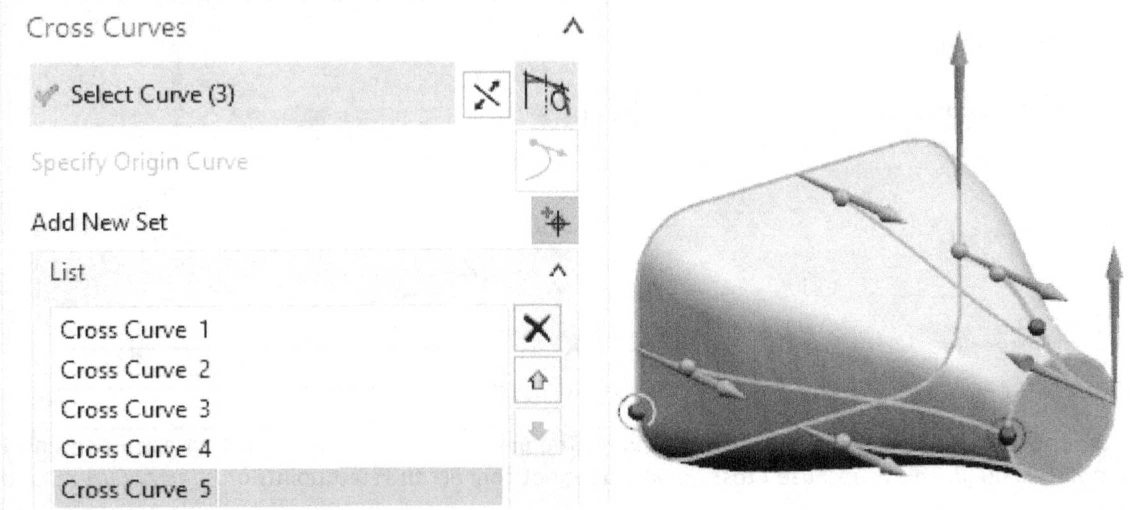

- Use the **Shell** tool to pierce the planar faces on the ends, to hollow out your body with a uniform wall thickness of *0.125 in*.

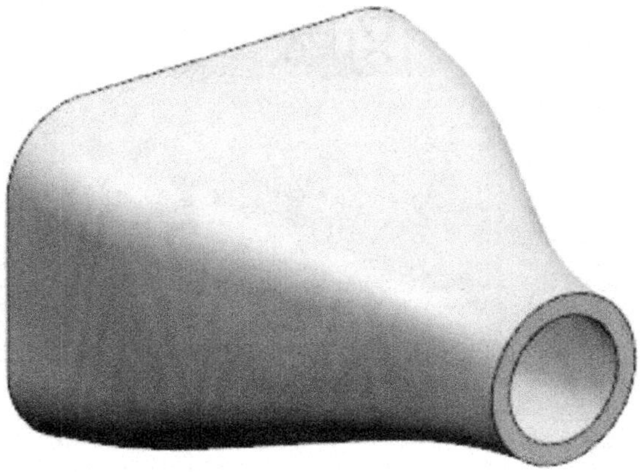

- **Save** your file!

19.4.3 Continuity

Our third exercise on the **Through Curve Mesh** command concerns continuity. The **Through Curves** command allows you to specify the degree of continuity along edges shared by surfaces adjacent to the **First Section** and **Last Section**. Since **Through Curve Mesh** requires at a minimum two primary and two cross curves, you can ask for higher degrees of continuity along each of those edges.

- Open your part file *"Bridge Surface.prt"*.
- Use the **Bridge Curve** command to create two splines joining the edges of the planar and cylindrical surfaces as shown below. Specify *G1* continuity at the **Start** and **End** of each, and a **Tangent Magnitude** of *1* at each end for both.

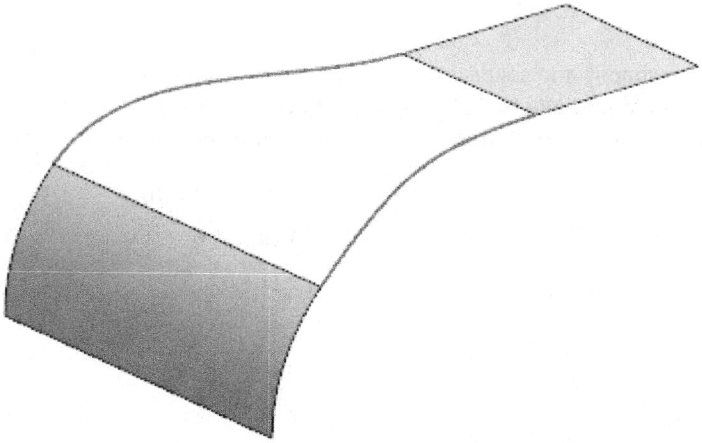

- Open the **Through Curve Mesh** tool. Select the linear edges of the planar and cylindrical surfaces as the **Primary Curves**, and the bridge curves as the cross curves. Set the **Continuity** as shown below.

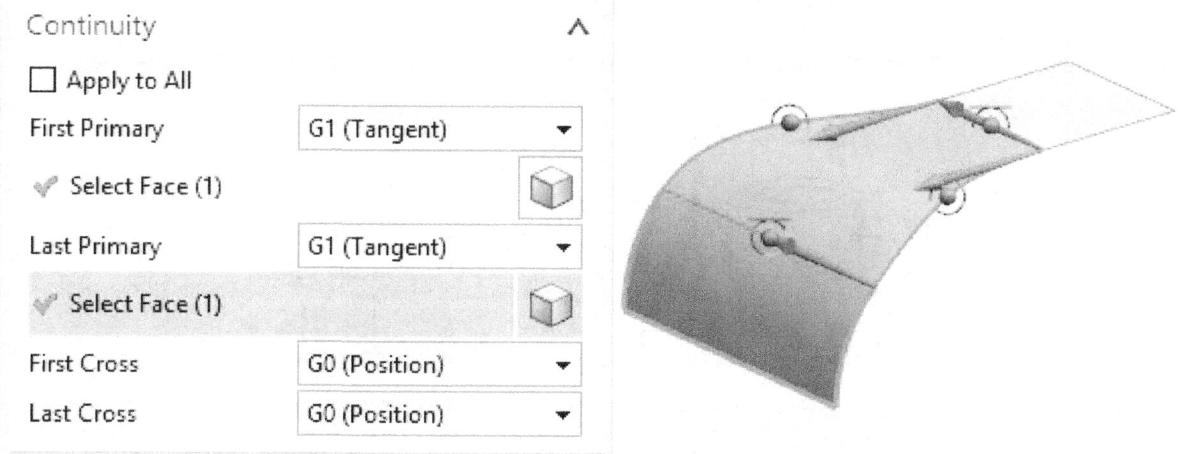

- If you attempt to change continuity to *G2* along either **Primary Curve**, you will encounter an error message – this is because the bridge curves are only *G1* continuous with the original surfaces.

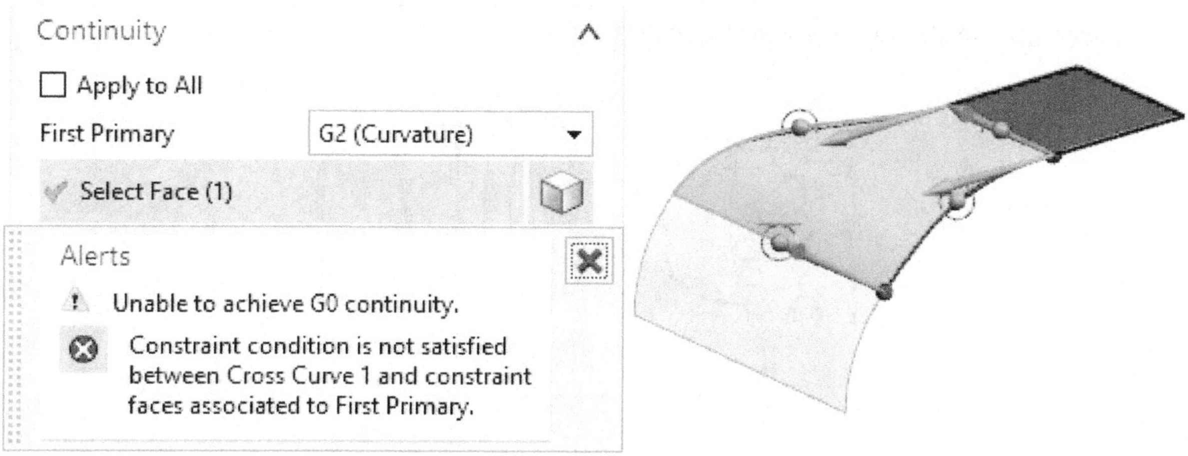

- The moral of the story here is that the degree of continuity your surface has with its neighbors along the **First Primary** and **Last Primary** is limited by the degree of continuity the **Cross Curves** have in relation to those surfaces.
- **Save** your file!

19.5 **Raster Image**

When you are building freeform shapes and solids you may want to translate curves from a hand sketch or a photograph into curves in your part file. The **Raster Image** tool allows you to import an image – usually a *JPEG*, *TIFF*, or *PNG* – and place it in your model. **Raster Image** is found in the **Datums** drop-down in the **Feature** group on the **Home** tab.

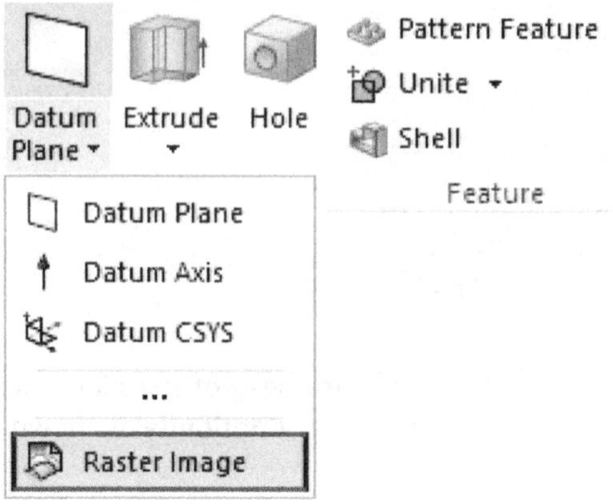

In the exercise below, you will build the handle of a device used to extract thermometers out of acid bottles.

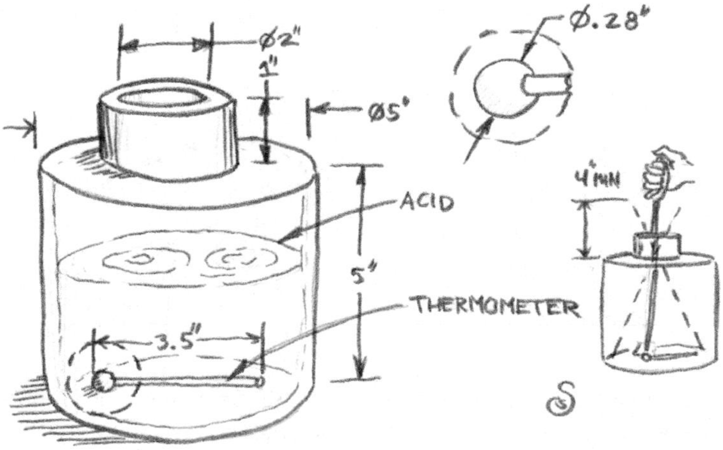

We can begin making our model from a hand sketch of the handle.

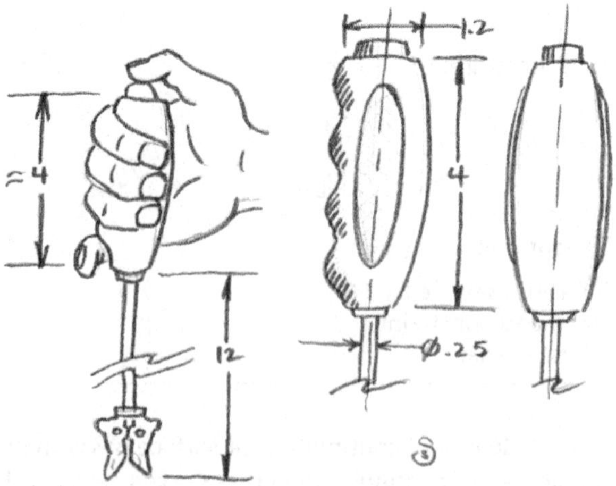

In your downloaded part files folder, you will find two *".png"* files that you will use in the exercise below.

- Create a new file (**Units**: *Inches*) called *"Freeform Handle.prt"*.

- Select the **Raster Image** tool. Find the *".png"* file called *"hand-sketchb.png"* in your downloaded part files folder.
- Position the image on the YZ plane of the datum coordinate system in your model, as shown below. Note that the datum coordinate system origin is aligned with the "center" of the bottom edge of the handle.

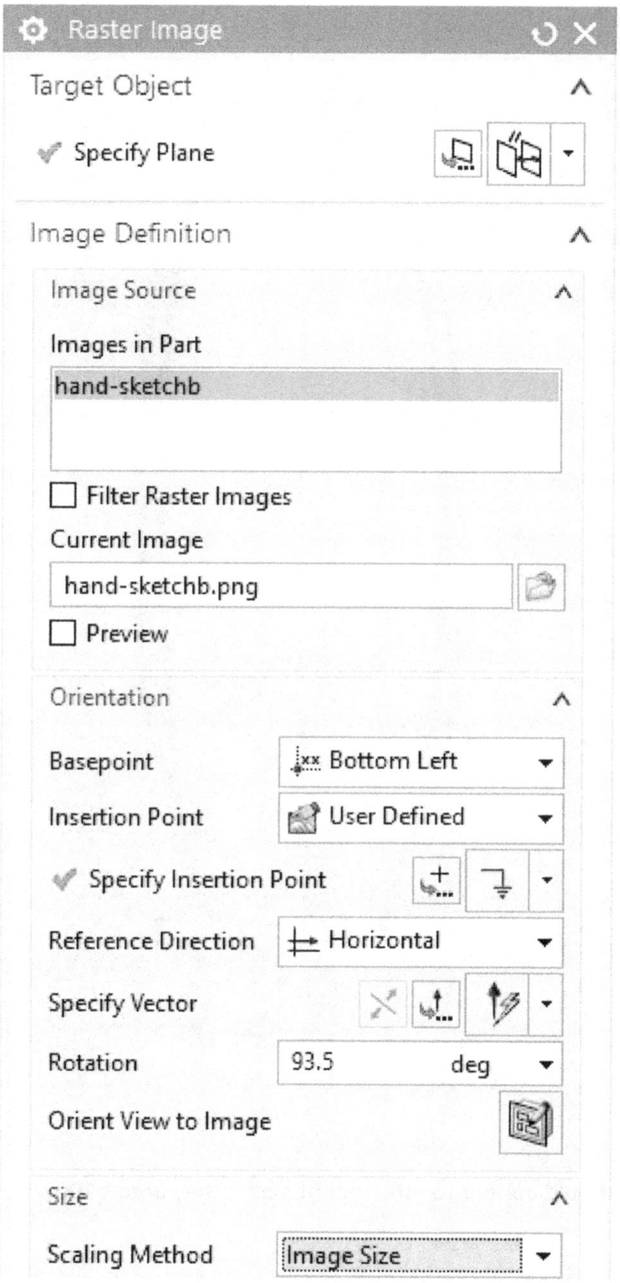

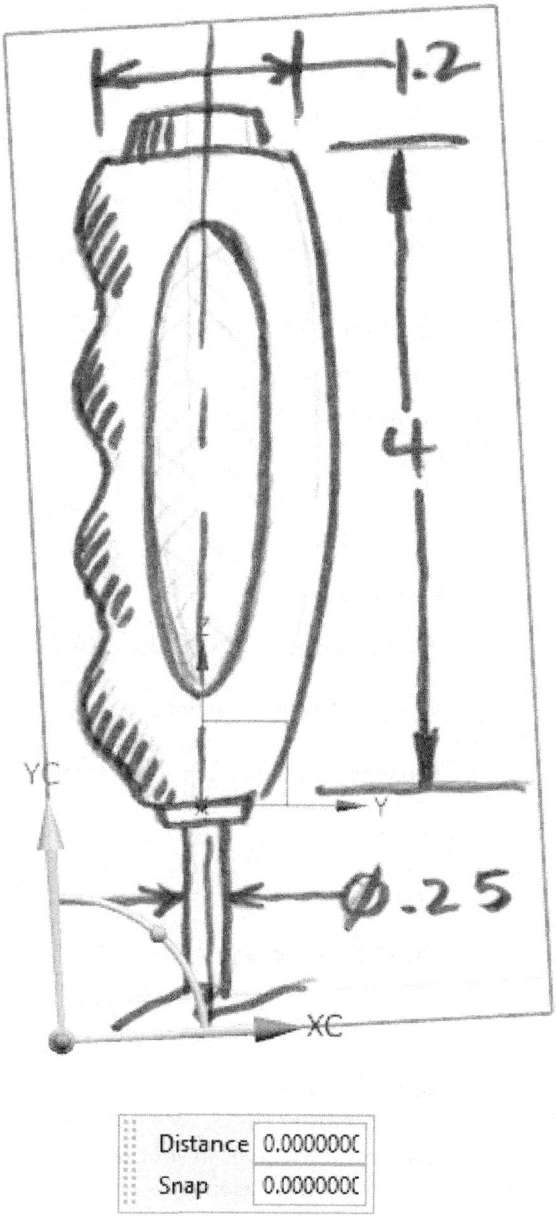

- Position the other hand sketch, *"hand-sketchc.png"*, in the same way, on the XZ plane of the datum coordinate system.

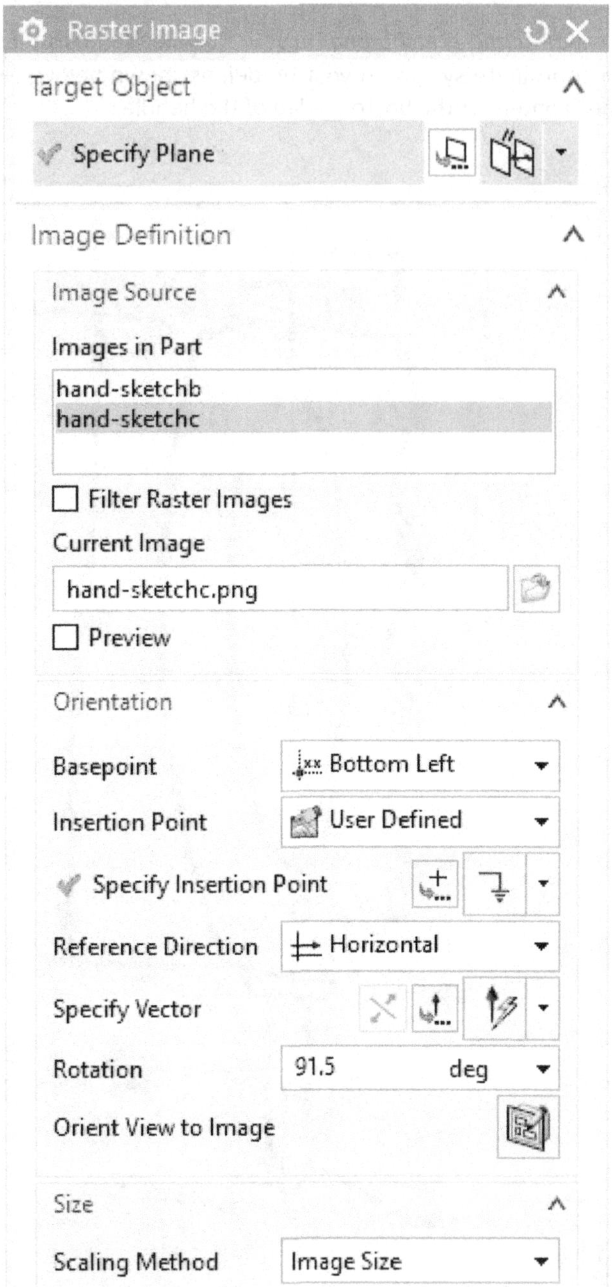

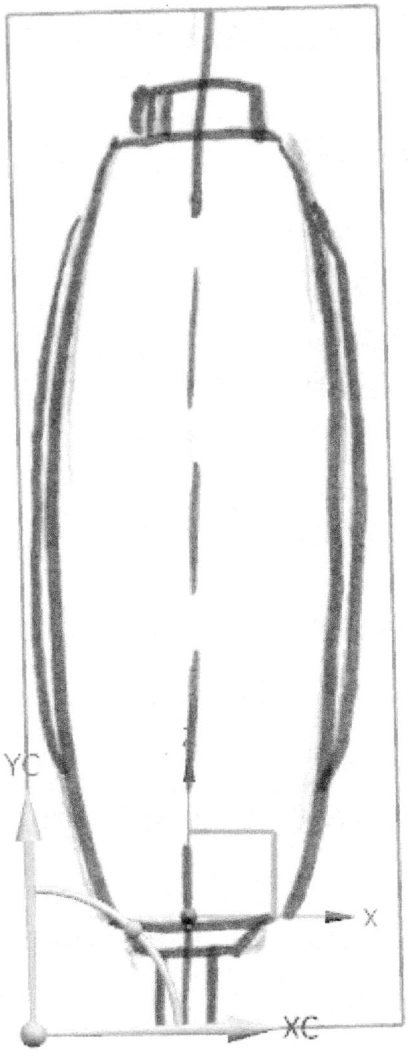

- From a diagonal view, the curves in your raster images should appear to intersect (it's ok if they aren't 100% perfect, this method is more art than science).

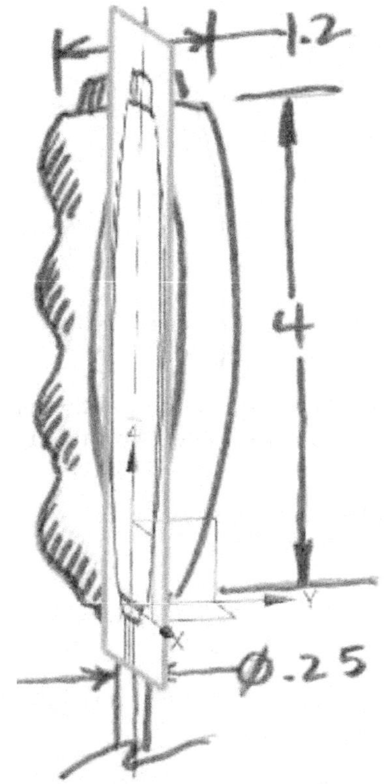

- Create two sketches and trace the curves from the hand sketches. One sketch consists of one spline and three arcs, and the other consists of four arcs, one of which is a mirror image about the sketch vertical axis.

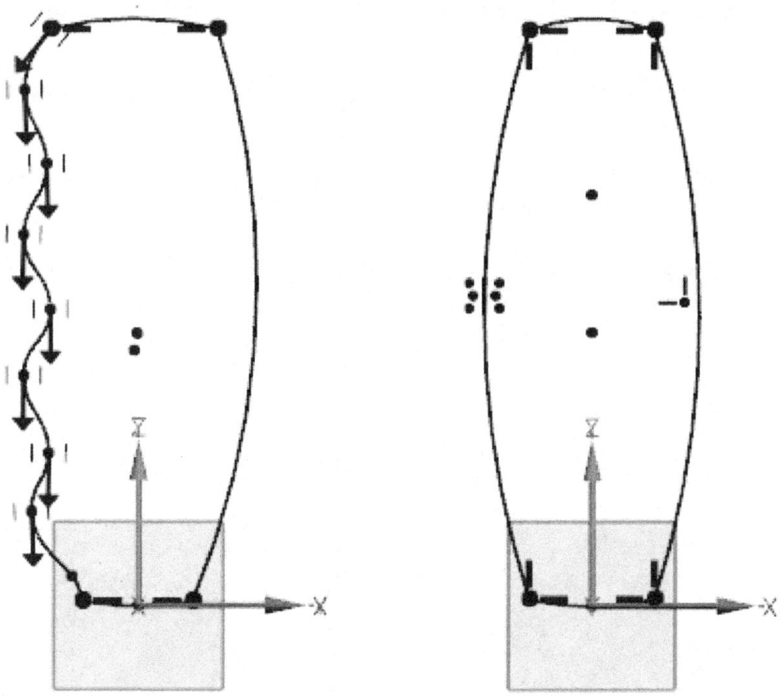

- In creating the second sketch, be sure to use the **Intersection Point** tool together with a *Point on Curve* constraint to constrain the arcs so that they actually intersect. Also, use *Horizontal Alignment* constraints to ensure that the "corners" are vertically aligned with the "corners" of the first sketch.

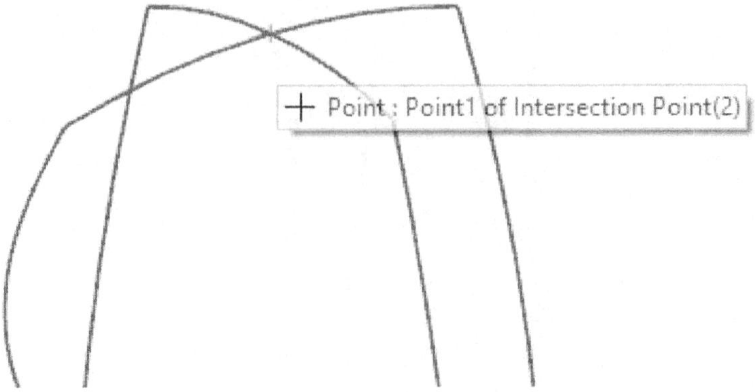

- Create a datum plane passing through three of the four corners at the top of your two sketches. It should necessarily pass through the fourth because of your horizontal alignment constraints. On that plane, sketch a closed cubic studio spline through the four corners.

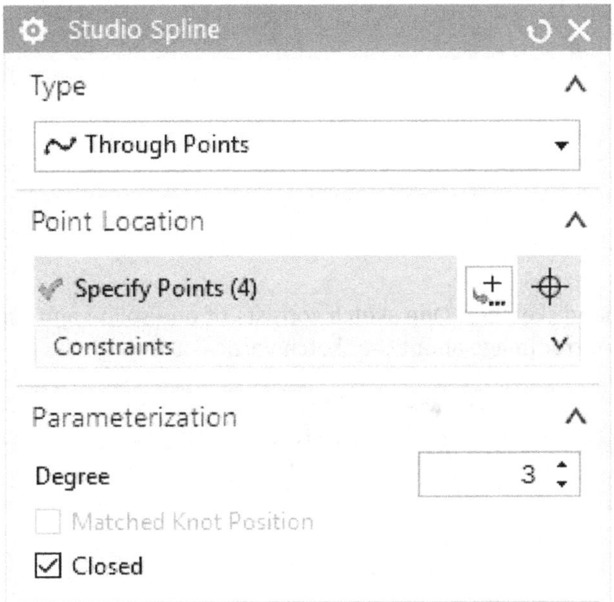

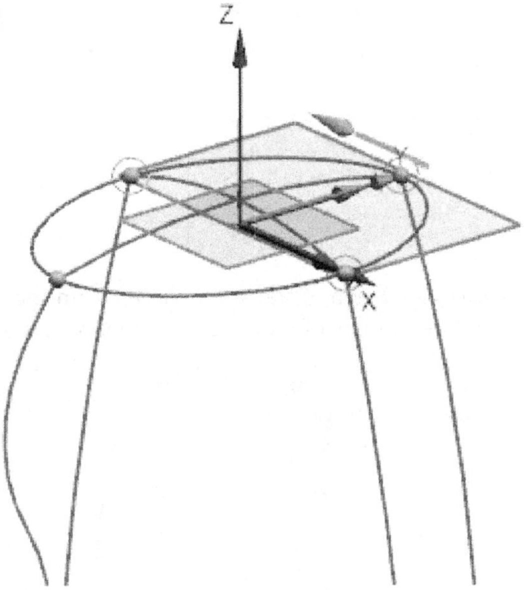

- Do the same on the bottom side.

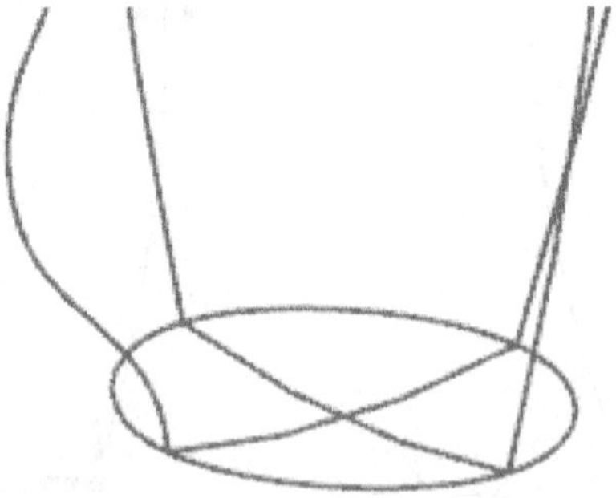

- **Save** your file!

19.6 **Studio Surface**

The **Studio Surface** is another mesh surfacing tool that will sometimes produce vastly superior results to the others. It works a lot like **Through Curve Mesh**, allowing you to specify both **Section (Primary) Curves**, and **Guide (Cross) Curves**, with the main difference being the number of required inputs. Where **Through Curve Mesh** requires at a minimum two

Primary and two **Cross** curves, **Studio Surface** will accept any combination of two at a minimum. This gives it flexibility, in that you can use it like the **Through Curves** tool, like the **Through Curve Mesh** tool, or like the **Swept** tool!

Studio Surface is found in the **Mesh Surface** drop-down in the **Surface** group on the **Surface** tab.

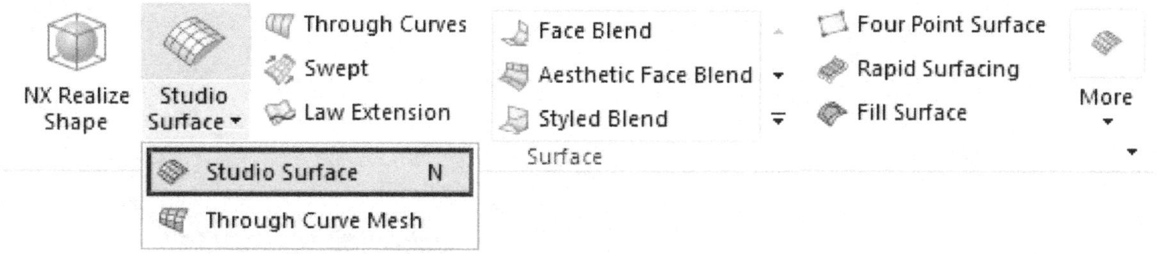

- Continue working with *"Freeform Handle.prt"*.
- Select the **Studio Surface** command.
- Begin by selecting the two closed splines you constructed last as the **Section (Primary) Curves**.

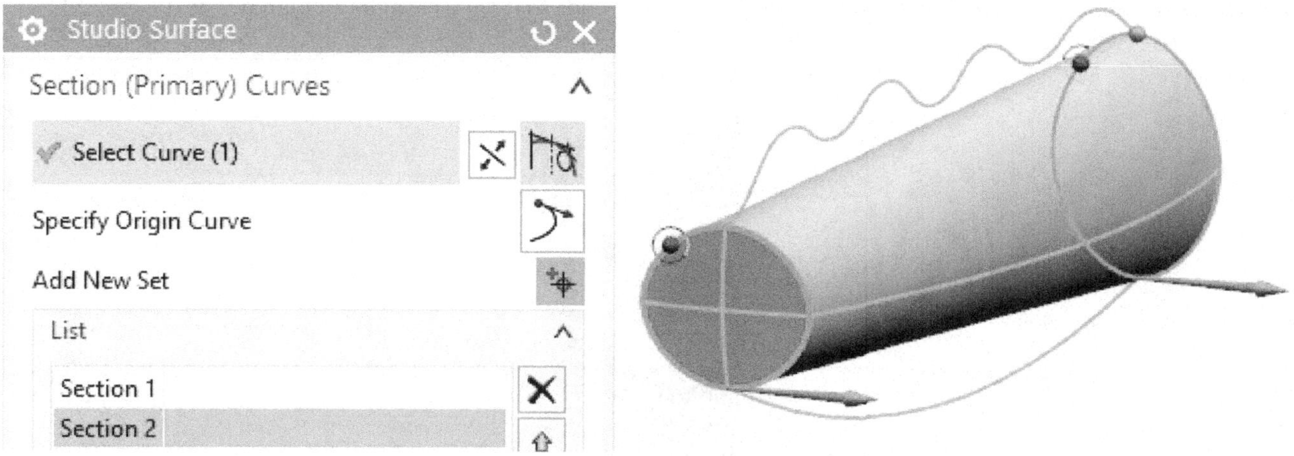

- For the **Guide (Cross) Curves**, start with the arc joining the start points of the two closed splines. You will get a preview right away, since the **Studio Surface** command doesn't need more than two inputs.

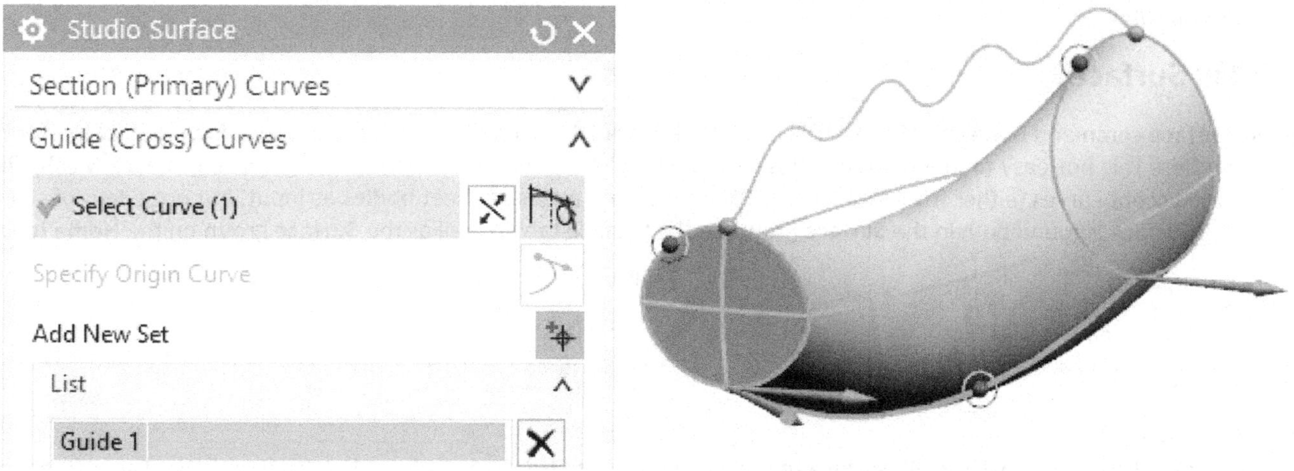

- Continue counterclockwise, making sure to conclude with the fifth **Guide Curve** the same as the first **Guide Curve**.

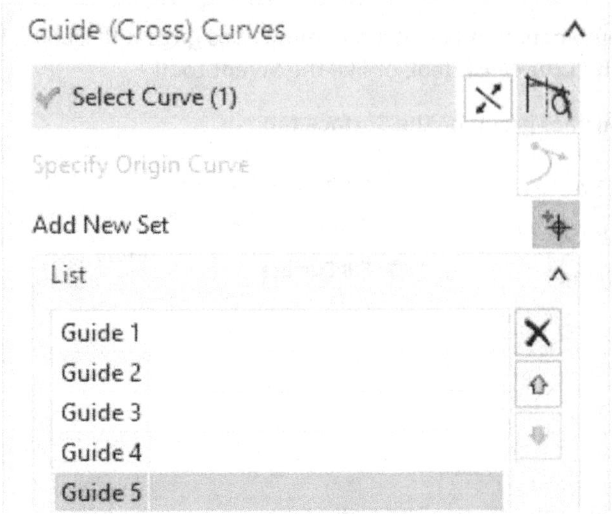

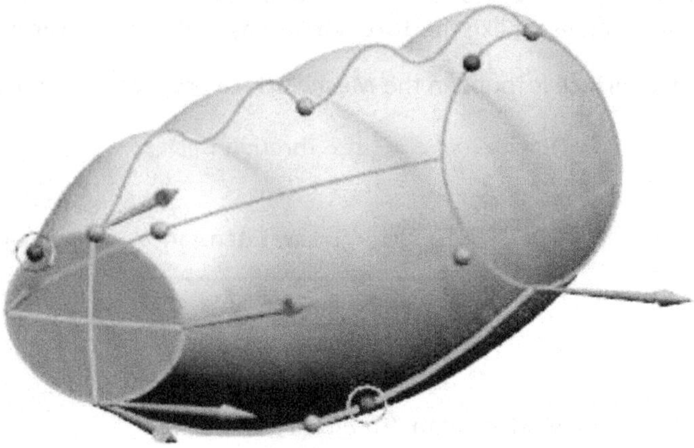

- Click **OK** to create your solid body. It has planar faces on the ends that we will smooth out in the next exercise!

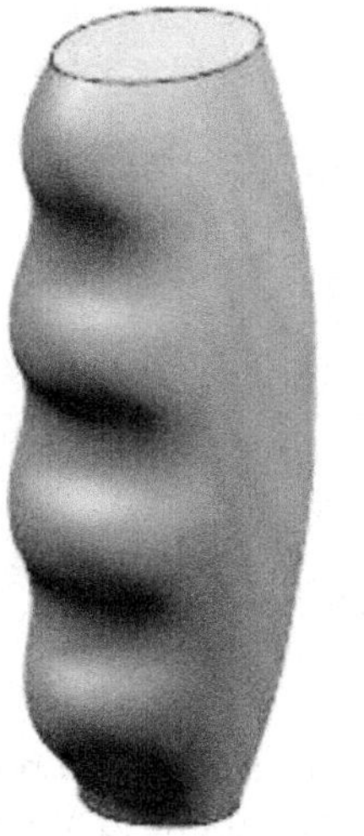

- **Save** your file!

19.7 Fill Surface

The **Fill Surface** command is a relatively new tool (introduced in NX 9) that makes it easy to produce high-quality freeform surfaces where it is not easy to identify four sides of a "rectangle" to build that surface using **Through Curve Mesh**. **Fill Surface** will accept curves (either from sketches, or 3D curves) and edges of sheet bodies as input, but not edges of solid bodies. **Fill Surface** is found both in the **Surface** group on the **Surface** tab, as well as the **Surface** group on the **Home** tab.

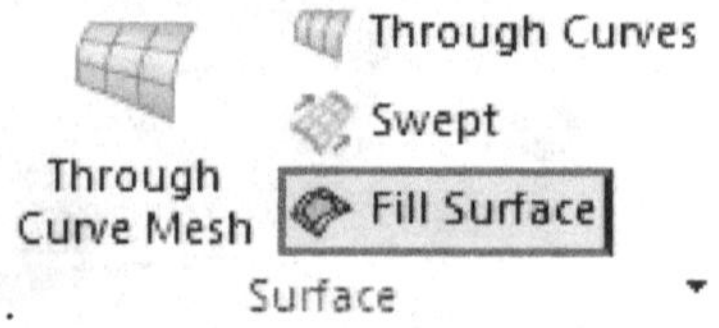

In the exercises below, you will use **Fill Surface** to build a high-quality freeform surface that "caps off" the planar faces on the solid body produced in the last exercise.

- Continue working with *"Freeform Handle.prt"*.
- Select the **Fill Surface** command and attempt to select the top edge of the solid body.

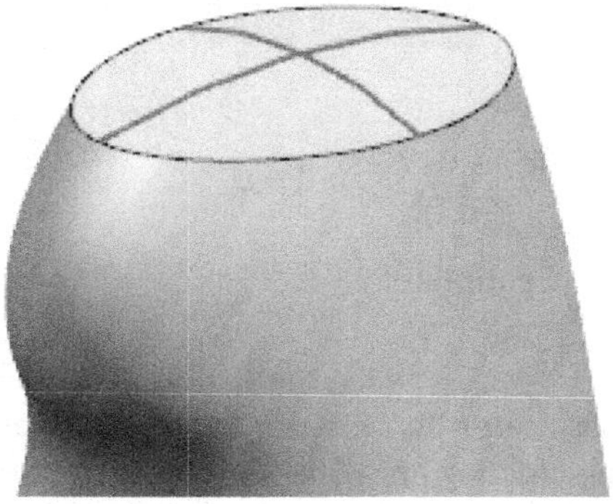

- No matter how hard you click, you cannot select this edge. Use **Delete Face** with the *Heal* checkbox unmarked to delete the planar face atop the freeform body.

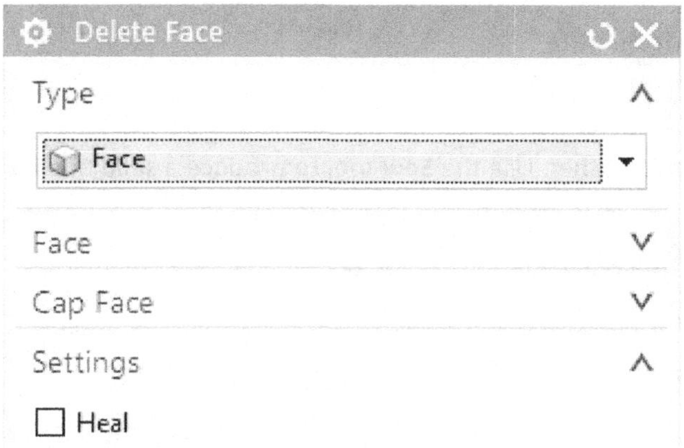

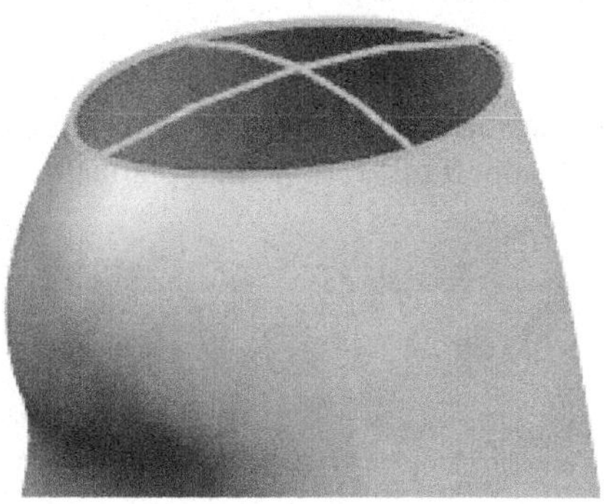

- Select the **Fill Surface** tool again. Now, you are able to select the edge of the sheet body that resulted from deleting the planar face!

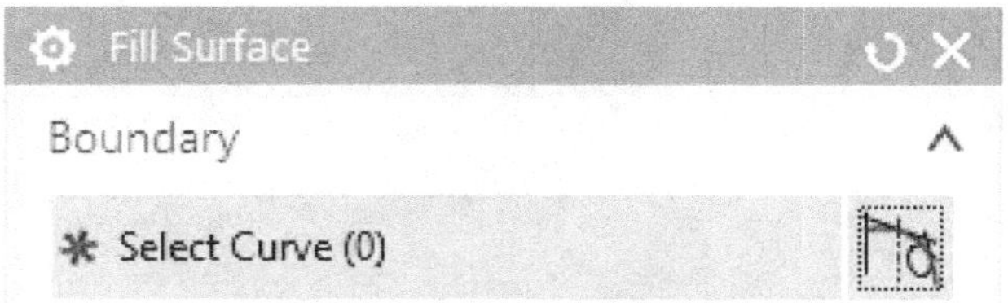

- Before actually selecting this edge, go into the **Settings** section of the **Fill Surface** dialog and uncheck *"Patch into Part"* and set the **Default Edge Continuity** to *G0*.

- Select the top edge of the sheet body. Note that the resulting surface is planar – this simply rebuilds the deleted face, since there are no higher continuity requirements on the edge.

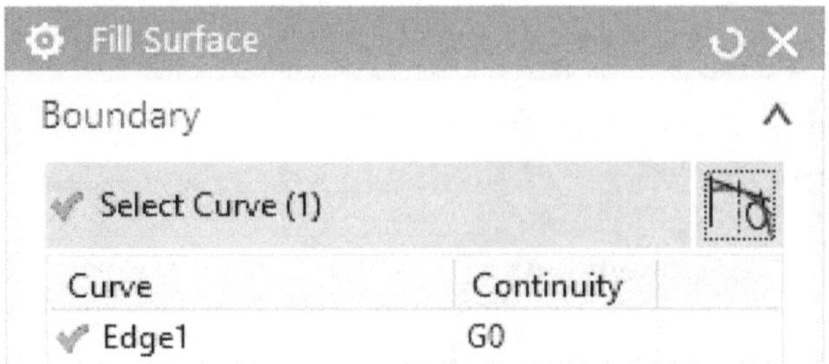

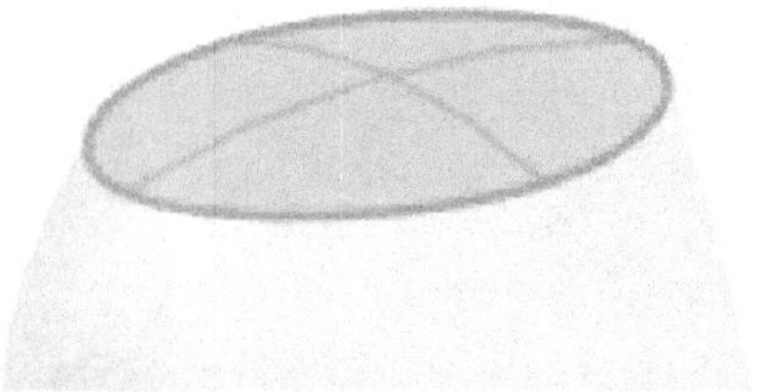

- You can influence the shape of the surface on its "interior" with the **Shape Control** section of the dialog. Set the **Method** to *Fit to Curves* and choose the crossing arcs. This will force the surface to try to contain the selected curves.

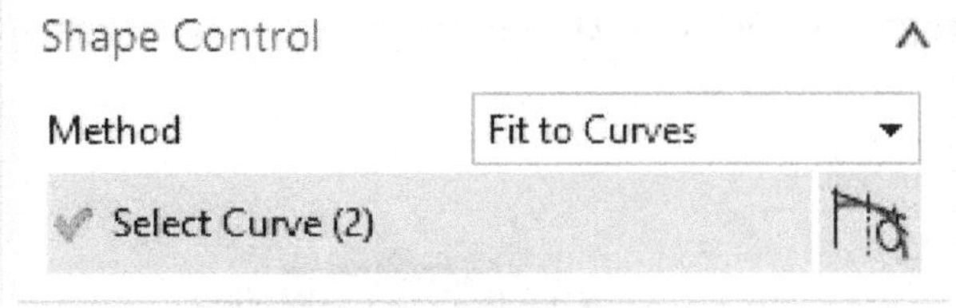

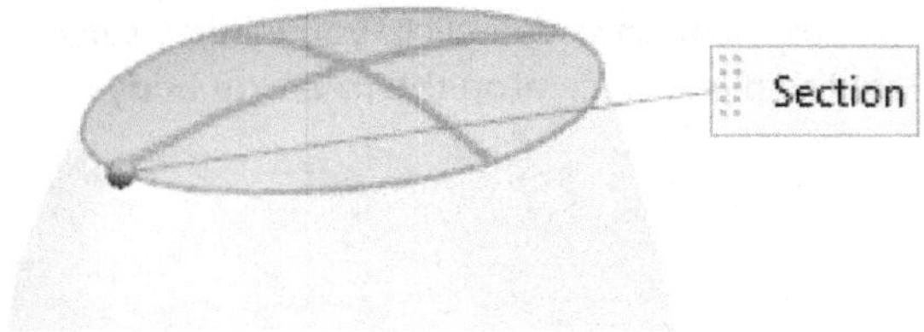

- The resulting surface is a sheet body unattached to the other. Use the **Sew** tool to produce a solid body with the top face now "capped off" rather than planar.

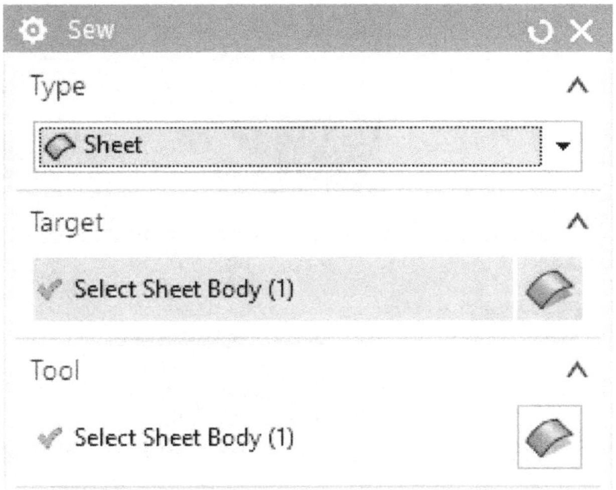

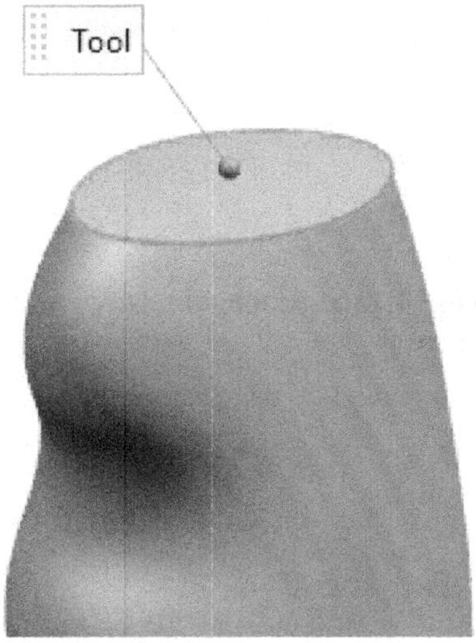

- **Save** your file!

19.7.1 Patch into Part

The *"Patch into Part"* checkbox allows you to skip the last step in the last exercise. In other words, if the **Boundary** curves for your **Fill Surface** operation are borrowed from a sheet body, the *Patch into Part* checkbox will automatically sew the new surface onto the existing sheet body. If the newly created surface results in a volume being enclosed, solid material will be filled in, just as with the **Sew** command.

- Continue working with *"Freeform Handle.prt"*.
- Use **Delete Face** with *Heal* unchecked to remove the planar face on the bottom of the handle.

- Use **Fill Surface**, again with **Method** set to *Fit to Curves*, and this time check the *Patch into Part* box.

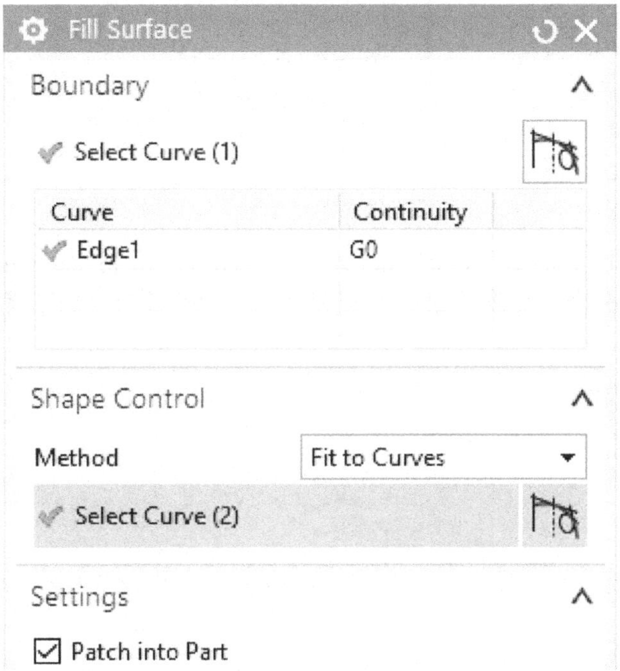

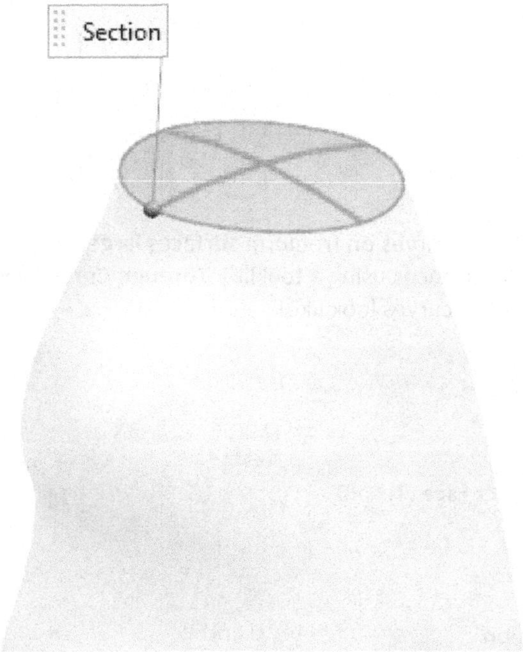

- *Patch into Part* will automatically perform the desired **Sew** operation for you so that the result is a solid body.

- **Save** your part file!

19.8 Isoparametric Curves

An important concept in NX is the rectangularity of all surfaces that you can model. Every surface has two unit parameters, U, and V, that vary in value between *0* and *1*. Every point on a surface is described by a unique *(U,V)* pair – you can think of U and V as "rectangular" coordinates on that surface. Many surfacing tools make mention of these *(U,V)* parameters somewhere on the dialog. Even "dumb", non-parametric surfaces have *(U,V)* parameters.

Isoparametric curves are the parameter curves with respect to this *(U,V)* parameterization. An isoparametric curve has constant *u* or *v* value. The **Isoparametric Curve** tool is found in the **Derived Curve** group on the **Curve** tab.

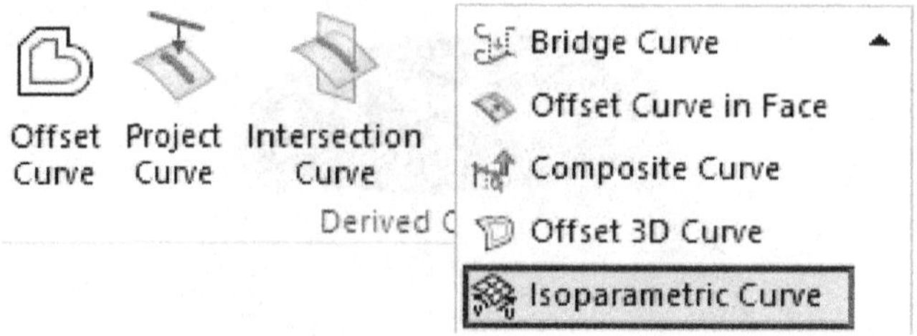

Isoparametric curves might sound abstract, but for familiar surfaces, they are also familiar.

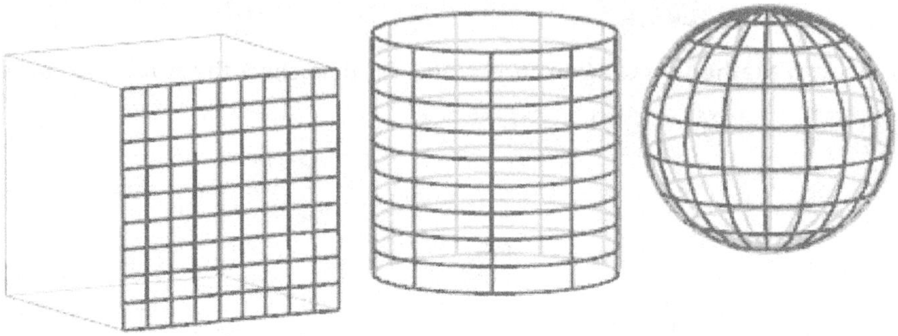

Isoparametric curves on freeform surfaces largely depend on the curves that your surface is built from. In fact, when you build mesh surfaces using a tool like **Through Curve Mesh** or **Studio Surface**, you are telling NX exactly what some of the isoparametric curves look like!

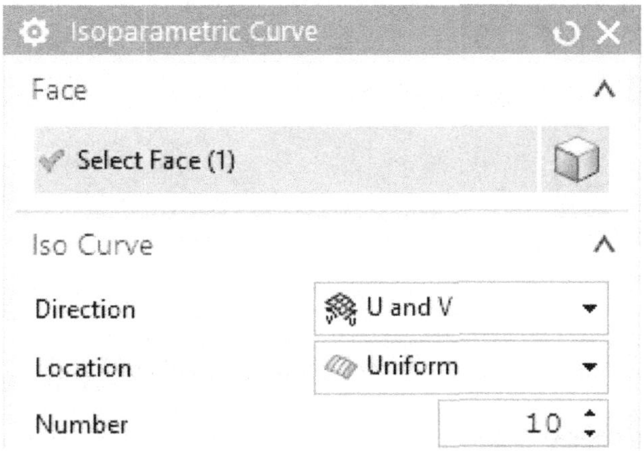

Every surface has this “rectangular” parameterization but not all surfaces appear to be rectangular. Many surfaces are in fact trimmed from their full rectangular form. For instance, the isoparametric curves on the face created by **Fill Surface** suggest that it was trimmed from a larger parametrical- rectangular surface, and indeed this is the case. This means that sometimes a (U,V) pair describes a point, not on the surface itself, but on its untrimmed form.

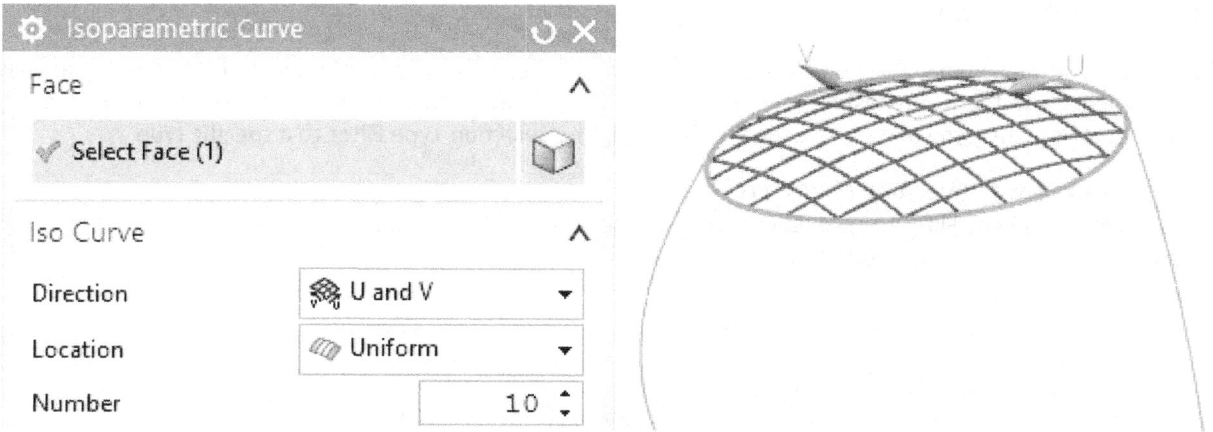

By the way, you can construct the "cap" surfaces on the end of the handle in this model using **Through Curve Mesh**. If you do so, the (U,V) parameterization takes on the appearance of "polar coordinates".

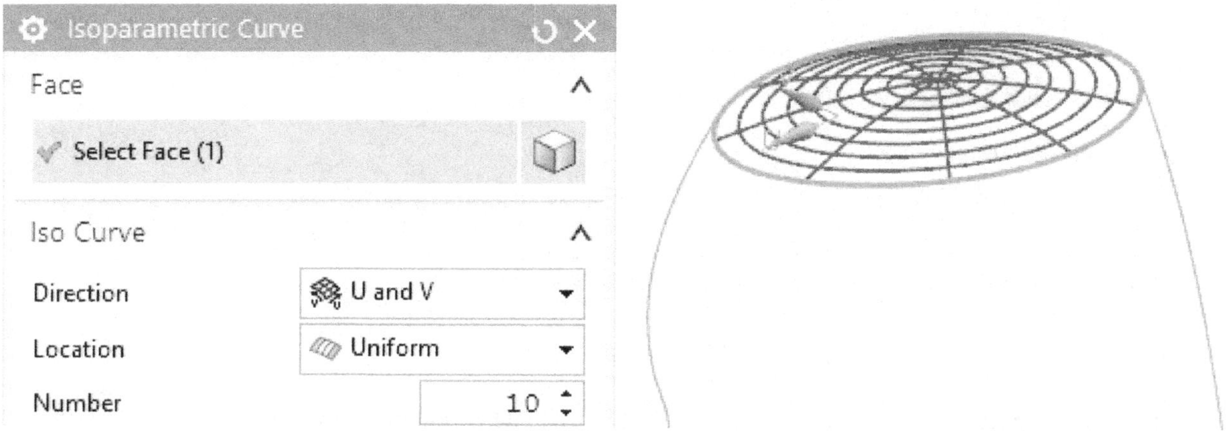

To build the "cap" surface in this way, you must make use of the fact that **Through Curve Mesh** will accept *points* as input for the **Primary Curves**. Here's a hint!

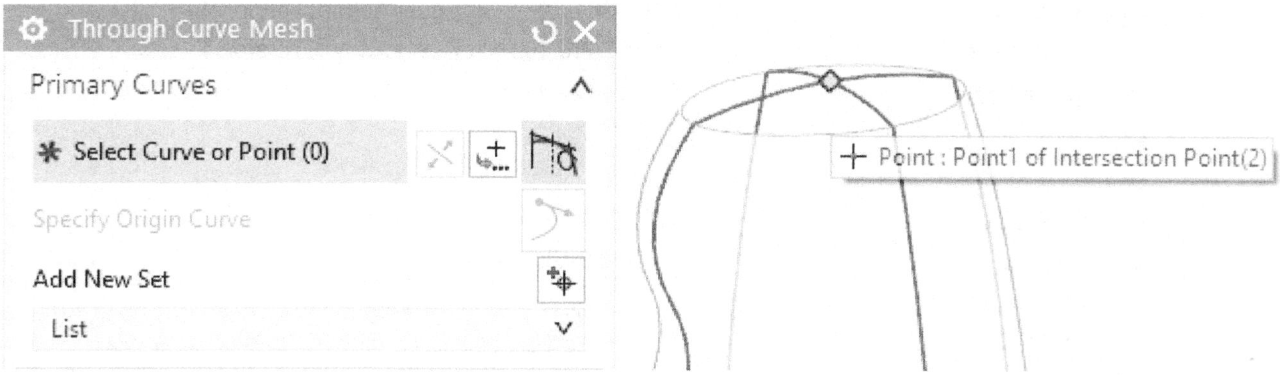

19.9 **B-Surfaces**

Throughout this chapter, we have been discussing "freeform" surfaces, and it is now time to be a bit more explicit about what a "freeform" surface is and what its important properties and features are. By a "freeform" surface, we mean what is known in NX as a *B-surface*. In some sense, you can think of a B-surface as the two-dimensional analog of a spline (which is one-dimensional). These surfaces are more widely known as NURBS (non-uniform rational B-spline) surfaces.

19.9.1 Object Information

The **Object Information** command is very useful when you are dealing with surfaces of many types and trying to diagnose potential problems with them. The fastest way to get to the **Object Information** command is from the type-specific pop-up toolbar that arises when you make a selection after setting the Selection Type Filter to a specific type.

- Continue working with *"Freeform Handle.prt"*.
- Set your **Selection Type Filter** to *Face*. Click on one of the "cap" surfaces created using **Fill Surface**, and select **Object Information** from the pop-up toolbar.

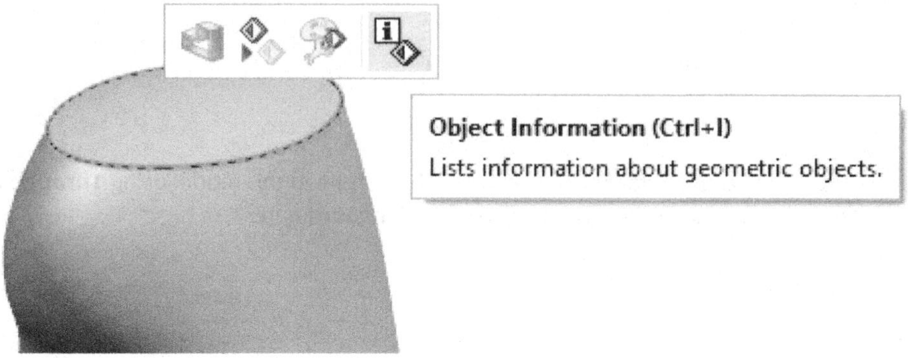

- An information window appears.

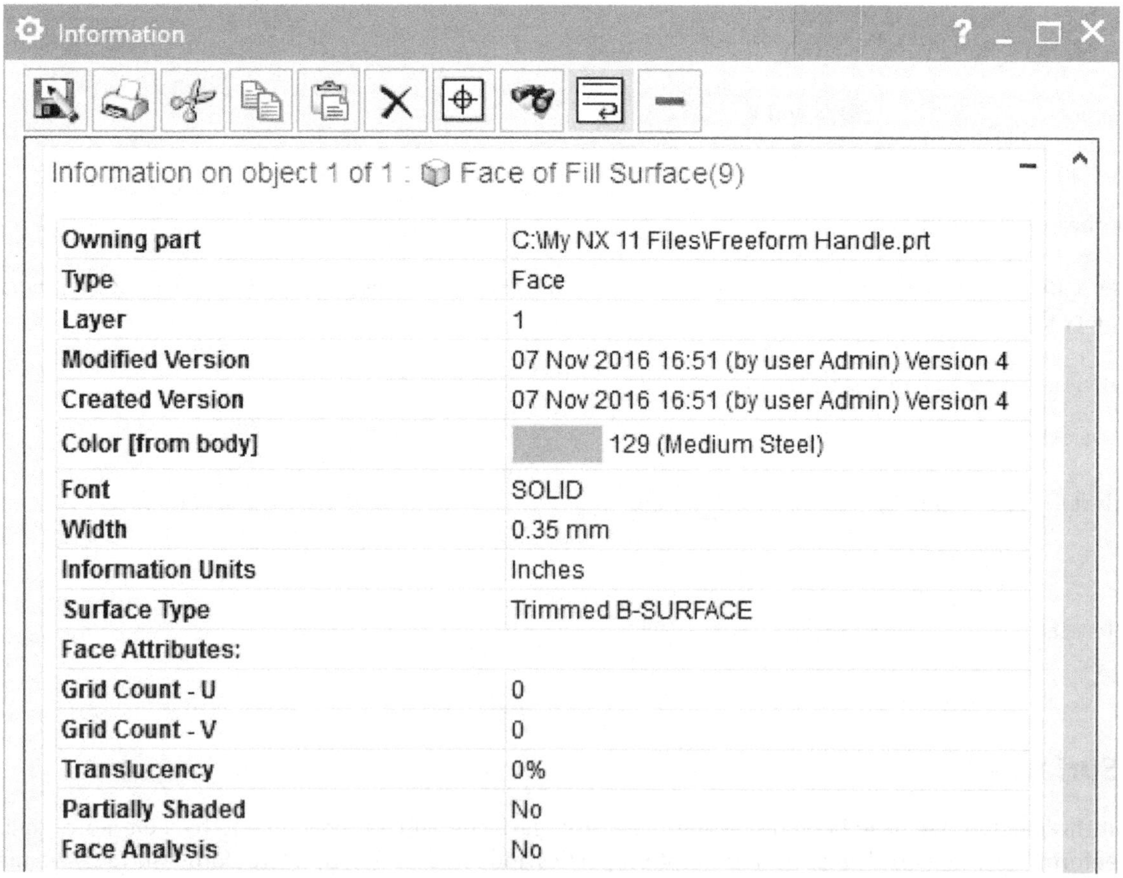

Owning part	C:\My NX 11 Files\Freeform Handle.prt
Type	Face
Layer	1
Modified Version	07 Nov 2016 16:51 (by user Admin) Version 4
Created Version	07 Nov 2016 16:51 (by user Admin) Version 4
Color [from body]	129 (Medium Steel)
Font	SOLID
Width	0.35 mm
Information Units	Inches
Surface Type	Trimmed B-SURFACE
Face Attributes:	
Grid Count - U	0
Grid Count - V	0
Translucency	0%
Partially Shaded	No
Face Analysis	No

- Amongst the information in this window is the **Surface Type** – in this case, the type is *"trimmed B-surface"*. Trimmed means that the surface has been trimmed back from its full rectangular form, as determined by its *(U,V)* parameterization. At the bottom of the information window, there is some information specific to B-surfaces.

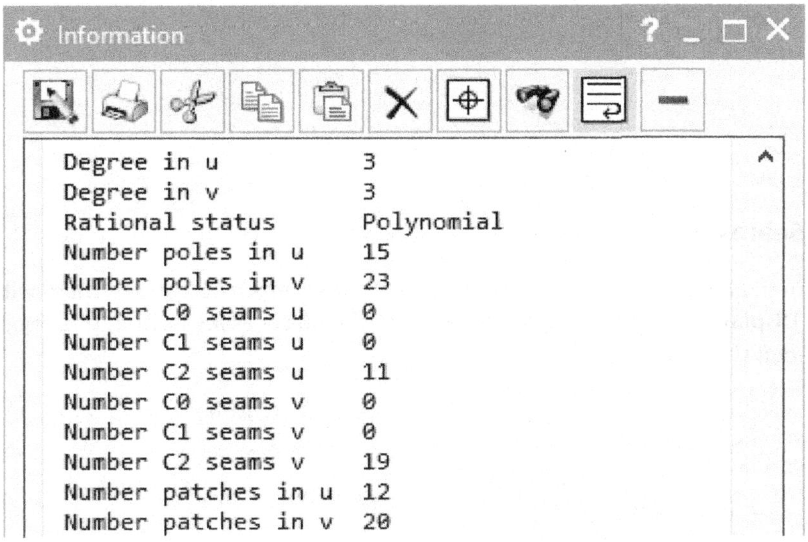

- The terminology here is inherited from the analogous terminology for splines explained in Chapter 13.2 – poles are poles, "seams" are knots, and "patches" are segments. The terms C0, C1, and C2 here mean essentially the same thing as G0, G1, and G2.
- **Close** the information window.

19.9.2 B-Surface Poles

In addition to the rectangular parameterization that every surface in NX has, B-surfaces also carry an additional special "rectangular" pole structure. The pole structure, degree, and patch information determine the full, rectangular, untrimmed version of every B-surface. The B-surface poles for the face created by **Fill Surface** are shown below.

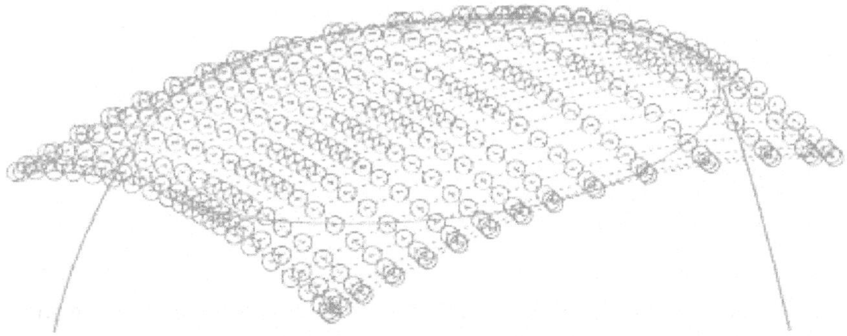

You can display the poles for a B-surface by using the **Show Poles** button found in the **Display** group on the **Analysis** tab. This button acts like a switch that toggles the display of poles on or off – there is no separate **Analysis Object** created as for the more sophisticated analysis tools discussed so far.

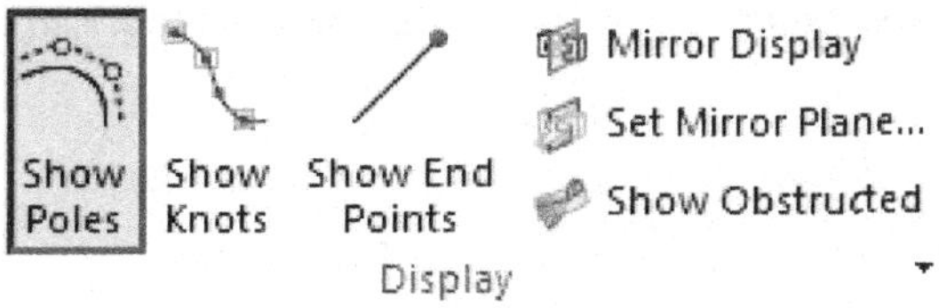

The poles of the "cap" face built with **Through Curve Mesh** appear as shown below.

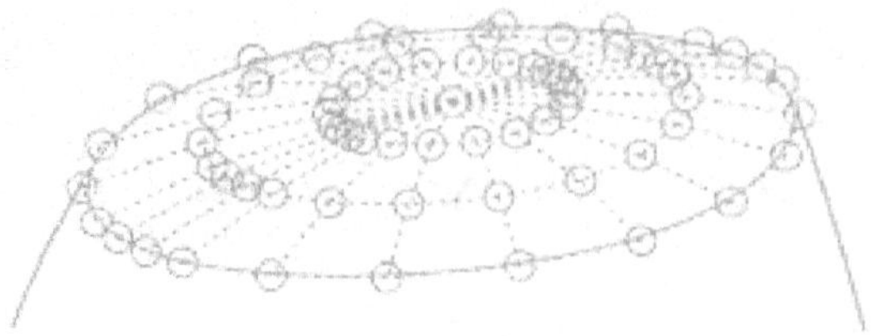

19.9.3 Patches and Seams

Patches and *seams*, as noted above, are like segments and knots of splines. You can see them with the help of the **Show Knots** button also in the **Display** group on the **Analysis** tab. Like the **Show Poles** button, this toggles the display on or off and does not create an *Analysis* object.

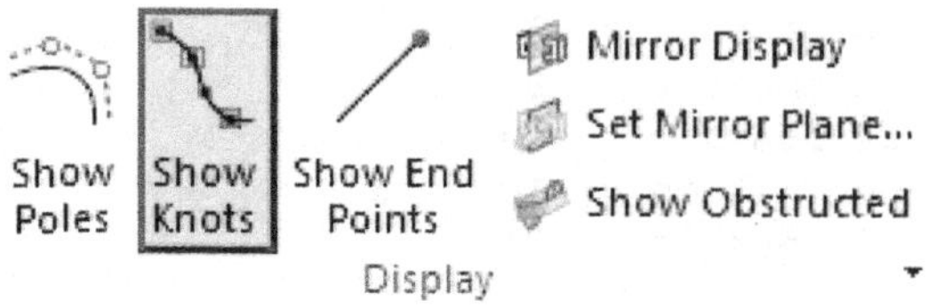

The patches and seams for the face created by **Fill Surface** are shown below.

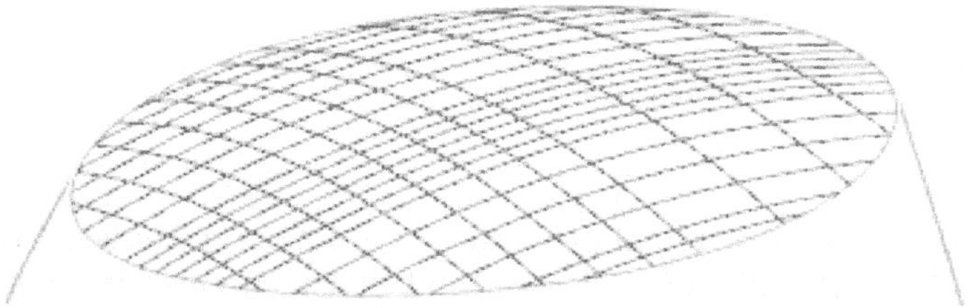

The patches and seams for the "cap" face created by **Through Curve mesh** are shown below.

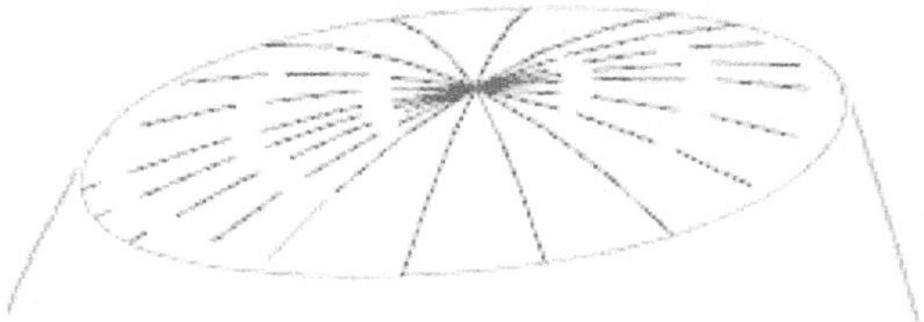

19.9.4 Curvature

As with splines, the curvature properties of B-surfaces are an important tool for understanding and improving them. There is a more robust notion of curvature for surfaces than for curves, as there are naturally many curves on a surface!

19.9.4.1 Section Analysis

The **Section Analysis** tool allows you to investigate the curvature combs of curves on the face in question. It is analogous to the **Curve Analysis** tool introduced in Chapter 13.3.1. The Section Analysis tool is in the Face Shape group on the Analysis tab.

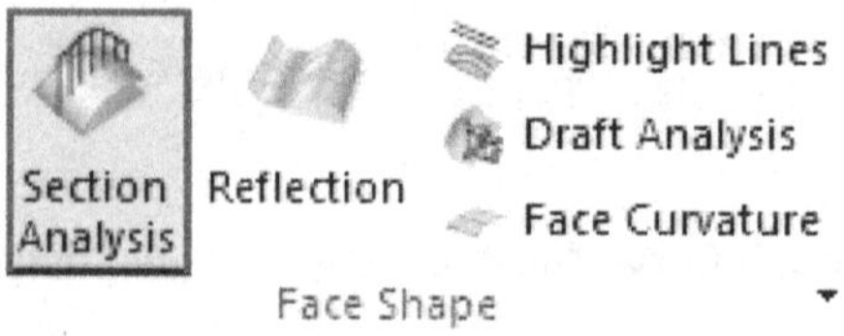

The **Section Analysis** tool has two checkboxes that are essential for controlling the dynamic display of curvature combs for curves on the selected face – *Show Comb* and *Suggest Scale Factor*.

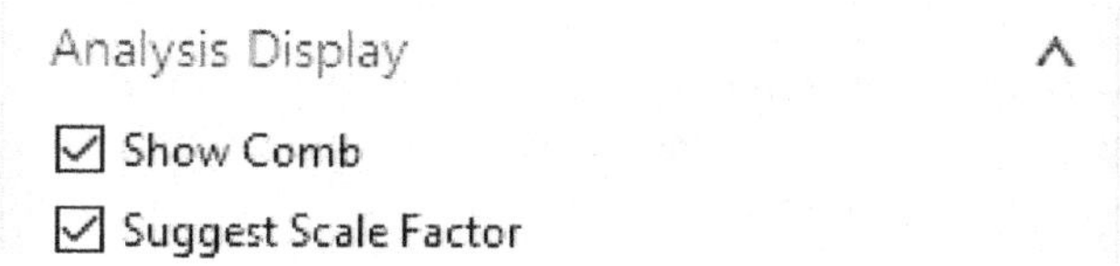

Below, the curvature combs for isoparametric curves in the U direction on the face created by **Fill Surface** are shown.

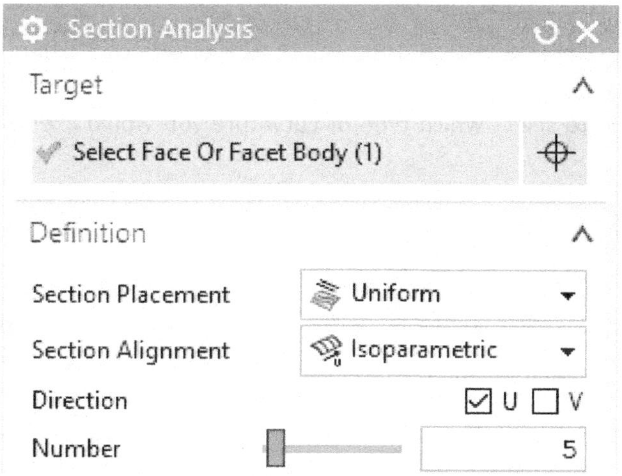

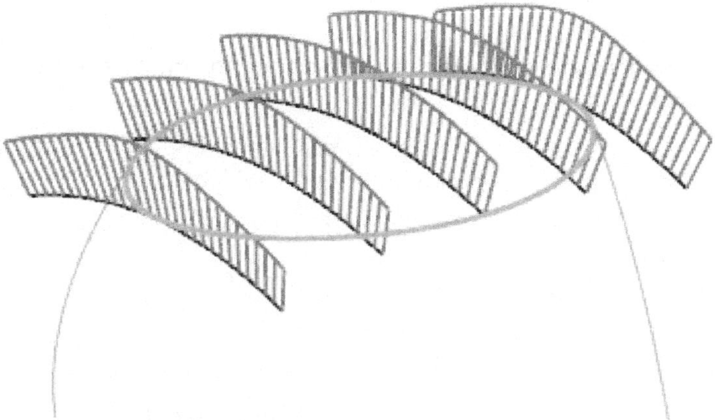

The output of **Section Analysis** is an *Analysis Object*. It behaves like **Deviation Gauge** in that the needle display will update dynamically as changes are made to the surface in question. Unlike **Draft Analysis** and **Reflection Analysis**, it is not necessary to use the *Face Analysis* rendering style to see the comb display, and it can be toggled on or off by simply checking or unchecking the red checkmark in the box adjacent to the name.

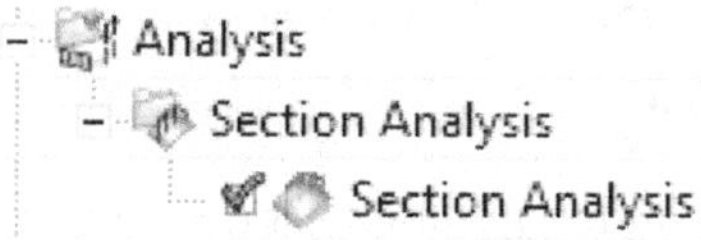

Here's what the curvature combs for isoparametric curves in the U direction look like for the "cap" surface created with **Through Curve Mesh**.

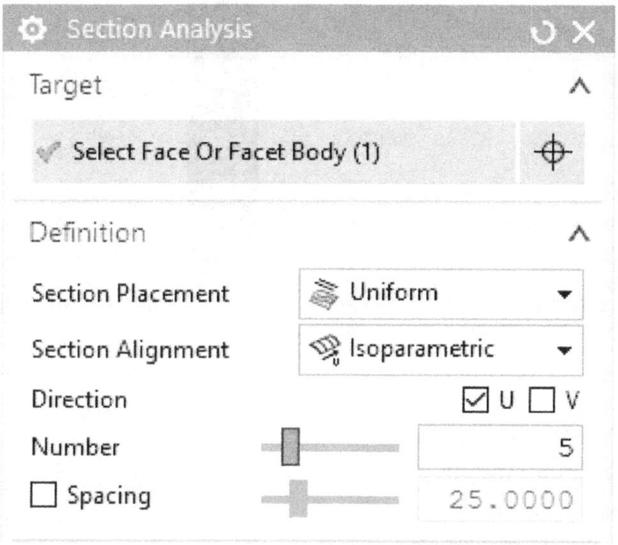

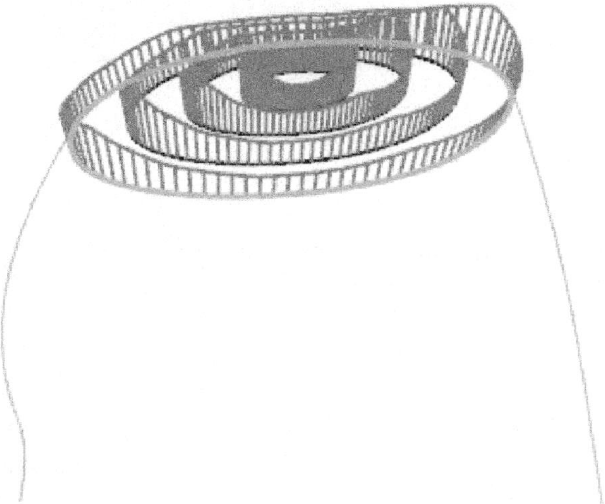

19.9.4.2 Face Curvature

Sometimes a color map gives you a better understanding of the curvature on a face, and indeed can be much more useful in identifying changes in concavity on a surface. The **Face Curvature Analysis** tool is used to produce such color maps and curves identifying where the curvature of a given type is zero on a surface (the zero contour).

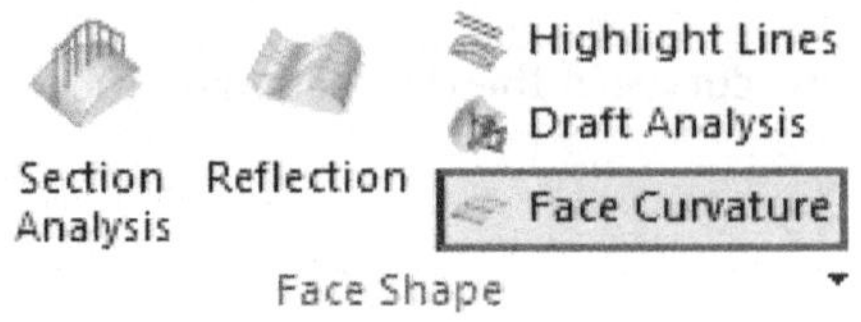

Since there are many notions of curvature for the tool allows you to select which type of curvature you would like to display.

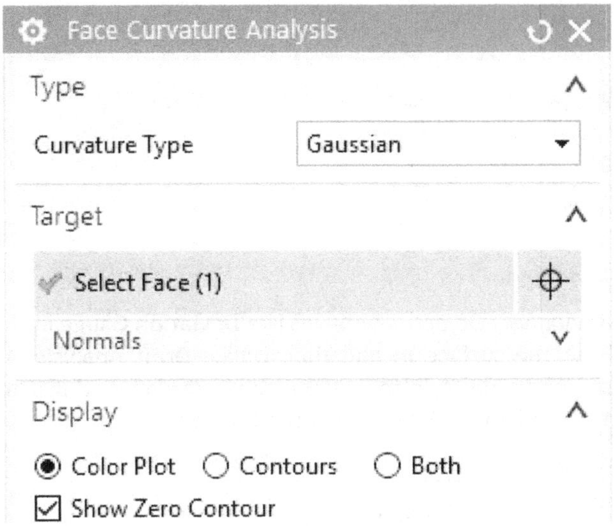

The color map comes with a legend on the right side of the graphics window. This map only appears while using the *Face Analysis* rendering style.

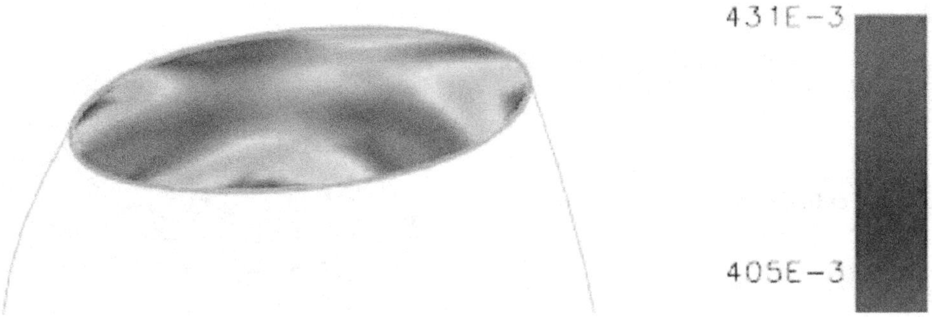

Note that no zero contours appear on this surface since it is convex! The output of **Face Curvature Analysis** is an Analysis Object.

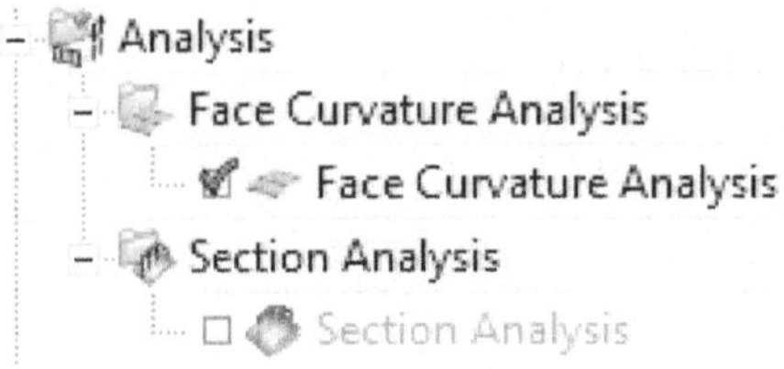

Here's what the color map looks like for the "cap" surface built using **Through Curve Mesh**. Which do you think looks better?

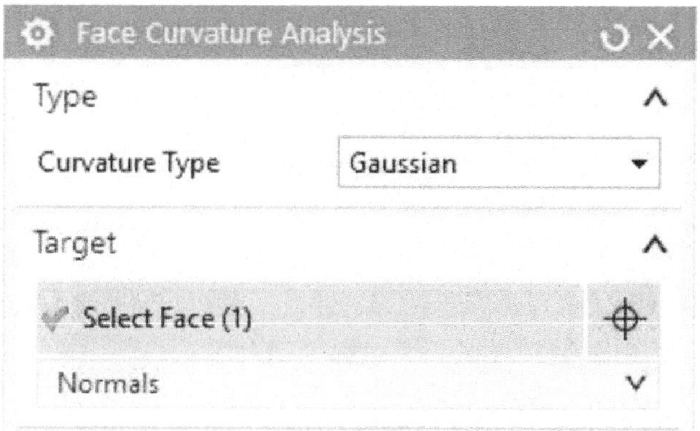

19.10 **Surface Operations**

The **Surface Operations** group contains tools for modifying surfaces or copies of surfaces. You have already used many of the tools in this group, so in this section the exercises will focus on the most useful remaining tools in this group. The **Surface Operations** group is found on the **Surface** tab.

19.10.1 Untrim

The **Untrim** command creates a copy of the untrimmed version of a given trimmed surface, meaning that it fills in any *(U,V)* pairs that might be missing. For ruled or extruded surfaces, the **Untrim** tool may also enlarge the surface by an unknown amount. For example, a surface that is linear in U will be enlarged in the U direction. The **Untrim** tool is found in the **More** gallery in the **Surface Operations** group on the **Surface** tab.

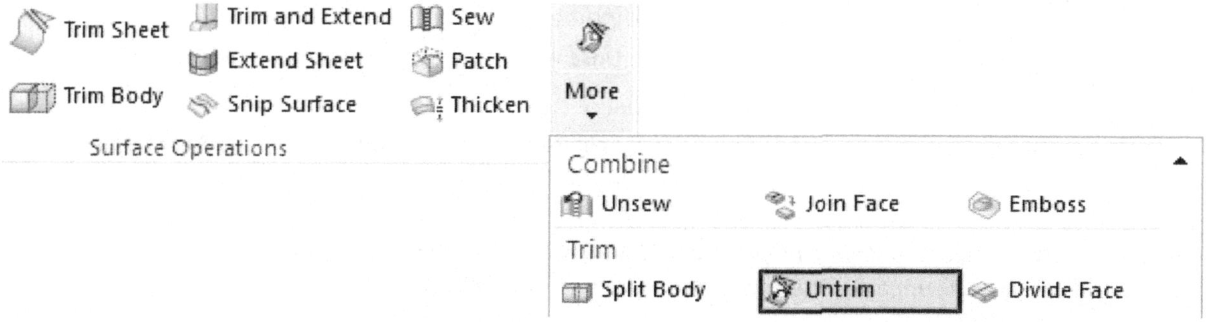

- Continue working with *"Freeform Handle.prt"*.
- Select the **Untrim** tool. Choose the "cap" face of the handle (the one created using **Fill Surface**, in case you also used **Through Curve Mesh**!) as the **Face to Untrim**. In **Settings**, check the *Associative* checkbox.

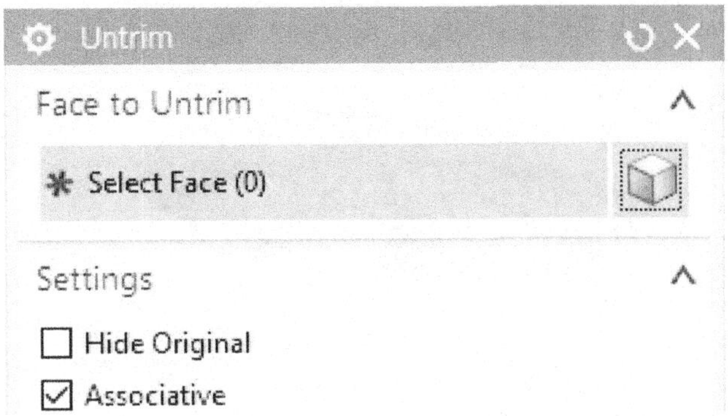

- The result is the full B-surface that was trimmed to meet the edge of the sheet body.

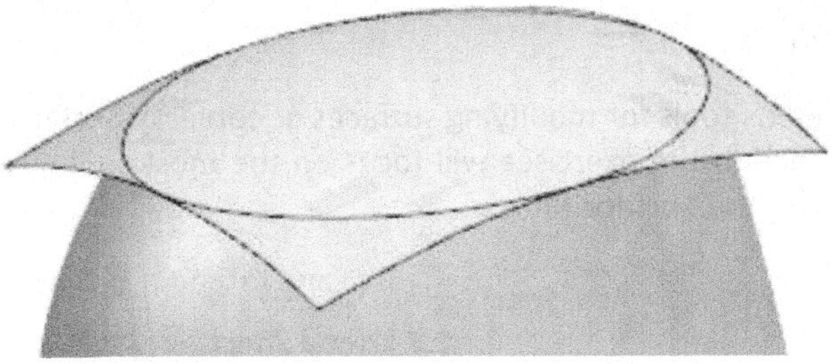

19.10.2 Delete Edge

In certain applications, you might not want to untrim a surface to its full rectangular form, but simply fill in a hole, or remove a rounded edge. For these types of operations, consider the **Delete Edge** command.

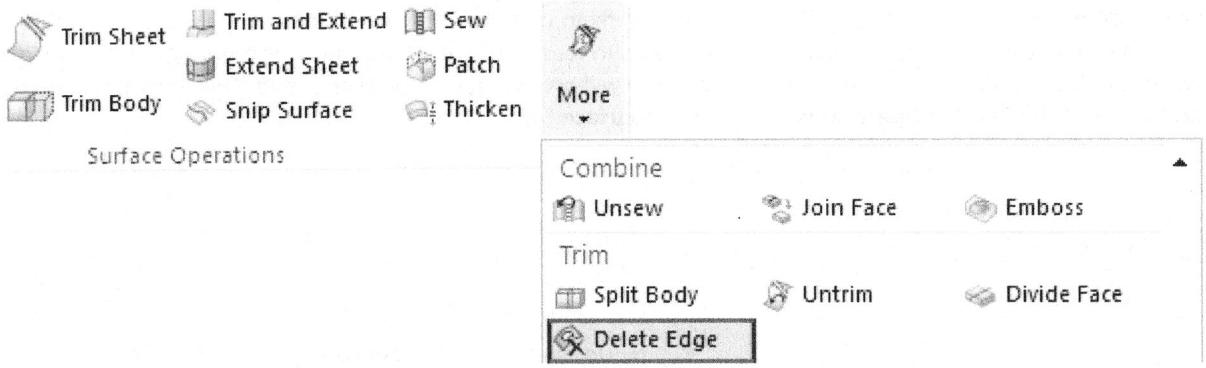

- Open the file *"Trimming and Untrimming.prt"* from *"C:\My NX 11 Files\Downloaded Files"*.
- Inside you will see a single unparameterized surface with two slots cut out of it.

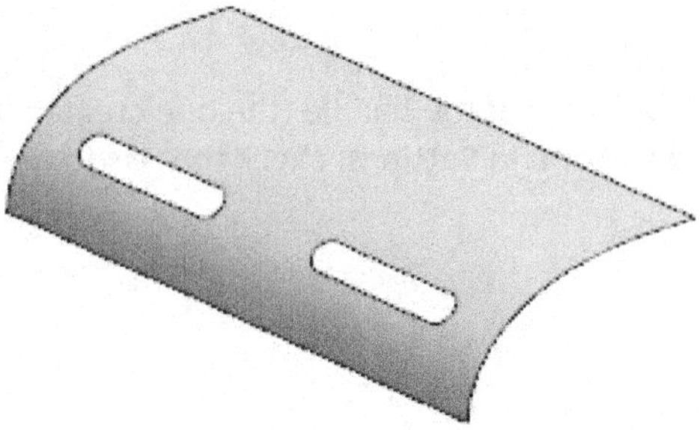

- Select the **Delete Edge** tool, and with **Selection Intent** set to *Tangent Curves*, select the left slot.

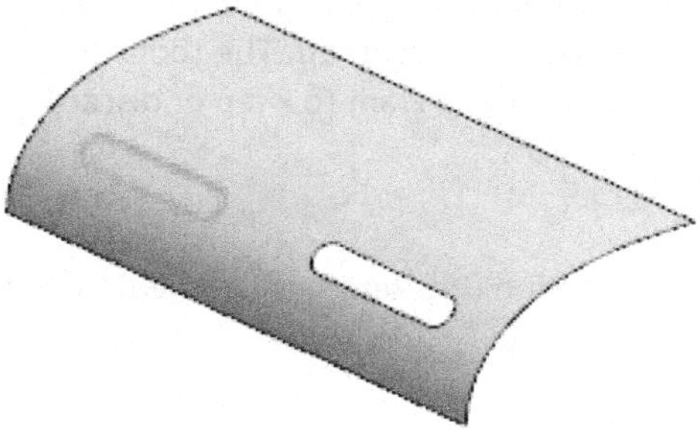

- This fills in all the missing *(U,V)* pairs in the slot.

19.10.3 Trim Sheet

In Chapter 8.4, you learned about the **Trim Body** command, which is capable of trimming a sheet body using a datum or another sheet body as trimming tool. **Trim Sheet** is another very useful tool for removing material from a sheet body, especially because it allows curves and edges of nearby bodies as trimming tools.

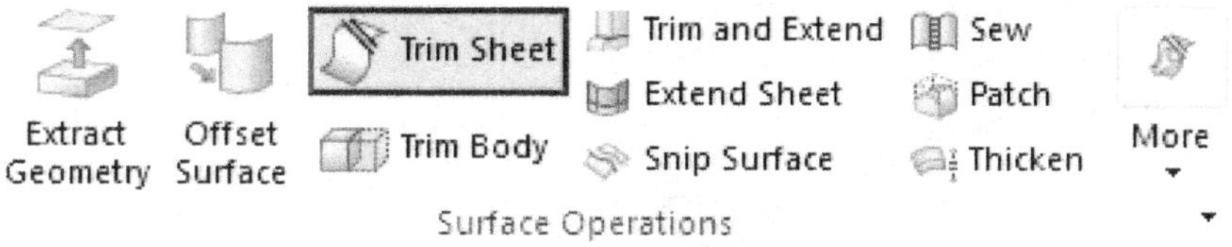

- Continue working with the file *"Trimming and Untrimming.prt"*.
- Make layer *41* visible. It contains some curves that we will use to trim the existing sheet body.

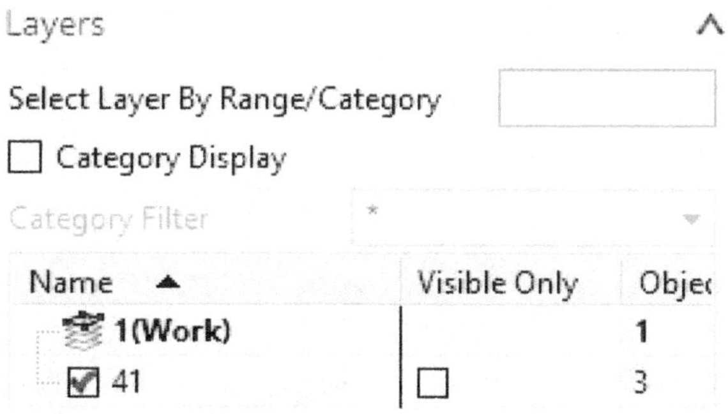

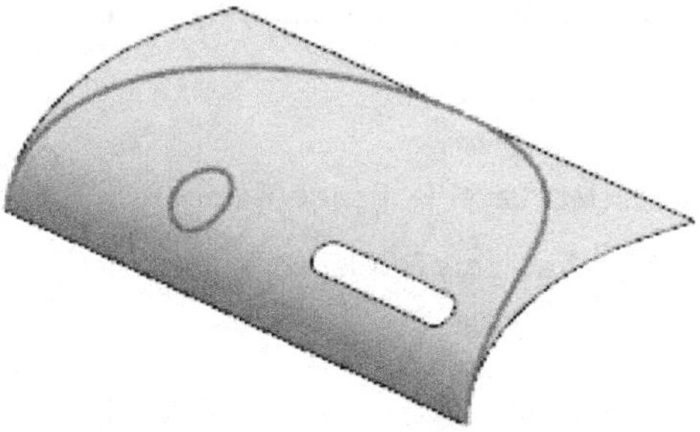

- Select **Trim Sheet**, and choose the sheet body as the **Target**. For the **Boundary**, choose the curves from layer *41*.

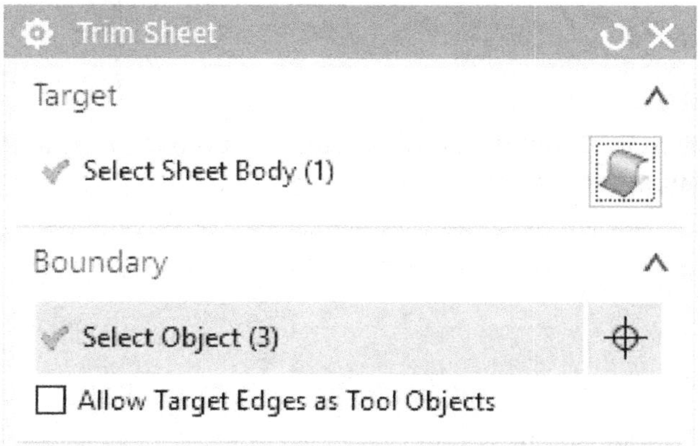

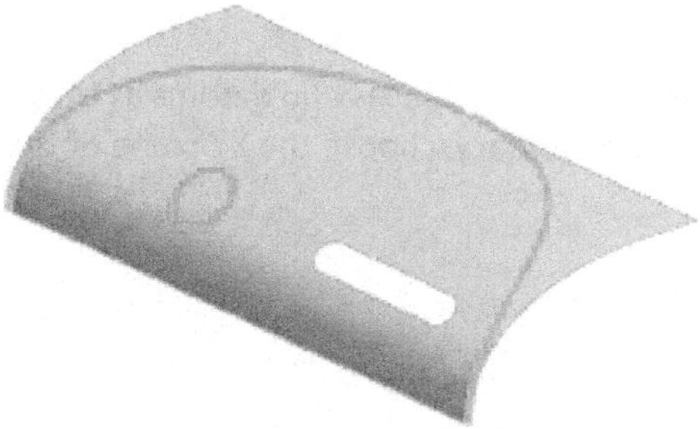

- An extremely important thing to do each time you use the **Trim Sheet** tool is to specify the **Region**, and say whether it should be kept or discarded. The tool will make an inference based on how you click, so if the highlighted region differs from what you want to keep or discard, you will need to deselect and reselect it here.

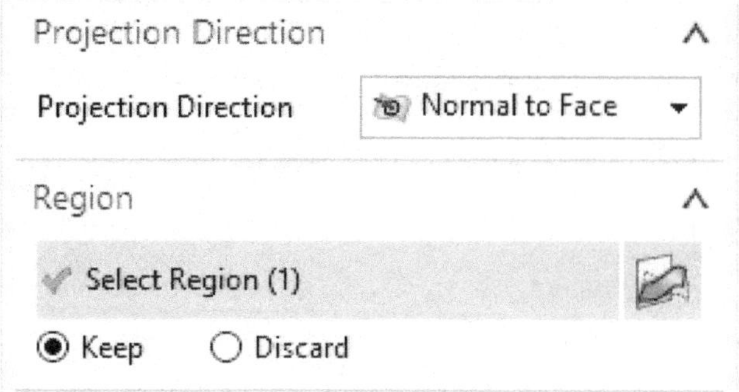

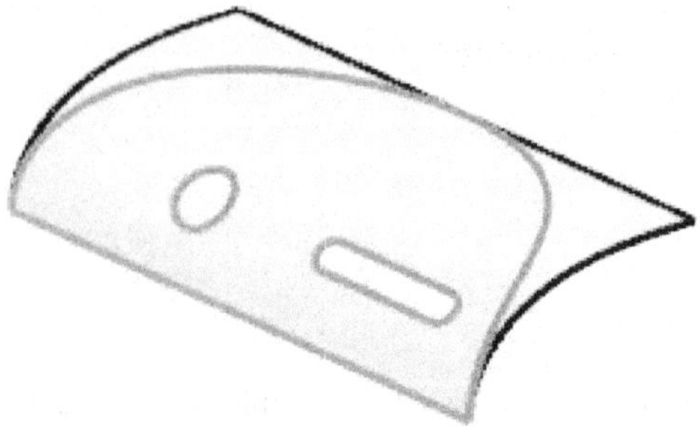

- Make layer *41* invisible, and **Extrude** the curved edge of the sheet body in the -ZC direction by *1 in*. (Note that this part is metric – you have to type *"1in"*!)

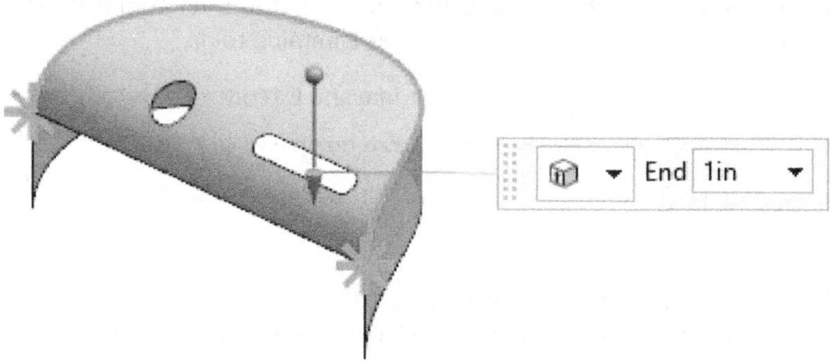

- Make layer *61* visible – it contains a datum coordinate system.

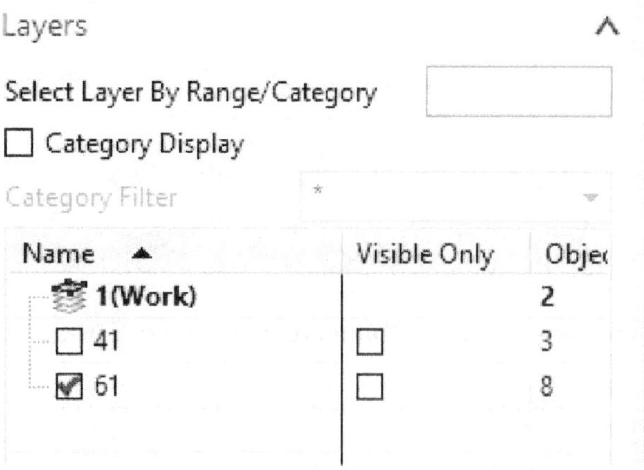

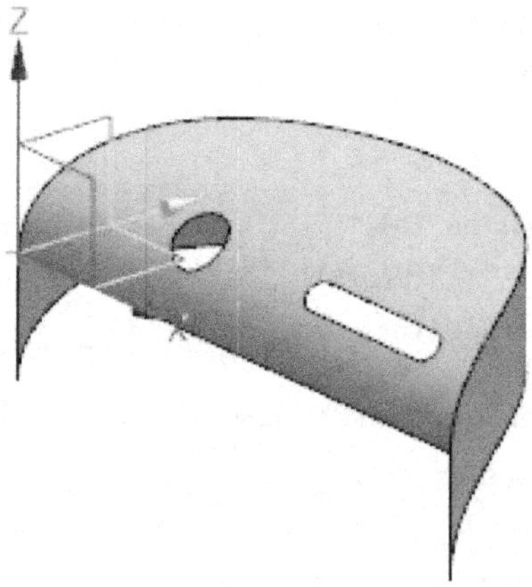

- **Trim Sheet** also allows faces and datums as trimming tools. Select the newly created surface of *Extrude* as the **Target**, and for the **Boundary**, choose the XY plane of the datum coordinate system.

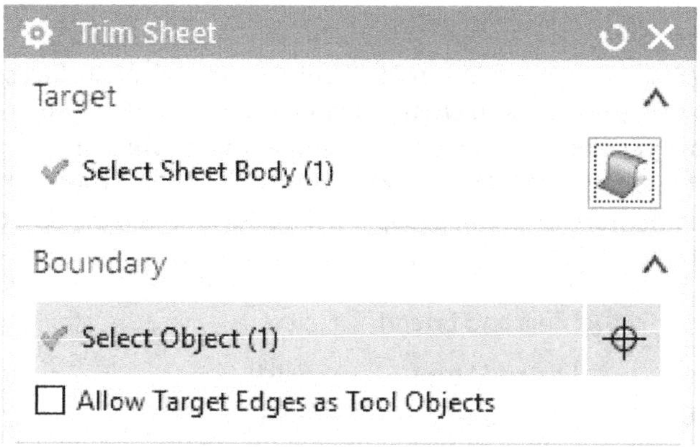

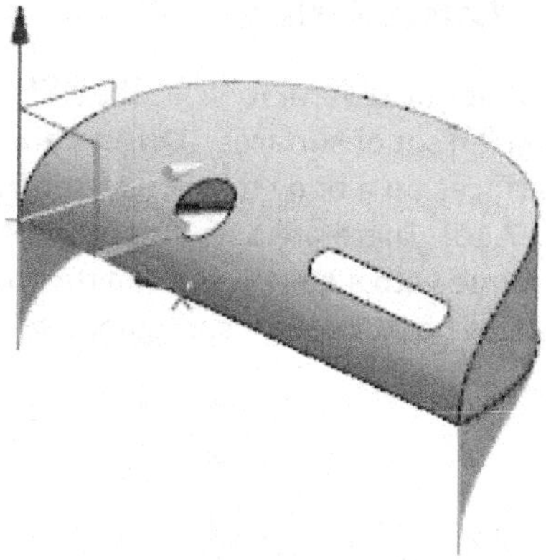

- Use the **Bounded Plane** tool to fill in the planar region at the bottom of your surfaces.

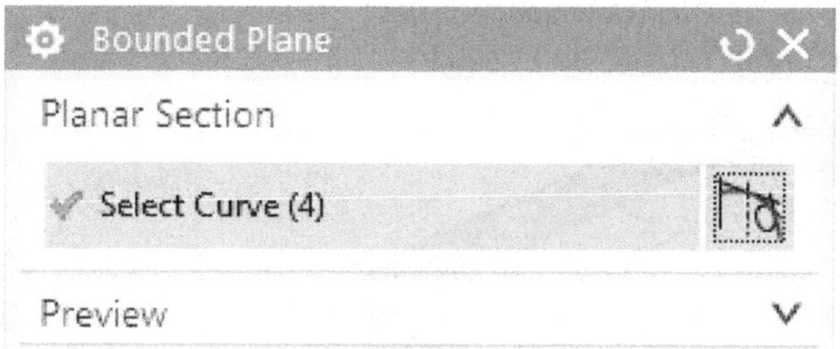

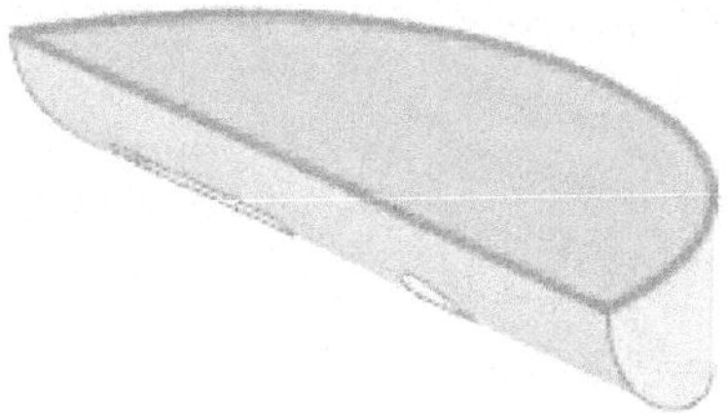

- Use **Thicken** to add thickness in the amount of *0.1 in* on the inside.

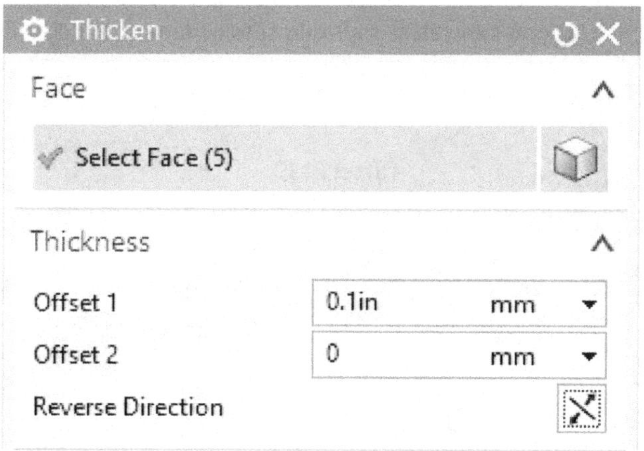

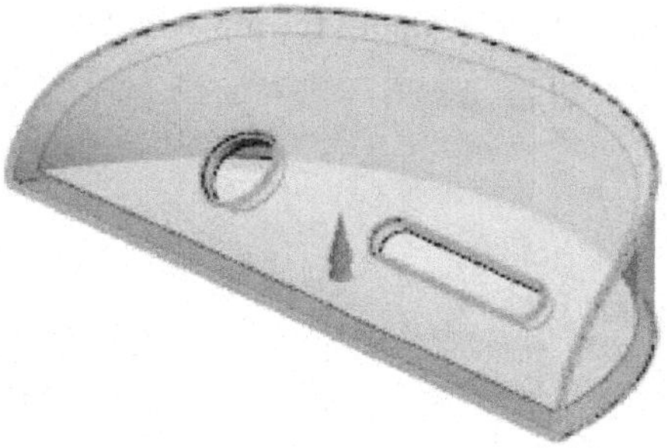

- Use **Delete Body** to associatively delete the sheet bodies, and enjoy the product housing you've built below!

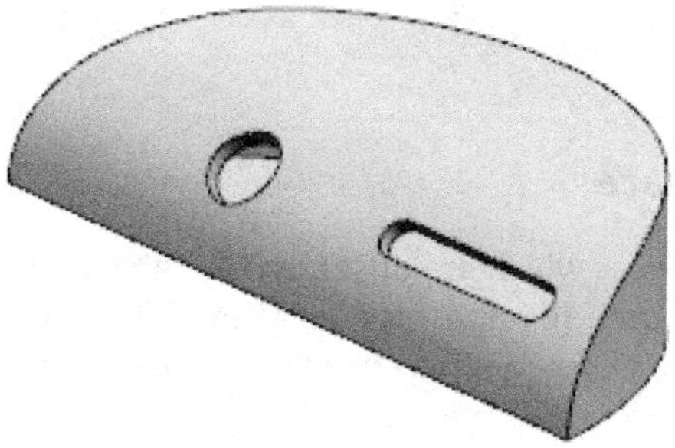

- **Save** your file!

19.10.4 Offset Surface

We have learned about several tools in the **Offset/Scale** group – some that offset a given set of surfaces by a uniform amount from a given set of surfaces (**Offset Face** in Chapter 7.11 and **Offset Region** in Chapter 16.3), and others yet that produce new surfaces on a body that are offset from specified surfaces by a uniform amount (**Thicken** in Chapter 7.8 and **Shell** in Chapter 7.10). There are a few more tools of interest in the **Offset/Scale** group, and to put them into their proper context, we continue with a quick lesson on the fundamental tool in this group – **Offset Surface**. Simply put, **Offset Surface** creates a new sheet body for each contiguous set of selected faces, by offsetting them a uniform amount.

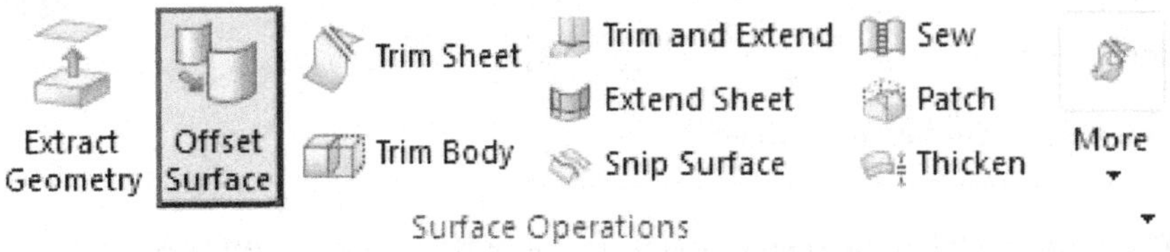

- Open the file *"Offset Surfaces.prt"* from *"C:\My NX 11 Files\Downloaded Files"*.
- Inside this file, you will find two sheet bodies, named *"FLOOR"* and *"WALL"*.

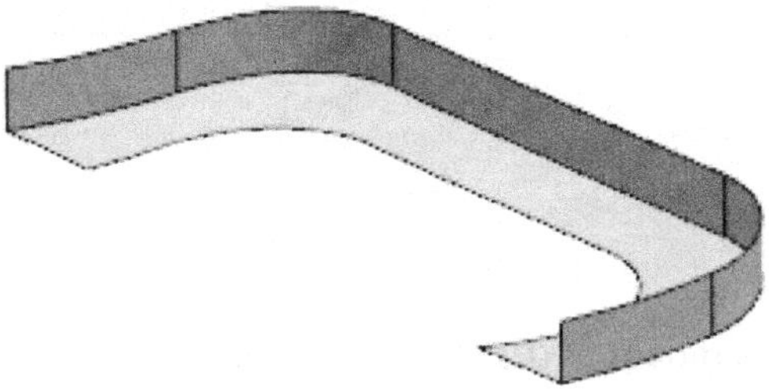

- Select the **Offset Surface** tool. Notice that it has a list, so you can select multiple sets of faces, and specify different directions and offset values for those faces.

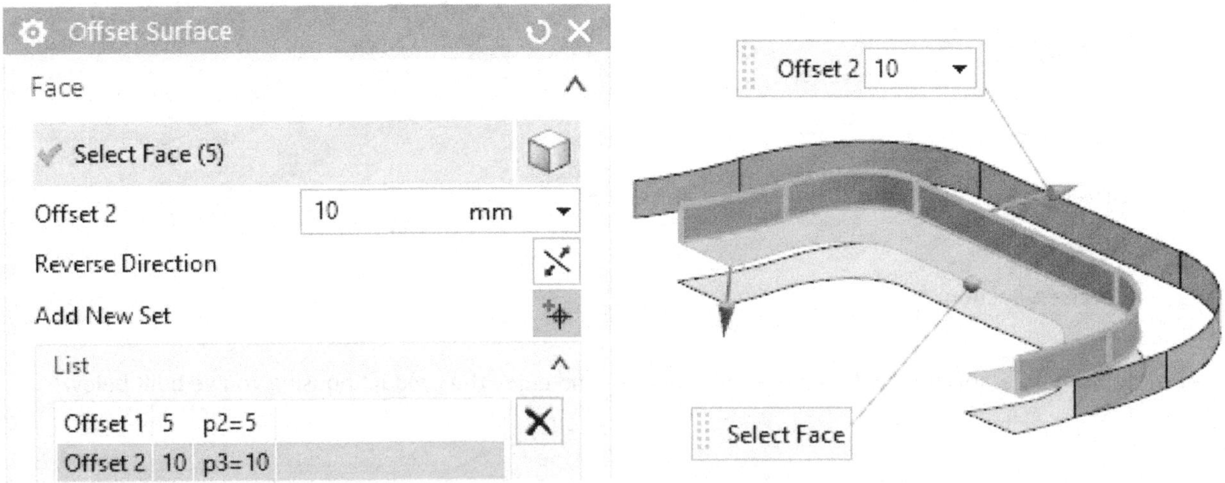

19.10.5 Variable Offset Surface

Occasionally you will want to produce surfaces which are offset by a variable amount over the parent surfaces. This is what the **Variable Offset Surface** command does, for one face at a time. This tool takes advantage of the "rectangular" parameterization of a surface in NX, and specifies a different offset at each of the four corners of the untrimmed form of a surface. As such, it only lets you offset one face at a time.

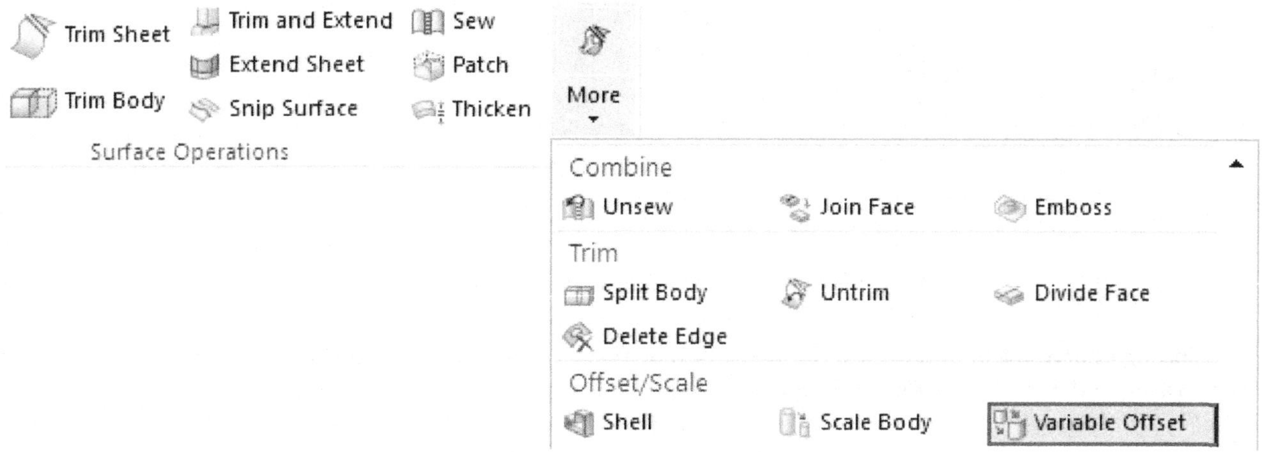

- Continue working with the file *"Offset Surfaces.prt"*.
- Hide the body *"WALL"*. Use **Show Poles** to see the edges of the untrimmed form of the surface.

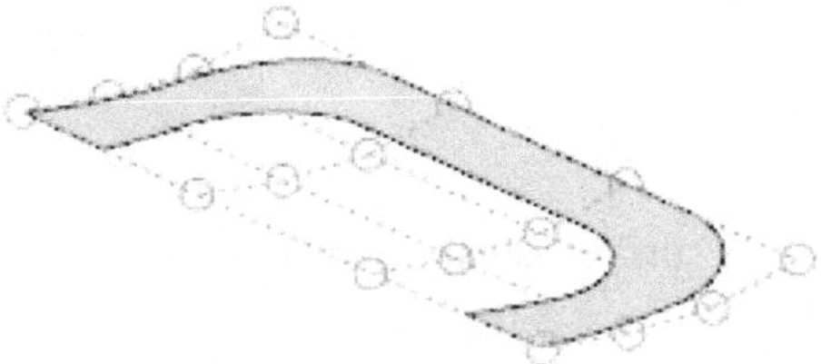

- Select the **Variable Offset Surface** tool, and choose *"FLOOR"* as the **Face to Offset**.

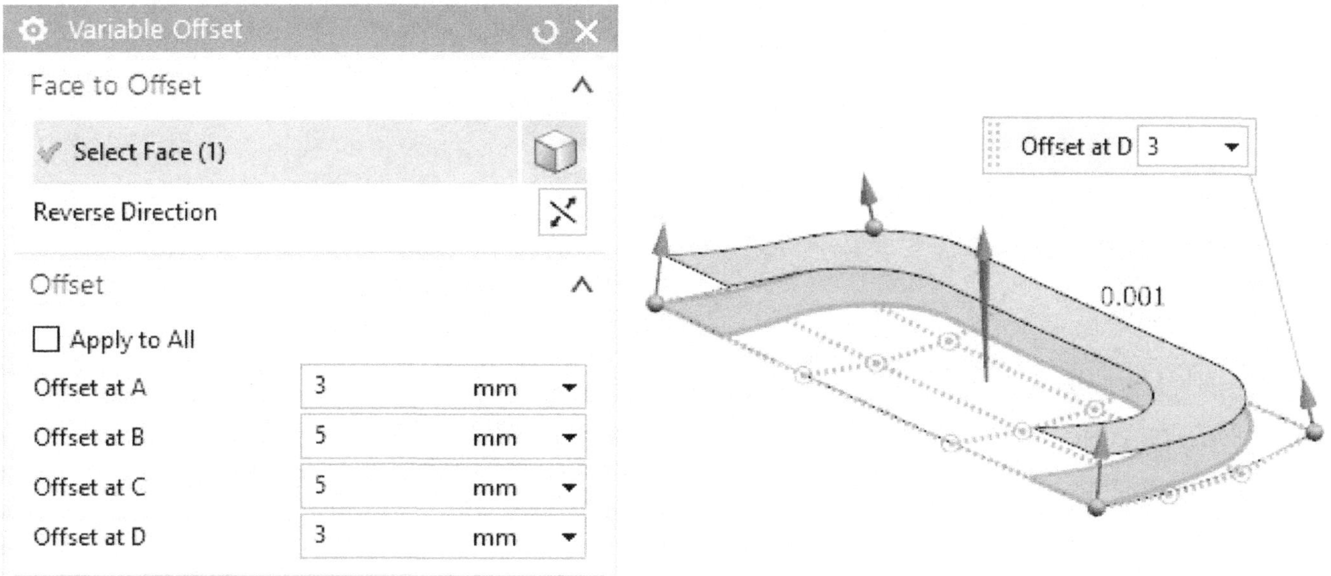

- Make layer *41* the Work Layer and layer *1* invisible. Layer *41* contains a copy of the edges of the surface *"FLOOR"*.

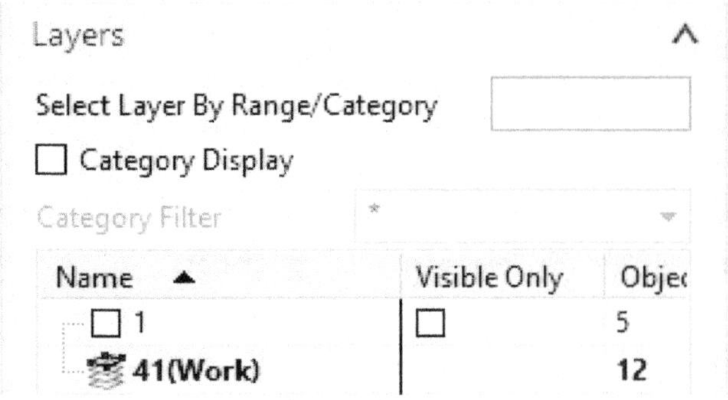

- Use **Through Curve Mesh** to build a surface enclosed by these curves. Move the surface to layer *2*, make layer *2* the Work Layer, and make layer *41* invisible again.

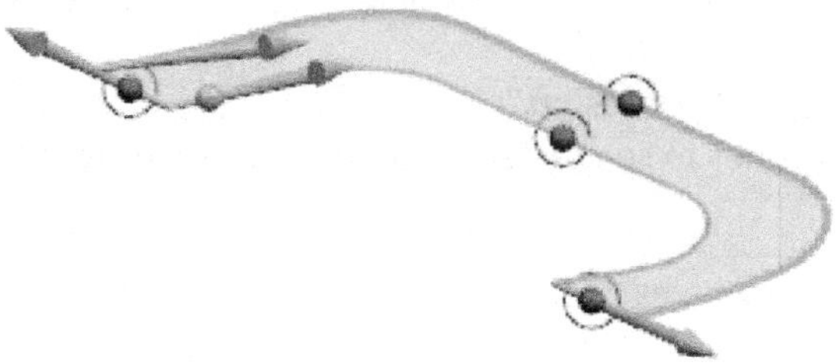

- The rectangular parameterization of this surface is very different from that of the original surface – now, the four corners are honestly the four corners!

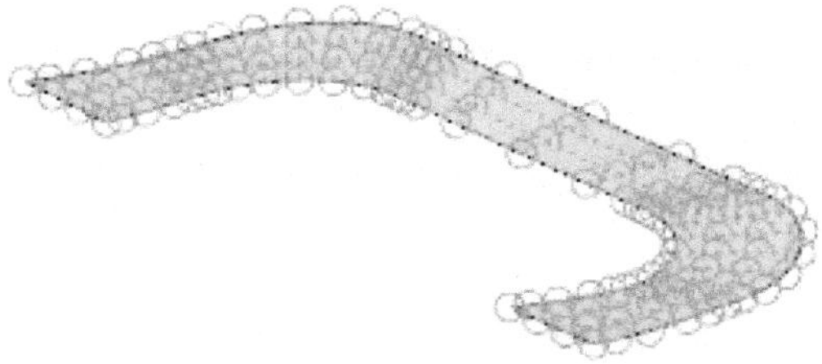

- Use the **Variable Offset Surface** tool once more.

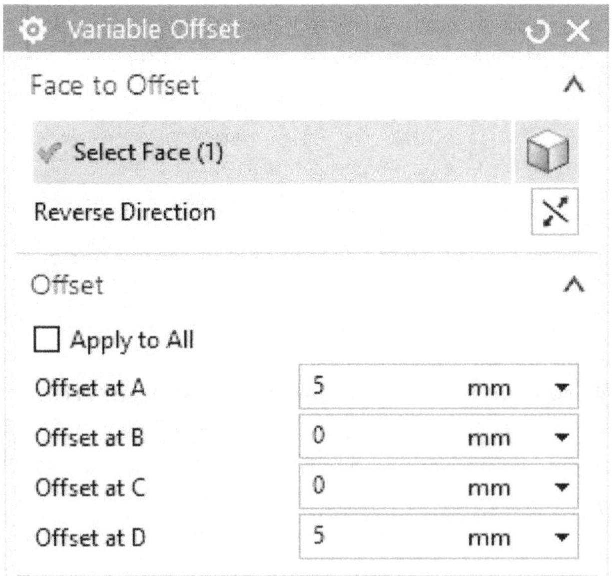

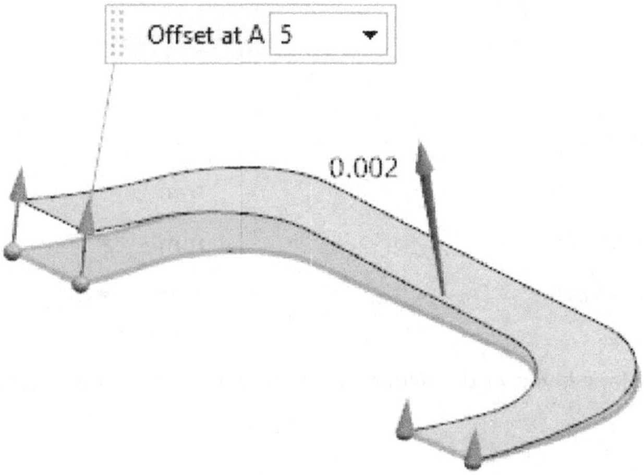

19.10.6 Variable Offset Face

When you need to create a body of non-uniform thickness, the **Offset Surface** and **Variable Offset Surface** tools can be moderately helpful, but they still require that you to perform surgery (e.g., trimming, blending) on the interior surfaces after offsetting them. The **Variable Offset Face** tool, new in NX 11, produces a single sheet body, offset by different amounts on different regions, with intermediate regions that transition continuously or smoothly.

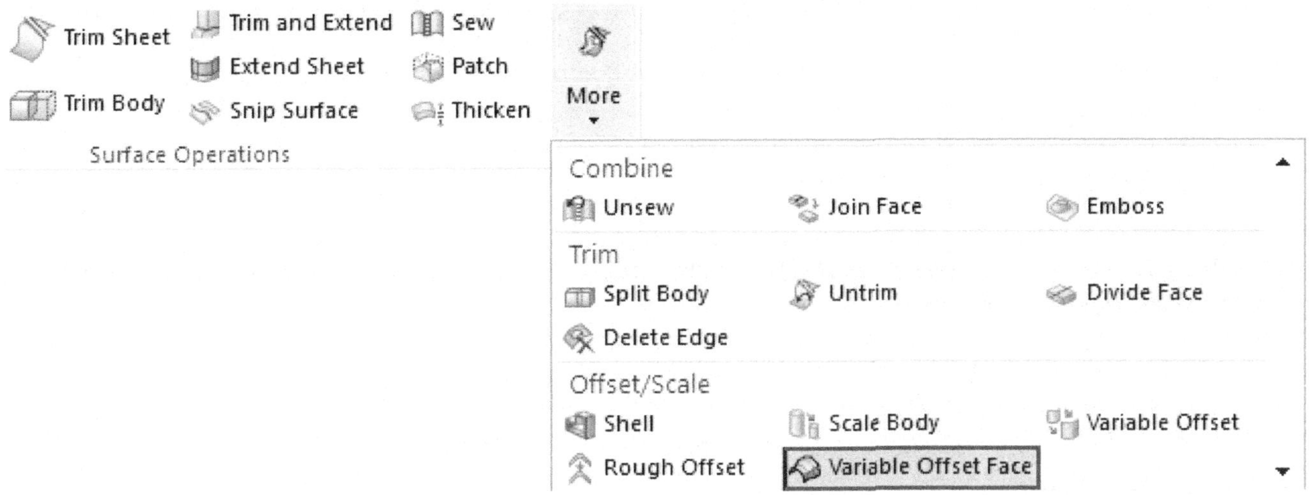

- Open the file *"Variable Offset Face.prt"* from *"C:\My NX 11 Files\Downloaded Files"*.
- Inside you will see two unparameterized sheet bodies, named *"EXTERIOR"* and *"MODIFIED EXTERIOR"*.
- The goal of this exercise is to produce a body of non-uniform thickness from the exterior, with smooth faces on the interior. Offsetting the faces of the body *"EXTERIOR"* results in a sharp edge on the interior, hence the need for the body *"MODIFIED EXTERIOR"*. The body *"MODIFIED EXTERIOR"* is simply the body *"EXTERIOR"* with an intermediate Face Blend.

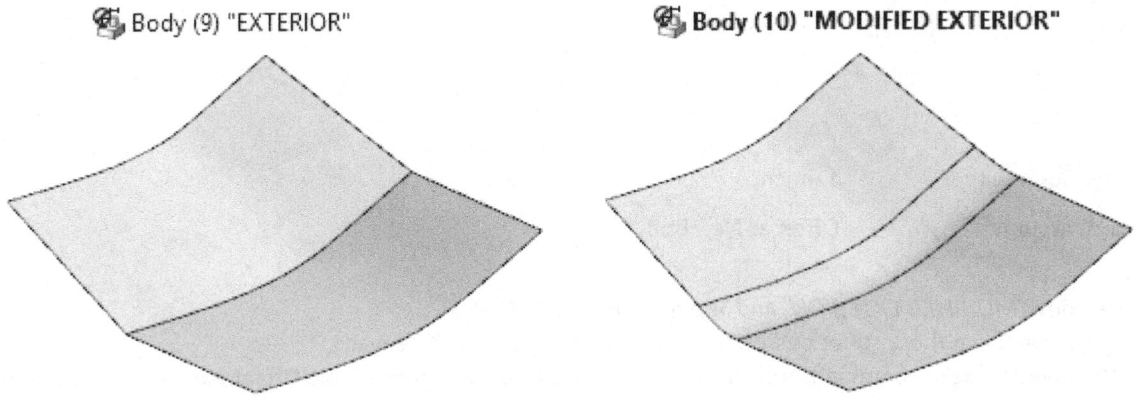

- **Hide** the body *"EXTERIOR"* and select the **Variable Offset Face** tool.
- For the Face, choose the body of *"MODIFIED EXTERIOR"*. The **Region Boundary** should consist of the two smooth edges on *"MODIFIED EXTERIOR"*. Use the **Reverse Direction** button if necessary to point the offset inward.

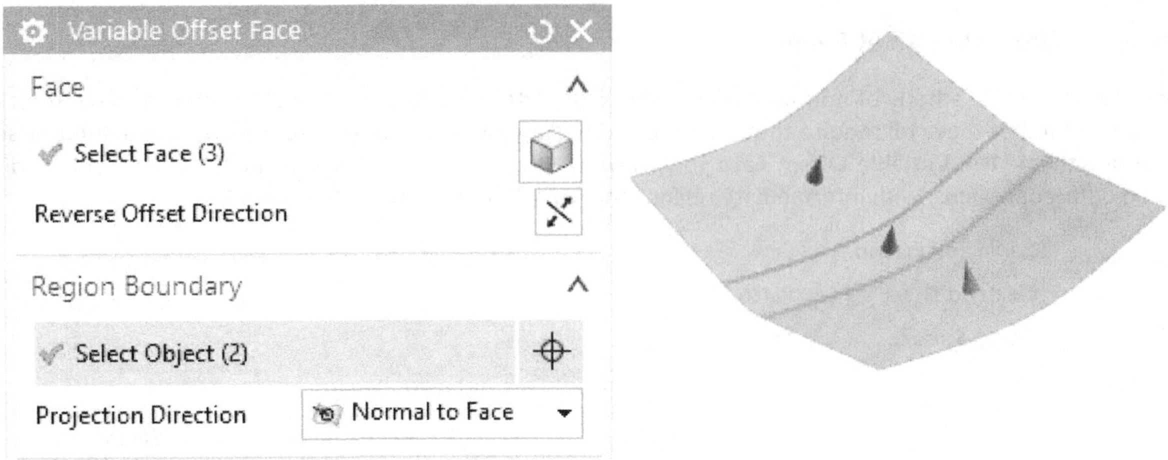

- This divides the faces of *"MODIFIED EXTERIOR"* into three regions. Enter the parameters shown below in the **Region** and **Settings** sections of the menu.

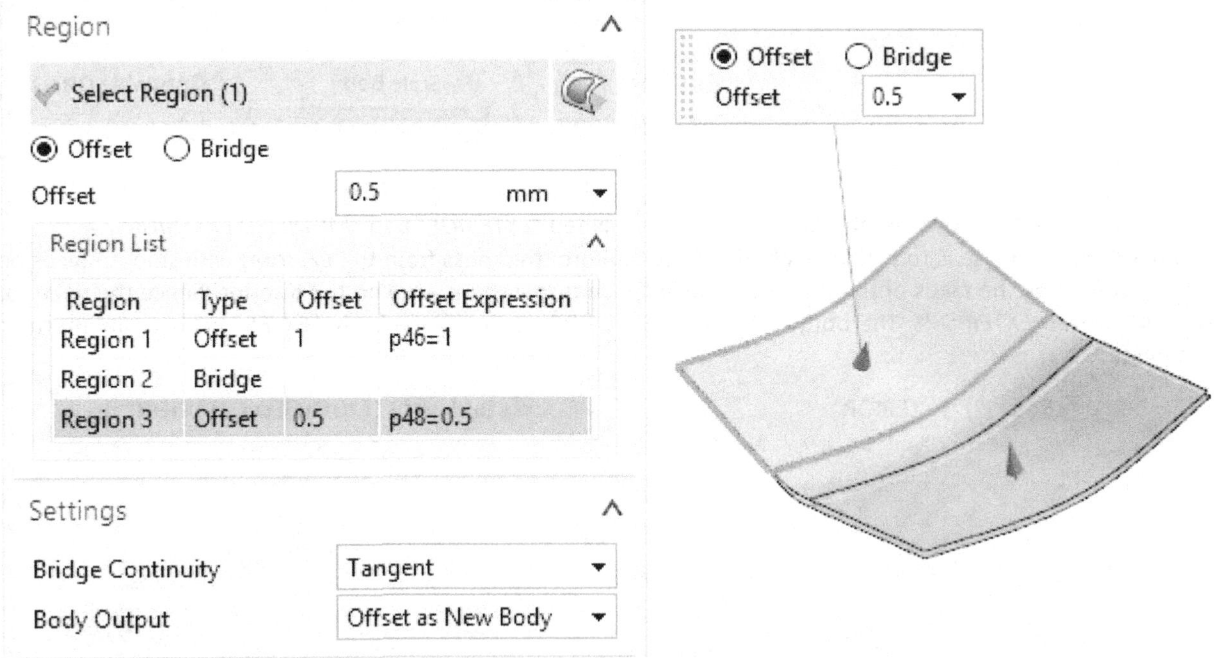

- **Hide** the body *"MODIFIED EXTERIOR"* and **Show** the body *"EXTERIOR"*.
- To create a solid from these sheet bodies, we will need to create the side walls.
- Select the **Ruled Surface** command, and choose the edges of the body of **Variable Offset Face** as **Section String 1**, and the edges of the body *"EXTERIOR"* as **Section String 2**.

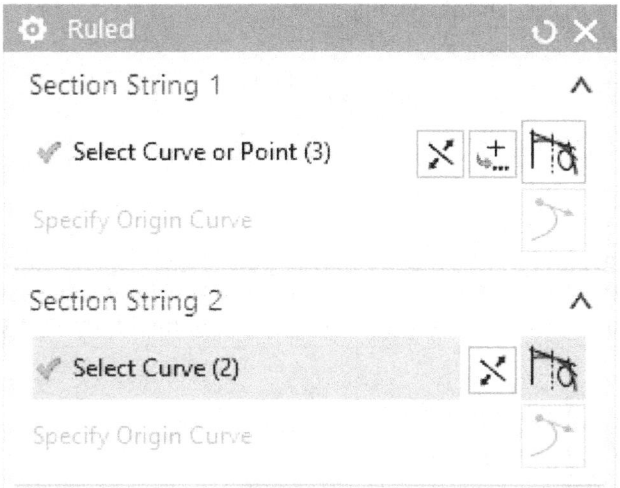

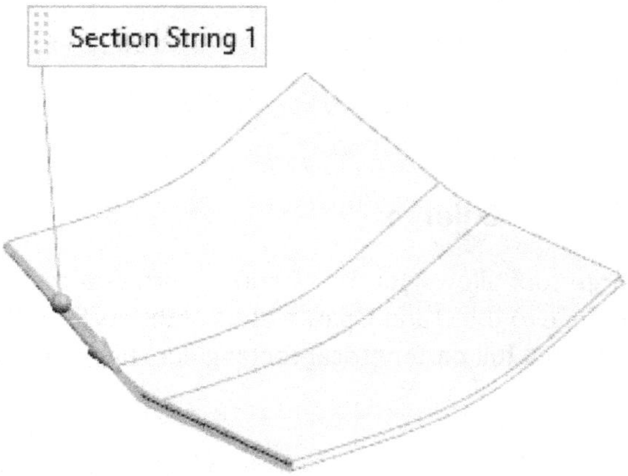

- This is one application of **Ruled Surface** where you want to <u>disable</u> the *Preserve Shape* checkbox – notice that when *Preserve Shape* is enabled, the ruled surface will inherit two undesirable edges.

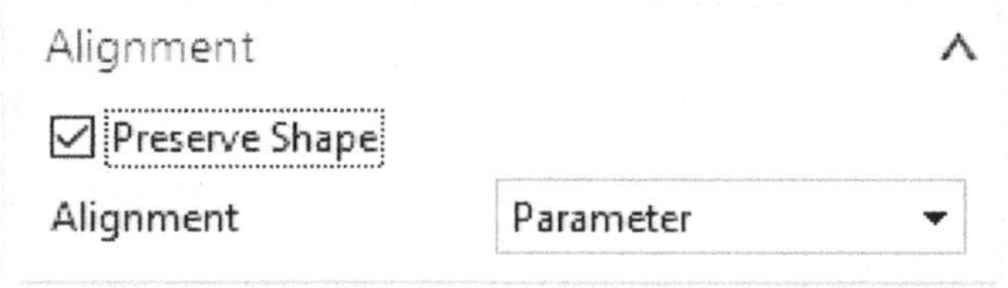

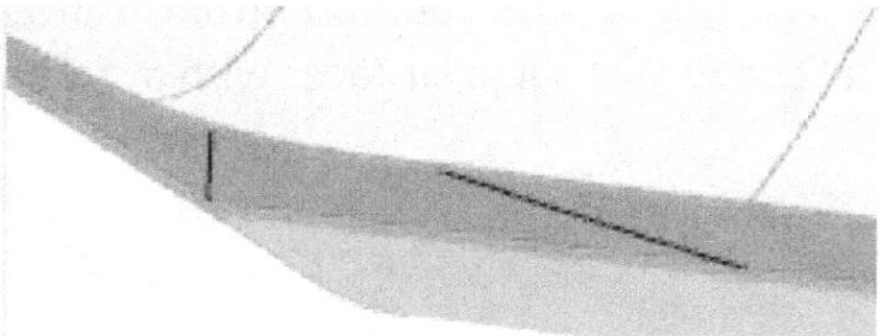

- Without *Preserve Shape*, the resulting surface is smooth.

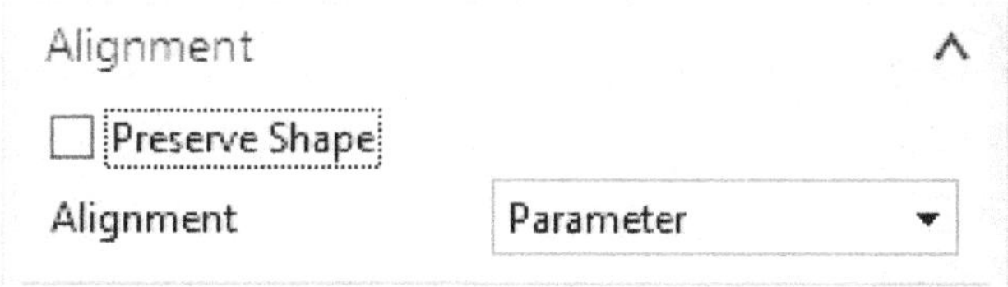

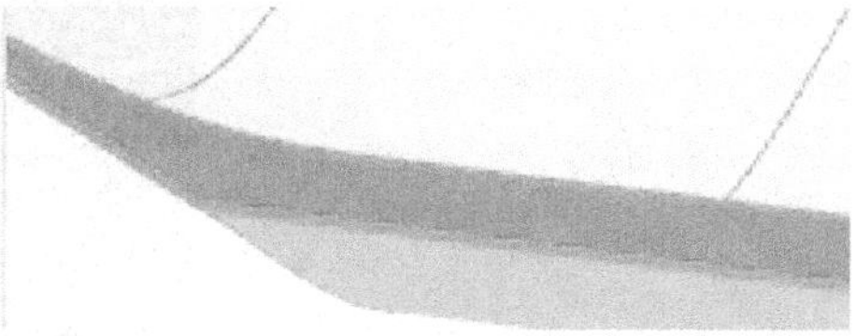

- Use **Ruled Surface** three more times, to create each of the other side walls, and then **Sew** all the sheet bodies together to produce the solid shown below.

- **Save** your file!

19.11 The Edit Surface Group

The tools in the **Edit Surface** group allow for no-frills direct editing of surfaces, and they excel at modifying both fully parametric and unparameterized surfaces. Several of these tools directly change the underlying *(U,V)* parameters of a surface, and in doing so, discard any model history for that surface. A deep understanding of the tools in this group will allow you to build robust, high-quality surfaces consistently in NX!

19.11.1 Enlarge

The **Enlarge** tool allows you to extend or retract a rectangular surface along each of its four edges, by quantifying the overall change to the *U* and *V* parameters of the surface. If your surface is missing *(U,V)* pairs, like **Untrim**, it will first extend the surface to a full parametrical-rectangular shape, although it may not be the same fully untrimmed form that **Untrim** yields.

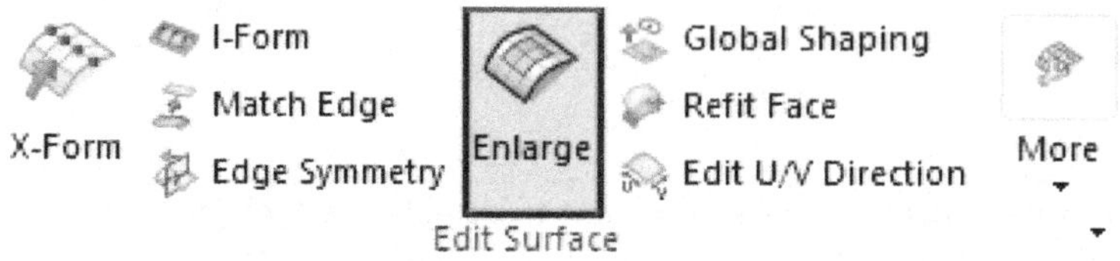

- Open the file *"Edit Surface Tools.prt"* from *"C:\My NX 11 Files\Downloaded Files"*.
- The model shows two ruled surfaces with a studio surface that transitions between them.

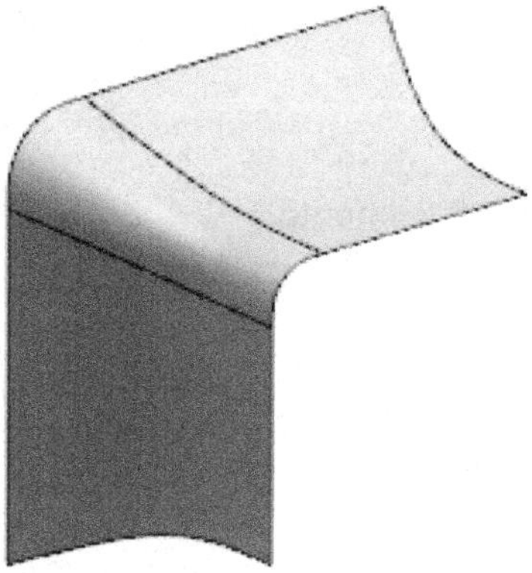

- Select the **Enlarge** tool and choose the vertical ruled surface. In **Settings**, the *Edit a Copy* checkbox automatically becomes marked and there is nothing you can do to disable it. Retract the **V Start** edge by *50%* as shown below.

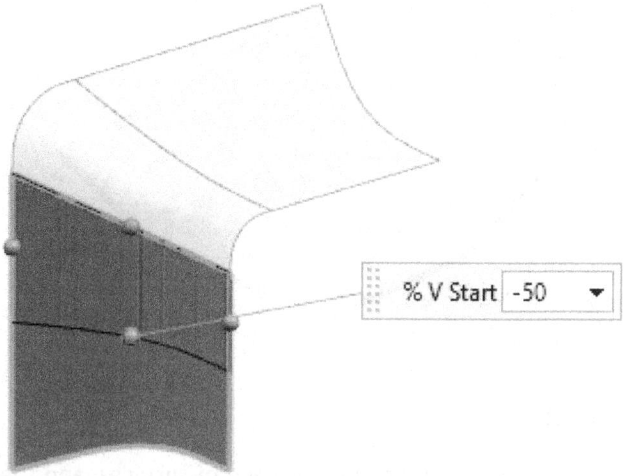

- Use **Delete Body** to remove the original surface associatively.
- Select the **Enlarge** tool again and choose the horizontal ruled surface. Push the **Show Poles** button to see the effect on the underlying parameters as you drag the **% V End Handle** to *50%*.

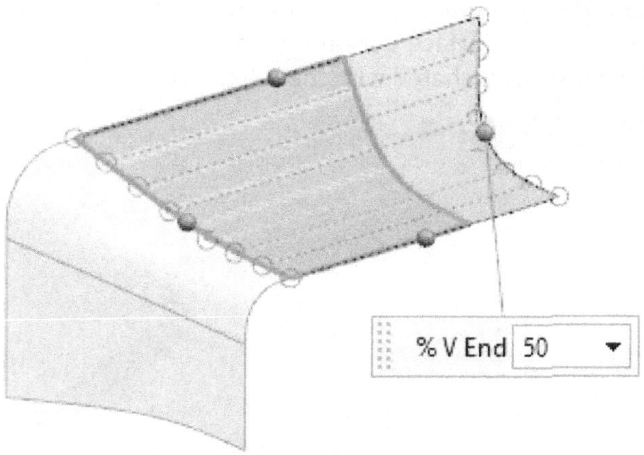

- Use **Delete Body** to delete the original surface associatively.

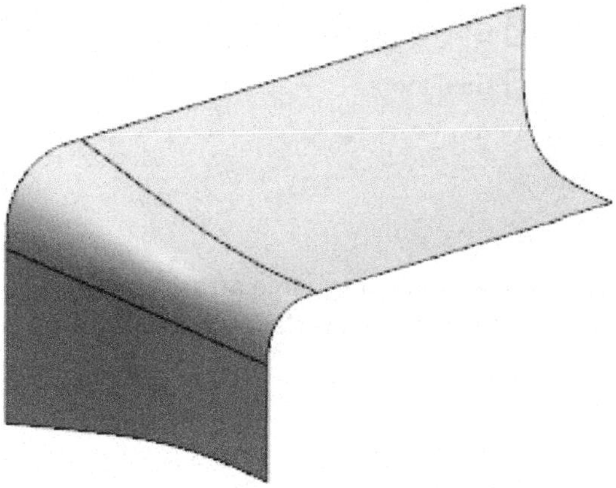

19.11.2 X-Form

The **X-Form** tool exposes the poles of a B-surface and allows you to manipulate them like the poles of a studio spline, as you practiced in Chapter 13.2.3. It can be quite challenging to use it in a controlled way, but it is the only tool that, with enough careful planning, can be used to smooth any eligible surface to meet its neighbors with *G3* continuity.

If the face you select to modify with **X-Form** is not a B-surface, **X-Form** will first convert it to a B-surface. This can result in real changes to the shape of the surface, so be careful when using **X-Form** to modify surfaces of other types.

X-Form can also be used to modify any curve – it will convert that curve to a spline, and allow you to reparametrize it and modify the poles as though it were a **Studio Spline** constructed *By Poles*.

X-Form is found both in the **Edit Surface** group on the **Surface** tab, as well as in the **Edit Curve** group on the **Curve** tab.

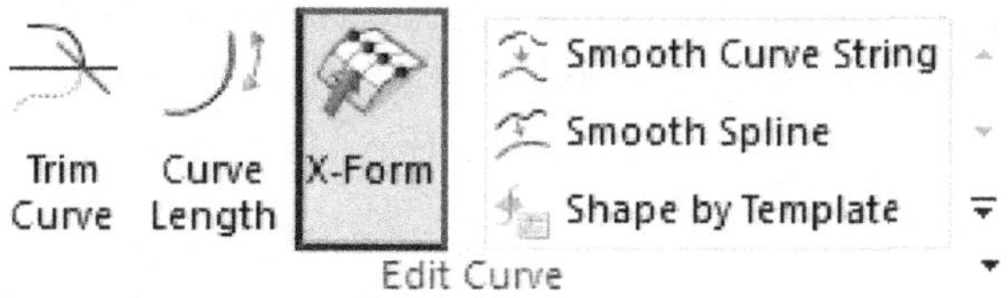

X-Form is arguably the deepest and most powerful surfacing tool in NX, but it can be very difficult to control it precisely in a parametric model. The exercise below shows how **X-Form** can be used to achieve *G2* continuity when other surfacing tools fail!

- Continue working with *"Edit Surface Tools.prt"*.

- The **Studio Surface** feature in this model was created with the intention of transitioning between the two ruled surfaces with *G2* continuity. However, a **Surface Continuity** analysis shows that the face of **Studio Surface** fails to achieve even *G1* continuity along the edges shared with the ruled surfaces.

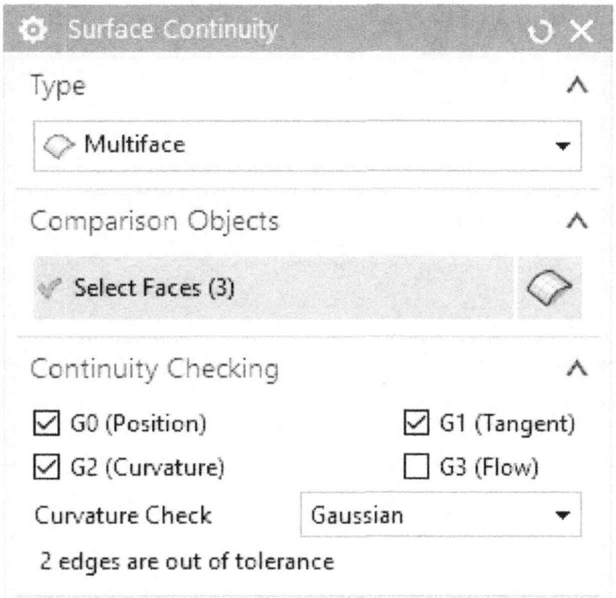

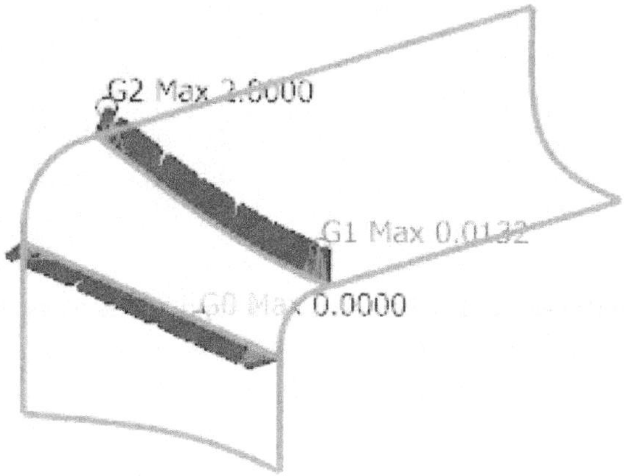

- On layer *61*, there is a set of points that represents a suitable arrangement of poles for the face of the studio surface to guarantee *G2* continuity with the neighboring ruled surfaces. Make layer *61* visible.

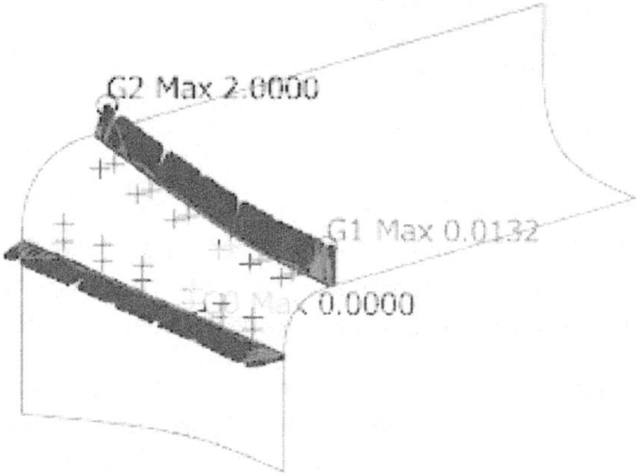

- Select the **X-Form** tool and choose the face of the *Studio Surface* feature.
- Enable the *Existing Point* snap point icon, and make sure that your **Move Method** is set to *View*. This will allow you to drag poles one-by-one freely in the view plane, and snap them onto the nearest point from layer *61*.
- As you drag poles, the needles of the surface continuity analysis will dynamically update.

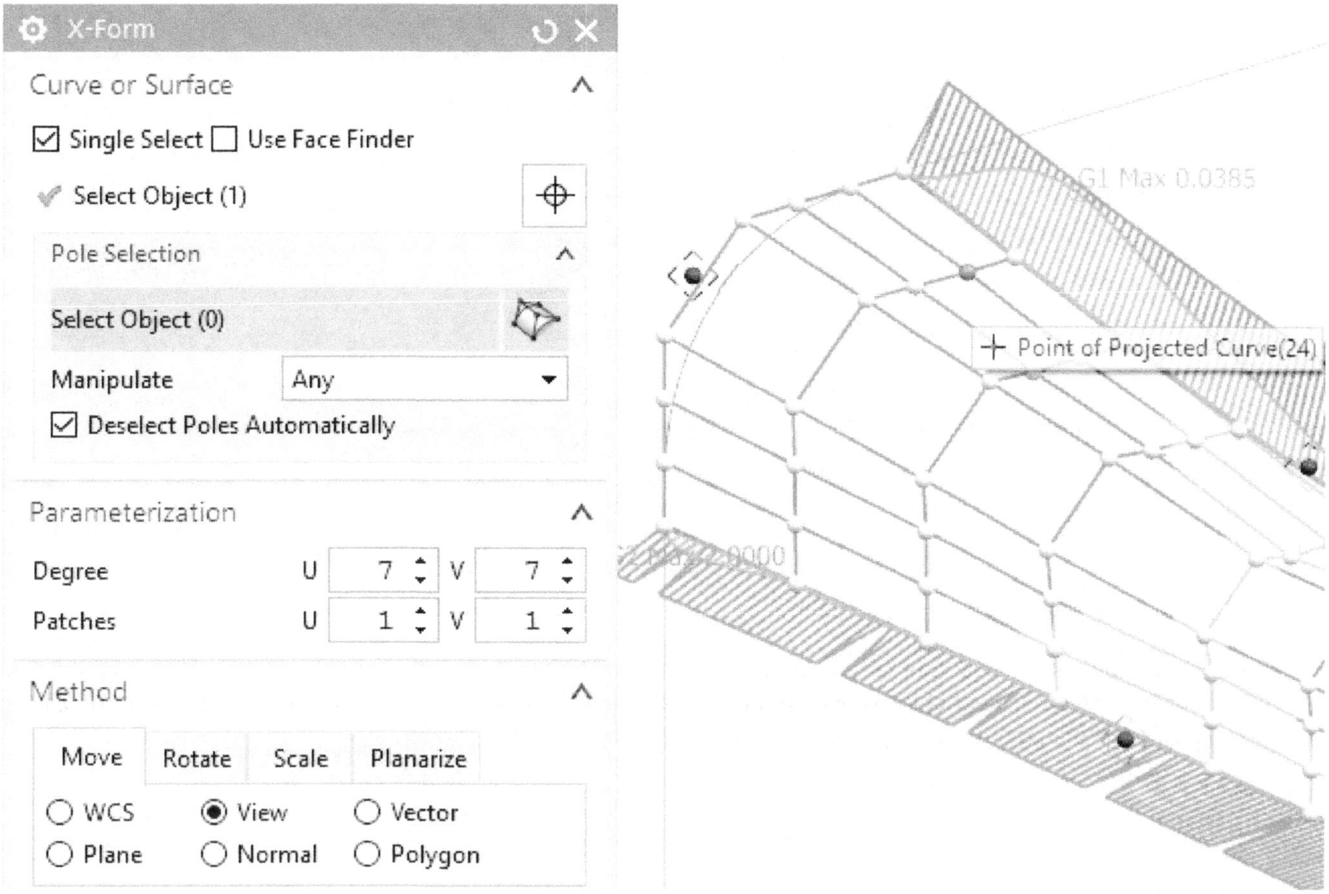

- When you finish the two rows adjacent to the horizontal ruled surface, the needles of the surface continuity analysis will vanish! Now you have achieved true *G2* continuity along the shared edge with that ruled surface.

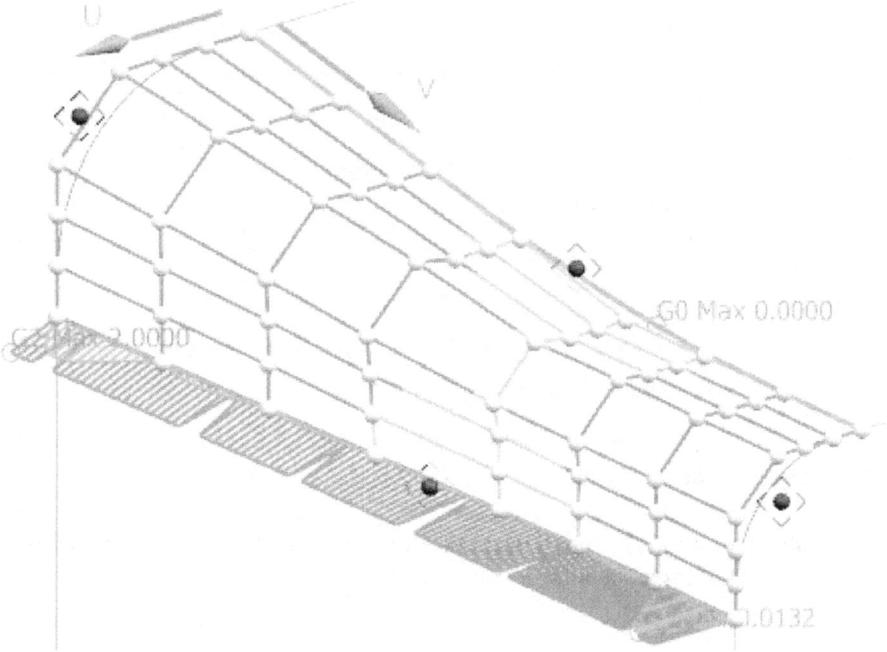

- After arranging the poles from the two rows adjacent to the vertical ruled surface, your needle display will reappear – NX detects some deviation between the tangent vectors to the surfaces along their shared edges, but if you look closely at the needles, you will see that the deviation is less than 0.0001 degrees! Deviation of this order is negligible.

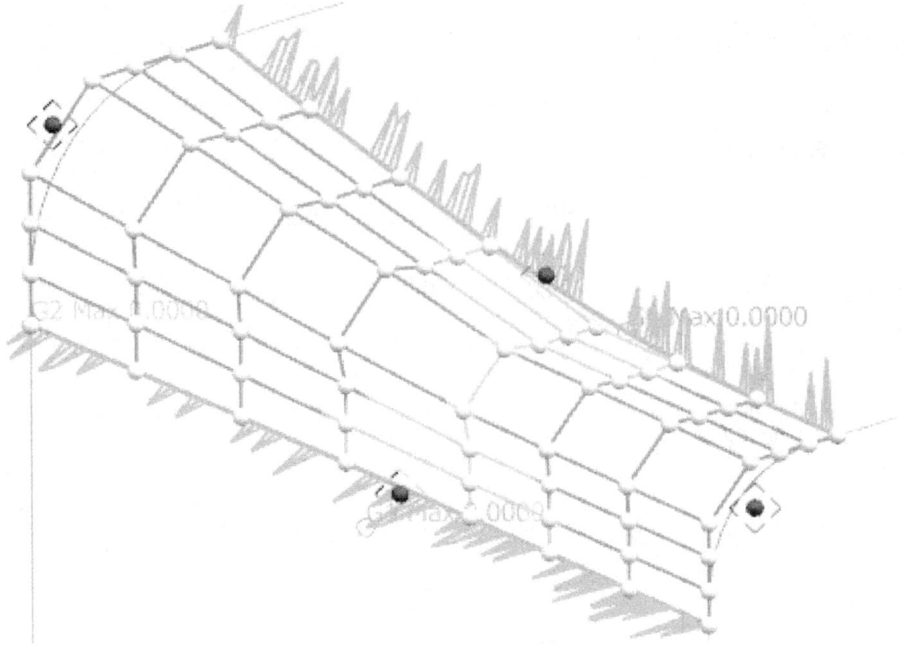

- You can also move sets of poles, or even rows of poles. Adjust the two inner rows of poles by selecting the polylines connecting the poles, and restrict the **Move Method** to only allow movement along the vector joining *(0,0,1)* to *(0,1,0)*. This will subtly change the surface while maintaining symmetry.

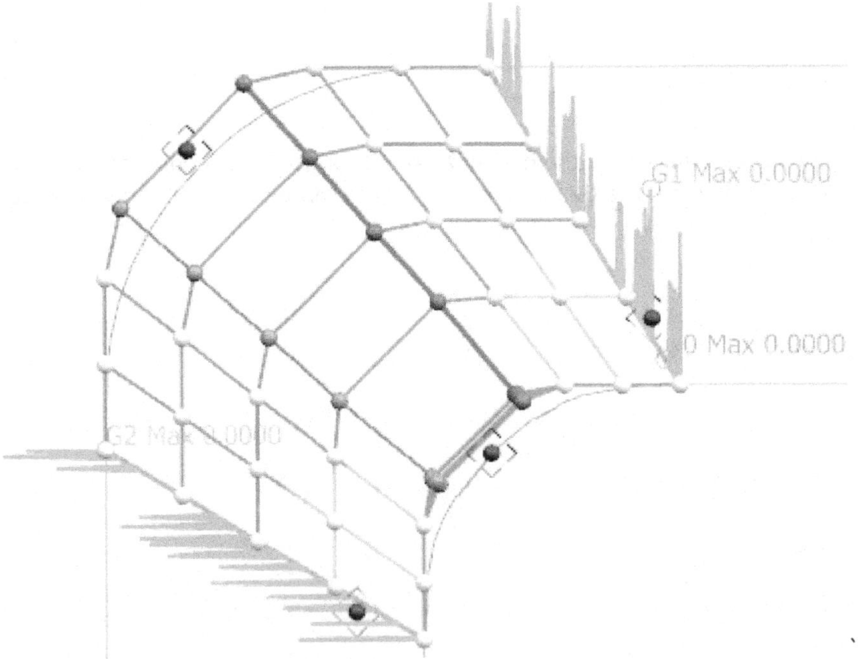

- Click **OK** to save the changes to your surface.
- **Hide** layer *61*.
- Use the **Reflection Analysis** tool to see the reflection, and compare the results with **X-Form** active vs. **X-Form** suppressed. There is a subtle, but tangible difference!

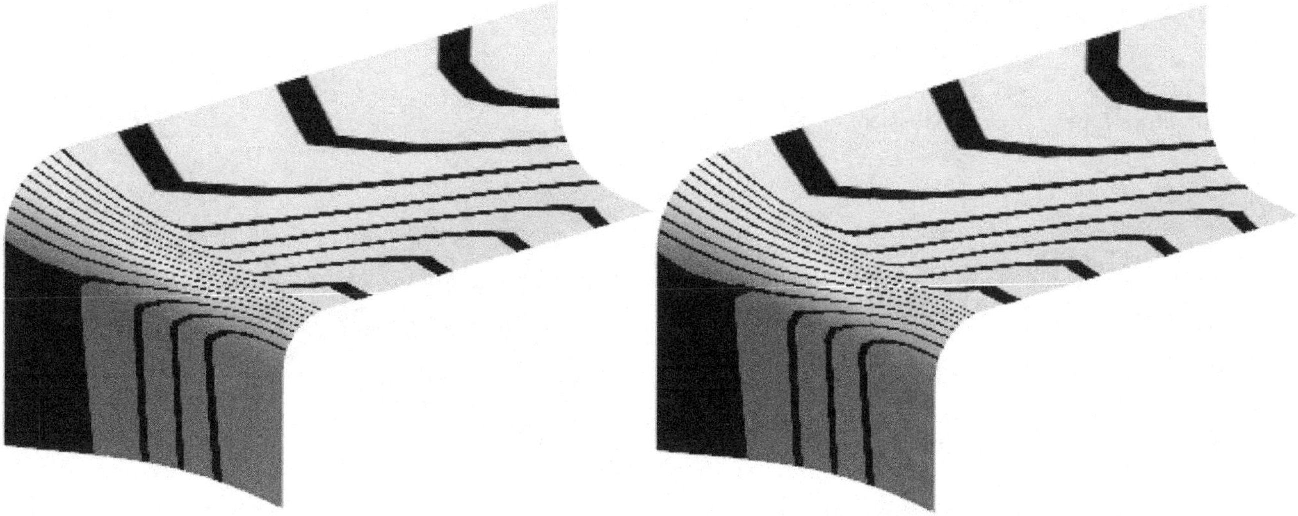

19.11.3 Flattening and Forming

The **Flattening and Forming** tool is brand new in NX 11, and it allows you to flatten non-developable surfaces, and then transform other geometry (e.g. curves) by reversing the flattening operation. This can be very useful for translating a chain of curves for which you know the precise shape when flat, but need to map onto the curved surface with as little deformation as possible.

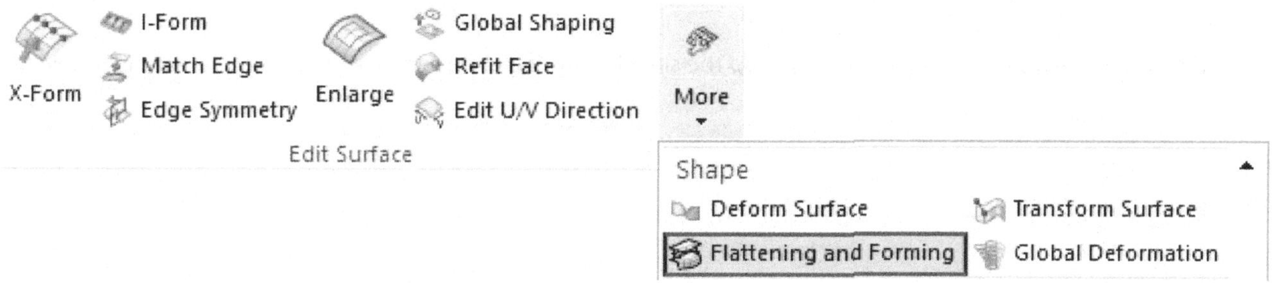

The main use of this tool is to "wrap" curves onto non-developable surfaces. The **Wrap/Unwrap Curve** command can be used on any developable surface, but to wrap curves onto non-developable surfaces, you must provide additional information about how to wrap – this is where this tool comes into play.

- Open the part file *"Flattening and Forming.prt"* from your downloaded part files folder.
- In this part file, you will find two sheet bodies – one obtained by extruding a sketch, and the other obtained by lofting from that sketch to another, and insisting on *G2* continuity with the extruded surface.
- The **Face Curvature Analysis** tool shows that the extruded surface is indeed developable, and that the lofted one is not. The precise definition of a developable surface is one with zero Gaussian curvature.

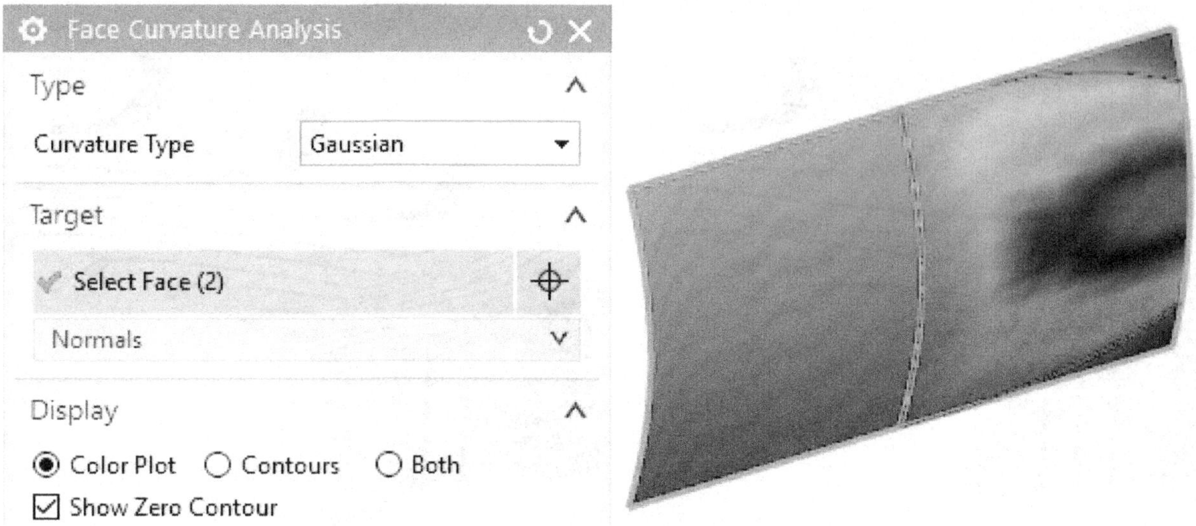

- There are two feature groups in this part file, each containing a point, a datum plane, and a sketch. Right-click on the feature group named *"Wrapping"* and select **Show**.

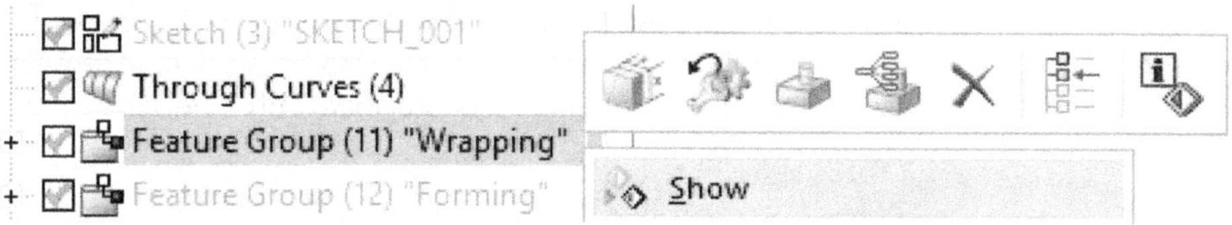

- Use the **Wrap/Unwrap Curve** command to wrap the sketch onto the extruded face, with the tangent plane selected.

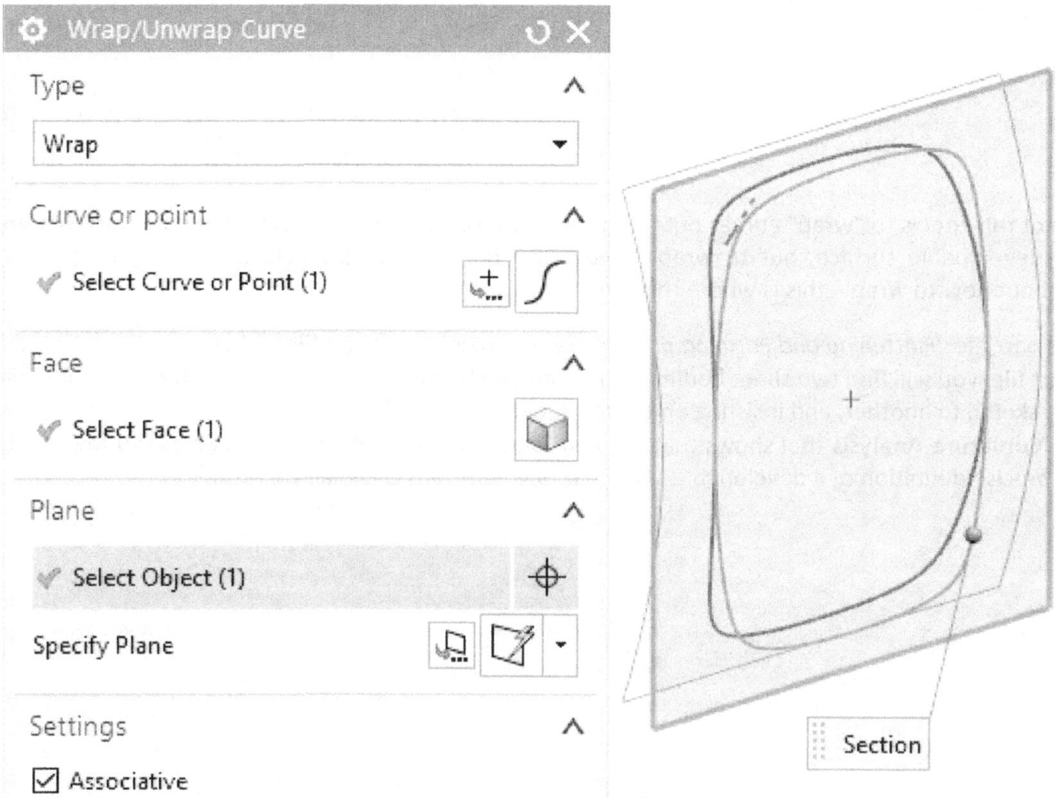

- **Show** the feature group named *"Forming"*. If you attempt to use **Wrap/Unwrap Curve** to wrap this second sketch onto the lofted surface, you will encounter an error, as shown below.

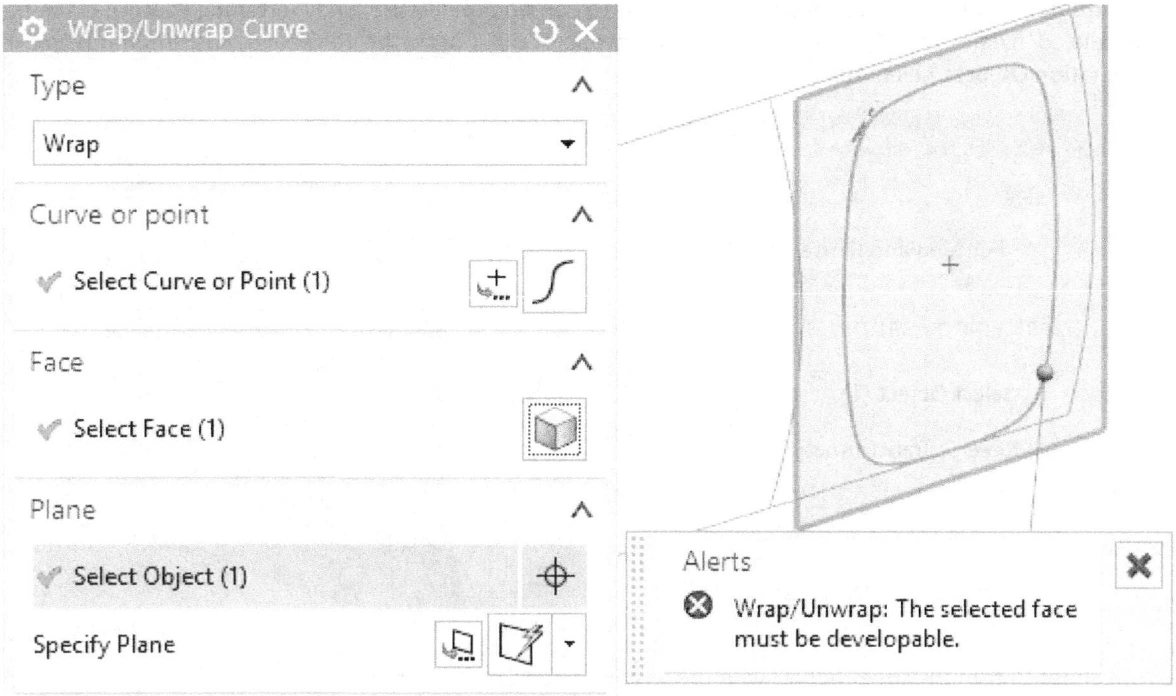

- Select the **Flattening and Forming** tool. For the **Source Face**, choose the face of *Through Curves*, and for the **Flattening Origin**, choose the point as **Origin**, and the normal vector to the tangent datum plane as the **Primary Direction**. Use the handles of the dynamic CSYS to modify the **Destination** until it appears as shown below. Set the **Distortion Map** to *Area*.

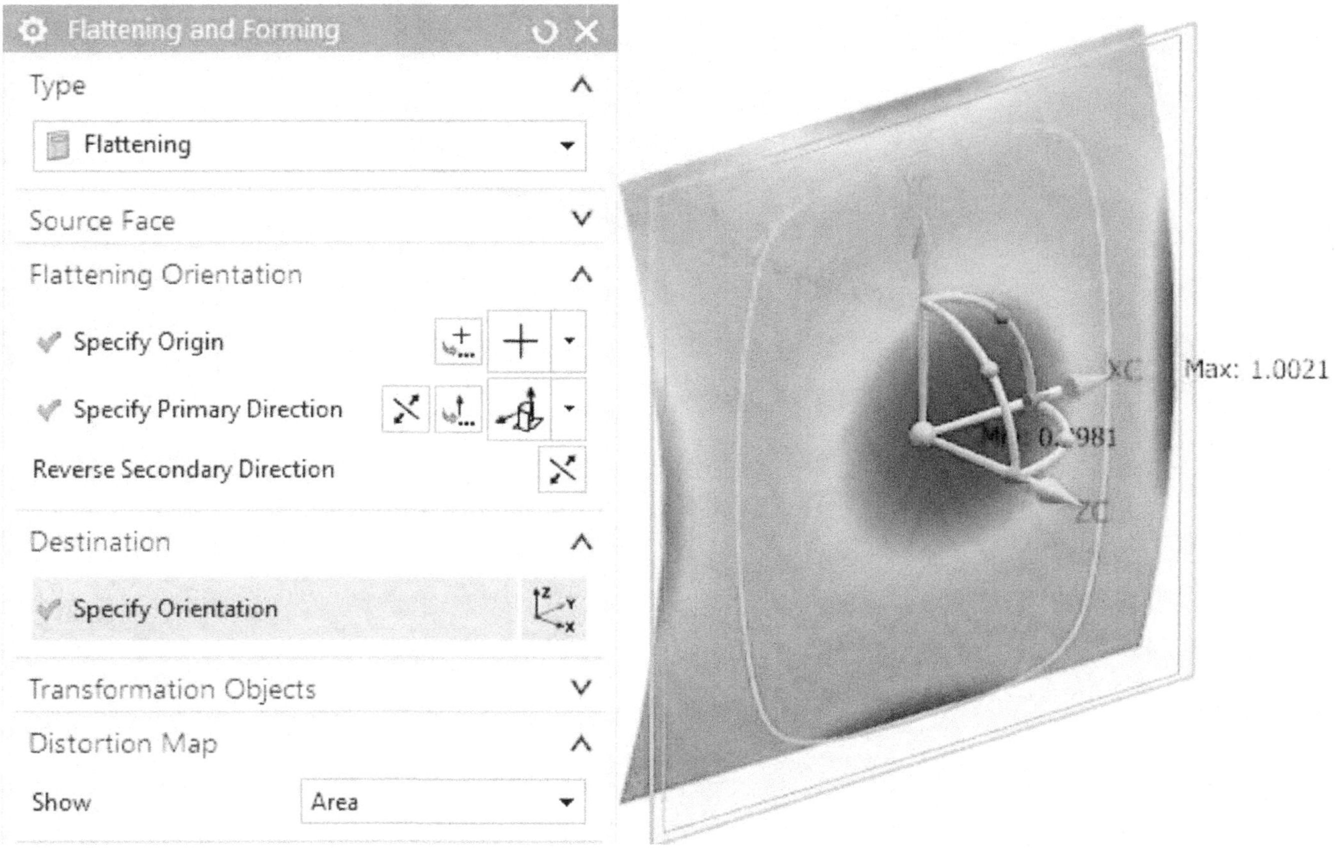

- Select **Flattening and Forming** again, and set the **Type** to *Forming and Reuse*. The **Flattening Feature** is the flattened body produced by the first **Flattening and Forming** operation. Check the *Reverse Transformation* box. For the **Transformation Objects**, select the sketch.

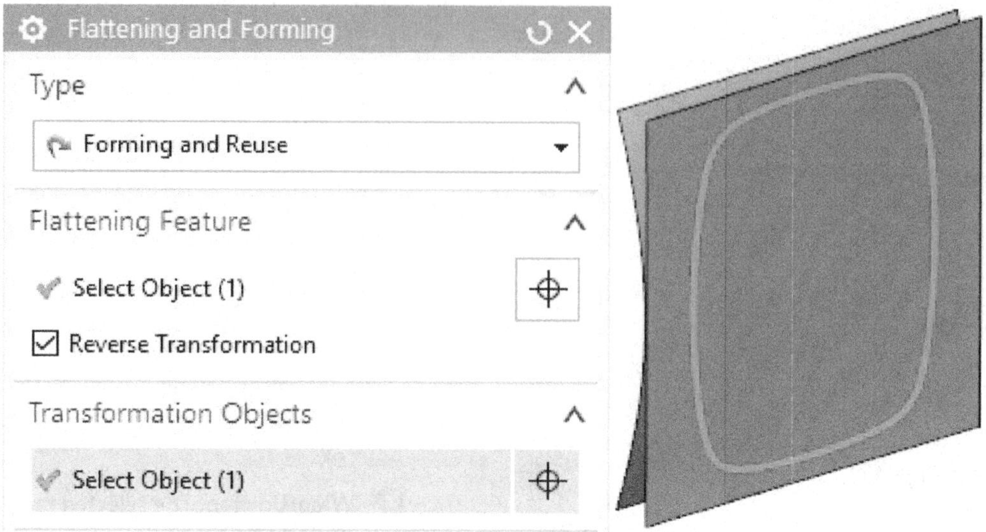

- Use **Trim Sheet** to trim the sheet bodies with the wrapped and formed curves.

- **Save** your part file!

20 Index

Made in United States
North Haven, CT
10 November 2026

22088263R00343